# LIFE

The Science of Biology

TENTH EDITION

# LIFE

## The Science of Biology

TENTH EDITION

DAVID
**SADAVA**
The Claremont Colleges

DAVID M.
**HILLIS**
University of Texas

H. CRAIG
**HELLER**
Stanford University

MAY R.
**BERENBAUM**
University of Illinois

SINAUER

MACMILLAN

THE COVER
The sea slug *Elysia crispata*. This animal is able to carry out photosynthesis using chloroplasts incorporated from the algae it feeds on (see back cover). Photograph © Alex Mustard/Naturepl.com.

THE FRONTISPIECE
Red-crowned cranes, *Grus japonensis*, gather on a river in Hokkaido, Japan. ©Steve Bloom Images/Alamy.

**LIFE:** The Science of Biology, Tenth Edition


ADDRESS EDITORIAL CORRESPONDENCE TO:
Sinauer Associates, Inc., 23 Plumtree Road, Sunderland, MA 01375 U.S.A.

www.sinauer.com
publish@sinauer.com

ADDRESS ORDERS TO:
MPS / W. H. Freeman & Co., Order Dept., 16365 James Madison Highway, U.S. Route 15, Gordonsville, VA 22942 U.S.A.

EXAMINATION COPY INFORMATION: 1-800-446-8923

Planet Friendly Publishing
✓ Made in the United States
✓ Printed on Recycled Paper
Text: 10% Cover: 10%
Learn more: www.greenedition.org

Courier Corporation, the manufacturer of this book, owns the *Green Edition* Trademark

**Library of Congress Cataloging-in-Publication Data**

Life : the science of biology / David Sadava ... [et al.]. -- 10th ed.
p. cm.
Includes bibliographical references and index.
ISBN 978-1-4292-9864-3 (casebound) — 978-1-4641-4122-5 (pbk. : v. 1) —
ISBN 978-1-4641-4123-2 (pbk. : v. 2) — ISBN 978-1-4641-4124-9 (pbk. : v. 3)
1. Biology--Textbooks. I. Sadava, David E.
QH308.2.L565 2013
570--dc23 2012039164

Printed in U.S.A.
First Printing December 2012
The Courier Companies, Inc.

*To all the educators who have worked tirelessly for quality biology education*

# The Authors

DAVID HILLIS MAY BERENBAUM CRAIG HELLER DAVID SADAVA

**DAVID SADAVA** is the Pritzker Family Foundation Professor of Biology, Emeritus at the Keck Science Center of Claremont McKenna, Pitzer, and Scripps, three of The Claremont Colleges. In addition, he is Adjunct Professor of Cancer Cell Biology at the City of Hope Medical Center in Duarte, California. Twice winner of the Huntoon Award for superior teaching, Dr. Sadava has taught courses on introductory biology, biotechnology, biochemistry, cell biology, molecular biology, plant biology, and cancer biology. In addition to *Life: The Science of Biology and Principles of Life*, he is the author or coauthor of books on cell biology and on plants, genes, and crop biotechnology. His research has resulted in many papers coauthored with his students, on topics ranging from plant biochemistry to pharmacology of narcotic analgesics to human genetic diseases. For the past 15 years, he has investigated multidrug resistance in human small-cell lung carcinoma cells with a view to understanding and overcoming this clinical challenge. At the City of Hope, his current work focuses on new anti-cancer agents from plants. He is the featured lecturer in "Understanding Genetics: DNA, Genes and their Real-World Applications," a video course for The Great Courses series.

**DAVID M. HILLIS** is the Alfred W. Roark Centennial Professor in Integrative Biology and the Director of the Dean's Scholars Program at the University of Texas at Austin, where he also has directed the School of Biological Sciences and the Center for Computational Biology and Bioinformatics. Dr. Hillis has taught courses in introductory biology, genetics, evolution, systematics, and biodiversity. He has been elected to the National Academy of Sciences and the American Academy of Arts and Sciences, awarded a John D. and Catherine T. MacArthur fellowship, and has served as President of the Society for the Study of Evolution and of the Society of Systematic Biologists. He served on the National Research Council committee that wrote the report *BIO 2010: Transforming Undergraduate Biology Education for Research Biologists*. His research interests span much of evolutionary biology, including experimental studies of viral evolution, empirical studies of natural molecular evolution, applications of phylogenetics, analyses of biodiversity, and evolutionary modeling. He is particularly interested in teaching and research about the practical applications of evolutionary biology.

**H. CRAIG HELLER** is the Lorry I. Lokey/Business Wire Professor in Biological Sciences and Human Biology at Stanford University. He has taught in the core biology courses at Stanford since 1972 and served as Director of the Program in Human Biology, Chairman of the Biological Sciences Department, and Associate Dean of Research. Dr. Heller is a fellow of the American Association for the Advancement of Science and a recipient of the Walter J. Gores Award for excellence in teaching and the Kenneth Cuthberson Award for Exceptional Service to Stanford University. His research is on the neurobiology of sleep and circadian rhythms, mammalian hibernation, the regulation of body temperature, the physiology of human performance, and the neurobiology of learning. He has done research on a huge variety of animals and physiological problems, including from sleeping kangaroo rats, diving seals, hibernating bears, photoperiodic hamsters, and exercising athletes. Dr. Heller has extended his enthusiasm for promoting active learning via the development of a two-year curriculum in human biology for the middle grades, through the production of Virtual Labs—interactive computer-based modules to teach physiology.

**MAY BERENBAUM** is the Swanlund Professor and Head of the Department of Entomology at the University of Illinois at Urbana-Champaign. She has taught courses in introductory animal biology, entomology, insect ecology, and chemical ecology and has received teaching awards at the regional and national levels from the Entomological Society of America. A fellow of the National Academy of Sciences, the American Academy of Arts and Sciences, and the American Philosophical Society, she served as President of the American Institute for Biological Sciences in 2009 and currently serves on the Board of Directors of AAAS. Her research addresses insect–plant coevolution and ranges from molecular mechanisms of detoxification to impacts of herbivory on community structure. Concerned with the practical application of ecological and evolutionary principles, she has examined impacts of genetic engineering, global climate change, and invasive species on natural and agricultural ecosystems. In recognition of her work, she received the 2011 Tyler Prize for Environmental Achievement. Devoted to fostering science literacy, she has published numerous articles and five books on insects for the general public.

# Contents in Brief

PART ONE **THE SCIENCE OF LIFE AND ITS CHEMICAL BASIS**

1 Studying Life 1
2 Small Molecules and the Chemistry of Life 21
3 Proteins, Carbohydrates, and Lipids 39
4 Nucleic Acids and the Origin of Life 62

PART TWO **CELLS**

5 Cells: The Working Units of Life 77
6 Cell Membranes 105
7 Cell Communication and Multicellularity 125

PART THREE **CELLS AND ENERGY**

8 Energy, Enzymes, and Metabolism 144
9 Pathways that Harvest Chemical Energy 165
10 Photosynthesis: Energy from Sunlight 185

PART FOUR **GENES AND HEREDITY**

11 The Cell Cycle and Cell Division 205
12 Inheritance, Genes, and Chromosomes 232
13 DNA and Its Role in Heredity 259
14 From DNA to Protein: Gene Expression 281
15 Gene Mutation and Molecular Medicine 304
16 Regulation of Gene Expression 328

PART FIVE **GENOMES**

17 Genomes 352
18 Recombinant DNA and Biotechnology 373
19 Differential Gene Expression in Development 392
20 Genes, Development, and Evolution 412

PART SIX **THE PATTERNS AND PROCESSES OF EVOLUTION**

21 Mechanisms of Evolution 427
22 Reconstructing and Using Phylogenies 449
23 Speciation 467
24 Evolution of Genes and Genomes 485
25 The History of Life on Earth 505

PART SEVEN **THE EVOLUTION OF DIVERSITY**

26 Bacteria, Archaea, and Viruses 525
27 The Origin and Diversification of Eukaryotes 549
28 Plants without Seeds: From Water to Land 569
29 The Evolution of Seed Plants 588
30 The Evolution and Diversity of Fungi 608
31 Animal Origins and the Evolution of Body Plans 629
32 Protostome Animals 651
33 Deuterostome Animals 678

PART EIGHT **FLOWERING PLANTS: FORM AND FUNCTION**

34 The Plant Body 708
35 Transport in Plants 726
36 Plant Nutrition 740
37 Regulation of Plant Growth 756
38 Reproduction in Flowering Plants 778
39 Plant Responses to Environmental Challenges 797

PART NINE **ANIMALS: FORM AND FUNCTION**

40 Physiology, Homeostasis, and Temperature Regulation 815
41 Animal Hormones 834
42 Immunology: Animal Defense Systems 856
43 Animal Reproduction 880
44 Animal Development 902
45 Neurons, Glia, and Nervous Systems 924
46 Sensory Systems 946
47 The Mammalian Nervous System 967
48 Musculoskeletal Systems 986
49 Gas Exchange 1005
50 Circulatory Systems 1025
51 Nutrition, Digestion, and Absorption 1048
52 Salt and Water Balance and Nitrogen Excretion 1071
53 Animal Behavior 1093

PART TEN **ECOLOGY**

54 Ecology and the Distribution of Life 1121
55 Population Ecology 1149
56 Species Interactions and Coevolution 1169
57 Community Ecology 1188
58 Ecosystems and Global Ecology 1207
59 Biodiversity and Conservation Biology 1228

# Preface

Biology is a constantly changing scientific field. New discoveries about the living world are being made every day, and more than 1 million new research articles in biology are published each year. Beyond the constant need to update the concepts and facts presented in any science textbook, in recent years ideas about how best to educate the upcoming generation of biologists have undergone dynamic and exciting change.

Although we and many of our colleagues had thought about the nature of biological education as individuals, it is only recently that biologists have come together to discuss these issues. Reports from the National Academy of Sciences, Howard Hughes Medical Institute, and College Board AP Biology Program not only express concern about how best to instruct undergraduates in biology, but offer concrete suggestions about how to design the introductory biology course—and by extension, our book. We have followed these discussions closely and have been especially impressed with the report "Vision and Change in Undergraduate Biology Education" (visionandchange.org). As participants in the educational enterprise, we have answered the report's call to action with this textbook and its associated ancillary materials.

The "Vision and Change" report proposes five core concepts for biological literacy:

1. Evolution
2. Structure and function
3. Information flow, exchange, and storage
4. Pathways and transformations of energy and matter
5. Systems

These five concepts have always been recurring themes in Life, but in this Tenth Edition we have brought them even more "front and center."

"Vision and Change" also advocates that students learn and demonstrate core competencies, including the ability to apply the process of science using quantitative reasoning. Life has always emphasized the experimental nature of biology. This edition responds further to these core competency issues with a new working with data feature and the addition of a statistics primer (Appendix B). The authors' multiple educational perspectives and areas of expertise, as well as input from many colleagues and students who used previous editions, have informed the approach to this new edition.

## Enduring Features

We remain committed to blending the presentation of core ideas with an emphasis on introducing students to the process of scientific inquiry. Having pioneered the idea of depicting important experiments in unique figures designed to help students understand and appreciate the way scientific investigations work, we continue to develop this approach in the book's 70 **Investigating Life** figures. Each of these figures sets the experiment in perspective and relates it to the accompanying text. As in previous editions, these figures employ the structure Hypothesis, Method, Results, and Conclusion. We have added new information focusing on the individuals who performed these experiments so students can appreciate more fully that science is a human and very personal activity. Each Investigating Life figure has a reference to BioPortal (*yourBioPortal.com*), where discussion and references to follow-up research can be found. A related feature is the **Research Tools** figures, which depict laboratory and field methods used in biology. These, too, have been expanded to provide more useful context for their importance.

Some 15 years ago, *Life's* authors and publishers pioneered the use of **balloon captions** in our figures. We recognized then that many students are visual learners, and this fact is even truer today. *Life*'s balloon captions bring the crucial explanations of intricate, complex processes directly into the illustration, allowing students to integrate information without repeatedly going back and forth between the figure, its legend, and the text.

We continue to refine our chapter organization. Our **opening stories** have always provide historical, medical, or social context to intrigue students and show how the subject of each chapter relates to the world around them. In the Tenth Edition, the opening stories all end with a question that is revisited throughout the chapter. At the end of each chapter the answer is presented in the light of material the student encountered in the body of the chapter.

A **chapter outline** asks questions to emphasize scientific inquiry, each of which is answered in a major section of the chapter. A **Recap** summarizes each section's key concepts and poses questions that help the student review and test their mastery of these concepts. The recap questions are similar in form to the learning objectives used in many introductory biology courses. The **Chapter Summaries** highlight each chapter's key figures and defined terms, while restating the major concepts presented in the chapter in a concise and student-friendly manner, with references to specific figures and to the activities and animated tutorials available in BioPortal.

At the end of the book, students will find a much-expanded glossary that continues *Life's* practice of providing Latin or Greek derivations for many of the defined terms. As students become gradually (and painlessly) more familiar with such root words, the mastery of vocabulary as they continue in their biological or medical studies will be easier. In addition, the popular **Tree of Life appendix** (Appendix A) presents the phylogenetic tree of life as a reference tool that allows students to place any group of organisms mentioned in the text into the context of the rest of life. The web-based version of Appendix A provides links to photos, keys, species lists, distribution maps, and other information (via the online database at DiscoverLife.org) to help students explore biodiversity in greater detail.

## New Features

The Tenth Edition of *Life* has a different look and feel from its predecessors. The new color palette and more open design will, we hope, be more accessible to students. And, in keeping with our heightened emphasis on scientific inquiry and quantitative analysis, we have added **Working with Data** exercises to almost all chapters. In these innovative exercises, we describe the context and approach of a research paper that provides the basis of the analysis. We then ask questions that require students to analyze data, make calculations, and draw conclusions. Answers (or suggested possible answers) to these questions are included in BioPortal and can be made available to students at the instructor's discretion.

Because many of the questions in the Working with Data exercises require the use of basic statistical methods, we have included a **Statistics Primer** as the book's Appendix B, describing the concepts and some methods of statistical analysis. We hope that the Working with Data exercises and statistics primer will reinforce students' skills and their ability to apply quantitative analysis to biology.

We have added links to **Media Clips** in the body of the text, with at least one per chapter. These brief clips are intended to enlighten and entertain. Recognizing the widespread use of "smart phones" by students, the textbook includes **instant access (QR) codes** that bring the Media Clips, Animated Tutorials, and Interactive Summaries directly to the screen in your hand. If you do not have a smart phone, never fear, we also provide direct web addresses to these features.

As educators, we follow current discussions of pedagogy in biological education. The chapter-ending **Chapter Reviews** now contain multiple levels of questions based on Bloom's taxonomy: Remembering, Understanding and Applying, and Analyzing and Evaluating. Answers to these questions appear at the end of the book.

For a detailed description of the media and supplements available for the Tenth Edition, please turn to "*Life's* Media and Supplements" on page xvii.

## The Ten Parts

**PART ONE, THE SCIENCE OF LIFE AND ITS CHEMICAL BASIS** Chapter 1 introduces the core concepts set forth in the "Vision and Change" report and continues the much-praised approach of focusing on a specific series of experiments that introduces students to biology as an experimentally based and constantly expanding science. Chapter 1 emphasizes the principles of biology that are the foundation for the rest of the book, including the unity of life at the cellular level and how evolution unites the living world. Chapters 2–4 cover the chemical principles and building blocks that underlie life. Chapter 4 also includes a discussion of how life could have evolved from inanimate chemicals.

**PART TWO, CELLS** The nature of cells and their role as the structural and functional basis of life is foundational to biology. These revised chapters include expanded explanations of how experimental manipulations of living systems have been used to discover cause and effect in biology. Students who are intrigued by the question "Where did the first cells come from?" will appreciate the updated discussion of ideas on the origin of cells and organelles, as well as expanded discussion of the evolution of multicellularity and cell interactions. In response to reviewer comments, the discussion of membrane potential has been moved to Chapter 45, where students may find it to be more relevant.

**PART THREE, CELLS AND ENERGY** The biochemistry of life and energy transformations are among the most challenging topics for many students. We have worked to clarify such concepts as enzyme inhibition, allosteric enzymes, and the integration of biochemical systems. Revised presentations of glycolysis and the citric acid cycle now focus, in both text and figures, on key concepts and attempt to limit excessive detail. There are also revised discussions of the ecological roles of alternate pathways of photosynthetic carbon fixation, as well as the roles of accessory pigments and reaction center in photosynthesis.

**PART FOUR, GENES AND HEREDITY** This crucial section of the book is revised to improve clarity, link related concepts, and provide updates from recent research results. Rather than being segregated into separate chapters, material on prokaryotic genetics and molecular medicine are now interwoven into relevant chapters. Chapter 11 on the cell cycle includes a new discussion of how the mechanisms of cell division are altered in cancer cells. Chapter 12 on transmission genetics now includes coverage of this phenomenon in prokaryotes. Chapters 13 and 14 cover gene expression and gene regulation, including new discoveries about the roles of RNA and an expanded discussion of epigenetics. Chapter 15 covers the subject of gene mutations and describes updated applications of medical genetics.

**PART FIVE, GENOMES** This extensive and up-to-date coverage of genomes expands and reinforces the concepts covered in Part Four. The first chapter of Part Five describes how genomes are analyzed and what they tell us about the biology of prokaryotes and eukaryotes, including humans. Methods of DNA sequencing and genome analysis, familiar to many students in a general way, are rapidly improving, and we discuss these advances as well as how bioinformatics is used. This leads to a chapter describing how our knowledge of molecular biology and genetics underpins biotechnology—the application of this knowledge to practical problems and issues such as stem cell research. Part Five closes with a unique sequence of two chapters that explore the interface of developmental processes with molecular biology (Chapter 19) and with evolution (Chapter 20), providing students with a link between these two crucial topics and a bridge to Part Six.

**PART SIX, THE PATTERNS AND PROCESSES OF EVOLUTION** Many students come to the introductory biology course with ideas about evolution already firmly in place. One common view, that evolution is only about Darwin, is firmly put to rest at the start of Chapter 21, which not only illustrates the practical value of fully understanding modern evolutionary biology, but briefly and succinctly traces the history of "Darwin's dangerous idea" through the twentieth century and up to the present syntheses of molecular evolutionary genetics and evolutionary developmental biology—fields of study that uphold and support the principles of evolutionary biology as the basis for comparing and comprehending all other aspects of biology. The remaining sections of Chapter 21 describe the mechanisms of evolution in clear, matter-of-fact terms. Chapter 22 describes phylogenetic trees as a tool not only of classification but also of evolutionary inquiry. The remaining chapters cover speciation and molecular evolution, concluding with an overview of the evolutionary history of life on Earth.

**PART SEVEN, THE EVOLUTION OF DIVERSITY** Continuing the theme of how evolution has shaped our world, Part Seven introduces the latest views on biodiversity and the evolutionary relationships among organisms. The chapters have been revised with the aim of making it easier for students to appreciate the major evolutionary changes that have taken place within the different groups of organisms. These chapters emphasize understanding the big picture of organismal diversity—the tree of life—as opposed to memorizing a taxonomic hierarchy and names. Throughout the book, the tree of life is emphasized as a way of understanding and organizing biological information.

**PART EIGHT, FLOWERING PLANTS: FORM AND FUNCTION** The emphasis of this modern approach to plant form and function is not only on the basic findings that led to the elucidation of mechanisms for plant growth and reproduction, but also on the use of genetics of model organisms. In response to users of earlier editions, material covering recent discoveries in plant molecular biology and signaling has been reorganized and streamlined to make it more accessible to students. There are also expanded and clearer explanations of such topics as water relations, the plant body plan, and gamete formation and double fertilization.

**PART NINE, ANIMALS: FORM AND FUNCTION** This overview of animal physiology begins with a sequence of chapters covering the systems of information—endocrine, immune, and neural. Learning about these information systems provides important groundwork and explains the processes of control and regulation that affect and integrate the individual physiological systems covered in the remaining chapters of the Part. Chapter 45, "Neurons and Nervous Systems," has been rearranged and contains descriptions of exciting new discoveries about glial cells and their role in the vertebrate nervous system. The organization of several other chapters has been revised to reflect recent findings and to allow the student to more readily identify the most important concepts to be mastered.

**PART TEN, ECOLOGY** Part Ten continues *Life*'s commitment to presenting the experimental and quantitative aspects of biology, with increased emphasis on how ecologists design and conduct experiments. New exercises provide opportunities for students to see how ecological data are acquired in the laboratory and in the field, how these data are analyzed, and how the results are applied to answer questions. There is also an expanded discussion of aquatic biomes and a more synthetic explanation of how aquatic, terrestrial, and atmospheric components integrate to influence the distribution and abundance of life on Earth. In addition there is an expanded emphasis on examples of successful strategies proposed by ecologists to mitigate human impacts on the environment; rather than an inventory of ways human activity adversely affects natural systems, this revised Tenth Edition provides more examples of ways that ecological principles can be applied to increase the sustainability of these systems.

## Exceptional Value Formats

We again provide *Life* both as the full book and as a set of paperback volumes. Thus, instructors who want to use less than the whole book can choose from these split volumes, each of which contains the book's front matter, appendices, glossary, and index.

- Volume I, *The Cell and Heredity*, includes: Part One, The Science of Life and Its Chemical Basis (Chapters 1–4); Part Two, Cells (Chapters 5–7); Part Three, Cells and Energy (Chapters 8–10); Part Four, Genes and Heredity (Chapters 11–16); and Part Five, Genomes (Chapters 17–20).
- Volume II, *Evolution, Diversity, and Ecology*, includes: Chapter 1, Studying Life; Part Six, The Patterns and Processes of Evolution (Chapters 21–25); Part Seven, The Evolution of Diversity (Chapters 26–33); and Part Ten, Ecology (Chapters 54–59).
- Volume III, *Plants and Animals*, includes: Chapter 1, Studying Life; Part Eight, Flowering Plants: Form and Function (Chapters 34–39); and Part Nine, Animals: Form and Function (Chapters 40–53).

Responding to student concerns, there also are two ways to obtain the entire book at a significantly reduced cost. The loose-leaf edition of *Life* is a shrink-wrapped, unbound, three-hole-punched version that fits into a three-ring binder. Students take only what they need to class and can easily integrate instructor handouts and other resources.

*Life* was the first comprehensive biology text to offer the entire book as a truly robust eBook, and we offer the Tenth Edition in this flexible, interactive format that gives students a different way to read the text and learn the material. The eBook integrates student media resources (animations, activities, interactive summaries, and quizzes) and offers instructors a powerful way to customize the textbook with their own text, images, web links, and, in BioPortal, quizzes, and other materials.

We are proud that our print edition is a greener *Life* that minimizes environmental impact. *Life* was the first introductory biology text to be printed on paper earning the Forest Stewardship Council label, the "gold standard in green paper," and it continues to be manufactured from wood harvested from sustainable forests.

## Many People to Thank

One of the wisest pieces of advice ever given to a textbook author is to "be passionate about your subject, but don't put your ego on the page." Considering all the people who looked over our shoulders throughout the process of creating this book, this advice could not be more apt. We are indebted to the many people who help to make this book what it is. First and foremost among these are our colleagues, biologists from over 100 institutions. Before we set pen to paper, we solicited the advice of users of *Life*'s Ninth Edition, as well as users of other books. These reviewers gave detailed suggestions for improvements. Other colleagues acted as reviewers when the book was almost completed, pointing out inaccuracies or lack of clarity. All of these biologists are listed in the reviewer credits, along with the dozens who reviewed all of the revised assessment resources.

Once we began writing, we had the superb advice of a team of experienced, knowledgeable, and patient biologists working as development and line editors. Laura Green of Sinauer Associates headed the team and coordinated her own fine work with that of Jane Murfett, Norma Roche, and Liz Pierson to produce a polished and professional text. We are especially indebted to Laura for her work on the important Investigating Life and new Working with Data elements. For the tenth time in ten editions, Carol Wigg oversaw the editorial process. Her positive influence pervades the entire book. Artist Elizabeth Morales again translated our crude sketches into beautiful new illustrations. We hope you agree that our art program remains superbly clear and elegant. Johannah Walkowicz effectively coordinated the hundreds of reviews described above. David McIntyre, photo editor extraordinaire, researched and provided us with new photographs, including many of his own, to enrich the book's content and visual statement. Joanne Delphia is responsible for the crisp new design and layout that make this edition of *Life* not just clear and readable but beautiful as well. Christopher Small headed Sinauer's production team and contributed in innumerable ways to bringing *Life* to its final form. Jason Dirks coordinated the creation of our array of media and instructor resources, with Mary Tyler, Mitch Walkowicz, and Carolyn Wetzel serving as editors for our expanded assessment supplements.

W. H. Freeman continues to bring *Life* to a wider audience. Associate Director of Marketing Debbie Clare, the regional specialists, regional managers, and experienced sales force are effective ambassadors and skillful transmitters of the features and unique strengths of our book. We depend on their expertise and energy to keep us in touch with how *Life* is perceived by its users. Thanks also to the Freeman media group for eBook and BioPortal production.

Finally, we thank our friend Andy Sinauer. Like ours, his name is on the cover of the book, and he truly cares deeply about what goes into it.

DAVID SADAVA

DAVID HILLIS

CRAIG HELLER

MAY BERENBAUM

# Reviewers for the Tenth Edition

### Between Edition Reviewers

Shivanthi Anandan, Drexel University
Brian Bagatto, The University of Akron
Mary Bisson, University at Buffalo, The State University of New York
Meredith Blackwell, Louisiana State University
Randy Brooks, Florida Atlantic University
Heather Caldwell, Kent State University
Jeffrey Carrier, Albion College
David Champlin, University of Southern Maine
Wesley Colgan, Pikes Peak Community College
Emma Creaser, Unity College
Karen Curto, University of Pittsburgh
John Dennehy, Queens College, The City University of New York
Rajinder Dhindsa, McGill University
James A. Doyle, University of California, Davis
Scott Edwards, Harvard University
David Eldridge, Baylor University
Joanne Ellzey, The University of Texas at El Paso
Douglas Gayou, University of Missouri
Stephen Gehnrich, Salisbury University
Arundhati Ghosh, University of Pittsburgh
Nathalia Glickman Holtzman, Queens College, The City University of New York
Elizabeth Good, University of Illinois at Urbana-Champaign
Harry Greene, Cornell University
Alice Heicklen, Columbia University
Albert Herrera, University of Southern California
David Hibbett, Clark University
Mark Holbrook, University of Iowa
Craig Jordan, The University of Texas at San Antonio
Walter Judd, University of Florida
John M. Labavitch, University of California, Davis
Nathan H. Lents, John Jay College of Criminal Justice, The City University of New York
Barry Logan, Bowdoin College
Barbara Lom, Davidson College
David Low, University of California, Davis
Janet Loxterman, Idaho State University
Sharon Lynn, The College of Wooster
Julin Maloof, University of California, Davis
Richard McCarty, Johns Hopkins University
Sheila McCormick, University of California, Berkeley
Marcie Moehnke, Baylor University
Roberta Moldow, Seton Hall University
Tsafrir Mor, Arizona State University
Alexander Motten, Duke University
Barbara Musolf, Clayton State University
Stuart Newfeld, Arizona State University
Bruce Ostrow, Grand Valley State University
Laura K. Palmer, The Pennsylvania State University, Altoona
Robert Pennock, Michigan State University
Kamini Persaud, University of Toronto, Scarborough
Roger Persell, Hunter College, The City University of New York
Matthew Rand, Carleton College
Susan Richardson, Florida Atlantic University
Brian C. Ring, Valdosta State University
Jay Rosenheim, University of California, Davis
Ben Rowley, University of Central Arkansas
Ann Rushing, Baylor University
Mikal Saltveit, University of California, Davis
Joel Schildbach, Johns Hopkins University
Christopher J. Schneider, Boston University
Paul Schulte, University of Nevada, Las Vegas
Leah Sheridan, University of Northern Colorado
Gary Shin, University of California, Los Angeles
Mitchell Singer, University of California, Davis
William Taylor, The University of Toledo
Sharon Thoma, University of Wisconsin, Madison
James F. A. Traniello, Boston University
Terry Trier, Grand Valley State University
Sara Via, University of Maryland
Curt Walker, Dixie State College
Fred Wasserman, Boston University
Alexander J. Werth, Hampden-Sydney College
Elizabeth Willott, University of Arizona

### Accuracy Reviewers

Rebecca Rashid Achterman, Western Washington University
Maria Ambrosetti, Emory University
Miriam Ashley-Ross, Wake Forest University
Felicitas Avendaño, Grand View University
David Bailey, St. Norbert College
Chhandak Basu, California State University, Northridge
Jim Bednarz, Arkansas State University
Charlie Garnett Benson, Georgia State University
Katherine Boss-Williams, Emory University
Ben Brammell, Asbury University

Christopher I. Brandon, Jr., Georgia Gwinnett College
Carolyn J. W. Bunde, Idaho State University
Darlene Campbell, Cornell University
Jeffrey Carmichael, University of North Dakota
David J. Carroll, Florida Institute of Technology
Ethan Carver, The University of Tennessee at Chattanooga
Peter Chabora, Queens College, The City University of New York
Heather Cook, Wagner College
Hsini Lin Cox, The University of Texas at El Paso
Douglas Darnowski, Indiana University Southeast
Stephen Devoto, Wesleyan University
Rajinder Dhindsa, McGill University
Jesse Dillon, California State University, Long Beach
James A. Doyle, University of California, Davis
Devin Drown, Indiana University
Richard E. Duhrkopf, Baylor University
Weston Dulaney, Nashville State Community College
David Eldridge, Baylor University
Kenneth Filchak, University of Notre Dame
Kerry Finlay, University of Regina
Kevin Folta, University of Florida
Douglas Gayou, University of Missouri
David T. Glover, Food and Drug Administration
Russ Goddard, Valdosta State University
Elizabeth Godrick, Boston University
Leslie Goertzen, Auburn University
Elizabeth Good, University of Illinois at Urbana-Champaign
Ethan Graf, Amherst College
Eileen Gregory, Rollins College
Julie C. Hagelin, University of Alaska, Fairbanks
Nathalia Glickman Holtzman, Queens College, The City University of New York
Dianne Jennings, Virginia Commonwealth University
Jamie Jensen, Bringham Young University
Glennis E. Julian
Erin Keen-Rhinehart, Susquehanna University
Henrik Kibak, California State University, Monterey Bay
Brandi Brandon Knight, Emory University
Daniel Kueh, Emory University
John G. Latto, University of California, Santa Barbara
Kristen Lennon, Frostburg State University
David Low, University of California, Santa Barbara
Jose-Luis Machado, Swarthmore College
Jay Mager, Ohio Northern University
Stevan Marcus, University of Alabama
Nilo Marin, Broward College
Marlee Marsh, Columbia College South Carolina
Erin Martin, University of South Florida, Sarasota-Manatee
Brad Mehrtens, University of Illinois at Urbana-Champaign
Michael Meighan, University of California, Berkeley
Tsafrir Mor, Arizona State University
Roderick Morgan, Grand Valley State University
Jacalyn Newman, University of Pittsburgh
Alexey Nikitin, Grand Valley State University
Zia Nisani, Antelope Valley College
Laura K. Palmer, The Pennsylvania State University, Altoona
Nancy Pencoe, State University of West Georgia
David P. Puthoff, Frostburg State University
Brett Riddle, University of Nevada, Las Vegas
Leslie Riley, Ohio Northern University
Brian C. Ring, Valdosta State University
Heather Roffey, McGill University
Lori Rose, Hill College
Naomi Rowland, Western Kentucky University
Beth Rueschhoff, Indiana University Southeast
Ann Rushing, Baylor University
Illya Ruvinsky, University of Chicago
Paul Schulte, University of Nevada, Las Vegas
Susan Sharbaugh, University of Alaska, Fairbanks
Jonathan Shenker, Florida Institute of Technology
Gary Shin, California State University, Long Beach
Ken Spitze, University of West Georgia
Bruce Stallsmith, The University of Alabama in Huntsville
Robert M. Steven, The University of Toledo
Zuzana Swigonova, University of Pittsburgh
Rebecca Symula, The University of Mississippi
Mark Taylor, Baylor University
Mark Thogerson, Grand Valley State University
Elethia Tillman, Spelman College
Terry Trier, Grand Valley State University
Michael Troyan, The Pennsylvania State University, University Park
Sebastian Velez, Worcester State University
Sheela Vemu, Northern Illinois University
Andrea Ward, Adelphi University
Katherine Warpeha, University of Illinois at Chicago
Fred Wasserman, Boston University
Michelle Wien, Bryn Mawr College
Robert Wisotzkey, California State University, East Bay
Greg Wray, Duke University
Joanna Wysocka-Diller, Auburn University
Catherine Young, Ohio Northern University
Heping Zhou, Seton Hall University

## Assessment Reviewers

Maria Ambrosetti, Georgia State University
Cecile Andraos-Selim, Hampton University
Felicitas Avendaño, Grand View University
David Bailey, St. Norbert College
Jim Bednarz, Arkansas State University
Charlie Garnett Benson, Georgia State University
Katherine Boss-Williams, Emory University
Ben Brammell, Asbury University
Christopher I. Brandon, Jr., Georgia Gwinnett College
Brandi Brandon Knight, Emory University

Ethan Carver, The University of Tennessee, Chattanooga
Heather Cook, Wagner College
Hsini Lin Cox, The University of Texas at El Paso
Douglas Darnowski, Indiana University Southeast
Jesse Dillon, California State University, Long Beach
Devin Drown, Indiana University
Richard E. Duhrkopf, Baylor University
Weston Dulaney, Nashville State Community College
Kenneth Filchak, University of Notre Dame
Elizabeth Godrick, Boston University
Elizabeth Good, University of Illinois at Urbana-Champaign
Susan Hengeveld, Indiana University Bloomington
Nathalia Glickman Holtzman, Queens College, The City College of New York
Glennis E. Julian
Erin Keen-Rhinehart, Susquehanna University
Stephen Kilpatrick, University of Pittsburgh
Daniel Kueh, Emory University
Stevan Marcus, University of Alabama
Nilo Marin, Broward College
Marlee Marsh, Columbia College
Erin Martin, University of South Florida, Sarasota-Manatee
Brad Mehrtens, University of Illinois at Urbana-Champaign
Darlene Mitrano, Christopher Newport University
Anthony Moss, Auburn University
Jacalyn Newman, University of Pittsburgh
Alexey Nikitin, Grand Valley State University
Zia Nisani, Antelope Valley College
Sabiha Rahman, University of Ottawa
Nancy Rice, Western Kentucky University
Brian C. Ring, Valdosta State University
Naomi Rowland, Western Kentucky University
Jonathan Shenker, Florida Institute of Technology
Gary Shin, California State University, Long Beach
Jacob Shreckengost, Emory University
Michael Smith, Western Kentucky University
Ken Spitze, University of West Georgia
Bruce Stallsmith, The University of Alabama in Huntsville
Zuzana Swigonova, University of Pittsburgh
William Taylor, The University of Toledo
Mark Thogerson, Grand Valley State University
Elethia Tillman, Spelman College
Michael Troyan, The Pennsylvania State University
Ximena Valderrama, Ramapo College of New Jersey
Sheela Vemu, Northern Illinois University
Suzanne Wakim, Butte College
Katherine Warpeha, University of Illinois at Chicago
Fred Wasserman, Boston University
Michelle Wien, Bryn Mawr College
Robert Wisotzkey, California State University, East Bay
Heping Zhou, Seton Hall University

# *LIFE's* Media and Supplements

yourBioPortal.com

BioPortal is the online gateway to all of *Life*'s digital resources, including the fully interactive eBook, a wide range of student and instructor media resources, and powerful assessment tools. BioPortal includes the following features and resources:

## *Life*, Tenth Edition eBook
(eBook also available stand-alone)

- Complete online version of the textbook
- Integration of all Media Clips, Activities, Animated Tutorials, and other media resources
- In-text links to all glossary entries, with audio pronunciations
- A flexible notes feature and easy text highlighting
- Searchable glossary and index
- Full-text search

### Additional eBook features for instructors:

- *Content Customization*: Instructors can easily hide chapters or sections that they don't cover in their course, re-arrange the order of chapters and sections, and add their own content directly into the eBook.
- *Instructor Notes*: Instructors can annotate the eBook with their own notes and content on any page. Instructor notes can include text, Web links, images, links to BioPortal resources, uploaded documents, and more.

## LearningCurve

New for the Tenth Edition, LearningCurve is a powerful adaptive quizzing system with a game-like format that engages students. Rather than simply answering a fixed set of questions, students answer dynamically-selected questions to progress toward a target level of understanding. At any point, students can view a report of how well they are performing in each topic area (with links to eBook sections and media resources), to help them focus on problem areas.

## Student BioPortal Resources

**DIAGNOSTIC QUIZZING.** The pre-built diagnostic quizzes assesses student understanding of each section of each chapter, and generates a Personalized Study Plan to effectively focus student study time. The plan includes links to specific textbook sections, animated tutorials, and activities.

**INTERACTIVE SUMMARIES.** For each chapter, these dynamic summaries combine a review of important concepts with links to all of the key figures, Activities, and Animated Tutorials.

**ANIMATED TUTORIALS.** In-depth tutorials that present complex topics in a clear, easy-to-follow format that combines a detailed animation or simulation with an introduction, conclusion, and brief quiz.

**MEDIA CLIPS.** New for the Tenth Edition, these short, engaging video clips depict fascinating examples of some of the many organisms, processes, and phenomena discussed in the textbook.

**ACTIVITIES.** A range of interactive activities that help students learn and review key facts and concepts through labeling diagrams, identifying steps in processes, and matching concepts.

**LECTURE NOTEBOOK.** New for the Tenth Edition, the Lecture Notebook is included online in BioPortal. The Notebook includes all of the textbook's figures and tables, with space for note-taking, and is available as downloadable PDF files.

**BIONEWS FROM SCIENTIFIC AMERICAN.** BioNews makes it easy for instructors to bring the dynamic nature of the biological sciences and up-to-the minute currency into their course, via an automatically updated news feed.

**BIONAVIGATOR.** A unique visual way to explore all of the Animated Tutorials and Activities across the various levels of biological inquiry—from the global scale down to the molecular scale.

**WORKING WITH DATA.** Online versions of the Working with Data exercises that are included in the textbook.

**FLASHCARDS AND KEY TERMS.** The Flashcards and Key Terms provide an ideal way for students to learn and review the extensive terminology of introductory biology, featuring a review mode and a quiz mode.

**INVESTIGATING LIFE LINKS.** For each Investigating Life figure in the textbook, BioPortal includes an overview of the experiment featured in the figure with links to the original paper(s), related research or applications that followed, and additional information related to the experiment.

**GLOSSARY.** The full glossary, with audio pronunciations for all terms.

**TREE OF LIFE.** An interactive version of the Tree of Life from Appendix A. The online Tree links to a wealth of information on each group listed.

**MATH FOR LIFE.** A collection of mathematical shortcuts and references to help students with the quantitative skills they need in the biology laboratory.

**SURVIVAL SKILLS.** A guide to more effective study habits, including time management, note-taking, effective highlighting, and exam preparation.

## Instructor BioPortal Resources

### Assessment

- LearningCurve and Diagnostic Quizzing reports provide instructors with a wealth of information on student comprehension, by textbook section, along with targeted lecture resources for those areas requiring the most attention.
- Comprehensive question banks include questions from the Test Bank, LearningCurve, Diagnostic Quizzes, Study Guide, and textbook Chapter Review.
- Question filtering allows instructors to select questions based on Bloom's category and/or textbook section, in order to easily select the desired mix of question types.
- Easy-to-use assessment tools allow instructors to create quizzes and many other types of assignments using any combination of publisher-provided questions and those created by the instructor.

### Media Resources

*(see Instructor's Media Library below for details)*

- Videos
- PowerPoint Presentations (Figures & Tables, Lecture, Editable Labels, Layered Art)
- Supplemental Photos
- Active Learning Exercises
- Instructor's Manual
- Lecture Notes
- Answers to Working with Data Exercises
- Course management features
- Complete course customization capabilities
- Custom resources/document posting
- Robust gradebook
- Communication Tools: Announcements, Calendar, Course Email, Discussion Boards

## Student Supplements

### *Life*, Tenth Edition Study Guide

*(Paper, ISBN 978-1-4641-2365-8)*

The *Life* Study Guide offers a variety of study and review resources to accompany each chapter of the textbook. The opening Big Picture section gives students a concise overview of the main concepts covered in the chapter. The Study Strategies section points out common problem areas that students may find more challenging, and suggests strategies for learning the material most effectively. The Key Concept Review section combines a detailed review of each section with questions that help students synthesize and apply what they have learned, including diagram questions, short-answer questions, and more open-ended questions. Each chapter concludes with a Test Yourself section that allows students to test their comprehension. All questions include answers, explanations, and references to textbook sections.

### *Life* Flashcards App

Available for iPhone/iPad and Android, the *Life* Flashcards App is a great way for students to learn and review all the key terminology from the textbook, whenever and wherever they want to study, in an intuitive flashcard interface. Available in the iTunes App Store and Google Play.

### *CatchUp Math & Stats*

*Michael Harris, Gordon Taylor, and Jacquelyn Taylor*
*(ISBN 978-1-4292-0557-3)*

Presented in brief, accessible units, this primer will help students quickly brush up on the quantitative skills they need to succeed in biology.

### *Student Handbook for Writing in Biology,* Third Edition

*Karen Knisely (ISBN 978-1-4292-3491-7)*

This book provides practical advice to students who are learning to write according to the conventions in biology, using the standards of journal publication as a model.

### *Bioethics and the New Embryology: Springboards for Debate*

*Scott F. Gilbert, Anna Tyler, and Emily Zackin*
*(ISBN 978-0-7167-7345-0)*

Our ability to alter the course of human development ranks among the most significant changes in modern science and has brought embryology into the public domain. The question that must be asked is: Even if we can do such things, should we?

### *BioStats Basics: A Student Handbook*

*James L. Gould and Grant F. Gould (ISBN 978-0-7167-3416-1)*

Engaging and informal, *BioStats Basics* provides introductory-level biology students with a practical, accessible introduction to statistical research.

### *Inquiry Biology: A Laboratory Manual,* Volumes 1 and 2

*Mary Tyler, Ryan W. Cowan, and Jennifer L. Lockhart (Volume 1 ISBN 978-1-4292-9288-7; Volume 2 ISBN 978-1-4292-9289-4)*

This introductory biology laboratory manual is inquiry-based—instructing in the process of science by allowing students to ask their own questions, gather background information, formulate hypotheses, design and carry out experiments, collect and analyze data, and formulate conclusions.

### Hayden-McNeil Life Sciences Lab Notebook

*(ISBN 978-1-4292-3055-1)*

This carbonless laboratory notebook is of the highest quality and durability, allowing students to hand in originals or copies, not entire composition books. Contains Hayden-McNeil's unique white paper carbonless copies and biology-specific reference materials.

## Instructor Media & Supplements

### Instructor's Media Library

*(Available both online via BioPortal and on disc; disc version ISBN 978-1-4641-2364-1)*

The *Life,* Tenth Edition Instructor's Media Library includes a wide range of electronic resources to help instructors plan their course, present engaging lectures, and effectively assess their students. The Media Library includes the following resources:

**TEXTBOOK FIGURES AND TABLES.** Every figure and table from the textbook (including all photos and all un-numbered figures) is provided in both JPEG (high- and low-resolution) and PDF formats, in multiple versions.

**UNLABELED FIGURES.** Every figure is provided in an unlabeled format, useful for student quizzing and custom presentations.

**SUPPLEMENTAL PHOTOS.** The supplemental photograph collection contains over 1,500 photographs, giving instructors a wealth of additional imagery to draw upon.

**ANIMATIONS.** An extensive collection of detailed animations, all built specifically for Life, and viewable in either narrated or step-through mode.

**VIDEOS.** Featuring many new segments for the Tenth Edition, the wide-ranging collection of video segments help demonstrate the complexity and beauty of life.

**POWERPOINT RESOURCES.** For each chapter of the textbook, many different PowerPoint presentations are available, providing instructors the flexibility to build presentations in the manner that best suits their needs, including the following:

- Textbook Figures and Tables
- Lecture Presentation
- Figures with Editable Labels
- Layered Art Figures
- Supplemental Photos
- Videos
- Animations
- Active Learning Exercises

**INSTRUCTOR'S MANUAL, LECTURE NOTES,** and **TEST BANK** are available in Microsoft Word format for easy use in lecture and exam preparation.

**MEDIA GUIDE.** A PDF version of the Media Guide from the Instructor's Resource Kit, convenient for searching.

**ACTIVE LEARNING EXERCISES.** Set up for easy integration into lectures, each exercise poses a question or problem for the class to discuss or solve during lecture. Each also includes a multiple-choice element, for easy use with clicker systems.

**ANSWERS TO WORKING WITH DATA EXERCISES.** Complete answers to all of the Working with Data exercises.

### Instructor's Resource Kit

*(Binder, ISBN 978-1-4641-4131-7)*

The *Life,* Tenth Edition Instructor's Resource Kit includes a wealth of information to help instructors in the planning and teaching of their course. The Kit includes:

**INSTRUCTOR'S MANUAL**

- *Chapter Overview*: A brief, high-level synopsis of the chapter.
- *What's New*: A guide to the revisions, updates, and new content added to the Tenth Edition.
- *Key Concepts & Learning Objectives*: New for the Tenth Edition, this section includes the major learning goals for the chapter, a detailed set of key concepts, and specific learning objectives for each key concept.
- *Chapter Outline*: All of the chapter's section headings and sub-headings.
- *Key Terms*: All of the important terms introduced in the chapter.

**LECTURE NOTES.** Detailed lecture outlines for each chapter, including references to relevant figures and media resources.

**MEDIA GUIDE.** A visual guide to the extensive media resources available with Life, including all animations, activities, videos, and supplemental photos.

### Overhead Transparencies

*(ISBN 978-1-4641-4127-0)*

The set of overheads includes over 1,000 transparencies—including all of the four-color line art and all of the tables from the text—in two convenient binders. All figures have been formatted and color-enhanced for clear projection in a wide range of conditions. Labels and images have been resized for improved readability.

### Test File

*(Paper, ISBN 978-1-4292-5579-0)*

The *Life*, Tenth Edition Test File includes over 5,000 questions and has been revised and reviewed for both accuracy and effectiveness. All questions are referenced to specific textbook headings and categorized according to Bloom's taxonomy. This allows instructors to easily build quizzes and exams with the desired mix of content, coverage, and question types (factual, conceptual, analyzing/applying, etc.). Each chapter includes a wide range of multiple choice and fill-in-the-blank questions, in addition to diagram questions that involve the student in working with illustrations of structures, graphs, steps in processes, and more.

### Computerized Test Bank

*(CD, ISBN 978-1-4641-4128-7)*

The entire Test File, plus the Diagnostic Quizzes, Learning-Curve questions, Study Guide questions, and Textbook End-of Chapter Review questions are all included in Wimba's easy-to-use Diploma program (software included). Designed for both novice and advanced users, Diploma allows instructors to quickly and easily create or edit questions, create quizzes or exams with a "drag-and-drop" feature (using any combination of publisher-provided and instructor-added questions), publish to online courses, and print paper-based assessments.

### Figure Correlation Tool

An invaluable resource for instructors switching to *Life,* Tenth Edition from another textbook or from *Life*, Ninth Edition, this online tool provides correlations between all of the figures in *Life,* Tenth Edition and figures in other majors biology textbooks and *Life*, Ninth Edition.

### Course Management System Support

As a service for *Life* adopters using Blackboard, WebCT, ANGEL, or other course management systems, full electronic course packs are available.

Faculty Lounge for Majors Biology is the first publisher-provided website for the majors biology community that lets instructors freely communicate and share peer-reviewed lecture and teaching resources. The Faculty Lounge offers convenient access to peer-recommended and vetted resources, including the following categories: Images, News, Videos, Labs, Lecture Resources, and Educational Research. **majorsbio.facultylounge.whfreeman.com**

iclicker

Developed for educators by educators, iclicker is a hassle-free radio-frequency classroom response system that makes it easy for instructors to ask questions, record responses, take attendance, and direct students through lectures as active participants. For more information, visit **www.iclicker.com**.

LabPartner is a site designed to facilitate the creation of customized lab manuals. Its database contains a wide selection of experiments published by W. H. Freeman and Hayden-McNeil Publishing. Instructors can preview, choose, and re-order labs, interleave their own original experiments, add carbonless graph paper and a pocket folder, customize the cover both inside and out, and select a binding type. Manuals are printed on-demand. **www.whfreeman.com/labpartner**

The Scientific Teaching Book Series is a collection of practical guides, intended for all science, technology, engineering and mathematics (STEM) faculty who teach undergraduate and graduate students in these disciplines. The purpose of these books is to help faculty become more successful in all aspects of teaching and learning science, including classroom instruction, mentoring students, and professional development. Authored by well-known science educators, the Series provides concise descriptions of best practices and how to implement them in the classroom, the laboratory, or the department. For readers interested in the research results on which these best practices are based, the books also provide a gateway to the key educational literature.

### Scientific Teaching

*Jo Handelsman, Sarah Miller, and Christine Pfund (ISBN 978-1-4292-0188-9)*

### *Transformations: Approaches to College Science Teaching*

*Deborah Allen and Kimberly Tanner (ISBN 978-1-4292-5335-2)*

### *Entering Research: A Facilitator's Manual*

*Workshops for Students Beginning Research in Science*

*Janet L. Branchaw, Christine Pfund, and Raelyn Rediske (ISBN 978-1-429-25857-9)*

### *Discipline-Based Science Education Research: A Scientist's Guide*

*Stephanie Slater, Tim Slater, and Janelle M. Bailey (ISBN 978-1-4292-6586-7)*

### *Assessment in the College Classroom*

*Clarissa Dirks, Mary Pat Wenderoth, Michelle Withers (ISBN 978-1-4292-8197-3)*

# Contents

## PART ONE
## The Science of Life and Its Chemical Basis

### 1 Studying Life 1

**1.1 What Is Biology? 2**
- Life arose from non-life via chemical evolution 3
- Cellular structure evolved in the common ancestor of life 3
- Photosynthesis allows some organisms to capture energy from the sun 4
- Biological information is contained in a genetic language common to all organisms 5
- Populations of all living organisms evolve 6
- Biologists can trace the evolutionary tree of life 6
- Cellular specialization and differentiation underlie multicellular life 9
- Living organisms interact with one another 9
- Nutrients supply energy and are the basis of biosynthesis 10
- Living organisms must regulate their internal environment 10

**1.2 How Do Biologists Investigate Life? 11**
- Observing and quantifying are important skills 11
- Scientific methods combine observation, experimentation, and logic 11
- Good experiments have the potential to falsify hypotheses 12
- Statistical methods are essential scientific tools 13
- Discoveries in biology can be generalized 14
- Not all forms of inquiry are scientific 14

**1.3 Why Does Biology Matter? 15**
- Modern agriculture depends on biology 15

- Biology is the basis of medical practice 15
- Biology can inform public policy 16
- Biology is crucial for understanding ecosystems 17
- Biology helps us understand and appreciate biodiversity 17

### 2 Small Molecules and the Chemistry of Life 21

**2.1 How Does Atomic Structure Explain the Properties of Matter? 22**
- An element consists of only one kind of atom 22
- Each element has a unique number of protons 22
- The number of neutrons differs among isotopes 22
- The behavior of electrons determines chemical bonding and geometry 24

**2.2 How Do Atoms Bond to Form Molecules? 26**
- Covalent bonds consist of shared pairs of electrons 26
- Ionic attractions form by electrical attraction 28
- Hydrogen bonds may form within or between molecules with polar covalent bonds 30
- Hydrophobic interactions bring together nonpolar molecules 30
- van der Waals forces involve contacts between atoms 30

**2.3 How Do Atoms Change Partners in Chemical Reactions? 31**

**2.4 What Makes Water So Important for Life? 32**
- Water has a unique structure and special properties 32
- The reactions of life take place in aqueous solutions 33
- Aqueous solutions may be acidic or basic 34

### 3 Proteins, Carbohydrates, and Lipids 39

**3.1 What Kinds of Molecules Characterize Living Things? 40**
- Functional groups give specific properties to biological molecules 40
- Isomers have different arrangements of the same atoms 41
- The structures of macromolecules reflect their functions 41

Most macromolecules are formed by condensation and broken down by hydrolysis 42

**3.2 What Are the Chemical Structures and Functions of Proteins? 42**

Amino acids are the building blocks of proteins 43

Peptide linkages form the backbone of a protein 43

The primary structure of a protein is its amino acid sequence 45

The secondary structure of a protein requires hydrogen bonding 45

The tertiary structure of a protein is formed by bending and folding 46

The quaternary structure of a protein consists of subunits 48

Shape and surface chemistry contribute to protein function 48

Environmental conditions affect protein structure 50

Protein shapes can change 50

Molecular chaperones help shape proteins 51

**3.3 What Are the Chemical Structures and Functions of Carbohydrates? 51**

Monosaccharides are simple sugars 52

Glycosidic linkages bond monosaccharides 53

Polysaccharides store energy and provide structural materials 53

Chemically modified carbohydrates contain additional functional groups 55

**3.4 What Are the Chemical Structures and Functions of Lipids? 56**

Fats and oils are triglycerides 56

Phospholipids form biological membranes 57

Some lipids have roles in energy conversion, regulation, and protection 57

**4 Nucleic Acids and the Origin of Life 62**

**4.1 What Are the Chemical Structures and Functions of Nucleic Acids? 63**

Nucleotides are the building blocks of nucleic acids 63

Base pairing occurs in both DNA and RNA 63

DNA carries information and is expressed through RNA 65

The DNA base sequence reveals evolutionary relationships 66

Nucleotides have other important roles 66

**4.2 How and Where Did the Small Molecules of Life Originate? 67**

Experiments disproved the spontaneous generation of life 67

Life began in water 68

Life may have come from outside Earth 69

Prebiotic synthesis experiments model early Earth 69

**4.3 How Did the Large Molecules of Life Originate? 71**

Chemical evolution may have led to polymerization 71

RNA may have been the first biological catalyst 71

**4.4 How Did the First Cells Originate? 71**

Experiments explore the origin of cells 73

Some ancient cells left a fossil imprint 74

## PART TWO Cells

**5 Cells: The Working Units of Life 77**

**5.1 What Features Make Cells the Fundamental Units of Life? 78**

Cell size is limited by the surface area-to-volume ratio 78

Microscopes reveal the features of cells 79

The plasma membrane forms the outer surface of every cell 79

Cells are classified as either prokaryotic or eukaryotic 81

**5.2 What Features Characterize Prokaryotic Cells? 82**

Prokaryotic cells share certain features 82

Specialized features are found in some prokaryotes 83

**5.3 What Features Characterize Eukaryotic Cells? 84**

Compartmentalization is the key to eukaryotic cell function 84

Organelles can be studied by microscopy or isolated for chemical analysis 84

Ribosomes are factories for protein synthesis 84

The nucleus contains most of the generic information 85

The endomembrane system is a group of interrelated organelles 88

Some organelles transform energy 91

There are several other membrane-enclosed organelles 93

The cytoskeleton is important in cell structure and movement 94

Biologists can manipulate living systems to establish cause and effect 98

**5.4 What Are the Roles of Extracellular Structures? 99**

The plant cell wall is an extracellular structure 99

The extracellular matrix supports tissue functions in animals 100

**5.5 How Did Eukaryotic Cells Originate? 101**

Internal membranes and the nuclear envelope probably came from the plasma membrane 101

Some organelles arose by endosymbiosis 102

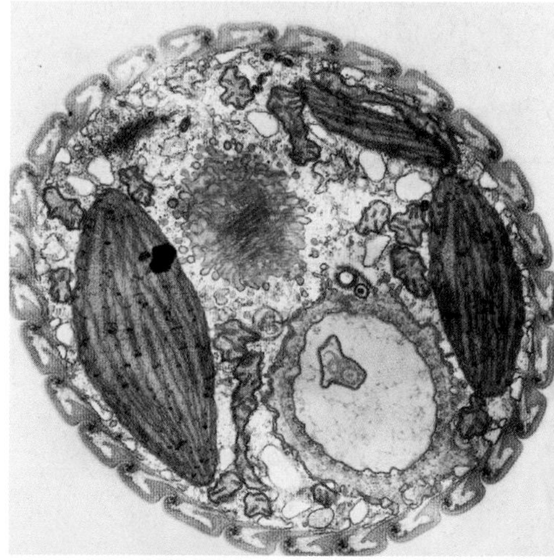

6 Cell Membranes 105

6.1 What Is the Structure of a Biological Membrane? 106
Lipids form the hydrophobic core of the membrane 106
Membrane proteins are asymmetrically distributed 107
Membranes are constantly changing 109
Plasma membrane carbohydrates are recognition sites 109

6.2 How Is the Plasma Membrane Involved in Cell Adhesion and Recognition? 110
Cell recognition and adhesion involve proteins and carbohydrates at the cell surface 111
Three types of cell junctions connect adjacent cells 111
Cell membranes adhere to the extracellular matrix 111

6.3 What Are the Passive Processes of Membrane Transport? 113
Diffusion is the process of random movement toward a state of equilibrium 113
Simple diffusion takes place through the phospholipid bilayer 114
Osmosis is the diffusion of water across membranes 114
Diffusion may be aided by channel proteins 115
Carrier proteins aid diffusion by binding substances 117

6.4 What are the Active Processes of Membrane Transport? 118
Active transport is directional 118


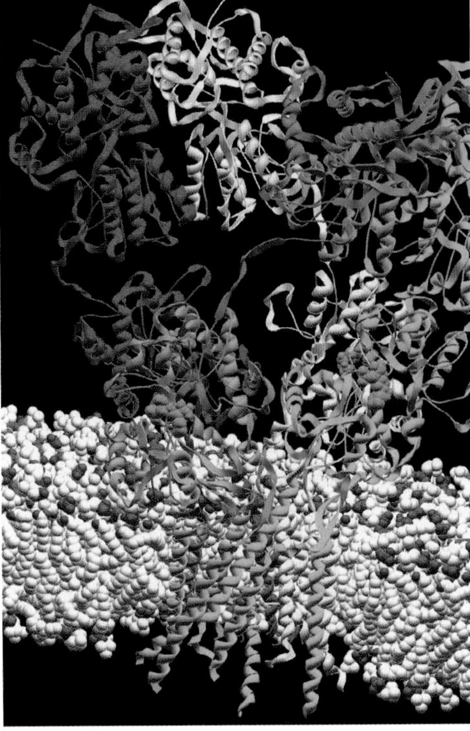


Different energy sources distinguish different active transport systems 118

6.5 How Do Large Molecules Enter and Leave a Cell? 120
Macromolecules and particles enter the cell by endocytosis 120
Receptor-mediated endocytosis is highly specific 121
Exocytosis moves materials out of the cell 122

7 Cell Communication and Multicellularity 125

7.1 What Are Signals, and How Do Cells Respond to Them? 126
Cells receive signals from the physical environment and from other cells 126
A signal transduction pathway involves a signal, a receptor, and responses 126

7.2 How Do Signal Receptors Initiate a Cellular Response? 127
Receptors that recognize chemical signals have specific binding sites 127
Receptors can be classified by location and function 128
Intracellular receptors are located in the cytoplasm or the nucleus 130

7.3 How Is the Response to a Signal Transduced through the Cell? 131
A protein kinase cascade amplifies a response to ligand binding 131
Second messengers can amplify signals between receptors and target molecules 132
Signal transduction is highly regulated 136

7.4 How Do Cells Change in Response to Signals? 137
Ion channels open in response to signals 137
Enzyme activities change in response to signals 138
Signals can initiate DNA transcription 139

7.5 How Do Cells in a Multicellular Organism Communicate Directly? 139
Animal cells communicate through gap junctions 139
Plant cells communicate through plasmodesmata 140
Modern organisms provide clues about the evolution of cell–cell interactions and multicellularity 140

PART THREE Cells and Energy

8 Energy, Enzymes, and Metabolism 144

8.1 What Physical Principles Underlie Biological Energy Transformations? 145
There are two basic types of energy 145
There are two basic types of metabolism 145
The first law of thermodynamics: Energy is neither created nor destroyed 146
The second law of thermodynamics: Disorder tends to increase 146
Chemical reactions release or consume energy 147
Chemical equilibrium and free energy are related 148

8.2 What Is the Role of ATP in Biochemical Energetics? 149
ATP hydrolysis releases energy 149

ATP couples exergonic and endergonic reactions 150

**8.3 What Are Enzymes? 151**

To speed up a reaction, an energy barrier must be overcome 151

Enzymes bind specific reactants at their active sites 152

Enzymes lower the energy barrier but do not affect equilibrium 153

**8.4 How Do Enzymes Work? 154**

Enzymes can orient substrates 154

Enzymes can induce strain in the substrate 154

Enzymes can temporarily add chemical groups to substrates 154

Molecular structure determines enzyme function 155

Some enzymes require other molecules in order to function 155

The substrate concentration affects the reaction rate 156

**8.5 How Are Enzyme Activities Regulated? 156**

Enzymes can be regulated by inhibitors 157

Allosteric enzymes are controlled via changes in shape 159

Allosteric effects regulate many metabolic pathways 160

Many enzymes are regulated through reversible phosphorylation 161

Enzymes are affected by their environment 161

**9 Pathways That Harvest Chemical Energy 165**

**9.1 How Does Glucose Oxidation Release Chemical Energy? 166**

Cells trap free energy while metabolizing glucose 166

Redox reactions transfer electrons and energy 167

The coenzyme $NAD^+$ is a key electron carrier in redox reactions 167

An overview: Harvesting energy from glucose 168

**9.2 What Are the Aerobic Pathways of Glucose Catabolism? 169**

In glycolysis, glucose is partially oxidized and some energy is released 169

Pyruvate oxidation links glycolysis and the citric acid cycle 170

The citric acid cycle completes the oxidation of glucose to $CO_2$ 170

Pyruvate oxidation and the citric acid cycle are regulated by the concentrations of starting materials 171

**9.3 How Does Oxidative Phosphorylation Form ATP? 171**

The respiratory chain transfers electrons and protons, and releases energy 172

Proton diffusion is coupled to ATP synthesis 173

Some microorganisms use non-$O_2$ electron acceptors 176

**9.4 How Is Energy Harvested from Glucose in the Absence of Oxygen? 177**

Cellular respiration yields much more energy than fermentation 178

The yield of ATP is reduced by the impermeability of mitochondria to NADH 178

**9.5 How Are Metabolic Pathways Interrelated and Regulated? 179**

Catabolism and anabolism are linked 179

Catabolism and anabolism are integrated 180

Metabolic pathways are regulated systems 181

**10 Photosynthesis: Energy from Sunlight 185**

**10.1 What Is Photosynthesis 186**

Experiments with isotopes show that $O_2$ comes from $H_2O$ in oxygenic photosynthesis 186

Photosynthesis involves two pathways 188

**10.2 How Does Photosynthesis Convert Light Energy into Chemical Energy? 188**

Light energy is absorbed by chlorophyll and other pigments 188

Light absorption results in photochemical change 190

Reduction leads to ATP and NADPH formation 191

Chemiosmosis is the source of the ATP produced in photophosphorylation 192

**10.3 How Is Chemical Energy Used to Synthesize Carbohydrates? 193**

Radioisotope labeling experiments revealed the steps of the Calvin cycle 193

The Calvin cycle is made up of three processes 194

Light stimulates the Calvin cycle 196

**10.4 How Have Plants Adapted Photosynthesis to Environmental Conditions? 197**

Rubisco catalyzes the reaction of RuBP with $O_2$ or $CO_2$ 197

$C_3$ plants undergo photorespiration but $C_4$ plants do not 198

CAM plants also use PEP carboxylase 200

**10.5 How Does Photosynthesis Interact with Other Pathways? 200**

# PART FOUR Genes and Heredity

## 11 The Cell Cycle and Cell Division 205


**11.1 How Do Prokaryotic and Eukaryotic Cells Divide? 206**

Prokaryotes divide by binary fission 206

Eukaryotic cells divide by mitosis or meiosis followed by cytokinesis 207

**11.2 How Is Eukaryotic Cell Division Controlled? 208**

Specific internal signals trigger events in the cell cycle 208

Growth factors can stimulate cells to divide 211

**11.3 What Happens during Mitosis? 211**

Prior to mitosis, eukaryotic DNA is packed into very compact chromosomes 211

Overview: Mitosis segregates copies of genetic information 212

The centrosomes determine the plane of cell division 212

The spindle begins to form during prophase 213

Chromosome separation and movement are highly organized 214

Cytokinesis is the division of the cytoplasm 216

**11.4 What Role Does Cell Division Play in a Sexual Life Cycle? 217**

Asexual reproduction by mitosis results in genetic constancy 217

Sexual reproduction by meiosis results in genetic diversity 218

**11.5 What Happens during Meiosis? 219**

Meiotic division reduces the chromosome number 219

Chromatid exchanges during meiosis I generate genetic diversity 219

During meiosis homologous chromosomes separate by independent assortment 220

Meiotic errors lead to abnormal chromosome structures and numbers 222

The number, shapes, and sizes of the metaphase chromosomes constitute the karyotype 224

Polyploids have more than two complete sets of chromosomes 224

**11.6 In a Living Organism, How Do Cells Die? 225**

**11.7 How Does Unregulated Cell Division Lead to Cancer? 227**

Cancer cells differ from normal cells 227

Cancer cells lose control over the cell cycle and apoptosis 228

Cancer treatments target the cell cycle 228


## 12 Inheritance, Genes, and Chromosomes 232


**12.1 What Are the Mendelian Laws of Inheritance? 233**

Mendel used the scientific method to test his hypotheses 233

Mendel's first experiments involved monohybrid crosses 234

Mendel's first law states that the two copies of a gene segregate 236

Mendel verified his hypotheses by performing test crosses 237

Mendel's second law states that copies of different genes assort independently 237

Probability can be used to predict inheritance 239

Mendel's laws can be observed in human pedigrees 240

**12.2 How Do Alleles Interact? 241**

New alleles arise by mutation 241

Many genes have multiple alleles 242

Dominance is not always complete 242

In codominance, both alleles at a locus are expressed 243

Some alleles have multiple phenotypic effects 243


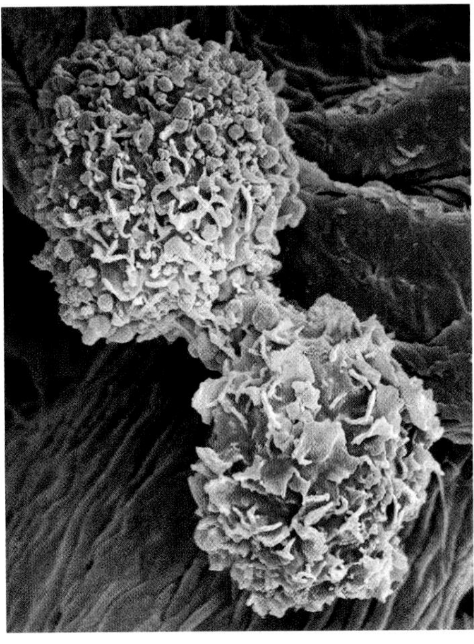


**12.3 How Do Genes Interact? 244**

Hybrid vigor results from new gene combinations and interactions 244

The environment affects gene action 245

Most complex phenotypes are determined by multiple genes and the environment 246

**12.4 What Is the Relationship between Genes and Chromosomes? 247**

Genes on the same chromosome are linked 247

Genes can be exchanged between chromatids and mapped 247

Linkage is revealed by studies of the sex chromosomes 249

**12.5 What Are the Effects of Genes Outside the Nucleus? 252**

**12.6 How Do Prokaryotes Transmit Genes? 253**

Bacteria exchange genes by conjugation 253

Bacterial conjugation is controlled by plasmids 254


## 13 DNA and Its Role in Heredity 259


**13.1 What Is the Evidence that the Gene Is DNA? 260**

DNA from one type of bacterium genetically transforms another type 260

Viral infection experiments confirmed that DNA is the genetic material 261

Eukaryotic cells can also be genetically transformed by DNA 263

**13.2 What Is the Structure of DNA? 264**

Watson and Crick used modeling to deduce the structure of DNA 264

Four key features define DNA structure 265

The double-helical structure of DNA is essential to its function 266

**13.3 How Is DNA Replicated? 267**

Three modes of DNA replication appeared possible 267

An elegant experiment demonstrated that DNA replication is semiconservative 268

There are two steps in DNA replication 268

DNA polymerases add nucleotides to the growing chain 269

Many other proteins assist with DNA polymerization 272

The two DNA strands grow differently at the replication fork 272

Telomeres are not fully replicated and are prone to repair 275

**13.4 How Are Errors in DNA Repaired? 276**

**13.5 How Does the Polymerase Chain Reaction Amplify DNA? 277**

The polymerase chain reaction makes multiple copies of DNA sequences 277

## 14 From DNA to Protein: Gene Expression 281

**14.1 What Is the Evidence that Genes Code for Proteins? 282**

Observations in humans led to the proposal that genes determine enzymes 282

Experiments on bread mold established that genes determine enzymes 282

One gene determines one polypeptide 283

**14.2 How Does Information Flow from Genes to Proteins? 284**

Three types of RNA have roles in the information flow from DNA to protein 285

In some cases, RNA determines the sequence of DNA 285

**14.3 How Is the Information Content in DNA Transcribed to Produce RNA? 286**

RNA polymerases share common features 286

Transcription occurs in three steps 286

The information for protein synthesis lies in the genetic code 288

**14.4 How Is Eukaryotic DNA Transcribed and the RNA Processed? 290**

Many eukaryotic genes are interrupted by noncoding sequences 290

Eukaryotic gene transcripts are processed before translation 291

**14.5 How Is RNA Translated into Proteins? 293**

Transfer RNAs carry specific amino acids and bind to specific codons 293

Each tRNA is specifically attached to an amino acid 294

The ribosome is the workbench for translation 294

Translation takes place in three steps 295

Polysome formation increases the rate of protein synthesis 297

**14.6 What Happens to Polypeptides after Translation? 298**

Signal sequences in proteins direct them to their cellular destinations 298

Many proteins are modified after translation 300

## 15 Gene Mutation and Molecular Medicine 304

**15.1 What Are Mutations? 305**

Mutations have different phenotypic effects 305

Point mutations are changes in single nucleotides 306

Chromosomal mutations are extensive changes in the genetic material 307

Retroviruses and transposons can cause loss of function mutations or duplications 308

Mutations can be spontaneous or induced 308

Mutagens can be natural or artificial 310

Some base pairs are more vulnerable than others to mutation 310

Mutations have both benefits and costs 310

**15.2 What Kinds of Mutations Lead to Genetic Diseases? 311**

Genetic mutations may make proteins dysfunctional 311

Disease-causing mutations may involve any number of base pairs 312

Expanding triplet repeats demonstrate the fragility of some human genes 313

Cancer often involves somatic mutations 314

Most diseases are caused by multiple genes and environment 314

15.3 How Are Mutations Detected and Analyzed? 315

Restriction enzymes cleave DNA at specific sequences 315

Gel electrophoresis separates DNA fragments 316

DNA fingerprinting combines PCR with restriction analysis and electrophoresis 317

Reverse genetics can be used to identify mutations that lead to disease 318

Genetic markers can be used to find disease-causing genes 318

The DNA barcode project aims to identify all organisms on Earth 319

15.4 How Is Genetic Screening Used to Detect Diseases? 320

Screening for disease phenotypes involves analysis of proteins and other chemicals 320

DNA testing is the most accurate way to detect abnormal genes 320

Allele-specific oligonucleotide hybridization can detect mutations 321

15.5 How Are Genetic Diseases Treated? 322

Genetic diseases can be treated by modifying the phenotype 322

Gene therapy offers the hope of specific treatments 323

16 Regulation of Gene Expression 328

16.1 How Is Gene Expression Regulated in Prokaryotes? 329

Regulating gene transcription conserves energy 329

Operons are units of transcriptional regulation in prokaryotes 330

Operator–repressor interactions control transcription in the *lac* and *trp* operons 330

Protein synthesis can be controlled by increasing promoter efficiency 332

RNA polymerases can be directed to particular classes of promoters 332

16.2 How Is Eukaryotic Gene Transcription Regulated? 333

General transcription factors act at eukaryotic promoters 333

Specific proteins can recognize and bind to DNA sequences and regulate transcription 335

Specific protein–DNA interactions underlie binding 335

The expression of transcription factors underlies cell differentiation 336

The expression of sets of genes can be coordinately regulated by transcription factors 336

16.3 How Do Viruses Regulate Their Gene Expression? 339

Many bacteriophages undergo a lytic cycle 339

Some bacteriophages can undergo a lysogenic cycle 340

Eukaryotic viruses can have complex life cycles 341

HIV gene regulation occurs at the level of transcription elongation 341

16.4 How Do Epigenetic Changes Regulate Gene Expression? 343

DNA methylation occurs at promoters and silences transcription 343

Histone protein modifications affect transcription 344

Epigenetic changes can be induced by the environment 344

DNA methylation can result in genomic imprinting 344

Global chromosome changes involve DNA methylation 345

16.5 How Is Eukaryotic Gene Expression Regulated after Transcription? 346

Different mRNAs can be made from the same gene by alternative splicing 346

Small RNAs are important regulators of gene expression 347

Translation of mRNA can be regulated by proteins and riboswitches 348

# PART FIVE Genomes

17 Genomes 352

17.1 How Are Genomes Sequenced? 353

New methods have been developed to rapidly sequence DNA 353

Genome sequences yield several kinds of information 355

17.2 What Have We Learned from Sequencing Prokaryotic Genomes? 356

Prokaryotic genomes are compact 356

The sequencing of prokaryotic and viral genomes has many potential benefits 357

Metagenomics allows us to describe new organisms and ecosystems 357

Some sequences of DNA can move about the genome 358

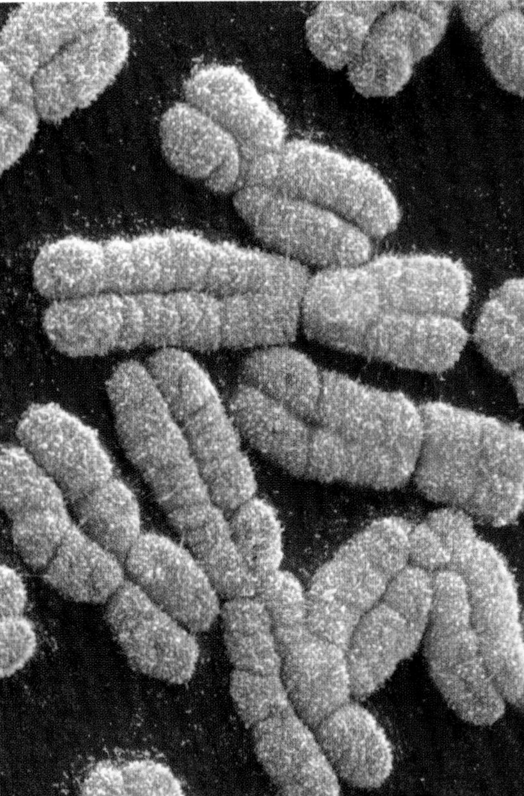

Will defining the genes required for cellular life lead to artificial life? 359

**17.3 What Have We Learned from Sequencing Eukaryotic Genomes? 361**

Model organisms reveal many characteristics of eukaryotic genomes 361

Eukaryotes have gene families 363

Eukaryotic genomes contain many repetitive sequences 364

**17.4 What Are the Characteristics of the Human Genome? 366**

The human genome sequence held some surprises 366

Comparative genomics reveals the evolution of the human genome 366

Human genomics has potential benefits in medicine 367

**17.5 What Do the New Disciplines of Proteomics and Metabolomics Reveal? 369**

The proteome is more complex than the genome 369

Metabolomics is the study of chemical phenotype 370

**18 Recombinant DNA and Biotechnology 373**

**18.1 What Is Recombinant DNA? 374**

**18.2 How Are New Genes Inserted into Cells? 375**

Genes can be inserted into prokaryotic or eukaryotic cells 376

A variety of methods are used to insert recombinant DNA into host cells 376

Reporter genes help select or identify host cells containing recombinant DNA 377

**18.3 What Sources of DNA Are Used in Cloning? 379**

Libraries provide collections of DNA fragments 379

cDNA is made from mRNA transcripts 379

Synthetic DNA can be made by PCR or by organic chemistry 380

**18.4 What Other Tools Are Used to Study DNA Function? 380**

Genes can be expressed in different biological systems 380

DNA mutations can be created in the laboratory 381

Genes can be inactivated by homologous recombination 381

Complementary RNA can prevent the expression of specific genes 382

DNA microarrays reveal RNA expression patterns 382

**18.5 What Is Biotechnology? 383**

Expression vectors can turn cells into protein factories 384

**18.6 How Is Biotechnology Changing Medicine and Agriculture? 384**

Medically useful proteins can be made using biotechnology 384

DNA manipulation is changing agriculture 386

There is public concern about biotechnology 388

**19 Differential Gene Expression in Development 392**

**19.1 What Are the Processes of Development? 393**

Development involves distinct but overlapping processes 393

Cell fates become progressively more restricted during development 394

**19.2 How Is Cell Fate Determined? 395**

Cytoplasmic segregation can determine polarity and cell fate 395

Inducers passing from one cell to another can determine cell fates 395

**19.3 What Is the Role of Gene Expression in Development? 397**

Cell fate determination involves signal transduction pathways that lead to differential gene expression 397

Differential gene transcription is a hallmark of cell differentiation 398

**19.4 How Does Gene Expression Determine Pattern Formation? 399**

Multiple proteins interact to determine developmental programmed cell death 399

Plants have organ identity genes 400

Morphogen gradients provide positional information 401

A cascade of transcription factors establishes body segmentation in the fruit fly 401

**19.5 Is Cell Differentiation Reversible? 405**

Plant cells can be totipotent 405

Nuclear transfer allows the cloning of animals 406

Multipotent stem cells differentiate in response to environmental signals 408

Pluripotent stem cells can be obtained in two ways 408

**20 Genes, Development, and Evolution 412**

**20.1 How Can Small Genetic Changes Result in Large Changes in Phenotype? 413**

Developmental genes in distantly related organisms are similar 413

**20.2 How Can Mutations with Large Effects Change Only One Part of the Body? 415**

Genetic switches govern how the genetic toolkit is used 415

Modularity allows for differences in the patterns of gene expression 416

**20.3 How Can Developmental Changes Result in Differences among Species? 418**

Differences in Hox gene expression patterns result in major differences in body plans 418

Mutations in developmental genes can produce major morphological changes 418

**20.4 How Can the Environment Modulate Development? 420**

Temperature can determine sex 420

Dietary information can be a predictor of future conditions 421

A variety of environmental signals influence development 421

**20.5 How Do Developmental Genes Constrain Evolution? 423**

Evolution usually proceeds by changing what's already there 423

Conserved developmental genes can lead to parallel evolution 423

# PART SIX
# The Patterns and Processes of Evolution

## 21 Mechanisms of Evolution 427

**21.1 What Is the Relationship between Fact and Theory in Evolution? 428**

Darwin and Wallace introduced the idea of evolution by natural selection 428

Evolutionary theory has continued to develop over the past century 430

Genetic variation contributes to phenotypic variation 431

**21.2 What Are the Mechanisms of Evolutionary Change? 432**

Mutation generates genetic variation 432

Selection acting on genetic variation leads to new phenotypes 432

Gene flow may change allele frequencies 433

Genetic drift may cause large changes in small populations 434

Nonrandom mating can change genotype or allele frequencies 434

**21.3 How Do Biologists Measure Evolutionary Change? 436**

Evolutionary change can be measured by allele and genotype frequencies 436

Evolution will occur unless certain restrictive conditions exist 437

Deviations from Hardy–Weinberg equilibrium show that evolution is occurring 438

Natural selection acts directly on phenotypes 438

Natural selection can change or stabilize populations 439

**21.4 How Is Genetic Variation Distributed and Maintained within Populations? 441**

Neutral mutations accumulate in populations 441

Sexual recombination amplifies the number of possible genotypes 441

Frequency-dependent selection maintains genetic variation within populations 441

Heterozygote advantage maintains polymorphic loci 442

Genetic variation within species is maintained in geographically distinct populations 443

**21.5 What Are the Constraints on Evolution? 444**

Developmental processes constrain evolution 444

Trade-offs constrain evolution 445

Short-term and long-term evolutionary outcomes sometimes differ 446

## 22 Reconstructing and Using Phylogenies 449

**22.1 What Is Phylogeny? 450**

All of life is connected through evolutionary history 451

Comparisons among species require an evolutionary perspective 451

**22.2 How Are Phylogenetic Trees Constructed? 452**

Parsimony provides the simplest explanation for phylogenetic data 454

Phylogenies are reconstructed from many sources of data 454

Mathematical models expand the power of phylogenetic reconstruction 456

The accuracy of phylogenetic methods can be tested 457

**22.3 How Do Biologists Use Phylogenetic Trees? 458**

Phylogenetic trees can be used to reconstruct past events 458

Phylogenies allow us to compare and contrast living organisms 459

Phylogenies can reveal convergent evolution 459

Ancestral states can be reconstructed 460

Molecular clocks help date evolutionary events 461

**22.4 How Does Phylogeny Relate to Classification? 462**

Evolutionary history is the basis for modern biological classification 463

Several codes of biological nomenclature govern the use of scientific names 463

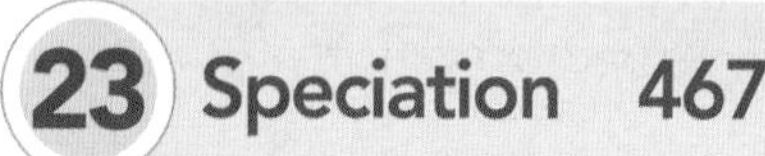

## 23 Speciation 467

**23.1 What Are Species? 468**

We can recognize many species by their appearance 468

Reproductive isolation is key 468

The lineage approach takes a long-term view 469

The different species concepts are not mutually exclusive 469

**23.2 What Is the Genetic Basis of Speciation? 470**

Incompatibilities between genes can produce reproductive isolation 470

Reproductive isolation develops with increasing genetic divergence 470

**23.3 What Barriers to Gene Flow Result in Speciation? 472**

Physical barriers give rise to allopatric speciation 472

Sympatric speciation occurs without physical barriers 473

**23.4 What Happens When Newly Formed Species Come into Contact? 475**

Prezygotic isolating mechanisms prevent hybridization 476

Postzygotic isolating mechanisms result in selection against hybridization 478

Hybrid zones may form if reproductive isolation is incomplete 478

**23.5 Why Do Rates of Speciation Vary? 480**

Several ecological and behavioral factors influence speciation rates 480

Rapid speciation can lead to adaptive radiation 481

## 24 Evolution of Genes and Genomes 485

**24.1 How Are Genomes Used to Study Evolution? 486**

Evolution of genomes results in biological diversity 486

Genes and proteins are compared through sequence alignment 486

Models of sequence evolution are used to calculate evolutionary divergence 487

Experimental studies examine molecular evolution directly 489

**24.2 What Do Genomes Reveal about Evolutionary Processes? 491**

Much of evolution is neutral 492

Positive and purifying selection can be detected in the genome 492

Genome size also evolves 494

**24.3 How Do Genomes Gain and Maintain Functions? 496**

Lateral gene transfer can result in the gain of new functions 496

Most new functions arise following gene duplication 496

Some gene families evolve through concerted evolution 498

**24.4 What Are Some Applications of Molecular Evolution? 499**

Molecular sequence data are used to determine the evolutionary history of genes 499

Gene evolution is used to study protein function 500

In vitro evolution is used to produce new molecules 500

Molecular evolution is used to study and combat diseases 501

## 25 The History of Life on Earth 505

**25.1 How Do Scientists Date Ancient Events? 506**

Radioisotopes provide a way to date fossils and rocks 507

Radiometric dating methods have been expanded and refined 507

Scientists have used several methods to construct a geological time scale 508

**25.2 How Have Earth's Continents and Climates Changed over Time? 508**

The continents have not always been where they are today 509

Earth's climate has shifted between hot and cold conditions 510

Volcanoes have occasionally changed the history of life 510

Extraterrestrial events have triggered changes on Earth 511

Oxygen concentrations in Earth's atmosphere have changed over time 511

**25.3 What Are the Major Events in Life's History? 514**

Several processes contribute to the paucity of fossils 514

Precambrian life was small and aquatic 515

Life expanded rapidly during the Cambrian period 516

Many groups of organisms that arose during the Cambrian later diversified 516

Geographic differentiation increased during the Mesozoic era 521

Modern biotas evolved during the Cenozoic era 521

The tree of life is used to reconstruct evolutionary events 522

# PART SEVEN
# The Evolution of Diversity

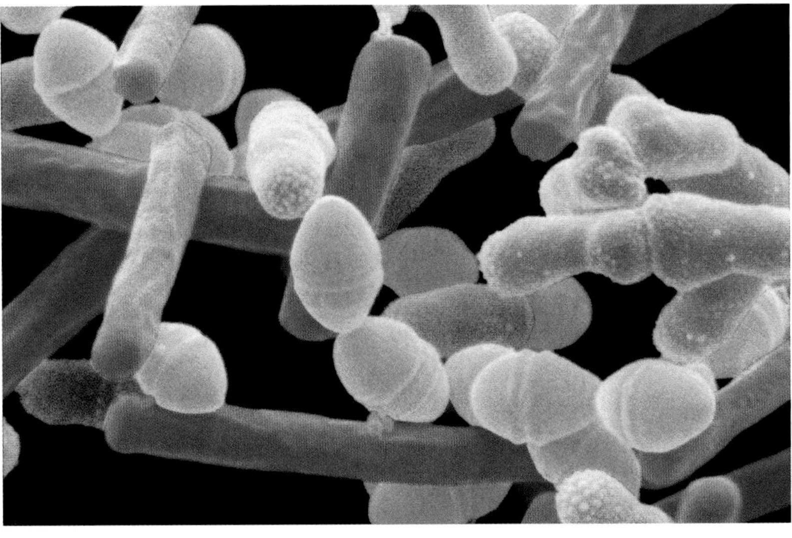

## 26 Bacteria, Archaea, and Viruses 525

**26.1 Where Do Prokaryotes Fit into the Tree of Life? 526**

The two prokaryotic domains differ in significant ways 526
The small size of prokaryotes has hindered our study of their evolutionary relationships 527
The nucleotide sequences of prokaryotes reveal their evolutionary relationships 528
Lateral gene transfer can lead to discordant gene trees 529
The great majority of prokaryote species have never been studied 530

**26.2 Why Are Prokaryotes So Diverse and Abundant? 530**

The low-GC Gram-positives include some of the smallest cellular organisms 530
Some high-GC Gram-positives are valuable sources of antibiotics 532
Hyperthermophilic bacteria live at very high temperatures 532
Hadobacteria live in extreme environments 532
Cyanobacteria were the first photosynthesizers 532
Spirochetes move by means of axial filaments 533
Chlamydias are extremely small parasites 533
The proteobacteria are a large and diverse group 534
Gene sequencing enabled biologists to differentiate the domain Archaea 534
Most crenarchaeotes live in hot or acidic places 536
Euryarchaeotes are found in surprising places 536
Korarchaeotes and nanoarchaeotes are less well known 537

**26.3 How Do Prokaryotes Affect Their Environments? 537**

Prokaryotes have diverse metabolic pathways 537
Prokaryotes play important roles in element cycling 538
Many prokaryotes form complex communities 539
Prokaryotes live on and in other organisms 539
Microbiomes are critical to human health 539
A small minority of bacteria are pathogens 541

**26.4 How Do Viruses Relate to Life's Diversity and Ecology? 543**

Many RNA viruses probably represent escaped genomic components of cellular life 544
Some DNA viruses may have evolved from reduced cellular organisms 544
Vertebrate genomes contain endogenous retroviruses 545
Viruses can be used to fight bacterial infections 545
Viruses are found throughout the biosphere 546

## 27 The Origin and Diversification of Eukaryotes 549

**27.1 How Did the Eukaryotic Cell Arise? 550**

The modern eukaryotic cell arose in several steps 550
Chloroplasts have been transferred among eukaryotes several times 551

**27.2 What Features Account for Protist Diversity? 552**

Alveolates have sacs under their plasma membranes 553
Stramenopiles typically have two flagella of unequal length 555
Rhizaria typically have long, thin pseudopods 557
Excavates began to diversify about 1.5 billion years ago 558
Amoebozoans use lobe-shaped pseudopods for locomotion 559

**27.3 What Is the Relationship between Sex and Reproduction in Protists? 562**

Some protists reproduce without sex and have sex without reproduction 562
Some protist life cycles feature alternation of generations 562

**27.4 How Do Protists Affect Their Environments? 563**

Phytoplankton are primary producers 563
Some microbial eukaryotes are deadly 563
Some microbial eukaryotes are endosymbionts 564
We rely on the remains of ancient marine protists 565

## 28 Plants without Seeds: From Water to Land 569

**28.1 How Did Photosynthesis Arise in Plants? 570**

Several distinct clades of algae were among the first photosynthetic eukaryotes 571

Two groups of green algae are the closest relatives of land plants 572
There are ten major groups of land plants 573

**28.2 When and How Did Plants Colonize Land? 574**
Adaptations to life on land distinguish land plants from green algae 574
Life cycles of land plants feature alternation of generations 574
Nonvascular land plants live where water is readily available 575
The sporophytes of nonvascular land plants are dependent on the gametophytes 575
Liverworts are the sister clade of the remaining land plants 577
Water and sugar transport mechanisms emerged in the mosses 577
Hornworts have distinctive chloroplasts and stalkless sporophytes 578

**28.3 What Features Allowed Land Plants to Diversify in Form? 579**
Vascular tissues transport water and dissolved materials 579
Vascular plants allowed herbivores to colonize the land 580
The closest relatives of vascular plants lacked roots 580
The lycophytes are sister to the other vascular plants 581
Horsetails and ferns constitute a clade 581
The vascular plants branched out 582
Heterospory appeared among the vascular plants 584

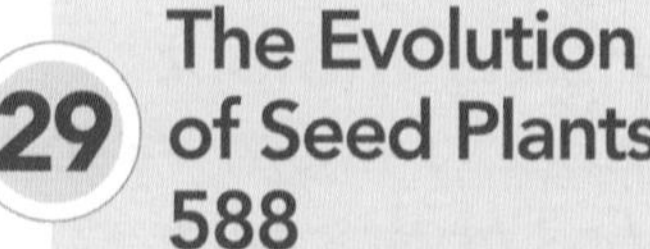

**29 The Evolution of Seed Plants 588**

**29.1 How Did Seed Plants Become Today's Dominant Vegetation? 589**
Features of the seed plant life cycle protect gametes and embryos 589
The seed is a complex, well-protected package 591

A change in stem anatomy enabled seed plants to grow to great heights 591

**29.2 What Are the Major Groups of Gymnosperms? 592**
There are four major groups of living gymnosperms 592
Conifers have cones and no swimming sperm 593

**29.3 How Do Flowers and Fruits Increase the Reproductive Success of Angiosperms? 596**
Angiosperms have many shared derived traits 596
The sexual structures of angiosperms are flowers 596
Flower structure has evolved over time 597
Angiosperms have coevolved with animals 598
The angiosperm life cycle produces diploid zygotes nourished by triploid endosperms 600
Fruits aid angiosperm seed dispersal 601
Recent analyses have revealed the phylogenetic relationships of angiosperms 601

**29.4 How Do Plants Benefit Human Society? 604**
Seed plants have been sources of medicine since ancient times 604
Seed plants are our primary food source 605

**30 The Evolution and Diversity of Fungi 608**

**30.1 What Is a Fungus? 609**
Unicellular yeasts absorb nutrients directly 609
Multicellular fungi use hyphae to absorb nutrients 609
Fungi are in intimate contact with their environment 610

**30.2 How Do Fungi Interact with Other Organisms? 611**
Saprobic fungi are critical to the planetary carbon cycle 611
Some fungi engage in parasitic or predatory interactions 611
Mutualistic fungi engage in relationships that benefit both partners 612
Endophytic fungi protect some plants from pathogens, herbivores, and stress 615

**30.3 How Do Major Groups of Fungi Differ in Structure and Life History? 615**
Fungi reproduce both sexually and asexually 616
Microsporidia are highly reduced, parasitic fungi 617
Most chytrids are aquatic 617
Some fungal life cycles feature separate fusion of cytoplasms and nuclei 619
Arbuscular mycorrhizal fungi form symbioses with plants 619
The dikaryotic condition is a synapomorphy of sac fungi and club fungi 620
The sexual reproductive structure of sac fungi is the ascus 620
The sexual reproductive structure of club fungi is the basidium 622

**30.4 What Are Some Applications of Fungal Biology? 623**
Fungi are important in producing food and drink 623
Fungi record and help remediate environmental pollution 624
Lichen diversity and abundance are indicators of air quality 624
Fungi are used as model organisms in laboratory studies 624
Reforestation may depend on mycorrhizal fungi 626

Fungi provide important weapons against diseases and pests 626

## 31 Animal Origins and the Evolution of Body Plans 629

**31.1 What Characteristics Distinguish the Animals? 630**

Animal monophyly is supported by gene sequences and morphology 630

A few basic developmental patterns differentiate major animal groups 633

**31.2 What Are the Features of Animal Body Plans? 634**

Most animals are symmetrical 634

The structure of the body cavity influences movement 635

Segmentation improves control of movement 636

Appendages have many uses 636

Nervous systems coordinate movement and allow sensory processing 637

**31.3 How Do Animals Get Their Food? 637**

Filter feeders capture small prey 637

Herbivores eat plants 637

Predators and omnivores capture and subdue prey 638

Parasites live in or on other organisms 638

Detritivores live on the remains of other organisms 639

**31.4 How Do Life Cycles Differ among Animals? 639**

Many animal life cycles feature specialized life stages 639

Most animal life cycles have at least one dispersal stage 640

Parasite life cycles facilitate dispersal and overcome host defenses 640

Some animals form colonies of genetically identical, physiologically integrated individuals 640

No life cycle can maximize all benefits 641

**31.5 What Are the Major Groups of Animals? 643**

Sponges are loosely organized animals 643

Ctenophores are radially symmetrical and diploblastic 644

Placozoans are abundant but rarely observed 645

Cnidarians are specialized predators 645

Some small groups of parasitic animals may be the closest relatives of bilaterians 648

## 32 Protostome Animals 651

**32.1 What Is a Protostome? 652**

Cilia-bearing lophophores and trochophores evolved among the lophotrochozoans 652

Ecdysozoans must shed their cuticles 654

Arrow worms retain some ancestral developmental features 655

**32.2 What Features Distinguish the Major Groups of Lophotrochozoans? 656**

Most bryozoans and entoprocts live in colonies 656

Flatworms, rotifers, and gastrotrichs are structurally diverse relatives 656

Ribbon worms have a long, protrusible feeding organ 658

Brachiopods and phoronids use lophophores to extract food from the water 658

Annelids have segmented bodies 659

Mollusks have undergone a dramatic evolutionary radiation 662

**32.3 What Features Distinguish the Major Groups of Ecdysozoans? 665**

Several marine ecdysozoan groups have relatively few species 665

Nematodes and their relatives are abundant and diverse 666

**32.4 Why Are Arthropods So Diverse? 667**

Arthropod relatives have fleshy, unjointed appendages 667

Jointed appendages appeared in the trilobites 668

Chelicerates have pointed, nonchewing mouthparts 668

Mandibles and antennae characterize the remaining arthropod groups 669

More than half of all described species are insects 671

## 33 Deuterostome Animals 678

**33.1 What Is a Deuterostome? 679**

Deuterostomes share early developmental patterns 679

There are three major deuterostome clades 679

Fossils shed light on deuterostome ancestors 679

**33.2 What Features Distinguish the Echinoderms, Hemichordates, and Their Relatives? 680**

Echinoderms have unique structural features 680

Hemichordates are wormlike marine deuterostomes 682

**33.3 What New Features Evolved in the Chordates? 683**

Adults of most lancelets and tunicates are sedentary 684

A dorsal supporting structure replaces the notochord in vertebrates 684

The phylogenetic relationships of jawless fishes are uncertain 685

Jaws and teeth improved feeding efficiency 686

Fins and swim bladders improved stability and control over locomotion 686

**33.4 How Did Vertebrates Colonize the Land? 689**

Jointed limbs enhanced support and locomotion on land 689
Amphibians usually require moist environments 690
Amniotes colonized dry environments 692
Reptiles adapted to life in many habitats 693
Crocodilians and birds share their ancestry with the dinosaurs 693
Feathers allowed birds to fly 695
Mammals radiated after the extinction of non-avian dinosaurs 696

**33.5 What Traits Characterize the Primates? 701**

Two major lineages of primates split late in the Cretaceous 701
Bipedal locomotion evolved in human ancestors 702
Human brains became larger as jaws became smaller 704
Humans developed complex language and culture 705

# PART EIGHT Flowering Plants: Form and Function

## 34 The Plant Body 708

**34.1 What Is the Basic Body Plan of Plants? 709**

Most angiosperms are either monocots or eudicots 709
Plants develop differently than animals 710
Apical–basal polarity and radial symmetry are characteristics of the plant body 711

**34.2 What Are the Major Tissues of Plants? 712**

The plant body is constructed from three tissue systems 712
Cells of the xylem transport water and dissolved minerals 714
Cells of the phloem transport the products of photosynthesis 714

**34.3 How Do Meristems Build a Continuously Growing Plant? 715**

Plants increase in size through primary and secondary growth 715
A hierarchy of meristems generates the plant body 715
Indeterminate primary growth originates in apical meristems 715
The root apical meristem gives rise to the root cap and the root primary meristems 716
The products of the root's primary meristems become root tissues 716
The root system anchors the plant and takes up water and dissolved minerals 718
The products of the stem's primary meristems become stem tissues 719
The stem supports leaves and flowers 720
Leaves are determinate organs produced by shoot apical meristems 720
Many eudicot stems and roots undergo secondary growth 721

**34.4 How Has Domestication Altered Plant Form? 723**

## 35 Transport in Plants 726

**35.1 How Do Plants Take Up Water and Solutes? 727**

Water potential differences govern the direction of water movement 727
Water and ions move across the root cell plasma membrane 728
Water and ions pass to the xylem by way of the apoplast and symplast 729

**35.2 How Are Water and Minerals Transported in the Xylem? 730**

The transpiration–cohesion–tension mechanism accounts for xylem transport 731

**35.3 How Do Stomata Control the Loss of Water and the Uptake of $CO_2$? 732**

The guard cells control the size of the stomatal opening 733
Plants can control their total numbers of stomata 734

**35.4 How Are Substances Translocated in the Phloem? 734**

Sucrose and other solutes are carried in the phloem 734
The pressure flow model appears to account for translocation in the phloem 735

## 36 Plant Nutrition 740

**36.1 What Nutrients Do Plants Require? 741**

All plants require specific macronutrients and micronutrients 741
Deficiency symptoms reveal inadequate nutrition 742
Hydroponic experiments identified essential elements 742

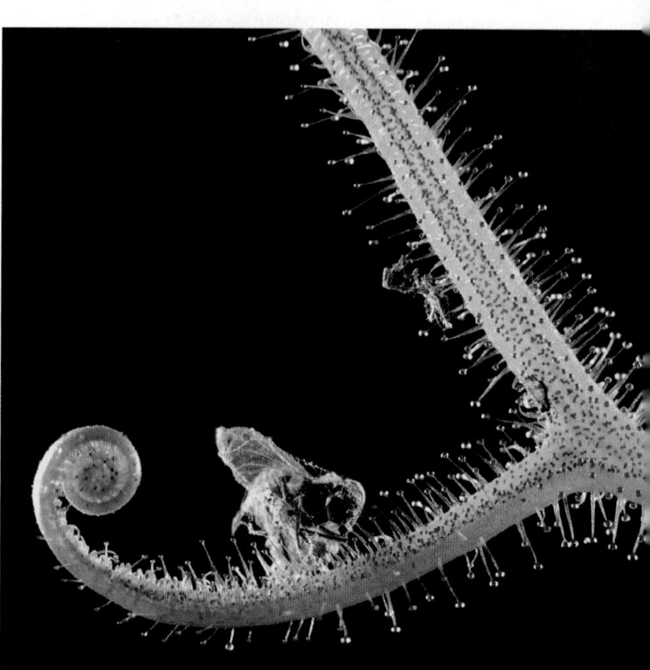

**36.2 How Do Plants Acquire Nutrients? 743**
Plants rely on growth to find nutrients 743
Nutrient uptake and assimilation are regulated 744

**36.3 How Does Soil Structure Affect Plants? 744**
Soils are complex in structure 745
Soils form through the weathering of rock 745
Soils are the source of plant nutrition 746
Fertilizers can be used to add nutrients to soil 746

**36.4 How Do Fungi and Bacteria Increase Nutrient Uptake by Plant Roots? 747**
Plants send signals for colonization 747
Mycorrhizae expand the root system 748
Soil bacteria are essential in getting nitrogen from air to plant cells 749
Nitrogenase catalyzes nitrogen fixation 749
Biological nitrogen fixation does not always meet agricultural needs 750
Plants and bacteria participate in the global nitrogen cycle 750

**36.5 How Do Carnivorous and Parasitic Plants Obtain a Balanced Diet? 751**
Carnivorous plants supplement their mineral nutrition 751
Parasitic plants take advantage of other plants 752
The plant–parasite relationship is similar to plant–fungus and plant–bacteria associations 753

## 37 Regulation of Plant Growth 756

**37.1 How Does Plant Development Proceed? 757**
In early development, the seed germinates and forms a growing seedling 757
Several hormones and photoreceptors help regulate plant growth 758
Genetic screens have increased our understanding of plant signal transduction 759

**37.2 What Do Gibberellins and Auxin Do? 760**
Gibberellins have many effects on plant growth and development 760
Auxin plays a role in differential plant growth 762
Auxin affects plant growth in several ways 765
At the molecular level, auxin and gibberellins act similarly 767

**37.3 What Are the Effects of Cytokinins, Ethylene, and Brassinosteroids? 768**
Cytokinins are active from seed to senescence 768
Ethylene is a gaseous hormone that hastens leaf senescence and fruit ripening 769
Brassinosteroids are plant steroid hormones 771

**37.4 How Do Photoreceptors Participate in Plant Growth Regulation? 771**
Phototropins, cryptochromes, and zeaxanthin are blue-light receptors 771
Phytochromes mediate the effects of red and far-red light 772
Phytochrome stimulates gene transcription 773
Circadian rhythms are entrained by light reception 774

## 38 Reproduction in Flowering Plants 778

**38.1 How Do Angiosperms Reproduce Sexually? 779**
The flower is an angiosperm's structure for sexual reproduction 779
Flowering plants have microscopic gametophytes 779
Pollination in the absence of water is an evolutionary adaptation 780
A pollen tube delivers sperm cells to the embryo sac 780

Many flowering plants control pollination or pollen tube growth to prevent inbreeding 782
Angiosperms perform double fertilization 783
Embryos develop within seeds contained in fruits 784
Seed development is under hormonal control 785

**38.2 What Determines the Transition from the Vegetative to the Flowering State? 785**
Shoot apical meristems can become inflorescence meristems 785
A cascade of gene expression leads to flowering 786
Photoperiodic cues can initiate flowering 787
Plants vary in their responses to photoperiodic cues 787
Night length is a key photoperiodic cue that determines flowering 788
The flowering stimulus originates in a leaf 788
Florigen is a small protein 790
Flowering can be induced by temperature or gibberellin 790
Some plants do not require an environmental cue to flower 792

**38.3 How Do Angiosperms Reproduce Asexually? 792**
Many forms of asexual reproduction exist 792
Vegetative reproduction has a disadvantage 793
Vegetative reproduction is important in agriculture 793

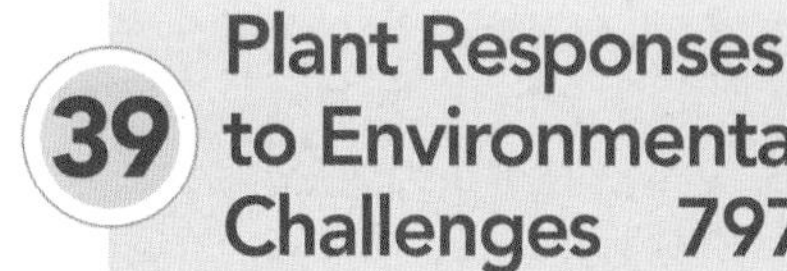


## 39 Plant Responses to Environmental Challenges 797

**39.1 How Do Plants Deal with Pathogens? 798**
- Physical barriers form constitutive defenses 798
- Plants can seal off infected parts to limit damage 798
- General and specific immunity both involve multiple responses 799
- Specific immunity involves gene-for-gene resistance 800
- Specific immunity usually leads to the hypersensitive response 800
- Systemic acquired resistance is a form of long-term immunity 801

**39.2 How Do Plants Deal with Herbivores? 801**
- Mechanical defenses against herbivores are widespread 801
- Plants produce constitutive chemical defenses against herbivores 802
- Some secondary metabolites play multiple roles 803
- Plants respond to herbivory with induced defenses 803
- Jasmonates trigger a range of responses to wounding and herbivory 805
- Why don't plants poison themselves? 805
- Plants don't always win the arms race 806

**39.3 How Do Plants Deal with Environmental Stresses? 806**
- Some plants have special adaptations to live in very dry conditions 806
- Some plants grow in saturated soils 808
- Plants can respond to drought stress 809
- Plants can cope with temperature extremes 810

**39.4 How Do Plants Deal with Salt and Heavy Metals? 810**
- Most halophytes accumulate salt 811
- Some plants can tolerate heavy metals 811

# PART NINE
# Animals: Form and Function

## 40 Physiology, Homeostasis, and Temperature Regulation 815

**40.1 How Do Multicellular Animals Supply the Needs of Their Cells? 816**
- An internal environment makes complex multicellular animals possible 816
- Physiological systems are regulated to maintain homeostasis 816

**40.2 What Are the Relationships between Cells, Tissues, and Organs? 817**
- Epithelial tissues are sheets of densely packed, tightly connected cells 817
- Muscle tissues generate force and movement 818
- Connective tissues include bone, blood, and fat 818
- Neural tissues include neurons and glial cells 819
- Organs consist of multiple tissues 820

**40.3 How Does Temperature Affect Living Systems? 820**
- $Q_{10}$ is a measure of temperature sensitivity 821
- Animals acclimatize to seasonal temperatures 821

**40.4 How Do Animals Alter Their Heat Exchange with the Environment? 822**
- Endotherms produce substantial amounts of metabolic heat 822
- Ectotherms and endotherms respond differently to changes in environmental temperature 822
- Energy budgets reflect adaptations for regulating body temperature 823
- Both ectotherms and endotherms control blood flow to the skin 824
- Some fish conserve metabolic heat 825
- Some ectotherms regulate metabolic heat production 825

**40.5 How Do Endotherms Regulate Their Body Temperatures? 826**
- Basal metabolic rates correlate with body size 826
- Endotherms respond to cold by producing heat and adapt to cold by reducing heat loss 827
- Evaporation of water can dissipate heat, but at a cost 829
- The mammalian thermostat uses feedback information 829
- Fever helps the body fight infections 830
- Some animals conserve energy by turning down the thermostat 830

## 41 Animal Hormones 834

**41.1 What Are Hormones and How Do They Work? 835**

Endocrine signaling can act locally or at a distance 835
Hormones can be divided into three chemical groups 836
Hormone action is mediated by receptors on or within their target cells 836
Hormone action depends on the nature of the target cell and its receptors 837

**41.2 What Have Experiments Revealed about Hormones and Their Action? 838**
The first hormone discovered was the gut hormone secretin 838
Early experiments on insects illuminated hormonal signaling systems 839
Three hormones regulate molting and maturation in arthropods 840

**41.3 How Do the Nervous and Endocrine Systems Interact? 842**
The pituitary is an interface between the nervous and endocrine systems 842
The anterior pituitary is controlled by hypothalamic neurohormones 844
Negative feedback loops regulate hormone secretion 844

**41.4 What Are the Major Endocrine Glands and Hormones? 845**
The thyroid gland secretes thyroxine 845
Three hormones regulate blood calcium concentrations 847
PTH lowers blood phosphate levels 848
Insulin and glucagon regulate blood glucose concentrations 848
The adrenal gland is two glands in one 849
Sex steroids are produced by the gonads 850
Melatonin is involved in biological rhythms and photoperiodicity 851
Many chemicals may act as hormones 851

**41.5 How Do We Study Mechanisms of Hormone Action? 852**
Hormones can be detected and measured with immunoassays 852
A hormone can act through many receptors 853

**42 Immunology: Animal Defense Systems 856**

**42.1 What Are the Major Defense Systems of Animals? 857**
Blood and lymph tissues play important roles in defense 857
White blood cells play many defensive roles 858
Immune system proteins bind pathogens or signal other cells 858

**42.2 What Are the Characteristics of the Innate Defenses? 859**
Barriers and local agents defend the body against invaders 859
Cell signaling pathways stimulate the body's defenses 860
Specialized proteins and cells participate in innate immunity 860
Inflammation is a coordinated response to infection or injury 861
Inflammation can cause medical problems 862

**42.3 How Does Adaptive Immunity Develop? 862**
Adaptive immunity has four key features 862
Two types of adaptive immune responses interact: an overview 863
Adaptive immunity develops as a result of clonal selection 865
Clonal deletion helps the immune system distinguish self from nonself 865
Immunological memory results in a secondary immune response 865
Vaccines are an application of immunological memory 866

**42.4 What Is the Humoral Immune Response? 867**
Some B cells develop into plasma cells 867
Different antibodies share a common structure 867
There are five classes of immunoglobulins 868
Immunoglobulin diversity results from DNA rearrangements and other mutations 868
The constant region is involved in immunoglobulin class switching 869
Monoclonal antibodies have many uses 871

**42.5 What Is the Cellular Immune Response? 871**
T cell receptors bind to antigens on cell surfaces 871
MHC proteins present antigen to T cells 872
T-helper cells and MHC II proteins contribute to the humoral immune response 872
Cytotoxic T cells and MHC I proteins contribute to the cellular immune response 874
Regulatory T cells suppress the humoral and cellular immune responses 874
MHC proteins are important in tissue transplants 874

**42.6 What Happens When the Immune System Malfunctions? 875**
Allergic reactions result from hypersensitivity 875
Autoimmune diseases are caused by reactions against self antigens 876
AIDS is an immune deficiency disorder 876

**43 Animal Reproduction 880**

**43.1 How Do Animals Reproduce without Sex? 881**
Budding and regeneration produce new individuals by mitosis 881
Parthenogenesis is the development of unfertilized eggs 881

**43.2 How Do Animals Reproduce Sexually? 882**
Gametogenesis produces eggs and sperm 882
Fertilization is the union of sperm and egg 884
Getting eggs and sperm together 887
Some individuals can function as both male and female 887
The evolution of vertebrate reproductive systems parallels the move to land 888
Animals with internal fertilization are distinguished by where the embryo develops 889

**43.3 How Do the Human Male and Female Reproductive Systems Work? 889**

Male sex organs produce and deliver semen 889
Male sexual function is controlled by hormones 892
Female sex organs produce eggs, receive sperm, and nurture the embryo 892
The ovarian cycle produces a mature egg 893
The uterine cycle prepares an environment for a fertilized egg 893
Hormones control and coordinate the ovarian and uterine cycles 894
FSH receptors determine which follicle ovulates 895
In pregnancy, hormones from the extraembryonic membranes take over 896
Childbirth is triggered by hormonal and mechanical stimuli 896

**43.4 How Can Fertility Be Controlled? 897**

Humans use a variety of methods to control fertility 897
Reproductive technologies help solve problems of infertility 897

**44 Animal Development 902**

**44.1 How Does Fertilization Activate Development? 903**

The sperm and the egg make different contributions to the zygote 903
Rearrangements of egg cytoplasm set the stage for determination 903

**44.2 How Does Mitosis Divide Up the Early Embryo? 904**

Cleavage repackages the cytoplasm 904
Early cell divisions in mammals are unique 905
Specific blastomeres generate specific tissues and organs 906
Germ cells are a unique lineage even in species with regulative development 908

**44.3 How Does Gastrulation Generate Multiple Tissue Layers? 908**

Invagination of the vegetal pole characterizes gastrulation in the sea urchin 908
Gastrulation in the frog begins at the gray crescent 909
The dorsal lip of the blastopore organizes embryo formation 910
Transcription factors and growth factors underlie the organizer's actions 911
The organizer changes its activity as it migrates from the dorsal lip 912
Reptilian and avian gastrulation is an adaptation to yolky eggs 913
The embryos of placental mammals lack yolk 914

**44.4 How Do Organs and Organ Systems Develop? 915**

The stage is set by the dorsal lip of the blastopore 915
Body segmentation develops during neurulation 916
Hox genes control development along the anterior–posterior axis 916

**44.5 How Is the Growing Embryo Sustained? 918**

Extraembryonic membranes form with contributions from all germ layers 918
Extraembryonic membranes in mammals form the placenta 919

**44.6 What Are the Stages of Human Development? 919**

Organ development begins in the first trimester 920
Organ systems grow and mature during the second and third trimesters 920
Developmental changes continue throughout life 920

**45 Neurons, Glia, and Nervous Systems 924**

**45.1 What Cells Are Unique to the Nervous System? 925**

The structure of neurons reflects their functions 925
Glia are the "silent partners" of neurons 926

**45.2 How Do Neurons Generate and Transmit Electric Signals? 927**

Simple electrical concepts underlie neural function 927
Membrane potentials can be measured with electrodes 928
Ion transporters and channels generate membrane potentials 928
Ion channels and their properties can now be studied directly 929
Gated ion channels alter membrane potential 930
Graded changes in membrane potential can integrate information 932
Sudden changes in $Na^+$ and $K^+$ channels generate action potentials 932
Action potentials are conducted along axons without loss of signal 934
Action potentials jump along myelinated axons 935

**45.3 How Do Neurons Communicate with Other Cells? 936**

The neuromuscular junction is a model chemical synapse 936
The arrival of an action potential causes the release of neurotransmitter 936
Synaptic functions involve many proteins 936
The postsynaptic membrane responds to neurotransmitter 936
Synapses can be excitatory or inhibitory 938
The postsynaptic cell sums excitatory and inhibitory input 938
Synapses can be fast or slow 938
Electrical synapses are fast but do not integrate information well 939

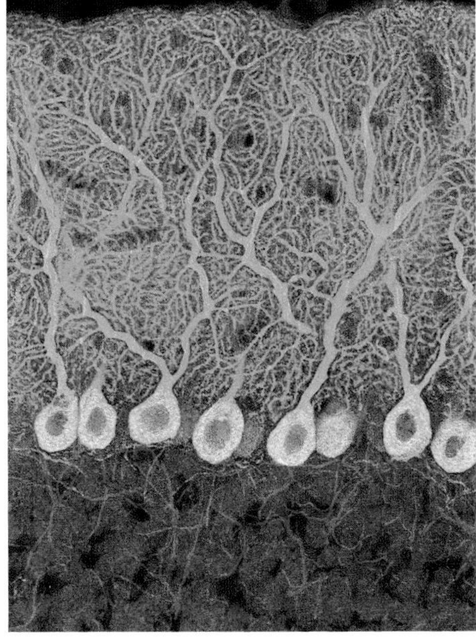

The action of a neurotransmitter depends on the receptor to which it binds 939
To turn off responses, synapses must be cleared of neurotransmitter 940
The diversity of receptors makes drug specificity possible 940

**45.4 How Are Neurons and Glia Organized into Information-Processing Systems? 940**
Nervous systems range in complexity 940
The knee-jerk reflex is controlled by a simple neural network 941
The vertebrate brain is the seat of behavioral complexity 943

## 46 Sensory Systems 946

**46.1 How Do Sensory Receptor Cells Convert Stimuli into Action Potentials? 947**
Sensory transduction involves changes in membrane potentials 947
Sensory receptor proteins act on ion channels 947
Sensation depends on which neurons receive action potentials from sensory cells 947
Many receptors adapt to repeated stimulation 948

**46.2 How Do Sensory Systems Detect Chemical Stimuli? 949**
Olfaction is the sense of smell 949
Some chemoreceptors detect pheromones 950
The vomeronasal organ contains chemoreceptors 950
Gustation is the sense of taste 951

**46.3 How Do Sensory Systems Detect Mechanical Forces? 952**
Many different cells respond to touch and pressure 952
Mechanoreceptors are also found in muscles, tendons, and ligaments 952
Hair cells are mechanoreceptors of the auditory and vestibular systems 953
Auditory systems use hair cells to sense sound waves 954
Flexion of the basilar membrane is perceived as sound 955
Various types of damage can result in hearing loss 956
The vestibular system uses hair cells to detect forces of gravity and momentum 956

**46.4 How Do Sensory Systems Detect Light? 957**
Rhodopsin is a vertebrate visual pigment 957
Invertebrates have a variety of visual systems 958
Image-forming eyes evolved independently in vertebrates and cephalopods 958
The vertebrate retina receives and processes visual information 959
Rod and cone cells are the photoreceptors of the vertebrate retina 960
Information flows through layers of neurons in the retina 962

## 47 The Mammalian Nervous System: Structure and Higher Functions 967

**47.1 How Is the Mammalian Nervous System Organized? 968**
Functional organization is based on flow and type of information 968
The anatomical organization of the CNS emerges during development 968
The spinal cord transmits and processes information 969
The brainstem carries out many autonomic functions 969
The core of the forebrain controls physiological drives, instincts, and emotions 970
Regions of the telencephalon interact to control behavior and produce consciousness 970
The size of the human brain is off the curve 973

**47.2 How Is Information Processed by Neural Networks? 973**
Pathways of the autonomic nervous system control involuntary physiological functions 974
The visual system is an example of information integration by the cerebral cortex 975
Three-dimensional vision results from cortical cells receiving input from both eyes 977

**47.3 Can Higher Functions Be Understood in Cellular Terms? 978**
Sleep and dreaming are reflected in electrical patterns in the cerebral cortex 978
Language abilities are localized in the left cerebral hemisphere 980
Some learning and memory can be localized to specific brain areas 981
We still cannot answer the question "What is consciousness?" 982

## 48 Musculoskeletal Systems 986

**48.1 How Do Muscles Contract? 987**
Sliding filaments cause skeletal muscle to contract 987
Actin–myosin interactions cause filaments to slide 988
Actin–myosin interactions are controlled by calcium ions 989
Cardiac muscle is similar to and different from skeletal muscle 991
Smooth muscle causes slow contractions of many internal organs 993

**48.2 What Determines Skeletal Muscle Performance? 994**

The strength of a muscle contraction depends on how many fibers are contracting and at what rate 994
Muscle fiber types determine endurance and strength 995
A muscle has an optimal length for generating maximum tension 996
Exercise increases muscle strength and endurance 996
Muscle ATP supply limits performance 997
Insect muscle has the greatest rate of cycling 997

**48.3 How Do Skeletal Systems and Muscles Work Together? 999**
A hydrostatic skeleton consists of fluid in a muscular cavity 999
Exoskeletons are rigid outer structures 999
Vertebrate endoskeletons consist of cartilage and bone 999
Bones develop from connective tissues 1001
Bones that have a common joint can work as a lever 1001

**49 Gas Exchange 1005**

**49.1 What Physical Factors Govern Respiratory Gas Exchange? 1006**
Diffusion of gases is driven by partial pressure differences 1006
Fick's law applies to all systems of gas exchange 1006
Air is a better respiratory medium than water 1007
High temperatures create respiratory problems for aquatic animals 1007
$O_2$ availability decreases with altitude 1007
$CO_2$ is lost by diffusion 1008

**49.2 What Adaptations Maximize Respiratory Gas Exchange? 1008**
Respiratory organs have large surface areas 1008
Ventilation and perfusion of gas exchange surfaces maximize partial pressure gradients 1009
Insects have airways throughout their bodies 1009
Fish gills use countercurrent flow to maximize gas exchange 1009
Birds use unidirectional ventilation to maximize gas exchange 1010
Tidal ventilation produces dead space that limits gas exchange efficiency 1012

**49.3 How Do Human Lungs Work? 1013**
Respiratory tract secretions aid ventilation 1013
Lungs are ventilated by pressure changes in the thoracic cavity 1015

**49.4 How Does Blood Transport Respiratory Gases? 1016**
Hemoglobin combines reversibly with $O_2$ 1016
Myoglobin holds an $O_2$ reserve 1017
Hemoglobin's affinity for $O_2$ is variable 1017
$CO_2$ is transported as bicarbonate ions in the blood 1018

**49.5 How Is Breathing Regulated? 1019**
Breathing is controlled in the brainstem 1019
Regulating breathing requires feedback 1020

**50 Circulatory Systems 1025**

**50.1 Why Do Animals Need a Circulatory System? 1026**
Some animals do not have a circulatory system 1026
Circulatory systems can be open or closed 1026
Open circulatory systems move extracellular fluid 1026
Closed circulatory systems circulate blood through a system of blood vessels 1026

**50.2 How Have Vertebrate Circulatory Systems Evolved? 1027**
Circulation in fish is a single circuit 1028
Lungfish evolved a gas-breathing organ 1028
Amphibians have partial separation of systemic and pulmonary circulation 1029
Reptiles have exquisite control of pulmonary and systemic circulation 1029
Birds and mammals have fully separated pulmonary and systemic circuits 1030

**50.3 How Does the Mammalian Heart Function? 1030**
Blood flows from right heart to lungs to left heart to body 1030
The heartbeat originates in the cardiac muscle 1032
A conduction system coordinates the contraction of heart muscle 1034
Electrical properties of ventricular muscles sustain heart contraction 1034
The ECG records the electrical activity of the heart 1035

**50.4 What Are the Properties of Blood and Blood Vessels? 1037**
Red blood cells transport respiratory gases 1038
Platelets are essential for blood clotting 1039
Arteries withstand high pressure, arterioles control blood flow 1039
Materials are exchanged in capillary beds by filtration, osmosis, and diffusion 1039
Blood flows back to the heart through veins 1041
Lymphatic vessels return interstitial fluid to the blood 1042
Vascular disease is a killer 1042

**50.5 How Is the Circulatory System Controlled and Regulated? 1043**
Autoregulation matches local blood flow to local need 1044

Arterial pressure is regulated by hormonal and neural mechanisms 1044

## 51 Nutrition, Digestion, and Absorption 1048

### 51.1 What Do Animals Require from Food? 1049

Energy needs and expenditures can be measured 1049
Sources of energy can be stored in the body 1050
Food provides carbon skeletons for biosynthesis 1051
Animals need mineral elements for a variety of functions 1052
Animals must obtain vitamins from food 1053
Nutrient deficiencies result in diseases 1054

### 51.2 How Do Animals Ingest and Digest Food? 1054

The food of herbivores is often low in energy and hard to digest 1054
Carnivores must find, capture, and kill prey 1055
Vertebrate species have distinctive teeth 1055
Digestion usually begins in a body cavity 1056
Tubular guts have an opening at each end 1056
Digestive enzymes break down complex food molecules 1057

### 51.3 How Does the Vertebrate Gastrointestinal System Function? 1058

The vertebrate gut consists of concentric tissue layers 1058
Mechanical activity moves food through the gut and aids digestion 1059
Chemical digestion begins in the mouth and the stomach 1060
The stomach gradually releases its contents to the small intestine 1061
Most chemical digestion occurs in the small intestine 1061
Nutrients are absorbed in the small intestine 1063
Absorbed nutrients go to the liver 1063
Water and ions are absorbed in the large intestine 1063
Herbivores rely on microorganisms to digest cellulose 1063

### 51.4 How Is the Flow of Nutrients Controlled and Regulated? 1064

Hormones control many digestive functions 1065
The liver directs the traffic of the molecules that fuel metabolism 1065
The brain plays a major role in regulating food intake 1067

## 52 Salt and Water Balance and Nitrogen Excretion 1071

### 52.1 How Do Excretory Systems Maintain Homeostasis? 1072

Water enters or leaves cells by osmosis 1072
Excretory systems control extracellular fluid osmolarity and composition 1072
Aquatic invertebrates can conform to or regulate their osmotic and ionic environments 1072
Vertebrates are osmoregulators and ionic regulators 1073

### 52.2 How Do Animals Excrete Nitrogen? 1074

Animals excrete nitrogen in a number of forms 1074
Most species produce more than one nitrogenous waste 1074

### 52.3 How Do Invertebrate Excretory Systems Work? 1075

The protonephridia of flatworms excrete water and conserve salts 1075
The metanephridia of annelids process coelomic fluid 1075
Malpighian tubules of insects use active transport to excrete wastes 1076

### 52.4 How Do Vertebrates Maintain Salt and Water Balance? 1077

Marine fishes must conserve water 1077
Terrestrial amphibians and reptiles must avoid desiccation 1077
Mammals can produce highly concentrated urine 1078
The nephron is the functional unit of the vertebrate kidney 1078
Blood is filtered into Bowman's capsule 1078
The renal tubules convert glomerular filtrate to urine 1079

### 52.5 How Does the Mammalian Kidney Produce Concentrated Urine? 1079

Kidneys produce urine and the bladder stores it 1080
Nephrons have a regular arrangement in the kidney 1081
Most of the glomerular filtrate is reabsorbed by the proximal convoluted tubule 1082
The loop of Henle creates a concentration gradient in the renal medulla 1082
Water permeability of kidney tubules depends on water channels 1084
The distal convoluted tubule fine-tunes the composition of the urine 1084
Urine is concentrated in the collecting duct 1084
The kidneys help regulate acid–base balance 1084
Kidney failure is treated with dialysis 1085

### 52.6 How Are Kidney Functions Regulated? 1087

Glomerular filtration rate is regulated 1087

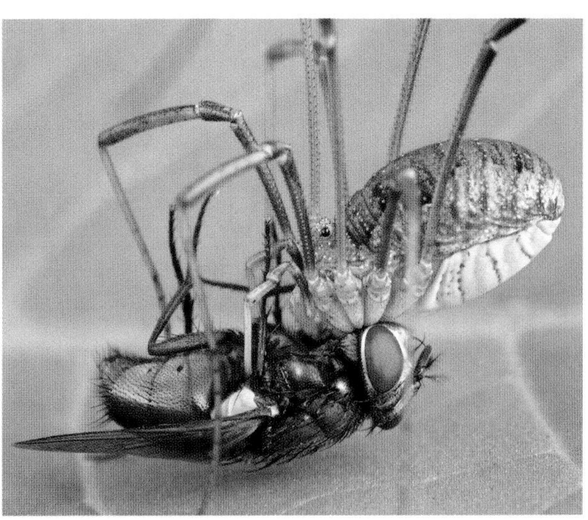

Regulation of GFR uses feedback information from the distal tubule 1087
Blood osmolarity and blood pressure are regulated by ADH 1088
The heart produces a hormone that helps lower blood pressure 1090

53 Animal Behavior 1093

53.1 What Are the Origins of Behavioral Biology? 1094
Conditioned reflexes are a simple behavioral mechanism 1094
Ethologists focused on the behavior of animals in their natural environment 1094
Ethologists probed the causes of behavior 1095

53.2 How Do Genes Influence Behavior? 1096
Breeding experiments can produce behavioral phenotypes 1096
Knockout experiments can reveal the roles of specific genes 1096
Behaviors are controlled by gene cascades 1097

53.3 How Does Behavior Develop? 1098
Hormones can determine behavioral potential and timing 1098
Some behaviors can be acquired only at certain times 1099
Birdsong learning involves genetics, imprinting, and hormonal timing 1099
The timing and expression of birdsong are under hormonal control 1101

53.4 How Does Behavior Evolve? 1102
Animals are faced with many choices 1103
Behaviors have costs and benefits 1103
Territorial behavior carries significant costs 1103
Cost–benefit analysis can be applied to foraging behavior 1104

53.5 What Physiological Mechanisms Underlie Behavior? 1106
Biological rhythms coordinate behavior with environmental cycles 1106
Animals must find their way around their environment 1109
Animals use multiple modalities to communicate 1110

53.6 How Does Social Behavior Evolve? 1113
Mating systems maximize the fitness of both partners 1113
Fitness can include more than your own offspring 1114
Eusociality is the extreme result of kin selection 1115
Group living has benefits and costs 1116
Can the concepts of sociobiology be applied to humans? 1116

PART TEN
Ecology

54 Ecology and the Distribution of Life 1121

54.1 What Is Ecology? 1122
Ecology is not the same as environmentalism 1122
Ecologists study biotic and abiotic components of ecosystems 1122

54.2 Why Do Climates Vary Geographically? 1122
Solar radiation varies over Earth's surface 1123
Solar energy input determines atmospheric circulation patterns 1124
Atmospheric circulation and Earth's rotation result in prevailing winds 1124
Prevailing winds drive ocean currents 1124
Organisms adapt to climatic challenges 1125

54.3 How Is Life Distributed in Terrestrial Environments? 1126
Tundra is found at high latitudes and high elevations 1128
Evergreen trees dominate boreal and temperate evergreen forests 1129
Temperate deciduous forests change with the seasons 1130
Temperate grasslands are widespread 1131
Hot deserts form around 30° latitude 1132
Cold deserts are high and dry 1133

Chaparral has hot, dry summers and wet, cool winters 1134
Thorn forests and tropical savannas have similar climates 1135
Tropical deciduous forests occur in hot lowlands 1136
Tropical rainforests are rich in species 1137

**54.4 How Is Life Distributed in Aquatic Environments? 1139**

The marine biome can be divided into several life zones 1139
Freshwater biomes may be rich in species 1140
Estuaries have characteristics of both freshwater and marine environments 1141

**54.5 What Factors Determine the Boundaries of Biogeographic Regions? 1141**

Geological history influences the distribution of organisms 1141
Two scientific advances changed the field of biogeography 1142
Discontinuous distributions may result from vicariant or dispersal events 1143
Humans exert a powerful influence on biogeographic patterns 1145

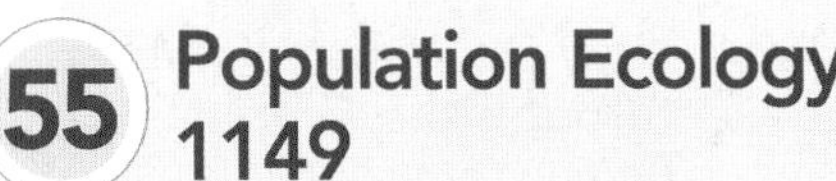

**55 Population Ecology 1149**

**55.1 How Do Ecologists Measure Populations? 1150**

Ecologists use a variety of approaches to count and track individuals 1150
Ecologists can estimate population densities from samples 1151
A population's age structure influences its capacity to grow 1151
A population's dispersion pattern reflects how individuals are distributed in space 1152

**55.2 How Do Ecologists Study Population Dynamics? 1153**

Demographic events determine the size of a population 1153
Life tables track demographic events 1154
Survivorship curves reflect life history strategies 1155

**55.3 How Do Environmental Conditions Affect Life Histories? 1156**

Survivorship and fecundity determine a population's growth rate 1156
Life history traits vary with environmental conditions 1156
Life history traits are influenced by interspecific interactions 1157

**55.4 What Factors Limit Population Densities? 1157**

All populations have the potential for exponential growth 1157
Logistic growth occurs as a population approaches its carrying capacity 1158
Population growth can be limited by density-dependent or density-independent factors 1159
Different population regulation factors lead to different life history strategies 1159
Several ecological factors explain species' characteristic population densities 1159
Some newly introduced species reach high population densities 1160
Evolutionary history may explain species abundances 1160

**55.5 How Does Habitat Variation Affect Population Dynamics? 1161**

Many populations live in separated habitat patches 1161
Corridors may allow subpopulations to persist 1162

**55.6 How Can We Use Ecological Principles to Manage Populations? 1163**

Management plans must take life history strategies into account 1163
Management plans must be guided by the principles of population dynamics 1163
Human population growth has been exponential 1164

**56 Species Interactions and Coevolution 1169**

**56.1 What Types of Interactions Do Ecologists Study? 1170**

Interactions among species can be grouped into several categories 1170
Interaction types are not always clear-cut 1171
Some types of interactions result in coevolution 1171

**56.2 How Do Antagonistic Interactions Evolve? 1172**

Predator–prey interactions result in a range of adaptations 1172
Herbivory is a widespread interaction 1175
Parasite–host interactions may be pathogenic 1176

**56.3 How Do Mutualistic Interactions Evolve? 1177**

Some mutualistic partners exchange food for care or transport 1178
Some mutualistic partners exchange food or housing for defense 1178
Plants and pollinators exchange food for pollen transport 1180
Plants and frugivores exchange food for seed transport 1181

**56.4 What Are the Outcomes of Competition? 1182**

Competition is widespread because all species share resources 1182
Interference competition may restrict habitat use 1183
Exploitation competition may lead to coexistence 1183
Species may compete indirectly for a resource 1184
Competition may determine a species' niche 1184

**57 Community Ecology 1188**

**57.1 What Are Ecological Communities? 1189**

Energy enters communities through primary producers 1189
Consumers use diverse sources of energy 1190
Fewer individuals and less biomass can be supported at higher trophic levels 1190
Productivity and species diversity are linked 1192

**57.2 How Do Interactions among Species Influence Communities? 1193**

Species interactions can cause trophic cascades 1193
Keystone species have disproportionate effects on their communities 1194

**57.3 What Patterns of Species Diversity Have Ecologists Observed? 1195**

Diversity comprises both the number and the relative abundance of species 1195
Ecologists have observed latitudinal gradients in diversity 1196
The theory of island biogeography suggests that immigration and extinction rates determine diversity on islands 1196

**57.4 How Do Disturbances Affect Ecological Communities? 1199**

Succession is the predictable pattern of change in a community after a disturbance 1199
Both facilitation and inhibition influence succession 1201
Cyclical succession requires adaptation to periodic disturbances 1201
Heterotrophic succession generates distinctive communities 1202

**57.5 How Does Species Richness Influence Community Stability? 1202**

Species richness is associated with productivity and stability 1202
Diversity, productivity, and stability differ between natural and managed communities 1202

**58 Ecosystems and Global Ecology 1207**

**58.1 How Does Energy Flow through the Global Ecosystem? 1208**

Energy flows and chemicals cycle through ecosystems 1208
The geographic distribution of energy flow is uneven 1208
Human activities modify the flow of energy 1210

**58.2 How Do Materials Move through the Global Ecosystem? 1210**

Elements move between biotic and abiotic compartments of ecosystems 1211
The atmosphere contains large pools of the gases required by living organisms 1211
The terrestrial surface is influenced by slow geological processes 1213
Water transports elements among compartments 1213
Fire is a major mover of elements 1214

**58.3 How Do Specific Nutrients Cycle through the Global Ecosystem? 1214**

Water cycles rapidly through the ecosystem 1215
The carbon cycle has been altered by human activities 1216

The nitrogen cycle depends on both biotic and abiotic processes 1218
The burning of fossil fuels affects the sulfur cycle 1219
The global phosphorus cycle lacks a significant atmospheric component 1220
Other biogeochemical cycles are also important 1221
Biogeochemical cycles interact 1221

**58.4 What Goods and Services Do Ecosystems Provide? 1223**

**58.5 How Can Ecosystems Be Sustainably Managed? 1224**

## 59 Biodiversity and Conservation Biology 1228

**59.1 What Is Conservation Biology? 1229**

Conservation biology aims to protect and manage biodiversity 1229
Biodiversity has great value to human society 1230

**59.2 How Do Conservation Biologists Predict Changes in Biodiversity? 1230**

Our knowledge of biodiversity is incomplete 1230
We can predict the effects of human activities on biodiversity 1231

**59.3 What Human Activities Threaten Species Persistence? 1232**

Habitat losses endanger species 1233
Overexploitation has driven many species to extinction 1234
Invasive predators, competitors, and pathogens threaten many species 1235
Rapid climate change can cause species extinctions 1236

**59.4 What Strategies Are Used to Protect Biodiversity? 1237**

Protected areas preserve habitat and prevent overexploitation 1237
Degraded ecosystems can be restored 1237
Disturbance patterns sometimes need to be restored 1239
Ending trade is crucial to saving some species 1240
Species invasions must be controlled or prevented 1241
Biodiversity has economic value 1241
Changes in human-dominated landscapes can help protect biodiversity 1243
Captive breeding programs can maintain a few species 1244
Earth is not a ship, a spaceship, or an airplane 1244

**APPENDIX A**
**The Tree of Life 1248**

**APPENDIX B**
**Statistics Primer 1255**

**APPENDIX C**
**Some Measurements Used in Biology 1264**

**ANSWERS TO CHAPTER REVIEW QUESTIONS A-1**

**GLOSSARY G-1**

**ILLUSTRATION CREDITS C-1**

**INDEX I-1**

# PART ONE
The Science of Life and Its Chemical Basis

# 1 Studying Life

## CHAPTER OUTLINE


1.1 What Is Biology?

1.2 How Do Biologists Investigate Life?

1.3 Why Does Biology Matter?

**What's Happening to the Frogs?** Tyrone Hayes grew up near the great Congaree Swamp in South Carolina collecting turtles, snakes, frogs, and toads. He is now a professor of biology at the University of California at Berkeley. In the laboratory and in the field, he is studying how and why populations of frogs are endangered by agricultural pesticides.

AMPHIBIANS—frogs, salamanders, and wormlike caecilians—have been around so long they watched the dinosaurs come and go. But for the last three decades, amphibian populations around the world have been declining dramatically. Today more than a third of the world's amphibian species are threatened with extinction. Why are these animals disappearing?

Tyrone Hayes, a biologist at the University of California at Berkeley, probed the effects of certain chemicals that are applied to croplands in large quantities and that accumulate in the runoff water from the fields. Hayes focused on the effects on amphibians of atrazine, a weed killer (herbicide) widely used in the United States and some other countries, where it is a common contaminant in fresh water (its use has been banned in the European Union). In the U.S., atrazine is usually applied in the spring, when many amphibians are breeding and thousands of tadpoles swim in the ditches, ponds, and streams that receive runoff from farms.

In his laboratory, Hayes and his associates raised frog tadpoles in water containing no atrazine and also in water with concentrations ranging from 0.01 parts per billion (ppb) up to 25 ppb. Concentrations as low as 0.1 ppb had a dramatic effect on tadpole development: it feminized the males. When these males became adults, their vocal structures—which are used in mating calls and thus are crucial for successful reproduction—were smaller than normal; in some, eggs were growing in the testes; some developed female sex organs. In other studies, normal adult male frogs exposed to 25 ppb had a tenfold reduction in testosterone levels and did not produce sperm. You can imagine the disastrous effects of such developmental and hormonal changes on the capacity of frogs to breed and reproduce.

But these experiments were performed in the laboratory, with a species of frog bred for laboratory use. Would the results be the same in nature? To find out, Hayes and his students traveled from Utah to Iowa, sampling water and collecting frogs. They analyzed the water for atrazine and examined the frogs. The only site where the frogs were normal was one where atrazine was undetectable. At all other sites, male frogs had abnormalities of the sex organs.

Like other biologists, Hayes made observations. He then made predictions based on those observations, and designed and carried out experiments to test his predictions.

Could atrazine in the environment affect species other than amphibians?

See answer on p. 18.

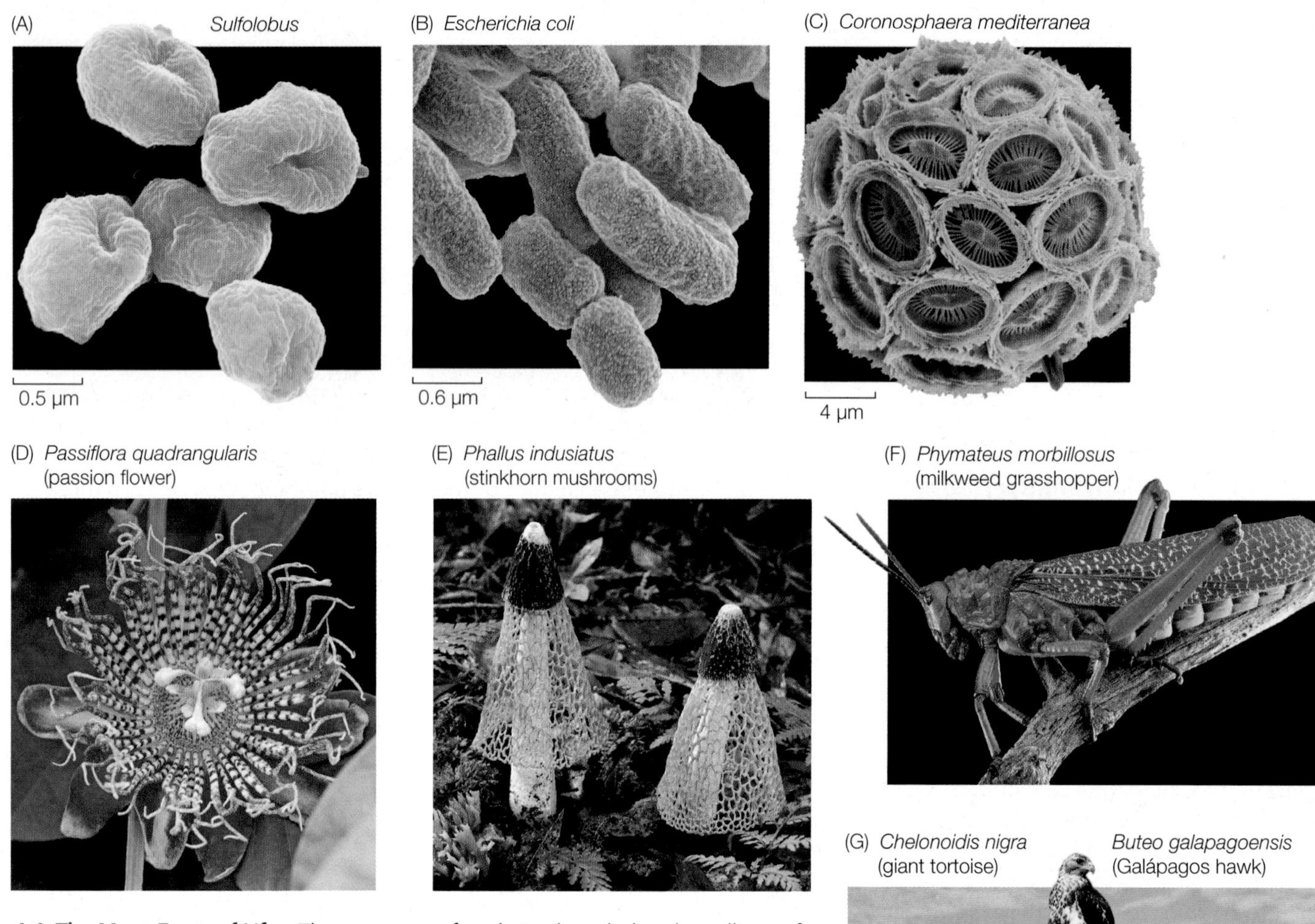


**1.1 The Many Faces of Life** The processes of evolution have led to the millions of diverse organisms living on Earth today. Archaea (A) and bacteria (B) are all single-celled, prokaryotic organisms, as described in Chapter 26. (C) Many protists are unicellular but, as discussed in Chapter 27, their cell structures are more complex than those of the prokaryotes. This protist has manufactured "plates" of calcium carbonate that surround and protect its single cell. (D–G) Most of the visible life on Earth is multicellular. Chapters 28 and 29 cover the green plants (D). The other broad groups of multicellular organisms are the fungi (E), discussed in Chapter 30, and the animals (F, G), covered in Chapters 31–33.

## 1.1 What Is Biology?

**Biology** is the scientific study of living things, which we call organisms (**Figure 1.1**). The living organisms we know about are all descended from a common origin of life on Earth that occurred almost 4 billion years ago. Living organisms share many characteristics that allow us to distinguish them from the nonliving world:

- Organisms are made up of a common set of chemical components, including particular carbohydrates, fatty acids, nucleic acids, and amino acids, among others.
- The building blocks of most organisms are **cells**—individual structures enclosed by plasma membranes.
- The cells of living organisms convert molecules obtained from their environment into new biological molecules.
- Cells extract energy from the environment and use it to do biological work.
- Organisms contain genetic information that uses a nearly universal code to specify the assembly of proteins.
- Organisms share similarities among a fundamental set of genes and replicate this genetic information when reproducing themselves.
- Organisms exist in populations that evolve through changes in the frequencies of genetic variants within the populations over time.
- Living organisms self-regulate their internal environments, thus maintaining the conditions that allow them to survive.

Taken together, these characteristics logically lead to the conclusion that all life has a common ancestry, and that the diverse organisms alive today all originated from one life form. If life had multiple origins, we would not expect to see the striking similarities across gene sequences, the nearly universal genetic code, or the common set of amino acids that characterizes every known living organism. Organisms from a separate origin of life—say, on another planet—might be similar in some ways to life on Earth. For example, such life forms would probably possess heritable genetic information that they could pass on to offspring. But we would not expect the details of their genetic code or the fundamental sequences of their genomes to be the same as or even similar to ours.

The list is necessarily simplified, and some forms of life may not display all of the listed characteristics all of the time. For example, the seed of a desert plant may go for many years without extracting energy from the environment, converting molecules, regulating its internal environment, or reproducing; yet the seed is alive. And there are viruses, which are not composed of cells and cannot carry out physiological functions on their own (they parasitize host cells to function for them). Yet viruses contain genetic information, and they mutate and evolve. So even though viruses are not independent cellular organisms, their existence depends on cells. In addition, it is highly probable that viruses evolved from cellular life forms. Thus most biologists consider viruses to be a part of life.

This book will explore the details of the common characteristics of life, how these characteristics arose, and how they work together to enable organisms to survive and reproduce. Not all organisms survive and reproduce with equal success, and it is through differential survival and reproduction that living systems evolve and become adapted to Earth's many environments. The processes of evolution have generated the enormous diversity of life on Earth, and evolution is a central theme of biology.

## Life arose from non-life via chemical evolution

Geologists estimate that Earth formed between 4.6 and 4.5 billion years ago. At first the planet was not a very hospitable place. It was some 600 million years or more before the earliest life evolved. If we picture the 4.6-billion-year history of Earth as a 30-day month, life first appeared some time around the end of the first week (**Figure 1.2**).

When we consider how life might have arisen from nonliving matter, we must take into account the properties of the young Earth's atmosphere, oceans, and climate, all of which were very different than they are today. Biologists postulate that complex biological molecules first arose through the random physical association of chemicals in that environment. Experiments simulating the conditions on early Earth have confirmed that the generation of complex molecules under such conditions is possible, even probable. The critical step for the evolution of life, however, was the appearance of **nucleic acids**—molecules that could reproduce themselves and also serve as templates for the synthesis of **proteins**, large molecules with complex but stable shapes. The variation in the shapes of these proteins enabled them to participate in increasing numbers and kinds of chemical reactions with other molecules. These subjects are covered in Part One of this book.

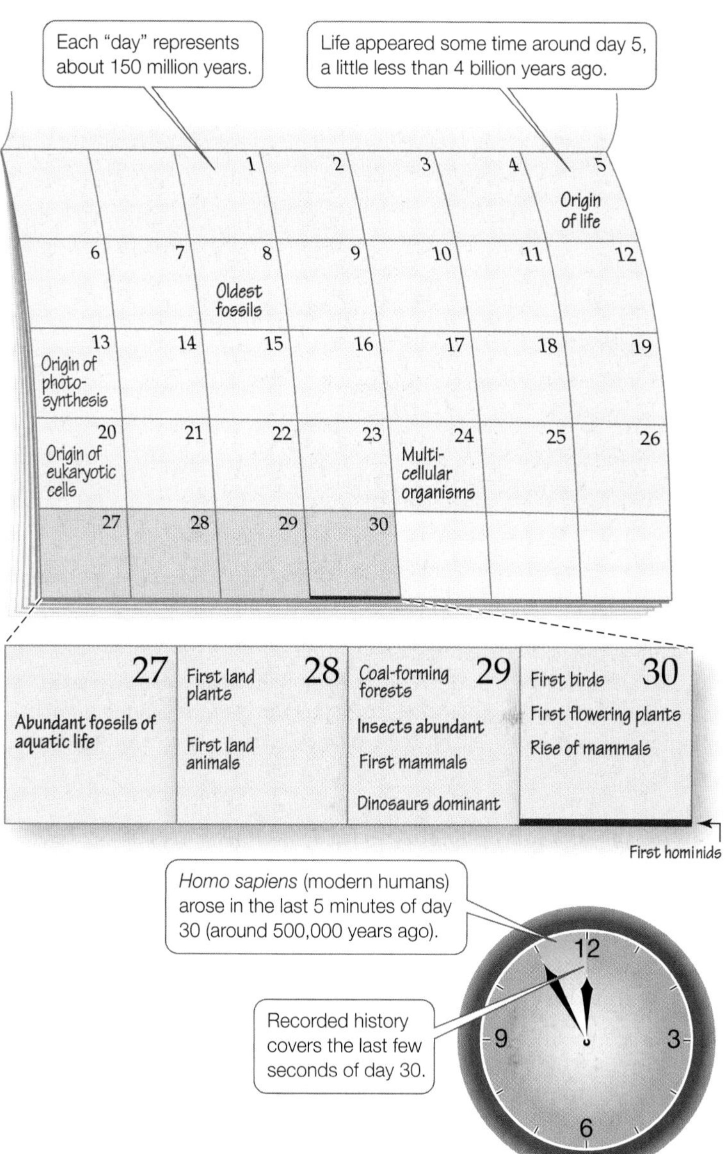


**1.2 Life's Timeline** Depicting the 4.6 billion years of Earth's history on the scale of a 30-day month provides a sense of the immensity of evolutionary time.

## Cellular structure evolved in the common ancestor of life

Another important step in the history of life was the enclosure of complex proteins and other biological molecules by membranes that contained them in a compact internal environment separate from the surrounding (external) environment. Molecules called fatty acids played a critical role because these molecules do not dissolve in water; rather they form membranous

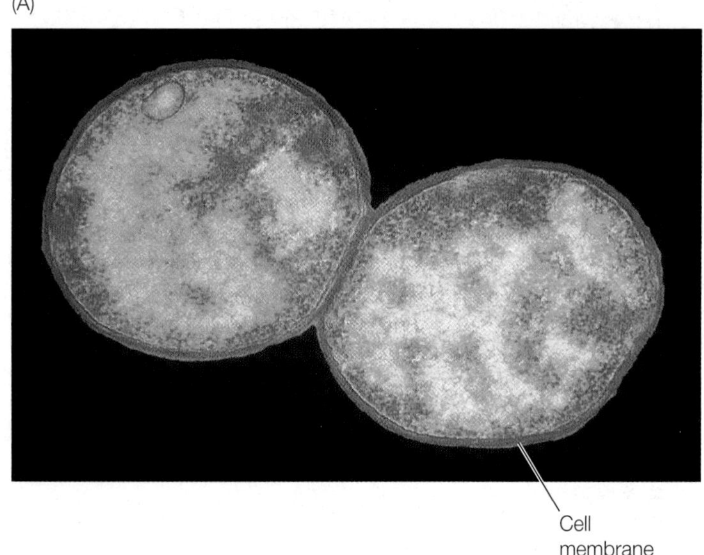


**1.3 Cells Are Building Blocks for Life** These photographs of cells were taken with a transmission electron microscope (see Figure 5.3) and enhanced with added color to highlight details. (A) Two prokaryotic cells of an *Enterococcus* bacterium that lives in the human digestive system. Prokaryotes are unicellular organisms with genetic and biochemical material enclosed inside a single membrane. (B) A human white blood cell (lymphocyte) represents one of the many specialized cell types that make up a multicellular eukaryote. Multiple membranes within the cell-enclosing outer membrane segregate the different biochemical processes of eukaryotic cells.

films that, when agitated, can form spherical structures. These membranous structures could have enveloped assemblages of biological molecules. The creation of an internal environment that concentrated the reactants and products of chemical reactions opened up the possibility that those reactions could be integrated and controlled within a tiny cell (**Figure 1.3**). Scientists postulate that this natural process of membrane formation resulted in the first cells with the ability to reproduce—that is, the evolution of the first cellular organisms.

For the first few billion years of cellular life, all the organisms that existed were unicellular and were enclosed by a single outer membrane. Such organisms, like the bacteria that are still abundant on Earth today, are called **prokaryotes**. Two main groups of prokaryotes emerged early in life's history: the **bacteria** and **archaea**. Some representatives of each of these groups began to live in a close, interdependent relationship with one another, and eventually merged to form a third major lineage of life, the **eukaryotes**. In addition to their outer membranes, the cells of eukaryotes have internal membranes that enclose specialized organelles within their cells. Eukaryote organelles include the nucleus that contains the genetic material and the mitochondria that power the cell. The structure of prokaryote and eukaryote cells and their membranes are the subjects of Part Two.

At some point, the cells of some eukaryotes failed to separate after cell division, remaining attached to each other. Such permanent colonial aggregations of cells made it possible for some of the associated cells to specialize in certain functions, such as reproduction, while other cells specialized in other functions, such as absorbing nutrients. This **cellular specialization** enabled multicellular eukaryotes to increase in size and become more efficient at gathering resources and adapting to specific environments.

## Photosynthesis allows some organisms to capture energy from the sun

Living cells require energy in order to function, and the biochemistry of the fundamental processes of energy conversion that drive life is covered in Part Three.

To fuel their cellular **metabolism** (energy transformations), the earliest prokaryotes took in small molecules directly from their environment and broke them down to their component atoms, thus releasing and using the energy contained in the chemical bonds. Many modern prokaryotes still function this way, and they function very successfully. But about 2.5 billion years ago, the emergence of **photosynthesis** changed the nature of life on Earth.

The chemical reactions of photosynthesis transform the energy of sunlight into a form of biological energy that powers the synthesis of large molecules. These large molecules can then be broken down to provide metabolic energy. Photosynthesis is the basis of much of life on Earth today because its energy-capturing processes provide food for other organisms. Early photosynthetic cells were probably similar to present-day prokaryotes called cyanobacteria (**Figure 1.4**). Over time, photosynthetic prokaryotes became so abundant that vast quantities of oxygen gas ($O_2$), which is a by-product of photosynthesis, began to accumulate in the atmosphere.

During the early eons of life, there was no $O_2$ in Earth's atmosphere. In fact, $O_2$ was poisonous to many of the prokaryotes living at that time. As $O_2$ levels increased, however, those

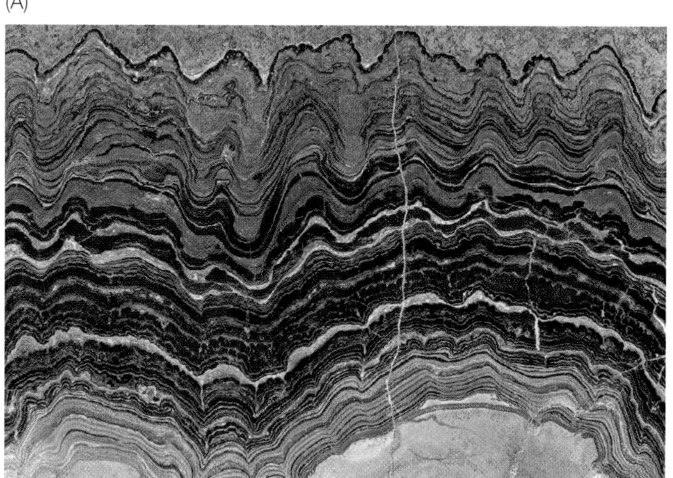

**1.4 Photosynthetic Organisms Changed Earth's Atmosphere** (A) Colonies of photosynthetic cyanobacteria and other microorganisms produced structures called stromatolites that were preserved in the ancient fossil record. This section of fossilized stromatolite reveals layers representing centuries of growth. (B) Living stromatolites can still be found in appropriate environments.

organisms that *did* tolerate $O_2$ were able to proliferate. The abundance of $O_2$ opened up vast new avenues of evolution because **aerobic metabolism**—a biochemical process that uses $O_2$ to extract energy from nutrient molecules—is far more efficient than **anaerobic metabolism** (which does not use $O_2$). Aerobic metabolism allows organisms to grow larger and is used by the majority of organisms today.

Oxygen in the atmosphere also made it possible for life to move onto land. For most of life's history, UV radiation falling on Earth's surface was so intense that it destroyed any organism that was not well shielded by water. But the accumulation of photosynthetically generated $O_2$ in the atmosphere for more than 2 billion years gradually produced a thick layer of ozone ($O_3$) in the upper atmosphere. By about 500 million years ago, the ozone layer was sufficiently dense and absorbed enough of the sun's UV radiation to make it possible for organisms to leave the protection of the water and live on land.

## Biological information is contained in a genetic language common to all organisms

The information that specifies what an organism will look like and how it will function—its "blueprint" for existence—is contained in the organism's **genome**: the sum total of all the DNA molecules contained in each of its cells. **DNA** (deoxyribonucleic acid) molecules are long sequences of four different subunits called **nucleotides**. The sequence of these four nucleotides contains genetic information. **Genes** are specific segments of DNA that encode the information the cell uses to create amino acids and form them into proteins (**Figure 1.5**). Protein molecules govern the chemical reactions within cells and form much of an organism's structure.

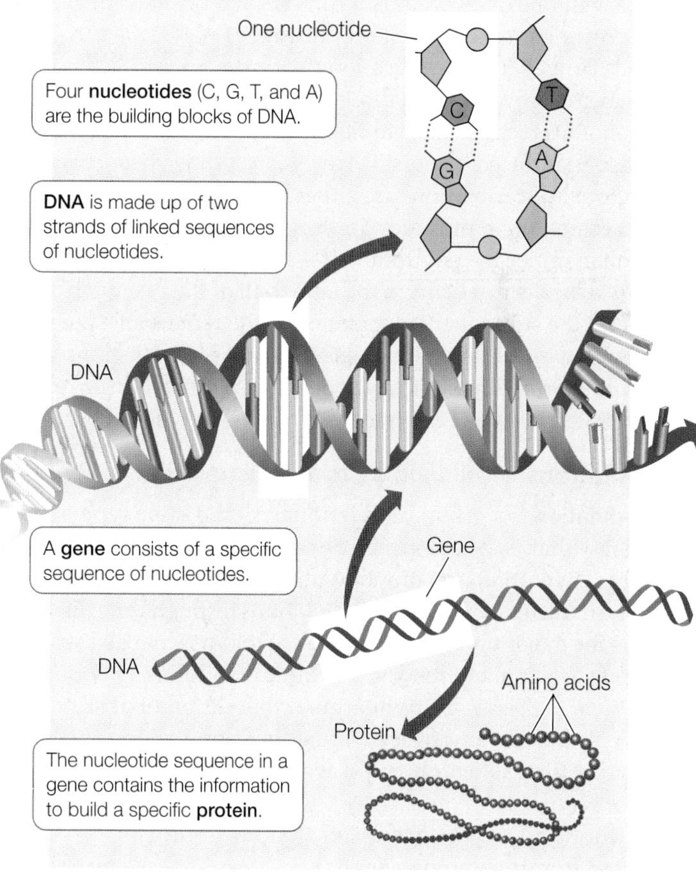


**1.5 DNA Is Life's Blueprint** The instructions for life are contained in the sequences of nucleotides in DNA molecules. Specific DNA nucleotide sequences comprise genes. The average length of a single human gene is 16,000 nucleotides. The information in each gene provides the cell with the information it needs to manufacture molecules of a specific protein.

By analogy with a book, the nucleotides of DNA are like the letters of an alphabet, and protein molecules are sentences. Combinations of proteins that form structures and control biochemical processes are the paragraphs. The structures and processes that are organized into different systems with specific tasks (such as digestion or transport) are the chapters of the book, and the complete book is the organism. If you were to write out your own genome using four letters to represent the four nucleotides, you would write more than 3 billion letters. Using the size type you are reading now, your genome would fill about 1,000 books the size of this one. The mechanisms of evolution are the authors and editors of all the books in the library of life.

All the cells of a multicellular organism contain essentially the same genome, yet different cells have different functions and form different structures—contractile proteins form in muscle cells, hemoglobin in red blood cells, digestive enzymes in gut cells, and so on. Therefore different types of cells in an organism must express different parts of the genome. How cells control gene expression in ways that enable a complex organism to develop and function is a major focus of current biological research.

The genome of an organism consists of thousands of genes. This entire genome must be replicated as new cells are produced. However, the replication process is not perfect, and a few errors, known as mutations, are likely to occur each time the genome is replicated. Mutations occur spontaneously; they can also be induced by outside factors, including chemicals and radiation. Most mutations are either harmful or have no effect, but occasionally a mutation improves the functioning of the organism under the environmental conditions it encounters.

The discovery of DNA in the latter half of the twentieth century and the subsequent elucidation of the remarkable mechanisms by which this material encodes and transmits information transformed biological science. These crucial discoveries are detailed in Parts Four and Five.

## Populations of all living organisms evolve

A **population** is a group of individuals of the same type of organism—that is, of the same **species**—that interact with one another. **Evolution** acts on populations; it is the change in the genetic makeup of biological populations through time. Evolution is the major unifying principle of biology. Charles Darwin compiled factual evidence for evolution in his 1859 book *On the Origin of Species*. Darwin argued that differential survival and reproduction among individuals in a population, which he termed **natural selection**, could account for much of the evolution of life.

Although Darwin proposed that all organisms are descended from a common ancestor and therefore are related to one another, he did not have the advantage of understanding the mechanisms of genetic inheritance and mutation. Even so, he observed that offspring resembled their parents; therefore, he surmised, such mechanisms had to exist. Part Six will describe how Darwin's theory of natural selection is both supported and explained by the massive body of molecular genetic data elucidated during the twentieth century, and how these elements coincide and mesh in the modern field of evolutionary biology.

If all the organisms on Earth today are the descendants of a single kind of unicellular organism that lived almost 4 billion years ago, how have they become so different? As mentioned earlier, organisms reproduce by replicating their genomes, and mutations are introduced almost every time a genome is replicated. Some of these mutations give rise to structural and functional changes in organisms. As individuals mate with one another, the genetic variants stemming from mutation can change in frequency within a population, and the population is said to evolve.

Any population of a plant or animal species displays variation, and if you select breeding pairs on the basis of some particular trait, that trait is more likely to be present in their offspring than in the general population. Darwin himself bred pigeons, and was well aware of how pigeon fanciers selected breeding pairs to produce offspring with unusual feather patterns, beak shapes, or body sizes (see Figure 21.5). He realized that if humans could select for specific traits in domesticated animals, the same process could operate in nature; hence the term "natural selection" as opposed to artificial (human-imposed) selection.

How does natural selection function? Darwin postulated that different probabilities of survival and reproductive success would do the job. He reasoned that the reproductive capacity of plants and animals, if unchecked, would result in unlimited growth of populations, but we do not observe such growth in nature; in most species, only a small percentage of an individual's offspring will survive to reproduce. Thus any trait that confers even a small increase in the probability that its possessor will survive and reproduce would spread in the population.

Because organisms with certain traits survive and reproduce best under specific sets of conditions, natural selection leads to **adaptations**: structural, physiological, or behavioral traits that enhance an organism's chances of survival and reproduction in its environment (**Figure 1.6**). In addition to natural selection, evolutionary processes such as sexual selection (for example, selection due to mate choice) and genetic drift (the random fluctuation of gene frequencies in a population due to chance events) contribute to the rise of biodiversity. These processes operating over evolutionary history have led to the remarkable diversity of life on Earth.

## Biologists can trace the evolutionary tree of life

As populations become geographically isolated from one another, they evolve differences. As populations diverge from one another, individuals in each population become less likely to reproduce with individuals of the other population. Eventually these differences between populations become so great that the two populations are considered different species. Thus species that share a fairly recent evolutionary history are generally more similar to each other than species

(A) *Dyscophus guineti*

(B) *Xenopus laevis*

(C) *Agalychnis callidryas*

(D) *Rhacophorus nigropalmatus*

**1.6 Adaptations to the Environment** The limbs of frogs show adaptations to the different environments of each species. (A) This terrestrial frog walks across the ground using its short legs and peglike digits (toes). (B) Webbed rear feet are evident in this highly aquatic species of frog. (C) This arboreal species has toe pads, which are adaptations for climbing. (D) A different arboreal species has extended webbing between the toes, which increases surface area and allows the frog to glide from tree to tree.

that share an ancestor in the more distant past. By identifying, analyzing, and quantifying similarities and differences between species, biologists can construct **phylogenetic trees** that portray the evolutionary histories of the different groups of organisms.

Tens of millions of species exist on Earth today; many times that number lived in the past but are now extinct. Biologists give each of these species a distinctive scientific name formed from two Latinized names—a **binomial**. The first name identifies the species' **genus** (plural *genera*)—a group of species that share a recent common ancestor. The second is the name of the species. For example, the scientific name for the human species is *Homo sapiens*: *Homo* is our genus, *sapiens* our species. *Homo* is Latin for "man," and *sapiens* is from the Latin word for "wise" or "rational." Our closest relatives in the genus *Homo* are the Neanderthals, *Homo neanderthalensis*. Neanderthals are now extinct and are known only from their fossil remains.

Much of biology is based on comparisons among species, and these comparisons are useful precisely because we can place species in an evolutionary context relative to one another. Our ability to do this has been greatly enhanced in recent decades by our ability to sequence and compare the genomes of different species. Genome sequencing and other molecular techniques have allowed biologists to augment evolutionary knowledge based on the fossil record with a vast array of molecular evidence. The result is the ongoing compilation of phylogenetic trees that document and diagram evolutionary relationships as part of an overarching tree of life, the broadest categories of which are shown in **Figure 1.7** and will be surveyed in more detail in Part Seven. (The tree is expanded in Appendix A, and you can also explore the tree interactively.)

Although many details remain to be clarified, the broad outlines of the tree of life have been determined. Its branching patterns are based on a rich array of evidence from fossils, structures, metabolic processes, behavior, and molecular

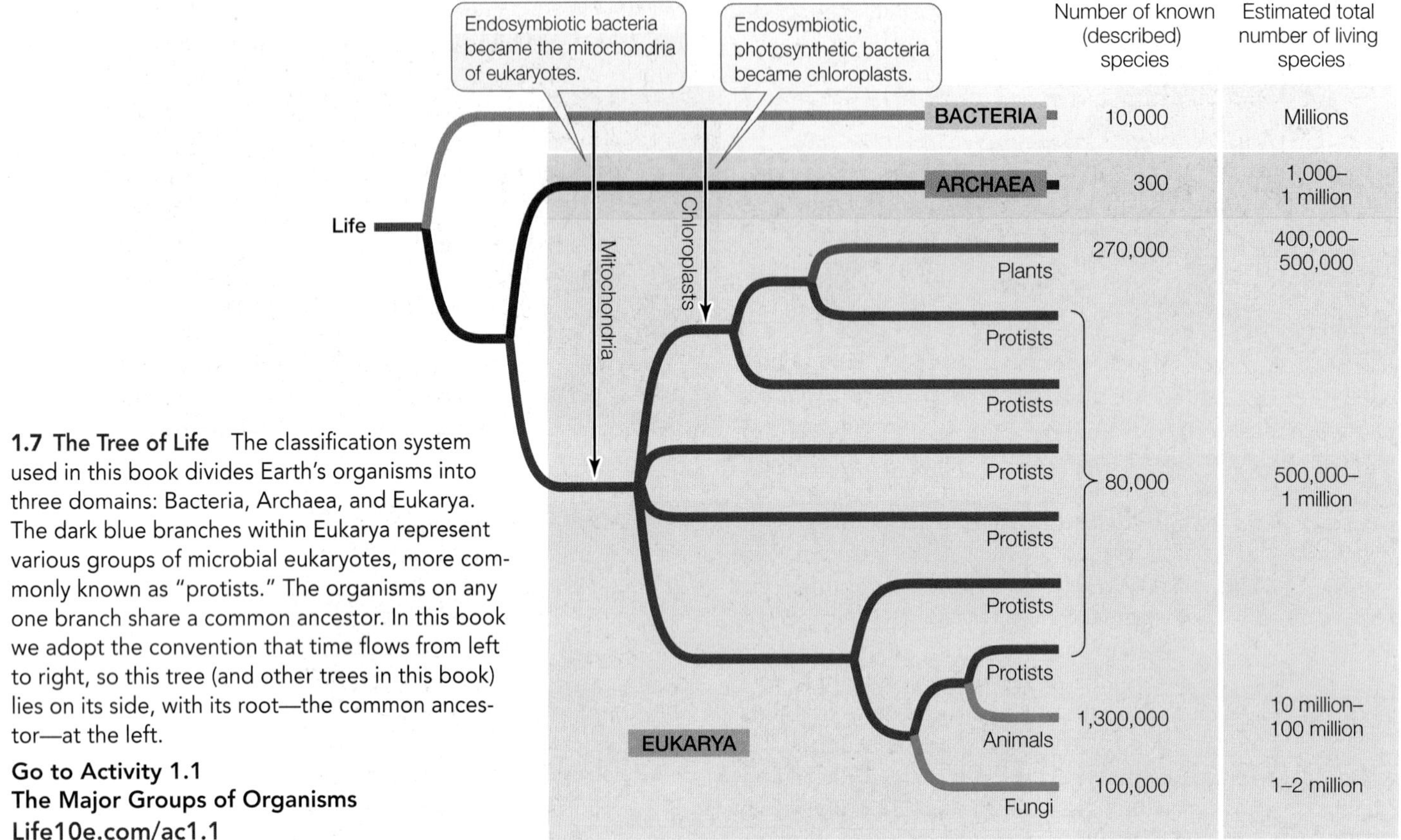


**1.7 The Tree of Life** The classification system used in this book divides Earth's organisms into three domains: Bacteria, Archaea, and Eukarya. The dark blue branches within Eukarya represent various groups of microbial eukaryotes, more commonly known as "protists." The organisms on any one branch share a common ancestor. In this book we adopt the convention that time flows from left to right, so this tree (and other trees in this book) lies on its side, with its root—the common ancestor—at the left.

**Go to Activity 1.1 The Major Groups of Organisms Life10e.com/ac1.1**

analyses of genomes. Two of the three main domains of life—Archaea and Bacteria—are single-celled prokaryotes, as mentioned earlier in this chapter. However, members of these two groups differ so fundamentally in their metabolic processes that they are believed to have separated into distinct evolutionary lineages very early. Species belonging to the third domain—Eukarya—have eukaryotic cells whose mitochondria and chloroplasts originated from endosymbioses of bacteria.

Plants, fungi, and animals are examples of familiar multicellular eukaryotes that evolved independently, from different groups of the unicellular eukaryotes informally known as protists. We know that plants, fungi, and animals had independent origins of multicellularity because each of these three groups is most closely

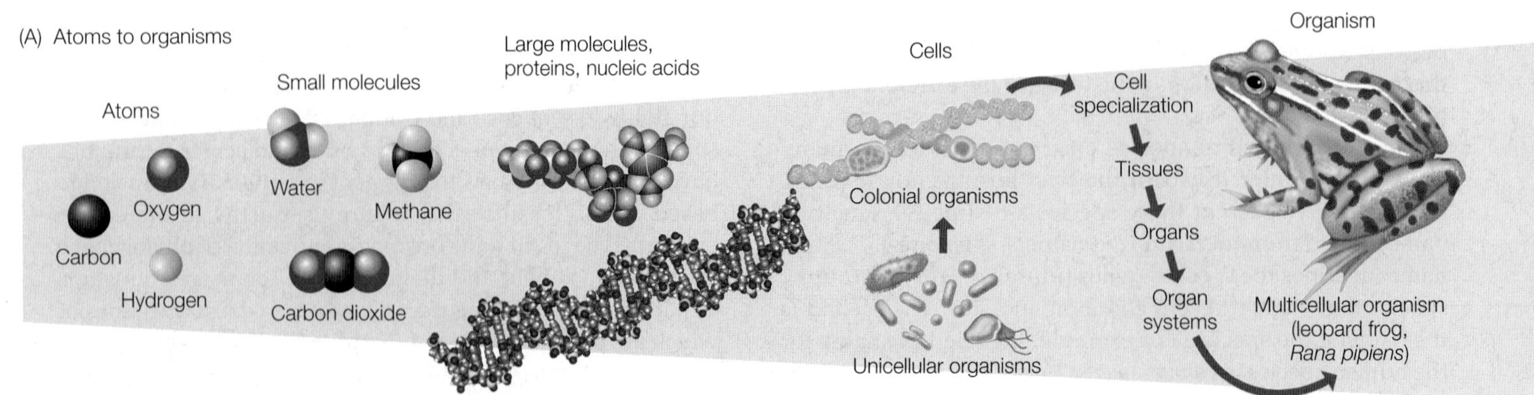


**1.8 Biology Is Studied at Many Levels of Organization** (A) Life's properties emerge when DNA and other molecules are organized in cells, which form building blocks for organisms. (B) Organisms exist in populations and interact with other populations to form communities, which interact with the physical environment to make up the many ecosystems of the biosphere.

**Go to Activity 1.2 The Hierarchy of Life Life10e.com/ac1.2**

related to different groups of unicellular protists, as can be seen from the branching pattern of Figure 1.7.

### Cellular specialization and differentiation underlie multicellular life

Looking back at Figure 1.2, you can see that for more than half of Earth's history, all life was unicellular. Unicellular species remain ubiquitous and highly successful in the present, even though the diverse multicellular organisms, owing to their much larger size, may seem to us to dominate the planet.

With the evolution of cells specialized for different functions within the same organism, these differentiated cells lost many of the functions carried out by single-celled organisms, and a **biological hierarchy** emerged (**Figure 1.8A**). To accomplish their specialized tasks, assemblages of differentiated cells are organized into **tissues**. For example, a single muscle cell cannot generate much force, but when many cells combine to form the tissue of a working muscle, considerable force and movement can be generated. Different tissue types are organized to form **organs** that accomplish specific functions. The heart, brain, and stomach are each constructed of several types of tissues, as are the roots, stems, and leaves of plants. Organs whose functions are interrelated can be grouped into **organ systems**; the esophagus, stomach, and intestines, for example, are all part of the digestive system. The physiology of two major groups of multicellular organisms (land plants and animals) is discussed in detail in Parts Eight and Nine, respectively.

### Living organisms interact with one another

Organisms do not live in isolation, and the internal hierarchy of the individual organism is matched by the external hierarchy of the biological world (**Figure 1.8B**). As mentioned earlier in this section, a group of individuals of the same species that interact with one another is a population. The populations of all the species that live and interact in a defined area (areas are defined in different ways and can be small or large) are called a **community**. Communities together with their abiotic (nonliving) environment constitute an **ecosystem**.

Individuals in a population interact in many different ways. Animals eat plants and other animals (usually members of another species) and compete with other species for food and other resources. Some animals prevent other individuals of their own species from exploiting a resource, be it food, nesting sites, or mates. Animals may also cooperate with members of their own species, forming social units such as a termite colony or a flock of birds. Such interactions have resulted in the evolution of social behaviors such as communication and courtship displays.

Plants also interact with their external environment, which includes other plants, fungi, animals, and microorganisms. All terrestrial plants depend on partnerships with fungi, bacteria, and animals. Some of these partnerships are necessary to obtain nutrients, some to produce fertile seeds, and still others to disperse seeds. Plants compete with each other for light and water and have ongoing evolutionary interactions with the animals that eat them. Through time, many adaptations have evolved in plants that protect them from predation (such as thorns) or that help then attract the animals that assist in their reproduction (such as sweet nectar or colorful flowers). The interactions of populations of plant and animal species in a community are major evolutionary forces that produce specialized adaptations.

Communities interacting over a broad geographic area with distinguishing physical features form ecosystems; examples

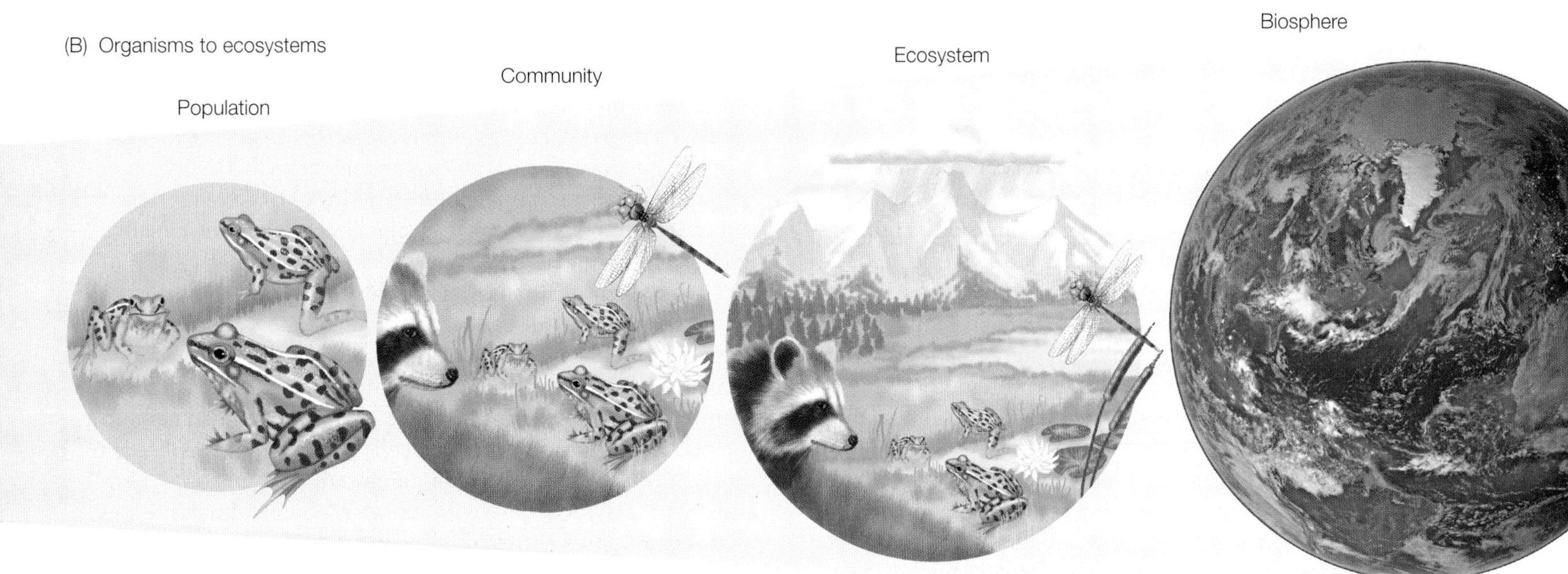

(B) *Spermophilus parryii*

(A) *Propithecus verreauxi*

**Go to Media Clip 1.1**
**Leaping Lemurs**
**Life10e.com/mc1.1**

**1.9 Energy Can Be Used Immediately or Stored** (A) Animal cells break down food molecules and use the energy contained in the chemical bonds of those molecules to do mechanical work, such as running and jumping. This composite image of a sifaka (a type of lemur from Madagascar) shows the same individual at five stages of a single jump. (B) The cells of this Arctic ground squirrel have broken down the complex carbohydrates in the plants it consumed and converted those molecules into fats. The fats are stored in the animal's body to provide an energy supply for the cold months.

include Arctic tundra, coral reef, and tropical rainforest. The ways in which species interact with one another and with their environment in populations, communities, and ecosystems is the subject of ecology, covered in Part Ten of this book.

## Nutrients supply energy and are the basis of biosynthesis

Living organisms acquire nutrients from the environment. Nutrients supply the organism with energy and raw materials for carrying out biochemical reactions. Life depends on thousands of biochemical reactions that occur inside cells. Some of these reactions break down nutrient molecules into smaller chemical units, and in the process some of the energy contained in the chemical bonds of the nutrients is captured by high-energy molecules that can be used to do different kinds of cellular work.

One obvious kind of work cells do is mechanical—moving molecules from one cellular location to another, moving whole cells or tissues, or even moving the organism itself, as muscles do (**Figure 1.9A**). The most basic cellular work is the building, or synthesis, of new complex molecules and structures from smaller chemical units. For example, we are all familiar with the fact that carbohydrates eaten today may be deposited in the body as fat tomorrow (**Figure 1.9B**). Still another kind of work is the electrical work that is the essence of information processing in nervous systems.

The myriad biochemical reactions that take place in cells are integrally linked in that the products of one reaction are the raw materials of the next. These complex networks of reactions must be integrated and precisely controlled; when they are not, the result is malfunction and disease.

## Living organisms must regulate their internal environment

The specialized cells, tissues, and organ systems of multicellular organisms exist in and depend on an **internal environment** that is made up of extracellular fluids. Because this environment serves the needs of the cells, its physical and chemical composition must be maintained within a narrow range of physiological conditions that support survival and function. The maintenance of this narrow range of conditions is known as **homeostasis**. A relatively stable internal (but extracellular) environment means that cells can function efficiently even when conditions outside the organism's body become unfavorable for cellular processes.

The organism's regulatory systems obtain information from sensory cells that provide information about both the internal and external conditions the organism is subject to at a given time. The cells of regulatory systems process and integrate this information and send signals to components of physiological systems, which can change in response to these signals so that the organism's internal environment remains reasonably constant.

The concept of homeostasis extends beyond the internal environment of multicellular organisms, however. In both unicellular and multicellular organisms, individual cells must regulate physiological parameters (such as acidity and salinity), maintaining them within a range that allows those cells to survive and function. Individual cells regulate these properties through actions of the plasma membrane that encloses them and are the cell's interface with its environment (either internal or external). Thus self-regulation to maintain a more or less constant internal environment is a general attribute of all living organisms.

RECAP 1.1

All organisms are related by common descent from a single ancestral form. They contain genetic information that encodes how they look and how they function. They also reproduce, extract energy from their environment, and use energy to do biological work, synthesize complex molecules to construct biological structures, regulate their internal environment, and interact with one another.

- Why did the evolution of photosynthesis so radically affect the course of life on Earth? **See pp. 4–5**
- Describe the relationship between evolution by natural selection and the genetic code. **See p. 6**
- What information have biologists used to construct a tree of life? **See pp. 6–8 and Figure 1.7**
- What do we mean by "homeostasis," and why is it crucial to living organisms? **See p. 10**

The preceding section briefly outlined the major features of life—features that will be covered in depth in subsequent chapters of this book. Before going into the details of what we know about life, however, it is important to understand how scientists obtain information and how they use that information in broadening our understanding of Earth's diverse living organisms and putting this understanding to practical use.

## 1.2 How Do Biologists Investigate Life?

Scientific investigations are based on observation, data, experimentation, and logic. Scientists use many different tools and methods in making observations, collecting data, designing experiments, and applying logic, but they are always guided by established principles that allow us to discover new aspects about the structure, function, evolution, and interactions of organisms.

### Observing and quantifying are important skills

Biologists have always observed the world around them, but today our ability to observe is greatly enhanced by technologies such as electron microscopes, rapid genome sequencing, magnetic resonance imaging, and global positioning satellites. These technologies allow us to observe everything from the distribution of molecules in the body to the movement of animals across continents and oceans.

Observation is a basic tool of biology, but as scientists we must also be able to quantify the information, or **data**, we collect as we observe. Whether we are testing a new drug or mapping the migrations of the great whales, applying mathematical and statistical calculations to the data we collect is essential. For example, biologists once classified organisms based entirely on qualitative descriptions of the physical differences among them. There was no way of objectively determining evolutionary relationships of organisms, and biologists had to depend on the fossil record for insight. Today our ability to quantify the molecular and physical differences among species, combined with explicit mathematical models of the evolutionary process, enables quantitative analyses of evolutionary history. These mathematical calculations, in turn, facilitate comparative investigations of all other aspects of an organism's biology.

### Scientific methods combine observation, experimentation, and logic

Textbooks often describe "*the* scientific method," as if there is a single, simple flow chart that all scientists follow. This is an oversimplification. Although flow charts such as the one shown in **Figure 1.10** incorporate much of what scientists do, you should not conclude that scientists necessarily progress through the steps of the process in one prescribed, linear order.

Observations lead to questions, and scientists make additional observations and often do experiments to answer those

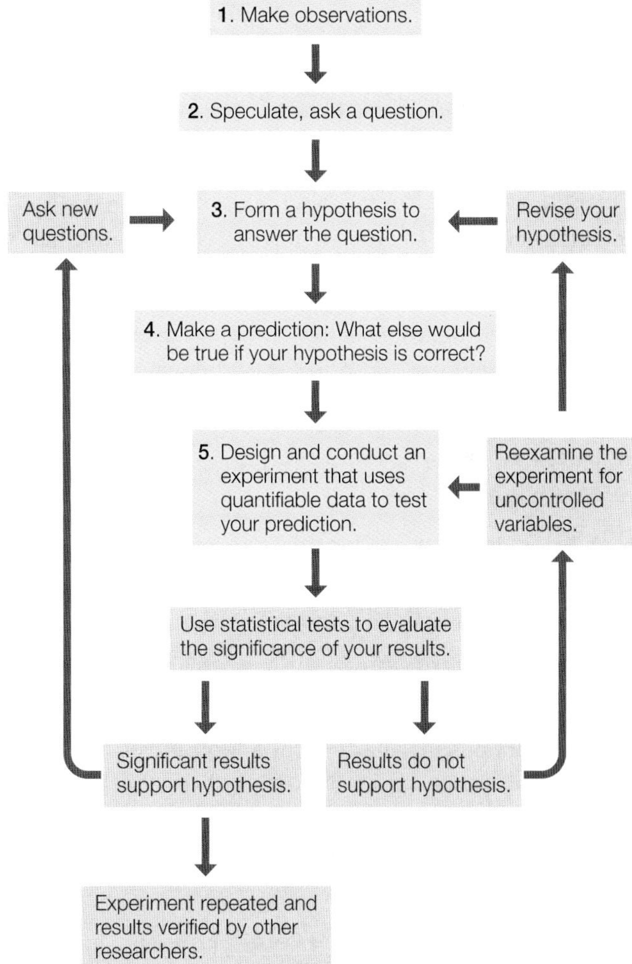


**1.10 Scientific Methodology** The process of observation, speculation, hypothesis, prediction, and experimentation is a cornerstone of modern science, although scientists may initiate their research at several different points. Answers gleaned through experimentation lead to new questions, more hypotheses, further experiments, and expanding knowledge.

questions. This hypothesis–prediction approach traditionally has five steps: (1) making observations; (2) asking questions; (3) forming hypotheses, which are tentative answers to the questions; (4) making predictions based on the hypotheses; and (5) testing the predictions by making additional observations or conducting experiments.

After posing a question, a scientist often uses **inductive logic** to propose a tentative answer. Inductive logic involves taking observations or facts and creating a new proposition that is compatible with those observations or facts. Such a tentative proposition is a **hypothesis** (plural *hypotheses*). In formulating a hypothesis, scientists put together the facts and data at their disposal to formulate one or more possible answers to the question. For example, at the opening of this chapter you learned that scientists have observed the rapid decline of amphibian populations worldwide and are asking why. Some scientists have hypothesized that a fungal disease is a cause; other scientists have hypothesized that increased exposure to ultraviolet radiation is a cause. Tyrone Hayes hypothesized that exposure to agricultural chemicals, specifically the widely used herbicide atrazine, could be a cause.

The next step in the scientific method is to apply a different form of logic—**deductive logic**—that starts with a statement believed to be true (the hypothesis) and then goes on to predict what facts would also have to be true to be compatible with that statement. Hayes knew that atrazine is commonly applied in the spring, when amphibians are breeding, and that atrazine is a common contaminant in the waters in which amphibians live as they develop into adults. Thus he predicted that frog tadpoles exposed to atrazine would show adverse effects of the chemical once they reached adulthood.

**Go to Animated Tutorial 1.1**
**Using Scientific Methodology**
Life10e.com/at1.1

## Good experiments have the potential to falsify hypotheses

Once predictions are made from a hypothesis, experiments can be designed to test those predictions. The most informative experiments are those that have the ability to show that the prediction is wrong. If the prediction is wrong, the hypothesis must be questioned, modified, or rejected.

There are two general types of experiments, both of which compare data from different groups or samples. A **controlled experiment** manipulates one or more of the factors being tested; **comparative experiments** compare unmanipulated data gathered from different sources. As described at the opening of this chapter, Tyrone Hayes and his colleagues conducted both types of experiments to test the prediction that the herbicide atrazine, a contaminant in freshwater ponds and streams throughout the world, affects the development of frogs.

INVESTIGATING**LIFE**

**1.11 Controlled Experiments Manipulate a Variable** The Hayes laboratory created controlled environments that differed only in the concentrations of atrazine in the water. Eggs from leopard frogs (*Rana pipiens*) raised specifically for laboratory use were allowed to hatch and the tadpoles were separated into experimental tanks containing water with different concentrations of atrazine.[a]

**HYPOTHESIS** Exposure to atrazine during larval development causes abnormalities in the reproductive tissues of male frogs.

*Method*

1. Establish 9 tanks in which all attributes are held constant except the water's atrazine concentration. Establish 3 atrazine conditions (3 replicate tanks per condition): 0 ppb (control condition), 0.1 ppb, and 25 ppb.
2. Place *Rana pipiens* tadpoles from laboratory-reared eggs in the 9 tanks (30 tadpoles per replicate).
3. When tadpoles have transitioned into adults, sacrifice the animals and evaluate their reproductive tissues.
4. Test for correlation of degree of atrazine exposure with the presence of abnormalities in the gonads (testes) of male frogs.

*Results*

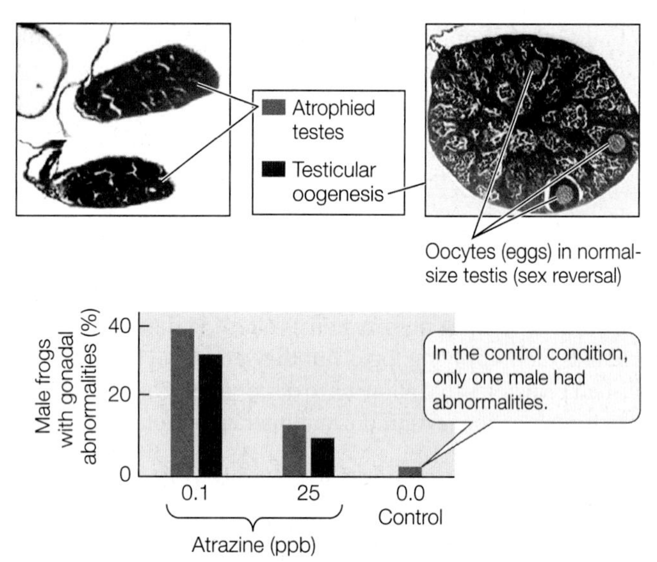


**CONCLUSION** Exposure to atrazine at concentrations as low as 0.1 ppb induces abnormalities in the gonads of male frogs. The effect is not proportional to the level of exposure.

Go to **BioPortal** for discussion and relevant links for all INVESTIGATING**LIFE** figures.

[a]Hayes, T. et al. 2003. *Environmental Health Perspectives III*: 568–575.

In a controlled experiment, we start with groups or samples that are as similar as possible. We predict on the basis of our hypothesis that some critical factor, or variable, has an effect on the phenomenon we are investigating. We devise some method to manipulate *only that variable* in an "experimental" group and compare the resulting data with data from an unmanipulated "control" group. If the predicted difference occurs, we then apply statistical tests to ascertain the probability that the manipulation created the difference (as opposed to the difference being the result of random

INVESTIGATING**LIFE**

**1.12 Comparative Experiments Look for Differences among Groups** To see whether the presence of atrazine correlates with testicular abnormalities in male frogs, the Hayes lab collected frogs and water samples from different locations around the U.S. The analysis that followed was "blind," meaning that the frogs and water samples were coded so that experimenters working with each specimen did not know which site the specimen came from.[a]

**HYPOTHESIS** Presence of the herbicide atrazine in environmental water correlates with gonadal abnormalities in frog populations.

*Method*

1. Based on commercial sales of atrazine, select 4 sites (sites 1–4) less likely and 4 sites (sites 5–8) more likely to be contaminated with atrazine.
2. Visit all sites in the spring (i.e., when frogs have transitioned from tadpoles into adults); collect frogs and water samples.
3. In the laboratory, sacrifice frogs and examine their reproductive tissues, documenting abnormalities.
4. Analyze the water samples for atrazine concentration (the sample for site 7 was not tested).
5. Quantify and correlate the incidence of reproductive abnormalities with environmental atrazine concentrations.

*Results*

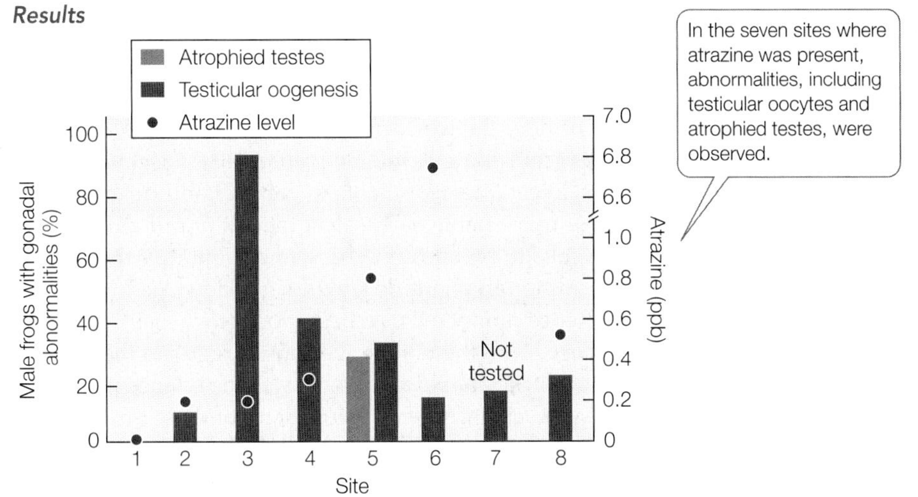


**CONCLUSION** Reproductive abnormalities exist in frogs from environments in which aqueous atrazine concentration is 0.2 ppb or above. The incidence of abnormalities does not appear to be proportional to atrazine concentration at the time of transition to adulthood.

Go to **BioPortal** for discussion and relevant links for all INVESTIGATING**LIFE** figures.

[a]Hayes, T. et al. 2003. *Nature* 419: 895–896.

chance). **Figure 1.11** describes one of the many controlled experiments performed by the Hayes laboratory to quantify the effects of atrazine on male frogs.

The basis of controlled experiments is that one variable is manipulated while all others are held constant. The variable that is manipulated is called the **independent variable**, and the response that is measured is the **dependent variable**. A good controlled experiment is not easy to design because biological variables are so interrelated that it is difficult to alter just one.

A comparative experiment starts with the prediction that there will be a difference between samples or groups based on the hypothesis. In comparative experiments, however, we cannot control the variables; often we cannot even identify all the variables that are present. We are simply gathering and comparing data from different sample groups.

When his controlled experiments indicated that atrazine indeed affects reproductive development in frogs, Hayes and his colleagues performed a comparative experiment. They collected frogs and water samples from eight widely separated sites across the United States and compared the incidence of abnormal frogs from environments with very different levels of atrazine (**Figure 1.12**). Of course, the sample sites differed in many ways besides the level of atrazine present.

The results of experiments frequently reveal that the situation is more complex than the hypothesis anticipated, thus raising new questions. In the Hayes experiments, for example, there was no clear direct relationship between the *amount* of atrazine present and the percentage of abnormal frogs: there were fewer abnormal frogs at the highest concentrations of atrazine than at lower concentrations. There are no "final answers" in science. Investigations consistently reveal more complexity than we expect, so scientists must design systematic approaches to identify, assess, and understand that complexity.

## Statistical methods are essential scientific tools

Whether we do comparative or controlled experiments, at the end we have to decide whether there is a difference between the samples, individuals, groups, or populations in the study. How do we decide whether a measured difference is enough to support or falsify a hypothesis? In other words, how do we decide in an unbiased, objective way that the measured difference is significant?

Significance can be measured with statistical methods. Scientists use statistics because they recognize that variation is always present in any set of measurements. Statistical tests calculate the probability that the differences observed in an experiment could be due to random variation. The results of statistical tests are therefore probabilities. A statistical test starts with a **null hypothesis**—the premise that any observed differences are simply the result of random differences that arise from drawing two finite samples from the same population. When quantified observations, or data, are collected, statistical methods are applied to those data to calculate the likelihood that the null hypothesis is correct.

More specifically, statistical methods tell us the probability of obtaining the same results by chance even if the null hypothesis were true. *We need to eliminate, insofar as possible, the chance that any differences showing up in the data are merely the result of random variation in the samples tested.* Scientists generally conclude that the differences they measure are significant if statistical tests show that the probability of error (that is, the probability that a difference as large as the one observed could be obtained by mere chance) is 5 percent or lower, although more stringent levels of significance may be set for some problems. Appendix B of this book is a short primer on statistical methods that you can refer to as you analyze data that will be presented throughout the text.

## Discoveries in biology can be generalized

Because all life is related by descent from a common ancestor, shares a genetic code, and consists of similar biochemical building blocks, knowledge gained from investigations of one type of organism can, with thought and care, be generalized to other organisms. Biologists use **model systems** for research, knowing that they can extend their findings from such systems to other organisms. For example, our basic understanding of the chemical reactions in cells came from research on bacteria but is applicable to all cells, including those of humans. Similarly, the biochemistry of photosynthesis—the process by which all green plants use sunlight to produce biological molecules—was largely worked out from experiments on *Chlorella*, a unicellular green alga. Much of what we know about the genes that control plant development is the result of work on *Arabidopsis thaliana*, a relative of the mustard plant. Knowledge about how animals, including humans, develop has come from work on sea urchins, frogs, chickens, roundworms, mice, and fruit flies. Being able to generalize from model systems is a powerful tool in biology.

## Not all forms of inquiry are scientific

Science is a unique human endeavor that has certain standards of practice. Other areas of scholarship share with science the practice of making observations and asking questions, but scientists are distinguished by *what they do with their observations* and *how they frame the answers*. Quantifiable data, subjected to appropriate statistical analysis, are critical in evaluating hypotheses (the Working with Data exercises you will find throughout this book are intended to reinforce this way of thinking). In short, scientific observation and evaluation is the most powerful approach humans have devised for learning about the world and how it works.

Scientific explanations for natural processes are objective and reliable because *a hypothesis must be testable* and *a hypothesis must have the potential of being rejected* by direct observations and experiments. Scientists must clearly describe the methods they use to test hypotheses so that other scientists can repeat their results. Not all experiments are repeated, but surprising or controversial results are always subjected to independent verification. Scientists worldwide share this process of testing and rejecting hypotheses, contributing to a common body of scientific knowledge.

If you understand the methods of science, you can distinguish science from non-science. Art, music, and literature all contribute to the quality of human life, but they are not science. They do not use scientific methods to establish what is fact. Religion is not science, although religions have historically attempted to explain natural events ranging from unusual weather patterns to crop failures to human diseases. Most such phenomena that at one time were mysterious can now be explained in terms of scientific principles. Fundamental tenets of religious faith, such as the existence of a supreme deity or deities, cannot be confirmed or refuted by experimentation and are thus outside the realm of science.

The power of science derives from strict objectivity and absolute dependence on evidence based on *reproducible and quantifiable observations*. A religious or spiritual explanation of a natural phenomenon may be coherent and satisfying for the person holding that view, but it is not testable and therefore it is not science. To invoke a supernatural explanation (such as a "creator" or "intelligent designer" with no known bounds) is to depart from the world of science. Science does not necessarily say that religious beliefs are wrong; they are simply not part of the world of science, and many religious beliefs are untestable using scientific methods.

Science describes how the world works. It is silent on the question of how the world "ought to be." Many scientific advances that contribute to human welfare also raise major ethical issues. Recent developments in genetics and developmental biology may enable us to select the sex of our children, to use stem cells to repair our bodies, and to modify the human genome. Although scientific knowledge allows us to do these things, science cannot tell us whether or not we *should* do so or, if we choose to do them, how we should regulate them. Such issues are as crucial to human society as the science itself, and a responsible scientist does not lose sight of these questions or neglect the contributions of the humanities or social sciences in attempting to come to grips with them.

### RECAP 1.2

Scientific methods of inquiry start with the formulation of hypotheses based on observations and data. Comparative and controlled experiments are carried out to test hypotheses.

- Explain the relationship between a hypothesis and an experiment. **See pp. 11–12 and Figure 1.10**
- What is controlled in a controlled experiment? **See pp. 11–12 and Figure 1.11**
- What features characterize questions that can be answered only by using a comparative approach? **See p. 13 and Figure 1.12**
- Explain why arguments must be supported by quantifiable and reproducible data in order to be considered scientific. **See pp. 13–14**
- Why can the results of biological research on one species often be generalized to very different species? **See p. 14**

The vast body of scientific knowledge accumulated over centuries of human civilization allows us to understand and manipulate aspects of the natural world in ways that no other species can. These abilities present us with challenges, opportunities, and above all, responsibilities.

## 1.3 Why Does Biology Matter?

Human beings exist in and depend on a world of living organisms. The oxygen in the air we breathe is produced by photosynthesis conducted by countless billions of individual organisms. The food that fuels our bodies comes from the tissues of other living organism. The fuels that drive our cars and power our electric plants are, for the most part, various forms of carbon molecules produced by living organisms—mostly millions of years ago. Inside and out, our bodies are covered in complex communities of living unicellular organisms, most of which help us maintain our health. There are also harmful species that invade our bodies and can cause mild to serious diseases, or even death. These interactions with other species are not limited to humans. Ecosystem function depends on thousands of complex interactions among the millions of species that inhabit Earth. In other words, understanding biological principles is essential to our lives and for maintaining the functioning of Earth as we know it and depend on it.

### Modern agriculture depends on biology

Agriculture represents some of the earliest human applications of biological principles. Even in prehistoric times, farmers selected the most productive or otherwise favorable plants and animals to use as seed stock for propagation, and over generations farmers continued and refined these practices. His knowledge of this kind of artificial selection helped Charles Darwin understand the importance of natural selection in evolution across all of life.

In modern times, increasing knowledge of plant biology has transformed agriculture in many ways and has resulted in huge boosts in food production (**Figure 1.13**), which in turn has allowed the planet to support a far larger human population than it once could have. Over the past few decades, detailed knowledge of the genomes of many domestic species and the development of technology for directly recombining genes have allowed biologists to develop new breeds and strains of animals, plants, and fungi of agricultural interest. For example, new strains of crop plants are being developed that are resistant to pests or can tolerate drought. Moreover, understanding evolutionary theory allows biologists to devise strategies for the application of pesticides that minimize the evolution of pest resistance. And better understanding of plant–fungus relationships results in better plant health and higher productivity. These are just a few of the many ways that biology continues to inform and improve agricultural practice.

**1.13 A Green Revolution** The agricultural advancements of the last 100 years have vastly increased yields and nutritional value of crops such as grains that sustain the expanding human population. In the last 30 years, these advancements have included genetic recombination techniques. Here a researcher with the U.S. Department of Agriculture works with a strain of "supernutritious" rice that provides high levels of the amino acid lysine.

### Biology is the basis of medical practice

People have speculated about the causes of diseases and searched for methods to combat them since ancient times. Long before the microbial causes of many diseases were known, people recognized that infections could be passed from one person to another, and the isolation of infected persons has been practiced as long as written records have been available.

Modern biological research informs us about how living organisms work, and about why they develop the problems and infections that we call disease. In addition to diseases caused by infection of other organisms, we now know that many diseases are genetic—meaning that variants of genes in our genomes cause particular problems in the way we function. Developing appropriate treatments or cures for diseases depends on understanding the origin, basis, and effects of these diseases, as well understanding the consequences of any changes that we make. For example, the recent resurgence of tuberculosis is the result of the evolution of bacteria that are resistant to antibiotics. Dealing with future tuberculosis epidemics requires understanding aspects of molecular biology, physiology, microbial ecology, and evolution—in other words, many of the general principles of modern biology.

Many of the microbial organisms that are periodically epidemic in human populations have short generation times and high mutation rates. For example, we need yearly vaccines for flu because of the high rate of evolution of influenza viruses, the causative agent of flu. Evolutionary principles help us understand how influenza viruses are changing, and can even help us predict which strains of influenza virus are likely to lead to future flu epidemics. This medical understanding—which combines an application of molecular biology, evolutionary

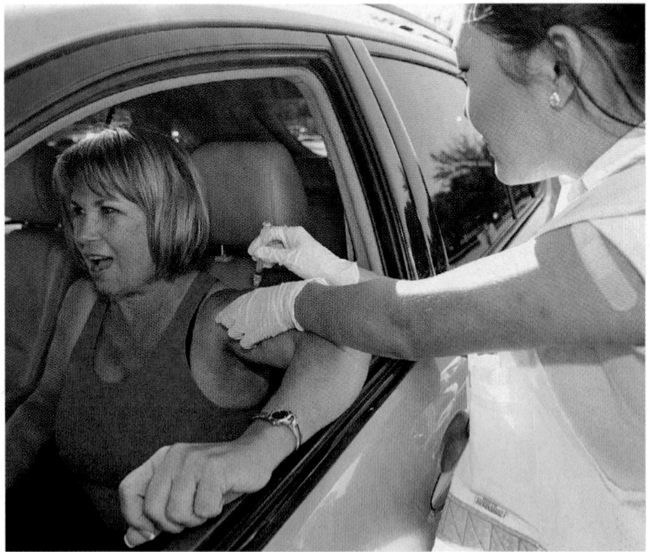

**1.14 Medical Applications of Biology Improve Human Health** Vaccination to prevent disease is a biologically based medical practice that began in the eighteenth century. Today evolutionary biology and genomics provide the basis for constant updates to vaccines that protect humans from virus-borne diseases such as flu. In the developed world, vaccinations have become so commonplace that some are offered on a "drive-through" basis.

theory, and basic principles of ecology—allows medical researchers to develop effective vaccines and other strategies for the control of major epidemics (**Figure 1.14**).

## Biology can inform public policy

Thanks to the deciphering of genomes and our newfound ability to manipulate them, vast new possibilities now exist for controlling human diseases and increasing agricultural productivity—but these capabilities raise ethical and policy issues. How much and in what ways should we tinker with the genes of humans and other species? Does it matter whether the genomes of our crop plants and domesticated animals are changed by traditional methods of controlled breeding and crossbreeding or by the biotechnology of gene transfer? What rules should govern the release of genetically modified organisms into the environment? Science alone cannot provide all the answers, but wise policy decisions must be based on accurate scientific information.

Biologists are increasingly called on to advise government agencies concerning the laws, rules, and regulations by which society deals with the increasing number of challenges that have a biological basis. As an example of the value of scientific knowledge for the assessment and formulation of public policy, consider a management problem. Scientists and fishermen have long known that Atlantic bluefin tuna (*Thunnus thynnus*) have a western breeding ground in the Gulf of Mexico and an eastern breeding ground in the Mediterranean Sea (**Figure 1.15**). Overfishing led to declining numbers of bluefin tuna,

(A)

(B)

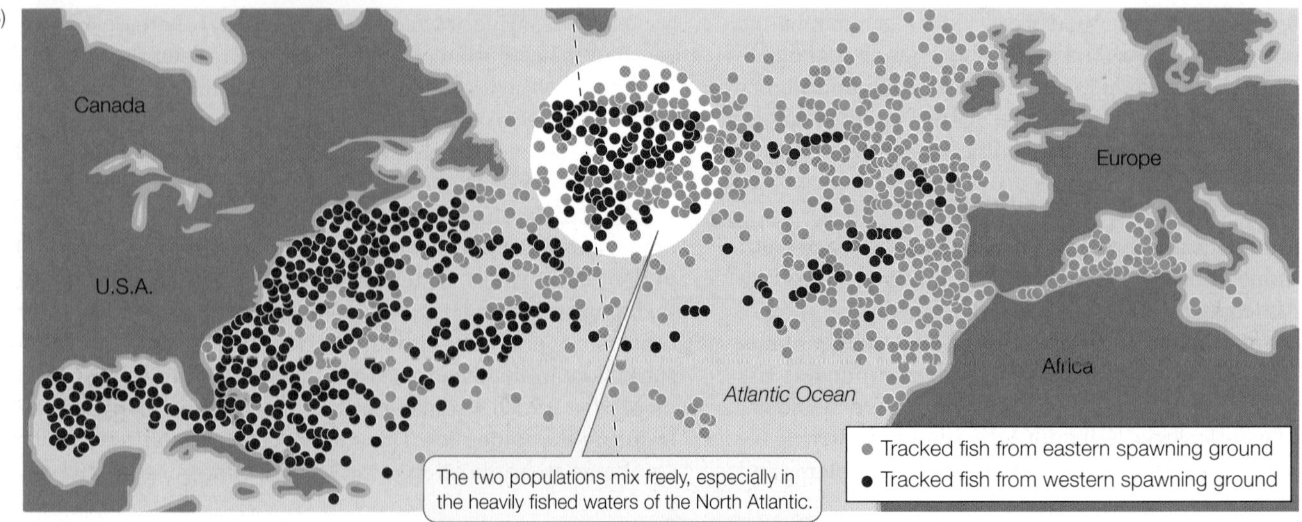


**1.15 Bluefin Tuna Do Not Recognize Boundaries** (A) Marine biologist Barbara Block attaches computerized data-recording tracking tags to a live bluefin tuna before returning it to the Atlantic Ocean, where its travels will be monitored. (B) At one time we assumed that bluefins from western- and eastern-breeding populations also fed on their respective sides of the Atlantic, so separate fishing quotas for each side (dashed line) in an attempt to speed recovery of the endangered western population. Now, however, tracking data have shown that the two populations *do not* remain separate after spawning, so in fact the arbitrary boundary and quotas do not protect the endangered population.

(A) 1941

(B) 2004

**1.16 A Warmer World** Earth's climate has been steadily warming for the last 150 years. The rate of this warming trend has also steadily increased, resulting in the rapid melting of polar ice caps, glaciers, and alpine (mountaintop) snow and ice. This photograph shows the effects of 63 years of climate change on two ancient, longstanding glaciers in Alaska. Over that time, Muir Glacier retreated some 7 kilometers and can no longer be seen from the original vantage point. Understanding how biological populations respond to such change requires integration of biological principles from molecular biology to ecosystem ecology.

especially in the western-breeding populations, to the point of these populations being endangered.

Initially it was assumed by scientists, fishermen, and policy makers alike that the eastern and western populations had geographically separate feeding grounds as well as separate breeding grounds. Acting on this assumption, an international commission drew a line down the middle of the Atlantic Ocean and established strict fishing quotas on the western side of the line, with the intent of allowing the western population to recover. Modern tracking data, however, revealed that in fact the eastern and western bluefin populations mix freely on their feeding grounds across the entire North Atlantic—a swath of ocean that includes the most heavily fished waters in the world. Tuna caught on the eastern side of the line could just as likely be from the western breeding population as the eastern; thus the established policy could not achieve its intended goal.

Policy makers take more things into consideration than scientific knowledge and recommendations. For example, studies on the effects of atrazine on amphibians have led one U.S. group, the Natural Resources Defense Council, to take legal action to have atrazine banned on the basis of the Endangered Species Act. The U.S. Environmental Protection Agency, however, must also consider the potential loss to agriculture that such a ban would create and thus has continued to approve atrazine's use as long as environmental levels do not exceed 30 to 40 ppb—which is 300 to 400 times the levels shown to induce abnormalities in the Hayes studies. Scientific conclusions do not always prevail in the political world. Some scientific conclusions may have more influence than others, however, especially when they indicate a strong possibility of negative effects on humans.

### Biology is crucial for understanding ecosystems

The world has been changing since its formation and continues to change with every passing day. Human activity, however, is resulting in an unprecedented *rate* of change in the world's ecosystems. For example, the mining and consumption of fossil fuels is releasing massive quantities of carbon dioxide into Earth's atmosphere. This anthropogenic (human-generated) increase in atmospheric carbon dioxide is largely responsible for the rapid rate of climate warming recorded over the last 50 years (**Figure 1.16**).

Our use of natural resources is putting stress on the ability of Earth's ecosystems to continue to produce the goods and services on which our society depends. Human activities are changing global climates at an unprecedented rate and are leading to the extinctions of large numbers of species (such as the amphibians featured in this chapter). The modern, warmer world is also experiencing the spread of new diseases and the resurgence of old ones. Biological knowledge is vital for determining the causes of these changes and for devising policies to deal with them.

### Biology helps us understand and appreciate biodiversity

Beyond issues of policy and pragmatism lies the human "need to know." Humans are fascinated by the richness and diversity of life, and most people want to know more about organisms and how they interact. Human curiosity might even be seen as an adaptive trait—it is possible that such a trait could have been selected for if individuals who were motivated to learn about their surroundings were likely to have survived and reproduced better, on average, than their less curious relatives.

**1.17 Discovering Life on Earth** These biologists are collecting insects in the top boughs of a spruce tree in the Carmanah Valley of Vancouver, Canada. Biologists estimate that the number of species discovered to date is only a small percentage of the number of species that inhabit Earth. To fill this gap in our knowledge, biologists around the world are applying thorough sampling techniques and new genetic tools to document and understand the Earth's biodiversity.

Far from ending the process, new discoveries and greater knowledge typically engender questions no one thought to ask before. There are vast numbers of questions for which we do not yet have answers, and the most important motivator of most scientists is curiosity.

Observing the living world motivates many biologists to learn more and to constantly collect new information (**Figure 1.17**). An intimate understanding of the **natural history** of a group of organisms—that is, how those organisms get their food, reproduce, behave, regulate their internal environments, and interact with other organisms—facilitates observations and provides a stronger basis for framing hypotheses about about those observations. The more information biologists have and the more the observer knows about general principles, the more he or she is likely to gain new insights from observing nature.

Most humans engage in activities that depend on biodiversity. You may be an avid birdwatcher, or enjoy gardening, or seek out particular species if you hunt or fish. Some people like to observe or collect butterflies, or mushrooms, or other groups of plants, animals, and fungi. Displays of spring wildflowers bring out throngs of human viewers in many areas of the world. Hiking and camping in natural areas full of diverse species are activities enjoyed by millions. All of these interests support the growing industry of eco-tourism, which depends on the observation of rare or unusual species. Learning about biology greatly increases our enjoyment of these activities.

**RECAP 1.3**

Biology informs us about the structure, processes, and interactions of the living organisms that make up our world. Informed decisions about food and energy production, health, and our environment depend on biological knowledge. Biology also addresses the human need to understand the world around us, and helps us appreciate the diverse planet we call home.

- Describe an example of how modern biology is applied to agriculture. **See p. 15**
- Why are some antibiotics not as effective for treating bacterial diseases as they were when the drugs were originally introduced? **See p. 15**
- What is an example of a biological problem that is directly related to global climate change? **See p. 17**

This chapter has provided a brief roadmap of the rest of the book. Thinking about the principles outlined here may help you to clarify and make sense of the pages of detailed description to come. At the end of the course you may wish to revisit Chapter 1 and see if you have a different perspective on the world of biology.

**Could atrazine in the environment affect species other than amphibians?**

**ANSWER**

An important aspect of the scientific process is the replication of experimental results. In some cases the exact same experiment is repeated in another laboratory by other investigators and the results are compared. In other cases the experiment is repeated on other species to test the generality of the findings.

Following the publications by Hayes and his students, other investigators tested the effects of atrazine on other species of amphibians as well as on vertebrates other than amphibians. Feminizing effects of atrazine have now been demonstrated in fish, reptiles, and mammals. These results are not surprising, because as you will learn in Chapters 41 and 43, the hormonal controls of sex development and function are the same, and therefore the effects of atrazine should generalize to other vertebrate species.

Biologists have now studied the molecular mechanisms of the effects of atrazine on the hormonal control of sex and found that very similar responses to atrazine are seen in fish and in cultures of human cells. So atrazine in the environment is increasingly a concern for the health of many other species—and that includes humans.

# CHAPTERSUMMARY 1

## 1.1 What Is Biology?

- **Biology** is the scientific study of living organisms, including their characteristics, functions, and interactions.
- All living organisms are related to one another through common descent. Shared features of all living organisms, such as specific chemical building blocks, a nearly universal genetic code, and sequence similarities across fundamental genes, support the common ancestry of life.
- Cells evolved early in the history of life. **Cellular specialization** allowed multicellular organisms to increase in size and diversity. **Review Figure 1.2**
- The instructions for a cell are contained in its **genome**, which consists of **DNA** molecules made up of sequences of **nucleotides**. Specific segments of DNA called **genes** contain the information the cell uses to make **proteins**. **Review Figure 1.5**
- **Photosynthesis** provided a means of capturing energy directly from sunlight and over time changed Earth's atmosphere.
- **Evolution**—change in the genetic makeup of biological **populations** through time—is a fundamental principle of life. Populations evolve through several different processes, including **natural selection**, which is responsible for the diversity of **adaptations** found in living organisms.
- Biologists use fossils, anatomical similarities and differences, and molecular comparisons of genomes to reconstruct the history of life. Three domains—**Bacteria**, **Archea**, and **Eukarya**—represent the major divisions, which were established very early in life's history. **Review Figure 1.7, ACTIVITY 1.1**
- Life can be studied at different levels of organization within a **biological hierarchy**. The specialized cells of multicellular organisms are organized into **tissues**, **organs**, and **organ systems**. Individual organisms form populations and interact with other organisms of their own and other species. The populations that live and interact in a defined area form a **community**, and communities together with their abiotic (nonliving) environment constitute an **ecosystem**. **Review Figure 1.9, ACTIVITY 1.2**
- Living organisms, whether unicellular or multicellular, must regulate their internal environment to maintain **homeostasis**, the range of physical conditions necessary for their survival and function.

## 1.2 How Do Biologists Investigate Life?

- Scientific methods combine observation, gathering information (**data**), experimentation, and logic to study the natural world. Many scientific investigations involve five steps: making observations, asking questions, forming hypotheses, making predictions, and testing those predictions. **Review Figure 1.10**
- **Hypotheses** are tentative answers to questions. Predictions made on the basis of a hypothesis are tested with additional observations and two kinds of experiments, **comparative** and **controlled experiments**. **Review Figures 1.11, 1.12, ANIMATED TUTORIAL 1.1**
- Quantifiable data are critical in evaluating hypotheses. Statistical methods are applied to quantitative data to establish whether or not the differences observed could be the result of chance. These methods start with the **null hypothesis** that there are no differences. **See Appendix B**
- Biological knowledge obtained from a **model system** may be generalized to other species.

## 1.3 Why Does Biology Matter?

- Application of biological knowledge is responsible for vastly increased agricultural production.
- Understanding and treatment of human disease requires an integration of a wide range of biological principles, from molecular biology through cell biology, physiology, evolution, and ecology.
- Biologists are often called on to advise government agencies on the solution of important problems that have a biological component.
- Biology is increasing important for understanding how organisms interact in a rapidly changing world.
- Biology helps us understand and appreciate the diverse living world.

**Go to the Interactive Summary to review key figures, Animated Tutorials, and Activities Life10e.com/is1**

# CHAPTERREVIEW

## REMEMBERING

1. Which of the following is *not* an attribute common to all living organisms?
   a. They are made up of a common set of chemical components, including particular nucleic and amino acids.
   b. They contain genetic information that uses a nearly universal code to specify the assembly of proteins.
   c. They share sequence similarities among their genes.
   d. They exist in populations that evolve over time.
   e. They extract energy from the sun in a process called photosynthesis.

2. In describing the hierarchy of life, which of the following descriptions of relationships is *not* accurate?
   a. An organ is a structure consisting of different types of cells and tissues.
   b. A population consists of all of the different animals in a particular type of environment.
   c. An ecosystem includes different communities.
   d. A tissue consists of a particular type of cells.
   e. A community consists of populations of different species.

3. Which of the following is a property of a good hypothesis?
   a. It is a statement of facts.
   b. It is general enough to explain a variety of possible experimental outcomes.
   c. It is independent of any observations.
   d. It explains things that are not addressable by experimentation.
   e. It can be falsified by experiments.

4. Which of the following events was most directly responsible for increasing oxygen in Earth's atmosphere?
   a. The cooling of the planet
   b. The origin of eukaryotes
   c. The origin of multicellularity
   d. The origin of photosynthesis
   e. The origin of prokaryotes

5. Which of the following is a reason to use statistics to evaluate data?
   a. It enables you to prove that your hypothesis is correct.
   b. It enables you to exclude data that do not fit your hypothesis.
   c. It makes it possible to exclude the null hypothesis.
   d. It enables you to predict experimental results.
   e. It accounts for variation in scientific measurements.

## UNDERSTANDING & APPLYING

6. Why is it important in science to design and perform experiments that are capable of falsifying a hypothesis?

7. What is the significance of the fact that mitochondria and chloroplasts contain the DNA that instructs their form and function?

8. The results in Dr. Hayes's comparative experiments were more variable than the results from his controlled experiments. How would you explain this?

## ANALYZING & EVALUATING

9. Biologists can now isolate genes from organisms and decode their DNA. When the nucleotide sequences from the same gene in different species are compared, differences are discovered. How could you use those data to deduce the evolutionary relationships among the organisms in your comparison?

10. Mitochondria are cell organelles that have their own DNA and replicate independently of the cell itself. In most organisms, mitochondria are inherited only from the mother. Based on this observation, when might it be advantageous or disadvantageous to use mitochondrial DNA rather than nuclear DNA for studying evolutionary relationships among populations?

Go to BioPortal at **yourBioPortal.com** for Animated Tutorials, Activities, LearningCurve Quizzes, Flashcards, and many other study and review resources.

# PART EIGHT
## Flowering Plants: Form and Function

# 34 The Plant Body

CHAPTER**OUTLINE**


**34.1** **What Is the Basic Body Plan of Plants?**

**34.2** **What Are the Major Tissues of Plants?**

**34.3** **How Do Meristems Build a Continuously Growing Plant?**

**34.4** **How Has Domestication Altered Plant Form?**

**Cassava** The roots of this plant supply food energy to hundreds of millions of people.

TO NORTH AMERICANS AND EUROPEANS, it is tapioca. To Central and South Americans, it is yuca. To Africans, it is manioc or bananku. The roots of the cassava plant, *Manihot esculenta*, are important in the diets of more than 800 million people. Cassava is grown mostly by family farmers for their own consumption. The root of the plant is a store of starch, which can be hydrolyzed and used by the plant as the stems and leaves grow. Other crop plants such as rice and wheat apportion about 35 percent of the total carbon fixed in photosynthesis into their storage organs (the grains), whereas cassava is a "starch factory," apportioning an astounding 80 percent of the plant's total photosynthate into the root. For humans, this is a convenient and concentrated source of food energy. Indeed, cassava has been nicknamed the "bread of the tropics" because, just as wheat bread is the major starchy food in the Western world, cassava supplies the starch in tropical areas.

Most cassava plants are clones. The plants have wide adaptability and grow well in dry soils, in both hot and cool climates. Typically, the farmer breaks off some pieces of stem and plants them. Some of the stem cells dedifferentiate and form roots, while others become growing shoots. A whole new plant develops, and the roots are ready to eat 6 months to 2 years later. It's a fairly easy process, much simpler than the work needed to grow rice, for example. As we pointed out in Chapter 19, totipotency is a remarkable property of plant cells, one that distinguishes them from animal cells.

There are countless recipes for preparing cassava as food. Cassava does not, however, offer complete nutrition. Although rich in carbohydrates, cassava is a relatively poor source of protein, a requirement for the human diet. A diet based on cassava presents other difficulties as well. Cassava roots and leaves contain cyanogenic compounds, which are converted to cyanide by digestive enzymes and gut flora. Cyanide is highly toxic and potentially lethal because it blocks electron transport in the mitochondria. Therefore it is essential to soak, cook, or ferment cassava, to break down the cyanogenic compounds and eliminate the cyanide, before it can be eaten safely. In all probability, the plant uses cyanide production as a protection against predators that eat it.

We open with the topic of cassava because it offers a preview of a wide range of studies embraced by the discipline of plant physiology. Plant physiology is a broad subject, covering photosynthesis, transport, plant nutrition, regulation of growth and development, reproduction, and the interactions between plants and their environments.

How might plant physiologists improve the cassava plant for human use?

See answer on p. 724.

## 34.1 What Is the Basic Body Plan of Plants?

Plants live by harvesting energy from sunlight and by collecting water and mineral nutrients from the atmosphere and the soil. Because these resources are sometimes limited, plants must collect them from large areas, both above and below ground. The plant is further challenged by its inability to move; a plant cannot, for example, relocate from a dry, shady location to one that is wet and sunny.

The plant body plan allows plants to respond to these challenges:

- Stems, leaves, and roots enable a plant anchored to one spot to capture scarce resources effectively, both above and below the ground.
- Plants can grow throughout their lifetimes, enabling them to respond to environmental cues. A plant can redirect its growth to exploit opportunities in its immediate environment; for example, it can extend its roots toward a water supply.

In Chapters 28 and 29 we saw how modern plants arose from aquatic ancestors, giving rise to simple land plants and then to vascular plants. Despite their obvious differences in size and form, all vascular plants have essentially the same simple structural organization. This chapter describes the basic architecture of the largest group of vascular plants, the angiosperms (flowering plants), and shows how so much diversity can literally grow out of such a simple, basic form.

As we saw in Figure 29.1, angiosperms first appeared about 145 million years ago. They radiated explosively over a period of about 60 million years and eventually became the dominant form of plant life on Earth. Today there are more than 250,000 angiosperm species. Flowers are the main distinguishing feature of angiosperms; they consist of modified leaves and stems and carry the organs for sexual reproduction. We will examine the structures and functions of flowers in detail in Section 38.1. In this chapter we'll focus on the three kinds of vegetative (nonsexual) organs that angiosperms possess: roots, stems, and leaves. Each of these vegetative organs can be understood in terms of its structure. By structure we mean both the overall form of the organ (its morphology) and the arrangement of its cells and tissues (its anatomy).

Plant organs are organized into two systems (**Figure 34.1**):

- The **root system** anchors the plant in place, absorbs water and dissolved minerals, and stores the products of photosynthesis from the shoot system. The extreme branching of plant roots and their high surface area-to-volume ratios allow them to absorb water and mineral nutrients from the soil efficiently.
- The **shoot system** of a plant consists of the stems, leaves, and flowers. Broadly speaking, the **leaves** are the chief organs of photosynthesis. The **stems** hold and display the leaves to the sun and provide connections for the transport of materials between roots and leaves.

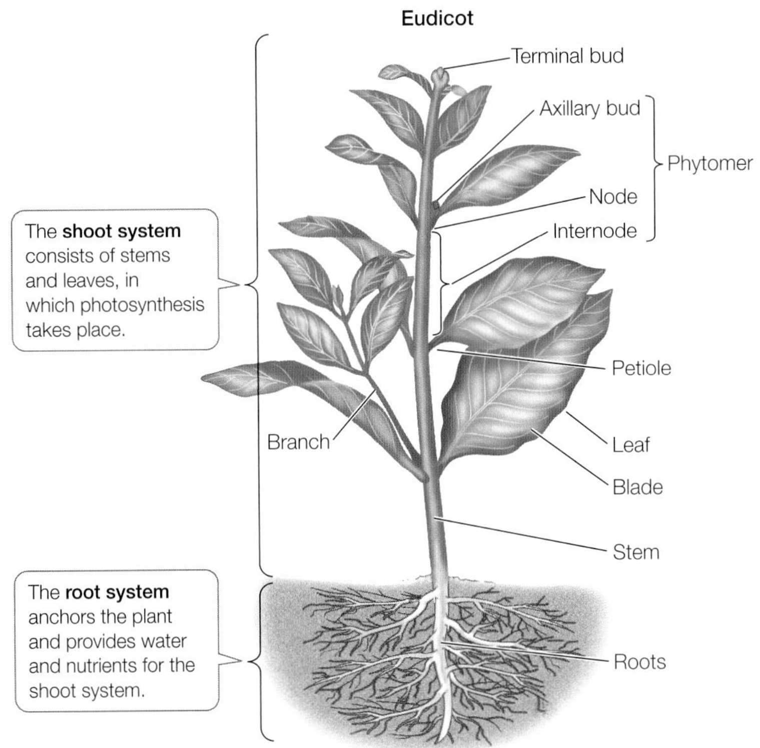


**34.1 Vegetative Organs and Systems** The basic plant body plan, with root and shoot systems, and the principal vegetative organs.

Shoots and roots are composed of repeating modules called **phytomers**. Each phytomer in the shoot consists of a **node** carrying one or more leaves; an **internode**, which is the interval of stem between two nodes; and one or more **axillary buds**, each of which forms in the angle (axil) where a leaf meets the stem. A **bud** is an undeveloped shoot that can develop further to produce another leaf, a phytomer, a flower, or a flowering stem. The axillary buds (also called lateral buds) are distinguished from the bud at the end of a stem or branch, which is called a **terminal bud**. If it becomes active, an axillary bud can develop into a new branch, or an extension of the shoot system. The arrangement of leaves along the stem (called the phyllotaxy) is characteristic of the plant species.

Plant roots also have a modular construction. In the roots, each phytomer consists of a root segment between two branches.

### Most angiosperms are either monocots or eudicots

As we saw in Section 29.3, most angiosperms belong to one of two major clades. *Monocots* are generally narrow-leaved flowering plants such as grasses, lilies, orchids, and palms. *Eudicots* are broad-leaved flowering plants such as soybeans, roses, sunflowers, and maples. These two clades, which account for 97 percent of flowering plant species, differ in several basic characteristics:

- Monocots have one **cotyledon** (leaf in the embryo), whereas eudicots have two.

- In eudicots, the vascular bundles in the stem are arranged in concentric circles; in monocots they are scattered.
- In monocots, the major leaf veins are usually parallel; in eudicots they are reticulate, meaning they form a network.
- Eudicots usually have taproot systems; monocots have fibrous root systems.
- Monocot flowers have parts (petals and sepals) that occur in threes; eudicots have floral parts that occur in fours or fives.
- Monocot pollen grains each have one furrow or pore; eudicot pollen grains have three.

We will discuss some of these differences in more detail in Section 34.3.

## Plants develop differently than animals

As we discussed in Chapter 19, the four processes that govern development in all multicellular organisms are **determination** (the commitment of cells to their ultimate fates); **differentiation** (cell specialization); **morphogenesis** (the organization of cells into tissues and organs); and **growth** (increase in body size). These processes govern plant development as well as animal development, but they are influenced by four unique properties of plants: apical meristems, totipotency, vacuoles, and cell walls.

APICAL MERISTEMS Animals have stem cells to replace tissues lost through damage or apoptosis. Plants have **meristems**: regions of undifferentiated cells where cell division occurs. **Apical meristems** are found at the tips of shoots and roots and allow plants to continue growing throughout their lives. We will discuss apical meristems in more detail below.

TOTIPOTENCY During normal animal development, only the early embryonic cells are totipotent: they can differentiate into any type of cell in the body (see Section 19.1). In contrast, some differentiated plant cells can dedifferentiate and become totipotent (see Figure 19.16). This means that a plant can readily repair damage wrought by the environment or herbivores.

VACUOLES Mature plant cells usually contain a single **central vacuole**, which may account for up to 90 percent of a cell's volume (see Figure 5.13). The vacuole is a watery sac containing a high concentration of solutes, including enzymes, amino acids, and sugars produced by photosynthesis. Many of these solutes are pumped into the vacuole by transporter proteins located in the **tonoplast**, the vacuolar membrane. This active accumulation of solutes provides the osmotic force for water uptake into the vacuole, as we will see in Section 35.1. As the vacuole expands, it exerts turgor pressure on the cell wall (see Chapter 6). Turgor pressure keeps plants upright and is essential for plant growth.

CELL WALLS Each plant cell is surrounded by a cell wall, which is interrupted by membrane-lined cytoplasmic channels called

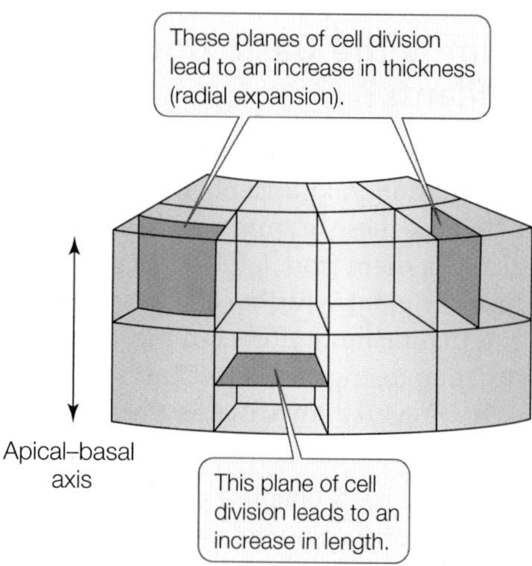


**34.2 Cytokinesis and morphogenesis** The plane of cell division can determine the growth pattern of a plant's organs, as in this section of a shoot.

plasmodesmata (see Section 5.4). This rigid extracellular matrix makes it impossible for cells to move from place to place, as they do in animal development. Instead, plant morphogenesis is controlled by the planes of cell division, which determine the direction in which a piece of tissue will grow (**Figure 34.2**). In addition, unequal cell division can occur when the cytoplasm contains differentiation signals that are localized in one part of a cell (see Section 19.2). Plant cytokinesis (the division of the cytoplasm) occurs along a cell plate laid down by membranous vesicles produced by the Golgi apparatus (see Figure 11.13B). Unlike animal cells, in which the location of cytokinesis depends on the location of the middle of the mitotic spindle, the location of the plant cell plate is determined earlier—as early as mitotic prophase.

One of the major ways that plants grow is by cell expansion. Some cells can increase in volume by 100,000 to 1,000,000 times! As a growing plant cell takes up water, it exerts turgor pressure on the cell wall, which resists cell expansion. For the cell to expand, the wall must expand too. Proteins called expansins reside in the cell wall and help loosen it by disrupting the noncovalent bonds between cellulose microfibrils and other polysaccharides in the cell wall. This is followed by the assembly of new polysaccharides and microfibrils, allowing the cell wall to grow:

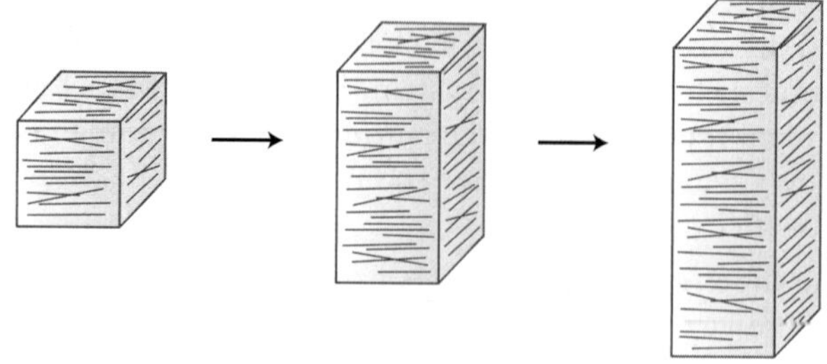

The wall of a growing plant cell is called the **primary cell wall**. When cell expansion stops, some types of plant cells deposit one or more additional cellulosic layers to form a thick, rigid **secondary cell wall** that is internal to the primary cell wall:

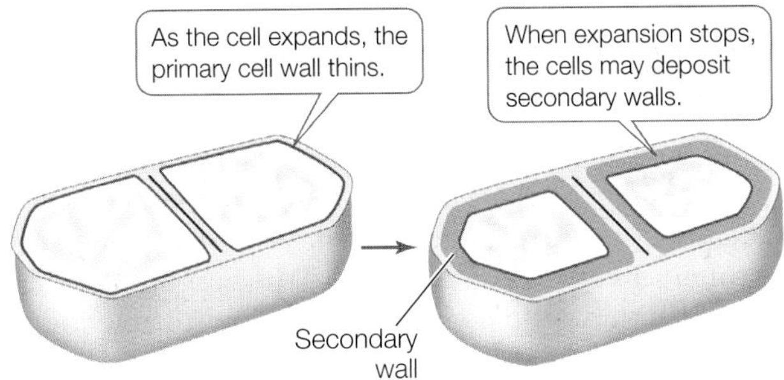


Secondary cell walls cannot expand. Instead they provide the mechanical support that allows some plants to produce large stems. The secondary wall contains layers of ordered cellulose microfibrils embedded in a remarkable substance called **lignin**. Lignin is a major component of wood; it is a complex, carbon-containing polymer that forms a hydrophobic matrix. This matrix is strong, waterproof, and resistant to digestion by animals. After cellulose, lignin is the most abundant biological polymer on Earth, accounting for 20 to 35 percent of the dry weight of wood.

## Apical–basal polarity and radial symmetry are characteristics of the plant body

Two basic patterns are established early in plant embryogenesis (embryo formation) (**Figure 34.3**):

- The *apical–basal axis*: the arrangement of cells and tissues along the main axis from root to shoot
- The *radial axis*: the concentric arrangement of the tissue systems. (which we will describe in Section 34.2)

Both axes are best understood in developmental terms. We will focus here on embryogenesis in *Arabidopsis thaliana*. As we have seen in previous chapters, *Arabidopsis* is a model eudicot

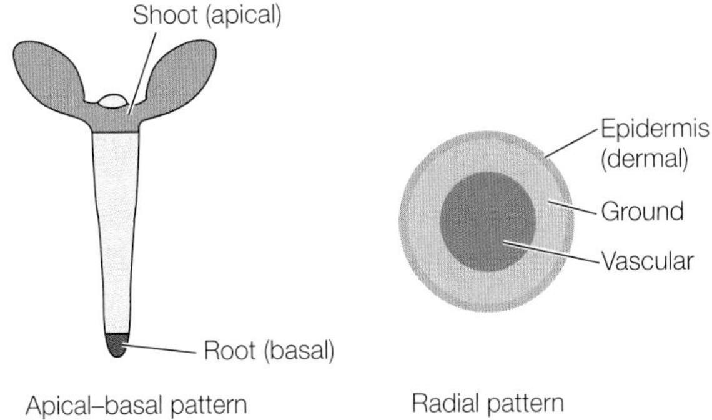


**34.3 Two Patterns for Plant Morphogenesis** (A) The apical–basal pattern is the arrangement of tissues along a main axis from root to shoot. (B) The radial pattern determines the concentric arrangement of tissues as organs grow in thickness.

that has been studied extensively by plant physiologists and geneticists.

The first step in the formation of an *Arabidopsis* embryo is a mitotic division of the zygote that gives rise to two daughter cells (**Figure 34.4, Step 1**). An asymmetrical plane of cell division results in an uneven distribution of cytoplasm between these two cells, and this determines their different fates. Signals in the smaller, upper daughter cell cause it to produce the embryo proper, and the other, larger daughter cell produces a supporting structure called the **suspensor** (**Figure 34.4, Step 2**). This division not only establishes the apical–basal axis of the new plant but also determines its polarity (which end is the tip, or apex, and which is the base). A long, thin suspensor and a more spherical or globular embryo are distinguishable after just four mitotic divisions (see Figure 19.1). The suspensor soon ceases to elongate.

In eudicots such as *Arabidopsis*, the initially globular embryo develops into the characteristic heart stage as the cotyledons start to grow (**Figure 34.4, Step 3**). Further elongation of the cotyledons and of the main axis of the embryo gives rise to the

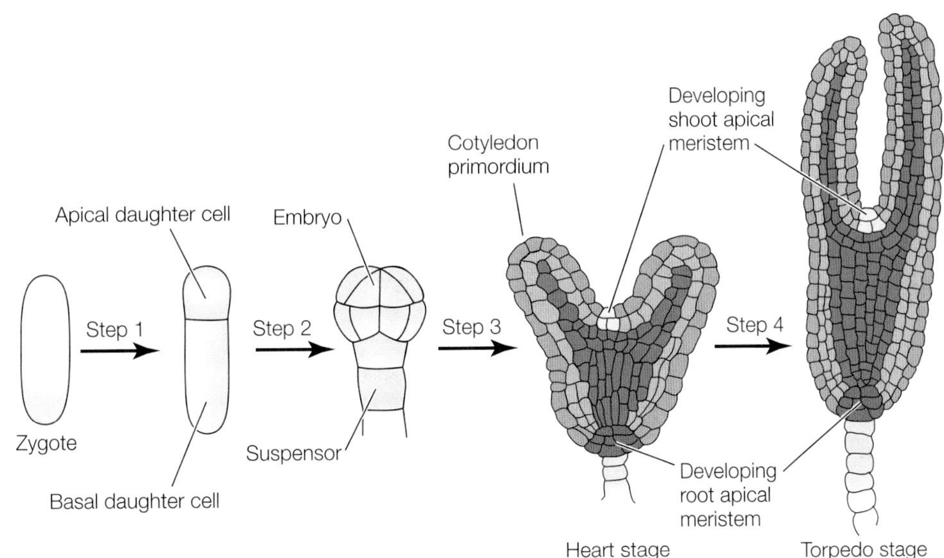


**34.4 Plant Embryogenesis** The basic body plant of the model eudicot *Arabidopsis thaliana* is established in several steps. By the heart stage, the three tissue systems are established: the dermal (gold), ground (light green), and vascular (blue) tissue systems.

torpedo stage, during which some of the internal tissues begin to differentiate (**Figure 34.4, Step 4**). Between the cotyledons is the **shoot apical meristem**; at the other end of the axis is the **root apical meristem**. Each of these meristems contains undifferentiated cells that will continue to divide to give rise to the organs that will develop during the life of the plant.

As shown in Step 2 of Figure 34.4, the plant embryo is first a sphere and later a cylinder. The root and stem retain this cylindrical shape throughout the plant's life. You can see this most easily in the trunk (mature stem) of a tree. By the end of embryogenesis, the radial symmetry of the plant has been established (see Figure 34.3). The embryonic plant contains three tissue systems, arranged concentrically, which will give rise to the tissues of the adult plant body. We will discuss these tissue systems in the next section.

### RECAP 34.1

The vegetative plant body consists of a root system and a shoot system. The plant body is modular, made up of repeated units called phytomers. Plant cells are characterized by apical meristems, totipotency, vacuoles, and cell walls. Plants have apical–basal and radial patterns. Most angiosperms are either monocots or eudicots, which differ in several basic ways.

- How do plants explore their environment for resources even though they cannot move? **See p. 709**
- How does plant development differ from animal development? **See pp. 710–711**
- How do apical–basal and radial patterns develop? **See pp. 711–712 and Figures 34.3 and 34.4**

The dramatic differences between plant and animal body plans are not surprising given that the multicellular forms of plants and animals evolved independently from entirely distinct protist ancestors (see Figure 1.7). In the next two sections we will look more closely at the unique characteristics of the plant body, by following its development from a zygote into an adult.

## 34.2 What Are the Major Tissues of Plants?

By the end of embryogenesis, the radial axis of the plant has been established. Unlike complex animals that can have dozens of different tissues (for example, in humans there are three kinds of muscle tissue alone), plants have just three major tissue systems, each of which has specialized cells. These three tissue systems are arranged concentrically in the embryo and give rise to the tissues of the adult plant body. In turn, these tissues form the major vegetative plant organs: roots, stems, and leaves.

### The plant body is constructed from three tissue systems

A tissue is an organized group of cells that have features in common and that work together as a structural and functional unit. Plant tissues are grouped into three **tissue systems**: dermal, ground, and vascular. Established during embryogenesis, these three tissue systems ultimately extend throughout the plant body in a concentric arrangement (**Figure 34.5**). Each tissue system has distinct functions and is composed of different mixtures of cell types.

**DERMAL TISSUE SYSTEM** The **dermal tissue system** forms the **epidermis** (outer covering) of a plant and usually consists of a single cell layer. The stems and roots of woody plants develop a dermal tissue called **periderm**.

During plant development, the epidermis grows to cover the expanding plant body. The cells of the epidermis are initially small and round and usually have a small central vacuole or none at all. Once cell division ceases in the epidermis of an organ, the epidermal cells expand. Some epidermal cells differentiate to form one of three specialized structures:

- Stomatal guard cells, which form stomata (pores) for gas exchange in leaves

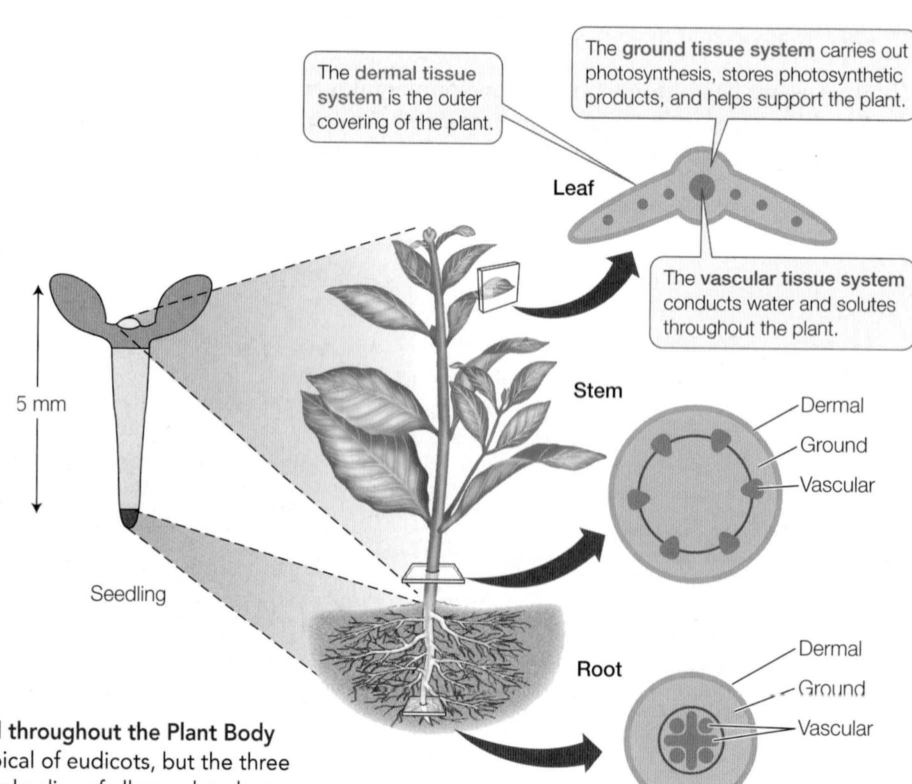


**34.5 Three Tissue Systems Extend throughout the Plant Body** The arrangement shown here is typical of eudicots, but the three tissue systems are continuous in the bodies of all vascular plants.

(A) Parenchyma cells

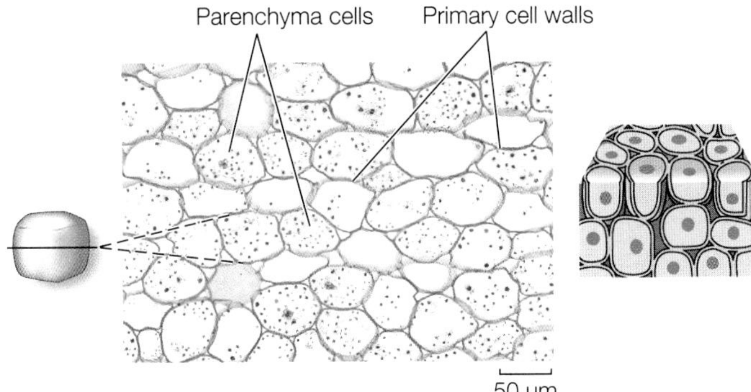


(B) Collenchyma cells

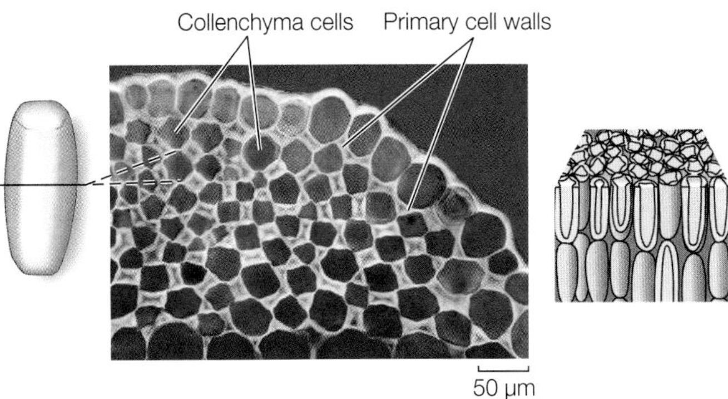


(C) Fibers

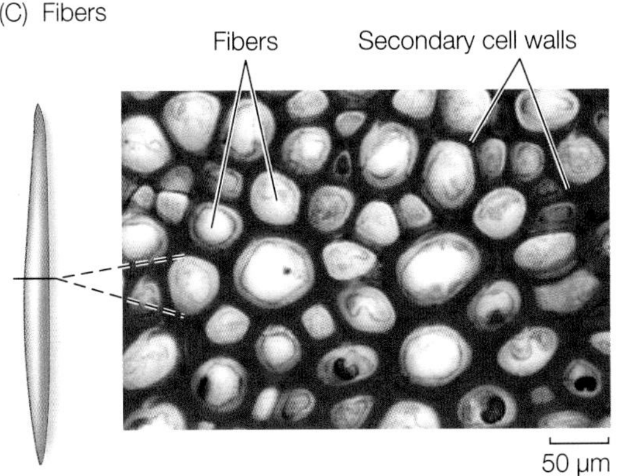


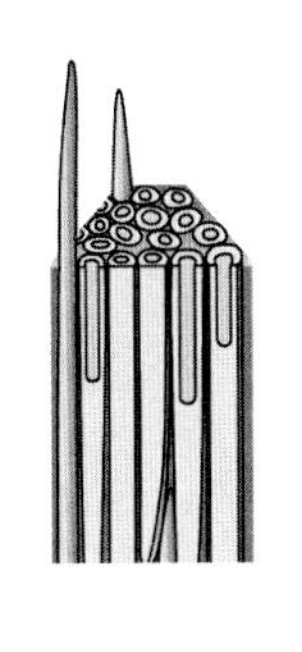

(D) Sclereids

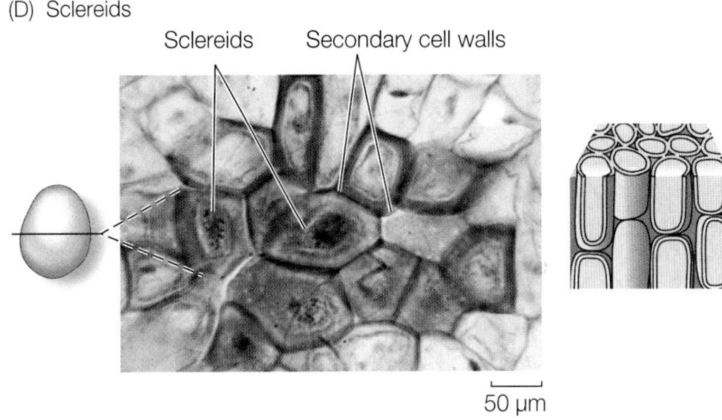


**34.6 Ground Tissue Cell Types** (A) Parenchyma cells in the petiole of *Coleus*. Note the thin, uniform cell walls. (B) Collenchyma cells make up the outer cell layers of this spinach leaf vein. Their walls are thick at the corners of the cells and thin elsewhere. (C) Sclerenchyma: fibers in a sunflower stem (*Helianthus*). The thick secondary walls are stained red. (D) Sclerenchyma: sclereids. The extremely thick secondary walls of sclereids are laid down in layers. They provide support and a hard texture to structures such as nuts and seeds.

- Trichomes, or leaf hairs, which provide protection against insects and damaging solar radiation
- Root hairs, which greatly increase root surface area, thus providing more surface for the uptake of water and mineral nutrients

The aboveground epidermal cells secrete a protective extracellular layer called a **cuticle**. The cuticle is made up of cutin (a polymer composed of long chains of fatty acids), a complex mixture of waxes, and cell wall polysaccharides. The cuticle limits water loss, reflects potentially damaging solar radiation, and serves as a barrier against pathogens.

GROUND TISSUE SYSTEM Virtually all the tissue lying between the dermal tissue and the vascular tissue in both shoots and roots is part of the **ground tissue system**. Therefore the ground tissues make up most of the plant body. The ground tissues function primarily in storage, support, and photosynthesis. To fulfill these diverse functions, ground tissues contain three cell types that are classified according to their cell wall structure: parenchyma, collenchyma, and sclerenchyma.

The most common cell type in plants is the **parenchyma** cell (**Figure 34.6A**). Parenchyma cells have large vacuoles and thin walls consisting only of a primary wall and the shared middle lamella. The **middle lamella** is a layer of pectin that cements adjacent plant cells together (see Figure 5.21). Parenchyma cells play important roles in photosynthesis (mainly in the leaves); proteins, starch, fats, or oils may be stored in parenchyma cells of the seeds and/or roots. Many retain the capacity to divide and give rise to new cells, as when a wound results in cell proliferation.

**Collenchyma** cells resemble parenchyma cells that have been modified to provide flexible support. They are generally elongated, and their primary walls are characteristically thick at the corners of the cells (**Figure 34.6B**). In these cells the primary wall thickens in part because of the deposition of pectins, but no secondary wall forms. Collenchyma cells provide support to leaf petioles, nonwoody stems, and growing organs. Tissue made of collenchyma cells is flexible, permitting stems and petioles to sway in the wind without snapping. The familiar "strings" in celery consist primarily of collenchyma cells.

**Sclerenchyma** cells have thickened secondary walls that enable the cells to perform their major function: support. Many sclerenchyma cells undergo programmed cell death (apoptosis; see

(A) Tracheids

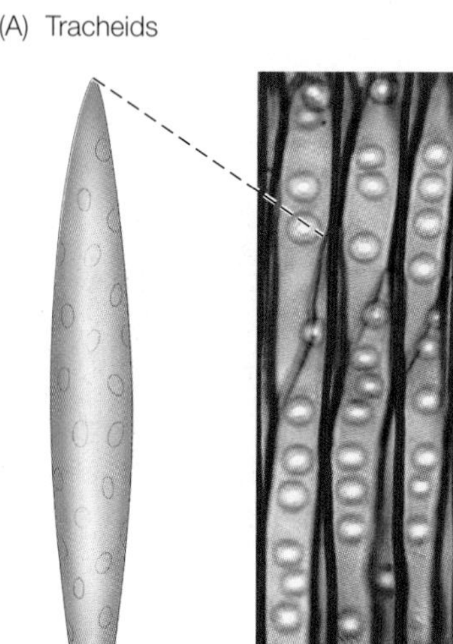


(B) Vessel elements

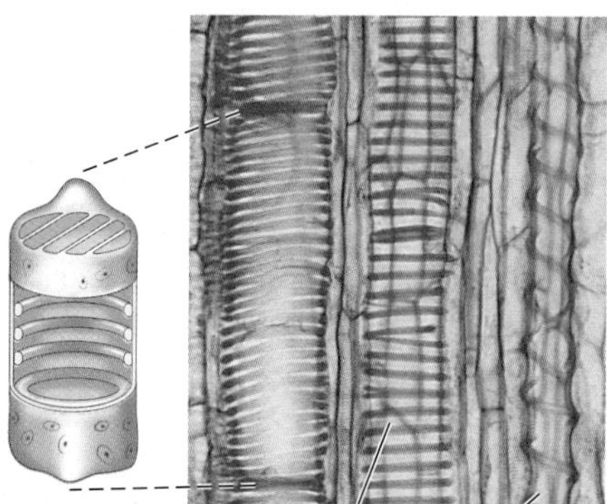


(C) Sieve tube elements

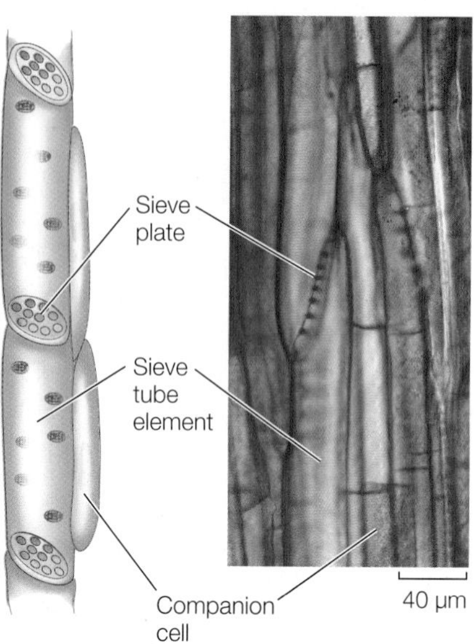


**34.7 Vascular Tissue Cell Types** (A, B) Tracheary elements: (A) Tracheids in pinewood. The thick secondary walls are stained dark red. (B) Vessel elements in the stem of a squash. The secondary walls are stained red; note the different patterns of thickening, including rings and spirals. (C) Sieve tube elements and companion cells in the stem of a cucumber.

Section 11.6) after developing the lignified secondary walls, and thus perform their supporting function when dead. There are two types of sclerenchyma cells: elongated **fibers** and variously shaped **sclereids**. Fibers are often organized into bundles and provide relatively rigid support to wood, bark, and other parts of the plant (**Figure 34.6C**). Sclereids may pack together densely, as in a nut's shell or in some seed coats (**Figure 34.6D**). Isolated clumps of sclereids, called stone cells, occur in pears and some other fruits and give them their characteristic gritty texture.

VASCULAR TISSUE SYSTEM The **vascular tissue system** is the plant's plumbing, or transport system—the distinguishing feature of vascular plants. Its two constituent tissues, the xylem and phloem, distribute materials throughout the plant. The **xylem** distributes water and mineral ions taken up by the roots to all the cells of the stems and leaves. **Phloem** performs a variety of functions, including transport, support, and storage. All the living cells of the plant body require a source of energy and chemical building blocks. The phloem meets these needs by transporting carbohydrates away from the sites of production, which are called **sources** (primarily leaves). The carbohydrates are transported to sites of utilization or storage, called **sinks**. Sinks include growing tissues, storage organs, and developing flowers.

Let's take a closer look at the structures of the diverse cell types that make up these vascular tissues. In Chapter 35 we will see how they transport water and materials throughout the plant body.

## Cells of the xylem transport water and dissolved minerals

Xylem tissue contains conducting cells called **tracheary elements** that have secondary cell walls and undergo apoptosis before assuming their function of transporting water and dissolved minerals. There are two types of tracheary elements: tracheids and vessel elements. The spindle-shaped **tracheids** (**Figure 34.7A**) are evolutionarily more ancient than vessel elements and are the major cell type in the wood of gymnosperms (see Section 29.2). When tracheids die, their internal components disintegrate and pits remain between the cells. Pits are cavities in the secondary cell walls spanned by porous structures that allow water and minerals to move between tracheids and thus through the xylem tissue.

Flowering plants have evolved a water-conducting system made up of vessels, which are formed from individual cells called **vessel elements**. These cells are laid down end-to-end. Like tracheids, vessel elements have pits in their cell walls, but their pits are generally larger in diameter than those of tracheids. Before they undergo apoptosis, the end walls of vessel elements partially break down, forming a continuous hollow tube that functions as an open pipeline for water conduction (**Figure 34.7B**). In the course of angiosperm evolution, vessel elements have become shorter and wider, and their end walls have become less oblique and less obstructed. These adaptations have presumably increased the efficiency of water transport through the vessels. The xylem of many angiosperms includes both tracheids and vessels.

## Cells of the phloem transport the products of photosynthesis

The transport cells of the phloem, unlike those of the mature xylem, are living cells. In flowering plants the characteristic cells of the phloem are called **sieve tube elements** (**Figure 34.7C**). Like vessel elements, these cells meet end-to-end. They form long sieve tubes, which transport carbohydrates and many other materials from their sources (usually leaves) to tissues that consume or store them (for example, roots).

Unlike in vessel elements, which break down their end walls, the end walls of sieve tube elements contain plasmodesmata, which enlarge to form pores. The end walls look and function like sieves and are thus called sieve plates. Although the sieve tube elements remain alive, some of their components, including the nucleus, ribosomes, and vacuole, break down during development. The sieve tube elements are, however, closely connected via plasmodesmata to **companion cells**—specialized parenchyma cells that retain all their organelles and function as "life support systems" for the sieve tube elements.

**RECAP 34.2**

The three concentric tissue systems of the plant embryo—dermal, ground, and vascular—give rise to the tissues and organs of the adult plant. These tissue systems have unique combinations of specialized cells that carry out the various functions necessary for plant life.

- What distinguishes the three tissue systems in terms of their location and functions? **See pp. 712–714 and Figure 34.5**
- What structural differences make tissues consisting of collenchyma cells more flexible than those consisting primarily of sclerenchyma cells? **See pp. 713–714**
- Outline the differences between tracheids and vessel elements. **See p. 714 and Figures 34.6 and 34.7**

After the plant embryo has formed, it is encased in a seed coat and is ready to germinate. We will discuss aspects of seed germination in the chapters that follow. For now, let's consider the processes by which the embryo grows into a mature plant.

## 34.3 How Do Meristems Build a Continuously Growing Plant?

As noted at the beginning of this chapter, plants and animals develop and function differently. While animals use their mobility to forage for food, plants are sessile (rooted in one place) and grow toward scarce resources, both above and below the ground. Therefore plants grow in two directions: the shoots grow toward sunlight, and the roots grow toward water and dissolved minerals in the soil.

In most animals, growth is **determinate**: it ceases when the adult state is reached. Determinate growth is also characteristic of some plant organs, such as leaves, flowers, and fruits. The growth of shoots and roots, however, is a lifelong process. Such open-ended growth is **indeterminate**.

### Plants increase in size through primary and secondary growth

Plant growth can occur in either of two ways:

- **Primary growth** is characterized by cell division followed by cell enlargement. It results in the proliferation and lengthening of shoots and roots. All seed plants have a primary plant body, which consists of all the nonwoody parts of the plant. Many herbaceous plants consist entirely of a primary plant body.
- **Secondary growth** increases plant thickness. Woody plants, such as trees and shrubs, have a secondary plant body consisting of wood and bark. As the tissues of the secondary plant body are laid down, the stems and roots thicken.

### A hierarchy of meristems generates the plant body

As we have already mentioned, meristems are localized regions of undifferentiated cells that are the sources of all new growth in the adult plant. Even before seed germination, the plant embryo has two meristems: a shoot apical meristem near the end of the embryonic shoot, and a root apical meristem at the end of the embryonic root (see Figure 34.4).

Meristematic cells are small and closely packed, with very small vacuoles and thin primary cell walls. They are undifferentiated and retain the ability to produce new cells indefinitely. The cells that perpetuate the meristems, called **initials**, are comparable to animal stem cells (discussed in Section 19.1). When the initials divide, some of the daughter cells develop into new initials, and some differentiate into more specialized cells.

Several types of meristem contribute to the growth and development of the adult plant:

- Apical meristems in the root and shoot (**Figure 34.8**) orchestrate primary growth, ultimately giving rise to every cell in the primary plant body.
- When the initials of apical meristems divide, some of their daughter cells differentiate and become the **primary meristems**. Three kinds of primary meristem (see below) give rise to the three major tissue systems (dermal, ground, and vascular) described in Section 34.2.
- **Lateral meristems** (also called secondary meristems) orchestrate secondary growth (see Figure 34.8). Two lateral meristems, vascular cambium and cork cambium, contribute to the secondary plant body.

### Indeterminate primary growth originates in apical meristems

Because apical meristems can perpetuate themselves indefinitely, a shoot or root can continue to lengthen and grow indefinitely; in other words, growth of the shoot or root is indeterminate. All plant organs arise ultimately from cell divisions in apical meristems, followed by cell expansion and differentiation. Several types of apical meristems play roles in organ formation:

- Shoot apical meristems supply the cells for new leaves and stems. In addition to the main stem of the plant, each branch has its own shoot apical meristem. Shoot apical meristems are also called **vegetative meristems**, because they give rise to vegetative tissues (leaves, stems, and roots).
- When the plant is ready to flower, one or more of its shoot apical meristems are transformed into **inflorescence meristems**, and these in turn develop **floral meristems**. See Section 38.2 for more on floral development.
- Root apical meristems supply the cells that extend roots, enabling the plant to penetrate and explore the soil for water and minerals. Each type of root (i.e., the taproot, a lateral root, or an adventitious root; see below) has its own root apical meristem.

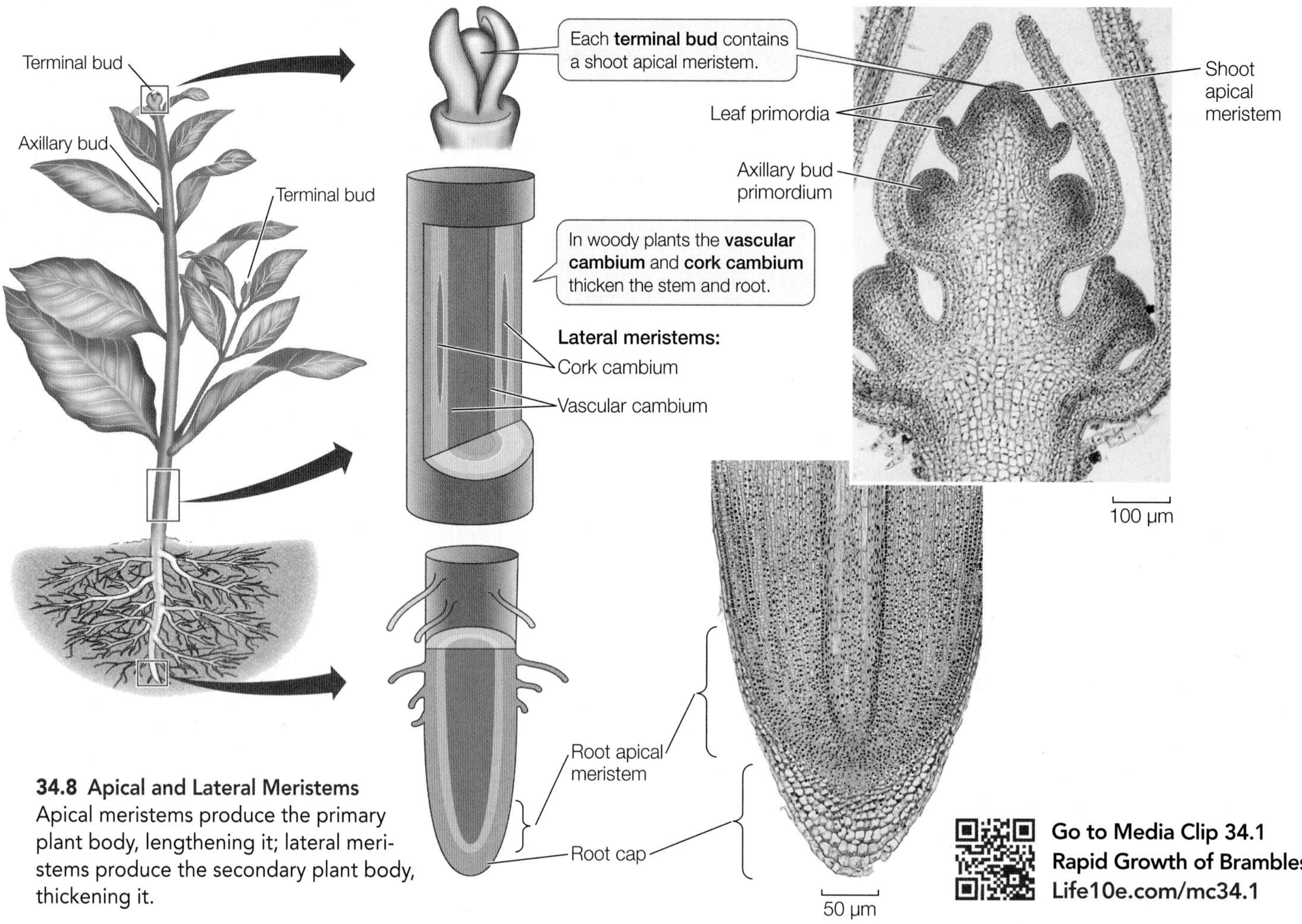

Terminal bud
Axillary bud
Terminal bud
Each **terminal bud** contains a shoot apical meristem.
Leaf primordia
Shoot apical meristem
Axillary bud primordium
In woody plants the **vascular cambium** and **cork cambium** thicken the stem and root.
**Lateral meristems:**
Cork cambium
Vascular cambium
100 μm
Root apical meristem
Root cap
50 μm

**34.8 Apical and Lateral Meristems**
Apical meristems produce the primary plant body, lengthening it; lateral meristems produce the secondary plant body, thickening it.

Go to Media Clip 34.1
Rapid Growth of Brambles
Life10e.com/mc34.1

Apical meristems in both the shoot and the root give rise to a set of **primary meristems**, which produce the tissues of the primary plant body. From the outside to the inside of the shoot or root, the primary meristems are the **protoderm**, the **ground meristem**, and the **procambium** (see Figure 34.8). These meristems, in turn, give rise to the three tissue systems:

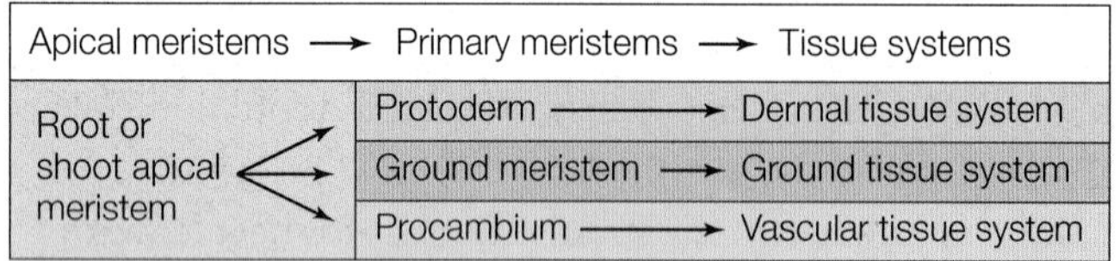

| Apical meristems → | Primary meristems → | Tissue systems |
|---|---|---|
| Root or shoot apical meristem | Protoderm → | Dermal tissue system |
| | Ground meristem → | Ground tissue system |
| | Procambium → | Vascular tissue system |

Because meristems can continue to produce new organs throughout the lifetime of the plant, the plant body is much more variable in form than the animal body, which produces each organ only once.

Let's look more closely at how the root apical meristem produces the root system.

## The root apical meristem gives rise to the root cap and the root primary meristems

The root apical meristem produces all the cells that contribute to the growth and development of the root (**Figure 34.9A**). Some of the daughter cells from the apical (tip) end of the root apical meristem contribute to a **root cap**, which protects the delicate growing region of the root as it pushes through the soil. The root cap secretes a mucopolysaccharide (slime) that acts as a lubricant. Even so, the cells of the root cap are often damaged or scraped away and must therefore be replaced constantly. The root cap is also the structure that detects the pull of gravity and thus controls the downward growth of roots.

In the middle of the root apical meristem is a quiescent center, in which cell divisions are rare. The quiescent center can become more active when needed—following injury, for example. The daughter cells produced above the quiescent center (that is, away from the root cap) become the three primary meristems.

The apical and primary meristems constitute the **zone of cell division**, the source of all the cells of the root's primary tissues. Just above this zone is the **zone of cell elongation**, where the newly formed cells are elongating and thus pushing the root farther into the soil. Above that zone is the **zone of maturation**, where the cells are differentiating, taking on specialized forms and functions. These three zones grade imperceptibly into one another; there is no abrupt line of demarcation.

## The products of the root's primary meristems become root tissues

The products of the three primary meristems (the protoderm, ground meristem, and procambium) are the tissue systems of the mature root. The differing arrangement of the three tissue

**34.9 Tissues and Regions of the Root Tip** (A) Extensive cell division creates the complex structure of the root. (B) Root hairs, seen with a scanning electron microscope.

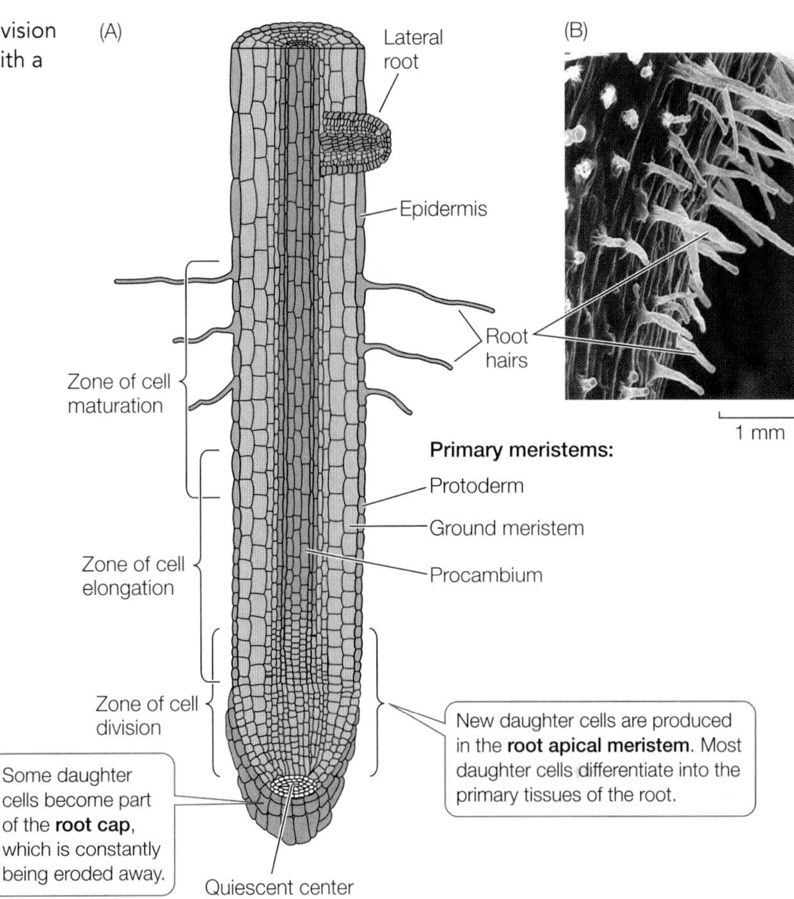


systems in the roots of eudicots and monocots is one of the ways in which the two clades are distinguished (**Figure 34.10**).

The protoderm gives rise to the epidermis, the outer layer of cells that is adapted for protection of the root and absorption of mineral ions and water. Many of the epidermal cells produce long, delicate **root hairs**, which vastly increase the surface area of the root (**Figure 34.9B**). Root hairs grow out among the soil particles, probing nooks and crannies and taking up water and minerals.

Internal to the epidermis, the ground meristem gives rise to a region of ground tissue that is many cells thick, called the **cortex**. The cells of the cortex are relatively unspecialized and often serve as storage depots. The innermost layer of the cortex is the **endodermis**. Unlike other cortical cells, the endodermal cells contain a waterproof substance called suberin in their primary cell walls. The suberin forms a cylindrical ring around the inside of the endodermis, which allows the endodermal cells to control the movement of water and dissolved mineral ions into and out of the vascular tissue system.

Within the endodermis is the vascular cylinder, or **stele**, produced by the procambium. The stele consists of three tissues: pericycle, xylem, and phloem. The **pericycle** consists of one or more layers of relatively undifferentiated cells. It has three important functions:

- It is the tissue within which lateral roots arise (**Figure 34.11**).
- It can contribute to secondary growth by giving rise to lateral meristems that thicken the root.
- Its cells contain membrane transport proteins that export nutrient ions into the cells of the xylem.

At the very center of the root of a eudicot lies the xylem. Seen in cross section, it typically has the shape of a star with a variable number of points (see Figure 34.10A). Between the points are bundles of phloem.

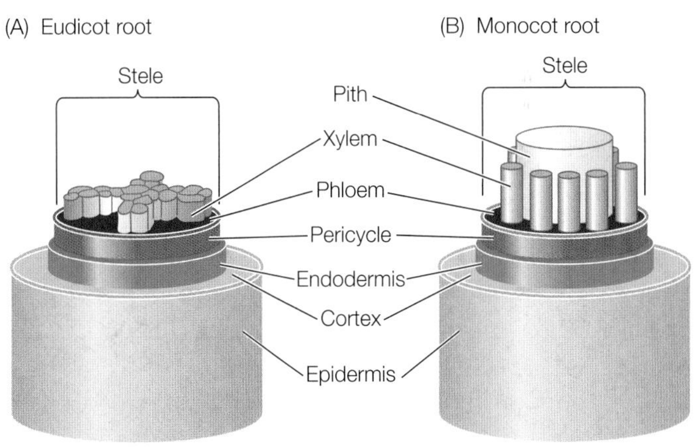


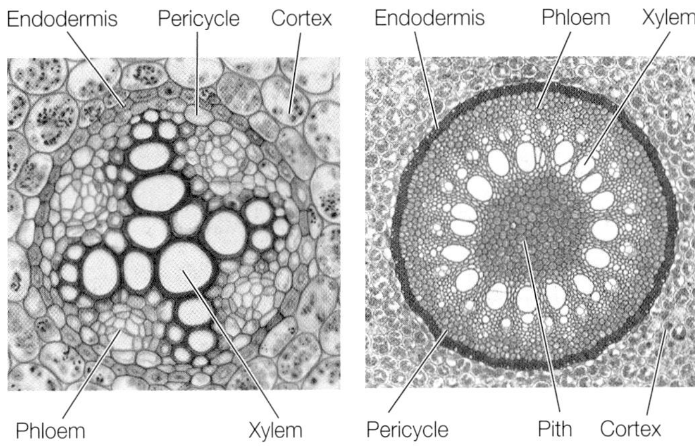


**34.10 Products of the Root's Primary Meristems** The protoderm gives rise to the outermost layer (epidermis). The ground meristem produces the cortex, the innermost layer of which is the endodermis. The primary vascular tissues of the root are found in the stele, which is the product of the procambium. The arrangement of tissues in the stele differs in the roots of (A) eudicots and (B) monocots. The photomicrographs show cross sections of the stele of a representative eudicot (the buttercup, *Ranunculus*) and a representative monocot (corn, *Zea mays*), showing the arrangement of the primary root tissues.

**Go to Activity 34.1 Eudicot Root Life10e.com/ac34.1**

**Go to Activity 34.2 Monocot Root Life10e.com/ac34.2**

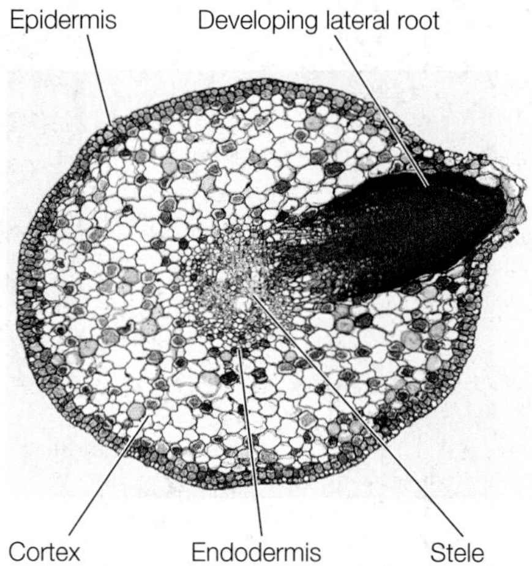


**34.11 Lateral Root Anatomy** Cross section through the tip of a lateral root in a willow tree. Cells in the pericycle divide and the products differentiate, forming the tissues of a lateral root.

In monocots, a region of parenchyma cells called the **pith** typically lies in the center of the root, surrounded by xylem and phloem (see Figure 34.10B). Pith, which often stores carbohydrate reserves, is also found in the stems of both eudicots and monocots.

### The root system anchors the plant and takes up water and dissolved minerals

Water and minerals enter most plants through the root system, which is located in the soil. Because light does not penetrate the soil, roots typically lack the capacity for photosynthesis. Although hidden from view, the root system is often larger than the visible shoot system. For example, the root system of a 4-month-old winter rye plant (*Secale cereale*) was found to be 130 times longer in total than the shoot system, with almost 13 million branches that had a cumulative length of more than 500 kilometers!

Angiosperm root systems develop from the embryonic root, called the **radicle**. From this common starting point, the root systems of monocots and eudicots develop differently. Following seed germination, the radicle of most eudicots develops as a primary root called the **taproot**, which extends downward by tip growth and outward by initiating **lateral roots**. The taproot and the lateral roots form a **taproot system**, which can take a variety of forms. For example, the taproot itself often functions as a nutrient storage organ, as in carrots (*Daucus carota*), sugar beets (*Beta vulgaris*), and sweet potato (*Ipomoea batatas*) (**Figure 34.12A**).

In contrast, the primary root of monocots (and some eudicots) is short-lived. Because they originate from the stem at ground level or just below, the roots of a typical monocot are called **adventitious** ("arriving from outside") **roots**, and they form a **fibrous root system** composed of numerous thin roots that are all roughly equal in diameter (**Figure 34.12B**). Many fibrous root systems have large surface areas for the absorption of water and minerals. A fibrous root system clings to soil very well. The fibrous root systems of grasses, for example, may protect steep hillsides where runoff from rain would otherwise cause erosion.

In some plants—corn, banyan, and pandanus trees, for example—adventitious roots grow down from above the ground and function as props to help support the shoot system (**Figure 34.12C**). These **prop roots** have evolved for different reasons in different species. For example, some monocots may develop prop roots because they are unable to support aboveground growth through the thickening of their stems. Pandanus trees often grow near coastal beaches, where their prop roots help provide support in very sandy soils. Banyans begin life as epiphytes (plants that grow on other plants) and then develop woody prop roots, which enable them to grow into huge trees.

(A) Taproots

(B) Fibrous root system

(C) Prop roots

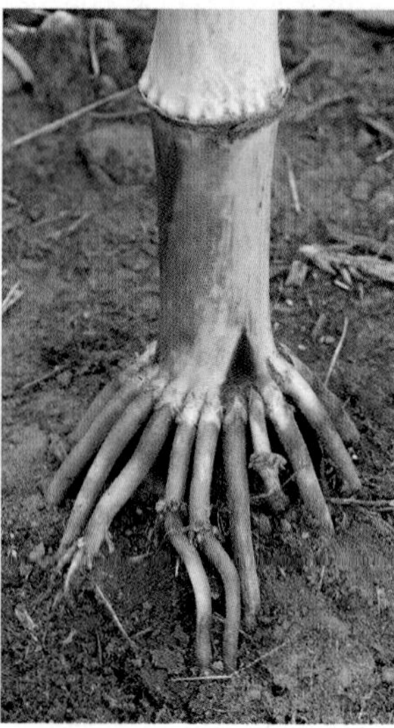

**34.12 Root Systems of Eudicots and Monocots** (A) The taproot systems of eudicots, such as carrots, sugar beets, and sweet potato, contrast with (B) the fibrous root system of a leek and (C) the adventitious prop roots of corn.

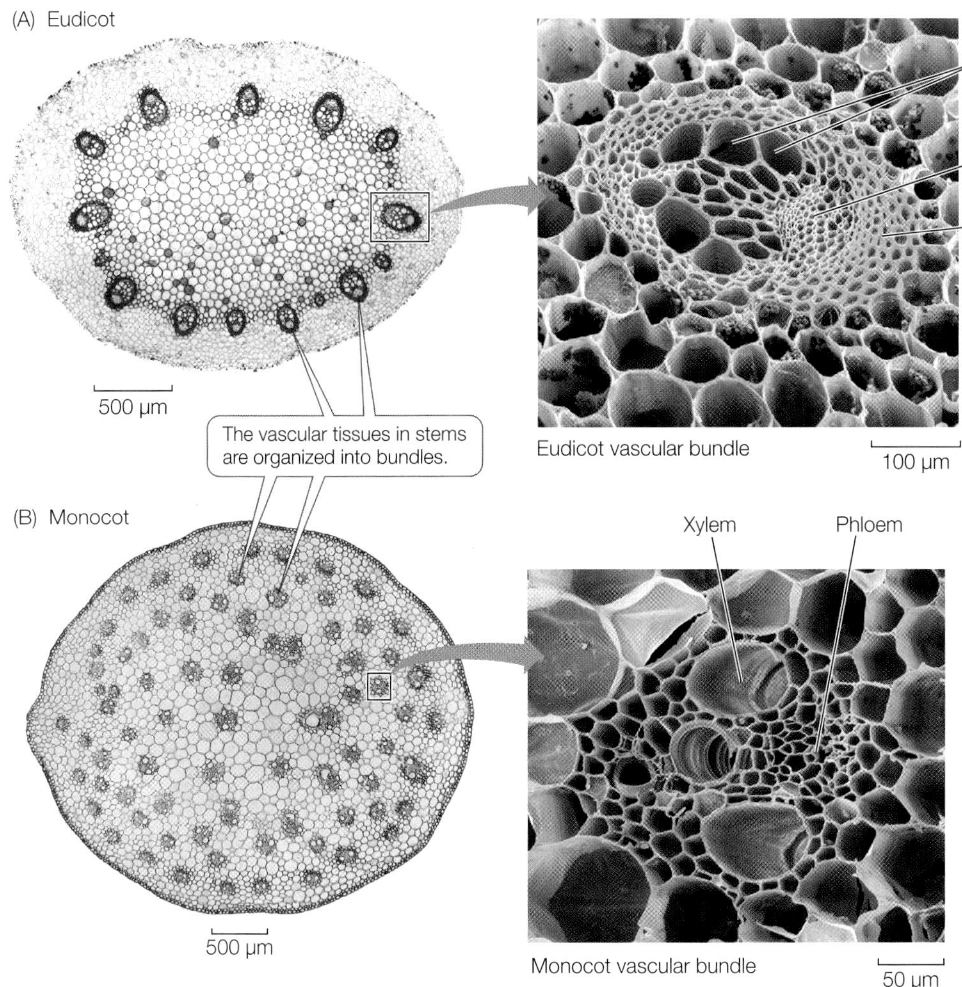


**34.13 Vascular Bundles in Stems** (A) In herbaceous eudicot stems, the vascular bundles are arranged in a cylinder, with pith in the center and the cortex outside the cylinder. (B) A scattered arrangement of vascular bundles is typical of monocot stems.

**Go to Activity 34.3 Eudicot Stem**
**Life10e.com/ac34.3**

**Go to Activity 34.4 Monocot Stem**
**Life10e.com/ac34.4**

## The products of the stem's primary meristems become stem tissues

Recall that shoots and roots are composed of repeating modules called phytomers. A phytomer in the shoot system consists of a node with its attached leaf or leaves, the internode below the node, and axillary buds, each of which forms in the angle between a leaf and the stem (see Figure 34.1). Shoots grow by adding new phytomers. Initially these form only on the primary stem of the plant, but if an axillary bud develops into a branch, it does so by producing new phytomers. The new phytomers originate from cells in the shoot apical meristems, which are present in the terminal buds of each branch and the main stem.

The shoot apical meristem, like the root apical meristem, forms three primary meristems: protoderm, ground meristem, and procambium. These primary meristems, in turn, give rise to the three shoot tissue systems.

As the shoot grows and extends, bulges called **leaf primordia** develop on the sides of the shoot apical meristem at regular intervals. These primordia are made up of primary meristematic tissues, which go on to develop into the mature tissues of the leaf. In addition, bud primordia form at the bases of the leaf primordia. These have the potential to become new apical meristems and initiate new shoots. The sites where leaf primordia form become nodes on the developing stem. The regions between the nodes (the internodes) lengthen initially via cell division in the primary meristematic tissues, and later by cell elongation in the mature stem tissues. The growing stem has no protective structure analogous to the root cap, but the leaf primordia can act as a protective covering for the shoot apical meristem.

The vascular systems of stems differ from those of roots. In a root, the vascular tissue lies deep in the interior, with the xylem at or near the center (see Figure 34.10). The vascular tissue of a young stem, however, is divided into discrete **vascular bundles** (**Figure 34.13**). Each vascular bundle contains both xylem and phloem. In eudicots the vascular bundles generally form a cylinder, whereas in monocots they are scattered throughout the stem.

In addition to the vascular tissues, the stem contains other important storage and supportive tissues. In eudicots the pith lies inside the ring of vascular bundles and also extends between them, forming regions called pith rays. To the outside lies the cortex, which may contain supportive collenchyma

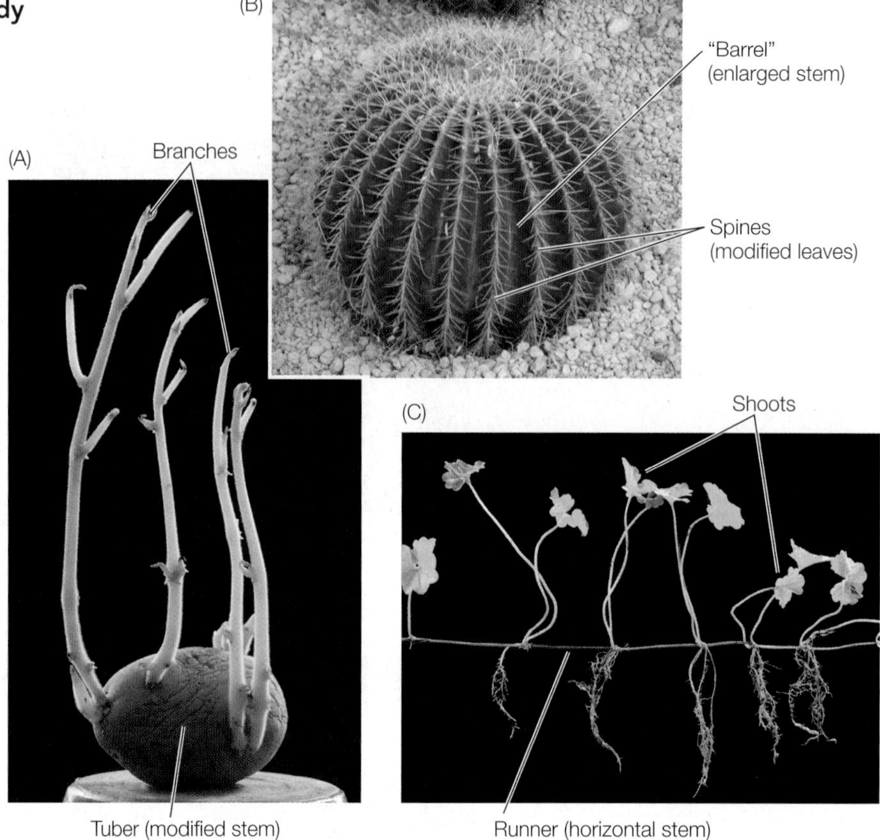


**34.14 Modified Stems** (A) A potato is a modified stem called a tuber; the sprouts that grow from its "eyes" are shoots, not roots. (B) The stem of this barrel cactus is enlarged to store water. Its highly modified leaves serve as thorny spines. Most of this plant's photosynthesis occurs in the stem. (C) The runners of ground ivy are horizontal stems that produce roots and shoots at intervals. Rooted portions of the plant can live independently if the runner is cut.

cells with thickened walls. The pith and cortex constitute the ground tissue system of the stem. The outermost cell layer of the young stem is the epidermis.

## The stem supports leaves and flowers

The central function of stems is to elevate and support the photosynthetic organs (leaves) as well as the reproductive organs (flowers). Various modifications of stems are seen in nature. The tuber of a potato, for example—the part of the plant eaten by humans—is not a root but rather an underground stem. The "eyes" of a potato are depressions containing axillary buds—in other words, a sprouting potato is just a branching stem (**Figure 34.14A**). Many desert plants have enlarged, water-retaining stems (**Figure 34.14B**). The runners of strawberry plants are horizontal stems that develop adventitious roots some distance from the main stem (**Figure 34.14C**); if the links between the rooted portions are broken, independent plants can develop on each side of the break—a form of vegetative (asexual) reproduction (see Section 38.3).

## Leaves are determinate organs produced by shoot apical meristems

For most of its life, a plant produces leaves from apical meristems. As we mentioned earlier, apical meristems that produce leaves are called vegetative meristems. Leaves originate from the edges of the apical meristem as initial cells that differentiate into leaf primordia. A highly simplified way to think of the development of a leaf from a leaf primordium is to imagine leaves as flattened stems. However, there are two important differences:

- Unlike the growth of the stem, which is indeterminate, the growth of a leaf is determinate.
- Whereas the tissues of the stem are arranged in a radial pattern, the leaf, as a flat organ, has a distinct top side and bottom side.

Leaf anatomy is beautifully adapted to carry out photosynthesis, and to support that process by exchanging the gases $O_2$ and $CO_2$ with the environment, while at the same time limiting evaporative water loss. Its extensive vascular system supplies the leaf with water and mineral nutrients and exports the products of photosynthesis to the rest of the plant. **Figure 34.15A** shows a section of a typical eudicot leaf in three dimensions.

The photosynthetic parenchyma tissue in leaves is called the **mesophyll** (which means "middle of the leaf"). Most eudicot leaves have two zones of mesophyll: an upper layer of elongated cells called the palisade mesophyll, and a lower layer of irregularly shaped cells called the spongy mesophyll. Within the mesophyll is a great deal of air space through which $CO_2$ can diffuse to photosynthesizing cells.

Vascular tissue branches extensively throughout the leaf, forming a network of veins (**Figure 34.15B**). Typically, monocot leaves have a parallel pattern of major veins; for an example, look at the grass blades on a nearby lawn. Dicot leaves, in contrast, have major veins in a netlike pattern, as in a maple or oak leaf. Veins extend to within a few cell diameters of all the cells of the leaf, ensuring that the mesophyll cells are well supplied with water and minerals. The products of photosynthesis are loaded into the veins for export to the rest of the plant.

The epidermis covers the entire surface of the leaf and is made up of nonphotosynthetic cells. The epidermal cells secrete a waxy cuticle that is impermeable to water. Although this impermeability prevents excessive water loss, it also poses

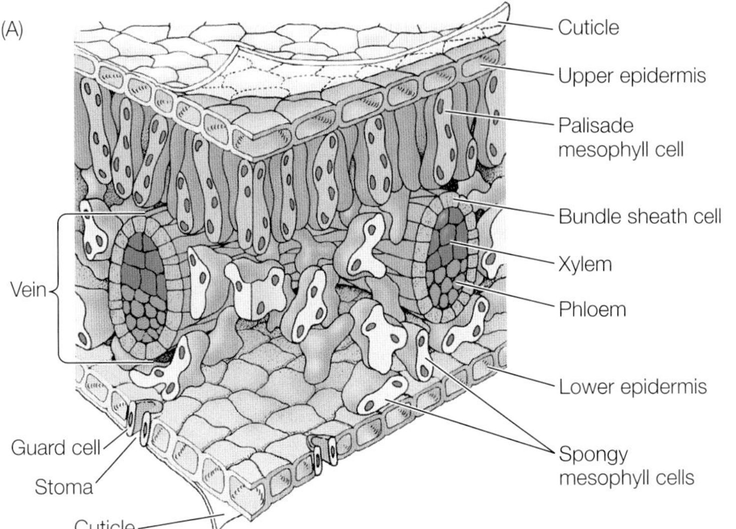


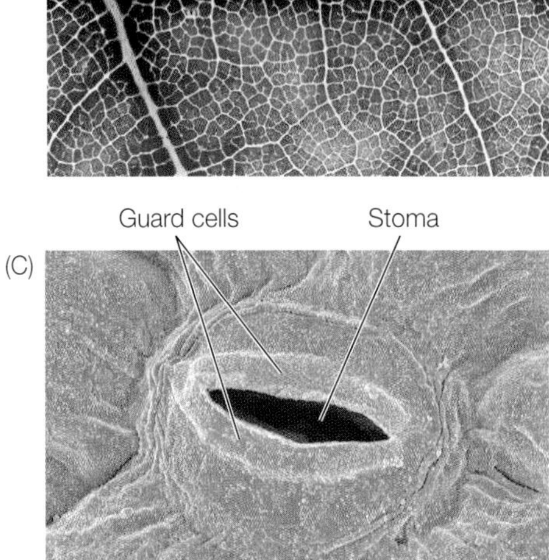


**34.15 The Eudicot Leaf** (A) This three-dimensional diagram shows a section of a eudicot leaf. (B) The network of fine veins in this maple leaf carries water to the mesophyll cells and carries photosynthetic products away from them. (C) Carbon dioxide enters the leaf through stomata like this one on the epidermis of a eudicot leaf.

**Go to Activity 34.5 Eudicot Leaf Life10e.com/ac34.5**

a problem: while the epidermis keeps water in the leaf, it also keeps out $CO_2$—the other raw material of photosynthesis.

The problem of balancing water retention and carbon dioxide availability is solved by an elegant regulatory system that will be discussed in more detail in Section 35.3. Stomatal guard cells are modified epidermal cells that can change their shape, thereby opening or closing pores called **stomata** (singular *stoma*). The stomata serve as passageways between the environment and the leaf's interior (**Figure 34.15C**). When the stomata are open, carbon dioxide can enter and oxygen can leave, but water can also be lost.

## Many eudicot stems and roots undergo secondary growth

As we have seen, the roots and stems of some eudicots develop a secondary plant body, the tissues of which we commonly refer to as wood and bark. These tissues are derived by secondary growth from the two lateral meristems, the vascular cambium and the cork cambium.

The **vascular cambium** is a cylindrical layer of tissue consisting predominantly of elongated cells that divide frequently. It supplies the cells of the secondary xylem and secondary phloem, which eventually become wood and bark. The **cork cambium** produces mainly waxy-walled protective cells. It supplies some of the cells that become bark.

Each year, deciduous trees lose their leaves and have bare branches and twigs over the winter (**Figure 34.16**). The apical meristems of the

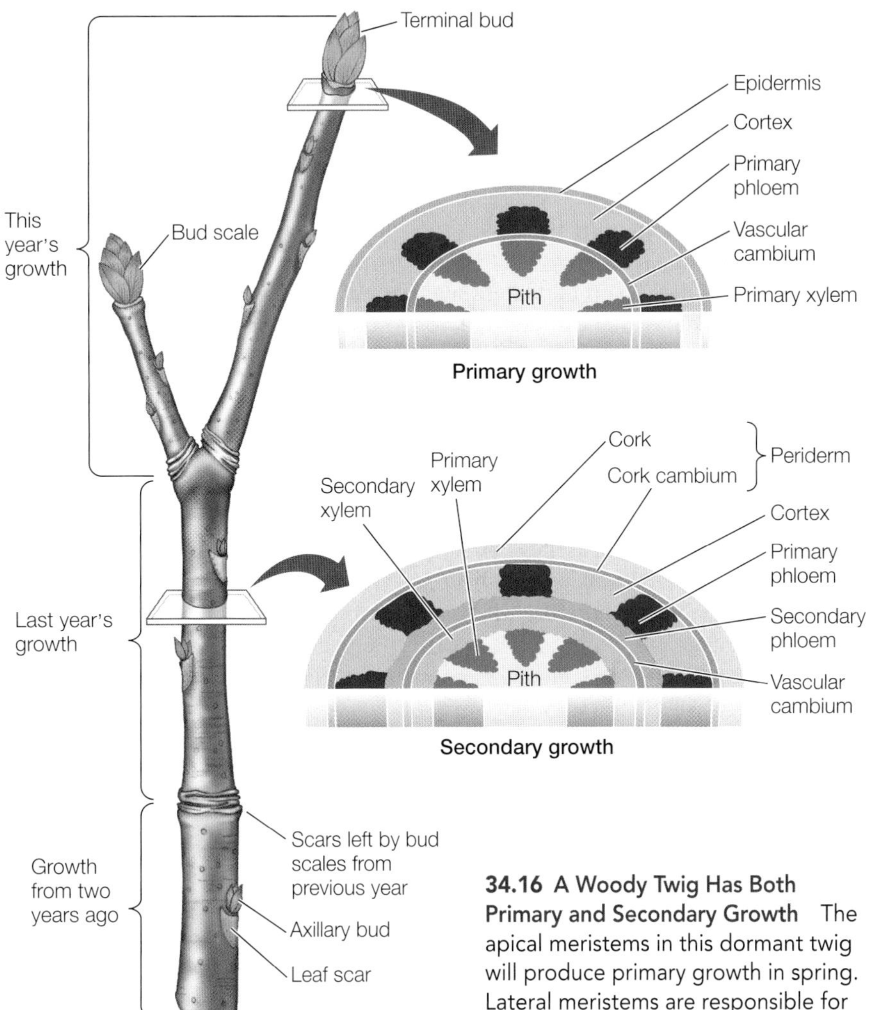


**34.16 A Woody Twig Has Both Primary and Secondary Growth** The apical meristems in this dormant twig will produce primary growth in spring. Lateral meristems are responsible for secondary growth.

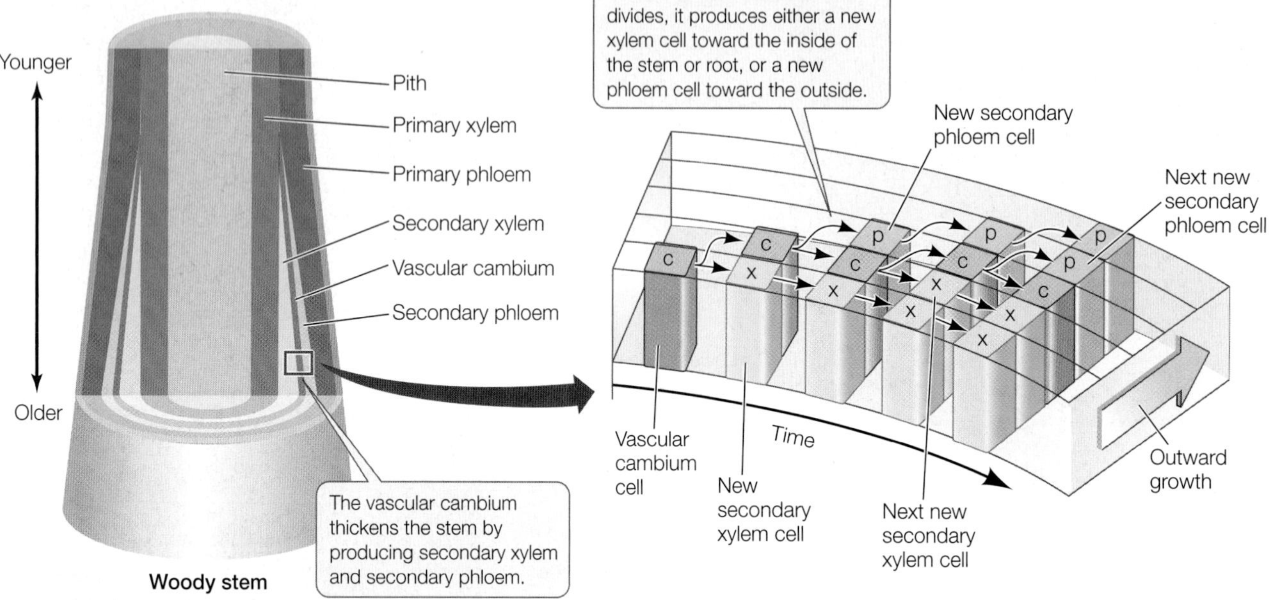


**34.17 Vascular Cambium Thickens Stems and Roots** Stems and roots grow thicker because a thin layer of cells, the vascular cambium, remains meristematic. These highly diagrammatic images emphasize the pattern of deposition of secondary xylem and phloem by the vascular cambium.

**Go to Animated Tutorial 34.1**
**Secondary Growth: The Vascular Cambium**
Life10e.com/at34.1

twigs are enclosed in buds protected by bud scales. When the buds begin to grow in spring, the scales fall away, leaving scars that show where the bud was. These scars allow us to identify the parts of the twig from each year's growth. The dormant twig shown in Figure 34.16 is the product of both primary and secondary growth. Only the buds consist entirely of primary tissues.

The vascular cambium is initially a single layer of cells lying between the primary xylem and the primary phloem within the vascular bundles. The root or stem increases in diameter when the cells of the vascular cambium divide, producing secondary xylem cells toward the inside of the root or stem and producing secondary phloem cells toward the outside (**Figure 34.17**). In the stem, cells in the pith rays between the vascular bundles also divide, forming a continuous cylinder of vascular cambium running the length of the stem. This cylinder, in turn, gives rise to complete cylinders of secondary xylem (the **wood**) and secondary phloem, which contributes to the bark. It also produces vascular rays for lateral transport, a structure not found in primary xylem and phloem. Therefore the vascular cambium produces vessel elements, tracheids, and supportive fibers in the secondary xylem; and sieve tube elements, companion cells, fibers, and parenchyma cells in the secondary phloem.

As secondary growth of stems or roots continues, the expanding vascular tissue stretches and breaks the epidermis and the outer layers of the cortex, which ultimately flake away. Before these dermal tissues are broken away, cells lying near the surface of the secondary phloem begin to divide, forming a cork cambium. This meristematic tissue produces layers of cork, a protective tissue composed of cells with thick walls waterproofed with suberin. The cork soon becomes the outermost tissue of the stem or root (see Figure 34.16). Without the activity of the cork cambium, the sloughing off of the outer primary tissues would expose the plant to potential damage, such as excessive water loss or invasion by microorganisms. Sometimes the cork cambium produces cells toward the inside as well as the outside; these cells constitute a tissue known as the phelloderm.

The cork cambium, cork, and phelloderm constitute the secondary dermal tissue called periderm. As the vascular cambium continues to produce secondary vascular tissue, these corky layers are lost, but the continuous formation of new cork cambia in the underlying secondary phloem gives rise to new corky layers. The periderm and the secondary phloem—that is, all the tissues external to the vascular cambium—constitute the **bark**.

When periderm forms on stems or roots, the underlying tissues still need to release carbon dioxide and take up oxygen for cellular respiration. Spongy regions in the periderm called lenticels allow such gas exchange (**Figure 34.18**).

Cross sections of most trunks (mature stems) of trees in temperate-zone forests show annual rings of wood (**Figure 34.19**), which result from seasonal environmental conditions.

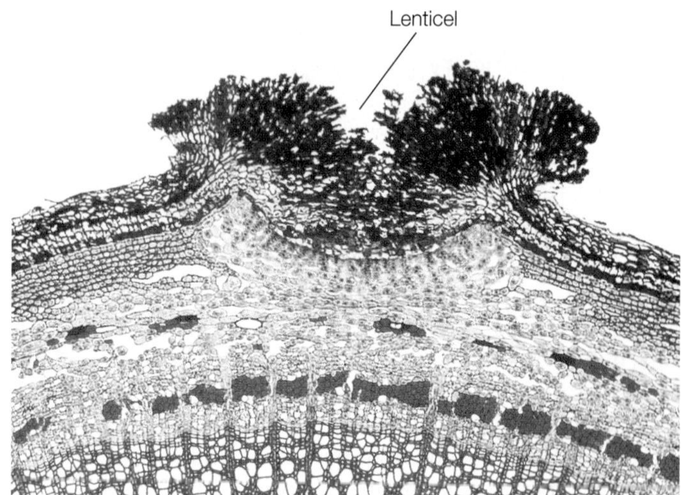


**34.18 Lenticels Allow Gas Exchange through the Periderm** The region of periderm that appears to be broken open is a lenticel in a year-old elderberry (*Sambucus*) twig; note the spongy tissue that constitutes the lenticel.

Secondary xylem (one year's growth)
Bark
Cork cambium
Cork
Pith
Spring wood
Summer wood
Secondary phloem

**34.19 Annual Rings** Rings of secondary xylem are the most noticeable feature of this cross section from a tree trunk.

In spring, when water is relatively plentiful, the tracheids or vessel elements produced by the vascular cambium tend to be large in diameter and thin-walled. Such wood is well adapted for transporting water and minerals. As water becomes less available during the summer, narrower cells with thicker walls are produced, making this summer wood darker and perhaps more dense than the wood formed in spring. Thus each growing season is usually recorded in a tree trunk by a clearly visible annual ring. Trees in the moist tropics do not undergo seasonal growth, so they do not lay down such obvious regular rings. Variations in temperature or water supply can lead to the formation of more than one ring in a single year, but most commonly a single annual ring is formed.

Only eudicots and other non-monocot angiosperms, along with many gymnosperms, have a vascular cambium and a cork cambium and thus undergo secondary growth. The few monocots that form thickened stems—palms, for example—do so without secondary growth. Palms have a very wide apical meristem that produces a wide stem, and dead leaf bases add to the diameter of the stem. All monocots grow in essentially this way, as do other angiosperms that lack secondary growth.

### RECAP 34.3

Meristems are localized regions of cell division that are the sources of all new organs in the adult plant. Apical meristems are responsible for primary growth, which is associated with the lengthening and branching of shoots and roots. Lateral meristems increase plant thickness and form wood and bark in many eudicots.

- Explain how an apical meristem can be maintained for years while continuing to form leaves. **See p. 715 and Figure 34.8**
- What cells are derived from the root apical meristem, and what is the general process of root growth? **See pp. 716–717 and Figures 34.9 and 34.10**
- How does the vascular cambium give rise to thicker stems and roots? **See pp. 721–722 and Figures 34.16 and 34.17**

The building of the plant body by meristems allows a plant to respond to its environment by redirecting its growth. Thus individual plants of the same species can vary greatly in form. What underlies this variation, and how have we humans used it to our advantage?

## 34.4 How Has Domestication Altered Plant Form?

We have seen in this chapter that a very simple, modular plant body plan underlies the remarkable diversity of the flowering plants that cover our planet. Differences in plant form among species are not surprising, given the high levels of genetic diversity among plant species. However, members of the same species can also show remarkable diversity in form. From a genetic perspective, this suggests that minor differences in gene content or gene regulation can underlie dramatic differences in plant form.

It is hard to believe that modern corn was domesticated from the wild grass teosinte, which still grows in the hills of Mexico (**Figure 34.20**). One of the most conspicuous differences is that teosinte, like other wild grasses, is highly branched, whereas domesticated corn has a single shoot. This morphological difference is due in large part to the activity of a single gene called *teosinte branched 1* (*tb1*). The protein product of *tb1* regulates the growth of axillary buds (see Figure 34.1). The allele of *tb1* in domesticated corn represses branching, whereas the allele in teosinte permits branching.

**34.20 Modern Corn Was Domesticated from the Wild Grass Teosinte** Each teosinte plant has multiple branches. Beginning more than 8,000 years ago in Mexico, farmers favored plants with minimal branching. Reducing the number of branches results in fewer ears per plant but allows each ear to grow larger and produce more seeds.

Even harder to believe is that a single species, *Brassica oleracea* (wild mustard), is the ancestor of many familiar and morphologically diverse crops, including kale, broccoli, Brussels sprouts, and cabbage (see Figure 21.4). But each of these familiar vegetable crops has the same basic body plan. Starting with morphologically diverse populations of the wild ancestor, humans selected and planted the seeds of variants with traits they found desirable. For example, Brussels sprout plants were selected for their enlarged axillary buds, cabbage plants were selected for their enlarged terminal buds and short internodes, and broccoli and cauliflower plants were selected for their large clusters of flower buds. Many generations of such artificial selection produced the crops that fill the produce section of the supermarket or the stands of the farmers' market.

Just as they were for ancient farmers, the genomes of plants are priceless resources today. The genetic variation in crop plants and their wild relatives can be used to improve our crop plants or adapt them to changing conditions. The improvement of crop plants is a work in progress that is being carried out in plant breeding programs worldwide. In fact, these programs are more important than ever. Increased human activity is dramatically changing our planet and leading to the extinction of more and more plant species. For this reason, various organizations around the world have developed seed banks, where seeds of diverse species, and variants within species, are stored.

**RECAP 34.4**

Crop domestication involves artificial selection of certain desirable traits found in wild plant populations. By understanding the basic body plan of plants, one can more easily understand the morphological relationship between a crop plant and its wild relatives.

- Why are the seeds from wild relatives of crop plants valuable? **See p. 725**

How might plant physiologists improve the cassava plant for human use?

**ANSWER**

Many people depend on cassava roots for food. Unfortunately, however, the varieties that are traditionally grown have inadequate levels of protein, iron, and β-carotene. (The human body converts β-carotene into vitamin A.) Furthermore, cassava plants have high levels of cyanogens that are converted to toxic cyanide when eaten. BioCassava Plus is an international consortium of scientists who are using biotechnology to develop cassava varieties with improved nutritional contents. Another team has developed cassava plants that contain reduced levels of cyanogens. The anatomy of the plant is also being changed. Cassava has been crossed with a treelike relative, *Manihot glaziovii*, and the resulting plants produce roots that are not only fleshy and edible but that grow deep into the soil, where they can tap into water supplies far below the surface in dry climates. This crop is being investigated for use in sub-Saharan Africa.

# CHAPTER**SUMMARY** 34

## 34.1 What Is the Basic Body Plan of Plants?

- The vegetative organs of flowering plants are roots, which form a **root system**, and stems and leaves, which form a **shoot system**. **Review Figure 34.1**
- Plant development differs from animal development in that plants have apical meristems, cell walls, vacuoles, and in some cases, totipotent cells.
- Plants have apical–basal and radial axes of symmetry. **Review Figures 34.3, 34.4**

## 34.2 What Are the Major Tissues of Plants?

- Three **tissue systems**, arranged concentrically, extend throughout the plant body: the vascular tissue, dermal tissue, and ground tissue systems. **Review Figure 34.5**
- The **dermal tissue system** protects the plant body surface. Dermal cells form the **epidermis** and, in woody plants, the **periderm**.
- The **ground tissue system** contains cells of three types. Some **parenchyma** cells carry out photosynthesis; others store starch. **Collenchyma** cells provide flexible support. **Sclerenchyma** cells include **fibers** and **sclereids** that provide strength and mechanical support. **Review Figures 34.6, 34.7**
- The **vascular tissue system** includes **xylem**, which conducts water and minerals absorbed by the roots, and **phloem**, which conducts the products of photosynthesis throughout the plant body.
- **Tracheary elements** include **tracheids** and **vessel elements**, which are the conducting cells of the xylem. **Sieve tube elements** are the conducting cells of the phloem.

## 34.3 How Do Meristems Build a Continuously Growing Plant?

- All seed plants possess a primary plant body consisting of nonwoody tissues. Woody plants also possess a secondary plant body consisting of wood and bark. **Apical meristems** generate the primary plant body, and **lateral meristems** generate the secondary plant body. **Review Figure 34.8**
- Apical meristems are responsible for **primary growth** (lengthening of roots and shoots). Apical meristems at the tips of stems and roots give rise to three **primary meristems** (**protoderm**, **ground meristem**, and **procambium**), which in turn produce the three tissue systems of the primary plant body.
- The root apical meristem gives rise to the **root cap** and to three primary meristems. Root tips have overlapping **zones of cell division**, **cell elongation**, and **cell maturation**. **Review Figure 34.9**

*continued*

- The vascular tissue of roots is contained within the **stele**. It is arranged differently in eudicot and monocot roots. **Review Figures 34.10, 34.11, ACTIVITIES 34.1, 34.2**
- In nonwoody stems, the vascular tissue is divided into **vascular bundles**, each containing both xylem and phloem. **Review Figure 34.13, ACTIVITIES 34.3, 34.4**
- Eudicot leaves have two zones of photosynthetic **mesophyll** that are supplied by veins with water and minerals. Veins also carry the products of photosynthesis to other parts of the plant body. A waxy **cuticle** limits water loss from the leaf. Guard cells control openings called **stomata** in the leaf that allow $CO_2$ to enter, but also allow some water to escape. **Review Figure 34.15, ACTIVITY 34.5**
- Two lateral meristems, the **vascular cambium** and **cork cambium**, are responsible for secondary growth. The vascular cambium produces secondary xylem (wood) and secondary phloem. The cork cambium produces a protective tissue called cork. **Review Figures 34.16, 34.17, ANIMATED TUTORAL 34.1**

## 34.4 How Has Domestication Altered Plant Form?

- The plant body plan is simple, yet it can be changed dramatically by minor differences in genes, as evidenced by the natural diversity of wild plants.
- Crop domestication involves artificial selection of certain desirable traits found in wild populations. **Review Figure 34.20**

**Go to the Interactive Summary to review key figures, Animated Tutorials, and Activities Life10e.com/is34**

# CHAPTER**REVIEW**

## REMEMBERING

1. Roots
   a. always form a fibrous root system that holds the soil.
   b. possess a root cap at their tip.
   c. form branches from axillary buds.
   d. are commonly photosynthetic.
   e. do not show secondary growth.

2. Which statement about parenchyma cells is *not* true?
   a. They are alive when they perform their functions.
   b. They typically lack a secondary wall.
   c. They often function as storage depots.
   d. They are the most common cell type in the plant body.
   e. They are found only in stems and roots.

3. Which statement about meristems is *not* true?
   a. They are formed during embryogenesis.
   b. They have secondary cell walls.
   c. Their cells have small central vacuoles.
   d. They are clusters of undifferentiated cells.
   e. They retain the ability to produce new cells indefinitely.

4. The pericycle
   a. is the innermost layer of the cortex.
   b. is the tissue within which lateral roots arise.
   c. consists of highly differentiated cells.
   d. forms a star-shaped structure at the very center of the root.
   e. is waterproofed by suberin.

5. Which statement about leaf anatomy is *not* true?
   a. Opening of stomata is controlled by guard cells.
   b. The cuticle is secreted by the epidermis.
   c. The veins contain xylem and phloem.
   d. The cells of the mesophyll are packed together, minimizing air space.
   e. The spines of cacti are actually modified leaves.

## UNDERSTANDING & APPLYING

6. Which of these statements is true of secondary growth but *not* of primary growth?
   a. It occurs in eudicots and monocots.
   b. It involves the proliferation of roots and shoots through branching.
   c. It derives from the vascular cambium and the cork cambium.
   d. It occurs in palms.
   e. It derives from the shoot apical meristem.

7. Which of the following is a difference between monocots and eudicots?
   a. Only eudicots have phytomers.
   b. Only monocots have shoot and root apical meristems.
   c. Monocot stems do not undergo secondary growth.
   d. The vascular bundles of monocot stems are commonly arranged as a cylinder.
   e. Eudicot embryos commonly have one cotyledon.

8. Compare sclerenchyma cells and collenchyma cells in terms of structure and function.

9. Compare primary and secondary growth. Do all angiosperms undergo secondary growth? Explain.

## ANALYZING & EVALUATING

10. When a young oak was 5 meters tall, a thoughtless person carved his initials in its trunk at a height of 1.5 meters above the ground. Today that tree is 10 meters tall. How high above the ground are those initials? Explain your answer in terms of plant growth.

11. Take a walk through a farmers' market or the produce section of a supermarket. Use your knowledge of plant growth and form to figure out what desirable trait was selected to produce some of your favorite salad vegetables.

Go to BioPortal at **yourBioPortal.com** for Animated Tutorials, Activities, LearningCurve Quizzes, Flashcards, and many other study and review resources.

# Transport in Plants

## CHAPTER OUTLINE


**35.1** How Do Plants Take Up Water and Solutes?

**35.2** How Are Water and Minerals Transported in the Xylem?

**35.3** How Do Stomata Control the Loss of Water and the Uptake of $CO_2$?

**35.4** How Are Substances Translocated in the Phloem?

**Thirsty Rice** Cultivation of rice, the most important food crop in Asia, requires large quantities of water.

EVERYONE KNOWS THAT PLANTS NEED WATER to grow. However, it may come as a surprise that the cultivation of crop plants consumes far more water than all other human activities combined. The worldwide demand for fresh water is increasing at a greater rate than the supply. This situation makes it imperative that we understand how plants use water so that we can select or breed plants that use it more efficiently. Much of the mass that plants acquire as they grow is due to their net fixation of atmospheric $CO_2$ into carbohydrates through photosynthesis. But plants need to take up a lot of water to grow. The ratio of net photosynthetic carbon fixation to water uptake is known as a plant's water-use efficiency.

Droughts and dwindling water supplies are challenging farmers all over the world. One of the least water-efficient of all crop plants is, unfortunately, one of our most important: rice. Rice plants use up to three times more water per unit of growth than crops such as wheat and corn. The precariousness of heavily water-dependent rice farming was dramatically demonstrated in eastern India between 1997 and 2003, when drought reduced rice production by more than 5 million tons—some farmers lost up to 50 percent of their crops.

A strain of rice requiring less water yet producing the same amount of grain would both make the world supply of rice less vulnerable to drought and help conserve water for other uses. A team of molecular biologists, plant physiologists, and crop scientists led by Andrew Pereira at Virginia Polytechnic began a quest for such a strain of rice by studying an entirely different plant—the model organism *Arabidopsis thaliana* (thale cress). They searched for genetic variants of *Arabidopsis* that had superior water-use efficiency. One variant they studied was particularly hard to pull out of the ground because of its extensive root system (indicating higher capacity for water uptake) and had thick leaves with abundant photosynthetic tissue (indicating prolific photosynthesis). Molecular and physiological characterization of this *Arabidopsis* strain showed that its improved water usage was linked to a mutation in a single gene that codes for a transcription factor. When this gene (called *HARDY*) was isolated and put into rice plants using recombinant DNA technology, the transformed rice plants not only had higher water-use efficiency than wild-type rice plants but were more tolerant of dry soil as well.

Many laboratories around the world are using *Arabidopsis* to isolate genes involved in water usage and other important physiological processes. The knowledge gained from these studies may lead to various improvements of crop plants.

What other methods are used to reduce water loss in agriculture?

See answer on p. 738.

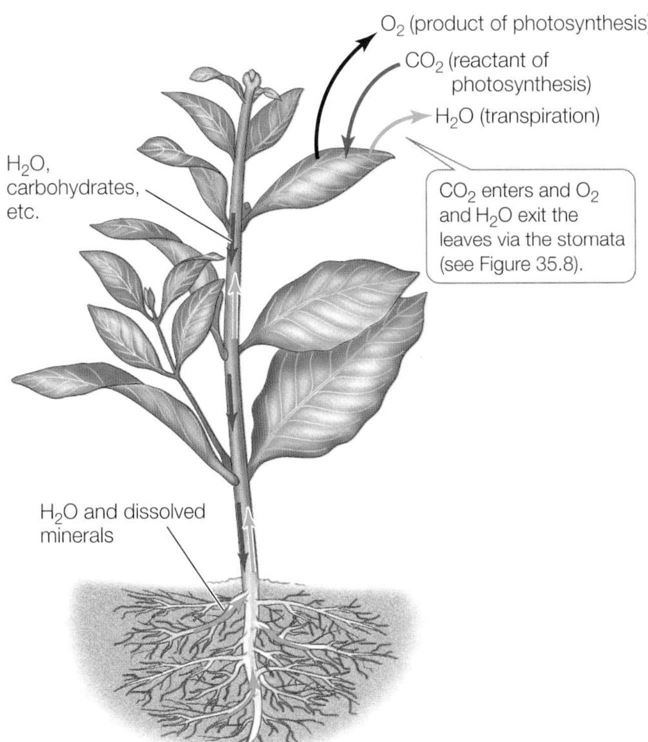


**35.1 The Pathways of Water and Solutes in a Plant**
Water travels from the soil to the atmosphere, with only a small fraction used within the plant.

## 35.1 How Do Plants Take Up Water and Solutes?

Terrestrial plants must obtain both water and mineral nutrients from the soil, usually through their roots. The roots, in turn, obtain carbohydrates and other important materials from the leaves (**Figure 35.1**). Water is required for photosynthesis in the leaves (see Section 10.1), for transporting solutes between plant organs, for cooling the plant, and for developing the internal pressure that supports the plant body.

The minerals that a plant needs are transported along with the water. Several steps in water and mineral transport will be considered in this chapter. In this section we will focus on the first part of the journey—the uptake of water and minerals into the roots and their transport into the xylem.

### Water potential differences govern the direction of water movement

To enter a root cell, a solution must pass through the cell's plasma membrane. In Section 6.3 we described osmosis, the movement of water through a selectively permeable membrane from a region of lower solute concentration (higher water potential) to a region of higher solute concentration (lower water potential). Plant biologists define **water potential** (**Ψ, psi**) as the tendency of a solution (water plus solutes) to take up water from pure water across a membrane. By definition, the water potential of pure water is zero. Any solution that has a water potential less than zero has a tendency to take up water from pure water. The lower (more negative) the water potential, the greater the driving force for water movement across the membrane.

Water potential has two major components:

- **Solute potential** ($\Psi_s$): Solutes affect the osmotic behavior of a solution. The solute potential of pure water is zero, so any solution will create a negative solute potential. The greater the concentration of solutes, the lower the water potential, and the lower (more negative) the solute potential.
- **Pressure potential** ($\Psi_p$): As plant cells take up water, they tend to swell. However, the presence of the cell wall provides resistance to swelling (see Figure 6.9). The result is an increase in pressure inside the cell (**turgor pressure**), which decreases the tendency of the cell to take up more water. Therefore the pressure potential within a plant cell is usually positive.

A solution's water potential is the sum of its (negative) solute potential ($\Psi_s$) and its (usually positive) pressure potential ($\Psi_p$):

$$\Psi = \Psi_s + \Psi_p$$

We measure solute potential, pressure potential, and water potential in megapascals (MPa), a unit of pressure. Atmospheric pressure, "one atmosphere," is about 0.1 MPa, or 14.7 pounds per square inch; a typical pressure in an automobile tire is about 0.2 MPa.

Whenever water moves by osmosis, the following important rule applies: *water always moves across a selectively permeable membrane toward the region of lower (more negative) water potential.* In the left-hand tube in **Figure 35.2A**, the solution has a negative solute potential and a pressure potential of zero (relative to the atmosphere), so water moves across the membrane into the solution. In the right-hand tube, a piston is used to increase the pressure potential in the solution. When the negative solute potential is balanced by the positive pressure potential there is no net movement of water across the membrane.

In a plant cell immersed in pure water (**Figure 35.2B**), turgor pressure is comparable to the pressure potential exerted by the piston in Figure 35.2A. Water enters the cell by osmosis until the pressure potential exactly balances the solute potential and the water potential is zero. At this point the cell is **turgid**—that is, it has a significantly positive pressure potential. The cells in a plant are not surrounded by pure water, and their water potential is dependent on the water potential in the soil. But because turgid cells have a positive pressure potential, there is no net movement of water into them. The physical structures of many plants are maintained by the positive pressure potential of the water in their cells. If the pressure potential drops (for example, if the plant does not have enough water), the plant wilts (**Figure 35.3**).

Within living plant tissues, the movement of water from cell to cell follows a gradient of water potential. Over long distances, in unobstructed tubes such as xylem vessels and phloem sieve tubes, the flow of water and dissolved solutes is driven by a *gradient of pressure potential*, not a gradient of water potential. The movement of a solution from a region of higher pressure

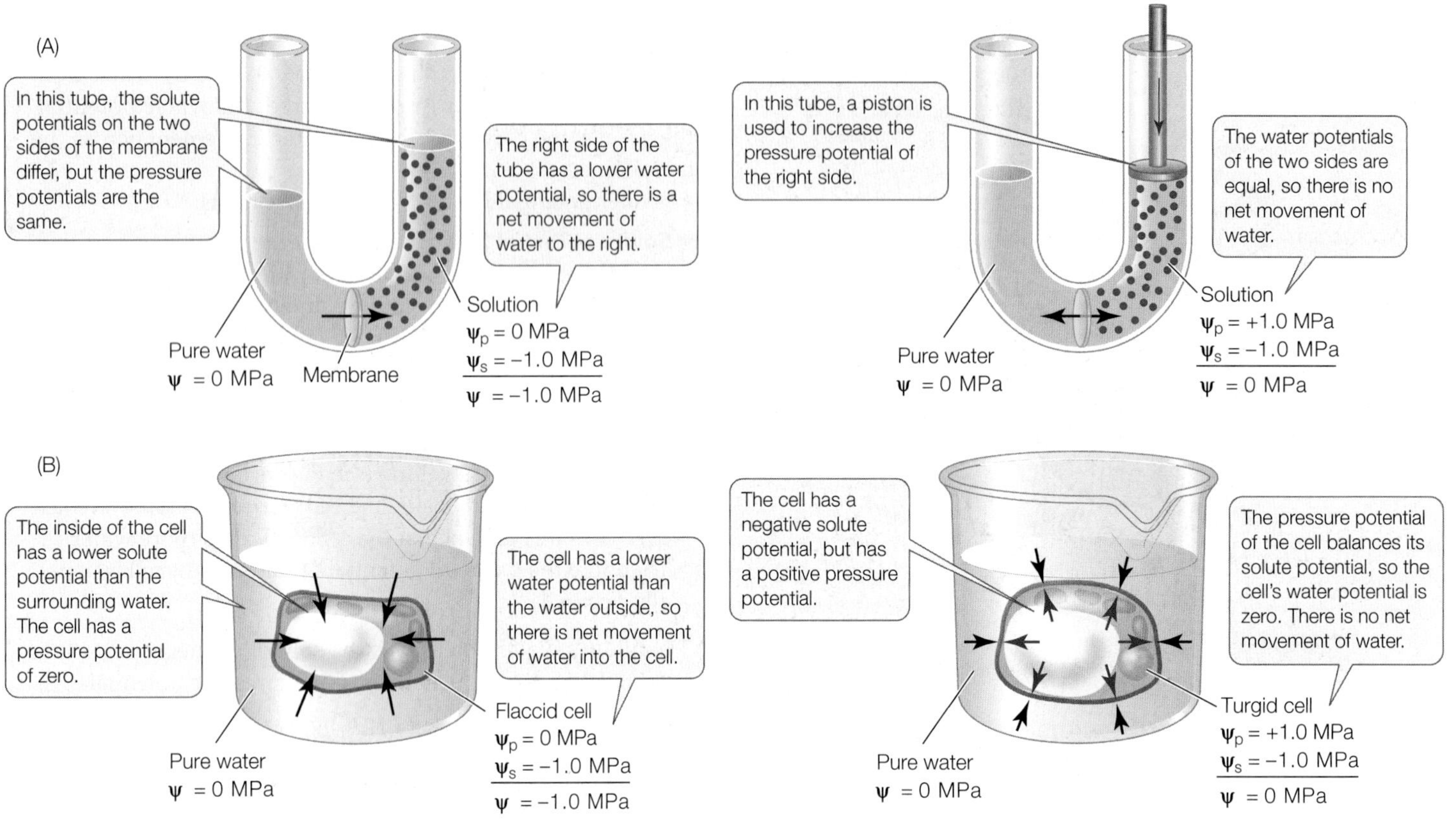
(A)
In this tube, the solute potentials on the two sides of the membrane differ, but the pressure potentials are the same.
The right side of the tube has a lower water potential, so there is a net movement of water to the right.
Pure water
$\psi$ = 0 MPa
Membrane
Solution
$\psi_p$ = 0 MPa
$\psi_s$ = −1.0 MPa
$\psi$ = −1.0 MPa
In this tube, a piston is used to increase the pressure potential of the right side.
The water potentials of the two sides are equal, so there is no net movement of water.
Pure water
$\psi$ = 0 MPa
Solution
$\psi_p$ = +1.0 MPa
$\psi_s$ = −1.0 MPa
$\psi$ = 0 MPa
(B)
The inside of the cell has a lower solute potential than the surrounding water. The cell has a pressure potential of zero.
The cell has a lower water potential than the water outside, so there is net movement of water into the cell.
Pure water
$\psi$ = 0 MPa
Flaccid cell
$\psi_p$ = 0 MPa
$\psi_s$ = −1.0 MPa
$\psi$ = −1.0 MPa
The cell has a negative solute potential, but has a positive pressure potential.
The pressure potential of the cell balances its solute potential, so the cell's water potential is zero. There is no net movement of water.
Pure water
$\psi$ = 0 MPa
Turgid cell
$\psi_p$ = +1.0 MPa
$\psi_s$ = −1.0 MPa
$\psi$ = 0 MPa

**35.2 Water Potential, Solute Potential, and Pressure Potential** A theoretical illustration of water potential. (B) The effect of differences in water potential on a plant cell.

**Go to Animated Tutorial 35.1**
**Water Uptake in Plants**
Life10e.com/at35.1

potential to a region of lower pressure potential is called **bulk flow**. As we will see later in this chapter, bulk flow in the xylem is between regions of differing *negative* pressure potentials (tension). By contrast, bulk flow in the phloem is between regions of differing *positive* pressure potentials (turgor pressure).

## Water and ions move across the root cell plasma membrane

The movement of water and mineral ions across a root cell plasma membrane can be impeded for two reasons:

- The membrane is hydrophobic, whereas water and mineral ions are polar.
- Some mineral ions must be moved against their concentration gradients.

However, as we saw in Chapter 6, membrane proteins assist with the movement of materials across membranes:

- *Aquaporins*. Aquaporins (see Figure 6.11) are located in both the plasma membrane and the tonoplast (vacuolar membrane) of a plant cell. Aquaporins allow water to diffuse rapidly across these membranes. The abundance of aquaporins in a particular cell depends on that cell's need to obtain and retain water, and can vary with environmental conditions. The permeability of some aquaporins also can be regulated. Alterations in aquaporin abundance and permeability change the *rate* of osmosis across the membrane. Note that water movement through aquaporins is always

The cells of this plant have low turgor pressure and the plant is wilted.
The water potential of cells of this plant is zero because the negative solute potential is balanced by an equally positive pressure potential. The plant is upright because its cells are turgid.

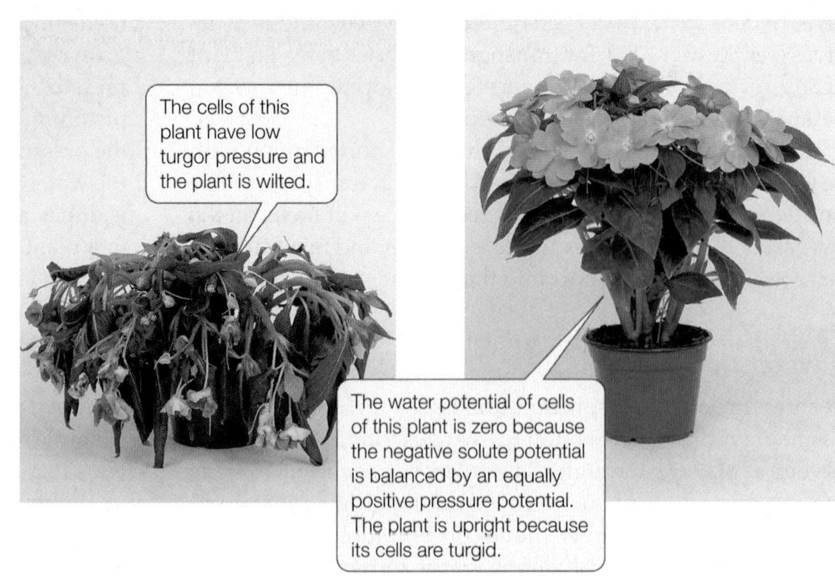

**35.3 A Wilted Plant** A plant wilts when the pressure potential in its cells (the turgor pressure) is low.

1 A proton pump generates differences in $H^+$ concentration and electric charge across the membrane.

2 The difference in electric charge causes cations such as $K^+$ to enter the cell.

3 Symport couples the diffusion of $H^+$ to the transport (against an electrochemical gradient) of anions such as $Cl^-$ into the cell.

Outside of cell

Plasma membrane

Symport protein

Proton pump

Potassium channel

ATP

ADP + $P_i$

Inside of cell

**35.4 The Proton Pump in Transport of $K^+$ and $Cl^-$** The active transport of hydrogen ions ($H^+$) out of the cell by the proton pump (1) drives the movement of both cations (2) and anions (3) into the cell.

passive: from a region of higher water potential to one of lower water potential.

- *Ion channels and pumps*. When the concentration of a charged ion in the soil is greater than that in the plant, transport proteins can move the ions into the plant by facilitated diffusion, which is a passive process (see Section 6.3). The concentrations of most ions in the soil solution, however, are lower than those inside the plant. In these cases the plant must actively take up ions *against* their concentration gradients—a process that requires energy (Section 6.4).

Electric charge differences also play a role in the uptake of mineral ions. For example, a negatively charged ion that moves into a negatively charged compartment is moving against an *electrical gradient*, and this requires energy. Concentration and electrical gradients combine to form an electrochemical gradient (see Section 45.2). Uptake against an electrochemical gradient involves active transport, which requires energy and specific transport proteins.

Unlike animals, plants do not have a sodium–potassium pump (see Section 6.4) to drive active transport. Rather, plants have a **proton pump**, which uses energy obtained from ATP to move protons out of the cell against a proton concentration gradient (**Figure 35.4, step 1**). Because protons ($H^+$) are positively charged, their accumulation outside the cell has two results:

- An electrical gradient is created, with the region outside the cell more positively charged than the inside.
- A proton concentration gradient develops, with more protons outside the cell than inside.

Both the electrical gradient and the concentration gradient assist with the movement of other ions into the cell. Because the inside of the cell is more negative than the outside, cations (positively charged ions) such as potassium ($K^+$) can move into the cell by facilitated diffusion through specific membrane channels (**Figure 35.4, step 2**). In addition, the proton concentration gradient can be harnessed to drive secondary active transport, in which anions (negatively charged ions) such as chloride ($Cl^-$) are moved into the cell. These ions can move against the electrochemical gradient because symport proteins couple their movement with that of $H^+$ (**Figure 35.4, step 3**).

## Water and ions pass to the xylem by way of the apoplast and symplast

The journey from the soil through the roots to the xylem occurs primarily by one of two pathways, either separately or simultaneously: the fast lane (called the apoplast) and the slow(er) lane (called the symplast) (**Figure 35.5**):

- The **apoplast** (Greek *apo*, "away from"; *plast*, "living material") consists of the cell walls, which lie outside the plasma membranes, and the intercellular spaces (spaces between cells) that are common in many plant tissues. The apoplast is a continuous meshwork through which water and dissolved substances can flow without ever having to cross a membrane. Movement of materials through the apoplast is thus unregulated and rapid.
- The **symplast** (Greek *sym*, "together with") is the continuous cytoplasm of the living cells, which are connected by plasmodesmata (see Figure 7.19). The selectively permeable plasma membranes of the root cells control access to the symplast, so movement of water and dissolved substances into the symplast is tightly regulated.

Water and minerals that pass from the soil solution through the apoplast can travel as far as the endodermis, the innermost

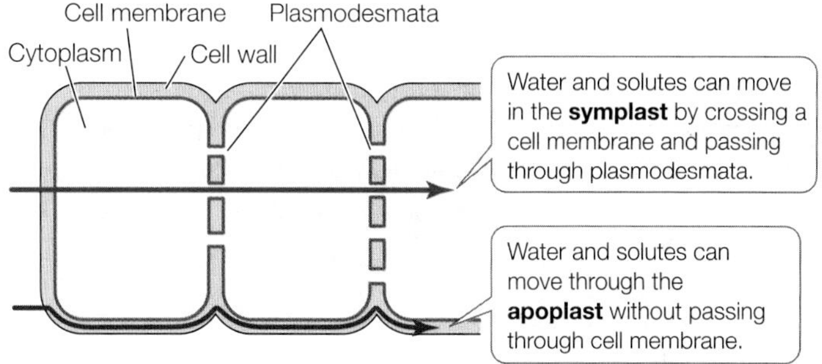


**35.5 Apoplast and Symplast** Plant cell walls and intercellular spaces constitute the apoplast. The symplast comprises the living cells, which are connected by plasmodesmata. To enter the symplast, water and solutes must pass through a plasma membrane. No such selective barrier limits movement through the apoplast.

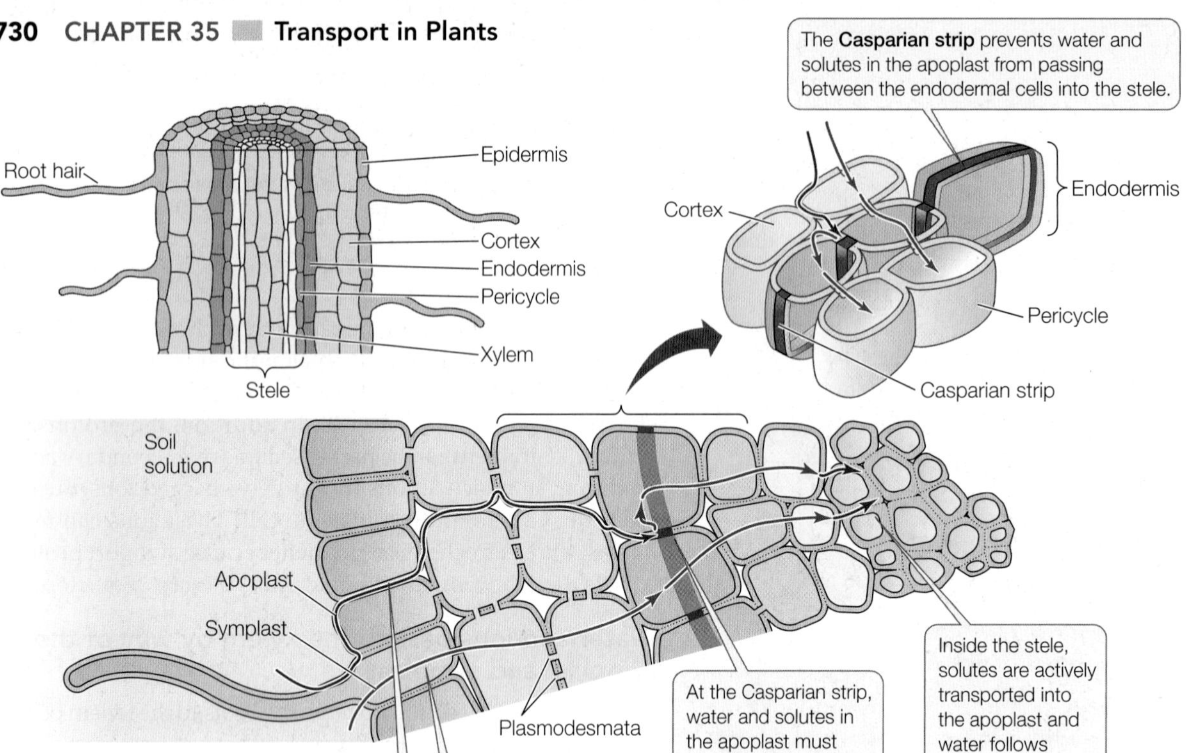


**35.6 Pathways to the Root Xylem** Water and solutes can move into the root through the symplast or the apoplast until they reach the endodermis (shown in dark green); there the water and solutes must enter the symplast to bypass the Casparian strip (red), a region of the endodermal cell wall that is impregnated with the water-repelling substance suberin. Inside the stele, the water and solutes enter the xylem (blue).

**Go to Activity 35.1 Apoplast and Symplast of the Root Life10e.com/ac35.1**

layer of the root cortex (**Figure 35.6**; see Section 34.3). The endodermis is distinguished from the rest of the ground tissue by the presence of the **Casparian strip**. This waxy, suberin-impregnated region of the endodermal cell wall forms a hydrophobic belt around each endodermal cell where it is in contact with other endodermal cells. The Casparian strip acts as a seal that prevents water and ions from moving through apoplastic spaces between the endodermal cells (see Figure 35.6). Therefore all water and ions must enter the symplast in order to cross the endodermis into the stele, which contains the vascular tissues of the root. The materials pass from the endodermal cells to cells in the stele via plasmodesmata.

Once they have passed the endodermal barrier, water and minerals remain in the symplast until they reach parenchyma cells in the pericycle or xylem. These cells then actively export mineral ions into the apoplast of the stele. As the concentrations of mineral ions in the apoplast increase, its water potential becomes more negative. Consequently, water moves out of the cells and into the apoplast by osmosis. In other words, ions are transported actively, and water follows passively. The end result is that water and minerals end up in the xylem, where they constitute the **xylem sap**.

**RECAP 35.1**

Differences in water potential govern the osmotic flow of water from the soil into the plant stele; this is a passive process. Uptake of minerals from the soil that occurs along an electrochemical gradient is an active process requiring energy and membrane transport proteins. Water and minerals can move into the root through either the apoplast or the symplast, but must enter and leave via the symplast to reach the xylem.

- What distinguishes water potential, solute potential, and pressure potential? **See p. 727 and Figure 35.2**
- Why is the cell wall important in determining the direction of water movement and plant form? **See pp. 727–728**
- What are aquaporins, and why are they needed? **See p. 728**
- What are the differences between the apoplast and the symplast? **See p. 729 and Figures 35.5 and 35.6**

So far we've described the movement of water and minerals into plant roots and their entry into the root xylem. How does the xylem sap move once it is in the xylem?

## 35.2 How Are Water and Minerals Transported in the Xylem?

Water has arrived in the xylem—and must travel "uphill" from there. Before considering the ascent of water and minerals to the leaves, reacquaint yourself with the cells that make up the xylem: the tracheids and vessel elements (see Figure 34.7A and B). Recall that these xylem cells are dead and lack all cell contents. When fused end to end, they form long tubular "straws" of lignified cell walls called **xylem vessels**. These vessels provide both structural support and the rigidity needed to maintain a gradient of pressure.

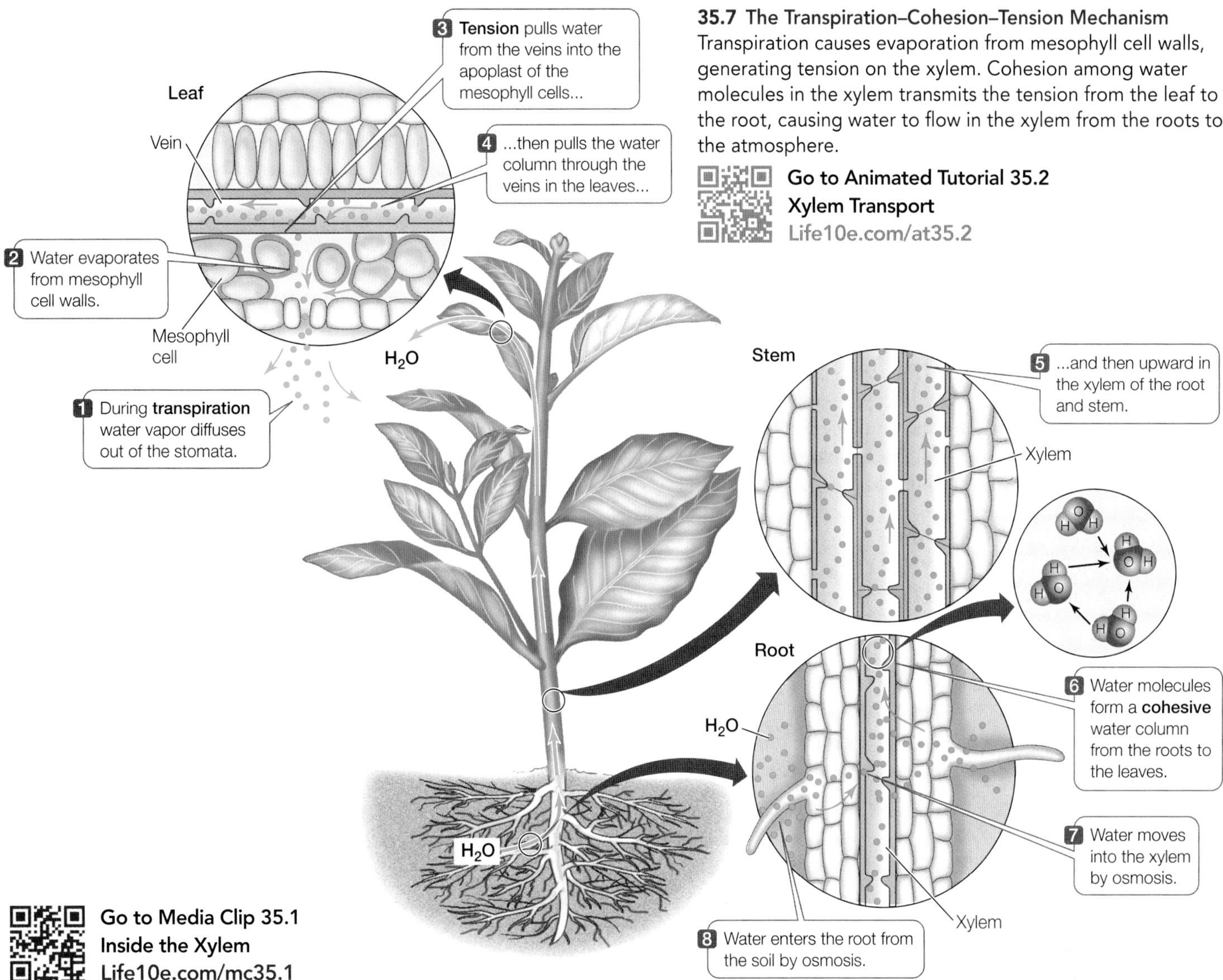


**35.7 The Transpiration–Cohesion–Tension Mechanism** Transpiration causes evaporation from mesophyll cell walls, generating tension on the xylem. Cohesion among water molecules in the xylem transmits the tension from the leaf to the root, causing water to flow in the xylem from the roots to the atmosphere.

**Go to Animated Tutorial 35.2**
**Xylem Transport**
Life10e.com/at35.2

**Go to Media Clip 35.1**
**Inside the Xylem**
Life10e.com/mc35.1

Consider the magnitude of what the xylem accomplishes when it transports large amounts of water over great distances within a plant. A single maple tree 15 meters tall was estimated to have some 177,000 leaves, with a total leaf surface area of 675 square meters—about one and a half times the area of a basketball court. During a summer day, that tree loses 220 liters of water *per hour* to the atmosphere by evaporation from the leaves. So to prevent wilting, the xylem needs to transport 220 liters of water up to 15 meters from the roots to the leaves every hour. (By comparison, a 50-gallon drum holds 189 liters.)

Until the twentieth century, scientists proposed two possible mechanisms for moving water through the xylem: upward pressure by living cells, and capillary action. Both of these possibilities were largely ruled out by experiments:

- A simple experiment in 1893 ruled out the hypothesis that root cells might initiate a *pumping mechanism* to propel water upward. A tree was cut at its base and the sawed-off part was placed in a vat containing a solution of poison that killed living cells. The poison rose up the trunk, killing any living cells it encountered along the way. The experiment demonstrated that a living pump in the root is not necessary to push the xylem sap up a tree. Because the roots were absent, it was clear that they are not involved in xylem movement. Furthermore, when the poison sap reached the leaves, they died and all upward movement of the solution stopped, showing that living leaves are necessary for water to move in the xylem.
- Because of its surface tension (see Section 2.4) and adhesive forces between water and its container, water will move up a narrow column by a mechanism called *capillary action*. Capillary action was ruled out as a primary mechanism for upward xylem sap transport when calculations showed that xylem vessels (at 100 micrometers in diameter) are too wide to get water to the top of a 15-meter tree in this fashion. In fact, the maximum height for a water column raised by capillary action alone in a 100-micrometer tube would be only 0.15 meters.

## The transpiration–cohesion–tension mechanism accounts for xylem transport

The current model of xylem transport involves three processes (**Figure 35.7**):

- *Transpiration* of water molecules from the leaves by evaporation
- *Tension* in the xylem sap resulting from transpiration from the leaves

- *Cohesion* of water molecules in the xylem sap, from the leaves to the roots

The relative amount of water vapor in the atmosphere is lower than that in the leaf. Because of this difference, water vapor diffuses from the intercellular spaces of the leaf to the outside air, in a process called **transpiration**. Within the leaf blade, water evaporates from the moist walls of the mesophyll cells and enters the intercellular spaces. As water evaporates from the aqueous film coating each cell, the film shrinks back into tiny spaces in the cell walls, increasing the curvature of the water surface and thus increasing its surface tension. This increased *tension* (negative pressure potential) in the surface film draws more water into the walls from the cells, replacing that which was lost. The resulting tension in the mesophyll draws water from the xylem of the nearest vein into the apoplast surrounding the mesophyll cells. The removal of water from the veins, in turn, establishes tension on the entire column of water contained in the xylem. *Cohesion* between water molecules in the column prevents the column from breaking. So these three forces—transpiration, tension, and cohesion—operate together to draw water up the xylem, all the way from the roots to the leaves.

Each part of this theory is supported by evidence:

- The difference in water potential between the soil solution and the air is huge, on the order of –100 MPa. This difference should generate more than enough tension to pull a water column up a tree.
- There is a continuous column of water in the xylem, which is caused by cohesion.
- Actual measurements of xylem pressures in cut stems show negative pressure potentials, indicating considerable tension in the xylem (**Figure 35.8**).

RESEARCH**TOOLS**

**35.8 Measuring the Pressure of Xylem Sap with a Pressure Chamber** Xylem sap pulls away from a cut stem because the pressure in the intact xylem is lower than that of the atmosphere. The negative pressure potential originally present in the plant can be measured in a pressure chamber in which the pressure can be raised. The cut surface remains outside the chamber. As gas pressure increases, the xylem sap is pushed back to the cut surface. When the sap first becomes visible again at the cut surface, the pressure in the chamber is recorded. This pressure is equal in magnitude but opposite in sign to the tension (negative pressure potential) originally present in the xylem.

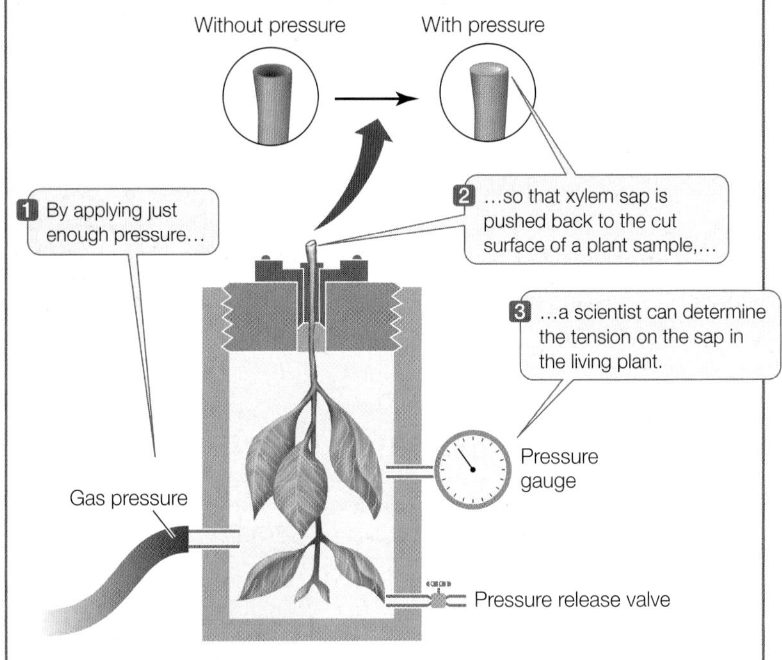


The transpiration–cohesion–tension mechanism accounts for the movement of water through the xylem. Dissolved mineral ions are carried along with the water to all of the plant's living tissues, where the ions are used for various cellular processes (see Chapter 36 for more on plant nutrition). In addition to promoting the transport of minerals, transpiration has an added benefit of cooling a plant's leaves. The evaporation of water from mesophyll cells consumes heat, thereby decreasing the leaf temperature. A farmer can hold a leaf between thumb and forefinger to estimate its temperature; if the leaf doesn't feel cool, that means transpiration is not occurring and it must be time to water.

RECAP 35.2

The transpiration–cohesion–tension mechanism explains the ascent of xylem sap. Transpiration draws water out of leaves, resulting in tension that pulls water from the xylem. Because of cohesion between water molecules, water is pulled passively through the xylem vessels in continuous columns, always toward a region with lower pressure potential.

- What are the roles of transpiration, cohesion, and tension in xylem transport? **See pp. 731–732 and Figure 35.7**
- What experiment rules out the role of pressure from the roots in the upward flow of water in xylem? **See p. 731**

Although transpiration provides the driving force for the transport of water and minerals in the xylem, it also results in the loss of tremendous quantities of water from the plant. How plants control this loss will be the subject of the next section.

## 35.3 How Do Stomata Control the Loss of Water and the Uptake of $CO_2$?

The epidermis of leaves and stems secretes a waxy cuticle, which is impermeable to water and thus helps minimize the loss of water from transpiration. However, the cuticle is also impermeable to carbon dioxide. The cuticle poses a dilemma: how can the plant balance its need to retain water with its need to obtain $CO_2$ for photosynthesis?

An elegant compromise has evolved in plants in the form of pores called **stomata** (singular *stoma*) in the epidermis of their leaves. A pair of specialized epidermal cells, called **guard cells**, controls the opening and closing of each stoma (**Figure 35.9A**). When the stomata are open, $CO_2$ can enter the leaf by diffusion—but water vapor diffuses out of the leaf at the same time. Closed stomata prevent water loss but also exclude $CO_2$ from the leaf.

Most plants open their stomata only when the light intensity is sufficient to maintain a moderate rate of photosynthesis. At night, when darkness precludes photosynthesis, their stomata are closed; no $CO_2$ is needed, and water is conserved. Even during the day, the stomata close if water is being lost at too rapid a rate.

Stomata are ancient structures; they have been found in plant fossils that are more than 400 million years old. For this reason they are thought to predate the evolution of leaves. Stomata are found in all vascular plants and in many non-vascular plants, including mosses (but not liverworts; see Chapter 28).

The stoma and guard cells seen in Figure 35.9A are typical of eudicots. Monocots typically have specialized epidermal cells associated with their guard cells. However, the principle of operation, which we will now describe in more detail, is the same for both monocot and eudicot stomata.

## The guard cells control the size of the stomatal opening

The opening and closing of stomata are regulated by several environmental factors, including light, $CO_2$ levels, temperature, and water availability. For example, light causes the stomata of most plants to open, admitting $CO_2$ for photosynthesis. A low level of $CO_2$ in the intercellular spaces of the leaf will trigger the opening of the stomata as well, allowing the uptake of more $CO_2$. However, to conserve water, plants will close their stomata on hot days and when water availability is limited.

Stomata can respond to these environmental stimuli in a matter of minutes. How does this important biological process happen so rapidly? Stomata open and close rapidly in response to changes in turgor pressure in the guard cells. These changes in turgor pressure arise in response to changes in $K^+$ concentrations in the guard cells. These concentration changes are triggered by environmental cues.

In sunlight, a pigment in the guard cell membrane absorbs blue light, which activates a proton pump that actively transports $H^+$ out of the guard cells and into the apoplast of the surrounding epidermis. The resulting electrochemical gradient drives $K^+$ into the guard cells, making the water potential of the guard cells more negative (**Figure 35.9B**). Negatively charged chloride ions and organic ions move into and out of the guard cells along with the potassium ions, functioning to maintain the electrical balance and contributing to the change in the solute potential of the guard cells. Water enters by osmosis (guard cell membranes are particularly rich in aquaporin protein channels), which increases the pressure potential of the guard cells. The cellulose microfibrils in the guard cell walls are arranged so that the shape of the cells changes in response to the increase in pressure potential, and a gap—the stoma—appears between them.

In the absence of sunlight, the stomata of most plants close. Lacking blue light, the proton pump becomes less active, potassium and chloride ions diffuse passively out of the guard cells, and water follows by osmosis (see Figure 35.9B, step 3). These changes lower the pressure potential in the guard cells so that they sag together, closing the stoma.

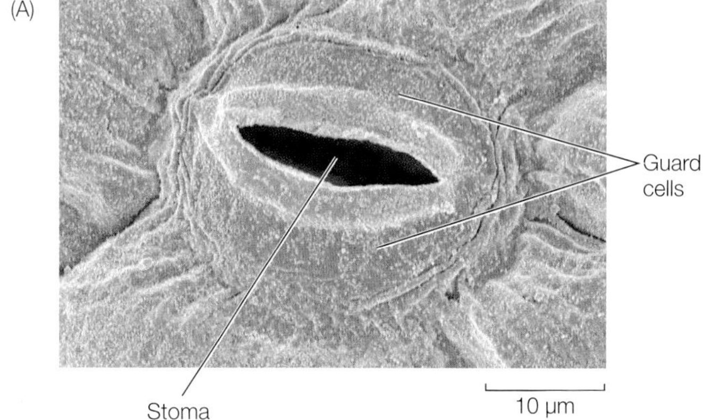


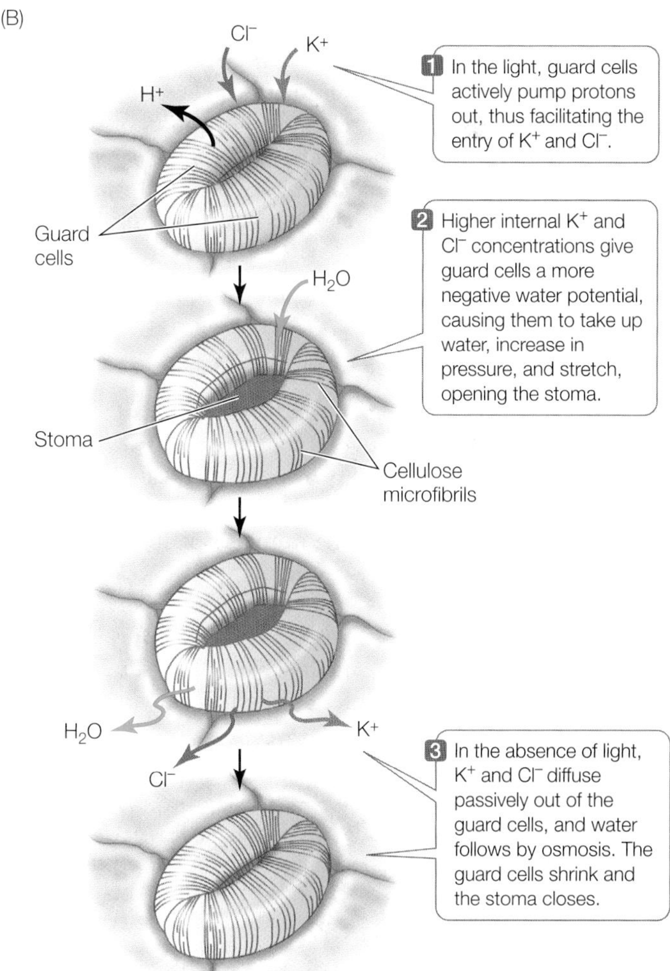


**35.9 Stomata** (A) Scanning electron micrograph of an open stoma formed by two sausage-shaped guard cells. (B) Potassium ion concentrations affect the water potential of the guard cells, controlling the opening and closing of stomata. Negatively charged ions (e.g., $Cl^-$) that accompany $K^+$ maintain electrical balance and contribute to the changes in water potential that open and close the stomata.

Stomata respond not only to light but to water availability. Water stress is a common problem for plants, especially on hot, windy days. On hot, dry days plants will close their stomata even when the sun is shining. The water potential of the mesophyll cells is the cue for this protective response. If the mesophyll is too dehydrated—that is, if its water potential is too negative—its cells release a plant hormone called abscisic acid. Abscisic acid causes the guard cells to close the stomata and prevent further drying of the leaf. Although this protective response reduces the rate of photosynthesis, it protects the plant from wilting.

Not all plants respond to environmental cues in the same way. CAM plants such as cacti (see Section 10.4) are often found in the desert. To avoid excess water loss, these plants close their stomata during the day and open them at night. They have a metabolic adaptation that allows them to accumulate $CO_2$ at night when their stomata are open, then release the $CO_2$ inside the plant for photosynthesis during the day.

### Plants can control their total numbers of stomata

Individual stomata are tiny, yet plants can lose large amounts of water. A single corn plant can lose 2 quarts of water per day. Plant water loss can be great because there can be a huge numbers of stomata on every leaf: up to 250,000 per square inch of leaf surface! Plants limit water loss by controlling stomata in two very different ways:

- By regulating stomatal opening and closing (described above)
- By controlling the total number of stomata

Whereas the process of opening and closing stomata can occur in minutes, as we saw above, the process of controlling the total number of stomata occurs over days or weeks. Trees can reduce their total number of stomata by shedding some of their leaves. Other plants reduce the numbers of stomata on new leaves that develop during a drought.

$CO_2$ levels can also trigger changes in stomatal density. If the model plant *Arabidopsis* is exposed to high $CO_2$ levels, its new leaves have fewer stomata than they would have under normal conditions. Why do you think this might be advantageous?

**RECAP 35.3**

$CO_2$, which is needed for photosynthesis, enters leaves via tiny pores called stomata. Stomata also permit the loss of water by transpiration. Guard cells open or close stomata in response to a variety of environmental cues. Plants control their total numbers of stomata by shedding leaves or by altering the density of stomata on new leaves.

- What is the role of $K^+$ ions in the functioning of guard cells? **See p. 733 and Figure 35.9**
- Describe how environmental cues (such as $CO_2$ level and water availability) can affect stomatal function and density during the life of a plant. **See p. 734**

Stomata are normally open during daylight hours, allowing photosynthesis—the production of carbohydrates from $CO_2$ and water. In the next section we'll see how the products of photosynthesis are delivered to other parts of the plant, supporting plant growth.

## 35.4 How Are Substances Translocated in the Phloem?

Photosynthesis occurs primarily in the leaf (see Figure 10.1). The carbohydrate products of photosynthesis (mainly sucrose) diffuse to the nearest small vein (composed of xylem and phloem), where they are actively transported into sieve tube elements of the phloem. The movement of carbohydrates and other solutes through the phloem is called **translocation**. The products of photosynthesis are called **photosynthates**, and the content of the phloem is called the **phloem sap**.

The photosynthates and other substances are translocated from sources to sinks.

- A **source** is an organ (such as a mature leaf or a storage root) that *produces*, by photosynthesis or by digestion of stored reserves, more sugars than it requires.
- A **sink** is an organ (such as a root, flower, developing fruit, or immature leaf) that *consumes* sugars for its own growth and storage needs of the plant.

A given organ can be a sink at some times and a source at others. An example is the roots of maple trees, which store sugars sent down from the leaves during one growing season, then send sugars back upward to support the emergence of new leaves the following spring.

### Sucrose and other solutes are carried in the phloem

Evidence that the phloem carries sucrose and other solutes was first obtained in the 1600s when the Italian scientist Marcello Malpighi removed a ring of bark from the trunk of a tree—that is, he "girdled" the tree. The bark contained the phloem, while the xylem in the underlying wood remained intact. Over time, the bark in the region above the girdle swelled:

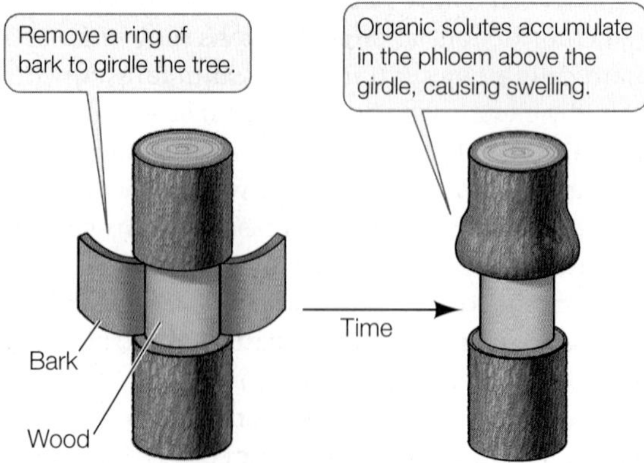


Malpighi correctly concluded that a solution coming from the leaves above the girdle was trapped in the bark. Later the bark below the girdle died, presumably because it no longer received nutrients from the leaves. Eventually the roots, and then

Pores of sieve plate
Sieve plate
Mature sieve tube elements do not have nuclei and have lost most of their organelles.
Sieve tube element
Phloem sap
The companion cell is a fully functional cell with a nucleus.
Companion cell
Sieve plate
Dr. R. Kessel & Dr. G. Shih/Visuals Unlimited
Pores

**35.10 Sieve Tubes** Individual sieve tube elements join together to form long tubes that transport carbohydrates and other nutrient molecules throughout the plant body in the phloem. Sieve plates form at the ends of each sieve tube element, and phloem sap passes through the pores in the sieve plate.

the entire tree, died—suggesting that sugar transport might occur in the phloem.

The cells that make up the phloem's conducting tubes are sieve tube elements (see Figure 34.7C). Like the vessel elements in the xylem, sieve tube elements meet end to end. However, unlike vessel elements, whose end walls are broken down as they mature, sieve tube elements retain their end walls. Communication between sieve tube elements is achieved by plasmodesmata in their end walls. During sieve tube development, the diameter of these plasmodesmata increases 10- to 100-fold, resulting in pores that allow the flow of phloem sap between neighboring cells. Because the end walls of sieve tube elements look and function like sieves, they are called **sieve plates** (**Figure 35.10**).

What happens next is truly remarkable and makes sieve tube elements among the most unusual cell types in nature. As the holes in the sieve plates expand, most of the cell contents are lost, including the nucleus, Golgi apparatus, and most of the ribosomes and cytoskeleton. Despite this, sieve tube elements live for an entire growing season in deciduous trees, and for decades in some other plants. How can sieve tube elements live for so long with no nucleus? The answer is that each sieve tube element has one or more **companion cells** (see Figure 35.10). Companion cells are produced as daughter cells along with the sieve tube elements when parent cells divide. Numerous plasmodesmata link a companion cell with its neighboring sieve tube element. Companion cells retain all their organelles and provide all the components needed to maintain the sieve tube elements—they may be thought of as the "life support systems" of the sieve tube elements.

Plant biologists in the twentieth century used aphids to precisely analyze the contents of the phloem. Aphids are insects that feed on plants by drilling into sieve tube elements with a specialized organ, the stylet. The pressure potential in the sieve tube is higher than that outside the plant, so the phloem contents are forced through the stylet into the aphid's digestive tract. So great is the pressure that some of the liquid is forced through the insect's body and out its anus. If an aphid is frozen in the act of feeding, its body can be chopped off the plant stem, leaving the stylet intact. Phloem sap continues to flow from the stylet for hours, and can be collected for analysis.

These and other experiments led to several important observations:

- Sucrose makes up 90 percent of the phloem sap solutes. The phloem sap also contains hormones, small molecules such as amino acids, mineral nutrients, and viruses.
- The flow rate can be very high, as much as 100 centimeters per hour.
- Different sieve tube elements conduct their contents in different directions—for example, up or down the stem. Therefore the overall movement in the phloem is bidirectional.
- The movement of phloem sap requires living cells, in contrast to movement in the xylem.

### The pressure flow model appears to account for translocation in the phloem

As noted above, phloem sap flows under positive pressure through the sieve tubes. It moves by bulk flow (i.e., from a region of higher pressure potential to a region of lower pressure potential) from one sieve tube element to the next. We need to

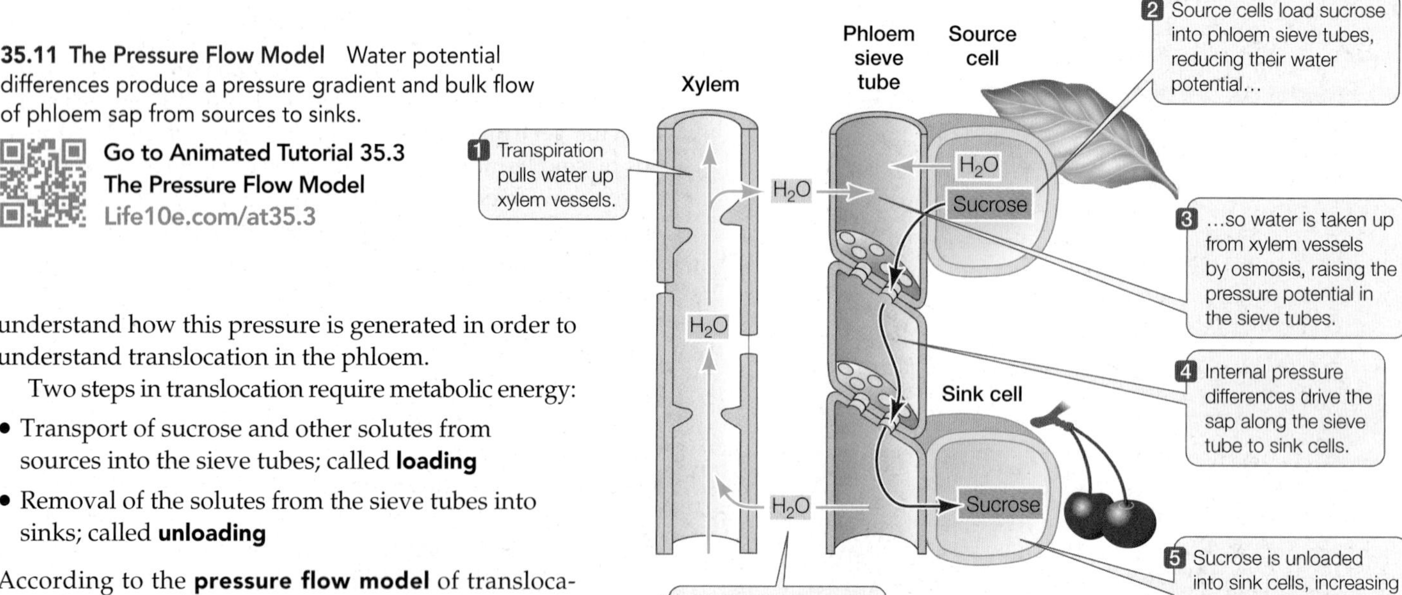


**35.11 The Pressure Flow Model** Water potential differences produce a pressure gradient and bulk flow of phloem sap from sources to sinks.

**Go to Animated Tutorial 35.3**
**The Pressure Flow Model**
Life10e.com/at35.3

understand how this pressure is generated in order to understand translocation in the phloem.

Two steps in translocation require metabolic energy:

- Transport of sucrose and other solutes from sources into the sieve tubes; called **loading**
- Removal of the solutes from the sieve tubes into sinks; called **unloading**

According to the **pressure flow model** of translocation in the phloem, sucrose is actively transported into sieve tube elements at a source, giving those cells a greater sucrose concentration than the surrounding cells. Therefore the cells develop a lower solute potential—a more negative $\Psi_s$—at which point water enters the sieve tube elements from xylem vessels by osmosis. The entry of this water causes a greater pressure potential (turgor pressure) at the source end of the sieve tube, so that the entire fluid content of the sieve tube is pushed toward the sink end of the tube—in other words, the sap moves by bulk flow in response to a pressure gradient (**Figure 35.11**). In the sink, the sucrose is unloaded both passively and by active transport, and water moves back to the xylem vessels. In this way the gradients of solute potential and pressure potential, which are needed for movement of the phloem sap, are maintained.

Sugars and other solutes move from the mesophyll cells in the leaf to the phloem by two general routes: apoplastic and symplastic. The exact details of these routes vary widely among plant species. In many plants, sugars and other solutes follow the **apoplastic pathway**: solutes move out of the mesophyll cells and then diffuse through the apoplast to the sieve tubes. Specific sugars and amino acids are then actively transported into the sieve tube elements. Active transport across membranes (see Section 6.4) allows plants to regulate which specific substances enter the phloem. In contrast, in the **symplastic pathway**, solutes remain within the symplast and pass through plasmodesmata all the way from the mesophyll cells to the sieve tube cells. Because no membranes are crossed in the symplastic pathway, the solutes are loaded into the phloem by mechanisms other than active membrane transport.

The pressure flow model of translocation in the phloem is contrasted with the transpiration–cohesion–tension model of xylem transport in **Table 35.1**.

Upon arriving in sink regions, solutes are actively transported *out* of the sieve tube elements and into the surrounding tissues. This unloading serves several purposes: it helps maintain the gradient of solute potential and hence of pressure potential in the sieve tubes; it supplies carbohydrates and amino acids to developing organs, where they are used for growth; and it helps build up high concentrations of proteins and carbohydrates in storage organs such as storage roots, fruits, and seeds. The demand for fixed carbon can be very large in some tissues, such as a rapidly growing stem or a developing potato tuber. These tissues receive a lot of phloem sap, while a slowly growing organ receives less.

An experiment by Lothar Willmitzer and colleagues at the Max Planck Institute in Germany demonstrated the importance of sink strength in potato tubers (**Figure 35.12**). Sink strength is the relative ability of an organ to attract photosynthates. A potato tuber is formed from an underground stem that changes its function from transport to synthesis and storage of starch. Because the source of glucose to make this starch is sucrose from the phloem, there is a high demand for phloem unloading into the developing potato tuber. The enzyme invertase catalyzes the hydrolysis of sucrose to glucose and fructose, and the fructose is quickly converted to glucose. Willmitzer and colleagues transformed potato plants by introducing a gene that caused invertase to accumulate in the apoplast of the tuber tissue, where the enzyme accumulated to high levels. In these plants, the sucrose that was transported out of the phloem into the apoplast was rapidly hydrolyzed. The resulting glucose was taken up by the tuber cells and converted to starch. The action of the invertase in the apoplast lowered the sucrose concentration in the potato tissue and led to a high rate of phloem unloading (see Figure 35.11). As a result, the potato tubers

TABLE **35.1**
**Mechanisms of Sap Flow in Plant Vascular Tissues**

| | Xylem | Phloem |
|---|---|---|
| Driving force for bulk flow | Transpiration from leaves | Active transport of sucrose at source and sink |
| Site of bulk flow | Nonliving vessel elements and tracheids | Living sieve tube elements |
| Pressure potential in sap | Negative (pull from top; tension) | Positive (push from source; pressure) |

INVESTIGATING**LIFE**

**35.12 Manipulating Sucrose Transport from the Phloem** Does a change in the concentration of sucrose in a sink tissue affect the unloading of sucrose from the phloem? Lothar Willmitzer and colleagues genetically modified potato plants so that sucrose in the developing tubers was immediately hydrolyzed. This lowered the effective sucrose concentration in the "sink" for phloem unloading. The researchers then tested whether this affected tuber development, a reflection of sucrose unloading from phloem.[a]

**HYPOTHESIS** Reducing the sucrose concentration in a sink organ will increase the transport and unloading of sucrose from the phloem.

*Method* Plants were transformed with a gene for invertase, an enzyme that hydrolyzes sucrose.

*Results*

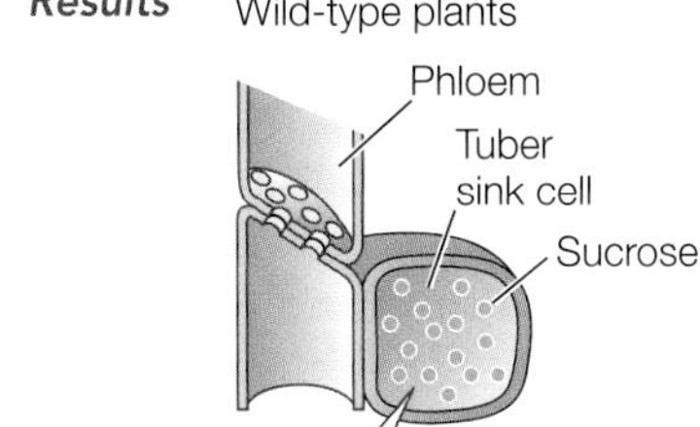


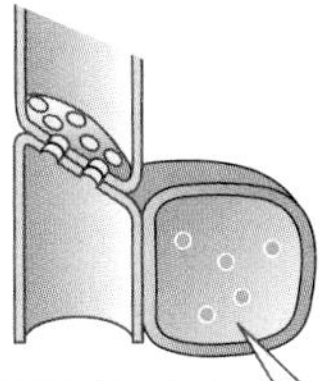


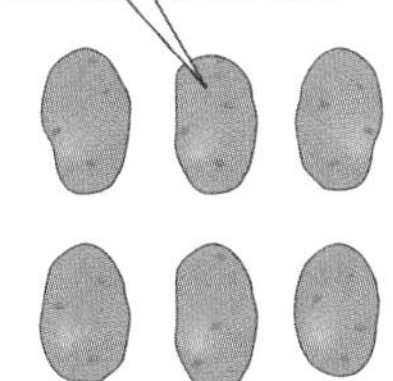

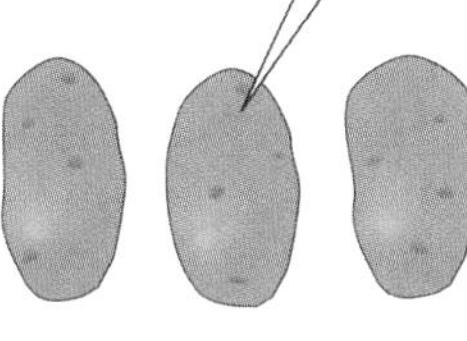

**CONCLUSION** Increasing sink strength increases sucrose transport into developing tissues.

Go to **BioPortal** for discussion and relevant links for all INVESTIGATING**LIFE** figures.

[a]Sonnewald, U. et al. 1997. *Nature Biotechnology* 15: 794–797.

that developed on the transformed plants were larger than those of wild-type plants. This experiment demonstrated that transport into the apoplast is an efficient way for phloem cells to unload sucrose, and that the sucrose concentration gradient between the phloem and the apoplast influences the rate of unloading.

WORKING WITH**DATA:**

## *Manipulating Sucrose Transport from the Phloem*

### Original Paper

Sonnewald, U., N.-R. Hajirezaei, J. Kossmann, A. Heyer, R. Tretheway, and L. Willmitzer. 1997. Increased potato tuber size from apoplastic expression of yeast invertase. *Nature Biotechnology* 15: 794–797.

### Analyze the Data

Plant tissues can be exporters (sources) or importers (sinks) of carbohydrates, mainly in the form of sucrose. Many of the foods grown for human consumption are sink tissues; we eat the crops' carbohydrates, mostly as starch. As a potato tuber develops from an underground stem, or stolon, it requires a lot of carbohydrate, which it obtains as sucrose from the phloem. Lothar Willmitzer and colleagues at the Max Planck Institute in Germany hypothesized that the rate of sucrose unloading from phloem to tuber is affected by the sucrose concentration difference (gradient) between the phloem and the developing tuber. To test their hypothesis, they used genetic engineering to express the gene for yeast invertase in developing tuber cells (see Figure 35.12).

Normal potato plants have invertase but do not express it at high levels in tubers. The yeast invertase gene used to transform the plants was coupled to a plant promoter that caused high expression of the enzyme in the developing tubers. Invertase hydrolyzes sucrose, so the researchers predicted that the concentration of sucrose would be lower in the tubers of transgenic plants (and the gradient steeper) than in wild-type plants. Experiments performed on the transgenic plants confirmed that increasing the sucrose concentration difference between the phloem and the tubers led to increased sucrose unloading from the phloem.

QUESTION 1

To test whether the introduced yeast invertase was having the predicted effects on sucrose levels, the researchers measured invertase activity and sugars in tubers of wild-type and transgenic plants. The results are in the first three rows of the table (mean ± standard deviation). Which group of plants had higher invertase activity? Carry out a statistical test to evaluate your conclusion (see Appendix B). How do the sugar content measurements relate to invertase activity?

QUESTION 2

The researchers estimated sucrose unloading from phloem by comparing the sizes of the tubers that developed on normal and transgenic plants, assuming that larger tubers must have unloaded more sucrose from the phloem. The tuber yields are shown in the last two rows of the table (mean ± standard deviation). What can you conclude about the distribution of sucrose to developing tubers?

| | Wild type | Transgenic |
|---|---|---|
| Invertase activity (units/mg protein) | 9 ± 4 | 598 ± 73 |
| Sucrose (mg/g tuber weight) | 15.8 ± 0.7 | 1.2 ± 0.2 |
| Glucose (mg/g tuber weight) | 2.1 ± 0.4 | 38.0 ± 4.3 |
| Tubers per plant | 6.4 ± 0.4 | 2.6 ± 0.1 |
| Tuber weight (g/tuber) | 18 | 57.3 |

Go to BioPortal for all WORKING WITH**DATA** exercises

The control of phloem sap travel from source to sink has great importance for humans. The parts of plants that we use for food, such as seeds (e.g., the rice in the opening story), fruits, and storage organs, are mostly sinks. Increasing the flow of sucrose into these organs as they develop can increase food production in crop plants. The experiment on potatoes, described above, is an important step toward understanding and manipulating phloem transport.

RECAP 35.4

Carbohydrates produced by photosynthesis are translocated from source to sink through the phloem by a pressure flow mechanism.

- Explain the difference between a source and a sink. See p. 734
- How does loading of sucrose at the source result in bulk flow toward the sink? See pp. 735–736 and Figure 35.11

What other methods are used to reduce water loss in agriculture?

ANSWER

Optimal plant growth depends on the presence of an appropriate amount of soil water near the roots during a plant's life cycle. In addition to using plants that are more water-efficient, plant growers can manipulate the environment to reduce water losses that are due to evaporation and soil runoff. Simply putting a pan out in the field and measuring the rate of evaporation can be a guide as to when to increase irrigation. Cheap sensors in the soil can measure water content and alert the grower to deficits or excess. Farmers can use laser technology to grade (shape) fields to reduce water runoff during irrigation.

# CHAPTER SUMMARY 35

## 35.1 How Do Plants Take Up Water and Solutes?

- Water moves through biological membranes by osmosis, always moving toward regions with a more negative water potential. The **water potential** ($\Psi$) of a cell or solution is the sum of the **solute potential** ($\Psi_s$) and the **pressure potential** ($\Psi_p$). **Review Figure 35.2, ANIMATED TUTORIAL 35.1**
- **Turgid** plant cells have significant positive pressure potential because the rigid cell wall limits expansion of the cell. This positive pressure (**turgor pressure**) maintains the physical structure of many plant cells; if the pressure potential drops, the plant wilts.
- The movement of a solution due to a difference in pressure potential between two parts of a plant is called **bulk flow**.
- Aquaporins are channel proteins that facilitate movement of water molecules through biological membranes.
- Mineral uptake requires transport proteins. Some minerals enter the plant passively by facilitated diffusion; others enter by active transport. A **proton pump** provides energy for the active transport of many mineral ions across membranes in plants. **Review Figure 35.4**
- Water and minerals pass from the soil into the root by way of the **apoplast** and **symplast**, but must pass through the symplast to cross the endodermis and enter the xylem. The **Casparian strip** in the endodermis blocks movement of water and minerals through the apoplast. **Review Figures 35.5, 35.6, ACTIVITY 35.1**

## 35.2 How Are Water and Minerals Transported in the Xylem?

- Experiments proved that neither a root pump nor capillary action can alone account for the ascent of xylem sap in trees.
- Water transport in the xylem results from the combined effects of **transpiration**, cohesion, and tension—the transpiration–cohesion–tension mechanism. Evaporation from the leaf produces tension in the mesophyll cells, which pulls a column of water—held together by cohesion—up through the xylem from the root. **Review Figure 35.7, ANIMATED TUTORIAL 35.2**
- Transport in the xylem is passive. It does not require the expenditure of energy by the plant.

## 35.3 How Do Stomata Control the Loss of Water and the Uptake of $CO_2$?

- The waxy cuticle of plant epidermis is impermeable to both water and carbon dioxide. **Stomata** allow for carbon dioxide uptake (when open) while minimizing transpirational water loss (when closed).
- A pair of **guard cells** controls the size of the stomatal opening. A light-activated proton pump moves protons out of the guard cells to the walls of surrounding epidermal cells, setting up an electrochemical gradient that drives the transport of potassium ions into the guard cells. Water follows osmotically, swelling the guard cells and opening the stomata. **Review Figure 35.9**
- When threatened by dehydration, mesophyll cells release abscisic acid, which causes guard cells to close the stomata, even in the light.

## 35.4 How Are Substances Translocated in the Phloem?

- Products of photosynthesis, as well as some minerals, are translocated through sieve tubes in the phloem by way of living sieve tube elements. **Review Figure 35.10**
- **Translocation** in the phloem can proceed in both directions in the stem. Translocation requires a supply of ATP.
- Translocation in the phloem is explained by the **pressure flow model**: the difference in solute concentration between **sources** and **sinks** creates a difference in (positive) pressure potential along the sieve tubes, resulting in bulk flow. **Review Figure 35.11, Table 35.1, ANIMATED TUTORIAL 35.3**

Go to the Interactive Summary to review key figures, Animated Tutorials, and Activities
Life10e.com/is35

# CHAPTERREVIEW

## REMEMBERING

1. Osmosis
   a. requires ATP.
   b. results in the bursting of plant cells placed in pure water.
   c. can cause a cell to become turgid.
   d. is independent of solute concentrations.
   e. continues until the pressure potential equals the water potential.

2. Water potential
   a. is the difference between the solute potential and the pressure potential.
   b. is analogous to the air pressure in an automobile tire.
   c. is the movement of water through a membrane.
   d. determines the direction of water movement between cells.
   e. is defined as 1.0 MPa for pure water under no applied pressure.

3. Which statement about proton pumping across the plasma membrane of plants is *not* true?
   a. It requires ATP.
   b. The region inside the membrane becomes positively charged with respect to the region outside.
   c. It enhances the movement of $K^+$ ions into the cell.
   d. It moves protons out of the cell against a proton concentration gradient.
   e. It can drive the secondary active transport of negatively charged ions.

4. Which statement is *not* true?
   a. The symplast consists of the interconnected cytoplasm of living cells.
   b. Water can enter the stele without entering the symplast.
   c. The Casparian strips prevent water from moving through the apoplast between endodermal cells.
   d. The endodermis is a cell layer in the cortex.
   e. Water can move freely in the apoplast without entering cells.

5. Which of the following is *not* part of the transpiration–cohesion–tension mechanism?
   a. Water evaporates from the walls of mesophyll cells.
   b. Removal of water from the xylem exerts a pull on the water column.
   c. Water is remarkably cohesive.
   d. The wider the tube, the greater the tension its water column can withstand.
   e. At each step, water moves to a region with a more strongly negative water potential.

6. Which statement about phloem transport is *not* true?
   a. It takes place in sieve tubes.
   b. It depends on mechanisms for loading solutes into the phloem at sources.
   c. It stops if the phloem is killed by heat.
   d. A high pressure potential is maintained in the sieve tubes.
   e. At sinks, solutes are actively transported into sieve tube elements.

## UNDERSTANDING & APPLYING

7. Shoot epidermal cells protect against excess water loss. How do they perform this function? What differences might you expect to find in the structure of the epidermis in stems, roots, and leaves?

8. Compare sources to sinks. Give examples of each. How might the distribution of sources and sinks change in the course of a year?

9. What is the minimum number of plasma membranes a water molecule would have to cross in order to get from the soil solution to the atmosphere by way of the stele? To get from the soil solution to a mesophyll cell in a leaf?

## ANALYZING & EVALUATING

10. In the story that opened this chapter we saw that a mutation in the *HARDY* gene resulted in *Arabidopsis* plants with a more extensive root system and thicker leaves than wild-type plants. When the *HARDY* gene was isolated, it was found to encode a transcription factor that stimulates expression of genes for increased water-use efficiency. What phenotype due to a mutation in the *HARDY* gene would cause *Arabidopsis* plants to use water more efficiently? How would you investigate the effect of the *HARDY* mutation on stomata? What results would you expect?

11. Here are measurements of water potential ($\Psi$) in a 100-meter-tall tree and its surroundings:

| Region | $\Psi$ (MPa) |
|---|---|
| Soil water | –0.3 |
| Xylem of root | –0.6 |
| Xylem of trunk | –1.2 |
| Inside of leaf | –2.0 |
| Outside air | –58.5 |

Gravity exerts a force of –0.01 MPa per meter of height above ground.
   a. Is the water potential in the leaf sufficiently low to draw water to the top of the tree?
   b. Would transpiration continue if soil water potential decreased to –1.0?
   c. What would you expect to happen to the xylem water potential if all of the stomata closed?

# 36 Plant Nutrition

## CHAPTER OUTLINE


**36.1** What Nutrients Do Plants Require?
**36.2** How Do Plants Acquire Nutrients?
**36.3** How Does Soil Structure Affect Plants?
**36.4** How Do Fungi and Bacteria Increase Nutrient Uptake by Plant Roots?
**36.5** How Do Carnivorous and Parasitic Plants Obtain a Balanced Diet?

**Nitrogen Sipper** Corn plants, such as those being fertilized here, extract a lot of nitrogen compounds from the soil. Excess nitrogen the corn plants do not take up gets left behind in the soil, where it can be a pollutant.

CROPS SUCH AS RICE, WHEAT, AND CORN supply more than half of the human diet. But like other organisms, the plants themselves require good nutrition in order to grow and produce our food. One of the nutrients most often in short supply in soils is nitrogen. Fertilizer or manure can be used to overcome this deficiency, producing a spectacular effect on plant growth and grain production. The production yield of many crops increased significantly in the past century, in part because of the application of more fertilizer.

But nitrogen fertilizer is expensive in two ways. First, the century-old process of manufacturing ammonium from hydrogen and nitrogen gases is very energy-intensive; when the price of oil goes up, the price of fertilizer also rises. In 2012 it cost farmers in the U.S. about $8 billion a year to spread nitrogen fertilizer on 90 million acres of cornfields. This was three times as much as it cost in 2000. Imagine what this means if you are a poor farmer without the money to pay the higher price for fertilizer needed for your crop to grow.

Nitrogen fertilizer is also environmentally expensive. When it rains excessively, nitrogen fertilizer can be lost from farm fields and end up in lakes, rivers, or groundwater. When nitrogen-laden rivers enter the sea, excessive growth of marine algae is likely to result. Eventually the algae die, and the organisms that decompose the algae use up so much oxygen that there is not enough left to support animal life. Nitrogen fertilizer runoff has thus resulted in vast "dead zones" in waters near the mouths of major rivers, including the Mississippi River Delta in the Gulf of Mexico. An additional environmental cost is the conversion of some nitrogen fertilizer to nitrous oxide gas ($N_2O$), which contributes to global warming.

Scientists are working on several strategies by which nitrogen fertilizer might be used efficiently. One strategy is to improve farming practices, to apply optimal rates of nitrogen to crops while reducing losses to the environment. The other strategy is to alter the genetics of crop plants to improve their uptake and assimilation of nitrogen.

Several companies in the U.S. are working to modify corn genetics to increase the nitrogen use efficiency. Many processes are involved in a plant's use of nitrogen from the soil, such as uptake into the roots, transport to other organs in the vascular system, and incorporation of nitrogen into molecules such as amino acids and nucleotides. Each process involves many genes, and so inheritance is complex. An understanding of these processes, however, will yield improvements in nitrogen use efficiency.

What progress has been made in improving the nitrogen use efficiency of corn?

See answer on p. 753.

## 36.1 What Nutrients Do Plants Require?

Every living thing—and a plant is no exception—must obtain raw materials from its environment. These **nutrients** include the major ingredients of macromolecules: carbon, hydrogen, oxygen, and nitrogen. Plants are autotrophs, and obtain both carbon and oxygen from atmospheric carbon dioxide through the reactions of photosynthesis (see Chapter 10). Hydrogen comes mainly from water, so it is plentiful when there is an adequate water supply. Nitrogen, as you will see later in this chapter, enters most plants from the soil. The activities of microorganisms are important in converting organic nitrogen and nitrogen gas into inorganic forms that are usable by plants.

In addition to nitrogen, organisms require other **mineral nutrients**: inorganic elements that are used for various cellular processes. For example, proteins contain sulfur (S), nucleic acids contain phosphorus (P), chlorophyll contains magnesium (Mg), cytochromes contain iron (Fe), and cellular signaling can involve calcium (Ca). Most plants obtain these nutrients from the soil. Within the soil, minerals dissolve in water as ions, forming a solution—called the **soil solution**—that contacts the roots of plants.

### All plants require specific macronutrients and micronutrients

A plant nutrient is called an **essential element** if the plant fails to complete its life cycle or grows abnormally when the element is absent or insufficient. In addition to needing the nonmineral elements carbon, oxygen, and hydrogen, plants require many mineral nutrients (**Table 36.1**). Essential elements fall roughly into two categories—macronutrients and micronutrients—based on the amounts required by plants.

- A plant needs **macronutrients** in concentrations of at least 1 gram per kilogram of the plant's dry matter.
- A plant needs **micronutrients** in concentrations of less than 100 milligrams per kilogram of the plant's dry matter.

How do we know if a plant is getting enough of a particular nutrient?

TABLE **36.1** Mineral Elements Required by Plants

| Element (Abbreviation; Absorbed Form) | Typical Amount in Plant (g/kg) | Major Functions | Deficiency Symptoms |
|---|---|---|---|
| MACRONUTRIENTS | | | |
| Nitrogen (N; $NO_3^-$ and $NH_4^+$) | 15 | In proteins, nucleic acids | Oldest leaves turn yellow and die prematurely; plant is stunted |
| Phosphorus (P; $H_2PO_4^-$ and $HPO_4^{2-}$) | 2 | In nucleic acids, ATP, phospholipids | Plant is dark green with purple veins and is stunted |
| Potassium (K; $K^+$) | 10 | Enzyme activation; water balance; ion balance; stomatal opening | Older leaves have dead edges |
| Sulfur (S; $SO_2^{4-}$) | 1 | In proteins and coenzymes | Young leaves are yellow to white with yellow veins |
| Calcium (Ca; $Ca^{2+}$) | 5 | Affects the cytoskeleton, membranes, and many enzymes; second messenger | Growing points die back; young leaves are yellow and crinkly |
| Magnesium (Mg; $Mg^{2+}$) | 2 | In chlorophyll; required by many enzymes; stabilizes ribosomes | Older leaves have yellow stripes between veins |
| MICRONUTRIENTS | | | |
| Iron (Fe; $Fe^{2+}$ and $Fe^{3+}$) | 0.1 | In active site of many redox enzymes and electron carriers; chlorophyll synthesis | Young leaves are white or yellow |
| Chlorine (Cl; $Cl^-$) | 0.1 | Photosynthesis; ion balance | Leaf tips wilt; leaves turn yellow and die |
| Manganese (Mn; $Mn^{2+}$) | 0.05 | Activation of many enzymes | Younger leaves are pale with green veins |
| Boron [B; $B(OH)_3$] | 0.02 | Required for proper cell wall formation and expansion | Poor growth of leaves and roots |
| Zinc (Zn; $Zn^{2+}$) | 0.02 | Enzyme activation; auxin synthesis | Young leaves are abnormally small; older leaves have many dead spots |
| Copper (Cu; $Cu^{2+}$) | 0.006 | In active site of many redox enzymes and electron carriers | New leaves are dark green, may have dead spots |
| Nickel (Ni; $Ni^{2+}$) | 0.00005 | Activation of the enzyme urease | Leaf tips die; deficiency is rare |
| Molybdenum (Mo; $MoO_4^{2-}$) | 0.0001 | Nitrate reduction | Leaves turn yellow between veins; older leaves die |

**36.1 Mineral Nutrient Deficiency Symptoms** The plants on the left were grown with a full complement of essential nutrients. The center plants were deprived of iron, whereas the plants on the right were deprived of nitrogen.

**Go to Animated Tutorial 36.1**
**Nitrogen and Iron Deficiences**
Life10e.com/at36.1

## Deficiency symptoms reveal inadequate nutrition

When a plant is deficient in an essential element, it displays characteristic **deficiency symptoms**. Table 36.1 lists some of these symptoms, and plants deficient in iron and nitrogen are also shown in **Figure 36.1**. Such symptoms help growers diagnose mineral nutrient deficiencies in plants. With proper diagnosis, the missing nutrient(s) can be provided in the form of a **fertilizer** (an added source of mineral nutrients).

We know that the elements listed in Table 36.1 are essential to the life of all plants. How did biologists discover which elements are essential?

## Hydroponic experiments identified essential elements

The essential elements for plants were identified by growing plants **hydroponically**—that is, with their roots suspended in nutrient solutions instead of soil. Growing plants in this manner allows for greater control of nutrient availability than is possible in a complex medium such as soil.

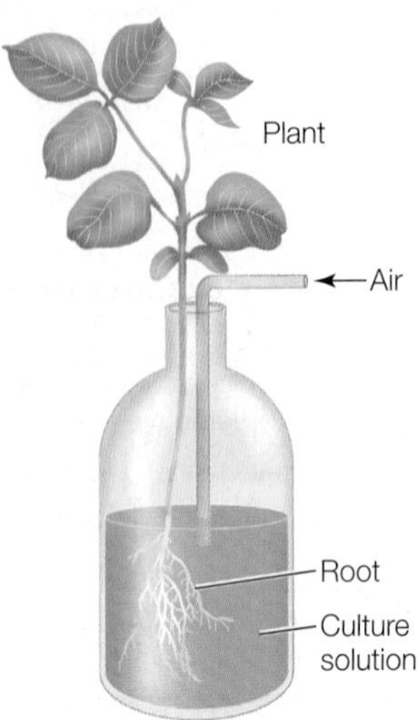


INVESTIGATING**LIFE**

**36.2 Is Nickel an Essential Element for Plant Growth?** Using highly purified salts in growth media, Patrick Brown and his colleagues tested whether barley can complete its life cycle in the absence of nickel.[a] Other investigators showed that no other element could substitute for nickel.

**HYPOTHESIS** Nickel is an essential element for a plant to complete its life cycle.

***Method***

1. Grow barley plants for 3 generations in nutrient solutions containing 0, 0.6, and 1.0 μM NiSO4.
2. Harvest seeds from 5–6 third-generation plants in each of the groups.
3. Determine the nickel concentration in seeds from each plant.
4. Germinate other seeds from the same plants on nickel-free medium and plot the success of germination against nickel concentration.

***Results*** There was a positive correlation between seed germination and seed nickel concentration. There was significantly less germination at the lowest nickel concentrations.

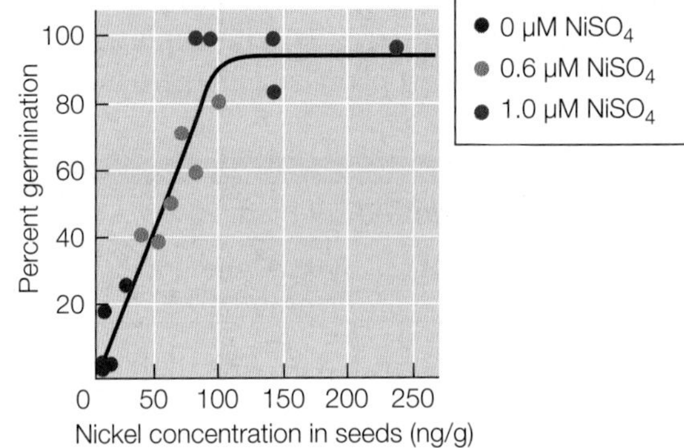


**CONCLUSION** Barley seeds require nickel in order to survive and germinate, and therefore nickel is an essential element for plant nutrition.

Go to **BioPortal** for discussion and relevant links for all INVESTIGATING**LIFE** figures.

[a]Brown, P. H., R. M. Welch, and E. E. Cary. 1987. *Plant Physiology* 85: 801–803.

In the first experiments of this type, performed a century and a half ago, plants seemed to grow normally in solutions containing only calcium nitrate [$Ca(NO_3)_2$], magnesium sulfate ($MgSO_4$), and potassium phosphate ($KH_2PO_4$). A solution missing any of these compounds could not support normal growth. Tests with other compounds that included various combinations of these elements soon established the existence of six essential elements: calcium, nitrogen, magnesium, sulfur, potassium, and phosphorus. These are now known as the essential mineral macronutrients.

Iron was the first micronutrient to be clearly established as essential, in the 1840s. The last micronutrient to be listed as essential was nickel, in 1983. That experiment is described in **Figure 36.2**. Identifying essential micronutrients proved to be more difficult than identifying macronutrients because of the small amounts involved. Sufficient amounts of micronutrients can be present in the environment used to grow plants or in the plants themselves. A seed may contain enough of a micronutrient to supply the embryo and the entire plant throughout its lifetime. There might even be enough left over to pass on to third-generation

WORKING WITH**DATA:**

## *Is Nickel an Essential Element for Plant Growth?*

### Original Paper

Brown, P. H., R. M. Welch, and E. E. Cary. 1987. Nickel: A micronutrient essential for higher plants. *Plant Physiology* 85: 801–803.

### Analyze the Data

For an element to be classified as an essential nutrient in plants, two criteria must be satisfied. Specifically, one must demonstrate that a plant cannot complete its life cycle in the absence of the element, and that no other element can substitute for the test element. To determine if nickel is an essential element for plants, Patrick Brown and colleagues tested whether barley could complete its life cycle in the absence of nickel. The researchers grew barley plants for three generations in nutrient solutions containing either 0, 0.6, or 1.0 μM $NiSO_4$, then tested the seeds for nickel content and germination. The results indicated a positive correlation between seed germination and nickel concentration (see Figure 36.2). Importantly, no germination was observed at the lowest nickel concentration. Together with results from another study showing that no other element could substitute for nickel, the investigators concluded that nickel is an essential micronutrient.

As shown in Figure 36.2, Ni depletion reduced the germination of barley grains. This could be caused by a reduced ability of the parental plant to produce grains, or, alternatively, a reduced germination ability of the grain itself. To investigate these two possibilities, Brown and colleagues measured grain production by twenty plants for each nickel treatment. The results are shown in **TABLE A**.

**QUESTION 1**

Did Ni depletion reduce the ability of the parental plants to produce grain? What data support your answer?

**QUESTION 2**

The data are expressed as mean ± standard deviation. What statistical test would you perform to determine whether the means are significantly different? Are they? (Refer to Appendix B if needed.)

**QUESTION 3**

To determine whether the reduced germination ability could be corrected by adding Ni, dry seeds from the three groups in the table were soaked in water with or without added 1.0 μM $NiSO_4$. The results are shown in **TABLE B**. Did adding Ni solution to dry seeds affect germination?

**QUESTION 4**

Plant biologists define the critical value as that concentration of a mineral nutrient in plant tissue that results in a 15 percent reduction in the optimum yield of the plant. Using maximum germination percentage in the graph in Figure 36.2 as the optimum yield, what is the critical value for nickel?

**TABLE A**

| Ni in nutrient solution (μM) | Total grain weight (g) | Grain number (per plant) |
|---|---|---|
| 0 | 7.3 ± 1.3 | 175 ± 26 |
| 0.6 | 7.5 ± 0.9 | 179 ± 35 |
| 1.0 | 8.4 ± 1.5 | 195 ± 41 |

**TABLE B**

| Ni supplied to maternal plant (μM) | Germination after soaking seed with or without nickel (%) | |
|---|---|---|
| | 0 μM $NiSO_4$ | 1 μM $NiSO_4$ |
| 0 | 9.1 ± 10 | 11.5 ± 10.3 |
| 0.6 | 55.2 ± 17 | 51.2 ± 20.3 |
| 1.0 | 98.2 ± 6.7 | 97.1 ± 9.1 |

**Go to BioPortal for all** WORKING WITH**DATA** **exercises**

plants. Because of such difficulties, nutrition experiments must be performed in tightly controlled laboratories with special air filters that exclude microscopic salt particles in the air, and must use only the purest available chemicals.

**RECAP 36.1**

Plants are autotrophs that obtain carbon and oxygen by photosynthesis, and mineral nutrients and water from the soil. Nutrients required by plants are classified as either macronutrients or micronutrients depending on the amount needed. Micronutrients are often needed in such minute amounts that only sophisticated chemical experiments can determine their essentiality.

- What are some specific mineral deficiency symptoms seen in plants? **See pp. 741–742 and Table 36.1**
- Why do plants need phosphorus? Why do they need nitrogen? **See p. 741**
- Outline an experimental method for determining whether an element is essential to a plant. **See Figure 36.2**

As we have seen, all plants require a specific set of nutrients for growth. Unlike many other organisms, however, plants can't move around to find nutrients. Let's look at how a plant finds and takes up nutrients from its environment.

## 36.2 How Do Plants Acquire Nutrients?

Many organisms can move from place to place to find the nutrients they need. But a plant cannot change its location (it is sessile), and so must obtain nutrients from its immediate environment. With the exception of the carbon and oxygen in $CO_2$, a plant's supply of nutrients is strictly local, and a plant may use up the water and mineral nutrients in its local environment as it grows. How does a plant cope with the problem of scarce nutrient supplies?

### Plants rely on growth to find nutrients

As discussed in Chapter 34, plants differ fundamentally from animals in that they grow throughout their lifetimes. In fact,

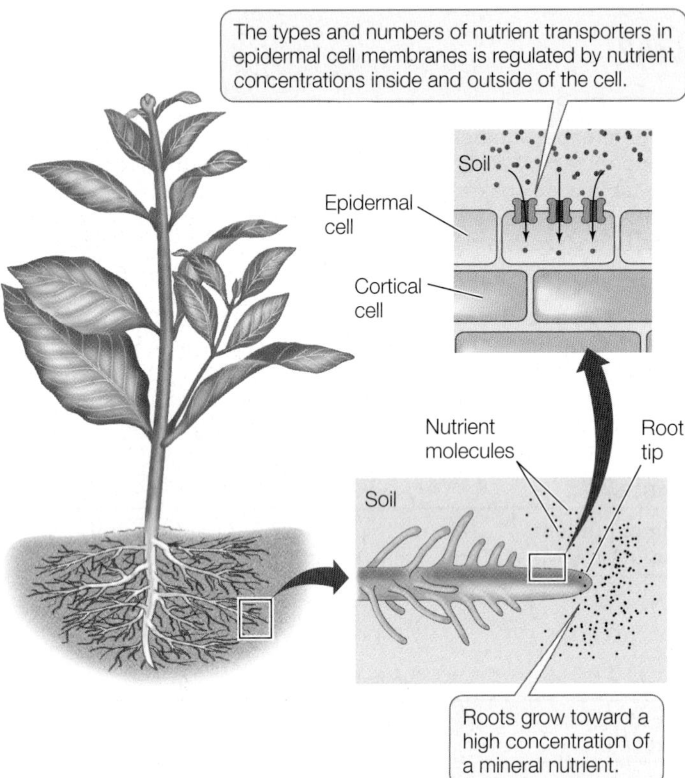


**36.3 Plants Regulate Their Nutrition** Plant roots branch and grow toward nutrients. Nutrients are taken up by transport proteins in the epidermal cell membrane. The number of transporters for a given nutrient can be regulated in response to nutrient availability.

growth is a plant's version of movement. For example, roots obtain most of the mineral nutrients plants need. By growing through the soil, roots mine it for new sources of mineral nutrients and water. The growth of stems and leaves helps a plant secure light and carbon dioxide, which in turn allows the roots to continue their growth through the soil. Deficiencies in water or specific mineral nutrients can stimulate plants to grow more roots, to improve the plants' chances of finding the nutrients they need.

As it grows, a plant—or even a single root—must deal with a variable environment. Animal droppings create high local concentrations of nitrogen. A particle of calcium carbonate may make a tiny area of the soil alkaline, while dead organic matter may make a nearby area acidic. Such microenvironments encourage or discourage the proliferation of a root system and help direct its growth. A major effort is underway to identify the signals in the soil and signaling pathways in the root that result in growth toward a source of nutrients (**Figure 36.3**).

### Nutrient uptake and assimilation are regulated

Nutrients must cross the plasma membranes of cells in order to be used or assimilated into larger molecules. Polar molecules, including mineral ions, cross the membrane via specialized transport systems. In Chapter 35 we discussed the movement of water and ions into plant roots by way of the apoplast (through cell walls and intercellular spaces) or symplast (directly through cells). The Casparian strip prevents water and ions from entering the xylem tissues of the roots (see Section 35.1); therefore these nutrients must enter the symplast before they can be transported to other tissues. In most cases, ions are actively transported across the plasma membrane of epidermal cells into the symplast because their concentrations in the soil solution are generally lower than their concentrations inside cells (see Figure 35.4).

Plants have specialized transport systems for the uptake of specific ions (see Figure 36.3). For example, *Arabidopsis thaliana* has more than 50 genes that encode nitrate ($NO_3^-$) transporters, 6 genes encoding ammonium ($NH_4^+$) transporters, and at least 4 genes for phosphate ($PO_4^{3-}$) transporters. Nutrient uptake is highly regulated because the levels of ions inside cells must be maintained at constant levels. The genes for ion transporters are regulated at the transcriptional level by the amounts of each nutrient inside cells: low nutrient levels stimulate transcription, whereas high levels repress transcription. In addition, the transporter proteins themselves are regulated (for example, by phosphorylation) to control their ion transport activity.

The assimilation of nutrients into more complex molecules is also regulated according to the plant's needs. The enzymes involved in assimilating nitrate and ammonium into amino acids are regulated at the transcriptional and posttranscriptional levels to increase assimilation when available nitrogen is abundant. The uptake and assimilation of nitrogen are also stimulated by photosynthesis, and this ensures that the nitrogen status in the plant is coordinated with its carbon status.

**RECAP 36.2**

Both the uptake and assimilation of nutrients are regulated according to a plant's needs.

- How does the ability to grow throughout their lifetime allow plants to seek out nutrients? **See p. 744**
- Describe how plants control the uptake and assimilation of mineral nutrients. **See Figure 36.3**

Plants acquire many essential elements from the soil. As we will see in the next section, soils have complex structures that affect the availability of nutrients for plants.

## 36.3 How Does Soil Structure Affect Plants?

Most terrestrial plants grow in soil. Soils provide:

- Mechanical support
- Mineral nutrients and water from the soil solution
- $O_2$ for root respiration

Soils also harbor many bacteria and other organisms; some of these are beneficial to plant life, but others are harmful. Some soils contain toxic levels of metal ions such as cadmium, chromium, and lead (see Section 39.4).

Soils are modified by natural phenomena such as rain, temperature extremes, and the activities of plants and animals. They are also modified by the activities of humans, particularly in agriculture. In this section we will examine the composition, structure, and formation of soils, as well as their role in plant nutrition.

## Soils are complex in structure

**Soils** have living and nonliving components (**Figure 36.4**). The living components include plant roots as well as populations of bacteria, fungi, protists, and animals such as earthworms and insects. The nonliving portion of the soil includes rock fragments ranging in size from large stones to sand to silt, and finally to tiny particles of clay that are 2 micrometers (μm) or less in diameter. Soil also contains water and dissolved mineral nutrients, air spaces, and dead organic matter. The air spaces in soil contain $O_2$.

Although soils vary greatly, almost all of them have a soil profile consisting of several recognizable horizontal layers, called **horizons**, lying on top of one another. Soil scientists recognize three major horizons—termed A, B, and C—in the profile of a typical soil (**Figure 36.5**).

- The **A horizon** is the **topsoil** that supports the plant's nutrient needs. It contains most of the soil's living and dead organic matter.
- The **B horizon** is the **subsoil**, which accumulates materials from the topsoil above it and from the parent rock below.
- The **C horizon** is the **parent rock**, also called bedrock, that is in the process of breaking down to form soil.

**Soil fertility** is a soil's ability to support plant growth. A topsoil's fertility is determined by several factors. Topsoils vary greatly in their proportions of sand, silt, and clay, and this influences their ability to support plant growth. For example, mineral nutrients tend to be **leached** from the upper soil horizons—dissolved in rain or irrigation water and carried to deeper horizons, where they are unavailable to plant roots. Dissolved minerals are readily leached from sandy soil because sand particles are relatively large and cannot hold water. Clay, by contrast, binds more water than sand does, and the charged surfaces of clay particles bind mineral ions that plant roots ultimately take up. But clay particles are tiny and pack tightly together, leaving little space for air. A **loam** is a soil that is an optimal mixture of sand, silt, and clay, and thus has sufficient levels of air, water, and available nutrients for plants. Loams also usually contain organic matter. Most of the best topsoils for agriculture are loams.

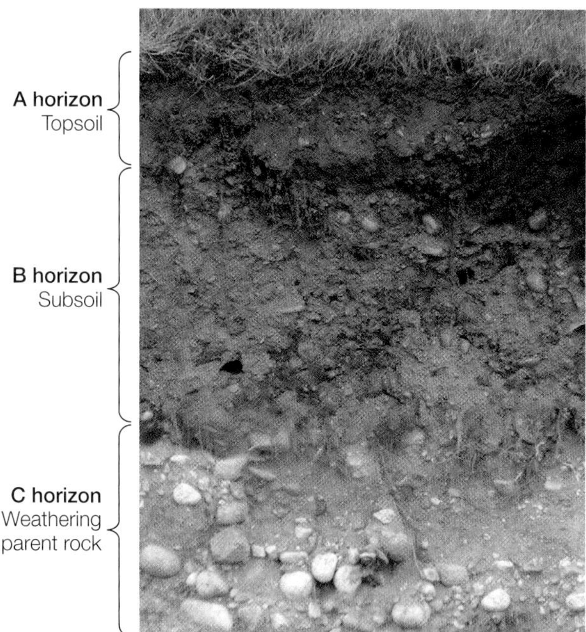


**36.5 A Soil Profile** The A, B, and C horizons can sometimes be seen at construction sites such as this one in Massachussets. The upper layer (the A horizon) is home to most of the living organisms in the soil.

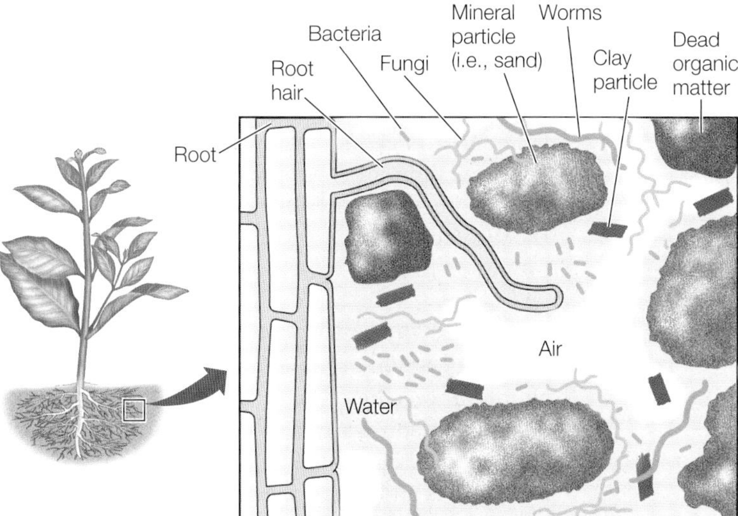


**36.4 The Complexity of Soil** Soils favorable for plant growth contain both clay and larger mineral particles, as well as water, air, and organic matter. Other organisms are also present.

In addition to mineral particles, soils contain dead organic matter, largely from plants. Soil organisms break down dead leaves and other plant organs on the ground into a substance called **humus**. This material is used as a food source by microbes that break down complex organic molecules and release simpler molecules into the soil solution. Humus also provides air spaces that increase $O_2$ availability to plant roots.

## Soils form through the weathering of rock

Rocks are broken down into soil particles—**weathered**—in two ways. First there is *mechanical weathering*, which is the physical breakdown of materials by wetting, drying, and freezing. Second there is *chemical weathering*, the alteration of the chemistry of the materials in the rocks. Several types of chemical weathering occur, all of which influence the availability of mineral nutrients:

- Oxidation by atmospheric oxygen
- Hydrolysis (reaction with water)
- Reaction with acids (particularly carbonic acid)

The parent rock and the weathering it undergoes determine the basic structure and chemical composition of a soil. However, a key soil characteristic for plants is the availability of nutrients, which must be dissolved in the soil solution for uptake by the plant. Chemical weathering often results in clay particles that are covered with negatively charged chemical groups, which bind positively charged mineral nutrients. How might roots obtain these mineral nutrients?

## Soils are the source of plant nutrition

Humus and clay particles often carry negative charges. These particles form ionic attractions (see Section 2.2) with the positively charged ions (cations) of many minerals that are important for plant nutrition, such as potassium ($K^+$), magnesium ($Mg^{2+}$), and calcium ($Ca^{2+}$). To become available to plants or other organisms, these cations must be detached from the clay particles.

Recall that the root surface is covered with root hair cells (see Figure 34.8). Transporters in the plasma membranes of these cells actively pump protons ($H^+$) out of the cell. In addition, cellular respiration in the roots releases $CO_2$, which dissolves in the soil water and reacts with it to form carbonic acid. This acid ionizes to form bicarbonate and free protons:

$$CO_2 + H_2O \rightleftharpoons H_2CO_3 \rightleftharpoons H^+ + HCO_3^-$$

Proton-pumping by the root and ionization of carbonic acid both act to increase the proton concentration (lower the pH) in the soil surrounding the root. The protons bind more strongly to clay particles than do mineral cations; in essence, they trade places with the cations in a process called **cation exchange** (**Figure 36.6**). Cation exchange releases important cations into the soil solution, where they are available to be taken up by the roots. Soil fertility is determined in part by the soil's ability to provide nutrients in this manner.

Some soil particles, such as ones containing oxides of iron or aluminium, are positively charged under acid conditions and can exchange anions in a process similar to cation exchange. However, soil pH is rarely low enough for anion exchange to occur. As a result, important anions such as nitrate ($NO_3^-$) and sulfate ($SO_4^{2-}$)—direct sources of nitrogen and sulfur, respectively—may leach rapidly from the A horizon.

As we have just seen, soil fertility is affected by soil pH. The proton concentration affects the binding of cations and anions to soil particles, and can affect the solubility of other nutrients, such as iron. In addition, soil pH affects the absorption of nutrients by plant roots. The pH level of a soil depends on its mineral and organic contents and can be altered by various factors, including rainfall, weathering, plant growth, and fertilizer applications. The optimal soil pH for most plants is in the range 6 to 7.5, but some plants, such as blueberries and cranberries, prefer pH levels of 4.5 to 5.

## Fertilizers can be used to add nutrients to soil

Leaching and the harvesting of crops can deplete a soil of its nutrients, so that new crops grow poorly on that soil. Soil fertility can be restored or increased in various ways, including shifting agriculture to another location or applying organic or chemical fertilizers.

**SHIFTING AGRICULTURE** In the past, when the soil could no longer support a level of plant growth sufficient for agricultural purposes, people simply moved to another location. The nutrients in the soil of a field allowed to lie fallow will be replenished gradually through the addition of organic matter from the growth and death of plants naturally present, and by the weathering of the parent rock. Both processes take a long time, which is not a problem as long as a lot of land is available. Today, however, the food needs of a large human population are too great to allow land to be left vacant for a long time, and people are disinclined to move away from settled homesteads. As a consequence, chemical fertilizers are now commonly used to improve soil fertility.

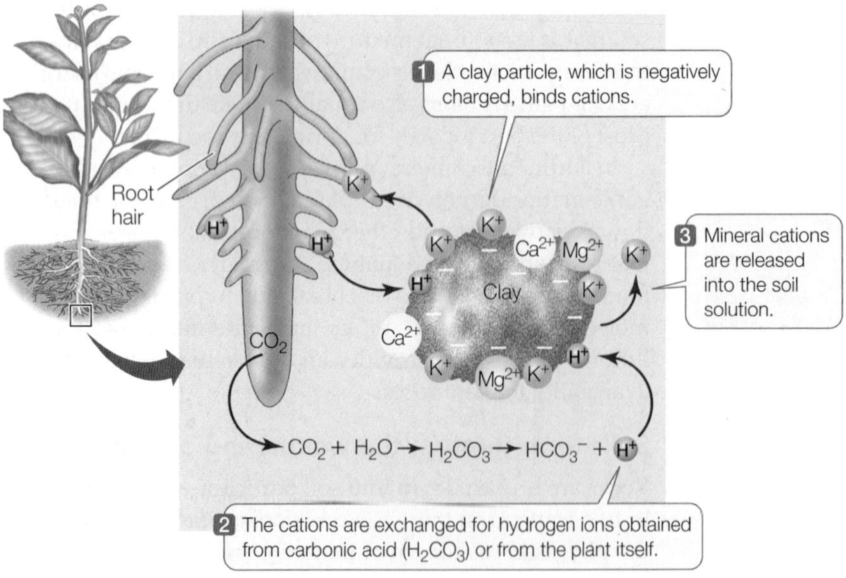

**36.6 Cation Exchange** Plants obtain mineral nutrients from the soil primarily in the form of positive ions; potassium ($K^+$) is the example shown here.

**ORGANIC FERTILIZERS** Microorganisms in the soil break down organic molecules into smaller, simpler molecules. These simpler molecules can dissolve in soil water and enter plant roots. For example, soil bacteria break down the proteins in dead leaves and produce ammonium ions ($NH_4^+$), which in turn are converted into nitrate ($NO_3^-$). Both ammonium and nitrate can be taken up and used by plants:

$$\text{Proteins in leaves} \xrightarrow{\text{Bacteria}} NH_4^+ \xrightarrow{\text{Bacteria}} NO_3^-$$

Farmers can increase the nutrient content of soil by adding organic materials such as compost (partially decomposed plant materials) or manure (waste from farm animals). Manure is a particularly good source of nitrogen. In either case, the addition of these **organic fertilizers** adds nutrients to the soil much more rapidly than weathering or the gradual addition of organic matter from natural vegetation. Organic fertilizers allow for a slow release of ions as the materials decompose.

INORGANIC FERTILIZERS Organic fertilizers may act too slowly to restore fertility if a soil is to be used every year. **Inorganic fertilizers** supply mineral nutrients in forms that can be taken up immediately by plants or that are rapidly converted to usable forms in the soil. Inorganic fertilizers are easily transported and handled, and allow farmers to control the amount of a particular nutrient that is supplied to each crop. Particular fertilizers are used in varying amounts, depending on the needs of the crop and the type of soil. For example, much higher amounts of nitrogen are applied to cornfields than to soybean fields. Inorganic fertilizers come in many forms. Common ones include ammonia ($NH_3$), urea ($NH_2$-CO-$NH_2$), and salts formed from positive and negative ions such as ammonium ($NH_4^+$), potassium ($K^+$), nitrate ($NO_3^-$), phosphate ($PO_4^{3-}$), and sulfate ($SO_4^{2-}$).

The use of inorganic fertilizers became widespread in many countries during the twentieth century. This contributed (along with improved genetic strains and other technological advances) to the "green revolution"—a massive increase in world food production during the second half of the century. However, as we saw in the opening story, excessive fertilizer use has led to environmental problems. Agricultural scientists are developing methods for determining the exact amount of a particular nutrient that a crop requires, and for applying fertilizer at an optimal time to maximize yield while reducing losses to the environment.

**RECAP 36.3**

Land plants live anchored in the soil and obtain water and mineral nutrients from it. Soils are complex in structure and vary in fertility. Farmers can add fertilizers to improve the nutrient contents of soils.

- Explain how mechanical and chemical weathering form soil from rock. **See pp. 745–746**
- How is soil fertility enhanced by the process of cation exchange? **See p. 746 and Figure 36.6**
- What are the differences between organic and inorganic fertilizers in terms of plant nutrition? **See pp. 746–747**

Thus far we have focused on the uptake of nutrients in the soil by plant roots. An understanding of how plants acquire nutrients from the soil would be incomplete, however, without taking into account the involvement of soil microbes, including fungi and bacteria. In the next section we will focus on the intimate interactions of plants with these organisms, which are essential to the success of most terrestrial plants.

## How Do Fungi and Bacteria Increase Nutrient Uptake by Plant Roots?

One gram of soil can contain 6,000 to 50,000 bacterial *species* and up to 200 *meters* of fungal hyphae (the long branching cells of fungi), although both are largely invisible to the naked eye. In Chapter 39 we will describe the strategies plants use to prevent infection by harmful soil microbes. But plants actively encourage a few species of fungi and bacteria to infect their roots and even invade root cells. In this section we will describe the mutually beneficial relationships in which products are exchanged between the plants and these special soil microbes.

### Plants send signals for colonization

In Chapter 30 we described mycorrhizae, the association of fungi with plant roots—an interaction that occurs in more than 90 percent of terrestrial plants. Our example in Chapter 30 was ectomycorrhizal fungi, which wrap themselves around a plant root (see Figure 30.9). In this chapter we will examine arbuscular mycorrhizae, in which fungal hyphae penetrate root cells. We will also describe here the close association between the roots of some plants and rhizobia, a group of nitrogen-fixing bacteria (bacteria that convert atmospheric $N_2$ into a more biologically useful form). We will see that mycorrhizal and rhizobial associations are both initiated by signals sent by plant roots that attract the soil organisms, and that the development of these associations involves similar genes and cellular pathways.

FORMATION OF MYCORRHIZAL ASSOCIATIONS The events in the formation of arbuscular mycorrhizae are shown in **Figure 36.7A**. Plant roots produce compounds called **strigolactones** that stimulate rapid growth of fungal hyphae toward the root. In response, the fungi produce signals that stimulate expression of plant symbiosis-related genes. The products of some of these genes give rise to the prepenetration apparatus (PPA), which guides the growth of the fungal hyphae into the root cortex. The sites of nutrient exchange between fungus and plant are the arbuscules, which form within root cortical cells. Despite the intimacy of this association, the plant and fungal cytoplasms never mix—they are separated by two membranes, the fungal plasma membrane and the periarbuscular membrane (PAM), which is continuous with the plant plasma membrane.

FORMATION OF NITROGEN-FIXING NODULES A group of plants called legumes (members of the plant family Fabaceae) can form symbioses with soil bacteria in several genera collectively known as rhizobia. The legume roots release flavonoids and other chemical signals that attract the rhizobia to the vicinity of the roots (**Figure 36.7B**). The flavonoids also trigger the transcription of bacterial *nod* genes, the products of which synthesize Nod (nodulation) factors. These factors, when secreted by the bacteria, cause cells in the root cortex to divide, leading to the formation of a primary nodule meristem. This meristem gives rise to the plant tissue that constitutes the **root nodule**.

**36.7 Roots Send Signals for Colonization** Plant roots send chemical signals to arbuscular mycorrhizal fungi (A) and nitrogen-fixing bacteria (B) to stimulate colonization.

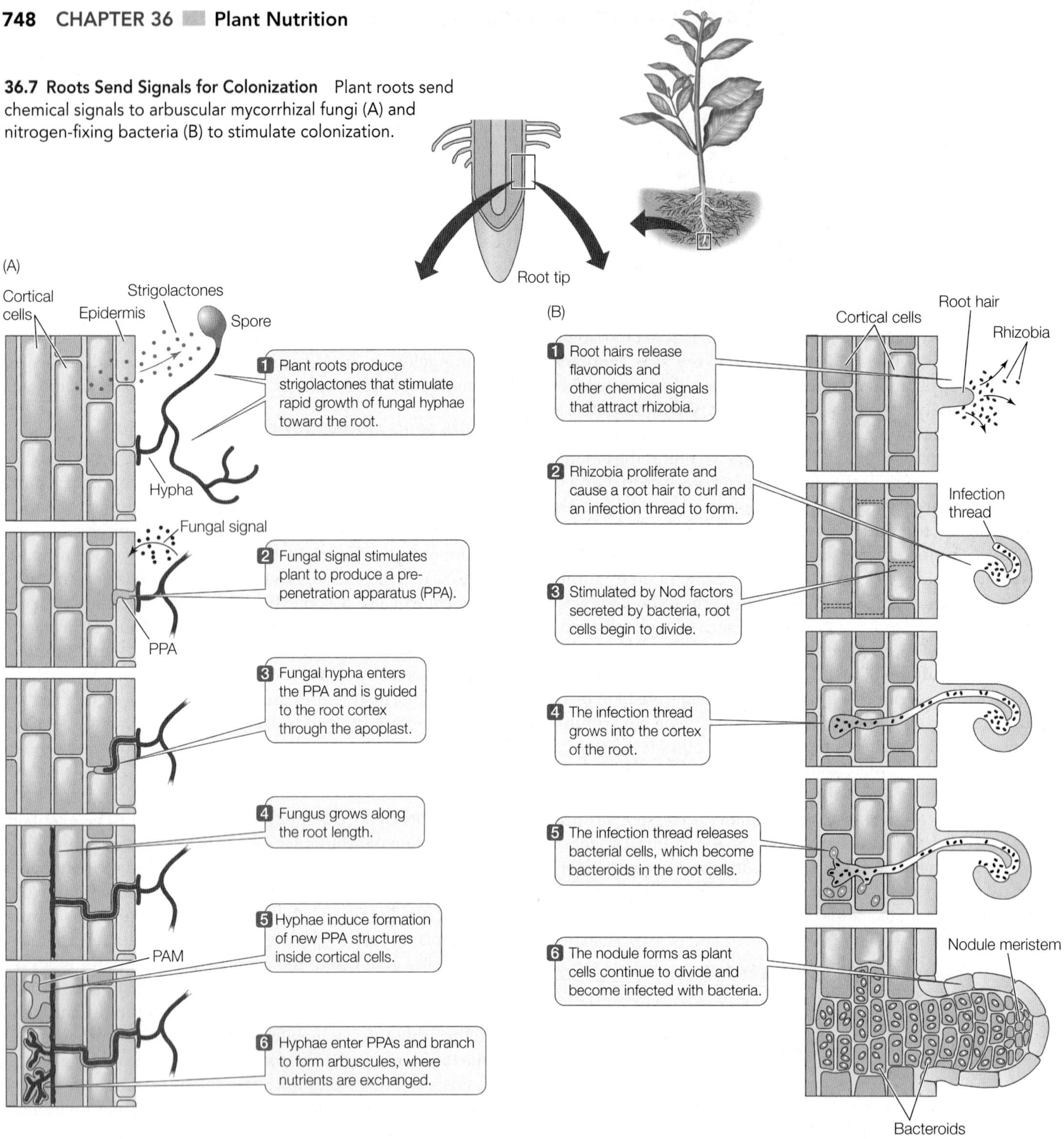

Root tip

(A)

Cortical cells
Epidermis
Strigolactones
Spore

1 Plant roots produce strigolactones that stimulate rapid growth of fungal hyphae toward the root.

Hypha
Fungal signal

2 Fungal signal stimulates plant to produce a pre-penetration apparatus (PPA).

PPA

3 Fungal hypha enters the PPA and is guided to the root cortex through the apoplast.

4 Fungus grows along the root length.

5 Hyphae induce formation of new PPA structures inside cortical cells.

PAM

6 Hyphae enter PPAs and branch to form arbuscules, where nutrients are exchanged.

(B)

Cortical cells
Root hair
Rhizobia

1 Root hairs release flavonoids and other chemical signals that attract rhizobia.

2 Rhizobia proliferate and cause a root hair to curl and an infection thread to form.

Infection thread

3 Stimulated by Nod factors secreted by bacteria, root cells begin to divide.

4 The infection thread grows into the cortex of the root.

5 The infection thread releases bacterial cells, which become bacteroids in the root cells.

6 The nodule forms as plant cells continue to divide and become infected with bacteria.

Nodule meristem
Bacteroids

Bacteria enter the root via an infection thread, analogous to the PPA in mycorrhizal associations, and eventually reach cells inside the root nodule. There the bacteria are released into the cytoplasm of the nodule cells, enclosed in membrane vesicles similar to the PAM. Inside the vesicles, the bacteria differentiate into **bacteroids**—the form of the bacteria that can fix nitrogen.

A COMMON MECHANISM There is increasing evidence that nodule formation depends on some of the same genes and mechanisms that allow mycorrhizae to develop. For example, both processes involve invagination of the plasma membrane to allow entry of the fungal hyphae or rhizobia. The similarities in the structures formed during the development of mycorrhizae and nodules are especially striking, considering that the symbioses involve members of two different kingdoms (fungi and bacteria) (**Figure 36.8**).

## Mycorrhizae expand the root system

In many cases the roots of vascular plants cannot nutritionally support plant growth alone—they simply cannot reach all the

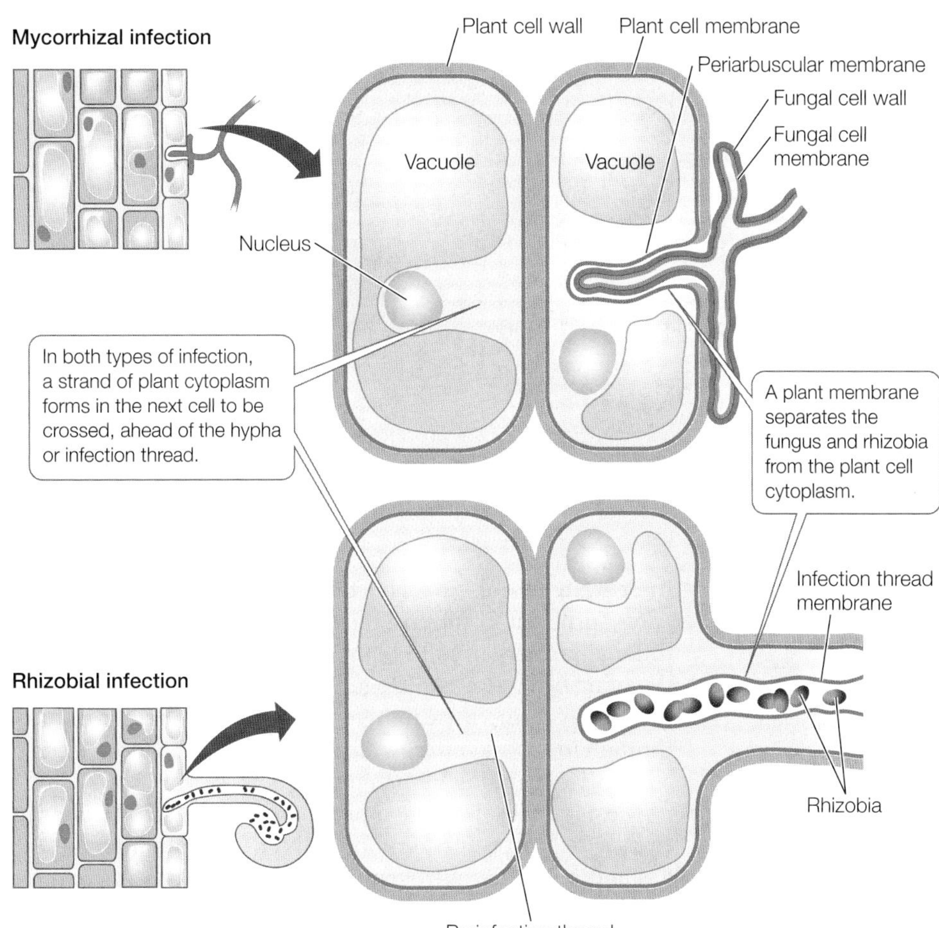


**36.8 Intracellular Structures in Plant–Fungus and Plant–Rhizobium Symbioses** Several steps in the development of mycorrhizae and nodules involve similar structures.

nutrients available in the soil. Mycorrhizae expand the root surface area 10-fold to 1,000-fold, increasing the amount of soil that can be mined for nutrients. In addition, because fungal hyphae are much finer than root hairs, they can get into pores in the soil that are inaccessible to roots. In this way, mycorrhizae probe a vast quantity of soil for nutrients and deliver them into root cortical cells.

The primary nutrient that the plant obtains from a mycorrhizal interaction is phosphorus. In exchange, the fungus obtains an energy source: the products of photosynthesis. In fact, up to 20 percent of the photosynthate of terrestrial plants is directed to and consumed by arbuscular mycorrhizal fungi. Such associations are excellent examples of mutualism, an interaction between two species in which both species benefit (further discussed in Chapter 56). Mutualism is a type of symbiosis, in which two different species live in close contact for a significant portion of their life cycles.

## Soil bacteria are essential in getting nitrogen from air to plant cells

The essential mineral nutrient most commonly in short supply, in both natural and agricultural situations, is nitrogen. This is surprising because elemental nitrogen ($N_2$) makes up almost four-fifths of Earth's atmosphere. However, plants cannot use $N_2$ directly as a nutrient. The triple bond linking the two nitrogen atoms is extremely stable, and a great deal of energy is required to break it; thus $N_2$ is a highly unreactive substance.

Some prokaryotes have an enzyme that enables them to convert $N_2$ into a more reactive and biologically useful form by a process called **nitrogen fixation**:

$$N_2 + 6\,H \rightarrow 2\,NH_3$$

**Nitrogen fixers**, including those present in root nodules, fix approximately 170 million metric tons of nitrogen per year. Humans use industrial methods to fix about 80 million metric tons per year. In addition, about 20 million metric tons per year are fixed in the atmosphere by nonbiological means such as lightning, volcanic eruptions, and forest fires. Rain brings these atmospherically formed products to the ground.

Two types of organisms can fix nitrogen:

- Free-living organisms living in soil and water (e.g., *Azotobacter* bacteria and *Nostoc* cyanobacteria [sometimes called blue-green algae])
- Symbiotic organisms living in other organisms (e.g., rhizobia in roots of legumes, and *Anabaena* cyanobacteria in aquatic ferns)

## Nitrogenase catalyzes nitrogen fixation

Nitrogen fixation is the reduction (see Section 9.1) of nitrogen gas. It proceeds by the stepwise addition of three pairs of

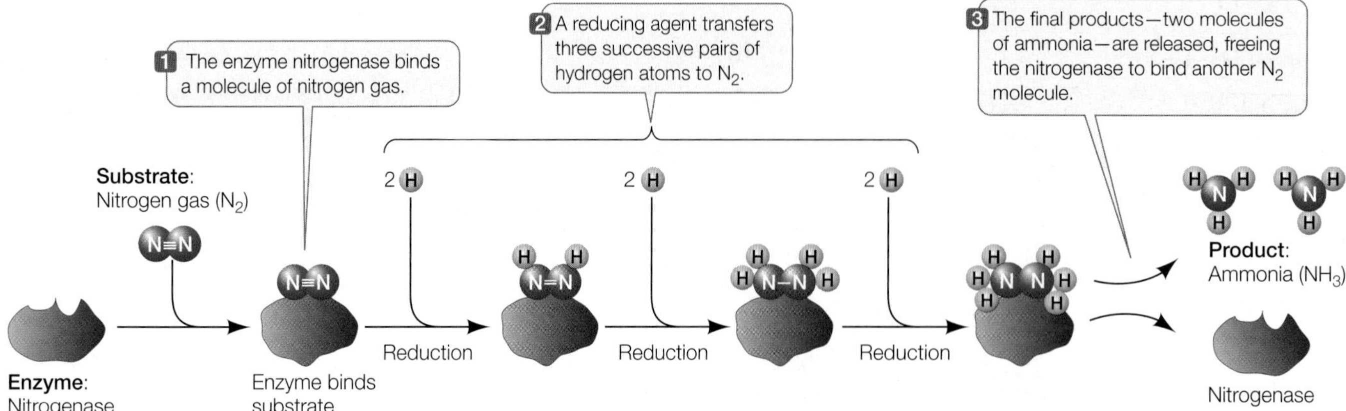


**36.9 Nitrogenase Fixes Nitrogen** Throughout the chemical reactions of nitrogen fixation, the reactants are bound to the enzyme nitrogenase. A reducing agent transfers hydrogen atoms to nitrogen, and eventually the final product—ammonia—is released. This reaction requires a large input of energy: about 16 ATPs are consumed per reaction.

hydrogen atoms to $N_2$ (**Figure 36.9**). In addition to $N_2$, these reactions require three things:

- A strong reducing agent to transfer hydrogen atoms (protons and electrons) to $N_2$ and to the intermediate products of the reaction
- A great deal of energy, which is supplied by ATP
- The enzyme **nitrogenase**, which catalyzes the reaction

Depending on the species of nitrogen fixer, either respiration or photosynthesis provides the necessary reducing agent and ATP.

Nitrogenase is strongly inhibited by oxygen, and many nitrogen fixers are anaerobes that live in environments with little or no $O_2$. But rhizobia are aerobic and fix nitrogen in aerobic plant roots. How can nitrogenase function under these circumstances?

Plants typically house nitrogen-fixing bacteria in root nodules. Within a nodule, $O_2$ is maintained at a low level that is sufficient to support respiration, but not so high as to inactivate nitrogenase. This is possible because the cytoplasm of nodule cells contains a plant-produced protein called **leghemoglobin**, which is an $O_2$ carrier. Leghemoglobin is a close relative of hemoglobin, the red, oxygen-carrying pigment of animals, and is thus an evolutionarily ancient molecule. Some plant nodules contain enough of it to be bright pink inside. Leghemoglobin, with its iron-containing heme groups, transports enough oxygen to the nitrogen-fixing bacteria to support their respiration, while keeping free oxygen concentrations low enough to protect nitrogenase.

## Biological nitrogen fixation does not always meet agricultural needs

Crop rotation systems have been used for hundreds or thousands of years by many human civilizations. In these systems, each field is used to grow different crops in different years, with legumes (such as alfalfa, clover, peas, and beans) included in the rotation. The rotation may also include periods of grazing by farm animals. Because of their association with nitrogen-fixing bacteria, legumes can replace all or some of the nitrogen removed by grain crops such as wheat and corn. Even with these systems, however, bacterial nitrogen fixation is not always sufficient to support the needs of agriculture. Some traditional farmers used to plant dead fish along with corn; the decaying fish released nitrogen that the developing corn could use. Today farmers use inorganic nitrogen fertilizers produced through industrial nitrogen fixation to meet the food needs of a rapidly expanding human population.

Most industrial nitrogen fixation is done by the Haber process, which produces ammonia ($NH_3$) from methane ($CH_4$) gas and molecular nitrogen ($N_2$). The Haber process requires a great deal of energy. (Biological nitrogen fixation also consumes lots of energy in the form of ATP—about 16 ATP per N fixed.) At present in the United States, the manufacture of nitrogen-containing fertilizer takes more energy—primarily natural gas and hydroelectric—than does any other aspect of crop production. The rising cost of energy sources raise serious questions about the sustainability of this approach to fertilizer production.

## Plants and bacteria participate in the global nitrogen cycle

Nitrogen moves through the biosphere in a **global nitrogen cycle** (**Figure 36.10**), which includes four key steps:

1. *Fixation* of atmospheric $N_2$ to $NH_3$ and $NH_4^+$ by bacteria and by abiotic processes
2. *Nitrification* of these molecules to nitrate by bacteria
3. *Nitrate reduction* by plants
4. *Denitrification* of nitrate by bacteria back to $N_2$, which is then released to the atmosphere to begin another cycle

Nitrogen fixation results in the formation of ammonia ($NH_3$), most of which is rapidly ionized to form ammonium ($NH_4^+$). The balance between ammonia and ammonium depends on pH; neutral and acidic pH levels favor the ionic form. Ammonia can be toxic to plants, but ammonium ions are taken up safely. Soil bacteria called **nitrifiers** oxidize ammonia to nitrate ions ($NO_3^-$)—another form that plants can take up—by the process of **nitrification**. Soil pH affects which form of nitrogen is taken up by plants: nitrate ions are taken up preferentially under more basic conditions, and ammonium

**36.10 The Nitrogen Cycle** Nitrogen fixation, nitrification, nitrate reduction, and denitrification are components of an essential chemical cycle that converts atmospheric nitrogen gas into ammonium ions and nitrate ions—forms of nitrogen that can be taken up by plants—and returns $N_2$ to the atmosphere.

**Go to Activity 36.1 The Nitrogen Cycle Life10e.com/ac36.1**

ions under more acidic ones. To use nitrate, a plant must first reduce it to ammonium in a process called **nitrate reduction**, which occurs in two enzyme-catalyzed steps. First, nitrate is converted to nitrite ($NO_2^-$) in the cytoplasm, and then nitrite is converted to ammonium in the plastids. The plant uses the ammonium to manufacture amino acids, from which the plant's proteins and all its other nitrogen-containing compounds are formed. Animals cannot reduce nitrogen and so depend on plants to supply them with reduced nitrogenous compounds.

The nitrogen cycle is essential for life on Earth: nitrogen-containing compounds constitute 5 to 30 percent of a plant's total dry weight. The nitrogen content of animals is even higher, and all of it arrives there by way of the plant kingdom.

**RECAP 36.4**

Two mutualistic interactions with soil microbes are critical to the success of terrestrial plants. Fungi and plants form mycorrhizae, which greatly increase the soil volume that roots can scavenge for nutrients. Bacteria in soils and root nodules fix atmospheric nitrogen into forms that plants and ultimately animals can use. Denitrification returns nitrogen from dead organisms and animal waste back to the atmosphere, completing the global nitrogen cycle.

- How is the formation of a root nodule on a legume similar to the formation of an arbuscular mycorrhiza? **See p. 747–749 and Figures 36.7 and 36.8**
- What is exchanged between plants and fungi in mycorrhizae? **See p. 749**
- What, besides nitrogenase, is required to reduce nitrogen gas to a form plants can use? **See p. 750 and Figure 36.9**

Let's turn now to some special mechanisms for obtaining nutrients that have evolved in plant species with unusual lifestyles.

## 36.5 How Do Carnivorous and Parasitic Plants Obtain a Balanced Diet?

Most plants obtain their mineral nutrients from the soil solution. Carnivorous and parasitic plants are examples of plants that obtain nutrients from other sources.

### Carnivorous plants supplement their mineral nutrition

Some plants augment their nitrogen supply by capturing and digesting flies and other insects. There are about 500 of these **carnivorous plant** species, the best known of which are Venus flytraps (genus *Dionaea*; **Figure 36.11A**), sundews (genus *Drosera*; **Figure 36.11B**), and pitcher plants (genus *Sarracenia*).

Carnivorous plants are typically found in boggy habitats that are acidic and nutrient deficient. To obtain extra nitrogen, these plants capture animals, digest their proteins, and absorb the amino acids. Pitcher plants have pitcher-shaped leaves that collect small amounts of rainwater. Insects and even small rodents are lured into the pitchers by bright colors or attractive scents and are prevented from leaving by stiff, downward-pointing hairs. The animals eventually die and are digested by a combination of plant enzymes and bacteria in the water. Sundews have leaves covered with hairs that secrete a sticky, sugary liquid. Insects become stuck to these hairs, and more hairs curve over to further entrap them. Enzymes secreted by the plant digest the insects. Venus flytraps have specialized leaves with two halves that fold together. When an insect touches special hairs called trigger hairs on the

(A) *Dionaea muscipula*

(B) *Drosera rotundifolia*

**36.11 Carnivorous Plants** Some plants have adapted to nitrogen-poor environments by becoming carnivorous. (A) The Venus flytrap obtains nitrogen from the bodies of insects trapped inside the plant when its hinges snap shut. (B) Sundews trap insects on sticky hairs. Secreted enzymes will digest the carcass externally.

**Go to Media Clip 36.1**
**A Venus Flytrap "Snaps to It"**
**Life10e.com/mc36.1**

leaf, its two halves snap together. The spiny margins interlock and imprison the insect before it can escape. The leaf then secretes enzymes that digest its prey.

The closing of the Venus flytrap's leaf is one of the fastest movements in the plant world, requiring only 0.1 seconds. To find out how this happens, Dr. Lakshminarayanan Mahadevan and colleagues at Harvard University painted fluorescent dots on the surface of the flytrap's leaf surface and used high-speed cameras to record the trap snapping shut when its trigger hairs were touched. They then used computer analysis of the recorded dot movements to generate a mathematical model to help explain the movement. The researchers found that the first step is the osmosis-driven elongation of cells on the outer surface of the leaf. The expansion of only one side of the leaf causes it to snap from a convex into a concave shape, much like a contact lens flipping inside out.

Carnivorous plants photosynthesize and extract soil nutrients just like other plants, but eating insects helps them grow faster in their natural habitats. They use the additional nitrogen from the insects to make more proteins, chlorophyll, and other nitrogen-containing compounds.

## Parasitic plants take advantage of other plants

Approximately 1 percent of flowering plant species derive some or all of their water, mineral nutrients, and sometimes even photosynthate from other plants. These **parasitic plants** have evolved absorptive organs called **haustoria**, which invade the host and tap into the vascular tissues in the root or stem.

Parasitic plants are divided into two broad classes based on their nutritional interactions with their hosts. **Hemiparasites** can still photosynthesize but derive water and mineral nutrients from the living bodies of other plants. Perhaps the most familiar hemiparasites are the several genera of mistletoes. Mistletoes are green and carry on some photosynthesis, but they parasitize other plants for water and mineral nutrients and may derive photosynthetic products from them as well. Dwarf mistletoe (*Arceuthobium americanum*) is a serious parasite in forests of the western United States, destroying more than 3 billion board feet of lumber per year.

**Holoparasites** are completely parasitic and do not perform photosynthesis. They are taxonomically and morphologically diverse. Some, such as members of the dodder family, are plantlike in appearance, with small leaf remnants and flowers (**Figure 36.12**). Some holoparasites do not have leaves or stems because they spend most of their life cycle underground and only break the surface to flower.

Several parasitic plant species lack many of the genes normally present in the chloroplast genome (which in turn is only a remnant of the genome of the original endosymbiont from which the chloroplast evolved; see Sections 5.5 and 27.1). These genes, which are needed for photosynthesis, have been lost because there is no evolutionary pressure to retain them. Thus while the parasitic lifestyle can be viewed as a free ride, for

**36.12 A Parasitic Plant** Tendrils of dodder (genus *Cuscuta*) wrap around a goldenrod (genus *Solidago*). The parasitic dodder obtains water, sugars, and other nutrients through tiny, rootlike protuberances that penetrate the surface of the host plant.

some plants it is also a one-way ticket, with no possibility of return to self-sufficiency.

### The plant–parasite relationship is similar to plant–fungus and plant–bacteria associations

Plant–bacteria and plant–fungus associations both involve reciprocal signaling between the two species (see Figure 36.7). Parasitic plants also need to detect signals from nearby plants so they can grow toward them and obtain their nutrients, but obviously this is to the disadvantage of the potential host plant. In one interesting case, a parasitic plant has evolved the ability to recognize the chemical signals produced by plants to attract beneficial fungi.

The holoparasite *Striga* (witchweed) is a serious pest of cereal crops in Africa. In Section 36.4 we saw that arbuscular fungi are attracted to plant roots by compounds called strigolactones. One of these same molecules induces the seed germination of some parasitic plants, including *Striga*. Scientists strongly suspect that this is no coincidence. The mycorrhizal interaction is ancient (more than 400 million years old) and predates the evolution of parasitic plants. For this reason scientists hypothesize that a mechanism evolved in the ancestors of modern *Striga* to recognize a compound that was already produced by plants to attract soil microbes.

In *Striga* we thus find an example of "opportunistic evolution"—that is, the repurposing of preexisting processes rather than the development of new processes. This is not the first time we have encountered this phenomenon in this chapter. Recall that the formation of nodules by rhizobia uses some of the same mechanisms used by arbuscular fungi to establish residence inside plant cells (see Figure 36.7). This implies an evolutionary connection between the two symbioses.

**RECAP 36.5**

Carnivorous plants supplement their nutrition by extracting materials from animals. Rapid reflexes have evolved in some of these plants for trapping their prey. Parasitic plants, by contrast, get some or all of their sustenance from other plants. Holoparasites cannot function as autotrophs, having lost chloroplast genes coding for photosynthetic machinery. At least one parasitic plant responds to the same signaling molecule that the host plant uses to attract beneficial fungi.

- Why do carnivorous plants eat animals? See p. 752
- How do the needs of holoparasitic plants differ from those of carnivorous plants? See p. 752
- What characteristics are shared among plant–parasite, plant–fungus, and plant–bacteria associations? See p. 753

What progress has been made in improving the nitrogen use efficiency of corn?

**ANSWER**

There has been significant progress in developing corn varieties with improved nitrogen use efficiency, using both conventional plant breeding (genetics) and biotechnology. The recent publication of the corn genome sequence will provide useful information about genes involved in nitrogen use. For example, the seed company Pioneer Hi-Bred is developing a strain of corn that produces a more efficient version of the enzyme glutamine synthetase (GS). GS adds ammonia to glutamate to form glutamine, and therefore plays an important role in nitrogen assimilation in plants. The new corn strain uses up to 20 percent less fertilizer to produce the same yields as other corn varieties.

## CHAPTER SUMMARY 36

### 36.1 What Nutrients Do Plants Require?

- Plants are photosynthetic autotrophs that can produce all their organic molecules from carbon dioxide, water, and minerals, including a nitrogen source.
- **Mineral nutrients** are obtained from the **soil solution**.
- Plants require 14 **essential elements**, 6 of which are **macronutrients** and 8 of which are **micronutrients**. **Deficiency symptoms** suggest what essential element a plant lacks. **Review Table 36.1, Figure 36.1, ANIMATED TUTORIAL 36.1**
- The essential elements were discovered by growing plants **hydroponically**, meaning in solutions that lacked individual elements. **Review Figure 36.2**

### 36.2 How Do Plants Acquire Nutrients?

- Root growth allows plants, which are sessile, to search for mineral resources.
- Plants can regulate the uptake of nutrients by increasing the number or activity of active transport proteins in root epidermal cells. **Review Figure 36.3**

### 36.3 How Does Soil Structure Affect Plants?

- **Soils** contain water, air, and inorganic and organic substances. Soils have living (biotic) and nonliving (abiotic) components. **Review Figure 36.4**
- A soil typically consists of two or three horizontal zones called **horizons**. **Topsoil** forms the uppermost or **A horizon**. Topsoil tends to lose mineral nutrients through **leaching**. **Loams** are excellent agricultural topsoils, with a good balance of sand, silt, clay, and organic matter. **Review Figure 36.5**
- Soils form by mechanical and chemical **weathering** of rock. Chemical weathering imparts mineral nutrients to clay particles. Plant litter and other organic matter decompose to form **humus**. Plants obtain some mineral nutrients through **cation exchange** between the soil solution and the surface of clay particles. **Review Figure 36.6**
- Farmers use **fertilizers** to make up for deficiencies in soil mineral nutrient content.

*continued*

## 36.4 How Do Fungi and Bacteria Increase Nutrient Uptake by Plant Roots?

- Mycorrhizae are symbiotic root–fungus associations that greatly increase a plant's absorption of water and minerals, especially phosphorus. They occur in more than 90 percent of terrestrial plant species.
- The arbuscules are the sites of nutrient exchange between the fungus and plant. **Review Figure 36.7**
- In the earliest stages of mycorrhiza formation, the hyphae of arbuscular fungi grow toward **strigolactones**, compounds that are produced by the plant roots.
- Some **nitrogen fixers** live free in soil or water; others live symbiotically as **bacteroids** within plant roots. The formation of a root nodule requires interaction between the root system of a legume and rhizobia. **Review Figure 36.7**
- Several steps in the formation of root nodules and arbuscules are similar and probably involve some of the same plant genes. **Review Figure 36.8**
- In **nitrogen fixation**, nitrogen gas ($N_2$) is reduced to ammonia ($NH_3$) or ammonium ions ($NH_4^+$) in a reaction catalyzed by **nitrogenase**. **Review Figure 36.9**
- Plants and bacteria interact in the **global nitrogen cycle**, which involves a series of reductions and oxidations of nitrogen-containing molecules. **Review Figure 36.10, ACTIVITY 36.1**

## 36.5 How Do Carnivorous and Parasitic Plants Obtain a Balanced Diet?

- **Carnivorous plants** are autotrophs that supplement a low nitrogen supply by feeding on insects or other small animals.
- **Parasitic plants** draw on other plants to meet their needs, which may include minerals, water, or the products of photosynthesis.
- **Hemiparasites**, such as mistletoes, can still photosynthesize. **Holoparasites** cannot function as autotrophs because they have lost chloroplast genes that code for components of the photosynthetic apparatus (which they no longer need).
- A strigolactone—a compound in the same category of compounds plants use to attract mycorrhizal fungi—also induces the germination of some parasitic plants, including *Striga*. Scientists hypothesize that a mechanism evolved in the ancestors of modern *Striga* to recognize a compound that was already produced by plants to attract arbuscular fungi.

**Go to the Interactive Summary to review key figures, Animated Tutorials, and Activities**
**Life10e.com/is36**

# CHAPTERREVIEW

## REMEMBERING

1. Macronutrients
   a. are so called because they are more essential than micronutrients.
   b. include manganese, boron, and zinc, among others.
   c. function as catalysts.
   d. are required in concentrations of at least 1 gram per kilogram of plant dry matter.
   e. are obtained by the process of photosynthesis.

2. Which of the following is *not* an essential mineral element for plants?
   a. Potassium
   b. Magnesium
   c. Calcium
   d. Lead
   e. Phosphorus

3. Fertilizers
   a. always have a defined chemical composition.
   b. are not required if crops are removed frequently enough.
   c. restore needed mineral nutrients to the soil.
   d. are needed to provide carbon, hydrogen, and oxygen to plants.
   e. are needed to destroy soil pests.

4. In a typical soil,
   a. the topsoil tends to lose mineral nutrients by leaching.
   b. there are four or more horizons.
   c. the C horizon consists primarily of loam.
   d. the dead and decaying organic matter gathers in the B horizon.
   e. more clay means more air space and thus more oxygen for roots.

5. Nitrogen fixation is
   a. performed only by plants.
   b. the oxidation of nitrogen gas.
   c. catalyzed by the enzyme nitrogenase.
   d. a single-step chemical reaction.
   e. possible because $N_2$ is a highly reactive substance.

6. Which of the following is true of the formation of *both* arbuscules and root nodules?
   a. Invasion of a plant root by a fungus
   b. Invasion of a plant root by a bacterium
   c. Strigolactones produced by the root are recognized by the microbe
   d. Root cells are invaded but there is no direct contact between plant and microbe cell contents
   e. Root cells are invaded and there is direct contact between plant and microbe cell contents

## UNDERSTANDING & APPLYING

7. Methods for determining whether a particular element is essential have been known for more than a century. Since these methods are so well established, why was the essentiality of some elements discovered only recently?

8. Soils are dynamic systems. What changes might result when land is subjected to heavy irrigation for agriculture after being relatively dry for many years? What changes in the soil might result when a virgin deciduous forest is cut down and replaced by crops that are harvested each year?

9. The biosphere of Earth as we know it depends on the existence of a few species of nitrogen-fixing prokaryotes. What do you think might happen if one of these species were to become extinct? If all of them were to disappear? (See also Figure 58.13.)

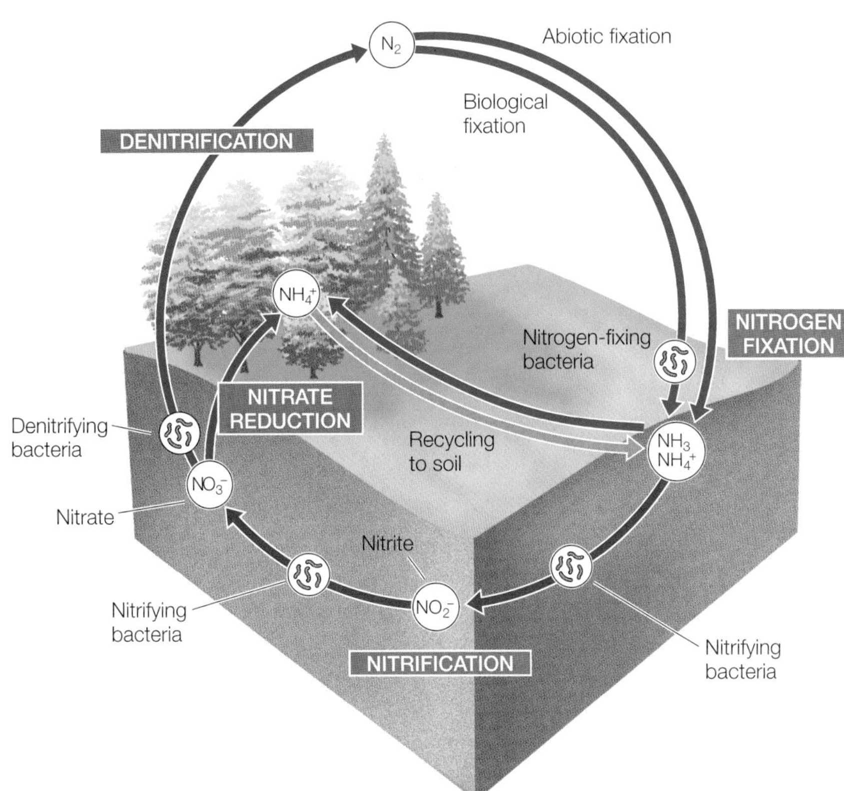


## ANALYZING & EVALUATING

10. Some mutant *Arabidopsis* plants that are very bushy (their shoots are more highly branched than wild-type plants) cannot make strigolactones because of a mutation in a gene necessary for strigolactone biosynthesis. If an investigator applies strigolactones to the plants, they grow normally. What does this experiment suggest about the role of strigolactones in plant growth? How does this add to the story of strigolactones as signals for arbuscules and parasitic plants?

11. Holoparasitic plants have lost many of the morphological and genetic traits necessary for an autotrophic lifestyle. From an evolutionary point of view, how do you think this happened? (Hint: think about selection pressures.)

Go to BioPortal at **yourBioPortal.com** for Animated Tutorials, Activities, LearningCurve Quizzes, Flashcards, and many other study and review resources.

# 37 Regulation of Plant Growth

## CHAPTER**OUTLINE**


**37.1 How Does Plant Development Proceed?**

**37.2 What Do Gibberellins and Auxin Do?**

**37.3 What Are the Effects of Cytokinins, Ethylene, and Brassinosteroids?**

**37.4 How Do Photoreceptors Participate in Plant Growth Regulation?**

**Norman Borlaug** Seen here in a field of semi-dwarf wheat, plant geneticist Norman Borlaug carried out a program of genetic crosses that led to high-yielding varieties and saved millions from starvation.

AGRICULTURAL SCIENTISTS are constantly searching for ways to help farmers produce more food for a growing population. One way is to breed crop plants whose physiology allows them to produce more grain per plant (resulting in higher yields). The drawback of this approach is that the sheer weight of the load of seeds may cause the stem to bend over. The problem is made worse when fertilizer makes the plants grow taller. Harvesting seeds on the ground is very difficult; think of how hard it would be to pick up seeds one by one, when some have already sprouted.

During World War II, the island nation of Japan was blockaded and could not import food or other supplies. Food was rationed and many people were hungry, but there were no major famines in Japan during that period. How were the Japanese able to produce enough grain to feed their population? One answer to this question lay in the fields: the Japanese had bred genetic strains of rice and wheat with short, strong stems that could bear high yields of grain without bending over. An agricultural advisor to the occupying American army sent samples of the grains to the U.S.

A decade later, the American plant geneticist Norman Borlaug, who was working in Mexico at the time, began making genetic crosses between the Japanese wheat and other varieties that had genes conferring rapid growth, adaptability to varying climates, and resistance to fungal diseases. The results were "semi-dwarf" wheat varieties that gave record yields. The varieties were grown first in Mexico, and later in India and Pakistan during the 1960s. At about the same time and using a similar strategy, scientists in the Philippines developed semi-dwarf rice with equally spectacular results. People who had lived on the edge of starvation now produced enough food. Countries that had relied on food from other countries were now able to grow more than enough grain, and export the surplus. The development of these semi-dwarf grains began what was called the "Green Revolution." Borlaug was awarded the Nobel Peace Prize for his research on wheat, which is estimated to have saved a billion lives.

?

What changes in growth patterns made the new strains of wheat and rice successful?

See answer on p. 775.

# 37.1 How Does Plant Development Proceed?

As described in Chapter 34, plants are sessile organisms that must seek out resources above and below the ground. A number of unique features enable plants to obtain the resources they need to grow and reproduce:

- *Meristems*. Plants have permanent collections of stem cells (undifferentiated, constantly dividing cells) that allow them to continue growing throughout their lifetimes (see Section 34.3).
- *Post-embryonic organ formation*. Unlike animals, plants can initiate development of new organs such as leaves and flowers throughout their lifetimes.
- *Differential growth*. Plants can allocate their resources so that they grow more of the organs that will benefit them most—for example, more leaves to harvest more sunlight or more roots to obtain more water and nutrients.

Plants must continuously monitor their ever-changing environments and redirect their growth appropriately. For example, the amount of available light changes from day to night and from season to season. In addition, other plants are often vying for what light there is, and plants modulate their growth to compete with their neighbors for this precious resource. As you will see in this chapter, several mechanisms have evolved in plants that enable them to sense changes in their environments, and trigger appropriate growth responses.

The development of a plant—the series of progressive changes that take place throughout its life—is regulated in many ways. Key factors involved in regulating plant growth and development are:

- *Environmental cues*, such as day length, water availability, and various chemicals in the environment
- *Receptors* that allow a plant to sense environmental cues, such as photoreceptors that absorb light, and chemoreceptors that signal the presence of pathogens (see Chapter 39)
- *Hormones*—chemical signals that mediate the effects of the environmental cues, including those sensed by receptors
- *Regulatory proteins and enzymes* that catalyze the biochemical reactions of development

We will explore these regulatory mechanisms in more detail later in this chapter. But first let's look at the initial steps of plant development—from seed to seedling—and the types of internal and external cues that guide them.

## In early development, the seed germinates and forms a growing seedling

Chapter 38 will describe the events of plant reproduction and development that lead to the formation of seeds. Here we begin with the seed, the structure that contains the early embryo. Unlike most animal embryos, plant seeds may be held in "suspended animation," with the development of the embryo halted, for long periods. If development stops even when external conditions (such as water supply) are adequate for development, the seed is said to be **dormant**.

**DORMANCY** Seed dormancy may last for weeks, months, or years. Plants use several mechanisms to maintain dormancy:

- *Exclusion of water or oxygen* from the embryo by an impermeable seed coat
- *Mechanical restraint* of the embryo by a tough seed coat
- *Chemical inhibition* of germination by growth regulators
- *Photodormancy*: some seeds need a period of light or dark before they can germinate
- *Thermodormancy*: some seeds need high or low temperatures to germinate

Dormancy can be broken by factors that overcome these mechanisms. For example, the seed coat may be damaged by passage through an animal's digestive system, or heavy rains may wash away chemical inhibitors. There are many unusual methods to overcome dormancy. One interesting example is the breaking of dormancy by components of smoke. *Emmenanthe penduliflora* is a common plant in dry chaparral of the southwestern U.S., an area that is prone to wildfires.

These plants germinate rapidly after a fire. John Keeley of Occidental College in Los Angeles found that dormancy in seeds of this plant is broken not by heat but by smoke—in particular, by the nitrogen oxides found in smoke. Other molecules in smoke have been identified that regulate seed germination.

Plant biologists distinguish between seed *dormancy*, which prevents germination under conditions that are suitable for plant growth, and seed *quiescence*, which occurs when a seed fails to germinate because conditions are unfavorable for growth. Some seeds may remain quiescent, yet viable, for centuries: botanists have germinated a 1,300-year-old lotus seed recovered from a dry lake bed in China.

Seed dormancy and quiescence are common, so they must provide selective advantages for plants. Dormancy ensures that the seed will germinate at a time suitable for the plant to complete its life cycle. For example, some seeds require exposure to a long cold period (winter) before they germinate in the spring; this ensures that the plant has the entire growing season to mature and set new seeds. Dormancy and quiescence also help seeds survive droughts or long-distance dispersal, allowing plants to colonize new territory.

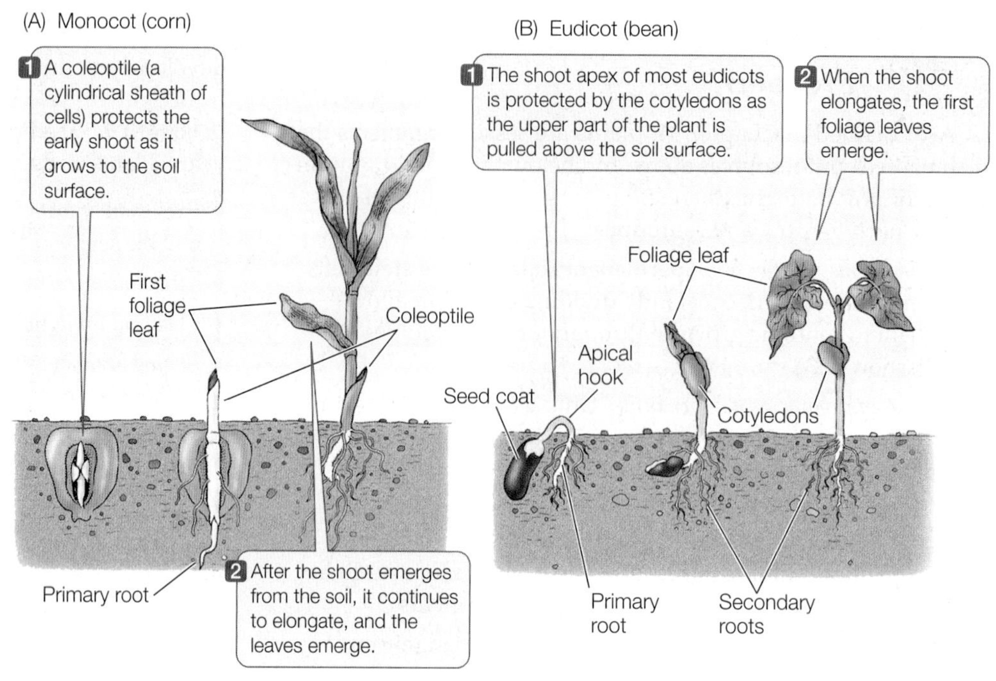


**37.1 Patterns of Early Shoot Development** (A) In grasses and some other monocots, growing shoots are protected by a coleoptile until they reach the soil surface. (B) In most eudicots, the growing point of the shoot is protected within the cotyledons.

**Go to Activity 37.1 Monocot Shoot Development Life10e.com/ac37.1**

**Go to Activity 37.2 Eudicot Shoot Development Life10e.com/ac37.2**

GERMINATION Seeds begin to **germinate**, or sprout, when dormancy is broken and environmental conditions are satisfactory. The first step in germination is the uptake of water, called **imbibition** (from *imbibe*, "to drink in"). Before germination, a seed contains very little water: only 5 to 15 percent of its weight is water, compared with 80 to 95 percent for most other plant parts. Seeds also contain polar macromolecules, such as cellulose and starch, which attract and bind polar water molecules. Consequently a seed has a very negative water potential (see Chapter 35) and will take up water if the seed coat is permeable to water. The force exerted by imbibing seeds, which expand several-fold in volume, demonstrates the magnitude of their water potential; for example, imbibing cocklebur seeds can exert a pressure of up to 1,000 atmospheres.

As a seed takes up water, it undergoes metabolic changes: enzymes are activated upon hydration, RNA and then proteins are synthesized, the rate of cellular respiration increases, and other metabolic pathways are activated. In many seeds, cell division is not initiated during the early stages of germination. Instead, growth results solely from the expansion of small, preformed cells.

As germination proceeds, starch, proteins, and lipids that are stored in the seed are hydrolyzed to provide metabolic energy and chemical building blocks—carbohydrates, amino acids, and lipid monomers—for the growing embryo. These reserves are stored in the cotyledons (the first leaf or leaves of the embryo) or in the endosperm (the non-embryonic storage tissue of the seed). Germination is completed when the **radicle** (embryonic root) emerges from the seed coat. The plant is then called a **seedling**.

If the seed germinates underground, the new seedling must elongate rapidly (in the right direction!) and cope with a period of life in darkness or dim light. Photoreceptors that sense light and specialized cells that sense gravity direct this stage of development and prepare the seedling for growth in the light.

The pattern of early shoot development varies among the flowering plants. **Figure 37.1** shows the shoot development patterns of monocots and eudicots. In monocots, the growing shoot is protected by a sheath of cells called the **coleoptile** as it pushes its way through the soil. In dicots, the shoot is protected by the cotyledons.

## Several hormones and photoreceptors help regulate plant growth

The above description of the early stages of plant development illustrates some of the environmental cues that influence plant growth. A plant's responses to these cues involve signal transduction pathways. Various mechanisms are used by the plant to sense changes in the environment, and these mechanisms activate signal transduction pathways that result in the synthesis and activation of specific plant hormones. In turn, these hormones act as signals that trigger pathways resulting in changes in plant growth. In some cases these changes involve alterations in the expression of specific genes.

**Hormones** are chemical signals that act at very low concentrations at sites often distant from where they are produced. Most plant hormones are very different from their counterparts in animals (**Table 37.1**). Each plant hormone plays multiple

TABLE **37.1**
Comparison of Plant and Animal Hormones

| Characteristic | Plant Hormones | Animal Hormones |
|---|---|---|
| Size, chemistry | Small organic molecules | Peptides, proteins, small molecules |
| Site of synthesis | Many locations in the plant | Specialized glands or cells |
| Site of action | Local or distant | Distant, transported |
| Effects | Often diverse | Often specific |
| Regulation | By biochemical feedback | By feedback and by central nervous system |

TABLE 37.2
Plant Growth Hormones

| Hormone | Structure | Typical Activities |
|---|---|---|
| Abscisic acid* | 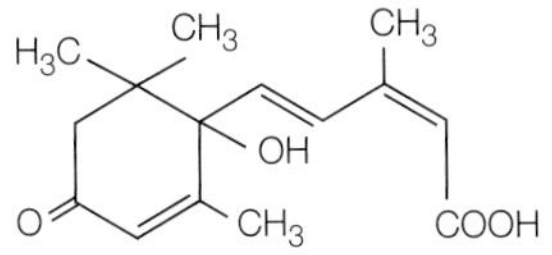 | Maintains seed dormancy; closes stomata |
| Auxins (mainly indole-3-acetic acid) | 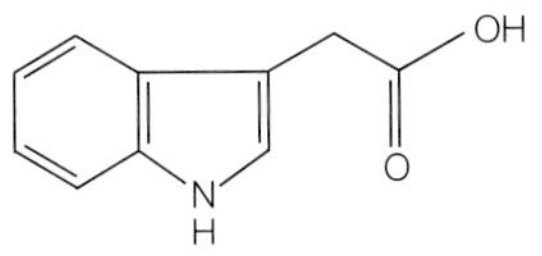 | Promote stem elongation, adventitious root initiation, and fruit growth; inhibit axillary bud outgrowth, leaf abscission, and root elongation |
| Brassinosteroids | 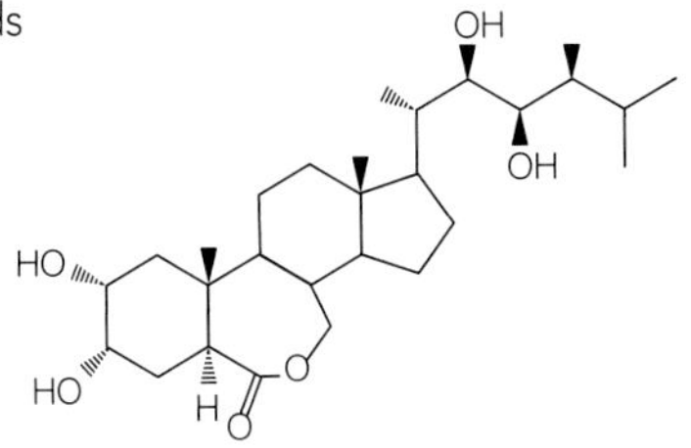 | Promote stem and pollen tube elongation; promote vascular tissue differentiation |
| Cytokinins | 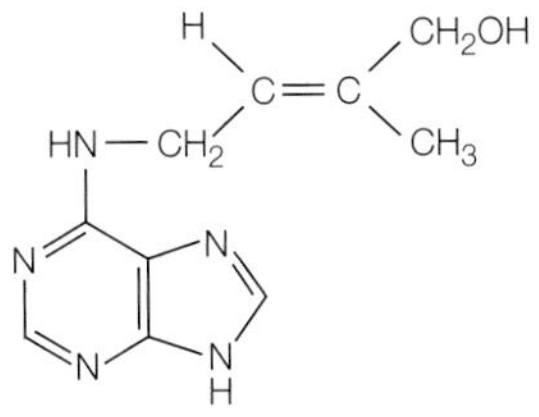 | Inhibit leaf senescence; promote cell division and axillary bud outgrowth; affect root growth |
| Ethylene | 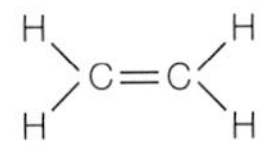 | Promotes fruit ripening and leaf abscission; inhibits stem elongation and gravitropism |
| Gibberellins (e.g., gibberellic acid) | 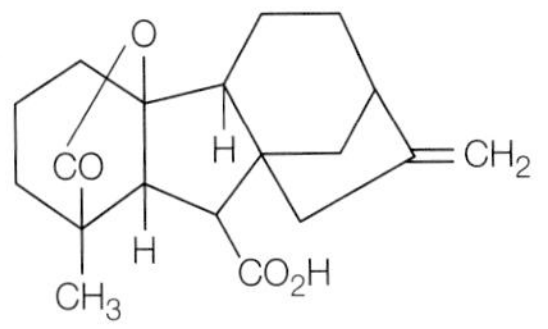 | Promote seed germination, stem growth, and ovule and fruit development; break winter dormancy; mobilize nutrient reserves in grass seeds |

*See Chapter 38.

regulatory roles, and interactions among them can be complex. Several hormones regulate the growth and development of plants from seedling to adult (**Table 37.2**). Other hormones (for example, jasmonic acid and salicylic acid) are involved in the plant's defenses against herbivores and microorganisms, as we will discuss in Chapter 39.

Perhaps the most important environmental cue for a plant is light: the source of energy for photosynthesis. Plants have an abundance of **photoreceptors** that detect changes in the quality and direction of light as well as the timing of light availability (daylength). Photoreceptors are often proteins associated with pigments. Light acts directly on photoreceptors, which in turn regulate developmental processes that need to be responsive to light, such as the many changes that occur as a seedling emerges from the soil.

## Genetic screens have increased our understanding of plant signal transduction

In Chapter 19 we described how genetic studies can be used to identify the steps along a developmental pathway. The reasoning behind these experiments is that if a mutation in a specific gene disrupts a developmental process, then the product of that gene must be involved in that process. Similarly, genetic studies can be used to analyze pathways of receptor activation and signal transduction in plants: if proper signaling does not occur in a mutant plant, then the mutant gene must be involved in the signal transduction process. Mapping the mutant gene and identifying its function is a starting point for understanding the signaling pathway. *Arabidopsis thaliana* has been a major model organism for plant biologists investigating signal transduction.

One technique for identifying the genes involved in a plant signal transduction pathway is illustrated in **Figure 37.2**. This technique, called a **genetic screen**, involves creating a large, random collection of mutant plants and identifying those individuals that are likely to have a defect in the pathway of interest. Plant genes can be randomly mutated in a variety of ways, including treatment with a chemical mutagen or the insertion of transposons (see Section 17.2) randomly in the genome. After treatment, the plants are grown and then examined for a specific phenotype, usually a characteristic that is influenced by the pathway of interest. Once mutant plants have been selected, their genotypes are compared with those of wild-type plants. *Arabidopsis* mutants with altered developmental patterns have provided a wealth of new information about the mechanisms of hormone and receptor (particularly photoreceptor) action.

### RECAP 37.1

Plant development is under the control of external cues in the environment and is mediated by hormones. These signals activate pathways that may result in changes in gene expression. Genetic screens have been useful in describing signal transduction pathways in the model plant *Arabidopsis thaliana*. Seed dormancy often precedes seed germination.

- Under what circumstances is seed dormancy advantageous? **See p. 757**
- Describe how monocots and eudicots differ in early development. **See p. 758 and Figure 37.1**
- What is a genetic screen, and how can it be used to analyze the regulation of plant development? **See pp. 759–760 and Figure 37.2**

RESEARCH**TOOLS**

**37.2 A Genetic Screen** Genetics of the model plant *Arabidopsis thaliana* can be used to identify the steps of a signal transduction pathway. If a mutant strain does not respond to a hormone (in this case, ethylene), the corresponding wild-type gene must be essential for the pathway (in this case, ethylene response). This method has been instrumental to scientists in understanding plant growth regulation.

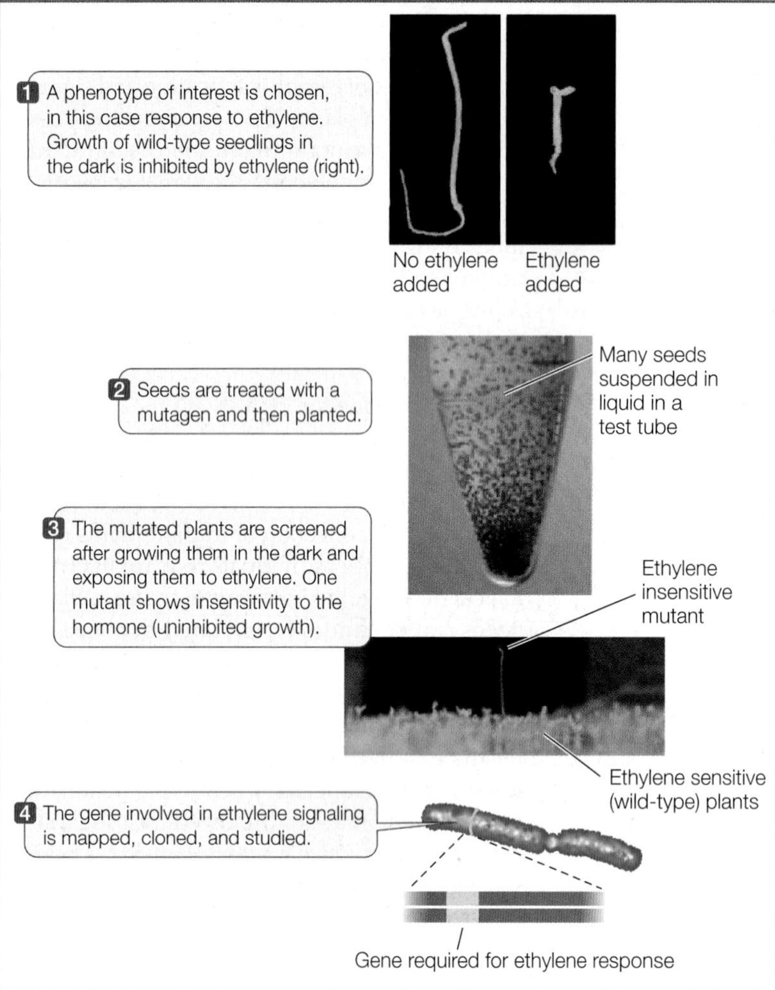


You have now seen the early stages of plant development and growth, and how the environment influences these processes. We will now turn to the subject of plant hormones, which are central to the internal regulation of development. We will describe how hormones were discovered and what physiological effects they have on plants. We will emphasize how genetic screens and other methods have led to a deeper molecular understanding of the action of plant hormones.

## 37.2 What Do Gibberellins and Auxin Do?

The discovery of two key plant hormones exemplifies the experimental approaches that plant biologists have used to investigate the mechanisms of plant development. **Gibberellins** (there are several active forms) and **auxin** (there are several forms, but one predominates) were the first plant hormones to be discovered, early in the twentieth century. Initially, the discoveries came from observations of natural phenomena:

- *Gibberellins*: In rice plants, a disease caused by the fungus *Gibberella fujikori* resulted in plants that grew overly tall and spindly.
- *Auxin*: Biologists and indoor gardeners noted that seedlings would bend toward the light when placed near a light source.

A chemical substance was then isolated that could cause each phenomenon:

- Gibberellic acid (see Table 37.2) made by the *G. fujikori* fungus caused rice plants to overgrow. Later it was found that plants make gibberellic acid as well, and that applying it to plants caused growth.
- Indole-3-acetic acid (see Table 37.2) applied asymmetrically to the growing tips of seedlings caused cell elongation on the side away from the light, which resulted in the shoot bending toward the light.

Finally, mutant plants that do not make each hormone exhibit a phenotype expected in the absence of the hormone, and adding the hormone reverses that phenotype (**Figure 37.3**):

- Tomato plants that do not make gibberellic acid are very short; supplying them with the hormone results in normal growth.
- *Arabidopsis thaliana* individuals that do not make auxin are also short; supplying them with that hormone reverses that phenotype.

Note that the phenotype involved—short stature, or dwarfism—is similar in both mutant plants. This observation exemplifies a concept that is important to keep in mind when studying plant hormones: their actions are not unique and specific, as is the case with animal hormones (see Table 37.1).

The approaches outlined above—observation, hormone isolation, hormone treatment, and analysis of mutants—are just some of the methods used to identify plant hormones and understand their roles in plant development. Plant biologists have also studied hormones using chemical inhibitors and using plant transformation experiments that alter hormone levels or the plants' responses to hormones.

### Gibberellins have many effects on plant growth and development

The functions of gibberellins can be inferred from the effects of experimentally decreasing gibberellins or blocking their action at various points in plant development. Such experiments reveal that gibberellins have multiple roles in regulating plant growth.

**37.3 Hormones Reverse a Mutant Phenotype** (A) The two mutant dwarf tomato plants in this photograph were the same size when the one on the right was treated with a gibberellin solution. (B) The short phenotype of this *Arabidopsis* mutant was reversed in the plant on the right by supplying auxin.

**STEM ELONGATION** The effects of gibberellins on wild-type plants are not as dramatic as those seen on dwarf plants. However, gibberellins are indeed active in wild-type plants, because inhibitors of gibberellin synthesis cause a reduction in stem elongation. Such inhibitors can be put to practical uses. For example, plants such as chrysanthemums that are grown in greenhouses tend to get tall, but leggy plants do not appeal to consumers. Flower growers spray such plants with gibberellin synthesis inhibitors to control their height. Some wheat crops are similarly sprayed to keep them short, so they do not fall over when they produce grain; the result is chemically produced semi-dwarfs similar to the genetically produced varieties described in the opening story of this chapter. In some plants, such as cabbage, the normal growth habit is to be a squat, leafy head near the ground. When environmental signals are right, however, the plant "bolts," quickly producing a tall stem with flowers—a response that can be mediated by gibberellins.

**FRUIT GROWTH** Gibberellins and other hormones regulate the growth of fruits. Grapevines that produce seedless grapes develop smaller fruits than varieties that produce seed-bearing grapes. Biologists wanting to explain this phenomenon removed seeds from immature seeded grapes and found that this prevented normal fruit growth, suggesting that the seeds are sources of a growth regulator. Biochemical studies showed that developing seeds produce gibberellins, which diffuse into the immature fruit tissue. Spraying young seedless grapes with a gibberellin solution causes them to grow as large as seeded ones, and this is now a standard commercial practice (**Figure 37.4**).

**MOBILIZATION OF SEED RESERVES** Early in seed germination, hydrolytic enzymes are produced to break down stored reserves of starch, proteins, and lipids. Just after imbibition in germinating seeds of barley and other cereals, the embryo secretes gibberellins. The hormones diffuse through the endosperm to a surrounding tissue called the **aleurone layer**, which lies underneath the seed coat. The gibberellins trigger a cascade of events in the aleurone layer, causing it to synthesize and secrete enzymes that hydrolyze proteins and starch stored in the endosperm (**Figure 37.5**). These observations have practical importance: in the beer brewing industry, gibberellins are used to enhance the "malting" (germination) of barley and the breakdown of its endosperm, producing sugar that is fermented to alcohol.

**37.4 Gibberellins and Fruit Growth** Spraying developing seedless grapes with gibberellins (right) increases their size compared with untreated fruit (left).

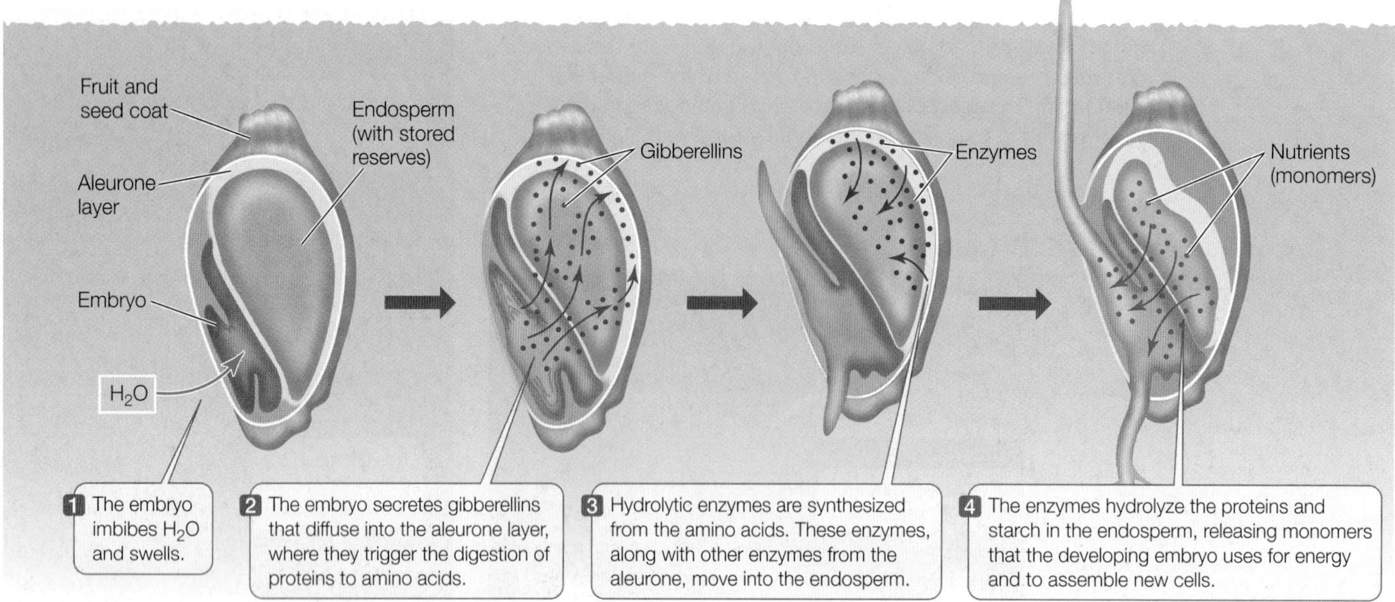


**37.5 Embryos Mobilize Their Reserves** During seed germination in cereal grasses, gibberellins trigger a cascade of events that result in the conversion of starch and protein reserves into monomers that can be used by the developing embryo.

**Go to Activity 37.3 Events of Seed Germination Life10e.com/ac37.3**

Another hormone, abscisic acid (ABA) (see Table 37.2) plays an antagonistic role with gibberellins in seed dormancy and germination. As we will describe in Section 38.1, ABA levels are high in dormant seeds and fall off during germination. Thus ABA plays a role in maintaining seed dormancy, whereas gibberellins function to break dormancy and promote germination.

## Auxin plays a role in differential plant growth

Auxin was first discovered for its role in the bending of plants toward light. Subsequent research has shown that auxin is involved in many other aspects of plant growth and development.

IDENTIFYING AUXIN AND ITS TRANSPORT Auxin (from the Latin "to increase") was discovered in the context of **phototropism**: a response to light in which plant stems bend toward a light source. This was a familiar observation by biologists and home gardeners when the ever-curious Charles Darwin and his son Francis investigated this phenomenon and concluded that a signal was made in the shoot apex and diffused down the shoot in a polar (unidirectional) fashion, stimulating cell elongation.

**Go to Animated Tutorial 37.1 Tropisms**
Life10e.com/at37.1

The Darwins worked with canarygrass (*Phalaris canariensis*) seedlings grown in the dark. While underground, a young grass seedling is protected by a coloeptile (see Figure 37.1). The coleoptiles of grasses are phototropic—they grow toward the light.

To find the light-receptive region of the coleoptile, the Darwins "blindfolded" the coleoptiles of dark-grown canarygrass seedlings in various places and then illuminated them from one side (**Figure 37.6**). The coleoptile grew toward the light whenever its tip was exposed. If the top millimeter or more of the coleoptile was covered, however, the coleoptile showed no phototropic response. The Darwins concluded that the tip contains the photoreceptor that responds to light. The actual bending toward the light, however, takes place in a growing region a few millimeters below the tip. Therefore, the Darwins reasoned, *some type of signal must travel from the tip of the coleoptile to the growing region.*

Others showed that placing the coleoptile tip (the source of the growth signal) on a decapitated coleoptile led to bending, even when a block of gelatin separated the tip and coleoptile.

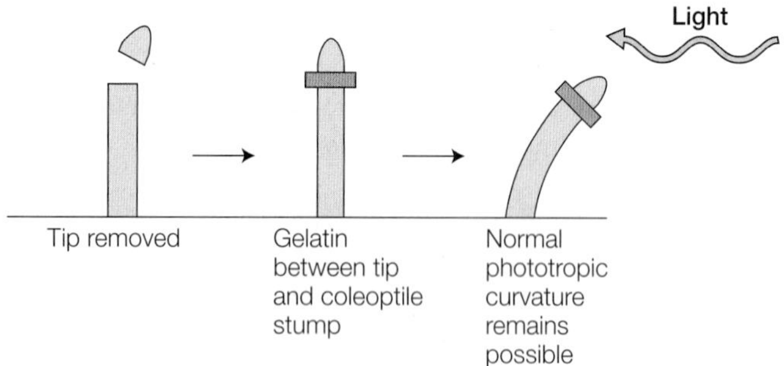


In these experiments, the growth signal moved from the coleoptile tip into the gelatin and then down into the decapitated coleoptile; later the growth signal was isolated from such gelatin blocks and identified as indole-3-acetic acid (see Table 37.2), the major form of auxin.

**Go to Animated Tutorial 37.2 Went's Experiment**
Life10e.com/at37.2

INVESTIGATING**LIFE**

**37.6 The Darwins' Phototropism Experiment** Charles Darwin and his son Francis wanted to know how plants bend toward the light. They grew canarygrass seedlings (coleoptiles) in the dark. To discover what part of the coleoptile responds to light, they covered up ("blindfolded") different regions of each coleoptile and then exposed the seedlings to light from one side. The Darwins discovered that the tip of the seedling senses the light and that growth occurs below the tip. Their observations led them to hypothesize the existence of a growth-promoting signal produced by the coleoptile tip.[a]

**HYPOTHESIS** Only part of the coleoptile senses the light that triggers phototropism.

*Method*

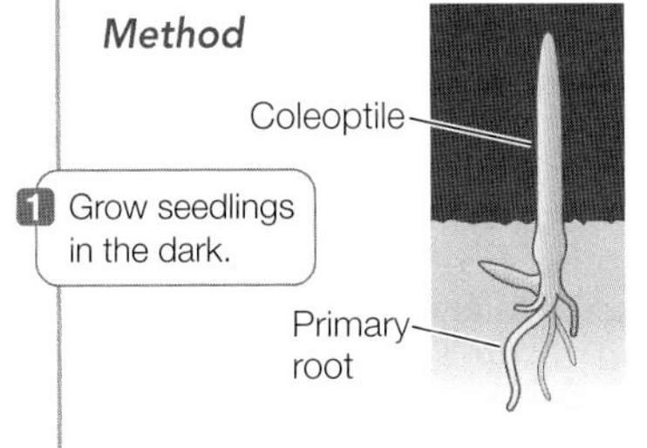


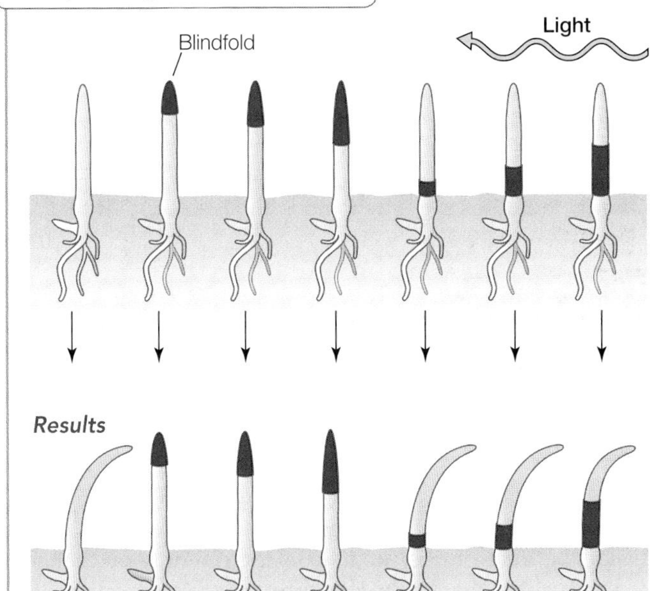


**CONCLUSION** The part of the coleoptile that senses light is in the tip, and it sends a signal from the tip to the growing region.

Go to **BioPortal** for discussion and relevant links for all INVESTIGATING**LIFE** figures.

[a]Darwin, C. R. 1880. *The Power of Movement in Plants.* London, John Murray.

**MECHANISM OF AUXIN TRANSPORT** The movement of auxin down a coleoptile is an example of polar (apical-to-basal) transport. Polar transport of auxin (**Figure 37.7**) depends on four biochemical processes that may be familiar from earlier chapters:

- *Diffusion across a plasma membrane.* Polar molecules (in the chemical sense) diffuse across plasma membranes less readily than nonpolar molecules (see Chapter 6).
- *Membrane protein asymmetry.* Active transport carriers (see Chapter 6) for auxin are located only in the portion of the plasma membrane at the basal (bottom) end of the cell.
- *Proton pumping/chemiosmosis.* A proton pump (see Chapter 36) moves $H^+$ from the cytoplasm to the cell wall, thereby increasing the intracellular pH and decreasing the pH in the cell wall. Proton pumping also sets up an electrochemical gradient, which provides potential energy to drive the transport of auxin by the carriers mentioned above (see Chapter 9).
- *Ionization of a weak acid.* The main form of auxin, indole-3-acetic acid, is a weak acid:

$$A^- + H^+ \rightleftharpoons HA$$

When the pH is low, this reaction is driven to the right, and HA (non-ionized auxin) is the predominant form. When the pH is higher, there is more $A^-$ (ionized auxin).

While polar auxin transport distributes the hormone along the *longitudinal* axis of the plant, *lateral* (side-to-side) redistribution

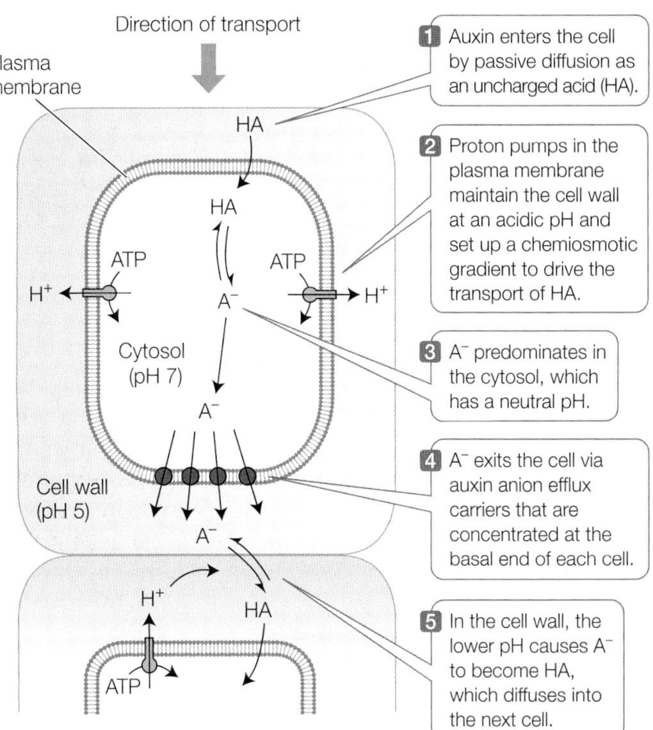


**37.7 Polar Transport of Auxin** Proton pumps set up a chemiosmotic gradient directing ionized auxin ($A^-$) toward the basally placed active transport carriers for auxin, which leads to a net movement of auxin in a basal direction.

## WORKING WITH**DATA:**

### *The Darwins' Phototropism Experiment*

#### Original Paper

Darwin, C. R. 1880. *The Power of Movement in Plants.* Chapter IX: Sensitiveness of plants to light: Its transmitted effects. London, John Murray.

#### Analyze the Data

Fascinated by experiments showing that light induces plants to bend toward it and that this is due to increased growth on the "dark side" of the stem, Charles Darwin and his son Francis performed experiments to learn more about how plants respond to light. Their experiments were done on canarygrass (*Phalaris canariensis*) seedlings grown in the dark to simulate events that occur when the seed germinates in the soil. These seedlings have a sheath—the coleoptile—that covers the developing leaf. When placed in a box with an opening at a window (unidirectional light), the seedlings bent toward the light source. This bending seemed to be led by the very tip of the coleoptile, and the Darwins hypothesized that the light-sensing mechanism might reside in the coleoptile, while growth occurs in tissue below the tip. Their hypothesis was validated when they repeated their experiments with the coleoptile tip cut off: bending did not occur. Because it was possible that the cut plants did not bend because they had been irreversibly damaged in some way, the Darwins repeated their initial experiment with undamaged seedlings whose coleoptile tips had been covered with a thin glass tube cap that was blacked out with India ink. In this case the plants did not bend. The Darwins' conclusion that the tip of the coloptile senses the light and transmits a message to tissue below the tip to grow was initially controversial but ultimately led to the identification of auxin, the first plant growth hormone.

**QUESTION 1**

The figure shows a drawing from the Darwins' book of the bending of coleoptiles after 8 hours of light exposure. From which direction was the light shone?

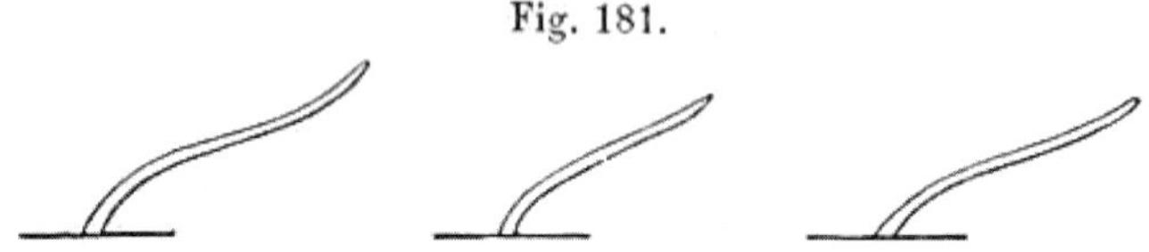


**QUESTION 2**

The Darwins reported:

> "Seven cotyledons [Note: The Darwins used the term "cotyledon" for what is now called a coleoptile.] had their tips cut off for lengths varying between 0.1 and 0.16 of an inch, and these, when left exposed all day to a lateral light, remained upright. In another set of 7 cotyledons, the tips were cut off for a length of only about 0.05 of an inch (1.27 mm) and these became bowed towards a lateral light, but not nearly so much as the many other seedlings in the same pots."

What do these data indicate about the possible role of the tip and about the possibility that injury in cutting blocks the bending response?

**QUESTION 3**

The Darwins describe their further experiments:

> "The summits of nine cotyledons, differing somewhat in height, were enclosed for rather less than half their lengths in uncoloured or transparent tubes; and these were then exposed before a south-west window on a bright day for 8 h. All of them became strongly curved towards the light, in the same degree as the many other free seedlings in the same pots; so that the glass-tubes certainly did not prevent the cotyledons from bending towards the light. Nineteen other cotyledons were, at the same time, similarly enclosed in tubes thickly painted with Indian ink. On five of them, the paint, to our surprise, contracted after exposure to the sunlight, and very narrow cracks were formed, through which a little light entered; and these five cases were rejected. Of the remaining 14 cotyledons, the lower halves of which had been fully exposed to the light for the whole time, 7 continued quite straight and upright; 1 was considerably bowed to the light, and 6 were slightly bowed, but with the exposed bases of most of them almost or quite straight."

What do these data indicate about the role of the tip? Can you explain why there was slight bending in the 6 coleoptiles that were covered with painted tubes?

**QUESTION 4**

The Darwins observed that the leaves of the insect-consuming Venus flytrap do not bend toward light. Why would this response not be important to an insectivorous plant?

**Go to BioPortal for all** WORKING WITH**DATA** **exercises**

of auxin is responsible for directional plant growth. This was shown in early experiments that followed the Darwins' when a coleoptile tip containing the growth hormone was placed asymmetrically on the decapitated coleoptile. The asymmetric distribution of growth hormone down the coleoptile resulted in excess growth on that side, and bending away from it, even in the absence of light:

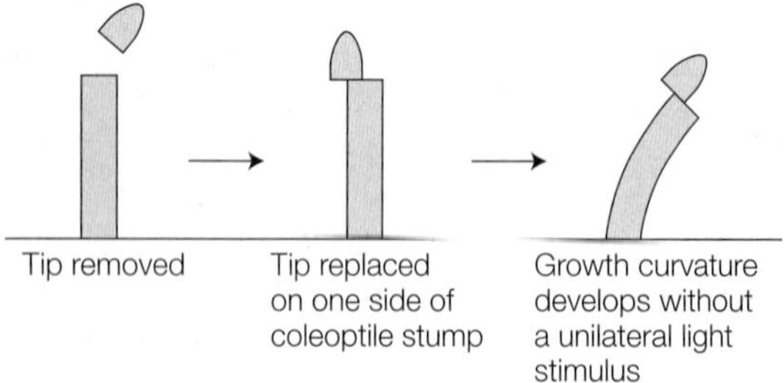


The redistribution of auxin to one side is carried out by auxin carrier proteins that move from the base of the cell to one side; because of this, auxin exits the cell only on that side of the cell, rather than at the base, and moves sideways within the tissue.

This lateral movement of auxin explains the bending of canarygrass seedlings toward light that the Darwins observed. When light strikes a canarygrass coleoptile on one side, auxin at the tip moves laterally toward the shaded side. The asymmetry thus established is maintained as polar transport moves auxin down the coleoptile, so that in the growing region below, the auxin concentration is highest on the shaded side. Cell elongation is thus speeded up on that side, causing the coleoptile to bend toward the light (**Figure 37.8A**).

Light is not the only signal that can cause the redistribution of auxin. Auxin moves to the lower side of a shoot that has been

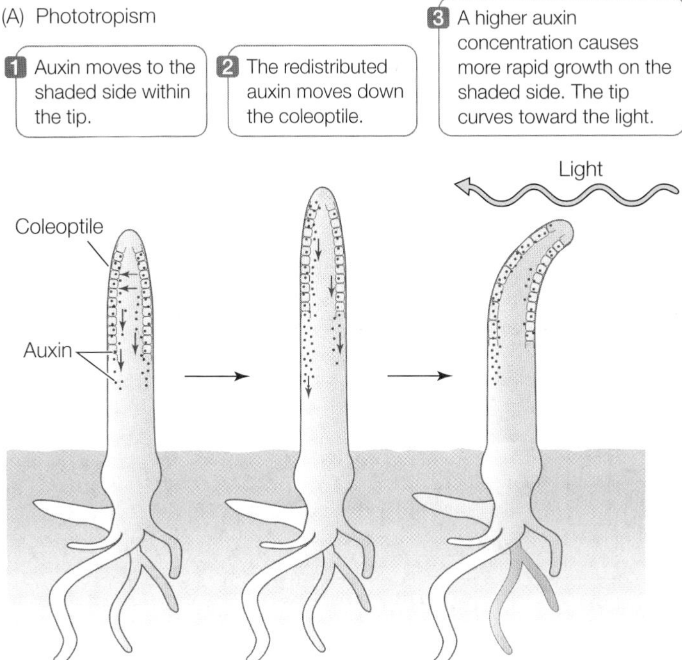


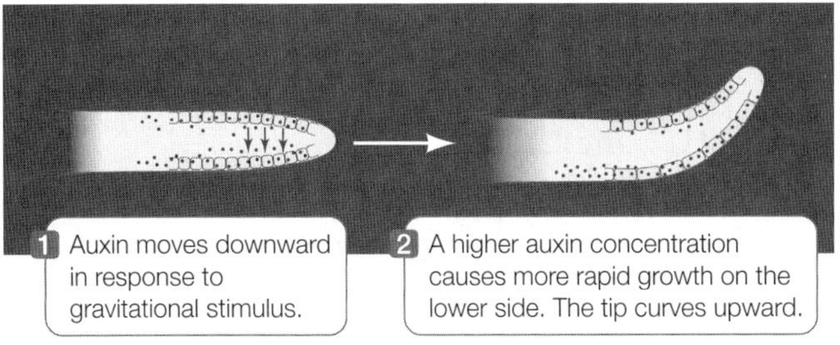


**37.8 Plants Respond to Light and Gravity** Phototropism (A) and gravitropism (B) occur in shoot apices in response to a redistribution of auxin.

tipped sideways, causing more rapid growth in the lower side and hence an upward bending of the shoot. Such growth in a direction determined by gravity is called **gravitropism** (**Figure 37.8B**). The upward gravitropic response of shoots is defined as negative gravitropism; that of roots, which bend downward, is positive gravitropism. Gravitropism in roots also involves differential growth caused by lateral movement of auxin, but the details of the mechanism differ between the root and the shoot.

## Auxin affects plant growth in several ways

Like the gibberellins, auxin has many roles in plant development. It affects the vegetative and reproductive growth of plants in several ways.

ROOT INITIATION Cuttings from the shoots of some plants can produce roots and develop into entire new plants. For this to occur, certain undifferentiated cells in the interior of the shoot, originally destined to function only in food storage, must set off on a new mission: they must change their cell fate and become organized into the apical meristem of a new root. These changes

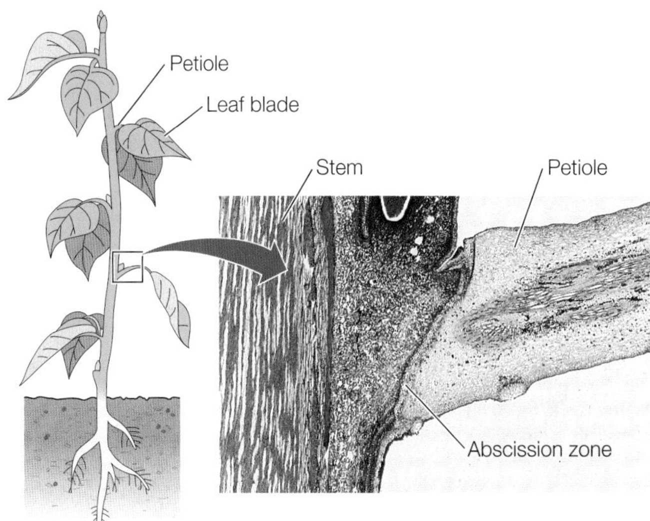


**37.9 Changes Occur When a Leaf Is about to Fall** The breakdown of cells in the abscission zone of the petiole causes the leaf to fall.

are similar to those that take place in the pericycle of a root when a lateral root forms (see Figure 34.10). Shoot cuttings of many species can be made to develop roots by dipping the cut surfaces into an auxin solution. These observations suggest that in an intact plant, the plant's own auxin plays a role in the initiation of lateral roots. Commercial preparations that enhance the rooting of plant cuttings typically contain synthetic auxins.

LEAF ABSCISSION In contrast to its stimulatory effect on root initiation, auxin inhibits the detachment of old leaves from stems. This detachment process, called **abscission**, is the cause of autumn leaf fall. Many leaves consist of a blade and a petiole that attaches the blade to the stem. Abscission results from the breakdown of a specific part of the petiole, the abscission zone (**Figure 37.9**). If the blade of a leaf is cut off, the petiole falls from the plant more rapidly than if the leaf had remained intact. If the cut surface is treated with an auxin solution, however, the petiole remains attached to the plant, often longer than an intact leaf would have. The timing of leaf abscission in nature appears to be determined in part by a decrease in the movement through the petiole of auxin produced in the blade.

APICAL DOMINANCE Auxin helps maintain **apical dominance**, a phenomenon in which apical buds inhibit the growth of axillary buds (see Figure 34.1), resulting in the growth of a single main stem with minimal branching. Apical dominance can be demonstrated by an experiment with a young seedling. If the plant remains intact, the stem elongates and the axillary buds remain inactive. Removal of the apical bud—the major site of auxin production—results in growth of the axillary buds. If the cut surface of the stem is treated with auxin, however, the axillary buds do not grow. The apical buds of branches also exert apical dominance: the axillary buds on the branch are inactive unless the apex of the branch is removed. That is why gardeners prune shrubs to encourage branching.

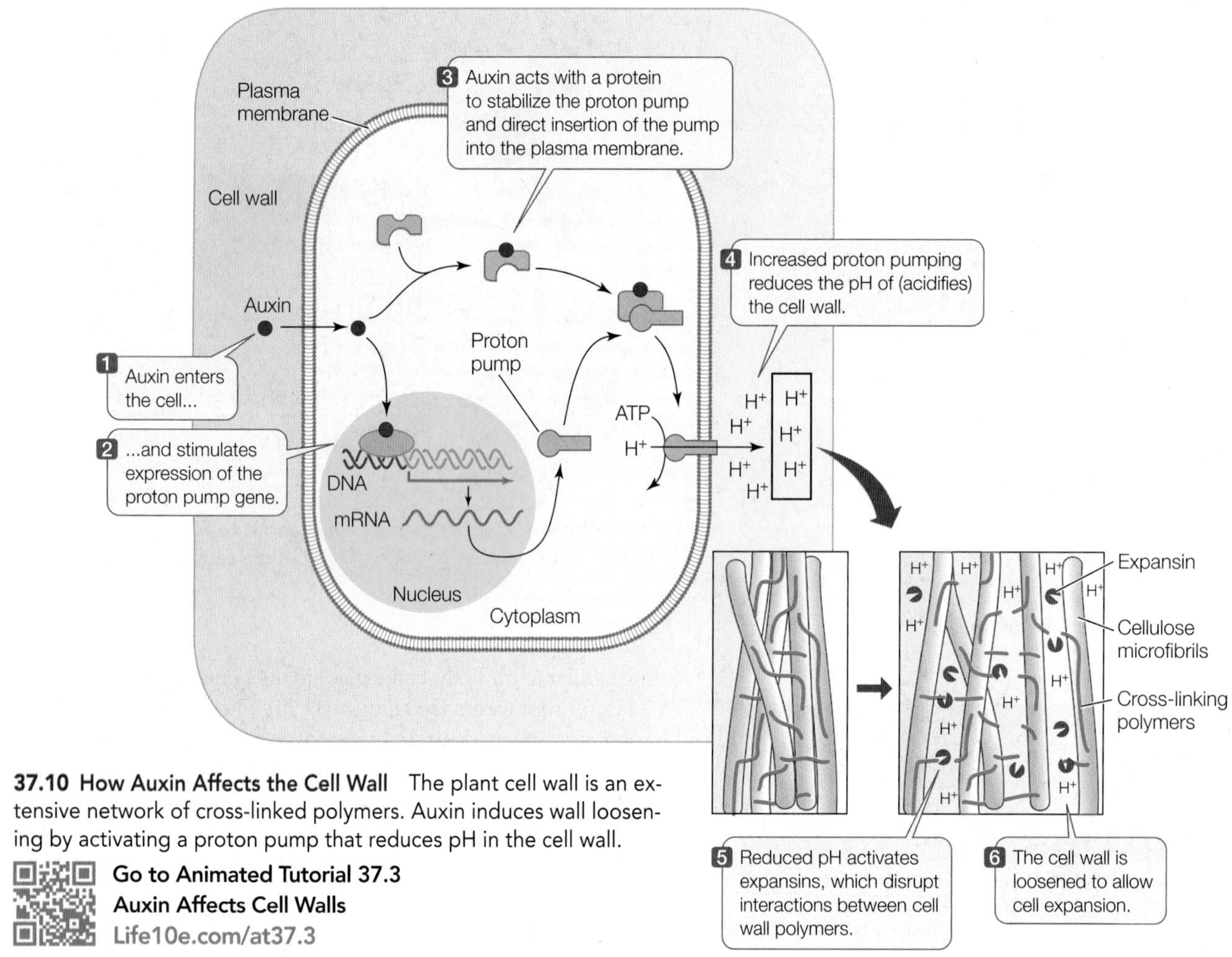


**37.10 How Auxin Affects the Cell Wall** The plant cell wall is an extensive network of cross-linked polymers. Auxin induces wall loosening by activating a proton pump that reduces pH in the cell wall.

**Go to Animated Tutorial 37.3**
**Auxin Affects Cell Walls**
Life10e.com/at37.3

In the experiments on leaves and stems just discussed, removal of a particular part of the plant elicits a response—abscission or loss of apical dominance—and that response is prevented by treatment with auxin. These results are consistent with other data showing that the excised part of the leaf or stem is an auxin source and that auxin in the intact plant delays the abscission of leaves and helps maintain apical dominance.

**FRUIT DEVELOPMENT** Fruit development normally depends on prior fertilization of the ovule (egg), but in many species treatment of an unfertilized ovary with auxin or gibberellins causes **parthenocarpy**—fruit formation without fertilization. Parthenocarpic fruits form spontaneously in some cultivated varieties of plants, including seedless grapes, bananas, and some cucumbers.

**CELL EXPANSION** Cell division followed by cell expansion is what causes plant growth. Because the plant cell wall normally prevents expansion of the cell contents inside the plasma membrane (see Section 34.1), the cell wall plays a key role in controlling the rate and direction of plant cell growth. Auxin acts on cell walls to regulate this process.

The expansion of a plant cell is driven primarily by the uptake of water, which enters the cytoplasm of the cell and accumulates in its central vacuole (see Section 35.1). Growth of the vacuole accounts for most of the increase in volume of a growing cell, and the vacuole often makes up more than 90 percent of the volume of a mature cell. As the vacuole expands, it presses the cytoplasm against the cell wall, and the wall resists this force (the basis of turgor pressure). The cell wall is an extensively cross-linked network of polysaccharides and proteins, dominated by cellulose microfibrils. If the cell is to expand, some adjustments must be made in the wall structure to allow the wall to "give" under turgor pressure. Think of a balloon (the cell surrounded by a membrane) inside a box (the cell wall). How does the cell wall "box" loosen to allow expansion?

The **acid growth hypothesis** explains auxin-induced cell expansion (**Figure 37.10**). The hypothesis holds that protons ($H^+$) are pumped from the cytoplasm into the cell wall, lowering the pH of the wall and activating enzymes called expansins that catalyze changes in the cell wall structure such that the polysaccharides adhere to each other less strongly. This loosens the cell wall, making it easier to stretch as the cell expands. Auxin has two roles in this process: to increase the synthesis of the proton

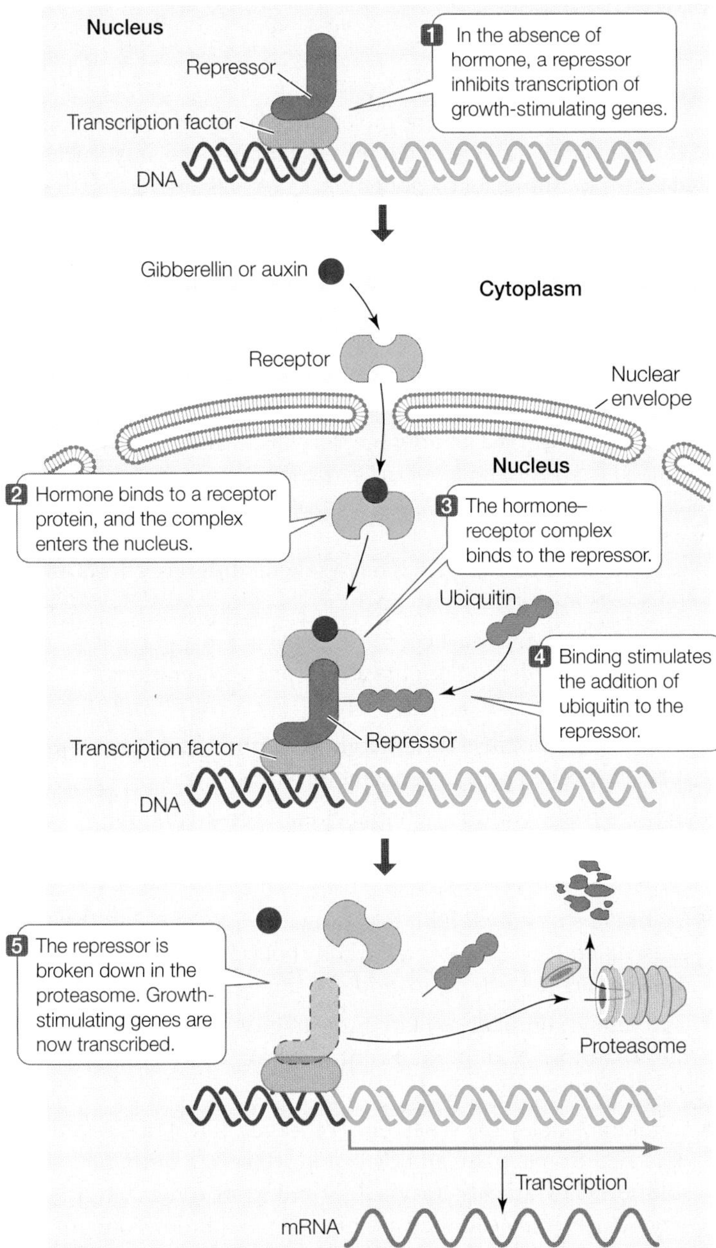


**37.11 Gibberellins and Auxin Have Similar Signal Transduction Pathways** Although the specific proteins involved are different, both hormones act to stimulate gene transcription by inactivating a repressor protein.

**Go to Media Clip 37.1**
**Gibberellin Binding to Its Receptor**
**Life10e.com/mc37.1**

pumps, and to guide their insertion into the plasma membrane. Auxin may also increase the activity of proton pump proteins already in the plasma membrane. Several lines of evidence support the acid growth hypothesis. For example, adding acid to the cell wall to lower the pH stimulates cell expansion even in the absence of auxin. Conversely, when a buffer is used to prevent the wall from becoming more acidic, auxin-induced cell expansion is blocked. The acid growth hypothesis holds more or less well depending on species; in some plants, auxin stimulates secretion of new cell wall components quickly enough to account for even rapid changes in growth rate.

## At the molecular level, auxin and gibberellins act similarly

The molecular mechanisms underlying both auxin and gibberellin action have been worked out with the help of genetic screens (see Figure 37.2). Biologists started by identifying mutant plants whose growth and development are insensitive to the hormones—that is, plants that are *not* affected by added hormone. Such mutants fall into two general categories:

- *Excessively tall plants.* These plants resemble wild-type plants given an excess of hormone and grow no taller when given extra hormone. They grow tall even when treated with inhibitors of hormone synthesis. Their hormone response is always "on," even in the absence of the hormone. In such cases, it is presumed that the normal allele for the mutant gene codes for an inhibitor of the hormone signal transduction pathway. In wild-type plants, that pathway is "off," but in the mutant plants, the pathway is "on" and the plant grows tall.
- *Dwarf plants.* These plants resemble dwarf plants that are deficient in hormone synthesis (see Figure 37.3), but they do not respond to added hormone. In these mutants the hormone response is always "off," regardless of the presence of the hormone.

Remarkably, some mutations of both types turned out to affect the same protein, which turns out to be a repressor of a transcription factor that stimulates the expression of growth-promoting genes. The repressor protein has two important domains, which explains how mutations in the same protein can have seemingly opposite effects:

- *One region of the repressor protein binds to the transcription complex to inhibit transcription of growth-promoting genes.* This is the region mutated in the excessively tall plants: the growth-promoting genes are always "on" because the repressor does not bind to the transcription complex.
- *Another region of the repressor protein causes it to be removed from the transcription complex.* This is the region mutated in the dwarf plants: the growth-promoting genes are always "off" because the repressor is always bound to the complex.

These observations allowed biologists to figure out how auxin and gibberellins work in wild-type plants. Of course, the repressor proteins involved in responding to the two hormones are different, but the actions of both hormones are similar: they *act by removing the repressor from the transcription complex* (**Figure 37.11**). The hormones do this by binding to a receptor protein, which in turn binds to the repressor. Binding of the hormone–receptor complex stimulates polyubiquitination of the repressor, targeting it for breakdown in the proteasome (see Figure 16.25). The receptors contain or associate with a region called an F-box that facilitates protein–protein interactions necessary for causing polyubiquitination of a target protein. Whereas animal genomes have few F-box-containing proteins, plant genomes have hundreds, an indication that this type of gene regulation is common in plants.

RECAP 37.2

Gibberellins are plant hormones that affect stem growth, fruit size, seed germination, and many other aspects of plant development; the effects vary from species to species. Auxin regulates cell expansion and thus mediates phototropism and gravitropism; it also plays roles in apical dominance, leaf abscission, fruit development, and root initiation. The acid growth hypothesis explains auxin-induced cell wall loosening. Similar molecular mechanisms explain the effects of auxin and gibberellins on gene expression.

- How were gibberellins and auxin shown to be plant hormones? **See p. 760**
- How do gibberellins contribute to the germination of barley seeds? **See Figure 37.5**
- What is the evidence for polar transport of auxin, and how does it occur? **See pp. 762–763 and Figures 37.6 and 37.7**
- Explain why, even though auxin moves *away* from the lighted side of a coleoptile tip, the coleoptile bends *toward* the light. **See p. 764 and Figure 37.8**
- How does auxin cause cell wall loosening? **See p. 766 and Figure 37.10**
- What is the general signal transduction pathway for auxin and gibberellin? **See p. 767 and Figure 37.11**

How can a single hormone, such as auxin or a gibberellin, have so many effects? As we have seen, a single signal transduction pathway may affect more than one gene. We will learn about other important plant hormones in the next section, and they too have multiple effects.

## 37.3 What Are the Effects of Cytokinins, Ethylene, and Brassinosteroids?

Like animal cells, plant cells differentiate after they form from undifferentiated stem cells (called meristem cells in plants). But unlike animal cells, which generally do not divide after differentiation, plant cells retain the ability to divide. For example, in leaf abscission (see Figure 37.9) differentiated parenchyma cells in the petiole resume division, forming a specialized, weak layer of cells. Also, cells of the phloem and cortex can resume division and form secondary meristems. What stimulates these cells to divide? An answer came from studies of cells isolated from plants and cultured in the laboratory.

### Cytokinins are active from seed to senescence

Like bacteria and yeasts, plant cells such as parenchyma cells can be grown in a liquid or solidified growth medium containing sugars and salts. The cells will divide continuously until they run out of nutrients. In the early days of plant cell culturing, scientists experimented with many supplements to determine the optimal chemical environment for growth. The best supplement was coconut milk, the fluid that surrounds the developing embryo in coconut fruit. Investigators suspected that a molecule in the fluid must stimulate plant cell division.

A clue to the identity of the molecule came when Folke Skoog at the University of Wisconsin tested various pure substances that might substitute for coconut milk. DNA was among the substances tested, and it did not work; however, heating DNA at high pressure in an autoclave produced a mixture that strongly promoted plant cell division. A derivative of adenine called kinetin was identified as the active ingredient. Because it stimulated cell division (cytokinesis), it was called a **cytokinin**.

Kinetin does not exist in cells, but it gave scientists a hint as to what type of molecule might be the active ingredient in coconut milk. In 1963 an adenine derivative called **zeatin** was extracted from corn endosperm. Since then, more than 150 different cytokinins have been isolated, and most are derivatives of adenine.

Cytokinins (see Table 37.2) have several different effects, in many cases interacting with auxin:

- Adding an appropriate combination of auxin and cytokinins to a growth medium induces rapid proliferation of cultured plant cells.
- Cytokinins can cause certain light-requiring seeds to germinate even when kept in constant darkness.
- In plant tissue cultures, a high cytokinin-to-auxin ratio promotes the formation of shoots; a low ratio promotes the formation of roots.
- Cytokinins usually inhibit the elongation of stems, but they cause lateral swelling of stems and roots (the fleshy roots of radishes are an extreme example).
- Cytokinins stimulate axillary buds to grow into branches; the auxin-to-cytokinin ratio controls the extent of branching (bushiness) of a plant.
- Cytokinins delay the senescence of leaves. If leaf blades are detached from a plant and placed in water or a nutrient solution, they quickly turn yellow and show other signs of senescence. If instead they are placed in a solution containing a cytokinin, they remain green and senesce much more slowly. Roots contain abundant cytokinins, and cytokinin transport to the leaves delays senescence.

Cytokinin signaling appears to act through a two-component pathway (a type of signal transduction pathway common in bacteria):

- A *receptor* that can act as a protein kinase, phosphorylating itself as well as a target protein

- A target protein, generally a transcription factor, that can act as an *effector*

Genetic screens in *Arabidopsis* for defects in the response to cytokinin have identified the receptor (AHK; *Arabidopsis* *h*istidine *k*inase) and target effector (ARR; *Arabidopsis* *r*esponse *r*egulator), the latter acting as a transcription factor when phosphorylated. The signal transduction pathway also includes a third protein (AHP; *Arabidopsis* *h*istidine *p*hosphotransfer protein), which transfers phosphates from the receptor to the effector (**Figure 37.12**). The *Arabidopsis* genome has more than 20 genes that are expressed in response to this signaling pathway.

## Ethylene is a gaseous hormone that hastens leaf senescence and fruit ripening

Whereas the cytokinins delay senescence, another plant hormone promotes it: the gas **ethylene** (see Table 37.2), which is sometimes called the senescence hormone. Ethylene can be produced by all parts of the plant, and like all plant hormones, it has several effects.

Back when streets were lit by gas rather than by electricity, leaves on trees near street lamps dropped earlier than those on trees farther from the lamps. We now know why: ethylene, a combustion product of the illuminating gas, caused the early abscission. Whereas auxin delays leaf abscission, ethylene strongly promotes it; thus the balance of auxin and ethylene controls abscission.

**FRUIT RIPENING** By promoting senescence, ethylene also speeds the ripening of fruit. As a fruit ripens, it loses chlorophyll and its cell walls break down; ethylene promotes both of these processes. Ethylene also causes an increase in its own production. Thus once ripening begins, more and more ethylene forms, and because it is a gas, it diffuses readily throughout the fruit and even to neighboring fruits on the same or other plants. The old saying "one rotten apple spoils the barrel" is true. That rotten apple is a rich source of ethylene, which speeds the ripening and subsequent rotting of the other fruit in a barrel or other confined space.

Farmers used to poke holes in developing figs to make them ripen faster. We now know that wounding causes an increase in ethylene production by the fruit, and that the raised ethylene level promotes ripening in many fruits, including apples, bananas, melons, apricots, and tomatoes. Today commercial shippers and storers of such fruit hasten ripening by adding ethylene to storage chambers. This use of ethylene is the single most important use of a natural plant hormone in agriculture and commerce.

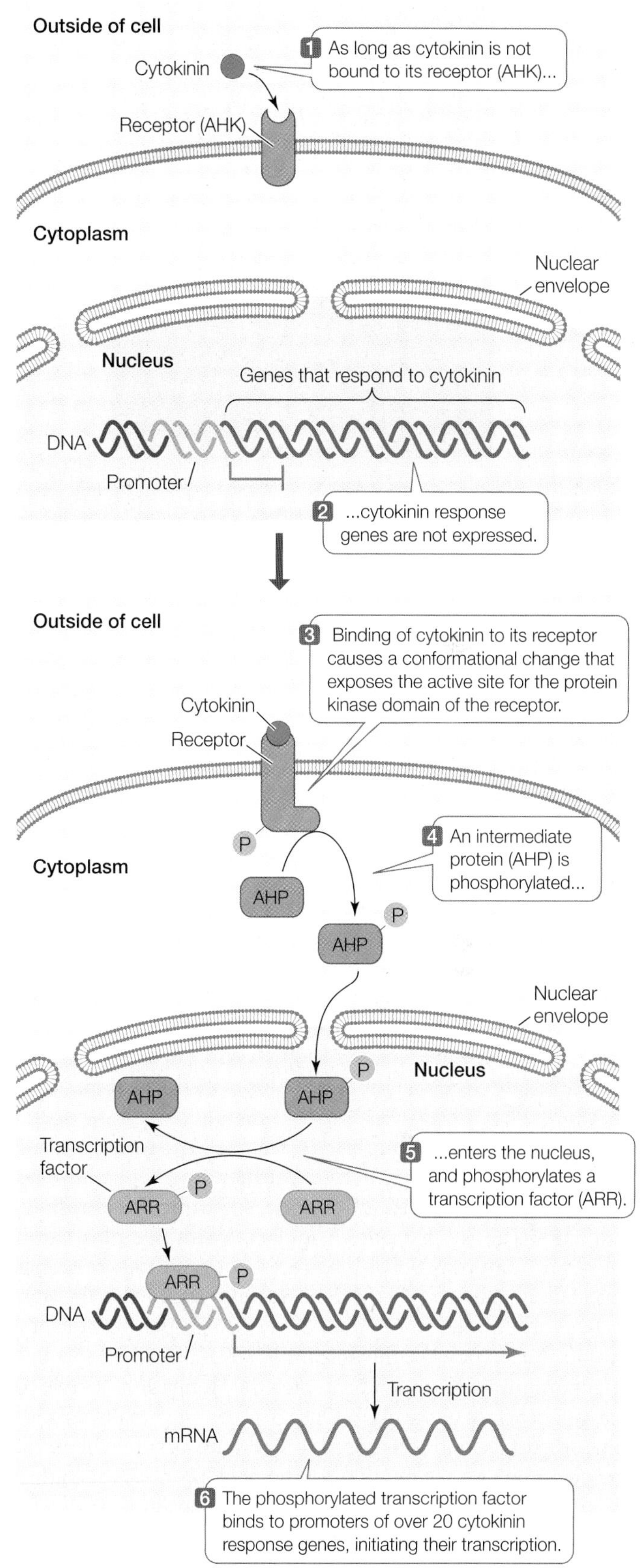


**37.12 The Cytokinin Response Pathway** Plant cells respond to cytokinins using a signal transduction pathway involving a receptor and an effector protein.

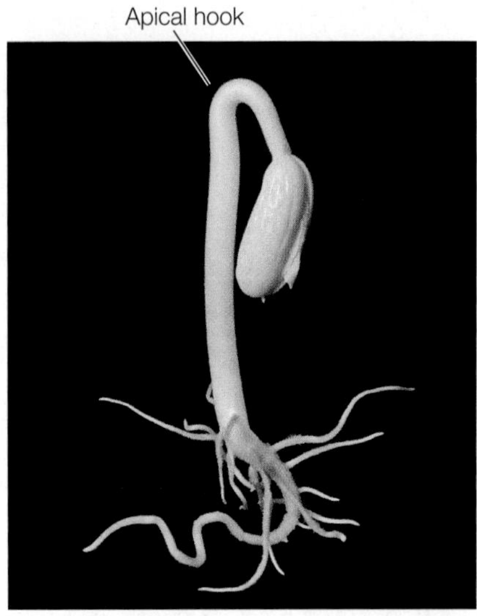


**37.13 The Apical Hook of a Eudicot** Asymmetrical production of auxin, controlled by ethylene, is responsible for the apical hook of this bean seedling. The ethylene concentration was highest on the right side, so more rapid growth on the left caused and maintained the hook.

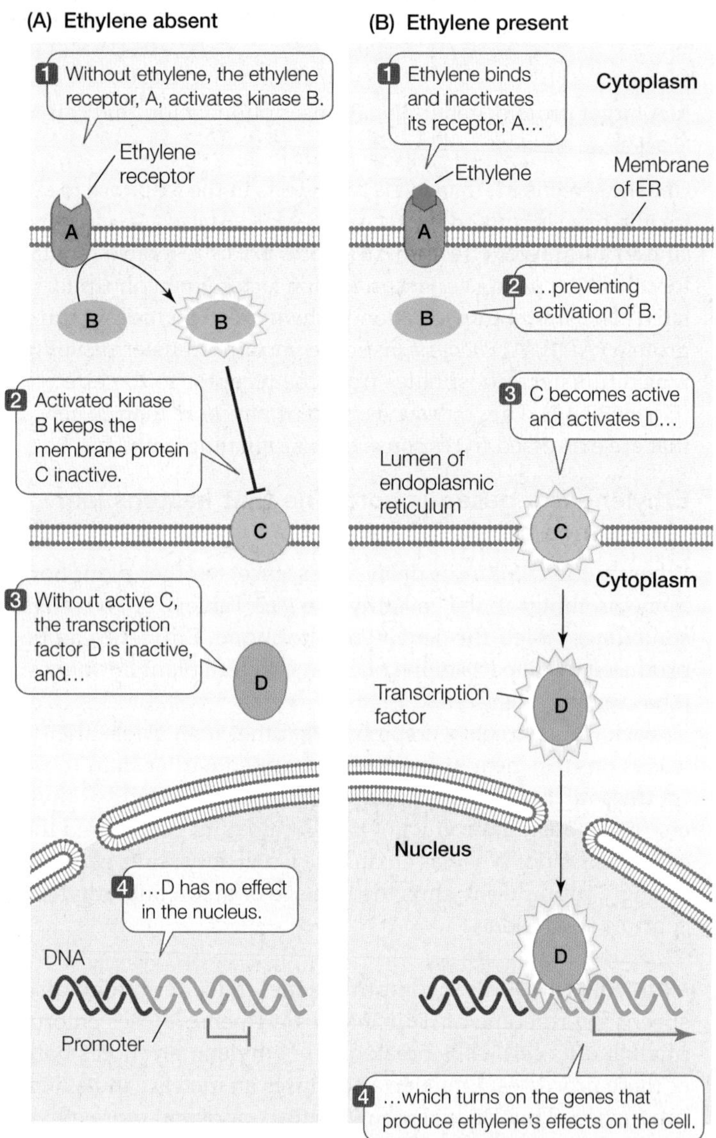


**37.14 The Signal Transduction Pathway for Ethylene** This diagram shows the roles of four proteins (A, B, C, and D) in the signal transduction pathway through which ethylene exerts its many effects.

Ripening can also be delayed by the use of "scrubbers" and adsorbents that remove ethylene from the atmosphere in fruit storage chambers. This strategy can even be used in the home. Many supermarkets sell plastic bags designed to keep fruits fresh; the bags are impregnated with a substance that binds ethylene.

As flowers senesce, their petals may abscise, decreasing their value in the cut-flower industry. Growers and florists often immerse the cut stems of ethylene-sensitive flowers in dilute solutions of silver thiosulfate before sale. Silver salts inhibit ethylene action by interacting directly with the ethylene receptor—thus they delay senescence, keeping flowers "fresh" for longer. An alternative product, used to delay fruit ripening and preserve cut flowers, is 1-methylcyclopropene, a gas that also binds ethylene receptors and blocks their function.

STEM GROWTH Although it is associated primarily with senescence, ethylene is active at other stages of plant development as well. Its effects on seedling development illustrate the interactions that occur among plant hormones. The hypocotyl of many eudicot seedlings forms an **apical hook** that protects the delicate shoot apex while the stem grows through the soil (**Figure 37.13**). As in phototropic and gravitropic responses, the apical hook is maintained through an auxin gradient, which promotes the elongation of cells on the outer surface of the hook. Once the seedling breaks through the soil surface and is exposed to light, the auxin level on the inside of the hook increases and the hook unfolds, raising the shoot apex and the expanding leaves into the sun.

There is evidence that ethylene controls the formation of the auxin gradient during seedling development. Treatment of dark-grown seedlings of some species with ethylene results in what is called the "triple response": an exaggeration of the apical hook and a thickening and shortening of the hypocotyl and root (this response was exploited in a genetic screen for ethylene response mutants; see Figure 37.2). It has been shown that ethylene affects both auxin synthesis and transport during apical hook development.

THE ETHYLENE SIGNAL TRANSDUCTION PATHWAY The mechanism of ethylene action has been worked out by analyzing *Arabidopsis* mutants that have ethylene-related defects. Some of these mutants do not respond to applied ethylene, and others act as if they have been exposed to ethylene even though they have not. Researchers studied the mutant genes and compared their protein products with other known proteins; thus they worked out some of the details of the signal transduction pathway through which ethylene acts (**Figure 37.14**).

The pathway includes two membrane proteins in the endoplasmic reticulum. The first is an ethylene receptor (labeled A in the figure), and the second is a channel-like protein (C). In the absence of ethylene, receptor A activates a protein kinase (B) that keeps C inactive by phosphorylation. When receptor A binds ethylene, it stops activating B. Without B to inactivate it, C activates a transcription factor (D), which then moves into the nucleus, where it turns on the genes that produce ethylene's effects in the cell. In other words, ethylene turns off the "off" signal.

### Brassinosteroids are plant steroid hormones

In animals, steroid hormones such as cortisol and estrogen are formed from cholesterol (see Figure 41.2). Initially, biologists isolated a plant steroid hormone from the pollen of rape, a member of the Brassicaceae (mustard family). When applied to various plant tissues, this **brassinosteroid** (see Table 37.2) stimulated cell elongation, pollen tube elongation, and vascular tissue differentiation, but it inhibited root elongation. Since then, dozens of chemically related, growth-affecting brassinosteroids have been found in most plants.

Mutant plants that either do not make brassinosteroids or have defects in brassinosteroid reception and signal transduction are usually dwarf, infertile, and slow to develop. These effects can be reversed by adding small amounts of brassinosteroids, indicating that brassinosteroids are true hormones. These hormones have diverse effects, which vary among plants. Brassinosteroids can:

- promote xylem differentiation
- promote growth of pollen tubes during reproduction
- promote seed germination
- promote apical dominance and leaf senescence
- enhance cell elongation and cell division in shoots

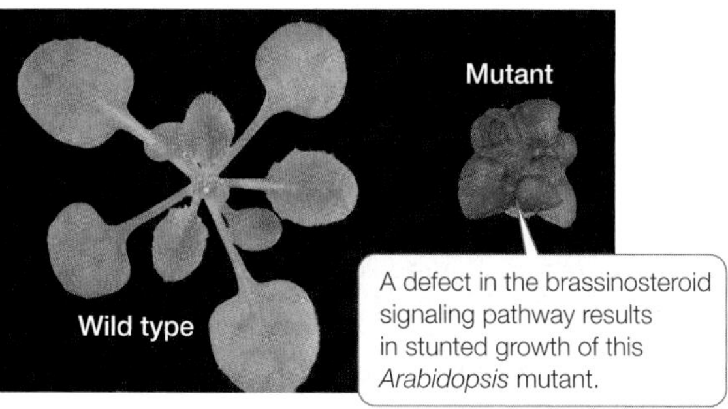


The signaling pathway for these plant steroids differs sharply from those for steroid hormones in animals. In animals, steroids diffuse through the plasma membrane and bind to receptors in the cytoplasm. In contrast, the receptor for brassinosteroids is an integral protein in the plasma membrane.

**RECAP 37.3**

Cytokinins, ethylene, and brassinosteroids work in concert with auxin and gibberellins to mediate plant development. Their signaling pathways vary from a simple two-component receptor–effector system (cytokinin) to inhibition of an inhibitor of an effector (ethylene).

- How do cytokinins interact with auxin to regulate a plant's development? **See p. 768**
- What is the role of ethylene in fruit ripening? How is this knowledge used commercially? **See p. 769**
- How do the signaling pathways for cytokinins and ethylene differ? **See pp. 769–770 and Figures 37.12 and 37.14**

A plant's response to light—the energy source for photosynthesis—is crucial to its survival. We saw how the Darwins' pioneering investigations of phototropism led to the discovery of auxin. Let's now look more closely at how plants sense and respond to light.

## 37.4 How Do Photoreceptors Participate in Plant Growth Regulation?

As we pointed out in Section 37.1, plants respond to many different environmental cues, and light is possibly the most important of these cues. Much has been learned about the receptors that plants use to sense light, and we will focus on those receptors here.

Plants respond to two aspects of light: (1) its *quality*—that is, the wavelengths of light that can be absorbed by molecules in the plant; and (2) its *quantity*—that is, the intensity and duration of light exposure.

Chapter 10 described photosynthesis: how chlorophyll and other pigments absorb light at certain wavelengths (quality), and how light intensity affects photosynthetic rate (quantity). Here we will consider how light affects plant development. Earlier in this chapter we described phototropism and how auxin mediates a plant stem's bending toward light. In addition to phototropism, light influences seed germination, shoot elongation, the initiation of flowering, and many other important aspects of plant development. Several photoreceptors take part in these processes. Three or more types of **blue-light receptors** mediate the effects of higher-intensity blue light, and phytochrome mediates the effects of red light.

### Phototropins, cryptochromes, and zeaxanthin are blue-light receptors

Charles and Francis Darwin showed that the apical tip of a growing coleoptile receives light as a signal and then redistributes auxin to stimulate cell elongation below the tip on the shaded side. You may recall from Chapter 10 that an action spectrum involves exposing plants to different wavelengths of light to determine what wavelengths are most effective in driving a given process (e.g., photosynthesis). For photosynthesis, such studies showed that the most effective wavelengths are those absorbed by chlorophylls (see Figure 10.5). When an action spectrum was obtained for phototropism of coleoptiles, blue light (peak 436 nm) was found to be the most effective at inducing the coleoptile to curve (**Figure 37.15**). What is the blue-light-absorbing receptor/pigment? Biologists have used a genetic approach to answer this question, once again employing the model plant *Arabidopsis*.

Researchers recovered blue-light-insensitive *Arabidopsis* mutants from a genetic screen and identified the gene for a blue-light receptor protein located in the plasma membrane called **phototropin**. Phototropin protein has a flavin mononucleotide associated with it that absorbs blue light, leading to a change in the shape of the protein. This change exposes an active site for a protein kinase, which in turn initiates a signal transduction cascade that ultimately results in the stimulation of cell elongation by auxin.

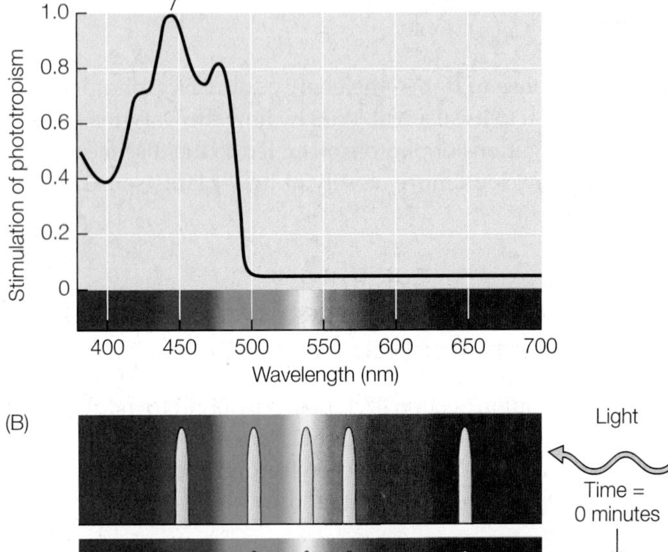


**37.15 Action Spectrum for Phototropism** (A) The action spectrum for bending of a coleoptile toward light is similar to the absorption spectrum for the receptor, phototropin. (B) After 90 minutes, only the coleoptiles exposed to blue light bend.

Phototropin is also involved in chloroplast movements in relation to light, and participates with another type of blue-light receptor, the plastid pigment **zeaxanthin**, in the light-induced opening of stomata (see Figure 35.8).

Yet another class of blue-light receptors is the **cryptochromes**, which absorb blue and ultraviolet light. These yellow pigments are located primarily in the plant cell nucleus and affect seedling development and flowering. The exact mechanism of cryptochrome action is not yet known. Strong blue light inhibits cell elongation through the action of cryptochromes, although the most rapid responses are mediated by phototropins.

## Phytochromes mediate the effects of red and far-red light

**Photomorphogenesis** refers to a number of physiological and developmental events in plants that are controlled by light. For example:

- A bean seedling germinating below ground has an elongated stem, a pale yellow, folded leaf, and a hook that protects the first leaves (see Figures 37.1 and 37.13)—it is **etiolated**. As the seedling reaches the surface of the soil, it undergoes several light-induced changes: the apical hook straightens, the rudimentary leaves unfold, and chlorophyll is made, so that photosynthesis can begin. Even very dim light will induce these changes.
- Lettuce seeds spread on the soil will germinate only in response to light. Even just a flash of dim light will suffice.
- Adult cocklebur plants flower when they are exposed to long nights. If there is a brief light flash in the middle of the night, they do not flower.

Action spectra of the above processes show that they are induced by red light (650–680 nm). This indicates that plants must have a photoreceptor pigment that absorbs red light and initiates photomorphogenesis.

What is especially remarkable about these red light responses is that *they are reversible by far-red light* (710–740 nm). For example, if lettuce seeds are exposed to brief, alternating periods of red and far-red light in close succession, they respond only to the final exposure. If it is red, they germinate; if it is far-red, they remain dormant (**Figure 37.16**). This reversibility of the effects of red and far-red light regulates many other aspects of plant development, including flowering and seedling growth.

The basis for the effects of red and far-red light resides in the bluish photoreceptor pigment protein in the cytosol of plants called **phytochrome**. Phytochrome exists in two

INVESTIGATING**LIFE**

**37.16 Sensitivity of Seeds to Red and Far-Red Light** Lettuce seeds will germinate if exposed to a brief period of light. An action spectrum indicated that red light was most effective in promoting germination, but far-red light would reverse the stimulation if presented right after the red light flash. Harry Borthwick and his colleagues asked what would be the effect of repeated alternating flashes of red and far-red light. In each case, the final exposure determined the germination response.[a] This observation led to the conclusion that a single, photoreversible molecule was involved. That molecule turned out to be phytochrome.

**HYPOTHESIS** The effects of red and far-red light on lettuce seed germination are mutually reversible.

***Method*** Expose lettuce seeds to alternate periods of red light R for 1 minute and far-red light FR for 4 minutes.

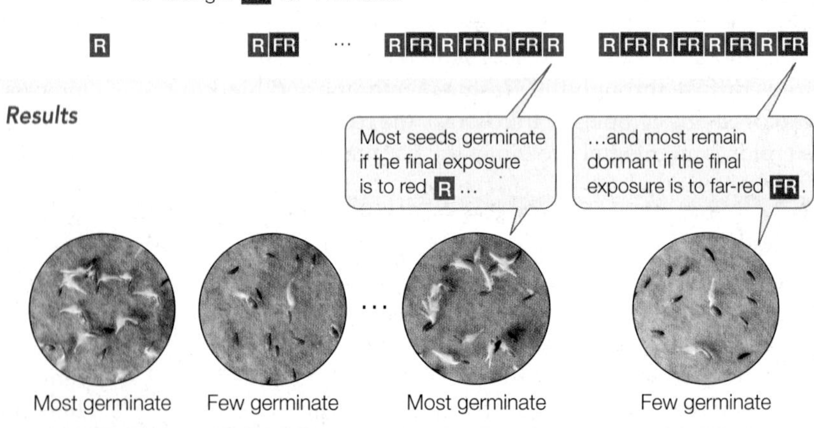


**CONCLUSION** Red light and far-red light reverse each other's effects.

Go to **BioPortal** for discussion and relevant links for all INVESTIGATING**LIFE** figures.

[a]Borthwick, H. A. et al. 1952. *Proceedings of the National Academy of Sciences USA* 38: 662–666.

interconvertible "isoforms," or states. The molecule undergoes a conformational change upon absorbing light at particular wavelengths. The default or "ground" state, which absorbs principally red light, is called **$P_r$**. When $P_r$ absorbs a photon of red light it is converted into **$P_{fr}$**. $P_{fr}$ is the active form of phytochrome—the form that triggers important biological processes in various plants.

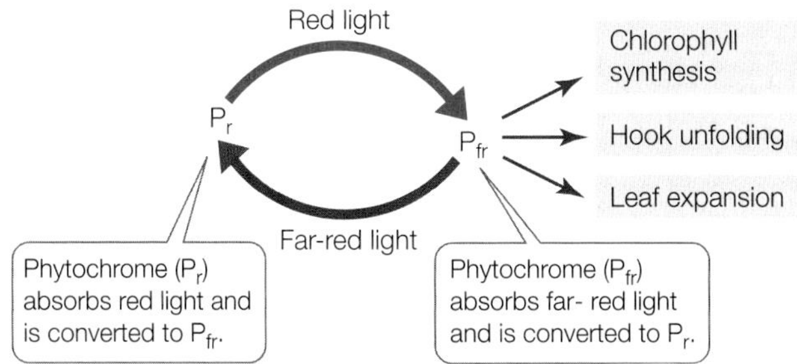


The part of phytochrome that absorbs red and far red light is a covalently attached pigment called a chromophore (**Figure 37.17A**). The chromophore of $P_r$ preferentially absorbs red light; when it does so, it changes conformation and the phytochrome is converted to the $P_{fr}$ form. When the chromophore of $P_{fr}$ absorbs far red light, the phytochrome is converted back to the $P_r$ form. If you know organic chemistry, this reaction is a familiar *cis-trans* isomerization.

The absorption spectra for $P_r$ and $P_{fr}$ correlate with their action spectra (**Figure 37.17B**). As we have seen, these processes include seed germination, shoot development after etiolation, and flowering. In *Arabidopsis* there is a gene family that encodes five slightly different phytochromes, each functioning in different photomorphogenic responses.

For a plant in nature, the ratio of red to far-red light determines whether a phytochrome-mediated response will occur. For example, during daylight the ratio is about 1.2:1; because there is more red than far-red light, the $P_{fr}$ form predominates. But for a plant growing in the shade of other plants, the ratio is as low as 0.13:1, and phytochrome is mostly in the $P_r$ form. The low ratio of red to far-red light in the shade results from absorption of red light by chlorophyll in the leaves overhead, so less of the red light gets through to the plants below. Shade-intolerant species respond by stimulating cell elongation in the stem and thus growing taller to escape the shade. Shade cast by other plants also prevents germination of seeds that require red light to germinate (see Figure 37.16). The reflective properties of the soil can also affect the red to far-red ratio—and thus plant behavior. For example, cotton seedlings grow more slowly on soils (such as clay) that reflect more red than far-red light.

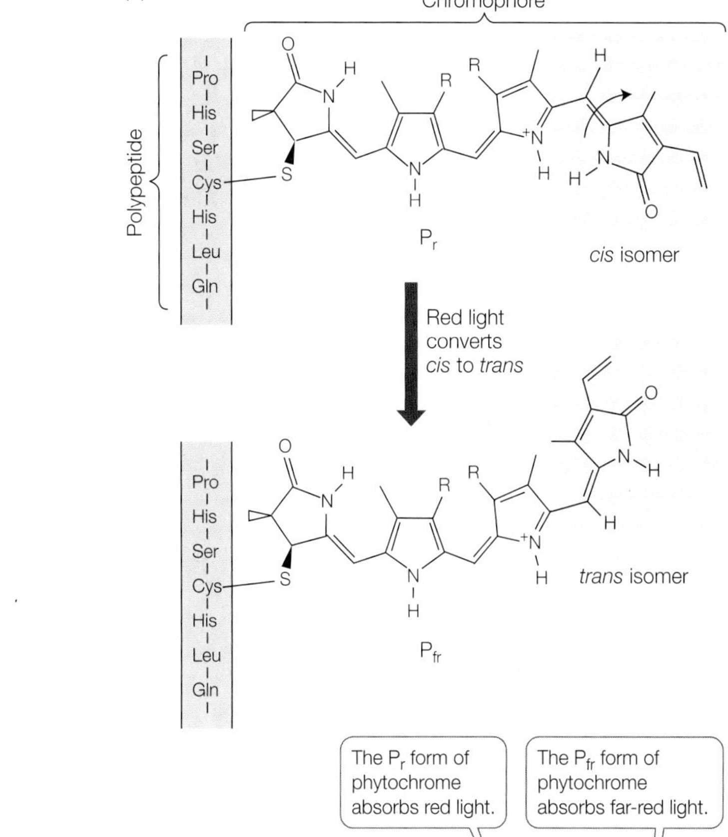


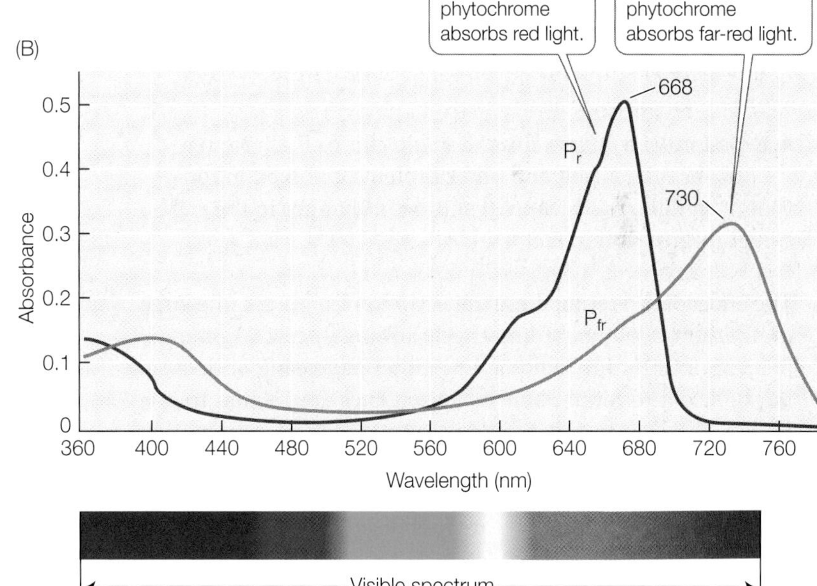


**37.17 Phytochrome Exists in Two Forms** Absorption spectra of phytochrome reveal two interconvertible forms. (A) The *cis* and *trans* isomers of phytochrome's chromophore. (B) The $P_r$ form absorbs red light; the $P_{fr}$ form absorbs far-red light.

## Phytochrome stimulates gene transcription

How does phytochrome, or more specifically $P_{fr}$, work? Phytochrome has two subunits (**Figure 37.18**), each composed of a protein chain and a chromophore. Gene transcription is stimulated when $P_r$ is converted to the $P_{fr}$ isoform. When $P_r$ absorbs red light and the chromophore changes shape (see Figure 37.17A), this leads to a change in the conformation of the protein itself, from the $P_r$ form to the $P_{fr}$ form. Conversion to the $P_{fr}$ form exposes two important regions of the phytochrome protein (see Figure 37.18), both of which affect transcriptional activity:

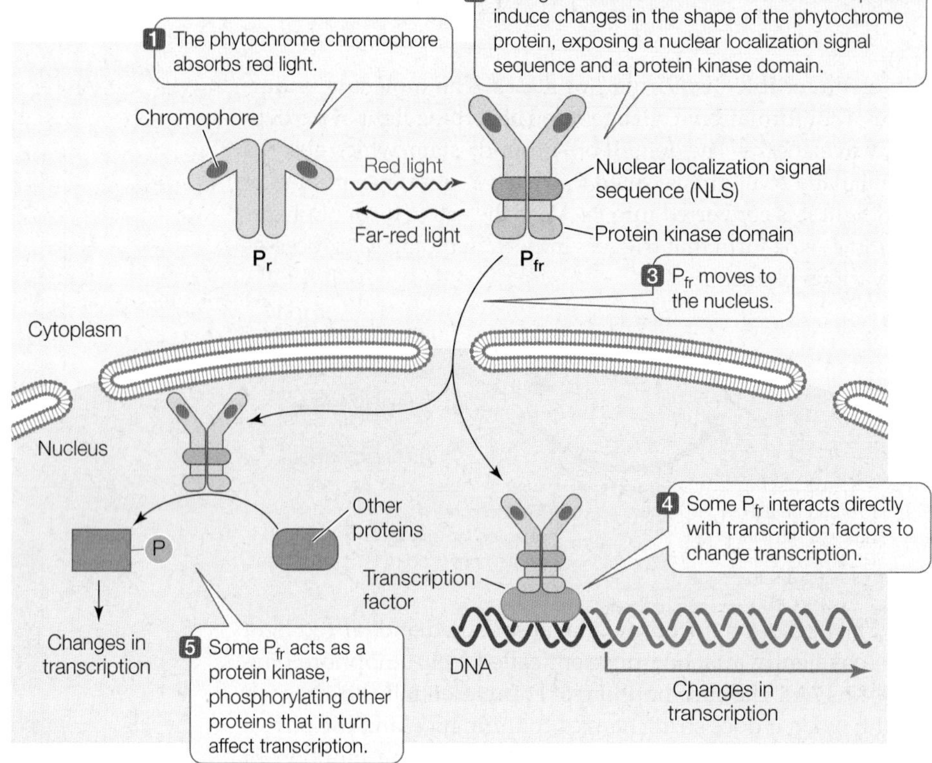


**37.18 Phytochrome Stimulates Gene Transcription** Phytochrome is composed of two polypeptide chains, each with a chromophore. This pair of polypeptides undergoes a conformational change upon absorbing light. When phytochrome absorbs red light, it converts to the $P_{fr}$ form, which activates transcription of phytochrome-responsive genes.

- Exposure of a *nuclear localization sequence* (see Figure 14.19) results in movement of $P_{fr}$ from the cytosol to the nucleus. Once in the nucleus, $P_{fr}$ binds to transcription factors and thereby stimulates expression of genes involved in photomorphogenesis.
- Exposure of a *protein kinase* domain causes $P_{fr}$ protein to phosphorylate itself and other proteins involved in red-light signal transduction, resulting in changes in the activity of transcription factors.

The effect of activating these transcription factors is quite large: in *Arabidopsis*, phytochrome affects an amazing 2,500 genes (10 percent of the entire genome!) by either increasing or decreasing their expression. Some of these genes are related to hormones. For example, when $P_{fr}$ is formed in seed germination, genes for gibberellin synthesis are activated and genes for gibberellin breakdown are repressed. As a result, gibberellins accumulate and seed reserves are mobilized.

## Circadian rhythms are entrained by light reception

The timing and duration of biological activities in living organisms are governed in all eukaryotes and some prokaryotes by what is commonly called a "biological clock"—an oscillator within cells that alternates back and forth between two states at roughly 12-hour intervals. The major outward manifestations of this clock are known as **circadian rhythms** (Latin *circa*, "about"; *dies*, "day"). Think of your own life: in all probability you sleep at night and are awake during the day. The circadian rhythms of animals are discussed in Section 53.3. In plants, circadian rhythms influence, for example, the opening (during the day) and closing (at night) of stomata in *Arabidopsis*, and the raising toward the sun (during the day) and lowering (at night) of leaves in bean plants. From these two examples, it is obvious that circadian rhythms are ecologically useful adaptations, in that they relate the plant's physiology to its environment.

Two qualities characterize circadian rhythms, as well as other regular biological cycles: the **period** is the length of one cycle, and the **amplitude** is the magnitude of the change over the course of a cycle. The circadian rhythms of plants have several noteworthy characteristics:

- The period of a circadian rhythm is remarkably insensitive to temperature, although lowering the temperature may drastically reduce the amplitude.
- Circadian rhythms are highly persistent; they may continue for days, even in the absence of environmental cues, such as light–dark periods.
- Circadian rhythms can be entrained, within limits, by light–dark cycles that do not exactly correspond to 24 hours. That is, the period of a rhythm can be made to coincide (within limits) with that of the light–dark cycle to which the organism is exposed.

Consider what happens when a person abruptly moves across many time zones: what was the night becomes the day, and gradually the person's sleep–wakefulness circadian rhythm entrains to the new environmental cues. Similar entrainment occurs in plants adapting to day length as the seasons progress during the year. The action spectrum for plant entrainment indicates that phytochrome (and to a lesser extent, blue-light receptors) is very likely involved. At sundown phytochrome is mostly in the active $P_{fr}$ form. But as the night progresses, $P_{fr}$ gradually gets converted back to the inactive $P_r$ form. By dawn phytochrome is mostly in the $P_r$ state, but as daylight begins, it rapidly converts to $P_{fr}$. The switch to the $P_{fr}$ state resets the plant's biological clock. However long the night, the clock is still reset at dawn every day. Thus while the total period measured by the clock is consistent, the clock adjusts to changes in day length over the course of the year.

RECAP 37.4

Light controls several physiological and developmental events in plants, a process called photomorphogenesis. Pigment photoreceptors such as phototropin, cryptochromes, and phytochrome mediate the effects of light on plant growth and development. Phytochrome exists in two interconvertible states; conversion from one state to the other is controlled by the ratio of red to far-red light. Circadian rhythms are influenced by light reception.

- Give the evidence for blue-light receptors in plants. **See pp. 771–772 and Figure 37.15**
- Why does red light affect seed germination differently than far-red light does? **See pp. 772–773 and Figure 37.16**
- What are circadian rhythms, and how are they related to photoreception? **See p. 774**

Photoreceptors also play a regulatory role in flowering. In addition to light, another environmental cue—temperature—regulates flowering. We will examine these topics and others in the next chapter, which focuses on reproduction in flowering plants.

What changes in growth patterns made the new strains of wheat and rice successful?

ANSWER

In normal wheat plants, gibberellins stimulate stem elongation. But in the semi-dwarf plants, a mutation affects the signal transduction mechanism for gibberellins so that the stem cells do not respond to it and growth is reduced. In rice, the mutation is in the gene for an enzyme in the biochemical pathway for the synthesis of gibberellins. Without the hormone, the stem does not elongate. The lives of countless people have been saved by intentional disruptions of hormone signaling.

## CHAPTER**SUMMARY** 37

### 37.1 How Does Plant Development Proceed?

- As sessile organisms, plants maximize their ability to grow by using meristems, forming new organs, and growing throughout life.
- The environment, photoreceptors, hormones, and the plant's genome all regulate plant development.
- Seed **dormancy**, which has adaptive advantages, is maintained by a variety of mechanisms. In nature, dormancy is broken by, for example, abrasion, fire, leaching, and low temperatures. When dormancy ends and the seed **imbibes** water, it **germinates** and develops into a **seedling**. **Review Figure 37.1, ACTIVITIES 37.1, 37.2**
- Plant hormones differ in structure and physiology from animal hormones. **Review Table 37.1**
- Plants have several hormones, each of which regulates multiple aspects of development. Interactions among these hormones are often complex. **Review Table 37.2**
- **Genetic screens** using the model organism *Arabidopsis thaliana* have contributed greatly to our understanding of signaling in plants. **Review Figure 37.2**

### 37.2 What Do Gibberellins and Auxin Do?

- Both **gibberellins** and **auxin** can induce growth in plants otherwise genetically destined to be dwarfs. **Review Figure 37.3**
- Gibberellins have many effects that vary among different plants, including cell elongation, fruit ripening, and mobilization of seed storage polymers. **Review Figures 37.3–37.5, ACTIVITY 37.3**
- Auxin was discovered in the context of stem and coleoptile growth, in particular **phototropism**. In the shoot, it is made in the growing tip and transported down to stimulate cell elongation. **Review Figures 37.6, 37.7, ANIMATED TUTORIALS 37.1, 37.2**
- According to the **acid growth hypothesis**, auxin stimulates cell elongation through the release of protons into the cell wall (acidification of the cell wall). **Review Figure 37.10, ANIMATED TUTORIAL 37.3**
- Both auxin and gibberellins act through the breakdown of transcriptional repressors.

### 37.3 What Are the Effects of Cytokinins, Ethylene, and Brassinosteroids?

- **Cytokinins** are adenine derivatives that promote plant cell division, promote seed germination in some species, inhibit stem elongation, promote lateral swelling of stems and roots, stimulate the growth of axillary buds, promote the expansion of leaf tissue, and delay leaf senescence.
- A balance between auxin and **ethylene** controls leaf abscission. Ethylene promotes senescence and fruit ripening. It indirectly causes the formation of a protective **apical hook** in eudicot seedlings. In stems, it inhibits elongation, promotes lateral swelling, and causes a loss of gravitropic sensitivity.
- Ethylene acts on cells by a protein kinase pathway located in the endoplasmic reticulum. **Review Figure 37.14**
- Dozens of different **brassinosteroids** affect cell elongation, pollen tube elongation, vascular tissue differentiation, and root elongation. These steroids act at a plasma membrane receptor.

*continued*

## 37.4 How Do Photoreceptors Participate in Plant Growth Regulation?

- **Phototropins** are blue-light photoreceptors for phototropism and chloroplast movements. **Zeaxanthin** acts in conjunction with the phototropins to mediate the light-induced opening of stomata. **Cryptochromes** are blue-light photoreceptors that control seedling development, stem elongation, and floral initiation.
- **Phytochrome** exists in the cytosol in two interconvertible forms, $\mathbf{P_r}$ and $\mathbf{P_{fr}}$. The relative amounts of these two forms are a function of the ratio of red to far-red light. Phytochrome affects seedling growth, flowering, and etiolation. **Review Figure 37.16**
- The phytochrome signal transduction pathway affects transcription in two different ways; the $P_{fr}$ form interacts directly with some transcription factors, and influences transcription indirectly through interactions with protein kinases. **Review Figure 37.18**
- **Circadian rhythms** are activities that occur on a near-24-hour cycle. Light can entrain these activities through photoreceptors such as phytochrome.

**Go to the Interactive Summary to review key figures, Animated Tutorials, and Activities Life10e.com/is37**

# CHAPTER**REVIEW**

## REMEMBERING

1. Which of the following is *not* an advantage of seed dormancy?
   a. It makes the seed more likely to be digested by birds that disperse it.
   b. It counters the effects of year-to-year variations in the environment.
   c. It increases the likelihood that a seed will germinate in the right place.
   d. It favors dispersal of the seed.
   e. It may result in germination at a favorable time of year.

2. The gibberellins
   a. are responsible for phototropism and gravitropism.
   b. are gases at room temperature.
   c. are produced only by fungi.
   d. cause flowering in plants.
   e. inhibit the synthesis of digestive enzymes by barley seeds.

3. In coleoptile tissue, auxin
   a. is transported from base to tip.
   b. is transported from tip to base.
   c. can be transported toward either the tip or the base, depending on the orientation of the coleoptile with respect to gravity.
   d. is transported by simple diffusion, with no preferred direction.
   e. is not transported, because auxin is used where it is made.

4. Signal transduction for both auxin and gibberellins involves
   a. binding of the hormone to a nuclear receptor.
   b. degradation of a repressor of gene transcription.
   c. production of a small molecule second messenger.
   d. light absorption followed by chemical changes.
   e. breakdown of the hormone.

5. Ethylene
   a. causes the triple response in seedlings growing underground.
   b. is liquid at room temperature.
   c. delays the ripening of fruits.
   d. generally promotes stem elongation.
   e. inhibits the swelling of stems, in opposition to cytokinin's effects.

6. Phytochrome
   a. is the only photoreceptor pigment in plants.
   b. exists in two forms that are interconvertible by light.
   c. is a pigment that is colored red or far-red.
   d. is a green-light receptor.
   e. is the photoreceptor for phototropism in coleoptiles.

## UNDERSTANDING & APPLYING

7. Describe the circumstances under which it would be advantageous for a species to have the dormancy of its seeds broken by fire.

8. Cocklebur fruits contain two seeds each that are kept dormant by two different mechanisms. Why might having two mechanisms of dormancy be advantageous to cockleburs?

9. Supermarkets sell plastic bags that are impregnated with activated charcoal, which binds gases. The bags are designed to keep fruit fresh. How do they work?

## ANALYZING & EVALUATING

10. Corn stunt spiroplasma (a bacterium) causes a great reduction in the growth rate of infected corn plants. Diseased plants take on a dwarfed form. Since their appearance is reminiscent of genetically dwarfed corn, you suspect that the bacterium may inhibit the synthesis of gibberellins by corn plants. Describe two experiments you might conduct to test this hypothesis, only one of which should require chemical measurement.

A corn plant infected with corn stunt spiroplasma

11. The semi-dwarf wheat and rice plants that led to the Green Revolution described in the chapter opening have mutations in the signal transduction pathway for gibberellins. You wish to use genetic engineering to make corn plants that are semi-dwarf.
    a. How would you do a genetic screen to identify the genes in corn involved in gibberellin signaling?
    b. Assuming that the signal transduction pathway is similar to that in *Arabidopsis*, what gene would you select for inactivation?
    c. Besides short stature, what other effects would you expect for the signal transduction mutant strain? How would you use other hormones to overcome them?

Go to BioPortal at **yourBioPortal.com** for Animated Tutorials, Activities, LearningCurve Quizzes, Flashcards, and many other study and review resources.

# 38 Reproduction in Flowering Plants

## CHAPTER OUTLINE


**38.1** How Do Angiosperms Reproduce Sexually?

**38.2** What Determines the Transition from the Vegetative to the Flowering State?

**38.3** How Do Angiosperms Reproduce Asexually?

**Dandelion** This plant can reproduce by forming seeds without sex.

YOU MAY NOT CONSIDER A BOTANIST specializing in plant reproduction to be "cool," but Arthur Hemmings, a fictional plant biologist at the famous Kew Botanical Gardens in London, might change your mind. In addition to being a plant biologist, Hemmings is an undercover agent for Britain's Office of Food Security. Hemmings is the protagonist in the recent novel *Day of the Dandelion* by Paul Pringle. The theft of a packet of seeds from the lab of an Oxford University geneticist, coupled with the disappearance of the scientist and his lab assistant, prompts the government to assign Hemmings to the case. The stolen seeds are unusual and highly valued. The seeds will be a hybrid between corn (*Zea mays*) and its close relative gamagrass (*Tripsacum dactyloides*), having the high grain production and disease-resistance characteristics of modern corn and an ability to reproduce asexually in a method called apomixis (thanks to a gene from gamagrass). The race to find the country (China? Russia?) or company (a multinational corporation that wants to dominate the food supply?) that has stolen the seeds takes the fictional scientist–spy around the world. It is an exciting tale, with a basis in reality because scientists at the U.S. Department of Agriculture have indeed bred a corn–*Tripsacum* hybrid that shows apomixis, although its productivity is low.

As we described in Chapter 12, modern farmers typically plant hybrid corn, created by seed companies from crosses between homozygous inbred strains. The heterozygous offspring of these crosses show hybrid vigor, with valuable characteristics such as high grain production. If a farmer allows these plants to reproduce among themselves, however, the result will be a collection of plants with some homozygosity and some heterozygosity (think of the results from this cross: AaBbCc × AaBbCc), and hybrid vigor will be lost. So crosses between inbred lines must be done every time hybrid seeds are needed.

Some species of plants, such as dandelions, blackberries, and gamagrass, have genes for apomixis, which prevents meiosis in the cells that form the female gametes during sexual reproduction. Instead of forming haploid egg cells, these plants can form diploid egg cells that don't need to be fertilized by male gametes. The egg cells go on to form clone diploid offspring: plants that are genetically identical to the mother plant. For these plant species, apomixis provides an evolutionary advantage in that it allows for reproduction without pollination, and for the propagation of well-adapted genotypes. However, a disadvantage is that the genetic variation that results from sexual reproduction is lost.

For the plant breeder, apomixis could be a boon. Apomixis would allow for the rapid propagation of hybrid plants without the need to make crosses and recreate the hybrids every time they are grown. The potential profits are staggering—thus the excitement of *Day of the Dandelion*.

By what genetic mechanism is apomixis brought about?

See answer on p. 794.

## 38.1 How Do Angiosperms Reproduce Sexually?

Most angiosperms (flowering plants) have evolved to reproduce sexually because this strategy has the selective advantage of producing the genetic diversity that is the raw material for evolution. Sexual reproduction in angiosperms involves mitosis, meiosis, and the alternation of haploid and diploid generations (see Figure 11.15). There are several important differences between sexual reproduction in angiosperms and in vertebrate animals (the latter also discussed in Chapter 43):

- Meiosis in plants produces spores, after which mitosis produces gametes; in animals, meiosis usually produces gametes directly.
- In most plants, there are multicellular diploid (sporophyte) and haploid (gametophyte) life stages (alternation of generations); in animals, there is no multicellular haploid stage.
- In plants, the cells that will form gametes are determined in the adult organism, usually in response to environmental conditions; in animals, the germline cells are determined before birth.

### The flower is an angiosperm's structure for sexual reproduction

The plant life cycle typically involves the alternation of haploid and diploid generations (see Chapter 11):

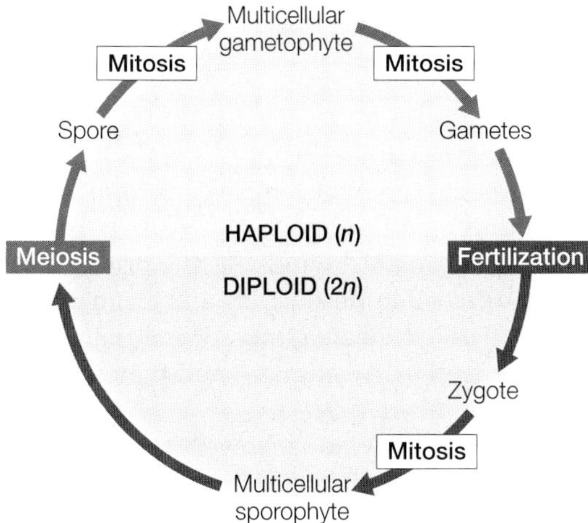


In angiosperms, the plant that we see in nature is a sporophyte, and the male and/or female gametophytes are contained in the flowers (see Section 29.3 for a description of flower parts and floral evolution). A complete flower consists of four concentric groups of organs arising from modified leaves: the carpels, stamens, petals, and sepals.

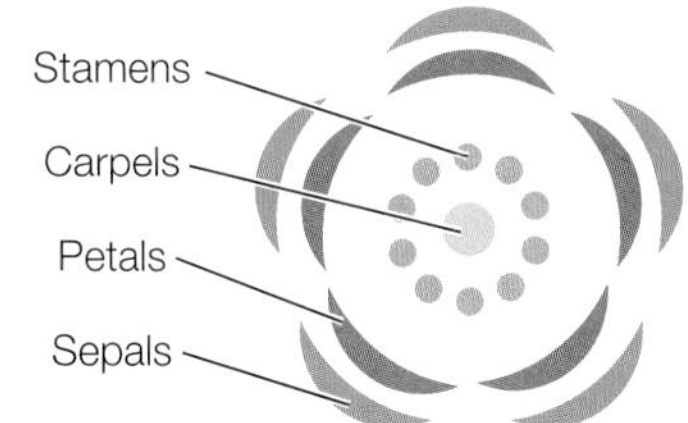


The parts of the flower are usually borne on a stem tip, and derive from a meristem.

- The carpels are the female sex organs that contain the developing female gametophytes.
- The stamens are the male sex organs that contain the developing male gametophytes.

The differentiation of the meristem into the various organs of the flower is controlled by specific transcription factors (see Figure 19.9).

Most angiosperm species are **hermaphroditic**, and have flowers with both stamens and carpels; such flowers are termed *perfect* (**Figure 38.1A**). *Imperfect* flowers, by contrast, are those with only male or only female sex organs. Male flowers have stamens but not carpels, and female flowers have carpels but not stamens. Some plants, such as corn, bear both male and female flowers on an individual plant; such species are called **monoecious** ("one house") (**Figure 38.1B**). In **dioecious** species, individual plants bear either male-only or female-only flowers; an example is American holly (**Figure 38.1C**).

### Flowering plants have microscopic gametophytes

**Figure 38.2** offers a detailed look at the gametophytes central to angiosperm reproduction. The haploid gametophytes—the gamete-producing structures—develop from haploid spores in the flower:

- Each female gametophyte (megagametophyte) is called an **embryo sac**, and it develops inside an ovule. One or more ovules are contained within the ovary, which is the lower part of the carpel.
- Male gametophytes (microgametophytes), which are called **pollen grains**, develop inside the anther, which is part of the stamen.

FEMALE GAMETOPHYTE Of the four haploid megaspores resulting from meiosis, three undergo apoptosis (programmed cell death). The remaining megaspore undergoes three mitotic divisions without cytokinesis, producing eight haploid nuclei, all initially contained within a single cell—three nuclei at one end, three at the other, and two in the middle. Subsequent cell wall formation leads to an elliptical, seven-celled megagametophyte with a total of eight nuclei:

- At one end of the megagametophyte are three small cells: the *egg cell* and two cells called *synergids*. The egg cell is the female gamete, and the synergids participate in fertilization by attracting the pollen tube. The pollen tube enters one of the synergids before the sperm cells are released for fertilization.
- At the opposite end of the megagametophyte are three antipodal cells, which eventually degenerate.
- In the large central cell are two **polar nuclei**.

The megagametophyte, or embryo sac, is the entire seven-cell, eight-nucleus structure.

(A) Perfect: lily (*Lilium* sp.)

(B) Imperfect monoecious: corn

(C) Imperfect dioecious: American holly (*Ilex opaca*)

Male flower with stamens

Female flower with carpels

**38.1 Perfect and Imperfect Flowers** (A) A lily is an example of a perfect flower, meaning one that has both male and female sex organs. (B) Imperfect flowers are either male or female. Corn is a monoecious species: both types of imperfect flowers are borne on the same plant. (C) American holly is a dioecious species; some American holly plants bear male imperfect flowers whereas others bear female imperfect flowers.

MALE GAMETOPHYTE The four haploid products of meiosis (the microspores) each develop a cell wall and undergo a single mitotic division, producing four two-celled pollen grains that are released into the environment. The two cells in a pollen grain have different roles:

- After pollination (see below) the generative cell divides by mitosis to form two sperm cells that participate in fertilization.
- The tube cell forms the elongating pollen tube that delivers the sperm to the embryo sac.

These events occur after the pollen grain is transferred to a stigma (part of the female reproductive organ)—a process called **pollination**.

## Pollination in the absence of water is an evolutionary adaptation

As Chapter 28 described, the union of gametes in aquatic plants is accomplished in the water. Fertilization of mosses and ferns also requires at least a film of water for the movement of gametes. While there are mechanisms to ensure that fertilization occurs if and when the two gametes meet, fertilization is clearly a low-probability event. The evolution of pollen made it possible for male gametes to reach the female gametophyte without an aqueous conduit.

In the first seed plants, wind was the primary vehicle by which pollen reached its destination, and the majority of gymnosperms are wind-pollinated today. Wind-pollinated flowers have sticky or featherlike stigmas, and they produce pollen grains in great numbers.

Pollen transport by wind is, however, a relatively chancy means of achieving pollination, explaining why about 90 percent of all angiosperms still rely on animals—including insects, birds, and bats—for pollen transport. Pollen transport by animals greatly increases the probability that pollen will get to the female gametophyte. As described in Sections 29.3 and 56.3, the structures of flowers have coevolved with their animal pollinators to enhance the plants' chances of successful pollination. Suitably pigmented, shaped, and scented flowers attract the pollinating animal, resulting in a pollen transfer from flower to flower within the same plant species (**Figure 38.3**).

## A pollen tube delivers sperm cells to the embryo sac

When a functional pollen grain lands on the stigma of a compatible stigma, it germinates. A key event is water uptake

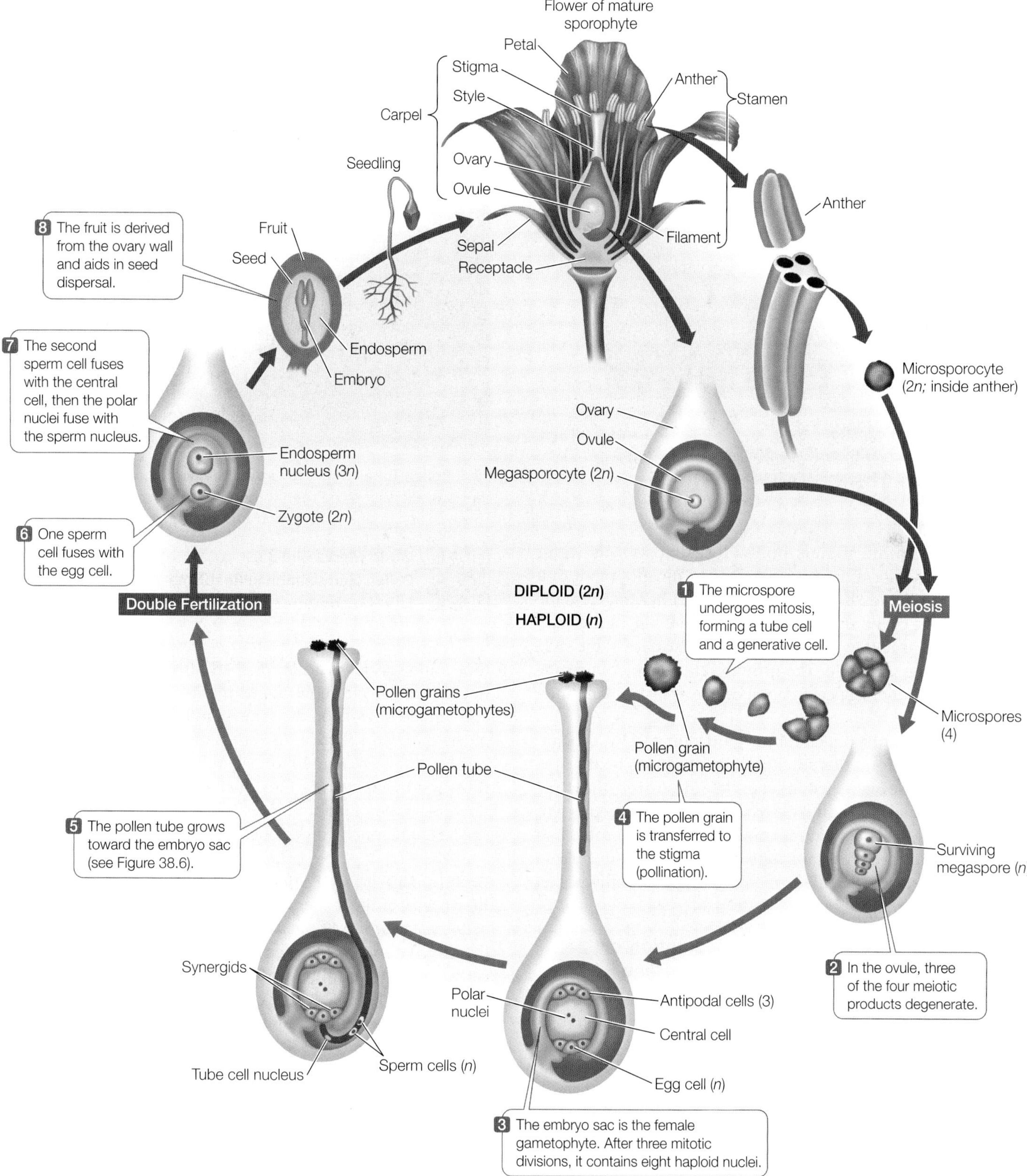


**38.2 Sexual Reproduction in Angiosperms** The embryo sac is the female gametophyte; the pollen grain is the male gametophyte. The male and female cells meet and fuse within the embryo sac. Angiosperms have double fertilization, in which a zygote and an endosperm form from separate fusion events. The zygote forms by the fusion of one sperm cell with the egg cell. The endosperm forms after the other sperm cell fuses with the central cell, which contains two nuclei. The three nuclei fuse, forming a triploid cell.

**Go to Activity 38.1**
**Sexual Reproduction in Angiosperms**
**Life10e.com/ac38.1**

**38.3 Flower and Pollinator**
Some flowers, such as these *Cavendishia* sp. flowers, have red pigments and a shape that attracts certain birds.

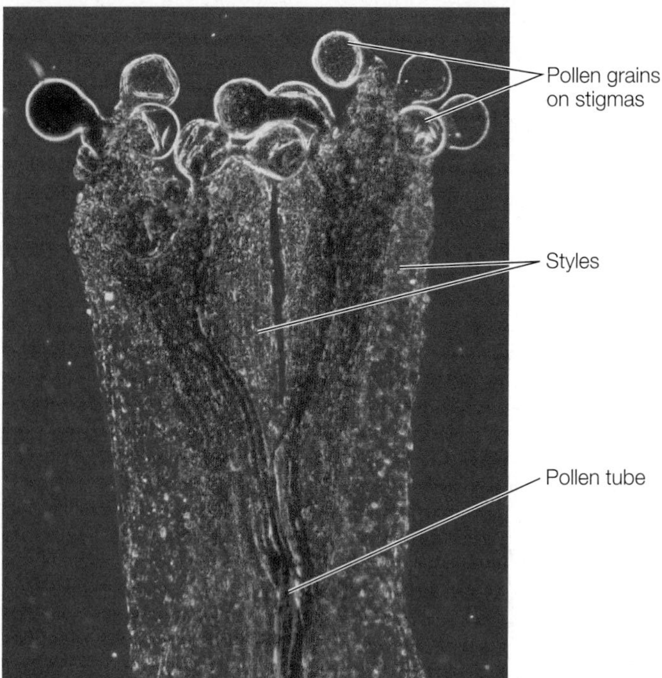


**38.4 Pollen Tubes Begin to Grow** Staining pollen with a fluorescent dye allows it to be seen through a fluorescence microscope. These pollen grains have landed on the stigmas of a crocus.

**Go to Media Clip 38.1**
**Pollen Germination in Real Time**
Life10e.com/mc38.1

by pollen from the stigma: pollen loses most of its water as it matures. Germination involves the development of a **pollen tube** (**Figure 38.4**). The pollen tube either traverses the spongy tissue of the style (part of the carpel; see Figure 38.2) or, if the style is hollow, grows on the inner surface of the style until it reaches an ovule. The growth rate of the pollen tube varies greatly among species, but it can be as fast as 3 millimeters an hour.

The growth of the pollen tube is guided in part by a chemical signal in the form of a small protein produced by the synergids within the ovule. If one synergid is destroyed, the ovule still attracts pollen tubes, but destruction of both synergids renders the ovule unable to attract pollen tubes, and fertilization does not occur. The attractant appears to be species-specific: in some cases, isolated female gametophytes attract only pollen tubes of the same species.

## Many flowering plants control pollination or pollen tube growth to prevent inbreeding

You may recall from discussions of Mendel's work (see Section 12.1) that some plants can reproduce sexually by both cross-pollination and self-pollination. Self-pollination increases the chances of successful pollination but leads to homozygosity, which reduces genetic diversity. Because diversity is the raw material of evolution by natural selection, homozygosity can be selectively disadvantageous. Most plants have evolved mechanisms that prevent self-fertilization. A multitude of strategies exist (for example, see Figure 29.13). Two primary means to prevent self-fertilization are (1) physical separation of male and female gametophytes and (2) genetic self-incompatibility.

SEPARATION OF MALE AND FEMALE GAMETOPHYTES Self-fertilization is prevented in dioecious species, which bear only male or female flowers on a particular plant. Pollination in dioecious species is accomplished only when one plant pollinates another. In monoecious plants, which bear both male and female flowers on the same plant, the physical separation of the male and female flowers is often sufficient to prevent self-fertilization. Some monoecious species prevent self-fertilization by staggering the development of male and female flowers so they do not bloom at the same time, making these species functionally dioecious.

GENETIC SELF-INCOMPATIBILITY A pollen grain that lands on the stigma of the same plant will fertilize the female gamete *only if the plant is self-compatible,* meaning capable of self-pollination. To prevent self-fertilization, many plants are genetically self-incompatible. **Self-incompatibility** depends on the ability of a plant to determine whether pollen is genetically similar or genetically different from itself. Rejection of "same-as-self" pollen prevents self-fertilization. How does it occur?

Self-incompatibility in plants is controlled by a cluster of tightly linked genes called the *S* locus (for self-incompatibility). The *S* locus encodes proteins in the pollen and style that interact during the recognition process. A self-incompatible species typically has many alleles of the *S* locus. The pollen phenotype may be determined by its own haploid genotype or by the diploid genotype of its parent plant. In either case, if the pollen expresses an allele that matches either of the alleles expressed in the recipient pistil, the pollen is rejected. Depending on the type of self-incompatibility system, the rejected pollen either fails to germinate or is prevented from growing through the style (**Figure 38.5**); either way, self-fertilization is prevented.

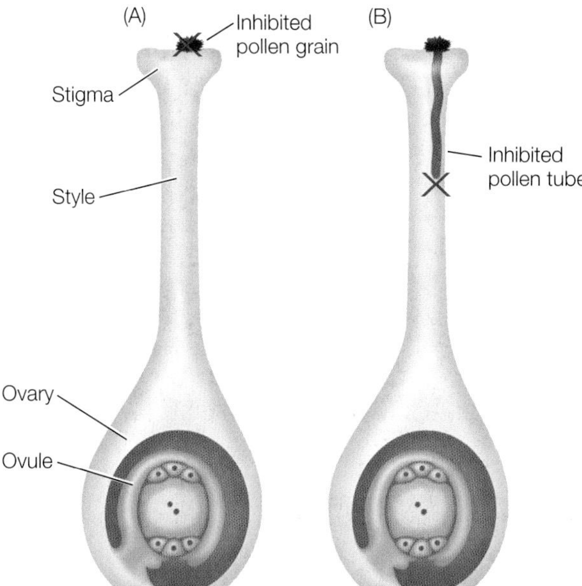


**38.5 Self-Incompatibility** In a self-incompatible plant, pollen is rejected if it expresses an *S* allele that matches one of the *S* alleles of the stigma and style. Self pollen may (A) fail to germinate or (B) its pollen tube may die before reaching an ovule. In either case, the egg cannot be fertilized by a sperm from the same plant.

## Angiosperms perform double fertilization

In most angiosperm species, the mature pollen grain consists of two cells: a large tube cell enclosing a much smaller generative cell. When a compatible pollen grain lands on the stigma of a plant of the same species, it germinates, and the pollen tube grows through the style tissue to the embryo sac. During this process, the generative cell undergoes one mitotic division and cytokinesis to produce two haploid **sperm cells** (**Figure 38.6, steps 1 and 2**).

Two fertilization events now occur. One of the two synergids degenerates when the pollen tube arrives and the two sperm cells are released into its remains. (**Figure 38.6, step 3**). Each sperm cell then fuses with a different cell of the embryo sac (**Figure 38.6, steps 4 and 5**). One sperm cell fuses with the egg cell, and the two nuclei fuse, producing the diploid zygote. The other sperm cell fuses with the central cell, and its nucleus fuses with the two polar nuclei, forming a **triploid** (**3n**) **cell**. Immediately after fertilization the triploid nucleus undergoes rapid mitotic divisions to form a specialized nutritive tissue, the **endosperm**. In most species the endosperm nucleus initially divides without cytokinesis, forming a large, multinucleate cell, and cell walls form later between the nuclei. After the endosperm begins developing, the zygote undergoes mitotic division to form the new sporophyte embryo. The developing embryo uses the endosperm tissue as a source of nutrients, energy, and carbon-based anabolic building blocks. In some cases the endosperm persists until germination and is used as a source of nutrients by the developing seedling. This source of nutrients is important because the seedling often begins its development underground and cannot perform photosynthesis right away.

The remaining cells of the male and female gametophytes—the antipodal cells, the remaining synergid, and the pollen tube nucleus—degenerate as the embryo begins to develop.

**Double fertilization** is so named because it involves two cell fusion events:

- One sperm cell fuses with the egg cell.
- The other sperm cell fuses with the central cell.

The fusion of a sperm cell with the central cell to form the triploid endosperm nucleus is one of the defining characteristics of angiosperms.

Go to Animated Tutorial 38.1
Double Fertilization
Life10e.com/at38.1

**38.6 Double Fertilization** Two sperm are involved in two cell fusion events, hence the term "double fertilization." One sperm is involved in the formation of the diploid zygote, and the other results in the formation of the triploid endosperm cell that divides to form endosperm. Double fertilization is a characteristic feature of angiosperm reproduction.

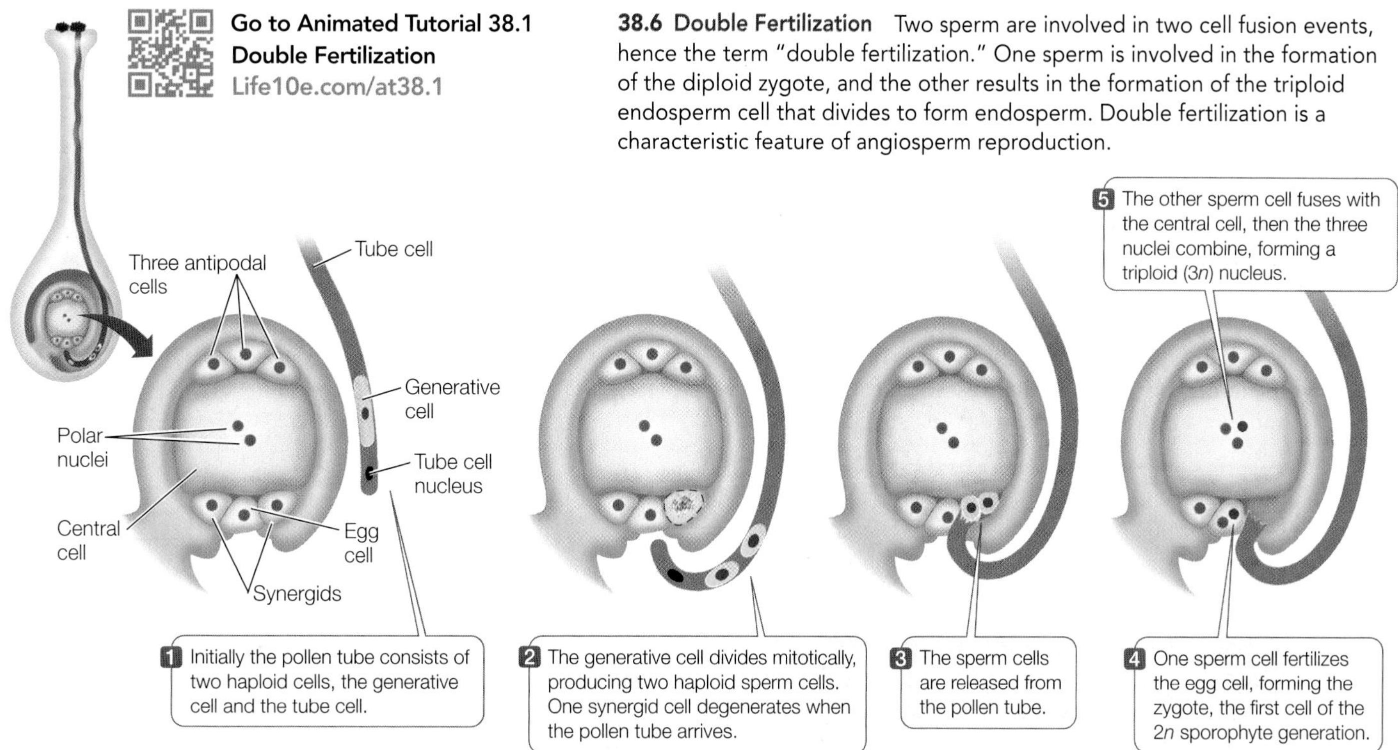

## Embryos develop within seeds contained in fruits

Fertilization initiates the highly coordinated growth and development of the embryo, endosperm, integuments, and carpel. The integuments—protective tissue layers surrounding the ovule—develop into the seed coat, and the ovary wall becomes the outer layers of the fruit that encloses the seed (see Figure 38.2).

In Chapter 37 we described the events in plant embryonic development and its hormonal control. As seeds develop, they prepare for dispersal and dormancy by losing up to 95 percent of their water content. You can see this desiccation by comparing corn grains (e.g., popcorn) with ripe corn from the cob or a can. A dry seed is still alive; it has protective proteins that keep its cells in a viscous state.

In angiosperms, the ovary—together with the seeds it contains—develops into a fruit after fertilization has occurred. Fruits have two main functions:

- They protect the seed from damage by animals and infection by microbial pathogens.
- They aid in seed dispersal.

A **fruit** may consist of only the mature ovary and seeds, or it may include other parts of the flower. Some species produce fleshy, edible fruits, such as peaches and tomatoes, whereas the fruits of other species are dry or inedible (**Figure 38.7**).

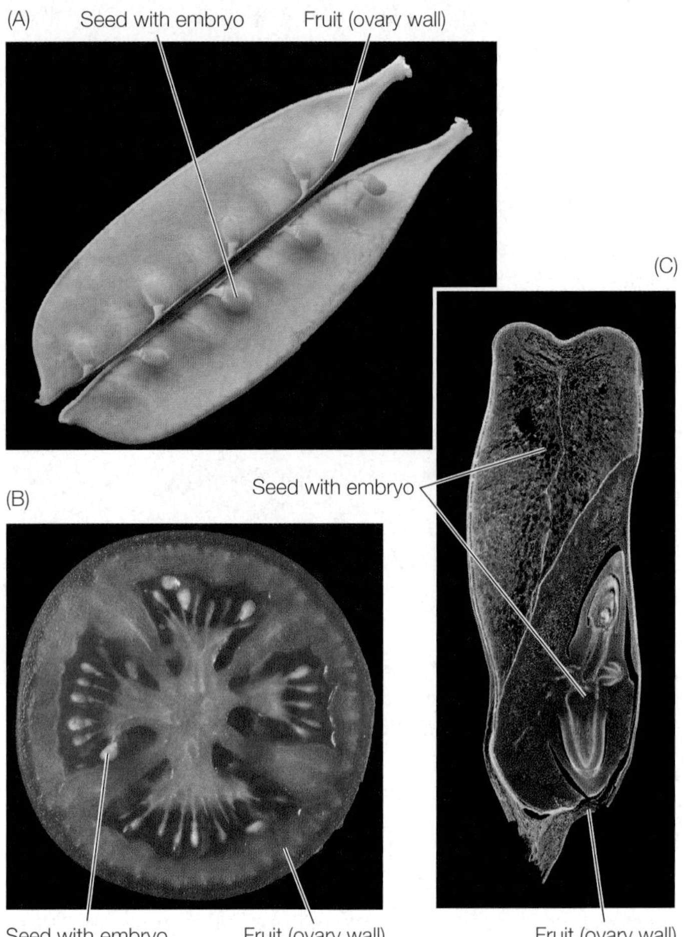


**38.7 Angiosperm Fruits** There are a variety of fruits, but all have seed containing the embryo, surrounded by a fruit that comes from the wall of the ovary. (A) Garden pea. (B) Tomato. (C) Corn.

The diverse forms of fruits reflect the varied strategies plants use to disperse their progeny. Because plants cannot move, their progeny need some mechanism for separating themselves from their parents. Wide dispersal of progeny may not always be advantageous, however. If a plant has successfully grown and reproduced, its location is likely to be favorable for the next generation too. Some offspring do indeed stay near their parents. This is the case in many tree species, whose seeds simply fall to the ground. However, this strategy has several potential disadvantages. If the species is a perennial, the offspring that germinate near their parents will be competing with their parents for resources, which may be too limited to support a dense population. Furthermore, even though local conditions were good enough for the parent to produce at least some seeds, there is no guarantee that conditions will still be good the next year, or that they won't be better elsewhere. Thus in many cases seed dispersal is vital to a species' survival.

Many fruits help disperse seeds over substantial distances, increasing the probability that at least a few of the seeds will find suitable conditions for germination and growth to sexual maturity. Some wind-dispersed seeds have fruits with "wings," like those of the familiar maple. In other cases, the fruits and seeds are tiny, and they include feathery structures, such as this thistle:

Other fruits, such as these burs, attach themselves to animals (or to your clothes and shoes):

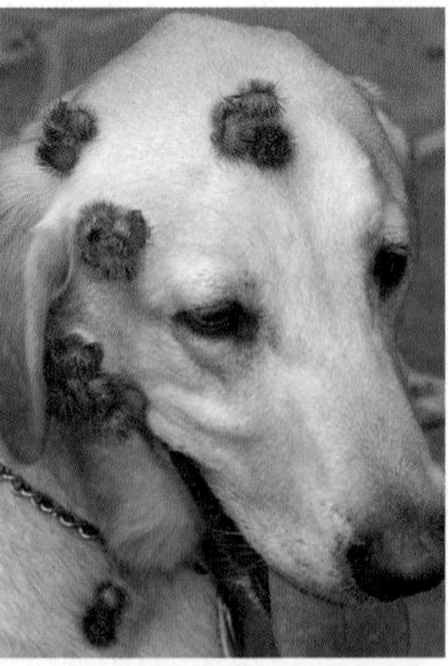

Water disperses some fruits; coconuts have been known to float thousands of miles between islands. Seeds swallowed whole by an animal along with fruits such as berries travel through the animal's digestive tract and are deposited some distance from the parent plant.

Biologists are beginning to understand the relationships between seed development (from ovules) and fruit development (from carpels). Some seedless fruits, such as varieties of watermelons and grapes, develop when fertilization occurs but the embryo then aborts. In other cases, such as bananas and pineapples, the fruit develops without fertilization. In most cases, however, fruit development does not occur in the absence of fertilization. Several years ago a farmer in Spain who grows sugar apples (*Annona squamosa*) noticed a seedless fruit and brought it to the attention of scientists. A single gene was subsequently identified whose mutated form results in this seedless phenotype. The wild-type version of this gene encodes a transcription factor important to the development of the outer tissues of the ovule. The gene is present in the genomes of all angiosperms examined. In the future, therefore, it may be possible to produce other seedless fruits by engineering mutations in this gene.

### Seed development is under hormonal control

Chapter 37 described the role of the gibberellin hormones in the mobilization of stored macromolecules in the seed endosperm during germination. The development of seeds is under the control of a different hormone, **abscisic acid** (**ABA**). (Unfortunately, its name is misleading, because it does not directly control leaf abscission.) Most plant tissues make ABA, and like other plant hormones, it has multiple effects (see Table 37.2). For example, ABA plays a role in stomatal closing under water stress conditions (see Section 35.3). During early seed development the ABA level is low, and it rises as the seed matures. This increase stimulates the endosperm to synthesize seed storage proteins. It also stimulates the synthesis of proteins that prevent cell death as the seeds dry.

ABA also keeps the developing seed from germinating on the plant before it dries. Premature germination, termed **vivipary**, is undesirable in seed crops (such as wheat) because the grain is damaged if it starts to sprout. Viviparous seedlings are also unlikely to survive if they remain attached to the parent plant and are unable to establish themselves in the soil. Mutants of corn that are insensitive to ABA have viviparous seeds, indicating the importance of ABA in preventing precocious germination.

The general effect of ABA in preventing germination extends to seed dormancy. Seeds stay dormant if their ABA levels are high and germinate when the levels go down. This usually occurs as dormancy is broken.

#### RECAP 38.1

Flowers contain the organs for sexual reproduction in angiosperms. Plants that use pollen for reproduction have several selective advantages, among them the ability to accomplish fertilization without water. After fertilization, the flower develops into seed(s) and fruit. The selective advantages of seeds and fruits include long-term viability and multiple modes of dispersal.

- What are the relationships between an ovule and an ovary, and between a fruit and a seed? **See p. 779 and Figure 38.2**
- How do plants prevent self-pollination? **See p. 782 and Figure 38.5**
- Describe the roles of the two sperm cells in double fertilization. **See p. 783 and Figure 38.6**
- How is plant development controlled by the hormone abscisic acid? **See p. 785**

We have now traced the sexual life cycle of angiosperms from the flower to the gametophytes, pollination, fertilization, and the dispersal of seeds. We discussed seed germination and seedling development in Chapter 37, and indeterminate, vegetative plant growth in Chapter 34. The next section will cover the rest of the angiosperm life cycle—the transition from the vegetative to the flowering state—and how this transition is regulated.

## 38.2 What Determines the Transition from the Vegetative to the Flowering State?

Flowering is one of the major events in a plant's life. Flowering requires a reallocation of energy and materials away from making more plant parts (vegetative growth) to making flowers and gametes (reproductive growth). Once a plant is old enough, it can respond to internal or external signals to initiate reproduction. Flowering can happen right at maturity as part of a predetermined developmental program (as in a dandelion plant in the summer) or in response to environmental cues such as light or temperature (as with most ornamental flowers).

Plants fall into three categories depending on when they mature and initiate flowering, and what happens after they flower:

- **Annuals** complete their lives in one year. This class includes many crops important to the human diet, such as corn, wheat, rice, and soybean. When the environment is suitable, these plants grow rapidly, with little or no secondary growth. After flowering, they use most of their materials and energy to develop seeds and fruits, and the rest of the plant withers away.
- **Biennials** take 2 years to complete their lives. They are much less common than annuals and include carrots, cabbage, and onions. Typically, biennials produce only vegetative growth during the first year and store carbohydrates in underground roots (carrot) and stems (onion). In the second year they use most of the stored carbohydrates to produce flowers and seeds rather than vegetative growth, and the plant dies after seeds form.
- **Perennials** live 3 or more—sometimes many more—years. Maple trees can live up to 400 years. Perennials include many trees and shrubs, as well as wildflowers. Typically these plants flower every year but stay alive and keep growing for another season; the reproductive cycle repeats each year. However, some perennials (e.g., century plant) grow vegetatively for many years, flower once, and die.

No matter what type of life cycle they have, angiosperms all make the transition to flowering. This transition entails significant developmental changes, to which we now turn.

### Shoot apical meristems can become inflorescence meristems

The first visible sign of a transition to the flowering state may be a change in one or more apical meristems in the shoot

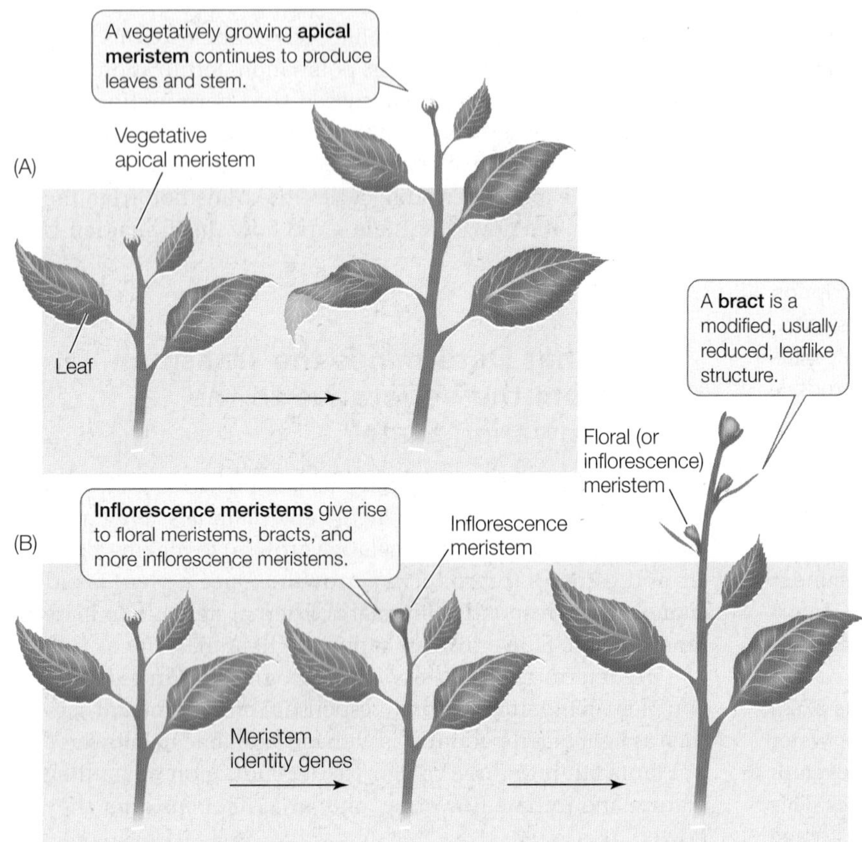


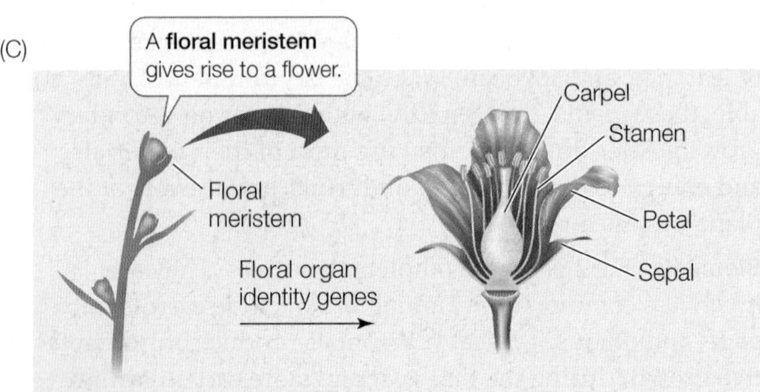


**38.8 Flowering and the Apical Meristem** A vegetative apical meristem (A) grows without producing flowers. Once the transition to the flowering state is made (B), inflorescence meristems give rise to bracts and to floral meristems (C), which become the flowers.

system. As described in Chapter 34, meristems have a pool of undetermined cells. During vegetative growth, a shoot apical meristem continually produces leaves, axillary buds, and stem tissues (**Figure 38.8A**) in a kind of unrestricted growth called *indeterminate growth* (see Section 34.3).

Flowers may appear singly or in an orderly cluster that constitutes an **inflorescence**. If a vegetative apical meristem becomes an **inflorescence meristem**, it ceases production of leaves and axillary buds and produces other structures: smaller leafy structures called bracts, as well as new meristems in the angles between the bracts and the stem (**Figure 38.8B**). These new meristems may also be inflorescence meristems, or they may be **floral meristems**, each of which gives rise to a flower.

Each floral meristem typically produces four consecutive whorls or spirals of organs—the sepals, petals, stamens, and carpels discussed earlier in the chapter—separated by very short internodes, keeping the flower compact (**Figure 38.8C**). In contrast to vegetative apical meristems and some inflorescence meristems, floral meristems are responsible for *determinate growth*—growth of limited duration.

## A cascade of gene expression leads to flowering

The genes that determine the transition from shoot apical meristems to inflorescence meristems and from inflorescence meristems to floral meristems have been studied in model organisms such as *Arabidopsis*.

**MERISTEM IDENTITY GENES** Expression of two **meristem identity genes** initiates a cascade of further gene expression that leads to flower formation. These genes encode the transcription factors LEAFY and APETALA1, which together are necessary and sufficient for determining the transition to flowering. Evidence for the roles of these factors comes from both genetic and plant transformation experiments. For example, a mutant allele of the *APETALA1* gene leads to continued vegetative growth, even if conditions are suitable for flowering. However, if the wild-type *APETALA1* gene is coupled with a constitutive (always on) promoter and used to transform *Arabidopsis* plants, the plants will flower prematurely, regardless of environmental conditions. This is powerful evidence that APETALA1 plays a role in switching meristem cells from a vegetative to a reproductive fate (see Figure 38.8B).

**FLORAL ORGAN IDENTITY GENES** The products of the meristem identity genes trigger the expression of **floral organ identity genes**. As described in Section 19.4, these genes work in concert to specify the successive whorls of the flower (see Figure 19.9). The floral organ identity genes encode transcription factors that determine whether cells in the floral meristem will be sepals, petals, stamens, or carpels. For example, *AGAMOUS* is a class C gene that causes florally determined cells to form stamens and carpels.

Depending on the species, plants initiate these gene expression changes, and the events that follow, in response to either internal or external cues. The most well studied external cues are photoperiod (day length) and temperature. We will begin with photoperiod.

**38.9 Mammoth Plant** Wild-type tobacco (left) is much smaller than the Maryland Mammoth mutant of the same age (right), which does not respond to an environmental cue to stop growing and flower.

## Photoperiodic cues can initiate flowering

The study of how light affects the transition to flowering began with two observations in the early twentieth century:

- Normally, tobacco grows to about 1.5 meters tall before flowering in the summer, but a variety called Maryland Mammoth grows to 5 meters (**Figure 38.9**). Farmers in Virginia were frustrated because they could not easily get seeds from this luxuriant plant for successive crops. Instead of flowering, it continued to grow until the late fall frost killed it.
- Because of improvements in agricultural techniques, soybean yields became so great that it was hard for farmers to harvest all the plants at once. Hoping to stagger the harvests, farmers tried planting the seeds in groups several weeks apart, but all resulting plants nevertheless formed flowers and seeds at the same time.

The explanation for both of these observations was the same: the signal that set the plants' shoot apical meristems on the path to flowering was the length of daylight, or **photoperiod**. When soybeans experience days of a certain length, they flower, regardless of how "old" they are. Maryland Mammoth tobacco *can* flower, but it doesn't do so in Virginia because it dies when the weather there gets cold. When the plant is grown in a greenhouse to prevent freezing, however, it flowers in December, when the days are short. Maryland Mammoth is now grown commercially in Florida.

Scientists used greenhouse experiments to measure the day length required for different plant species to flower. Maryland Mammoth tobacco did not flower if exposed to more than 14 hours of light per day; flowering was only initiated once day length became shorter than 14 hours, as it does in December. Other plants (such as soybeans and henbane) flowered only when the days were long (**Figure 38.10**). Control of an organism's responses by the length of day or night is called **photoperiodism**.

## Plants vary in their responses to photoperiodic cues

Plants that flower in response to photoperiodic stimuli fall into two main classes:

- **Short-day plants (SDPs)** flower only when the day is shorter than a critical maximum. They include poinsettias and chrysanthemums, as well as Maryland Mammoth tobacco. Thus, for example, we see chrysanthemums in nurseries in the fall and poinsettias in winter.
- **Long-day plants (LDPs)** flower only when the day is longer than a critical minimum. Spinach and clover are examples of LDPs. For example, spinach tends to flower in the summer, so it is normally planted in early spring.

While there are variations on these two patterns, photoperiodic control of flowering serves an important role: it synchronizes the flowering of plants of the same species in a local population, and this promotes cross-pollination and successful reproduction.

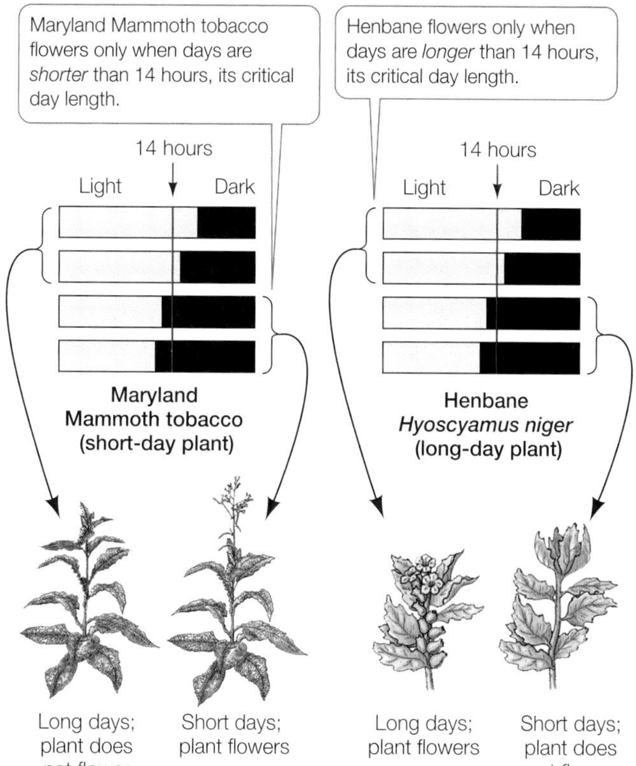


**38.10 Day Length and Flowering** Flowering of Maryland Mammoth tobacco is initiated when the days become shorter than a critical length. Maryland Mammoth tobacco is thus called a short-day plant. Henbane, a long-day plant, shows an inverse pattern of flowering.

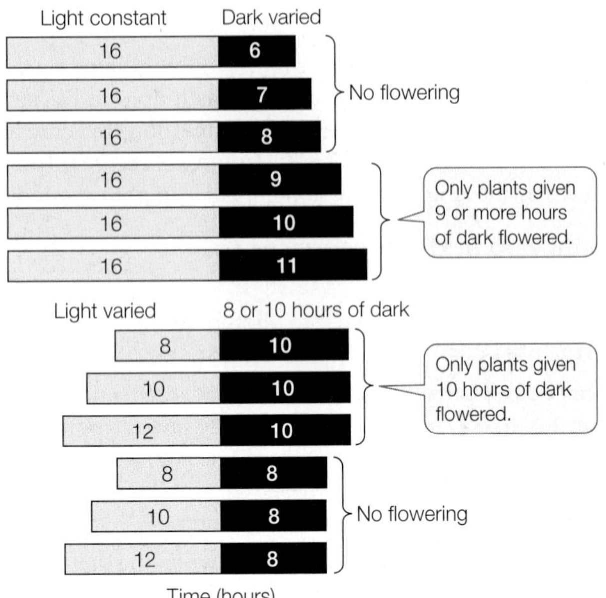


**38.11 Night Length and Flowering** Greenhouse experiments using cocklebur, a short-day plant, showed that night length, not day length, is the environmental cue that initiates flowering.

## Night length is a key photoperiodic cue that determines flowering

The terms "short-day plant" and "long-day plant" imply that *day length* is the environmental cue that triggers flowering. Actually, the important cue is *night length*, as a series of greenhouse experiments confirmed (**Figure 38.11**). In a greenhouse, the overall length of a day or night can be varied, irrespective of the 24-hour natural cycle. For example, if cocklebur, an SDP, is exposed to several long periods of light (16 hours each), it will still flower as long as the dark period between them is 9 hours or longer. This 9-hour inductive dark period also induces flowering even if the light period varies from 8 hours to 12 hours.

Biologists noticed that when the inductive dark period was interrupted by a brief period of light, the flowering signal generated by the long night disappeared. It took several days of long nights for the plant to recover and initiate flowering. Interrupting the day with a dark period had no effect on flowering. A clue as to what occurred in the plant when the flash of light was given came when biologists determined the action spectrum for the wavelengths of light that were effective. As with lettuce seed germination (see Figure 37.16), red light was most effective at breaking the "night" stimulus, and its effect was reversible by far-red light (**Figure 38.12**). Later, the photoreceptor involved was identified as a phytochrome.

One way that plants sense night length is through phytochrome activity. As explained in Section 37.4, during the day there is more light in the red than in the far-red range, and the inactive $P_r$ isoform of phtyochrome is converted to the active $P_{fr}$ isoform. At night there is a gradual, spontaneous conversion of this $P_{fr}$ back to $P_r$, so at the end of the night the phytochrome is predominantly in its inactive form. But a short burst of red light in the middle of the night will disrupt this rhythm by converting the $P_r$ to $P_{fr}$ again.

INVESTIGATING**LIFE**

**38.12 Interrupting the Night** Knowing that plants measure night duration, the question became whether the dark hours to which a plant is exposed must be continuous. Using SDPs and LDPs as test subjects, Karl Hamner and James Bonner demonstrated that this was the case by interrupting the night with short bursts of light. Sterling Hendricks and William Siegelman repeated the experiments using light of different wavelengths to gain information about the photoreceptor involved.[a]

**HYPOTHESIS** Red light participates in the photoperiodic timing mechanism.

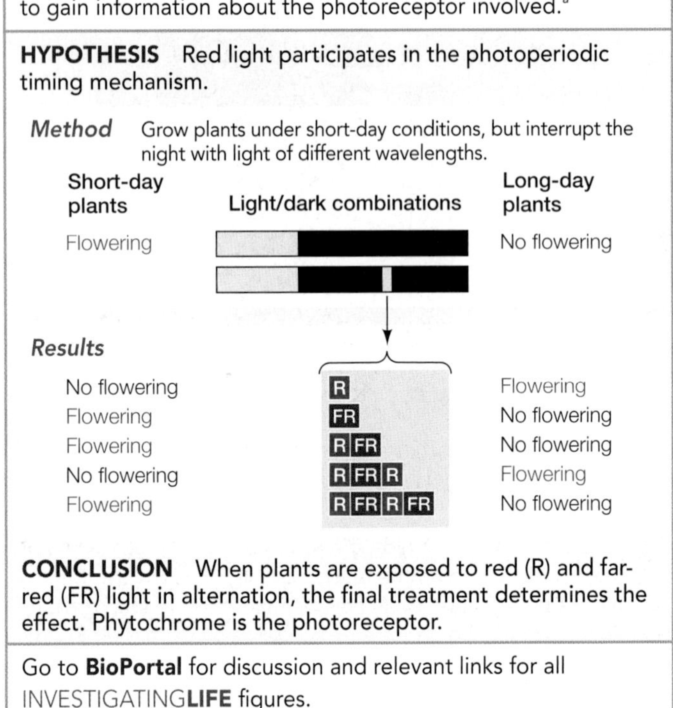


**CONCLUSION** When plants are exposed to red (R) and far-red (FR) light in alternation, the final treatment determines the effect. Phytochrome is the photoreceptor.

Go to **BioPortal** for discussion and relevant links for all INVESTIGATING**LIFE** figures.

[a]Hendricks, S. B. and H. W. Siegelman. 1967. *Comparative Biochemistry* 27: 211–235.

**Go to Animated Tutorial 38.2**
**The Effect of Interrupted Days and Nights**
Life10e.com/at38.2

As described in Section 37.4, phytochromes and a blue-light receptor function together to "entrain" a circadian rhythm in plants. These photoreceptors cycle through active and inactive phases over repeated 24-hour periods, and in doing so they activate signaling pathways that result in regular cycles in the expression of specific genes. One gene whose expression follows a circadian rhythm is *CONSTANS* (*CO*), which encodes a transcriptional regulator that controls the expression of flowering genes.

Experiments with *Arabidopsis*, an LDP, have shown that photoperiodic flowering times are determined by interactions between photoreceptors and the CO protein. *CO* gene expression goes through regular 24-hour cycles with peak expression late in the day. On long days, this occurs in daylight, late in the afternoon, but on short days the peak expression occurs after dark. On long days, the active forms of phytochrome and a blue-light receptor activate pathways that stabilize the CO protein, which promotes flowering. This process does not occur on short days.

## The flowering stimulus originates in a leaf

Early experiments indicated that reception of the photoperiodic stimulus occurs within the leaf. For example, in the LDP spinach, flowering occurred if the leaves were exposed to long-day periods of light, while the shoot apical meristem was masked to simulate short days. Flowering could *not* occur when the

INVESTIGATING**LIFE**

**38.13 The Flowering Signal Moves from Leaf to Bud** The receptors for photoperiod are in the leaf, but the transition to flowering occurs in the shoot apical meristem. To investigate whether there is a diffusible substance that travels from leaf to bud, Hamner and Bonner exposed only the leaf to the photoperiodic stimulus[a]

**HYPOTHESIS** The leaves measure the photoperiod.

*Method* Grow cocklebur plants (short-day) under long days and short nights. Mask a leaf on some plants and see if flowering occurs.

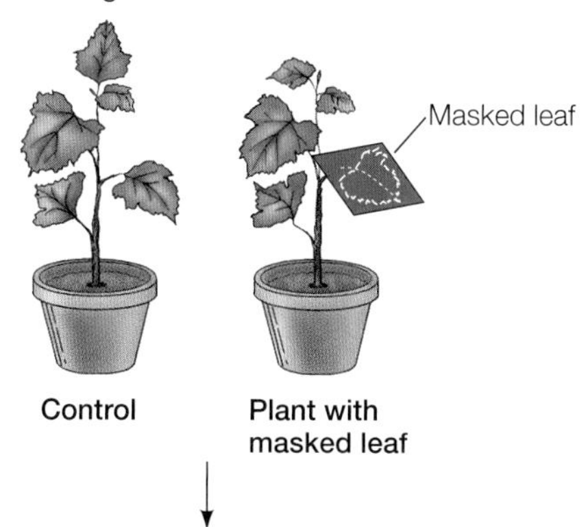


*Results* If even one leaf is masked for part of the day—thus shifting that leaf to short days and long nights—the plant will flower.

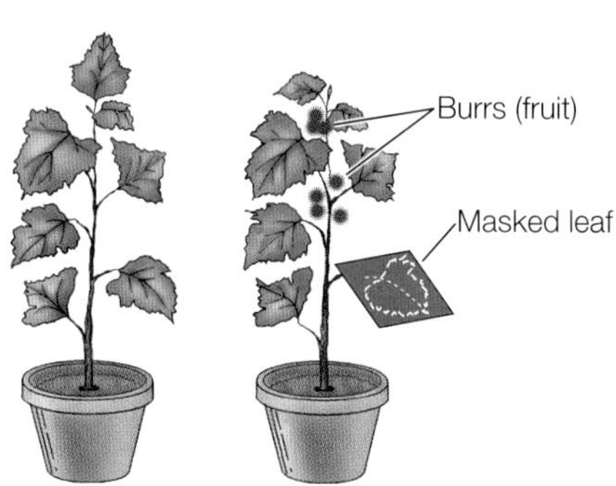


**CONCLUSION** The leaves measure the photoperiod. Therefore some signal must move from the induced leaf to the flowering parts of the plant.

Go to **BioPortal** for discussion and relevant links for all INVESTIGATING**LIFE** figures.

[a]Hamner, K. C. and J. Bonner. 1938. *Botanical Gazette* 100: 388–431.

leaves were masked to simulate short days, while the bud was exposed to long-day periods of light.

These "masking" experiments were extended to SDP plants as well (**Figure 38.13**). Because the receptor of the stimulus (in the leaf) is physically separated from the tissue on which the stimulus acts (the bud meristem), the inference can be drawn that a systemic signal travels from the leaf through the plant's tissues to the bud meristem. Other evidence that a diffusible chemical travels from the leaf to the bud meristem signal includes the following:

- If a photoperiodically induced leaf is immediately removed from a plant after the inductive dark period, the plant does not flower. If, however, the induced leaf remains attached to the plant for several hours, the plant will flower. This

WORKING WITH**DATA:**

## *The Flowering Signal Moves from Leaf to Bud*

### Original Paper

Hamner, K. C. and J. Bonner. 1938. Photoperiodism in relation to hormones as factors in floral initiation and development. *Botanical Gazette* 100: 388–431.

### Analyze the Data

In 1938 Karl Hamner at the University of Chicago was working on the role of plant nutrition in flowering. The plant he studied, cocklebur, is a short-day plant that requires 16 hours of darkness to flower. When the plants were kept in 6 hours darkness (16 hours light) in a greenhouse, they did not flower. One day Hamner came to the lab to find all the plants flowering. It turned out that there had been a power outage, and the plants had received a single inductive short day (long night). Realizing that this provided a simple system to study flowering, Hamner invited a major scientist in the field, James Bonner from Caltech, to join him for the summer. The two biologists carried out a series of experiments using the single inductive period that showed that flowering is induced by night length as opposed to day length (see Figure 38.12) and that the flowering signal is received by the leaf from which it travels to a bud, inducing flowering. A Russian plant physiologist, Mikhail Chailakhyan, named this signal florigen. More recently, the molecular nature of this signal was described.

QUESTION 1

Intact cocklebur plants (6) or plants with their leaves removed (6) were placed in the inductive (short day) photoperiod. After 14 days, the researchers obtained the results in **TABLE A**. Based on these data, which part of the plant senses the photoperiod?

QUESTION 2

Cocklebur plants were treated so that a single leaf was exposed to the inductive (short) photoperiod while the rest of the plant received the long photoperiod. After 18 days, the results were as shown in **TABLE B**. What do these results indicate about the location of the receptor for flowering?

QUESTION 3

What do the data tell you about the signal generated by the plant in response to photoperiod and that induces flowering in the apical meristem?

TABLE A

| Treatment | Number of plants that flowered |
|---|---|
| No inductive period, intact plant | 0 |
| Inductive period, intact plant | 6 |
| Inductive period, leaves removed | 0 |

TABLE B

| Treatment | Number of plants | Result |
|---|---|---|
| Untreated | 6 | Vegetative |
| Treated, one leaf | 32 | Flower |

**Go to BioPortal for all** WORKING WITH**DATA** **exercises**

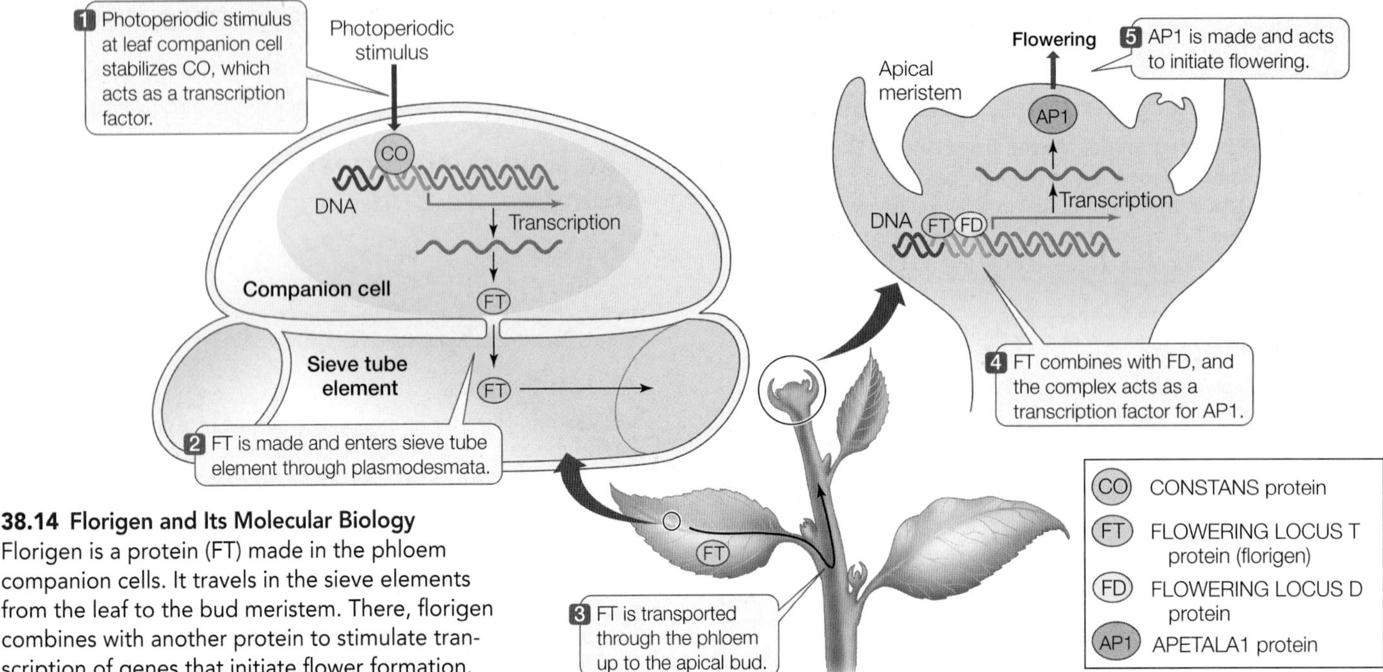


**38.14 Florigen and Its Molecular Biology** Florigen is a protein (FT) made in the phloem companion cells. It travels in the sieve elements from the leaf to the bud meristem. There, florigen combines with another protein to stimulate transcription of genes that initiate flower formation.

result suggests that something is synthesized in the leaf in response to the inductive dark period, and then moves out of the leaf to induce flowering.

- If two cocklebur plants are grafted together, and if only one of the plants is exposed to inductive long nights, both plants flower.
- In several species, if an induced leaf from one species is grafted onto another, noninduced plant of a different species, the recipient plant flowers.

The transmissible signal was given a name, **florigen** ("flower inducing"), in 1937, but its chemical nature has been described only in the past decade.

## Florigen is a small protein

The characterization of florigen was made possible by genetic and molecular studies of the model plant *Arabidopsis*. Three genes are involved in the signaling response for flowering (**Figure 38.14**).

- ***FT* (*FLOWERING LOCUS T*)** *codes for florigen.* FT is a small (20 kDa) protein that can travel through plasmodesmata. FT is synthesized in the phloem companion cells of the leaf and then diffuses into the adjacent sieve elements. It then is carried through the phloem to the apical meristem. If the *FT* gene is coupled to an active promoter and expressed at high levels in the shoot meristem, flowering is induced even in the absence of an appropriate photoperiodic stimulus.
- ***CO* (*CONSTANS*)** *codes for the transcription factor that activates the synthesis of FT.* As described above, *CO* expression follows a circadian rhythm, and stabilization of the CO protein by photoreceptors allows it to function. Like *FT*, *CO* is expressed in leaf companion cells. If *CO* is experimentally overexpressed in the leaf, flowering is induced. However, if *CO* is overexpressed in the apical meristem, flowering is not induced, indicating that *CO* functions in the leaf.
- ***FD* (*FLOWERING LOCUS D*)** *encodes a protein that binds to FT protein when it arrives in the apical meristem.* The FD protein is a transcription factor that, when bound to FT, activates promoters for meristem identity genes, such as *APETALA1* (see Figure 38.8). The expression of FD primes meristem cells to change from a vegetative fate to a reproductive fate once FT arrives.

Before FT was isolated, grafting experiments indicated that many different plant species could be induced to flower by the same chemical signal. Results of molecular experiments confirmed that the *FT* gene is involved in photoperiod signaling in many species:

- Transgenic plants (e.g., tobacco and tomato) that express the *Arabidopsis FT* gene at high levels flower regardless of day length.
- Transgenic *Arabidopsis* plants that express *FT* homologs from other plants (e.g., rice and tomato) flower regardless of day length.

While the molecular basis of the action of florigen has been elucidated, commercial applications of this knowledge have been harder to realize. It was hoped that florigen might be a very small molecule, like an auxin or gibberellin, that could be sprayed on economically important plants to induce flowering at will. The fact that florigen is a protein that cannot readily enter cells from the outside environment makes the development of commercial florigen treatments unlikely.

We have considered the photoperiodic regulation of flowering, from photoreceptors in the leaf to florigen that travels from the induced leaf to the sites of flower formation. In some plants, however, flowering is induced by other stimuli. These additional stimuli can function with photoperiodism or independently of it.

## Flowering can be induced by temperature or gibberellin

Whereas some plants use the environmental cue of day length to induce flowering, other plants use different mechanisms. These include temperature (another environmental cue) and gibberellin (an internal, hormonal cue).

**38.15 Vernalization** A genetic strain of *Arabidopsis* (winter-annual *Arabidopsis*) requires vernalization for flowering. Without it, the plant is large and vegetative (left), but with the cold period it is smaller and flowers (right).

TEMPERATURE In some plant species, notably certain cereal grains, the environmental signal for flowering is cold temperature, a phenomenon called **vernalization** (Latin *vernus*, "spring"). In both wheat and rye, we distinguish two categories of flowering behavior. Spring wheat, for example, is a typical annual plant: it is sown in the spring and flowers in the same year. Winter wheat is sown in the fall, grows to a seedling, overwinters (often covered by snow), and flowers the following summer. If winter wheat is not exposed to cold in its first year, it will not flower normally the next year.

How vernalization leads to flowering has been elucidated using model organisms such as *Arabidopsis*. In strains of *Arabidopsis* that require vernalization to flower (**Figure 38.15**), a gene called *FLC* (*FLOWERING LOCUS C*) encodes a transcription factor that blocks the FT–FD florigen pathway (see Figure 38.14) by inhibiting expression of FT and FD. Cold temperature inhibits the synthesis of FLC protein, allowing the FT and FD proteins to be expressed, and flowering to proceed. Similar proteins control some steps in vernalization in cereals.

Epigenetics (see Section 16.4) plays an important role in the inhibition of *FLC* gene expression by cold temperature. Before vernalization, the chromatin at the promoter of the *FLC* gene is in a relaxed configuration, with histone protein acetylation lowering the ionic attraction of these proteins for DNA, which allows transcription (**Figure 38.16A**). During vernalization, a gene is expressed whose protein product is involved in the deacetylation of histones on the *FLC* gene. Deactylation causes the chromatin to be more compact, which blocks *FLC* gene expression (**Figure 38.16B**).

GIBBERELLIN *Arabidopsis* plants do not flower if they are genetically deficient in the hormone gibberellin, or if they are treated with an inhibitor of gibberellin synthesis. These observations implicate gibberellins in flowering. Direct application of gibberellins to *Arabidopsis* buds results in activation of the

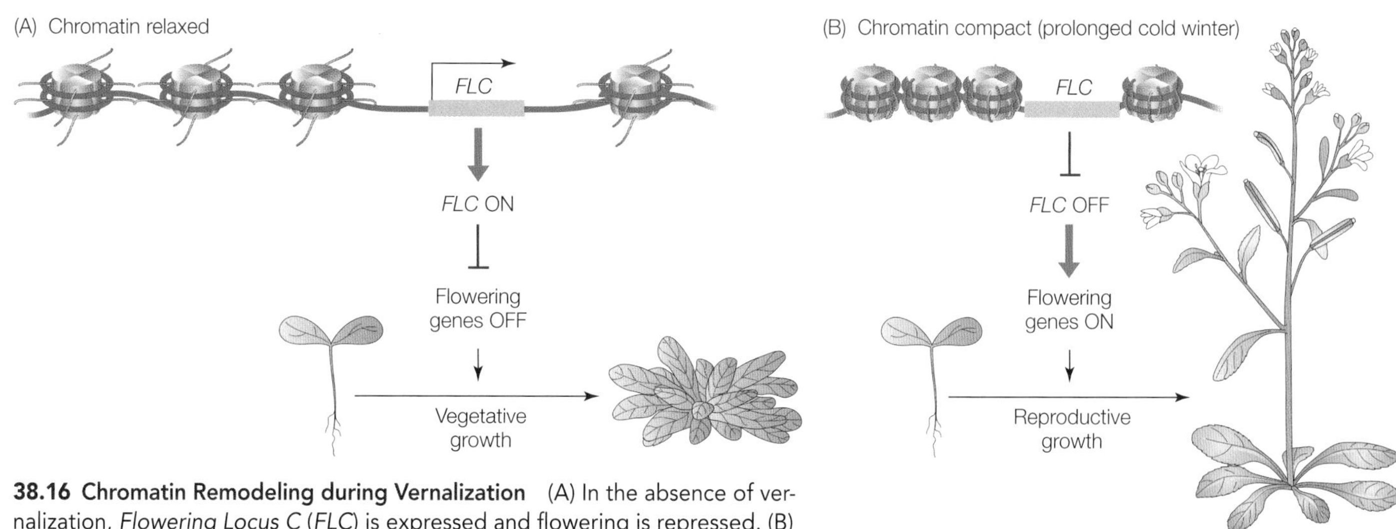


**38.16 Chromatin Remodeling during Vernalization** (A) In the absence of vernalization, *Flowering Locus C* (*FLC*) is expressed and flowering is repressed. (B) Prolonged cold weather leads to chromatin remodeling that represses expression of *FLC*. The absence of FLC protein allows flowering genes to be expressed.

meristem identity gene *LEAFY*, which in turn promotes the transition to flowering.

### Some plants do not require an environmental cue to flower

Several plant species and strains do not require a photoperiod or vernalization to flower, but instead flower on cue from an "internal clock." For example, flowering in some strains of tobacco will be initiated in the terminal bud when the stem has grown four phytomers in length (recall that stems are composed of repeating units called phytomers; see Figure 34.1). If such a bud with a single adjacent phytomer is removed and planted, the cutting will flower because the bud has already received the cue for flowering. But the rest of the shoot below the bud that has been removed will not flower because it is only three phytomers long. After it grows an additional phytomer, it will flower. These results suggest that there is something about the *position* of the bud (atop four phytomers of stem) that determines its transition to flowering.

The bud might "know" its position by the concentration of some substance that forms a positional gradient along the length of the plant. Such a gradient could be formed if the root makes a diffusible inhibitor of flowering whose concentration diminishes with plant height. When the plant reaches a certain height, the concentration of the inhibitor would become sufficiently low at the tip of the shoot to allow flowering. What this inhibitor might be is unclear, but there is evidence that it acts by decreasing the amount of FLC, allowing the FT–FD pathway to proceed (just as cold acts on FLC in vernalization). A positional gradient that acts on FLC would be consistent with other mechanisms affecting flowering, which all converge on *LEAFY* and *APETALA1*:

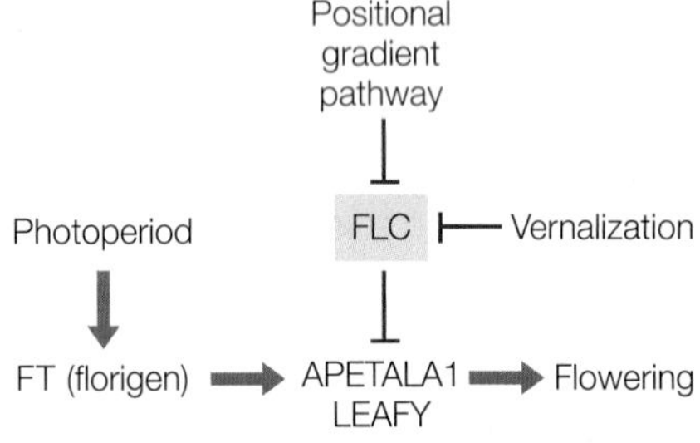


**RECAP 38.2**

Flowering of some angiosperms is controlled by night length, a phenomenon called photoperiodism. Low temperatures can induce flowering in some species (vernalization). Some species flower when their stems have grown by a certain amount, independent of environmental cues. All pathways to flowering converge on the meristem identity genes.

- What are the differences between apical meristems, inflorescence meristems, and floral meristems? What genes control the transitions between them? **See p. 786 and Figure 38.8**
- Explain why "short-day plant" is a misleading term. **See p. 788 and Figure 38.11**
- What is the evidence for florigen? What is its molecular mechanism of action? **See pp. 788–790 and Figures 38.13 and 38.14**

We have seen how environmental factors interact with genes to control flowering in angiosperms. The function of flowers is sexual reproduction, which maintains beneficial genetic variation in a population. Many angiosperms, however, also benefit from being able to reproduce asexually.

## 38.3 How Do Angiosperms Reproduce Asexually?

Although sexual reproduction takes up most of the space in this chapter, asexual reproduction accounts for many of the individual plants present on Earth. This fact suggests that in some circumstances asexual reproduction must be advantageous.

We have noted that genetic recombination is one of the advantages of sexual reproduction. Self-fertilization is a form of sexual reproduction, but offers fewer opportunities for genetic recombination than does cross-fertilization. A diploid, self-fertilizing plant that is heterozygous for a certain locus can produce both kinds of homozygotes for that locus plus the heterozygote among its progeny, but it cannot produce any progeny carrying alleles that it does not itself possess. Nevertheless many self-fertilizing plant species produce viable and vigorous offspring.

Asexual reproduction eliminates genetic recombination altogether. A plant that reproduces asexually produces progeny genetically identical to the parent (clones). What, then, is the advantage of asexual reproduction? If a plant is well adapted to its environment, asexual reproduction allows it to pass on to all its progeny a superior combination of alleles, which might otherwise be separated by sexual recombination.

### Many forms of asexual reproduction exist

Stems, leaves, and roots are considered vegetative organs and are distinguished from flowers, the reproductive parts of the plant. Asexual reproduction is often accomplished through the modification of a vegetative organ, which is why the term **vegetative reproduction** is sometimes used to describe asexual reproduction in plants. Another type of asexual reproduction, apomixis (see the opening story), involves flowers but no fertilization.

Often the stem is the organ that is modified for vegetative reproduction. Strawberries, for example, produce horizontal stems, called stolons or runners, which grow along the soil surface, form roots at intervals, and establish potentially independent plants. Asexual reproduction by shoot tips is accomplished when the tips of upright branches sag to the ground and develop roots, as in blackberry and forsythia.

Some plants, such as potatoes, form enlarged fleshy tips of underground stems, called tubers, that can produce new plants (from the "eyes"). Rhizomes are horizontal underground stems that can give rise to new shoots. Bamboo is a striking example of a plant that reproduces vegetatively by means of rhizomes. A single bamboo plant can give rise to a stand—even a forest—of plants constituting a single, physically connected entity (**Figure 38.17A**).

(A)

(B)

(C)

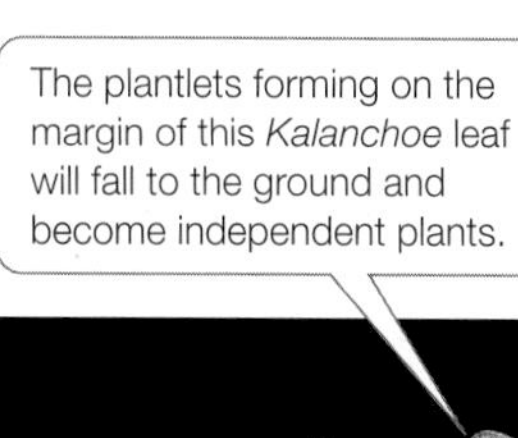


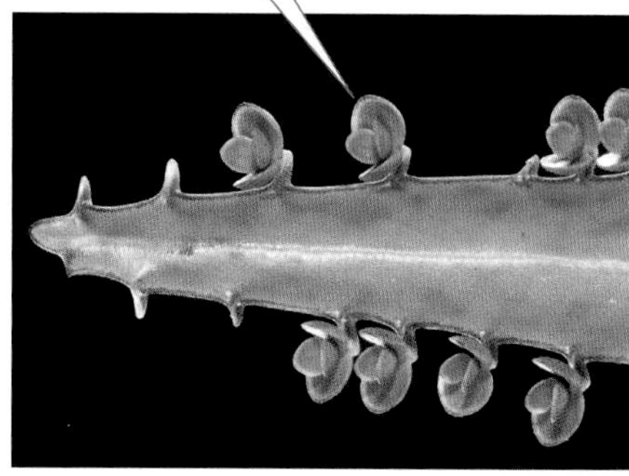

**38.17 Vegetative Organs Modified for Reproduction** (A) The rhizomes of bamboo are underground stems that produce plants at intervals. (B) Bulbs are short stems with large leaves that store nutrients and can give rise to new plants. (C) In *Kalanchoe*, new plantlets can form on leaves.

Whereas stolons and rhizomes are horizontal stems, bulbs and corms are short, vertical, underground stems. Lilies and garlic form bulbs (**Figure 38.17B**), short stems with many fleshy, highly modified leaves that store nutrients. These storage leaves make up most of the bulb. They can give rise to new plants by dividing or by producing new bulbs from axillary buds. Crocuses, gladioli, and many other plants produce corms, underground stems that function very much as bulbs do. Corms are disclike and consist primarily of stem tissue; they lack the fleshy modified leaves that are characteristic of bulbs.

Stems are not the only vegetative organs modified for asexual reproduction. Leaves may also be the source of new plantlets, as in some succulent plants of the genus *Kalanchoe* (**Figure 38.17C**). Many kinds of angiosperms, ranging from grasses to trees such as aspens and poplars, form interconnected, genetically homogeneous populations by means of suckers—shoots produced by roots. What appears to be a whole stand of aspen trees, for example, may be a clone derived from a single tree by suckers. This is why the leaves of a whole stand of aspens typically turn yellow at the same time.

Plants that reproduce vegetatively often grow in physically unstable environments such as eroding hillsides. Plants with stolons or rhizomes, such as beach grasses, rushes, and sand verbena, are common pioneers on coastal sand dunes. Rapid vegetative reproduction enables these plants, once introduced, not only to multiply but also to survive burial by the shifting sand; in addition, the dunes are stabilized by the extensive network of rhizomes or stolons that develops. Vegetative reproduction is also common in some deserts, where the environment is often not suitable for seed germination and the establishment of seedlings.

## Vegetative reproduction has a disadvantage

Vegetative reproduction is highly efficient in an environment that is stable over the long term. A change in the environment, however, can leave an asexually reproducing species at a disadvantage.

A striking example is provided by the demise of the English elm, *Ulmus procera*, which was apparently introduced into England as a clone by the ancient Romans. This tree reproduces asexually by suckers and is incapable of sexual reproduction. In 1967 Dutch elm disease first struck the English elms. After two millennia of clonal growth, the population lacked genetic diversity, and no individuals carried genes that would protect them against the disease. Today the English elm is all but gone from England.

## Vegetative reproduction is important in agriculture

One of the oldest methods of vegetative reproduction used in agriculture consists of simply making cuttings of stems, inserting them in soil, and waiting for them to form roots and become autonomous plants. The cuttings are usually encouraged to root by treatment with auxin, a plant hormone.

Woody plants can be propagated asexually by **grafting**: attaching a bud or a piece of stem from one plant to a root or root-bearing stem of another plant. The part of the resulting plant that comes from the root-bearing "host" is called the **stock**; the part grafted on is the **scion** (**Figure 38.18**). The vascular cambium of the scion associates with that of the stock, forming a continuous cambium that produces xylem and phloem. The cambium allows the transport of water and minerals to the scion and of photosynthate to the stock. Much of the fruit grown for market in the United States is produced on grafted trees, as are wine grapes.

Another method widely used for asexual plant propagation is **meristem culture**, in which pieces of shoot apical meristem are cultured on growth media to generate plantlets, which can then be planted in the field. This strategy is vital when uniformity is desired, as in forestry, or when virus-free plants are the goal, as with strawberries and potatoes.

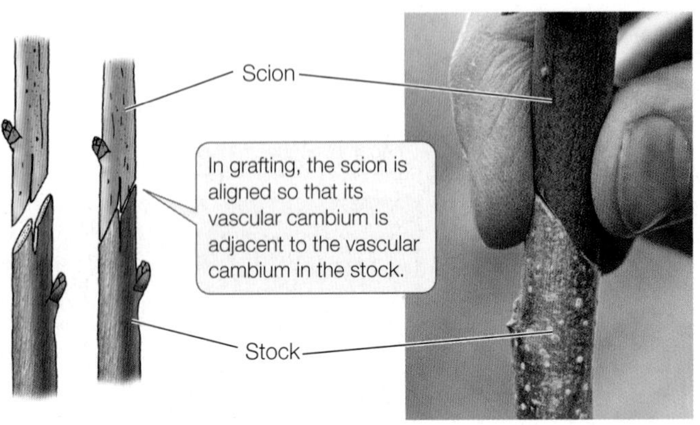


**38.18 Grafting** Grafting—attaching a piece of a plant to the root or root-bearing stem of another plant—is a common horticultural technique. The "host" root or stem is the stock; the upper grafted piece is the scion. In the photo, a scion of one apple variety is being grafted onto a stock of another variety.

**RECAP 38.3**

Angiosperms may reproduce asexually by means of modified stems, roots, or leaves. Asexual reproduction is advantageous when a plant has a superior genotype well adapted to its environment, but it decreases the genetic diversity of plant populations.

- What are the advantages and disadvantages of asexual reproduction for a plant? See pp. 792–793
- Explain how vegetative reproduction of plants is advantageous to humans. See pp. 793–794

We have seen how angiosperms reproduce sexually and asexually. A disadvantage of asexual reproduction is that its genetic inflexibility may leave a population unable to cope with new challenges. In the next chapter we will focus on the mechanisms that have evolved in plants to cope with biological and physical challenges in their environment.

By what genetic mechanism is apomixis brought about?

**ANSWER**

Apomixis involves several processes, including abnormal meiosis, followed by the development of an embryo within a seed. *Arabidopsis* plants with a mutation in the *SWI1* gene show apomixis but no viable seeds. In normal plants, the expression of *SWI1* is essential for chromosome pairing during meiosis I (see Figure 11.16). But in the mutant strain, meiosis I resembles mitosis, and the chromosomes replicate again before what would be meiosis II. The search is now on for genes that promote embryo and seed development in plants that exhibit apomixis. Scientists are trying to isolate and transfer such genes into corn and other cereal crops with the hope that plant breeders can use apomixis to propagate plants with desirable traits (such as high yields and disease- and insect-resistance) without compromising their hybrid vigor.

# CHAPTER SUMMARY 38

## 38.1 How Do Angiosperms Reproduce Sexually?

- Sexual reproduction promotes genetic diversity in a population. The flower is an angiosperm's structure for sexual reproduction.
- Flowering plants have microscopic gametophytes. The megagametophyte is the **embryo sac**, which typically contains eight nuclei in a total of seven cells. The microgametophyte is the **pollen grain**, which usually contains two cells. **Review Figure 38.2, ACTIVITY 38.1**
- Following **pollination**, the pollen grain delivers **sperm cells** to the embryo sac by means of a **pollen tube**. **Review ANIMATED TUTORIAL 38.1**
- Plants have both physical and genetic methods of preventing inbreeding. Physical separation of the gametophytes and genetic **self-incompatibility** prevent self-pollination. **Review Figure 38.5**
- Most angiosperms exhibit **double fertilization**: one sperm cell fertilizes the egg cell, forming a zygote, and the other sperm cell fertilizes the central cell, where its nucleus unites with the two **polar nuclei** to form a triploid **endosperm**. **Review Figure 38.6**
- Ovules develop into seeds, and the ovary wall and the enclosed seeds develop into a **fruit**.
- The hormone **abscisic acid** promotes seed development and dormancy.

## 38.2 What Determines the Transition from the Vegetative to the Flowering State?

- In **annuals** and **biennials**, flowering and seed formation usually leads to death of the rest of the plant. **Perennials** live a long time and typically reproduce repeatedly.
- For a vegetatively growing plant to flower, an apical meristem in the shoot system must become an **inflorescence meristem**, which in turn must give rise to one or more **floral meristems**. These events are under the influence of **meristem identity genes** and **floral organ identity genes**. **Review Figure 38.8**

*continued*

- Some plants flower in response to **photoperiod**. **Short-day plants** flower when the nights are longer than a critical night length specific to each species; **long-day plants** flower when the nights are shorter than a critical night length. **Review Figure 38.11**
- The mechanism of photoperiodic control involves phytochromes and a biological clock. **Review Figure 38.12, ANIMATED TUTORIAL 38.2**
- A flowering signal, called **florigen**, is formed in a photoperiodically induced leaf and is translocated to the sites where flowers will form. **Review Figures 38.13, 38.14**
- In some angiosperm species, exposure to low temperatures—**vernalization**—is required for flowering; in others, internal signals (one of which is gibberellin in some plants) induce flowering. **Review Figures 38.15, 38.16**

### How Do Angiosperms Reproduce Asexually?

- Asexual reproduction allows rapid multiplication of organisms that are well suited to their environment.
- **Vegetative reproduction** involves the modification of a vegetative organ—usually the stem—for reproduction. **Review Figure 38.17**
- Horticulturists often **graft** different plants together to take advantage of favorable properties of both **stock** and **scion**. **Review Figure 38.18**

**Go to the Interactive Summary to review key figures, Animated Tutorials, and Activities Life10e.com/is38**

# CHAPTER**REVIEW**

## REMEMBERING

1. The typical angiosperm female gametophyte
   a. is called a microspore.
   b. has eight nuclei.
   c. has eight cells.
   d. is called a pollen grain.
   e. is carried to the male gametophyte by wind or animals.

2. Pollination in angiosperms
   a. always requires wind.
   b. never occurs within a single flower.
   c. always requires help by animal pollinators.
   d. is also called fertilization.
   e. makes most angiosperms independent of external water for reproduction.

3. Which statement about double fertilization is *not* true?
   a. It is found in most angiosperms.
   b. It includes fusion between the microsporocyte and the megasporocyte.
   c. One of its products is a triploid nucleus.
   d. One sperm cell fuses with the egg cell.
   e. One sperm cell fuses with the central cell.

4. Which statement about photoperiodism is *not* true?
   a. It is related to the biological clock.
   b. Phytochrome plays a role in the timing process.
   c. It is based on measurement of the length of the night.
   d. Some plants do not flower in response to photoperiod.
   e. It is limited to plants.

5. Florigen is
   a. produced in the leaves and transported to the apical bud.
   b. produced in the roots and transported to the shoots.
   c. produced in the apical meristem of the stem and transported to the base.
   d. the same as gibberellin.
   e. activated by prolonged (more than a month) high temperature.

6. Which statement about vernalization is *not* true?
   a. It decreases the abundance of an inhibitor of flowering.
   b. It involves exposure to cold temperatures.
   c. It only occurs in crop plants such as cereals.
   d. It inhibits synthesis of the FLC protein.
   e. If winter wheat is not exposed to cold, it will not flower.

## UNDERSTANDING & APPLYING

7. Thompson Seedless grapes are produced by vines that are triploid. Think about the consequences of this chromosomal condition for meiosis in the flowers. Why are these grapes seedless? Describe the role played by the flower in fruit formation when no seeds are being formed. How do you suppose Thompson Seedless grapes are propagated?

8. Poinsettias are popular ornamental plants that typically bloom just before Christmas. Their flowering is photoperiodically controlled. Are they long-day or short-day plants? Explain.

9. You plan to induce the flowering of a crop of long-day plants in the field by using artificial light. Is it necessary to keep the lights on continuously from sundown until the point at which the critical day length is reached? Why or why not?

## ANALYZING & EVALUATING

10. Describe the proteins and mutations that could be involved in the following observations:
    a. A mutant plant flowers without its normal inductive dark period. When a leaf from the mutant plant is grafted onto an unexposed wild-type plant, the recipient plant flowers.
    b. A mutant plant does not flower when exposed to the normal inductive dark period. When a leaf from a mutant plant that has been exposed to the inductive dark period is grafted onto an unexposed wild-type plant, the recipient plant flowers.
    c. A plant flowers only after exposure to cold.
    d. If a gene is experimentally overexpressed in the leaf, flowering is induced. Overexpression of the gene in the shoot apical meristem does not, however, induce flowering.
11. The isolation of a mutation in the *Arabidopsis SWII* gene that results in abnormal meiosis has offered insights into apomixis. How would you try to identify other genes or mutations that function along with the gene product of *SWII*, with the goal of producing fully fertile apomictic plants?

# 39 Plant Responses to Environmental Challenges

## CHAPTER OUTLINE


**39.1** How Do Plants Deal with Pathogens?

**39.2** How Do Plants Deal with Herbivores?

**39.3** How Do Plants Deal with Environmental Stresses?

**39.4** How Do Plants Deal with Salt and Heavy Metals?

**A New Way to Fight Malaria** Sweet wormwood (*Artemisia annua*) grows in forests throughout the world. It synthesizes a defensive chemical, called artemisinin, that is now being used to treat people with malaria.

*ARTEMISIA ANNUA*, or sweet wormwood, is a fernlike shrub about 2 meters tall that is native to Asia but grows all over the world. Unlike animals, which can sometimes escape their enemies, plants must confront their enemies in place. Over time, plants have evolved elaborate mechanisms to fight off attackers. One such mechanism is the production of defensive chemicals. *Artemisia* produces a chemical called artemisinin that is toxic to certain cells, including those of the parasite that causes malaria.

Malaria has infected humans for more than 50,000 years. Today about 400 million cases arise worldwide and close to 1 million people die from malaria each year. The disease is caused by a mosquito-borne parasite that infects liver and blood cells in the human body.

Many cultures have long histories of extracting medicines from plants growing in their environments. About 1,700 years ago the Chinese herbalist Ge Hong noted that drinking a tea made from *Artemisia* was useful in treating malaria. Meanwhile, the indigenous people of Peru were using tinctures from the bark of a different plant—cinchona—to treat malaria. In the 1600s Jesuit priests noted how effective the cinchona treatment was, and brought it back to Europe. During the nineteenth century the active ingredient of cinchona was identified as quinine, and quinine became the most widely used medicine for both the prevention and treatment of malaria. But during the twentieth century, the supply of quinine could not keep up with demand, especially for soldiers fighting in the world wars. This led to the development of various synthetic antimalarial drugs.

As more modern drugs were developed to treat malaria, traditional herbs such as *Artemisia* fell into disuse. But gradually the parasite developed resistance to the drugs, and doctors turned back to *Artemisia*. Indeed, during the Vietnam War, drinking *Artemisia* tea helped Vietnamese soldiers cope with the quinine-resistant malaria that struck American soldiers. Chinese scientists isolated the active ingredient of the antimalarial tea and called it artemisinin.

The exact mechanism of action of artemisinin is still unclear. It appears that the drug reacts with iron in red blood cells, forming free radicals that damage lipids and DNA in the infecting parasite. Because of political differences between China and the rest of the world, news of the discovery of artemisinin, and access to the drug, were restricted until the early 1980s. Now it is widely available.

Until recently, hundreds of such plant chemicals were contemplated only in the context of plant biochemistry (how plants make them). Today we view them as adaptations arising from a plant's interactions with its environment.

What is the current status of artemisinin therapy for malaria?

See answer on p. 812.

## 39.1 How Do Plants Deal with Pathogens?

Botanists know of thousands of diseases that can affect plants. Each is caused by a different strain of pathogen. Plant pathogens—which include bacteria, fungi, protists, nematodes, and viruses—are part of nature, and for that reason alone they merit our study in biology. For example, many diseases affect tomato plants, some of which may be familiar to you from growing tomatoes in your backyard or shopping for them (**Figure 39.1**). Like some human infectious diseases such as pneumonia, which may be bacterial or viral, plant diseases are named for the symptoms rather than the agent that causes them. For example, the term "blight," characterized by browning and death of plant tissues, can be caused by bacteria, fungi, or oomycetes. Just as medical schools have departments of pathology, many universities in agricultural regions have departments of plant pathology.

Successful infection by a pathogen can have significant effects on a plant, reducing photosynthesis and causing massive cell and tissue death. Like the responses of the human immune system (see Chapter 42), the responses by which plants fight off disease are varied and fascinating. Plants and pathogens have evolved together in a continuing "arms race": pathogens have evolved ways to attack plants, and plants have evolved ways to defend themselves against those attacks.

What determines the outcome of a battle between a plant and a pathogen? The key to success for the plant is to respond to information from the pathogen quickly and massively. Plants use both mechanical and chemical defenses in this effort. These defenses can either be:

- **Constitutive**, always present in the plant, or
- **Induced**, produced in reaction to damage or stress.

### Physical barriers form constitutive defenses

A plant's first line of defense is its outer surfaces, which can prevent the entry of pathogens. As Chapter 34 described, the organs of a growing plant that are exposed to the outside environment are largely covered with cutin, suberin, and waxes. These substances not only prevent water loss by evaporation but can also prevent fungal spores and bacteria from entering the underlying tissues. Some fungi get around this defense by secreting enzymes that hydrolyze these substances, breaking them down to gain entry.

Much more important to the plant are the induced resistance mechanisms, summarized in **Figure 39.2**, that are initiated when a pathogen lands on a plant.

### Plants can seal off infected parts to limit damage

Whereas animals generally repair tissues that have been damaged by pathogens, plants do not. Instead, plants seal off and sacrifice damaged tissues so that the rest of the plant does not

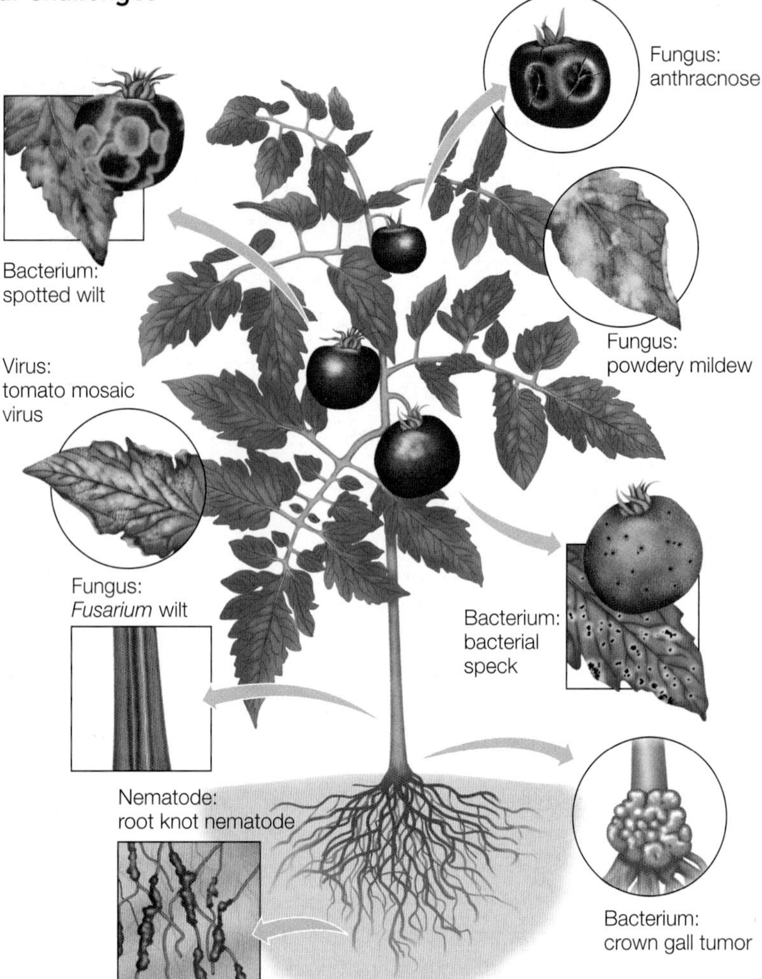


**39.1 Diseases of Tomato Plants** A wide variety of disease agents cause a variety of symptoms.

become infected. Plants have the option of discarding damaged tissues because most plants, unlike most animals, can replace damaged parts by growing new ones.

Before we look at the details of the defensive process, note that a key response by plant cells to invasion by pathogens is the rapid deposition of additional polymers on the inside of the cell wall. These macromolecules not only reinforce the mechanical barrier formed by the cell wall, but also block the plasmodesmata, limiting the ability of viral pathogens to move from cell to cell. The polysaccharides also serve as a base on which lignin may be laid down. Lignin enhances the mechanical barrier, and the toxicity of lignin precursor chemicals makes the cell inhospitable to some pathogens. These lignin building blocks are only one example of the toxic substances that plants use as chemical defenses.

Induced responses to pathogens are controlled by receptors. Plant pathogens cause the host plant to activate various chemical defense responses. A wide range of molecules called **elicitors** have been identified that trigger these defenses. These molecules vary in character, from peptides made by bacteria to cell wall fragments from fungi. Elicitors can also be derived from fragments of plant cell wall components broken down by pathogens.

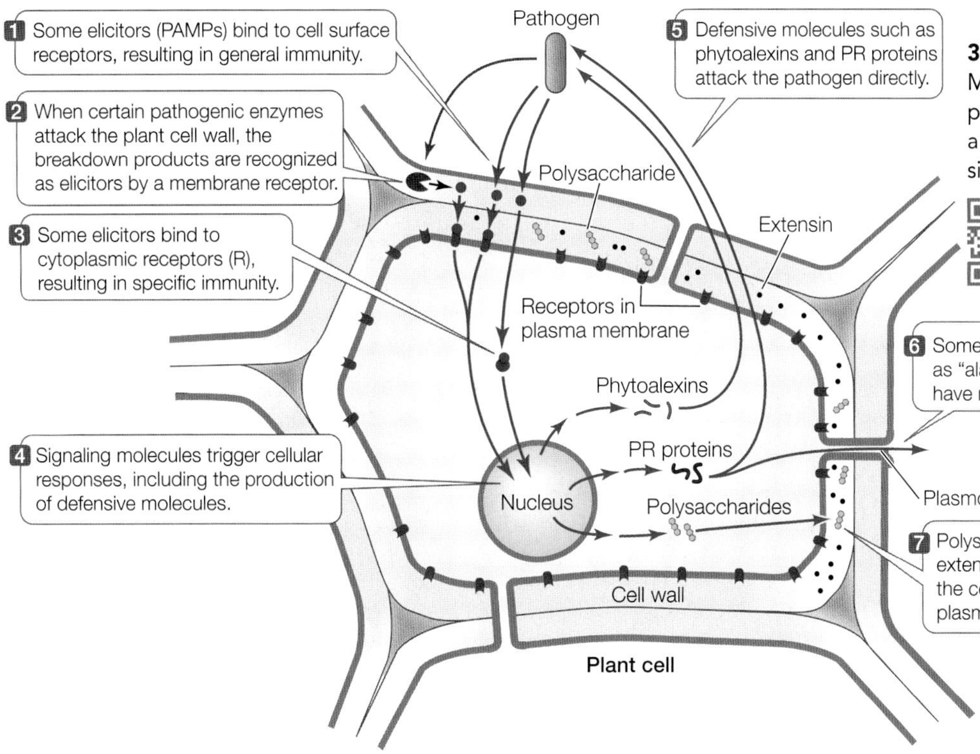

1 Some elicitors (PAMPs) bind to cell surface receptors, resulting in general immunity.

2 When certain pathogenic enzymes attack the plant cell wall, the breakdown products are recognized as elicitors by a membrane receptor.

3 Some elicitors bind to cytoplasmic receptors (R), resulting in specific immunity.

4 Signaling molecules trigger cellular responses, including the production of defensive molecules.

5 Defensive molecules such as phytoalexins and PR proteins attack the pathogen directly.

6 Some PR proteins serve as "alarm signals" to cells that have not yet been attacked.

7 Polysaccharides and extensin strengthen the cell wall and block plasmodesmata.

Pathogen

Polysaccharide

Extensin

Receptors in plasma membrane

Phytoalexins

PR proteins

Nucleus

Polysaccharides

Plasmodesma

Cell wall

Plant cell

**39.2 Signaling between Plants and Pathogens** Molecular interactions between plants and pathogens are highly coevolved. The presence of a pathogen stimulates the plant to produce defensive molecules that work in many different ways.

**Go to Animated Tutorial 39.1 Signaling between Plants and Pathogens** Life10e.com/at39.1

The responses of plants to elictors can be described in terms of the "plant immune system." Two forms of immunity are recognized (see Figure 39.2):

- General immunity is triggered by general elicitors called pathogen associated molecular patterns (PAMPs). PAMPs are usually molecules that are produced by entire classes of pathogens, such as flagellin (found in bacterial flagella) or chitin (found in fungal cell walls). Thus general immunity is an overall response rather than a response that is triggered by a specific pathogen in a particular plant. PAMPs are recognized by transmembrane receptors called pattern recognition receptors, which activate signaling pathways that lead to general immunity (also called PAMP-triggered immunity, or PTI).
- Specific immunity is triggered by specific elicitors called effectors. Effectors include a wide variety of specific pathogen-produced molecules that enter the plant cell. Once inside the cell, effectors bind to cytoplasmic receptors called R proteins that trigger the specific immunity response (also called effector-triggered immunity, or ETI).

## General and specific immunity both involve multiple responses

Many of the signaling pathways associated with general and specific immunity are the same, although the latter is specific for particular pathogens in particular plants, and is much stronger than general immunity. Both forms of immunity involve signaling pathways that are triggered by binding between the elicitors (PAMPs or effectors) with their receptors. These pathways lead to various responses:

- *Formation of reactive oxygen and NO*: Receptor binding triggers the rapid production of nitric oxide (NO) and reactive oxygen species such as superoxide and hydrogen peroxide. These reactive molecules are toxic to some pathogens, and they are components of signal transduction pathways leading to local and systemic (plantwide) defenses.
- *Callose deposition*: The β-1,3-glucan polymer callose is deposited on the inside of the cell wall to strengthen the wall and seal off the cell.
- *Hormone signaling*: Some pathways result in the production of plant hormones, including salicylic acid and jasmonic acid. We will describe the roles of these hormones in immunity later in the chapter.
- *Changes in gene expression*: Signal transduction cascades lead to changes in gene expression. The upregulated genes include pathogenesis-related (PR) genes and genes encoding the production of antimicrobial substances called phytoalexins.

PHYTOALEXINS **Phytoalexins** are antibiotics that are produced by infected plants and are toxic to many fungi and bacteria. Most are small molecules, and each is made by only a few plant species. They are produced by infected cells and their immediate neighbors within hours of the onset of infection. Because their antimicrobial activity is nonspecific, phytoalexins can destroy many species of fungi and bacteria in addition to the one that originally triggered their production. Some phytoalexins can also kill the plant cells that produced them, thus sealing off the infection site (the hypersensitive response; see below).

Phytoalexins are an example of an induced plant defense: they are not normally present in plants but are synthesized rapidly when a bacterial or fungal infection occurs. Physical injuries and viral infections can also induce the production of phytoalexins.

Camalexin, a phytoalexin made by the model organism *Arabidopsis thaliana*, appears to function by disrupting the cell membranes of invading fungal or bacterial pathogens. Its production is induced by a conserved protein kinase cascade (see Section 7.3) that is triggered by receptor binding in either general or specific immunity. This protein kinase cascade results in the upregulation of genes encoding enzymes that convert the amino acid tryptophan to camalexin:

Tryptophan → Camalexin

**PATHOGENESIS-RELATED PROTEINS** Plants produce several types of **pathogenesis-related (PR) proteins**. Some are enzymes that break down the cell walls of pathogens. Chitinase, for example, is a PR protein that breaks down chitin, which is found in many fungal cell walls. In some cases the breakdown products of the pathogen cell walls serve as elicitors that trigger further defensive responses.

Another class of PR proteins are the plant defensins, which are similar to defensins produced by animals (see Section 42.2). These small peptides bind to fungal membranes and are toxic to a wide range of fungal targets, but they are not toxic to plant or animal cells.

Other PR proteins may serve as alarm signals to plant cells that have not yet been attacked. In general, PR proteins appear not to be rapid-response weapons; rather, they act more slowly, perhaps after other, faster responses have blunted the pathogen's attack.

## Specific immunity involves gene-for-gene resistance

As mentioned above, pathogen effectors are molecules that are secreted inside plant cells by fungal and bacterial pathogens. These molecules are often proteins whose function is to inhibit some aspect of plant immunity. For example, many effectors are proteases that break down specific plant proteins involved in immune responses. Thus effectors enable pathogens to overcome general immunity and invade the plant, causing disease. The genes that encode effectors have evolved as part of the "arms race" between plants and their pathogens.

In response to the evolution of effectors, plants have evolved intracellular receptors that recognize specific effectors. These receptors are called R (for *r*esistance) proteins. When an R protein binds its ligand (a specific effector), it activates the signal transduction pathways of the specific immunity response. As we mentioned above, specific immunity is stronger than general immunity, and it enables the plant to prevent growth of the pathogen and remain healthy.

R proteins are encoded by **resistance (*R*) genes**. During the middle of the twentieth century, the plant pathologist Harold

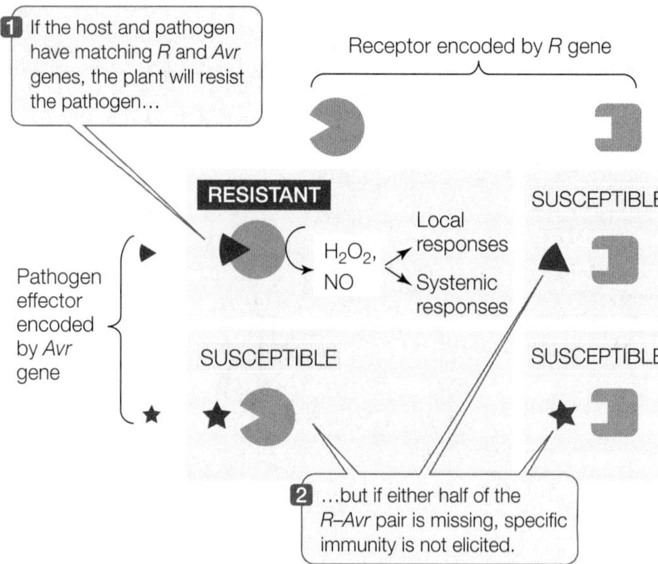


**39.3 Gene-for-Gene Resistance** If a gene in a pathogen that codes for an elicitor "matches" a gene in a plant that codes for a receptor, the receptor binds the elicitor, and a defensive response results.

Henry Flor, at North Dakota State University, realized that there is a special relationship between *R* genes and certain pathogen genes named ***Avirulence* (*Avr*) genes**. Flor studied strains of the rust fungus *Melampsora lini* and the flax plant (*Linum usitatissimum*). He found that specific strains of rust fungus (with particular *Avr* genes) were able to cause disease only on some varieties of flax—those that *didn't* carry specific *R* genes. If a flax variety carried the right *R* gene, it was resistant to those strains of rust fungus. Thus a particular *Avr* gene rendered the fungus *avirulent* on flax plants with the corresponding *R* gene. Flor named this the **gene-for-gene concept**. We now know that *Avr* genes encode effector molecules that bind to receptors encoded by the *R* genes, resulting in the specific immunity response (**Figure 39.3**). Hundreds of *R* genes and their corresponding *Avr* genes have been identified. A major goal of plant breeders for the past 50 years has been to breed new *R* genes into crops to make them more resistant to pathogens.

## Specific immunity usually leads to the hypersensitive response

Many signaling pathways and plant responses are shared between general and specific immunity, although in the latter these responses are accelerated and amplified. Furthermore, specific immunity usually leads to a form of programmed cell death called the **hypersensitive response**. When this occurs, plant cells at and near the site of pathogen infection die, forming a necrotic lesion (**Figure 39.4**). This cell death deprives the pathogen of nutrients and prevents the spread of the infection.

As described above, specific immunity is triggered by binding between specific pathogen effectors (*Avr* gene products) and R receptor proteins. The pathways that lead to cell death have not been worked out in detail, but the generation of nitric oxide and reactive oxygen species is an important component.

**39.4 Sealing Off the Pathogen and the Damage** These necrotic lesions on the leaves of a broad bean plant are a response to "chocolate spot" fungus, *Botrytis fabae*.

These reactive molecules may contribute to cell death directly—for example, by disrupting cellular membranes. They also function as signaling molecules that activate pathways leading to cell death.

### Systemic acquired resistance is a form of long-term immunity

Thus far we have described the events that occur in plant cells at or near the site of invasion by a pathogen. In both general and specific immunity, the infected cells also send hormonal signals to the rest of the plant, stimulating a systemic response. **Systemic acquired resistance** is a general increase in the resistance of the entire plant to a wide range of pathogens. It is not limited to the pathogen that originally triggered it, or to the site of the original infection, and its effect may last as long as an entire growing season.

This defensive response is initiated by the plant hormone salicylic acid.

Salicylic acid

Salicylic acid production is triggered by receptor binding in both general and specific immunity; salicylic acid then functions as a signal that mediates various defense responses. For example, salicylic acid triggers the production of reactive oxygen species and the induction of genes that encode PR proteins. There are many classes of PR proteins, which function in various ways to protect plants against insect attack and against invasion by fungi, bacteria, viruses, and nematodes.

These responses occur at the site of infection and, to a lesser extent, throughout the plant. A derivative of salicylic acid can form a gas that can travel through the air, carrying the defense signal not only to other parts of the same plant but to other nearby plants.

Another type of systemic acquired resistance is a more specific defense against viruses with RNA genomes. The plant uses its own enzymes to convert some of the single-stranded RNA of the invading virus into double-stranded RNA (dsRNA) and to chop that dsRNA into small interfering RNAs (siRNAs) (see Section 16.5). Simultaneously, some of the viral RNA is transcribed, forming mRNAs that advance the infection. However, the siRNAs interact with another cellular component to degrade those mRNAs, blocking viral replication. The siRNAs spread quickly throughout the entire plant through plasmodesmata, providing systemic resistance.

**RECAP 39.1**

Plants protect themselves against pathogens with constitutive and induced defenses. General and specific immunity involve several common signaling pathways, but specific immunity is a stronger response that usually leads to hypersensitive cell death. Systemic acquired resistance provides a longer-lasting, more general immunity throughout the plant.

- Name two types of defensive compounds that plant cells produce when they are infected by bacteria or fungi. **See pp. 799–800 and Figure 39.2**
- How do *R* and *Avr* genes determine which pathogens a plant can resist? **See p. 800 and Figure 39.3**
- How do infected plant cells signal infection to other parts of the plant, or to other plants? **See p. 801**

Not all biological threats to plants come from pathogens. Another threat comes from the many animals that eat plants.

## 39.2 How Do Plants Deal with Herbivores?

Herbivores—animals that eat plants—depend on plants for energy and nutrients. Their foraging activities cause physical damage to plants, and they often spread disease among plants as well. While the majority of herbivores are insects (**Figure 39.5**), every major class of vertebrates includes at least a few herbivores (see also Section 56.2, which discusses herbivory in the ecological context of species interactions). Plants cannot evade their consumers by running away, but they have many other ways of protecting themselves against herbivory.

### Mechanical defenses against herbivores are widespread

Plants have both constitutive and induced mechanical defenses against herbivores. Constitutive anatomical barriers include trichomes (specialized hairs; see p. 807) and thorns, spines that are specialized for defense. An example of an induced mechanical defense is the production of latex. Some plants, such as *Euphorbia* species, produce a thick, white aqueous suspension of cellular debris, oils, and resins called latex when they are injured by an herbivore. Insects trapped by this sticky substance starve to death.

**TABLE 39.1**
**Secondary Metabolites Used in Defense**

| Class | Type | Role | Example |
|---|---|---|---|
| Nitrogen-containing | Alkaloids | Neurotoxin | Nicotine in tobacco |
| | Glycosides | Inhibit electron transport | Dhurrin in sorghum |
| Ephedrine (an alkaloid) | Nonprotein amino acids | Disrupt protein structure | Canavanine in jack bean |
| Nitrogen–sulfur-containing<br>Methylglucosinolide | Glucosinolates | Inhibit respiration | Methylglucosinolate in cabbage |
| Phenolics | Coumarins | Block cell division | Umbelliferone in carrots |
| | Flavonoids | Phytoalexins | Capsidol in peppers |
| Umbelliferone | Tannins | Inhibit enzymes | Gallotannin in oak trees |
| Terpenes | Monoterpenes | Neurotoxins | Pyrethrin in chrysanthemums |
| | Diterpenes | Disrupt reproduction and muscle function | Gossypol in cotton |
| | Triterpenes | Inhibit ion transport | Digitalis in foxglove |
| | Sterols | Block animal hormones | Spinasterol in spinach |
| Pyrethrin | Polyterpenes | Deter feeding | Latex in *Euphorbia* |

**39.5 Insect Herbivores** The great majority of herbivores are insects. (A) Some herbivores, such as this locust, are generalists that will attack nearly any plant. (B) Others are specialists, like this tobacco hornworm, which feeds only on tobacco and related plants.

### Plants produce constitutive chemical defenses against herbivores

Plants attract, resist, and inhibit other organisms with a wide range of chemicals known as secondary metabolites. You learned about one of these chemicals, artemisinin, in the opening of this chapter.

Primary metabolites are substances such as proteins, nucleic acids, carbohydrates, lipids, and their building blocks, which are produced and used by all living organisms, including plants. Primary metabolites are used in basic cellular processes such as photosynthesis, respiration, and nutrient uptake. **Secondary metabolites** are substances that are not used for basic cellular processes. Each is found in only certain organisms or groups of organisms.

The more than 10,000 known plant secondary metabolites range in molecular mass from about 70 to more than 390,000 daltons, but most have a low molecular mass (**Table 39.1**). Some are produced by only a single plant species, whereas

INVESTIGATING**LIFE**

**39.6 Nicotine Is a Defense against Herbivores** The secondary metabolite nicotine, made by tobacco plants, is an insecticide, yet most commercial varieties of tobacco are susceptible to insect attack. Ian Baldwin demonstrated that a tobacco strain with a reduced nicotine concentration was more susceptible to insect damage.[a]

**HYPOTHESIS** Nicotine helps protect tobacco plants against insects.

*Method*

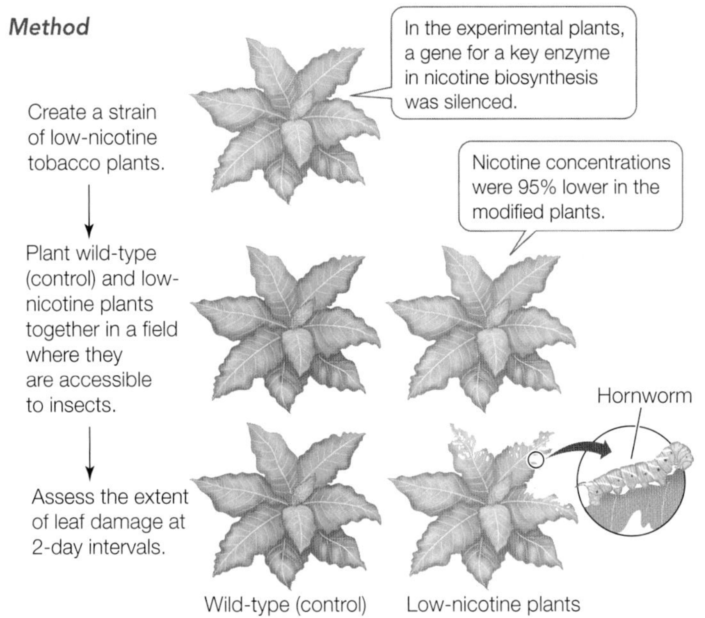


*Results* The low-nicotine plants suffered more than twice as much leaf damage as did the wild-type controls.

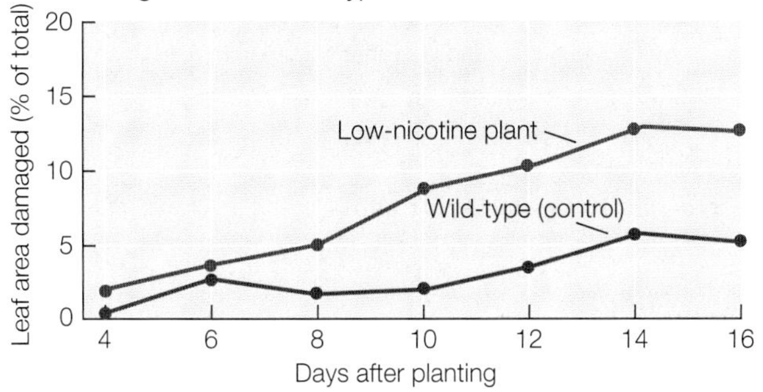


**CONCLUSION** Nicotine provides tobacco plants with at least some protection against insects.

Go to **BioPortal** for discussion and relevant links for all INVESTIGATING**LIFE** figures.

[a]Steppuhn, A. et al. 2004. *PLoS Biology* 2(8): e217.

others are characteristic of entire genera or even families. The effects of defensive secondary metabolites on animals are diverse. Some act on the nervous systems of herbivorous insects, mollusks, or mammals. Others mimic the natural hormones of insects, causing some larvae to fail to develop into adults. Still others damage the digestive tracts of herbivores. Some secondary metabolites are toxic to fungal pathogens. For example, as we saw at the opening of this chapter, humans make use of secondary metabolites as medicines.

The secondary metabolite nicotine was one of the first insecticides to be used by farmers and gardeners. This molecule kills insects by inhibiting the functioning of the nervous system. Yet commercial varieties of tobacco and related plants that produce nicotine are still attacked, with moderate damage, by pests such as the tobacco hornworm (see Figure 39.5B). Given that observation, does nicotine really deter herbivores? Biologists answered this question conclusively with a study that used tobacco plants in which an enzyme involved in nicotine biosynthesis had been silenced, lowering the nicotine concentration in the plants by more than 95 percent. These low-nicotine plants suffered much more damage from insect herbivory than did normal plants (**Figure 39.6**).

## Some secondary metabolites play multiple roles

Canavanine is a secondary metabolite whose role is defensive and is based on its chemical structure. Canavanine is an amino acid that is not found in proteins, but is very similar to the amino acid arginine, which is found in almost all proteins:

A seemingly slight chemical difference...

| Arginine | Canavanine |
| --- | --- |
| $NH_2$ | $NH_2$ |
| C=NH | C=NH |
| N—H | N—H |
| H—C—H | O |
| H—C—H | H—C—H |
| H—C—H | H—C—H |
| $H_2N$—C—H | $H_2N$—C—H |
| C—OH | C—OH |
| ‖ O | ‖ O |

...results in an inactive protein.

When an insect larva consumes canavanine-containing plant tissue, the canavanine is incorporated into the insect's proteins in some of the places where the mRNA codes for arginine, because the enzyme that charges the tRNA specific for arginine fails to discriminate accurately between the two amino acids (see Section 14.5). The structure of canavanine, however, is different enough from that of arginine that some of the resulting proteins end up with a modified tertiary structure, and hence reduced biological activity. These defects in protein structure and function lead to developmental abnormalities that kill the insect.

In plants that produce them, canavanine and other secondary metabolites are constitutive defenses—that is, they are present regardless of whether the plant is under attack. Other chemical defenses come into play only when an herbivore strikes.

## Plants respond to herbivory with induced defenses

In Section 39.1 we described the defenses that are induced in plants in response to pathogen attack. Plants also respond to wounding and herbivory with induced defenses involving signal transduction pathways. Less is known about the elicitors and receptors involved in these responses, but several classes of chemical elicitors have been identified. These elicitors are either derived from the herbivores themselves or are products of the digestion of plant tissues. For example, the enzymes

## WORKING WITH**DATA**:

### *Nicotine Is a Defense against Herbivores*

#### Original Paper

Steppuhn, A., K. Gase, B. Krock, R. Halitschke, and I. T. Baldwin. 2004. Nicotine's defensive function in nature. *PLoS Biology* 2(8): e217.

#### Analyze the Data

In 2004 Anke Steppuhn, Ian Baldwin, and colleagues at the Max Planck Institute for Chemical Ecology in Germany tested the hypothesis that nicotine helps protect tobacco plants against insects. They generated a line of low-nicotine transgenic plants by modifying the gene for putrescine *N*-methyl transferase, a key regulatory enzyme in the nicotine biosynthesis pathway. Both the low-nicotine and wild-type tobacco plants were transplanted into a field plantation where they were accessible to naturally occurring herbivores. The extent of leaf damage by insects was then measured at 2-day intervals for a period of 16 days. Results showed that the low-nicotine plants lost more than twice as much of their total leaf area as did the wild-type controls, indicating that nicotine provides tobacco plants with protection against insects (see Figure 39.6).

In a separate experiment, Baldwin and his colleagues showed that treatment with jasmonic acid (jasmonate) increased the concentration of nicotine in wild-type tobacco plants but not in the low-nicotine transgenic plants. The researchers then planted a group of wild-type and low-nicotine tobacco plants and treated them with jasmonate, a plant hormone, 7 days after planting. The plants were assessed for herbivore damage every 2 days after being planted. The results are shown in the figure.

**QUESTION 1**

Compare these data to those in Figure 39.6 for the untreated plants. What was the effect of jasmonate treatment on the resistance of wild-type plants to herbivore damage? What was the effect on low-nicotine plants?

**QUESTION 2**

What do these data reveal about the role of nicotine in preventing herbivore damage?

**QUESTION 3**

Explain how jasmonate could have had the effect it did on the transgenic low-nicotine plants, even though their nicotine levels were still low after jasmonate treatment.

**QUESTION 4**

What statistical test would you use to determine the possible significance of differences between the jasmonate-treated and untreated plants? At 10 days, the mean damage ± standard deviation for untreated low-nicotine plants was 6.0 ± 1.5 percent ($n$ = 36); for treated low-nicotine plants it was 2.2 ± 0.6 percent ($n$ = 28). Run a statistical test comparing the two results, calculate the $P$ value, and comment on significance (see Appendix B).

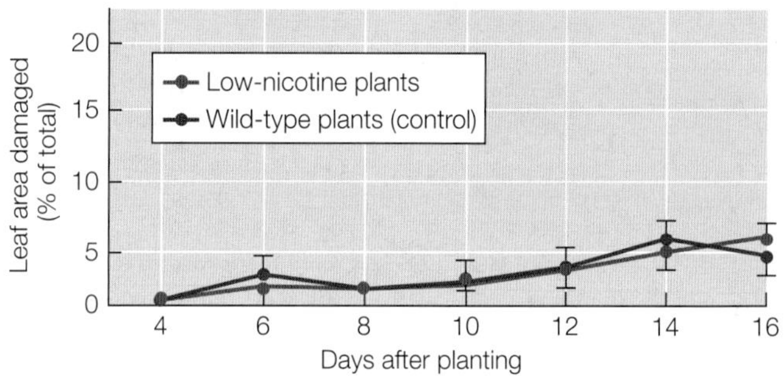


Go to BioPortal for all WORKING WITHDATA exercises

that insects use to digest plant carbohydrates and lipids can elicit defensive responses. Some elicitors are produced when plant material passes through an herbivore's digestive tract; these elicitors are composed of a fatty acid derived from the plant and an amino acid derived from the insect gut. One such elicitor, produced by insects feeding on corn plants, has been named volicitin for its ability to induce production of volatile signals that can travel to other plant parts—and to neighboring corn plants—and stimulate their defense responses. In addition, many herbivorous insects lay their eggs on plants, and some components of the fluids secreted during egg laying have been identified as elicitors.

The signal transduction pathways that are activated by herbivory or wounding involve several key components, some of which are shared by the pathways that are induced by pathogens:

- *Membrane signaling:* The plasma membrane is the part of the plant cell that is in contact with the environment. Within the first minute after an herbivore strikes, changes in the electric potential of the plasma membrane occur in the damaged area. As we will describe in our chapter on the animal nervous system (see Section 45.2), such changes can be rapidly transmitted as a signal along the plasma membrane. In the case of plants responding to herbivory, the continuity of the symplast (see Figure 35.5) ensures that the signal travels over much of the plant within 10 minutes.
- *Reactive oxygen species:* Both wounding and herbivory trigger the production of reactive oxygen species (such as superoxide and hydrogen peroxide), which act as signaling molecules in pathways that lead to changes in gene expression.

*Hormone signaling:* Herbivory induces the production of several hormones that stimulate various plant responses. The most important of these is **jasmonic acid** (**jasmonate**), which triggers systemic defenses against herbivores (**Figure 39.7**).

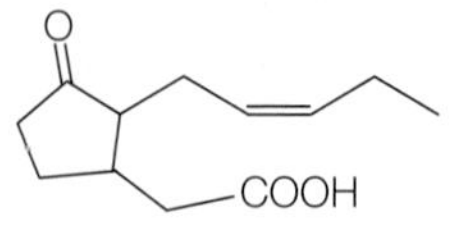

O

COOH

Jasmonic acid

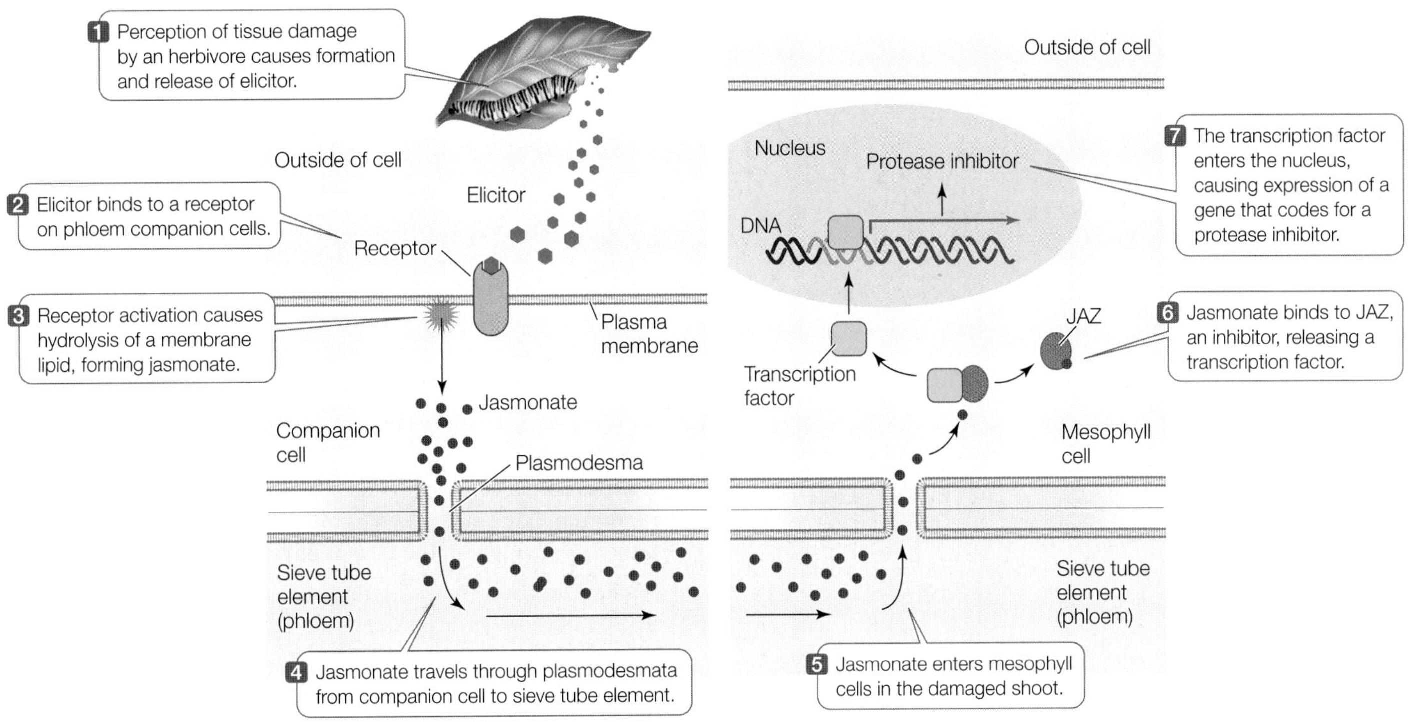

**39.7 A Signal Transduction Pathway for Induced Defenses** The chain of events initiated by herbivory that leads to the production of a defensive chemical can consist of many steps. These steps may include the synthesis of one or two hormones, binding of receptors, gene activation, and finally, synthesis of defensive compounds.

### Jasmonates trigger a range of responses to wounding and herbivory

When the plant senses an herbivore-produced elicitor, it makes jasmonate and a variety of jasmonate derivatives. These molecules trigger many plant defenses both at the site of herbivore attack and throughout the plant (see Figure 39.7). These defenses include the production of specific secondary metabolites and defensive proteins. Jasmonates induce changes in gene expression by binding a transcriptional inhibitor called a JAZ protein. After binding by jasmonate, the JAZ protein is targeted for degradation, and the previously inhibited genes can be expressed.

Protease inhibitors are an important group of defensive proteins that are synthesized in response to insect attack. Once inside an insect's gut, these inhibitors interfere with the digestion of proteins and thus stunt the insect's growth.

Jasmonates can also "call for help" by triggering the formation of volatile compounds that attract insects that prey on the herbivores attacking the plant.

### Why don't plants poison themselves?

Why don't the defensive chemicals that are so toxic to herbivores and pathogens kill the plants that produce them? In some cases the defensive chemicals are directed at organs or systems that are not found in plants, such as the nervous, digestive, or endocrine systems of animals. In addition, plants that produce toxic defensive chemicals use one or more other measures to protect themselves.

**COMPARTMENTALIZATION** Isolation of the toxic substance is a common means of avoiding exposure. Plants store their toxins in vacuoles if the toxins are water-soluble. If they are hydrophobic, the toxins may be dissolved in latex (see p. 806) and stored in a specialized compartment, or they may be dissolved in waxes on the epidermal surface. Such compartmentalized storage keeps the toxins away from the mitochondria, chloroplasts, and other parts of the plant's metabolic machinery.

**STORAGE OF PRECURSORS** Some plants store the precursors of toxic substances in one type of tissue, such as the epidermis, and store the enzymes that convert those precursors into the active toxin in another type, such as the mesophyll. When an herbivore chews part of the plant, cells are ruptured, the enzymes come into contact with the precursors, and the toxin is produced. The only part of the plant that is damaged by the toxin is that which was already damaged by the herbivore. Plants such as sorghum and some legumes, which respond to herbivory by producing cyanide (a potent inhibitor of cellular respiration), are among those that use this type of protective measure.

**MODIFIED PROTEINS** The plant has modified proteins that do not react with the toxin. As noted above, canavanine resembles arginine and therefore plays havoc with protein synthesis in insect larvae. In plants that make canavanine, the plant enzyme that charges the arginine tRNA discriminates correctly between arginine and canavanine, so canavanine is not incorporated into the plant's proteins.

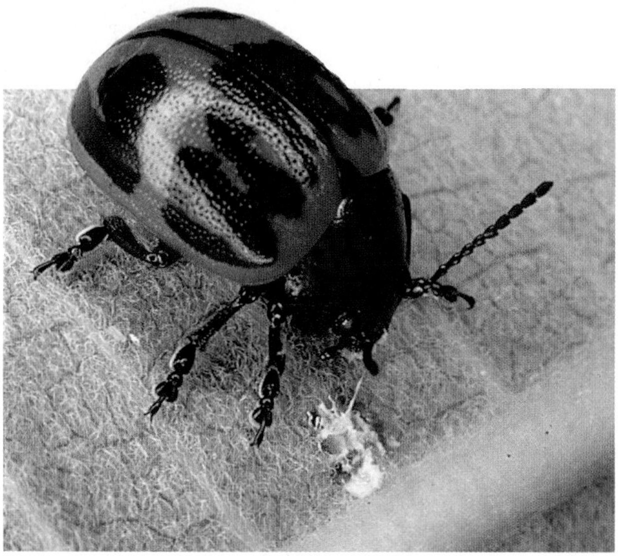

**39.8 Disarming a Plant's Defenses** This beetle is inactivating a milkweed's defense system by cutting its laticifer supply lines.

## Plants don't always win the arms race

Milkweeds such as *Asclepias syriaca* store their defensive chemicals in latex in specialized tubes called **laticifers**, which run alongside the veins in the leaves. When damaged, a milkweed releases copious amounts of toxic latex from its laticifers. Field studies have shown that most insects that feed on neighboring plants of other species do not attack laticiferous plants, but there are exceptions. One population of beetles that feeds on *A. syriaca* exhibits a remarkable prefeeding behavior: these beetles cut a few veins in the leaves before settling down to dine. Cutting the veins causes massive latex leakage from the adjacent laticifers and interrupts the latex supply to a downstream portion of the leaf. The beetles then move to the relatively latex-free portion and eat their fill (**Figure 39.8**).

We have described just one of many ways that herbivores circumvent plant defenses. A successful plant defense exerts strong selection pressure on herbivores to get around it somehow; a successful herbivore, in turn, exerts strong selection pressure on plants to develop new defensive strategies. One can imagine, for example, that over time *A. syriaca* might evolve to have thicker walls around the base of its laticifers, such that the beetles can no longer cut them, or to produce a different toxin that does not depend on laticifers.

**RECAP 39.2**

Many plants use secondary metabolites as constitutive defenses against herbivory. Other defenses are induced by herbivory through signal transduction pathways. The plant hormone jasmonate and its derivatives stimulate local and systemic responses to herbivores.

- Describe one example of a secondary metabolite and how it affects herbivores. **See pp. 802–803**
- What role do jasmonates play in plant defense? **See p. 805 and Figure 39.7**
- What are three ways in which a plant avoids being poisoned by its own defensive chemicals? **See p. 805**

A plant's survival depends not only on successful defenses against pathogens and herbivores but also on coping with a sometimes hostile physical environment. In the next section we will consider how plants deal with environmental stresses.

## 39.3 How Do Plants Deal with Environmental Stresses?

Plants are threatened by many aspects of the physical environment, such as drought, waterlogged soils, and extreme temperatures. These conditions are biologically harmful (**Table 39.2**). Plants cope with environmental stresses through adaptation or acclimation.

- **Adaptation** is genetically encoded resistance to stress. A plant may have structures or biochemical properties that aid in its survival in the face of environmental challenges.
- **Acclimation** is increased tolerance for environmental extremes because of prior exposure to them. An individual plant previously exposed to extreme cold, for example, may be more likely to survive the subsequent winter.

**Go to Media Clip 39.1**
**Leaves for Every Environment**
**Life10e.com/mc39.1**

### Some plants have special adaptations to live in very dry conditions

Many plants, especially those living in deserts, must cope with extremely limited water supplies. A variety of anatomical and life-cycle adaptations allow plants to survive under these conditions. Many of these adaptations are ways to avoid, reduce, or cope with the inevitable water loss through transpiration that occurs during active photosynthesis. Other adaptations help plants tolerate the excessive light and heat that are often found in deserts.

**DROUGHT AVOIDERS** Some desert plants have no special structural adaptations for water conservation. Instead, these desert annuals, called drought avoiders, simply evade periods of drought. Drought avoiders carry out their entire life cycle—from seed to seed—during a brief period in which rainfall has

**TABLE 39.2** Environmental Stresses on Plants

| Condition | Effect on Plants |
|---|---|
| Drought | Reduced water potential, dehydration |
| Flooding | Reduced $O_2$ and respiration |
| High temperature | Changes in membrane fluidity and in proteins |
| Low temperature | Changes in membrane fluidity, damage by ice crystals |
| Salinity | Reduced water potential, dehydration |
| Metal element toxicity | Disruption of metabolism |

39.9 **Desert Annuals Avoid Drought** The seeds of many desert annuals lie dormant for long periods, awaiting conditions appropriate for germination. When they do receive enough moisture to germinate, they grow and reproduce rapidly before the short wet season ends. During the long dry spells, only dormant seeds remain alive.

made the surrounding desert soil sufficiently moist for growth and reproduction (**Figure 39.9**). A different drought avoidance strategy is seen in some African and South American deciduous perennial plants, which shed their leaves in response to drought as a way to conserve water. These plants remain dormant until conditions are again favorable for growth, much as deciduous trees in temperate climates shed their leaves in fall and are dormant until the following spring.

LEAF STRUCTURES Most desert plants are not drought avoiders, but rather grow in their dry environment year-round. Plants adapted to dry environments are called **xerophytes** (Greek *xeros*, "dry"). Three structural adaptations are found in the leaves of many xerophytes:

- Specialized leaf anatomy that reduces water loss
- A thick cuticle and a profusion of hairs over the leaf epidermis, which retard water loss
- Trichomes that diffract and diffuse sunlight, thereby decreasing the intensity of light impinging on the leaves and the risk of damage to the photosynthetic apparatus by excess light

In some xerophytes the stomata are strategically located in sunken cavities below the leaf surface (known as **stomatal crypts**), where they are sheltered from the drying effects of air currents (**Figure 39.10**). Trichomes surrounding the stomata slow air currents further. Cacti and similar plants have spines rather than typical leaves, and photosynthesis is confined to the fleshy stems. The spines may help the plants cope with desert conditions by reflecting solar radiation or by dissipating heat. The spines may also deter herbivores.

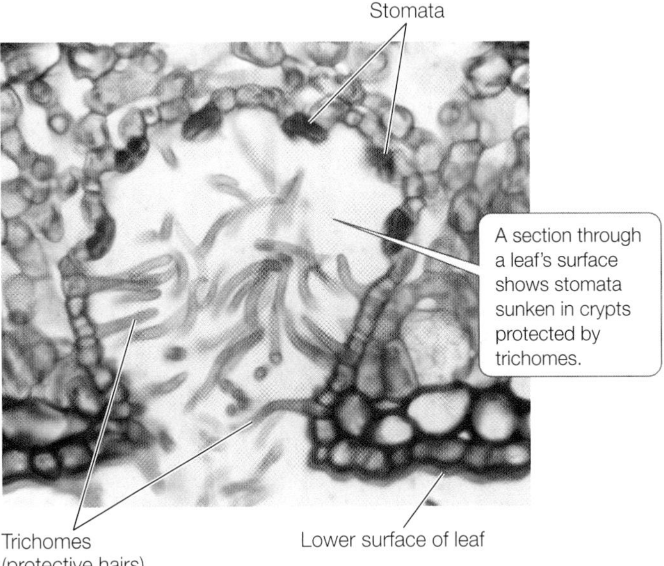


39.10 **Stomatal Crypts** Stomata in the leaves of some xerophytes are located in sunken cavities called stomatal crypts. The trichomes (hairs) covering these crypts trap moist air.

WATER-STORING STRUCTURES **Succulence**—the possession of fleshy, water-storing leaves or stems—is another adaptation to dry environments (**Figure 39.11**). This adaptation allows plants to take up large amounts of water when it is available (such as after a brief thunderstorm) and then draw on the stored water during subsequent dry periods. Other adaptations of succulents include a reduced number of stomata and a variant form of photosynthesis (the CAM pathway; see Section 10.4), both of which reduce water loss.

39.11 **Succulence** The *Aloe* plant stores water in its fleshy leaves.

**39.12 Mining Water with Deep Taproots** In Death Valley, California, the root of this mesquite tree must reach far beneath the dunes for its water supply.

**ROOT SYSTEMS THAT MAXIMIZE WATER UPTAKE** Roots may also be adapted to dry environments. Cacti have shallow but extensive fibrous root systems that effectively intercept water at the soil surface following even light rains. Mesquite (*Prosopis*) (**Figure 39.12**) obtains water through taproots that grow to great depths, reaching water supplies far underground, as well as from condensation on its leaves. The Atacama Desert in northern Chile often goes for several years without measurable rainfall, but the landscape there has many surprisingly large mesquite shrubs.

**SOLUTE ACCUMULATION** Xerophytes and other plants that must cope with inadequate water supplies may accumulate high concentrations of the amino acid proline or of secondary metabolites in their vacuoles. This solute accumulation lowers the water potential in the plant's cells below that of the soil, which allows the plant to take up water via osmosis. Plants living in saline environments share this and several other adaptations with xerophytes, as we will see shortly.

## Some plants grow in saturated soils

For some plants, the environmental challenge is too much water, the opposite of that faced by xerophytes. They live in environments so wet that the diffusion of oxygen to their roots is severely limited. These plants have shallow root systems that grow slowly; oxygen levels are likely to be highest near the surface of the soil, and slow growth decreases the roots' need for oxygen.

The root systems of some plants adapted to swampy environments, such as cypresses and some plants that grow in coastal mangrove habitats, have **pneumatophores**, which are extensions that grow out of the water and up into the air (**Figure 39.13A**). Pneumatophores contain lenticels (openings; see Figure 34.17) that allow oxygen to diffuse through them, aerating the submerged parts of the root system.

Many submerged or partly submerged aquatic plants have large air spaces in the leaf and stem parenchyma and in the petioles. Tissue containing such air spaces is called **aerenchyma** (**Figure 39.13B**). Aerenchyma stores oxygen produced by photosynthesis and permits its ready diffusion to parts of the plant where it is needed for cellular respiration. Aerenchyma also imparts buoyancy. Furthermore, because aerenchyma contains far fewer cells than most other plant tissues, metabolism in aerenchyma proceeds at a lower rate, so the need for oxygen is much reduced.

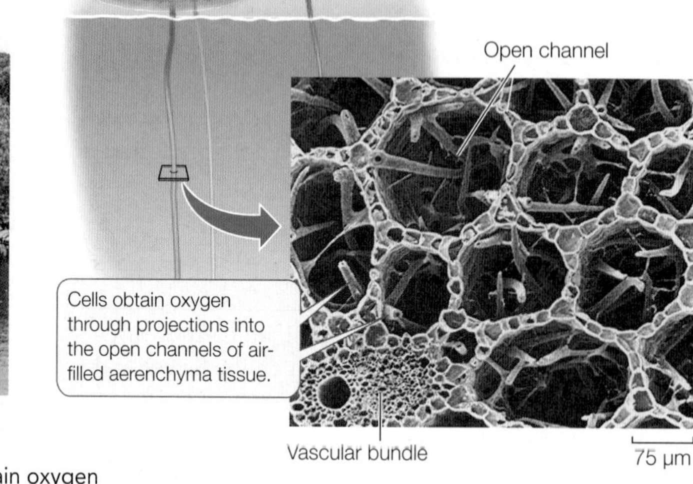


**39.13 Coming Up for Air** (A) The roots of these mangroves obtain oxygen through pneumatophores. (B) This scanning electron micrograph of a cross section of a petiole of the yellow water lily shows the structure of the air-filled channels that make up aerenchyma tissue.

## Plants can respond to drought stress

The adaptations of xerophytes for coping with dry environments are generally constitutive—they are always present—and under normal conditions they prevent the plants from experiencing drought stress. When conditions become so dry that even xerophytes are stressed, however, the plants turn to inducible responses. The same responses are found in many other plants, including those that are not adapted to grow in dry climates.

When the weather is abnormally dry, the water content of the soil is reduced and less water is available to plants. Water deficits in plant cells have two major biochemical effects:

- A reduction in membrane integrity as the polar–nonpolar forces that orient the lipid bilayer are reduced
- Changes in the three-dimensional structures of proteins

Plant growth is reduced when the structure of plant cells is compromised in these ways. Indeed, inadequate water supply is the single most important factor that limits production of our most important food crops.

When plants sense a water deficit in their roots, a signaling pathway is set in motion that initiates several measures to conserve water and maintain cellular integrity. This pathway begins with the production of the hormone abscisic acid in the roots. This hormone travels from the roots to the shoot, where it results in stomatal closure and initiates gene transcription that leads to other physiological events that conserve water and cellular integrity (**Figure 39.14**).

Many plant genes whose expression is altered by drought stress have been identified, largely through research using DNA microarrays, proteomics, and other molecular approaches (see Chapters 17 and 18). One group of proteins whose production is upregulated during drought stress is the *l*ate *e*mbryogenesis *a*bundant (LEA; pronounced "lee-yuh") group of proteins. These hydrophobic proteins also accumulate in maturing seeds as they dry out (hence their name). LEA proteins bind to membrane proteins and other cellular proteins to stabilize them, preventing their aggregation during desiccation. The importance of LEA proteins in coping with drought stress was shown by Ray Wu and colleagues at Cornell University, who found that transgenic rice plants expressing a high level of LEA protein in leaves and roots grew better than normal plants under drought conditions (**Figure 39.15**). Genes that encode LEA proteins occur in many plants and confer drought tolerance.

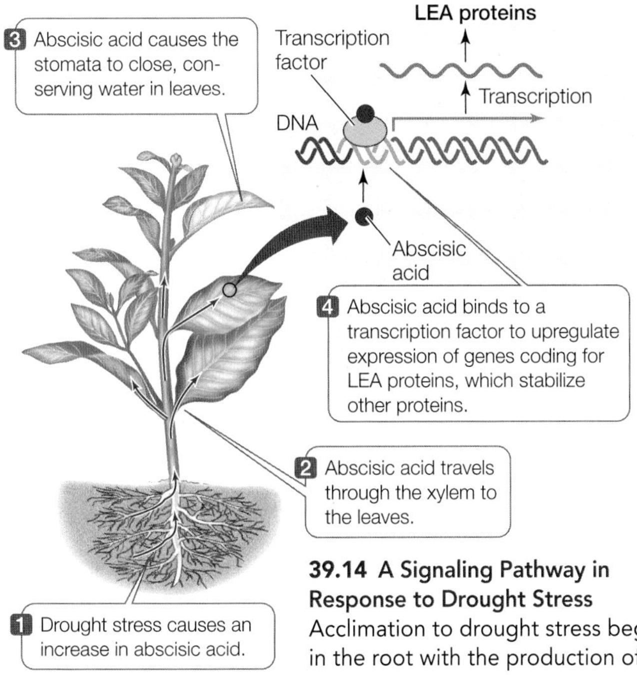

**39.14 A Signaling Pathway in Response to Drought Stress** Acclimation to drought stress begins in the root with the production of the hormone abscisic acid.

INVESTIGATING**LIFE**

**39.15 A Molecular Response to Drought Stress** Understanding the responses of plants to drought conditions is vital for agriculture. Ray Wu and colleagues transformed rice cells with a gene that codes for a LEA protein that is expressed in seeds as they mature and dry out. The investigators then measured the response of the transgenic rice plants to 28 days of drought.[a]

**HYPOTHESIS** LEA proteins protect plants from the effects of drought stress.

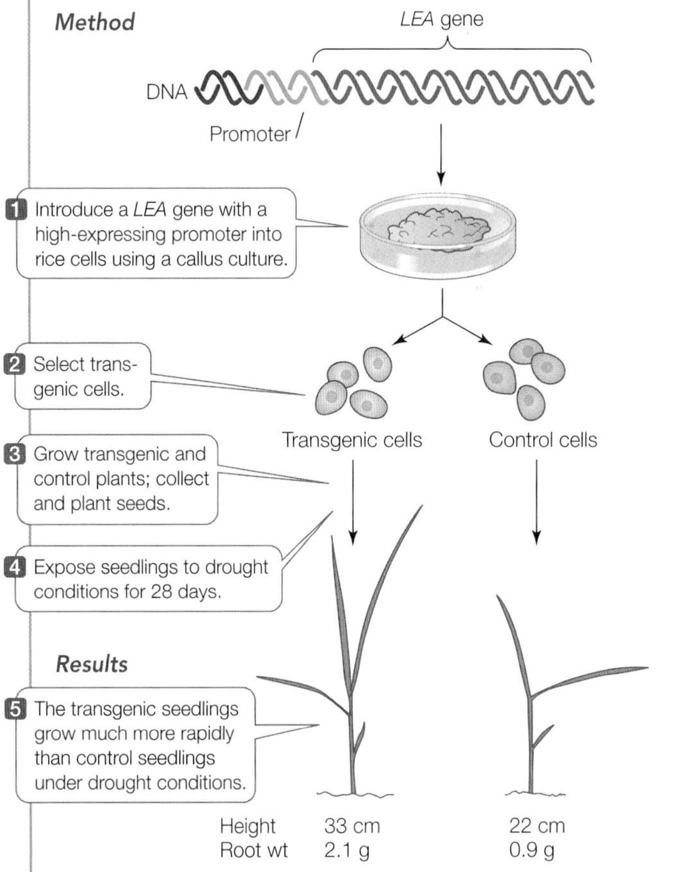

**CONCLUSION** Plants with higher *LEA* gene expression grow better under drought conditions.

Go to **BioPortal** for discussion and relevant links for all INVESTIGATING**LIFE** figures.

[a]Xu, D. et al. 1996. *Plant Physiology* 110:249–257.

## Plants can cope with temperature extremes

Temperatures that are too high or too low can stress plants and even kill them. Plant species differ in their sensitivity to heat and cold, but all plants have their limits. Any temperature extreme can damage cellular membranes.

- High temperatures destabilize membranes and denature many proteins, especially some of the enzymes of photosynthesis.
- Low temperatures cause membranes to lose their fluidity and alter their permeabilities to solutes.
- Freezing temperatures may cause ice crystals to form, damaging membranes.

Plants have both constitutive adaptations and inducible responses for coping with temperature extremes.

ANATOMICAL ADAPTATIONS Many plants living in hot environments have constitutive adaptations similar to those of xerophytes. These adaptations include hairs and spines that dissipate heat and leaf forms that intercept less direct sunlight.

HEAT SHOCK RESPONSE The plant inducible response to heat stress is similar to the response to drought stress in that new proteins are made, often under the direction of an abscisic acid–mediated signaling pathway. Within minutes of experimental exposure to raised temperatures (typically a 5°C–10°C increase), plants synthesize several kinds of **heat shock proteins**. Among these proteins are chaperonins, which help other proteins maintain their structures and avoid denaturation. Threshold temperatures for the production of heat shock proteins vary, but 39°C is sufficient to induce them in most plants.

COLD-HARDENING Low temperatures above the freezing point can cause chilling injury in many plants, including crops such as rice, corn, and cotton as well as tropical plants such as bananas. Many plant species can acclimate to cooler temperatures through a process called **cold-hardening**, which requires repeated exposure to cool temperatures over many days. A key change during hardening is an increase in the proportion of unsaturated fatty acids in cell membranes, which allows them to retain their fluidity and function normally at cooler temperatures (see Figure 3.21). Plants have a greater ability to modify the degree of saturation of their membrane lipids than animals do. In addition, low temperatures induce the formation of proteins similar to heat shock proteins, which protect against chilling injury.

If ice crystals form within plant cells, they can kill the cells by puncturing organelles and plasma membranes. Furthermore, the growth of ice crystals outside the cells can draw water from the cells and dehydrate them. Freeze-tolerant plants have a variety of adaptations to cope with these problems, including the production of antifreeze proteins that slow the growth of ice crystals.

RECAP 39.3

Plants that live in continually dry or water-saturated environments have structural adaptations to cope with those conditions. Mechanisms that protect plants from drought stress are initiated by a signaling pathway involving abscisic acid. Heat shock proteins help plants acclimate to high and low temperatures.

- Describe two structural adaptations for growth in water-saturated soils. See p. 808
- What is the role of abscisic acid in acclimation to drought stress? See p. 809 and Figure 39.14
- What environmental conditions induce the formation of heat shock proteins, and what functions do those proteins serve? See p. 810

Just as climatic extremes can limit plant growth, the presence of certain substances, such as salt and heavy metals, can make an environment inhospitable to plant growth.

## 39.4 How Do Plants Deal with Salt and Heavy Metals?

A number of toxic solutes are found in soils, but worldwide, no toxic substance restricts angiosperm growth more than ordinary salt (sodium chloride). Saline—salty—habitats support, at best, limited types of vegetation. Saline habitats are found in diverse locales, from hot, dry deserts to moist, cool coastal marshes. Along the seashore, saline environments are created by ocean spray. The ocean itself is a saline environment, as are estuaries, where fresh and salt water meet and mingle. Salinization of agricultural land is an increasing global problem (**Figure 39.16**). Even where crops are irrigated with fresh water, sodium ions from the water accumulate in the soil to ever greater concentrations as the water evaporates.

Saline environments pose an osmotic challenge for plants. Because of its high salt concentration, a saline environment

**39.16 Salty Soil** Accumulation of salt from irrigation water with inadequate drainage has caused this soil in central California to become unsuitable for most plant growth.

has an unusually negative water potential (see Figure 35.2). To obtain water from such an environment, a plant must have an even more negative water potential; otherwise water will diffuse out of its cells, and the plant will wilt and die. Plants in saline environments are also challenged by the potential toxicity of sodium, which inhibits enzymes and protein synthesis.

### Most halophytes accumulate salt

**Halophytes**—plants adapted to saline habitats—are found in a wide variety of flowering plant groups. Most halophytes share one adaptation: they take up sodium and, usually, chloride ions and transport those ions to their leaves. The accumulated ions are stored in the central vacuoles of leaf cells, away from more sensitive parts of the cells. Nonhalophytes accumulate relatively little sodium, even when placed in a saline environment; of the sodium that is absorbed by their roots, very little is transported to the shoot. The increased salt concentration in the tissues of halophytes lowers their water potential and allows them to take up water from their saline environment.

Some halophytes have other adaptations to life in saline environments. Some, for example, have **salt glands** in their leaves. These glands excrete salt, which collects on the leaf surface until it is removed by rain or wind (**Figure 39.17**). This adaptation, which reduces the danger of poisoning by accumulated salt, is found in some desert plants, such as *Frankenia palmeri*, and in some mangroves growing in seawater.

Salt glands can play multiple roles, as in the desert shrub *Atriplex halimus*. This shrub has glands that secrete salt into small bladders on the leaves. By lowering the water potential of the leaves, this salt not only helps them obtain water from the roots but also reduces their transpirational loss of water to the atmosphere.

The adaptations we have just discussed are specific to halophytes. Several other adaptations are shared by halophytes and xerophytes, including thick cuticles, succulence, and CAM photosynthesis.

**39.17 Excreting Salt** This saltwater mangrove plant has special salt glands that excrete salt, which appears here as crystals on the leaves.

### Some plants can tolerate heavy metals

Salt is not the only toxic solute found in soils. High concentrations of some heavy metal ions, such as chromium, mercury, lead, and cadmium, are toxic to most plants; many of these ions are more toxic than sodium at equivalent concentrations.

Some geographic sites are naturally rich in heavy metals as a result of normal geological processes. In other places, acid rain leads to the release of toxic aluminum ions in the soil. Human activities, notably the mining of metallic ores, leave localized areas—known as tailings—with high concentrations of heavy metals and low concentrations of nutrients. Such sites are hostile to most plants, and seeds falling on them generally do not produce adult plants.

Most mine tailings rich in heavy metals, however, are not completely barren. They may support healthy plant populations that differ genetically from populations of the same species on the surrounding normal soils. How do these plants survive?

Initially, botanists believed that some plants were able to tolerate heavy metals by excluding them: that by not taking up the metal ions, the plants avoided being poisoned. Further investigations have shown, however, that tolerant plants growing on mine tailings do take up heavy metals, accumulating concentrations that would kill most plants. More than 200 plant species have been identified as **hyperaccumulators** that store large quantities of metals such as arsenic (As), cadmium (Cd), nickel (Ni), aluminum (Al), and zinc (Zn).

Perhaps the best-studied hyperaccumulator is alpine pennycress (*Thlaspi caerulescens*). Before the advent of chemical analysis, miners used to use the presence of this plant as an indicator of mineral-rich deposits. A *Thlaspi* plant may accumulate as much as 30,000 ppm Zn (most plants contain 100 ppm) and 1,500 ppm Cd (most plants contain 1 ppm). Studies of *Thlaspi* and other hyperaccumulators have revealed the presence of several common mechanisms:

- Increased ion transport into the roots
- Increased rates of translocation of ions to the leaves
- Accumulation of ions in vacuoles in the shoot
- Resistance to the ions' toxicity

Knowledge of these hyperaccumulation mechanisms and the genes underlying them has led to the emergence of **phytoremediation**, a form of bioremediation (see Section 18.6) that uses plants to clean up environmental pollution. Some phytoremediation projects use natural hyperaccumulators, whereas others use genes from hyperaccumulators to create transgenic plants that grow more rapidly and are better adapted to a particular polluted environment. In either case, the plants are grown in the contaminated soil, where they act as natural "vacuum cleaners" by taking up the contaminants (**Figure 39.18**). The plants are then harvested and disposed of to remove the contaminants. Perhaps the most dramatic use of phytoremediation occurred after an accident at the nuclear power plant at Chernobyl, Ukraine (then part of the Soviet Union), in 1986, when sunflower plants were used to remove uranium from the nearby soil. Phytoremediation is now widely used in cleaning up land after strip mining.

**39.18 Phytoremediation** Plants that accumulate heavy metals can be used to clean up contaminated soils. Here, poplars are being used to remove contaminants from an air force base.

After finding plants that accumulate valuable metals such as Ni, cobalt (Co), and silver (Ag), some scientists have proposed using those plants for *phytomining*. As in phytoremediation, the plants would be used to take up metals from the soil, but the metals would be extracted from the plants after they are harvested.

### RECAP 39.4

Halophytes have several adaptations to saline habitats, most of which involve mechanisms that lower their water potential. Some plants can tolerate heavy-metal-rich soils that are toxic to most other plants.

- What are some of the roles of salt glands in halophyte leaves? **See p. 811**
- How are plants used for phytoremediation? **See p. 811**

What is the current status of artemisinin therapy for malaria?

**ANSWER**

Since 2000, artemisinin has become a mainstay of malaria treatment worldwide. Millions of people take it every day. Because the solubility of the natural molecule is quite limited, chemists have synthesized derivatives of it that are more soluble and easier to take. There is, however, a serious danger that the malaria parasite will develop resistance to artemisinin and its derivatives, so these drugs are used in combination with other synthetic drugs.

## CHAPTER**SUMMARY** 39

### 39.1 How Do Plants Deal with Pathogens?

- Plants and pathogens have evolved together in a continuing "arms race": pathogens have evolved mechanisms for attacking plants, and plants have evolved mechanisms for defending themselves against those attacks.
- **Constitutive** defenses include plants' ability to strengthen their cell walls and block plasmodesmata when attacked, limiting the ability of viral pathogens to move from cell to cell.
- **Induced** defenses are triggered by a wide range of molecular **elicitors** and fall into two main categories: general immunity and specific immunity. **Review Figure 39.2, ANIMATED TUTORIAL 39.1**
- The **gene-for-gene concept** depends on a match between a plant's **resistance (*R*) genes** and a pathogen's ***Avirulence* (*Avr*) genes**. **Review Figure 39.3**
- In the **hypersensitive response** to infection by bacteria or fungi, cells produce two kinds of defensive molecules: **phytoalexins** and **pathogenesis-related (PR) proteins**. Some cells around the infected area die, sealing off the pathogens and the damage they have caused.
- The hypersensitive response is often followed by **systemic acquired resistance**, in which salicylic acid activates further synthesis of defensive compounds.
- Plants use RNA interference to develop specific immunity to invading RNA viruses.

### 39.2 How Do Plants Deal with Herbivores?

- Some plants produce **secondary metabolites** as defenses against herbivores. **Review Table 39.1, Figure 39.6**
- Hormones, including **jasmonates**, participate in signal transduction pathways leading to the production of defensive compounds. **Review Figure 39.7**
- Plants protect themselves against their own toxic defensive chemicals by isolating them in specialized compartments, by producing them only after the plant has already been damaged, or by having modified enzymes or receptors that are not affected by the toxic substance.

### 39.3 How Do Plants Deal with Environmental Stresses?

- Plants cope with environmental stresses by **adaptation** (genetically encoded resistance) or **acclimation** (increased tolerance) **Review Table 39.2**
- **Xerophytes** are plants that are adapted to dry environments.

*continued*

- Some xerophytic adaptations are structural, including thickened cuticles, specialized trichomes, **stomatal crypts**, **succulence**, and long taproots.
- Some plants accumulate solutes, making their water potential lower so they can tolerate drought.
- Adaptations to water-saturated habitats include **pneumatophores**, extensions of roots that allow oxygen uptake from the air, and **aerenchyma**, tissue in which oxygen can be stored and ready for diffusion throughout the plant.
- A signaling pathway involving abscisic acid initiates a plant's response to drought stress. **Review Figures 39.14, 39.15**
- Membranes and proteins can be damaged by extremely high or low temperatures. Plants respond to extreme temperatures by producing **heat shock proteins**.
- Some plants undergo **cold-hardening**, an acclimation process that includes changes in membrane lipids and production of heat shock proteins.
- Some plants resist freezing by producing antifreeze proteins.

## How Do Plants Deal with Salt and Heavy Metals?

- Most **halophytes** accumulate salt. Some have **salt glands** that excrete salt to the leaf surface.
- Some plants living in soils that are rich in heavy metals are **hyperaccumulators** that take up large amounts of those metals into their tissues.
- **Phytoremediation** is the use of hyperaccumulating plants or their genes to clean up environmental pollution.

**See ACTIVITY 39.1 for a concept review of this chapter.**

**Go to the Interactive Summary to review key figures, Animated Tutorials, and Activities Life10e.com/is39**

# CHAPTER**REVIEW**

## REMEMBERING

1. Plants sometimes protect themselves from their own toxic secondary metabolites by
   a. producing special enzymes that destroy the toxin.
   b. storing precursors of the toxic substances in one compartment and the enzymes that convert those precursors to toxic products in another compartment.
   c. storing the toxic substances in mitochondria or chloroplasts.
   d. distributing the toxic substances to all cells of the plant.
   e. performing crassulacean acid metabolism.

2. Which statement about secondary metabolites is *not* true?
   a. They may be used in defense against fungi.
   b. Some are poisonous to herbivores.
   c. Some are amino acids that are normally part of proteins.
   d. Water-soluble molecules are stored in vacuoles.
   e. Some mimic the hormones of animals.

3. Which of the following is *not* an adaptation to dry environments?
   a. Increased solute concentration in the vacuoles
   b. Hairy leaves
   c. A heavier cuticle over the leaf epidermis
   d. Sunken stomata
   e. A root system that grows each rainy season and dies back when it is dry

4. Some plants adapted to swampy environments meet the oxygen needs of their roots by means of a specialized tissue called
   a. parenchyma.
   b. aerenchyma.
   c. collenchyma.
   d. sclerenchyma.
   e. chlorenchyma.

5. Halophytes
   a. may accumulate abscisic acid in their vacuoles.
   b. may have water potentials that are lower than those of other plants.
   c. only accumulate sodium.
   d. have low root-to-shoot ratios.
   e. rarely accumulate sodium.

6. Which of the following is *not* true of the general immunity response of a plant to a pathogen?
   a. It can result in changes in gene expression.
   b. It can result in the production of plant hormones.
   c. It is a weak response in comparison with specific immunity.
   d. It involves sealing off damaged tissues to contain the infection.
   e. It is produced when a plant receptor recognizes a molecular pattern in the pathogen.

## UNDERSTANDING & APPLYING

7. How might the adaptations of herbivores affect plant evolution? How might plant adaptations affect the evolution of herbivores?

8. A tomato plant can be infected with the fungus *Cladosporium*. The host plant and pathogen can have various genes involved in the hypersensitive response. Fill in the table that describes the fungal strains and the results of infection of plants:

| | *Cladosporium* genotype | | |
|---|---|---|---|
| Tomato genotype | *R1R2* | *R3R4* | *R1R4* |
| *Avr1Avr2* | Healthy | Diseased | Healthy |
| *Avr2Avr3* | | | |
| *Avr1Avr4* | | | |

## ANALYZING & EVALUATING

9. In the coming decades, climate change may have significant effects on the growth and productivity of plants, in particular the crops on which we depend for our food. Discuss the physiological effects, and possible genetic responses in terms of plant breeding, of the following:
   a. In Pakistan, reduced rainfall causes a reduction in wheat yields.
   b. In the Mekong Delta of Vietnam, rising sea level inundates rice fields, causing a drastic reduction in yields.
   c. Increased temperature and humidity in western Canada causes an increase in wheat rust.

10. The tobacco hornworm (*Manduca sexta*) is adapted to feeding on nicotine-producing plants. Using the genetically modified tobacco plants described in Figure 39.6, how might you test the hypothesis that dietary nicotine protects the tobacco hornworm against its parasite *Cotesia congregata*?

Go to BioPortal at **yourBioPortal.com** for Animated Tutorials, Activities, LearningCurve Quizzes, Flashcards, and many other study and review resources.

# 40 Physiology, Homeostasis, and Temperature Regulation

## CHAPTER OUTLINE


**40.1** How Do Multicellular Animals Supply the Needs of Their Cells?

**40.2** What Are the Relationships between Cells, Tissues, and Organs?

**40.3** How Does Temperature Affect Living Systems?

**40.4** How Do Animals Alter Their Heat Exchange with the Environment?

**40.5** How Do Endotherms Regulate Their Body Temperatures?

**Limits to Performance** Paula Radcliffe, photographed here during her winning performance at the 2008 New York City Marathon, collapsed from heat stress during the 2004 Olympic marathon. When the body is subjected to extreme heat, its homeostatic mechanisms may fail.

THE 2008 NEW YORK CITY MARATHON took place on a cold, clear, windy day in November. For the third time, the first-place woman in this 41-km race was world record holder Paula Radcliffe. Radcliffe had also been expected to win the women's marathon in the 2004 Olympics. But that race took place on an extremely hot (a high of 34°C), humid day in Athens. Overcome by heat stress, Radcliffe collapsed 6 km from the finish line. In contrast, the average temperature for the three New York marathons Radcliffe won was 7°C.

Based on a survey of many marathons, elite runners have their best times when temperatures are below 10°C; higher temperatures can mean serious problems. The 2012 Boston Marathon coincided with an unseasonable April heat wave, with temperatures exceeding 27°C. During the course of the race, 120 runners were rushed to hospitals with severe heat stress.

When a person's internal body temperature rises above 40°C, major organs begin to fail, a condition known as heat stroke. Every year some athletes suffer heat stroke, which leads to death in a high percentage of cases. Soldiers in desert environments are at extreme risk of heat stroke, as are workers in many occupations, including firefighting, agriculture, and construction.

Why is heat stroke a particular danger for those who must be active in the heat? The short answer is that working muscles generate heat. That heat leaves the muscles in the blood and is circulated around the body, raising the temperature of the body's internal tissues. Although some of the heated blood flows to the skin, where heat can be lost to the environment, humans are subject to the problems faced by all mammals in losing excess heat. First, their normal internal temperatures are not far from the environmental temperatures that cause heat stress, so they don't have much of a safety zone. Second, most mammalian skin surfaces are covered with an insulating layer of fur—great for conserving body heat in cold environments, but an impediment to heat loss in warm ones.

Evolutionary adaptation in mammals has resulted in the efficient heat-loss portals of non-furred areas such as the nose, tongue, and footpads. In these areas, specialized blood vessels can open up and act like radiators to disperse heat (conversely, these portals can close down to conserve heat). Humans are not furred, but our evolutionary ancestors were, and we retain these general mammalian blood vessel adaptations in our hands, feet, and face (which is why we blush).

Can we increase heat loss from our natural heat portals to protect against heat stress?

See answer on p. 831.

## 40.1 How Do Multicellular Animals Supply the Needs of Their Cells?

All animal cells must obtain nutrients and oxygen from the environment and must eliminate carbon dioxide and other waste products of metabolism to the environment. The cells of very small or very thin aquatic animals meet these needs by direct exchanges with the external environment. In such animals, no cell is far from direct contact with the water it lives in; the water contains nutrients, absorbs wastes, and provides a relatively unchanging physical environment. Most cells of larger animals do not have direct contact with the external environment, and their needs must be served by an environment that is wholly internal to the animal.

### An internal environment makes complex multicellular animals possible

The cells of multicellular animals exist within an **internal environment** of extracellular fluid. A human, for example, is about 60 percent water. Two-thirds of that water is contained within our cells, while one-third is the extracellular fluid (ECF) that is our internal environment. About 20 percent of the ECF (about 3 liters) is the **plasma** that circulates in our blood vessels. The remaining 80 percent (about 11 liters) is the **interstitial fluid** that bathes every cell of the body (**Figure 40.1**). Individual cells get their nutrients from this interstitial fluid and dump their waste products into it. As long as conditions in the body's internal environment are held within certain limits, cells are protected from the changes and harsh conditions of the external environment. A stable internal environment makes it possible for an animal to occupy habitats that would kill its cells if they were directly exposed to the external conditions. How is the internal environment kept constant?

As multicellular organisms evolved, cells became specialized for maintaining specific aspects of the internal environment. In turn, the internal environment enabled these specializations, since each cell no longer had to provide for all of its own needs. Some cells evolved to be the body's interface between the internal and the external environments and to provide the necessary transport functions to get nutrients in and move wastes out. Other cells became specialized for internal functions such as circulation of the extracellular fluids, energy storage, movement, and information processing. The evolution of physiological systems to maintain the internal environment made it possible for multicellular animals to become larger, thicker, and more complex, and allowed them to occupy many different habitats.

The composition of the internal environment is constantly being challenged by the external environment and by the metabolic activity of the cells of the body. Organisms must maintain their internal environment in a state of **homeostasis**—a narrow range of stable physical and biochemical conditions under which the body functions optimally. If a physiological system fails to function properly, homeostasis is compromised, and cells are damaged and can die. To avoid the loss of homeostasis, physiological systems must be controlled and regulated in response to changes in both the external and internal environments. The maintenance of homeostasis is a central theme of physiology.

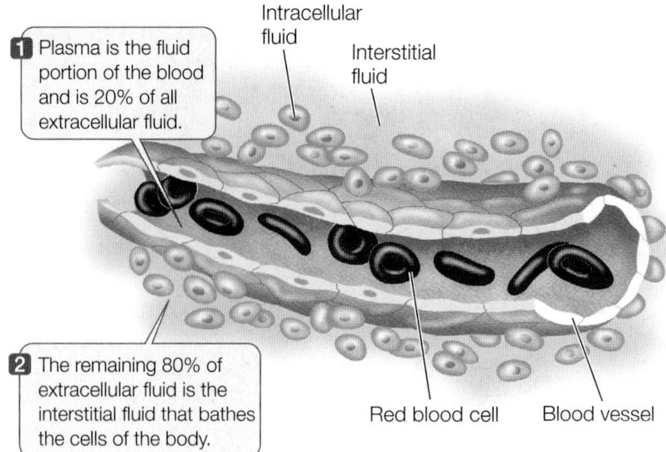


**40.1 The Internal Environment** The "internal environment" is the extracellular fluid, or ECF, which accounts for about one-third of total body water. The ECF is made up of the blood plasma and the interstitial fluid. The physiological state of the ECF must remain stable within narrow limits, and maintaining that stability is the job of the body's organ systems.

### Physiological systems are regulated to maintain homeostasis

The activities of physiological systems are *controlled*—speeded up or slowed down—by actions of the nervous and endocrine systems. But to *regulate* these systems and maintain homeostasis, information is required. As an analogy, think of the thermostat that controls the furnace or air conditioner to regulate the temperature of a house (**Figure 40.2**). The desired temperature is a **set point**, or reference point on the thermostat. The thermostat sensor responds to the temperature of the air, providing **feedback**—information that is compared with the set point. Any difference between the set point and the air temperature is an **error signal**. The error signal is converted into a corrective action—turning the air conditioner or furnace on or off.

Some components of physiological systems are called **effectors** because they *effect* changes in the internal environment. Muscles are a notable example of effectors. Another example are cells in the stomach that secrete digestive juices. Effectors are **controlled systems** because their activities are controlled by neural or hormonal signals from regulatory systems. **Regulatory systems** obtain, process, and integrate information, then issue commands to controlled systems. Important components of any regulatory system are the **sensors** (e.g., light-, temperature-, and pressure-sensitive cells) that provide feedback information to be compared with the internal set point.

**Negative feedback** is information used to counteract the influence that created an error signal. Whatever force is pushing the system away from its set point must be "negated." In our thermostat analogy, an air temperature below the set point causes the furnace to be turned on, which then reverses the direction of change in the air temperature.

Although not as common as negative feedback, **positive feedback** also exists in physiological systems. Rather than returning a system to a set point, positive feedback amplifies a response (i.e., it increases the deviation from the set point). An

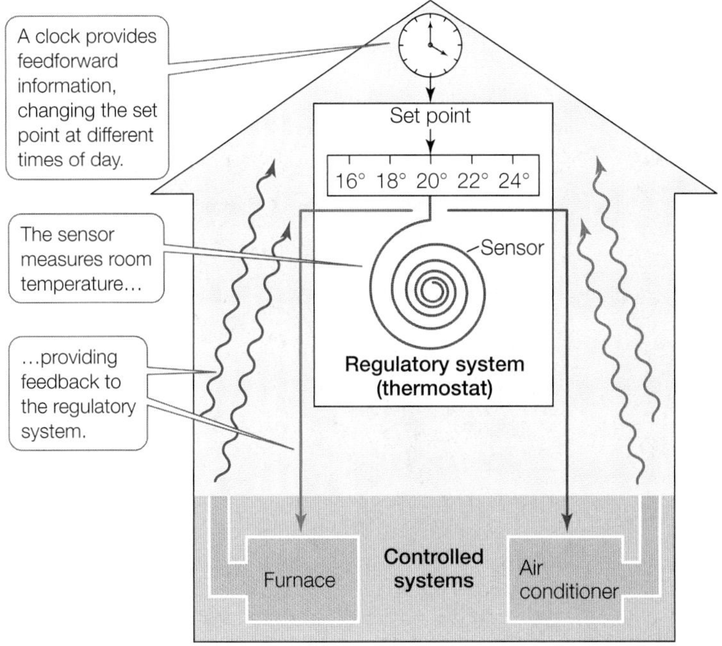


**40.2 Thermostat Regulates Temperature** A thermostat regulates the temperature of a room by turning the furnace or air conditioner on or off in response to the difference between feedback information (room temperature) and set points that are programmed into the thermostat.

example is sexual behavior, in which a little stimulation causes more behavior, which causes more stimulation, and so on. Positive feedback responses tend to reach a limit and terminate rapidly. The birth process is a good example. Contractions of the uterus stretch the birth canal, and that stretching stimulates more and stronger contractions until the baby is delivered, at which time contractions cease.

**Feedforward information** is another feature of regulatory systems. Its function is to change the set point in anticipation of a change in conditions. The timer on a thermostat provides feedforward information by changing the system's temperature set point, usually lowering it in the evening and raising it in the morning. Hearing the words "on your mark" before a race is feedforward information that raises your heart rate in anticipation of running. Feedforward information anticipates a change in the internal environment before that change occurs.

### RECAP 40.1

The internal environment provides for the needs of all the cells that make up a complex multicellular animal. Organs and organ systems control the composition of the internal environment so as to maintain homeostasis.

- What are the three water-containing compartments in the human body? Why is the ECF crucial to survival in multicellular animals? **See p. 816 and Figure 40.1**
- Explain the difference between negative and positive feedback control mechanisms. **See pp. 816–817**
- Thinking about the thermoregulatory system in your own home, what is the set point, what is the feedback, and what would be feedforward information? **See pp. 816–817 and Figure 40.2**

Principles of control and regulation help organize our thinking about physiological systems. Once we understand how a system works, we can then ask how it is regulated. Part 9 of this book describes various physiological systems, how they function, and how they are regulated. But first we summarize some important structural features of the physiological systems that link the functioning of individual cells with the systemic functioning of multicellular animals.

## 40.2 What Are the Relationships between Cells, Tissues, and Organs?

Control, regulation, and the resulting maintenance of homeostasis is the work of a hierarchy of interacting physiological systems that make up a multicellular animal (see Figure 1.9A). Each system is composed of discrete **organs**, such as the stomach, heart, lungs, and kidneys. Organs are made up of assemblages of similar cells called **tissues**; tissues in turn are made up cells of different types. There are many specialized cell types, but there are only four kinds of tissues: epithelial, muscle, connective, and nervous tissues. The word "tissue" is often used in a general way to refer to a piece of an organ, such as "lung tissue" or "kidney tissue." An organ, however, always consists of more than one of the four tissue types. Here we will provide only brief descriptions of the four tissues types because each will be dealt with in more detail in the context of an organ or organ system in which it plays a major role.

### Epithelial tissues are sheets of densely packed, tightly connected cells

**Epithelial tissues** are formed by sheets of cells, creating boundaries between the inside and the outside of the body and between body compartments. Epithelial tissues comprise the outer layer of the skin, line the blood vessels, and make up various ducts and tubules (**Figure 40.3**). They control which molecules and ions can move between different body compartments, such as the blood and the interstitial fluid. They selectively transport ions and molecules from one side of an epithelial membrane to the other; for example, the gut epithelium absorbs nutrients from the gut and secretes digestive juices into the gut.

The cells that make up epithelial tissues are of different types and configurations. The multilayered epithelial tissues of the skin are subject to a lot of wear and tear. Accordingly, epithelial cells in the deepest layer of the skin have a high rate of cell division, producing new cells that move progressively to the skin surface, die, and are shed (see Figure 40.3A). In contrast, gut epithelium consists of a single layer of tall, closely packed cells.

Epithelial cells have many specialized roles. Some secrete hormones, milk, mucus, digestive enzymes, or sweat. Others have cilia that move substances over surfaces or through tubes (see Figure 40.3B). Epithelial cells can also provide information to the nervous system. Smell- and taste-sensitive cells, for example, are epithelial cells that detect specific chemicals.

(A) Squamous cells
Stratified epithelium
30 µm

(B) Columnar epithelium
Cilia
20 µm

(C) Cuboidal epithelial cells
25 µm

**40.3 Epithelial Tissues** (A) Epithelial cells make up the outer layers of skin. This epithelium is stratified, from extremely thin (squamous) older cells at the surface to rapidly dividing new cells that will rise to the surface as older cells are shed. (B) Ciliated columnar epithelium from the male reproductive duct (the vas deferens). (C) A single layer of cuboidal epithelial cells forms a tubule in the kidney. These cells have many molecular transport functions.

## Muscle tissues generate force and movement

Muscle tissues are the most abundant tissues in the animal body. They are the most prominent example of effectors (see p. 816).

The cells that make up muscle tissues contain long filaments of proteins called actin and myosin, which interact to cause muscle cells to contract and exert force. There are three types of muscle tissues (**Figure 40.4**):

- *Skeletal muscle* (so named because most of it is attached to bone) is responsible for locomotion and other body movements such as facial expressions, shivering, and breathing.
- *Cardiac muscle* makes up the heart and is responsible for the beating of the heart and the pumping of blood.
- *Smooth muscle* makes up the walls of many hollow internal organs such as the gut, bladder, and blood vessels.

## Connective tissues include bone, blood, and fat

The cells that make up connective tissues are generally dispersed in an extracellular matrix that these cells themselves secrete. The composition and properties of the matrix differ among the different types of connective tissues, but protein fibers are always an important component of the matrix. The dominant protein in the extracellular matrix is collagen (see Figure 5.22), which makes up about 25 percent of total body protein. Collagen fibers are strong and resistant to stretch, giving strength to the skin and to the connections between bones and between bones and muscles. These fibers also provide a netlike framework for organs, giving them shape and structural strength. Elastin is another protein fiber found in the extracellular matrix of connective tissues. It recoils after

(A) Skeletal muscle

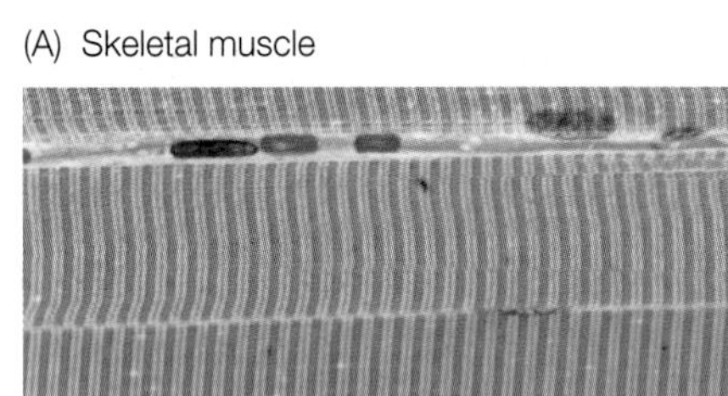

15 µm

(B) Cardiac muscle

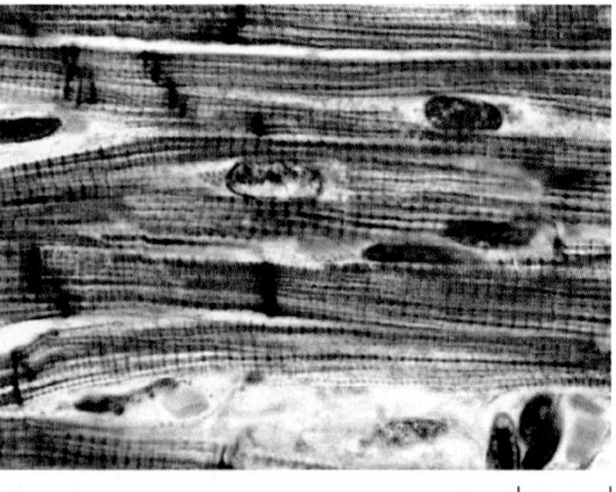

15 µm

(C) Smooth muscle

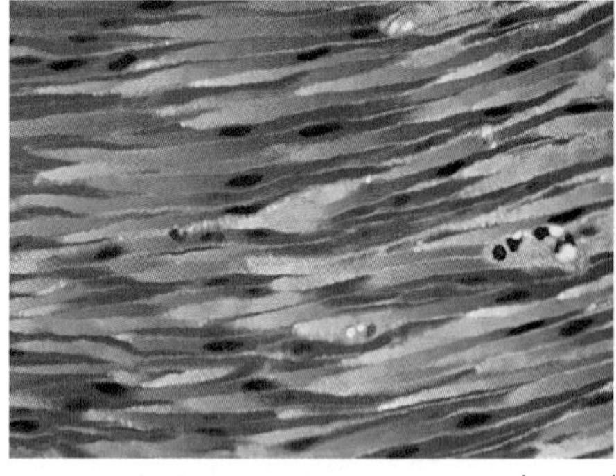

30 µm

**40.4 Muscle Cells Contain Protein Filaments** Filaments of two specific proteins—actin and myosin—interact to cause contraction and generate force in muscle tissue. (A) The regular arrangement of actin and myosin filaments results in the striated (striped) appearance of skeletal muscle. (B) The individual cells of cardiac muscle are branched and form a strong structural meshwork. (C) The actin and myosin filaments of smooth muscle are not regularly arranged and thus it does not have a striated appearance.

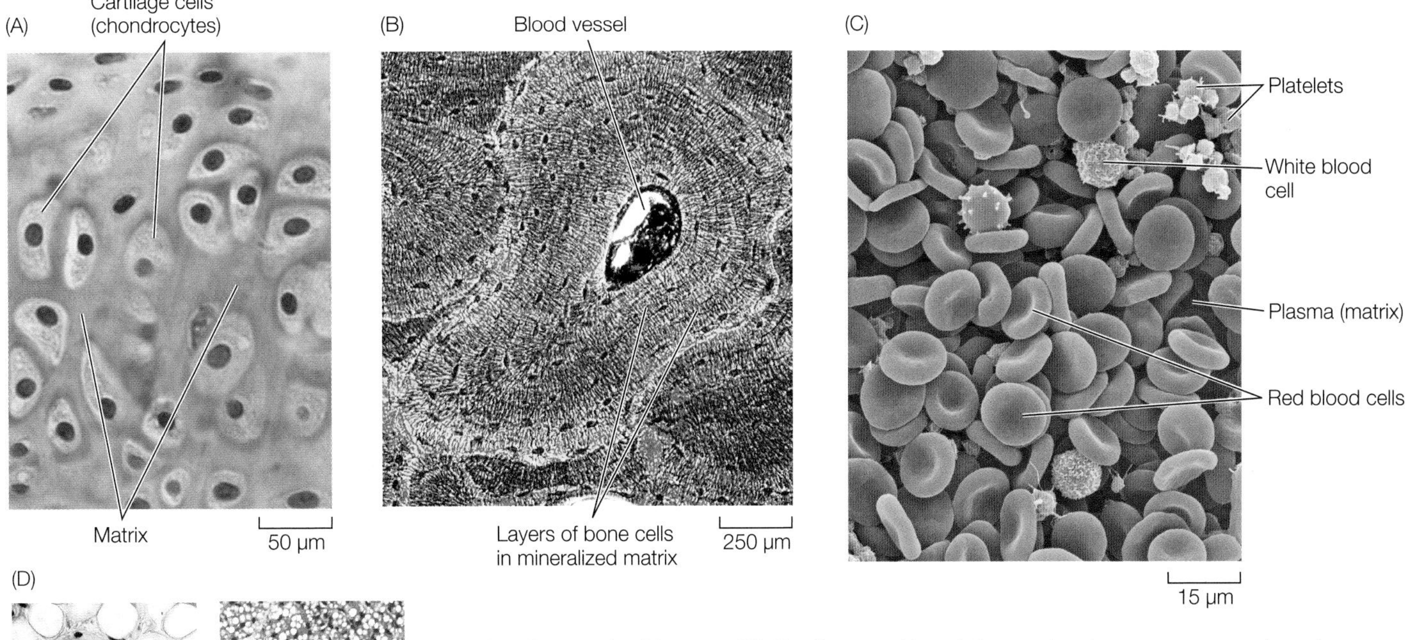


**40.5 Connective Tissues** (A) Cartilage cushions joints and makes structures such as the ear stiff but flexible. Cartilage cells, or chondrocytes, secrete an extracellular matrix rich in collagen and chondroitin sulfate. (B) Bone is the mineral-rich connective tissue of the vertebrate skeleton. (C) Blood is unique among the connective tissues, consisting of blood cells floating in an extracellular matrix of plasma. (D) "White" fat (left) is typical adipose tissue, with large droplets of energy-storing lipids and limited blood supply. In some mammals, specialized "brown" fat produces heat; this tissue is packed with mitochondria and blood vessels.

being stretched, so elastin fibers are abundant in tissues that are regularly stretched, such as the walls of the lungs and the large arteries.

- *Cartilage* and *bone* (**Figure 40.5A,B**) are connective tissues that provide firm structural support.
- *Blood* (**Figure 40.5C**) is a connective tissue consisting of cells dispersed in an extensive liquid extracellular matrix, the blood plasma.
- *Adipose* cells (**Figure 40.5D**) form loose connective tissue that stores lipids. Adipose tissue, or "fat," is a major source of stored energy. It also cushions organs, and layers of adipose tissue under the skin can provide a barrier to heat loss.

## Neural tissues include neurons and glial cells

The many different cells of neural tissues are specialized for processing information. Nerve cells fall into one of two categories, neurons or glial cells.

- *Neurons* (**Figure 40.6A**) come in many shapes and sizes, but all neurons encode and conduct information as electrical signals. Most neurons release chemical signals that are received by target cells; the target cells can be other neurons,

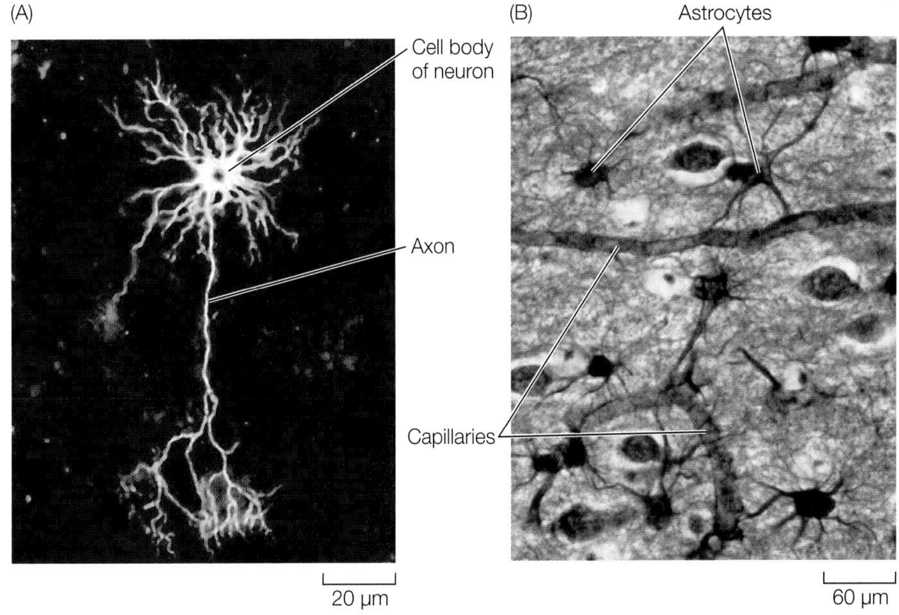


**40.6 Neural Tissue Includes Neurons and Glial Cells** (A) This human neuron consists of a cell body, several processes that receive input from other neurons, and one long axon that sends information to other cells. (B) A section through human brain tissue shows astrocytes, a type of glial cell.

muscle cells, or cells that secrete hormones and other molecules and substances, such as saliva.

- *Glial cells*, or *glia*, provide a variety of support functions for neurons (**Figure 40.6B**). One class of glia creates a barrier between the blood vessels and the neural tissue that protects the nervous system from potentially harmful chemicals circulating in the blood. Although glia do not generate electrical signals, they can communicate information through the release of chemical signals, and their roles in regulatory processes are becoming increasingly known.

Chapters 45, 46, and 47 detail the properties of nervous tissues.

## Organs consist of multiple tissues

Organs are composed of an epithelium and one or more other kinds of tissue. Indeed, most organs include all four tissue types. The wall of the gut is a good example (**Figure 40.7**). Its inner surface is lined with a sheet of columnar epithelial cells. Different types of epithelial cells in this lining secrete hormones or digestive juices or absorb nutrients from the gut. Beneath the epithelial lining is a layer of connective tissue called the mucosa. Within this connective tissue are blood vessels, neurons, and glands (clusters of secretory epithelial cells). Concentric layers of smooth muscle tissue enable the gut to contract to mix food with digestive juices. A network of neurons between the muscle layers controls these movements.

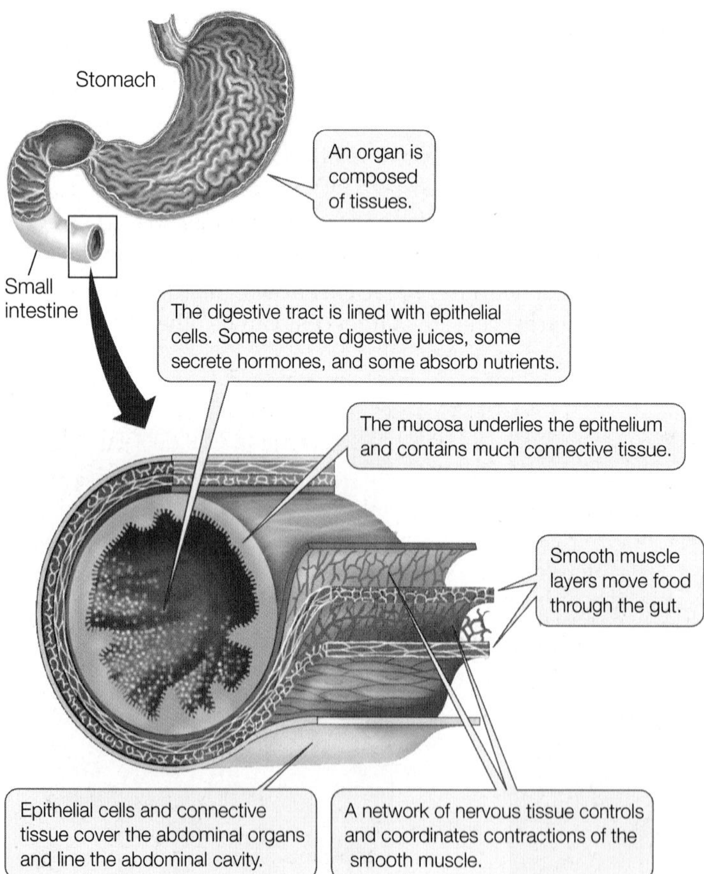


**40.7 Tissues Form Organs** The organs of the human digestive system, such as the stomach and small intestine, are made up of all four tissue types.

An individual organ is usually part of an **organ system,** a group of organs that work together to carry out certain functions. The stomach, small intestine, liver, and pancreas, for example, are parts of the digestive system. Thus we see an organizational hierarchy, with cells forming tissues that become part of organs, which in turn are organized in the functioning physiological systems of an individual organism.

**RECAP 40.2**

There are four tissue types in the animal body: epithelial, connective, nervous, and muscle. Individual organs include tissues of multiple types and are part of organ systems

- Why is it crucial that the cells of epithelial tissue be densely packed into tight sheets? **See p. 817 and Figure 40.3**
- What is the key difference between neurons and glia, the two cell types that make up nervous tissue? **See pp. 819–820**
- What is meant by the hierarchical organization of physiological function? **See p. 820 and Figure 40.7**

Subsequent chapters will describe each of the organ systems mentioned above in much greater detail. The remainder of this chapter focuses on the mechanisms of homeostasis, using one important variable of the internal environment—temperature—as our example.

## 40.3 How Does Temperature Affect Living Systems?

Temperatures where organisms live vary enormously, from the boiling hot springs of Yellowstone National Park to the interior of Antarctica, where the temperature can fall below –80°C. Cells can function over only a narrow range of temperatures. If cells cool below 0°C, ice crystals form and damage cell structures. Some animals have adaptations, such as antifreeze molecules in their blood that help them resist freezing; others can survive freezing. Generally, however, cells must remain above 0°C to stay alive. The upper temperature limit for survival in most cells is about 45°C (although some specialized algae can grow in hot springs at 70°C, and some archaea live at near 100°C).

In general, proteins begin to denature and lose their function as temperatures rise above 40°C. Most cellular functions are limited to the range between 0°C and 40°C, which approximates the thermal limits for most organisms. Most species, however, have much narrower thermal limits. To stay within those limits in spite of environmental conditions, animals have evolved thermoregulatory adaptations. Those adaptations give them certain thermal tolerances, which determine their geographic ranges. When environments change rapidly, as may be happening as the global climate warms, animals may find themselves in situations that exceed their thermal tolerances.

## $Q_{10}$ is a measure of temperature sensitivity

Even between 0°C and 40°C, changes in tissue temperature create problems for animals. Most physiological processes, like the biochemical reactions that constitute them, are temperature-sensitive, going faster at higher temperatures (see Figure 8.20). The temperature sensitivity of a reaction or process is described in terms of **$Q_{10}$**, a factor calculated by dividing the rate of a process or reaction at a certain temperature, $R_T$, by that rate at a temperature 10°C lower, $R_{T-10}$:

$$Q_{10} = \frac{R_T}{R_{T-10}}$$

$Q_{10}$ can be measured for a simple enzymatic reaction or for a complex physiological process, such as rate of oxygen consumption. If a reaction or process is not temperature-sensitive, it has a $Q_{10}$ of 1. Most biological $Q_{10}$ values are between 2 and 3. A $Q_{10}$ of 2 means that the reaction rate doubles as temperature increases by 10°C, and a $Q_{10}$ of 3 indicates a tripling of the rate over a 10°C temperature range (**Figure 40.8**).

You will notice that the $Q_{10}$ values (except for $Q_{10} = 1$) plotted in Figure 40.8 produce curves rather than straight lines. This is because the temperatures increase in additive intervals (10, 20, 30, etc.) but the reaction rates increase in a multiplicative fashion (2, 4, 8, 16, 32, etc.). Such curvilinear plots are common in biological data.

Changes in body temperature can disrupt an animal's physiology because not all of the biochemical reactions that constitute the metabolism of an animal have the same $Q_{10}$. These biochemical reactions are linked together in complex networks: the products of one reaction are the reactants for other reactions. Because different reactions have different $Q_{10}$'s, changes in tissue temperature will shift the rates of some reactions more than others, disrupting the overall network. Therefore, to maintain homeostasis, organisms must be able to compensate for or prevent changes in body temperature.

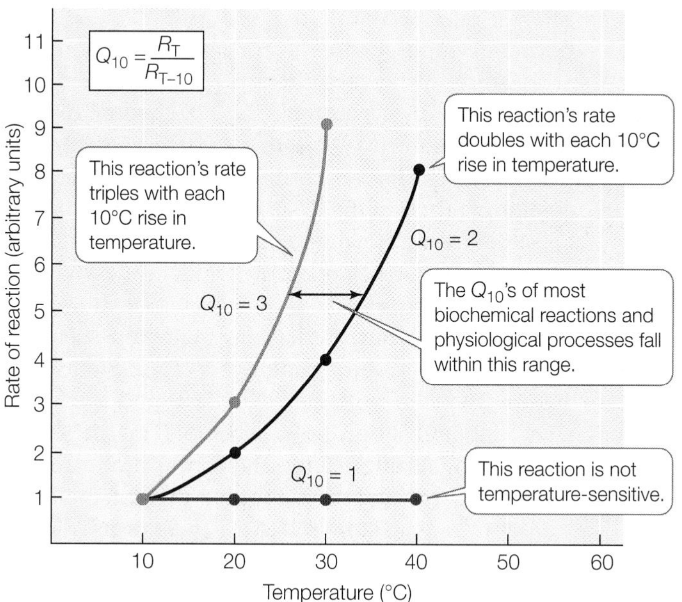


**40.8 $Q_{10}$ and Reaction Rate** The larger the $Q_{10}$ of a reaction or process, the faster its rate rises in response to an increase in temperature.

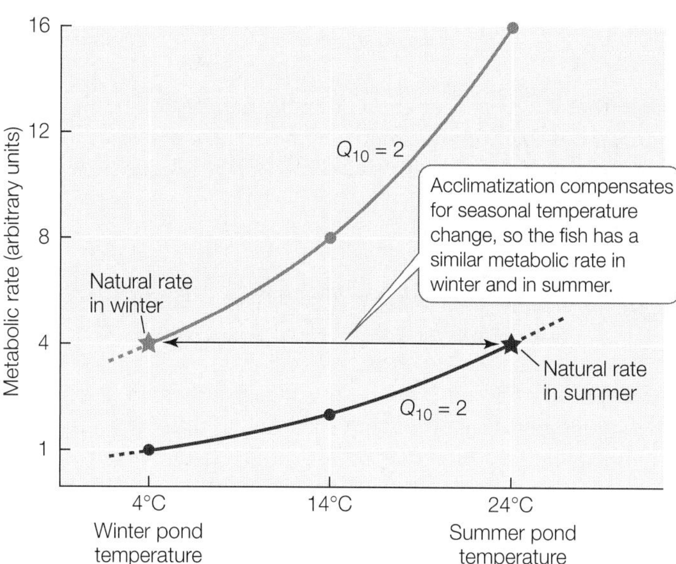


**40.9 Metabolic Compensation** In its natural environment, a fish's metabolic rate readjusts, or acclimatizes, to compensate for seasonal changes in temperature.

## Animals acclimatize to seasonal temperatures

The body temperature of some animals (especially aquatic animals) is coupled to environmental temperature. The body temperature of a fish in a pond, for example, will be the same as the water temperature, which might range from 4°C in winter to 24°C in summer. If we bring that fish into the laboratory in the summer and measure the rates of any of its physiological or biochemical processes such as oxygen consumption, we will demonstrate a $Q_{10}$ relationship. On the basis of that relationship, we can predict what the fish's metabolic rate will be in its pond in the winter. However, if we bring that fish back into the laboratory in the winter and measure its metabolic rate at winter pond temperature, we will find that rate to be higher than we predicted. The fish's biochemistry and physiology will have **acclimatized** to the seasonal change in water temperature so that it can remain active at winter temperatures (**Figure 40.9**). What could be the mechanism of this acclimatization? One possibility is that the fish may express isozymes with different temperature optima in summer and in winter. The ability of animals to acclimatize means that their metabolic functions are less sensitive to long-term changes in temperature than to short-term changes.

**RECAP 40.3**

Cells can survive only within a narrow range of temperatures, but even changes within that range can be disruptive because different physiological processes have different temperature sensitivities.

- Plot a $Q_{10} = 2.5$ curve for a physiological process. **See Figure 40.8**
- Explain how a change in body temperature can disrupt physiological processes. **See p. 821**

Now that we have seen how animals are affected by the temperature of their environments, we next take a look at the adaptations that allow animals to control and regulate their body temperatures.

## 40.4 How Do Animals Alter Their Heat Exchange with the Environment?

Animals have different ways of dealing with changes in environmental temperature. Many of us learned to think of animals as being either "cold-blooded" or "warm-blooded," which implies a comparison with our own body temperature and sets mammals and birds apart from other animals. This simple classification breaks down when we realize that mammals that hibernate become cold, and that many reptiles and insects can have body temperatures similar to ours when they are active. Physiologists sometimes classify animals according to whether they have a constant body temperature (homeotherms) or a variable body temperature (poikilotherms). But there are situations where applications of these descriptive terms also lead to strange conclusions. A deep-sea fish has a constant body temperature. Should it be classified with mammals? And the hibernator's body temperature in winter varies between a typical high mammalian level and a level that is close to environmental temperatures.

Another classification system is based on the source of heat that predominantly influences the body temperature of the animal. **Ectotherms** are animals whose body temperatures are determined primarily by external sources of heat. **Endotherms** can regulate their body temperatures by producing heat metabolically. Mammals and birds are endotherms—most of the time; other animals are ectotherms—most of the time. Therefore we have a third category: a **heterotherm** is an animal that sometimes behaves as an endotherm and at other times as an ectotherm. For example, a mammal that hibernates is an endotherm over the summer, but during its winter bouts of hibernation its internal heat production falls and it behaves much like an ectotherm. And some ectotherms can produce substantial amounts of metabolic heat, thus behaving like endotherms.

### Endotherms produce substantial amounts of metabolic heat

Transfers of energy in biological systems are always inefficient, as Section 8.1 explained. With every transfer of energy—from food molecules to ATP, from ATP to biological work—some of the energy is lost as heat. This is true for both ectotherms and endotherms. In both cases, working muscles produce heat, as do the metabolic activities of all tissues. So why do endotherms produce more heat? The answer is that the cells of endotherms are less efficient at using energy than are the cells of ectotherms.

The cells of endotherms are more "leaky" to ions than are the cells of ectotherms. Therefore $Na^+$ ions are constantly diffusing into the cells, and $K^+$ ions are constantly diffusing out. Even an endotherm at rest must spend considerable amounts of energy to transport $Na^+$ out of the cells and transport $K^+$ back in. Because of their leaky membranes, endotherms expend more energy (and thus release more heat) than do ectotherms just to maintain the ion concentration gradients across their cell membranes. This situation is analogous to a leaky rowboat: the faster water comes in (i.e., the faster ions leak), the more metabolic energy has to be expended to bail the water out (i.e., pump ions) just to remain afloat.

We can speculate that a mutation resulting in seemingly faulty or leaky ion channels may underlie the evolution of endothermy. Such a mutation in a small ectotherm would have increased its energy expenditure and therefore its heat production. Increased heat production would allow the animal to be active for a longer time after sunset. Being active in the evening would open up a new world of ecological opportunities—a world in which there was less competition from similar-sized ectotherms.

Two major differences between endotherms and ectotherms are their resting metabolic rates—the sum total of all energy expenditures in their bodies when at rest—and their responses to changes in environmental temperature.

### Ectotherms and endotherms respond differently to changes in environmental temperature

How do two similar-size animals, a lizard (an ectotherm) and a mouse (an endotherm), respond to changes in environmental temperature? We put each animal in a closed chamber and measure its body temperature and its resting metabolic rate as we change the temperature of the chamber from 37°C to 0°C. The body temperature of the lizard equilibrates with that of the chamber, whereas the body temperature of the mouse remains stable (**Figure 40.10A**). The metabolic rate of the lizard (already lower than that of the mouse) decreases as the temperature drops (**Figure 40.10B**). In contrast, the mouse's metabolic rate increases as the chamber temperature falls below 25°C. This increase in metabolism produces enough heat to prevent the mouse's body temperature from falling. In other words, the mouse can regulate its body temperature by increasing its metabolic rate; the lizard cannot.

This experiment might lead us to conclude that the ectotherm cannot regulate its body temperature, but observations of the lizard in nature do not support this conclusion. In nature, unlike in the laboratory, the lizard's body temperature is sometimes considerably different than the environmental temperature. Air temperature in the desert can fluctuate by 40°C in a few hours; the lizard, however, maintains a fairly stable body temperature by using behavior to alter its heat exchange with the environment (**Figure 40.11**). Its behavioral strategies include spending time in a burrow, basking in the sun, seeking shade, climbing vegetation, and changing its orientation with respect to the sun. Thus the lizard can regulate its body temperature quite well, although it does so by behavioral mechanisms rather than by altering its internal metabolic heat production.

Behavioral thermoregulation is not the exclusive domain of ectotherms. Endotherms usually select the most comfortable thermal environment possible. They may change posture, orient to the sun, move between sun and shade, and move between still air and moving air. Examples of more complex thermoregulatory behaviors include nest construction and social behaviors such as huddling. Humans put on or remove clothing and burn fossil fuels to generate the energy to heat or cool buildings.

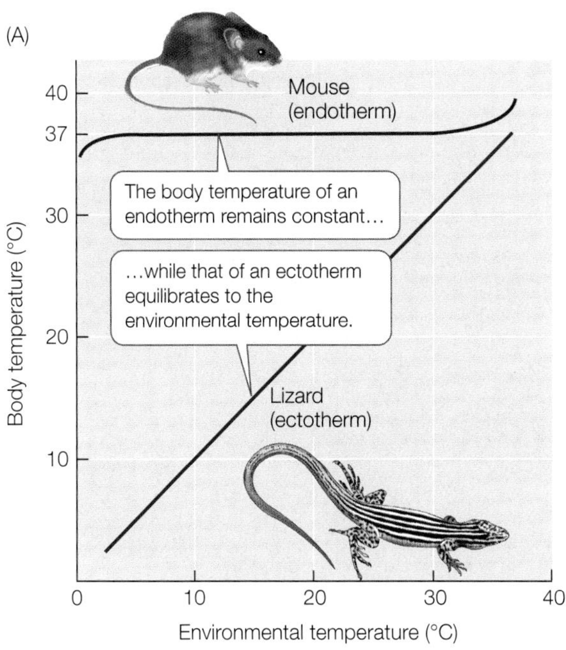


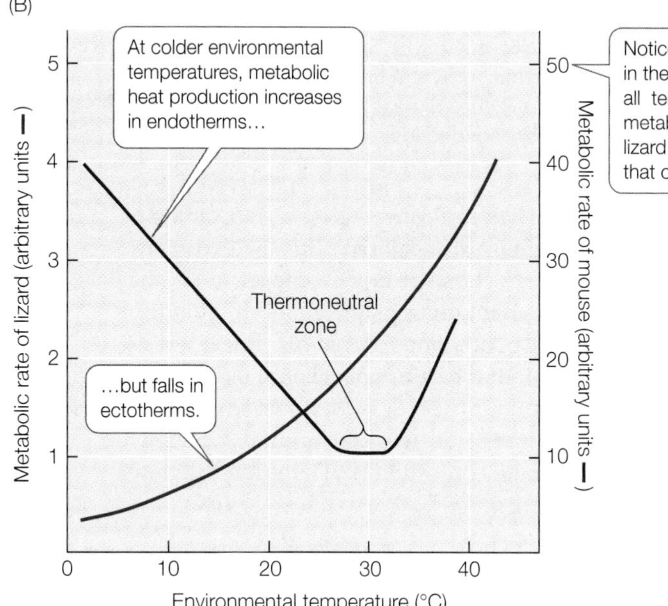


**40.10 Ectotherms and Endotherms React Differently to Environmental Temperatures** (A) At the same environmental temperature, an ectotherm and an endotherm of approximately the same body size (here, a lizard and a mouse) have different body temperatures. (B) The metabolic rates of the lizard and mouse react in opposite ways to cooler temperatures. (The mouse's metabolic rate rises again at higher temperatures because, after a certain point, it takes metabolic energy to dissipate heat by sweating or panting.)

## Energy budgets reflect adaptations for regulating body temperature

Both ectotherms and endotherms can influence their body temperatures by altering four avenues of heat exchange between their bodies and the environment (**Figure 40.12**):

- **Radiation** Heat moves from warmer objects to cooler ones via the exchange of infrared radiation (what you feel when you stand in front of a fire).
- **Convection** Heat transfers to a surrounding medium such as air or water as that medium flows over a surface (the wind-chill factor).
- **Conduction** Heat transfers directly between two objects at different temperatures when they come into contact (e.g., an icepack on a sprained ankle).
- **Evaporation** Heat transfers away from a surface when water evaporates on that surface (the effect of sweating).

The total balance of heat production and heat exchange can be expressed as an **energy budget**, based on the simple fact that if the body temperature of an animal is to remain constant, the heat entering the animal must equal the heat leaving it. The heat coming in is usually from metabolism and radiation ($R_{abs}$, for radiation absorbed). Heat leaves the body via the four mechanisms listed above—radiation emitted ($R_{out}$), convection, conduction, and evaporation. The energy budget takes the mathematical form

$$\underbrace{\text{metabolism} + R_{abs}}_{\text{heat}_{in}} = \underbrace{R_{out} + \text{convection} + \text{conduction} + \text{evaporation}}_{\text{heat}_{out}}$$

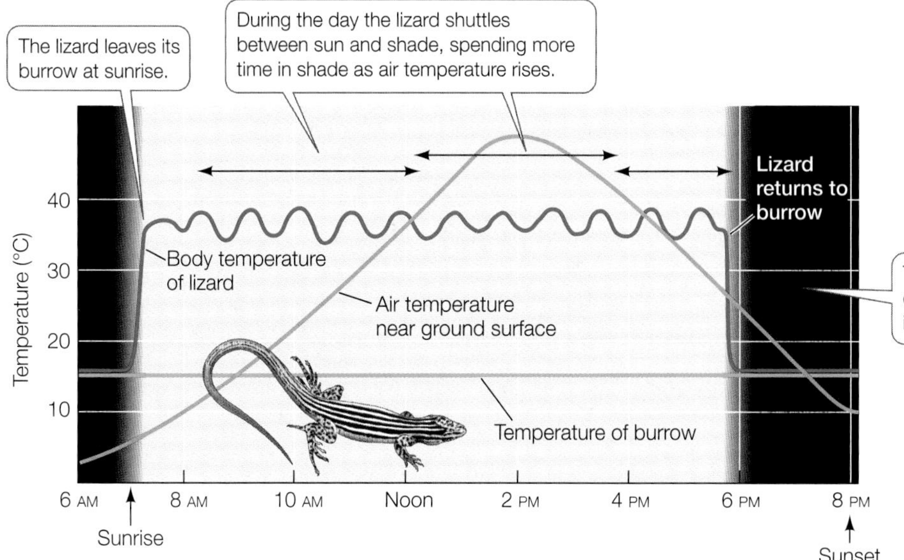


**40.11 Using Behavior to Regulate Body Temperature** The body temperature of a lizard (an ectotherm) depends on environmental temperature, but the lizard can regulate its temperature by moving from place to place within its environment.

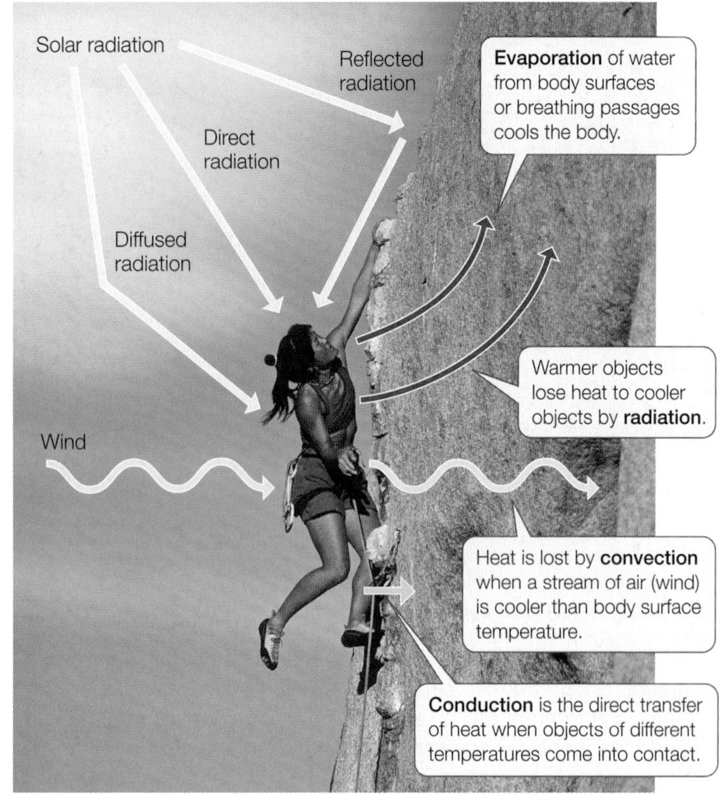


**40.12 Animals Exchange Heat with the Environment** An animal's body temperature is determined by the balance between internal heat production and four avenues of heat exchange with the environment: radiation, convection, conduction, and evaporation.

The energy budget is a useful concept because any adaptation that influences the ability of an animal to deal with its thermal environment must affect one or more components of the budget. So the energy budget gives us the ability to quantify and compare the thermal adaptations of animals. One interesting observation is that all of the components on the right side of the energy budget equation—that is, the heat-loss side—depend on the surface temperature of the animal. One way surface temperature can be controlled is by altering the flow of blood to the skin.

## Both ectotherms and endotherms control blood flow to the skin

Heat exchange between the internal environment and the skin occurs largely through blood flow. As described at the beginning of this chapter, when body temperature rises because of exercise, blood flow to the skin increases, and the skin surface becomes warm. The heat that the blood brings from the body core to the skin is lost to the environment through the four avenues listed above, which helps bring the body temperature back to normal. In contrast, when body temperature is too low or the environment is too cold, the blood vessels supplying the skin constrict, reducing heat loss to the environment.

The ability to control blood flow to the skin can be an important adaptation for an ectotherm such as the marine iguana (a reptile) of the Galápagos archipelago (**Figure 40.13**). The Galápagos are volcanic islands that lie on the equator but are bathed by cold ocean currents. The iguanas bask on hot black lava rocks on the shore, but periodically they enter the cold ocean water to feed on seaweed. While the iguanas are feeding, they cool to the temperature of the sea. This cooling lowers their metabolism, making them slower, more vulnerable to predators, and incapable of efficient digestion. They therefore alternate between feeding in the cold seawater and basking on the hot rocks. It is advantageous for iguanas to retain body heat as long as possible while swimming and to warm up as fast as possible when basking. They can accomplish these results by changing their heart rate and thus the rate of blood flow to their skin.

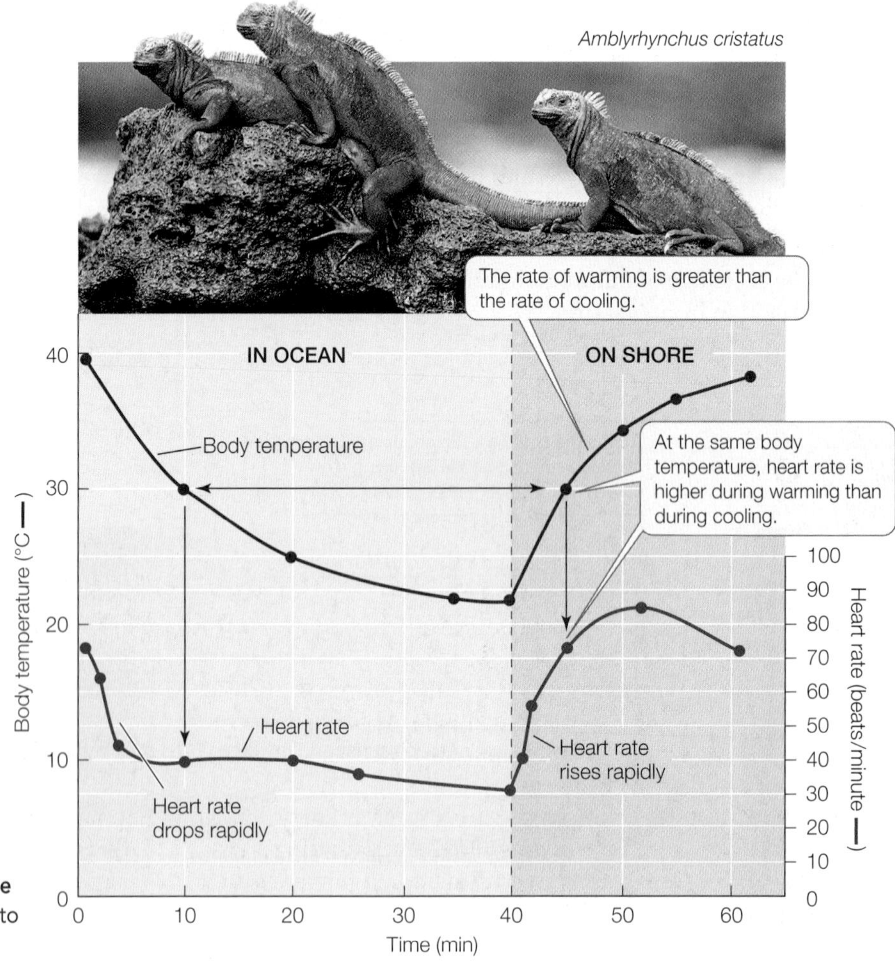


**40.13 Some Ectotherms Regulate Blood Flow to the Skin** Galápagos marine iguanas control blood flow to the skin to alter their heating and cooling rates.

(A) "Cold" fish

1 In the gills, blood is oxygenated and cooled to seawater temperature.

2 Cold blood flows through the center of the fish in the large dorsal aorta.

3 Arteries carry blood to the tissues.

4 Veins return blood to the heart.

5 The heart pumps blood to the gills.

Heart

(B) "Hot" fish

1 Blood is oxygenated in the gills.

2 Cold blood flows from the gills to the body in arteries just under the skin.

3 In the countercurrent heat exchangers, arterial blood flowing into the muscle is warmed by venous blood flowing out of the muscle.

4 Veins under the skin return blood to the heart.

5 The heart pumps blood to the gills.

The major vein and artery just under the skin are at the same cool temperature.

Capillary bed (muscles generate heat)

Countercurrent heat exchanger

Vein

Artery

**40.14 "Cold" and "Hot" Fish** (A) In most fish species, the circulatory system conducts cool, oxygenated blood (red) from the gills through a large dorsal aorta to the rest of the body. By the time it returns to the heart, deoxygenated venous blood (blue) has been warmed by the metabolism of the muscles. (B) The anatomy of "hot" fish species includes a mechanism that allows heat to pass into cold arterial blood from the warmed venous blood so that both are at the same temperature. (C) The principle of this countercurrent heat exchange allows heat (orange) to be transferred with maximum efficiency.

What about furred mammals? Fur acts as insulation to keep body heat in, making it possible for mammals to function in cold environments. When they are active, however, mammals still must get rid of excess heat, and it does little good to transport that heat to the skin under the fur. Thus, as mentioned at the opening of this chapter, mammals have specialized blood vessels for transporting heat to their hairless skin surfaces. Heat loss from these areas is tightly controlled by the opening and closing of these blood vessels. When you are cold, the blood flow to your hands and feet decreases and they feel cold, but when you exercise, these same surfaces can get hot quickly.

**Go to Media Clip 40.1**
**Thermoregulation in Animals**
**Life10e.com/mc40.1**

## Some fish conserve metabolic heat

The muscles of active fish produce substantial amounts of metabolic heat, but they have difficulty retaining any of that heat. Blood pumped from the fish heart goes directly to the gills, where it must come very close to the surrounding water to exchange respiratory gases. So any heat that the blood picks up from metabolically active muscles is lost to the water flowing across its gills. It is thus surprising that some large, rapidly swimming fish, such as giant bluefin tuna and great white sharks, can maintain temperature differences as great as 10°C to 15°C between their bodies and the surrounding water. The heat comes from their powerful swimming muscles, and the ability of these "hot" fish to conserve that heat is based on the remarkable arrangement of their blood vessels.

In the typical ("cold") fish circulatory system, oxygenated blood from the gills collects in a large dorsal vessel, the aorta, and travels through the center of the fish, distributing blood to all organs and muscles (**Figure 40.14A**). "Hot" fish have a smaller central dorsal aorta; most of their oxygenated blood is transported in large side vessels just under their skin (**Figure 40.14B**). The cold blood leaving the gills is thus kept close to the surface of the fish as the blood flows posteriorly to the swimming muscles. Smaller vessels transport this cold blood into the muscle mass, and these small vessels run parallel to the vessels carrying warm blood from the swimming muscles back toward the heart. Because the vessels carrying the cold blood into the muscles are in close contact with the vessels carrying warm blood away, heat flows from the warm to the cold blood by conduction and is therefore retained in the muscle mass.

Because heat is exchanged between blood vessels carrying blood in opposite directions, this adaptation is called **countercurrent heat exchange** (**Figure 40.14C**). It keeps the heat within the muscles, enabling these fish to have an internal body temperature considerably higher than the water temperature. Each 10°C rise in muscle temperature increases the fish's sustainable power output almost threefold, giving it a faster swimming capability.

## Some ectotherms regulate metabolic heat production

Some ectotherms raise their body temperature by producing metabolic heat. For example, the powerful flight muscles of many insects must reach 35°C to 40°C before the insects can fly, and they must maintain these high temperatures during flight. Such insects warm up to fly by contracting their flight

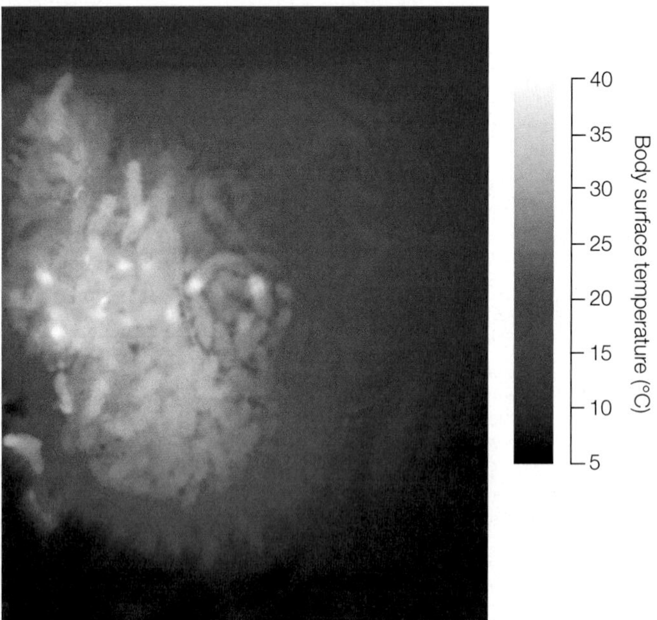


**40.15 Bees Keep Warm in Winter** Honey bee colonies survive winter cold because workers generate metabolic heat. In this infrared photograph of the center of an overwintering hive, individual bees are discernible by the heat their bodies produce as they cluster around their queen.

muscles in a manner analogous to shivering in mammals. The heat-producing ability of insects can be quite remarkable. Probably the most impressive case is a species of scarab beetle that lives mostly underground in mountains north of Los Angeles, California. These beetles come aboveground to mate, with the males flying in search of females. They undertake this mating ritual at night, in winter, and only during snowstorms.

Honey bees regulate temperature as a group. They live in large colonies consisting mostly of female worker bees that maintain the hive and rear the larval offspring of the single queen bee. During winter, worker bees cluster around the brood (eggs and larvae). They adjust their individual metabolic heat production and density of clustering so that the brood temperature remains remarkably constant, at about 34°C, even as the outside air temperature drops below freezing (**Figure 40.15**).

### RECAP 40.4

Animals that can maintain constant high body temperatures because of their metabolic heat production are called endotherms. Those whose body temperatures are determined primarily by environmental sources of heat are called ectotherms. Heat exchange between an animal and its environment occurs via radiation, convection, conduction, and evaporation.

- Explain why the two curves in Figure 40.10A are different from the two curves in Figure 40.10B. **See p. 822**
- In terms of the energy budget, why is the control of blood flow to the skin so important for thermoregulation? **See pp. 823–824**
- Explain how countercurrent heat exchange makes it possible for some fish to have a body temperature higher than that of the surrounding water. **See p. 825 and Figure 40.14**

Endotherms respond to changes in environmental temperature by changing their rates of metabolic heat production. They also have other adaptations for controlling their rates of heat exchange with their environments. How do they regulate these various avenues of heat exchange to achieve a constant internal body temperature?

## 40.5 How Do Endotherms Regulate Their Body Temperatures?

Physiologists can determine an animal's metabolic rate by measuring its consumption of $O_2$ or production of $CO_2$. Within a narrow range of environmental temperatures, called the **thermoneutral zone** (see Figure 40.10B), the metabolic rates of endotherms (birds and mammals) are at low levels and independent of temperature. The metabolic rate of a resting animal at a temperature within the thermoneutral zone is known as the **basal metabolic rate**, or **BMR**. It is usually measured in animals that are quiet but awake and not using energy for digestion, reproduction, or growth. Thus the BMR is the rate at which a resting animal is consuming just enough energy to carry out its minimal body functions.

### Basal metabolic rates correlate with body size

As you might expect, the BMR of an elephant is greater than that of a mouse. After all, the elephant is more than 100,000 times larger than the mouse. However, the BMR of the elephant is only about 7,000 times greater than that of the mouse. That means that a gram of mouse tissue uses energy at a rate 15 times greater than a gram of elephant tissue (**Figure 40.16**). Across all of the endotherms, BMR per gram of tissue increases as animals get smaller.

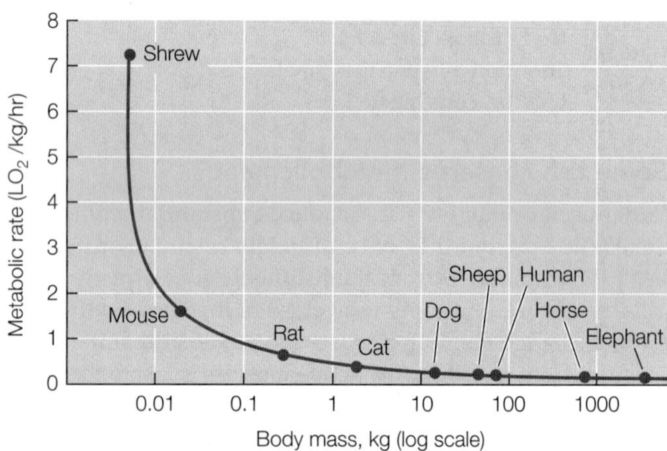


**40.16 The Mouse-to-Elephant Curve** On a weight-specific basis, the metabolic rate of small endotherms is much greater than that of larger endotherms. This classic illustration originally published in the 1930s, plots $O_2$ consumption per kilogram of body mass (a measure of metabolic rate) against a logarithmic plot of body mass.

WORKING WITH**DATA:**

## *A Mammal's BMR Is Proportional to Its Body Size*

### Original Paper

White, C. R. and R. S. Seymour. 2003. Mammalian basal metabolic rate is proportional to body mass. *Proceedings of the National Academy of Sciences USA* 100: 4046–4049.

### Analyze the Data

Studies conducted in the late 1800s concluded that BMR varied as a function of body mass to the ⅔ power. This conclusion made sense because metabolism is a function of body *mass* (a cube function of a linear dimension), whereas heat loss is a function of body *surface area* (a square function of a linear dimension). Thus, because of this limitation on heat loss, BMR would have to decrease as body size increased.

Subsequent studies in the early 1900s, however, concluded that BMR varies as a function of body mass to the ¾ power. Recent theoretical considerations of fractal geometry led to support for this ¾ power function. The table at right gives a small subset of the data from ten species presented in the paper cited above.

QUESTION 1

Do you think these data support the ⅔ power function or the ¾ power function?

QUESTION 2

You know from the "mouse-to-elephant curve" in Figure 40.16 that these data relate to each other as an exponential function and therefore produce a curvilinear plot. You can linearize the presentation of these data by conducting a logarithmic transformation. Using a calculator, create a new data table that expresses the values below as $\log_{10}$ values. Plot these values with body size ($\log_{10}$ body mass) on the *x* axis and metabolic rate ($\log_{10}$ BMR) on the *y* axis. Then, using a linear regression program on your calculator or computer, determine the equation that describes the graph you created in Question 2.

QUESTION 3

Remember that you transformed your data into logarithmic values, so in your equation, $y = \log_{10}$ BMR and $x = \log_{10}$ body mass. Convert your equation to a non-logarithmic form. Does your answer support the conclusion expressed in the title of the paper listed above?

| Species | Body size (g) | BMR (ml $O_2$/hr) |
|---|---|---|
| Bat | 16 | 40 |
| Mouse | 30 | 63 |
| Rat | 400 | 146 |
| Muskrat | 1,000 | 640 |
| Marmot | 4,000 | 1,550 |
| Beaver | 10,000 | 4,500 |
| Coyote | 10,000 | 2,690 |
| Baboon | 17,000 | 5,150 |
| Antelope | 37,800 | 9,300 |
| Moose | 325,000 | 51,400 |

**Go to BioPortal for all** WORKING WITH**DATA** **exercises**

Why should this disproportionate difference exist? There are several possible reasons. As animals get bigger, they have a smaller ratio of surface area to volume (see Figure 5.2). Since heat production is related to the volume (i.e., mass) of the animal, but its capacity to dissipate heat is related to its surface area, it has been reasoned that larger animals evolved lower metabolic rates to avoid overheating. However, this explanation alone is insufficient because the relationship between body mass and metabolic rate holds for even very small organisms and for ectotherms, in which overheating is not a problem. Other hypotheses have also been proposed. For example, a larger animal has a greater proportion of support tissues (e.g., skin and bone), which are not as metabolically active as other tissue types. The real explanation is probably a mixture of different factors, but the relationship holds over a very broad range of species.

For an endotherm, a metabolic rate versus environmental temperature curve represents the integrated response of all the animal's thermoregulatory adaptations (**Figure 40.17**). The thermoneutral zone is bounded by a lower and an upper **critical temperature**. When the environmental temperature is within its thermoneutral zone, an endotherm's thermoregulatory responses do not require much energy and could be considered passive; such responses include changing posture, fluffing fur, and controlling blood flow to the skin. Outside its thermoneutral zone, however, an endotherm's thermoregulatory responses are active and require considerable metabolic energy.

### Endotherms respond to cold by producing heat and adapt to cold by reducing heat loss

When environmental temperatures fall below the lower critical temperature, endotherms must produce heat to compensate for the heat they lose to the environment. Mammals can accomplish this by shivering and/or nonshivering heat production. Birds use only shivering heat production. Shivering uses the contractile machinery of skeletal muscles to convert ATP to ADP, with the energy from this process released as heat. Shivering muscles pull against each other so that little movement other than a tremor results. "Shivering heat production" is perhaps too narrow a term, however; increased muscle tone and increased body movements also contribute to increased heat production in cold environments.

Most nonshivering heat production occurs in specialized adipose tissue called **brown fat** (see Figure 40.5D). This tissue looks brown because of its abundant mitochondria and rich blood supply. In brown fat cells, a protein called thermogenin uncouples proton movement from ATP production,

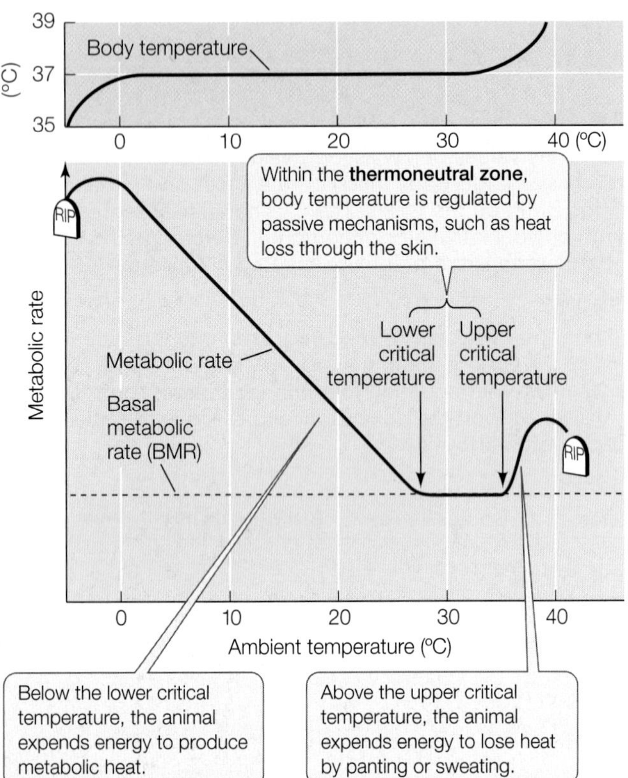


**40.17 Environmental Temperature and Mammalian Metabolic Rates** A plot of an endotherm's metabolic rate versus environmental temperature represents the integrated response of all of its thermoregulatory mechanisms. Outside the thermoneutral zone, maintaining a constant body temperature requires expending energy. Outside extreme limits (0°C and 40°C in this instance), the animal cannot maintain its body temperature and dies.

allowing protons to leak across the inner mitochondrial membrane rather than having to pass through the ATP synthase and generate ATP (review the discussion of brown fat and the chemiosmotic mechanism in Chapter 9). As a result, metabolic fuels are consumed without producing ATP, but heat is still released. Pads of brown fat are found in newborns of many mammalian species (including humans), in some adult mammals that are small and acclimatized to cold, and in mammals that hibernate. Recently it has been discovered that adult humans have small amounts of brown fat distributed around the body and that its metabolic activity is stimulated by cold exposure. One study found less brown fat activity in obese than in lean individuals, leading to the hypothesis that individual differences in propensity for weight gain may be related to the amount of brown fat in an individual, as described in the opening stories of Chapters 9 and 41.

In addition to their ability to produce heat, endotherms that live in cold climates have evolved adaptations to reduce their heat loss. Heat is lost from the body surface, and cold-climate species have anatomical adaptations that give them smaller surface-to-volume ratios than their warm-climate relatives (**Figure 40.18**). These adaptations include rounder body shapes and shorter appendages.

The most common and important means of decreasing heat loss is to increase thermal insulation. Animals adapted to cold climates have much thicker layers of fur, feathers, or fat than do their warm-climate relatives. Fur and feathers are good insulators because they trap a layer of still, warm air close to the skin surface. If that air is displaced by water, insulation is drastically reduced. In many species, oil secretions spread through fur or feathers by grooming are critical for resisting wetting and maintaining a high level of insulation.

The ability to decrease blood flow to the skin is an important thermoregulatory adaptation for cold-climate endotherms. Constriction of blood vessels in the skin, and especially in the appendages, greatly improves an animal's ability to conserve heat. Countercurrent heat exchange like that we saw in "hot" fish is also an important adaptation in the appendages of endotherms. Blood flowing out to the paw of a wolf, the hoof of a caribou, or the foot of a bird parallels the flow of the blood returning to the body core. Heat is transferred from the outgoing to the returning blood, thus retaining heat in the animal's core.

**Go to Activity 40.1 Thermoregulation in an Endotherm Life10e.com/ac40.1**

**40.18 Adaptations to Cold and Hot Climates** (A) The antelope jackrabbit is found in the Sonoran Desert of Arizona. Its large ears serve as heat exchangers, passing heat from the animal's blood to the surrounding air. (B) The thick fur of the Arctic hare provides insulation and its rounded body shape lowers its surface-to-volume ratio. The ears and extremities are smaller than those of its warm-climate relatives so less heat is lost to the environment.

## Evaporation of water can dissipate heat, but at a cost

As environmental temperature rises within an endotherm's thermoneutral zone, the animal dissipates more of its metabolic heat by increasing blood flow to the skin, seeking shade and cool breezes, and decreasing activity. When the temperature exceeds the upper critical temperature, however, overheating becomes a problem. For exercising animals (including athletes), overheating can occur even at low environmental temperatures. Large mammals, especially those in hot habitats such as elephants, rhinoceroses, and water buffaloes, have little or no insulating fur and seek out water to wallow in when the air temperature is high. Having water in contact with the skin greatly increases heat loss because the heat-absorbing capacity of water is much greater than that of air.

Evaporation from external or internal body surfaces through sweating or panting can also cool an endotherm. A gram of water absorbs about 580 calories of heat when it evaporates. If this evaporation occurs on the skin, most of the heat comes from the skin. However, sweat or saliva that falls off the body provides no cooling. Thus when the need for heat loss is greatest, water from the internal environment can be squandered with no cooling benefit. Water is heavy, so animals do not carry an excess supply of it, and many hot environments are also arid. In habitats that are both hot and dry, sweating and panting are cooling adaptations of last resort.

Sweating and panting are *active* processes that require expending metabolic energy. That is why the metabolic rate increases when the upper critical temperature is exceeded (see Figure 40.17). A sweating or panting animal is generating heat in the process of dissipating heat, which can be a losing battle.

## The mammalian thermostat uses feedback information

The thermoregulatory mechanisms and adaptations of endotherms that we have just discussed are controlled by neural regulatory systems that integrate information from environmental and physiological sources and then issue commands to the effectors that alter the heat content of the body. These regulatory systems are similar in principle in birds and mammals but differ in many details. Here we focus on the nervous system thermostats of mammals.

The major thermoregulatory integrative center of mammals is at the base of the brain in a structure called the **hypothalamus**. As we will see in upcoming chapters, the hypothalamus is a key player in many animal regulatory systems. Experiments demonstrating its role have shown that slight cooling of the hypothalamus stimulates constriction of skin blood vessels and that stronger cooling increases metabolic heat production. As a result, cooling of the hypothalamus in an unchanging, thermoneutral environment will cause body temperature to rise. Conversely, hypothalamic heating causes the overall body temperature to fall (**Figure 40.19**).

In mammals, the temperature of the hypothalamus itself is the major feedback signal. The hypothalamus generates set points for thermoregulatory responses. When the temperature of the hypothalamus exceeds or drops below those set points, thermoregulatory responses are activated to reverse the direction of temperature change (**Figure 40.20**). The system integrates other sources of information in addition to hypothalamic temperature. For example, temperature sensors in the skin register environmental temperature. A change in skin temperature is feedforward information that shifts hypothalamic set points; the set point for metabolic heat production is higher when the skin is cold and lower when the skin is warm.

INVESTIGATING**LIFE**

**40.19 The Hypothalamus Regulates Body Temperature**
A mammal's hypothalamus was subjected directly to temperature manipulation. The body's responses to the manipulations were as expected if the hypothalamus is the mammalian "thermostat."[a]

**HYPOTHESIS** Heating or cooling the mammalian hypothalamus results in predictable changes in body temperature.

***Method*** 1. Implant a probe into the hypothalamus of a living ground squirrel's brain. Use the probe to heat or cool the hypothalamus directly (i.e., without affecting the ambient temperature).

2. Manipulate the hypothalamic temperature $T_H$.

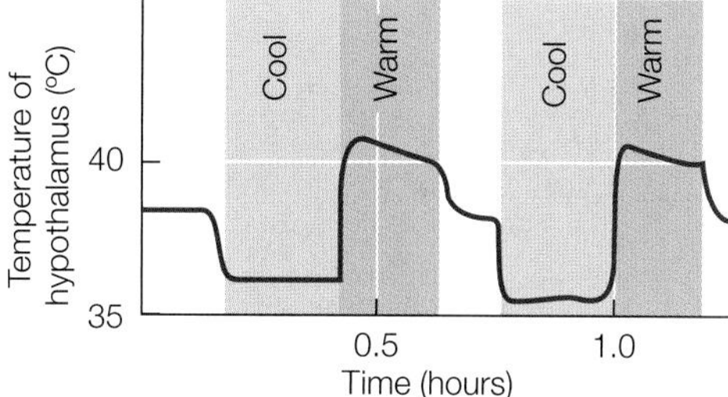


3. Measure the animal's metabolic rate and body temperature throughout the period of hypothalamic manipulation.

***Results***

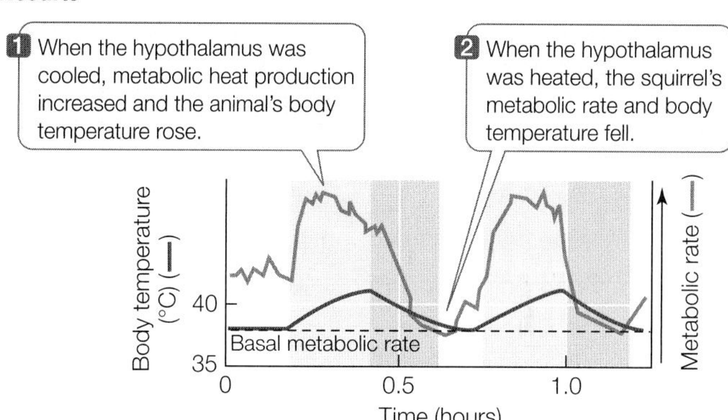


**CONCLUSION** The ground squirrel's hypothalamus acts as a thermostat. When cooled it activates metabolic heat production; when warmed, it suppresses metabolic heat production and favors heat loss.

Go to **BioPortal** for discussion and relevant links for all INVESTIGATING**LIFE** figures.

[a]Heller, H. C. and G. W. Colliver. 1974. *American Journal of Physiology* 227: 583–589.

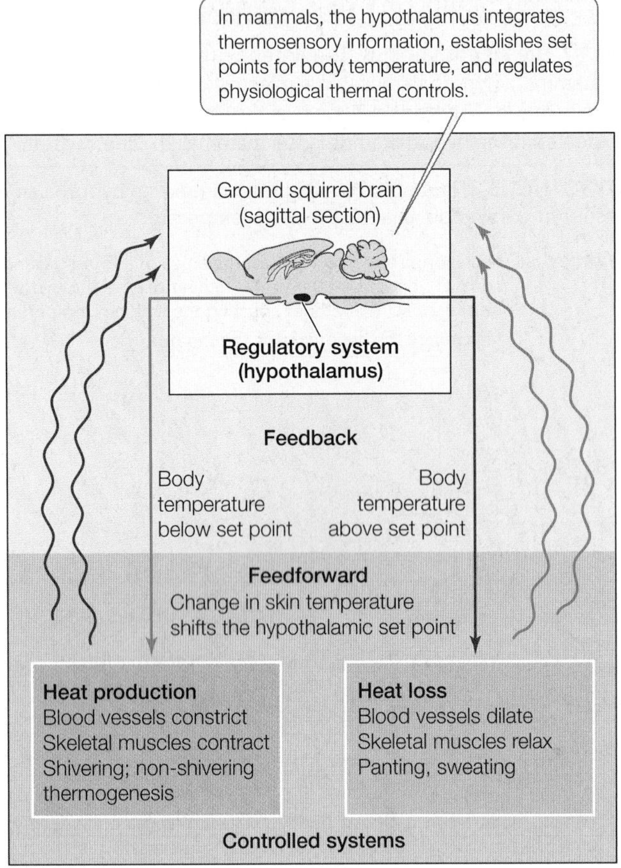


**40.20 The Mammalian Thermostat** Like the home thermostat in Figure 40.2, the mammalian hypothalamus is the regulatory system that controls the body's heating and cooling mechanisms.

**Go to Animated Tutorial 40.1**
**The Hypothalamus**
Life10e.com/at40.1

Hypothalamic set points are higher during wakefulness than during sleep, and they are higher during the active part of the daily cycle than the inactive part, even if the animal is awake at both times. Even when an endotherm is kept under constant environmental conditions, its body temperature displays a daily cycle of changes in set point. This kind of cycle is controlled by an internal biological clock, as we will discuss in Chapter 53.

## Fever helps the body fight infections

Fever is an adaptive response that helps the body fight pathogens. A fever is a rise in body temperature in response to molecules called **pyrogens**. Pyrogens can be exogenous, such as foreign proteins produced by bacteria or viruses that invade the body, or endogenous, including substances produced by cells of the immune system in response to infection.

The presence of a pyrogen causes a rise in the body's hypothalamic set point for metabolic heat production. As a result, you shiver, put on a sweater, or crawl under a blanket, and your body temperature rises until it matches the new set point. At the higher body temperature you no longer feel cold, and you may not feel hot, but someone touching your forehead will say that you are "burning up." Taking aspirin lowers your set point to normal. Now you feel hot, take off clothes, and even sweat until your elevated body temperature returns to normal. Although modest fevers help the body fight infections, extreme fevers can be dangerous and must be controlled, usually with fever-reducing drugs.

## Some animals conserve energy by turning down the thermostat

**Hypothermia** is a below-normal body temperature. It can result from starvation (lack of metabolic fuel), exposure to extreme cold, serious illness, or anesthesia. In each of these cases, the drop in body temperature is unregulated. However, many birds and mammals undergo *regulated* hypothermia to survive periods of cold and food scarcity.

Hummingbirds, for example, are very small endotherms with a high metabolic rate, and going even a single day without food could exhaust their metabolic reserves. Hummingbirds and other small endotherms can extend the period over which they can survive without food by dropping their body temperature 10°C to 20°C during the portion of day or night when they are normally inactive, thus lowering their metabolic rate and conserving energy. This adaptive hypothermia is called **daily torpor**.

Regulated hypothermia that lasts for days or even weeks, during which the body temperature falls close to the ambient temperature, is called **hibernation** (**Figure 40.21**). Many species of mammals, including bats, bears, marmots, and ground squirrels, hibernate, but only one species of bird (the poorwill) has been shown to hibernate. The metabolic rate needed to sustain a hibernating animal may be only one-fiftieth its basal metabolic rate, and many hibernating animals maintain body temperatures close to the freezing point. Arousal from hibernation occurs when the hypothalamic set point returns to the normal level. The ability of animals to enter daily torpor or deep hibernation to reduce their thermoregulatory set point so dramatically probably evolved as an extension of the set point decrease that accompanies sleep in all mammals and birds.

**RECAP 40.5**

Within the thermoneutral zone, an endotherm controls its body temperature principally by passive means. Above or below the thermoneutral zone, an endotherm must expend considerable metabolic energy to control its body temperature. Thermoregulatory responses in mammals are regulated by the hypothalamus.

- Describe how endotherms produce heat. **See pp. 827–828 and Figure 40.17**
- Why is dependence on evaporative water loss a dangerous strategy for dealing with hot environments? **See p. 829**
- What is the nature of the negative feedback information and feedforward information used by the mammalian thermostat? **See pp. 829–830 and Figure 40.20**

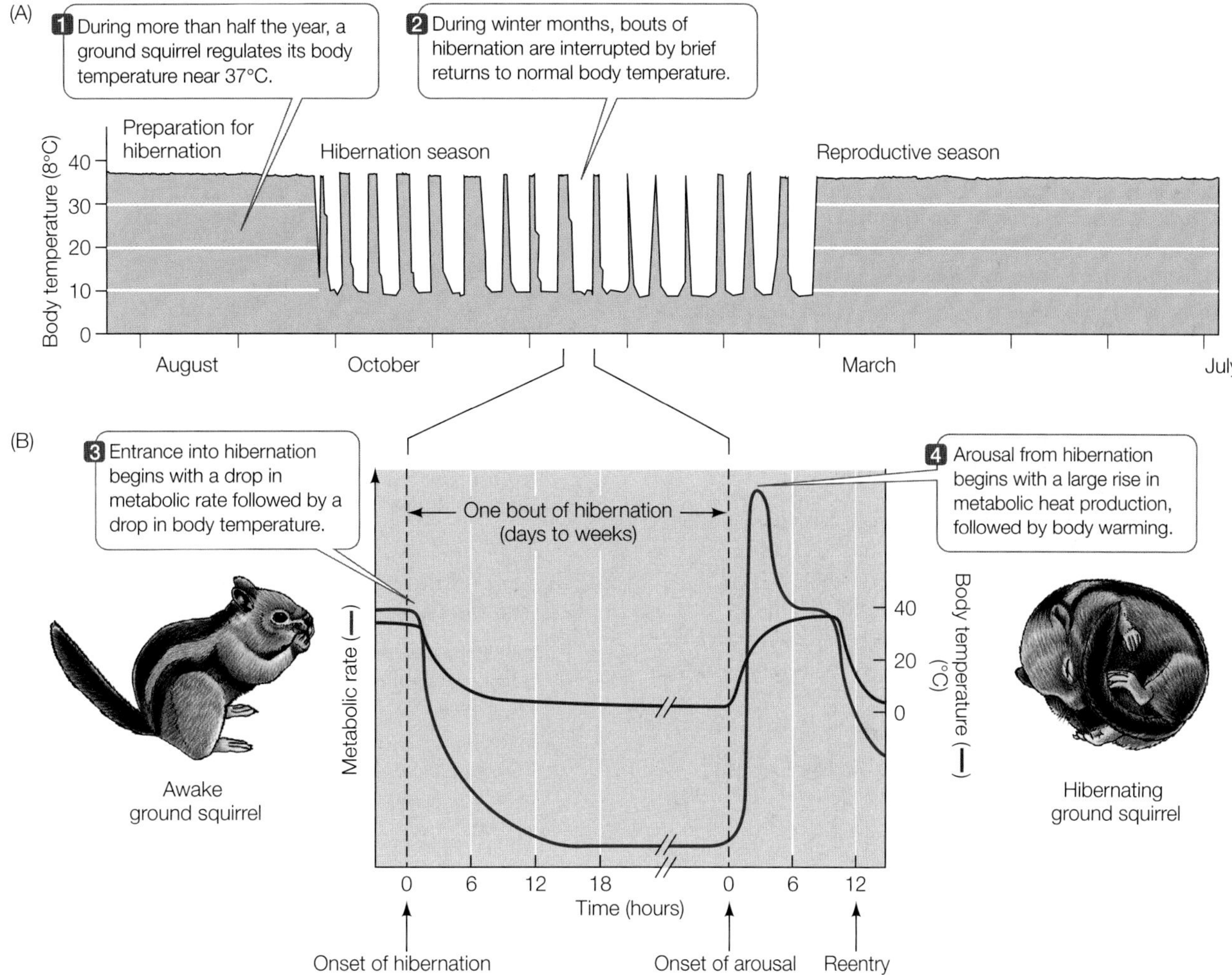


**40.21 Hibernation Patterns in a Ground Squirrel** During most of the year, the ground squirrel regulates its body temperature around 37°C. During winter months, however, these animals hibernate in underground burrows, living off stored fuel (in the form of either fat or cached food). The metabolic demands for these stored fuels are decreased by bouts of torpor, during which body temperature drops close to that of the environment for long periods of time.

**Can we increase heat loss from our natural heat portals to protect against heat stress?**

**ANSWER**

Normally, blood flows from the heart through large arteries, through the tiny capillaries, and then into larger veins that carry the blood back to the heart. The evolutionary adaptations of the "heat portals" in non-hairy mammalian skin, however, includes gated shunts that can deliver arterial blood directly to veins, bypassing the slower flow of the capillaries. The veins in non-hairy skin form networks that can accommodate a large volume of blood when the shunts are open. Based on this knowledge, biologists at Stanford University developed a "rapid-cooling" technology that maximizes heat extraction from non-hairy skin. An area such as the palm of the hand is placed in contact with a cooled surface and a mild vacuum is used to pull more blood into the large, heat-exchanging blood vessels. When this device is used, body temperature rises more slowly during exercise and cools more rapidly during rest after exercise.

An unexpected by-product of the Stanford discovery was enhanced athletic performance. Because muscle fatigue is partly due to increased muscle temperature, enhanced cooling can reduce fatigue and increase exercise capacity, which in turn can lead to conditioning gains. In one study, first-year college students in a conditioning program improved their push-up performance at a rate of 5 push-ups a day without cooling, but 9 push-ups a day with cooling. Some men and women in the study achieved more than 800 push-ups or hundreds of pull-ups in a 45-minute workout session.

## CHAPTER**SUMMARY** 40

### 40.1 How Do Multicellular Animals Supply the Needs of Their Cells?

- Multicellular animals provide for the needs of all their cells by maintaining a stable **internal environment.** That environment consists of two extracellular fluid compartments: the **interstitial fluid** and the blood **plasma**. **Review Figure 40.1**
- Regulation of physiological systems is mostly through **negative feedback**. **Feedforward information** functions to change **set points**. **Review Figure 40.2**

### 40.2 What Are the Relationships between Cells, Tissues, and Organs?

- The cells of the body are organized into assemblages called **tissues**.
- Although there are many cell types, there are only four tissue types: **epithelial**, **muscle**, **connective**, and **neural** tissues.
- **Organs** are made up of tissues, and most organs contain all four types of tissues. Organs are grouped into **organ systems**. **Review Figure 40.7**

### 40.3 How Does Temperature Affect Living Systems?

- Life is possible only within a narrow range of environmental temperatures. $\mathbf{Q_{10}}$ is a measure of the sensitivity of a life process to temperature. A $Q_{10}$ of 2 means that the reaction rate of that process doubles as temperature increases by 10°C. **Review Figure 40.8**
- Animals can acclimatize to seasonal changes in temperature through biochemical and physiological adaptations. **Review Figure 40.9**

### 40.4 How Do Animals Alter Their Heat Exchange with the Environment?

- The body temperatures of **ectotherms** are determined primarily by external sources of heat. **Endotherms** regulate their body temperatures by producing heat metabolically. **Review Figure 40.10**
- The four avenues of heat exchange with the environment are **radiation**, **convection**, **conduction**, and **evaporation**. The balance between heat production and heat exchange can be expressed as an **energy budget**. **Review Figure 40.12**
- Control of blood flow to the skin is an important means of temperature regulation. Circulatory system adaptations such as **countercurrent heat exchange** can conserve metabolic heat. **Review Figures 40.13, 40.14**

### 40.5 How Do Endotherms Regulate Their Body Temperatures?

- Within the **thermoneutral zone**, resting endotherms have a **basal metabolic rate** (**BMR**) that scales with body size. **Review Figures 40.16, 40.17, ACTIVITY 40.1**
- In mammals, control of body temperature relies on commands from a regulatory center in the **hypothalamus**. This thermostat uses its own temperature as negative feedback information and skin temperature as feedforward information. **Review Figure 40.20, ANIMATED TUTORIAL 40.1**

**Go to the Interactive Summary to review key figures, Animated Tutorials, and Activities Life10e.com/is40**

## CHAPTER**REVIEW**

### REMEMBERING

1. Which of the following best describes the limits to life on earth?
   a. The freezing point and the boiling point of water
   b. The effects of temperature on biochemical rates of reaction
   c. The freezing point of water and the temperature at which proteins denature
   d. The ability of the organism to actively produce and dissipate heat
   e. The effect of temperature on the oxygen content of the environment

2. If the $Q_{10}$ of an animal's metabolic rate is 2, then
   a. the animal is better acclimatized to a cold environment than if its $Q_{10}$ were 3.
   b. the animal is an ectotherm.
   c. the animal consumes half as much oxygen per hour at 20°C as it does at 30°C.
   d. the animal's metabolic rate is not at basal levels.
   e. the animal produces twice as much heat at 20°C as it does at 30°C.

3. Which statement about brown fat is true?
   a. It produces heat without producing much ATP.
   b. It insulates animals acclimatized to cold.
   c. It is a major source of heat production for birds.
   d. It is found only in hibernators.
   e. It provides fuel for muscle cells.

4. Which of the following is the most important and most general characteristic of endotherms adapted to cold climates compared with endotherms adapted to warm climates?
   a. Higher basal metabolic rates
   b. Higher $Q_{10}$ values
   c. Brown fat
   d. Greater insulation
   e. Ability to hibernate

5. Which of the following would cause a decrease in the hypothalamic temperature set point for metabolic heat production?
   a. Entering a cold environment
   b. Taking an aspirin when you have a fever
   c. Arousing from hibernation
   d. Getting an infection that causes a fever
   e. Cooling the hypothalamus

## UNDERSTANDING & APPLYING

6. What is the advantage of feedforward information for homeostasis? Can you suggest what some sources of feedforward information could be for regulation of breathing, blood pressure, and secretion of digestive juices?

7. Newton's law of cooling describes how a physical object comes into thermal equilibrium with its environment. This law can be expressed as

$$HL = K(T_o - T_a)$$

where $HL$ is the rate of heat loss, K is the thermal conductance constant (how easily an object loses heat), $T_o$ is the temperature of the object, and $T_a$ is the ambient temperature. Apply this expression to the metabolic rate/environmental temperature curve for endotherms (see Figure 40.17). What would be the equivalent of $HL$? In Newton's law of cooling, K is a constant reflecting the properties of the object. What would K represent for an endotherm? What would 1/K represent? Using a version of Newton's law that replaces $T_o$ with $T_b$ (body temperature), explain why the curve projects to zero at an ambient temperature that equals body temperature.

8. What do you expect to be the effect of temperature on the ability of a heart to pump blood and the ability of skeletal muscle to contract? How does this create a physiological challenge for a great white shark or a giant bluefin tuna?

## ANALYZING & EVALUATING

The following data from the White and Seymour paper cited in the Working with Data exercise (p. 827) should be used to answer Questions 9 and 10.

| Species | Body mass (kg) | Heart size (g) | BMR (ml $O_2$/hr) |
|---|---|---|---|
| Mouse | 0.03 | 0.2 | 63 |
| Rat | 0.4 | 1.5 | 146 |
| Muskrat | 1.0 | 8 | 640 |
| Marmot | 4 | 24 | 1,550 |
| Coyote | 10 | 60 | 2,690 |
| Baboon | 17 | 100 | 5,150 |

9. Plot the data for (1) BMR versus body mass and (2) heart mass versus body size. Describe the differences between these two curves.

10. The BMR is supported by the amount of blood pumped by the heart, and the amount of blood the heart can pump depends on its size (volume). What other factor would you propose to explain the differences between these two curves?

11. The observations on the Galápagos marine iguana in Figure 40.13 showed that this animal's body temperature rose faster in air than it fell in water. The inference was that the iguana was influencing its gain or loss of heat by altering the blood flow to its skin. However, the thermal properties of air and water are different, and the animal was breathing when in air, but not while swimming. In terms of the energy budget

$$\underbrace{\text{metabolism} + R_{abs}}_{\text{heat}_{in}} = \underbrace{R_{out} + \text{convection} + \text{conduction} + \text{evaporation}}_{\text{heat}_{out}}$$

what factors other than blood flow to the skin could influence these rates of heat exchange, and what experiment could you do to strengthen the argument that changes in blood flow to the skin were critical variables?

Go to BioPortal at **yourBioPortal.com** for Animated Tutorials, Activities, LearningCurve Quizzes, Flashcards, and many other study and review resources.

# 41 Animal Hormones

## CHAPTER**OUTLINE**


**41.1** **What Are Hormones and How Do They Work?**

**41.2** **What Have Experiments Revealed about Hormones and Their Action?**

**41.3** **How Do the Nervous and Endocrine Systems Interact?**

**41.4** **What Are the Major Endocrine Glands and Hormones?**

**41.5** **How Do We Study Mechanisms of Hormone Action?**

**Exercised Muscles Secrete a Hormone** Research on mice has shown that irisin, a hormone produced by exercising muscles, travels to fat cells in the body, where it stimulates molecular changes that make the adipose tissue more metabolically active.

IF YOU ARE a person who eats a lot but never puts on weight, it is likely that you also exercise regularly. But do you stay thin simply because exercising burns the excess calories that would otherwise be stored as fat? We learned in earlier chapters that typical adipose tissue, or "white" fat, stores lipids, whereas "brown" fat metabolizes lipids to produce heat without producing ATP. Brown fat is known to be present in cold-acclimated rodents, hibernators, and newborn humans, but it was not thought to be present in adult humans. Recently, however, imaging techniques revealed brown fat activity in cold-exposed adults. What was interesting was that the amount of brown fat was inversely proportional to total body mass—lean people had more brown fat and obese people had less. It was suggested that the excess calories burned by the brown fat were responsible for low body mass.

A recently discovered signaling molecule may explain the difference in individual propensity to put on weight. The molecule has been classified as a hormone and given the name irisin, after the mythological character Iris. Irisin was discovered in a strain of mice bred for increased exercise endurance capacity, and it was shown that the molecule is released from muscles. It has been well documented that exercise training causes numerous structural and metabolic changes in muscle. But exercise training also improves many other aspects of health. How are these changes mediated? Analysis of fat tissue from the super athletic mice showed a remarkable finding: their white fat had properties of brown fat. It was then shown that this "browning" of the white fat was triggered by a blood-borne chemical signal produced by the exercising muscles. Thus the muscle is telling the white fat to change its properties to a tissue that is more metabolically active, burns more calories, and produces more heat. So exercise improves the condition of muscles, and it also causes the muscles to talk to adipose tissues, telling them to "shape up."

Why did the scientists call their signaling molecule a hormone and choose to name it after Iris? As we will discuss in this chapter, a hormone is a chemical message that circulates in the blood and activates distant target cells. In Greek mythology, Iris was a messenger of the gods, traveling the world with the speed of the wind.

How can we demonstrate that a molecule found in the blood is a hormone?

See answer on p. 853.

## 41.1 What Are Hormones and How Do They Work?

In multicellular animals, physiological regulatory systems require information and cell-to-cell communication. Most intercellular communication is by means of chemical signals that bind to receptors, as described in Chapter 7. Examples of chemical signals include hormones, growth factors and morphogens, cytokines, and neurotransmitters. These signals operate in different contexts—the endocrine system, growth and development, the immune system, and the nervous system—but the principles of their function are the same: one cell releases a chemical signal that travels to and binds to a receptor on a second cell (the "target"), causing a response in the target cell.

Not all signaling is chemical; keep in mind that there are receptors in the nervous system that encode information from external physical sources, such as temperature, pressure, and light. And the nervous system uses electric signals called action potentials to get information from place to place in the body. Regardless of the system, the processing of information depends on which cells have receptors for the signals and how those target cells respond to the signals.

Some analogies might help distinguish how the immune, nervous, and endocrine informational systems operate. The immune system (the topic of Chapter 42) operates like an army of private security guards. The various cellular agents make their rounds of the body, and if they detect a security breach, they sound their alarms—cytokines—that activate the body's defenses. The nervous system (see Chapters 45–47) operates like a landline telephone system, with a central integration and command center that sends signals along specific wires to specific receivers. This chapter focuses on the endocrine system, which is more like a radio or television network, broadcasting signals that can be picked up by anyone who has an appropriate receiver that is turned on and tuned in.

### Endocrine signaling can act locally or at a distance

The many and varied types of **endocrine cells** produce and release chemical signals directly into the extracellular fluid (ECF). These molecules may then diffuse through the ECF and enter the bloodstream. Endocrine signals that enter the blood are called **hormones**, and hormones activate target cells far from their site of release (**Figure 41.1A**). Irisin is an example of a hormone. You probably are familiar with several other hormones, such as testosterone, estrogen, adrenaline, and insulin.

Some endocrine signals are released in such tiny quantities, or are so rapidly inactivated by enzymes, or are taken up so efficiently by local cells that they never diffuse into the blood in sufficient amounts to act on distant cells (**Figure 41.1B**). Because these signals affect only target cells near their release site, they are called **paracrines** (*para*, "near"). An example of a paracrine signal is histamine, one of the mediators of inflammation. The most local action an endocrine signal can have is when it binds to receptors on or in the same cell that secreted it. When a chemical signal influences the cell that secreted it, it is called an **autocrine**. Hormones and paracrines can have

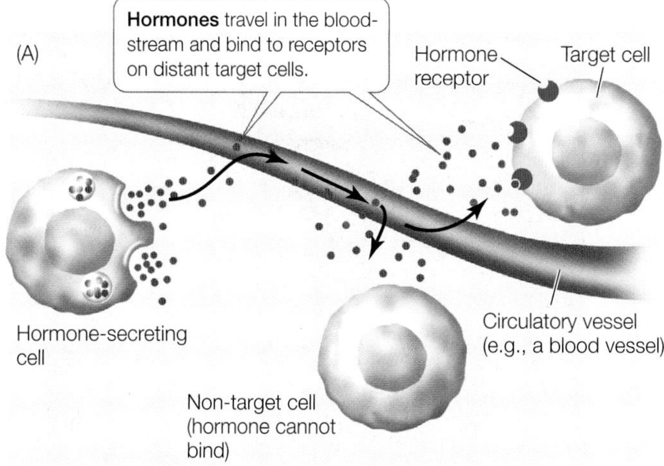


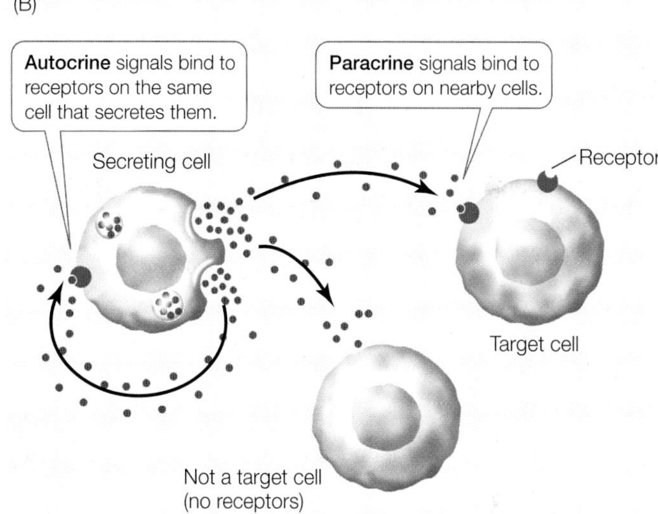


**41.1 Chemical Signaling Systems** (A) Hormones are distributed throughout the body in the bloodstream. (B) Paracrines and autocrines do not enter the blood. Paracrines simply diffuse to nearby cells, and autocrines influence the same cells that release them.

autocrine functions as a means of providing negative feedback to control their own rates of secretion.

Some endocrine cells exist as single cells within a tissue. Hormones of the digestive tract, for example, are secreted by isolated endocrine cells in the walls of the stomach and small intestine. Many hormones are secreted by aggregations of endocrine cells in secretory organs called **endocrine glands**. A single endocrine gland may secrete multiple hormones.

The name "endocrine" (Greek, "separated within") reflects the fact that the substances secreted by these cells are secreted directly into the "internal environment," the extracellular fluid (see Section 40.1). From the ECF they can diffuse locally or enter the bloodstream. **Exocrine glands**, in contrast, have ducts through which the products of exocrine cells are carried to an *external* environment such as the surface of the skin or a body cavity such as the gut (sweat and salivary glands are examples).

(A) Protein hormones

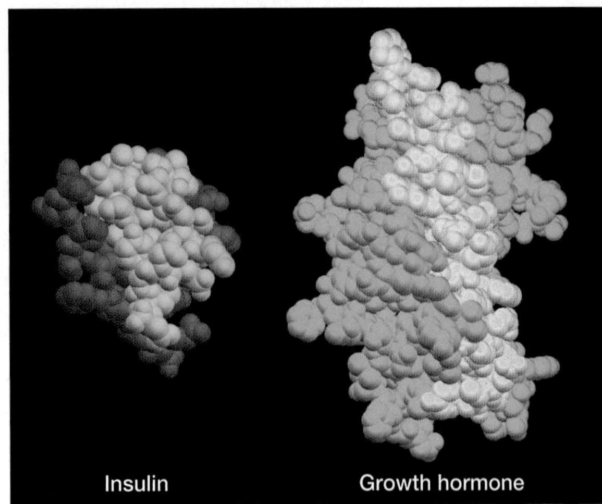


(B) Steroid hormones

Sterol backbone

Cholesterol

Corticosteroids

Sex steroids

Aldosterone

Cortisol

Testosterone

Estrogen

(C) Amine hormones

Tyrosine

Epinephrine

Thyroxine

**41.2 Three Classes of Hormones** (A) The largest hormone molecules are peptides and proteins. This class includes insulin and growth hormone. (B) Steroid hormones are modified from cholesterol molecules. They include the corticosteroids produced by the adrenal gland and the sex steroids produced primarily by the gonads. (C) Amine hormones are tiny molecules synthesized from a single amino acid. Both epinephrine and thyroxine are synthesized from tyrosine units, but thyroxine is lipid-soluble and epinephrine is water-soluble. Their modes of release and transport and the locations of their receptors differ accordingly.

To complete our overview of intercellular chemical communication, we must mention neurotransmitters, which we will discuss in detail in Chapters 45 through 47. Neurons, the cells of the nervous system, conduct information over long distances as electric signals, but where a neuron communicates that information to another cell, it does so by releasing chemical signals called neurotransmitters. Most neurotransmitters act very locally (frequently on the same neuron that released them). Some, however, diffuse into the blood and act on distant targets; neurotransmitters that act in this manner are referred to as **neurohormones**.

## Hormones can be divided into three chemical groups

There is enormous diversity in the chemical structure of hormones, but by and large they can be classified into three groups:

- The majority of hormones are peptides or proteins. These hormones (insulin is an example; **Figure 41.2A**) are water-soluble and thus easily transported in the blood without carrier molecules. Peptide and protein hormones can be packaged in vesicles within the cells that make them, and then released by exocytosis. Their receptors are on cell surfaces.
- **Steroid hormones** (such as estrogen and testosterone) are synthesized from the steroid cholesterol (**Figure 41.2B**), are lipid-soluble, and pass easily through cell membranes. Steroid hormones diffuse out of the cells that make them and are usually bound to carrier molecules in the blood. Their receptors are mostly intracellular.
- **Amine hormones** are mostly synthesized from the amino acid tyrosine (thyroxine is one example; **Figure 41.2C**). Some amine hormones are water-soluble and others are lipid-soluble; their modes of release differ accordingly.

## Hormone action is mediated by receptors on or within their target cells

Water-soluble hormones cannot pass readily through plasma membranes, so their receptors must be located on the surfaces of target cells. These receptors are large transmembrane glycoprotein complexes with three domains: a binding domain that projects outside the plasma membrane, a transmembrane domain that anchors the receptor in the membrane, and a cytoplasmic domain that extends into the cytoplasm of the cell. When a hormone binds to the binding domain, the cytoplasmic domain initiates the target cell's response. Second messengers activate a cascade of intracellular events, eventually activating protein kinases or protein phosphatases (see Figures 7.6 and 7.7). In most cases these protein kinases and phosphatases activate or inactivate enzymes in the cytoplasm that lead to the cell's response, but the signaling cascade initiated by the receptor can also generate signals that enter the nucleus and alter gene expression (see Figure 7.10).

Lipid-soluble hormones can diffuse through plasma membranes, and therefore their receptors are usually inside cells, in either the cytoplasm or the nucleus (although some membrane-bound receptors for lipid-soluble hormones have recently been described). In most cases, the complex formed by the lipid-soluble hormone and its receptor acts by altering gene expression in the cell's nucleus (see Figure 7.8).

## Hormone action depends on the nature of the target cell and its receptors

Wherever a hormone encounters a cell with an appropriate receptor, it can bind to that receptor and trigger a response. The nature of the response depends on the responding cell and its receptors. Thus the same hormone can cause different responses in different types of cells.

Consider the amine hormone adrenaline, or **epinephrine**. Suppose you are walking in the forest and almost step on a rattlesnake. You jump back, your heart starts to thump, and protective reactions are set in motion. The jump and the heart thumping are driven by your rapidly responding nervous system. Simultaneously with these muscular responses, your nervous system stimulates endocrine cells in the adrenal glands just above your kidneys to secrete epinephrine. Within seconds, epinephrine is diffusing into your blood and circulating around your body to activate the many components of the **fight-or-flight response** (Figure 41.3).

Epinephrine binds to receptors in your heart, causing a faster and stronger heartbeat. Your heart is now pumping more blood. Epinephrine also binds to receptors in certain blood vessels. By causing constriction of blood vessels supplying your skin, kidneys, and digestive tract (digesting lunch can wait!), the hormone diverts more blood to the muscles needed for your escape from danger.

Epinephrine binds to cells in the liver, stimulating them to break down glycogen and release its glucose units into the blood as a quick energy supply (see Figure 7.18). In fatty tissue, epinephrine stimulates the breakdown of fats to yield fatty acids—another source of energy. These are just some of the actions triggered by one hormone. In each case the cellular response depends on the cell's receptors and associated intracellular signaling cascade, but they all contribute to increasing your chances of surviving a dangerous situation.

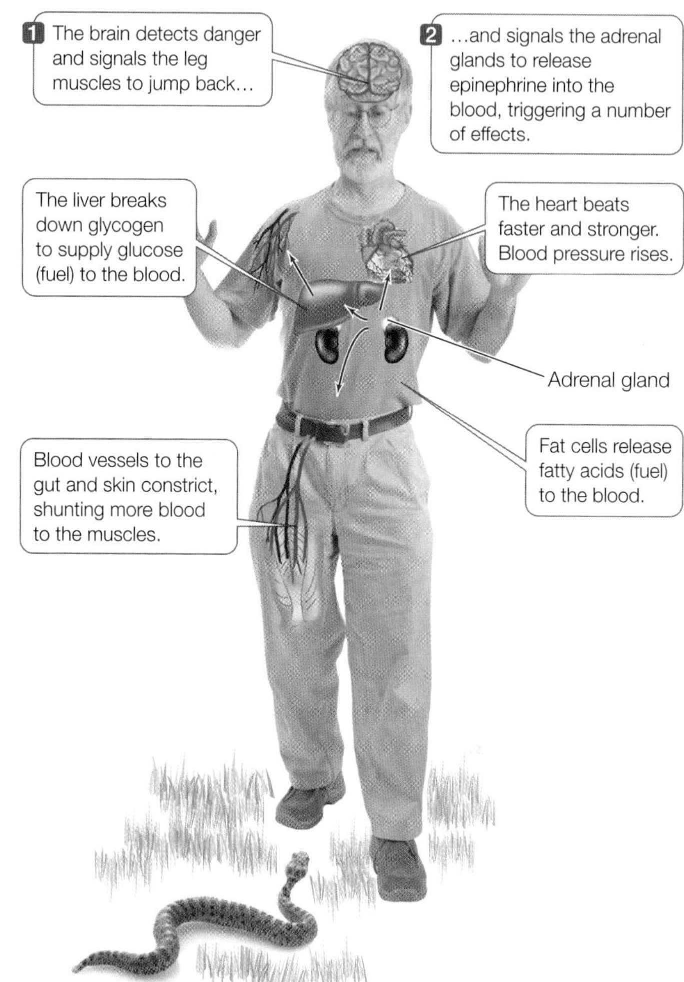


**41.3 The Fight-or-Flight Response** The brain of a person suddenly faced with a threatening situation sends a signal to the adrenal glands, which almost instantaneously release the hormone epinephrine. Epinephrine circulates around the body and induces the various components of the fight-or-flight response in different tissues.

**RECAP 41.1**

Hormones are chemical signals released by endocrine cells into the extracellular fluid, where they can diffuse into the blood and travel to distant target cells. The receptors for water-soluble hormones are on the surfaces of target cells; receptors for most lipid-soluble hormones are inside the target cells.

- What are the three general chemical categories of hormones? **See p. 836 and Figure 41.2**
- Describe the different methods by which water-soluble and lipid-soluble hormones reach their receptors. **See p. 837**
- How can a single hormone have diverse effects in the body? **See p. 837 and Figure 41.3**

Scientists discovered and elucidated the transformative actions of many hormones before their specific molecular structures were known. Ingenious experimental approaches have been used to show that a chemical released by one cell has effects on other distant cells. Subsequent studies have focused on the nature of the signal and the mechanisms of its actions.

## 41.2 What Have Experiments Revealed about Hormones and Their Action?

Intercellular chemical signaling was critical for the evolution of multicellularity. A protist, the slime mold *Dictyostelium*, uses a chemical signal (cAMP) to coordinate the aggregation of individual cells to form a multicellular fruiting structure (see Figure 27.18). The least complex of the multicellular animals—the sponges—do not have nervous systems, but they do have intercellular chemical communication. Molecules that diffuse between cells and communicate information have a long evolutionary history in the multicellular lineages. As we discussed in Chapter 37, plant growth is regulated by a variety of hormones, and hormones are identifiable throughout the animal kingdom.

Studying the evolution of hormonal signaling reveals an interesting generalization: the signal molecules themselves are highly conserved. We find the same chemical compounds over broad groups of organisms, although their functions may differ. As organisms have evolved to occupy different environments and have different lifestyles, the same hormone–receptor systems have diversified to serve different functions. A good example of this evolutionary diversification is seen in the hormone prolactin, described in **Figure 41.4**.

The existence and functions of hormones were experimentally demonstrated many years before specific molecules were isolated and identified chemically. That is not surprising when you consider their small size and the tiny amounts of certain hormones that exist in an organism.

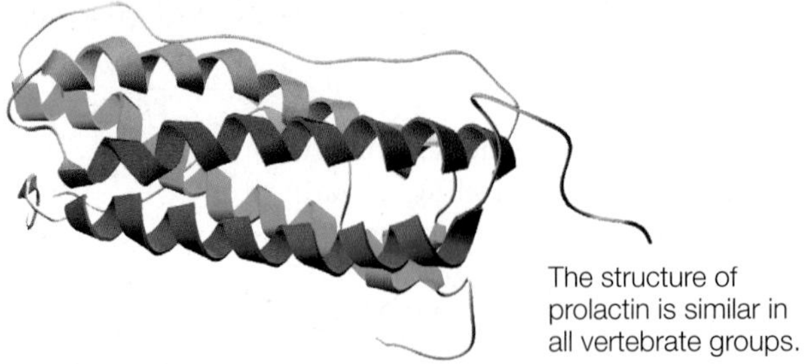
The structure of prolactin is similar in all vertebrate groups.

**Fish**
Required for osmoregulation in freshwater species. In saltwater species that return to fresh water to spawn (e.g., salmon), prolactin production in adults may play a role in generating the drive to return to natal streams.

**Amphibians**
In some species, creates a "water drive" that returns adults to breeding locations. Stimulates oviduct development and production of egg jelly in females. In some species, controls development of sexual characteristics.

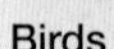

**Birds**
In some species, stimulates nesting activity, incubation behavior, and parental care in both sexes. Stimulates the epithelial cells of the upper GI tract to proliferate and slough off to form "crop milk" to nourish the young.

**Mammals**
In females, stimulates growth of the mammary glands and milk production. In humans, it is responsible for the sensation of sexual gratification as well as the male refractory period following sexual intercourse.

**41.4 Prolactin's Structure Is Conserved, but Its Functions Have Evolved** The hormone prolactin is found in all vertebrate groups and has a long evolutionary history. Its probable function in early vertebrates was in regulating the body's salt and water balance (osmoregulation). It maintains this function in some species, and has evolved in others to control a number of physiological processes, most of which are associated with reproduction.

### The first hormone discovered was the gut hormone secretin

Secretin, a protein released from cells in the gut, stimulates the pancreas to secrete digestive fluids into the gut. At the start of the twentieth century, the prevailing view was that the secretion of digestive juices was controlled by the nervous system. This view was in part the result of the famous work of the Russian physiologist Ivan Pavlov, who discovered the neural control of salivary secretions of the mouth (see Figure 53.1). Pavlov failed, however, when he tried similar experiments with the secretions of the pancreas.

In 1902 William Bayliss and Ernest Starling, working on anesthetized dogs, surgically removed all of the nerves to the pancreas. They then showed that the denervated pancreas still produced secretions when stomach acid was injected into the gut. Then they took tissue from the gut, ground it up, and injected an extract from this preparation into the bloodstream; this injected fluid stimulated pancreatic secretions. Bayliss and Starling's work proved that a chemical extracted from one tissue could travel in the blood to cause a reaction in a different tissue. They called this type of chemical a "hormone," from the Greek word for "impetus."

INVESTIGATING**LIFE**

**41.5 Muscle Cells Can Produce a Hormone** One effect of exercise is the "browning" of white fat (see Figure 40.5D), which increases the fat's metabolic activity and leads to other health benefits. Bruce Spiegelman and his colleagues investigated the possibility that exercised muscles secrete a hormone that changes the characteristics of white fat.[a]

**HYPOTHESIS** Exercised muscle cells produce a hormone that stimulates browning of fat cells.

*Method*

1. Two types of muscle cell cultures were prepared. One culture received a treatment that mimicked the effects of exercise on muscle cells.
2. After culture, the muscle cells were removed and their used media (the culture fluid) was added to cultures of developing fat cells.

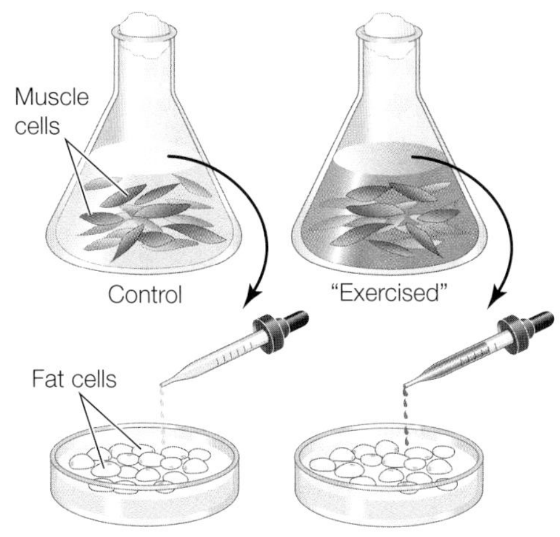


*Results*

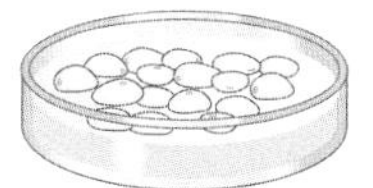
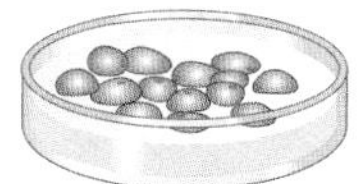
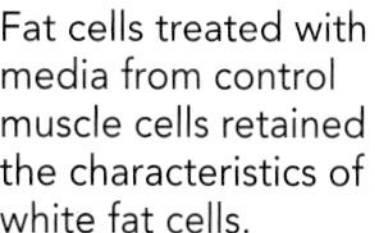

Fat cells treated with media from control muscle cells retained the characteristics of white fat cells.

Fat cells treated with media from "exercised" cells developed properties of brown fat.

**CONCLUSION** A substance secreted by exercised muscle cells stimulates "browning" of cultured fat.

Go to **BioPortal** for discussion and relevant links for all INVESTIGATING**LIFE** figures.

[a]Boström, P. et al. 2012. *Nature* 481: 463–469.

WORKING WITH**DATA:**

## *Identifying a Hormone Secreted by Exercised Muscles*

### Original Paper

Boström, P. and 17 others. 2012. A PGC1-α-dependent myokine that drives brown-fat-like development of white fat and thermogenesis. *Nature* 481: 463–469.

### Analyze the Data

Over the course of many experiments, Bruce Spiegelman and his colleagues demonstrated that exercising mammalian muscle cells secrete a substance (identified as irisin) that they hypothesized was a hormonal signal stimulating white fat cells to develop some of the characteristics of brown fat cells. They cultured muscle cells from wild-type mice (the equivalent of sedentary mice) and from transgenic mice that overexpress irisin (the equivalent of exercising mice). They then took the fluid media from the two culture types and added that media to cultures of white fat cells (see Figure 41.5). They compared the properties of the white fat cells exposed to "exercised" muscle cell media and to "sedentary" muscle cell media.

**QUESTION 1**

In Experiment 1 (left-hand bars in the figure below), the investigators measured the expression of uncoupling protein 1 (UCP1), a mitochondrial inner membrane protein abundant in brown fat that allows the energy in fat to be converted to heat (see Chapter 9). They then extracted mRNA from treated fat cells and quantified the amount of UCP1 mRNA. Looking at the results, what effect did the "exercised" muscle cell media have on the amount of UCP1 mRNA? Why does this support the idea that exercised muscle cells stimulate "browning" of white fat? In testing the hypothesis that a hormone was involved in stimulating UCP1 expression, why was it important to remove the muscle cells from the conditioned media before adding the media to the fat cells? (See Figure 41.5.)

**QUESTION 2**

In a second experiment (right-hand bars in the figure), the investigators repeated the procedure in Figure 41.5 but pretreated the culture media with an antibody that recognizes and *inactivates* the molecule of interest (i.e., irisin). What was the effect of pretreating the culture media with antibody? What additional information did this experiment provide?

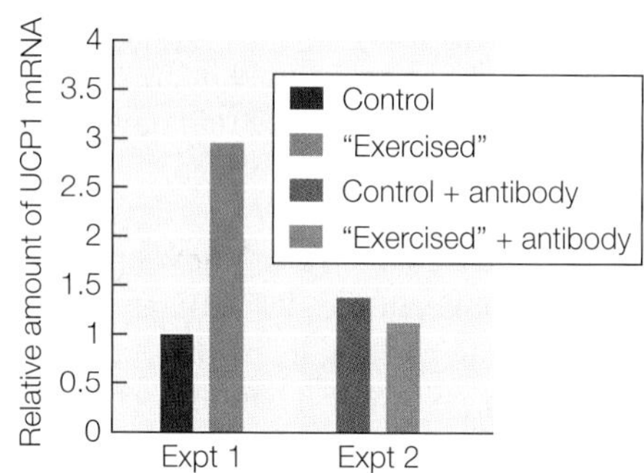


Go to BioPortal for all WORKING WITH**DATA** exercises

The discovery of irisin described at the start of this chapter was achieved using modern methods that did not rely on animal subjects (**Figure 41.5**). However, these experiments were similar in principle to those of Bayliss and Starling.

## Early experiments on insects illuminated hormonal signaling systems

The British physiologist Sir Vincent Wigglesworth was a pioneer in the study of hormonal action and control in insects. Insects, like all arthropods, have rigid exoskeletons (see Chapter 32). Their growth is therefore episodic, punctuated with molts (shedding of the exoskeleton). Each growth stage between two molts is called an instar. In the 1930s and 1940s Wigglesworth studied the phenomenon of insect molting in a series of experiments on the bloodsucking bug *Rhodnius prolixus*. Newly hatched *Rhodnius* look like miniature adults but lack certain adult features. A juvenile bug molts five times before developing into a mature adult; a blood meal triggers each episode of molting and growth.

*Rhodnius* is an amazingly hardy experimental animal—it survives for quite a long time after its head is cut off. Wigglesworth's studies revealed that, if decapitated within an hour after a blood meal, *Rhodnius* can survive for up to a year, but it never molts. If decapitated a week after its blood meal, however, it does molt. Wigglesworth hypothesized that the time lag meant that the substance that triggers molting diffuses slowly from the head. He tested this hypothesis with the experiment described in **Figure 41.6**, which showed that molting was triggered by a substance that diffused throughout the body from a point of origin in the bug's head.

Wigglesworth's experiments yielded another curious result. Regardless of which instar was used, decapitated *Rhodnius* that molted always molted directly into adults, bypassing the usual juvenile instars. Additional experiments demonstrated that a different substance (distinct from that identified by the experiment in Figure 41.6) determines whether a bug molts into another juvenile instar or into an adult.

Because the head of *Rhodnius* is long, it is possible to remove just the front part of it, which contains the brain, while leaving the rear part intact. When fourth-instar bugs that had had a blood meal a week earlier were partly decapitated in this way, they molted into fifth-instar juveniles, not into adults.

This experiment was followed by more experiments in which Wigglesworth used glass tubes to connect individual bugs. When an unfed, completely decapitated fifth-instar bug was connected to a blood-fed, partly decapitated fourth-instar bug (i.e., with only the front part of its head removed), both bugs molted into juvenile forms. A substance from the rear part of the head of the fourth-instar bug prevented both bugs from molting into adults.

INVESTIGATING**LIFE**

**41.6 A Diffusible Substance Triggers Molting** The blood-sucking bug *Rhodnius prolixus* develops from hatchling to adult in a series of five molts (instars) that are triggered by ingesting blood. Sir Vincent Wigglesworth's experiments demonstrated that a blood meal stimulates production of some molt-inducing substance in the insect's head.[a]

**HYPOTHESIS** The substance that controls molting in *R. prolixus* is produced in the head segment and diffuses slowly through the body.

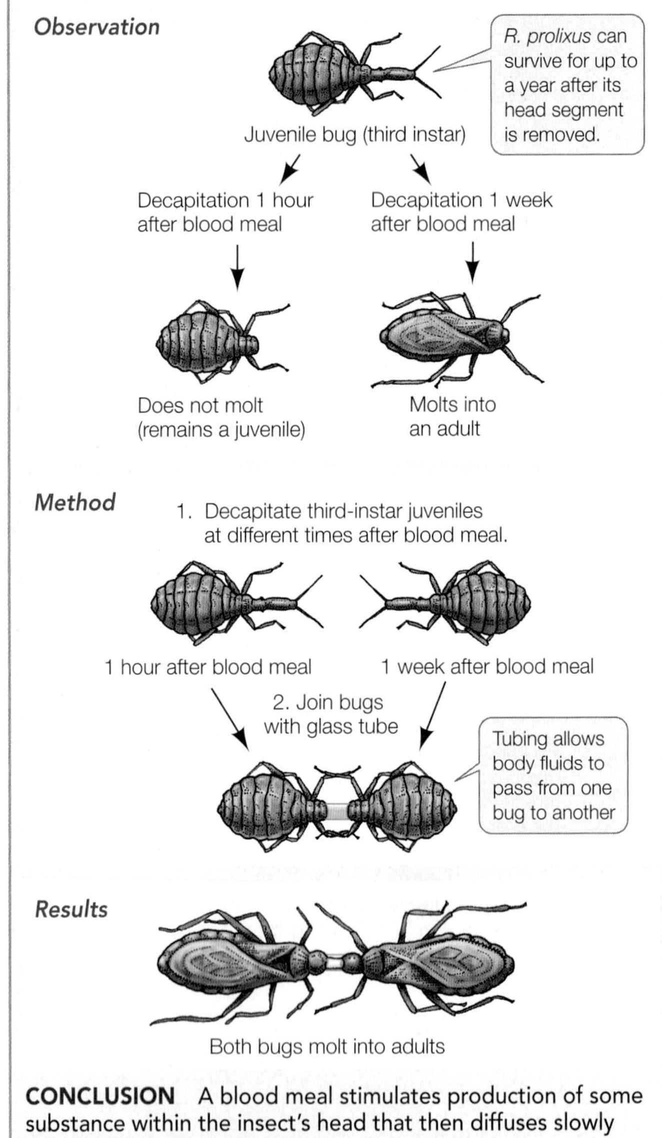


**CONCLUSION** A blood meal stimulates production of some substance within the insect's head that then diffuses slowly through the body, triggering a molt.

Go to **BioPortal** for discussion and relevant links for all INVESTIGATING**LIFE** figures.

[a]Wigglesworth, V. B. 1934. *Quarterly Journal of Microscopical Science* 77: 191–223.

## Three hormones regulate molting and maturation in arthropods

An arthropod's nervous system receives various types of information about the environment (e.g., day length, temperature, social cues, and nutrition) that help determine the optimal timing for the stages of growth and development. When conditions are right, the brain signals the prothoracic gland to produce the hormones that orchestrate physiological processes involved in development and molting.

**PTTH AND ECDYSONE** We now know that two hormones, prothoracicotropic hormone (PTTH) and ecdysone, work in sequence to regulate molting in arthropods. Cells in the brain produce PTTH, which is why it has also been called "brain hormone." PTTH is transported to and stored in paired structures called the corpora cardiaca attached to the brain (see Figure 41.7) After appropriate stimulation (which for *Rhodnius* is a blood meal), PTTH is released and diffuses through the extracellular fluid to an endocrine gland, the prothoracic gland. PTTH stimulates the prothoracic gland to secrete the hormone ecdysone. Ecdysone diffuses to target tissues and stimulates molting.

Ecdysone is a steroid hormone (see Figure 41.2B) and thus is related to the vertebrate hormones estrogen and testosterone (which also play roles in controlling growth and development). Ecdysone is lipid-soluble and readily passes through the plasma membrane of its target cells (mostly cells of the

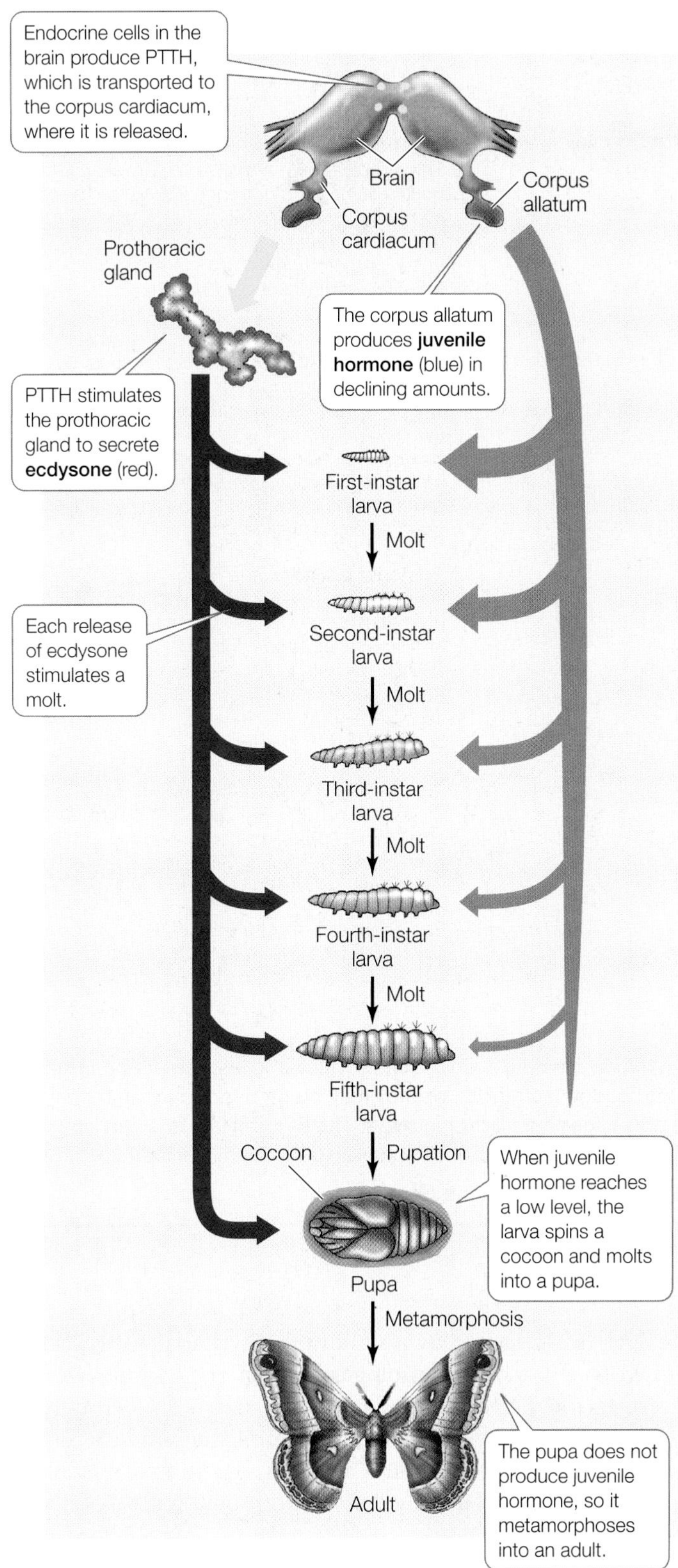


**41.7 Hormonal Control of Metamorphosis** Three hormones control molting and metamorphosis in the silkworm moth *Hyalophora cecropia*.

**Go to Animated Tutorial 41.1 Complete Metamorphosis**
Life10e.com/at41.1

epidermis). In the target cells, ecdysone binds to a receptor that is probably related to the vertebrate testosterone receptor. The resulting hormone–receptor complex induces expression of the genes encoding enzymes involved in digesting the old cuticle and secreting a new one.

**JUVENILE HORMONE** What about the other substance noted in Wigglesworth's experiments—the one that produced sequential juvenile instars rather than adults? The substance responsible for preventing maturation is now known to be **juvenile hormone**, a molecule that is released continuously from the corpora allata (structures that are attached to the corpora cardiaca, which release PTTH). As long as juvenile hormone is present, *Rhodnius* molts into another juvenile instar. Normally *Rhodnius* stops producing juvenile hormone during the fifth instar, then molts into an adult, a life cycle called incomplete metamorphosis (see Section 32.4).

The role of juvenile hormone is more complex in insects that undergo complete metamorphosis (such as butterflies). These animals undergo dramatic developmental changes in their life cycles. The fertilized egg hatches into a larva, which feeds and molts several times, becoming bigger each time. After a fixed number of molts, it enters an inactive stage called pupation. The pupa undergoes major body reorganization and finally emerges as an adult.

An example of complete metamorphosis is provided by the silkworm moth *Hyalophora cecropia* (**Figure 41.7**). As long as juvenile hormone is present in high concentrations, larvae molt into larger larvae. When the level of juvenile hormone falls, larvae spin cocoons and molt into pupae. Because no juvenile hormone is produced in pupae, they molt into adults. Many modern pesticides use juvenile hormone analogs to prevent larvae developing into adults.

## RECAP 41.2

The chemical structure of hormones is highly conserved across many organisms. Early experiments identifying secretin as a hormone defined the characteristics of hormonal signaling in mammals. Early studies on insects revealed much about the nature of hormone action, although the chemical structures of the signaling molecules involved were not established until much later.

- What is meant when we say that hormone molecules are highly conserved? **See p. 838 and Figure 41.4**
- Why did decapitation of *Rhodnius* prevent molting when done 1 day after feeding but not when done 1 week after feeding? **See pp. 840–841 and Figure 41.6**
- What is the role of juvenile hormone in metamorphosis? **See p. 841 and Figure 41.7**

The hormonal control of molting and maturation described above is a general arthropod hormonal control mechanism and exemplifies how the endocrine system works with the nervous system to integrate information and induce long-term effects. The next section will describe similar links between the nervous system and endocrine glands in mammals.

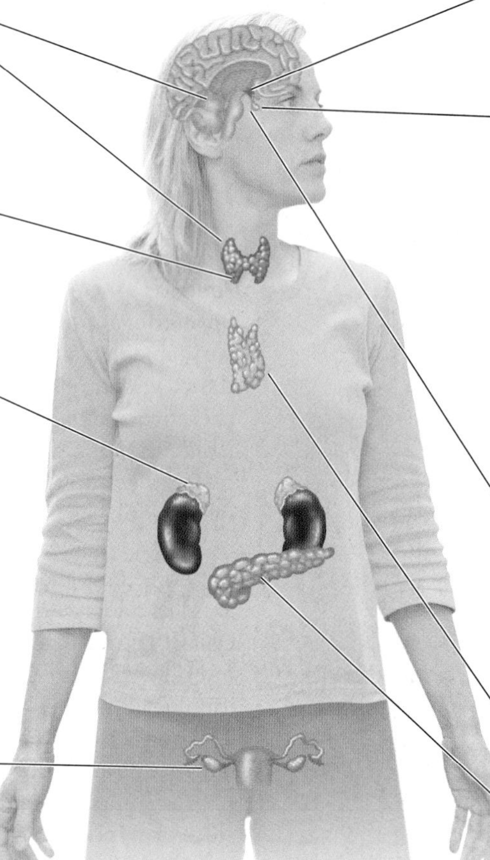


**Other organs include cells that produce and secrete hormones**

| Organ | Hormone |
|---|---|
| Adipose tissue | Leptin |
| Heart | Atrial natriuretic peptide |
| Kidney | Erythropoietin |
| Stomach | Gastrin, ghrelin |
| Intestine | Secretin, cholecystokinin |
| Liver | Somatomedins, insulin-like growth factors |

**41.8 The Endocrine System of Humans** Cells that produce and secrete hormones may be organized into discrete endocrine glands, or they may be embedded in the tissues of other organs, such as the digestive tract or kidneys. The hypothalamus is part of the brain, but it includes cells that secrete neurohormones into the extracellular fluid.

**Go to Activity 41.1 The Human Endocrine Glands**
**Life10e.com/ac41.1**

## How Do the Nervous and Endocrine Systems Interact?

The list of hormones known to exist is long and growing longer (as shown by the recent identification of irisin). To make the subject manageable, we will focus primarily on information about the endocrine system of humans (**Figure 41.8**), much of which is applicable to other mammals. We will begin by considering the hormones involved in the integration of nervous system and endocrine system functions.

### The pituitary is an interface between the nervous and endocrine systems

The **pituitary gland** sits in a depression at the bottom of the skull, just over the back of the roof of the mouth (**Figure 41.9A**). It is attached by a stalk to the hypothalamus, which is involved in many physiological regulatory systems (we detailed its role in thermoregulation in Section 40.5). Through its close connection with the hypothalamus, the pituitary serves as the interface between the nervous system and the endocrine system and is involved in the hormonal control of many physiological processes.

The pituitary has two parts with different developmental origins. The **anterior pituitary** originates as an outpocketing of the roof of the embryonic mouth cavity, whereas the **posterior pituitary** originates as an outpocketing of the floor of the developing brain. Thus the anterior pituitary originates from gut epithelial tissue and the posterior pituitary from neural tissue. Both parts interact with the nervous system but in different ways. The anterior pituitary contains endocrine cells controlled by neurohormones secreted by the hypothalamus. The posterior pituitary contains axons from hypothalamic neurons.

Go to Animated Tutorial 41.2
The Hypothalamic–Pituitary–Endocrine Axis
Life10e.com/at41.2

THE POSTERIOR PITUITARY Long axons extend into the posterior pituitary from neurons in the hypothalamus. The ends, or terminals, of those axons release two neurohormones, antidiuretic hormone and oxytocin (**Figure 41.9B**). These neurohormones are packaged in vesicles that are transported down the axons. The vesicles are stored in the nerve terminals until an action potential stimulates their release.

The main action of **antidiuretic hormone** (**ADH**) in mammals and birds is to increase the amount of water conserved by the kidneys. When ADH secretion is high, the kidneys produce only a small volume of highly concentrated urine. When ADH secretion is low, the kidneys produce a large volume of dilute urine. The posterior pituitary increases its release of ADH when blood pressure falls or the blood becomes too salty. ADH is also known as **vasopressin** because at high concentrations it causes the constriction of peripheral blood vessels as a means of elevating blood pressure.

When a woman is about to give birth, her posterior pituitary releases **oxytocin**, which stimulates the uterine contractions that deliver the baby. Oxytocin also brings about the flow of milk from the mother's breasts. The baby's suckling stimulates neurons in the mother's brain that cause the secretion of oxytocin. Even the sound of a baby can cause a nursing mother to secrete oxytocin and release breast milk—a good example of how the nervous system integrates information that regulates hormonally mediated processes.

Hormones, in turn, can influence the nervous system. Oxytocin, for example, promotes bonding (see the story that opens Chapter 7). If oxytocin release is experimentally blocked, mammalian mothers, from rats to sheep, will reject their newborn offspring, but if a virgin rat is given a dose of oxytocin, she will adopt strange pups as if they were her own. Oxytocin promotes pair bonding and trust in a variety of animals. In humans its secretion rises with intimate sexual contact, and it has been nicknamed the "cuddle hormone."

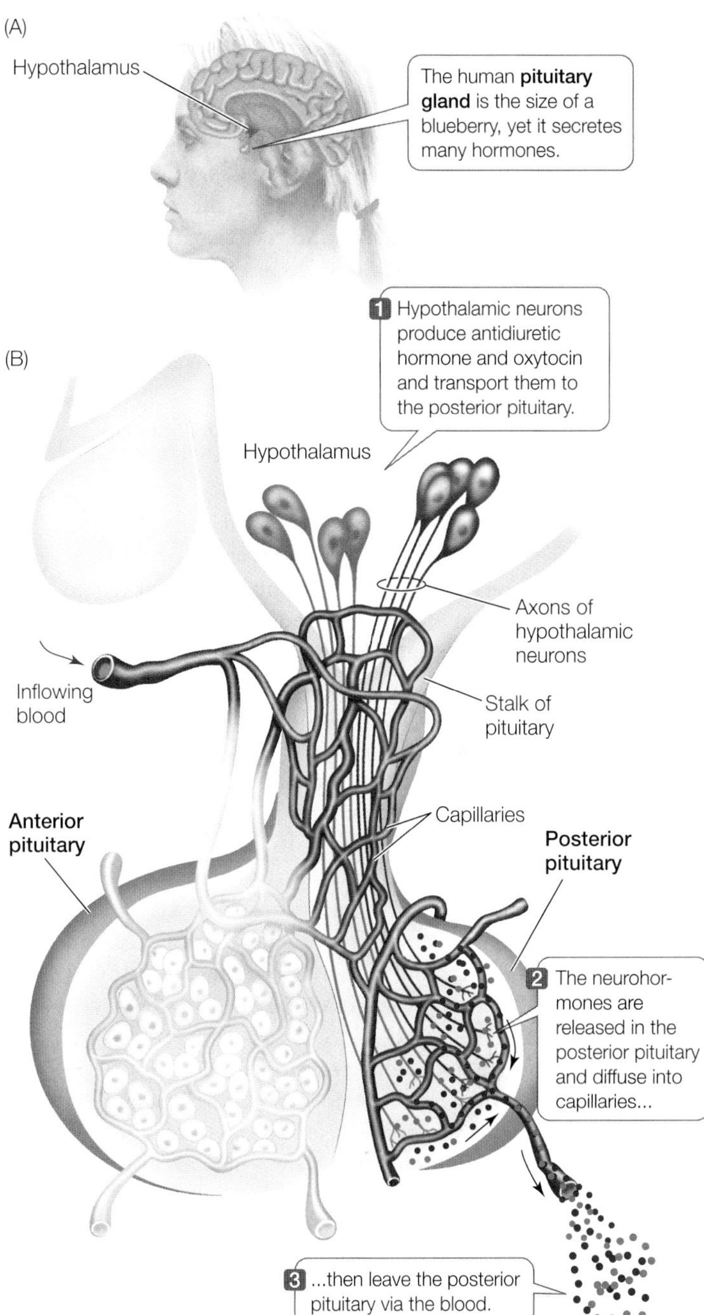


**41.9 The Posterior Pituitary Releases Neurohormones** Neurons in the hypothalamus produce two peptide neurohormones, which are stored and released by the posterior pituitary.

THE ANTERIOR PITUITARY The anterior pituitary produces and releases four peptide and protein hormones that act as **tropic hormones**, meaning they control the activities of other endocrine glands. These four tropic hormones are thyrotropin (thyroid-stimulating hormone), luteinizing hormone, follicle-stimulating hormone, and corticotropin. Each is produced by a different type of pituitary cell. We will say more about the tropic hormones when we describe their target glands—the thyroid, testes, ovaries, and adrenal cortex—later in this chapter and in Chapter 43. Other peptide and protein hormones produced by the anterior pituitary are prolactin (see Figure 41.4), growth hormone, enkephalins, and endorphins.

**Growth hormone** (**GH**) acts on a wide variety of tissues to promote growth. One of its important effects is to stimulate cells to take up amino acids. Growth hormone also promotes growth by stimulating the liver to produce chemical signals that stimulate the growth of bone and cartilage. Overproduction of growth hormone in children causes gigantism, in which affected individuals may grow to nearly 8 feet tall. Underproduction causes pituitary dwarfism, in which individuals fail to reach normal adult height.

**Endorphins** and **enkephalins** are the body's natural painkillers. In the brain, these molecules act as neurotransmitters in pathways that control pain. Their production in the anterior pituitary is normally quite small and probably has little significant effect.

### The anterior pituitary is controlled by hypothalamic neurohormones

In contrast to the posterior pituitary, the anterior pituitary makes and secretes its own hormones, but its secretion of these hormones is controlled by the hypothalamus. The hypothalamus senses and receives information about conditions in the body and in the external environment and communicates that information to the anterior pituitary by releasing neurohormones. If the connection between the hypothalamus and the pituitary is experimentally cut, the release of pituitary hormones no longer changes when conditions in the internal or external environment change.

Hypothalamic neurons do not extend into the anterior pituitary as they do into the posterior pituitary. Remember that the posterior pituitary develops from neural tissue, whereas the anterior pituitary develops from gut tissue. Instead, a special set of **portal blood vessels** bridges the gap between the hypothalamus and the anterior pituitary (**Figure 41.10**). Secretions from neurons in the hypothalamus enter the blood and are conducted down the portal vessels to the anterior pituitary, where they stimulate the release of anterior pituitary hormones.

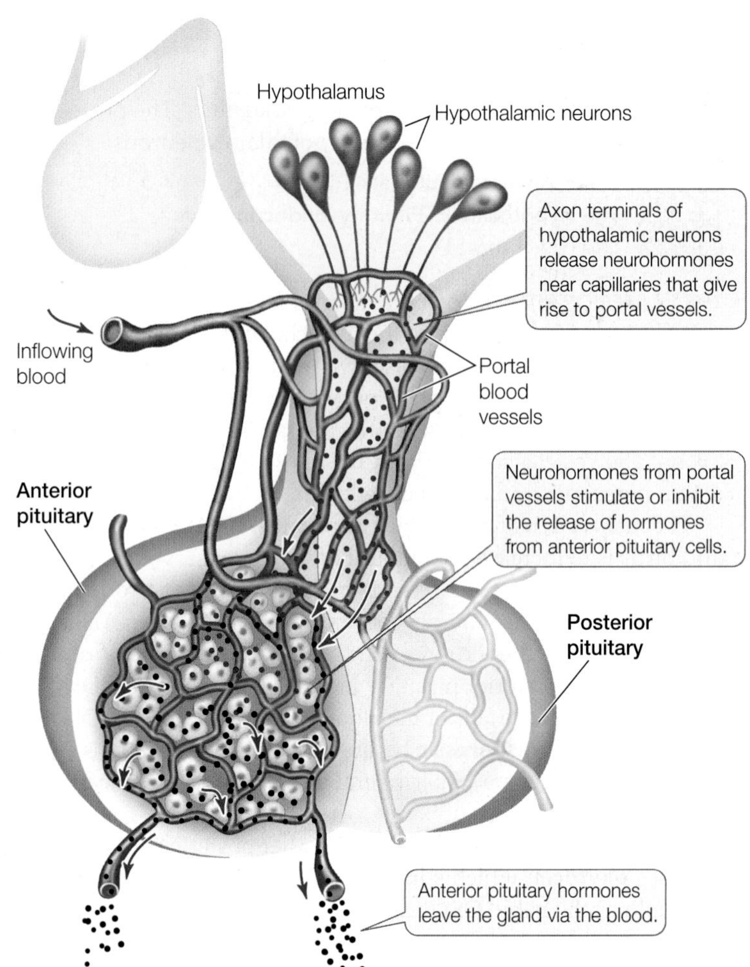


**41.10 The Anterior Pituitary Is Controlled by the Hypothalamus** Cells of the anterior pituitary produce four tropic hormones that control other endocrine glands, as well as several other peptide and protein hormones. These cells are controlled by neurohormones produced in the hypothalamus and delivered through portal blood vessels that run between the hypothalamus and the anterior pituitary through the pituitary stalk.

In the 1960s two large teams of scientists led by Roger Guillemin and Andrew Schally initiated the search for these hypothalamic secretions. Because the amounts of such neurohormones in any individual mammal would be tiny, massive numbers of hypothalami from pigs and sheep were collected from slaughterhouses and shipped to laboratories. One extraction effort began with the hypothalami from 270,000 sheep and yielded only 1 milligram of purified **thyrotropin-releasing hormone** (**TRH**). TRH was the first hypothalamic release-stimulating hormone to be isolated and characterized. It turned out to be a simple tripeptide consisting of glutamine, histidine, and proline. It causes certain anterior pituitary cells to release the tropic hormone thyrotropin, which in turn stimulates the activity of the thyroid gland.

Soon after discovering TRH, Guillemin's and Schally's teams identified **gonadotropin-releasing hormone** (**GnRH**), which stimulates certain anterior pituitary cells to release the tropic hormones that control the activity of the gonads (the ovaries and the testes). For these discoveries, Guillemin and Schally shared the 1977 Nobel Prize in Medicine with Rosalyn Yalow, who invented a technique that allows measurement of miniscule amounts of specific molecules (see Figure 41.19).

Many other hypothalamic neurohormones, including both releasing and release-inhibiting hormones, are now known. The major hypothalamic neurohormones that control anterior pituitary function are:

- Thyrotropin-releasing hormone
- Gonadotropin-releasing hormone
- Prolactin-releasing and release-inhibiting hormones
- Growth hormone-releasing hormone and growth hormone release-inhibiting hormone (somatostatin)
- Corticotropin-releasing hormone

### Negative feedback loops regulate hormone secretion

In addition to being controlled by hypothalamic releasing and release-inhibiting hormones, the endocrine cells of the anterior pituitary are also under direct and indirect negative feedback control by the hormones of the target glands they stimulate (**Figure 41.11**). For example, cortisol, produced by the adrenal gland in response to corticotropin secreted by the anterior pituitary, reaches the pituitary in the circulating blood and inhibits further release of corticotropin. Cortisol also acts as a negative feedback signal to the hypothalamus, inhibiting the release of corticotropin-releasing hormone. In some cases a tropic hormone also exerts negative feedback control on the hypothalamic cells that produce the corresponding releasing hormone.

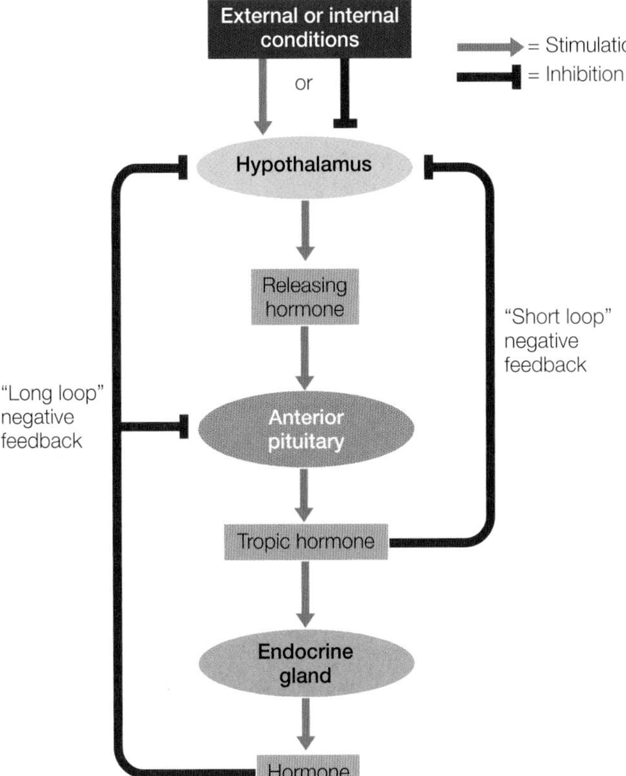


**41.11 Multiple Feedback Loops Control Hormone Secretion** Multiple negative feedback loops regulate the chain of command from hypothalamus to anterior pituitary to endocrine glands.

**RECAP 41.3**

The pituitary is the interface between the nervous system and the endocrine system. The posterior pituitary releases two neurohormones. The anterior pituitary, under the control of other neurohormones from the hypothalamus, releases hormones that control other endocrine glands.

- Describe the anatomical and functional relationships between the brain and the two parts of the pituitary. **See pp. 842–843**
- Describe the role of hypothalamic neurohormones and the portal blood vessels in the secretion of hormones from the anterior pituitary. **See p. 844 and Figure 41.10**

Now that we have described some of the mechanisms by which endocrine systems are controlled, we will take a more detailed look at the functions of the major endocrine glands of mammals, as exemplified by humans.

## 41.4 What Are the Major Endocrine Glands and Hormones?

Hormones help regulate functions in all mammalian physiological systems. In this section we will examine a few major examples of hormonal action in physiological processes. Although we will focus on humans, these systems are very similar in all mammals. Be aware that these are only a few examples; hormonal actions play an important role in virtually all the physiological systems that will be described in the chapters that follow.

### The thyroid gland secretes thyroxine

The **thyroid gland** wraps around the front of the windpipe (trachea) and expands into a lobe on either side (see Figures 41.8 and 41.14). There are two cell types in the thyroid gland, each of which produces a specific hormone. **Thyroxine** is produced by epithelial cells that make up round, colloid-containing structures called follicles (**Figure 41.12A**). **Calcitonin** is produced by cells in the spaces between the follicles and is involved in blood calcium regulation (which we will describe shortly).

Thyroxine, a crucial signal in the regulation of cellular energy metabolism, begins as the glycoprotein thyroglobulin, which is synthesized by the follicle cells and packaged in secretory vesicles. The follicle cells actively take up iodide from the blood and move it into the lumen of the follicle. Each thyroglobulin molecule contains about 100 tyrosine units. When the secretory vesicles release thyroglobulin into the lumen of the follicle, they also release an enzyme that catalyzes the iodination of the tyrosine units in the thyroglobulin. When the thyroid gland is stimulated to release thyroxine, the follicle cells take up thyroglobulin from the follicle by endocytosis. These bits of thyroglobulin are then cleaved to form smaller molecules consisting of only two tyrosine units, and these molecules leave the follicle cells and enter the blood (**Figure 41.12B**). If these molecules are iodinated at the maximum of four sites on the tyrosine units, the hormone is tetraiodothyronine, or $T_4$:

and if they are iodinated at only three sites, they are triiodothyronine, or $T_3$:

The thyroid usually releases about ten times as much $T_4$ as $T_3$. However, $T_3$ is a much more active hormone than $T_4$ is, so when you read about the effects of thyroxine, keep in mind that the actions discussed are primarily those of $T_3$. The difference in the activities of $T_3$ and $T_4$ makes it possible to control the effects of thyroxine in different tissues. Within target cells, $T_4$ can be converted to $T_3$ by an enzyme called a deiodinase. Another deiodinase can convert $T_4$ into an inactive hormone called reverse $T_3$. Deiodinase can also inactivate $T_3$ by converting it into $T_2$ or $T_1$. Thus each target cell can set a unique sensitivity to thyroid hormones using these enzymes to control the conversion of $T_4$ to $T_3$ or to reverse $T_3$.

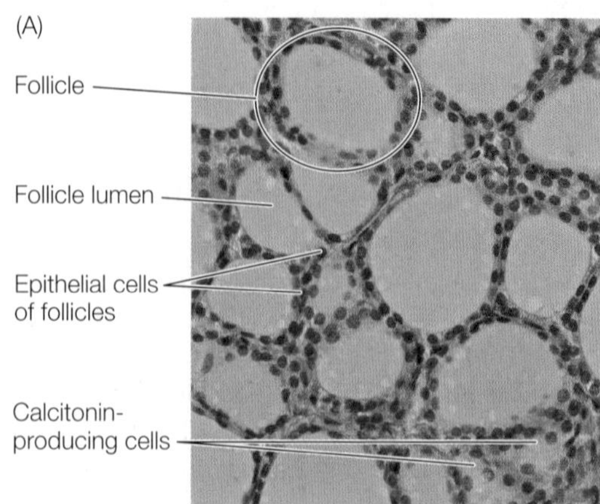


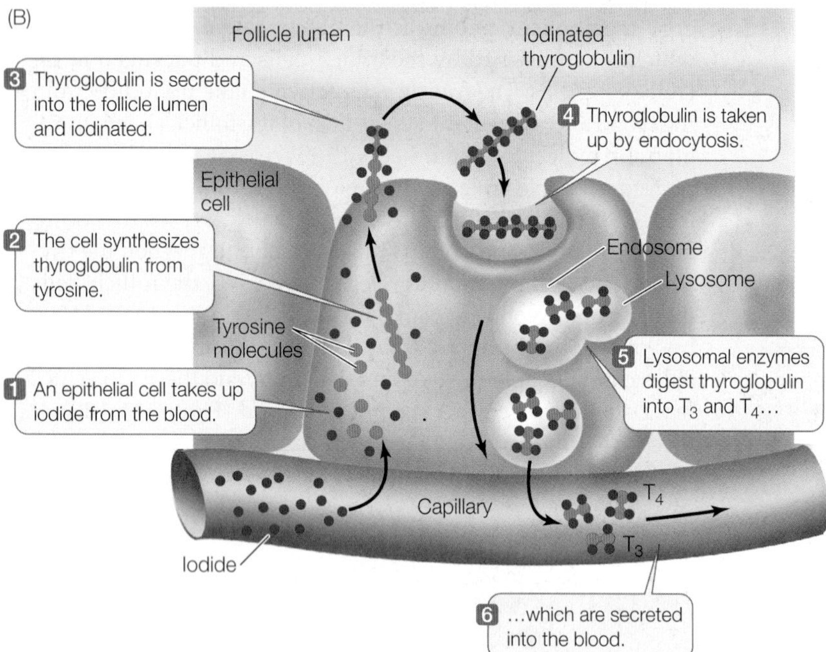


**41.12 The Thyroid Gland Consists of Many Follicles** (A) Cross section through a thyroid gland, showing numerous follicles bounded by epithelial cells. Calcitonin-secreting cells are located in the spaces between the follicles. (B) The epithelial cells of the follicle synthesize thyroglobulin and secrete it into the lumen of the follicle, where it is iodinated and stored until it is processed by the epithelial cells to generate $T_3$ and $T_4$.

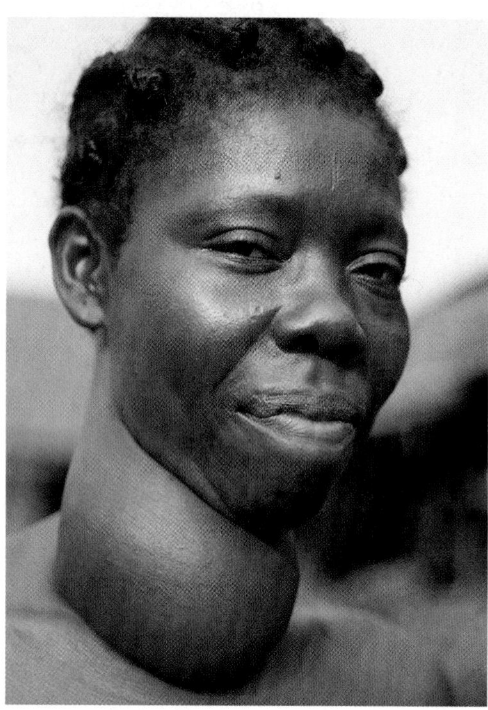

**41.13 A Hypothyroid Goiter** In this condition, dietary iodide deficiency leads to a lack of functional thyroxine, resulting in the oversynthesis of thyroglobulin and subsequent enlarged follicles.

**TSH AND TRH REGULATE THYROXINE PRODUCTION** The tropic hormone **thyroid-stimulating hormone** (**TSH**, also known as **thyrotropin**), produced by the anterior pituitary, activates the thyroxine-producing follicle cells in the thyroid. Thyrotropin-releasing hormone (TRH), produced in the hypothalamus and transported to the anterior pituitary through the portal blood vessels, activates the TSH-producing pituitary cells. The hypothalamus uses environmental information, such as temperature or day length, to determine whether to increase or decrease its secretion of TRH. This sequence of steps is regulated by a negative feedback loop (see Figure 41.11). Circulating thyroxine inhibits the response of pituitary cells to TRH, so less TSH is released when thyroxine levels are high, and more TSH is released when thyroxine levels are low. Circulating thyroxine also exerts negative feedback on the production and release of TRH by the hypothalamus.

Because thyroxine is lipid-soluble, it enters cells readily and binds to receptors in the nucleus. When combined with thyroxine, these receptors (which are found in most cells of the body) stimulate the transcription of numerous genes whose products are transport proteins, structural proteins, and enzymes involved in metabolic pathways; thus thyroxine elevates the metabolic rates of most cells and tissues. Exposure to cold for several days leads to an increased release of thyroxine, an increased conversion of $T_4$ to $T_3$, and therefore an increased basal metabolic rate (see Section 40.5).

During development and growth, thyroxine promotes amino acid uptake and protein synthesis. Insufficient thyroxine in a human fetus or growing child greatly retards physical and mental development, resulting in a condition known as cretinism.

**GOITER** A **goiter** is an enlarged thyroid gland (**Figure 41.13**) that can be associated with either hyperthyroidism (excess production of thyroxine) or hypothyroidism (thyroxine deficiency). The negative feedback loop whereby thyroxine controls TSH release helps explain how two seemingly opposite conditions can result in the same symptom.

- The most common cause of *hyperthyroid* goiter is Graves' disease, an autoimmune disease involving an antibody to the TSH receptor. This antibody binds to and activates the TSH receptors on the follicle cells, causing uncontrolled production and release of thyroxine. Blood levels of TSH are low due to negative feedback from high levels of thyroxine, but the thyroid remains maximally stimulated and grows bigger. People with hyperthyroidism have high metabolic rates, usually feel hot, and may develop a buildup of fat behind the eyeballs that causes their eyes to bulge.

- *Hypothyroid* goiter results when there is not enough circulating thyroxine to turn off TSH production. The most common cause is a deficiency of dietary iodide, without which the follicle cells cannot make thyroxine. Without sufficient thyroxine, TSH levels remain high and the thyroid continues to produce large amounts of thyroglobulin. Because sufficient iodine is not available, however, the thyroglobulin is poorly iodinated. When it is broken down by the follicle cells, it produces little functional $T_3$ or $T_4$. TSH levels remain high and stimulate more and more synthesis of thyroglobulin, and the thyroid gets bigger. The symptoms of hypothyroidism are low metabolism, intolerance of cold, and general physical and mental sluggishness.

Goiter affects about 5 percent of the world's population. The addition of iodide to table salt has greatly reduced the incidence of hypothyroid goiter in industrialized nations, but the condition is still common in the other parts of the world and is a leading cause of intellectual impairment.

## Three hormones regulate blood calcium concentrations

The regulation of calcium concentration in the blood is crucial, and shifts in blood calcium concentration above or below a narrow range can cause serious problems. When blood calcium falls below this range, the nervous system becomes overly excited, resulting in muscle spasms and even seizures. When blood calcium rises above this range, the nervous system becomes depressed and muscles—including the heart—weaken. Regulation of blood calcium is difficult because only about 0.1 percent of the calcium in the body is located in the extracellular fluid. About 1 percent is in cells, and almost 99 percent is in the bones. Therefore the body must maintain a tiny pool of calcium in the blood at a precise concentration, and that tiny pool can be influenced greatly by relatively small shifts in the much larger pools of calcium in the cells and bones.

The body has multiple mechanisms for changing blood calcium levels, including:

- Deposition or absorption of bone
- Excretion or retention of calcium by the kidneys
- Absorption of calcium from the digestive tract

Go to Animated Tutorial 41.3
Hormonal Regulation of Calcium
Life10e.com/at41.3

These mechanisms are controlled by three hormones: calcitonin, parathyroid hormone, and calcitriol (synthesized from vitamin D).

**CALCITONIN REDUCES BLOOD CALCIUM** Calcitonin is released by the thyroid and lowers the concentration of calcium in the blood, mainly by regulating bone turnover (**Figure 41.14**). Bone is continuously remodeled through a dynamic process that involves both resorption of old bone and synthesis of new bone, as we will discuss in Section 48.3. Cells called osteoclasts break down bone and release calcium into the blood, and cells called osteoblasts take up calcium from the blood and deposit it in new bone. Calcitonin decreases the activity of osteoclasts and thereby favors removal of calcium from the blood and its deposition in bone by osteoblasts. The turnover of bone in adult humans is not very high, so calcitonin does not play a major role in calcium homeostasis in adults. It is probably more important in young individuals whose bones are actively growing.

**PARATHYROID HORMONE INCREASES BLOOD CALCIUM** The **parathyroid glands** are four tiny structures embedded in the

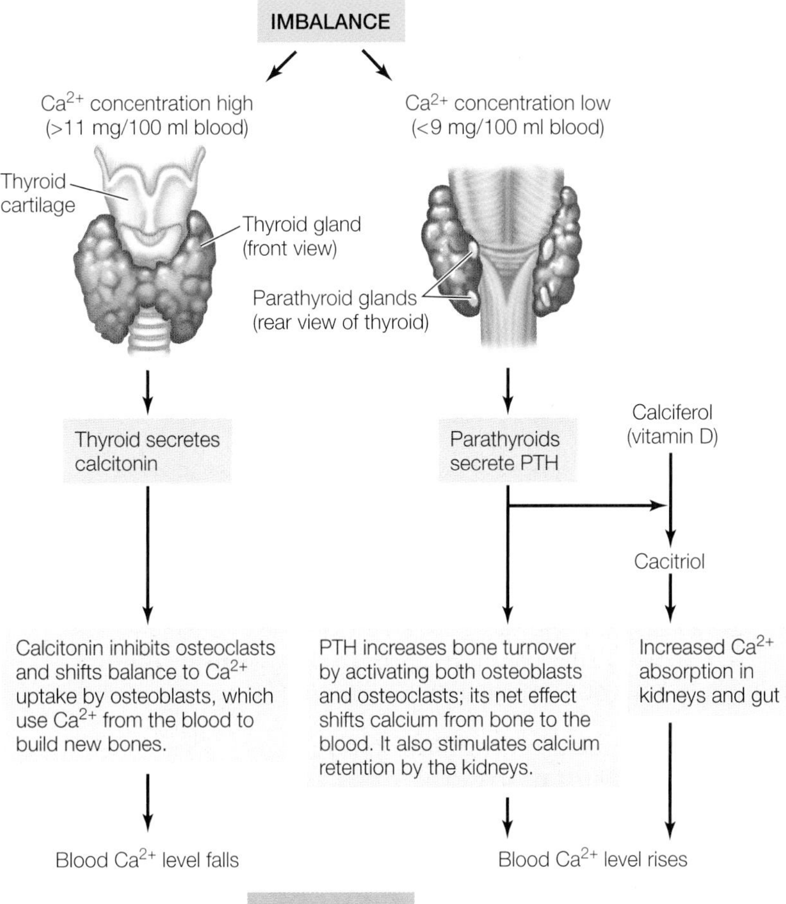


**41.14 Hormonal Regulation of Calcium** Calcitonin, parathyroid hormone (PTH), and calcitriol (the active form of calciferol, or vitamin D) regulate $Ca^{2+}$ levels in the blood.

posterior surface of the thyroid gland (see Figure 41.14). Their single hormone product, **parathyroid hormone** (**PTH**, also called parathormone), is the most important hormone in the regulation of blood calcium levels. Circulating calcium activates receptors in the plasma membrane of the parathyroid cells. When these receptors are active, they inhibit the synthesis and release of PTH. A fall in blood calcium removes this inhibition and triggers the synthesis and release of PTH. PTH stimulates bone turnover by actions on both osteoclasts and osteoblasts. The end result of these actions of PTH is a net increase of calcium in the blood. PTH also raises blood concentration of calcium by stimulating the kidneys to reabsorb it rather than excrete it in the urine.

CALCITRIOL INCREASES BLOOD CALCIUM It had long been known that fragile bones were common among people living at high latitudes, where winter days are short and the winter diet often lacks fish, dairy products, and fresh vegetables. Since the condition could be reversed by taking cod-liver oil, it was assumed that a vitamin deficiency was involved. That vitamin was vitamin D, but when its chemical identity was established, this molecule turned out not to be a vitamin at all.

A vitamin is a substance that the body requires in small quantities but cannot synthesize for itself and must therefore obtain from food (or from supplements such as vitamin pills). However, vitamin D—now more accurately named calciferol—is synthesized naturally from cholesterol when skin cells receive ultraviolet light. Calciferol is not an active hormone, but through actions of the liver and kidneys it is converted into the active form called **calcitriol**, which circulates in the blood and acts on distant cells (and therefore *is* a hormone). The conversion of calciferol to calcitriol is activated by PTH. Calcitriol promotes the absorption of calcium from food in the gut. Thus the combined actions of PTH and calcitriol raise blood calcium levels.

## PTH lowers blood phosphate levels

Bones are made of phosphate as well as calcium, and when PTH stimulates the release of calcium from bone it also releases phosphate. Normal blood concentrations of calcium and phosphate are just below the levels at which they precipitate out of solution as calcium phosphate salts. Even a small rise may cause such precipitation, leading to maladies such as kidney stones and calcium deposits in the arteries (hardening of the arteries). To reduce this risk, PTH acts on the kidneys to increase the elimination of phosphate via the urine.

## Insulin and glucagon regulate blood glucose concentrations

Before the 1920s, the disease diabetes mellitus was fatal. Characterized by weakness, lethargy, and a dramatic loss of body mass, this condition was known to be connected somehow with the pancreas—a large gland located just below the stomach (see Figure 41.8)—and with abnormal glucose metabolism. The exact links, however, were not clear.

Today we know that diabetes mellitus is caused by a lack of the protein hormone **insulin** (in type I diabetes) or by a lack of insulin responsiveness in target tissues (in type II diabetes). Glucose enters cells by diffusion, but cell membranes are not very permeable to glucose. Glucose transporter proteins in cell membranes facilitate the movement of glucose into cells, and the glucose transporters most common in muscle and adipose tissue are controlled by insulin. When insulin binds to its receptor on the cell membrane, it causes these glucose transporters to move from cytoplasmic vesicles to the cell membrane, thus making the cell more permeable to glucose. When insulin is not present, these transporters are returned to the cytoplasmic pool through endocytosis.

In the absence of insulin or insulin responsiveness, glucose entry into cells is impaired, resulting in so much glucose accumulating in the blood that it starts to spill over into the urine. A high concentration of glucose in the blood increases urine output by two mechanisms. First, it causes water to move from cells into the blood by osmosis, and this increase in blood volume results in increased urine production. Second, the increased glucose in the tubules of the kidneys pulls more water into the urine by osmosis. Diabetic individuals thus can become dehydrated, but more importantly they suffer from a lack of metabolic fuel. Because glucose uptake by muscle and adipose tissue is impaired in the absence of insulin, muscle cells must depend on fat and protein for fuel and adipose tissue cannot replenish its stores of triglycerides. If the condition is not treated, the body can waste away.

For centuries, the prospects for people suffering with diabetes were bleak. A change came almost overnight in 1921, when the physician Frederick Banting and a medical student, Charles Best, at the University of Toronto, discovered they could reduce the symptoms of diabetes by injecting an extract prepared from pancreatic tissue. The active component of this extract was found to be insulin, a small protein consisting of just 51 amino acids. In the United States today, insulin replacement therapy using manufactured insulin allows more than 1.5 million people with type I diabetes to lead almost normal lives.

ISLETS OF LANGERHANS Insulin is produced in clusters of endocrine cells in the pancreas. These clusters are called **islets of Langerhans** after the German medical student who discovered them. They contain three types of cells, each of which produces a specific hormone:

- Beta (β) cells produce and secrete insulin.
- Alpha (α) cells produce and secrete **glucagon**, a hormone that has effects mostly opposite from those of insulin.
- Delta (δ) cells produce the hormone **somatostatin**.

The rest of the pancreas is made up of exocrine tissue, which produces enzymes and other secretions that travel through ducts to the gut, where they participate in digestion.

After a meal, the concentration of glucose in the blood rises, stimulating the β cells of the islets to release insulin. Insulin causes target cells throughout the body to use circulating glucose as fuel and convert it into storage products such as glycogen and fat. When the gut is empty of food, blood glucose concentration falls and the islets stop releasing insulin. As a result, most cells shift to using glycogen and fat rather than glucose as fuel. If blood glucose concentration falls substantially below normal, the islet $\alpha$ cells release glucagon, which stimulates the liver to break down stored glycogen and release glucose into the blood. These actions will be discussed in greater detail in Section 51.4.

**SOMATOSTATIN** Somatostatin is released from the $\delta$ cells of the pancreas in response to rapid increases of glucose and amino acids in the blood. This hormone has paracrine functions within the islets, where it inhibits the release of both insulin and glucagon. Outside the pancreas it acts as a hormone, slowing the digestive activities of the gut and extending the period during which nutrients are absorbed. Somatostatin is also produced in very small amounts by cells in the hypothalamus. Hypothalamic somatostatin is transported in the portal blood vessels to the anterior pituitary, where it acts as a neurohormone to inhibit the release of growth hormone and thyrotropin.

## The adrenal gland is two glands in one

An **adrenal gland** sits above each kidney, just below the middle of your back. Functionally and anatomically, each adrenal gland consists of a gland within a gland (**Figure 41.15**). The core, or **adrenal medulla**, produces: **epinephrine** (also known as adrenaline) and, to a lesser degree, **norepinephrine** (noradrenaline). The medulla develops from nervous tissue and is under the control of the nervous system. Surrounding the medulla is the **adrenal cortex**, which produces steroid hormones. The cortex is under hormonal control, largely by corticotropin produced by the anterior pituitary.

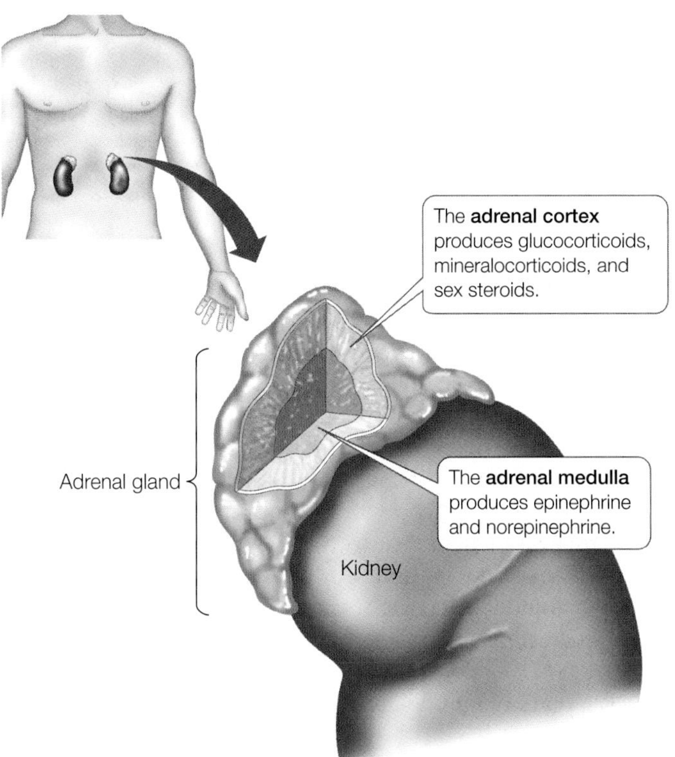


**41.15 The Adrenal Is Really Two Glands** An adrenal gland, consisting of an outer cortex and an inner medulla, sits above each kidney. The medulla and the cortex produce different hormones.

**THE ADRENAL CORTEX** The cells of the adrenal cortex use cholesterol to produce three classes of steroid hormones (see Figure 41.2B), collectively called **corticosteroids**:

- *Mineralocorticoids* influence the salt and water balance of the extracellular fluid.
- *Glucocorticoids* influence blood glucose concentrations as well as other aspects of fat, protein, and carbohydrate metabolism.
- *Sex steroids* play roles in sexual development, sexual behavior, and anabolism (tissue-building).

In adult humans, the adrenal cortex secretes only negligible amounts of sex steroids. The major producers of sex steroids are the gonads, as we will see in the following section.

**Aldosterone**, the primary mineralocorticoid, stimulates the kidneys to conserve sodium and excrete potassium, as we will discuss in Chapter 51. If the adrenal glands are removed from an animal, sodium must be added to its diet, or its sodium will be depleted and it will die.

The main glucocorticoid in humans is **cortisol**, which is critical for mediating the body's metabolic responses to stress. Within minutes of a stressful stimulus (one provoking fear or anger, for example), blood cortisol levels begin to rise. This response is much slower than the neurally mediated epinephrine and norepinephrine response to stress, but it lasts longer. Cells not critical for a sustained stress response are stimulated by cortisol to decrease their use of blood glucose and shift instead to using fats and proteins for energy. Cortisol also inhibits the immune system (because dealing with the immediate stressor is more important than feeling sick, having allergic reactions, or healing wounds). This explains why cortisol and drugs that mimic its action are useful for reducing inflammation and allergic responses.

Cortisol release is controlled from the anterior pituitary by **corticotropin** (also called **adrenocorticotropic hormone**, or **ACTH**), whose release is controlled in turn by **corticotropin-releasing hormone (CRH)** from the hypothalamus. The action of ACTH on the adrenal cortex is to stimulate the synthesis of cortisol. Like other steroid hormones, cortisol is not stored in vesicles and therefore is available for immediate release. As cortisol or other steroid hormones diffuse into the blood, they combine with carrier proteins, and their release from these proteins can have a long time course, thus stretching out their actions. Also, many of their actions stimulate gene expression in target cells, which also takes time but has a long-lasting effect.

Turning off the stress responses activated by cortisol is as important as turning them on. A study of stress in rats showed that old rats could turn on these stress responses as effectively as young rats, but they had lost the ability to turn them off as rapidly. As a result, they suffered from the well-known consequences of stress seen in humans: digestive system problems, cardiovascular problems, strokes, impaired immune system function, and increased susceptibility to cancers and other diseases. Acute stress responses are controlled by negative feedback from cortisol on both the ACTH-secreting cells of the anterior pituitary and CRH-secreting cells of the hypothalamus. With chronic or prolonged stress, these control mechanisms

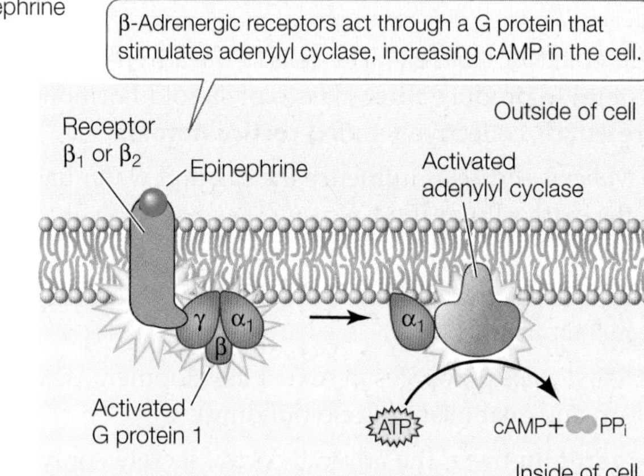


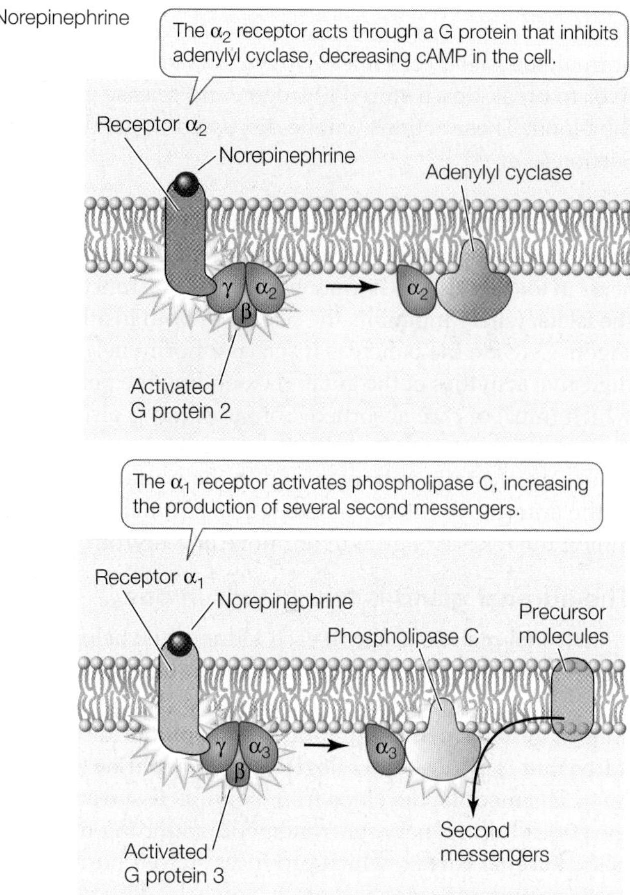


**41.16 Hormones Can Activate a Variety of Signal Transduction Pathways** Epinephrine and norepinephrine bind to G protein-linked adrenergic receptors that act through different signal transduction pathways. Epinephrine (A) acts equally on both α- and β-adrenergic receptors; norepinephrine (B) acts mostly on α-adrenergic receptors.

become insufficient and cortisol must exert negative feedback through another brain region, the hippocampus. Prolonged exposure to cortisol, however, causes trauma to and loss of hippocampal cells, resulting in the decreased ability to turn off the stress response.

**THE ADRENAL MEDULLA** The adrenal medulla produces epinephrine and norepinephrine in response to stressful situations, arousing the body to action. As we saw earlier in this chapter, epinephrine increases heart rate and blood pressure and diverts blood flow to active muscles and away from the gut and skin. Norepinephrine has similar functions, but since it is also a neurotransmitter involved in many physiological regulatory processes, it has many ongoing functions in addition to its involvement in flight-or-fight reactions.

Epinephrine and norepinephrine are both water-soluble, and both bind to the same set of receptors on the surfaces of target cells. These **adrenergic receptors** are of two general types, α-adrenergic and β-adrenergic (**Figure 41.16**). The α-adrenergic receptors respond more strongly to norepinephrine than to epinephrine, whereas β-adrenergic receptors respond about equally to both epinephrine and norepinephrine. Because of this difference in receptor affinities, it is possible for drugs to blunt the flight-or-fight responses without disrupting physiological regulatory processes. Such drugs are called "beta blockers" because, by inhibiting β-adrenergic receptors, they can reduce the fight-or-flight response to epinephrine without disrupting the physiological regulatory functions of norepinephrine mediated through the α-adrenergic receptors. Beta blockers are commonly prescribed to reduce symptoms of anxiety such as dry mouth and elevated heart rate (palpitations).

## Sex steroids are produced by the gonads

The **gonads**—the testes of the male and the ovaries of the female—produce hormones as well as sperm and ova. The male steroid hormones are collectively called **androgens**, and the dominant hormone is testosterone. The female steroids are **estrogens** and **progesterone**. The dominant estrogen is estradiol, which is synthesized from testosterone. Males and females both synthesize testosterone, but females have an enzyme (aromatase) that converts testosterone to estradiol.

Go to Media Clip 41.1
The Testosterone Factor
Life10e.com/mc41.1

**PHENOTYPIC SEX DETERMINATION** The sex steroids determine whether a mammalian embryo develops into a phenotypic female or male. In humans, the gonads of an early embryo are undifferentiated. Beginning in about the seventh week of development, the expression of genes on the Y chromosome of an XY individual normally causes the undifferentiated gonads to produce androgens. In response to androgens, the reproductive system develops the male phenotype. If no Y chromosome is present (i.e., the individual is genotype XX), androgens are not produced at this time and female structures develop (**Figure 41.17**). After birth, the sex steroids control the maturation of the reproductive organs and the development and maintenance of secondary sexual characteristics, such as breasts and facial hair.

**PUBERTY** In humans, the sex steroids are produced at low levels by juvenile gonads, but their production increases rapidly at puberty (around age of 12 or 13). Why does this sudden increase occur? In both juvenile and adult humans, the activities of the gonads are controlled by the tropic hormones luteinizing hormone (LH) and follicle-stimulating hormone (FSH), which

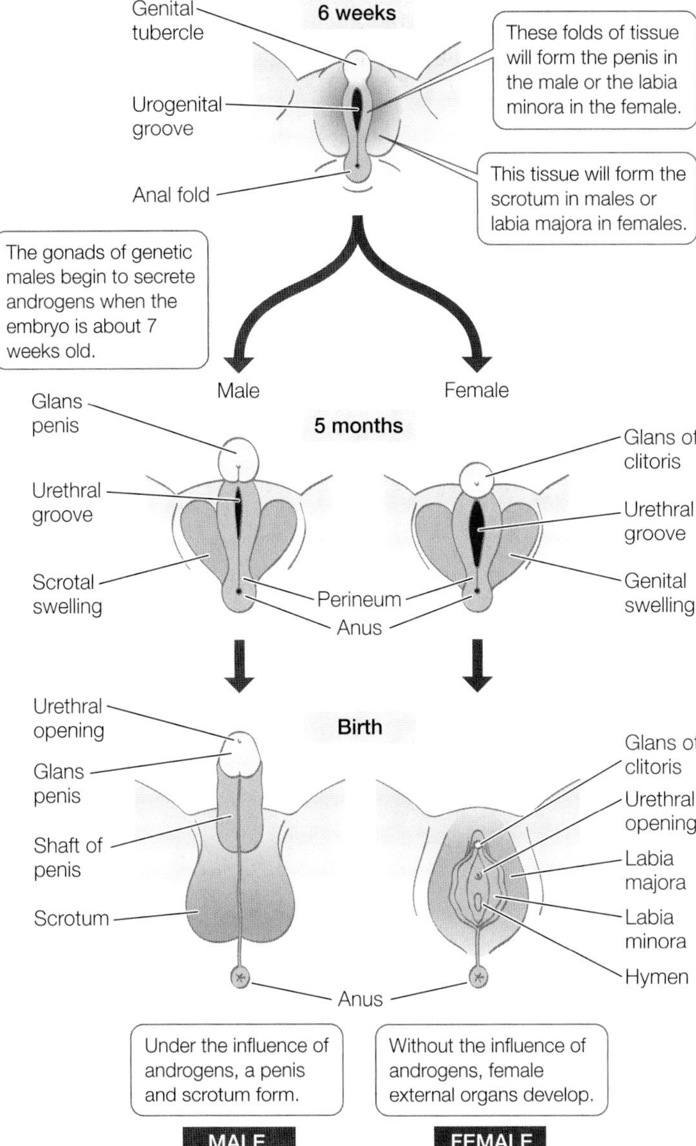


**41.17 Sex Steroids Direct the Development of Human Sex Organs** The sex organs of early human embryos are undifferentiated. Androgens promote the development of male sex organs. In the absence of androgens, female sex organs form.

together are the **gonadotropins**. The production of gonadotropins by the anterior pituitary is under the control of gonadotropin-releasing hormone (GnRH) produced by the hypothalamus. Before puberty, the hypothalamus produces only low levels of GnRH. Puberty is initiated by a reduction in the sensitivity of hypothalamic GnRH-producing cells to negative feedback from sex steroids and from gonadotropins. As a result, GnRH production increases, stimulating increased production of gonadotropins and hence increased production of sex steroids.

In females, increasing levels of LH and FSH at puberty stimulate the ovaries to increase their production of the female sex hormones. The increased circulating levels of these hormones initiate the development of the traits of a sexually mature woman: enlarged breasts, vagina, and uterus; broadened hips; increased subcutaneous fat; pubic hair; and initiation of the menstrual and ovarian cycles (see Figure 43.13). In males, an increasing level of LH stimulates groups of cells in the testes to increase their synthesis of testosterone, which in turn initiates the physiological, anatomical, and psychological changes associated with adolescence. The voice deepens, hair begins to grow on the face and body, and the testes and penis grow larger. Testosterone also stimulates bone and skeletal muscle growth. FSH in males stimulates production of sperm.

The roles that sex steroids play in adult sexual behavior and reproduction will be described in Chapter 43.

## Melatonin is involved in biological rhythms and photoperiodicity

The **pineal gland** is situated between the two hemispheres of the brain and is connected to the brain by a stalk. It synthesizes the amine hormone **melatonin** from the amino acid tryptophan. The pineal gland releases melatonin in the dark, and therefore melatonin levels indicate the length of the night. Exposure to light inhibits the release of melatonin.

In vertebrates, melatonin is involved in biological rhythms, including **photoperiodicity**—the phenomenon whereby seasonal changes in day length cause physiological changes. Many species, for example, come into reproductive condition when the days begin to lengthen (**Figure 41.18**). Humans are not strongly photoperiodic, but melatonin in humans may play a role in synchronizing daily biological rhythms to the daily cycle of light and dark.

## Many chemicals may act as hormones

We have discussed the major mammalian endocrine glands and their hormones in this chapter, but many more hormones

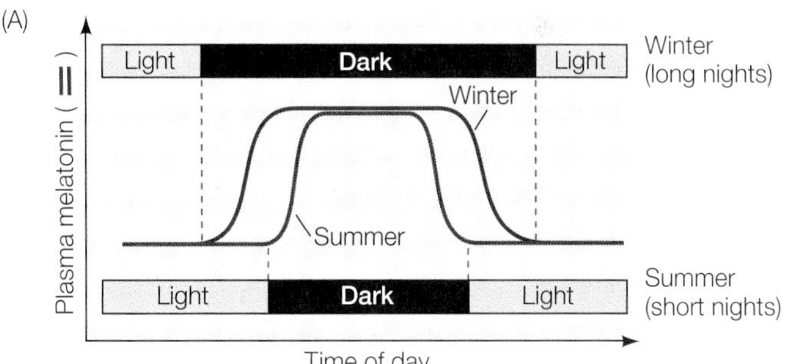


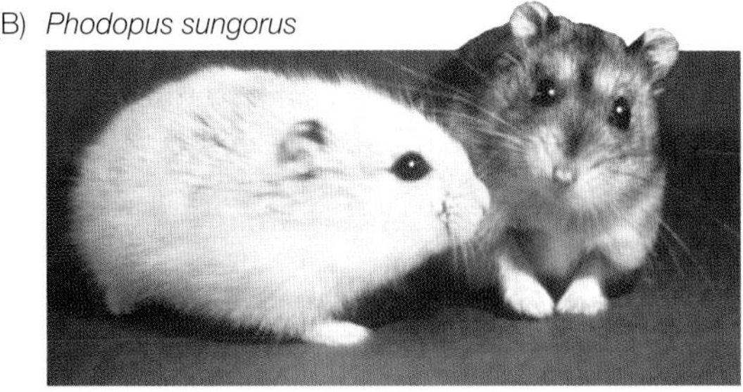


**41.18 Melatonin Regulates Seasonal Changes** (A) Melatonin release occurs in the dark and is inhibited by exposure to light. The duration of daily melatonin release thus changes as day length (photoperiod) changes, inducing dramatic seasonal physiological changes in some animals. (B) In winter, Siberian hamsters are white and do not reproduce. In summer, they are mottled brown and breed.

exist. As we discuss the organ systems of the body in the chapters that follow, we will frequently describe hormones that their tissues produce as well as hormones that control their functions.

### RECAP 41.4

The major endocrine glands of mammals include the hypothalamus, pituitary gland, thyroid gland, parathyroid glands, pancreas, adrenal glands, gonads, and pineal gland. Each of these glands secretes and responds to hormones that play crucial roles in controlling physiology and development.

- Describe how thyroxine is produced and how its production and release are controlled. **See pp. 845–846 and Figure 41.12**
- How is the concentration of calcium in the blood regulated? **See pp. 847–848 and Figure 41.14**
- How does insulin control the rate of glucose uptake by cells? **See p. 848**
- What changes in the feedback control of sex steroids result in puberty? **See pp. 850–851**

Many hormones are released in very small quantities, and some disappear from the extracellular fluid rapidly. A hormone's receptors may be found on diverse cells around the body, and those cells can respond in different ways to the same hormone. How have we overcome these difficulties to learn how hormones work?

RESEARCH**TOOLS**

**41.19 An Immunoassay Allows Measurement of Small Concentrations** An immunoassay uses labeled and unlabeled samples of a purified hormone (or other substance to be measured) and an antibody to that hormone to develop a standard curve against which a sample of unknown concentration can be measured.

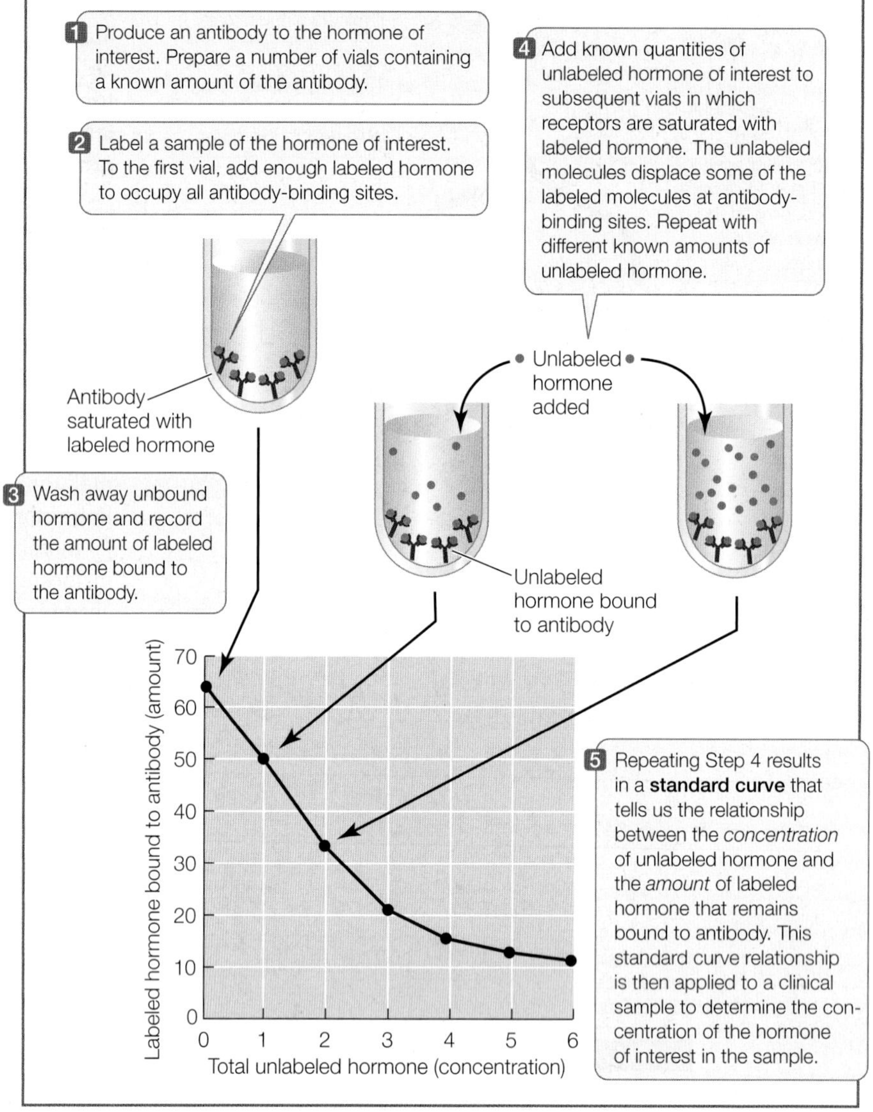

## 41.5 How Do We Study Mechanisms of Hormone Action?

In the current age of molecular biology, we can break the study of hormone actions into different sets of problems. First we must be able to detect, identify, and measure hormones. Next we must be able to identify and characterize hormone receptors. Finally, we must understand the signal transduction pathways activated by hormones in different tissues.

### Hormones can be detected and measured with immunoassays

As we have seen, testosterone has many dramatic and diverse effects, yet its concentration in the blood of adult human males is only about 30 to 100 *billionths* of a gram per milliliter. Measuring hypothalamic neurohormones requires calibrations in the range of *trillionths* of a gram per milliliter.

The ability to detect and measure minute quantities of hormones was an important breakthrough. Rosalyn Yalow developed a method she named radioimmunoassay because it used radioactive labels (she used radioactive isotopes of iodine) to track interactions between an antigen (the hormone or other substance of interest to be measured) and an antibody made to that antigen. Today we are more likely to use nonradioactive labels, so the technique is called simply **immunoassay** (**Figure 41.19**). Being able to measure hormones in the blood made it possible to study many important hormonal mechanisms.

An important characteristic of a hormone is the time course over which it acts. This time course can be measured by the hormone's half-life in the blood (defined as the length of time it takes for one-half of the hormone molecules to disappear). Soon after endocrine cells are stimulated to secrete their hormone, the hormone reaches its maximum concentration in the blood. By examining a series of blood samples using immunoassays, researchers can determine how long it takes for the circulating hormone to drop to half its maximum concentration. The fight-or-flight response to epinephrine is fast both in its onset and its termination; the half-life of epinephrine in the

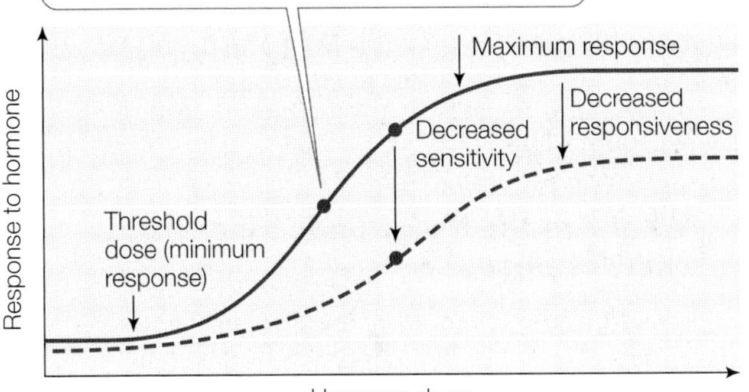


**41.20 Dose–Response Curves Quantify the Body's Response to a Hormone** Between the threshold and maximum doses, a dose–response curve frequently has an S shape. Anything that changes the responsiveness of a system—such as an increase or decrease in the number of receptors in target cells—affects the position of the curve.

blood is 1 to 3 minutes. The effects of other hormones, such as cortisol and thyroxine, are expressed over much longer periods, with half-lives measured in days.

Immunoassays have also facilitated the measurement of dose–response relationships. To evaluate a drug or a natural hormone for therapeutic use, it is critical to know the sensitivity of the body to that drug or hormone. Being able to measure the concentrations of drugs or hormones in the blood makes it possible to construct dose–response curves that help physicians adjust dosages appropriately (**Figure 41.20**).

## A hormone can act through many receptors

Different receptors may be involved in mediating the actions of a single hormone. Because there are slight differences among the receptors for a particular hormone, it is possible to create drugs that are selective in blocking or stimulating specific responses. A number of receptors have been identified, isolated, and purified through biochemical separation techniques. For example, a hormone can be bound to a substrate such as resin beads packed into a glass column. When an extract of cells suspected of containing receptors to that hormone is added to the column, the receptors bind to the hormone molecules on the beads. The hormone–receptor complexes can then be washed off the beads and the receptors isolated. This technique is called **affinity chromatography**.

As more receptors are isolated and characterized, researchers discover that they frequently exist in families with common structural features. These common features result from common nucleotide sequences in the receptor genes. Genomic analyses have led to the discovery of many new receptors. Investigators "scan" the genome for sequences that bear homologies to known receptor gene sequences; when they get a "hit," they have found a candidate gene for a new receptor. They can then identify the molecule to which that receptor binds (its ligand), describe where the receptor occurs within the body, and characterize the receptor's physiological effects.

Knowing the molecular identity of a receptor and being able to measure its concentration makes it possible to study that receptor's regulation. We saw above that the release of hormones can be under negative feedback control. Similarly, the abundance of *receptors* for a hormone can be under feedback control. In some cases, continuous high concentrations of a hormone can decrease the number of its receptors, a process known as **downregulation**. An increased number of receptors can occur when hormone secretion is suppressed, resulting in **upregulation**. The regulation of receptor abundance is an important mechanism controlling the sensitivity of the body to hormonal signaling.

An example of downregulation occurs in type II diabetes, which is characterized by elevated insulin concentrations in the blood and a loss of insulin receptors. Although genetic factors are probably involved, a possible immediate cause of the disease is an overstimulation of pancreatic release of insulin by excessive carbohydrate intake, which leads to downregulation of the insulin receptors. An example of upregulation may be seen in people who have been on a regular dose of beta blockers (see p. 850). As the activity of the β-adrenergic receptors is blocked over time, more of these receptors are produced. If the person goes off the medication suddenly, the effects of the receptors are amplified, resulting in heightened anxiety. Changes in dosage in the long-term use of such medications thus are usually gradual and carefully supervised.

**RECAP 41.5**

Studying the mechanisms of hormone action requires the ability to measure hormone concentrations and to identify and characterize hormone receptors.

- Describe how an immunoassay is performed. **See p. 852 and Figure 41.19**
- How are receptors for a particular hormone identified? **See p. 853**

How can we demonstrate that a molecule found in the blood is a hormone?

**ANSWER**

In order for a molecule to be considered a hormone, we have to show that the molecule is released from cells in a certain tissue and that it has an effect on cells in another, distant tissue. We also have to show that the molecule is *necessary* to stimulate its presumed target cells and that it is *sufficient* to stimulate those target cells. In the experiment shown in Figure 41.6, when the bugs were decapitated soon after a blood meal, molting did not occur, but when they were decapitated a week after a blood meal, molting did occur. These results showed that some substance from the head, but not the head itself, was necessary for the molting response. Sufficiency was demonstrated by connecting two bugs with a glass tube allowing diffusion of substances between them. In this experiment, both bugs molted. In the irisin experiments (Figure 41.5), it was shown that the molecule came from muscle cells and acted on fat cells; that the molecule could stimulate "browning" of white fat cells in culture; (sufficiency); and that when the molecule was inactivated it blocked this "browning" of the white fat (necessity).

## CHAPTER SUMMARY 41

### 41.1 What Are Hormones and How Do They Work?

- **Endocrine cells** secrete chemical signals that induce responses in other cells that have receptors for those molecules. In some cases endocrine cells are aggregated into **endocrine glands**.
- **Hormones** are endocrine signals that are secreted from a cell, circulate in the blood, and bind to target cells distant from the secreting cell. **Review Figure 41.1**
- Hormones fall into three general categories: proteins and peptides, steroids, and amines. Peptide and protein hormones and some amines are water-soluble; steroids and some amines are lipid-soluble. **Review Figure 41.2**
- Receptors for water-soluble hormones are located on the cell surface. Receptors for most lipid-soluble hormones are inside the cell.
- Hormones can cause different responses in different target cells. **Review Figure 41.3**

### 41.2 What Have Experiments Revealed about Hormones and Their Action?

- The chemical structures of hormones are highly conserved. Through evolution, however, hormones acquire different functions in different animal groups. **Review Figure 41.4**
- Early experiments identifying secretin as a hormone defined the characteristics of hormonal signaling. Modern experiments demonstrate these characteristics in order to identify hormone molecules. **Review Figure 41.5**
- Pioneering experiments illustrating hormonal action showed that two hormones, PTTH and ecdysone, control molting in arthropods. A third hormone, **juvenile hormone**, prevents maturation. **Review Figures 41.6, 41.7, ANIMATED TUTORIAL 41.1**

### 41.3 How Do the Nervous and Endocrine Systems Interact?

- In humans, the major endocrine glands are distributed around the body. **Review Figure 41.8, ACTIVITY 41.1**
- The **pituitary gland** is the interface between the nervous and endocrine systems. The **anterior pituitary** develops from embryonic mouth tissue; the **posterior pituitary** develops from the developing brain. **Review Figures 41.9, 41.10**
- The posterior pituitary secretes two **neurohormones**: **antidiuretic hormone** (**ADH**) and **oxytocin**. The anterior pituitary secretes **tropic hormones** (thyrotropin, corticotropin, luteinizing hormone, and follicle-stimulating hormone) as well as **growth hormone**, **prolactin**, **endorphins**, and **enkephalins**.
- The anterior pituitary is controlled by neurohormones produced by cells in the hypothalamus and transported through **portal blood vessels** to the anterior pituitary. **See ANIMATED TUTORIAL 41.2**
- Hormone release is controlled in part by negative feedback loops. **Review Figure 41.11**

### 41.4 What Are the Major Endocrine Glands and Hormones?

- The **thyroid gland** is controlled by **thyrotropin** and secretes **thyroxine**, which controls cell metabolism. **Review Figure 41.12**
- The level of calcium in the blood is regulated by three hormones. **Calcitonin** from the thyroid lowers blood calcium by promoting bone deposition. **Parathyroid hormone** (**PTH**) raises blood calcium by promoting bone turnover and decreasing calcium excretion. **Calcitriol** promotes calcium absorption from the digestive tract. **Review Figure 41.14, ANIMATED TUTORIAL 41.3**
- The pancreas secretes three hormones. **Insulin** stimulates glucose uptake by cells and lowers blood glucose, **glucagon** raises blood glucose, and **somatostatin** slows the rate of nutrient processing.
- The **adrenal gland** has two portions, one within the other. The inner portion, the **adrenal medulla**, releases **epinephrine** and **norepinephrine** in response to stress. The outer portion, the **adrenal cortex,** produces three classes of **corticosteroids**: glucocorticoids, mineralocorticoids, and small amounts of sex steroids. **Review Figure 41.15**
- **Aldosterone** is a mineralocorticoid that stimulates the kidneys to conserve sodium and excrete potassium. **Cortisol** is a glucocorticoid that is released in response to stressful stimuli but acts more slowly than the hormones of the adrenal medulla.
- Sex hormones (**androgens** in males, **estrogens** and **progesterone** in females) control sexual development, secondary sexual characteristics, and reproductive functions. **Review Figure 41.17**
- The **pineal gland** releases **melatonin**, a hormone involved in controlling biological rhythms. **Review Figure 41.18**

### 41.5 How Do We Study Mechanisms of Hormone Action?

- **Immunoassay** techniques are used to measure concentrations of hormones and other substances. **Review Figure 41.19**
- The body's sensitivity to a hormone is measured by a dose–response curve. **Review Figure 41.20**
- The sensitivity of a cell to a hormone can be altered by **downregulation** or **upregulation** of the hormone's receptors in that cell.

**See ACTIVITY 41.2 for a concept review of this chapter**

**Go to the Interactive Summary to review key figures, Animated Tutorials, and Activities Life10e.com/is41**

# CHAPTER**REVIEW**

## REMEMBERING

1. Prior to puberty
   a. the pituitary secretes luteinizing hormone and follicle-stimulating hormone, but the gonads are unresponsive.
   b. the hypothalamus does not secrete much gonadotropin-releasing hormone.
   c. males can stimulate massive muscle development through a vigorous training program.
   d. testosterone plays no role in development of the male sex organs.
   e. genetic females will develop male genitals unless estrogen is present.

2. Both epinephrine and cortisol are secreted in response to stress. Which of the following statements is also true for *both* of these hormones?
   a. They act to increase blood glucose availability.
   b. Their receptors are on the surfaces of target cells.
   c. They are secreted by the adrenal cortex.
   d. Their secretion is stimulated by corticotropin.
   e. They are secreted into the blood within seconds of the onset of stress.

3. The posterior pituitary
   a. synthesizes oxytocin.
   b. is under the control of hypothalamic releasing hormones.
   c. secretes tropic hormones.
   d. secretes neurohormones.
   e. is under feedback control by thyroxine.

4. PTH
   a. stimulates osteoblasts to lay down new bone.
   b. reduces blood calcium levels.
   c. stimulates calcitonin release.
   d. is produced by the thyroid gland.
   e. is released when blood calcium levels fall.

5. Steroid hormones
   a. are produced only by the adrenal cortex.
   b. have only cell-surface receptors.
   c. are water-soluble.
   d. act by altering the activity of proteins in the target cell.
   e. act by altering gene expression in the target cell.

## UNDERSTANDING & APPLYING

6. Compare the characteristics you would expect of a hormone signaling system that controls a short-term process, such as digestion, with the characteristics you would expect of a hormone signaling system that controls a long-term process, such as embryonic development.

7. Some body builders, males and females, take high doses of synthetic male steroid hormones to enhance the growth of their muscles. Among other side effects, the ovarian cycles of the females stop and the males become sterile. Explain these consequences in terms of the hypothalamic/pituitary/gonadal hormonal axis.

8. Explain how both hyperthyroidism and hypothyroidism can both be associated with goiter.

## ANALYZING & EVALUATING

9. Honey bees build honeycombs from beeswax that they produce. After extracting honey from the combs, beekeepers store the combs until they are needed again as the bees build up their honey reserves. Wax moths lay their eggs in honeycombs. The larvae eat the wax as they go through cycles of molting and growth until they metamorphose into adult moths. A beekeeper decided to store his honeycombs in a cold room. Months later when he went to get the spare combs out of the cold room, he found them full of large holes and gigantic moth larvae, but no adult moths. Hypothesize as to what aspect of the moth endocrine system was altered by the low temperature.

10. Neurons (the cells of the nervous system) do not require insulin. Why might this be so, and why is it important?

Go to BioPortal at **yourBioPortal.com** for Animated Tutorials, Activities, LearningCurve Quizzes, Flashcards, and many other study and review resources.

# 42 Immunology: Animal Defense Systems

CHAPTER**OUTLINE**


**42.1 What Are the Major Defense Systems of Animals?**

**42.2 What Are the Characteristics of the Innate Defenses?**

**42.3 How Does Adaptive Immunity Develop?**

**42.4 What Is the Humoral Immune Response?**

**42.5 What Is the Cellular Immune Response?**

**42.6 What Happens When the Immune System Malfunctions?**

**Fighting the Pox** This 1905 illustration depicts Parisians being vaccinated against smallpox with serum from a cow infected with cowpox, a disease similar to smallpox that causes mild symptoms in humans.

TWO TINY VIALS IN DEEP FREEZERS, one in Atlanta and the other in Siberia, are all that is left of the smallpox virus, long a scourge of humanity. It last occurred as a human pathogen in 1978, after killing more than 300 million people in the twentieth century alone. Smallpox was eliminated by the human immune system with the help of vaccination. A vaccine is usually an inactive form of a pathogen or toxin that nevertheless provokes the immune system to produce antibodies: specific proteins directed against the target. The immune system destroys whatever is bound to the antibodies.

The eradication of smallpox was a spectacular international accomplishment. One might think that vaccination for other potentially lethal diseases such as the flu (influenza) would be widely accepted by the public. But in the U.S. and some other Western countries, this is not necessarily so. In fact, some surveys show that more than one-third of Americans refuse flu shots, and a significant number of parents refuse vaccination for their children.

Unfortunately, those who refuse vaccination may harm people other than themselves. A vaccination program can control or eradicate a disease only if a high percentage (typically above 80 percent) of people are vaccinated, thus disrupting the chain of infection from person to person. This level of vaccination results in "herd immunity," meaning that even those who cannot be vaccinated or who have weak immune systems are protected from infection. This protection is lost if the vaccination rate falls below the level needed for herd immunity. Those who are old or sick, and infants whose immune systems have not yet fully developed, are most at risk.

The best way to develop herd immunity in a population is compulsory vaccination. In many countries, vaccination is a prerequisite for school enrollment and military enlistment and is required during epidemics. For example, in the periodic smallpox epidemics during the twentieth century in the U.S., doctors accompanied by police would go into neighborhoods where the disease raged, vaccinating all those who were uninfected and removing infected people to quarantine. You can imagine the reaction of parents whose children were taken away, often to die. When Henning Jacobson, a Swedish immigrant in Massachusetts, refused vaccination during a smallpox epidemic, he was arrested and took his case to the U.S. Supreme Court. It ruled in 1905 that while personal freedom is important, each state was entitled to protect its citizens. This provided a legal framework for compulsory vaccination that continues to this day. But opposition and court challenges continue. These cases of "freedom v. immunity" are a political dilemma.

Why do many people resist vaccination?

See answer on p. 877.

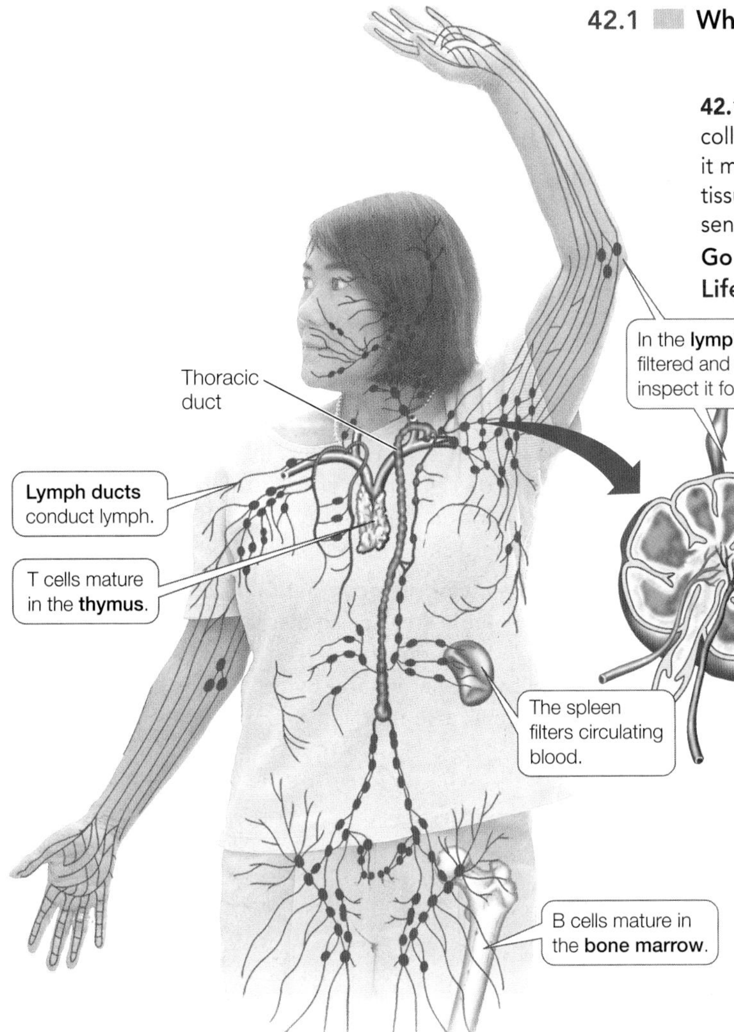


**42.1 The Human Lymphatic System** A network of ducts and vessels collects lymph from body tissues and carries it toward the heart, where it mixes with blood to be pumped back to the tissues. Other lymphoid tissues, including the thymus, spleen, and bone marrow, are also essential to the body's defense system.

**Go to Activity 42.1 The Human Defense System Life10e.com/ac42.1**

## 42.1 What Are the Major Defense Systems of Animals?

Animals have several ways of defending themselves against **pathogens**—harmful organisms and viruses that can cause disease. These defense systems are based on the distinction between *self*—the animal's own molecules—and *nonself*, or foreign, molecules. The defensive response involves three phases:

- *Recognition phase*. The organism must be able to discriminate between self and nonself.
- *Activation phase*. The recognition event leads to a mobilization of cells and molecules to fight the invader.
- *Effector phase*. The mobilized cells and molecules destroy the invader.

There are two general types of defense mechanisms:

- **Innate defenses**, or nonspecific defenses, provide the first line of defense against pathogens. They typically act very rapidly and include barriers such as the skin, molecules that are toxic to invaders, and phagocytic cells that ingest invaders. (Recall from Section 6.5 that phagocytosis is a form of endocytosis, in which a cell engulfs a large particle or another cell.) This system recognizes broad classes of organisms or molecules and responds quickly, within minutes or hours. All animals have some level of innate immunity, and certain innate immunity mechanisms are shared between animals and plants.
- **Adaptive defenses** are aimed at specific pathogens and are activated by the innate immune system. For example, an adaptive defense system can make an antibody protein that will recognize, bind to, and aid in the destruction of a specific virus, if that virus ever enters the body. Adaptive defenses are typically slower to develop than the innate response and longer-lasting. Adaptive immunity evolved in vertebrate animals.

Mammals have both kinds of defense mechanism and are the focus of this chapter. In mammals and other vertebrates, the innate and adaptive mechanisms operate together as a coordinated defense system. **Table 42.1** gives an overview of these processes during the course of an infection. Innate immunity is the body's first line of defense because the adaptive defenses often require days or even weeks to become effective.

### Blood and lymph tissues play important roles in defense

The components of the mammalian defense system are dispersed throughout the body and interact with almost all of its other tissues and organs. The lymphoid tissues, which include the thymus, bone marrow, spleen, and lymph nodes, are essential parts of the defense system (**Figure 42.1**). The blood and lymph are complex systems with nondefensive functions that will be discussed in Chapter 50. They each have central roles in defense as well.

The blood and lymph both consist of liquids in which cells are suspended:

- **Blood plasma** is a yellowish solution containing ions, small molecule solutes, and soluble proteins. Suspended in the plasma are red blood cells, white blood cells, and platelets (cell fragments essential to blood clotting). Whereas red blood cells are normally confined to the closed circulatory system (the heart, arteries, capillaries, and veins), white blood cells and platelets are also found in the lymph.

TABLE 42.1
Innate and Adaptive Immune Responses to an Infection

| Response (Time after Infection by a Pathogen) | System | Mechanisms |
|---|---|---|
| Early (0–4 hr) | Innate, nonspecific (first line) | Barrier (skin and lining of organs)<br>Dryness, low pH<br>Mucus<br>Lysozyme, defensins |
| Middle (>4–96 hr) | Innate, nonspecific (second line) | Inflammation<br>Phagocytosis<br>Natural killer cells<br>Complement system<br>Interferons |
| Late (>96 hr) | Adaptive, specific | Humoral immunity (antibodies from B cells)<br>Cellular immunity (T cells) |

- **Lymph** is a fluid that is derived from the blood (but lacking red blood cells) and other tissues and accumulates in intercellular spaces throughout the body. From these spaces, the lymph moves slowly into the vessels of the lymphatic system. Tiny lymph capillaries conduct this fluid to larger ducts that eventually join together, forming one large vessel, the thoracic duct, which joins a major vein (the left subclavian vein) near the heart. By this system of vessels, the lymph is eventually returned to the blood and the circulatory system.

At many sites along the lymph vessels are small, roundish structures called **lymph nodes**, which contain a type of white blood cell called a **lymphocyte**. As lymph passes through a lymph node, the lymphocytes encounter foreign cells and molecules that have entered the body, and if they are recognized as nonself, an immune response is initiated.

## White blood cells play many defensive roles

One milliliter of human blood typically contains about *5 billion* red blood cells and *7 million* of the larger **white blood cells** (also called leukocytes). All of these cells originate from multipotent stem cells (constantly dividing undifferentiated cells that can form several different cell types; see Section 19.2) in the bone marrow. There are two major families of white blood cells: lymphocytes and phagocytes (**Figure 42.2**). Lymphocytes include the B cells and T cells; they are smaller than other white blood cells and are not phagocytic. **Phagocytes** include most of the other cells shown in Figure 42.2, and as their name suggests, they are phagocytic. Each kind of white blood cell has specialized functions. Some phagocytes are also referred to collectively as granulocytes because they contain numerous granules (vesicles containing defensive molecules). Defensive proteins and signals play fundamental roles in the interactions and functioning of these cells.

## Immune system proteins bind pathogens or signal other cells

The cells that defend mammalian bodies work together, interacting with one another and with the cells of invading pathogens. These cell–cell interactions are accomplished by a variety of key proteins, including receptors, other cell surface proteins, and signaling molecules. Four of the major players are listed here, and will be discussed in more detail later in the chapter.

- **Antibodies** are proteins that bind specifically to certain substances identified by the immune system as nonself. They recognize and bind specific configurations of atoms. The molecules that bind antibodies are called **antigens**. This binding can directly inactivate viruses and toxins; on nonself cells, antibody–antigen complexes can act as tags, making the cells easier for the immune system cells to recognize and attack. Antibodies are produced by B cells.
- **Major histocompatibility complex** (**MHC**) proteins are used to display antigens on the surfaces of self cells, so that the antigens can be detected by antibodies and by cells of the immune system. MHC proteins also function as important self-identifying

| TYPE OF CELL | FUNCTION |
|---|---|
| Basophils (I, A) | Release histamine; may promote development of T cells |
| Eosinophils (A) | Kill antibody-coated parasites |
| Neutrophils (I) | Stimulate inflammation; engulf and digest microorganisms |
| Mast cells (I) | Release histamine when damaged |
| Monocytes (I, A) | Develop into macrophages and dendritic cells |
| Macrophages (I, A) | Engulf and digest microorganisms; activate T cells |
| Dendritic cells (A) | Present antigens to T cells |
| Natural killer cells (I) | Attack and lyse virus-infected or cancerous body cells |
| B lymphocytes (A) | Differentiate to form antibody-producing cells and memory cells |
| T lymphocytes (A) | Kill virus-infected cells or cancer cells; regulate activities of other white blood cells |

**42.2 White Blood Cells** White blood cells have key roles in both innate (I) and adaptive (A) immunity. The lymphocytes are the B cells and T cells; the other cell types are phagocytes.

**Go to Animated Tutorial 42.1**
**Cells of the Immune System**
Life10e.com/at42.1

labels. There are two major classes of MHC proteins: MHC I proteins are found on the surfaces of most cells in the mammalian body, whereas MHC II proteins are found on immune system cells.

- **T cell receptors** are integral membrane proteins on the surfaces of T cells. They recognize and bind to antigens presented by the MHC proteins on the surfaces of other cells.
- **Cytokines** are soluble signaling proteins released by many cell types. They bind to cell surface receptors and alter the behavior of their target cells. Various cytokines activate or inactivate B cells, macrophages (see Figure 42.1), and T cells.

**RECAP 42.1**

All animals have innate defenses against pathogens, and vertebrates have innate and adaptive defenses. Both kinds of mechanisms are based on the ability to differentiate self from nonself. Innate defenses target a broad range of molecules and organisms, whereas adaptive defenses target specific pathogens.

- List the differences between innate and adaptive defenses. **See p. 857**
- What are the two classes of white blood cells, and how do they function in innate or adaptive immunity? **See p. 858 and Figure 42.2**

The outcome of a disease—the life or death of the host—often depends on the success of both rapid, innate responses and long-lasting, adaptive responses to invading pathogens. We will turn now to the innate defenses that protect vertebrates from disease.

## 42.2 What Are the Characteristics of the Innate Defenses?

Innate defenses are general protection mechanisms that attempt to stop pathogens from invading the body or to quickly eliminate those that do manage to invade. They are genetically programmed (innate) and "ready to go," in contrast to adaptive responses, which take time to develop after a pathogen or toxin has been recognized as nonself. In mammals, innate defenses include physical barriers as well as cellular and chemical defenses (**Figure 42.3**).

### Barriers and local agents defend the body against invaders

The first line of innate defense is encountered by a potential pathogen as soon as it lands on the surface of an animal. Consider a pathogenic bacterium that lands on human skin. The challenges faced by the bacterium just to reach its target are formidable:

- The *physical barrier of the skin*: Bacteria rarely penetrate intact skin; by the same token, broken skin increases the risk of infection.
- The *saltiness and dryness of skin*: This environment may not be hospitable to the growth of the bacterium.
- The *presence of normal flora*: Bacteria and fungi that normally live and sometimes reproduce in great numbers on our body surfaces without causing disease will compete with pathogens for space and nutrients.

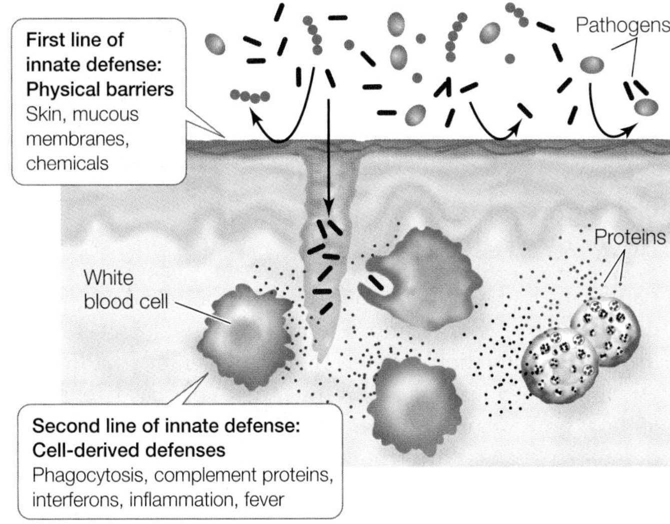


**42.3 Innate Immunity** Physical barriers, cells, and proteins (complement and interferons) provide nonspecific defenses against invading pathogens.

**Go to Media Clip 42.1**
**The Chase Is On: Phagocyte versus Bacteria**
**Life10e.com/mc42.1**

If a pathogen lands inside the nose or another internal organ, it faces other innate defenses:

- **Mucus** is a slippery secretion produced by mucous membranes found at the inner surfaces of the nose (as well as the digestive, respiratory, and urogenital systems). Mucus traps microorganisms so they can be removed by the beating of cilia (see Figure 5.17), which continuously move the mucus and its trapped debris away.
- **Lysozyme** is an enzyme made by mucous membranes that attacks the cell walls of many bacteria, causing them to lyse (burst open).
- **Defensins**, also made by mucous membranes, are peptides of 18 to 45 amino acids that contain hydrophobic domains. They are toxic to a wide range of pathogens, including bacteria, microbial eukaryotes, and enveloped viruses. Defensins insert themselves into the plasma membranes of these organisms and make the membranes permeable, thus killing the invaders. Defensins are also produced in phagocytes, where they kill pathogens ingested by phagocytosis. Plants also produce defensins in response to pathogen exposure (see Chapter 39).

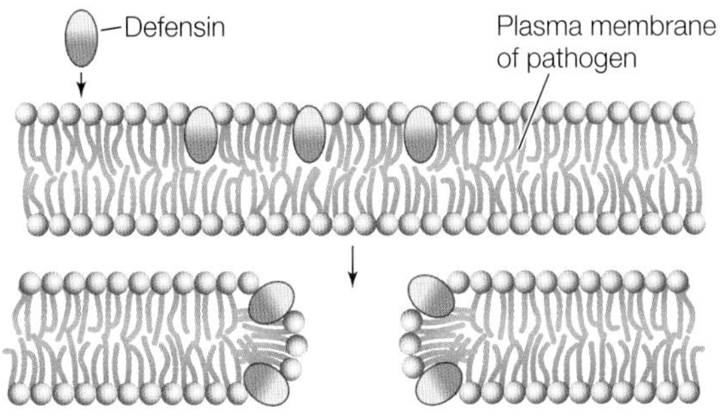

- Harsh conditions in an animal's internal environment can also kill pathogens. For example, gastric juice in the stomach is a deadly environment for many bacteria because of the hydrochloric acid and proteases that are secreted into it.

## Cell signaling pathways stimulate the body's defenses

Pathogens that are able to penetrate the body's outer and inner surfaces encounter more complex innate defenses. These include the activation of defensive cells. Over the past 15 years, much has been learned about the innate cellular defense systems of both plants and animals. A critical feature is that the defense responses are triggered by nonself rather than self molecules. A class of receptors called **pattern recognition receptors** (**PRRs**) plays an important role in distinguishing self from nonself.

PRRs are present in cells that play roles in the innate immune systems of both plants and animals. In mammals these include macrophages, dendritic cells, and natural killer cells. The molecules recognized by PRRs are called **pathogen associated molecular patterns** (**PAMPs**). As we described in Chapter 39, these are molecules that are unique to large classes of microbes, such as bacterial lipopolysaccharides, which are found in bacterial cell membranes.

An invading pathogen can be regarded as a signal. In response to that signal, the body produces molecules (complement proteins, interferons, and other cytokines) that regulate phagocytosis and other defense processes. Not surprisingly, the link between signal and response is a signal transduction pathway, similar to the ones we considered in Section 7.3. A key group of PRRs in mammals is the **toll-like receptors**, which activate signal transduction pathways involved in both innate and adaptive defenses (**Figure 42.4**). The toll protein was first identified in insects, where it is involved in development and in sensing infection. Comparative genomics has revealed at least ten similar receptors in humans. Bruce Beutler, the scientist who first described these receptors, won the Nobel Prize in 2011. Binding of a PAMP to the receptor sets in motion a cascade of molecular changes, including the activation of the transcription factor NF-κB. (NF-κB stands for *n*uclear *f*actor *kappa* light chain enhancer of activated *B* cells.) The activated NF-κB enters the nucleus, where it activates the transcription of genes encoding defensive proteins.

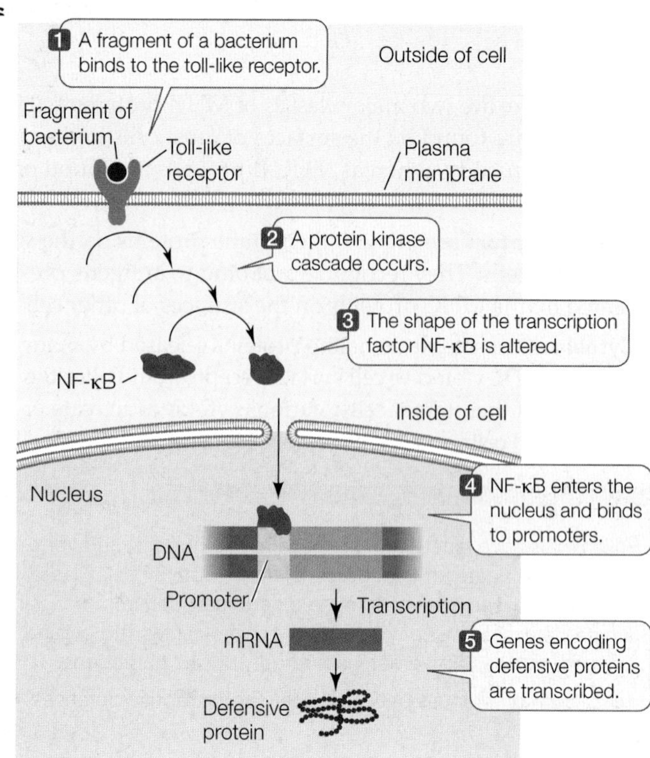


**42.4 Cell Signaling and Defense** Binding of a pathogenic molecule or fragment to the toll-like receptor initiates a signal transduction pathway that results in the transcription of genes whose products are involved in adaptive and innate defenses.

## Specialized proteins and cells participate in innate immunity

Several proteins are produced by the body either before an infection occurs or in response to invasion by pathogens. Two important groups are the complement and interferon proteins.

**COMPLEMENT PROTEINS** Vertebrate blood contains more than 20 different proteins that make up the antimicrobial **complement system**. This system can be activated by various mechanisms, including both innate and adaptive defense responses. The proteins act in a characteristic sequence, or cascade, with each protein activating the next:

- First, the proteins attach to specific components on the surface of a microbe or to an antibody that has already bound to the microbe's surface. In either case, binding helps phagocytes recognize and destroy the microbe.
- Then, complement proteins activate the inflammatory response (see next page) and attract phagocytes to the site of infection.
- Finally, complement proteins lyse invading cells (such as bacteria).

**INTERFERONS** When a cell is infected by a pathogen, it produces small amounts of signaling proteins called **interferons** that increase the resistance of neighboring cells to infection. Interferons are a class of cytokines and have been found in many vertebrates. Various molecules, including double-stranded (viral) RNA, induce the production of interferons. Thus interferons are particularly important as a first line of defense against viruses. Interferons bind to receptors on the plasma membranes of uninfected cells, stimulating a signaling pathway that inhibits viral reproduction if the cells are subsequently infected. In addition, interferons stimulate the cells to hydrolyze bacterial or viral proteins to peptides, an initial step in adaptive immunity (see Section 42.3).

**PHAGOCYTES** Some phagocytes travel freely in the circulatory and lymphatic systems; others can move out of blood vessels and adhere to certain tissues. Pathogenic cells, viruses, or fragments of these invaders are recognized by phagocytes, which then ingest them by phagocytosis. Defensins, nitric oxide, and

**42.5 Interactions of Cells and Chemical Signals Result in Inflammation** Histamine and other signals are released from mast cells to initiate the inflammatory response. The chemical signals associated with inflammation attract phagocytes, which digest the pathogens and damaged cells.
**Go to Activity 42.2 Inflammatory Response**
**Life10e.com/ac42.2**

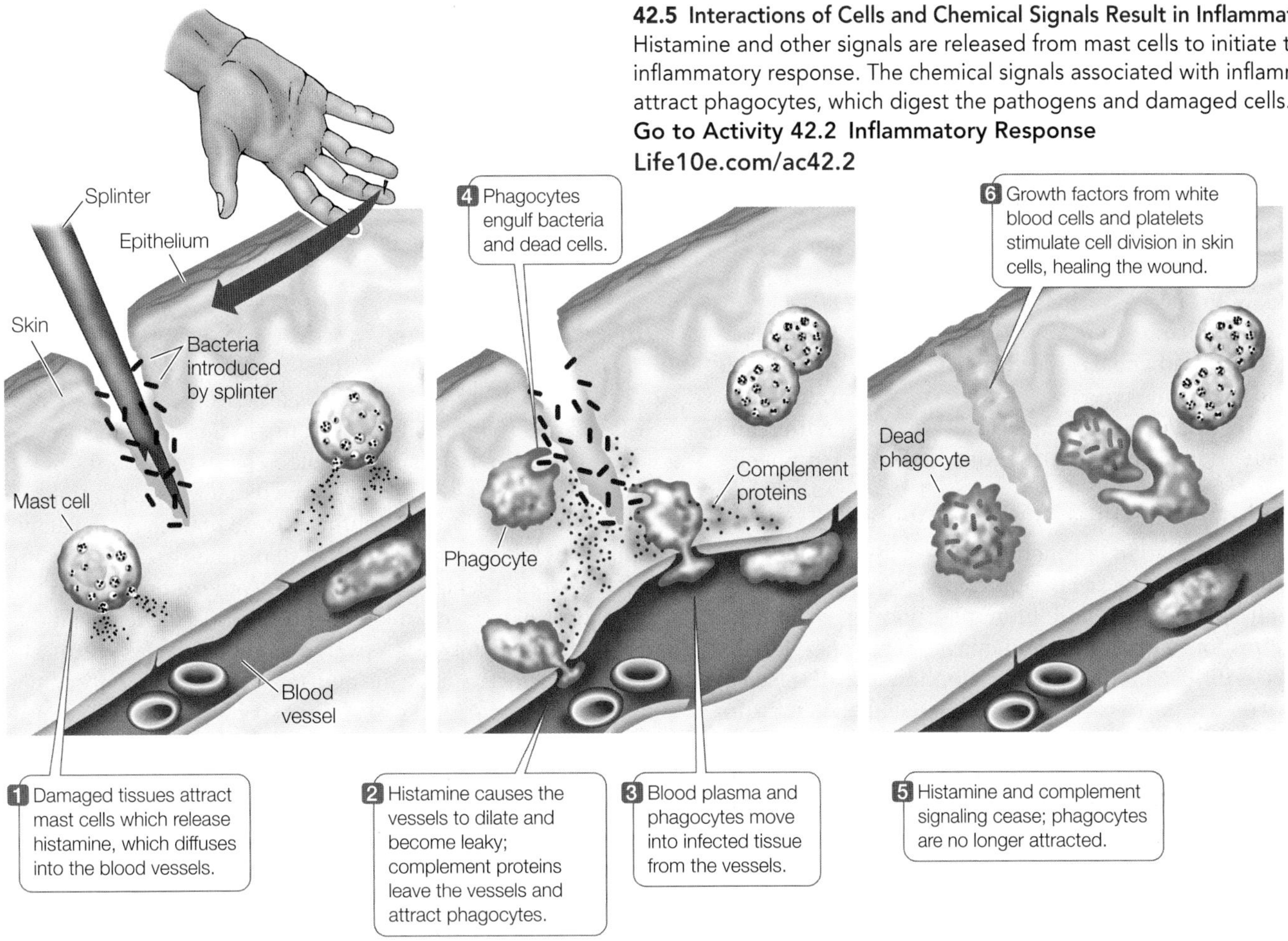

reactive oxygen intermediates (see also Section 39.1) inside these phagocytes then kill the pathogens.

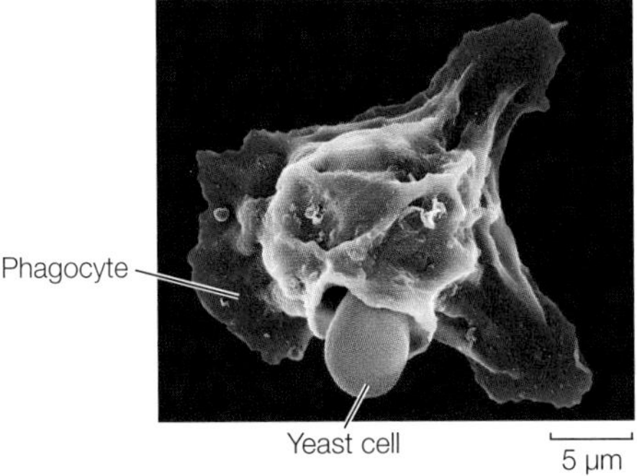


NATURAL KILLER CELLS One class of lymphocytes, known as **natural killer cells**, can distinguish virus-infected cells and some tumor cells from their normal counterparts and initiate the apoptosis of these target cells. In addition to this innate defense action, natural killer cells interact with the adaptive defense mechanisms by lysing antibody-labeled target cells.

DENDRITIC CELLS These phagocytes act as messengers between the innate and adaptive systems. They can endocytose microbes, viruses, and even virus-infected host cells. Once inside a dendritic cell, these particles are digested to fragments, and if fragments have PAMPs, the dendritic cell "presents" an antigenic fragment on its surface, along with class II MHC proteins. In addition, the dendritic cell secretes signals that activate cells of the adaptive immune system.

## Inflammation is a coordinated response to infection or injury

When mammalian tissue is damaged because of infection or injury, the body responds with **inflammation**. This response can happen almost anywhere in the body, internally as well as on the surface. Inflammation is an important phenomenon: it isolates the damaged area to stop the spread of the damage; it recruits cells and molecules to the damaged location to kill the invader; and it promotes healing. The first responders to tissue damage are **mast cells**, which adhere to the skin and the linings of organs and release numerous chemical signals, including:

- **Tumor necrosis factor**, a cytokine protein that kills target cells and activates immune cells.
- **Prostaglandins**, fatty acid derivatives involved in various responses, including the widening of blood vessels. Prostaglandins interact with nerve endings and are partly responsible for the pain caused by inflammation.
- **Histamine**, an amino acid derivative that leads to itchy, watery eyes and rashes seen with some types of allergic reactions.

The redness and heat of inflammation result from the dilation and leakiness of blood vessels in the infected or injured area (**Figure 42.5**). Phagocytes enter the inflamed area, where they engulf the invaders and dead tissue cells. Phagocytes are responsible for most of the healing associated with inflammation. They produce several cytokines, which (among other functions) can signal the brain to produce a fever. This rise in body temperature accelerates lymphocyte production and phagocytosis, thereby speeding the immune response. In some cases, pathogens are temperature-sensitive and their growth is inhibited. The pain of inflammation results from increased pressure due to swelling, the action of leaked enzymes on nerve endings, and the action of prostaglandins, which increase the sensitivity of the nerve endings to pain.

Following inflammation, pus may accumulate. Pus is a mixture of leaked fluid and dead cells: bacteria, neutrophils (the most abundant white blood cells—see Figure 42.2), and damaged body cells. Pus is a normal result of inflammation and is gradually consumed and further digested by macrophages.

### Inflammation can cause medical problems

Although inflammation is generally a good thing, sometimes the inflammatory response is inappropriately strong, resulting in some allergies, cases of autoimmunity, and sepsis. In these cases the response causes more damage than was originally there. We will discuss allergy and autoimmune diseases in Section 42.6. In some cases of severe bacterial infection, the inflammatory response does not remain local. Instead it extends throughout the bloodstream in a condition called sepsis. As in a local infection or injury, blood vessels dilate, but they do so throughout the body. The lowering of blood pressure that results is a medical emergency and can be lethal.

The symptoms of swelling, pain, and fever caused by excessive inflammation can be bothersome to the point of incapacitation. Diseases such as rheumatoid arthritis and chronic obstructive pulmonary disease, and accidents such as athletic injuries, result in tissue damage and an inflammatory response. In order to manage excessive inflammation, drugs have been developed that act on the various cytokines and signal transduction pathways to reduce inflammation and its symptoms. For example, aspirin works by inhibiting an enzyme in the pathway for the synthesis of prostaglandins. Other anti-inflammatory drugs act on the prostaglandin pathway, on the actions of tumor necrosis factor, and on the actions of histamine.

**RECAP 42.2**

Innate immunity is the first line of defense against pathogens. Innate immunity includes physical barriers such as the skin, and cellular responses involving the recognition of self and nonself molecules. Recognition of nonself molecules by white blood cells leads to coordinated responses such as the production of defensive proteins and inflammation.

- How do complement proteins and interferons defend the body against microbes? **See p. 860**
- What are the roles of pathogen-associated patterns (PAMPs) and pattern recognition receptors (PRRs) in innate defenses? **See p. 860 and Figure 42.4**
- Describe the inflammatory response. **See pp. 861–862 and Figure 42.5**

Often the innate immune system, with its nonspecific defenses, is adequate to prevent or fight off a pathogenic infection. But in many cases this system works together with adaptive immunity, which detects and responds to specific pathogens. We will now turn to the development and functioning of adaptive immunity.

## 42.3 How Does Adaptive Immunity Develop?

Before the twentieth century, scientists had long suspected that blood was somehow involved in immunity against pathogens. More than a century ago, Emil von Behring and Shibasaburo Kitasato at the University of Marburg in Germany performed a key experiment that pointed to blood as an important factor in immunity (**Figure 42.6**). They showed that guinea pigs injected with a sublethal dose of diphtheria toxin or bacteria developed in their blood serum (the noncellular fluid that remains after blood is clotted) a factor that protected other guinea pigs from a lethal dose of the same toxin. In other words, the recipients had developed **immunity**. Moreover, the immunity was specific: the immune factor made by the guinea pigs protected only against the specific toxin, from one strain of diphtheria-causing bacteria, with which they had been injected.

In this section we outline the main features of the adaptive immune system, much of which does indeed occur in blood serum. We will consider the two major types of adaptive responses: the humoral immune response, which produces antibodies; and the cellular immune response, which destroys infected cells.

### Adaptive immunity has four key features

Four important features of the adaptive immune system are:

- specificity
- the ability to distinguish self from nonself
- the ability to respond to an enormous diversity of nonself molecules
- immunological memory

**SPECIFICITY** Lymphocytes (B and T cells) are crucial components of adaptive immunity. T cell receptors and the antibodies produced by B cells recognize and bind to specific nonself substances (antigens), and this interaction initiates an adaptive immune response. The specific sites on antigens that the immune system recognizes are called **antigenic determinants**, or **epitopes**:

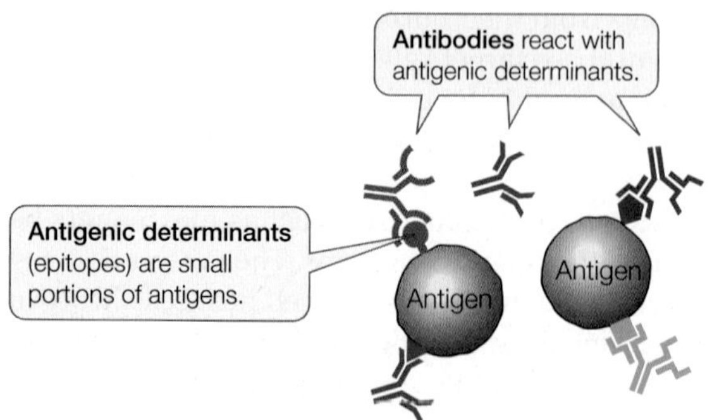


An antigenic determinant is a specific portion of a large molecule, such as a certain sequence of amino acids that may be present in a protein. Antigens are usually proteins or

INVESTIGATING**LIFE**

**42.6 The Discovery of Adaptive Immunity** Until the twentieth century, most people did not survive an attack of the bacterium that causes diphtheria, but a few did. Emil von Behring and Shibasaburo Kitasato performed a key experiment using an animal model, and demonstrated that the factor(s) responsible for immunity against diphtheria were in blood serum.[a]

**HYPOTHESIS** Serum from guinea pigs injected with a sublethal dose of diphtheria toxin protects other guinea pigs that are exposed to a lethal dose of the same toxin.

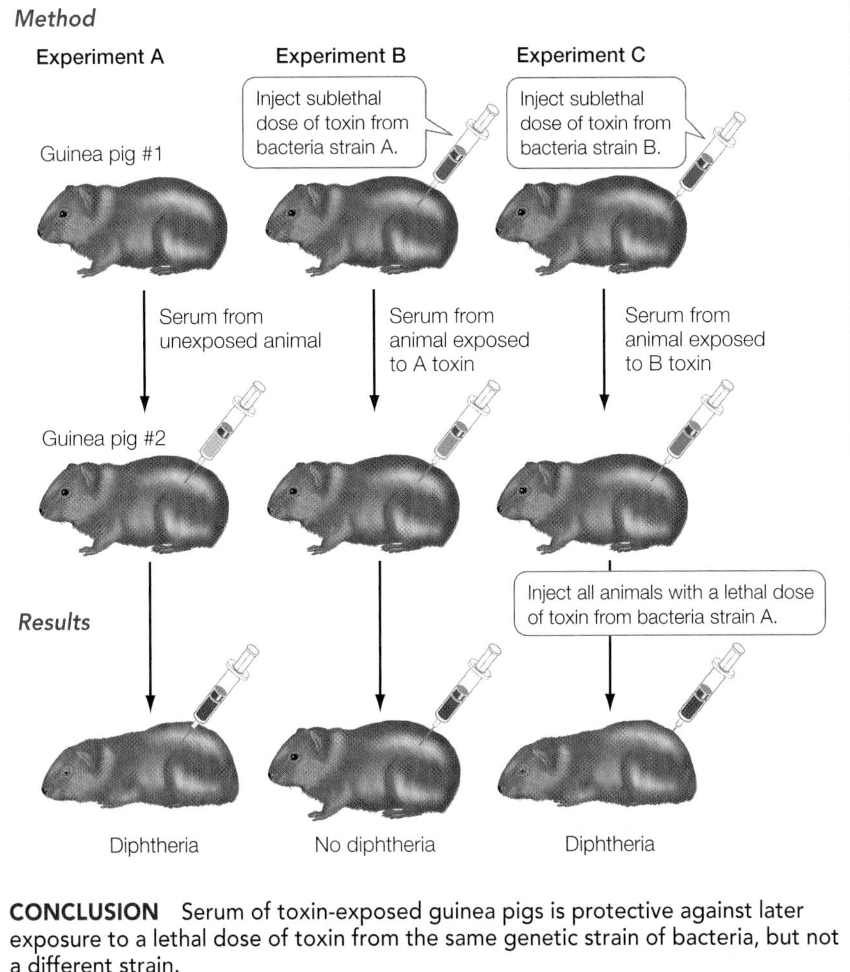


**CONCLUSION** Serum of toxin-exposed guinea pigs is protective against later exposure to a lethal dose of toxin from the same genetic strain of bacteria, but not a different strain.

Go to **BioPortal** for discussion and relevant links for all INVESTIGATING**LIFE** figures.

[a]Behring, E. and S. Kitasato. 1890. *Deustche medizinische Wochenschrift* 16: 1113–1114.

polysaccharides, and there can be multiple antigens on a single invading bacterium. A single antigenic molecule can have multiple, different antigenic determinants. The host animal responds to the presence of an antigen with highly specific defenses involving T cell receptors and antibodies. These receptors and soluble proteins bind to the antigenic determinants. Each T cell and each antibody is specific for a single antigenic determinant. For the remainder of the chapter, we will refer to antigenic determinants simply as "antigens."

DISTINGUISHING SELF FROM NONSELF We have seen how the innate immune system distinguishes between self and nonself molecules. The adaptive immune system has another set of mechanisms for distinguishing self from nonself. The human body contains tens of thousands of different proteins, each with a specific three-dimensional structure capable of generating immune responses. Thus every cell in the body bears a tremendous number of antigens. A crucial requirement of an individual's adaptive immune system is that it recognize the body's own antigens and not attack them. This is accomplished by clonal deletion, negative selection, and the action of Treg cells (we will discuss these mechanisms later in the chapter.)

DIVERSITY Challenges to the immune system are numerous. Pathogens take many forms: viruses, bacteria, protists, fungi, and multicellular parasites. Furthermore, each pathogenic species usually exists as many subtly different genetic strains, and each strain possesses multiple surface features. Estimates vary, but a reasonable guess is that humans can respond specifically to 10 million different antigens. Upon recognizing an antigen, the adaptive immune system responds by activating lymphocytes of the appropriate specificity. This capacity is accomplished by a special genetic recombination mechanism that we will describe in Section 42.4.

IMMUNOLOGICAL MEMORY After the innate immune system responds to a particular type of pathogen once, the adaptive immune system "remembers" that pathogen and can usually respond more rapidly and powerfully to the same threat in the future. This **immunological memory** usually saves us from repeats of childhood diseases such as chicken pox. Vaccination against specific diseases works because the adaptive immune system "remembers" the antigens that were introduced into the body.

All four of these features of adaptive immune defense characterize both the humoral immune response and the cellular immune response.

## Two types of adaptive immune responses interact: an overview

The adaptive immune system mounts two types of responses against invaders: the humoral immune response and the cellular immune response. B cells that make antibodies are the workhorses of the humoral immune response, and cytotoxic (killer) T cells are the workhorses of the cellular immune response. These two responses operate simultaneously and cooperatively, sharing many mechanisms. A key event early in these two processes is the exposure of the nonself antigen's three-dimensional structure to the immune system. This occurs

## WORKING WITH**DATA**:

### *The Discovery of Adaptive Immunity*

#### Original Paper

Behring, E. and S. Kitasato. 1890. Uber das Zustandekommen der Diptherie-Immunitat und der Tetanus-Immunitat bel thieren. *Deustche medizinische Wochenschrift* 16: 1113–1114. In *Milestones in Microbiology: 1556–1940*, translated and edited by T. D. Brock, 1998, p. 138. ASM Press, Washington, D.C.

#### Analyze the Data

Until the early twentieth century, diphtheria, an infectious disease called the "strangling angel" for the way it disrupts breathing, exacted a terrible death toll, particularly on children. In the late 1800s a team of scientists led by Robert Koch, a leader in the then new field of microbiology, focused on diseases such as diphtheria and the equally lethal tetanus. Both diseases are caused by bacteria that kill their victims by secreting toxins into the bloodstream. A member of Koch's team from Japan, Shibasaburo Kitasato isolated the bacteria that cause diphtheria, and a recent medical school graduate, Emil Behring, set out to see how the disease could be treated. They discovered that animals injected with the disease-causing bacteria had substances in their blood that could prevent productive infections in naïve (previously unexposed) animals and even increase survival of animals already infected (see Figure 42.6). They called these substances anti-toxins (we know them as antibodies) and soon found that they were not species-specific: anti-toxin from a horse could be used to treat people. This and similar work on tetanus led quickly to the development of anti-diphtheria and anti-tetanus medicines. Behring shared the first Nobel Prize for Medicine in 1901.

In their experiments, Behring and Kitasato used different doses of toxin to test the immunity of guinea pig #2 in each experiment (see Figure 42.6). The results are shown in the table.

**QUESTION 1**

What can you conclude from these data in terms of the level of protection afforded by the serum?

**QUESTION 2**

These experiments could be performed with the same results with either intact bacteria that cause diphtheria or a bacteria-free filtrate of a 10-day-old culture of the bacteria. Explain.

| | Symptoms | |
|---|---|---|
| Experiment | 0.5 ml dose | 10 ml dose |
| A | Diphtheria | Diphtheria |
| B | No diphtheria | No diphtheria |
| C | Diphtheria | Diphtheria |

Go to BioPortal for all WORKING WITHDATA exercises

when an antigenic molecule, or a fragment of the molecule, is displayed on the surface of a cell and the unique epitope structure protrudes from the cell, where it is exposed to nearby T or B cells. Cells that can "present" the antigen to the immune system in this way are collectively referred to as **antigen-presenting cells**. Dendritic cells play a key role as antigen-presenting cells, although other cells can also perform this function. **Figure 42.7** provides a simplified overview of antigen presentation and the roles of T and B cells in the adaptive immune response.

The key player integrating the humoral and cellular immune responses is the T-helper ($T_H$) cell. By binding to the antigen on a presenting cell, the $T_H$ cell stimulates events in both responses.

HUMORAL IMMUNE RESPONSE In the **humoral immune response** (from the Latin *humor*, "fluid"), antibodies react with antigens on pathogens in blood, lymph, and tissue fluids. An animal can produce a staggering diversity of antibodies capable of binding to almost any conceivable antigen the animal encounters. Antibodies are secreted by B cells and travel freely in the blood and lymph. A particular B cell also possesses receptors on its surface with the same specificity as the antibodies it produces.

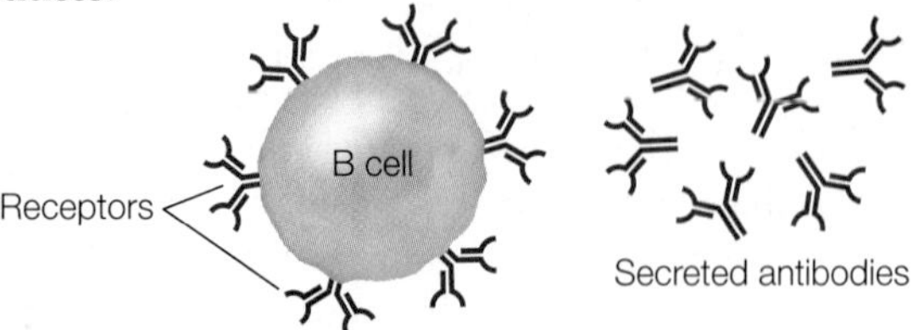


The first time a specific antigen invades the body, it may be presented and then detected by binding to a T cell receptor. This binding activates a B cell with the appropriate antibody; this cell proliferates, and its daughter cells make and secrete multiple copies of the antibody.

Go to Animated Tutorial 42.2
Pregnancy Test
Life10e.com/at42.2

CELLULAR IMMUNE RESPONSE The **cellular immune response** is directed against antigens that have become established within a cell of the host animal. It detects and destroys virus-infected or mutated cells, such as cancer cells expressing unique proteins caused by mutations.

T cells in the lymph nodes, bloodstream, and intercellular spaces carry out the cellular immune response. These T cells have integral membrane proteins—T cell receptors—that recognize and bind to antigens. T cell receptors are rather similar to antibodies in structure and function, each including specific molecular configurations that bind to specific antigens. Once a T cell is bound to an antigen, it initiates an immune response that typically results in the total destruction of the antigen-containing cell.

T cell
T cell receptor

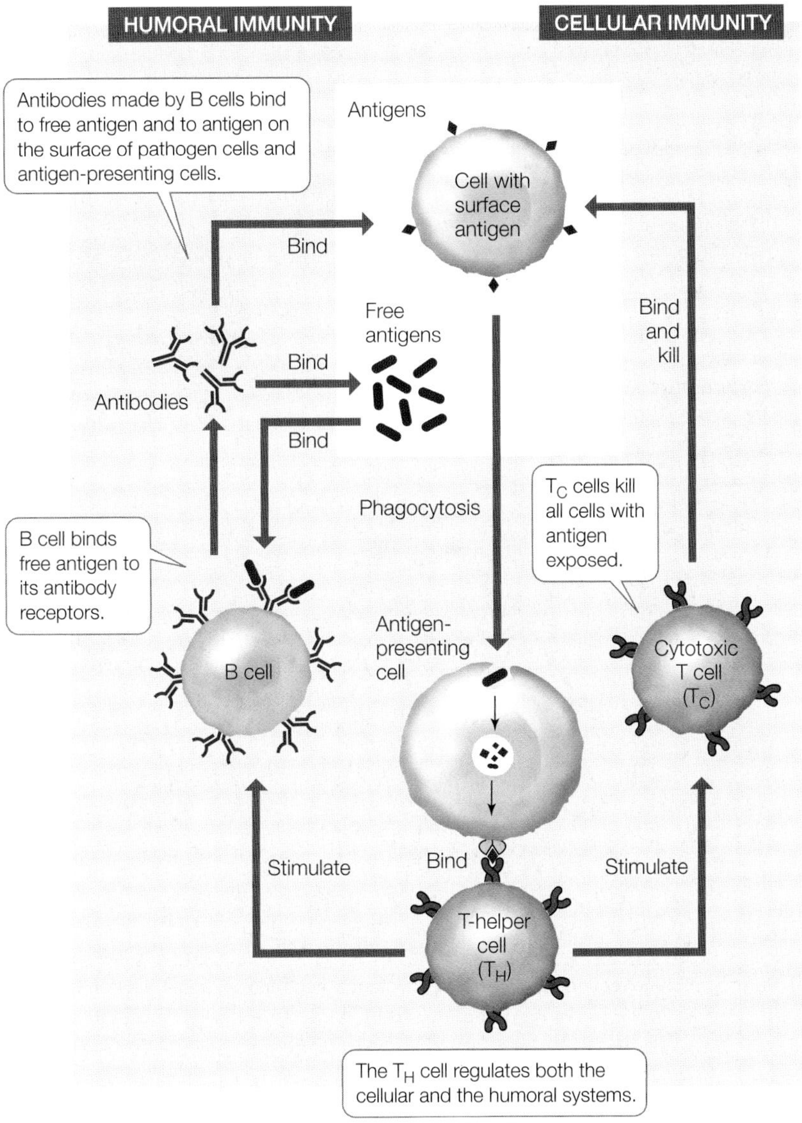

**42.7 The Adaptive Immune System** Humoral immunity involves the production of antibodies by B cells. Cellular immunity involves the activation of cytotoxic T cells that bind to cells expressing the antigen. For further details, see Figure 42.16.

## Adaptive immunity develops as a result of clonal selection

Before the reactions just described for the humoral and cellular immune responses can take place, the body needs to generate a vast diversity of lymphocytes that have the ability to bind different antigens. How does this tremendous diversity arise? As we will discuss in Section 42.4, this diversity is generated primarily by DNA changes—chromosomal rearrangements and other mutations—that occur just after the B and T cells are formed in the bone marrow. Millions of different B cells develop, each of which can produce only one kind of antibody. Similarly, there are millions of different T cells, each with one specific kind of T cell receptor. Thus the adaptive immune system is "predeveloped"—*all of the machinery available to respond to an immense diversity of antigens is already there, even before the antigens are ever encountered.*

As we have described, when a pathogen enters the vertebrate body it stimulates the innate immune system. In addition to triggering its own defensive responses, the innate immune system triggers adaptive defensive responses via specific antigens that are presented on the surfaces of antigen-presenting cells, particularly dendritic cells. This triggers the proliferation of lymphocytes that are specific for those particular antigens. How does this proliferation occur? The answer lies in the process of **clonal selection**: *antigen binding "selects" a particular B or T cell for proliferation*. When an antigen fits the surface receptor on a B or T cell and binds to it, that cell is activated. It divides to form a clone of cells (a genetically identical group derived from a single cell), all of which recognize and react to the same antigen. This process is illustrated for B cells in **Figure 42.8**. Binding and activation select a particular lymphocyte, while proliferation generates the clone, hence the term "clonal selection."

## Clonal deletion helps the immune system distinguish self from nonself

Normally, the body is tolerant of its own molecules—the same molecules that would generate an immune response in another individual. One way that the immune system does this is through the process of **clonal deletion**. This occurs primarily in the thymus, during the early differentiation of T and B cells, when these cells encounter self antigens. Any immature B or T cell that shows the potential to mount an immune response against self antigens undergoes programmed cell death (apoptosis) within a short time.

## Immunological memory results in a secondary immune response

The first time a vertebrate animal is exposed to a particular antigen there is a time lag (usually several days) before the B cell–produced antibody molecules and T cells specific to that antigen slowly increase. But for years afterward—sometimes for life—the immune system "remembers" that particular antigen, allowing the body to mount a faster response the next time it encounters the antigen. How does this happen?

The answer lies in the fact that activated lymphocytes divide and differentiate to produce *two types* of daughter cells: effector cells and memory cells.

- **Effector cells** carry out the attack on the antigen. Effector B cells, called **plasma cells**, secrete antibodies. Effector T cells release cytokines and other molecules that initiate reactions that destroy nonself or altered cells. Effector cells live only a few days.
- **Memory cells** (see Figure 42.8) are long-lived cells that retain the ability to start dividing on short notice to produce more effector and more memory cells. Memory B and T cells may survive in the body for decades, rarely dividing.

These two types of lymphocytes can respond to an antigen in two different ways:

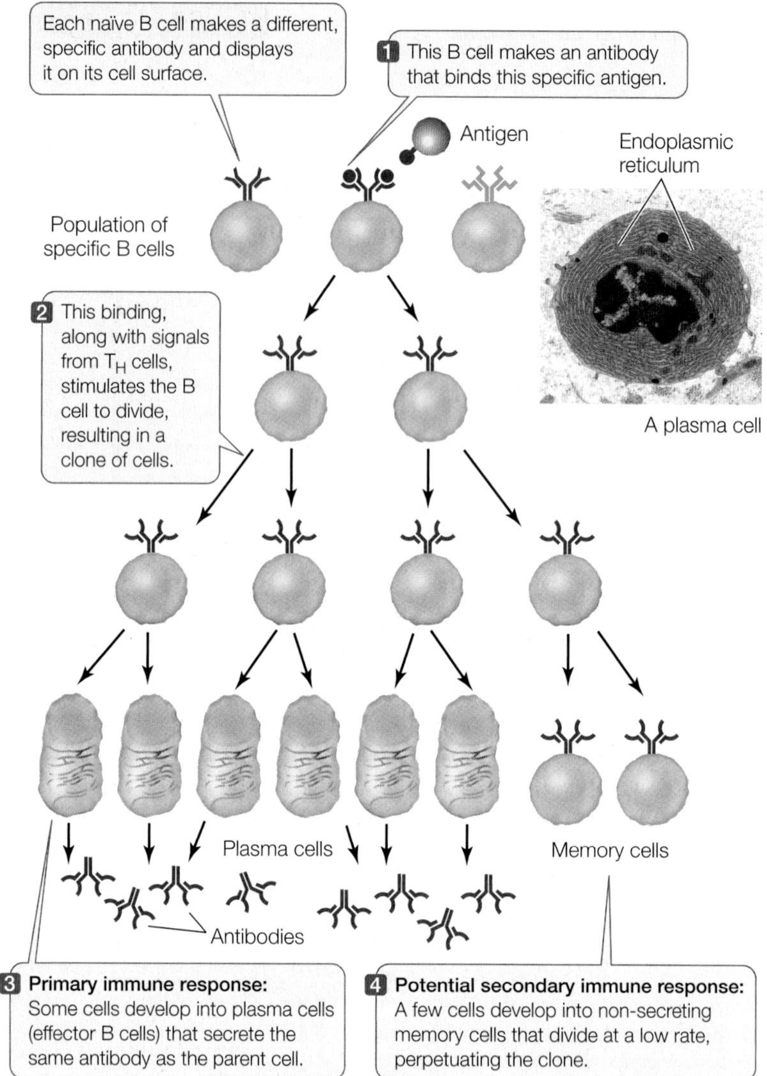


**42.8 Clonal Selection in B Cells** The binding of an antigen to a specific receptor on the surface of a B cell stimulates that cell to divide, producing a clone of genetically identical cells to fight that invader. Plasma cells have extensive endoplasmic reticulum for synthesizing antibodies.

- When the body first encounters a particular antigen, a **primary immune response** is activated, in which the "naïve" (previously unexposed) lymphocytes that recognize that antigen proliferate to produce clones of effector and memory cells.
- After a primary immune response to a particular antigen, subsequent encounters with the same antigen will trigger a much more rapid and powerful **secondary immune response**. The memory cells that bind with that antigen proliferate, launching a huge army of plasma cells and effector T cells.

## Vaccines are an application of immunological memory

You will recall Behring's experiment on using the serum of diphtheria-exposed animals to protect other animals from the disease (see Figure 42.6). The animals that survived and donated serum had developed an adaptive immune response, including memory cells that lead to long-term protection.

Thanks to immunological memory, exposure to many diseases (including childhood diseases such as chicken pox) provides a natural immunity to those diseases. Furthermore, it is possible to provide artificial immunity against many life-threatening diseases by **vaccination**: the introduction of antigen into the body in a form that does not cause disease.

Vaccination initiates a primary immune response, generating memory cells without making the person ill. Later, if a pathogen carrying the same antigen attacks, specific memory cells already exist. They recognize the antigen and quickly overwhelm the invaders with a massive production of lymphocytes and antibodies (**Figure 42.9**).

Because the antigens used for immunization or vaccination are produced by pathogenic organisms, they must be altered so that they cannot cause disease but are still able to provoke an immune response. There are three principal ways to do this:

- *Inactivation* involves killing the pathogen with heat or chemicals.
- *Attenuation* involves reducing the virulence of a virus by repeatedly infecting cells with it in the laboratory; this results in mutations in the virus that render it nonpathogenic but still recognized as nonself.
- *Recombinant DNA technology* can be used to produce peptide fragments that bind to and activate lymphocytes but do not have the harmful part of a protein toxin.

For most of the 70 or so bacteria, viruses, fungi, and parasites known to cause serious human diseases, vaccines are already available or will be in the next few years. As you saw in the opening story, vaccination has completely or almost completely wiped out some deadly diseases, including smallpox, diphtheria, and polio, in industrialized countries.

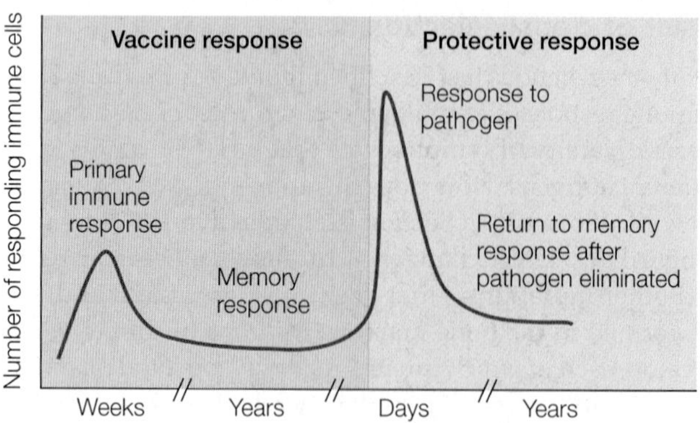


**42.9 Vaccination** Immunological memory from exposure to an antigen that does not cause disease can result in a massive response to the disease agent when it appears later.

RECAP 42.3

The adaptive immune system reacts against nonself or mutated self molecules called antigens. The system generates amazing diversity in both antibodies (produced by B cells) and in T cell receptors. In the primary immune response, B cells and T cells that recognize a particular pathogen proliferate by clonal selection. Immunological memory prepares the body for a much stronger secondary immune response.

- How does an antigen initiate an adaptive immune response? **See pp. 864–865 and Figure 42.7**
- Describe clonal selection. How does it contribute to immunological memory? **See pp. 865–866 and Figure 42.8**
- How do vaccines make use of immunological memory? **See p. 866 and Figure 42.9**

Now that we have discussed some general features of the adaptive immune system, let's focus in more detail on the B lymphocytes and the humoral immune response.

## 42.4 What Is the Humoral Immune Response?

Every day, billions of B cells survive the test of clonal deletion and are released from the bone marrow into the circulation. B cells are the basis for the humoral immune response.

**Go to Animated Tutorial 42.3**
**Humoral Immune Response**
Life10e.com/at42.3

### Some B cells develop into plasma cells

A B cell begins by making a receptor protein on its cell surface. As we have seen, if a B cell is activated by antigen binding to this receptor, it gives rise to clones of plasma cells and memory cells. The plasma (effector B) cells secrete antibodies into the bloodstream (see Figure 42.8).

Usually, for a naïve B cell to develop into an antibody-secreting plasma cell, a T-helper ($T_H$) cell with the same specificity must also bind to the antigen (see Figure 42.7). The division and differentiation of the B cell is stimulated by chemical signals from the $T_H$ cell.

As plasma cells develop, the number of ribosomes and the amount of endoplasmic reticulum in their cytoplasms increase greatly. These increases allow the cells to synthesize and secrete large amounts of antibody proteins—up to 2,000 molecules per second! *All the plasma cells arising from a given B cell produce antibodies that are specific for the antigen that originally bound to the parent B cell.* Thus antibody specificity is maintained as B cells proliferate.

### Different antibodies share a common structure

Antibodies belong to a class of proteins called **immunoglobulins**. There are several types of immunoglobulins, but all contain a tetramer consisting of four polypeptide chains (**Figure 42.10**). In each immunoglobulin molecule, two of these polypeptides are identical light chains, and two are identical heavy chains. Disulfide bonds hold the chains together. Each polypeptide chain has a constant region and a variable region:

- The amino acid sequences of the **constant regions** are similar among the immunoglobulins. They determine the destination and function—the class—of each immunoglobulin.
- The amino acid sequences of the **variable regions** are different for each specific immunoglobulin. Their three-dimensional antigen-binding sites are determined by their secondary structures and are responsible for antibody specificity.

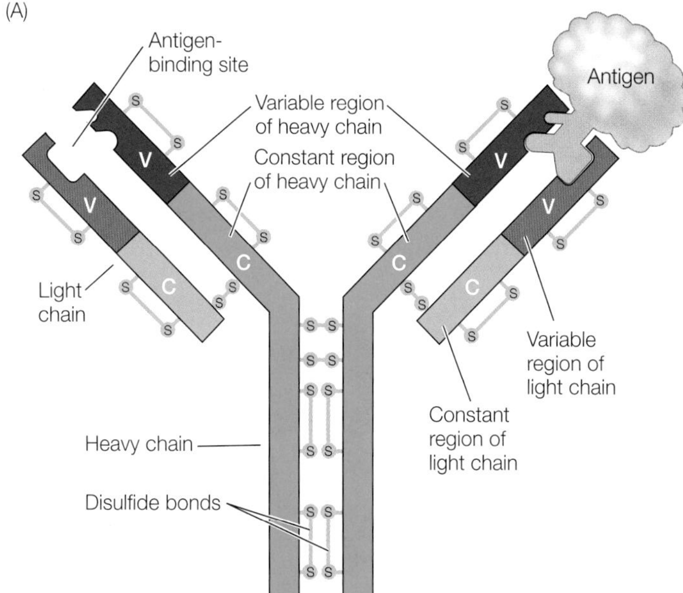


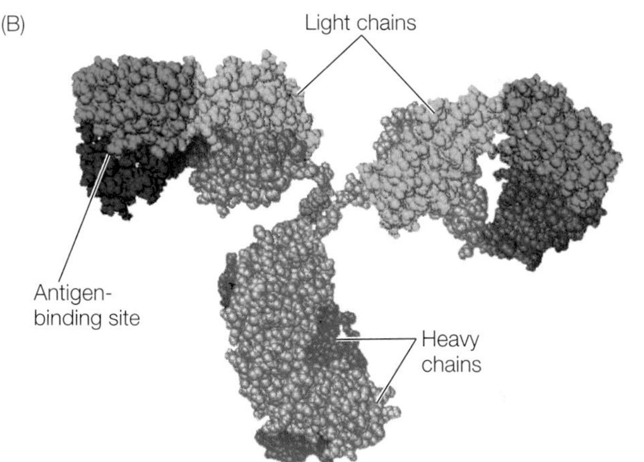


**42.10 The Structure of an Immunoglobulin** Four polypeptide chains (two light, two heavy) make up an immunoglobulin molecule. Both diagrammatic (A) and space-filling (B) representations of immunoglobulin are shown here.

**Go to Activity 42.3 Immunoglobulin Structure**
**Life10e.com/ac42.3**

The two antigen-binding sites on each immunoglobulin molecule are identical, making the antibody bivalent (*bi*, "two"; *valent*, "binding"). This ability to bind two antigen molecules at once, along with the presence of multiple epitopes on the surfaces of many antigens (including large proteins, viruses, and bacteria) permits antibodies to form large complexes with the antigens. These complexes are easy targets for ingestion and breakdown by phagocytes.

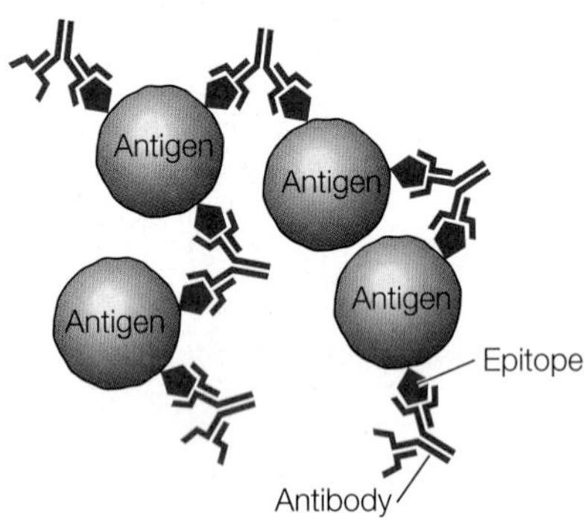


## There are five classes of immunoglobulins

While the variable regions are responsible for the specificity of an immunoglobulin, the constant regions of the heavy chain determine the class of the immunoglobulin—for example, whether it will be an integral membrane receptor (e.g., on the surface of a B cell) or a soluble antibody that is secreted into the bloodstream. The five immunoglobulin classes are described in **Table 42.2**. The most abundant class is IgG; these soluble antibody proteins make up about 80 percent of the total immunoglobulin content of the bloodstream. They are made in greatest quantity during a secondary immune response. IgG molecules defend the body in several ways. For example, after some IgG molecules bind to antigens, they become attached by their heavy chains to macrophages. This attachment permits the macrophages to destroy the antigens by phagocytosis.

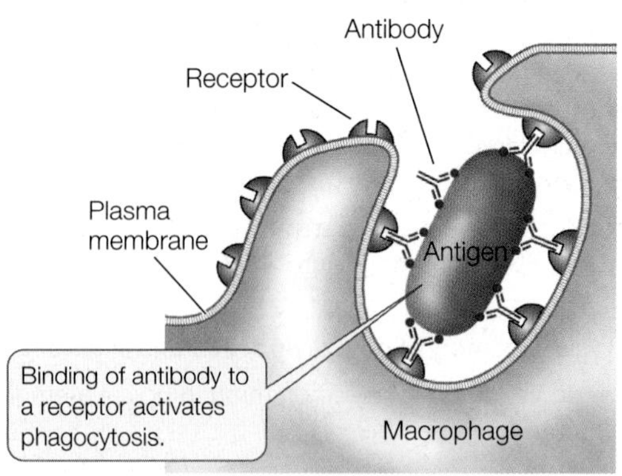


## Immunoglobulin diversity results from DNA rearrangements and other mutations

Each mature B cell makes one—and only one—specific antibody targeted to a single epitope. As we have seen, there are millions of possible epitopes to which a human is exposed or can be exposed. A simple calculation using approximate numbers shows that it would be impossible to have a unique gene for each of these epitopes:

$$\text{One antibody gene} \approx 2{,}100 \text{ base pairs (bp) DNA}$$
$$10 \text{ million different antibodies} \approx 21 \text{ billion bp DNA}$$

This is seven times the size of the entire human genome! There must be another way to generate antibody diversity.

It turns out that instead of a single gene encoding each immunoglobulin, the genome of the differentiating B cell has a limited number of alleles for each of several regions (domains) of the protein, and that combinations of these alleles generate

**TABLE 42.2** Antibody Classes

| Class | General Structure | Location | Function |
|---|---|---|---|
| IgG | Monomer | Free in blood plasma; about 80 percent of circulating antibodies | Most abundant antibody in primary and secondary immune responses; crosses placenta and provides passive immunization to fetus |
| IgM | Pentamer | Surface of B cell; free in blood plasma | Antigen receptor on B cell membrane; first class of antibodies released by B cells during primary response |
| IgD | Monomer | Surface of B cell | Cell surface receptor of mature B cell; important in B cell activation |
| IgA | Dimer | Saliva, tears, milk, and other body secretions | Protects mucosal surfaces; prevents attachment of pathogens to epithelial cells |
| IgE | Monomer | Secreted by plasma cells in skin and tissues lining gastrointestinal and respiratory tracts | Binds to mast cells and basophils to sensitize them to subsequent binding of antigen, which triggers release of histamine that contributes to inflammation and some allergic responses |

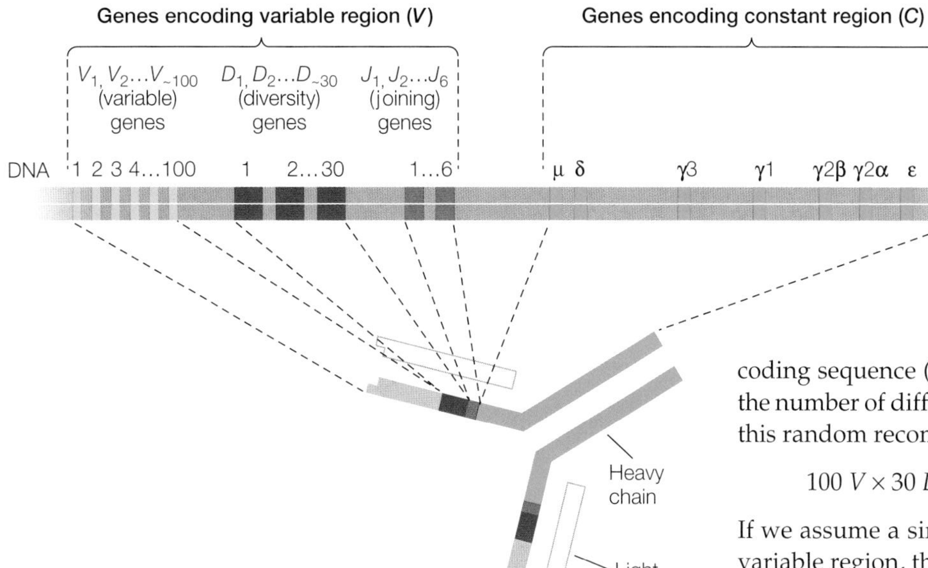


**42.11 Heavy-Chain Genes** Mouse immunoglobulin heavy chains have four domains, each of which is coded for by one of several possible genes selected from a cluster of similar genes.

diversity. First let's look at the unusual process of shuffling this genetic deck to generate the enormous immunological diversity that characterizes each individual mammal. This process affects the variable region, the part of the immunoglobulin that recognizes a particular epitope. Next we will see how similar events involving the constant region produce the five classes of immunoglobulins, which have different cellular locations or functions in the body.

Each gene encoding an immunoglobulin chain is in reality a "supergene" assembled by means of genetic recombination from several clusters of smaller genes scattered along part of a chromosome. Such a region in the mouse genome is shown in **Figure 42.11**. Every cell in the body has hundreds of immunoglobulin genes located in separate clusters that are potentially capable of participating in the synthesis of both the variable and constant regions of immunoglobulin chains. In most body cells and tissues, these genes remain intact and separated from one another. But during B cell development, these genes are cut out, rearranged, and joined together in DNA recombination events. One gene from each cluster is chosen randomly for joining, and the others are deleted.

In this manner, a unique immunoglobulin supergene is assembled from randomly selected "parts." Each B cell precursor assembles two supergenes, one for a specific heavy chain and the other, assembled independently, for a specific light chain. In humans and mice, two families of genes encode the variable region of the light chain, and three families encode the variable region of the heavy chain. For example, in mice the variable region of the heavy chain is assembled from 100 *V*, 30 *D*, and 6 *J* genes (see Figure 42.11). Each B cell randomly selects one gene from each of these clusters to make the final coding sequence (*VDJ*) of the heavy-chain variable region. So the number of different heavy chains that can be made through this random recombination process is quite large:

$$100\ V \times 30\ D \times 6\ J = 18{,}000 \text{ possible combinations}$$

If we assume a similar amount of diversity in the light chain variable region, the number of possible combinations of light- and heavy-chain variable regions is:

$$18{,}000 \text{ different light chains} \times 18{,}000 \text{ different heavy chains} = 324 \text{ million possibilities!}$$

Even more diversity is generated by various kinds of mutation that occur during the recombination events. These mutations can occur through imprecise recombination and the high spontaneous mutation rates in immunoglobulin genes.

These genetic events are irreversible—once the final coding sequence has been assembled for the variable regions of a B cell's light and heavy chains, that B cell's epitope specificity cannot change. This remarkable example of irreversible cell differentiation generates an enormous diversity of immunoglobulins from the same starting genome. A similar process results in the diversity of T cell receptors, which we will discuss in more detail in Section 42.5.

Once the pre-transcriptional processing is completed, each supergene is transcribed and then translated to produce an immunoglobulin light chain or heavy chain. These combine to form an active immunoglobulin protein, as shown in **Figure 42.12**. This figure shows the variable region joining with a μ constant region segment, which forms a gene for an immunoglobulin in the IgM class (see below). However, this genetic system is capable of still other kinds of changes. The B cell or plasma cell can switch the immunoglobulin class it produces while retaining its antigen specificity.

**Go to Animated Tutorial 42.4**
**A B Cell Builds an Antibody**
Life10e.com/at42.4

## The constant region is involved in immunoglobulin class switching

Table 42.2 describes the different classes of immunoglobulins and their functions. Generally, a B cell makes only one class

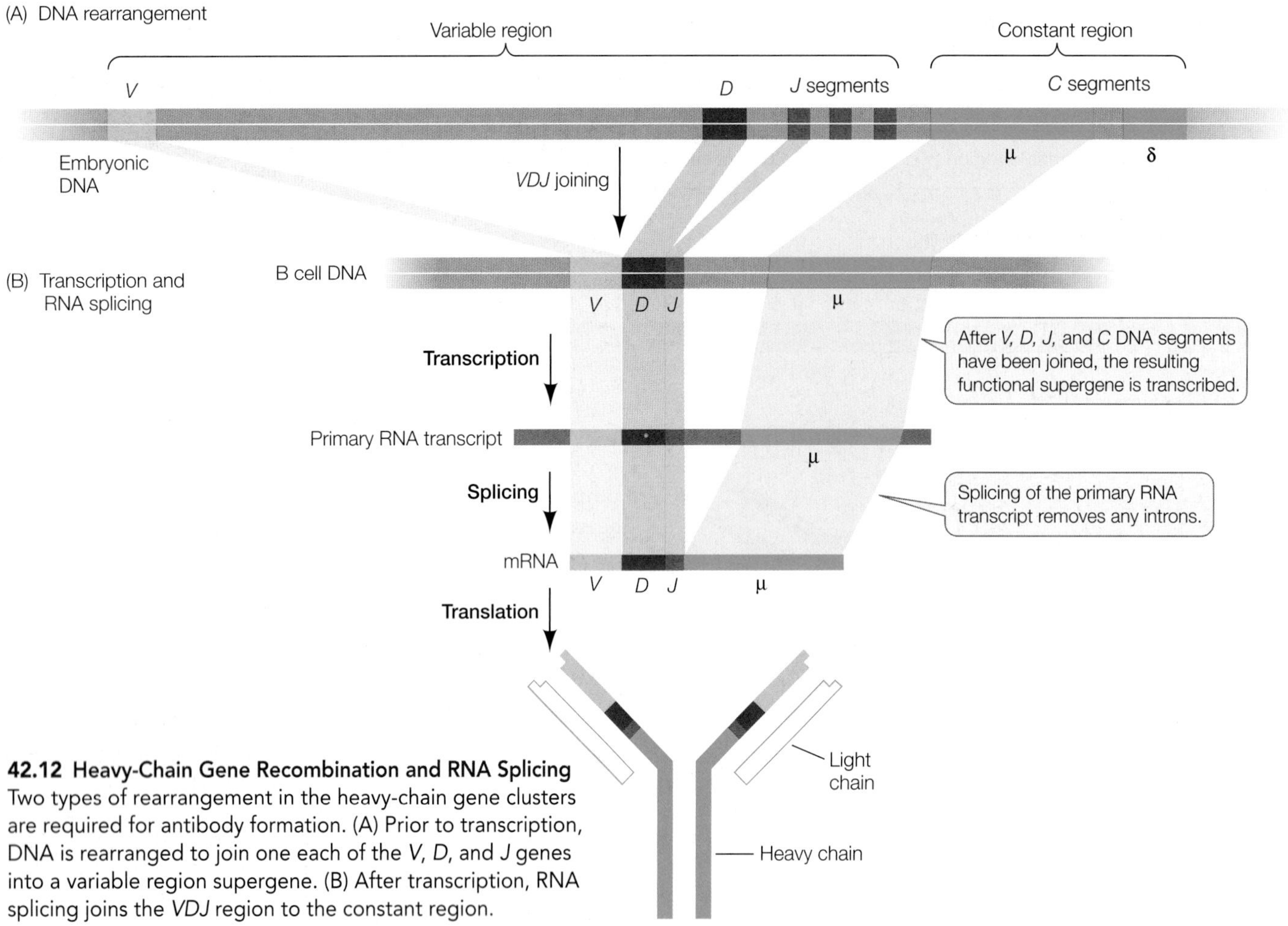

(A) DNA rearrangement

Variable region

Constant region

V

D

J segments

C segments

Embryonic DNA

μ

δ

*VDJ* joining

(B) Transcription and RNA splicing

B cell DNA

V D J

μ

Transcription

After *V*, *D*, *J*, and *C* DNA segments have been joined, the resulting functional supergene is transcribed.

Primary RNA transcript

μ

Splicing

Splicing of the primary RNA transcript removes any introns.

mRNA

V D J μ

Translation

Light chain

Heavy chain

**42.12 Heavy-Chain Gene Recombination and RNA Splicing** Two types of rearrangement in the heavy-chain gene clusters are required for antibody formation. (A) Prior to transcription, DNA is rearranged to join one each of the *V*, *D*, and *J* genes into a variable region supergene. (B) After transcription, RNA splicing joins the *VDJ* region to the constant region.

at a time. But **class switching** can occur, in which a B cell changes the immunoglobulin class it synthesizes. For example, a B cell making IgM can switch to making IgG.

Early in its life, a B cell produces IgM molecules, which are the receptors responsible for its recognition of a specific antigen. At this time, the constant region of the heavy chain is encoded by the first constant region gene, the μ gene (see Figures 42.11 and 42.12). If the B cell later becomes a plasma cell during a humoral immune response, another deletion occurs in the cell's DNA, positioning the variable region genes (consisting of the same *V*, *D*, and *J* genes) next to a constant region gene farther away on the original DNA molecule (**Figure 42.13**). Such a DNA deletion results in the production of a new immunoglobulin with a different constant region of the heavy chain, and therefore a different function (see Table 42.2). However, this immunoglobulin has the same variable regions—and therefore the same antigen specificity—as the IgM produced by the parent B cell. The new immunoglobulin protein falls into one of the other four classes (IgA, IgD, IgE, or IgG), depending on which of the constant region genes is placed adjacent to the variable region genes.

What triggers class switching? $T_H$ cells direct the course of an immune response and determine the nature of the attack on the antigen. These T cells induce class switching by sending cytokine signals. The cytokines bind to receptors

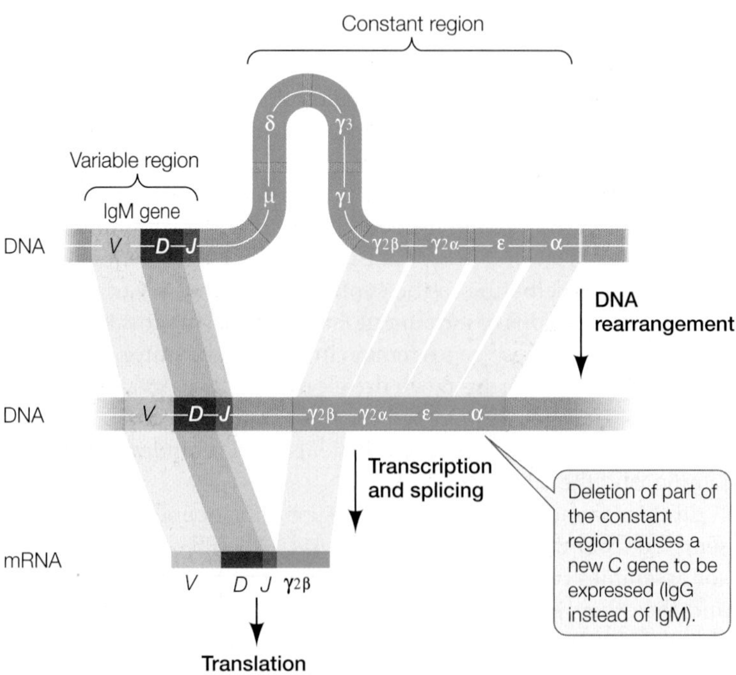


**42.13 Class Switching: Exchanging C Regions** The supergene produced by joining *V*, *D*, *J*, and *C* genes (see Figure 42.12) may later be modified, causing a different *C* region to be transcribed. This modification, known as class switching, is accomplished by deletion of part of the constant region gene cluster. Shown here is class switching from IgM to IgG.

on the target B cells, generating signal transduction cascades that result in recombination and altered expression of the immunoglobulin genes.

### Monoclonal antibodies have many uses

The specificity of antibodies suggested to scientists that they might be useful for detecting specific substances in the laboratory. However, the immune response to a complex antigen is polyclonal—that is, most antigens carry many different antigenic determinants and will produce a complex mixture of antibodies, each made by a different clone of B cells. Furthermore, as emphasized in our study of biochemistry, many biological molecules share regions of similar structure—all human steroid hormones, for example, have a similar multi-ring structure (see Figure 41.2). A polyclonal group of antibodies targeted to estrogen might be uninformative because some of the antibodies would bind to any steroid hormone present in the blood sample. More useful would be a clone of B cells that produce large amounts of an antibody that binds to only one specific epitope—a **monoclonal antibody**. Various methods can be used to produce monoclonal antibodies in the laboratory.

Monoclonal antibodies have many applications:

- *Immunoassays* involve the use of monoclonal antibodies to detect tiny amounts of molecules in tissues and fluids. For example, this technique is used in pregnancy tests to detect human chorionic gonadotropin, the hormone made by the developing embryo.
- *Immunotherapy* involves the use of monoclonal antibodies targeted against antigens on the surfaces of cancer cells. The coupling of a radioactive ligand or a toxin to the antibody makes it into a medical "smart bomb." In a related approach, binding of the antibody itself is enough to trigger a cellular immune response that destroys the cancer. This is the case with trastuzumab (Herceptin), a monoclonal antibody that binds to a growth factor receptor on some breast cancer cells.

**RECAP 42.4**

The humoral immune response is based on the synthesis by B cells of specific immunoglobulins. The specificity of an immunoglobulin derives from the amino acid sequence of its variable regions. B cells can make millions of immunoglobulins with different specificities by rearranging the genes that encode the variable regions of the heavy and light chains. Monoclonal antibodies are specific to one epitope and can be produced artificially for use in diagnostics and therapy.

- How does a B cell respond to an antigen? **See p. 867**
- How is the structure of an antibody molecule related to its function? **See pp. 867–868 and Figure 42.10**
- How can millions of antibodies with different specificities be generated from a relatively small number of genes? **See pp. 869–870 and Figures 42.11 and 42.12**
- What is the role of the constant region of the immunoglobulin in class switching? **See p. 870 and Figure 42.13**
- What are monoclonal antibodies, and how are they used? **See p. 871**

By making antibodies, B cells are the major players in the humoral immune response. We will now turn to the cellular immune response, where T cells are active at all stages.

## 42.5 What Is the Cellular Immune Response?

Two types of effector T cells (T-helper cells and cytotoxic T cells) are involved in the cellular immune response. They work along with proteins of the major histocompatibility complex (the MHC proteins), which present antigens on the surfaces of cells and contribute to the immune system's tolerance for the body's own cells.

**Go to Animated Tutorial 42.5**
**Cellular Immune Response**
Life10e.com/at42.5

### T cell receptors bind to antigens on cell surfaces

Like B cells, T cells possess specific membrane receptors. The T cell receptor is not an immunoglobulin, however, but a glycoprotein with a molecular weight of about half that of an IgG. It is made up of two polypeptide chains, each encoded by a separate gene (**Figure 42.14**). The two chains have distinct regions with constant and variable amino acid sequences. As in the immunoglobulins, the variable regions provide the site for specific binding to antigens. But there is one major difference: whereas an antibody can bind to any antigen, whether it is present on the surface of a cell or not, a T cell receptor binds only to an antigen displayed by an MHC protein on the surface of an antigen-presenting or target cell.

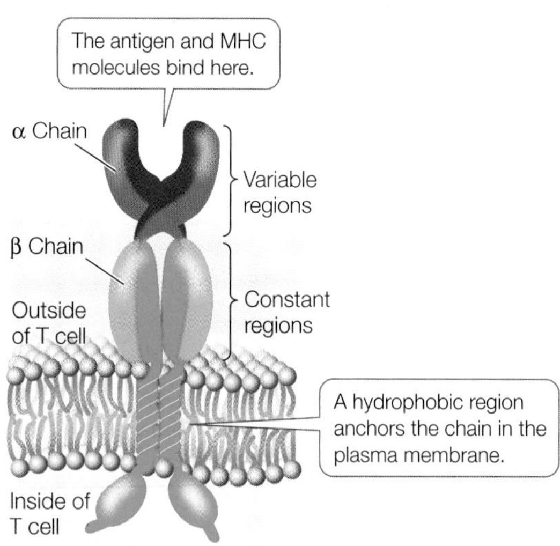


**42.14 A T Cell Receptor** The receptors on T lymphocytes are smaller than those on B lymphocytes, but their two polypeptides contain both variable and constant regions. As with the B cell receptors, the constant region fixes the receptor in the plasma membrane, while the variable regions establish the specificity for binding to antigen.

When a T cell is activated by contact with a specific antigen, it proliferates and forms a clone. Its descendants form clones of two types of effector T cells:

- **Cytotoxic T cells**, or **$T_C$ cells**, recognize virus-infected or mutated cells and kill them by inducing lysis (see p. 874).
- **T-helper cells ($T_H$ cells**, also called helper T cells) assist both the cellular and the humoral immune responses.

## MHC proteins present antigen to T cells

T cell receptors do not bind directly to antigens. Instead, they bind to antigens that are bound to a cell surface glycoprotein, an MHC protein. With several gene families and hundreds of alleles, the MHC protein is a cell surface marker of genetic individuality. The diversity of MHC proteins also means that there are many possibilities for presenting an antigen to the T cell receptor.

There are two classes of MHC proteins. Both function to present antigens to the different T lymphocytes:

- **Class I MHC** proteins are present on the surface of every nucleated cell in the vertebrate body. They enable $T_C$ cells to recognize virus-infected cells and kill them. Viral protein fragments that are antigenic are complexed with MHC I inside the cell and then the complex is carried to the plasma membrane. A $T_C$ cell with the appropriate T cell receptor then binds to the MHC–antigen complex. To ensure binding, the $T_C$ cell also has a cell surface protein called CD8 that recognizes and binds to MHC I.
- **Class II MHC** proteins are found mostly on the surfaces of B cells, macrophages, and other antigen-presenting cells, including dendritic cells (see Figure 42.2). When one of these cells ingests a pathogen such as a bacterium, the bacterial antigens are broken down in a phagosome. An MHC II molecule may bind to one of the fragments and carry it to the cell surface, where it is presented to a $T_H$ cell (**Figure 42.15**). $T_H$ cells have a surface protein called CD4 that recognizes and binds to MHC II.

**Table 42.3** summarizes the information on MHC proteins, the cellular origins of antigens, and T lymphocytes. To accomplish its role in antigen presentation, each kind of MHC protein has an antigen-binding site that can hold a peptide of about 10 to 20 amino acids. The T cell receptor recognizes not just the antigenic fragment but the MHC I or II protein to which the fragment is bound.

MHC proteins play a vital role in the selection of T cells during their development in the thymus gland:

- *Binding to MHC proteins.* A T cell receptor should bind not to an antigen alone but to an antigen–MHC complex, because T cells are activated by antigen presented on the surface of cells, not free antigen. Here there is positive selection for T cells that bind to MHC proteins. Any T cells that do not recognize MHC proteins (and thus would not bind to antigen-presenting cells) are eliminated soon after they develop; the rest of the T cells go on to the next selection step.
- *Binding to self peptides bound to self MHC proteins.* In this case there is negative selection of T cells that bind to self antigens presented on MHC proteins. This eliminates the further production of T cells that react to self antigens. Negative selection through clonal deletion (see p. 874) is a mechanism that prevents the adaptive immune system from reacting to self molecules.

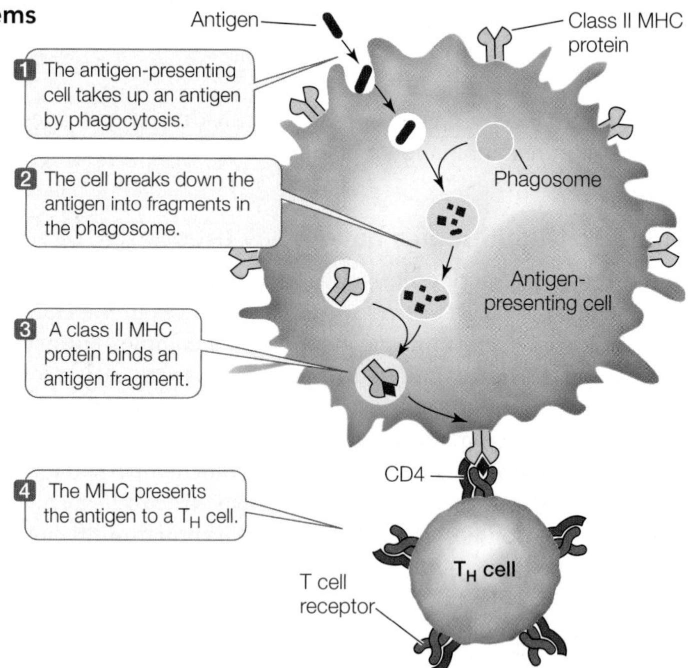


**42.15 Macrophages Are Antigen-Presenting Cells** A fragment of an antigen is displayed by MHC II on the surface of a macrophage. T cell receptors on a specific T-helper cell can then bind to and interact further with the antigen–MHC II complex.

## T-helper cells and MHC II proteins contribute to the humoral immune response

When a $T_H$ cell survives the selection processes and binds to an antigen-presenting cell, it releases cytokines that activate the $T_H$ cell to proliferate, producing a clone of $T_H$ cells with the same specificity. The steps to this point constitute the activation phase of the humoral immune response, and they occur in the lymphoid tissues. Next comes the effector phase, in which the $T_H$ cells activate naïve B cells with the same specificity to produce antibodies.

B cells are also antigen-presenting cells. B cells take up antigens bound to their surface immunoglobulin receptors by endocytosis, break them down, and display antigenic fragments on class II MHC proteins. When a $T_H$ cell binds to the displayed antigen–MHC II complex, it releases cytokines that cause the B cell to produce a clone of plasma cells (**Figure 42.16A**). Finally, the plasma cells secrete antibodies, completing the effector phase of the humoral immune response.

TABLE **42.3**
**The Interaction between T Cells and Antigen-Presenting Cells**

| Presenting Cell Type | Antigen Presented | MHC Class | T Cell Type | T Cell Surface Protein |
|---|---|---|---|---|
| Any cell | Intracellular protein fragment | Class I | Cytotoxic T cell ($T_C$) | CD8 |
| Macrophages, dendritic cells, and B cells | Fragments from extracellular proteins | Class II | Helper T cell ($T_H$) | CD4 |

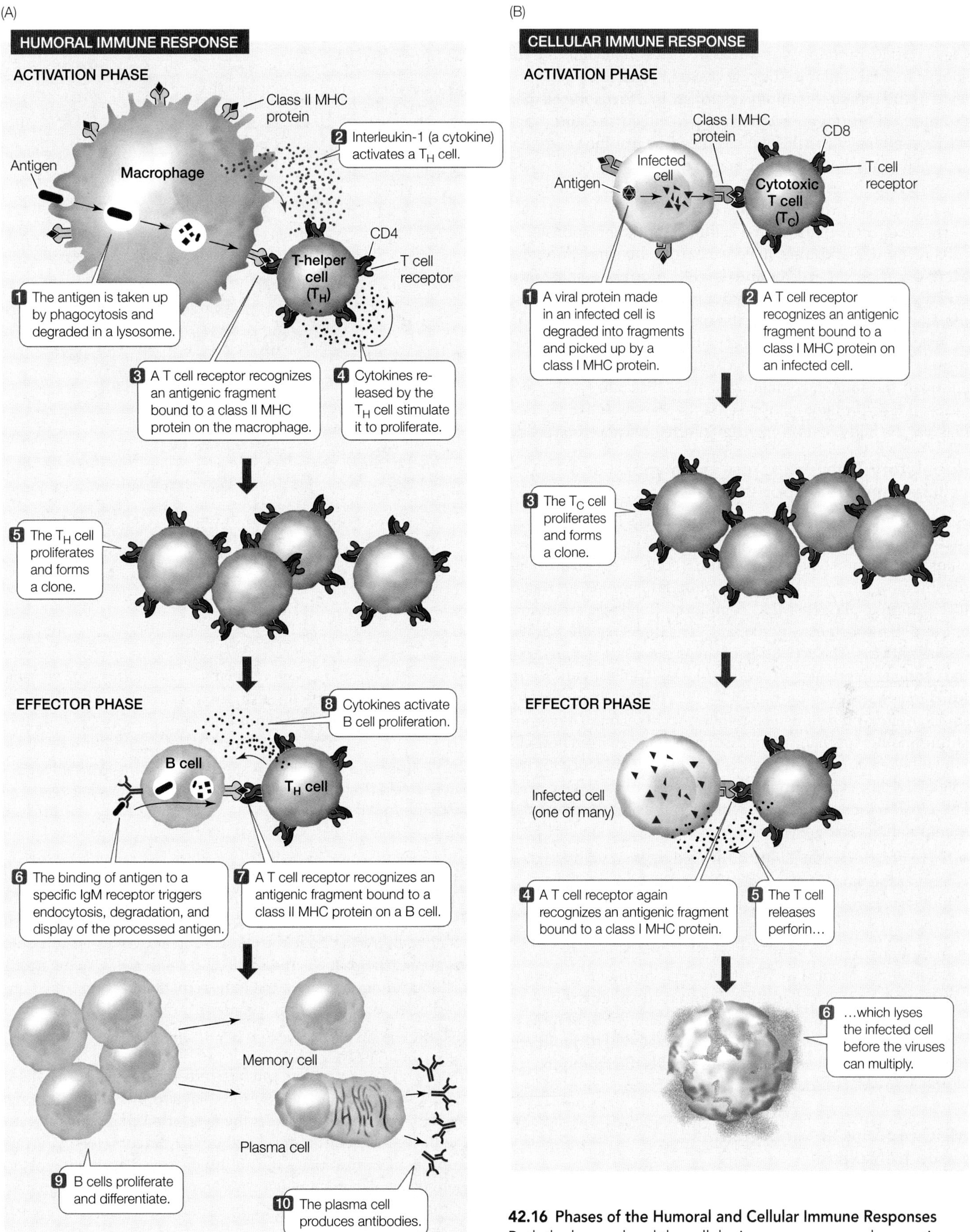


**42.16 Phases of the Humoral and Cellular Immune Responses** Both the humoral and the cellular immune responses have activation and effector phases, all of which involve T cells.

## Cytotoxic T cells and MHC I proteins contribute to the cellular immune response

Class I MHC proteins play a role in the cellular immune response that is similar to the role played by class II MHC proteins in the humoral immune response. In a virus-infected or mutated cell, foreign or abnormal proteins or peptide fragments combine with MHC I molecules. The resulting complex is displayed on the cell surface and presented to $T_C$ cells. When a $T_C$ cell recognizes and binds to this antigen–MHC I complex, it is activated to proliferate (**Figure 42.16B**).

In the effector phase of the cellular immune response, $T_C$ cells recognize and bind to cells bearing the same antigen–MHC I complex. These bound $T_C$ cells produce a substance called perforin, which lyses the bound target cell. In addition, the $T_C$ cells can bind to a specific receptor (called Fas) on the target cell that initiates apoptosis in that cell. These two mechanisms, cell lysis and programmed cell death, work in concert to eliminate the antigen-containing host cell. Because $T_C$ cells recognize MHC proteins complexed with nonself antigens, they help rid the body of its own virus-infected or cancerous cells.

## Regulatory T cells suppress the humoral and cellular immune responses

A third class of T cells called **regulatory T cells** (**Tregs**) ensures that the immune system does not attack self cells and molecules indiscriminately. Like $T_H$ and $T_C$ cells, Tregs mature in the thymus gland, carry T cell receptors, and become activated if they bind to antigen–MHC complexes. But Tregs are different in one important way: the antigens that Tregs recognize are *self antigens*. The activation of Tregs causes them to secrete the cytokine interleukin-10, which blocks T cell activation and leads to apoptosis of the $T_C$ and $T_H$ cells that are bound to the same antigen-presenting cell (**Figure 42.17**). Thus Tregs constitute another mechanism for distinguishing self from nonself. How do we know this? There are two lines of evidence for the role of Tregs. As in many other biological studies, the cause-and-effect relationships were worked out using experimental manipulations and genetics:

- If Tregs are experimentally destroyed in the thymus of a mouse, the mouse grows up with an out-of-control immune system, mounting strong immune responses to self antigens (autoimmunity—see Section 42.6).
- In humans, a rare X-linked hereditary disease occurs when a gene critical to Treg function is mutated. An infant with this disease, called IPEX (*i*mmune dysregulation, *p*olyendocrinopathy and *e*nteropathy, *X*-linked), mounts an immune response that attacks the pancreas, thyroid, and intestines. Most affected individuals die within the first few years of life.

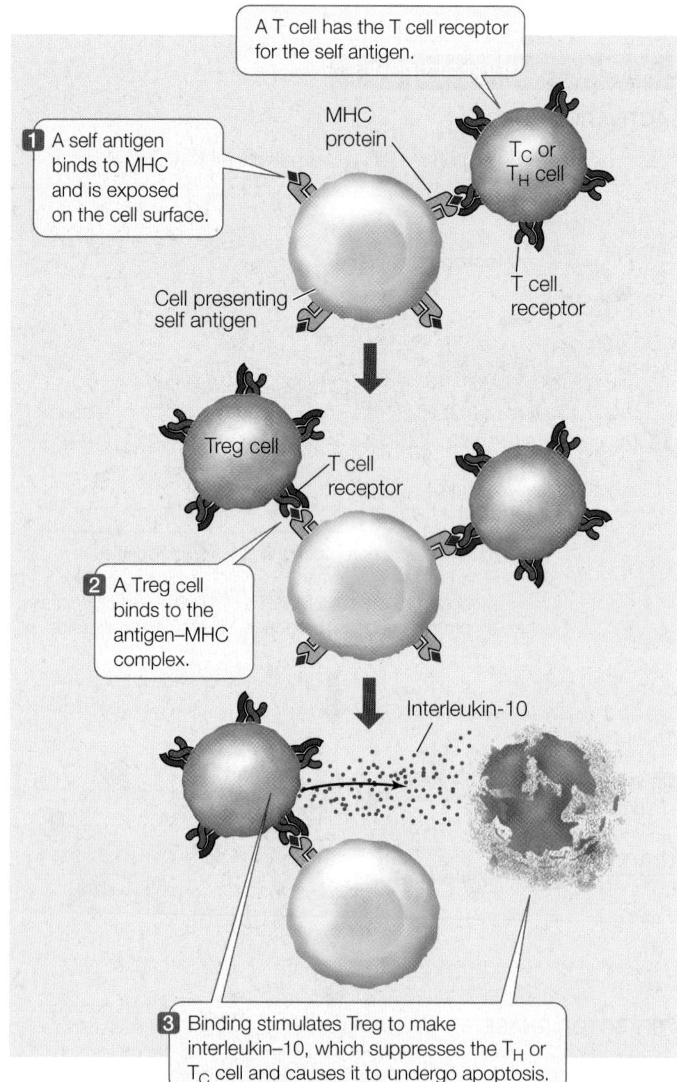


**42.17 Tregs and Tolerance** A special class of T cells called regulatory T cells (Tregs) inhibits the activation of the immune system in response to self antigens.

## MHC proteins are important in tissue transplants

In humans, one consequence of the major histocompatibility complex became important with the development of organ transplant surgery. Because the proteins produced by the MHC are specific to each individual, they act as nonself antigens if transplanted into another individual. An organ or a piece of tissue transplanted from one person to another is recognized as nonself by the host body and soon provokes an immune response; the tissue is then killed, or "rejected," by the host's cellular immune system. But if the transplant is performed immediately after birth, or if it comes from a genetically identical person (an identical twin), the material is recognized as self and is not rejected.

The rejection problem can be overcome by treating a patient with a drug, such as cyclosporin, that suppresses the immune system. Cyclosporin blocks the activation of a transcription factor that is essential for T cell development. However, this approach compromises the ability of transplant recipients to defend themselves against pathogens. This problem must be managed by the use of antibiotics and other drugs to combat infections that develop.

RECAP 42.5

Specific receptors on T cells bind to antigen–MHC complexes displayed on cell surfaces. During development, T cells are selected that recognize MHC proteins. T-helper cells bind antigens on MHC II proteins and contribute to the humoral immune response. Cytotoxic T cells bind antigens on MHC I proteins and contribute to the cellular immune response. The cellular immune response acts against virus-infected or mutated body cells. Tregs suppress immune responses to self antigens.

- What are the roles of a T cell receptor in cellular immunity? **See p. 871 and Figure 42.14**
- What parts do MHC proteins play in the cellular immune response? **See p. 872 and Figure 42.15**
- What occurs during the cellular immune response to a virus-infected cell? **See pp. 873–874 and Figure 42.16**

Given the numerous and complex cellular interactions that activate the immune system and generate antibody diversity, you may have perceived many points at which the immune system could fail. We will now turn to several situations in which one or more components of this complex system malfunction.

## 42.6 What Happens When the Immune System Malfunctions?

Sometimes the immune system fails us in one way or another. It may overreact, as in an allergic reaction; it may attack self antigens, as in an autoimmune disease; or it may function weakly or not at all, as in an immune deficiency disease.

### Allergic reactions result from hypersensitivity

An **allergic reaction** arises when the human immune system overreacts to (is hypersensitive to) a dose of antigen. Although the antigen itself may present no danger to the host, the inappropriate immune response may produce inflammation and other symptoms, which can cause serious illness or even death. Allergic reactions are the most familiar examples of this phenomenon. Allergic reactions may involve immediate hypersensitivity or delayed hypersensitivity.

IMMEDIATE HYPERSENSITIVITY **Immediate hypersensitivity** arises when an allergic individual is exposed to an antigen (in this case referred to as an allergen) from the environment, such as a food, pollen, or the venom of an insect. In response to the allergen, the individual makes large amounts of IgE. When this happens, mast cells in tissues and basophils in the blood bind the constant end of the IgE. If that individual is exposed to the same allergen again, binding of the allergen to the IgE causes the mast cells and basophils to rapidly release a large amount of histamine (**Figure 42.18**). This results in symptoms such as dilation of blood vessels, inflammation, and difficulty breathing. If not treated with antihistamines, a severe allergic reaction can lead to death. It is not known why some people produce excessive amounts of IgE in response to allergens. There is some evidence for genetic factors predisposing people to allergic responses.

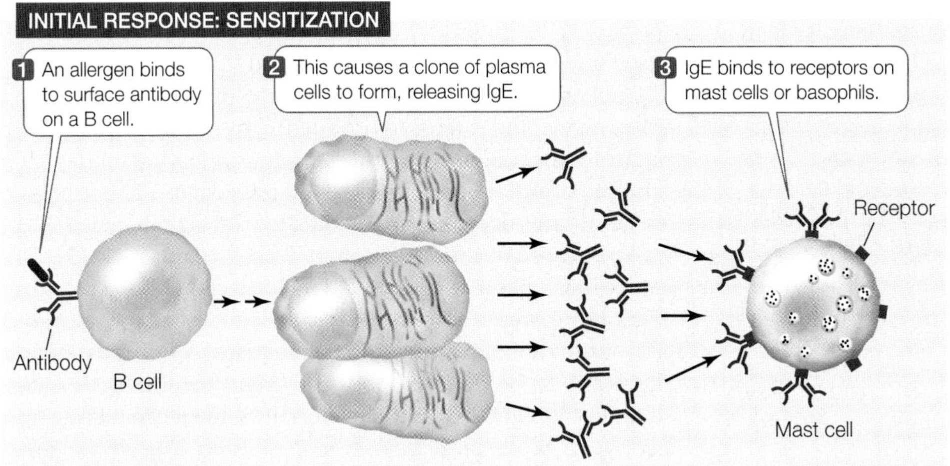

**42.18 An Allergic Reaction** An allergen is an antigen that stimulates B cells to make large amounts of IgE antibodies, which bind to mast cells and basophils. When the body encounters the allergen again, these cells produce large amounts of histamine, which has harmful physiological effects.

Allergy to pollen can be treated using a process called desensitization. The process involves injecting small amounts of the allergen (typically just an extract of the offending plant tissue) into the skin—enough to stimulate IgG production but not enough to stimulate IgE production. The next time the person is exposed to the allergen, IgG binds to it, tying it up before IgE can bind it and exert its harmful effects.

Desensitization does not work for food allergens because the IgE response to those substances is so strong that even a small amount of antigen provokes it. The best approach for those with food allergies—there are an estimated 3 million people in the U.S.—is to avoid foods containing the allergens. This can be difficult, but food labels listing all the ingredients are helpful. Molecular biologists are beginning to identify the antigens that act as allergens, with the hope of developing vaccines or genetically modified foods that lack the allergenic epitopes.

DELAYED HYPERSENSITIVITY **Delayed hypersensitivity** is an allergic reaction that does not begin until hours after exposure to an antigen. In this case the antigen is taken up by antigen-presenting cells and a T cell response is initiated. A $T_H$ cell produces a clone of cells that secrete various cytokines, which cause such reactions as inflammation and rash. These events take time (hence the term "delayed"). An example is the rash that develops after exposure to poison ivy.

## Autoimmune diseases are caused by reactions against self antigens

Errors in the selection of T cells in the thymus can result in T cells that bind to antigen–MHC complexes that carry self antigens. Although the precise origin of **autoimmunity** is not known, there are several hypotheses:

- *Failure of negative selection*. A clone of lymphocytes making antibodies against self antigens that should have been destroyed by clonal deletion is not destroyed.
- *Molecular mimicry*. T cells that recognize a nonself antigen, such as a virus, also recognize something on a self antigen that has a similar structure.

Autoimmunity does not always result in disease, but several autoimmune diseases are common:

- People with *systemic lupus erythematosis* (SLE) have antibodies to many cellular components, including DNA and nuclear proteins released from dying cells. These antinuclear antibodies can cause serious damage when they bind to normal tissue antigens and form large circulating antigen–antibody complexes, which become stuck in tissues and provoke inflammation.
- People with *rheumatoid arthritis* have difficulty in shutting down a T cell response to self antigens. These patients may have low activity of CTLA4, an inhibitory protein that blocks T cells from reacting to self antigens. This results in inflammation of the joints and other tissue damage.
- *Hashimoto's thyroiditis* is the most common autoimmune disease in women over 50. Immune cells attack thyroid tissue, resulting in fatigue, depression, weight gain, and other symptoms.
- *Insulin-dependent diabetes mellitus*, or type I diabetes, occurs most often in children. It is caused by an immune reaction against several proteins in the cells of the pancreas that manufacture the protein hormone insulin. This reaction kills the insulin-producing cells, so people with type I diabetes must take insulin daily in order to survive.

## AIDS is an immune deficiency disorder

There are several inherited and acquired immune deficiency disorders. In some individuals, T or B cells never form; in others, B cells lose the ability to give rise to plasma cells. In either case, the affected individual is unable to mount an adaptive immune response and thus lacks a major line of defense against pathogens. The $T_H$ cell is perhaps the most central component of the immune system because of its essential roles in both the humoral and cellular immune responses (see Figure 42.7). This cell is the target of **human immunodeficiency virus** (**HIV**), the retrovirus that results in **acquired immune deficiency syndrome** (**AIDS**).

HIV can be transmitted from person to person in body fluids containing the virus (such as blood, semen, vaginal fluid, or breast milk). The recipient tissue is either blood (by transfusion) or a mucous membrane lining an organ (the mucus contains a high concentration of lymphocytes). HIV initially infects macrophages, $T_H$ cells, and antigen-presenting dendritic cells in the blood and tissues. At first there is an immune response to the viral infection, and $T_H$ cells are activated. But because HIV infects the $T_H$ cells, they are killed both by HIV itself and by $T_C$ cells that lyse infected $T_H$ cells. Consequently $T_H$ cell numbers decline after the first month or so of infection. Meanwhile, the extensive production of HIV by infected cells activates the humoral immune system. Antibodies bind to HIV, and the complexes are removed by phagocytes. The HIV level in blood goes down. There is still a low level of infection, however, because of the depletion of $T_H$ cells (**Figure 42.19**). This process reaches a low, steady-state level called the "set point." This point varies among individuals and is a strong predictor of the rate of progression of the disease. For most people it takes 8 to 10 years without treatment for the more severe manifestations of AIDS to develop. In some it can take as little as a year; in others, 20 years.

During this dormant period, people carrying HIV generally feel fine, and their $T_H$ cell levels are adequate for them to mount immune responses. Eventually, however, the virus destroys the $T_H$ cells, and their numbers fall to the point where the infected person is susceptible to infections that the $T_H$ cells would normally eliminate. These infections result in conditions such as Kaposi's sarcoma, a skin tumor caused by a herpes virus; pneumonia caused by the fungus *Pneumocystis jirovecii*; and lymphoma tumors caused by the Epstein–Barr virus. These conditions result from opportunistic infections because

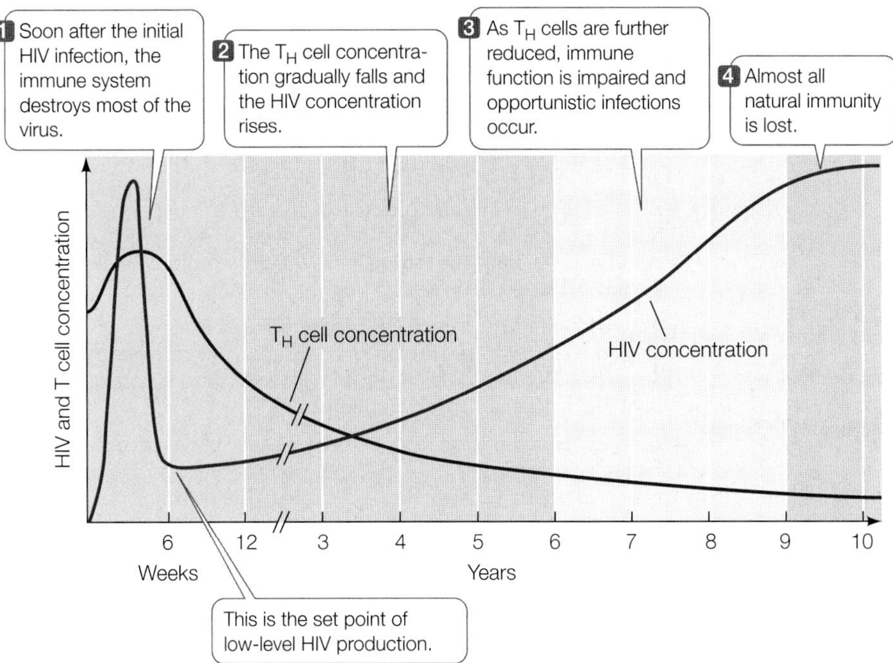


**42.19 The Course of an HIV Infection** An HIV infection may be carried, unsuspected, for many years before the onset of symptoms.

the pathogens take advantage of the crippled immune system of the host. They lead to death within a year or two.

The molecular biology of HIV and its life cycle have been intensively studied (see Figure 16.16). This has resulted in the development of drugs targeted to HIV proteins, such as the reverse transcriptase that makes cDNA from the viral RNA, and the viral protease that cuts the large precursor viral protein into its final active proteins. Treatment with combinations of such drugs has had spectacular success. Getting AIDS before the 1990s was a death sentence, with few sufferers surviving beyond a year or two. Today the survival of a treated, infected person is not much different from that of an uninfected person. Unfortunately, like many medical treatments, HIV drugs are not available to all who need them—particularly in poor regions of the world where AIDS is prevalent. As a result, there are about 1.7 million deaths per year worldwide from AIDS.

## RECAP 42.6

Failures of the immune system include allergic reactions (caused by hypersensitivity to antigens), autoimmune diseases (caused by reactions against self antigens), and immune deficiency disorders.

- How does immediate hypersensitivity develop? **See p. 875 and Figure 42.18**
- What is an autoimmune disease? Give an example. **See p. 876**
- Describe the course of events in the human immune system during HIV infection. **See pp. 876–877 and Figure 42.19**

Why do many people resist vaccination?

### ANSWER

There seem to be three main reasons why a person might refuse a vaccine. The first is complacency. Measles, which used to kill thousands of children every year in the U.S. and still does in poor countries, is no longer a highly visible threat to public health. Second, some people believe that vaccines, although exhaustively tested for safety, are actually unsafe and cause disease. The internet is full of such assertions. Third, people are wary of experts because of past false alarms. The discovery of the H1N1 flu virus (swine flu) in Mexico in 2009 led to a high alert and a mass vaccination program that turned out not to be necessary. Vaccination is a scientific success in terms of bolstering immunity and eradicating disease. But like any technology, its acceptance is a political issue.

## 42.1 What Are the Major Defense Systems of Animals?

- Animal defenses against **pathogens** are based on the body's ability to distinguish between self and nonself.
- **Innate** (nonspecific) **defenses** are inherited mechanisms that protect the body from many kinds of pathogens. They typically act rapidly.
- **Adaptive** (specific) **defenses** respond to specific pathogens. They develop more slowly than innate defenses but are long-lasting.
- Many defenses are implemented by cells and proteins carried in the blood plasma and **lymph**. **Review Figure 42.1, ACTIVITY 42.1**
- **White blood cells** fall into two broad groups. **Phagocytes** engulf pathogens by phagocytosis. **Lymphocytes**, which include B cells and T cells, participate in adaptive responses. **Review Figure 42.2, ANIMATED TUTORIAL 42.1**

## 42.2 What Are the Characteristics of the Innate Defenses?

- An animal's innate defenses include physical barriers such as the skin, and competing resident microorganisms known as normal flora. **Review Figure 42.3**
- The **complement system** consists of more than 20 different antimicrobial proteins that act to alter membrane permeability and kill targeted cells.
- Circulating defensive cells, such as phagocytes and **natural killer cells**, eliminate invaders.
- A cell signaling pathway involving the **toll-like receptor** stimulates the body's defenses. **Review Figure 42.4**
- **Inflammation** involves activation of several types of cells and proteins that act against invading pathogens. **Mast cells** release **histamines**, which cause blood vessels to dilate and become "leaky." **Review Figure 42.5, ACTIVITY 42.2**

## 42.3 How Does Adaptive Immunity Develop?

- The adaptive immune response recognizes specific **antigens**, responds to an enormous diversity of **antigenic determinants**, distinguishes self from nonself, and remembers the antigens it has encountered. **Review ANIMATED TUTORIAL 42.2**
- Each antibody and each T cell is specific for a single antigenic determinant. **T cell receptors** bind to antigens on the surfaces of virus-infected cells and abnormal cells.
- The **humoral immune response** is directed against pathogens in the blood, lymph, and tissue fluids. The **cellular immune response** is directed against an antigen established within a host cell. Both responses are mediated by antigenic fragments being presented on a cell surface along with the proteins of the **major histocompatibility complex** (MHC). **Review Figure 42.7**
- **Clonal selection** accounts for the specificity and diversity of the immune response and for **immunological memory**. **Review Figure 42.8**
- An activated B or T lymphocyte produces **effector cells** that attack the antigen, and **memory cells** that are long-lived and rarely divide. Effector B cells are called **plasma cells** and secrete specific **antibodies**.
- **Vaccination** is inoculation with modified pathogens or antigens that provoke an immune response but are not pathogenic. **Review Figure 42.9**

## 42.4 What Is the Humoral Immune Response?
**See ANIMATED TUTORIAL 42.3**

- B cells are the basis of the humoral immune response. Naïve B cells are activated by binding of antigen and by stimulation by $T_H$ cells with the same specificity, and then form plasma cells. These cells synthesize and secrete specific antibodies.
- An antibody is an **immunoglobulin**, a tetramer of four polypeptides: two identical light chains and two identical heavy chains, each consisting of a **constant region** and a **variable region**. **Review Figure 42.10, ACTIVITY 42.3**
- The variable regions determine the specificity of an immunoglobulin, and the constant regions of the heavy chain determine its class. There are five classes of immunoglobulins with different body locations and functions. **Review Table 42.2**
- B cell genomes undergo random recombination of genes coding for regions of the immunoglobulin polypeptide chains so that each cell can produce a specific antibody protein. The immunoglobulin chains derive from "supergenes" that are constructed from different combinations of *V*, *D*, *J*, and *C* genes. This DNA rearrangement and rejoining yields millions of different immunoglobulin chains. **Review Figures 42.11, 42.12, ANIMATED TUTORIAL 42.4**
- Once a B cell becomes a plasma cell, it may undergo **class switching**, in which a deletion of one or more constant region genes results in the production of an immunoglobulin with a different constant region and a different function. **Review Figure 42.13**
- A **monoclonal antibody** can be used in diagnosis and therapy.

## 42.5 What Is the Cellular Immune Response?
**See ANIMATED TUTORIAL 42.5**

- T cells are the effectors of the cellular immune response. T cell receptors are somewhat similar in structure to the immunoglobulins, having variable and constant regions. **Review Figure 42.14**
- There are three types of T cells. **Cytotoxic T cells ($T_C$ cells)** recognize and kill virus-infected cells or mutated cells. **T-helper cells ($T_H$ cells)** direct both the cellular and humoral immune responses. **Regulatory T cells (Tregs)** inhibit the other T cells from mounting an immune response to self antigens.
- The genes of the major histocompatibility complex (MHC) encode membrane proteins that bind antigenic fragments and present them to T cells. **Review Figures 42.15, 42.16**
- Organ transplants are rejected when the host's immune system recognizes MHC proteins on transplanted tissue as nonself and initiates an immune defense attacking the foreign tissue.

## 42.6 What Happens When the Immune System Malfunctions?

- An **allergic reaction** is an inappropriate immune response caused by **immediate hypersensitivity** or **delayed hypersensitivity** to certain antigens. **Review Figure 42.18**
- Autoimmune diseases result when the immune system produces B and T cells that attack self antigens.
- Immune deficiency disorders result from failure of one or another part of the immune system. **Acquired immune deficiency syndrome (AIDS)** is a disorder that arises from depletion of the $T_H$ cells as a result of infection with **human immunodeficiency virus (HIV)**. **Review Figure 42.19**

**Go to the Interactive Summary to review key figures, Animated Tutorials, and Activities**
**Life10e.com/is42**

# CHAPTER**REVIEW**

## REMEMBERING

1. Phagocytes kill harmful bacteria by
   a. endocytosis.
   b. producing antibodies.
   c. complement proteins.
   d. T cell stimulation.
   e. inflammation.

2. Which statement about an antigenic determinant is *not* true?
   a. It is a specific chemical grouping.
   b. It may be part of many different molecules.
   c. It is the part of an antigen to which an antibody binds.
   d. It may be part of a cell.
   e. A single protein has only one on its surface.

3. T cell receptors
   a. are the primary receptors for the humoral immune system.
   b. are carbohydrates.
   c. cannot function unless the animal has previously encountered the antigen.
   d. are produced by plasma cells.
   e. are important in combating viral infections.

4. According to the clonal selection theory,
   a. an antibody changes its shape to match the antigen it meets.
   b. an individual animal contains only one type of B cell.
   c. an individual animal contains many types of B cells, each producing one kind of antibody.
   d. each B cell produces many types of antibodies.
   e. many clones of antiself lymphocytes appear in the bloodstream.

5. The extraordinary diversity of antibodies results in part from
   a. the action of monoclonal antibodies.
   b. the splicing of protein molecules.
   c. the action of cytotoxic T cells.
   d. the rearrangement of genes.
   e. their remarkable nonspecificity.

6. The major histocompatibility complex
   a. codes for specific proteins found on the surfaces of cells.
   b. plays no role in T cell immunity.
   c. plays no role in antibody responses.
   d. plays no role in skin graft rejection.
   e. is encoded by a single locus with multiple alleles.

## UNDERSTANDING & APPLYING

7. Describe the part of an antibody molecule that interacts with an antigen. How is it similar to the active site of an enzyme? How does it differ from the active site of an enzyme?

8. Contrast immunoglobulins and T cell receptors with respect to their structures and functions.

9. The gene family (genes A, B, and D) determining MHC on the cell surface in humans is on a single chromosome. A father's MHC genotype is A1, A3, B5, B7, D9, D11. A mother's genotype is A2, A4, B6, B7, D11, D12. Their child's is A1, A4, B6, B7, D11, D12. What are the parents' haplotypes—that is, which alleles are linked on each of the two chromosomes of each parent? Assuming there is no recombination among the genes determining the MHC type, can these same two parents have a child who has the genotype A1, A2, B7, B8, D9, D11?

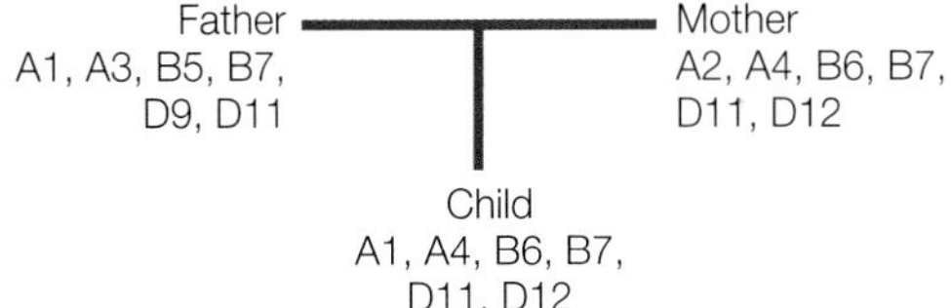


## ANALYZING & EVALUATING

10. Discuss the diversity of antibody specificities in an individual in relation to the diversity of enzymes. Does every cell in an animal contain genetic information for all the organism's enzymes? Does every cell contain genetic information for all the organism's immunoglobulins?

11. Development of an effective HIV vaccine requires that the person being vaccinated develop both cellular and humoral immunity against HIV. What experiments would you do to test whether both types of immunity developed in people given a potential new vaccine?

Go to BioPortal at **yourBioPortal.com** for Animated Tutorials, Activities, LearningCurve Quizzes, Flashcards, and many other study and review resources.

# 43 Animal Reproduction

CHAPTER**OUTLINE**


**43.1** **How Do Animals Reproduce without Sex?**

**43.2** **How Do Animals Reproduce Sexually?**

**43.3** **How Do the Human Male and Female Reproductive Systems Work?**

**43.4** **How Can Fertility Be Controlled?**

**A Unique Reproductive Strategy** Among honey bees and other hymenopteran insects, the only reproductive female is the queen, seen in the center. The female workers attending her are sterile.

THE HONEY BEE *Apis mellifera* has been studied throughout recorded history. Sex and reproduction in these social insects are intriguing. A honey bee hive contains a single reproductive female, the queen, and thousands of infertile female worker bees. There are few if any male bees in a hive. The queen lays eggs continuously while the worker bees build the honeycomb, forage, clean and defend the hive, and feed and care for the young. In honey bees, fertilized eggs develop into females and unfertilized eggs develop into males. The queen controls which of her eggs are fertilized. She carries sperm from males she has mated with and occasionally produces males by *not* releasing sperm onto her eggs. It is not in her best interest to produce many males because they don't contribute to raising young or maintaining the hive; they just hang around until they have a very rare chance to mate.

Eventually a hive needs a new queen. The old queen might die, or she and a retinue of workers might leave an overcrowded hive to start a new hive (a phenomenon known as swarming). Whether the queen dies or swarms, the remaining worker bees enlarge a few cells in the honeycomb that contain fertilized eggs laid by the old queen. The larvae from those eggs are fed special food that stimulates their growth and development into prospective queens.

The first queen to pupate and emerge from her royal chamber kills the other developing queens. She then leaves the hive for her mating flight. Males from all around sense a chemical message (a pheromone) that a virgin queen is available and congregate around her. While in flight, she mates with 15 to 20 males, and each coupling is an event. After a male manages to insert his penis into the queen's vagina, he literally explodes, leaving behind not only his sperm but also his sex organs (the latter will drop out later). The males die and the queen returns to the hive with a lifetime supply of sperm. The queen lives for about 2 years and lays as many as 3,000 eggs each day. Most of those eggs develop into female worker bees that devote their lives to feeding her and raising their sisters.

Natural selection has resulted in some amazing adaptations, none more so than those involved in reproduction. Sexual or asexual, bizarre or otherwise, the anatomy, physiology, biochemistry, and behavior surrounding the urge to propagate are fascinating.

How can a queen be so different from her worker sisters when they all share the same genome?

See answer on p. 899.

## 43.1 How Do Animals Reproduce without Sex?

Sexual reproduction is a nearly universal trait in animals, although many species can also reproduce asexually and some reproduce only asexually. Offspring produced asexually are genetically identical to one another and to their parents. Asexual reproduction is efficient because no time or energy is wasted on mating and every member of the population can convert resources into offspring. However, asexual reproduction does not generate genetic diversity in a population as sexual reproduction does, and this diversity is the raw material that enables natural selection to shape adaptations in response to environmental change. When environmental changes occur, lack of genetic diversity can be disadvantageous to a population.

A variety of animals, mostly invertebrates, reproduce asexually. They tend to be species that are attached to their substrates and cannot search for mates, or species that live in sparse populations and rarely encounter potential mates. Asexually reproducing species are likely to be found in relatively constant environments where genetic diversity is less important for species success. In fact, asexual reproduction is a good way to preserve a successful genotype in a particular environment—as long as that environment does not change.

Three common modes of asexual reproduction are budding, regeneration, and parthenogenesis.

### Budding and regeneration produce new individuals by mitosis

Many simple multicellular animals produce offspring by **budding**. New individuals form as outgrowths or buds from the bodies of older animals. A bud grows by mitotic cell division, and the cells differentiate before the bud breaks away from the parent (**Figure 43.1A**). The bud is genetically identical to the parent, and it may grow as large as the parent before it becomes independent.

**Regeneration** is usually thought of as the replacement of damaged tissues or lost limbs, but in some cases pieces of an organism can regenerate complete individuals. Echinoderms, for example, have remarkable abilities to regenerate. If sea stars (starfish) are cut into pieces, each piece that includes an arm and a portion of the central disc can grow into a new animal (**Figure 43.1B**). In the early 1900s oyster fishermen in Narragansett Bay tried to eliminate the sea stars that were preying on their oysters. Whenever they encountered sea stars, they chopped them up with their knives and threw them back into the water. As a result, the sea star population increased explosively.

Regeneration can occur when an animal is broken by an outside force such as wave action in the intertidal zone. In some cases, breakage occurs in the absence of external forces. Some species of segmented marine worms develop segments with rudimentary heads bearing sensory organs. The segments then break apart and each one forms a new worm.

### Parthenogenesis is the development of unfertilized eggs

Not all eggs must be fertilized to develop. A common mode of asexual reproduction in arthropods is the development of offspring from unfertilized eggs. This phenomenon, called **parthenogenesis**, also occurs in some species of fishes, amphibians, and reptiles. Most species that reproduce parthenogenetically also engage in sexual reproduction or at least sexual behavior at other times. In some species, parthenogenesis is part of the mechanism that determines sex. As we saw at the beginning of this chapter, in honey bees (as well as in most ants and wasps), males develop from unfertilized eggs and are haploid. Females develop from fertilized eggs and are diploid.

Parthenogenetic reproduction in some species requires sexual behavior even though sperm are not delivered to the female reproductive tract and eggs are not fertilized. David Crews and his students at the University of Texas extensively investigated one such case, that of parthenogenetic reproduction in a species of whiptail lizard. There are no males of this species. Females can act as males, engaging in all aspects of courtship display and mating, although no sperm are produced or transferred

(A) *Hydra* sp.

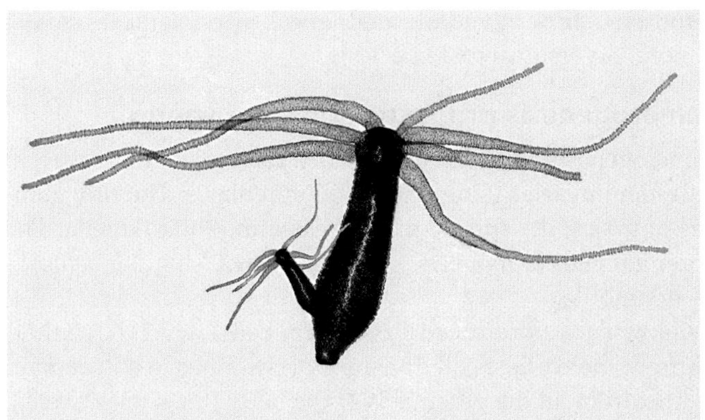

(B) *Fromia* sp.

**43.1 Two Forms of Asexual Reproduction** (A) Budding: A new individual forms as an outgrowth from an adult hydra. (The hydra in this photo was stained with a blue dye to make it easier to see.) (B) Regeneration: A single severed arm and a piece of the central disc of a mature sea star can regenerate into an entire animal.

(A)

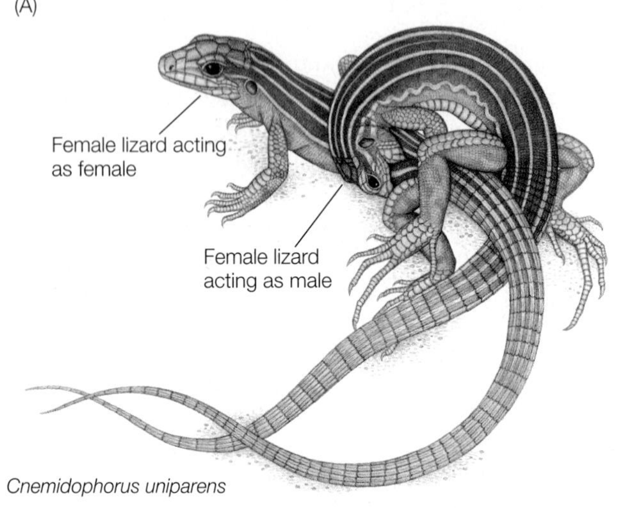


(B)

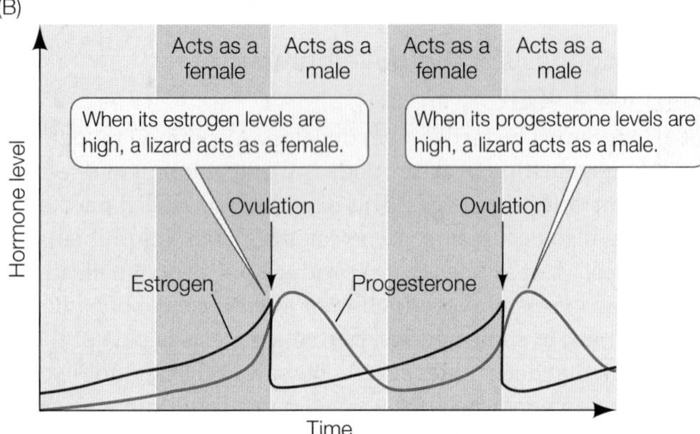


**43.2 Asexual Reproduction May Require Sexual Behavior** (A) Parthenogenetic whiptail lizards are all females, but they take turns acting the male role in reproductive behavior. The stimulation from sexual behavior is necessary for ovulation to occur. (B) The ovarian cycle determines the role an individual whiptail plays.

(**Figure 43.2**). Whether a specific female acts as a female or as a male depends on cyclical hormonal states. When estrogen levels are high, she acts as a female. When her progesterone levels peak, she acts as a male. The stimulation resulting from the sexual activity triggers the release of eggs from the ovaries of the acting female.

**RECAP 43.1**

Most animals reproduce sexually, but many can also or can only reproduce asexually, through budding, regeneration, or parthenogenesis.

- Explain why asexual reproduction might be disadvantageous for an animal living in a changing environment. **See p. 881**
- How is parthenogenesis related to sex determination in honey bees? **See p. 881**

Asexual reproduction is an efficient way to use resources. Since most animals reproduce sexually, however, the genetic diversity produced by sexual reproduction must confer a tremendous advantage.

## 43.2 How Do Animals Reproduce Sexually?

Given the efficiency of asexual reproduction in perpetuating an organism's genome, the prevalence of sexual reproduction is somewhat surprising. Because one gender (i.e., males) cannot produce offspring, it is a much less efficient strategy. And mating behaviors involve costs and risks. Costs include time and energy spent finding, attracting, and competing for a mate, as well as the "opportunity costs" of detracting from other activities such as feeding and caring for existing offspring. Risks include increased exposure to predation and the potential for physical damage. Despite these disadvantages, most eukaryotic organisms reproduce sexually. Thus it would seem that the production of genetic diversity is an evolutionary advantage that overwhelms the cost of sex (see Sections 21.2 and 21.4).

Sexual reproduction requires the joining of two haploid sex cells to form a diploid individual. These haploid cells, or **gametes**, are produced through gametogenesis, a process that involves meiotic cell divisions. Two events in meiosis contribute to genetic diversity: crossing over between homologous chromosomes and the independent assortment of chromosomes (see Sections 11.5 and 12.1). Sexual reproduction itself also contributes to genetic diversity. The genetic variation among the gametes of a single individual and the genetic variation between any two parents produce an enormous potential for genetic variation between any two offspring of a sexually reproducing pair of individuals.

Sexual reproduction in animals consists of three fundamental steps:

- **Gametogenesis**: making gametes
- **Spawning** or **mating**: bringing gametes together
- **Fertilization**: fusing gametes

The process of gametogenesis is similar across sexually reproducing animal species. Processes of fertilization are also quite similar in widely different species. Therefore, while our discussion of gametogenesis will focus generally on mammals, and our discussion of fertilization will feature sea urchins, the facts would not be dramatically different were we to consider many other animal groups. Adaptations for spawning and mating, in contrast, show incredible anatomical, physiological, and behavioral diversity across species.

### Gametogenesis produces eggs and sperm

Gametogenesis occurs in the gonads: **testes** (singular *testis*) in males and **ovaries** (singular *ovary*) in females. The tiny gametes of males, the **sperm**, move by beating their flagella. The larger gametes of females—the eggs or **ova** (singular *ovum*)—are nonmotile.

Gametes are produced from **germ cells**, which have their origin in the earliest cell divisions of the embryo and remain distinct from all the other cells of the body (the somatic cells). Germ cells are sequestered in the body of the embryo until its gonads begin to form. The germ cells then migrate to the developing gonads, where they take up residence and proliferate by

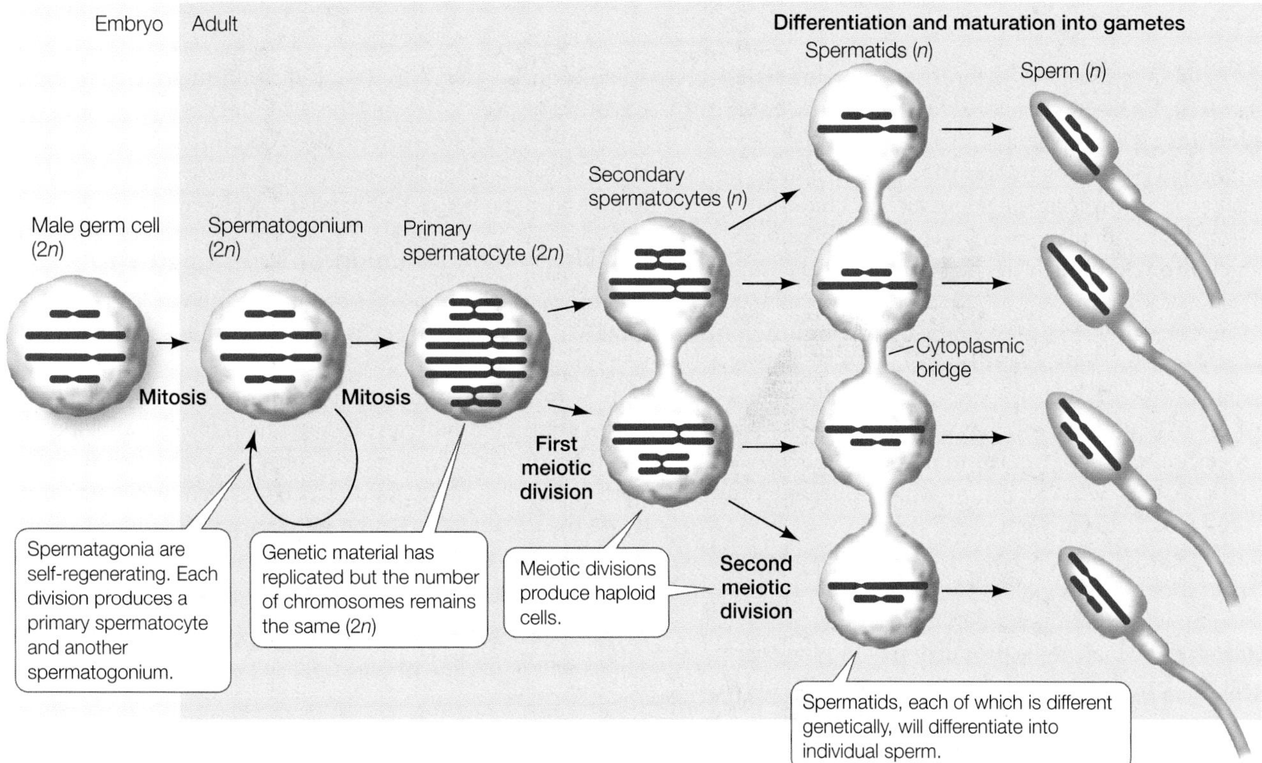


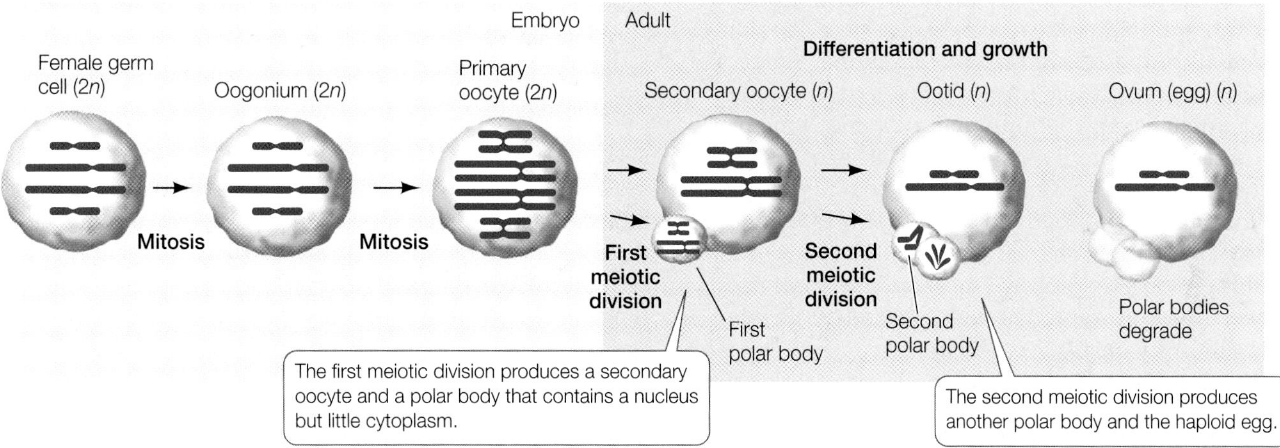


**43.3 Gametogenesis** Male and female germ cells proliferate by mitosis and produce diploid spermatogonia and oogonia that mature into primary spermatocytes and oocytes before entering meiosis. (A) Spermatogonia continue to divide by mitosis in adults, producing a steady supply of spermatocytes that divide meiotically to produce haploid spermatids, which differentiate into sperm. In many species, the progeny of spermatocytes remain in contact through cytoplasmic bridges until the sperm mature. (B) In mammals, oogonia cease division in the embryo, and primary oocytes remain arrested in prophase I of meiosis until they are ovulated and fertilized. Each oocyte will produce one haploid ootid, which matures into an ovum.

mitosis, producing **spermatogonia** (singular *spermatogonium*) in males and **oogonia** (singular *oogonium*) in females (**Figure 43.3**). Spermatogonia and oogonia are diploid, multiply by mitosis, and are stem cells. They are self-regenerating, and they also produce progeny—spermatocytes and oocytes—that will enter the next stage of gametogenesis.

In this next stage of gametogenesis, meiotic cell division reduces the chromosomes to the haploid number (see Section 11.5). The progeny of the spermatogonia and oogonia that enter meiosis are **primary spermatocytes** and **primary oocytes**. The steps of meiosis are similar in males and females, but there are important sex differences in gametogenesis.

**SPERMATOGENESIS** The initial proliferation of male germ cells into spermatogonia proceeds by mitosis in the embryo. But throughout the male life span, spermatogonia continue to divide by mitosis, with one daughter cell retaining the self-regenerating capacity of a spermatogonium while the other daughter cell becomes a primary spermatocyte. As illustrated in Figure 43.3A, primary spermatocytes then undergo the first meiotic division—the reduction division—to form two haploid **secondary spermatocytes**. The second meiotic division produces four haploid **spermatids** for each primary spermatocyte that enters meiosis. In mammals, the progeny of primary spermatocytes remain connected by cytoplasmic bridges after each division.

One reason that mammalian spermatocytes remain in cytoplasmic contact throughout their development is the asymmetry of sex chromosomes in males. Half the secondary spermatocytes receive an X chromosome, the other half a Y chromosome. The Y chromosome contains fewer genes than the X chromosome, and some of the products of genes found only on the X chromosome are essential for spermatocyte development. By remaining in cytoplasmic contact, all four spermatocytes can share the gene products of the X chromosomes, although only half of them have an X chromosome.

A spermatid bears little resemblance to a mature sperm. Through further differentiation (spermiogenesis), the spermatid becomes compact, streamlined, and grows a flagellum to become motile. We will look at the production of human sperm in Section 43.3.

OOGENESIS Oogonia, like spermatogonia, proliferate through mitosis (see Figure 43.3B). The resulting primary oocytes immediately enter prophase of the first meiotic division. In many species, including humans, the oocyte experiences developmental arrest at this point and may remain in that state for days, months, or years. In the human female, this period of arrest is at least 10 years (i.e., until puberty), and some primary oocytes remain in prophase I for up to 50 years (i.e., until menopause). In contrast, spermatogenesis continues, uninterrupted, to completion once the primary spermatocyte has differentiated.

During this prolonged prophase I, or shortly before it ends, the primary oocyte grows larger through increased production of ribosomes, RNA, cytoplasmic organelles, and energy stores. At this point the primary oocyte acquires all the energy, raw materials, and RNA that the ovum will need to survive its first cell divisions after fertilization. In fact, the nutrients in the egg must maintain the embryo until it is either nourished by the maternal circulatory system or can feed on its own.

When a primary oocyte resumes meiosis, its nucleus completes the first meiotic division near the surface of the cell. The daughter cells of this division receive grossly unequal shares of cytoplasm. This asymmetry represents another major difference from spermatogenesis, in which cytoplasm is apportioned equally. The daughter cell that receives almost all the cytoplasm becomes the **secondary oocyte**, and the one that receives almost none forms the **first polar body** (see Figure 43.3B).

The second meiotic division—that of the large, secondary oocyte—is also accompanied by an asymmetrical division of the cytoplasm. One daughter cell forms the large, haploid **ootid**, which eventually differentiates into a mature egg, and the other forms the **second polar body**. Polar bodies degenerate, so the end result of oogenesis is only one mature egg for each primary oocyte that entered meiosis. However, that egg is a large, well-provisioned cell.

A second period of arrested development occurs after the first meiotic division forms the secondary oocyte. The egg may be expelled from the ovary in this condition. In many species, including humans, the second meiotic division is not completed until the egg is fertilized by a sperm.

## Fertilization is the union of sperm and egg

The union of the haploid sperm and the haploid egg in fertilization creates a single diploid cell, called a **zygote**, which will develop into an embryo. Fertilization does more than just restore the full genetic complement of the animal. The processes associated with fertilization help the egg and sperm get together, prevent the union of the sperm and egg of different species, and guarantee that only one sperm will enter and activate the egg metabolically. Fertilization involves a complex series of events:

1. The sperm and the egg chemically recognize each other.
2. The sperm is activated, enabling it to gain access to the plasma membrane of the egg.
3. The plasma membrane of the egg fuses with the plasma membrane of a single sperm.
4. The egg blocks entry of additional sperm.
5. The egg is metabolically activated and stimulated to start development.
6. The egg and sperm nuclei fuse to create the diploid nucleus of the zygote.

SPECIFICITY IN SPERM–EGG INTERACTIONS Specific recognition molecules mediate interactions between sperm and eggs. These molecules ensure that the activities of sperm are directed toward eggs and not other cells, and they help prevent eggs from being fertilized by sperm from the wrong species. The latter function is particularly important in aquatic species that release eggs and sperm into the surrounding water, because the eggs of such animals may readily be exposed to sperm of other species. The sea urchin is a good example of such a species, and its mechanisms of fertilization have been well studied.

Sea urchin eggs release chemical attractants that increase the motility of sperm and cause them to swim toward the egg. These chemical attractants are species-specific. For example, eggs of one species of sea urchin release a specific peptide consisting of 14 amino acids. This peptide binds to receptors present on sperm of the same species. The sperm respond by increasing their mitochondrial respiration and motility. Before exposure to the peptide, the sperm swim in tight little circles, but after binding to the peptide, they swim energetically up the concentration gradient of the peptide until they reach the egg that is releasing it.

When sperm reach an egg, they must get through two protective layers before they can fuse with the egg plasma membrane. The eggs of sea urchins are covered with a jelly coat, which surrounds a proteinaceous **vitelline envelope** (**Figure 43.4A**). The success of a sperm's assault on these protective layers depends on a membrane-enclosed structure at the front of the sperm head called an **acrosome**.

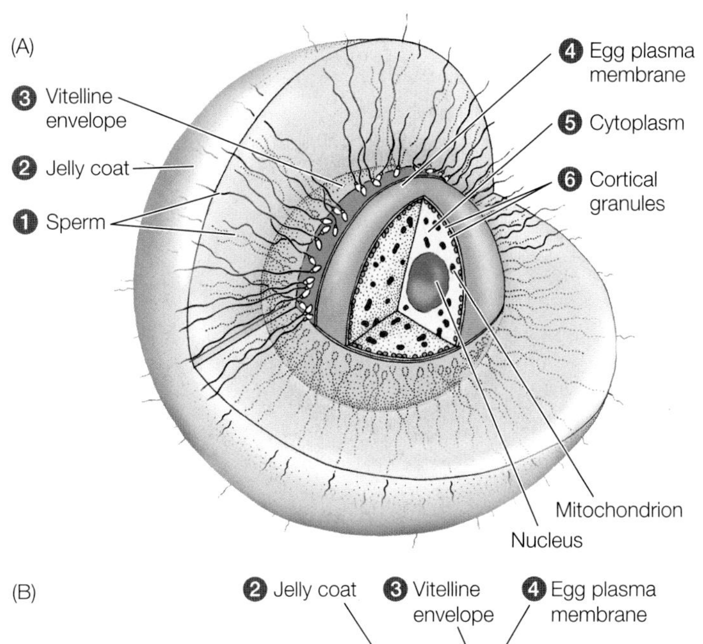

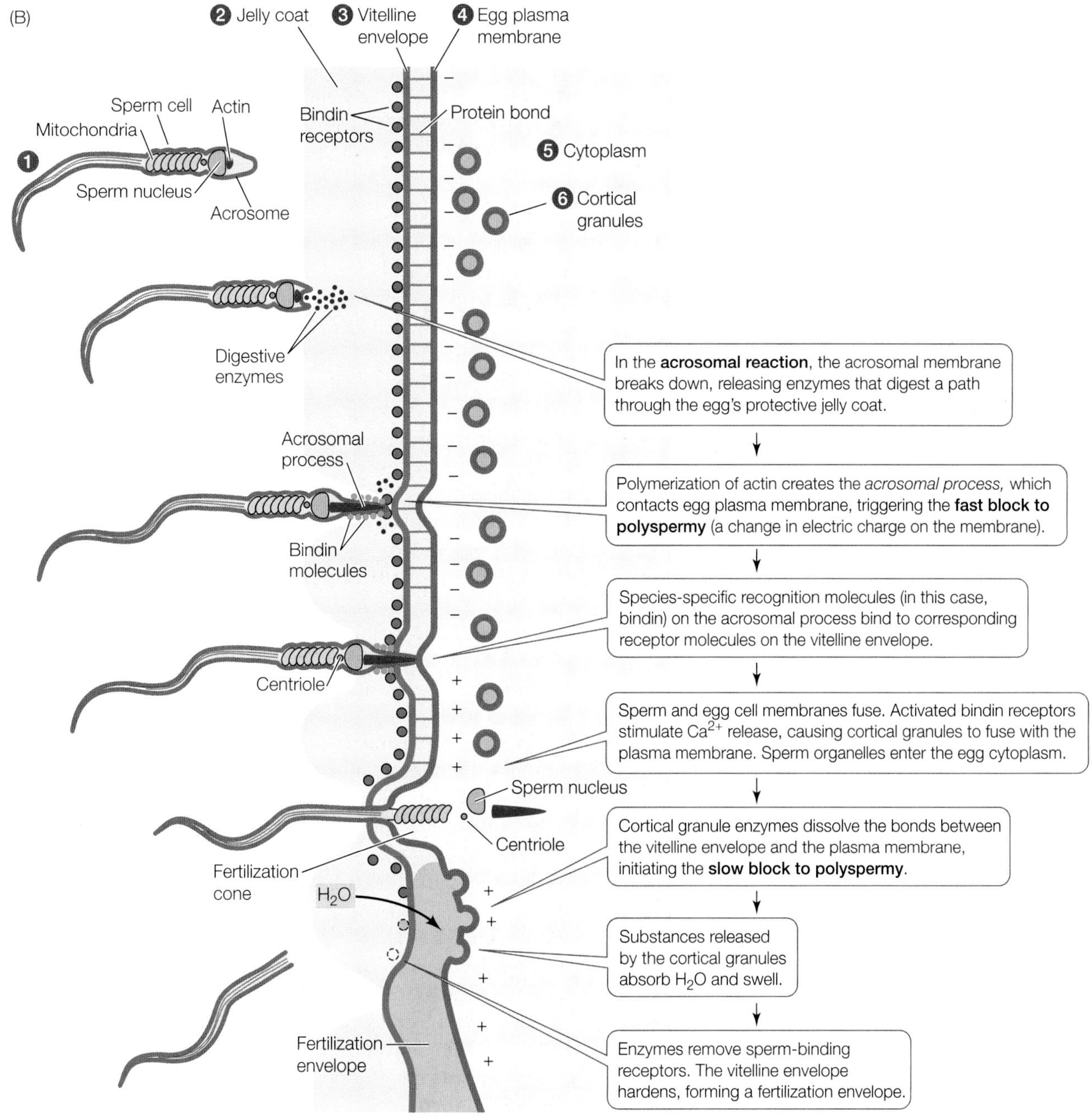

**43.4 Fertilization of the Sea Urchin Egg** (A) Sea urchin eggs are protected by a jelly layer and a proteinaceous vitelline envelope. Sperm must penetrate both to reach the egg plasma membrane. Many sperm attach to the vitelline envelope, but only one penetrates the egg cell membrane and achieves fertilization. Circled numbers match structures with the events shown in panel B. (B) The acrosomal reaction allows a sea urchin sperm to recognize an egg of the same species and pass through its protective layers. Enzymes from the egg's cortical granules trigger the slow block to polyspermy.

**Go to Animated Tutorial 43.1**
**Fertilization in a Sea Urchin**
Life10e.com/at43.1

(B)

2 Jelly coat
3 Vitelline envelope
4 Egg plasma membrane
Sperm cell
Actin
Mitochondria
1
Sperm nucleus
Acrosome
Bindin receptors
Protein bond
5 Cytoplasm
6 Cortical granules
Digestive enzymes

In the **acrosomal reaction**, the acrosomal membrane breaks down, releasing enzymes that digest a path through the egg's protective jelly coat.

Acrosomal process
Bindin molecules

Polymerization of actin creates the *acrosomal process,* which contacts egg plasma membrane, triggering the **fast block to polyspermy** (a change in electric charge on the membrane).

Species-specific recognition molecules (in this case, bindin) on the acrosomal process bind to corresponding receptor molecules on the vitelline envelope.

Centriole

Sperm and egg cell membranes fuse. Activated bindin receptors stimulate $Ca^{2+}$ release, causing cortical granules to fuse with the plasma membrane. Sperm organelles enter the egg cytoplasm.

Sperm nucleus
Centriole
Fertilization cone
$H_2O$

Cortical granule enzymes dissolve the bonds between the vitelline envelope and the plasma membrane, initiating the **slow block to polyspermy**.

Substances released by the cortical granules absorb $H_2O$ and swell.

Fertilization envelope

Enzymes remove sperm-binding receptors. The vitelline envelope hardens, forming a fertilization envelope.

The acrosome contains enzymes and other proteins. When the sperm makes contact with an egg of its own species, substances in the jelly coat trigger an acrosomal reaction, which begins with the breakdown of the plasma membrane covering the sperm head and the underlying acrosomal membrane (**Figure 43.4B**). The acrosomal enzymes are released and digest a hole through the jelly coat.

As a result of the polymerization of actin triggered by the acrosomal reaction, an acrosomal process extends out of the head of the sperm. The acrosomal process is coated with species-specific recognition molecules called bindin, and there are bindin receptors on the vitelline envelope of the egg. The interaction of bindin and bindin receptors enables the sperm to contact the egg plasma membrane. That contact results in fusion of the sperm and egg plasma membranes and the formation of a fertilization cone that engulfs the sperm head, bringing it into the egg cytoplasm. The sperm mitochondria, which largely constitute the midpiece of the sperm, are also drawn into the egg cytoplasm, but they degrade and disappear; this means that the mitochondria and mitochondrial genes of the new urchin are derived only from the egg.

In animals that practice internal fertilization, mating behaviors help guarantee species specificity, but egg–sperm recognition mechanisms still exist. The mammalian egg is surrounded by a thick layer called the **cumulus**, which consists of a loose assemblage of maternal cells in a gelatinous matrix (**Figure 43.5**). Beneath the cumulus is a glycoprotein envelope called the **zona pellucida**, which is functionally similar to the vitelline envelope of sea urchin eggs. When mammalian sperm are deposited in the female reproductive tract, they are metabolically activated and made capable of an acrosomal reaction should they encounter an egg. An activated sperm can penetrate the cumulus and interact with the zona pellucida.

Unlike the jelly coat of sea urchin eggs, the cumulus of mammalian eggs does not trigger the acrosomal reaction. When sperm make contact with the zona pellucida, a species-specific glycoprotein binds to recognition molecules on the head of the sperm. This binding triggers the acrosomal reaction, releasing acrosomal enzymes that digest a path through the zona pellucida. When the sperm head reaches the egg plasma membrane, other proteins facilitate its adhesion to and fusion with the egg plasma membrane.

The importance of the zona pellucida and its sperm-binding molecules as a species-specific recognition mechanism was revealed in experiments on mammalian eggs and sperm in culture dishes. When the zona pellucida was stripped from human eggs and the eggs were exposed to hamster sperm, fertilization took place, resulting in a hamster–human hybrid zygote. The hybrid zygote did not survive its first cell division, but the experiment demonstrated that a recognition mechanism in mammalian species resides in the zona pellucida.

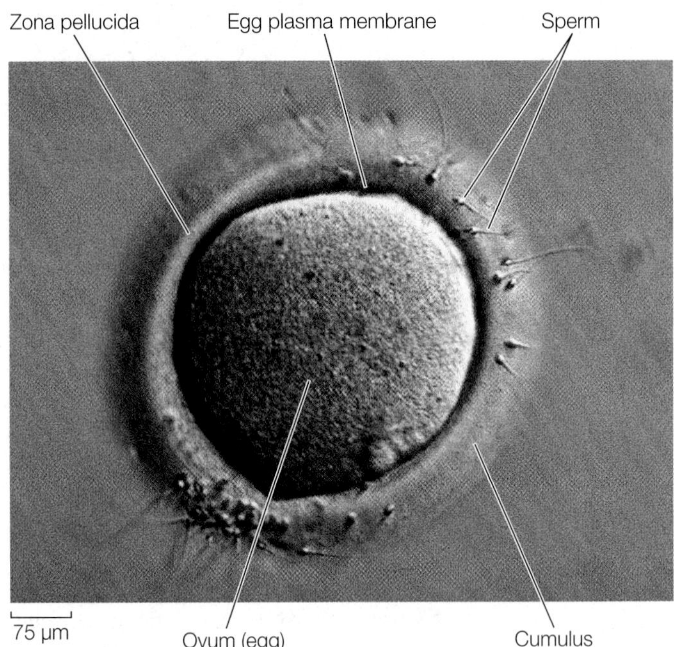


**43.5 Barriers to Sperm** This human egg, like other mammalian eggs, is surrounded by the cumulus and zona pellucida. Sperm must penetrate both to fertilize the egg. Only one sperm will penetrate the zona pellucida and fuse with the plasma membrane.

**BLOCKS TO POLYSPERMY** The fusion of the sperm and egg plasma membranes and the entry of the sperm into the egg initiate a programmed sequence of events. The first responses to sperm entry are **blocks to polyspermy**. These blocks are mechanisms that prevent more than one sperm from entering the egg. If more than one sperm enters the egg, the embryo is unlikely to survive.

Blocks to polyspermy have been studied extensively in sea urchin eggs, which can be fertilized in a dish of seawater. Within seconds after the sperm membrane contacts the egg membrane, an influx of sodium ions changes the electric charge difference across the egg's plasma membrane. This fast block to polyspermy prevents the fusion of any other sperm with the egg plasma membrane, but it is transient. The change in membrane electric charge lasts only about a minute, but that is enough time to allow a slower block to sperm entry to develop.

The slow block to polyspermy involves converting the vitelline envelope to a physical barrier that sperm cannot penetrate. Before fertilization, the vitelline envelope is bonded to the egg plasma membrane. Just under the plasma membrane are vesicles called cortical granules (see Figure 43.4), which contain enzymes and other proteins.

The sea urchin egg, like all animal cells, sequesters calcium in its endoplasmic reticulum. Sperm entry into the egg stimulates the release of calcium ions from the endoplasmic reticulum and into the egg cytoplasm. This increase in cytosolic calcium causes the egg's cortical granules to fuse with the plasma membrane and release their contents. Cortical granule enzymes break the bonds between the vitelline envelope and the plasma membrane, and other proteins released from the cortical granules attract water into the space between them. As a result, the vitelline envelope rises to form a fertilization envelope. Cortical granule enzymes also degrade sperm-binding molecules on the surface of the fertilization envelope and cause it to harden, thus preventing additional sperm from contacting the egg's plasma membrane.

In mammalian eggs, sperm entry does not cause a rapid change in membrane potential, but it does trigger a release of calcium from the endoplasmic reticulum. As in the sea urchin egg, increased calcium causes the cortical granules to fuse egg with the egg plasma membrane. A fertilization envelope does not form around the mammalian egg, but the cortical granule enzymes destroy the sperm-binding molecules in the zona pellucida. The rise in cytosolic calcium also signals the egg to complete meiosis. The stage is set for the first cell division.

## Getting eggs and sperm together

As we have just seen, sexual reproduction requires the production of haploid gametes (gametogenesis) and the joining together of those gametes to form a diploid zygote (fertilization). Spawning and mating behaviors get eggs and sperm close enough together that fertilization can occur. Fertilization can occur externally or internally.

EXTERNAL FERTILIZATION In an aquatic environment, animals can bring their gametes together by simply releasing them into the water. This practice, called **spawning**, results in **external fertilization**. Many aquatic animals are not very mobile, but they produce huge numbers of gametes that can travel far from the point of release. A female oyster, for example, will release millions of eggs when she spawns, and the number of sperm produced by a male oyster is astronomical.

Numbers alone, however, do not guarantee that gametes will meet. The reproductive activities of the males and females of a population must be synchronized, since released gametes have a limited life span. Seasonal breeders may use day length, changes in temperature, or changes in weather to time the production and release of their gametes. Mutual stimulation is also important. Release of gametes into the water by one individual can stimulate others to spawn.

Behavior can play an important role in bringing gametes together even when fertilization is external. Many species travel great distances to congregate with potential mates and release their gametes at the same time in a suitable environment. Salmon are an extreme example. They hatch and develop in freshwater streams and then migrate to the ocean, where they remain for years. When they are mature, they travel hundreds of miles back in the stream where they hatched to spawn. Males and females expend great amounts of energy to swim up the streams to the spawning grounds, where they pair up, prepare a depression in the streambed gravel, and together release their sperm and eggs. As the gametes drift down into the gravel, fertilization occurs.

INTERNAL FERTILIZATION Terrestrial animals cannot simply release their gametes into the environment. Sperm can move only through liquid, and delicate gametes released into air would dry out and die. Terrestrial animals avoid these problems by **internal fertilization**, the release of sperm into the female reproductive tract. Some aquatic animals also practice internal fertilization, but it is ubiquitous in terrestrial animals.

Animals have evolved an astonishing diversity of behavioral and anatomical adaptations for internal fertilization. As we saw above, gametogenesis occurs in the gonads, which are the **primary sex organs**. All additional anatomical components of an animal's reproductive system are called **accessory sex organs**. An obvious accessory sex organ in males of many species is the **penis**, which enables the male to deposit sperm in the female's **vagina**, the entry to her reproductive tract. Accessory sex organs include a variety of glands, tubules, ducts, and other structures.

**Copulation** is the physical joining of male and female accessory sex organs. Most male insects copulate and transfer sperm to the female's vagina through a penis. The **genitalia**—external sex organs—of insects often have species-specific shapes that ensure that the male and female genitalia match in a lock-and-key fashion. This mechanism ensures a tight, secure fit between the mating pair during the prolonged period of sperm transfer. In some insect species in which females mate with more than one male, the males have elaborate structures on their penises that can scoop sperm deposited by other males out of a female's reproductive tract, replacing it with their own.

Transfer of sperm in internal fertilization can also be indirect. Males of many invertebrate species (e.g., mites and scorpions) and a few vertebrates (e.g., salamanders) deposit spermatophores—packets of sperm protected from desiccation—in the environment. When a female mite encounters a spermatophore from a potential mate, she straddles it and opens a pair of plates in her abdomen so that the tip of the spermatophore enters her reproductive tract and allows the sperm to enter.

Male squids and spiders play a more active role in spermatophore transfer. The male spider secretes a drop containing sperm onto a bit of web, then uses a special structure on his foreleg to pick up the sperm-containing web and insert it through the female's genital opening. Male squids use one specialized tentacle to pick up a spermatophore and insert it into the female's genital opening.

## Some individuals can function as both male and female

In most species, gametes are produced by individuals that are either male or female. Species that have separate male and female members are referred to as **dioecious** (Greek, "two houses"). In some species, however, a single individual may produce both sperm and eggs. Such species are **monoecious** ("one house") or **hermaphroditic**.

Almost all invertebrate groups contain some hermaphroditic species. An earthworm is an example of a *simultaneous* hermaphrodite, meaning an individual is both male and female at the same time. When two earthworms mate, they exchange sperm, and as a result, the eggs of each are fertilized (see Figure 32.12C). Some vertebrates are *sequential* hermaphrodites, meaning that an individual may function as a male or a female at different times in its life. An example is the anemone fish, or clown fish, a species that lives in small groups within large sea anemones (**Figure 43.6**). All anemone fish are born male. The largest fish in a group becomes a functional female. If that fish is removed from the group, the next-largest male becomes a female. The second-largest fish in the group is the only male in breeding condition.

What is the evolutionary advantage of hermaphroditism? Some simultaneous hermaphrodites, such as parasitic

*Amphiprion* sp.

**43.6 When Size Determines Sex** Anemone fish (also known as clown fish) live in groups of about a dozen centered on a single sea anemone. All anemone fish are born male; the largest fish in the group becomes a functional female. Thus a fish may function as a male and as a female at different times in its life.

tapeworms, have a low probability of meeting a potential mate—an individual tapeworm may be the only one in its host. Tapeworms can fertilize their own eggs, but most simultaneous hermaphrodites must mate with another individual. Because every simultaneous hermaphrodite is both male and female, however, the probability of encountering a possible mate doubles. In some sequential hermaphrodites, siblings are all male or all female at the same time, thus reducing the incidence of inbreeding.

## The evolution of vertebrate reproductive systems parallels the move to land

The earliest vertebrates evolved in aquatic environments. The closest living relatives of those earliest vertebrates are modern-day fishes. They remain exclusively aquatic, and most practice external fertilization. The most primitive of the fishes, the lampreys and hagfishes, simply release their gametes into the environment. In most fishes, however, mating behaviors bring females and males into close proximity at the time of gamete release. In sharks and rays, fins have evolved into claspers that hold the male and female together and enable sperm to be transferred directly into the female reproductive tract.

Amphibians were the first vertebrates to live in terrestrial environments. They dealt with the challenge of a dry environment by returning to water to reproduce, as most amphibians still do today.

Reptiles were the first vertebrate group to solve the problem of reproduction in the terrestrial environment (**Figure 43.7**). Their solution was the amniote egg (see Figure 33.19A). A hard or leathery shell protects the embryo and impedes water loss while allowing the diffusion of oxygen into the egg and carbon dioxide out of the egg. The eggshell creates an obvious problem for fertilization. Sperm cannot penetrate the shell, so they must reach the egg before the shell forms. Hence internal fertilization and the evolution of accessory sex organs were necessary for the evolution of the amniote egg.

Male snakes and lizards have paired hemipenes, which can be filled with blood and thereby extruded from the male's body. Only one hemipenis is inserted into the female's reproductive tract at a time. It is usually rough or spiny at the end to achieve a secure hold while sperm are transferred down a groove on its surface. Retractor muscles pull the hemipenis back into the male's body when mating is completed. Some evolutionarily ancient bird species have erectile penises that channel sperm along a groove into the female's reproductive tract. Birds with more recent evolutionary origins, however, do not have erectile penises; instead, the male and female simply bring their genital openings close together to transfer sperm.

(A) *Chelonia mydas*

(B) *Merops apiaster*

**43.7 The Shelled Egg** The shelled egg was a major evolutionary step that allows reptiles and birds to reproduce in the terrestrial environment. (A) A female green sea turtle has deposited her eggs in the sand. (B) The shelled egg requires that sperm meet egg before the shell forms. Terrestrial animals thus must practice internal fertilization, as these European bee-eaters are doing.

Usually this involves the male standing on the female's back (see Figure 43.7B).

All mammals practice internal fertilization. With the exception of the prototherian mammals, the developing embryo is retained for some time in the female reproductive tract. Prototherian mammals (the monotremes; see Figure 33.24) lay eggs. The other mammals (the therians) vary enormously as to the developmental stage of their offspring at the time of birth.

### Animals with internal fertilization are distinguished by where the embryo develops

Two patterns of care and nurture of the embryo have evolved in animals: oviparity (egg laying) and viviparity (live bearing).

**Oviparous** animals lay eggs in the environment, and their embryos develop outside the mother's body. Oviparous terrestrial animals such as insects, reptiles, and birds protect their eggs from desiccation with waterproof membranes or shells. Oviparity is possible because eggs are stocked with abundant nutrients to supply the needs of the embryo. Some oviparous animals engage in various forms of parental behavior to protect their eggs, but until the eggs hatch, the embryos depend entirely on the nutrients stored in the egg.

**Viviparous** animals retain the embryo within the mother's body during its early developmental stages. Although examples of viviparity exist in all vertebrate groups except the crocodiles, turtles, and birds (even some sharks retain fertilized eggs in their bodies and give birth to free-living offspring), there is a big difference between viviparity in mammals and in other species.

All mammals except the prototherians are viviparous and have a specialized portion of the female reproductive tract, the **uterus**, or womb, that holds the embryo and interacts with it to produce a **placenta**, which enables the exchange of nutrients and wastes between the blood of the mother and that of the embryo. Very few non-mammalian species have evolved such a connection between the embryo and the mother.

In most non-mammalian viviparous animals, such as garter snakes and the well-known aquarium fish the guppy, fertilized eggs are retained in the mother's body until they hatch. These embryos still receive nutrition from stores in the egg, so this reproductive adaptation is called **ovoviviparity**.

RECAP  43.2

Sexual reproduction involves gametogenesis, mating, and fertilization. Fertilization can be external or internal and involves mechanisms for ensuring that only one sperm from the right species enters the egg.

- Describe the steps by which a sea urchin sperm penetrates the egg. **See Figure 43.4**
- Explain how polyspermy is prevented and why it is crucial to do so. **See pp. 886–887 and Figure 43.4**
- What reproductive adaptations made life on land possible? **See p. 888**

Now that we have covered some of the general aspects of gametogenesis and fertilization and have briefly discussed the great diversity of mating systems, we will next consider the human reproductive systems in detail.

##  43.3 How Do the Human Male and Female Reproductive Systems Work?

In this section we describe the structures and functions of male and female reproductive systems in mammals, and we will use humans as our prime example. We will also discuss the hormonal regulation of both male and female systems. Our discussion covers:

- *The primary sex organs* (testes in males and ovaries in females) that produce gametes and serve endocrine functions.
- *The accessory sex organs*, which include the ducts through which the gametes pass, the various glands that empty into those ducts, and the external genitalia.
- *Secondary sexual characteristics*, which are not directly involved in reproduction but are responsible for the major differences in external appearance of men and women and are important in mating and in rearing offspring.

### Male sex organs produce and deliver semen

**Semen** is the product of the male reproductive system. Semen contains sperm and a complex mixture of fluids and molecules that support the sperm and facilitate fertilization. Sperm make up less than 5 percent of the volume of the semen.

The male reproductive organs are diagrammed in **Figure 43.8**. Sperm are produced in the testes, the paired male gonads. The testes of most mammals are located outside the body cavity in a pouch of skin called the **scrotum**.

Why should the testes be located outside the body cavity? The optimal temperature for spermatogenesis in most mammals is slightly lower than the normal body temperature. The scrotum keeps the testes at this optimal temperature. Muscles in the scrotum contract in a cold environment, bringing the testes closer to the warmth of the body; in a hot environment they relax, cooling the testes by suspending them farther from the body.

Spermatogenesis takes place within the **seminiferous tubules** tightly coiled in each testis (**Figure 43.9A**). Between the seminiferous tubules are clusters of **Leydig cells** that produce testosterone (**Figure 43.9B**). Spermatogonia reside in the outer regions of the seminiferous tubules, just under the basement membrane. Moving inward toward the lumen of the tubule are germ cells in successive stages of spermatogenesis (**Figure 43.9C**). The germ cells are intimately associated with **Sertoli cells** that provide nutrients for the developing sperm.

When the second meiotic division is complete, each primary spermatocyte has produced four spermatids (see Figure 43.3A). The spermatids develop into spermatozoa as they migrate toward the lumen of the seminiferous tubule. The nucleus becomes compact, and the surrounding cytoplasm is lost. A flagellum—the sperm tail—develops. The mitochondria that provide the energy for sperm motility become condensed into a midpiece between the head and tail. An acrosome forms over the nucleus in the head of the sperm. Immature sperm are shed into the lumen of the seminiferous tubule.

(A) Posterior view

Urinary bladder
Ureter
Seminal vesicle
Ejaculatory duct
Prostate gland
Bulbourethral gland
Erectile tissue
Urethra
Penis
Glans penis

Seminal fluids are produced by the **seminal vesicles**, **prostate gland**, and **bulbourethral glands**.

Sperm are delivered to the urethra by the vas deferens which joins the seminal vesicle duct in the prostate gland to form the ejaculatory duct.

Sperm are stored and mature in the **epididymis**.

Sperm are produced in the **testes**.

Semen is ejaculated through the male copulatory organ, the **penis**.

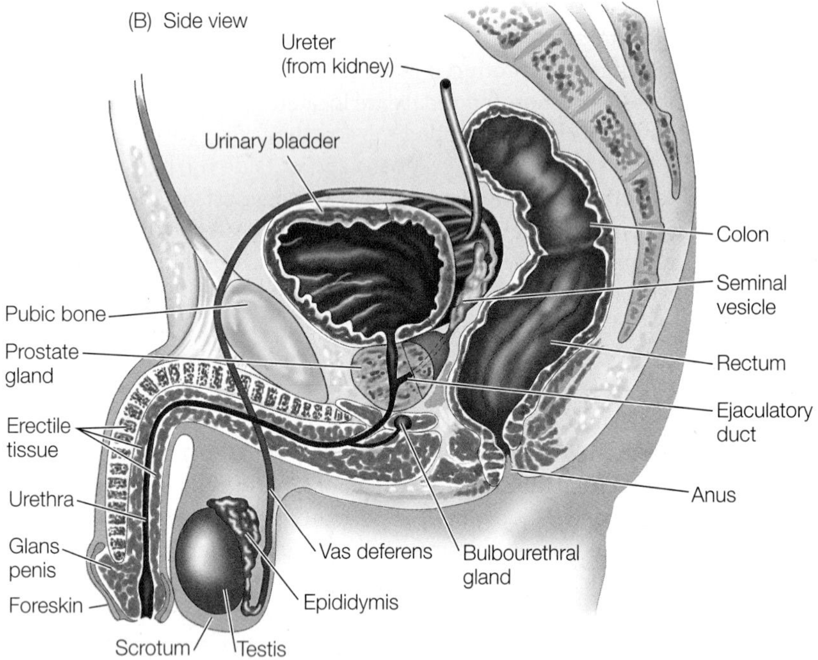


**43.8 Reproductive Tract of the Human Male** The male reproductive organs are shown (A) from the rear and (B) from the side.

**Go to Activity 43.1 The Human Male Reproductive Tract Life10e.com/ac43.1**

From the seminiferous tubules, sperm move into the **epididymis** (see Figure 43.8), where they mature, become motile, and are stored. The epididymis connects to the **urethra** via the **vas deferens** (plural *vasa deferentia*) and the **ejaculatory duct.**

The urethra originates in the bladder, runs through the penis, and opens to the outside of the body at the tip of the penis. It serves as the common final duct for the urinary and reproductive systems.

The components of the semen other than sperm come from several accessory glands. About 60 percent of the volume of semen is secreted by the paired **seminal vesicles**, which empty into the vas deferens just before it joins the urethra. Seminal fluid is thick because it contains mucus and fibrinogen, a protein also found in the blood, where it can polymerize to form blood clots. Seminal fluid also contains the monosaccharide fructose, an energy source for the sperm.

The **prostate gland** contributes about 30 percent of the volume of the semen. Prostate fluid is alkaline, so it neutralizes the acidity in the male and female reproductive tracts and makes these environments more hospitable to sperm. The prostate also secretes a clotting enzyme that causes fibrinogen from the seminal vesicles to convert the semen into a clotted, gelatinous mass, facilitating the semen's propulsion into the upper regions of the female reproductive tract. Another prostate enzyme, fibrinolysin, then dissolves the clotted semen and liberates the sperm. Prostaglandins—hormones produced by the prostate—stimulate contractions of the female reproductive tract.

The **bulbourethral glands** produce a small volume of an alkaline, mucoid secretion that helps neutralize acidity in the urethra and lubricate it to facilitate the passage of semen immediately preceding the climax of sexual intercourse. These secretions can carry with them residual sperm from prior sexual activity. It is therefore possible for pregnancy to occur even if the penis is withdrawn just before climax (an ancient but ineffective birth control practice known as coitus interruptus).

The penis and the scrotum are the male genitalia. The shaft of the penis is covered with normal skin, but the highly sensitive tip, the **glans penis**, is covered with thinner, more sensitive skin that is especially responsive to sexual stimulation. A fold of skin called the foreskin covers the glans of the human penis. The procedure known as circumcision removes a portion of the foreskin.

Sexual stimulation triggers responses in the nervous system that result in penile **erection**. Nerve endings release a neurotransmitter that causes the endothelial cells lining the penile blood vessels to release a gaseous neurotransmitter, nitric oxide (NO). NO diffuses into the muscle cells that control the diameter of the penile arteries and stimulates them to produce the second messenger cGMP (see Figure 7.15). Increased cGMP in these muscle cells causes them to relax; the arteries dilate and carry more blood into the penis. Increased blood flow swells

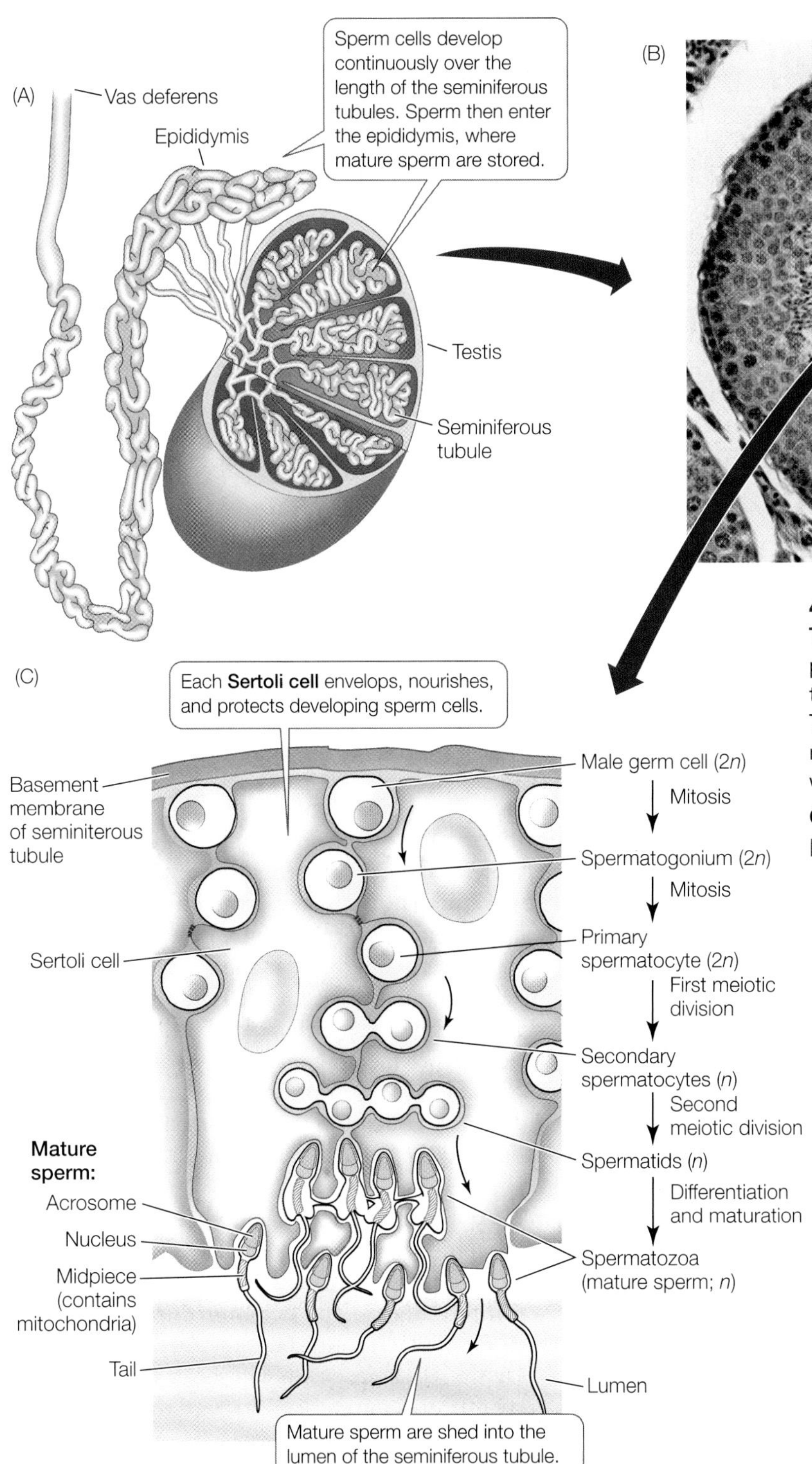


**43.9 Spermatogenesis Takes Place in the Seminiferous Tubules** (A) Seminiferous tubules fill the testes, continuously producing millions of sperm. (B) Cross section of seminiferous tubules and the Leydig cells in the spaces between them. (C) This longitudinal diagram shows how, as sperm mature, they move from the outer layer of the tubule toward the center, where they are shed into the lumen of the tubule.

**Go to Activity 43.2 Spermatogenesis Life10e.com/ac43.2**

the shafts of spongy erectile tissue located along the length of the penis. The enlargement of these blood-filled cavities compresses the vessels that carry blood out of the penis, and the erectile tissue becomes engorged with blood. The penis becomes hard and erect, facilitating its insertion into the vagina. Many species of mammals (though not humans) have a bone in the penis, but these species still depend on erectile tissue for copulation.

At the climax of copulation, 2 to 6 milliliters of semen are propelled through the vasa deferentia and the urethra in two steps, emission and ejaculation. During emission, rhythmic contractions of smooth muscles in the vasa deferentia and accessory glands move the semen into the urethra at the base of the penis. Ejaculation is caused by contractions of other muscles at the base of the penis surrounding the urethra. These contractions force the coagulum of semen through the urethra and out of the penis. The muscle contractions of ejaculation are accompanied by feelings of intense pleasure known as orgasm. They are also accompanied by transient increases in heart rate, blood pressure, breathing, pupil dilation and skeletal muscle contractions throughout the body.

After ejaculation, NO release decreases and enzymes break down cGMP, causing the blood vessels flowing into the penis to constrict. The blood pressure in the erectile tissue decreases, relieving the compression of the blood vessels leaving the penis, and the erection declines.

Erectile dysfunction (ED), or impotence, is the inability to achieve or sustain an erection. ED may have different causes, including cardiovascular disease. Drugs used to treat ED act by inhibiting the breakdown of cGMP, thus enhancing the effect of NO released in the penis, which improves the ability to achieve and maintain an erection.

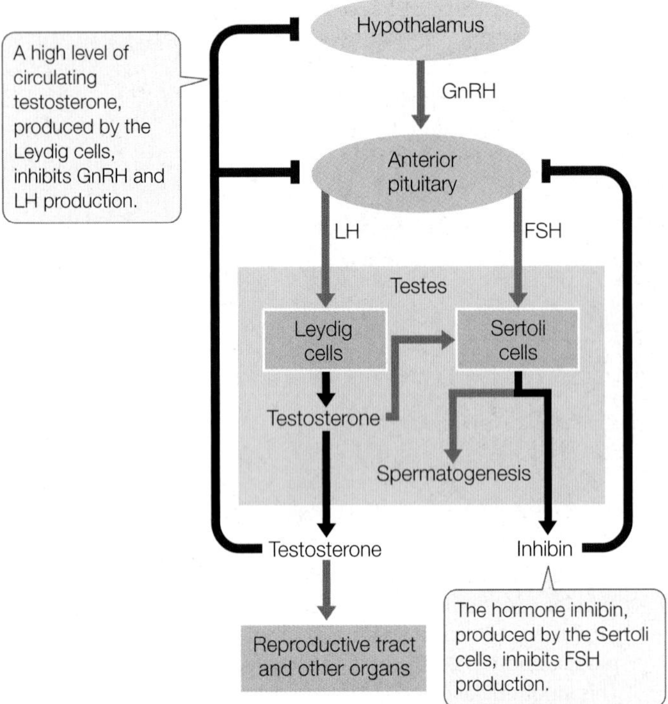


**43.10 Male Reproductive Hormones** The male reproductive system is under hormonal control by the hypothalamus and the anterior pituitary. Red lines indicate inhibition; green lines indicate stimulation.

## Male sexual function is controlled by hormones

Spermatogenesis and maintenance of male secondary sexual characteristics such as facial hair and a deep voice depend on testosterone produced by the Leydig cells of the testes. As described in Section 41.3, increased production of testosterone at puberty results from an increased release of gonadotropin-releasing hormone (GnRH) by the hypothalamus. GnRH stimulates anterior pituitary cells to secrete luteinizing hormone (LH) and follicle-stimulating hormone (FSH) (**Figure 43.10**). Higher levels of LH stimulate the Leydig cells to increase their production and release of testosterone. Testosterone exerts negative feedback on the anterior pituitary and the hypothalamus. At the time of puberty, the sensitivity of the hypothalamus to negative feedback from testosterone declines; as a result, the level of circulating testosterone increases.

Increased testosterone in pubertal boys causes the development of pubic and facial hair, a deeper voice, enlarged genitals, and an increased growth rate. Testosterone also promotes increased muscle mass and maturation of the testes. Continued production of testosterone after puberty is essential for the maintenance of secondary sexual characteristics and the production of sperm.

Spermatogenesis is controlled by the influence of FSH and testosterone on the Sertoli cells in the seminiferous tubules. The Sertoli cells also produce a hormone called inhibin that exerts negative feedback on the anterior pituitary cells producing and secreting FSH.

## Female sex organs produce eggs, receive sperm, and nurture the embryo

The human female reproductive system is shown in **Figure 43.11**. Eggs are produced in and released from the ovaries located on either side of the lower abdominal cavity. When an egg leaves the ovary (**ovulation**), it enters the abdominal cavity, but it does not go far. The ovaries are close to the openings of the **oviducts** (also known as the Fallopian tubes). The openings are surrounded by undulating, fringed tissues called fimbria that sweep the egg into the oviduct. Cilia lining the oviduct propel the egg toward the **uterus**, a muscular, thick-walled cavity shaped in humans like an upside-down pear. The uterus is where the embryo will develop if the egg is fertilized. At its bottom, the uterus narrows into a region called the **cervix**, which opens into the vagina.

Externally, the vagina is enclosed by two sets of skin folds. The inner, more delicate folds are the labia minora; the outer, thicker folds are the labia majora. At the anterior tip of the labia minora is the clitoris, a small bulb of highly sensitive erectile tissue that has the same developmental origins as the male glans penis. The labia minora and the clitoris become engorged with blood in response to sexual stimulation. Between the vagina and the clitoris is the opening of the urethra through which urine flows.

The external opening of an infant's vagina is usually, but not always, partly covered by a thin membrane, the hymen. Eventually the hymen can be torn by vigorous physical activity or by first sexual intercourse; it can sometimes make first intercourse difficult or painful for the woman.

To fertilize an egg, sperm deposited in the vagina swim and are propelled by contractions of the female reproductive tract through the cervical opening, across the uterus, and most of the way up an oviduct. Before meeting a sperm, the egg is still a secondary oocyte, and fertilization stimulates it to complete its second meiotic division. Following that division, the haploid nuclei of the sperm and the egg can fuse to produce a diploid zygote nucleus.

While still in the oviduct, the zygote undergoes its first few cell divisions to become a **blastocyst**. The blastocyst moves down the oviduct to the uterus, where it attaches itself to the epithelial lining of the uterus—the **endometrium**. Once attached, the blastocyst implants in the endometrium and interacts with it to form the placenta, as we will see in Chapter 44. The mother's body nurtures the embryo through the placenta, which also produces hormones that help sustain pregnancy.

Just as the maturation of eggs and ovulation in the ovaries is a cyclical process, so are critical events in the uterus. In anticipation of receiving a blastocyst, the endometrium thickens and develops lots of blood vessels. If a blastocyst does not arrive within a certain window of time, the endometrium regresses. Thus female reproductive functions consist of two linked cycles: an **ovarian cycle** that produces eggs and hormones; and a **uterine**, or **menstrual**, **cycle** that prepares the endometrium for the arrival of a blastocyst. The two cycles must be synchronized so that a blastocyst arrives in the uterus at the optimal time to embed in the endometrium and continue its development.

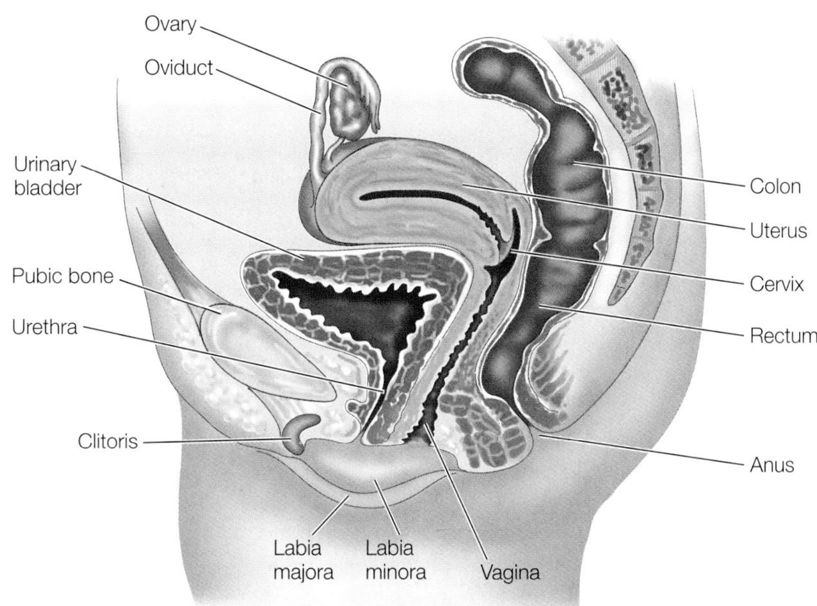


**43.11 Reproductive Tract of the Human Female** The female reproductive organs are shown (A) from the front and (B) from the side.

**Go to Activity 43.3 The Human Female Reproductive Tract Life10e.com/ac43.3**

## The ovarian cycle produces a mature egg

A newborn baby girl has about a million primary oocytes in each ovary. By the time she reaches puberty, she has only about 200,000; the rest have degenerated. During a woman's fertile years, her ovaries go through about 450 ovarian cycles (**Figure 43.12A**). During each cycle, 10 to 20 primary oocytes begin to mature, but usually only one matures completely and is ovulated; the others degenerate. At around the age of 50, a woman reaches **menopause**—the end of fertility—and may have few if any oocytes left in each ovary.

A primary oocyte surrounded by a layer of ovarian cells is the functional unit of the ovary—the **follicle** (**Figure 43.12B**). The follicle cells surrounding the oocyte supply it with nutrients and growth factors. The follicle cells are also the main site of production of the female hormones estrogen and progesterone. Usually only one follicle matures completely and reaches the stage of ovulation, releasing its egg at midcycle.

The second half of the ovarian cycle is called the luteal phase because the follicle cells left in the ovary after the egg is ovulated develop into an endocrine gland—the **corpus luteum** (yellow body)—secreting estrogen and progesterone. If the egg is not fertilized and does not embed into the endometrium, the corpus luteum becomes inactive and degenerates in 12 to 14 days.

## The uterine cycle prepares an environment for a fertilized egg

The uterine cycle parallels the ovarian cycle and consists of a buildup and then a breakdown of the endometrium (**Figure 43.13**). About 5 days into the ovarian cycle, the endometrium starts to thicken in preparation for receiving a blastocyst. The uterus attains its maximal state of preparedness about 5 days after ovulation and remains in that state for another 9 days. If a blastocyst has not arrived by that time, the endometrium breaks down and the sloughed-off tissue, including blood, flows from the body through the vagina—the process of **menstruation** (Latin *menses*, "months").

The uterine cycles of most mammals other than humans do not include menstruation; instead, the uterine lining typically is resorbed. In these species, the most obvious correlate of the ovarian cycle is a state of sexual receptivity called **estrus** ("heat") around the time of ovulation. You may be aware of the bloody discharge that occurs in dogs at the time of estrus. This discharge is not the same as menstruation—in fact it is exactly the opposite. Bleeding in dogs occurs during the *proliferation* of the uterine lining, which occurs just before ovulation. When the female mammal comes into estrus, she actively solicits male attention and may be aggressive to other females. Humans are unusual among mammals in that females are potentially sexually receptive throughout their ovarian cycles and at all seasons of the year.

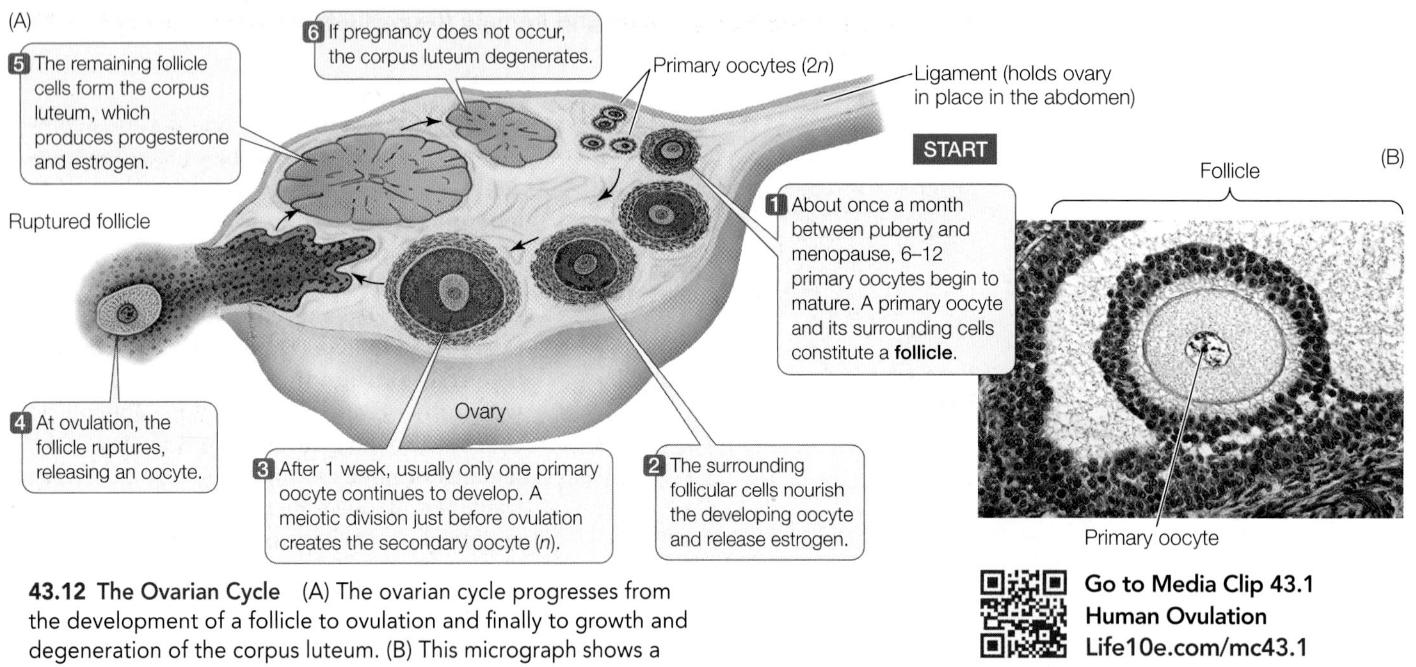


**43.12 The Ovarian Cycle** (A) The ovarian cycle progresses from the development of a follicle to ovulation and finally to growth and degeneration of the corpus luteum. (B) This micrograph shows a mature mammalian follicle; the oocyte is in the center.

**Go to Media Clip 43.1**
**Human Ovulation**
**Life10e.com/mc43.1**

## Hormones control and coordinate the ovarian and uterine cycles

The ovarian and uterine cycles are coordinated and timed by the same hormones that initiate sexual maturation. Gonadotropins (FSH and LH) secreted by the anterior pituitary are the central elements of this control. Before puberty (that is, before about 11 years of age), the secretion of FSH and LH is low and the ovaries are inactive. At puberty the hypothalamus increases its release of GnRH, stimulating the anterior pituitary to secrete FSH and LH. In response to FSH and LH, ovarian tissue grows and produces estrogen. The rise in estrogen causes the maturation of the accessory sex organs and the development of female secondary sexual characteristics. Between puberty and menopause, interactions of GnRH, gonadotropins, and sex steroids control the ovarian and uterine cycles.

Menstruation marks the beginning of each uterine and ovarian cycle. A few days before menstruation begins, the anterior pituitary begins to increase its secretion of FSH and LH. In response, several follicles begin to mature in the ovaries, and these follicles steadily increase

(A) Gonadotropins (from anterior pituitary)

FSH and LH are under control of GnRH from the hypothalamus and the ovarian hormones estrogen and progesterone (part C).

Estrogen inhibits LH and FSH release — Estrogen stimulates LH and FSH release — Estrogen inhibits LH and FSH release

Luteinizing hormone (LH)

LH surge triggers ovulation.

Follicle-stimulating hormone (FSH)

(B) Events in ovary (ovarian cycle)

FSH stimulates the development of follicles; the LH surge causes ovulation and then the development of the corpus luteum.

Oocyte maturation — Developing follicle — Ovulation (day 14) — Corpus luteum — Developing oocyte

(C) Ovarian hormones and the uterine cycle

Estrogen and progesterone stimulate the development of the endometrium in preparation for pregnancy.

Estrogen — Progesterone

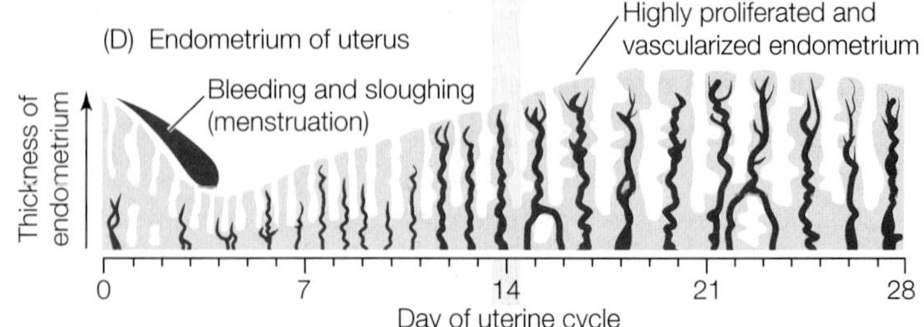


**43.13 The Ovarian and Uterine Cycles** During a woman's ovarian and uterine cycles, coordinated changes occur in (A) gonadotropin release by the anterior pituitary, (B) the ovary, (C) the release of female sex steroids, and (D) the uterus. The cycles begin with the onset of menstruation; ovulation is at midcycle (yellow bar).

**Go to Animated Tutorial 43.2**
**The Ovarian and Uterine Cycles**
Life10e.com/at43.2

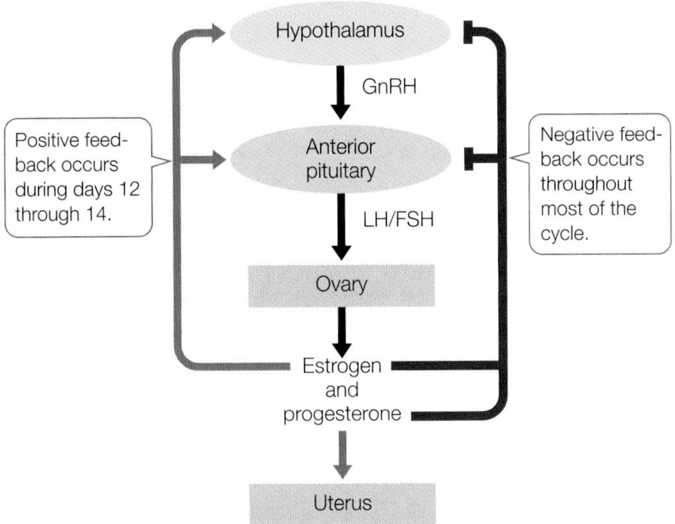


**43.14 Hormones Control the Ovarian and Uterine Cycles** The ovarian and uterine cycles are under a complex series of positive and negative feedback controls involving several hormones.

their production of estrogen. After about a week, all but one of the follicles wither away.

Estrogen exerts negative feedback control on gonadotropin release by the anterior pituitary during the first 12 days of the ovarian cycle. Then, on about day 12, estrogen exerts positive rather than negative feedback control on the pituitary (**Figure 43.14**). As a result, a surge of LH and a lesser surge of FSH occur (see Figure 43.13A). The LH surge triggers the mature follicle to rupture and release its egg, and it stimulates the cells of the ruptured follicle to develop into a corpus luteum.

Estrogen and progesterone secreted by the corpus luteum following ovulation are crucial to growth and maintenance of the endometrium. These sex steroids also exert negative feedback control on the pituitary, inhibiting gonadotropin release and thus preventing new follicles from maturing.

If the egg is not fertilized, the corpus luteum degenerates on about day 26 of the cycle. Without production of progesterone by the corpus luteum, the endometrium sloughs off and menstruation occurs. The decrease in circulating steroids also releases the hypothalamus and pituitary from negative feedback control, so GnRH, FSH, and LH all begin to increase. The increase in these hormones induces the next round of follicle development, and the ovarian cycle begins again.

### FSH receptors determine which follicle ovulates

Early in the ovarian cycle several follicles begin to develop, but only one reaches full maturity. What determines which one will survive? There are two types of follicle cells, arranged in two layers. The inner layer immediately surrounding the ovum is made up of granulosa cells, and the outer layer contains thecal cells. Thecal cells are stimulated by LH to produce testosterone

## WORKING WITH**DATA:**

### *Circadian Timing, Hormone Release, and Labor*

#### Original Paper

Olcese, J., S. Lozier, and C. Paradise. 2012. Melatonin and the circadian timing of human parturition. *Reproductive Sciences*, epub before print, May 3, 2012.

#### Analyze the Data

Pregnant women are more likely to go into labor during the night than in the daytime. Frequently, however, when women in labor are taken to a well-lit hospital, their contractions decrease. James Olcese and his colleagues thought that the timing of labor contractions might be controlled by a signal from the brain's circadian timing system—specifically, the release of the hormone melatonin from the pineal gland. Melatonin is always released at night, and its release is inhibited by light (see Section 41.5).

The researchers hypothesized that rising levels of melatonin potentiate contractility of the uterine muscles. To test this hypothesis, they collected two types of data from pregnant women close to term. They collected saliva samples from 5 women at 30-minute intervals overnight and measured the melatonin levels in the samples. They also recorded the frequency of contractions of two women who were having regular but premature contractions at night. In both situations the women were exposed to bright light between 11 P.M. and midnight. The results are shown in the table.

**QUESTION 1**

Plot the data in the tables. How would you interpret these data? Do they support the authors' hypothesis?

**QUESTION 2**

It is known that the posterior pituitary hormone oxytocin stimulates the uterine contractions leading to expulsion of the fetus from the uterus. If you cultured some uterine muscle tissue, how would you investigate the possible interaction of oxytocin and melatonin? What would your hypothesis be?

| | Data Set 1 | Data Set 2 | |
|---|---|---|---|
| | Salivary melatonin (percent of maximum; mean for 5 subjects) | Contractions per hour (2 subjects) | |
| Time | | Subject A | Subject B |
| 7 P.M. | 5 | | |
| 8 P.M. | 5 | 2 | 3 |
| 9 P.M. | 12 | 4 | 5 |
| 10 P.M. | 30 | 5 | 5 |
| 11 P.M. } Bright light | 55 | 6 | 2 |
| Midnight } Bright light | 30 | 1 | 1 |
| 1 A.M. | 50 | 2 | 0 |
| 2 A.M. | 60 | 2 | 0 |
| 3 A.M. | 65 | 2 | 11 |
| 4 A.M. | 90 | 5 | 18 |
| 5 A.M. | 85 | 5 | 15 |
| 6 A.M. | 70 | 4 | 12 |
| 7 A.M. | 60 | 2 | 4 |

Go to BioPortal for all WORKING WITH**DATA** exercises

(similar to Leydig cells in the male). Testosterone diffuses into the granulosa cells, where the enzyme aromatase converts the testosterone to estrogen. Estrogen, along with FSH, stimulates the growth and maturation of the granulosa cells (similar to the Sertoli cells in the male). The estrogen plays two important roles in the selection of the follicle that will ovulate: (1) estrogen stimulates the granulosa cells to express more FSH and LH receptors, and (2) estrogen entering the circulation feeds back on the pituitary to decrease the production of FSH. The granulosa cells also produce inhibin (similar to Sertoli cells in the male), and inhibin also decreases the production of FSH. As FSH and LH levels fall, the follicle with the most FSH and LH receptors survives and matures while the others regress.

## In pregnancy, hormones from the extraembryonic membranes take over

If the egg is fertilized and a blastocyst arrives in the uterus and implants in the endometrium, a new hormone comes into play. A layer of cells covering the blastocyst begins to secrete **human chorionic gonadotropin**, or **hCG** (**Figure 43.15A**). This gonadotropin, a molecule similar to LH, stimulates the corpus luteum to continue to produce estrogen and progesterone to support the growth and maintenance of the endometrium and thereby prevent menstruation. Because it is present only in the blood of pregnant women, the presence of hCG is the basis for pregnancy testing. Pregnancy tests use an antibody to detect hCG in urine; they take only minutes and can be done at home.

Blastocyst and endometrial tissues form the placenta, which nourishes the embryo. The placenta also replaces the corpus luteum as the major producer of estrogen and progesterone. Continued high levels of estrogen and progesterone prevent the pituitary from secreting gonadotropins; thus the ovarian cycle ceases for the duration of pregnancy. This mechanism underlies the action of birth control pills, which contain synthetic hormones resembling estrogen and progesterone that exert negative feedback control on the hypothalamus and pituitary.

## Childbirth is triggered by hormonal and mechanical stimuli

Throughout pregnancy, the muscles of the uterine wall periodically undergo slow, weak, rhythmic contractions called Braxton Hicks contractions. These contractions become stronger during the third trimester of pregnancy and are sometimes called false labor. The contractions of true labor mark the beginning of childbirth. Both hormonal and mechanical stimuli contribute to the onset of labor.

Progesterone inhibits and estrogen stimulates contractions of uterine muscle. Toward the end of the third trimester, the estrogen–progesterone ratio shifts in favor of estrogen. The onset of labor is marked by increased secretion of the hormone oxytocin by the posterior pituitaries of both mother and fetus. Oxytocin is a powerful stimulant of uterine muscle contraction. Manufactured oxytocin is used to induce labor when that is necessary.

Mechanical stimuli come from the stretching of the uterus by the fully grown fetus and the pressure of the fetal head on the cervix. These mechanical stimuli increase the release of oxytocin by the mother's posterior pituitary, which in turn

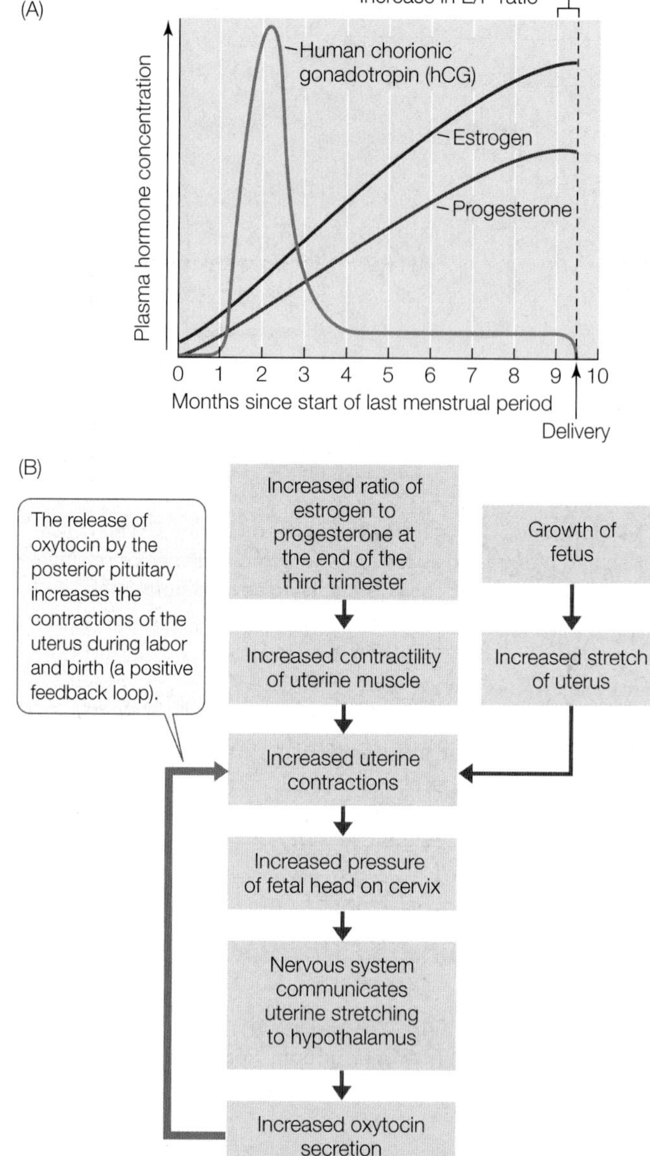


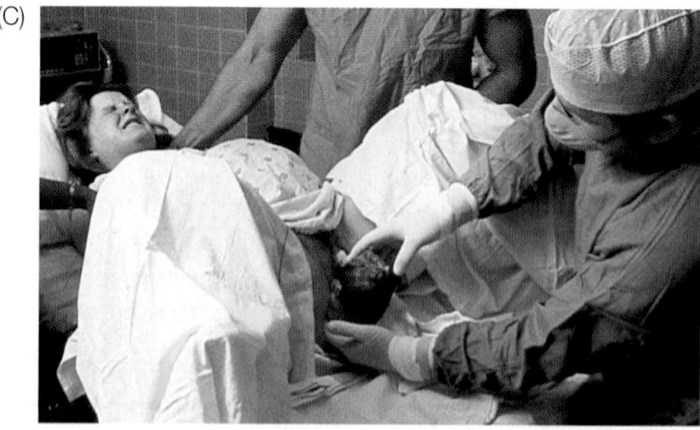

**43.15 Pregnancy and Childbirth** (A) When a fertilized ovum implants in the uterus, cells surrounding it produce human chorionic gonadotropin, which acts like LH and keeps the corpus luteum functioning as an endocrine gland. The ovarian and uterine cycles are put on hold for the duration of pregnancy. (B) Both mechanical and hormonal signals are involved in stimulating the uterine contractions of labor and delivery. (C) A new person comes into the world head first.

increases the activity of uterine muscle, causing even more pressure on the cervix. This positive feedback loop converts the weak, slow, rhythmic Braxton Hicks contractions into stronger labor contractions (**Figure 43.15B**).

In the early stage of labor, hormonal changes and pressure created by the contractions cause the cervix to dilate (expand) until it is large enough to allow the baby to pass through. Gradually the contractions become more frequent and more intense. This stage of labor lasts an average of 12 to 15 hours in a first pregnancy, but is usually 8 hours or less in subsequent ones.

The second stage of labor begins when the cervix is fully dilated to a diameter of about 10 cm (**Figure 43.15C**). The baby's head can now move into the vagina. Passage of the fetus through the vagina is assisted by the mother's bearing down ("pushing") with her abdominal and other muscles. Once the head and shoulders of the baby clear the cervix, the rest of its body eases out rapidly, but it is still connected to the placenta by the umbilical cord. Once the baby clears the birth canal, it starts breathing and is independent of its mother's circulation. The umbilical cord may then be clamped and cut. The segment still attached to the baby dries up and sloughs off in a few days, leaving behind its distinctive signature, the belly button—more properly called the umbilicus. The third stage of labor, the detachment and expulsion of the placenta and fetal membranes, takes from a few minutes to an hour, and may be accompanied by uterine contractions. A baby's suckling at the breast immediately following birth stimulates additional secretion of oxytocin, which augments uterine contractions, reduces the size of the uterus, and helps stop bleeding.

**RECAP 43.3**

The reproductive systems of men and women produce gametes and hormones, and these functions are controlled by hypothalamic and anterior pituitary hormones. In women, the hormonal control of reproductive functions produces linked ovarian and uterine cycles.

- Describe the path the human sperm and ovum take in moving from their respective gonads to the point at which fertilization occurs. **See Figures 43.8 and 43.11**
- In males, increased production of GnRH at puberty stimulates the release of what two hormones of the anterior pituitary? What effect do these hormones have? **See p. 892**
- Explain the events in the ovarian cycle that result in release of a single ovum each month. What events prepare the uterus to receive the egg? **See Figures 43.12 and 43.13**

Understanding the physiology of human reproduction has led to numerous methods and technologies for controlling it, either to prevent unwanted pregnancies or to overcome infertility.

## 43.4 How Can Fertility Be Controlled?

Sexual issues and sexual behavior are dominant aspects of our society, and reproductive technologies have had huge impacts on our sexual and reproductive lives.

### Humans use a variety of methods to control fertility

According to a recent study, almost half of the more than 6 million pregnancies that occur in the United States each year are unintended. For women of college age, a single act of unprotected intercourse in the two days prior to ovulation carries a chance of conception as high as 50 percent.

The only failure-proof methods of preventing pregnancy are either complete abstinence from sexual activity or surgical removal of the gonads. Those options are not acceptable to most people, and they turn to other methods to prevent pregnancy. Many of these methods prevent fertilization or implantation (conception) and are therefore referred to as **contraception**. **Table 43.1** lists some of the most commonly used contraceptive methods and their relative failure rates; note that these methods vary enormously in their effectiveness. Most methods are used by the woman, although some are used by the man.

Once a fertilized egg is successfully implanted in the uterus, any termination of the pregnancy is called an **abortion**. A spontaneous abortion is the medical term for what is commonly called a miscarriage. Spontaneous abortions frequently occur early in pregnancy and are usually the result of either a chromosomal abnormality in the fetus or a breakdown in the process of implantation. Many spontaneous abortions occur before the woman even realizes she is pregnant.

Abortions that result from medical intervention may be performed either for therapeutic purposes or for fertility control. A therapeutic abortion may be necessary to protect the health of the mother, or it may be performed because prenatal testing reveals that the fetus has a severe defect. In a medical abortion, the cervix is dilated and some of the endometrium, along with the implanted fetus, is removed. When performed in the first trimester of a pregnancy, a medical abortion carries less risk of death to the mother than a full-term pregnancy. The risk rises after the first 12 weeks of pregnancy but remains less than that of a full-term pregnancy through the second trimester.

### Reproductive technologies help solve problems of infertility

About 15 percent of couples in the United States are infertile—that is, they can't have children. The reasons for infertility are many, and are about equally distributed between men and women. Several technologies have been developed to overcome barriers to both conceiving and bearing a child. The simplest of these is **artificial insemination**, in which the physician places sperm in the female's reproductive tract. This technique is useful if the male's sperm count is low, if his sperm lack motility, or if problems in the female's reproductive tract prevent the normal movement of sperm up to and through the oviducts. Artificial insemination is also widely used in the production of domesticated animals such as cattle.

**Assisted reproductive technologies**, or **ARTs**, involve procedures that remove unfertilized eggs from the ovary, combine them with sperm outside the body, and then place fertilized eggs or egg–sperm mixtures in the appropriate location in the female's reproductive tract for development to take place. The

TABLE **43.1**
Methods of Contraception

| Method | Mode of Action | Failure Rate[a] | Comments |
|---|---|---|---|
| Unprotected | No form of birth control. | 85 | High risk of pregnancy, especially for women 15–30. |
| **Nontechnological methods** | | | |
| Rhythm method | The couple abstains from intercourse between days 10 and 20 of the ovarian cycle (peak fertility). | 15–35 | High failure rate due to miscalculation and/or variation of individual cycles. |
| Coitus interruptus | The man withdraws his penis prior to ejaculation with the intention of not depositing sperm into the vagina. | 20–40 | Requires self-control, especially by the man. Very high failure rate. |
| **Barrier methods**[b] | | | |
| Condom | A sheath of impermeable material (often latex) is fitted over the erect penis. Semen is trapped in the condom, so no sperm are deposited in the vagina. | 15 | If fitted correctly, an intact condom can prevent pregnancy and provide protection against sexually transmitted diseases (STDs), including HIV (AIDS). |
| Spermicidal jellies | Applied inside the vagina, these chemical compounds kill or immobilize sperm. | 25 | Used alone, spermicidal compounds have a fairly high failure rate. |
| Diaphragms, cervical caps | Inserted by the woman prior to intercourse, these devices work by blocking the cervix so that sperm cannot pass into the uterus. | 10–15 | Approximately the same failure rate as condom use by men, but do not protect against STDs. Can be used in conjunction with spermicidal jelly for extra protection. |
| **Hormone-based contraceptives** | | | |
| Oral hormones ("the pill") | A daily pill for women containing a combination of synthetic estrogens and progesterone. These hormones mimic pregnancy to the extent that the ovarian cycle and ovulation are suspended. The uterine cycle is allowed to continue by including a week of non-hormone administration every 21–28 days. | 0–3 | Requires medical consultation and prescription. Taken correctly, oral contraceptives are extremely effective. In the U.S., more than 12 million women use them each year; they are sometimes prescribed to treat menstrual disorders. |
| Non-orally administered hormones | Making use of same hormonal actions as the pill, these methods include long-acting injections, patches that release hormones transdermally (through the skin), and a hormone-containing vaginal ring. | <1 | Same as oral hormones. A slightly lower failure rate because the woman does not have to remember to take a daily pill. |
| Progestin-only pill (Plan B®) | An oral contraceptive meant to be taken within 72 hours after unprotected sex. A high dose of progestin in two pills prevents ovulation in the same manner birth control pills do. | 5–40 | Not an "abortion pill," this drug will not terminate an existing pregnancy. Currently available to women over 17 without a prescription. Failure rate varies widely depending on when taken. |
| **Implantation blockers** | | | |
| Intrauterine device (IUD) | A medical professional inserts a small plastic or metal device into the uterus. The resulting inflammation reaction (see Chapter 42) releases prostaglandins, which prevent implantation of the fertilized egg. | 0.5–5 | A highly effective contraceptive, it is the most widely used birth control device in China (and hence the world). With medical monitoring, can remain in place for several years. |
| Mifepristone (RU-486) | This drug blocks progesterone receptors necessary to maintain the endometrium during implantation and pregnancy. | 0.5–6 | Prevents implantation when taken up to several days after unprotected intercourse. Can terminate a pregnancy up to the time of the first missed menstrual period. In the U.S., available from specialized providers. |
| **Sterilization** | | | |
| Vasectomy | The vasa deferentia (see Figure 43.8) are cut and tied off so that sperm can no longer pass into the urethra. Sperm continue to be produced but are reabsorbed by the man's body. Male hormone levels and sexual responses are not affected. | 0–0.15 | A simple surgical procedure performed under local anesthetic in a doctor's office. Although theoretically it can be reversed, vasectomy should be considered permanent. |
| Tubal ligation | The oviducts (see Figure 43.11A) are tied off so that eggs cannot reach the uterus and sperm cannot reach the egg. As with vasectomy, hormone levels and sexual responses are not affected. | 0–0.05 | This surgical procedure is somewhat more complex than vasectomy. It is often performed in conjunction with childbirth when a woman has decided that her family is complete. |

[a]Failure rate refers to the number of pregnancies per 100 women per year
[b]All of these barrier methods are routinely available without medical prescription.

Pipette holding egg

Egg

Pipette injecting sperm

**43.16 Intracytoplasmic Sperm Injection** In this procedure, a sperm is injected directly into a mature egg cell. The fertilized egg is then placed in the female reproductive tract, where it can implant and develop into a fetus.

first successful ART was in vitro fertilization (IVF). In IVF the female is treated with hormones that stimulate many follicles in her ovaries to mature. Eggs are collected from these follicles, and sperm are collected from the male. Eggs and sperm are combined in a culture medium outside the body, where fertilization takes place. The resulting embryos can be injected into the mother's uterus in the blastocyst stage or kept frozen for implantation later. The first "test-tube baby" resulting from IVF was born in England in 1978. Since then, millions of babies have resulted from this ART.

A major cause of IVF failure is failure of sperm to gain access to the egg plasma membrane (see Figure 43.4). To solve this problem, methods have been developed to inject a sperm cell directly into the cytoplasm of an egg. In intracytoplasmic sperm injection (ICSI), an egg is held in place by suction applied to a polished glass pipette. A slender, sharp pipette is then used to penetrate the egg and inject a sperm (**Figure 43.16**). This ART was used successfully for the first time in 1992 by researchers in Belgium; now thousands of these procedures are performed in U.S. clinics each year, with a success rate of about 25 percent.

IVF, coupled with techniques of genetic analysis, can eliminate the risk that adults who are carriers of genetic diseases will produce affected children. It is now possible to take a cell from a human embryo at the 4- or 8-cell stage (see Figure 44.4) without damaging its developmental potential. The sampled cell can be subjected to molecular analysis to determine whether it carries the harmful gene. This procedure, called preimplantation genetic diagnosis (PGD), makes it possible to determine whether an embryo produced by IVF carries the genetic defect of concern.

### RECAP 43.4

Controlling fertility is an important aspect of modern human life. Decreasing the probability of pregnancy is achieved through methods that prevent sperm and egg from meeting and from preventing implantation. Pregnancies can be facilitated through medical technology.

- Which method of contraception is the only one to offer protection against sexually transmitted diseases (STDs)? **See Table 43.1**
- Explain what a couple who are both carriers of a genetic disease could do to ensure that their offspring would not have the disease. **See p. 899**

The fertilized egg of a sexually reproducing organism is a single cell containing all the genetic information needed to create a new organism. Chapters 19 and 20 introduced some of the molecular aspects of the process of development in multicellular animals. Chapter 44 will describe the physiological and anatomical events of animal development.

How can a queen be so different from her worker sisters when they all share the same genome?

### ANSWER

When a hive loses its queen, the workers create new queens from a few eggs that the old queen laid by feeding a substance called "royal jelly" to the larvae that hatch from the chosen eggs. This special food stimulates growth and has recently been discovered to have powerful epigenetic effects on honey bee development. As described in Chapter 16, the expression of a gene can be more or less permanently altered by the chemical modification of histones—proteins that are closely associated with the DNA—or by methylation or demethylation of the DNA itself (see Section 16.4). Because these changes do not alter the gene's nucleotide sequence, they are referred to as epigenetic ("outside the gene"). One component of royal jelly is phenyl butyrate, an inhibitor of histone deacetylation. Furthermore, the queen bee has well over 500 genes that are methylated differently than those same genes in workers. It is therefore likely that the royal jelly fed to a future queen dramatically alters the expression of her genome.

# CHAPTER SUMMARY 43

## 43.1 How Do Animals Reproduce without Sex?

- Asexual reproduction produces offspring that are genetically identical to their parent and to one another; it produces no genetic diversity.
- Means of asexual reproduction include **budding**, **regeneration**, and **parthenogenesis**. **Review Figures 43.1, 43.2**

## 43.2 How Do Animals Reproduce Sexually?

- Sexual reproduction involves three basic steps: **gametogenesis**, **spawning** or **mating**, and **fertilization**.
- **Gametogenesis** and fertilization are similar in all animals, but spawning and mating include a great variety of anatomical, physiological, and behavioral adaptations.
- Gametogenesis occurs in **testes** and **ovaries**. In spermatogenesis (the production of **sperm**) and oogenesis (the production of eggs), the **germ cells** proliferate mitotically, undergo meiosis, and mature into gametes.
- Each **primary spermatocyte** can produce four haploid sperm through the two divisions of meiosis. **Review Figure 43.3A**
- **Primary oocytes** immediately enter prophase of the first meiotic division, and in many species, including humans, their development is arrested at this point. Each **oogonium** produces only one egg. **Review Figure 43.3B**
- Fertilization involves sperm activation, species-specific binding of sperm to egg, the acrosomal reaction, digestion of a path through the protective coverings of the egg, and fusion of sperm and egg plasma membranes. Fusion of these two membranes triggers **blocks to polyspermy**, which prevent additional sperm from entering the egg and, in mammals, signal the egg to complete meiosis and begin development. **Review Figure 43.4, ANIMATED TUTORIAL 43.1**
- **External fertilization** is common in aquatic species. **Internal fertilization** is necessary in terrestrial species and usually involves **copulation**.
- **Hermaphroditic**, or **monoecious**, species have both male and female reproductive systems in the same individual, either sequentially or simultaneously. **Dioecious** species have separate male and female individuals.
- Animals can be classified as **oviparous** or **viviparous**, depending on whether the early stages of development occur outside or inside the mother's body.

## 43.3 How Do the Human Male and Female Reproductive Systems Work?

- Men produce **semen** consisting of sperm suspended in seminal fluid (which nourishes the sperm and facilitates fertilization).
- Sperm are generated in the **seminiferous tubules** of the testes, mature in the **epididymis**, and are delivered to the **urethra** through the **vasa deferentia**. Other components of semen are produced in the **seminal vesicles**, **prostate gland**, and **bulbourethral gland**. **Review Figures 43.8, 43.9, ACTIVITIES 43.1, 43.2**
- All components of the semen join in the urethra at the base of the **penis** and are ejaculated through the erect penis by muscle contractions at the culmination of copulation.
- Spermatogenesis depends on testosterone secreted by the **Leydig cells** of the testes, which are under the control of hormones produced in the anterior pituitary and the hypothalamus. The production of these hormones is controlled by negative feedback from testosterone and from inhibin, a hormone produced by the **Sertoli cells** of the testes. **Review Figure 43.10**
- Eggs mature in the woman's ovaries and are released into the **oviducts**. Sperm deposited in the **vagina** during **copulation** move up through the **cervix** and **uterus** into the oviducts. Fertilization occurs in the upper regions of the oviducts. **Review Figure 43.11, ACTIVITY 43.3**
- The maturation and release of eggs constitute an **ovarian cycle**. The **uterine cycle** prepares the uterus for receipt of a blastocyst. If no blastocyst is implanted, the lining of the uterus sloughs off in the process of **menstruation**. **Review Figure 43.13, ANIMATED TUTORIAL 43.2**
- Both the ovarian and the uterine cycles are under the control of hypothalamic and pituitary hormones, which in turn are under the feedback control of estrogen and progesterone. **Review Figure 43.14**
- Childbirth is initiated by hormonal and mechanical stimuli that increase the contraction of uterine muscle. **Review Figure 43.15**

## 43.4 How Can Fertility Be Controlled?

- Methods of **contraception** include abstention from copulation and the use of technologies that decrease the probability of fertilization. **Review Table 43.1**
- **Assisted reproductive technologies** (**ARTs**) have been developed to increase fertility.

**Go to the Interactive Summary to review key figures, Animated Tutorials, and Activities Life10e.com/is43**

# CHAPTERREVIEW

## REMEMBERING

1. Which statement about human oocytes is true?
   a. By birth, a human female infant has produced her lifetime supply of oocytes.
   b. At the onset of puberty, ovarian follicles produce new oocytes in response to hormonal stimulation.
   c. A woman stops producing oocytes at the onset of menopause.
   d. A woman produces oocytes throughout adolescence.
   e. Oocytes are stored in the oviducts.

2. Spermatogenesis and oogenesis differ in that
   a. spermatogenesis produces gametes with greater stores of raw materials than those produced by oogenesis.
   b. spermatocytes remain in prophase of the first meiotic division longer than oocytes.
   c. oogenesis produces four equally functional haploid cells per meiotic event, and spermatogenesis does not.
   d. spermatogenesis produces many gametes with meager energy reserves, whereas oogenesis produces relatively few, well-provisioned gametes.
   e. spermatogenesis begins before birth in humans, whereas oogenesis does not start until the onset of puberty.

3. Semen contains all of the following except
   a. fructose.
   b. mucus.
   c. clotting enzymes.
   d. testosterone.
   e. an active clot-dissolving enzyme.

4. Which of the following statements about the ovarian and uterine cycles is *not* true?
   a. Falling estrogen and progesterone levels induce menstruation.
   b. A sudden rise in LH induces ovulation.
   c. Estrogen levels reach highest levels in the follicular phase and progesterone reaches highest levels in the luteal phase of the ovarian cycle.
   d. If fertilization occurs, the corpus luteum secretes hCG.
   e. Estrogen is produced by follicle cells.

5. Contractions of muscles in the uterine wall are stimulated by
   a. progesterone.
   b. estrogen.
   c. prolactin.
   d. oxytocin.
   e. human chorionic gonadotropin.

6. Which of the following methods of contraception is *most likely* to fail?
   a. Rhythm method
   b. Birth control pills
   c. Diaphragm
   d. Vasectomy
   e. Condom

## UNDERSTANDING & APPLYING

7. In terms of characteristics and functions, explain how you would pair Leydig cells, Sertoli cells, thecal cells, and granulosa cells, and why.

8. A drug called RU-486 blocks the actions of progesterone. It is called a contragestational drug because it can terminate a pregnancy after fertilization has occurred. How does RU-486 do this?

## ANALYZING & EVALUATING

9. Females of the marine worm *Bonellia viridis* release fertilized eggs into the water. The eggs hatch into larvae that swim and then settle on a substrate. If the larvae land on sand, they burrow in and develop into females with large bodies and a proboscis that extends into the water above to collect food. If a larva lands on a female proboscis, it burrows into the female body and becomes a tiny male whose only function is to produce sperm to fertilize the female's eggs. What selection pressures could contribute to this extreme sexual dimorphism?

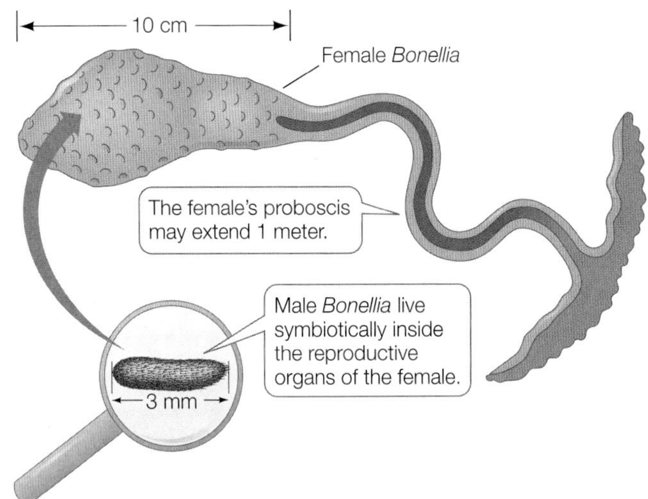


10. If a man carries a genetic mutation that prevents the maintenance of cytoplasmic bridges between his secondary spermatocytes and his spermatids, would you expect that he would father only sons or only daughters? Why?

Go to BioPortal at **yourBioPortal.com** for Animated Tutorials, Activities, LearningCurve Quizzes, Flashcards, and many other study and review resources.

# 44 Animal Development

## CHAPTER**OUTLINE**


**44.1** How Does Fertilization Activate Development?

**44.2** How Does Mitosis Divide Up the Early Embryo?

**44.3** How Does Gastrulation Generate Multiple Tissue Layers?

**44.4** How Do Organs and Organ Systems Develop?

**44.5** How Is the Growing Embryo Sustained?

**44.6** What Are the Stages of Human Development?


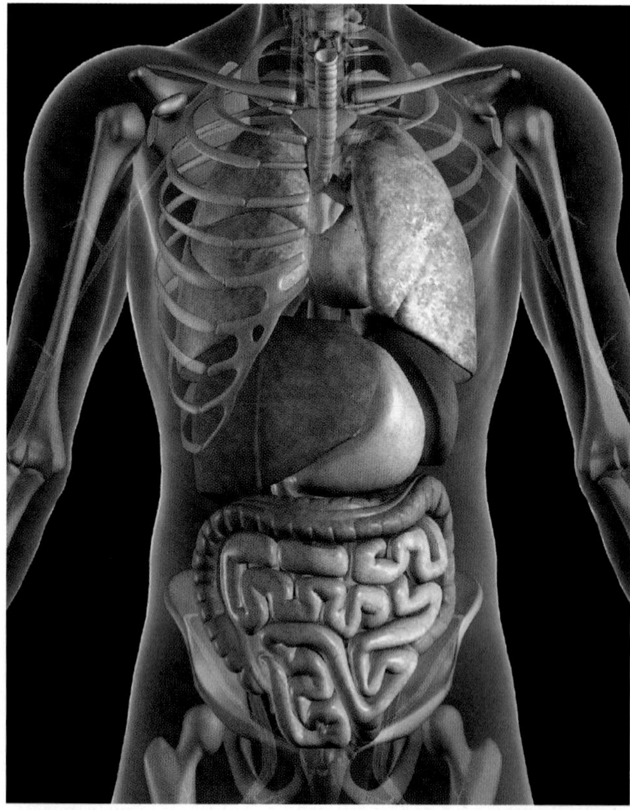

**Go with the Flow** The internal organs of humans are not all symmetrical, and some individuals are born with the mirror-image pattern of what is seen in most people—a condition called situs inversus. The left–right asymmetry of the internal organs is initiated by asymmetrical stimulation of primary cilia at a very early stage in development.

PLACE YOUR HAND over your heart. Next point to your appendix. Surely you first put your hand on the left side of your chest and then pointed to the right side of your lower abdomen. Like other vertebrates, humans are bilaterally symmetrical—but our symmetry is not absolute. Some of our internal organs are oriented differently with respect to the left and right sides of the body. In about 1 out of every 7,000 people, the arrangement of the internal organs is reversed, a condition known as situs inversus ("location inverted"). The difference arises from events very early in the development of the embryo, and most people with situs inversus lead normal lives.

As you will learn in this chapter, to get from an embryo with a single layer of cells to the next stage with two layers of cells, a midline slit forms as cells in one area of the embryo migrate inward. The place where the inward movement of cells starts is called the node. Cells at the node have motile cilia that beat in a clockwise motion and move extracellular fluid across the surface of the node. That fluid movement is always leftward—that is, from right to left. Why?

Imagine the movement of the cilia describes a clock face oriented so that 12 o'clock and 6 o'clock lie along the embryo's midline slit, with 12 o'clock closest to the future head. The cilia protrude from the cell surfaces at an angle. Thus when the cilia are rotating through the 6 o'clock position, they are closer to the cell surface and experience greater shear forces (resistance) than when they are rotating through the 12 o'clock position. In the bilaterally symmetrical early embryo, the circular beating of cilia at this critical spot—the node—creates left–right information.

How do we know the information generated by beating nodal cilia is critical for the left–right asymmetrical patterns of gene expression and developmental processes that follow? In mutant strains of mice that have no cilia or nonmotile cilia in the node, about half of the mice have reversed organ symmetry. Similarly, a spectrum of rare genetic disorders in humans, collectively called Kartagener's syndrome, are characterized by nonmotile cilia, and about 50 percent of these individuals also have situs inversus.

How does the directional flow of extracellular fluid across the node stimulate a left–right asymmetry in gene expression and development?

See answer on p. 921.

## 44.1 How Does Fertilization Activate Development?

Fertilization is the joining of sperm and egg to produce a diploid zygote. You might therefore think of it as the event that begins development. Keep two things in mind, however. First, in animals that reproduce asexually, development proceeds without fertilization. And second, in animals where fertilization does occur, it is preceded by critical events in the maturing egg that will influence subsequent development but are suspended in developmental arrest. Thus in studying fertilization we are asking how it activates or restarts multicellular development in sexually reproducing animals.

Fertilization does more than restore a full diploid complement of maternal and paternal genes. The fusion of sperm and egg plasma membranes accomplishes several things:

- It stimulates ion fluxes across the egg membrane.
- It sets up blocks to the entry of additional sperm into the egg.
- It changes the pH of the egg cytoplasm.
- It increases egg metabolism and stimulates protein synthesis.
- It initiates the rapid series of cell divisions that produce a multicellular embryo.

The mechanisms of fertilization were described in Section 43.2. Here we will take a closer look at the cellular and molecular interactions of sperm and egg that initiate the first steps of development.

### The sperm and the egg make different contributions to the zygote

In most species, eggs are much larger than sperm. Egg cytoplasm is well stocked with organelles, nutrients, and a variety of molecules, including cytoplasmic determinants such as transcription factors and mRNAs (see Section 19.2). Nearly everything the embryo needs during its early stages of development comes from the mother, including its mitochondria (and therefore all of its mitochondrial DNA). In addition to its haploid nucleus, the sperm makes one other crucial contribution to the zygote in most species—the centriole. The centriole contributes to the zygote's centrosome, which organizes the mitotic spindles for subsequent cell divisions (see Figure 11.10). Centrioles are also the origin of the microtubules of the primary cilia, which are important in cell signaling, as we saw in the opening story about situs inversus.

Cytoplasmic determinants in the egg play important roles in setting up the signaling cascades that orchestrate the major events of development: determination, differentiation, morphogenesis, and growth.

### Rearrangements of egg cytoplasm set the stage for determination

The unique attributes of amphibian eggs make them ideal models for illustrating how rearrangements of egg cytoplasm set the stage for determination. The molecules in the cytoplasm of the amphibian egg are not homogeneously distributed. The entry of the sperm into the egg stimulates rearrangements of the egg cytoplasm that introduce additional organization to the egg. Sperm entry establishes the polarity of the zygote, and informational molecules in the egg cytoplasm are organized with respect to that polarity. Therefore, when cell divisions begin, these informational molecules—which guide subsequent development—are not divided equally among daughter cells.

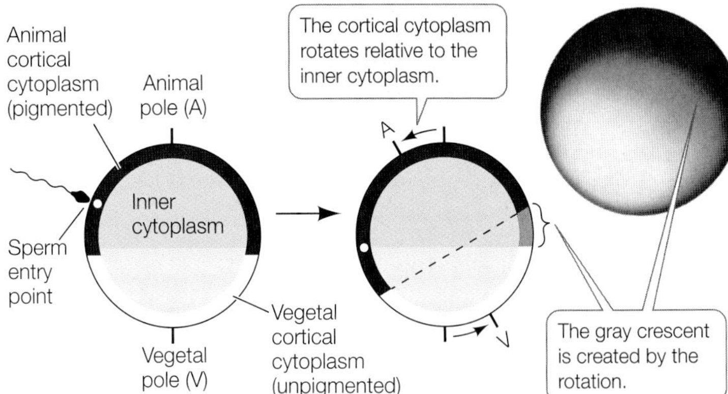


**44.1 The Gray Crescent** In amphibian eggs, cortical rotation and rearrangement of the cytoplasm after fertilization create the gray crescent opposite the point of sperm entry. These events are important for specifying the body axes and other important events in later development.

Rearrangement of egg cytoplasm following fertilization is easily observed in some frog species because of pigments in the cytoplasm. The nutrients in an unfertilized frog egg are dense yolk granules that are concentrated by gravity in the lower half of the egg, called the **vegetal hemisphere**. The haploid nucleus of the egg is located at the opposite end, in the **animal hemisphere**. The outermost (cortical) cytoplasm of the animal hemisphere is heavily pigmented, and the underlying cytoplasm has more diffuse pigmentation. The vegetal hemisphere is not pigmented. Because of these differences, it is easy to observe how the cytoplasm rearranges when the egg is fertilized.

The frog egg is radially symmetrical. You can turn it on its vegetal–animal pole axis, and all sides are the same. Sperm-binding sites are localized on the surface of the animal hemisphere. When a sperm binds to and enters the egg, the egg's radial symmetry is converted to bilateral symmetry and an anterior–posterior axis is created. Cortical cytoplasm rotates toward the site of sperm entry (**Figure 44.1**). This rotation brings the animal and vegetal regions of cytoplasm into contact with each other, producing a band of pigmented cytoplasm on the side opposite the site of sperm entry. This band, called the **gray crescent**, marks the location of important developmental events in some species of amphibians.

The centriole that was the sperm's contribution to the egg initiates the cytoplasmic reorganization that coincides with the appearance of the gray crescent. The centriole organizes the microtubules in the vegetal hemisphere cytoplasm into a parallel array that guides the movement of the cortical cytoplasm. These

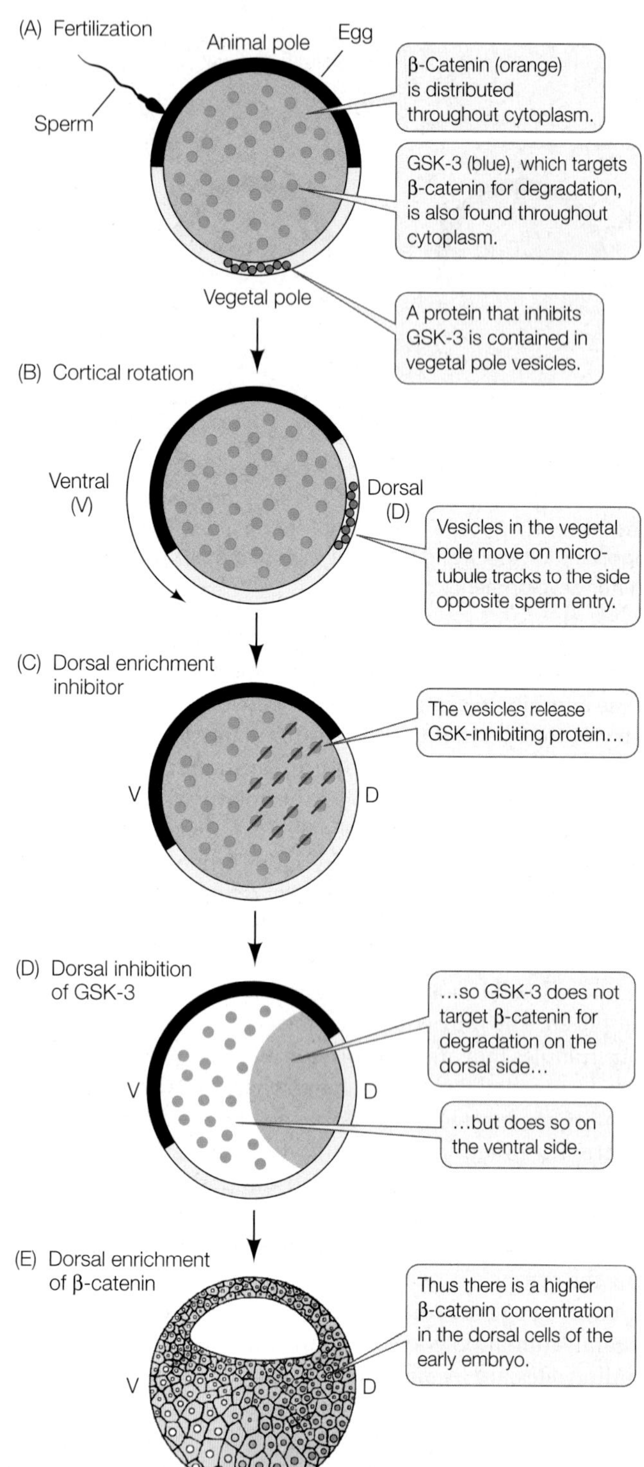


**44.2 Cytoplasmic Factors Set Up Signaling Cascades** Cytoplasmic movement changes the distributions of critical developmental signals. In the frog zygote, the interaction of the protein kinase GSK-3, its inhibitor, and the protein β-catenin are crucial in specifying the dorsal–ventral axis of the embryo.

microtubules also appear to be directly responsible for movement of specific organelles and proteins, because these organelles and proteins move from the vegetal hemisphere to the gray crescent region even faster than the cortical cytoplasm rotates.

The movement of cytoplasm, proteins, and organelles changes the distribution of critical signals. A key transcription factor in early development is β-catenin, which is produced from maternal mRNA (mRNA produced and stored in the egg while it was maturing in the ovary). β-Catenin is found throughout the egg cytoplasm (**Figure 44.2**). Also present throughout the egg cytoplasm is a protein kinase, glycogen synthase kinase-3 (GSK-3), that phosphorylates β-catenin and thereby targets it for degradation. An inhibitor of GSK-3 is localized in the vegetal cortex of the egg. After sperm entry, this inhibitor moves along microtubules to the gray crescent, where it prevents the degradation of β-catenin. As a result, the concentration of β-catenin is higher on the dorsal than on the ventral side of the developing embryo.

**RECAP 44.1**

The egg is stocked with nutrients and informational molecules that power and direct the early stages of development. Fertilization activates the egg and stimulates rearrangement of the cytoplasm, setting up the body axes and organizing positional information that will control determination and differentiation.

- What are the contributions of the sperm and of the egg to the zygote? **See p. 903**
- Explain how β-catenin becomes more concentrated on the dorsal side of the embryo. **See p. 904 and Figure 44.2**

The uneven distribution of informational molecules in the cytoplasm of the fertilized egg is essential to later events in development. In the next section we will see how these informational molecules end up in different cells of the developing embryo.

## 44.2 How Does Mitosis Divide Up the Early Embryo?

β-Catenin plays a major role in the cell–cell signaling cascade that begins the process of cell determination and the formation of the embryo. But cell–cell signaling requires more than one cell. The single-celled zygote must become a multicellular embryo.

### Cleavage repackages the cytoplasm

**Cleavage** is the sequence of early cell divisions that transforms the diploid zygote into a mass of undifferentiated cells that will develop as the embryo. Because the cytoplasm of the zygote is not homogeneous, these first cell divisions result in the differential distribution of nutrients and cytoplasmic determinants in the early embryo.

In most animals, cleavage proceeds with rapid DNA replication and mitosis but with no cell growth and little gene expression. The embryo becomes a solid ball of smaller and smaller cells. Eventually this ball forms a central fluid-filled cavity called a **blastocoel**, at which point the embryo is called a **blastula**. Its individual cells are called **blastomeres**. The pattern of cleavage in different species influences the form of their blastulas.

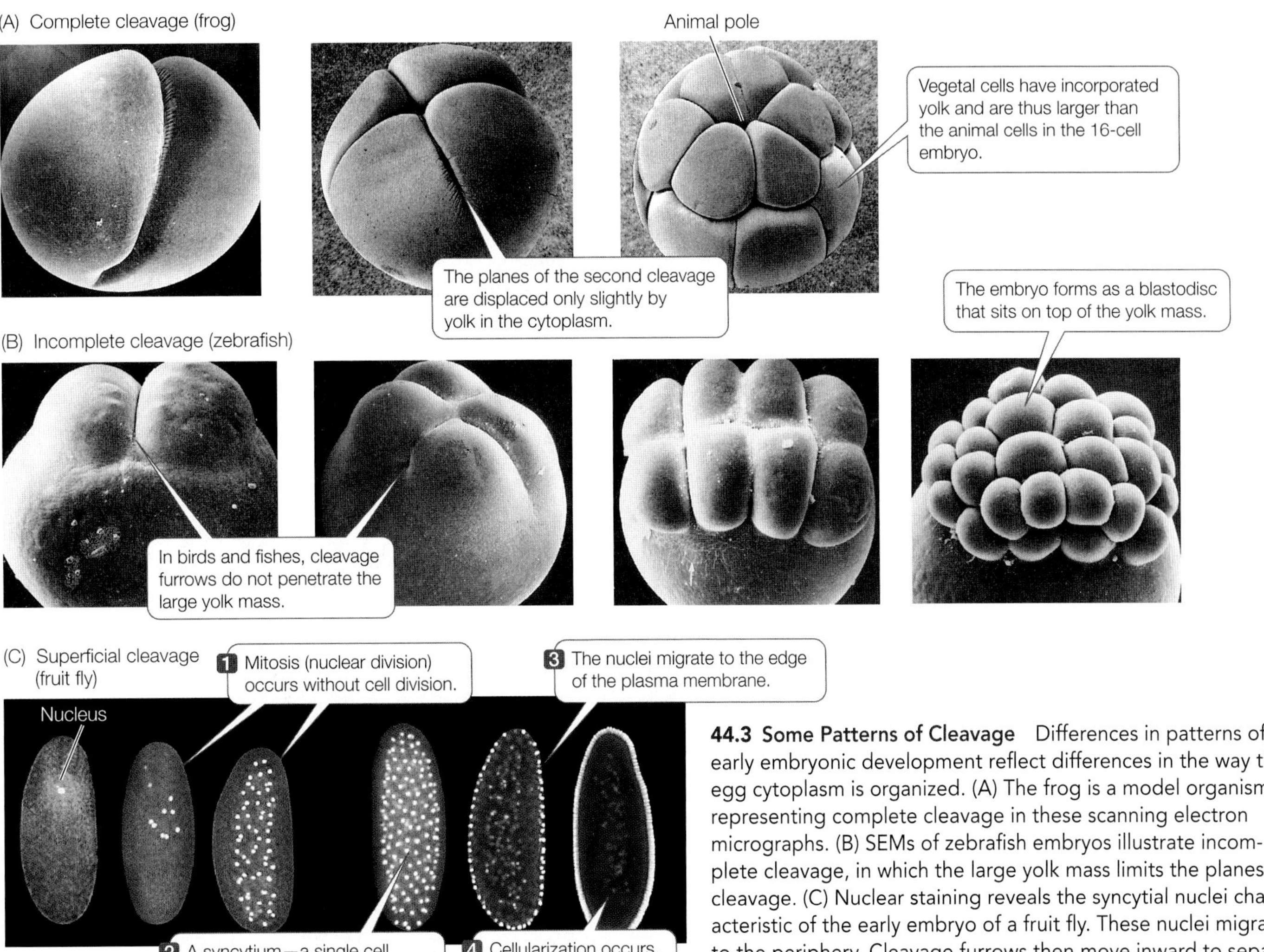


**44.3 Some Patterns of Cleavage** Differences in patterns of early embryonic development reflect differences in the way the egg cytoplasm is organized. (A) The frog is a model organism representing complete cleavage in these scanning electron micrographs. (B) SEMs of zebrafish embryos illustrate incomplete cleavage, in which the large yolk mass limits the planes of cleavage. (C) Nuclear staining reveals the syncytial nuclei characteristic of the early embryo of a fruit fly. These nuclei migrate to the periphery. Cleavage furrows then move inward to separate the nuclei into individual cells, forming the blastoderm.

- **Complete cleavage** occurs in most eggs that have little yolk. Early cleavage furrows divide the egg completely. The frog egg undergoes complete cleavage, but because its vegetal pole contains more yolk, the division of the cytoplasm is unequal and the blastomeres in the animal hemisphere are smaller than those in the vegetal hemisphere (**Figure 44.3A**).
- **Incomplete cleavage** occurs in species in which the egg contains a lot of yolk and the cleavage furrows do not penetrate it all. **Discoidal cleavage** is a type of incomplete cleavage common in fishes and in birds and other reptiles, in which the embryo forms as a disc of cells, or **blastodisc**, that sits on top of the dense yolk mass (**Figure 44.3B**).
- **Superficial cleavage** is a variation of incomplete cleavage that occurs in insects such as the fruit fly (*Drosophila*). Early in development, cycles of mitosis occur without cell division, producing a **syncytium**—a single cell with many nuclei (**Figure 44.3C**). The nuclei eventually migrate to the periphery of the egg, after which the plasma membrane of the egg grows inward, creating a **blastoderm** by partitioning the nuclei into individual cells surrounding a core of yolk.

The positions of the mitotic spindles during cleavage are not random but are defined by cytoplasmic determinants produced from the maternal genome and stored in the egg (see Section 19.2). The orientation of the mitotic spindles can determine the planes of cleavage and the arrangement of the blastomeres.

In complete cleavage, if the mitotic spindles of successive cell divisions form parallel or perpendicular to the animal–vegetal axis of the zygote, a pattern of **radial cleavage** occurs. The first two cell divisions are parallel to the animal–vegetal axis, and the third is perpendicular to it (see Figure 44.4A). **Spiral cleavage** results when the mitotic spindles are at oblique angles to the animal–vegetal axis. In spiral cleavage, each new cell layer is shifted to the left or right, depending on the orientation of the mitotic spindles. Most mollusks have spiral cleavage, reflected in some species by a coiling shell pattern (as seen in snails).

## Early cell divisions in mammals are unique

Several features of early cell divisions in placental mammals (eutherians) are very different from those seen in other animal groups. First, this process in mammals is very slow. Cell divisions are 12 to 24 hours apart, compared with tens of minutes to a few hours in non-mammalian species. Also, the cell

(A) Parallel plane of second division
A
Plane of first division
V
Perpendicular plane of second division

**44.4 Becoming a Blastocyst** (A) Mammals have rotational cleavage, in which the plane of the first cleavage is parallel to the animal–vegetal (A–V) axis, but the second cell division involves two planes (beige) at right angles to each other. (B) Scanning electron micrographs (color added) of early cleavage (leading to the formation of the blastocyst) in a human embryo. The cells' outer surfaces are covered with cilia (bright yellow). The small spheres, or "blebs," of cytoplasmic material, prominent at the 8-cell stage, disintegrate as cleavage progresses. (C) Seen in cross section under a light microscope, a mammalian blastocyst consists of an inner cell mass adjacent to a fluid-filled blastocoel and surrounded by trophoblast cells.

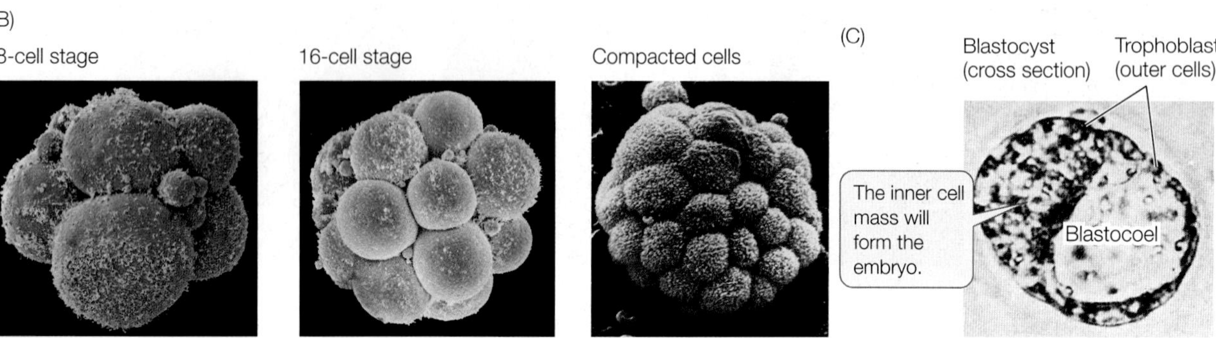


divisions of mammalian blastomeres are not in synchrony with each other. Because the blastomeres do not undergo mitosis at the same time, the number of cells in the embryo does not increase in the regular progression (2, 4, 8, 16, 32, etc.) typical of other species. This slow mammalian cleavage means that genes expressed during cleavage play roles in cleavage. In animals such as sea urchins and frogs, very little if any gene transcription occurs in the blastomeres, with cleavage being directed by molecules that were present in the egg before fertilization.

The pattern of mammalian cleavage is unique and is called rotational cleavage. The first cell division is parallel to the animal–vegetal axis as in radial cleavage, but in the second cell division, the two blastomeres divide at right angles to one other. One blastomere divides parallel to the animal–vegetal axis, while the other divides perpendicular to this axis (**Figure 44.4A**). As in other animals that have complete cleavage, the early cell divisions in a mammalian zygote produce a loosely associated ball of cells. After the 8-cell stage, however, the behavior of the mammalian blastomeres changes. They change shape to maximize their surface contact with one another, form tight junctions (see Figure 6.7), and become a compact mass of cells (**Figure 44.4B**).

Soon after the transition to the 32-cell stage, the cells separate into two groups. The **inner cell mass** will develop as the embryo, while the surrounding outer cells become an encompassing sac called the **trophoblast**. Trophoblast cells secrete fluid, creating a cavity—the blastocoel—with the inner cell mass at one end. At this stage the mammalian embryo is called a **blastocyst**, distinguishing it from the blastulas of other animal groups (**Figure 44.4C**). The pluripotent cells of the inner cell mass are known as **embryonic stem cells** and are the subject of much research because of their therapeutic potential (see Section 19.5).

Why is mammalian cleavage so different? A key factor is that mammalian eggs contain little or no yolk and must derive all nutrients from the mother. To support the developing embryo, a connection develops between the circulatory systems of the embryo and the mother. As we will see later in this chapter, the structures that provide this connection are the placenta and the umbilical cord. Thus the blastocyst of placental mammals must produce both the embryo (from the inner cell mass) and its support structures (from the trophoblast).

Fertilization in mammals occurs in the upper reaches of the oviduct, and cleavage occurs as the zygote travels down the oviduct to the uterus (**Figure 44.5**). When the blastocyst arrives in the uterus, the trophoblast adheres to the lining of the uterus (the **endometrium**), beginning the process of **implantation**. In humans, implantation begins about 6 days after fertilization and is aided by adhesion molecules and enzymes secreted by the trophoblast.

As the blastocyst moves down the oviduct to the uterus, it must not embed itself in the oviduct (Fallopian tube) wall, or the result will be an ectopic, or tubal, pregnancy—a very dangerous condition. Early implantation is prevented by the zona pellucida, which surrounded the egg (see Figure 43.5) and remains around the cleaving ball of cells. At about the time the blastocyst reaches the uterus, it hatches from the zona pellucida, and implantation can occur.

## Specific blastomeres generate specific tissues and organs

Cleavage results in a repackaging of the egg cytoplasm into a large number of small cells surrounding the fluid-filled blastocoel. Except in mammals, there is little gene expression during cleavage. Nevertheless, cells in different regions of the blastula possess different complements of the nutrients and cytoplasmic determinants that were present in the egg. For example, Figure 44.2 illustrated the processes by which β-catenin becomes localized in the region of the zygote that will become the dorsal side of the embryo.

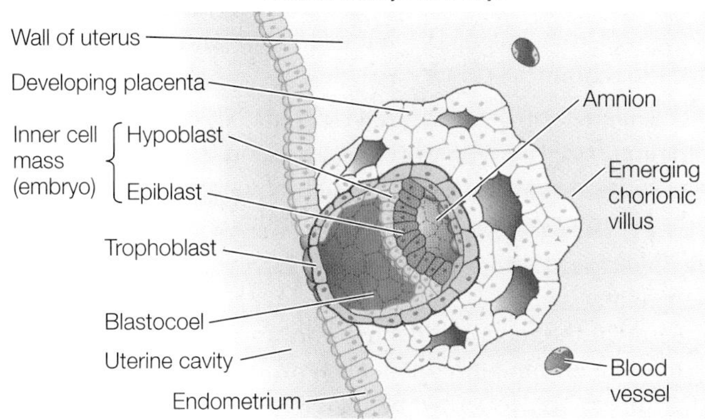


**44.5 A Human Blastocyst at Implantation** Adhesion molecules and proteolytic enzymes secreted by trophoblast cells allow the blastocyst to burrow into the endometrium. Once the blastocyst is implanted in the wall of the uterus, the trophoblast cells send out numerous projections—the chorionic villi—which increase the embryo's area of contact with the mother's bloodstream.

The blastocoel prevents cells from different regions of the blastula from coming into contact and interacting, but that will soon change. During the next stage of development, the cells of the blastula will move around and come into new associations with one another, communicate instructions to one another, and begin to differentiate. In many animals, these movements of the blastomeres are so regular and well orchestrated that it is possible to label each specific blastomere with a dye, thus producing **fate maps** that identify the tissues and organs formed from each blastomere's progeny (**Figure 44.6**).

Blastomeres become **determined**—committed to specific fates—at different times in different species. In some species, such as roundworms, the fates of blastomeres are restricted as early as the two-cell stage. If one of these blastomeres is experimentally removed, a particular portion of the embryo will not form. This type of development has been called **mosaic development** because each blastomere seems to contribute a specific set of "tiles" to the final "mosaic" that is the adult animal.

In contrast to mosaic development, the loss of some cells during cleavage in **regulative development** does not affect the developing embryo, because the remaining cells compensate for the loss. Regulative development is typical of many vertebrate species. Because development is regulative in humans, a single blastomere can be removed from early embryos without harming the remaining blastomeres or disrupting normal development. Cells removed from embryos produced by in vitro fertilization can be used for preimplantation genetic diagnosis to ensure that healthy embryos are selected for implantation in the mother.

If some blastomeres can change their fate to compensate for the loss of other cells during cleavage and blastula formation, can those cells form an entire embryo? To a certain extent, yes. During cleavage or early blastula formation in mammals, for example, if the blastomeres are physically separated into two groups, both groups can produce complete embryos. Since the two embryos come from the same zygote, they will be monozygotic twins—genetically identical.

Non-identical twins occur when two separate eggs are fertilized by two separate sperm. Thus, although identical twins are always of the same sex, non-identical twins have a 50 percent chance of being the same sex (that is, the same as two non-twin siblings). In about 1 out of 50,000 human pregnancies, genetic or environmental factors cause the inner cell mass to split partially. The result is twins who are conjoined at some point on their bodies and usually share some of their organs and limbs.

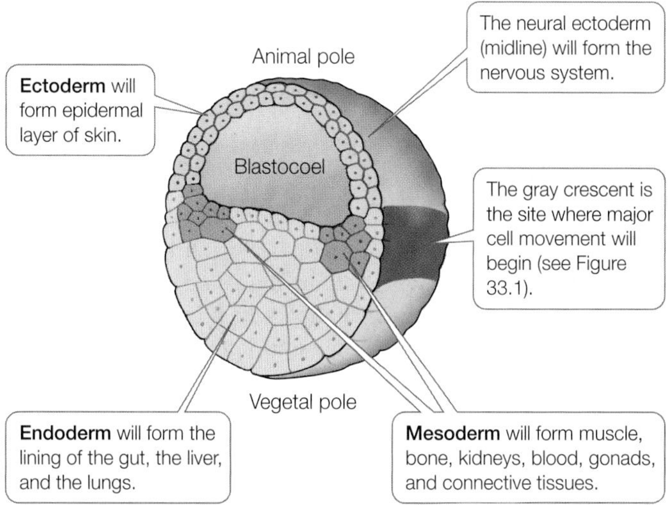


**44.6 Fate Map of a Frog Blastula** Colors indicate the portions of the blastula that will form the three germ layers and subsequently the frog's tissues and organs. This cutaway view shows the inside of the blastula. The cells fated to become mesoderm are in the deeper layer of cells, whereas the superficial layers of cells will mostly form ectoderm and endoderm.

### Germ cells are a unique lineage even in species with regulative development

Molecules present in the egg cytoplasm determine which lineage of cells will eventually populate the gonads and become the reproductive stem cells—oogonia and spermatogonia. In fruit flies, at the ninth nuclear division (recall that the egg is a multinucleate syncytium at this stage), a group of nuclei migrate to the posterior pole of the egg where they become surrounded by **pole plasm**—cytoplasm containing a complex mixture of fibrils, mitochondria, and specific proteins and mRNAs. As the cellularization of the blastoderm proceeds, the nuclei within the pole plasm give rise to the lineage of cells that will eventually migrate to the gonads (when they form) and produce germ cells (eggs and sperm).

As in fruit flies, the germ cell lineage in frogs starts with a special type of cytoplasm—the germ cell plasm—localized to one part of the egg. As a result of cleavage, the germ cell plasm becomes enclosed within some of the cells in the vegetal hemisphere; descendants of these cells will eventually migrate to the gonads once those structures form. The components of germ cell plasm have not been fully characterized, but one hypothesis is that they include inhibitors of transcription and translation that prevent these cells from differentiating into anything other than germ cells.

**RECAP 44.2**

Cleavage divides up the cytoplasm of the zygote such that different blastomeres contain different combinations of informational molecules. The amount of nutrients stored in the egg influences the pattern of cell cleavage that produces the blastula. Blastulation in mammals is different than in other vertebrates. The cells that will give rise to germ cell lineages are set aside very early in development.

- In general terms, describe the difference between complete and incomplete cleavage. **See p. 905 and Figure 44.3**
- What does a fate map tell us? How are fate maps constructed? **See p. 907 and Figure 44.6**
- Explain the statement that "the germ cell lineage exists outside of the processes of determination and differentiation." **See p. 908**

Of the next stage of development—gastrulation—the developmental biologist Lewis Wolpert once said, "It is not birth, marriage, or death, but gastrulation which is the most important time in your life." During gastrulation, cell movements create new cell-to-cell contacts, which in turn sets up signaling cascades that initiate the differentiation of cells and tissues and set the stage for the emergence of the body plan.

## 44.3 How Does Gastrulation Generate Multiple Tissue Layers?

The blastula is typically a fluid-filled ball of cells. How does this simple ball of cells become an embryo made up of multiple tissue layers with head and tail ends and dorsal and ventral sides? **Gastrulation** is the process whereby the blastula is transformed by massive movements of cells into an embryo with multiple tissue layers and distinct body axes. The resulting spatial relationships between tissues make possible the inductive interactions between cells that trigger differentiation and organ formation (see Figure 19.4).

In the triploblastic animals (see Section 33.1), three **germ layers** (also called cell layers or tissue layers, and not to be confused with germ cells) form during gastrulation:

- The **endoderm** is the innermost germ layer, created as some blastomeres move to the inside of the embryo. The endoderm gives rise to the lining of the digestive tract, respiratory tract, pancreas, thyroid, and liver.
- The **ectoderm** is the outer germ layer, formed from those cells remaining on the outside of the embryo. The ectoderm gives rise to the nervous system, including the eyes and ears; and to the epidermal layer of the skin and structures derived from skin, such as hair, feathers, nails or claws, sweat glands, oil glands, and even teeth and other tissues of the mouth.
- The **mesoderm** is the middle layer and is made up of cells that migrate between the endoderm and the ectoderm. The mesoderm contributes tissues to many organs, including the heart, blood vessels, muscles, and bones.

The three germ layers are illustrated for a very early embryo in the fate map shown in Figure 44.6.

Some of the most interesting and important challenges in animal development have dealt with two related questions: what directs the cell movements of gastrulation, and what is responsible for the resulting patterns of cell differentiation and organ formation? Scientists have made significant progress in answering both these questions at the molecular level. In the following discussion we will begin with sea urchin gastrulation because it is the simplest to conceptualize in spatial terms. We will then describe the more complex pattern of gastrulation in frogs, and then to the still more complex patterns in reptiles, birds, and mammals.

**Go to Animated Tutorial 44.1**
**Gastrulation**
Life10e.com/at44.1

### Invagination of the vegetal pole characterizes gastrulation in the sea urchin

The sea urchin blastula is a hollow ball of cells only one cell layer thick. The end of blastulation is marked by slowing of the rate of mitosis, and the beginning of gastrulation is marked by a flattening of the vegetal hemisphere (**Figure 44.7**). Some cells at the vegetal pole break away from neighboring cells and migrate into the cavity. These cells become **mesenchyme**—cells of the middle germ layer, the mesoderm. Mesenchymal cells are not organized in tightly packed sheets or tubes like epithelial cells are; they act as independent units, migrating into and among the other tissue layers.

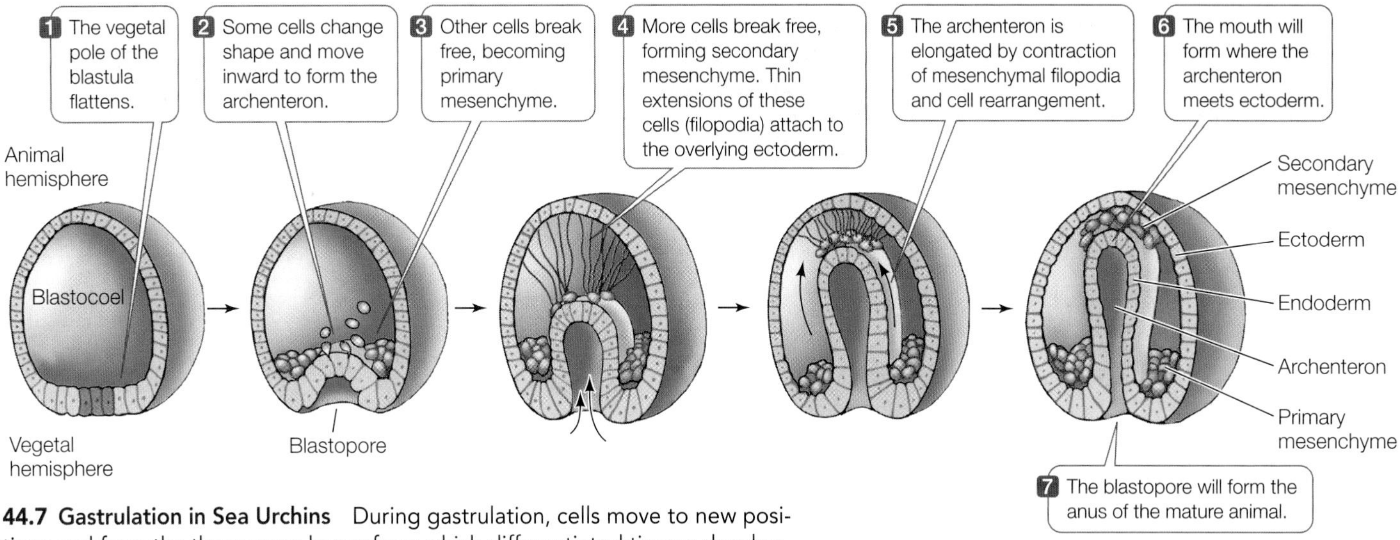

**44.7 Gastrulation in Sea Urchins** During gastrulation, cells move to new positions and form the three germ layers from which differentiated tissues develop.

The flattening at the vegetal pole results from changes in the shape of individual blastomeres. These cells, which are originally rather cuboidal, become wedge-shaped, with smaller outer edges and larger inner edges. As a result, the vegetal pole bulges inward, or invaginates, as if someone were poking a finger into a hollow ball (see Figure 44.7). The invaginating cells become endoderm and form the primitive gut, called the **archenteron**. At the tip of the archenteron, more cells enter the blastocoel to form mesoderm.

Changes in cell shapes cause the initial invagination of the archenteron, but eventually the archenteron is pulled inward by the mesenchyme cells. These cells, attached to the tip of the archenteron, send out extensions called filopodia that adhere to the overlying ectoderm. When the filopodia contract, they pull the archenteron toward the ectoderm at the opposite end of the embryo from where the invagination began. The mouth of the animal forms where the archenteron makes contact with this overlying ectoderm. The opening created by the invagination of the vegetal pole is called the **blastopore**, and it will become the anus of the animal.

What mechanisms control the various cell movements of sea urchin gastrulation? The immediate answer is that specific properties of particular blastomeres change. For example, some vegetal cells change shape and bulge into the blastocoel, and these cells become mesenchyme. Once they lose contact with their neighboring cells on the surface of the blastula, they send out filopodia that then move along an extracellular matrix of proteins laid down by the cells lining the blastocoel.

A deeper understanding of gastrulation requires that we discover the molecular mechanisms whereby different blastomeres develop different properties. Cleavage systematically divides up the cytoplasm of the egg. The sea urchin blastula at the 64-cell stage is radially symmetrical, but it has polarity, as described in Section 19.2. It consists of tiers of cells. As in the frog blastula, the top is the animal pole and the bottom the vegetal pole.

If different tiers of blastula cells are separated experimentally, they show different developmental potentials; only cells from the vegetal pole are capable of initiating the development of a complete larva. It has been proposed that these differences are due to uneven distribution of various transcriptional regulatory proteins in the egg cytoplasm. As cleavage progresses, these proteins end up in different groups of cells. Therefore specific sets of genes are activated in different cells, determining their different developmental capacities.

Next we will turn to gastrulation in the frog and to the key signaling molecules involved.

## Gastrulation in the frog begins at the gray crescent

Amphibian blastulas have considerable yolk and are more than one cell layer thick; gastrulation is therefore more complex in amphibians than in sea urchins. Variation is considerable across different species of amphibians, so this brief account describes results from studies done on different species to produce a generalized picture of amphibian development.

Amphibian gastrulation begins when certain cells in the gray crescent region (see Figure 44.1) change their shapes and cell-adhesion properties. These cells bulge inward toward the blastocoel while they remain attached to the outer surface of the blastula by slender necks; because of their shape, they are called bottle cells. Bottle cells mark the spot where the **dorsal lip** of the blastopore will form (**Figure 44.8**).

As the bottle cells move inward, the dorsal lip is created, and a sheet of cells moves over it into the blastocoel. This process is called **involution**. One group of involuting cells is the prospective endoderm; these cells form the primitive gut, or archenteron. Another group will move between the endoderm and the outermost cells to form the mesoderm. These rearrangements are due to changes in cell properties called **convergent extension**. The cells elongate in the direction of movement, but they also intercalate (move in between each other). If they just elongated, the migrating group of cells would become much

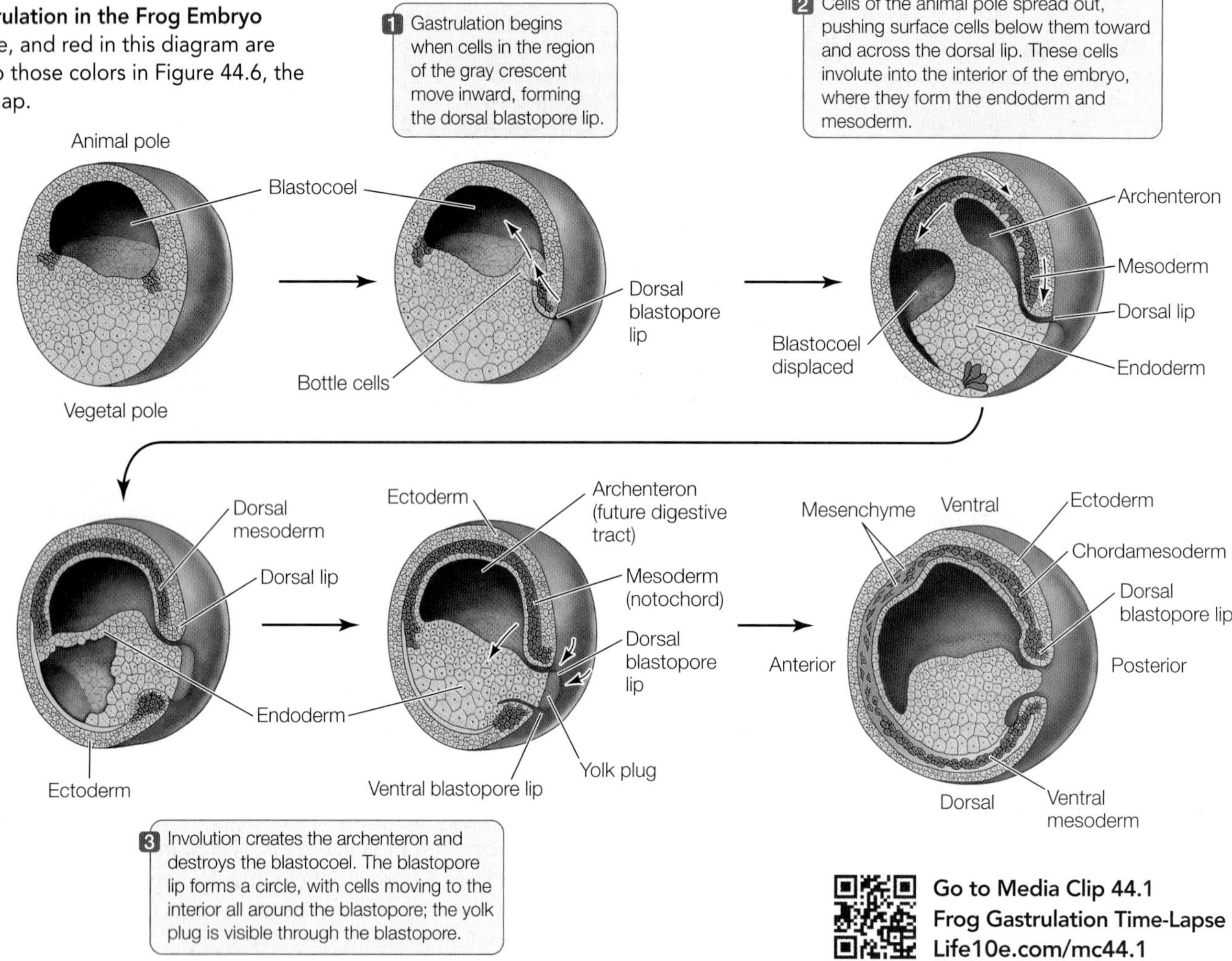


**44.8 Gastrulation in the Frog Embryo** Yellow, blue, and red in this diagram are matched to those colors in Figure 44.6, the frog fate map.

Go to Media Clip 44.1
Frog Gastrulation Time-Lapse
Life10e.com/mc44.1

narrower; by intercalating, they maintain the width of the migrating cell group.

As gastrulation proceeds, cells from the animal hemisphere flatten and move toward the site of involution in a process called **epiboly**. The blastopore lip widens and eventually forms a complete circle surrounding a "plug" of yolk-rich cells. As cells continue to move inward through the blastopore, the archenteron grows, gradually displacing the blastocoel.

As gastrulation comes to an end, the amphibian embryo consists of three germ layers: ectoderm on the outside, endoderm on the inside, and mesoderm in between. The embryo also has a dorsal–ventral and anterior–posterior organization. Most importantly, the fates of specific regions of the endoderm, mesoderm, and ectoderm have been determined. The beautiful experiments revealing how determination takes place in the amphibian embryo are an old but exciting story.

## The dorsal lip of the blastopore organizes embryo formation

In the early 1900s the German biologist Hans Spemann was studying the development of salamander eggs. He was interested in finding out whether the nuclei of blastomeres remain capable of directing the development of complete embryos or whether these nuclei lose some developmental potential. With great patience and dexterity, he formed loops from single hairs taken from a baby (in fact, his daughter) and tied them around fertilized eggs along the plane of the first cell division, effectively dividing the eggs in half, with the nucleus restricted to one side. That side went through cell divisions and developed into a salamander; the other half simply degenerated. Up until the 16-cell stage, if one nucleus escaped to the other side of the constriction, twin salamanders could develop. Thus each of the nuclei of the blastula (at least up to the 16-cell stage) was capable of directing and supporting development of the whole organism.

As often happens in science, Spemann's bisection experiments revealed a new phenomenon. Sometimes the half of the blastula receiving an escaped nucleus did not develop. When his loops bisected the gray crescent, both halves of the zygote developed into a complete embryo. When he tied the loops so the gray crescent was on only one side of the constriction, however, only that half of the zygote developed into a complete embryo (**Figure 44.9**). The half lacking gray crescent material underwent cell division, but even if it contained a nucleus, it became a clump of undifferentiated cells that Spemann called a "belly piece." Spemann hypothesized that cytoplasmic factors unequally distributed in the fertilized egg were necessary for gastrulation and the development of a normal salamander.

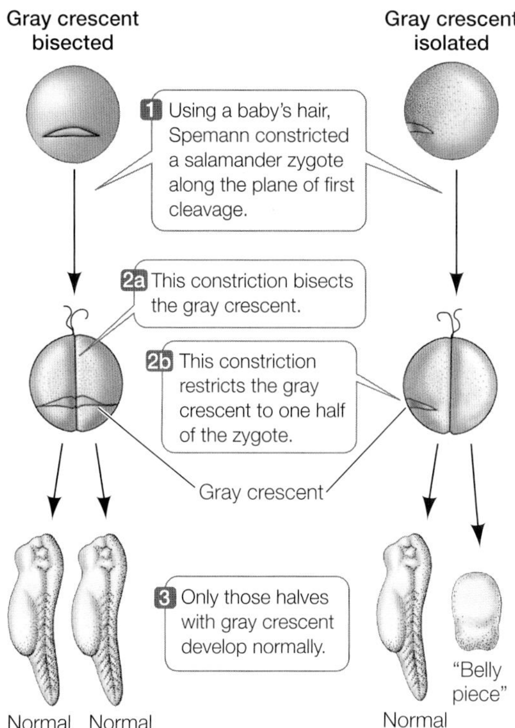


**44.9 Gastrulation and the Gray Crescent** Spemann's research revealed that gastrulation and subsequent normal development in salamanders depend on cytoplasmic determinants localized in the gray crescent.

To further test the hypothesis that cells receiving different complements of cytoplasmic factors had different developmental fates, Spemann transplanted pieces of early gastrulas to various locations on other gastrulas. Guided by fate maps (see Figure 44.6), he was able to take a piece of ectoderm he knew would develop into the epidermis of the skin and transplant it to a region that normally becomes part of the nervous system, and vice versa.

When he performed these transplants in early gastrulas—when the blastopore was just beginning to form—the transplanted pieces always developed into tissues that were appropriate for the location where they were placed. Transplanted cells destined to become epidermis in their original location developed into nervous system tissue, and transplanted cells destined to become nervous system tissue in their original location developed into host epidermis. Thus Spemann learned that the fates of the transplanted cells had not been determined before the transplantation.

In late gastrulas, however, the same experiment yielded opposite results. Transplanted cells destined to become epidermis in their original location produced patches of skin cells in the host nervous system, and the transplanted cells from regions that would develop into nervous system tissue produced neural tissue in the skin of the recipient. At some point during gastrulation, the fates of the embryonic cells had become determined.

Spemann's next experiment, done with his student Hilde Mangold, produced momentous results: they transplanted the dorsal lip of the blastopore (**Figure 44.10**). When this small piece of tissue was transplanted into the presumptive belly area of another gastrula, it stimulated a second site of gastrulation—and a second complete embryo formed belly-to-belly with the original embryo. Because the dorsal lip of the blastopore was apparently capable of inducing the host tissue to form an entire embryo, Spemann and Mangold dubbed the dorsal lip tissue the **primary embryonic organizer**, or simply the **organizer**. For more than 90 years, the organizer has been an active area of research.

## Transcription factors and growth factors underlie the organizer's actions

With the advent of modern molecular methods, the primary embryonic organizer has been studied intensively to discover the molecular mechanisms involved in its action. The distribution of the transcription factor β-catenin in the late blastula corresponds to the location of the organizer in the early gastrula, so β-catenin is a candidate for the initiator of organizer activity. To prove that a protein is an inductive signal, it has to be shown that it is both *necessary* and *sufficient* for the proposed effect. In other words, the effect should not occur if the candidate protein is not present (necessity), and the candidate protein should be capable of inducing the effect where it would otherwise not occur (sufficiency).

The criteria of necessity and sufficiency have been satisfied for β-catenin. If β-catenin mRNA transcripts are depleted by injections of antisense RNA into the egg (see Section 18.4), gastrulation does not occur. If β-catenin is experimentally overexpressed in another region of the blastula, it can induce a second axis of embryo formation, as the transplanted dorsal lip did in the Spemann–Mangold experiments. Thus β-catenin appears to be both necessary and sufficient for the formation of the primary embryonic organizer—but it is only one component of a complex signaling process.

How the presence of β-catenin creates the organizer, and how the organizer then induces the beginnings of the body plan, involves a complex series of interactions between transcription factors and growth factors that control gene expression. What follows is only a portion of this complex and still emerging story. What you should take from this description is not the names of the genes and gene products involved. Rather, we hope you will gain a basic appreciation for how signaling molecules interact to produce different combinations of signals that convey spatial and temporal information. This information guides cells into different paths of determination and differentiation.

Studies of early gastrulas revealed that primary embryonic organizer activity is generated by the interaction of β-catenin with signals coming from the vegetal cells. Together they activate the expression of the transcription factor Goosecoid. Expression of the *goosecoid* gene depends on two signaling pathways.

The first of these pathways involves a *goosecoid*-promoting transcription factor called Siamois. The *siamois* gene is normally

INVESTIGATING**LIFE**

**44.10 The Dorsal Lip Induces Embryonic Organization** In a classic experiment, Hans Spemann and Hilde Mangold transplanted the dorsal blastopore lip mesoderm of an early gastrula stage salamander embryo.[a] The results showed that the cells of this embryonic region, which they dubbed "the organizer," could direct the formation of an entire embryo.

**HYPOTHESIS** The early dorsal blastopore lip organizes cell differentiation in amphibian embryos.

*Method*

1. Excise a patch of mesoderm tissue from above the dorsal blastopore lip of an early gastrula stage salamander embryo (the donor).
2. Transplant the donor tissue onto a recipient embryo at the same stage. The donor tissue is transplanted onto a region of ectoderm that should become epidermis (skin).

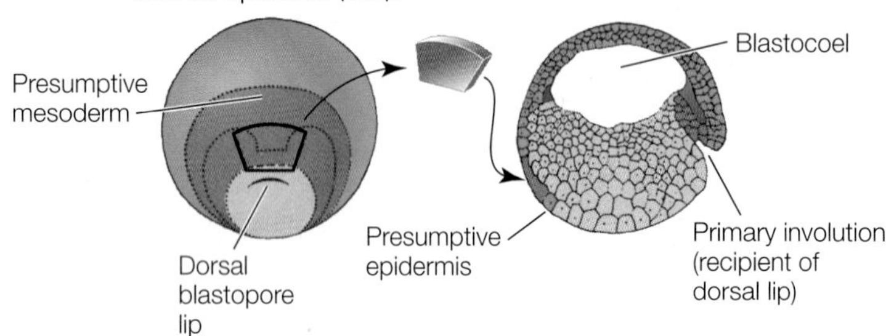


*Results*

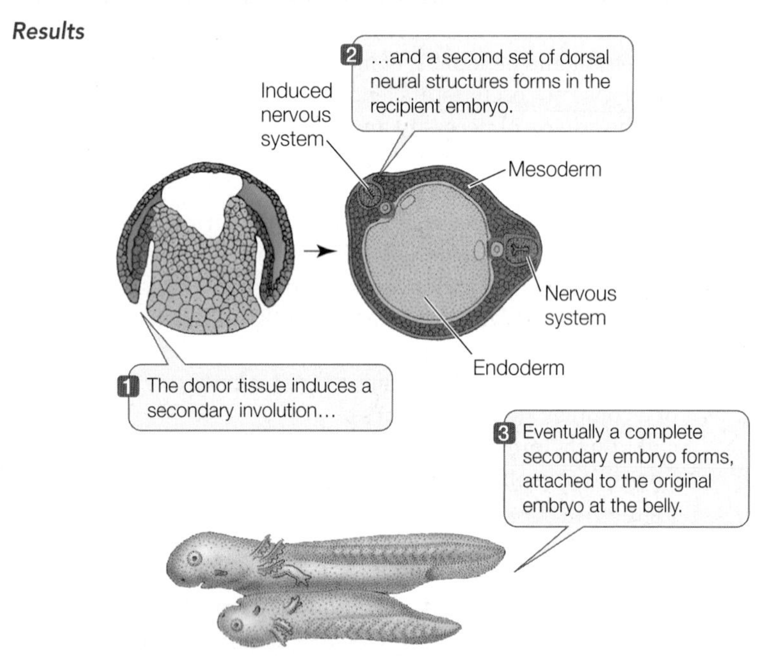


**CONCLUSION** The cells of the dorsal blastopore lip can induce other cells to change their developmental fates.

Go to **BioPortal** for discussion and relevant links for all INVESTIGATING**LIFE** figures.

[a]Spemann, H. and H. Mangold. 1924. *Roux' Arch. Entw. Mech.* 100: 599–638. Viktor Hamburger's translation appeared in *Foundations of Experimental Embryology, 1964*, (B.H. Willier and J.M. Oppenheimer, eds.), pp. 146–184.

**Go to Animated Tutorial 44.2**
**Tissue Transplants Reveal the Process of Determination**
Life10e.com/at42.2

repressed by a ubiquitous transcription factor called Tcf-3, but in cells in which β-catenin is present, an interaction between Tcf-3 and β-catenin induces *siamois* expression (**Figure 44.11**). But Siamois protein alone is not sufficient for *goosecoid* expression.

The second pathway involves mRNAs from the original egg cytoplasm for a family of proteins called transforming growth factor-β (TGF-β). TGF-β interacts with the Siamois protein to turn on *goosecoid* transcription. Thus you can see it is a complex combination of factors that determines which cells become the primary organizer.

## The organizer changes its activity as it migrates from the dorsal lip

Organizer cells begin the process of formation of the dorsal lip of the blastopore. Specifically, these cells are at the center of the dorsal lip and involute, moving forward on the midline (i.e., the middle of the anterior–posterior axis). The first organizer cells to enter the embryo migrate anteriorly to become the head endoderm and head mesoderm. Here they induce neighboring cells to participate in making structures of the head. Organizer cells that involute into the later embryo will induce structures of the trunk, and the last of the organizer cells to move inward from the dorsal lip will induce structures of the tail. How does the nature of the organizer cells change to enable them to induce head, trunk, or tail structures?

Inductive tissue interactions can suppress as well as activate. As we learned above, the early organizer cells express the transcription factor Goosecoid, which activates genes encoding soluble signals. As these cells move forward in the blastocoel, they come into contact with new populations of cells that produce several different growth factors. For head structures to form, certain of these growth factors have to be suppressed. The most anterior organizer cells, under the influence of Goosecoid, produce and release antagonists to those growth factors.

The induction of trunk structures requires suppression of a different set of growth factors. In organizer cells that involute later than the head organizers, Goosecoid is no longer the dominant transcription factor, and these cells express different growth factor antagonists. The induction of tail structures requires still different activities of the organizer cells that involute last. Thus the organizer cells express appropriate sets of growth factor antagonists at the right times to achieve different patterns of differentiation on the anterior–posterior axis.

The initiation of nervous system development also involves a suppressive tissue interaction. For a long time it was thought that the involuting organizer cells actively induced the overlying ectoderm to form neural tissue rather than becoming epidermis. We now

Gray crescent
1 Repression of *siamois* by Tcf-3 proteins prevents expression of organizer-specific genes.
2 β-Catenin in vegetal cells below the gray crescent blocks Tcf-3 repression of *siamois* gene expression.
No β-catenin
Tcf-3 proteins
*siamois* gene repressed
DNA
No transcription
β-Catenin proteins
*siamois* gene activated
Transcription
Siamois protein
3 TGF-β-related signaling pathway acts synergistically with Siamois to activate the *goosecoid* gene.
*goosecoid*
Transcription
4 Goosecoid protein activates numerous genes in the organizer.

**44.11 Molecular Mechanisms of the Organizer** In amphibians, the organizing potential of the gray crescent depends on the activity of the *goosecoid* gene, which in turn is activated by signaling pathways set up in the vegetal cells below the gray crescent.

know, however, that epidermis is not the default state of the dorsal ectoderm. Rather, the underlying mesoderm secretes factors called BMP proteins that induce the ectoderm to become epidermis. The role of the involuting organizer cells is to block that induction, allowing the overlying ectodermal cells to follow what is really their default pathway—differentiation into neural tissue (**Figure 44.12**).

## Reptilian and avian gastrulation is an adaptation to yolky eggs

The eggs of reptiles and birds contain a mass of yolk, and the blastulas of these groups develop as a disc of cells on top of the yolk (see Figure 44.3B). We will use the chicken egg to show how gastrulation proceeds in a flat disc of cells rather than in a ball of cells.

Cleavage in the chick results in a flat, circular layer of cells called a blastodisc (**Figure 44.13**). Between the blastodisc and the yolk mass is a fluid-filled space. Some cells from the blastodisc break free and move into this space. These cells come together to form a continuous layer called the **hypoblast**, which will later contribute to extraembryonic membranes that will support and nourish the developing embryo. The overlying cells make up the **epiblast**, from which the embryo will form.

INVESTIGATING**LIFE**

**44.12 Differentiation Can Be Due to Inhibition of Growth Factors** When organizer cells involute to underlie dorsal ectoderm along the embryo midline, that overlying ectoderm becomes neural tissue rather than skin (epidermis). But do the organizer cells cause dorsal ectoderm to become neural tissue, or do they prevent this ectoderm from becoming skin?[a]

**HYPOTHESIS** The default state of amphibian dorsal ectoderm is neural; it is induced by underlying mesoderm to become epidermis.

***Method*** 1. Excise the animal caps of late-stage frog blastulas and disperse the cells in culture medium so there is no cell-to-cell contact.

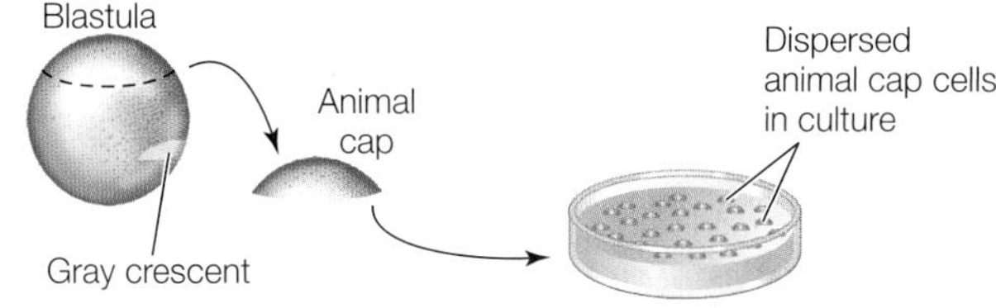


2. Prepare four separate cultures of embryonic ectodermal cells. Incubate with no additions (control); with BMP4 (isolated from mesodermal cells) with BMP4 inhibitor (isolated from organizer cells) and with both molecules.

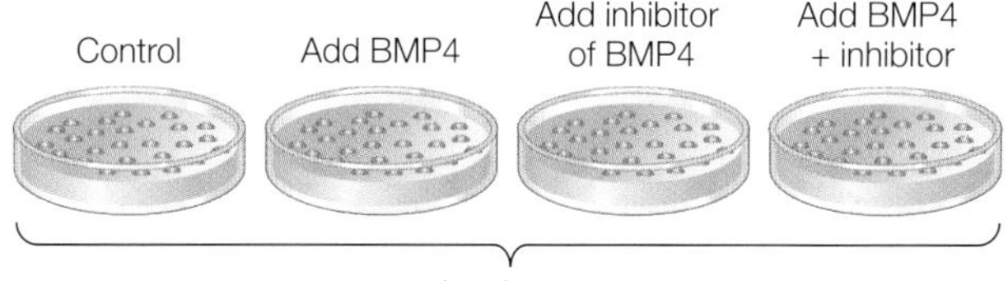


3. After incubation, extract mRNAs from the ectodermal cells and analyze for the presence of mRNAs for marker proteins NCAM (neural cell adhesion molecule, a neural protein) and/or keratin (an epidermal protein).

***Results*** The control ectoderm (no inductive factors added) expresses the neural marker. In the presence of mesodermal BMP4, ectoderm expresses the epidermal marker. If BMP4 is inhibited, ectoderm expresses the neural marker.

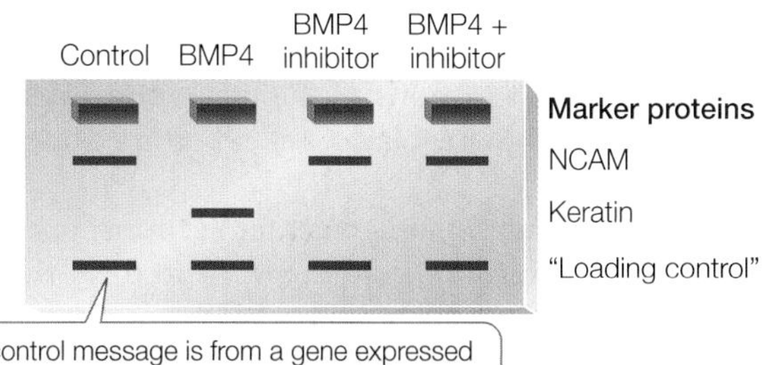


**CONCLUSION** The default state of amphibian dorsal ectoderm is neural. MP4 protein from mesoderm can induce ectoderm cells to differentiate into epidermis. Thus the organizer cells must secrete an inhibitor of BMP4.

Go to **BioPortal** for discussion and relevant links for all INVESTIGATING**LIFE** figures.

[a]Wilson, P. A. and A. Hemmati-Brivanlou. 1995. *Nature* 376: 331–333.

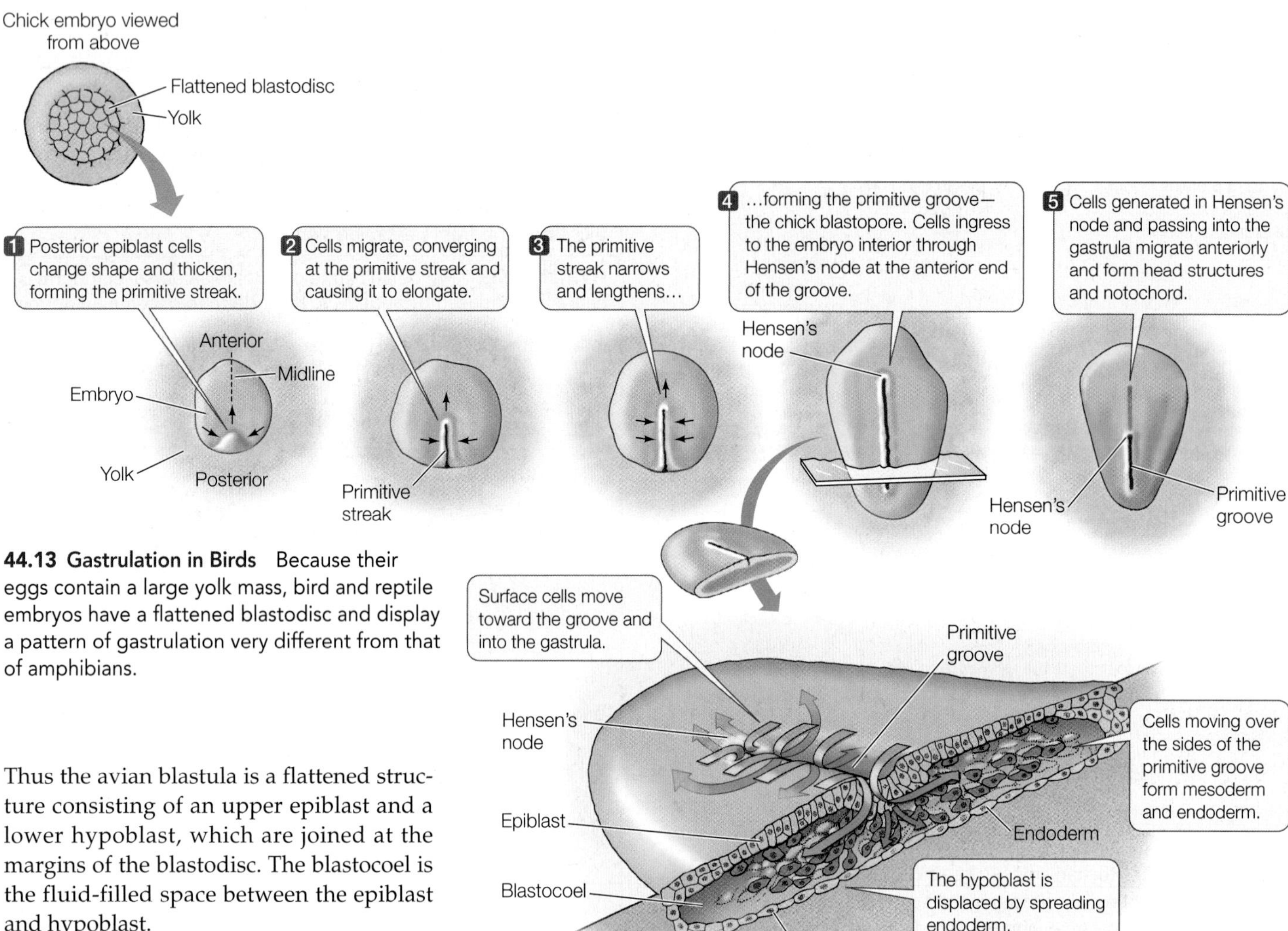


**44.13 Gastrulation in Birds** Because their eggs contain a large yolk mass, bird and reptile embryos have a flattened blastodisc and display a pattern of gastrulation very different from that of amphibians.

Thus the avian blastula is a flattened structure consisting of an upper epiblast and a lower hypoblast, which are joined at the margins of the blastodisc. The blastocoel is the fluid-filled space between the epiblast and hypoblast.

Gastrulation begins with a thickening in the posterior region of the epiblast, caused by the movement of cells toward the midline and then forward along the midline (see Figure 44.13). The result is a midline ridge called the **primitive streak**. A depression called the primitive groove forms along the length of the primitive streak. The primitive groove functions as the blastopore, and cells migrate through it into the blastocoel to become endoderm and mesoderm.

In the chick embryo, no archenteron forms, but the endoderm and mesoderm migrate forward to form the gut and other structures. At the anterior end of the primitive groove is a thickening called **Hensen's node**, which in reptiles, birds, and mammals is the equivalent of the dorsal lip of the amphibian blastopore. Many signaling molecules that have been identified in the frog organizer are also expressed in Hensen's node. Cells moving into the blastocoel and moving anteriorly from Hensen's node become the notochord and organize the chick embryo in a manner similar to that of the frog embryo.

## The embryos of placental mammals lack yolk

Mammalian embryos (with the exception of monotremes) derive their nourishment from the maternal circulation, and therefore mammalian eggs do not have large amounts of yolk constraining their cleavage and early development. Nevertheless, mammals evolved from reptilian ancestors, so it is not surprising that mammals, birds, and reptiles share certain patterns of early development. Earlier we described the development of the mammalian inner cell mass (the equivalent of the avian blastodisc) and the outer trophoblast.

As in avian development, in placental mammals the inner cell mass splits into an upper layer called the epiblast and a lower layer called the hypoblast. The embryo forms from the epiblast, while the hypoblast contributes to the extraembryonic membranes that will encase the developing embryo and help form the placenta (see Figure 44.5). The epiblast also contributes to the extraembryonic membranes; specifically, it splits off an upper layer of cells that will form the amnion. The amnion will grow to surround the developing embryo as a membranous sac filled with amniotic fluid. Gastrulation occurs in the mammalian epiblast just as it does in the avian epiblast. A primitive groove forms, and epiblast cells migrate through the groove to become layers of endoderm and mesoderm. At the top of the groove is the node, which, as we learned at the start of this chapter, is where the beating of nodal cell cilia creates an asymmetrical flow of extracellular fluid. The asymmetrical

WORKING WITH**DATA:**

## *Nodal Flow and Inverted Organs*

### Original Paper

Nonaka, S., H. Shiratori, and H. Hamata. 2002. Determination of left–right patterning of the mouse embryo by artificial nodal flow. *Nature* 418: 96–99.

### Analyze the Data

The phenotype of the mutant mouse strain *inversus viscerum* (*iv/iv*) mimics human situs inversus. These mice have nonmotile primary cilia in the ventral (bellyward) node of the early embryo. Left alone, roughly half of all *iv/iv* mouse embryos will develop with normal left–right organ asymmetry while the other half show reversed asymmetry. Researchers in Japan used *iv/iv* embryos to test the hypothesis that the leftward flow of extracellular fluid created by the beating of nodal primary cilia is the stimulus for breaking bilateral symmetry in organ development.

The researchers anchored very early mouse embryos with their nodal ends pointed upward in a chamber filled with culture medium. Culture medium was then artificially pumped through the chamber either from the left (normal) or from the right (reversed) and at slow or fast flow rates. After 4 days the embryos were assessed as to the direction of looping (normal or reversed) of their developing heart tubes. The researchers compared the effects of speed and direction culture medium flow on wild-type and mutant embryos that were at the presomite stage when they were placed in the chamber.

A second set of experiments explored the the effects of fast, reversed flow on wild-type embryos that were at later stages of development (1, 2, or 3 somites) when placed in the chamber. The results are shown in the table below.

**QUESTION 1**

Do the data support the hypothesis that nodal flow is a stimulus that determines left–right organ asymmetry? Why or why not?

**QUESTION 2**

How would you explain the different results for the slow rightward flow in the presomite wild-type and *iv/iv* mice?

**QUESTION 3**

What do you conclude from the results on the 1-, 2-, and 3-somite wild-type embryos?

| | Speed and direction of culture medium flow | | | |
|---|---|---|---|---|
| Genotype, stage | Fast left | Slow left | Slow right | Fast right |
| *iv/iv*, presomite | 10N, 0R, 10T[a] | 8N, 2R, 10T | 3N, 25R, 34T | 1N, 11R, 12T |
| Wild-type, presomite | 9N, 0R, 10T | 12N, 4R, 16T | 13N, 3R, 16T | 2N, 21R, 24T |
| Wild-type, 1-somite | — | — | — | 9N, 0R, 12T |
| Wild-type, 2-somite | — | — | — | 22N, 0R, 22T |
| Wild-type, 3-somite | — | — | — | 14N, 0R, 14T |

[a]The normal direction for fluid flow is left. Developmental outcomes indicated as follows: N, normal; R, reversed; T, total embryos in sample. Note that some embryos were neither normal nor reversed but were included in the total.

Go to BioPortal for all WORKING WITH**DATA** exercises

flow stimulates nonmotile cilia, generating signaling cascades that determine the left–right asymmetry of the internal organs.

RECAP 44.3

The cell movements of gastrulation convert the blastula into an embryo with three tissue layers. New contacts between cells set up inductive signaling interactions that determine cell fates. Dorsal lip tissue is the source of organizer cells that induce development of preliminary head, trunk, and tail structures.

- Describe and compare the cell movements that occur during gastrulation in a sea urchin, a frog, and a bird. **See Figures 44.7, 44.8, and 44.13**
- Explain the molecular basis for the inductive capabilities of the organizer. **See pp. 912–913 and Figures 44.11 and 44.12**

We have described how the fertilized egg develops into an embryo with three germ layers and how cellular signals trigger different patterns of differentiation. In the next section we will describe how organs and organ systems develop.

## 44.4 How Do Organs and Organ Systems Develop?

Gastrulation produces an embryo with three germ layers that are positioned to influence one another through inductive tissue interactions. During the next phase of development, called **organogenesis**, organs and organ systems develop simultaneously and in coordination with each other. In the chordates (see Section 33.1), an early process of organogenesis is **neurulation**, the initiation of the nervous system. We will examine neurulation in the amphibian embryo, but it occurs in a similar fashion in reptiles, birds, and mammals.

### The stage is set by the dorsal lip of the blastopore

As we learned in the previous section, one group of cells that passes over the dorsal lip of the blastopore moves anteriorly and becomes the endodermal lining of the digestive tract. Another group of cells that involutes over the dorsal lip on the midline becomes chordamesoderm, so named because it forms a rod of mesoderm—the **notochord**—that extends down the

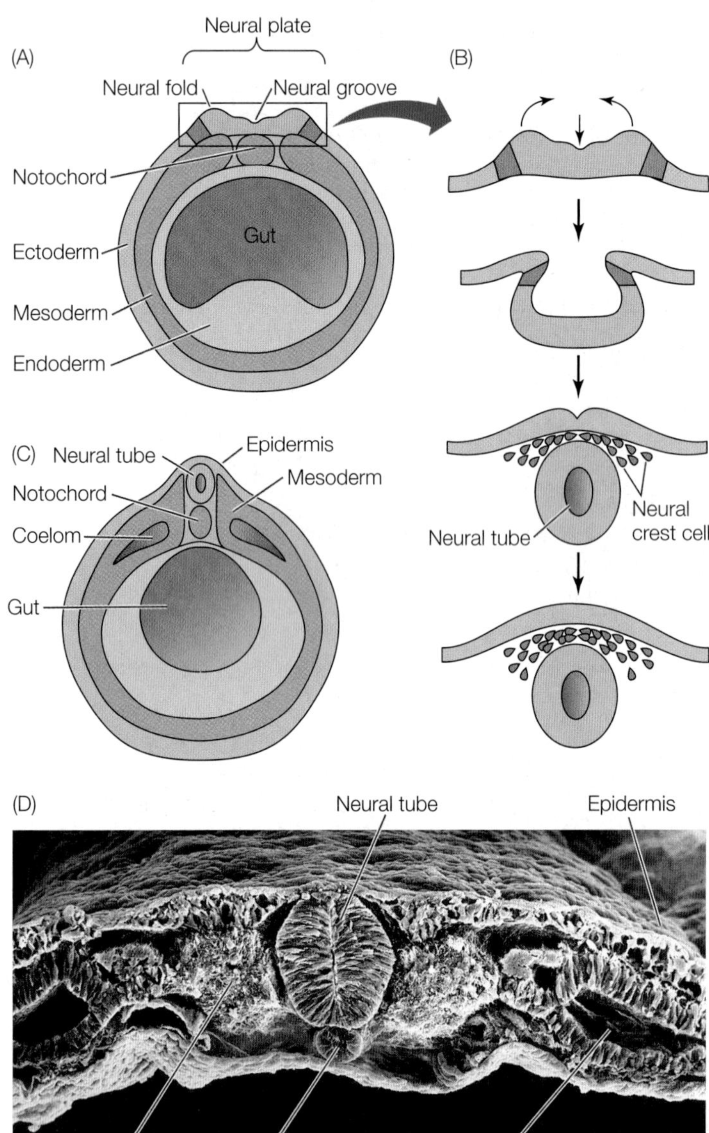


**44.14 Neurulation in a Vertebrate** (A) At the start of neurulation, the ectoderm of the neural plate (green) is flat. (B) The neural plate invaginates and folds, forming a tube. (C,D) The completely formed neural tube seen (C) in diagrammatic form and (D) in a scanning electron micrograph of a chick embryo.

center of the embryo. These cells also have important organizer functions (see Figure 44.8). The notochord gives structural support to the developing embryo and in vertebrates is replaced by the vertebral column. The organizing capacity of the chordamesoderm enables the overlying ectoderm to become neural ectoderm (see Figure 44.12). It does this by expressing signaling molecules (one appropriately called Noggin and another one called Chordin) that initiate differentiation of the different divisions of the nervous system.

Neurulation involves the formation of an internal neural tube from an external sheet of cells. The first signs of neurulation are flattening and thickening of the ectoderm overlying the notochord; this thickened area forms the neural plate (**Figure 44.14A**). The edges of the neural plate that run in an anterior–posterior direction continue to thicken to form ridges or folds. Between these neural folds, a groove forms and deepens as the folds roll over it to converge on the midline. The folds fuse, forming a cylinder, the **neural tube**, and a continuous overlying layer of epidermal ectoderm (**Figure 44.14B–D**).

Cells from the most lateral portions of the neural plate do not become part of the neural tube, but disassociate from it and come to lie between the neural tube and the overlying epidermis. These **neural crest cells** migrate outward to lead the development of the connections between the central nervous system (brain and spinal cord) and the rest of the body.

The neural tube develops bulges at the anterior end, which become the major divisions of the brain; the rest of the tube becomes the spinal cord. In humans, failure of the neural folds to fuse in this posterior region results in spina bifida, a birth defect in which the spinal cord is exposed because the vertebrae do not fuse. If the folds fail to fuse at the anterior end, an infant can develop without a forebrain (a condition called anencephaly). Although several genetic factors can cause these defects, other factors are environmental, including maternal diet. The incidence of neural tube defects in the United States in the early 1900s was as high as 1 in 300 live births; today it is less than 1 in 1,000. A major factor in this improvement has been the inclusion of folic acid (a B vitamin, also known as folate) in the mother's diet. It is essential for pregnant women to ingest sufficient folic acid.

## Body segmentation develops during neurulation

The vertebrate body plan, like that of arthropods, consists of repeating segments that are modified during development. These segments are most evident as the repeating patterns of vertebrae, ribs, nerves, and muscles along the anterior–posterior axis.

As the neural tube forms, mesodermal tissues gather along the sides of the notochord to form separate, segmented blocks of cells called **somites** (**Figure 44.15**). Somites produce cells that will become the vertebrae, ribs, muscles of the trunk, and limbs.

Nerves that connect the brain and spinal cord with tissues and organs throughout the body are also arranged segmentally. The somites help guide the organization of these peripheral nerves, but the nerves are not of mesodermal origin. As we saw above, when the neural tube fuses, the neural crest cells break loose and migrate inward between the epidermis and the somites and through the somites. These neural crest cells have diverse fates, including the development of peripheral nerves.

As development progresses, the different segments of the body change. Regions of the spinal cord differ, regions of the vertebral column differ in that some vertebrae grow ribs of various sizes and others do not, forelegs arise in the anterior part of the embryo, and hind legs arise in the posterior region.

## Hox genes control development along the anterior–posterior axis

How is mesoderm in the anterior part of a mouse embryo programmed to produce forelegs rather than hind legs? In Section 19.4, we saw how homeotic genes control body segmentation in *Drosophila*. We also learned that all homeotic genes contain a

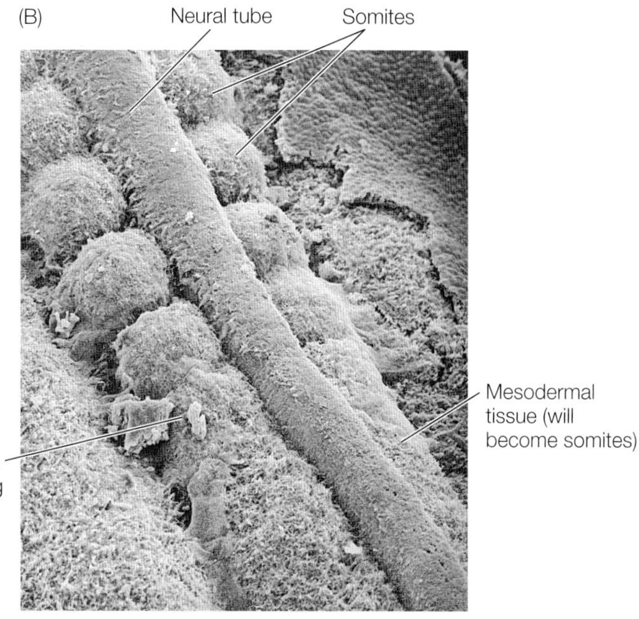


**44.15 Developing Body Segmentation** (A) Repeating blocks of tissue called somites form on either side of the neural tube. Muscle, cartilage, bone, and the inner layer of the skin form from the somites. (B) In this scanning electron micrograph of somite formation in a chick embryo, the overlying ectoderm has been removed and the neural tube and somites are seen from above.

DNA sequence called the **homeobox**. Some of the genes directing gastrulation in the frog are homeobox genes—for example, *goosecoid* and *siamois*. In vertebrates, the homeotic genes that control differentiation along the anterior–posterior body axis are called **Hox genes**.

In mammals, four Hox gene complexes reside on different chromosomes in clusters of about 10 genes each. Remarkably, the temporal and spatial expression of these genes follows the same pattern as their linear order on their chromosome. That is, the Hox genes closest to the 3′ end of each gene complex are expressed first and in the anterior of the embryo. The Hox genes at the 5′ end of the gene complex are expressed later and in a more posterior part of the embryo. As a result, different segments of the embryo receive different combinations of Hox gene products, which serve as transcription factors (**Figure 44.16**; see also Figure 20.2).

Whereas Hox genes give cells information about their position on the anterior–posterior body axis, other genes provide information about their dorsal–ventral position. Tissues in each segment of the body differentiate according to their dorsal–ventral location. The notochord provides many of these signals. One example of a dorsal–ventral difference is seen in the spinal cord; sensory nerve connections develop in the dorsal region, and motor nerve connections in the ventral region. The protein Sonic hedgehog (named for the video-game character), which is expressed in the mammalian notochord, induces cells in the overlying neural tube (i.e., the ventralmost cells of the tube) to become motor neurons.

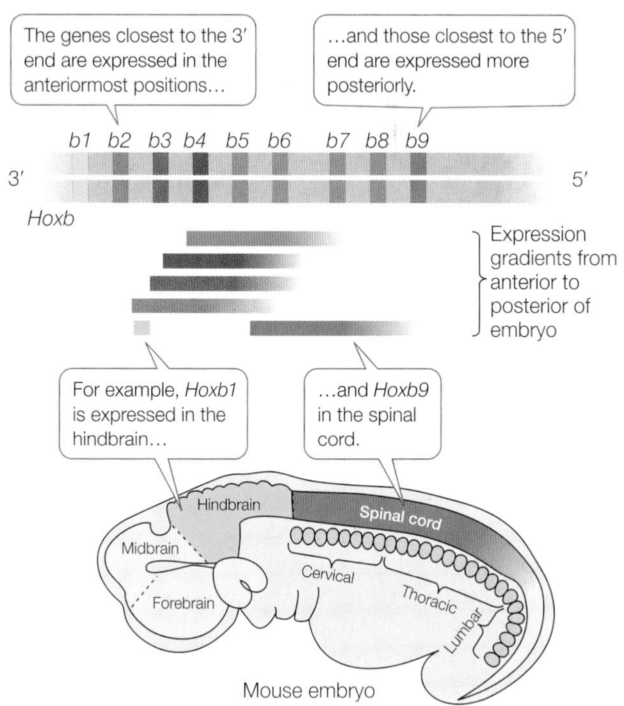


**44.16 Hox Genes Control Body Segmentation** Hox genes are expressed along the anterior–posterior axis of the embryo in the same order as their arrangement between the 3′ and 5′ ends of the gene complex. As a result of gene duplication during evolution, vertebrates have four copies of the Hox gene complex shown.

After body segmentation develops, the formation of organs and organ systems progresses rapidly. The development of an organ involves extensive inductive interactions of the kind we saw in the example of the vertebrate eye (see Figure 19.4). These inductive interactions are a current focus of study for developmental biologists.

RECAP 44.4

Gastrulation sets up tissue interactions that initiate organogenesis. Neurulation is initiated by organizer mesoderm that forms the notochord.

- Describe the formation of the neural tube in vertebrates. **See p. 916 and Figure 44.14**
- How do somites relate to segmentation of the body axis? **See p. 916 and Figure 44.15**
- Explain what Hox genes are and how they instruct patterns of differentiation along the body axis. (You may want to refer back to Section 19.4 for more information on Hox genes.) **See pp. 916–917 and Figure 44.16**

We have seen how the basic structure of the developing embryo arises, through the establishment of the anterior–posterior and dorsal–ventral axes, the formation of the neural tube, and the emergence of a segmented body plan. In the next section we will examine the developmental events that result in the formation of structures that support the developing embryo: the extraembryonic membranes and the placenta.

## 44.5 How Is the Growing Embryo Sustained?

There is more to a developing reptile, bird, or mammal than the embryo itself. As mentioned earlier, the embryos of these vertebrates are surrounded by several **extraembryonic membranes** that originate from the embryo but are not part of it. Extraembryonic membranes function in nutrition, gas exchange, and waste removal. In mammals they interact with tissues of the mother to form the placenta. The evolutionary relationships between the extraembryonic membranes of birds and mammals were discussed in Section 33.4.

### Extraembryonic membranes form with contributions from all germ layers

The chicken provides a good example of how extraembryonic membranes form from the germ layers created during gastrulation. In the chick, four membranes form—the yolk sac, the allantoic membrane, the amnion, and the chorion. The **yolk sac** is the first to form, and it does so by extension of the hypoblast layer along with some adjacent mesoderm. The yolk sac grows to enclose the entire body of yolk in the egg (**Figure 44.17**). It constricts at the top to create a tube that is continuous with the gut of the embryo. However, yolk does not pass through this tube. Yolk is digested by the cells of the yolk sac, and the nutrients are transported to the embryo through blood vessels that form from mesoderm and line the outer surface of the yolk sac. The **allantoic membrane** is also an outgrowth of the extraembryonic endoderm plus adjacent mesoderm. It forms the **allantois**, a sac for storage of metabolic wastes.

Ectoderm and mesoderm combine and extend beyond the limits of the embryo to form the other extraembryonic membranes. Two layers of cells extend all along the inside of the eggshell, both over the embryo and below the yolk sac. Where they meet, they fuse, forming two membranes, the inner **amnion** and the outer **chorion**. The amnion surrounds the embryo, forming the amniotic cavity. The amnion secretes fluid into the cavity, providing a protective environment for the embryo. The outer membrane, the chorion, forms a continuous membrane just under the eggshell (see Figure 44.17). It limits water loss from the egg and also works with the enlarged allantoic membrane to exchange respiratory gases between the embryo and the outside world.

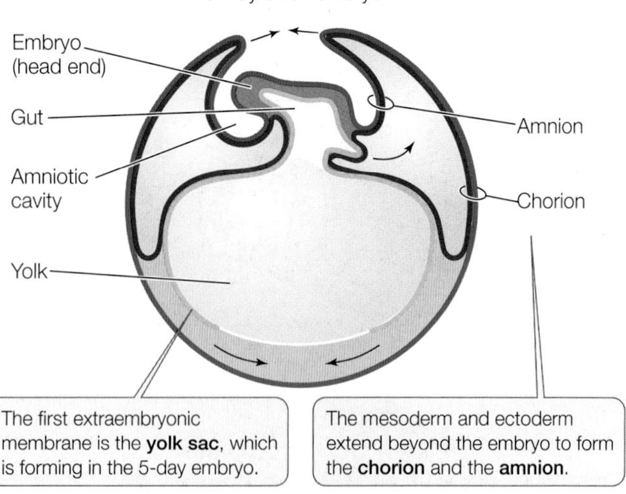


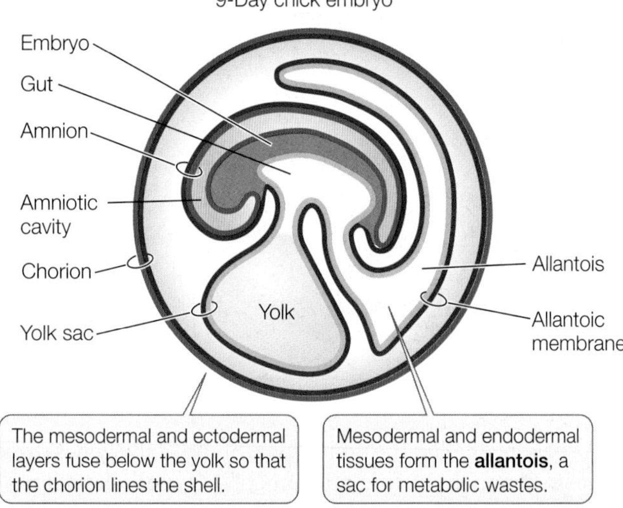


**44.17 The Extraembryonic Membranes** In birds and mammals, the embryo constructs four extraembryonic membranes. In birds, the yolk sac encloses the yolk, and the amnion and chorion enclose the embryo. Fluids secreted by the amnion fill the amniotic cavity, providing an aqueous environment for the embryo. The chorion, along with the allantoic membrane, mediates gas exchange between the embryo and its environment. The allantois stores the embryo's waste products. (See also Figure 33.19.)

**Go to Activity 44.1 Extraembryonic Membranes Life10e.com/ac44.1**

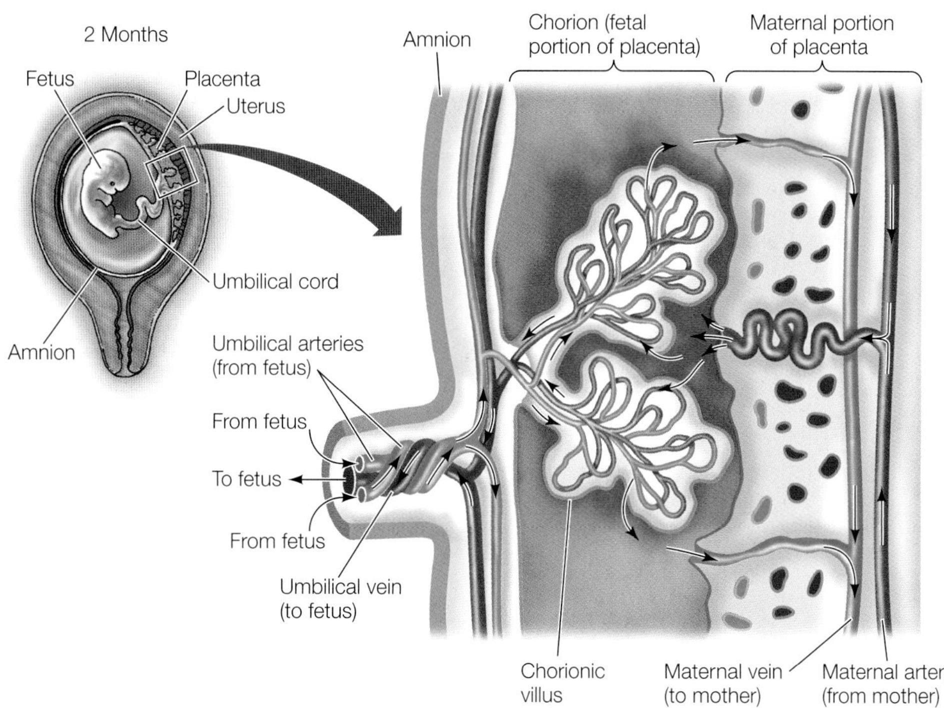


**44.18 The Mammalian Placenta** In humans and most other mammals, nutrients and wastes are exchanged between maternal and fetal blood in the placenta, which forms from the chorion and tissues of the uterine wall. The embryo is attached to the placenta by the umbilical cord. Embryonic blood vessels invade the placental tissue to form fingerlike chorionic villi. Maternal blood flows into the spaces surrounding the villi, and placental blood flows through the villi so nutrients and respiratory gases can be exchanged between the maternal and fetal blood.

### Extraembryonic membranes in mammals form the placenta

In placental mammals, the first extraembryonic membrane to form is the trophoblast (see Figures 44.4C and 44.5). When the blastocyst reaches the uterus and hatches from its encapsulating zona pellucida, trophoblast cells interact directly with the endometrium. Adhesion molecules expressed on the surfaces of these cells attach them to the uterine wall. By secreting proteolytic enzymes, the trophoblast burrows into the endometrium, beginning the process of implantation. Eventually the entire trophoblast is within the wall of the uterus. The trophoblast cells send out numerous projections, or villi, to increase the surface area of contact with maternal blood.

Meanwhile, the hypoblast cells proliferate to form what in the bird would be the yolk sac. But there is virtually no yolk in eggs of placental mammals, so the yolk sac contributes mesodermal tissues that interact with trophoblast tissues to form the chorion. The chorion, along with tissues of the uterine wall, produces the **placenta**, the organ that exchanges nutrients, respiratory gases, and metabolic wastes between the mother and the embryo (**Figure 44.18**).

At the same time the yolk sac is forming from the hypoblast, the epiblast produces the amnion, which grows to enclose the entire embryo in a fluid-filled amniotic cavity. The rupturing of the amnion and chorion and the loss of the **amniotic fluid** (the "water breaks") herald the onset of labor in humans.

An allantois also develops in mammals, but its importance depends on how well nitrogenous wastes can be transferred across the placenta. The human placenta deals effectively with the fetal nitrogenous wastes, so the human allantoic sac is small. In contrast, the pig placenta is not very good at clearing nitrogenous wastes from the fetus, so the pig's allantoic sac is large. In humans and other placental mammals, allantoic tissues contribute to the formation of the umbilical cord, by which the embryo is attached to the chorionic placenta (see Figure 33.19B). It is through the blood vessels of the umbilical cord that nutrients and oxygen from the mother reach the developing fetus, and wastes, including carbon dioxide and urea, are removed.

**RECAP 44.5**

The extraembryonic membranes of reptiles, birds, and mammals sustain the growing embryo. In reptiles and birds these membranes surround the embryo within the shelled egg. In mammals the extraembryonic membranes form the placenta, an organ that exchanges nutrients, respiratory gases, and metabolic wastes between the mother and the embryo.

- Describe each of the four extraembryonic membranes and their functions in the developing chick egg. **See p. 918 and Figure 44.17**
- Explain the role of the trophoblast in the early development of a mammalian embryo. **See p. 919 and Figure 44.5**

## 44.6 What Are the Stages of Human Development?

In humans, **gestation**, or pregnancy, lasts about 266 days, or 9 months. Gestation is shorter in smaller mammals—21 days in mice, for example—and in larger mammals it is longer—330 days in horses and 600 days in elephants. The events of human gestation can be divided into three **trimesters** of roughly 3 months each.

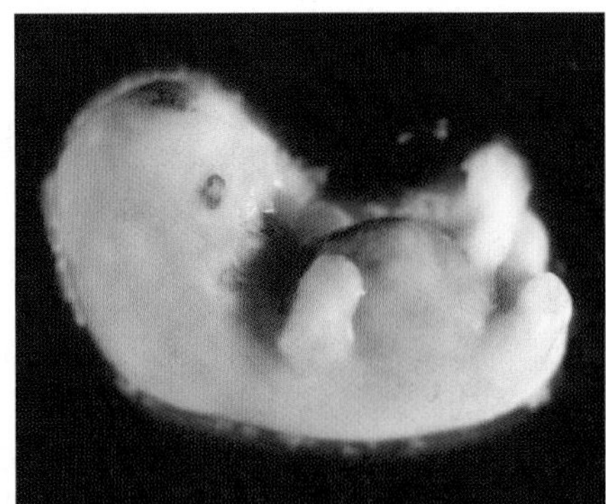


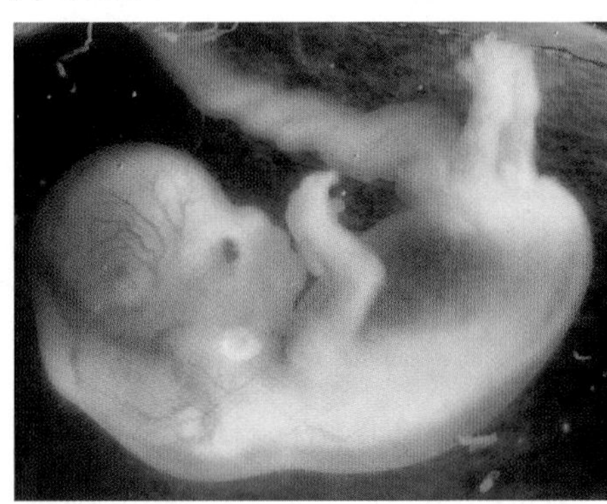

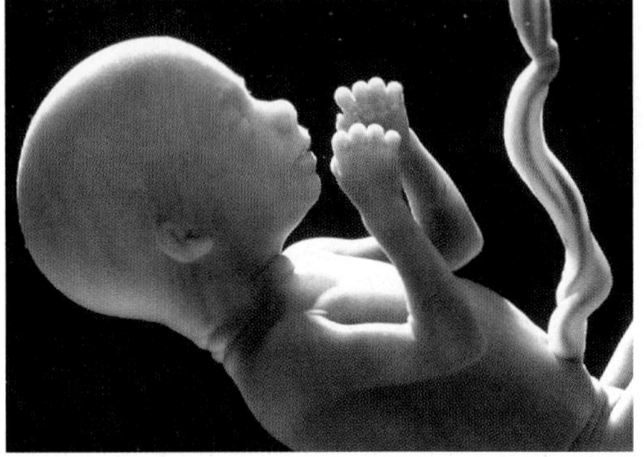


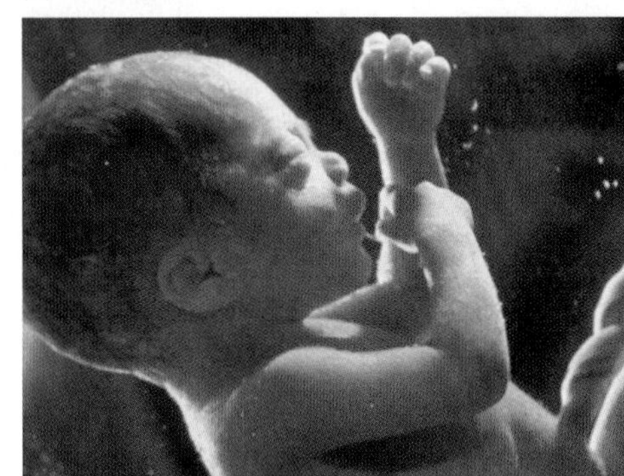


**44.19 Stages of Human Development** (A) At 4 weeks, most of the embryo's organ systems have been formed and the heart is beating. (B) The body structures of this 8-week-old embryo are forming rapidly, and it is visibly a male. The umbilical cord attaches the embryo to the placenta (upper left). (C) At 4 months, the fetus has fully formed limbs with fingers and toes and moves freely within the amniotic cavity. (D) This fetus is well along in its ninth month. Soon its lungs will be mature enough to trigger the onset of contractions and birth.

## Organ development begins in the first trimester

Implantation of the human blastocyst begins about 6 days after fertilization. After implantation, gastrulation occurs, tissues differentiate, the placenta forms, and organs begin to develop. The heart begins to beat during week 4, and limbs are formed by week 8 (**Figure 44.19A,B**). By the end of the first trimester, most organs have started to form. The embryo is about 8 centimeters long and weighs about 40 grams (less than 2 ounces). At about this point in time, the human embryo is medically and legally referred to as a **fetus**. (The distinction between embryo and fetus is not made for other mammals; developing mice, for example, remain embryos until they are born.)

The first trimester is a time of rapid cell division and tissue differentiation. Signal transduction cascades and the resulting branching sequences of developmental processes are in their early stages. Therefore the first trimester is the period during which the embryo is most sensitive to damage from radiation, drugs, chemicals, and pathogens that can cause birth defects. An embryo can be damaged before the mother even realizes she is pregnant. A classic case is that of thalidomide, a drug widely prescribed in Europe in the late 1950s to treat nausea. Women who took this drug in the fourth and fifth weeks of pregnancy, when the embryo's limbs are beginning to form, gave birth to children with missing or severely malformed arms and legs.

## Organ systems grow and mature during the second and third trimesters

During the second trimester the fetus grows rapidly to a weight of about 600 grams. The limbs of the fetus elongate, and the fingers, toes, and facial features become well formed (**Figure 44.19C**). Eyebrows and fingernails grow, and the fetus's nervous system develops rapidly. Fetal movements are first felt by the mother early in the second trimester and become progressively stronger and more coordinated.

The fetus grows rapidly during the third trimester (**Figure 44.19D**). As this final stage approaches its end, the internal organs mature. The digestive system begins to function, the liver stores glycogen, the kidneys produce urine, and the brain undergoes cycles of sleep and waking. A human infant is born when the last of its critical organs—the lungs—mature.

Although the first-trimester embryo is the most susceptible to adverse effects of drugs, chemicals, and diseases, the potential for serious effects from exposure to environmental factors exists throughout pregnancy. Protein malnutrition and exposure to alcohol and cigarette smoke are examples of factors that can result in low birth weight, mental retardation, and other developmental complications.

## Developmental changes continue throughout life

Development does not end with birth. Growth continues until adult size is reached, and even when growth stops, organs of the body continue to repair and renew themselves through cycles of cell replacement by the progeny of undifferentiated stem cells. In humans especially, enormous developmental changes occur in the brain in the years between birth and adolescence. Especially in the early years, there is a great deal of plasticity in the organization of the nervous system as the connections between neurons develop.

For example, a child born with misaligned eyes (a condition known as strabismus) will use mostly one eye. The connections to the brain from one eye will become strong while connections from the other eye remain weak, and the child will develop with reduced visual acuity and depth perception. If eye alignment is corrected in the first 3 years of life, the connections between the eyes and the brain can improve and the child is likely to develop normal vision. After the age of 3, corrective measures are less likely to result in improvement and visual impairments may persist. Thus plasticity in human visual system development declines during early childhood. However, recent data indicate that it is not lost entirely and may be reactivated even in adulthood.

RECAP 44.6

Human gestation lasts 9 months and is divided into three trimesters. At the end of the first trimester, the fetus is very small but most of its organs have begun to form. In the second trimester, limbs elongate and the fetus moves. By the end of the third trimester, most organs have begun to function.

- Why is a first-trimester embryo particular sensitive to environmental risks? **See p. 920**

Having briefly outlined how multicellular tissues and organ systems emerge from a single cell (the fertilized egg), the remainder of this unit will discuss the physiological functioning of organ systems. The next three chapters will describe the workings of perhaps the most complicated of these, the vertebrate nervous system.

How does the directional flow of extracellular fluid across the node stimulate a left–right asymmetry in gene expression and development?

ANSWER

Surrounding the node are cells with nonmotile cilia that are believed to sense the direction of fluid flow across the node. There are two hypotheses about how these cilia function, one chemical and the other mechanical. Both sensory mechanisms are used by cilia elsewhere in the body. In the nose, cilia of the olfactory cells sense chemicals. In the ear, cilia are bent by sound waves, which opens ion channels. The leftward flow of fluid across the node certainly can create mechanical forces on the nonmotile cilia that differ on the two sides of the node. Other research, however, supports the possibility of chemical signaling. Research on the mouse node shows that the beating of the cilia causes proteins of a certain size range to form a concentration gradient across the node. More important, it was discovered that nodal cells secrete small vesicles that are swept to the left side of the node by the flow and burst when they contact the nonmotile cilia of surrounding cells. The contents of these vesicles could be the chemical signals that initiate left–right asymmetry in gene expression and development.

## CHAPTER SUMMARY 44

### 44.1 How Does Fertilization Activate Development?

- The sperm and the egg contribute differentially to the zygote. The sperm contributes a haploid nucleus and, in most species, a centriole. The egg contributes a haploid nucleus, nutrients, ribosomes, mitochondria, mRNAs, and proteins.
- In amphibians, the cytoplasmic contents of the egg are not distributed homogeneously, and they are rearranged after fertilization to set up the major axes of the future embryo. The nutrient molecules are generally found in the **vegetal hemisphere**, whereas the nucleus is found in the **animal hemisphere**. **Review Figures 44.1, 44.2**

### 44.2 How Does Mitosis Divide Up the Early Embryo?

- **Cleavage** is a period of rapid cell division. Except in mammals, little if any gene expression occurs during cleavage. Cleavage can be complete or incomplete, and the pattern of cell divisions depends on the orientation of the mitotic spindles. The result of cleavage is a ball or mass of cells called a **blastula**. **Review Figure 44.3**
- Early cell divisions in mammals are unique in being slow and allowing for gene expression early in the process. These cell divisions produce a **blastocyst** composed of an **inner cell mass** that becomes the embryo and an outer cell mass that develops as the **trophoblast**. At the time of **implantation**, the trophoblast secretes molecules that help the blastocyst implant in the uterine wall. **Review Figures 44.4, 44.5**
- A **fate map** can be created by labeling specific **blastomeres** and observing what tissues and organs are formed by their progeny. **Review Figure 44.6**
- Some species undergo **mosaic development**, in which the fate of each cell is determined during early divisions. Other species, including vertebrates, undergo **regulative development**, in which remaining cells can compensate for cells lost in early cleavages.

### 44.3 How Does Gastrulation Generate Multiple Tissue Layers?

- **Gastrulation** involves massive cell movements that produce three **germ layers** and place cells from various regions of the blastula into new associations with one another. **Review Figure 44.7, ANIMATED TUTORIAL 44.1**
- The initial step of sea urchin and amphibian gastrulation is inward movement of certain blastomeres. The site of inward movement becomes the **blastopore**. Cells that move into the blastula become the **endoderm** and **mesoderm**; cells remaining on the outside become the **ectoderm**. Cytoplasmic factors in the vegetal pole cells are essential to initiate development. **Review Figures 44.7, 44.8**
- The **dorsal lip** of the amphibian blastopore is a critical site for cell determination. It has been called the **primary embryonic organizer** because it induces determination in cells that pass over it during gastrulation. **Review Figures 44.8, 44.9, 44.10, ANIMATED TUTORIAL 44.2**

*continued*

- The protein β-catenin activates a signaling cascade that induces the primary embryonic organizer and sets up the anterior–posterior body axis. **Review Figures 44.2, 44.11**
- Gastrulation in reptiles and birds differs from that in sea urchins and frogs because the large amount of yolk causes the blastula to form a flattened disc of cells. **Review Figure 44.13**
- Although their eggs have no yolk, placental mammals have a pattern of gastrulation similar to that of reptiles and birds.

## 44.4 How Do Organs and Organ Systems Develop?

- Gastrulation is followed by **organogenesis**, the process whereby tissues interact to form organs and organ systems.
- In the formation of the vertebrate nervous system, one group of cells that migrates over the blastopore lip is determined to become the **notochord**. The notochord organizes the overlying ectoderm to thicken, form parallel ridges, and fold in on itself to form a **neural tube** below the epidermal ectoderm. The nervous system develops from this neural tube. **Review Figure 44.14**
- The notochord and **neural crest cells** participate in the segmental organization of mesoderm into structures called **somites** along the body axis. Rudimentary organs and organ systems form during these stages. **Review Figure 44.15**
- In vertebrates, **Hox genes** determine the pattern of anterior–posterior differentiation along the body axis. Other genes, such as *Sonic hedgehog*, contribute to dorsal–ventral differentiation. **Review Figure 44.16**

## 44.5 How Is the Growing Embryo Sustained?

- The embryos of reptiles, birds, and mammals are protected and nurtured by four **extraembryonic membranes**. In birds and reptiles the **yolk sac** surrounds the yolk and provides nutrients to the embryo, the **chorion** lines the eggshell and participates in gas exchange, the **amnion** surrounds the embryo and encloses it in an aqueous environment, and the **allantois** stores metabolic wastes. **Review Figure 44.17, ACTIVITY 44.1**
- In mammals the chorion and the trophoblast cells interact with the maternal uterus to form a **placenta**, which provides the embryo with nutrients and gas exchange. The amnion encloses the embryo in an aqueous environment. **Review Figure 44.18**

## 44.6 What Are the Stages of Human Development?

- Human pregnancy, or **gestation**, can be divided into three trimesters. The embryo forms in the first trimester; during this time, it is most vulnerable to environmental factors that can lead to birth defects. During the second and third trimesters the **fetus** grows, the limbs elongate, and the organ systems mature.
- Development continues throughout childhood and throughout life.

**Go to the Interactive Summary to review key figures, Animated Tutorials, and Activities Life10e.com/is44**

# CHAPTER**REVIEW**

## REMEMBERING

1. How does cleavage in mammals differ from cleavage in frogs?
   a. Slower rate of cell division
   b. Formation of tight junctions
   c. Expression of the embryo's genome
   d. Early separation of cells that will not contribute to the embryo
   e. All of the above

2. Which statement about gastrulation is *true*?
   a. In frogs, gastrulation begins in the vegetal hemisphere.
   b. In sea urchins, gastrulation produces the notochord.
   c. In birds, cells from the surface of the blastodisc move down through the primitive groove to form the hypoblast.
   d. In mammals, gastrulation occurs in the hypoblast.
   e. In sea urchins, gastrulation produces only two germ layers.

3. Which of the following was a conclusion from the experiments of Spemann and Mangold?
   a. Cytoplasmic determinants of development are homogeneously distributed in the amphibian zygote.
   b. In the late blastula, certain regions of cells are determined to form skin or nervous tissue.
   c. The dorsal lip of the blastopore can be isolated and will form a complete embryo.
   d. The dorsal lip of the blastopore can initiate gastrulation.
   e. The dorsal lip of the blastopore gives rise to the neural tube.

4. Which of the following characterizes neurulation?
   a. The notochord forms a neural tube.
   b. The neural tube is formed from ectoderm.
   c. A neural tube forms around the notochord.
   d. The neural tube forms somites.
   e. In birds, the neural tube forms from the primitive groove.

5. Which statement about trophoblast cells is *true*?
   a. They are capable of producing monozygotic twins.
   b. They are derived from the hypoblast of the blastocyst.
   c. They are endodermal cells.
   d. They secrete proteolytic enzymes.
   e. They prevent the zona pellucida from attaching to the oviduct.

## UNDERSTANDING & APPLYING

6. The glycogen synthase kinase-3 (GSK-3)-inhibiting protein in the amphibian egg is a product of the *Disheveled* gene. If you had both Disheveled protein and an inhibitor of Disheveled protein, what experiments might you do to test whether β-catenin was both necessary and sufficient to initiate gastrulation?

7. If you used a laser to kill a small number of cells on the midline of the dorsal blastopore lip of an amphibian embryo, what defects would you expect to see during subsequent development?

## ANALYZING & EVALUATING

8. During gastrulation in birds, the *Sonic hedgehog* gene is expressed only on the left side of Hensen's node. What might be the cause of this expression pattern, and what is its significance? How could you test your hypotheses?

9. When oogonia or spermatogonia divide by mitosis, one daughter cell remains a germinal stem cell and one becomes a primary oocyte or spermatocyte. What mechanism could possibly account for these different cell fates?

10. There is controversy over therapeutic cloning as a way of obtaining embryonic stem cells to treat diseases. Given that human development is regulative, suggest a way to produce a source of isogenic (i.e., identically matching a person's own body) stem cells for an individual without resorting to therapeutic cloning.

Go to BioPortal at **yourBioPortal.com** for Animated Tutorials, Activities, LearningCurve Quizzes, Flashcards, and many other study and review resources.

# Neurons, Glia, and Nervous Systems

## CHAPTER OUTLINE


**45.1** What Cells Are Unique to the Nervous System?

**45.2** How Do Neurons Generate and Transmit Electric Signals?

**45.3** How Do Neurons Communicate with Other Cells?

**45.4** How Are Neurons and Glia Organized into Information-Processing Systems?

**Trying and Learning** To varying degrees, individuals with Down syndrome are unable to learn basic concepts and tasks despite intense effort on their own part and that of their teachers. Understanding the cellular basis of this learning disability may lead to new ways of alleviating it.

YOUR BRAIN ENABLES YOU to learn the material in this chapter. It enables you to read the words, understand the illustrations, and store that information so you can use it to answer the questions at the end of the chapter. You have to spend time studying in order to master this chapter; it may be hard work, but eventually you will have learned something about how the brain receives and processes information.

Imagine what it would be like if you could not learn any of this material—no matter how much you poured over the book, no matter how hard you tried. This is the situation faced by individuals with a learning disability that is part of the condition known as Down syndrome, which affects 1 out of every 700 children born in the United States.

Individuals with Down syndrome are born with three copies of chromosome 21. This smallest of the human somatic chromosomes contains only some 250 genes, but having an extra copy of these genes causes numerous developmental and functional problems, including a learning disability in which the brain does not properly take in and store new information. How can we understand the cause of this disability and perhaps find a way to remedy it?

One productive way of investigating the causes and possible treatments for a human disease or deficit is to develop an animal model. Through genetic engineering, researchers created a "Down syndrome mouse" that has most of the same genes triplicated as those in humans with Down syndrome. Using this mouse model, biologists found out that the learning disability of these mice was due to overinhibition in the brain, and that when this inhibition was reduced with drugs, the ability of these mice to learn was increased. But what do we mean by "overinhibition in the brain"?

We often think of the brain as a puppet master, pulling the strings that activate the muscles and organs of the body. In fact the brain is more like an orchestra conductor, making some sections louder, some softer, speeding up, slowing down. The brain must constantly maintain a delicate balance of excitation and inhibition, acting on some signals and ignoring others. In the brains of "Down syndrome mice," there is consistently too much inhibition. If inhibition is reduced using certain drugs, the mice appear able to learn. Research like this is a first step toward bringing potential therapies for humans into clinical trials in the hope of someday alleviating suffering.

What causes overinhibition in the nervous system, and how can it be reduced?

See answer on p. 943.

# 45.1 What Cells Are Unique to the Nervous System?

Nervous systems are informational systems. They encode, process, and store a wide variety of information from the external and internal environments, and they use that information to control and regulate the physiological processes and behavioral actions of the organism. Nervous systems are able to carry out these functions because of the properties of two unique types of cells: nerve cells, or **neurons,** and glial cells, or **glia**. The many different types of neurons vary enormously in structure and appearance, but all neurons are **excitable**, meaning they can generate and transmit electric signals; the electric signals generated by neurons are known as **action potentials**. Generally glia do not generate action potentials, but they interact intimately with neurons, modulating their activity and supporting them in many ways.

## The structure of neurons reflects their functions

All neurons have a basic structure that includes four regions (**Figure 45.1**).

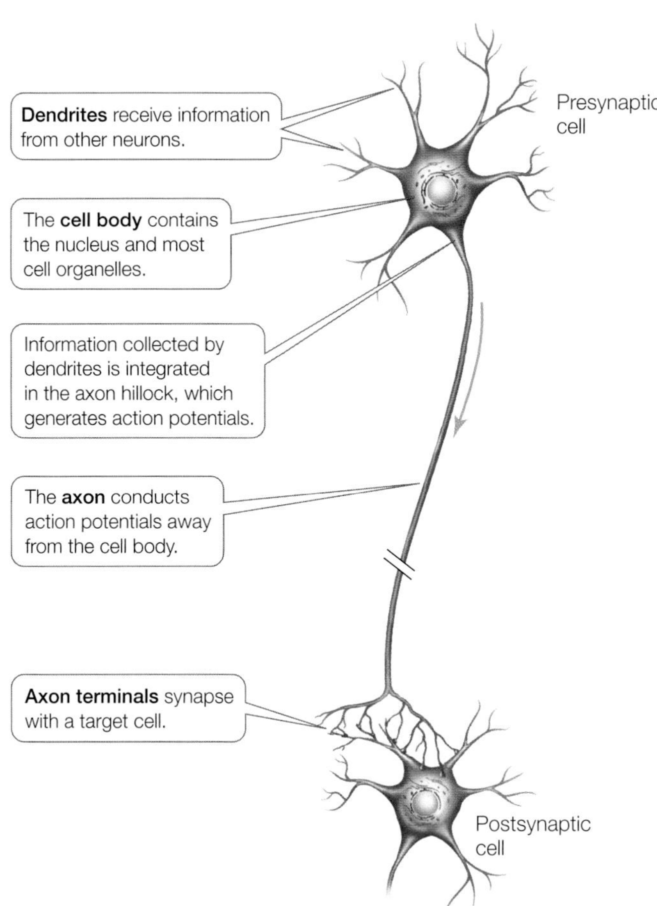


**45.1 A Generalized Neuron** The diagram shows the features typical of most neurons. The forms of these features, including the length of the single axon and the density and branching patterns of the dendrites, vary greatly across the many different types of neurons.

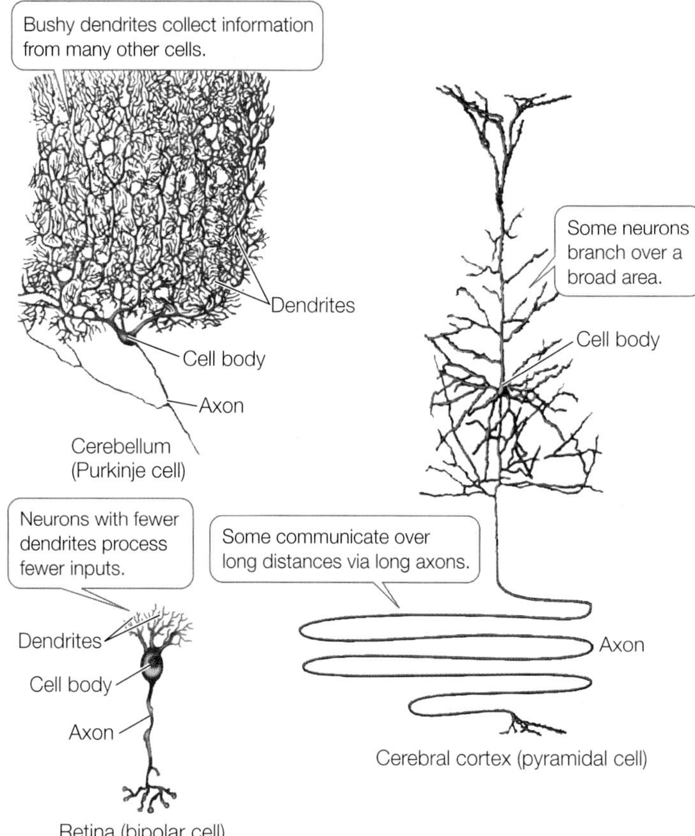


**45.2 Neurons Have Many and Varied Forms** The morphological differences in neurons from different parts of the mammalian nervous system are related to their specific functional adaptations. The small sample here shows two neuronal types from the human brain (a Purkinje cell and a pyramidal cell) and a sensory neuron from the retina of the human eye.

- A **cell body** contains the nucleus and most of the cell's organelles.
- Shrublike projections called **dendrites** (Greek *dendron*, "tree") may extend from the cell body. Dendrites bring information from other neurons or sensory cells to the cell body. Neurons with few dendrites are receiving information from specific and limited sources, whereas neurons with large arrays of dendrites can collect and integrate information from a wider range of sources.
- In most neurons, one projection—the **axon**—is much longer than the others. Axons carry information in the form of action potentials away from the originating cell body (the **presynaptic cell**) to the receiving target cell (the **postsynaptic cell**).
- At the postsynaptic cell, the axon divides into a spray of fine nerve endings. At the tip of each of these tiny nerve endings is a swelling called the **axon terminal**.

A wide variety of forms can be seen among the many different types of neurons (**Figure 45.2**), but all neurons share the mechanisms whereby their plasma membranes generate and conduct action potentials. Information received by dendrites is integrated by the cell body, and the result of that integration can generate an action potential that is conducted down the axon toward its terminals at the target cell. Action potentials can travel at speeds up to 100 m/sec (360 km/hr), making it possible for an individual to sense, process, and act on information very quickly. The axon terminal comes extremely close to (apposes) the membrane of the target cell, forming a **synapse** at which the information conveyed by the action potential is communicated from the presynaptic cell to the postsynaptic cell (see Figure 45.1).

As we will discuss in Section 45.3, synapses can be either chemical or electrical. Electrical synapses allow the action potential to pass directly between two neurons. In vertebrates, however, most synapses are chemical. At chemical synapses a space about 25 nanometers wide (about 1/2000th the width of a human hair) separates the presynaptic and postsynaptic membranes. An action potential arriving at an axon terminal causes it to release chemical messenger molecules called **neurotransmitters**. The neurotransmitters diffuse across the space and bind to receptors on the plasma membrane of the postsynaptic (target) cell. This binding alters the activity of the postsynaptic neuron. Some neurotransmitter–receptor combinations inhibit activity of the postsynaptic neuron, and other neurotransmitter–receptor combinations excite it. Neurons integrate information by summing excitatory and inhibitory inputs.

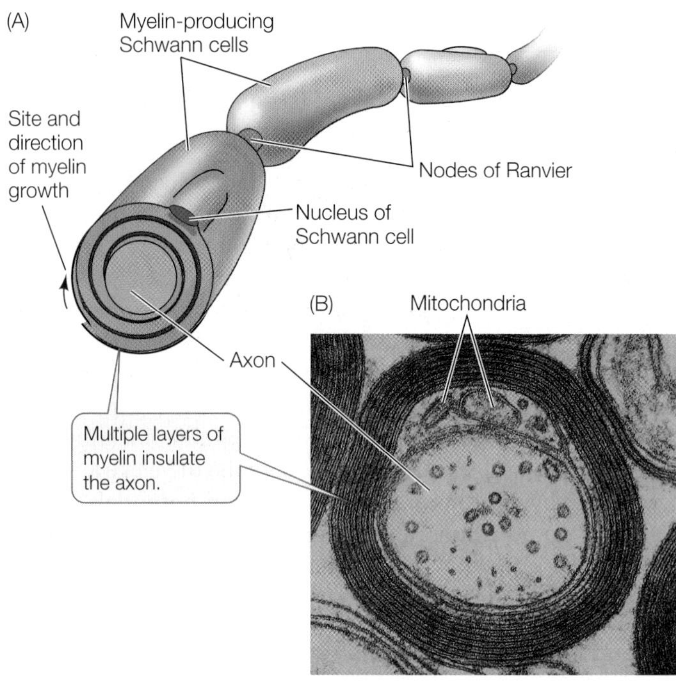


**45.3 Wrapping Up an Axon** (A) Schwann cells produce layers of myelin, a type of plasma membrane that provides electrical insulation to the axon. At the intervals between Schwann cells—the nodes of Ranvier—the axon is exposed. Action potentials travel along the axon by "jumping" from node to node. (B) A myelinated axon, seen in cross section through an electron microscope.

## Glia are the "silent partners" of neurons

The human brain has about 10 times more glial cells than neurons. A neurobiologist once said that "flashy neurons get all of the attention, but glial cells do most of the brain's work and are the cause of many of its diseases." It has been easier to study the functions of neurons because their action potentials can be observed. The mostly silent glia have been more difficult to study, and therefore we know much less about them. Our knowledge of glia will grow enormously in the years to come, and there are likely to be many surprises.

Like neurons, glia come in several forms and have diverse functions. In the brain and spinal cord, glia called **oligodendrocytes** wrap around the axons of neurons, covering them with concentric layers of insulating plasma membrane. Outside the brain and spinal cord, glia called **Schwann cells** wrap axons (**Figure 45.3**). **Myelin** is the covering produced by oligodendrocytes and Schwann cells, and it gives many parts of the nervous system a glistening white appearance. Not all axons are myelinated, but those that are can conduct action potentials more rapidly than can axons that are not myelinated, for reasons we will describe in Section 45.2.

Diseases that affect myelin can be devastating because they impair conduction of action potentials. The most common of these demyelinating diseases is multiple sclerosis—literally "multiple scars"—which occurs in about 1 in 700 people in the United States. Individuals with this autoimmune disease produce antibodies to proteins in the myelin in the brain and spinal cord. The symptoms and damage from the disease depend on where in the nervous system the antibody attacks occur. Motor impairment is common. An example of a demyelinating disease that attacks myelin outside the brain and spinal cord is Guillain–Barre syndrome, which is usually the result of a severe infection. Environmental factors such as pesticide exposure can also damage myelin. There are no known cures for demyelinating diseases.

Glia called **astrocytes** (because they look like stars) contribute to the **blood–brain barrier**, which protects the brain from toxic chemicals in the blood. Blood vessels throughout the body are very permeable to many chemicals, including toxic ones, which would reach the brain if this barrier did not exist. Astrocytes help form the blood–brain barrier by surrounding the smallest, most permeable blood vessels in the brain. The barrier is not perfect, however. Because it consists of plasma membranes, it is permeable to fat-soluble substances such as anesthetics and alcohol (which explains why these substances have such rapid and marked effects on the nervous system).

In addition to their role in the blood–brain barrier, astrocytes have several known functions at the synapse:

- They can take up neurotransmitter that has been released into the synapse and thereby control communication between the pre- and postsynaptic cells.
- They can supply neurons with nutrients. Neurons have no energy reserves, but astrocytes store glycogen that they can break down to supply the neurons with fuel.
- They have signaling properties. Even though most astrocytes do not generate action potentials, they do release neurotransmitters that can alter the activities of neurons.

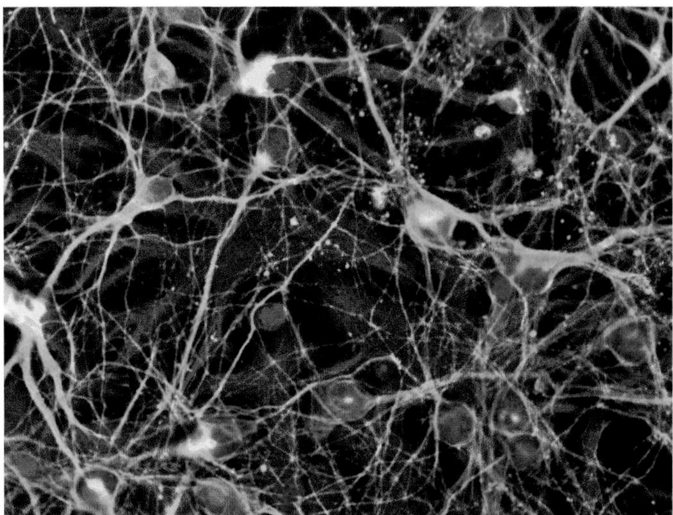

**45.4 Astrocytes Communicate with Many Synapses** Cell-type-specific antibodies have been used to label the astrocytes and their processes (red) and the neurons and their processes (green) in this fluorescence micrograph. The blue label visualizes the nuclei of both cell types.

- They aid in the repair and regeneration of neurons.
- They make contact with both blood vessels and neurons and can therefore signal changes in the composition of the blood.

Astrocytes play crucial yet poorly understood roles in modulating synapse activity. The projections of a single astrocyte may make contact with more than 100,000 synapses (**Figure 45.4**). The contact of the astrocyte with the neuronal components of the synapse is so intimate that it has inspired the concept of the **tripartite synapse**—the idea that a synapse includes not only the pre- and postsynaptic neurons but also connections from astrocytes.

**Microglia** are another type of glial cell**.** The blood–brain barrier typically prevents antibodies in the general circulation from entering the brain and spinal cord. Microglia that originate during development from stem cells in the bone marrow come to reside in the nervous system and act as macrophages and mediators of inflammatory responses, thus providing the nervous system with immune defenses.

**RECAP 45.1**

Nervous systems have two unique types of cells: neurons and glia. There are many types of neurons, but they all generate action potentials. A neuron has four regions: cell body, dendrites, axon, and axon terminals. Neurons communicate with target cells at synapses, which can be chemical or electrical. Although glia do not generate action potentials, they have a wide variety of functions that are only beginning to be elucidated.

- Describe the different parts of a neuron and their functions. **See pp. 925–926 and Figure 45.1**
- What are some types of glia, and what are their functions? **See pp. 926–927**
- What is meant by the "tripartite synapse"? **See p. 927**

The one feature common to all neurons is that they process information in the form of action potentials. In the next section we will focus on how action potentials are generated and transmitted.

## 45.2 How Do Neurons Generate and Transmit Electric Signals?

Sodium–potassium pumps create concentration gradients across all animal cell membranes, with $Na^+$ concentration higher outside the cell and $K^+$ concentration higher inside. Both inside and outside the cell, the positive charges of these ions are balanced by negatively charged ions. But *across the plasma membrane* there is an electric charge difference, with the inside of the cell being negative relative to the outside. This is because there are "leak channels" that allow only certain ions—usually $K^+$—to passively leak across the cell membrane.

Because there is a higher concentration of $K^+$ inside the cell than in the extracellular fluid outside, $K^+$ diffuses out of the cell down its concentration gradient. But when $K^+$ leaks out of the cell, it leaves behind an unbalanced negative electric charge that tends to pull $K^+$ back into the cell. An equilibrium is reached when the tendency for $K^+$ to diffuse out is countered by the electric charge pulling $K^+$ back in. The result is a charge difference—a **membrane potential**—across the membrane, with the inside of the cell negative relative to the outside.

Membrane potentials exist in all cells. In neurons the steady state membrane potential is called the **resting potential**. An action potential is a sudden, large, transient reversal in the resting potential that is generated by sudden openings and rapid closings of ion channels. Before describing the properties of ion channels and action potentials in detail, a review of some simple concepts of electricity may be useful.

### Simple electrical concepts underlie neural function

**Voltage** (electric potential difference) is a force that causes electrically charged particles to move between two points. Voltage is to the flow of electrically charged particles as pressure is to the flow of water. If the negative and positive poles of a battery are connected by a wire, an electric current will flow through the wire because there is a voltage difference between the two poles. This flow of electric current can be used to do work, just as a current of water can be used to do work.

In wires, electric current is carried by electrons, but in solutions and across cell membranes, electric current is carried by ions. The major ions that carry electric charges across the plasma membranes of neurons are sodium ($Na^+$), potassium ($K^+$), calcium ($Ca^{2+}$), and chloride ($Cl^-$). Recall that ions with opposite charges attract one another and that those with like charges repel one another. How do these basic principles of bioelectricity establish the resting potential of the neural plasma membrane? And how is the flow of ions through membrane channels turned on and off to generate action potentials? We will address these questions next.

RESEARCH**TOOLS**

**45.5 Measuring the Membrane Potential** An electrode can be made from a glass pipette with a very sharp tip filled with a solution that conducts electric charges. If one electrode is placed inside the plasma membrane of an axon and another is placed just outside the axon, the difference in voltage can be measured.

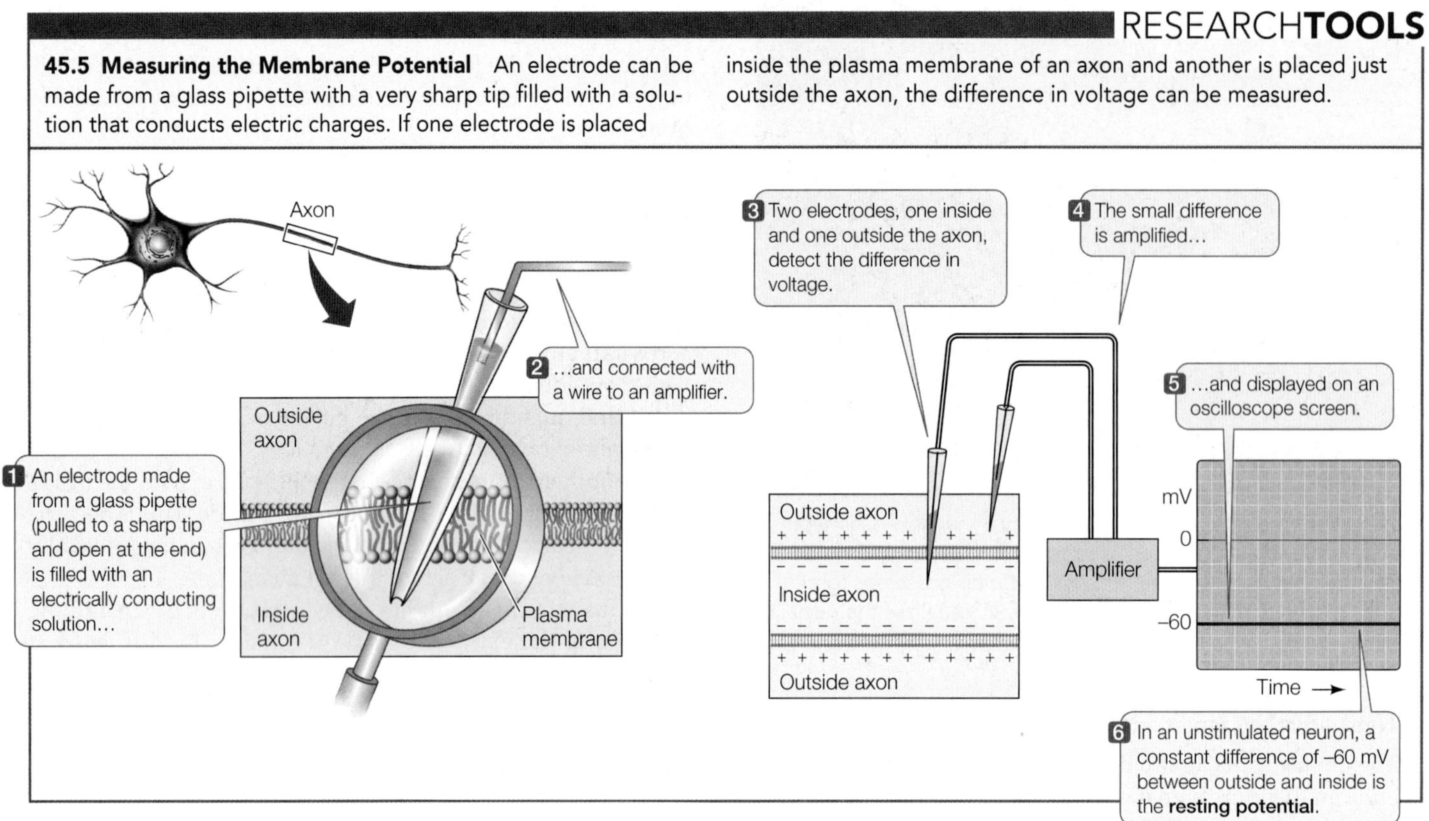


## Membrane potentials can be measured with electrodes

We can record electrical events in a cell using electrodes. **Figure 45.5** shows how this technique is applied across an unstimulated axon to measure the resting potential, which is usually between –60 and –70 millivolts (mV). The minus sign indicates that the inside of the cell is electrically negative compared with the outside.

The resting potential provides a means for neurons to respond to a stimulus. Because of the voltage difference across the membrane, and the different ion concentrations on either side of the membrane, ions would cross the membrane if they could. For example, $Na^+$ ions are more abundant outside the cell than inside, and the inside of the resting cell is negatively charged. Therefore if the membrane suddenly became permeable to $Na^+$, those positively charged ions would rush into the cell. Any chemical or physical stimulus that changes the permeability of the plasma membrane to ions will produce a change in the cell's membrane potential. The most extreme change in membrane potential is the action potential, a sudden and rapid reversal in the voltage across a portion of the plasma membrane. For 1 or 2 milliseconds, positively charged ions flow into the cell, making the inside of the cell *more positive* than the outside.

**Go to Animated Tutorial 45.1**
**The Resting Membrane Potential**
Life10e.com/at45.1

## Ion transporters and channels generate membrane potentials

The plasma membranes of neurons, like those of all other cells, are lipid bilayers that are impermeable to ions but contain many protein molecules that serve as ion transporters and channels. Ion transporters and channels are responsible for the distribution of charges across the membrane that create resting and action potentials.

Ion transporters require energy to move ions against their concentration or electrical gradients and are therefore called ion pumps. A major ion transporter in the plasma membranes of neurons (and all other cells) is the **sodium–potassium pump**, so called because it actively expels $Na^+$ ions from inside the cell, exchanging them for $K^+$ ions from outside the cell (**Figure 45.6A**). The $Na^+$–$K^+$ pump is also known as sodium–potassium ATPase, a term emphasizing that it is an enzyme complex requiring ATP to do its work. The $Na^+$–$K^+$ pump keeps the concentration of $K^+$ inside the cell greater than the $K^+$ concentration of the extracellular fluid, and the concentration of $Na^+$ inside the cell less than that of the extracellular fluid. The concentration differences established by this active transporter means that $K^+$ would diffuse out of the cell and $Na^+$ would diffuse in if the ions could cross the lipid bilayer. How do these concentration gradients relate to the electrical gradients we discussed above?

Ion channels permit the diffusion of ions across membranes. These channels are water-filled pores formed by proteins that cross the lipid bilayer and are generally selective,

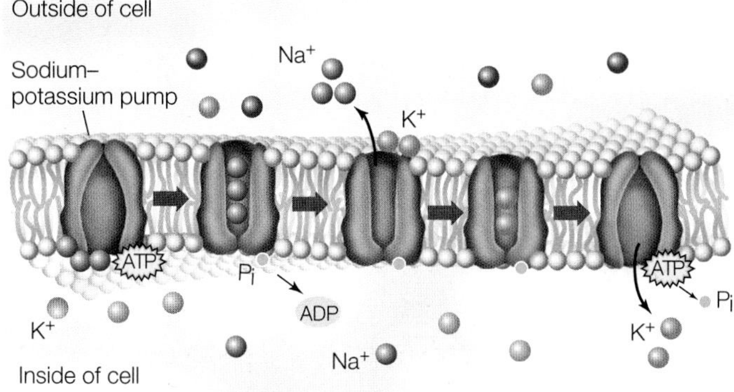


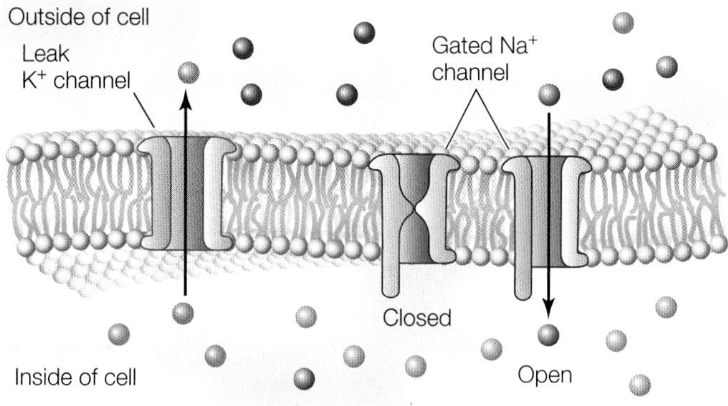


**45.6 Ion Transporters and Channels** (A) The sodium–potassium pump is an active transporter that moves $K^+$ to the inside of a neuron and $Na^+$ to the outside. (B) Ion channels allow specific ions to diffuse down their concentration gradients; $K^+$ tends to leave neurons when potassium channels are open, and $Na^+$ tends to enter neurons when sodium channels are open. Leak channels like the $K^+$ channel shown are always open and create the resting membrane potential. Gated channels like the $Na^+$ channels shown are opened by chemical or electrical stimulation.

allowing some types of ions to pass through more easily than others (**Figure 45.6B**). Thus there are potassium channels, sodium channels, chloride channels, and calcium channels, and there are different kinds of channels for each ion. Ions can diffuse through these channels in either direction. The direction and magnitude of the net movement of ions through a channel depend on the concentration gradient of that ion type across the plasma membrane, as well as on the voltage difference across that membrane. These two motive forces acting on an ion are termed its **electrochemical gradient**. Although the electrochemical gradient drives the movement of ions through channels, that movement is modified by gates that open and close the channels.

Potassium channels are the most common open, or leak, channels in the plasma membranes of resting (nonstimulated) neurons. As a consequence, resting neurons are more permeable to $K^+$ than to any other ion. Thus open potassium channels are largely responsible for the resting membrane potential. Because the potassium channels make the plasma membrane permeable to $K^+$, and because the $Na^+$–$K^+$ pump keeps the concentration of $K^+$ inside the cell much higher than that outside the cell, $K^+$ tends to diffuse down its electrochemical gradient, out of the cell, through the channels. As these positively charged potassium ions diffuse out of the cell, they leave behind unbalanced negative charges, generating an electric potential across the membrane that tends to pull $K^+$ back into the cell.

The membrane potential at which the net diffusion of $K^+$ out of the cell ceases (that is, the point at which $K^+$ diffusion out due to the concentration gradient is balanced by its movement in due to the negative electric potential) is the **potassium equilibrium potential**, or $\boldsymbol{E_K}$. The value of $E_K$ can be calculated from the concentrations of $K^+$ on the two sides of the membrane using the **Nernst equation** (**Figure 45.7**). This equation, developed in the late 1800s, shows that the existence of ion channels in neural membranes was hypothesized long before their specific structures and properties were described.

In the late 1940s, A. L. Hodgkin and A. F. Huxley at the University of Cambridge set out to study the electrical properties of axonal membranes. With the techniques available at that time, the necessary measurements could be made only if you had a very large axon to work with. Such an axon exists in nature, in the huge neuron that controls the escape response of squid. Hodgkin and Huxley used electrodes to measure the voltage across the plasma membrane of this large axon, as seen in Figure 45.7, and to pass electric current into it to change its membrane potential. They also changed the concentrations of $Na^+$ and $K^+$ both inside and outside the squid axon and measured the resulting changes in membrane potential. On the basis of their many careful experiments, Hodgkin and Huxley developed virtually all of our basic concepts about the electrical properties of neurons, and shared a Nobel Prize in 1963.

We now know that, in general, the resting potential is less negative than the $E_K$ calculated from the Nernst equation. This means that the resting potential is not due solely to leak $K^+$ channels. The neuronal membrane is slightly permeable to other ions, especially $Na^+$ and $Cl^-$, and movements of these ions influence the resting potential. A different equation takes into account (1) all of the ions that can cross the membrane and (2) the relative permeability of the membrane to those ions. This equation, called the **Goldman equation**, predicts the membrane potential more accurately than does the Nernst equation (see p. 931).

## Ion channels and their properties can now be studied directly

Because Hodgkin and Huxley were working long before there were laboratory techniques that could investigate ion channels, they could only hypothesize their properties. These hypotheses could not be tested until the late 1970s, when B. Sakmann and E. Neher developed a technique called **patch clamping**, for which they shared the Nobel Prize in 1991. Patch clamping, described in **Figure 45.8**, is widely used by neurobiologists, enabling them to record in real time the tiny electric currents caused by the openings and closings of single ion channels.

RESEARCH**TOOLS**

**45.7 Using the Nernst Equation** The Nernst equation calculates membrane potential when only one type of ion can cross a membrane that separates solutions with different concentrations of that ion.

**1. Measure concentrations of ions inside and outside a neuron.**

To measure the concentration of ions in a neuron, the neuron (and its axon) must be big. Squid have giant neurons that control their escape response (see Figure 45.17C). It is possible to sample the cytoplasm of these axons, which are about 1 mm in diameter.

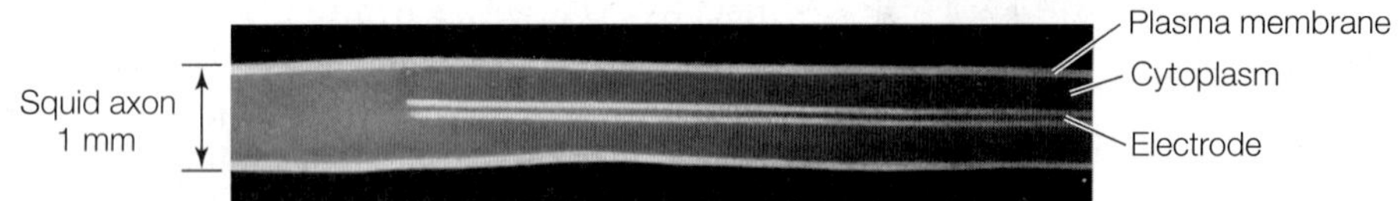


**2. Use the Nernst equation to calculate what the membrane potential would be if it were permeable to each of the ions that are differently concentrated on the two sides of the membrane: $Na^+$, $K^+$, $Ca^{2+}$, and $Cl^-$.**

The Nernst equation predicts the membrane potential resulting from membrane permeability to a single type of ion that differs in concentration on the two sides of the membrane. The equation is written

$$E_{ion} = 2.3\frac{RT}{zF}\log\frac{[ion]_o}{[ion]_i}$$

where $E$ is the equilibrium (resting) membrane potential (the voltage across the membrane in mV), $R$ is the universal gas constant, $T$ is the absolute temperature, $z$ is the charge on the ion (+1, +1, +2, or –1), and $F$ is the Faraday constant. The subscripts o and i indicate the ion concentrations outside and inside the cell, respectively.

***At this point you could just "plug and play," but do you understand this equation?***

A concentration difference of ions across a membrane creates a *chemical* force that pushes the ions across the membrane; however, the resulting unbalanced *electric charges* will pull the ions back the other way. At *equilibrium*, the work done moving ions in each direction will be the same.

The *chemical* energy pushing the ions will equal $2.3\ RT \log\ [ion]_o/[ion]_i$
The *electric* energy pulling the ions will equal $zEF$. So, at equilibrium:

$$zEF = 2.3\ RT\log\frac{[ion]_o}{[ion]_i}$$

Rearranging the equation to solve for $E$, we get the Nernst equation:

$$E_{ion} = 2.3\frac{RT}{zF}\log\frac{[ion]_o}{[ion]_i}$$

We can simplify the equation by picking a temperature—let's use "room temperature," or 20°C—and solving for $2.3\ RT/F$. At 20°C, $2.3\ RT/F$ equals 58. Thus:

$$E_{ion} = 58/z\ \log\frac{[ion]_o}{[ion]_i}$$

**3. Measuring ion concentrations in squid giant axon cytoplasm and in seawater, then solving the Nernst equation for each ion, we find:**

| Ion | Ion concentration (mM) in squid axon | in seawater | Predicted membrane potential (mV) |
|---|---|---|---|
| $K^+$ | 400 | 20 | –75 |
| $Na^+$ | 50 | 460 | +56 |
| $Ca^{2+}$ | 0.5 | 10 | +38 |
| $Cl^-$ | 50 | 560 | –60 |

**4. Since the measured membrane potential is –66 mV, it is clear that the resting potential of the axon is due to permeability of the membrane to more than just one type of ion.**

## Gated ion channels alter membrane potential

The ion channels called leak channels are always open, but other ion channels in the plasma membranes of neurons behave as if they contain "gates"; they are open under some conditions and closed under other conditions. **Voltage-gated channels** open or close in response to a change in the voltage across the plasma membrane. **Chemically gated channels** open or close depending on the presence or absence of a specific molecule that binds to the channel protein, or to a separate receptor that in turn alters the channel protein. **Mechanically gated channels** open or close in response to mechanical force applied to the plasma membrane. Gated channels play important roles in neural function.

Openings and closings of gated channels alter the membrane potential. Imagine what happens, for example, if sodium channels in the plasma membrane suddenly open. $Na^+$ diffuses into the neuron down its electrochemical gradient to approach

WORKING WITH**DATA:**

## *Equilibrium Membrane Potential: The Goldman Equation*

### Original Papers

Goldman, D. E. 1943. Potential, impedence and rectification in membranes. *Journal of General Physiology* 27: 37–60.

Hodgkin, A. L. and B. Katz. 1949. The effect of sodium ions on the electrical activity of the giant axon of the squid. *Journal of Physiology* 108: 37–77.

### Analyze the Data

Figure 45.7 presented the Nernst equation by which the membrane potential for a single ion can be determined. But we also saw (at the end of Figure 45.7) that the equilibrium membrane potential is the product of more than one ion. The Goldman equation (sometimes called the Goldman, Hodgkin, Katz equation) calculates the equilibrium membrane potential by taking into account all of the ions that can diffuse across a given membrane and the relative permeabilities of the membrane to those ions. The ions involved in mammalian neurons here are $K^+$, $Na^+$, and $Cl^-$, and the Goldman equation is

$$V_m = \frac{RT}{F}\ln\left(\frac{p_K[K^+]_o + p_{Na}[Na^+]_o + p_{Cl}[Cl^-]_i}{p_K[K^+]_i + p_{Na}[Na^+]_i + p_{Cl}[Cl^-]_o}\right)$$

Relative permeabilities (*p*) are expressed as percentages. The membrane's permeability to potassium ions is the highest, so $p_K = 1.0$. Then $p_{Na} = 0.05$ and $p_{Cl} = 0.45$. Bracketed elements refer to the inside and outside ion concentrations, as in the Nernst equation.

**QUESTION**

The table gives the intra- and extracellular ion concentrations for a mammalian neuron. Use these values and the Goldman equation to calculate the membrane potential. Refer to Figure 45.7 for a comparison with calculations based on the Nernst equation. (Hint: In redrafting the equation, you can substitute "2.3*RT*/*F* log" for "*RT*/*F* ln".)

| | Ion concentration (mM) | |
|---|---|---|
| | Intracellular | Extracellular |
| $K^+$ | 140 | 5 |
| $Na^+$ | 10 | 145 |
| $Cl^-$ | 20 | 110 |

**Go to BioPortal for all** WORKING WITH**DATA** **exercises**

RESEARCH**TOOLS**

**45.8 Patch Clamping** The patch clamp is a glass micropipette filled with an electrically conductive solution that has the same composition as extracellular fluids. When this pipette/electrode is positioned against the membrane of a cell and slight suction is applied, a seal forms. If a single ion channel (or a few ion channels) are within the patch of membrane bounded by the seal, the openings and closings of individual channels can be recorded by the electrode. If the pipette is retracted, it can tear the patched membrane away from the cell, and the activities of the ion channels in the patch can continue to be recorded.

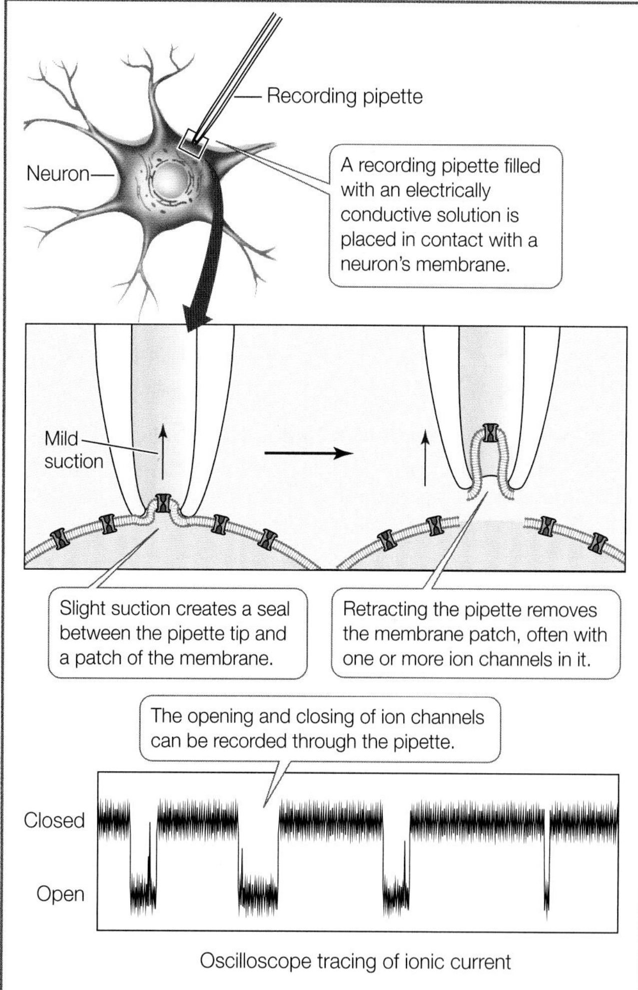


the equilibrium potential for $Na^+$ ($E_{Na}$). Therefore the inside of the cell becomes less negative. When the inside of a neuron becomes less negative (or more positive) in comparison to its resting condition, its plasma membrane is **depolarized** (**Figure 45.9**).

An opposite change in the membrane potential occurs if gated $K^+$ channels open. When $K^+$ efflux from the neuron increases over the normal leak current (the movement of $K^+$ through the leak channels), the membrane potential becomes even more negative, and the plasma membrane is **hyperpolarized**.

The openings and closings of ion channels that result in changes in the voltage across the plasma membrane are the basic mechanisms by which neurons respond to stimuli, be they electrical, chemical, or mechanical. How do such local changes in membrane potential get communicated to other parts of the cell?

A local change in membrane potential causes a flow of ions that spreads the change in membrane potential to adjacent regions of the membrane. For example, when $Na^+$ enters a neuron through open sodium channels at one location, those positively charged ions are attracted to adjacent areas on the inside of the membrane that are more negative, and thus there is a rapid flow of ionic electric current (movement of charged ions) away from the site of the open $Na^+$ channels. However,

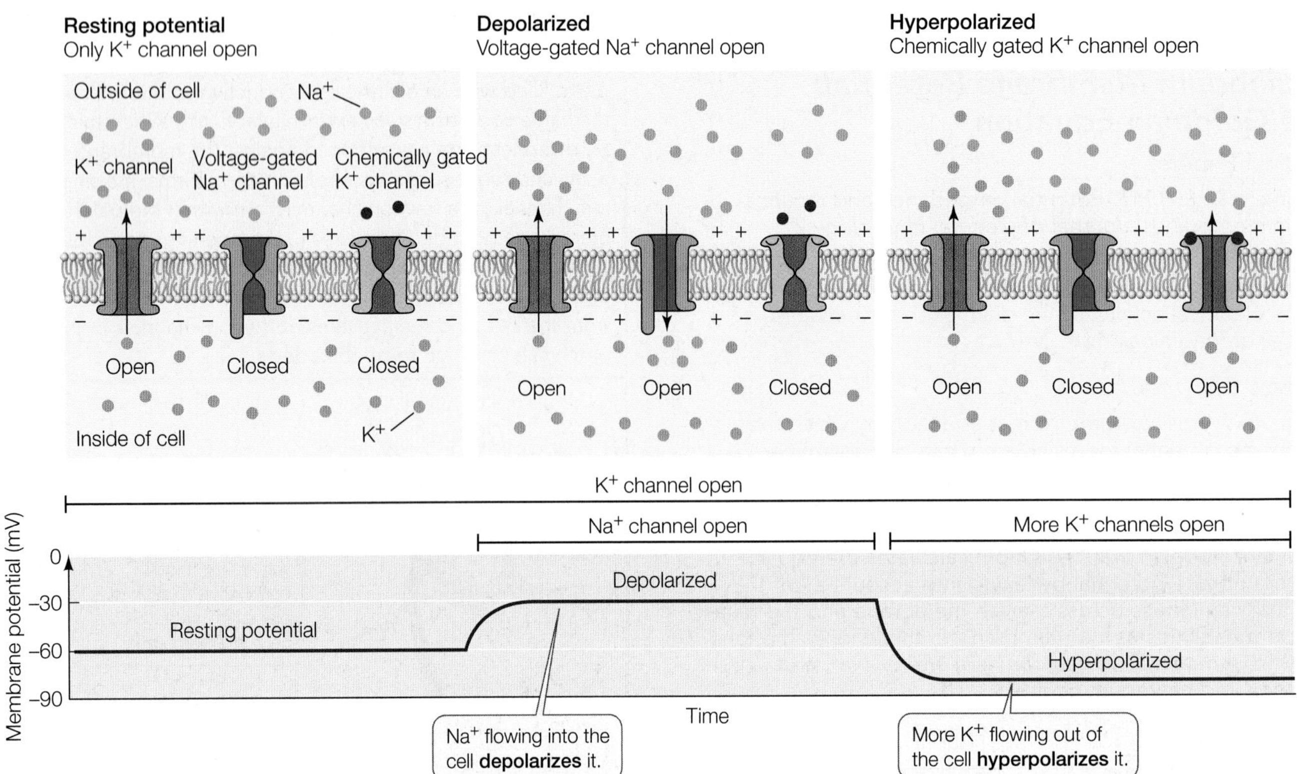


**45.9 Membranes Can Be Depolarized or Hyperpolarized** The resting potential is produced by leak $K^+$ channels. A shift from the resting potential to a less negative membrane potential, as occurs when $Na^+$ enters the cell through a gated sodium channel, is called depolarization. Hyperpolarization occurs when the membrane potential becomes more negative, as when additional $K^+$ leaves the cell through gated $K^+$ channels, which occurs extensively in your brain when you fall asleep.

this local flow of ionic electric current decays as it spreads and therefore does not spread very far. The small number of electrically charged ions that enter are rapidly diluted by the large number of oppositely charged particles, and the $Na^+/K^+$ ATPases are continuously pumping them out.

## Graded changes in membrane potential can integrate information

Even though the flow of ionic electric current along plasma membranes can only extend over short distances, it can cause graded changes in membrane potentials locally. A **graded membrane potential** is a change from the resting potential. Such changes can be due to chemical or mechanical influences on ion channels. Graded potentials are a means of integrating inputs to a cell because the membrane can respond to those inputs with proportional amounts of depolarization or hyperpolarization.

Graded potentials can transmit signals over very short distances and play an important role at the neuromuscular junction (see Section 45.3). In the next chapter we will learn how they play important roles in sensory systems. However, axons are too long to transmit information as a continuous flow of ionic electric current. Therefore axons code information as discrete action potentials that travel along their membranes. Graded potentials, however, play an important role in the generation of action potentials.

## Sudden changes in $Na^+$ and $K^+$ channels generate action potentials

Action potentials are sudden, transient, large changes in membrane potential. In unmyelinated axons (those not wrapped in myelin by oligodendrocytes or Schwann cells), they can be conducted at speeds of up to 2 meters per second, but in myelinated axons the conduction velocity can be 100 meters per second. Think of running the 100-meter dash—the world record is slightly under 10 seconds.

If we place the tips of a pair of electrodes on either side of the plasma membrane of a resting axon and measure the voltage difference, the reading might be about –60 mV, as we saw in Figure 45.5. If these electrodes are in place when an action potential travels down the axon, they register a rapid change in membrane potential, from –60 mV to about +50 mV. The membrane potential then rapidly returns to its resting level of –60 mV as the action potential passes (**Figure 45.10**).

The action potential is generated by the actions of voltage-gated $Na^+$ and $K^+$ channels in the plasma membrane of the axon. At the resting potential, most of these channels are closed (balloon 1 in Figure 45.10). A slight depolarization of the membrane causes them to open. For example, if a neuron is stimulated sufficiently to cause the plasma membrane of its cell body to depolarize slightly, that graded potential can spread by local current flow to the **axon hillock**, the region of the cell body at the base of the axon (see Figure 45.1). Voltage-gated

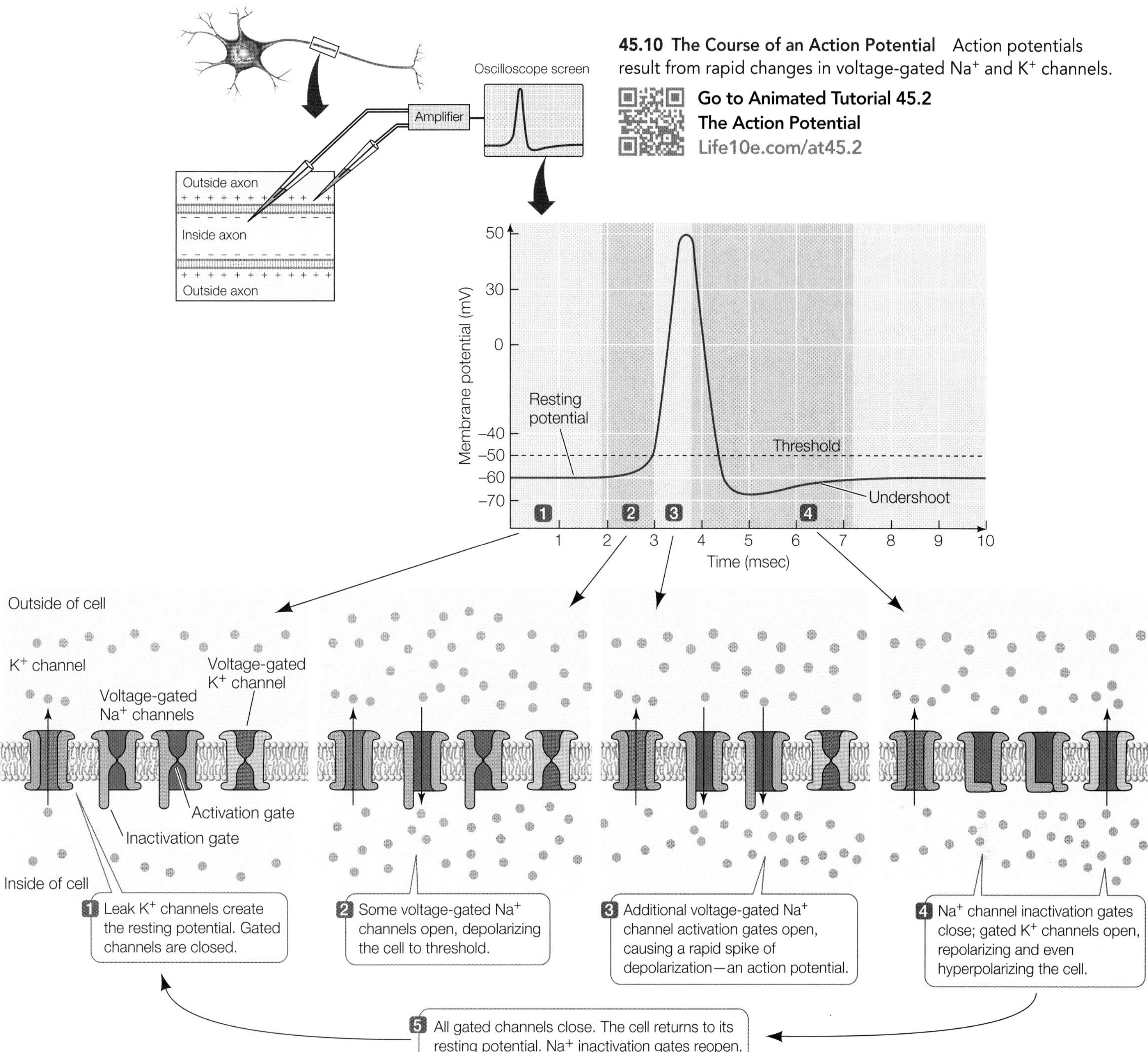


**45.10 The Course of an Action Potential** Action potentials result from rapid changes in voltage-gated $Na^+$ and $K^+$ channels.

**Go to Animated Tutorial 45.2**
**The Action Potential**
Life10e.com/at45.2

$Na^+$ channels are concentrated in the axon hillock. A slight depolarization of the plasma membrane in this area causes some of these voltage-gated channels to open briefly—for less than a millisecond (balloon 2 in Figure 45.10). When these channels open, $Na^+$ rushes into the axon and depolarizes the membrane even more, causing more $Na^+$ channels to open—a positive feedback effect. When the membrane is depolarized about 5 to 10 mV above the resting potential, a **threshold** is reached; a large number of sodium channels open (balloon 3 in Figure 45.10), and the membrane potential becomes positive—an action potential. The rising phase of the action potential halts abruptly in 1 to 2 milliseconds, and the membrane potential rapidly becomes negative once again.

What causes the axon to return to resting potential? There are two contributing factors: the voltage-gated $Na^+$ channels close, and voltage-gated $K^+$ channels open (balloon 4 in Figure 45.10). Voltage-gated $K^+$ channels open more slowly than the $Na^+$ channels and stay open longer, allowing $K^+$ to carry excess positive charges out of the axon. As a result, the membrane potential returns to a negative value and usually becomes even more negative than the resting potential until the voltage-gated $K^+$ channels close (balloon 5 in Figure 45.10).

Another feature of the voltage-gated $Na^+$ channels is that once they open and close, they have a **refractory period** of 1 to 2 milliseconds during which they cannot open again. This property can be explained by the channels having two gates, an

**activation gate** and an **inactivation gate** (see Figure 45.10). Under resting conditions, the activation gate is closed and the inactivation gate is open. Depolarization of the membrane to the threshold level causes both gates to change state, but the activation gate responds faster. As a result, the channel is open for a brief time between the opening of the activation gate and the closing of the inactivation gate. Inactivation gates remain closed for 1 to 2 milliseconds before they spontaneously open again, thus explaining why the membrane has a refractory period before it can fire another action potential. By the time the inactivation gate reopens, the activation gate is closed, and the membrane is poised to generate another action potential. Another contribution to the refractory period is the duration of the opening of the voltage-gated $K^+$ channels, as we saw above. The dip in the membrane potential following an action potential is called the **after-hyperpolarization** or **undershoot**.

The difference in the concentration of $Na^+$ across the plasma membrane and the negative resting potential constitute the "battery" that drives action potentials. How rapidly does the battery run down? It might seem that a substantial number of ions would have to cross the membrane for the membrane potential to change from –60 mV to +50 mV and back to –60 mV again. In fact, only a vanishingly small percent of the $Na^+$ concentrated just outside the plasma membrane moves through the channels during the passage of an action potential. Thus the effect of a single action potential on the concentration gradients of $Na^+$ and $K^+$ is very small, and it is possible in most cases for the sodium–potassium pump to keep the "battery" charged, even when the neuron is generating many action potentials every second.

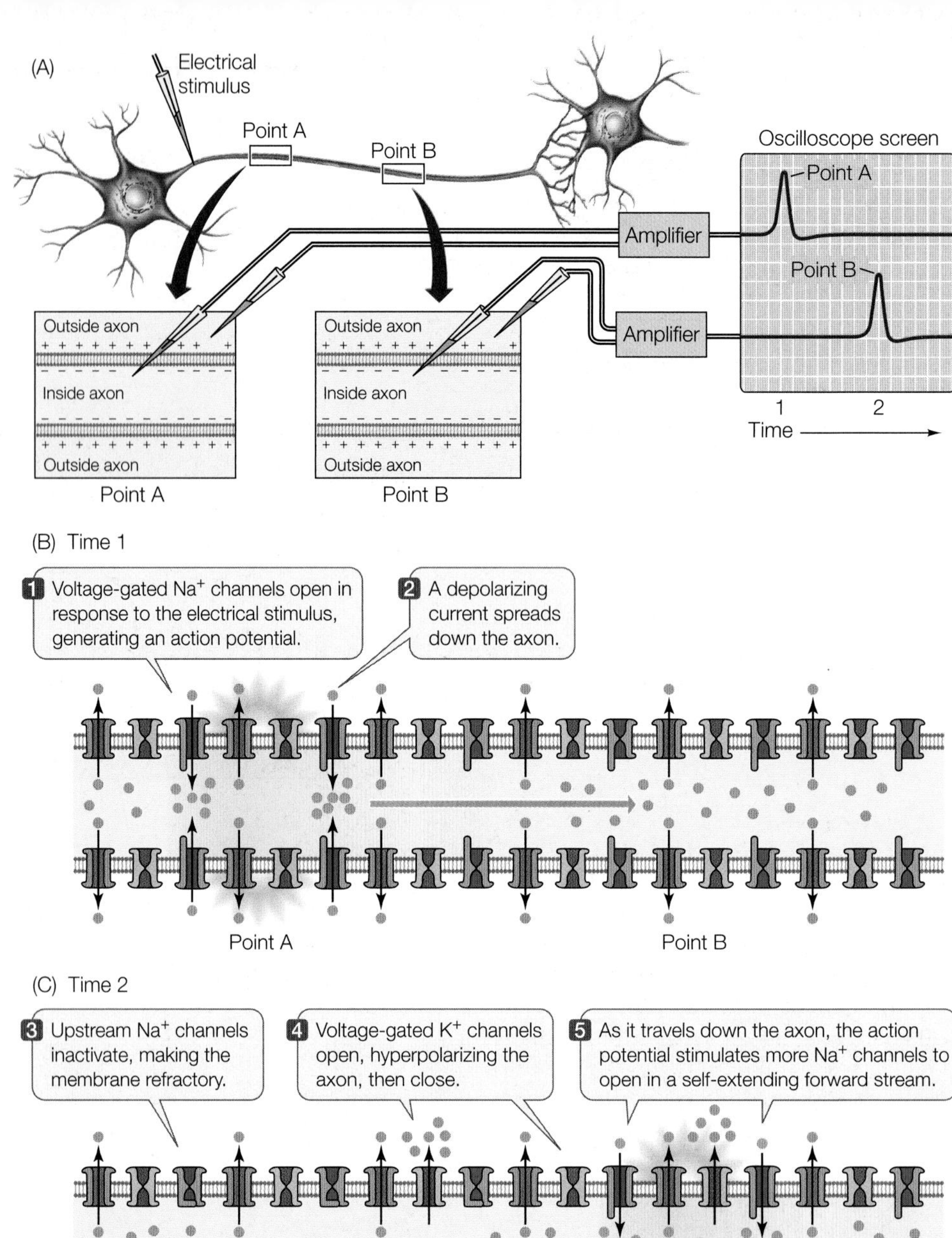


**45.11 Action Potentials Travel along Axons** (A) There is no loss of signal as an action potential travels along an axon. (B) When an action potential is stimulated in one region of membrane, electric current flows to adjacent areas of membrane and depolarizes them. (C) The advancing wave of depolarization causes more $Na^+$ channels to open, and the action potential is generated anew in the next section of membrane. Meanwhile, in the region where the action potential has just fired, the $Na^+$ channels are inactivated and the voltage-gated $K^+$ channels are still open, rendering this section of the axon incapable of generating an action potential. Hence the action potential cannot "back up," but moves continuously forward, regenerating itself as it goes.

## Action potentials are conducted along axons without loss of signal

Action potentials can travel over long distances with no loss of signal. If we place two pairs of electrodes at two different locations along an axon, we can record an action potential at those two locations as it travels along the axon (**Figure 45.11A**). The magnitude of the action potential does not change between the two recording sites. This constancy is possible because an action potential is an all-or-none, self-regenerating event.

- An action potential is *all-or-none* because of the interaction between the voltage-gated $Na^+$ channels and the membrane potential. If the membrane is depolarized slightly, some voltage-gated $Na^+$ channels open. Some sodium ions cross the plasma membrane and depolarize it even more, opening more voltage-gated $Na^+$ channels, and so on, generating an action potential. This positive feedback mechanism ensures that action potentials always rise to their maximum value.

- An action potential is *self-regenerating* because it spreads by local current flow to adjacent regions of the plasma membrane. The resulting depolarization brings those neighboring areas of membrane to threshold. So when an action potential occurs at one location on an axon, it stimulates the adjacent region of axon to generate an action potential, and so on down the length of the axon.

We can use an electrode to stimulate an axon, causing it to depolarize and to fire an action potential that is then conducted along the axon. **Figure 45.11B** shows the changes in the ion channels in the membrane that are responsible for conducting the action potential along the axon without a reduction in amplitude. Normally an action potential is propagated in only one direction—away from the cell body. It cannot reverse itself because the voltage-gated $Na^+$ channels in the region of the membrane it came from are in their refractory period (**Figure 45.11C**).

Action potentials do not travel along all axons at the same speed. They travel faster in large-diameter axons than in small-diameter axons because the resistance to ionic current flow decreases as an axon's diameter gets bigger. They travel faster in myelinated than in nonmyelinated axons because they can move down the axon in short "jumps" (think of a kangaroo jumping down a path) (**Figure 45.12**). Invertebrates mostly depend on increased axon diameter for fast conduction, but vertebrates mostly depend on myelination of axons to increase conduction velocity.

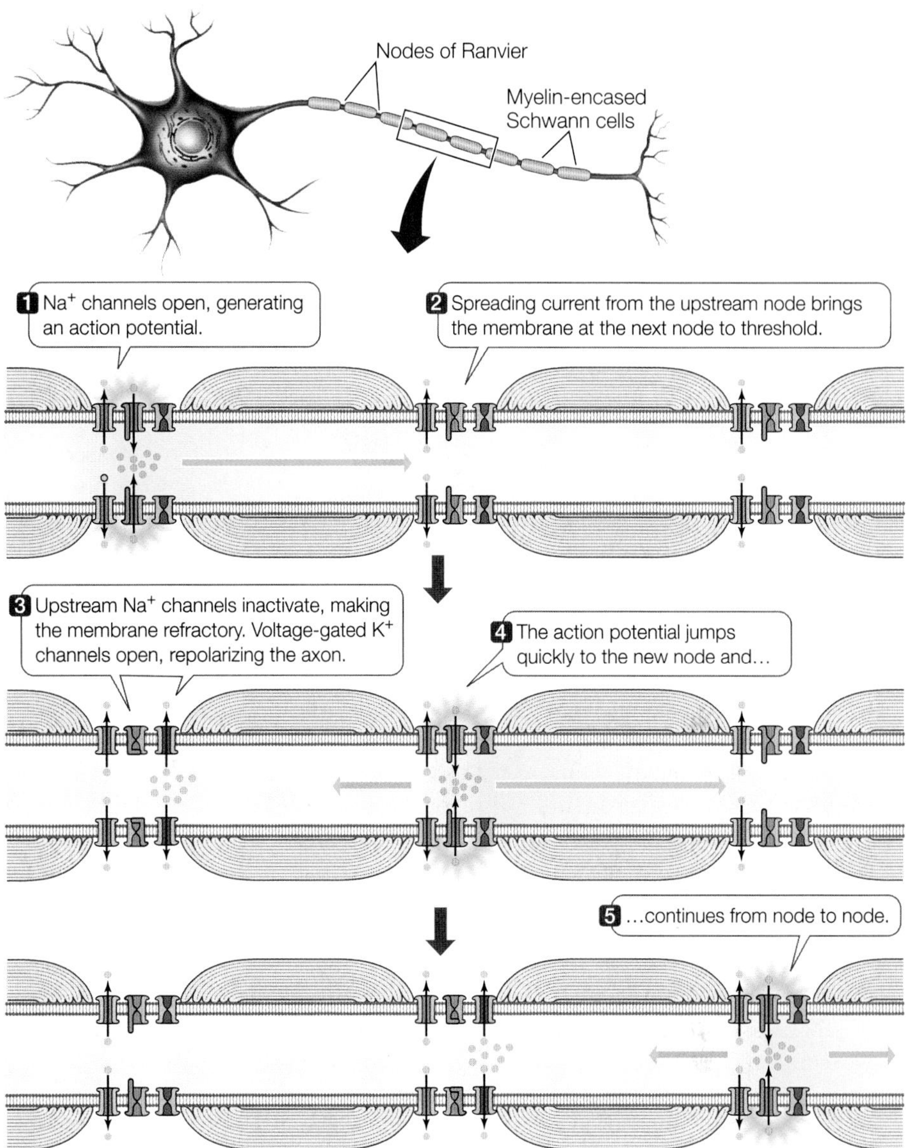


**45.12 Saltatory Action Potentials** Action potentials "jump" from node to node in myelinated axons, allowing faster transmission of information.

## Action potentials jump along myelinated axons

In vertebrate nervous systems, increasing the speed of action potentials by increasing the diameter of axons is not feasible because of the huge number of axons involved. Each of our eyes, for example, has about a million axons connecting it to the brain. These axons conduct action potentials at about the same speed as does the squid giant axon—about 20 meters per second—yet the diameter of each is 200 times smaller than the squid axon's diameter. Imagine having optic tracts 200 times bigger. A different way of increasing conduction velocity of axons has evolved in vertebrates, and that adaptation is myelination.

When glia wrap around axons, they cover the axons with concentric layers of myelin (see Figure 45.3). However, they leave regularly spaced gaps called **nodes of Ranvier**, where the axon is not covered (see Figure 45.12). The leakage of ions across the regions of the plasma membrane that are wrapped in myelin is reduced, so ionic electric current can spread farther along the inside of a myelinated axon than it can along a nonmyelinated axon. Additionally, voltage-gated ion channels are clustered at the nodes of Ranvier. Thus an axon can fire action potentials only at nodes, and those action potentials cannot be propagated through the adjacent patch of membrane covered with myelin. The positive charges that flow into the axon at the node do, however, flow down the inside of the axon in the form of ionic electric current. When the current reaches the next node, the plasma membrane at that node is depolarized to threshold and fires another action potential. Action potentials therefore jump from node to node along the axon.

The speed of conduction is increased in these myelin-wrapped axons because ionic electric current flows much faster through the cytoplasm than ion channels can open and close. This form of rapid impulse propagation is called **saltatory conduction** (Latin *saltare*, "to jump").

**RECAP 45.2**

Neurons have membrane potentials due to ionic concentration differences across their membranes and because leak channels make the membrane differentially permeable to ions. Changes in ion channel permeabilities cause graded changes in membrane potentials. Sudden openings and closings of gated ion channels in the membrane produce action potentials. Action potentials are rapid, all-or-none changes in membrane potential that are conducted along axons from the cell body to the axon terminals.

- How are membrane resting potentials generated and altered? **See pp. 928–929 and Figures 45.6 and 45.9**
- What does the Nernst equation calculate, and why does that calculation not equal the measured resting potential of a membrane? **See p. 929 and Figure 45.7**
- How are action potentials generated? **See pp. 932–934 and Figure 45.10**
- How are action potentials transmitted along axons? **See pp. 934–936 and Figures 45.11 and 45.12**

Having described how action potentials are generated and transmitted along axons, we will next address the question of what happens when an action potential reaches the axon terminal. How is its signal communicated to the next cell—which could be another neuron, a muscle cell, or a secretory cell?

## 45.3 How Do Neurons Communicate with Other Cells?

Neurons communicate with each other and with other cells at synapses. In **electrical synapses**, which are common among invertebrates, the action potential spreads directly from presynaptic to postsynaptic cell. The most common type of synapse in the vertebrate nervous system is the **chemical synapse**, in which neurotransmitters released from a presynaptic cell induce changes in a postsynaptic cell. We will begin this section with a discussion of the synapses between neurons and muscle cells. We will then consider the diversity in synapses and how they integrate information.

### The neuromuscular junction is a model chemical synapse

**Neuromuscular junctions** are synapses between neurons and skeletal muscle cells. They are excellent models for how chemical synaptic transmission works. Like other neurons, a motor neuron has only one axon, but that axon can branch into numerous axon terminals that form many synapses with muscle cells. At each axon terminal an enlarged knob or buttonlike structure contains vesicles filled with neurotransmitter molecules. The neurotransmitter used by all vertebrate neuromuscular synapses is **acetylcholine** (**ACh**). ACh is released by exocytosis when the membrane of a vesicle fuses with the presynaptic membrane of the axon terminal.

Where does the neurotransmitter come from? Some neurotransmitters, such as ACh, are synthesized in the axon terminal and packaged in vesicles. The enzymes required for ACh biosynthesis, however, are produced in the cell body of the motor neuron and are transported along microtubules down the axon to the terminals. In contrast, peptide neurotransmitters are produced in the cell body and packaged into membrane-bound vesicles by the Golgi apparatus. These vesicles are transported down the axon to the terminals.

The postsynaptic membrane of the neuromuscular junction is a modified part of the muscle cell plasma membrane called a **motor end plate**. It appears as a depression in the muscle cell membrane, and the terminals of the motor neuron sit in the depression. The space between the presynaptic membrane and the postsynaptic membrane is the **synaptic cleft**, which in chemical synapses is about 20 to 40 nanometers wide. ACh released into the cleft by the presynaptic cell diffuses across to the postsynaptic membrane (**Figure 45.13**).

### The arrival of an action potential causes the release of neurotransmitter

Neurotransmitter is released when an action potential arrives at the axon terminal and causes the opening of voltage-gated $Ca^{2+}$ channels in the presynaptic membrane. Because the $Ca^{2+}$ concentration is greater outside the cell than inside, $Ca^{2+}$ enters the axon terminal. This increase in $Ca^{2+}$ inside the axon terminal causes the vesicles containing neurotransmitter to fuse with the presynaptic membrane and empty their contents into the synaptic cleft.

In neuromuscular synapses, vesicle fusion and emptying is all-or-none. The vesicle membrane is incorporated into the presynaptic membrane, which actually gets larger as a result—at least until the extra membrane is recycled through endocytosis. The membrane is reprocessed by the cell into new vesicles that are refilled with neurotransmitter.

### Synaptic functions involve many proteins

The description above of the release of neurotransmitter from the presynaptic membrane may seem simple, but it involves hundreds of proteins that are responsible for various aspects of the process: vesicle formation, transport of neurotransmitter into vesicles, anchoring of vesicles to cytoskeletal elements, docking of the vesicles with the presynaptic membrane, fusion of the vesicular and cell membranes, and endocytosis of the vesicle membrane for recycling.

Some of these proteins are the targets of toxins. For example, botulinum and tetanus toxins from bacteria of the genus *Clostridium* act on several of the proteins necessary for the docking of vesicles to the presynaptic membrane, resulting in diseases that are frequently fatal. Botulinum toxin impairs muscle contraction, whereas tetanus toxin causes uncontrolled muscle contraction. Poisons can become medicines, however. Botulinum toxin (in the form of Botox®) is used therapeutically to subdue muscle spasms and cosmetically to subdue wrinkles.

### The postsynaptic membrane responds to neurotransmitter

When acetylcholine is released at a synapse, some of it diffuses across the synaptic cleft and binds to ACh receptors on the postsynaptic membrane (**Figure 45.14**). The postsynaptic

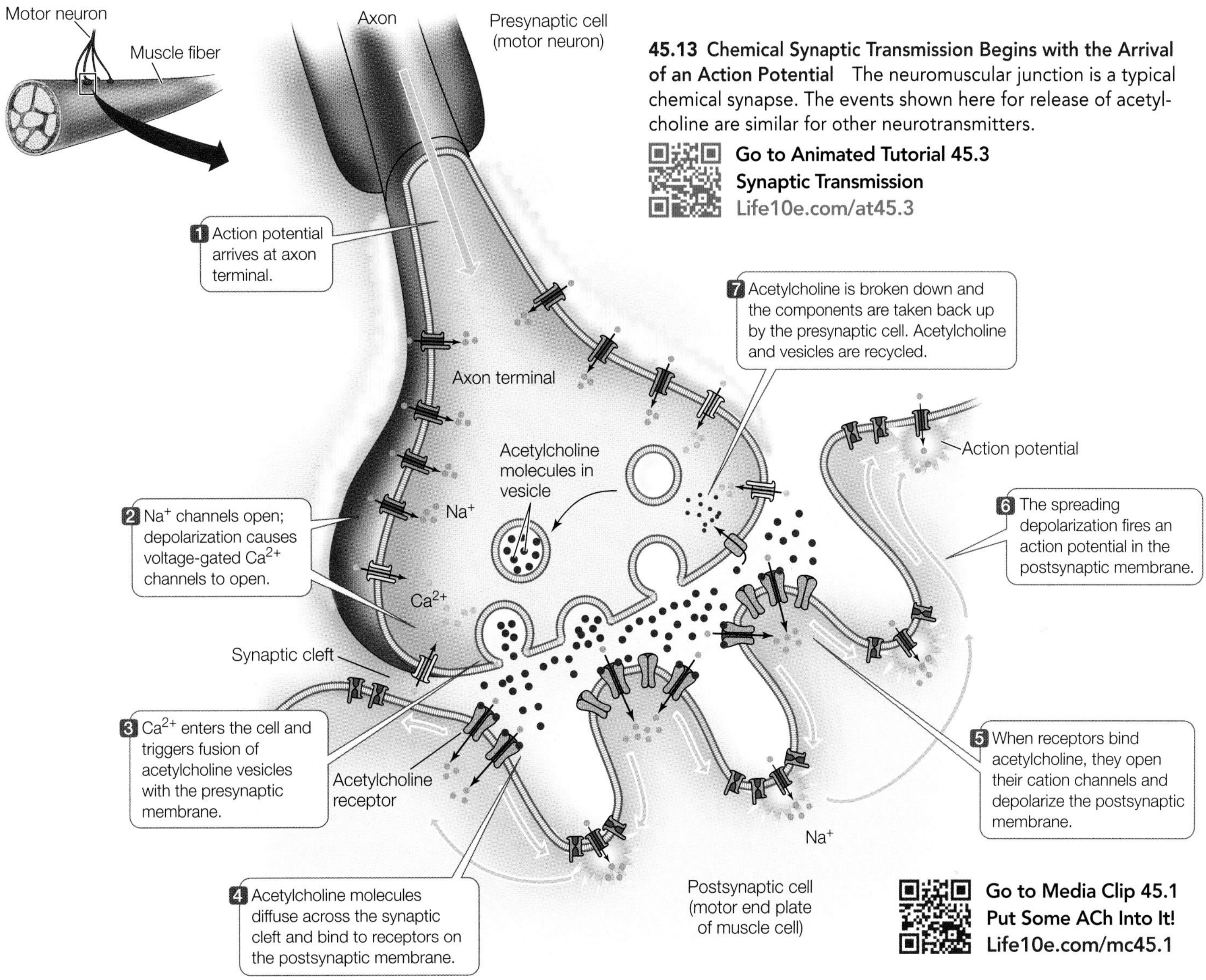

Motor neuron
Muscle fiber
Axon
Presynaptic cell (motor neuron)

1 Action potential arrives at axon terminal.

Axon terminal
Acetylcholine molecules in vesicle
$Na^+$
$Ca^{2+}$

2 $Na^+$ channels open; depolarization causes voltage-gated $Ca^{2+}$ channels to open.

Synaptic cleft

3 $Ca^{2+}$ enters the cell and triggers fusion of acetylcholine vesicles with the presynaptic membrane.

Acetylcholine receptor

4 Acetylcholine molecules diffuse across the synaptic cleft and bind to receptors on the postsynaptic membrane.

5 When receptors bind acetylcholine, they open their cation channels and depolarize the postsynaptic membrane.

6 The spreading depolarization fires an action potential in the postsynaptic membrane.

7 Acetylcholine is broken down and the components are taken back up by the presynaptic cell. Acetylcholine and vesicles are recycled.

Action potential
$Na^+$
Postsynaptic cell (motor end plate of muscle cell)

**45.13 Chemical Synaptic Transmission Begins with the Arrival of an Action Potential** The neuromuscular junction is a typical chemical synapse. The events shown here for release of acetylcholine are similar for other neurotransmitters.

**Go to Animated Tutorial 45.3**
**Synaptic Transmission**
Life10e.com/at45.3

**Go to Media Clip 45.1**
**Put Some ACh Into It!**
**Life10e.com/mc45.1**

membrane of the motor end plate is highly folded. ACh receptors are on the crests of the folds, and voltage-gated cation channels are at the bottoms of the folds and in the surrounding muscle cell membrane (see Figure 45.13). The ACh receptors are chemically gated channels that allow both $Na^+$ and $K^+$ to flow through, but since the electrochemical gradients favor a net influx of $Na^+$, the response of the motor end plate to ACh is to depolarize. That graded potential reflecting the number of receptors activated spreads to the depths of the folds of the motor end plate membrane and to surrounding muscle cell membrane, which contain voltage-gated $Na^+$ channels.

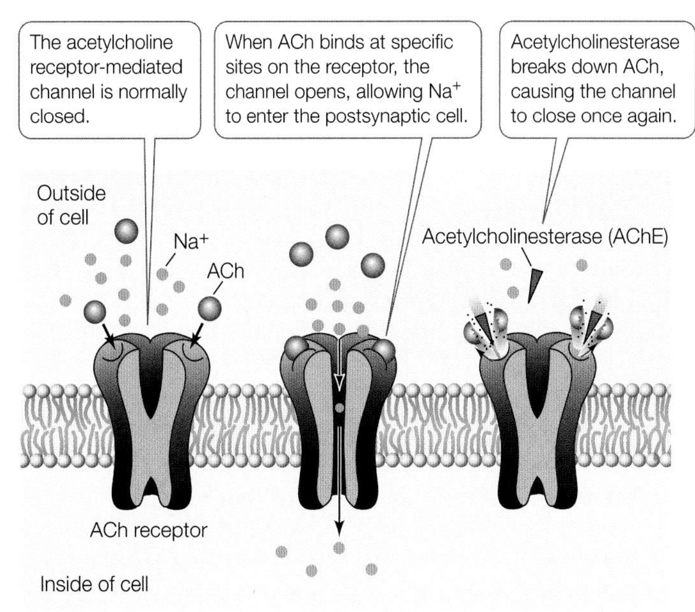


**45.14 Acetylcholine Receptors** ACh receptors are chemically gated ion channels found in the motor end plate and other types of postsynaptic membranes. When one of these receptors binds ACh, its channel pore opens and $Na^+$ moves into the postsynaptic cell, depolarizing its membrane. The enzyme acetylcholinesterase (AChE) breaks down ACh in the synapse, closing the channel. The breakdown products (acetate and choline) are then taken up by the presynaptic membrane and resynthesized into more ACh.

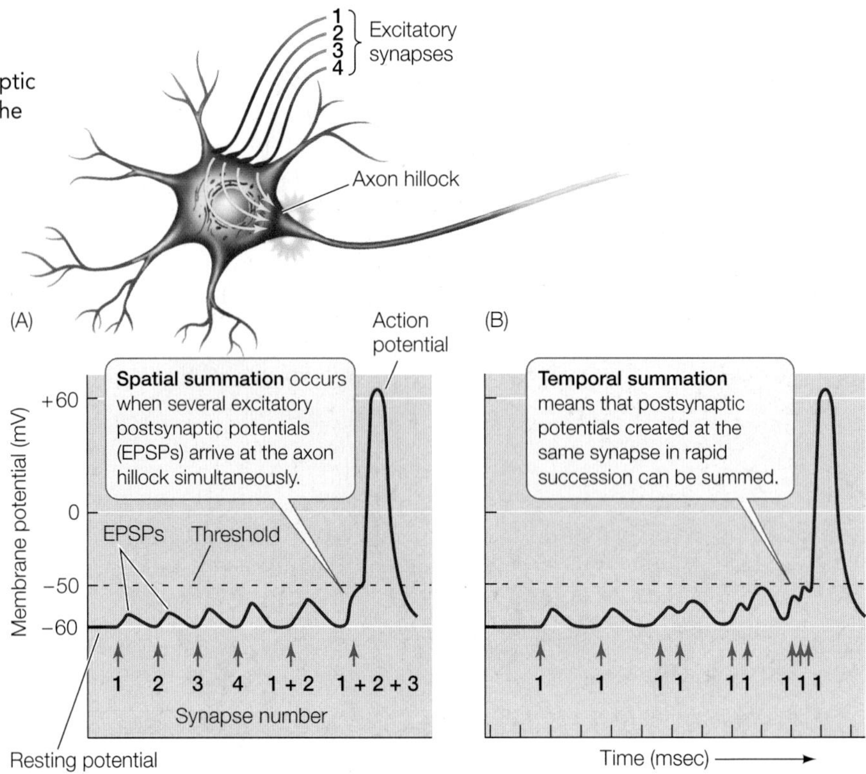


**45.15 The Postsynaptic Neuron Sums Information** Individual neurons sum excitatory and inhibitory postsynaptic potentials over space (A) and time (B). When the sum of the potentials depolarizes the axon hillock to threshold, the neuron generates an action potential.

If the axon terminal of a motor neuron releases sufficient amounts of ACh to depolarize a motor end plate, that spreading depolarization activates the voltage-gated $Na^+$ channels and causes the firing of an action potential. This action potential is then conducted throughout the muscle cell's system of membranes, causing the cell to contract. We will discuss the contraction of muscle cells in greater detail in Section 48.1.

How much neurotransmitter is enough? Neither a single ACh molecule nor the contents of an entire vesicle (about 10,000 ACh molecules) will bring the membrane of a muscle cell to threshold. However, a single action potential in an axon terminal releases the contents of about 100 vesicles—more than enough to fire an action potential in the muscle cell and cause it to contract.

**Go to Animated Tutorial 45.4**
**Neurons and Synapses**
**Life10e.com/at45.4**

## Synapses can be excitatory or inhibitory

In vertebrates, the synapses between motor neurons and muscle cells are always excitatory; that is, motor end plates always respond to ACh with a graded potential that is less negative than the resting potential (depolarization). However, synapses between neurons are frequently inhibitory; such a synapse causes hyperpolarization of the postsynaptic membrane. For example, recall that there are more chloride ions ($Cl^-$) outside the cell than inside it. If the receptors on the postsynaptic membrane open $Cl^-$ channels, $Cl^-$ ions enter the postsynaptic cell and hyperpolarize it. Hyperpolarization takes the membrane farther from the threshold potential for the voltage-gated $Na^+$ channels, and therefore makes it less likely that the cell will fire action potentials.

Recall that most neurons have many dendrites. Axon terminals from many other neurons can form synapses with those dendrites and with the cell body. The axon terminals of different presynaptic neurons can store and release different neurotransmitters, and the plasma membrane of the dendrites and cell body of a postsynaptic neuron can have receptors for a variety of neurotransmitters. The mix of synaptic activity impinging on a cell will cause it to have a graded membrane potential that can be either more positive or more negative than its resting potential.

## The postsynaptic cell sums excitatory and inhibitory input

What determines when an individual neuron will fire an action potential? As we just learned, the sum of excitatory and inhibitory postsynaptic potentials creates a graded membrane potential in the postsynaptic cell body. This summation ability is the major mechanism by which the nervous system integrates information. Each neuron receives 1,000 or more synaptic inputs, but it has only one output: an action potential in a single axon. At any one time, the information from all of the active inputs is translated into the rate at which that neuron generates action potentials in its axon.

For most neurons, summation takes place in the axon hillock at the base of the axon. The plasma membrane of the axon hillock is not insulated by glia and has many voltage-gated $Na^+$ channels. Excitatory and inhibitory postsynaptic potentials from synapses anywhere on the dendrites or the cell body spread to the axon hillock by local current flow. If the resulting graded potential depolarizes the axon hillock to threshold, it fires an action potential. Because postsynaptic potentials decrease in strength as they spread from the site of the synapse, a synapse at the tip of a dendrite has less influence than a synapse on the cell body, near the axon hillock.

Excitatory and inhibitory postsynaptic potentials are summed over space and over time. **Spatial summation** adds up the simultaneous influences of synapses at different sites on the postsynaptic cell (**Figure 45.15A**). **Temporal summation** adds up postsynaptic potentials generated at the same site in a rapid sequence (**Figure 45.15B**).

## Synapses can be fast or slow

Most neurotransmitter receptors induce changes in postsynaptic cells by opening or closing ion channels. How they do so is the basis for grouping receptors into two general categories:

- **Ionotropic receptors** are themselves ion channels. Neurotransmitter binding to an ionotropic receptor causes a direct change in ion movement across the plasma membrane of the postsynaptic cell. These proteins enable fast, short-lived responses.

The ACh receptor of the motor end plate is an example of an ionotropic receptor. Each of its five subunits extends through the plasma membrane. When assembled, the subunits create a central pore that allows ions to pass through (see Figure 45.14). Of several different kinds of subunit, only one kind has the ability to bind ACh. Each functional receptor has two ACh-binding subunits and three other subunits.

- **Metabotropic receptors** are not ion channels, but they induce signaling cascades in the postsynaptic cell that secondarily lead to changes in ion channels (see Figure 7.10A). Postsynaptic cell responses mediated by metabotropic receptors are generally slower and longer-lived than those induced by ionotropic receptors.

Metabotropic receptors are also transmembrane proteins, but instead of acting as ion channels, they initiate an intracellular signaling process that results in the opening or closing of an ion channel and other changes in the postsynaptic cell.

## Electrical synapses are fast but do not integrate information well

Electrical synapses are different from chemical synapses because they couple neurons electrically. Electrical synapses contain numerous gap junctions. At these synapses, the presynaptic and postsynaptic cell membranes are separated by a space of only 2 or 3 nanometers, and membrane proteins called connexins link the two neurons by forming pores that connect the cytoplasm of the two cells (see Figure 7.19A). Ions and small molecules can pass directly from cell to cell through these pores. Transmission at electrical synapses is very fast and can proceed in either direction, whereas transmission at chemical synapses is slower and unidirectional.

Electrical synapses are less common in the nervous systems of vertebrates than are chemical synapses for several reasons. First, electrical continuity between neurons does not allow temporal summation of synaptic inputs. Second, an effective electrical synapse requires a large area of contact between the presynaptic and postsynaptic cells. This condition rules out the possibility of thousands of synaptic inputs to a single neuron—which is the norm in complex nervous systems. Third, electrical synapses cannot be inhibitory. Thus electrical synapses are useful for rapid communication, but they are less useful for processes of integration and learning.

## The action of a neurotransmitter depends on the receptor to which it binds

More than 100 neurotransmitters are now recognized, and more will surely be discovered. ACh, as we have seen, is an important neurotransmitter because it is how the nervous system commands muscles to contract. ACh also plays roles in certain synapses between neurons in the brain, but it accounts for only a small percent of the total neurotransmitter content of the brain.

The workhorse neurotransmitters of the brain are three simple amino acids: glutamate, which is excitatory, and glycine and $\gamma$-aminobutyric acid (GABA), which are inhibitory. The integration of information at the cellular level is a balance between excitation and inhibition, so it is understandable that excessive inhibition such as we saw in the Down syndrome mice described at the start of this chapter could impair an animal's ability to process and integrate new information. Indeed, when GABA's inhibitory action in these mice was reduced by a drug that blocks the GABA receptor, the learning ability of the mice increased (**Figure 45.16**).

Another important group of neurotransmitters in the brain is the monoamines, which are derivatives of amino acids. They

INVESTIGATING**LIFE**

**45.16 Reducing Neuronal Inhibition May Enhance Learning** Learning and memory in mice can be studied using the "novel object recognition task" described in the method below. Genetically engineered mice that are models for Down syndrome (i.e., the genes known to be present on human chromosome 21 have been triplicated in the mouse) cannot perform this task. Treatment of these model mice with a drug that partially blocks GABA receptors (GABA being an inhibitor of synaptic transmission) appears to increase their ability to learn.[a]

**HYPOTHESIS** Excessive inhibition of neurons by GABA impairs learning ability in a mouse model of Down syndrome.

*Method*

1. Place the mouse in arena with two objects and allow them to explore and investigate; remove mouse.
2. After 24 hours, change one of the two objects and return the mouse to the arena. Compare the amount of time the mouse spends with the old versus the novel object. If the mouse spends more time with the novel object, it remembers the old object (has "learned").
3. Repeat the experiment with normal and Down syndrome mice before and after the subject mice are treated with a drug that blocks GABA receptors (GABAr).

*Results*

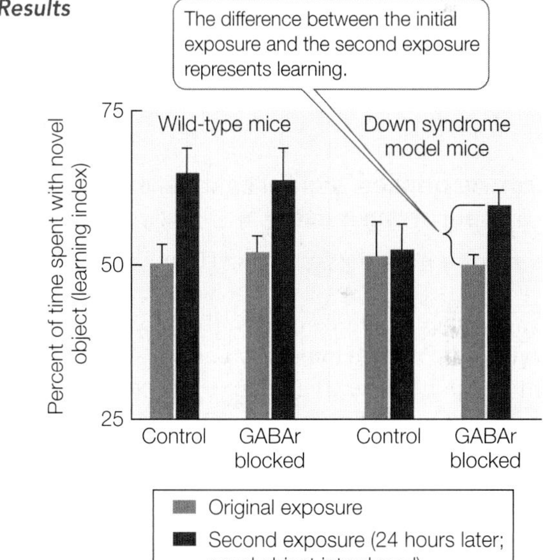


**CONCLUSION** Blocking GABA receptors improved learning capacity in the Down syndrome model mice.

Go to **BioPortal** for discussion and relevant links for all INVESTIGATING**LIFE** figures.

[a]Fernandez, F. et al. 2007. *Nature Neuroscience* 10: 411–413.

include dopamine and norepinephrine (derivatives of tyrosine) and serotonin (a derivative of tryptophan). Peptides also function as neurotransmitters; for example, endorphins and enkephalins are the body's opiates and modulate the sensation of pain. Another peptide, substance P, transmits pain sensations. Even a gas, nitric oxide, is used by neurons as an intercellular messenger (see Figure 7.15).

Neurotransmission is complex in part because each neurotransmitter has multiple receptor types. ACh, for example, has two receptor types: nicotinic receptors, which are ionotropic; and muscarinic receptors, which are metabotropic. Both types of ACh receptors are found in the brain and spinal cord, where nicotinic receptors tend to be excitatory and muscarinic receptors tend to be inhibitory. ACh actions can differ outside the brain and spinal cord as well. ACh acting through nicotinic receptors causes the smooth muscle of the gut to increase its motility, but ACh acting through muscarinic receptors causes cardiac muscle to hyperpolarize and therefore to slow down. There are many more examples of neurotransmitters that have different effects in different tissues, but the important thing to remember is that the action of a neurotransmitter depends on the receptor to which it binds. In addition, keep in mind that turning neurotransmitter action off is as important as turning it on.

**Go to Activity 45.1 Neurotransmitters**
**Life10e.com/ac45.1**

### To turn off responses, synapses must be cleared of neurotransmitter

If released neurotransmitter molecules simply remained in the synaptic cleft, the postsynaptic membrane would become saturated and receptors would be constantly activated. The postsynaptic cell would remain hyperpolarized or depolarized and would be unresponsive to short-term changes in the presynaptic cell. The more rapidly neurons can respond to input, the more information they can process in a given amount of time. Thus neurotransmitter must be cleared from the synaptic cleft shortly after it is released by the axon terminal.

Neurotransmitter action can be terminated in several ways. First, enzymes may destroy the neurotransmitter. Acetylcholine, for example, is rapidly destroyed by the enzyme acetylcholinesterase (AChE), which is present in the synaptic cleft in close association with ACh receptors on the postsynaptic membrane (see Figure 45.14). When AChE is inhibited, ACh lingers in the synaptic cleft, causing spastic (contracted) muscle paralysis and usually resulting in death. Some of the most deadly nerve gases developed for chemical warfare work by inhibiting AChE. Some agricultural insecticides, such as malathion, also inhibit AChE and can poison farm workers if used without safety precautions.

Neurotransmitter can also simply diffuse away from the cleft, or be taken up via active transport by nearby cell membranes, most notably glial cell membranes. The antidepressant drug commonly prescribed under the brand name Prozac slows the reuptake of the neurotransmitter serotonin, thus enhancing serotonin's activity at the synapse.

### The diversity of receptors makes drug specificity possible

Many drugs used to treat the nervous system act by modulating specific synaptic interactions. Drugs that mimic or potentiate the effect of a neurotransmitter are called **agonists**; those that block the actions of neurotransmitters are called **antagonists**. For example, morphine is an agonist at the endorphin receptor and therefore blocks pain. Propranolol, a widely used β-blocker, is an antagonist of certain adrenergic receptors and therefore decreases panic attacks and anxiety. A major emphasis in neurobiology is to identify neurotransmitter receptor subtypes and design drugs that selectively bind to them to have highly specific effects on nervous system activity.

**RECAP 45.3**

Chemical synapses involve the release of neurotransmitter molecules stored in vesicles in the presynaptic terminal. Action potentials reaching that terminal cause the fusion of vesicles with the presynaptic membrane, releasing neurotransmitter that can then bind to receptors on the postsynaptic membrane and influence its membrane potential. There is a great diversity of neurotransmitters and their receptors.

- Describe the role of $Ca^{2+}$ channels in synaptic events. **See p. 936 and Figure 45.13**
- How can some synapses be excitatory and others inhibitory? **See p. 938**
- How do neurons integrate the input from various synapses? **See p. 938 and Figure 45.15**

## 45.4 How Are Neurons and Glia Organized into Information-Processing Systems?

Nervous systems can process information because their neurons are organized into **neural networks**. These networks include three functional categories of neurons, which can be thought of as being involved with input, output, and integration:

- **Afferent neurons** carry sensory information into the nervous system. That information comes from specialized sensory cells that transduce (convert) various kinds of sensory stimuli (e.g., light, heat, pressure) into action potentials.
- **Efferent neurons** carry commands to physiological and behavioral effectors such as muscles and glands.
- **Interneurons** integrate and store information and communicate between afferent and efferent neurons.

### Nervous systems range in complexity

Simple animals such as cnidarians (sea anemones, for example) process information with a limited number of simple neural networks that do little more than provide direct lines of communication from sensory cells to effectors; there is little or no integration or processing of signals (**Figure 45.17A**). The

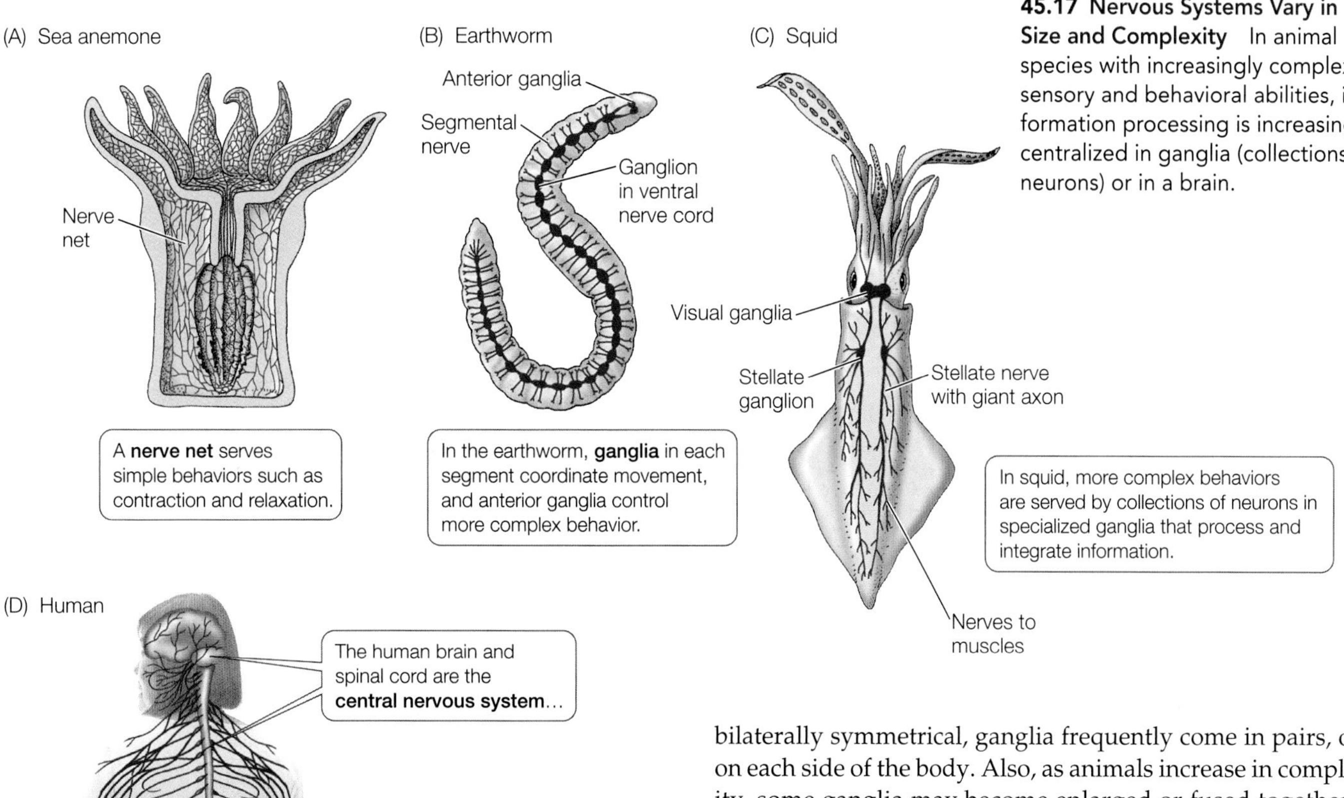


**45.17 Nervous Systems Vary in Size and Complexity** In animal species with increasingly complex sensory and behavioral abilities, information processing is increasingly centralized in ganglia (collections of neurons) or in a brain.

cnidarian's **nerve net** is most developed around the tentacles and the oral opening, where it facilitates detection of food or danger and causes tentacles to extend or retract.

Animals that are more complex and move about in search of food and mates must process and integrate larger amounts of information. Even earthworms fit this description, and their increased need for information processing is met by higher numbers of neurons organized into clusters of neuronal cell bodies called **ganglia** (singular *ganglion*). Ganglia serving different functions may be distributed around the body, as in earthworms and squid (**Figure 45.17B,C**). In animals that are bilaterally symmetrical, ganglia frequently come in pairs, one on each side of the body. Also, as animals increase in complexity, some ganglia may become enlarged or fused together at the anterior end, forming a **brain**. Small nervous systems of invertebrates can be remarkably complex. Consider the nervous systems of spiders, which have programmed within them the thousands of precise movements necessary to construct an intricate web without prior experience or opportunities to learn the specific web architecture of their species.

In vertebrates, most cells of the nervous system are found in the **central nervous system** (**CNS**). The CNS includes the brain and the spinal cord, which are the sites of most information processing, storage, and retrieval (**Figure 45.17D**). Information is transmitted from sensory cells to the CNS and from the CNS to effectors via neurons that extend or reside outside the brain and the spinal cord. These sensory and effector neurons and their supporting cells are the **peripheral nervous system** (**PNS**).

The spinal cord conducts information in both directions between the brain and the peripheral neurons of the body's organs. Neural networks integrate information coming from the peripheral nervous system and issue motor commands. An example of an information-processing network in humans is the spinal knee-jerk reflex.

## The knee-jerk reflex is controlled by a simple neural network

A cross section of the human spinal cord reveals a central area of gray matter in the shape of a butterfly, surrounded by an area of white matter (see Figure 45.18). In the nervous system, **gray matter** is rich in neural cell bodies, and **white matter** contains myelinated axons. The gray matter of the spinal cord contains the cell bodies of the spinal neurons; the white matter contains myelinated axons that conduct information up and down the spinal cord.

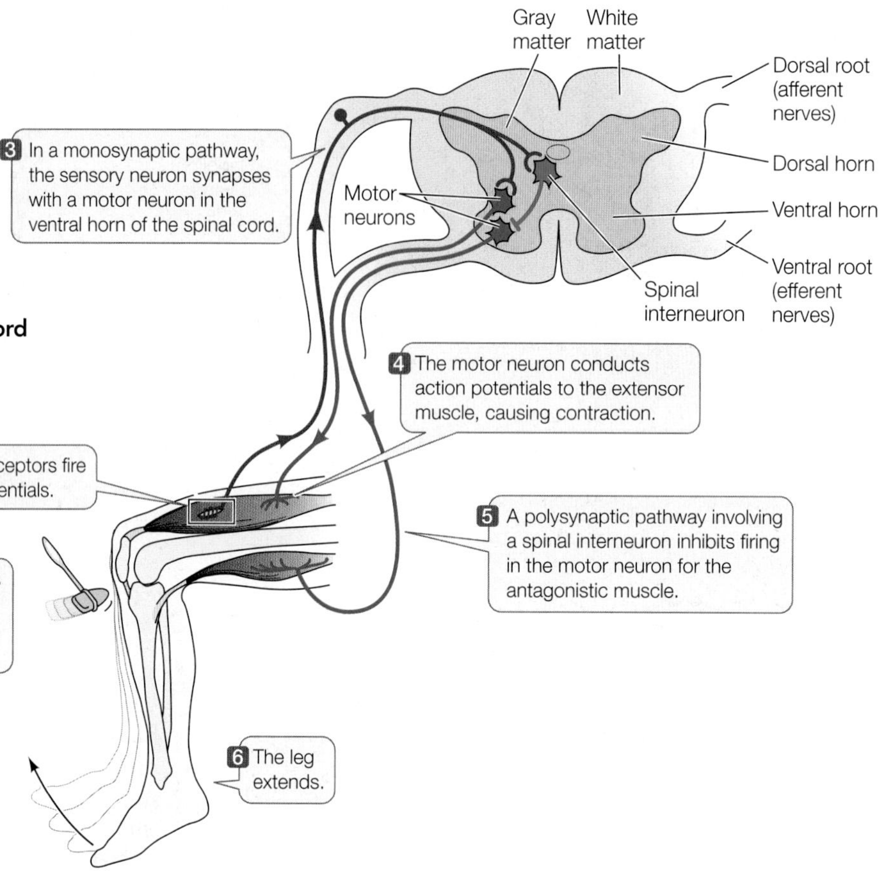

**45.18 A Neural Network in the Spinal Cord Generates the Knee-Jerk Reflex** Sensory (afferent) information enters the spinal cord through the dorsal horns (red pathway), and motor (efferent) output leaves it via the ventral horns (blue pathways). Information travels to the brain in white matter tracts. Interneurons make connections within the spinal cord that result in a complex, coordinated behavior pattern.

**Go to Animated Tutorial 45.5 Information Processing in the Spinal Cord**
Life10e.com/at45.5



Spinal nerves extend from the spinal cord at regular intervals on each side. Each spinal nerve has two roots, one connecting with the dorsal horn of the gray matter, the other with the ventral horn. The afferent (sensory) axons in a spinal nerve enter the spinal cord through the dorsal root, and the efferent (motor) axons leave through the ventral root. The conversion of afferent to efferent information in the spinal cord without participation of the brain is called a **spinal reflex**. The neural network involved in one such spinal reflex—the "knee-jerk"—is diagrammed in **Figure 45.18.** The stimulus for this particular reflex is a tap on the knee with a small mallet. That tap stretches the tendon going over the knee, which attaches the muscle of the upper leg to bone in the lower leg. Stretching the tendon stretches muscle fibers in the upper leg, and stretch receptors in that muscle transduce the physical stimulus into action potentials. The action potentials are then conducted by a sensory neuron into the dorsal horn of the spinal cord. That sensory neuron goes all the way to the ventral horn and synapses onto a motor neuron, which fires action potentials. The axon of that motor neuron travels out through the ventral horn of the spinal cord and extends all the way to the same muscle that initially was stretched, causing that muscle to contract.

The function of this simple circuit is to sense an increased load on the muscle and to increase the strength of its contraction to compensate for that additional load. Because there is only one synapse between the afferent and efferent neurons in this simple network, it is called a monosynaptic reflex; most spinal networks are more complex. For example, limb movement is controlled by antagonistic sets of muscles that work against each other. When one member of an antagonistic set of muscles contracts, it bends (flexes) the limb; it is therefore called a flexor. The antagonist muscle, the extensor, straightens (extends) the limb. For a limb to move, one muscle of the pair must relax while the other contracts. Thus sensory input that activates the motor neuron of one muscle also inhibits its antagonist. This coordination is achieved by an interneuron, which makes an inhibitory synapse onto the motor neuron of the antagonistic muscle (see Figure 45.18). Thus the reciprocal inhibition of antagonistic muscles involves an interneuron between the sensory cell and the motor neuron of the inhibited muscle, and therefore at least two synapses.

The withdrawal reflex is an example of a polysynaptic spinal reflex that involves many interneurons. When you step on a tack, you immediately pull back your foot: the tack stimulates pain receptors in the foot, and the sensory neurons transmit action potentials into the dorsal horn of the spinal cord on the same side of the body. In the dorsal horn, these neurons synapse with interneurons that send information through their axons to the brain, resulting in the conscious sensation of pain. Before the brain is aware of the pain, however, synapses of the sensory neurons with other interneurons stimulate and inhibit a variety of different motor neurons in the spinal cord. Interneurons on the same side of the spinal cord coordinate the activity of the muscles that withdraw the foot and leg. To

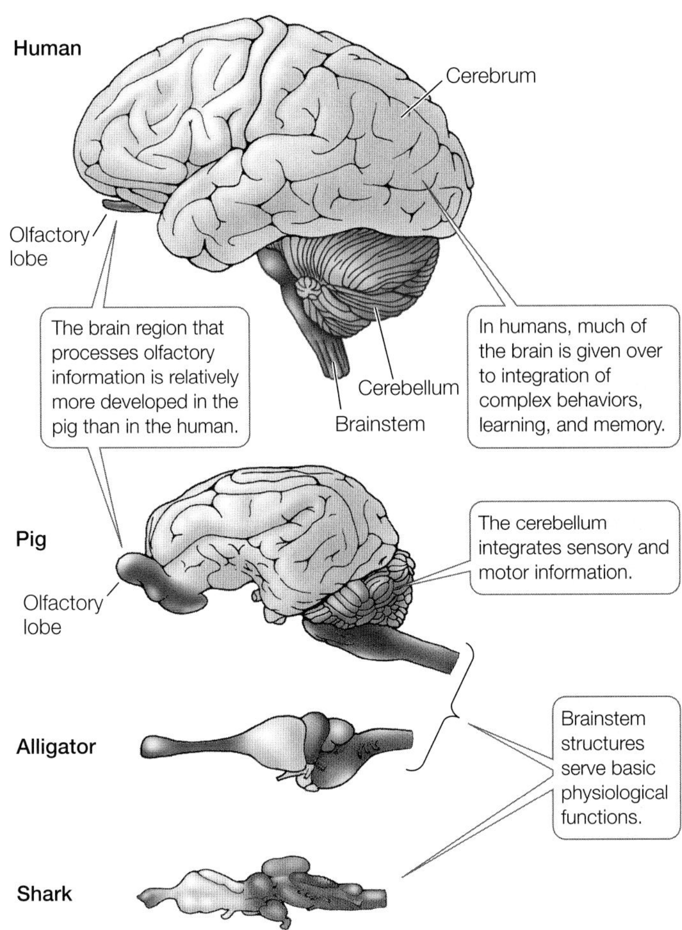


**45.19 Brains Vary in Size and Complexity** The brains of four vertebrate species—all of which may have a similar body mass—show immense differences. Note that the brainstem, which is involved in physiological regulation and stereotypic behavior, differs less among these species than does the cerebrum, which is responsible for complex behavior and learning.

pull away, however, the other leg has to extend and balance must be shifted. The coordination of these actions involves interneurons that make connections across the spinal cord to motor neurons on the opposite side. Thus a rather complex suite of movements is coordinated by a network in the spinal cord. By extension, you can appreciate how much more complex a neural network is required to enable you to execute complex movements in time with music and coordinated with another person—in other words, to dance.

### The vertebrate brain is the seat of behavioral complexity

Vertebrates differ greatly in their behavioral complexity and in their physiological specializations, and their neural networks reflect this diversity. **Figure 45.19** shows the brains of four vertebrate species of similar body mass drawn to the same scale.

The human nervous system contains an estimated $10^{11}$ neurons. A given neuron in the brain can have 1,000 or more synapses. Thus the human brain can contain $10^{14}$ synapses ($10^{11}$ neurons $\times$ $10^3$ synapses per neuron). Then there are the glia. A single astrocyte might participate in 100,000 synapses while at the same time monitoring signals in the extracellular fluid and the blood. And beyond these numbers, synapses are not constant but can be highly plastic. They can increase or decrease in number and size, and they can become more or less sensitive.

This astronomical number of neurons and synapses is divided into thousands of distinct but interacting networks that function in parallel. The possible number of informational networks in the brain is almost infinite, and therein lies the incredible ability of the human brain to process information, to learn, to do complex tasks, to remember, and to have emotions.

#### RECAP 45.4

Nervous systems are composed of neural networks that include afferent neurons, interneurons, and efferent neurons. Nervous systems range in complexity from simple nerve nets to the human brain. The spinal knee-jerk reflex is an example of a simple neural network, but networks that control more complex behavior become increasingly complicated.

- What are ganglia, and why are they concentrated in the anterior region of many invertebrates? **See p. 941 and Figure 45.17**
- Explain the components of a spinal reflex network. How can the same stimulus cause contraction in one muscle and relaxation in another? **See p. 942 and Figure 45.18**

What causes overinhibition in the nervous system, and how can it be reduced?

#### ANSWER

Some neurotransmitter–receptor combinations excite activity in the postsynaptic neuron, depolarizing it and making it likely to fire an action potential. Other neurotransmitter–receptor combinations inhibit the responses of the postsynaptic cell. Neurons integrate information by summing excitatory and inhibitory synaptic inputs. Overinhibition in the nervous system can arise when too much inhibitory neurotransmitter (such as GABA) is released, or if postsynaptic cells have too many receptors for inhibitory neurotransmitters. Drugs that decrease the synthesis and release of the inhibitory neurotransmitter can reduce the inhibition, as can drugs that block the receptors for such neurotransmitters. In the case of the Down syndrome model mice, drugs that blocked the GABA receptors that are also $Cl^-$ channels reduced the level of inhibition in the mice nervous systems, apparently increasing the ability of the mice to learn and form memories.

# CHAPTER SUMMARY 45

## 45.1 What Cells Are Unique to the Nervous System?

- The cells of the nervous systems include many types of **neurons** and **glia**.
- All neurons are **excitable**, which means they can generate and conduct electric signals called **action potentials**. Glia support and modulate the activities of neurons but do not generate action potentials.
- A neuron generally receives information via its **dendrites**, of which there can be many, and transmits information via its single **axon**, which ends in **axon terminals**. **Review Figure 45.1**
- Where neurons and their target cells meet, information is transmitted across specialized junctions called **synapses**.
- Glia include **Schwann cells** and **oligodendrocytes,** both of which generate **myelin** sheets on axons. Glia also include **astrocytes**, which support neurons metabolically, modulate synaptic signaling, and contribute to the **blood–brain barrier**. **Review Figures 45.3, 45.4**

## 45.2 How Do Neurons Generate and Transmit Electric Signals?

- Neurons have an electric charge difference across their plasma membranes, the **membrane potential**. The membrane potential is created by ion transporters and channels. When a neuron is not firing action potentials, its membrane potential is referred to as the **resting potential**. **Review Figures 45.5, 45.6, ANIMATED TUTORIAL 45.1**
- The **sodium–potassium pump** concentrates $K^+$ on the inside of a neuron and $Na^+$ on the outside. Potassium leak channels allow $K^+$ to diffuse out of the neuron, leaving behind unbalanced negative charges. **Review Figures 45.6, 45.7**
- **Patch clamping** allows the study of single ion channels. **Review Figure 45.8**
- The resting potential is perturbed when ion channels open or close, changing the permeability of the plasma membrane to charged ions. Through this mechanism, the plasma membrane can become **depolarized** or **hyperpolarized** and therefore have a **graded membrane potential** response to input. **Review Figure 45.9**
- An action potential results from a rapid reversal in charge across a portion of the plasma membrane resulting from the sequential opening and closing of voltage-gated $Na^+$ and $K^+$ channels. These changes in voltage-gated channels occur when the plasma membrane depolarizes to a **threshold** level. **Review Figure 45.10, ANIMATED TUTORIAL 45.2**
- Action potentials are all-or-none, self-regenerating events. They are conducted down axons because local current flow depolarizes adjacent regions of membrane and brings them to threshold. **Review Figure 45.11**
- In myelinated axons, action potentials appear to jump between **nodes of Ranvier**, areas of axonal plasma membrane that are not covered by myelin. **Review Figure 45.12**

## 45.3 How Do Neurons Communicate with Other Cells?

- Neurons communicate with each other and with other cell types by transmitting information over **electrical synapses** or by the transmission of molecular signals called **neurotransmitters** over **chemical synapses**.
- The **neuromuscular junction** is a well-studied chemical synapse between a motor neuron and a skeletal muscle cell. Its neurotransmitter is **acetylcholine** (**ACh**), which causes depolarization of the postsynaptic membrane when it binds to its receptor at the **motor end plate**. **Review Figure 45.13, ANIMATED TUTORIAL 45.3**
- When an action potential reaches an axon terminal, it causes the release of neurotransmitters, which diffuse across the **synaptic cleft** and bind to receptors on the postsynaptic membrane. **Review Figures 45.13, 45.14, ANIMATED TUTORIAL 45.4**
- Synapses between neurons can be either excitatory or inhibitory. A postsynaptic neuron integrates information by summing excitatory and inhibitory postsynaptic potentials in both spatially and temporally. **Review Figure 45.15**
- **Ionotropic receptors** are ion channels or directly influence ion channels. **Metabotropic receptors** influence the postsynaptic cell through various signal transduction pathways and can result in the opening or closing of ion channels. The actions of ionotropic synapses are generally faster than those of metabotropic synapses.
- There are many different neurotransmitters and even more types of receptors. The action of a neurotransmitter depends on the type of receptor to which it binds. **See ACTIVITY 45.1**
- Neurotransmitter molecules cannot be allowed to accumulate in a synapse but must be cleared in order to turn off responses in the postsynaptic cell. This may be done by enzymatic degradation, simple diffusion, or reuptake of the neurotransmitter.

## 45.4 How Are Neurons and Glia Organized into Information-Processing Systems?

- In vertebrates, the brain and spinal cord form the **central nervous system** (**CNS**), which communicates with the rest of the body via the **peripheral nervous system** (**PNS**). The CNS increases in complexity from invertebrates to vertebrates and from fish to mammals. **Review Figures 45.17, 45.19**
- **Neural networks** include **afferent neurons** and **efferent neurons**, generally connected through **interneurons**.
- A **spinal reflex** is an example of a simple neural network that integrates information and controls a response. **Review Figure 45.18, ANIMATED TUTORIAL 45.5**

**Go to the Interactive Summary to review key figures, Animated Tutorials, and Activities Life10e.com/is45**

# CHAPTER**REVIEW**

## REMEMBERING

1. The rising phase of an action potential is due to the
   a. closing of $K^+$ channels.
   b. opening of chemically gated $Na^+$ channels.
   c. closing of voltage-gated $Ca^{2+}$ channels.
   d. opening of voltage-gated $Na^+$ channels.
   e. spread of positive current along the plasma membrane.

2. Which statement about synaptic transmission is *not* true?
   a. The synapses between neurons and skeletal muscle cells use ACh as their neurotransmitter.
   b. A single vesicle of neurotransmitter can cause a muscle cell to contract.
   c. The release of neurotransmitter at the neuromuscular junction causes the motor end plate to depolarize.
   d. In vertebrates, the synapses between motor neurons and muscle fibers are always excitatory.
   e. Inhibitory synapses cause the resting potential of the postsynaptic membrane to become more negative.

3. Which statement accurately describes an action potential?
   a. Its magnitude increases along the axon.
   b. Its magnitude decreases along the axon.
   c. All action potentials in a single neuron are of the same magnitude.
   d. During an action potential, the membrane potential of a neuron remains constant.
   e. An action potential permanently shifts a neuron's membrane potential away from its resting value.

4. Graded membrane potentials
   a. can be hyperpolarizing.
   b. can be depolarizing.
   c. integrate the many synaptic inputs to a cell.
   d. are important means of summing sensory inputs.
   e. All of the above

5. The binding of an inhibitory neurotransmitter to the postsynaptic receptors of a neuron at its resting potential results in
   a. depolarization of the membrane.
   b. generation of an action potential.
   c. hyperpolarization of the membrane.
   d. increased permeability of the membrane to sodium ions.
   e. increased permeability of the membrane to calcium ions.

## UNDERSTANDING & APPLYING

6. If you stimulate an axon in the middle, action potentials are conducted in both directions. Yet when an action potential is generated at the axon hillock, it goes only toward the axon terminals and does not backtrack. Explain why action potentials are bidirectional in the first example and unidirectional in the second.

7. How does a neuron integrate excitatory and inhibitory synapses?

## ANALYZING & EVALUATING

8. Benzodiazepines are drugs that potentiate the effects of GABA on its receptor. What effects might you expect to observe in a person taking these drugs?

9. The language of the nervous system consists of one "word," the action potential. How can this single word convey a diversity of information? How can that information be quantified, and how can it be integrated?

Go to BioPortal at **yourBioPortal.com** for Animated Tutorials, Activities, LearningCurve Quizzes, Flashcards, and many other study and review resources.

# 46 Sensory Systems

## CHAPTER**OUTLINE**


**46.1** How Do Sensory Receptor Cells Convert Stimuli into Action Potentials?

**46.2** How Do Sensory Systems Detect Chemical Stimuli?

**46.3** How Do Sensory Systems Detect Mechanical Forces?

**46.4** How Do Sensory Systems Detect Light?

**Sensing Infrared Radiation** The "hole" to the right of this diamondback rattlesnake's eye is one of its bilateral pit organs. Pit organs detect infrared radiation from the snake's preferred prey—small rodents—with unerring precision, even in total darkness. The forked tongue also provides positional information, picking up molecular signals that are transmitted to the brain by a specialized organ in the roof of the snake's mouth.

A RATTLESNAKE can see to strike a running rodent in complete darkness. How can this be, when "seeing" means using the eyes to detect light waves, and "complete darkness" means no light? It is possible because these definitions are based on human capabilities. What we call "light" is actually only a small portion (red, orange, yellow, green, blue, indigo, and violet) of the spectrum of electromagnetic radiation. Other animals see wavelengths humans cannot. In Chapter 27 we saw that insects perceive patterns on flowers that reflect ultraviolet wavelengths invisible to humans. Similarly, rattlesnakes "see" infrared wavelengths that we cannot (although at high enough levels of intensity, humans feel infrared wavelengths as heat).

It is not the snake's eyes that perceive infrared light. Rattlesnakes and their relatives have pit organs located between the nostril and the eye on each side of the skull that contain high densities of infrared-sensitive neurons. The two pits are positioned in such a way that sensory receptor cells in the pits receive directional information. The fields of "view" of the bilateral pits are overlapping and thus convey a three-dimensional perspective. Information from the pit organs goes to the same region of the brain as information from the eyes, so rattlesnakes actually do "see" the world in a range of electromagnetic radiation that is different from the human visual spectrum.

Our definition of silence is as arbitrary as our definition of darkness. "Sound" is actually pressure waves in the environment, and many animals are sensitive to pressure waves with frequencies, or pitches, we cannot hear. Elephants communicate using sound waves that are below human hearing range; such long waves travel great distances, an advantage to large animals that roam over extensive areas. Bats emit incredibly loud, brief sound pulses that are above our range of hearing. A flying bat hears echoes of these pulses bouncing off objects in the environment. The pulses are so loud and the echoes so weak that it is rather like a construction worker trying to overhear a whispered conversation while using a pneumatic drill.

"Reality" is what our eyes see, our ears hear, our noses smell, and what we touch and taste. But human beings sense only a limited range of the information available. Animals with different ranges of sensitivity process different sources of information and may perceive "reality" quite differently.

How can bats emit loud pulses of sound and not be deaf to the faint echoes that return within milliseconds?

See answer on p. 963.

## 46.1 How Do Sensory Receptor Cells Convert Stimuli into Action Potentials?

Sensory receptor cells, usually simply called **sensors** or **receptors**, transduce (convert) physical and chemical stimuli such as light and sound waves, pressure (touch), and odorant and taste molecules into neural signals. These signals are then transmitted to the central nervous system (CNS) for processing and interpretation. The first step in this process of **sensory transduction** is a change in the membrane potential of the receptor cell in response to a specific type of stimulus.

### Sensory transduction involves changes in membrane potentials

Sensory transduction typically begins with a **receptor protein** that opens or closes ion channels in response to a specific stimulus such as heat, light, chemicals, mechanical force (including sound waves), or electric fields. The resulting change in ion flow alters the receptor cell's membrane potential. A change in the membrane potential of a receptor cell in response to a stimulus is called a **receptor potential**. Receptor potentials are *graded membrane potentials* that spread over only short distances. In order to signal over long distances in the nervous system, receptor potentials must generate action potentials, which they can do in two ways:

- The receptor potential may trigger action potentials in the receptor cell itself.
- The receptor potential may cause the receptor cell to release neurotransmitters that can induce a postsynaptic neuron to generate action potentials.

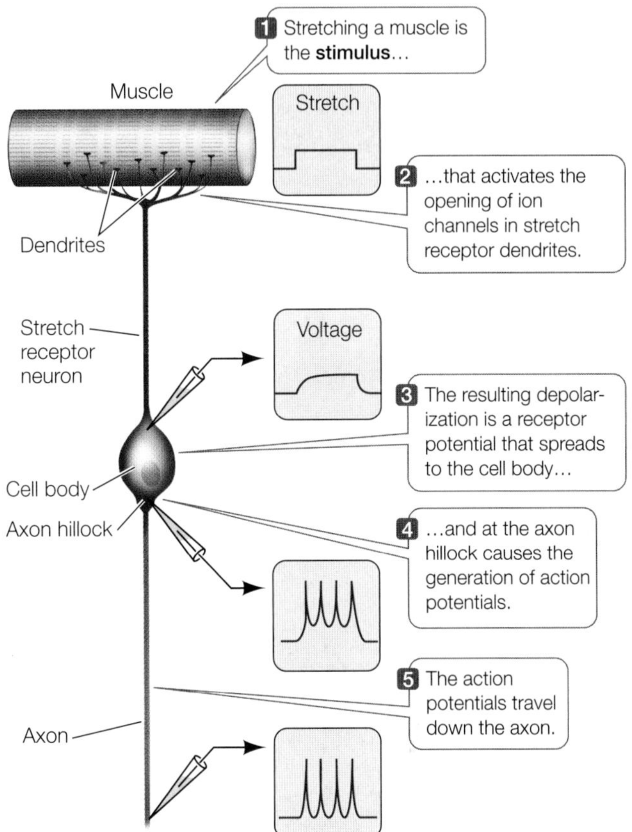


**46.1 Stimulating a Sensory Cell Produces a Receptor Potential** Signal transduction in the stretch receptor of a crayfish can be investigated by measuring the membrane potential at different places on the stretch receptor neuron while stretching the muscle innervated by that sensory neuron.

A good model of how a receptor cell generates action potentials is the **stretch receptor** of a crayfish (**Figure 46.1**). Stretching the muscle to which the stretch receptor is attached causes receptor potentials. These receptor potentials spread to the base of the cell's axon, where they generate action potentials that travel down the axon to the CNS. The rate at which action potentials are fired depends on the magnitude of the receptor potential; that magnitude, in turn, depends on how much the muscle is stretched.

In a receptor cell that does not fire action potentials (such as the photoreceptors in the vertebrate eye), the spreading receptor potential reaches a presynaptic patch of plasma membrane and induces the release of a neurotransmitter. The intensity of the stimulus influences how much neurotransmitter is released. That neurotransmitter binds to receptor proteins on an associated sensory neuron, altering its membrane potential and causing it to increase or decrease its rate of firing action potentials. In a few cases, this second cell also does not generate action potentials, but simply changes the rate at which it releases neurotransmitter onto another neuron. Eventually, however, the stimulation of a sensory cell is always coded as a change in firing of action potentials in a sensory circuit.

### Sensory receptor proteins act on ion channels

Sensory receptor proteins respond to stimuli by directly or indirectly opening or closing ion channels in the sensory cell (**Figure 46.2**), leading either to an action potential or to the release of neurotransmitter. Section 45.3 noted that synaptic receptor proteins are either ionotropic or metabotropic; the same distinction can be applied to sensory receptor proteins. Ionotropic sensory receptor proteins are either ion channels themselves or directly affect the opening of an ion channel. Examples are receptors that respond to physical force (mechanoreceptors) and those that respond to temperature (thermoreceptors). Electrosensors most likely do not have receptor proteins, but they are grouped with the ionotropic receptors; the plasma membrane of electrosensory cells is sensitive to the voltage across it and releases neurotransmitter in response to slight changes in membrane potential.

Metabotropic sensory receptor proteins influence ion channels indirectly, through G proteins and second messengers as described in Chapter 7. Examples are most chemoreceptors and photoreceptors.

### Sensation depends on which neurons receive action potentials from sensory cells

All sensory systems process information in the form of action potentials. But the sensations we perceive—heat, pressure, light, smell, sound—differ because the messages from different kinds of sensory cells arrive at different places in the central nervous system. Action potentials arriving in the visual cortex

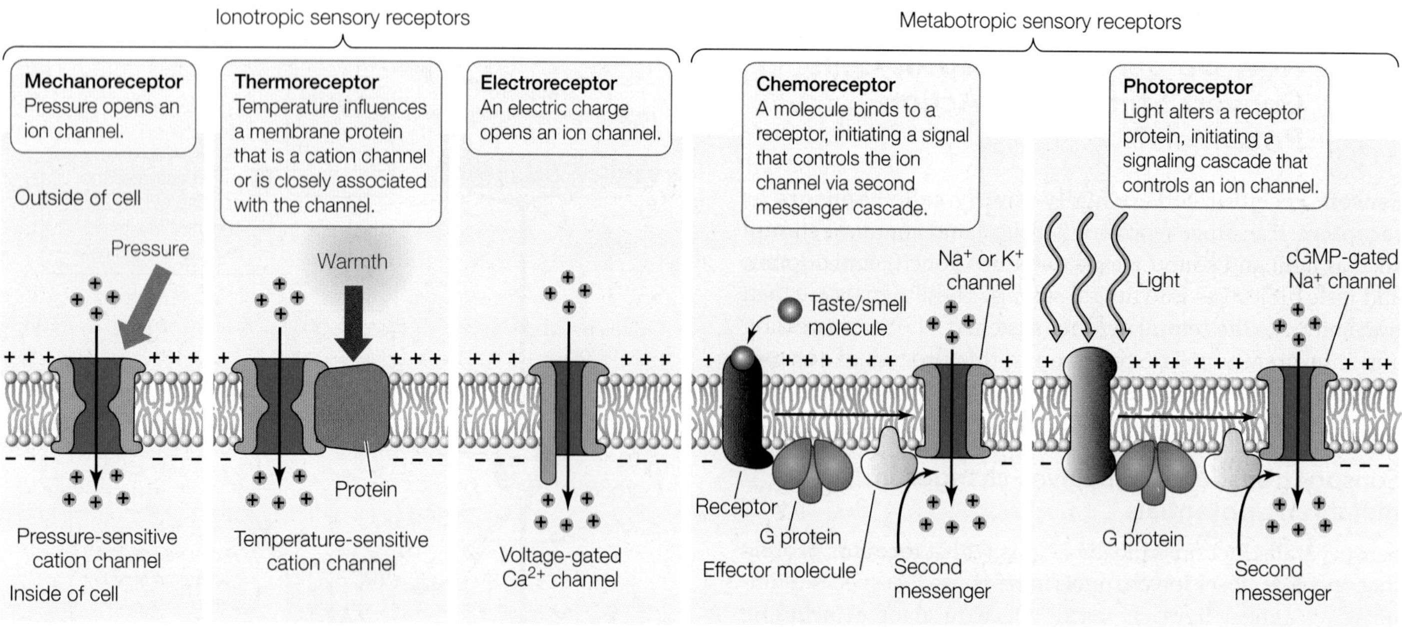


**46.2 Sensory Cell Membrane Receptor Proteins Respond to Stimuli** The receptor proteins in mechanoreceptors are ion channels. The activated receptor proteins of metabotropic chemoreceptors and photoreceptors initiate signal transduction cascades that eventually open or close ion channels.

of the brain, for example, are interpreted as light, whereas those that arrive in the olfactory bulb are perceived as smells.

A small patch of skin on your arm contains some sensory receptor cells that increase their firing rates when the skin is warmed and others that increase their activity when the skin is cooled. Other sensory cells in the same patch of skin respond to touch, irritants such as insect bites, and painful stimuli. These receptor cells transmit their messages through axons that enter the CNS at the spinal cord. The synapses made by those axons in the spinal cord and the subsequent pathways of transmission determine whether the stimulation of the skin on your arm is perceived as warmth, cold, touch, itch, or pain; even though the action potentials carried by all of these sensory axons look the same, the connectivity of each axon is specific for a given sensory modality. The *intensity* of a given sensation is coded by the frequency of the action potentials.

Some sensory cells transmit information about the body's internal conditions of which we may not be consciously aware. The brain continuously receives information about body temperature, the concentrations of carbon dioxide and oxygen in the blood, arterial pressure, muscle tension, and the position of the limbs. All of this information is important for homeostasis but does not necessarily result in conscious sensation.

Some sensory receptor cells are assembled with other types of cells into **sensory organs**, such as eyes, ears, and noses, that enhance the ability of sensory cells to collect, filter, and amplify stimuli. **Sensory systems** include the sensory cells, their associated organs, and the neural networks that process the information.

## Many receptors adapt to repeated stimulation

Some sensory cells give gradually diminishing responses to maintained or repeated stimulation. This phenomenon, known as **adaptation**, enables an animal to ignore background or unchanging conditions while remaining sensitive to changes and new information. (Note that this use of the term "adaptation" is different from its application in an evolutionary context.) When you get dressed, you feel each item of clothing touch your skin, but the sensation of clothes touching your skin is not constantly on your mind throughout the day. You are immediately aware of new sensations, however, such as your shoe coming untied or someone touching your back.

Some sensory cells adapt very little or very slowly; examples are some types of pain receptors and the mechanoreceptors that control balance. You do not want to ignore pain, which usually is signaling that something is wrong in your body, and to maintain equilibrium you must continuously know (albeit unconsciously) the tensions and forces on all of your joints and muscles.

### RECAP 46.1

Sensory receptor cells have receptor proteins that respond to specific stimuli from the external or internal environment by opening or closing ion channels, which results in the generation of action potentials in sensory neurons.

- What is the difference between ionotropic and metabotropic sensory receptor proteins? **See p. 947 and Figure 46.2**
- How are we able to perceive action potentials—all of which are essentially the same—as different sensations? **See pp. 947–948**

Now that we have a general view of how sensory systems code and process information, we will next discuss how sensory systems gather and filter stimuli, transduce specific stimuli into action potentials, and transmit action potentials to the central nervous system, which perceives them in many different forms.

## 46.2 How Do Sensory Systems Detect Chemical Stimuli?

A colony of corals responds to a small amount of meat extract in seawater by extending their bodies and tentacles in search of food; a solution of a single amino acid can stimulate this response. Conversely, a small amount of seawater in which corals were crushed will stimulate a defensive retraction of the coral polyps. Humans also react strongly to certain chemical stimuli. When we smell freshly baked bread we salivate and feel hungry, whereas when we smell rotting meat we feel nauseated.

All animals receive information about chemical stimuli through **chemoreceptors**, which are receptor proteins that bind to specific molecules—their **ligands**—and are responsible for smell and taste. Chemoreceptors are also responsible for monitoring some aspects of the internal environment, such as the level of carbon dioxide in the blood.

### Olfaction is the sense of smell

The sense of smell, **olfaction**, depends on chemoreceptors. In vertebrates the olfactory sensors are neurons embedded in a layer of epithelial tissue in the uppermost region of the nasal cavity. Axons from these neurons extend into the **olfactory bulb** (the olfactory integration area of the brain), whereas their dendrites end in olfactory cilia on the surface of the nasal epithelium. A protective layer of mucus covers the epithelium. Molecules from the environment must diffuse through this mucus to reach the receptor proteins on the olfactory cilia. When you have a cold, the amount of mucus in your nose increases, and the epithelium swells. With this in mind, study **Figure 46.3**, and you will easily understand why respiratory infections can cause you to lose your sense of smell.

An **odorant** is a molecule in the environment that binds to and activates an olfactory receptor protein on the cilia of **olfactory receptor neurons** (**ORNs**). Different olfactory receptor proteins bind specific subsets of odorant molecules. An odorant molecule binding to its receptor on an ORN activates a G protein. The G protein then activates an enzyme that causes an increase of a second messenger (cAMP in vertebrates) in the cytoplasm (see Figure 7.17). The second messenger binds to and opens cation channels in the ORN's plasma membrane, causing an influx of $Na^+$ into the ORN—which then depolarizes to threshold and fires action potentials. An interesting feature of ORNs is that they are continuously regenerated. Because they are embedded in nasal epithelium that, like other epithelial linings, is regularly shed, ORNs have to be constantly replaced.

The olfactory world has an enormous number of odors and a correspondingly large number of olfactory receptor proteins. In the 1990s Linda Buck and Richard Axel discovered a family of about 1,000 genes in mice (about 3 percent of the mouse genome) that code for olfactory receptor proteins. Each receptor protein that is expressed is found in a limited number of ORNs in the olfactory epithelium, and each ORN expresses just one receptor type. Using a combination of patch clamping (see Figure 45.8) and molecular techniques, the investigators were able to match specific gene products with the odorants they detect. For their discoveries of the molecular nature of the olfactory system, Buck and Axel received the Nobel Prize in 2004.

Olfactory sensitivity enables discrimination of many more odorants than there are olfactory receptors. An odorant molecule can be quite complex, and different regions of that molecule may bind to different receptor proteins. The next stage of processing olfactory information is in the olfactory bulb, where axons from ORNs expressing the same receptor protein cluster

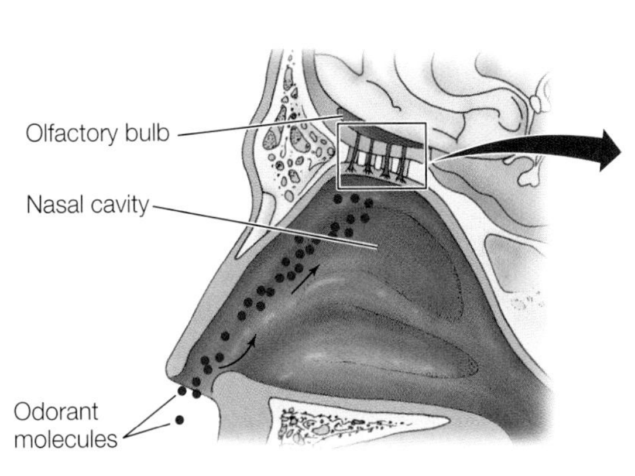


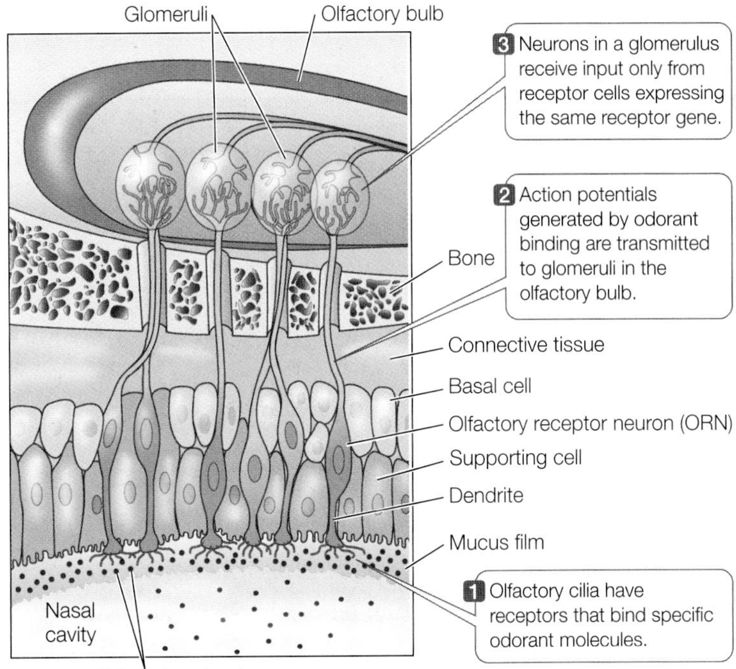


**46.3 Olfactory Receptors Communicate Directly with the Brain** The receptor cells of the human olfactory system are embedded in epithelial tissues lining the nasal cavity and send their axons to the olfactory bulb of the brain.

together on olfactory bulb neurons, forming structures called glomeruli (see Figure 46.3). A complex odorant molecule can activate a unique combination of glomeruli in the olfactory bulb, so an olfactory system with hundreds of different receptor proteins can discriminate an astronomically large number of smells. The more odorant molecules that bind to ORNs, the greater the frequency of action potentials and thus the greater the intensity of the perceived smell.

Humans have a sensitive olfactory system, but in comparison with most mammals we depend far more on vision than on olfaction. The nasal epithelium of a typical dog is 15 to 20 times larger than a human's and has about *1 billion* odorant receptors, compared with about 20 million in the average human. For some scents, the threshold sensitivity of the dog is 100 million times lower; a dog's nose reveals a huge amount of information not available to people.

## Some chemoreceptors detect pheromones

A specialized type of chemical signal used for communication among conspecifics (individuals of the same species) is called a **pheromone**. Individual animals secrete pheromone molecules into the environment, triggering behavioral responses in other individuals of the species. Pheromones may communicate alarm signals, mark food trails, or define territories, among many other uses. Their function in mating and mate attraction especially illustrates the remarkable sensitivity of chemosensory systems, and the silkworm moth *Bombyx mori* is an extensively studied example.

To attract a mate, the female silkworm moth releases a pheromone called bombykol from a gland at the tip of her abdomen (**Figure 46.4A**). The male silkworm moth has about 10,000 receptors for this molecule on each of his feathery antennae (**Figure 46.4B**). A single molecule of bombykol may be sufficient to generate action potentials in the antennal nerve, which transmits the signal to the male's CNS. The extreme sensitivity of the male bombykol receptors ensures that the sexual message sent by a female moth is likely to reach any male within a huge downwind area. When approximately 200 hairs per second are activated, a male orients upwind in search of the female. Because the rate of firing in the male's sensory nerves is proportional to the concentration of bombykol in the air, he can follow an airborne concentration gradient to "home in" on the signaling female.

## The vomeronasal organ contains chemoreceptors

The **vomeronasal organ** (**VNO**) is a small, paired tubular structure embedded in the nasal epithelium of amphibians, reptiles, and many mammals (although not humans). In mammals the VNO is located on the septum dividing the two nostrils (see Figure 53.3).

The vomeronasal organ has a pore that opens into the nasal cavity. When the animal sniffs, the VNO pulsates and draws a sample of nasal fluid over the chemoreceptors embedded in its walls. The information from these chemoreceptors goes to an accessory olfactory bulb in the brain, and from there to brain regions involved in sexual and other instinctive behaviors.

In snakes the VNO opens into the roof of the mouth cavity. Each time the snake's forked tongue darts in and out, the forks fit into the VNO openings and present the chemoreceptors located there with a sample of molecules from the surrounding air (see the chapter-opening photo). Thus the snake uses its tongue to smell its environment, not to taste it. Why doesn't the snake simply use the flow of air to and from its lungs, as we do, to smell the environment? In reptiles, air flows to and from

**46.4 Pheromones Can Communicate over Great Distances** Mating in silkworm moths of the genus *Bombyx* is coordinated by a pheromone called bombykol.

the lungs slowly (and can even stop entirely for long periods of time), but the tongue can dart in and out rapidly. It is a quick source of olfactory information.

Studies on mice have led to the hypothesis that the mammalian VNO is a specialized olfactory organ that detects pheromones. Lawrence Katz and his colleagues at Duke University recorded the activity of neurons in the mouse accessory olfactory bulb, which receives input from chemosensors in the VNO. These accessory olfactory neurons were activated when a mouse attached to recording electrodes sniffed another mouse placed in the same cage. However, the neurons fired differentially, depending on the gender and strain of the "intruder" mouse. Other studies on mouse behavior have supported a role for the VNO in gender identification and sexutal behaviors that are linked to pheromone perception (see Section 53.2).

## Gustation is the sense of taste

In humans and other vertebrates, the sense of taste, **gustation**, depends on clusters of chemoreceptors called **taste buds**. The taste buds of terrestrial vertebrates are confined to the mouth cavity, but some fish have taste buds in the skin that enhance their ability to sense their environment. Some fish living in murky water are very sensitive to small amounts of amino acids in the water and can find food without the use of vision. The duck-billed platypus, a prototherian mammal (see Figure 33.26), has similar talents as a result of taste buds on the sensitive skin of its bill.

The human tongue has 5,000 to 10,000 taste buds embedded in the epithelium. Most of them are found on the sides of the papillae (**Figure 46.5**). (Look at your tongue in a mirror—the papillae make it look fuzzy.) The outer surface of a taste bud has a pore that exposes the tips of the sensory receptor cells. Microvilli (tiny hairlike projections) increase the surface area of these cells where their tips converge at the pore. These chemosensory cells generate action potentials and release neurotransmitter at their bases, where they form synapses with sensory neurons that convey the signals into the central nervous system.

The tongue does a lot of hard work, so its epithelium, along with cells of its taste buds, are shed and replaced at a rapid rate. Individual taste bud cells last about 10 days before they are replaced, but the sensory neurons associated with them live on, constantly forming new synapses as new taste buds form.

Humans perceive five taste classes: sweet, salty, sour, bitter, and umami. However, taste buds can distinguish among a variety of sweet-tasting molecules and a variety of bitter-tasting molecules, and small families of genes for receptor proteins responding to sweet and bitter tastes have been discovered. Umami is a savory, meaty taste that originates from receptors for amino acids, including monosodium glutamate (MSG), a commonly used flavor enhancer. In addition, spicy/hot tastes involve the activation of heat sensors, and minty tastes activate cold sensors. The full complexity of the chemosensitivity that enables us to enjoy the subtle flavors of food comes from the combined activation of gustatory and olfactory receptors, which is why you may lose your sense of taste when you have a cold.

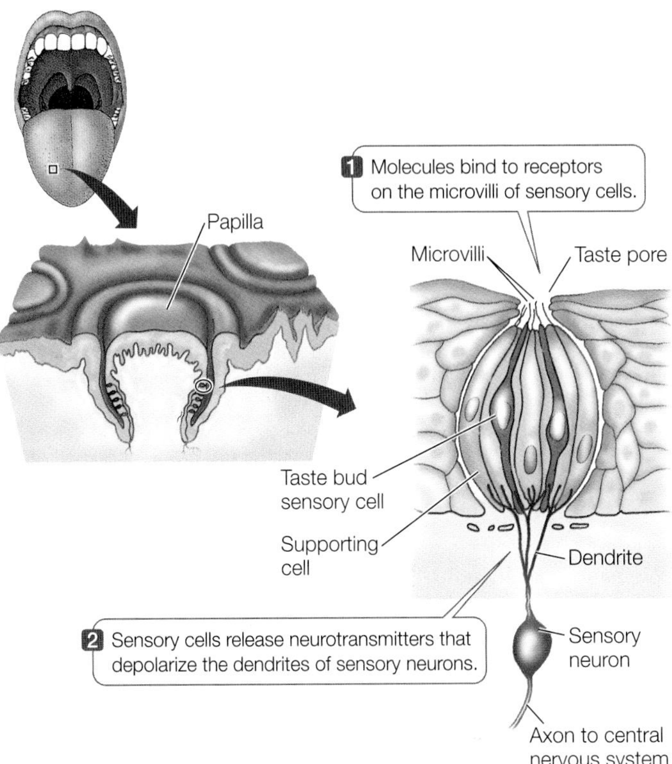


**46.5 Taste Buds Are Clusters of Sensory Cells** A human tongue has as many as 10,000 taste buds, most of which are found on the papillae.

Gustation begins with receptor proteins in the membranes of the microvilli of the taste bud sensory cells (see Figure 46.5). The nature of these proteins and the mechanisms by which they depolarize sensory receptor cells differ for the different tastes. Saltiness receptor proteins are ionotropic and allow $Na^+$ to diffuse into taste bud sensory cells through open $Na^+$ channels, depolarizing the sensory cell. Sourness receptors probably are ionotropic as well, and their depolarization is also due to a direct effect of $H^+$ ions on $Na^+$ channels. In contrast, sweet, bitter, and umami taste reception is metabotropic, involving families of G protein-coupled receptor proteins. The bitter taste may have evolved as a protective mechanism, since it enables animals to detect toxic plant compounds such as quinine, caffeine, and nicotine. Because many such toxic compounds occur in plants, where they have evolved in response to herbivorous predators, a variety of receptors is essential. Similarly, a large number of molecules in food could indicate nutritional value, so a variety of receptors is of value. The diversity of sweet receptors helps explain why it has been possible to invent many chemically distinct artificial sweeteners.

Regardless of the mechanism of taste transduction by the sensory cells of the taste buds, all these cells release neurotransmitter onto sensory neurons. These neurons then generate action potentials that are conducted to the CNS, where they are interpreted as specific taste sensations.

RECAP 46.2

All animals receive information about chemical stimuli through chemoreceptors, which have diverse structures and bind to a tremendous variety of stimulus molecules. Chemoreceptors are the basis of the sensations of olfaction and gustation and the reception of pheromones. They also monitor some aspects of an animal's internal environment.

- Why are we able to distinguish so many different smells? Why do some animals experience more or different odors than others? **See pp. 949–950 and Figure 46.3**
- What is distinctive about pheromones as opposed to odorant molecules? **See p. 950**
- Describe how different substances in food are transduced into action potentials in taste buds. **See p. 951 and Figure 46.5**

We have now seen how chemoreceptors give rise to the sensations of smell and taste, and how some animals use chemoreception to communicate with others of their species. Next we will describe the sensory cells that respond to mechanical forces, including the vibrations we perceive as sound.

## 46.3 How Do Sensory Systems Detect Mechanical Forces?

**Mechanoreceptors** respond to mechanical (physical) forces. Physical distortion of a mechanoreceptor's plasma membrane causes ion channels to open, altering the membrane potential of the cell to create a graded receptor potential, which in turn leads to either the release of neurotransmitter or the generation of action potentials. The rate of action potentials tells the CNS the strength of the stimulus to the mechanoreceptor. A considerable diversity of mechanosensory cells involved in many sensory systems has evolved. The functions of these cells range from interpreting skin sensations to sensing blood pressure to hearing and maintaining balance.

### Many different cells respond to touch and pressure

Human skin (and that of other mammals) is packed with diverse mechanoreceptors that generate varied sensations (**Figure 46.6**). The most important tactile receptors, found in both hairy and nonhairy skin, are **Merkel's discs**, which adapt rather slowly and provide continuous information about anything touching the skin. **Meissner's corpuscles**, found primarily in nonhairy skin, are very sensitive but adapt rapidly; they provide information about *changes* in things touching the skin. The rapid adaptation of Meissner's corpuscles is why you roll a small object between your fingers (rather than holding it still) to discern its shape and texture: as you roll it, the object continually stimulates Meissner's corpuscles.

Deeper in the skin, **Ruffini endings** adapt slowly and are good at providing information about vibrating stimuli of low frequencies, while **Pacinian corpuscles**, which adapt rapidly, provide information about vibrating stimuli of higher frequencies. Even deeper in the skin, the dendrites of sensory neurons wrap around hair follicles. When the surface hairs are displaced, those neurons are stimulated.

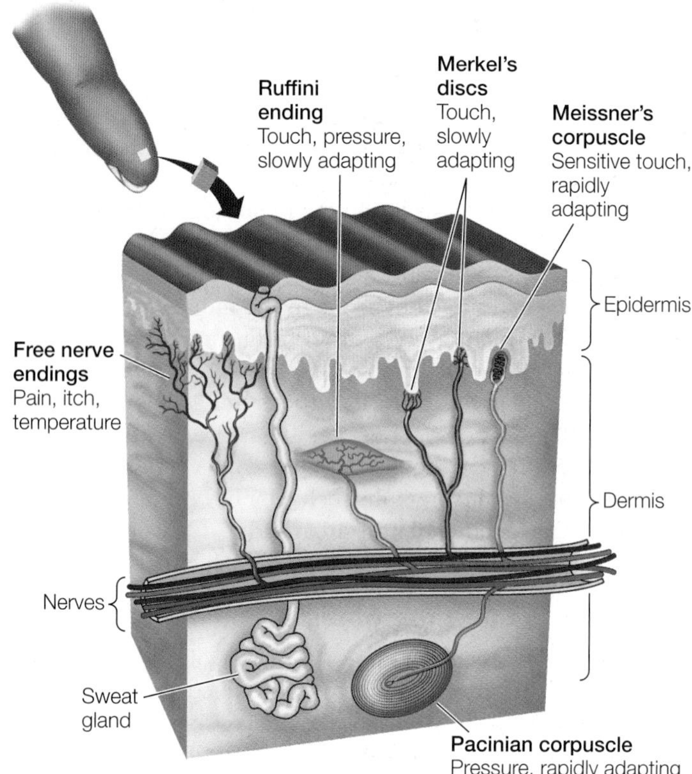


**46.6 The Skin Feels Many Sensations** Even a very small patch of skin contains a variety of sensory cells, making the skin a multimodal receptor that can sense temperature, pressure, texture, pain, touch, and itch.

The density of tactile mechanoreceptor cells varies across the body's surface. By touching the skin with two toothpicks simultaneously, you can determine how far apart two stimuli have to be before a person can tell whether the sensations are produced by one toothpick or by two. On a person's back, for example, stimuli have to be relatively far apart before they are perceived as two discrete stimuli. But when this same "two-point spatial discrimination test" is applied to the lips or fingertips, a person can identify two stimuli as separate even when they are quite close together, meaning that receptor density is much greater in these regions.

### Mechanoreceptors are also found in muscles, tendons, and ligaments

An animal receives information from mechanoreceptors about the position of its limbs and the stresses on its muscles and joints. These mechanoreceptors supply information continuously to the CNS, and this information is essential for postural control and the coordination of movements.

The mechanoreceptors in skeletal muscle are the **muscle spindles**. These stretch receptors are modified muscle cells embedded in connective tissue inside muscles and innervated by sensory neurons (**Figure 46.7A**). When a muscle is stretched, the muscle spindles are stretched as well, signaling the spindle neurons transmit action potentials to the central nervous system. Figure 46.1 showed how crayfish stretch receptors transduce physical force into action potentials; the actions of muscle spindles are similar. The CNS uses the information from

**46.7 Stretch Receptors** Stretch receptors provide information about the stresses on muscles and joints in an animal's limbs. (A) Signals from muscle spindles to the CNS initiate muscle contraction. (B) Golgi tendon organs in tendons and ligaments inhibit a contraction that becomes too forceful, triggering a reduction in muscle tension and protecting the muscle from tearing.

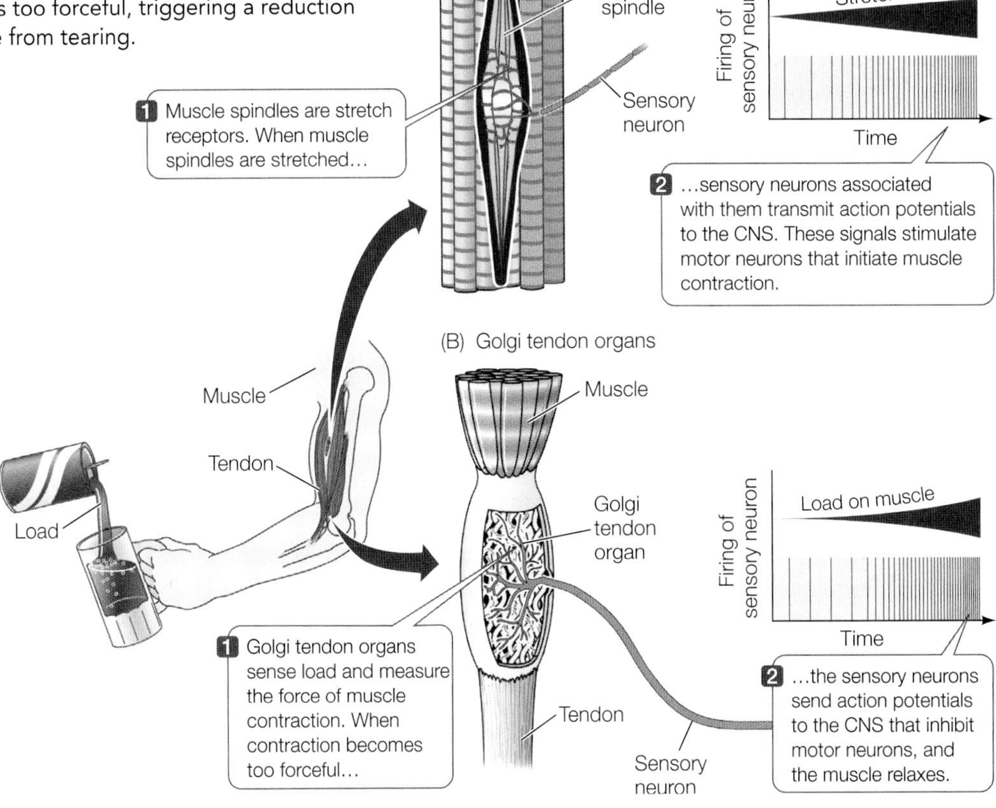


muscle spindles to adjust the strength of the muscle contraction so that it matches the load put on the muscle; thus a person can hold a mug steady while it is being filled.

Another type of mechanoreceptor, the **Golgi tendon organ**, is found in tendons and ligaments and provides information about the force generated by a contracting muscle (**Figure 46.7B**). When a contraction becomes too forceful, action potentials from the Golgi tendon organ inhibit the spinal cord motor neurons innervating that muscle, causing the muscle to relax and protecting it from tearing. (You may recall a cell organelle called the Golgi apparatus. What these two very different structures have in common is their discovery in the late nineteenth century by the Italian anatomist Camillo Golgi.)

## Hair cells are mechanoreceptors of the auditory and vestibular systems

**Hair cells** are the mechanoreceptors for the vertebrate auditory system (sound-perceiving) and vestibular (equilibrium-maintaining) systems. Both of these systems are housed in the complex structures of the vertebrate ear. **Stereocilia**—fingerlike extensions of the cell membrane stiffened by cross-linked actin filaments—project from the surface of each hair cell like a set of organ pipes (**Figure 46.8A**). Stereocilia bend in response to waves of pressure; bending of the stereocilia in one direction depolarizes the hair cell, and bending in the other direction hyperpolarizes it (**Figure 46.8B**).

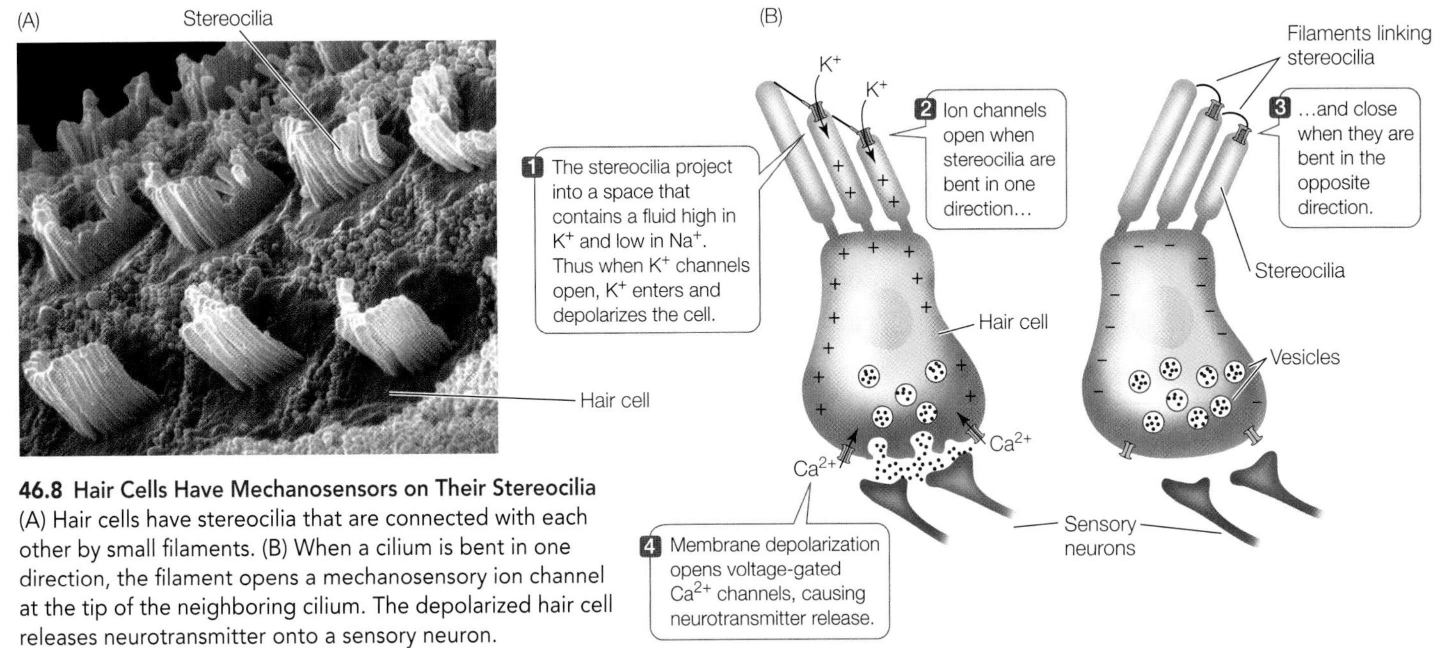


**46.8 Hair Cells Have Mechanosensors on Their Stereocilia** (A) Hair cells have stereocilia that are connected with each other by small filaments. (B) When a cilium is bent in one direction, the filament opens a mechanosensory ion channel at the tip of the neighboring cilium. The depolarized hair cell releases neurotransmitter onto a sensory neuron.

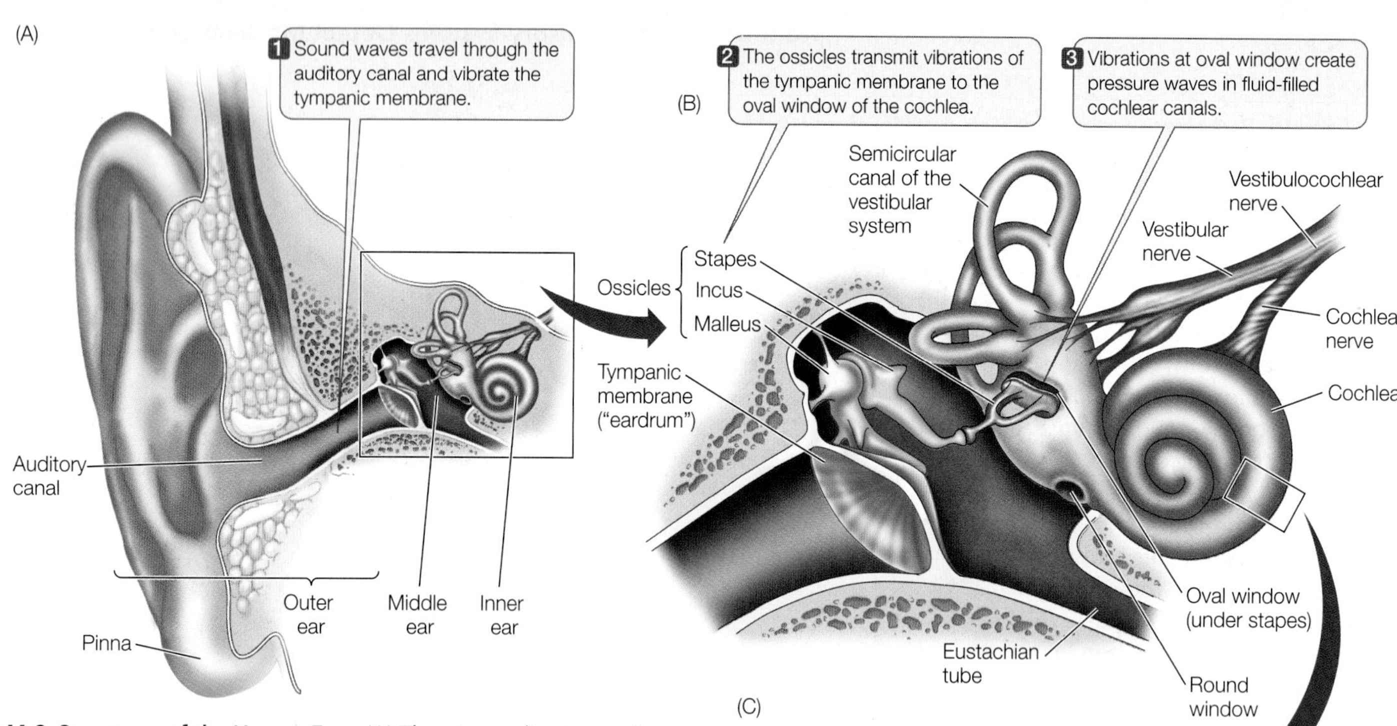


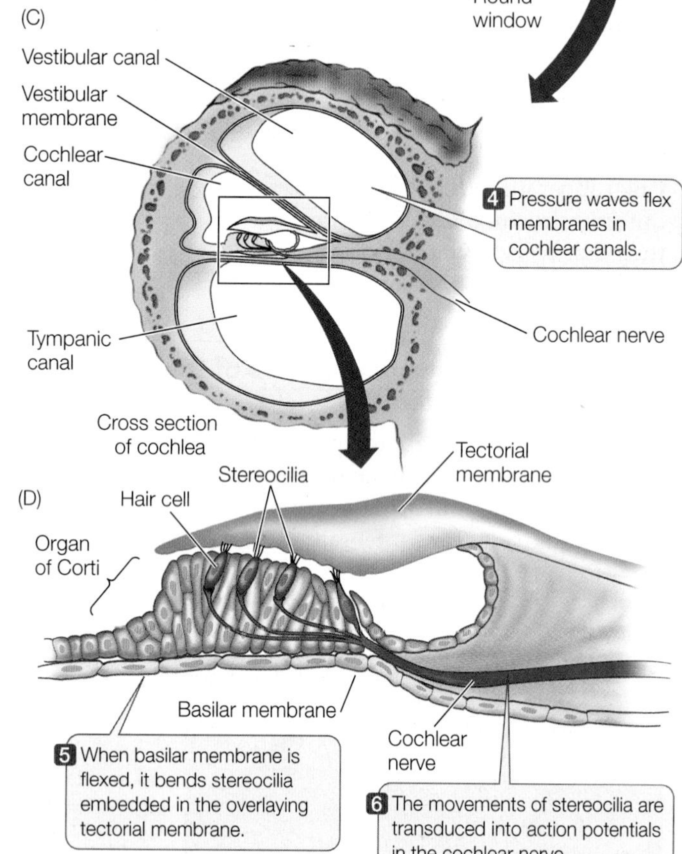


**46.9 Structures of the Human Ear** (A) The pinnae direct sound waves down the auditory canal to impinge on the tympanic membrane. The tympanic membrane mechanically transmits these pressure waves into movements of the ossicles in the middle ear. (B) The ossicles transmit their movement into pressure waves in the fluid of the cochlea at the oval window. (C) The cochlea is divided into fluid-filled chambers; pressure waves from the ossicles cause the membranes between the chambers to flex. (D) Flexing of the basilar membrane bends stereocilia on hair cells in the organ of Corti.

**Go to Activity 46.1 Structures of the Human Ear**
**Life10e.com/ac46.1**

Measurements with microelectrodes have shown that the bending of stereocilia creates local electric currents near their tips, indicating that ion channels near the tips must be opening or closing. Electron microscope images reveal minute filaments that connect the tip of each stereocilium to its taller neighbor. It is hypothesized that these filaments are fine molecular attachments to the ion channels, and that they act like springs that open the channels. If the taller neighboring stereocilium is bent away, the spring tightens and the ion channel is opened. If the taller neighbor bends toward its shorter neighbor, the spring is relaxed and the channel closes (see Figure 46.8B).

## Auditory systems use hair cells to sense sound waves

The stimuli that animals perceive as sounds are pressure waves. Auditory systems use mechanoreceptors to convert pressure waves into receptor potentials. Auditory systems include special structures that gather sound waves, direct them to the sensory organ, and amplify their effect on the mechanoreceptors. A good example of an auditory system is the human ear, which can be divided into three major areas: the outer, middle, and inner ear (**Figure 46.9A**).

**OUTER EAR** The outer ear consists of the pinnae (singular, pinna) and the auditory canal. The pinnae collect sound waves and direct them into the auditory canals; watch a dog change the orientation of its ears to focus on a particular sound to get the idea of the role pinnae play in hearing. Covering the end of the auditory canal is the **tympanic membrane** (commonly called the eardrum), which vibrates in response to pressure waves traveling down the canal, thus converting the pressure waves to physical forces in the middle ear.

**MIDDLE EAR** On the other side of the tympanic membrane is the middle ear, an air-filled cavity connected to the throat at the back of the mouth through the **eustachian tube** (also called the pharyngotympanic tube). Because the eustachian tube is also filled with air, pressure equilibrates between the middle ear and the environment. When you have a cold, the tube can become blocked by mucus or tissue swelling and you have difficulty equilibrating the pressure in the middle ear with the outside air pressure (as you have to do when changing altitude in an airplane).

The middle ear contains the **ossicles**, three delicate bones individually named the malleus ("hammer"), incus ("anvil"), and stapes ("stirrup") (**Figure 46.9B**). The ossicles transmit the vibrations of the tympanic membrane to another flexible membrane, the **oval window**. The ossicles act as a lever (like a hammer pulling out a nail), translating a large movement of the tympanic membrane into a smaller movement of the oval window, but a movement of greater force. Because the oval window is much smaller than the tympanic membrane, the pressure the stapes transmits to the oval window is more than 20 times greater than the pressure exerted by a sound wave on the tympanic membrane.

Behind the oval window lies the fluid-filled inner ear. Movements of the oval window impart pressure changes to that enclosed fluid. These pressure waves are transduced into action potentials.

**INNER EAR** The inner ear is a bony structure consisting of two sets of canals. One is the organ of balance, the **vestibular system**, and the other is the organ of hearing, the **cochlea**. The cochlea ("snail" or "spiral shell") is a long, tapered, coiled structure. A cross section of the cochlea reveals that it is composed of three parallel canals separated by two membranes, the vestibular membrane and the basilar membrane (**Figure 46.9C**). Sitting on the basilar membrane is the **organ of Corti**, which transduces pressure waves into action potentials. The organ of Corti contains hair cells with stereocilia (see Figure 46.8). The tips of the hair cells are embedded in a gelatinous overhanging shelf called the tectorial membrane (**Figure 46.9D**).

Stereocilia are not motile, but because their tips are attached to the more rigid tectorial membrane, stereocilia bend when the basilar membrane flexes. The response of the hair cell is a graded membrane potential. The hair cells do not fire action potentials, but the changes in their membrane potential alter the rate at which the hair cells release neurotransmitter onto the sensory neurons whose axons make up the auditory nerve and transmit action potentials to the brain.

The vestibular and tympanic canals are separate until they reach the distal end of the cochlea (the end farthest from the oval window), where they join; thus they form one continuous canal that turns back on itself (see Figure 46.10). Just as the oval window is a flexible membrane at the beginning of the vestibular canal, the **round window** is a flexible membrane at the end of the tympanic canal. When the oval window vibrates, the waves of fluid pressure create traveling waves, or flexions, in the basilar membrane.

## Flexion of the basilar membrane is perceived as sound

What causes the basilar membrane to flex, and how does this mechanism distinguish sounds of different frequencies? Air is highly compressible but fluids are not; therefore a pressure wave can travel through air without displacing much air, whereas a pressure wave in fluid displaces that fluid. When the stapes pushes on the oval window, the fluid in the vestibular canal is displaced. If the movement of the oval window occurs slowly, the cochlear fluid pressure wave travels down the vestibular canal, around the bend, and back through the tympanic canal (**Figure 46.10**). At the end of the tympanic canal, the displacement pressure is dissipated by the outward bulging of the round window.

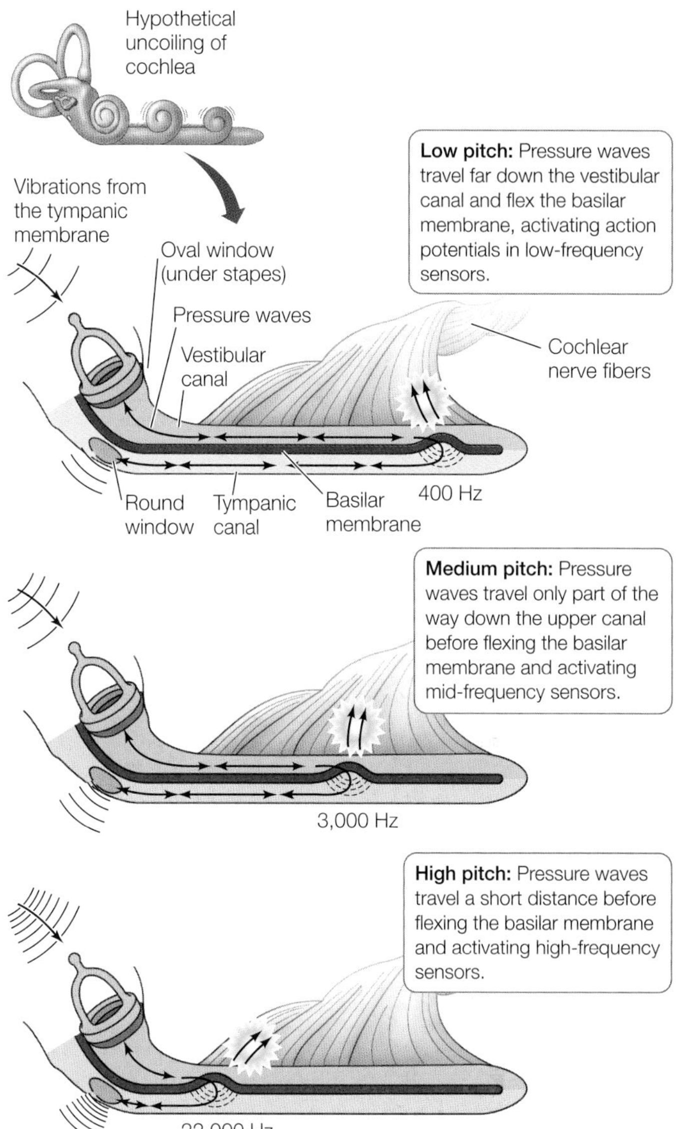


**46.10 Sensing Pressure Waves in the Inner Ear** Pressure waves of different frequencies flex the basilar membrane at different locations. Information about sound frequency is specified by which hair cells are activated. For simplicity, this representation illustrates the cochlea as uncoiled, and leaves out the middle ear.

**Go to Animated Tutorial 46.1**
**Sound Transduction in the Human Ear**
Life10e.com/at46.1

The basilar membrane is not uniform—it is thicker and stiffer at its base and wider and thinner at its apical end.

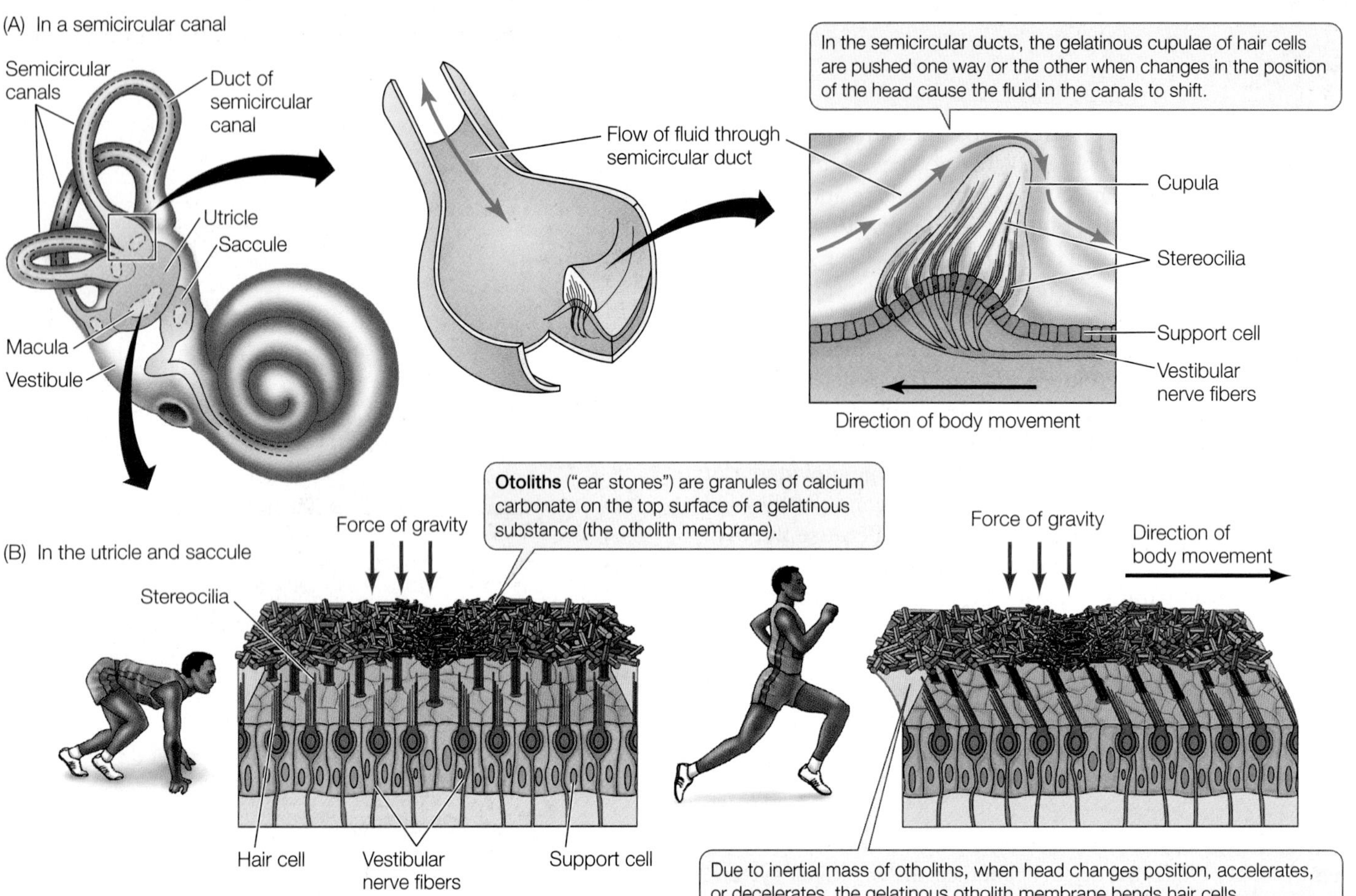


**46.11 Organs of Equilibrium** The vestibular system consists of bony chambers and fluid-filled canals. (A) Each semicircular duct has a cupula containing stereocilia. When fluid moves against the cupula, the stereocilia bend. (B) In the saccule and utricle, stereocilia are bent by gravitational forces on the otoliths.

Pressure waves in the cochlear fluid have different frequencies and set up different patterns of traveling waves. High-frequency waves cause maximal flexion at the basal end of the basilar membrane, whereas low-frequency pressure waves result in maximal flexion at the apical end (see Figure 46.10). Thus different pitches of sound flex the basilar membrane at different locations and activate different sets of hair cells. Action potentials stimulated by the mechanoreceptors at different positions along the organ of Corti travel along the cochlear nerve and are transmitted to different regions of the brain's auditory cortex by the vestibulocochlear nerve.

## Various types of damage can result in hearing loss

There are two general types of acquired hearing loss, or deafness. Conduction deafness is caused by the loss of function of the tympanic membrane and/or the ossicles of the middle ear. Repeated infections of the middle ear can cause scarring of the tympanic membrane and stiffening of the connections between the ossicles. The consequence is less efficient conduction of sound waves from the tympanic membrane to the oval window. With increasing age, the ossicles inevitably stiffen, resulting in a gradual loss of the ability to hear high-frequency sounds.

Nerve deafness is caused by damage to the inner ear or the auditory pathways. A common cause of nerve deafness is damage to the hair cells of the delicate organ of Corti by exposure to loud sounds such as jet engines, pneumatic drills, or highly amplified music. Consistent exposure to sounds above 85 decibels can damage hearing; this damage is cumulative and irreversible. Even using earphones can put you at risk for hearing loss because they generate high-pressure sound waves close to the tympanic membrane. Personal stereo earphones can reach 120 decibels, and people commonly use them at 100 decibels (equivalent to being at a rock concert).

## The vestibular system uses hair cells to detect forces of gravity and momentum

Hair cells in the vestibular system of the inner ear detect the position and movement of the head—information that is essential for maintaining equilibrium (balance). Information from the vestibular system is also crucial for the control of eye movements. When you look at something, you can move your head while staying focused on the object because of your vestibulo-ocular reflex.

In the mammalian inner ear, the vestibular system consists of three bony **semicircular canals** and a bony chamber called the **vestibule**. Within each canal is a membranous semicircular duct, and within the vestibule are the membranous saccule and the utricle. The ducts and the saccule and utricle are filled with fluid. In the semicircular ducts, the fluid shifts when the head changes position (**Figure 46.11A**). Since the three semicircular canals have different orientations, the fluid in their ducts responds differentially to the direction of movement. Projecting into the base of each duct is a **cupula**, a gelatinous swelling enclosing a cluster of hair cell stereocilia. When the shifting fluid pushes on the cupula, it bends the stereocilia and causes a graded potential in their hair cell plasma membranes.

The stereocilia in the saccule and utricle are bent in a different way. These stereocilia are embedded gelatinous membranes that contain otoliths ("ear stones") that are crystals of calcium carbonate. When the head changes position or when it accelerates or decelerates, gravitational forces are exerted on the otoliths and the stereocilia bend (**Figure 46.11B**).

As in the cochlea, the hair cells of the vestibular system do not fire action potentials, but they release neurotransmitter at synapses with sensory neurons, which in turn fire action potentials.

**Go to Animated Tutorial 46.2**
**Mechanoreceptors**
Life10e.com/at46.2

RECAP  46.3

Sensations that derive from mechanoreceptors include touch, tickle, pressure, joint position, muscle load, hearing, and equilibrium.

- Describe some of the different mechanoreceptors in the skin and their properties. **See p. 952 and Figure 46.6**
- How do hair cells transduce force into action potentials? **See pp. 953–954 and Figure 46.8**
- How do different frequencies of sound result in action potentials being fired in different acoustic neurons? **See pp. 954–955 and Figures 46.9 and 46.10**

We turn next to another example of metabotropic sensory reception, one in which light is the stimulus. We will see how light energy is converted into action potentials that in higher vertebrates are perceived as vision, perhaps the most elaborate of the senses.

## 46.4 How Do Sensory Systems Detect Light?

Sensitivity to light—**photosensitivity**—confers on the simplest animals the ability to orient to the sun and sky, and it gives more complex animals rapid and extremely detailed information about objects in their environment. Photosensitivity is ubiquitous in the animal kingdom, and the molecular basis for that sensitivity is a family of visual that has been evolutionarily conserved.

In this section we will learn how a visual pigment molecule responds when stimulated by light energy and how that response is transduced into neural signals. We will also examine the structures of eyes, the organs that gather light energy and focus it onto **photoreceptor cells**, the metabotropic sensory receptors that transform light energy into action potentials, and the routes those impulses travel to the brain.

### Rhodopsin is a vertebrate visual pigment

Photosensitivity depends on the ability of visual pigments to absorb photons of light and to undergo a change in conformation. One such pigment is **rhodopsin**, a well-studied vertebrate visual pigment. A rhodopsin molecule consists of the protein **opsin** (which by itself is not photosensitive) and 11-*cis*-retinal, a nonprotein, light-absorbing functional group cradled in the center of the opsin protein and covalently bound to it (**Figure 46.12**). The entire rhodopsin molecule sits within the plasma membrane of a photoreceptor cell, such as the rod cells of humans (see p. 960).

When 11-*cis*-retinal absorbs a photon of light energy, it changes into a different isomer of retinal, called all-*trans*-retinal. This change puts a strain on the bonds between retinal and opsin, changing the conformation of opsin and thus signaling the detection of light. In vertebrate eyes, the retinal and the opsin eventually separate from each other (a process called bleaching), which causes the molecule to lose its photosensitivity. A series of enzymatic reactions returns the all-*trans*-retinal to the 11-*cis* isomer, which then recombines with opsin so that it once again becomes the photosensitive pigment rhodopsin.

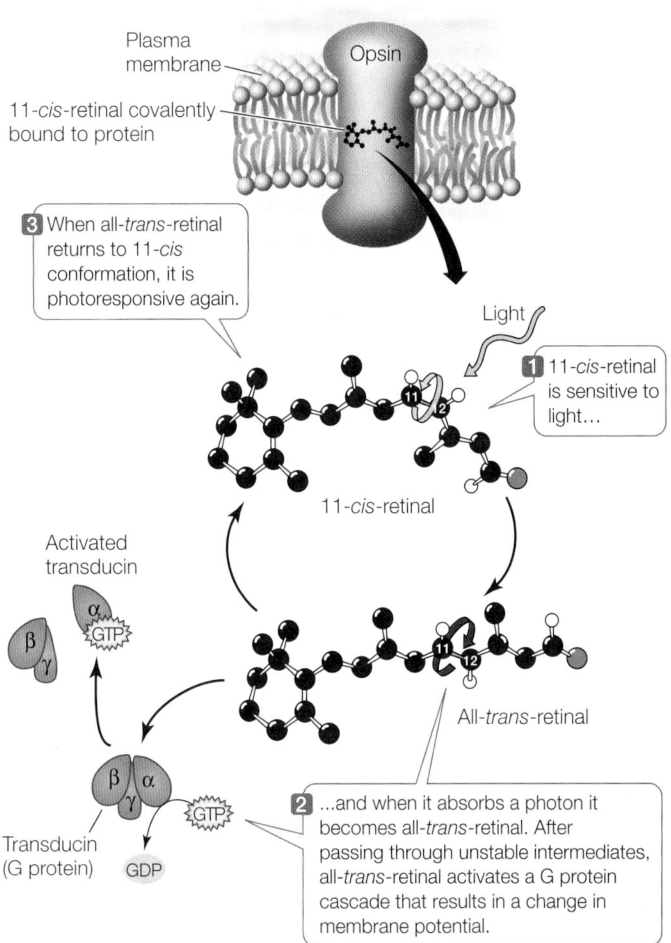


**46.12 Light Changes the Conformation of Rhodopsin** The light-absorbing molecule 11-*cis*-retinal bonds with the protein opsin to form the vertibrate visual pigment rhodopsin.

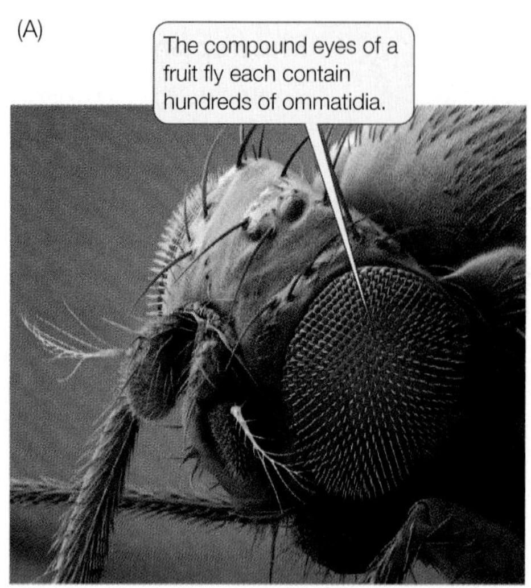


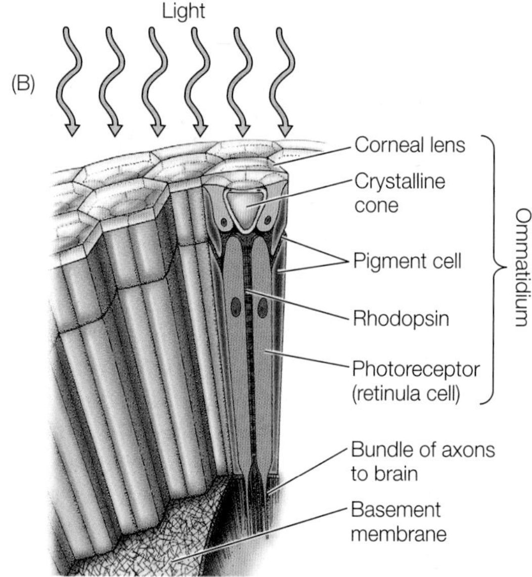


**46.13 Ommatidia: The Functional Units of Insect Eyes** (A) The micrograph shows the compound eye of a fruit fly. (B) The rhodopsin-containing retinula cells are the photoreceptors in ommatidia.

How does the conformational change of rhodopsin transduce light into a cellular response? After retinal is converted from the 11-*cis* to the all-*trans* form, its interactions with opsin pass through several unstable intermediate stages. One of these stages triggers a cascade of reactions involving a G protein signaling mechanism that results in the alteration of membrane potential that is the photoreceptor cell's response to light.

**Go to Animated Tutorial 46.3**
**Photosensitivity**
Life10e.com/at46.3

## Invertebrates have a variety of visual systems

Photoreceptors and visual pigments are incorporated into a variety of visual systems, from simple to complex. Flatworms obtain directional information about light from photoreceptor cells that are organized into **eye cups**. The eye cups are paired, bilateral structures, each partly shielded from light by a layer of pigmented cells lining the cup. The photoreceptors on the two sides of the animal are unequally stimulated unless the animal is facing directly toward or away from a light source. The flatworm generally uses directional information from the eye cups to move away from light.

Arthropods have **compound eyes** that provide them with information about patterns or images in the environment. These eyes are called compound because each eye consists of many optical units called **ommatidia** (singular *ommatidium*), each with its own narrow-angle lens (**Figure 46.13**). In contrast, a vertebrate eye consists of just one optical unit with a wide-angle lens. The number of ommatidia in a compound eye varies from only a few in some ants to 800 in fruit flies and to 30,000 in some dragonflies.

Each ommatidium has a lens that directs light onto photoreceptor cells. Flies, for example, have eight elongated photoreceptors in each ommatidium. The inner borders of the photoreceptors are covered with microvilli that contain rhodopsin and trap light. Axons from the photoreceptors send the light information to the nervous system. Since each ommatidium of a compound eye is directed at a slightly different part of the visual world, only a low-resolution (pixillated) image can be communicated from the compound eye to the CNS.

## Image-forming eyes evolved independently in vertebrates and cephalopods

Both vertebrates and cephalopod mollusks have eyes with exceptional abilities to form detailed images of the visual world. Like cameras, both of these eye types focus inverted images on an internal surface that is sensitive to light. Considering that they evolved completely independently of each other, their degree of similarity is remarkable (**Figure 46.14**).

The vertebrate eye (see Figure 46.14A) is a spherical, fluid-filled structure bounded by a tough connective tissue layer called the **sclera**. At the front of the eye, the sclera forms the transparent **cornea**, through which light passes to enter the eye. Just inside the cornea is the pigmented **iris**, which gives the eye its color. The iris controls the amount of light that reaches the photoreceptor cells at the back of the eye, just as the diaphragm of a camera controls the amount of light reaching the film. The central opening of the iris is the **pupil**. The iris is under neural control. In bright light, the iris constricts and the pupil is very small. As light levels fall, the iris relaxes and the pupil enlarges.

Behind the iris is the crystalline protein **lens**, which makes fine adjustments in the focus of images falling on the photosensitive layer—the **retina**—at the back of the eye. The cornea and the fluids within the eye bend light rays passing through them so that they are focused on the retina. The lens makes fine adjustments to the focus and allows the eye to accommodate—that is, to focus on objects at various locations in the near visual field. To focus a camera on objects close at hand, you adjust the distance between the lens and the internal surface sensitive to light. Fishes, amphibians, and non-avian reptiles accommodate in a similar manner, moving the lenses of their eyes closer to or farther from their retinas. Mammals and birds use a different method; they alter the shape of the lens.

The mammalian lens is contained in a connective tissue sheath that tends to keep it in a spherical shape, but it is attached to suspensory ligaments that pull it into a flatter shape. Circular ciliary muscles counteract the pull of the suspensory ligaments, permitting the lens to round up. When the ciliary muscles are at rest, the flatter lens has the correct optical properties to focus distant images on the retina. Contracting the

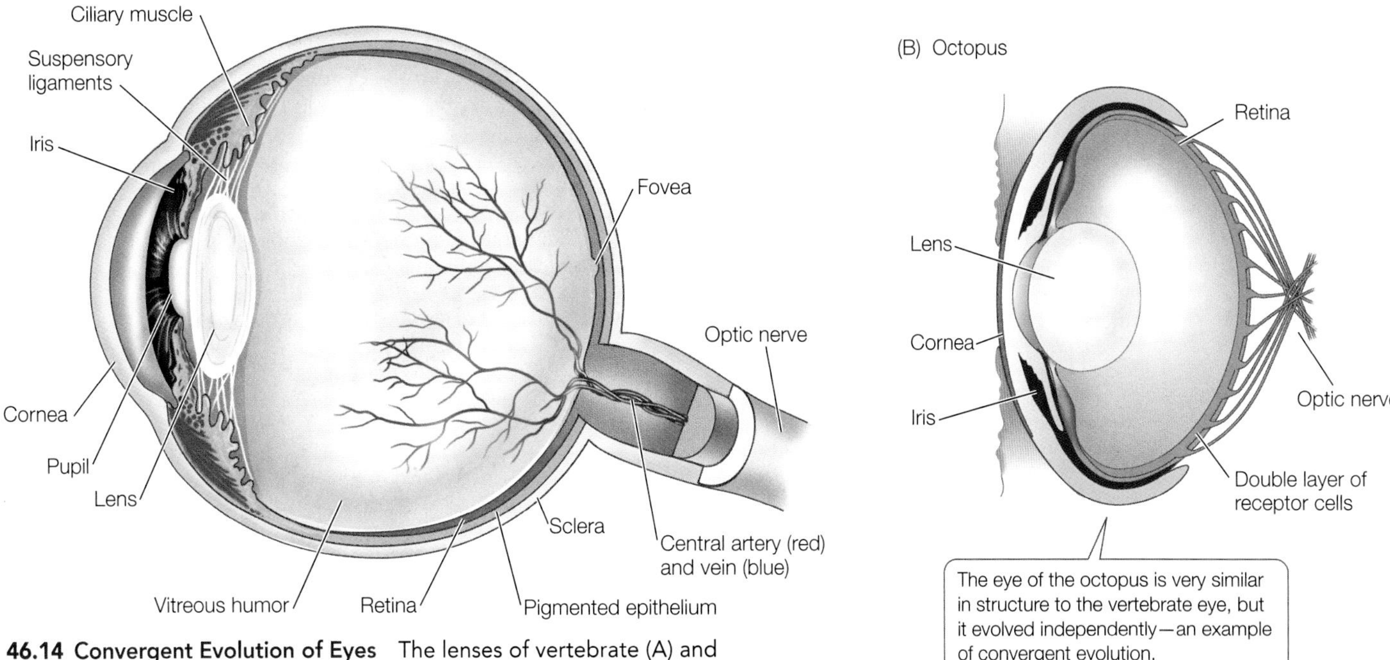


**46.14 Convergent Evolution of Eyes** The lenses of vertebrate (A) and cephalopod (B) eyes focus images on layers of photoreceptor cells.

ciliary muscles rounds up the lens, changing its light-bending properties to bring close images into focus (**Figure 46.15**).

Lenses become less elastic with age, so we lose the ability to focus on objects close at hand without the help of corrective lenses. Most people over the age of 45 need the assistance of reading glasses or bifocal lenses.

**Go to Activity 46.2 Structure of the Human Eye Life10e.com/ac46.2**

## The vertebrate retina receives and processes visual information

During embryonic development, neural tissue grows out from the brain to form the retina. In addition to a layer of photoreceptor cells, the retina includes four additional types of cells that process visual information from the photoreceptors (see Figure 46.20). Light must pass through all the layers of retinal cells before being captured by photosensors. In humans and other day-active animals, the light that is not captured is absorbed by a black-pigmented epithelial tissue layer behind the retina. In contrast, nocturnal animals such as deer and raccoons have an iridescent reflective layer behind their retinas, which maximizes the capture of photons by reflecting them back onto the photoreceptors. This is why a deer in the headlights appears to have bright white eyes. Because humans do not have a white reflective layer in the retina, photographs taken indoors often have a "red-eye effect," caused by the flash of light from the camera being reflected by the abundant blood vessels in the retina.

The pigmented epithelium also plays a role in the renewal of the photoreceptors. The photoreceptor cells are always shedding discs from their distal ends as new ones are being generated by the inner segments of those cells. The pigmented epithelial cells phagocytose the shed discs. Each outer segment is totally renewed about every two weeks.

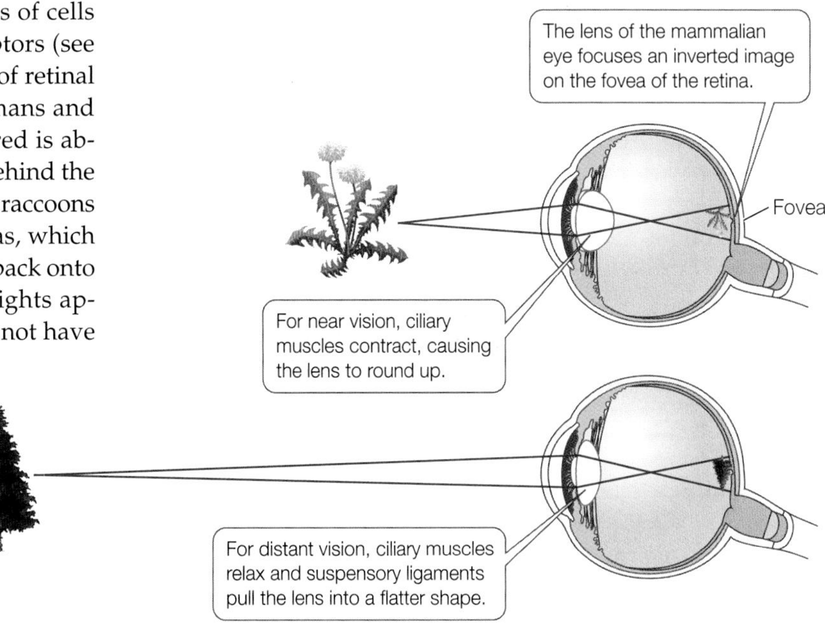


**46.15 Staying in Focus** Mammals and birds focus their eyes by changing the shape of the lens depending on the eye's distance from the object of focus.

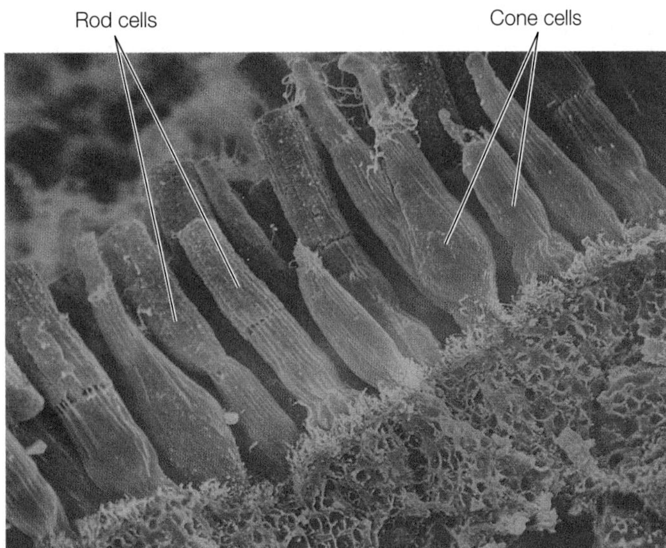


**46.16 Rods and Cones** This scanning electron micrograph of photoreceptors in the retina of a mud puppy (an amphibian) shows cylindrical rods and tapered cones.

## Rod and cone cells are the photoreceptors of the vertebrate retina

The photoreceptors of the vertebrate retina are two types of modified neurons called **rod cells** and **cone cells** based on their shapes (**Figure 46.16**). Rod cells are highly light-sensitive and perceive shades of gray in dim light. Cone cells function at high light levels and are responsible for high-acuity color vision. The human retina has about 5 million cones and 100 million rods, but the density of each varies across the retina.

ROD CELLS Rod cells do not produce action potentials, but they release neurotransmitter from their bases, where they form synapses with the next neurons in the visual pathway. Each rod cell has an outer segment, an inner segment, and a synaptic terminal (**Figure 46.17**). The highly specialized outer segment contains a stack of discs made up of plasma membrane densely packed with the visual pigment rhodopsin (see p. 957). The function of these discs is to capture photons of light. The inner segment contains the cell nucleus, mitochondria, and other organelles. The synaptic terminal is where the rod cell communicates with other neurons.

To see how a rod cell responds to light, we can penetrate a single rod cell with an electrode and record its membrane potential in the dark and in the light, as shown in Figure 46.17. From what we have learned about other types of sensory receptors, we might expect that stimulation of the rod cell by light would make its membrane potential less negative, but the opposite is true—it becomes more negative.

When a rod cell is kept in the dark, it has a relatively depolarized resting potential compared with other neurons. In fact, the plasma membrane of the rod cell is almost as permeable to $Na^+$ as to $K^+$. In the dark, $Na^+$ continually enters the outer segment of the cell—the dark current. When light is flashed on the dark-adapted rod cell, its membrane potential becomes more negative—it hyperpolarizes (see Figure 46.17). The rate of neurotransmitter release changes as membrane potential changes. As the rod cell hyperpolarizes, its release of neurotransmitter decreases.

INVESTIGATING**LIFE**

**46.17 A Rod Cell Responds to Light** The plasma membrane of a rod cell hyperpolarizes—becomes more negative—in response to a flash of light. Rod cells do not fire action potentials, but in response to the absorption of light energy, the neuron experiences a change in membrane potential.[a]

**HYPOTHESIS** When a rod cell absorbs photons (light energy), its membrane potential changes in proportion to the strength of the light stimulus.

*Method*

1. Record membrane potentials from the inner segment of a rod cell.
2. Stimulate the rod cells with light flashes of varying intensity and record the results.

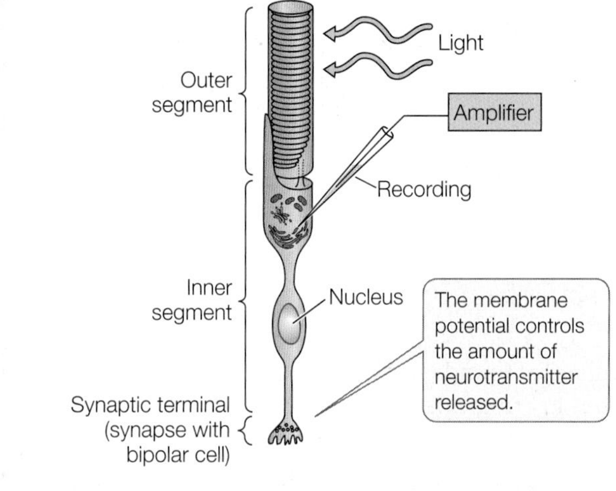


*Results*

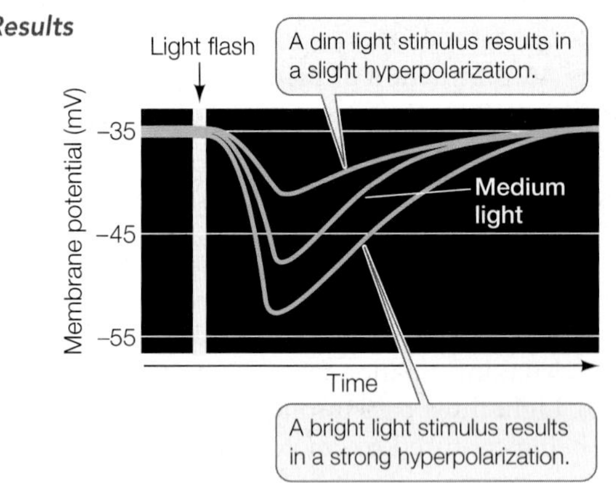


**CONCLUSION** The membrane potential of rod cells is depolarized in the dark and hyperpolarizes (becomes more negative) in response to light.

Go to **BioPortal** for discussion and relevant links for all INVESTIGATING**LIFE** figures.

[a]Baylor, D. A. et al. 1979. *Journal of Physiology* 288: 589–634.

WORKING WITH**DATA:**

## *Membrane Currents and Light Intensity in Rod Cells*

### Original Paper

Baylor, D. A., T. D. Lamb, and K.-W. Yau. 1979. The membrane current of single rod outer segments. *Journal of Physiology* 288: 589–611.

### Analyze the Data

In a slightly different set of experiments than that described in Figure 46.17, researchers measured the effect of light on the current across the rod cell membrane. The figure at right shows a series of recordings of the membrane currents (i.e., inward currents of positive ions) generated when rod cells were illuminated by light flashes of varying intensities. The initial values for the membrane currents represent the condition of a rod cell in total darkness. The light flash was given at time 0, and the intensity of the flashes is indicated on the right side of the response curves.

**QUESTION 1**

Why is there no difference between the *maximum* currents induced by flashes of light at 7.8 and 16 photons per $\mu m^2$?

**QUESTION 2**

Why does a rod maintain its maximum current for longer in response to a flash of light at 16 photons per $\mu m^2$ than it does to one at 7.8 photons per $\mu m^2$?

**QUESTION 3**

If you measured membrane potential instead of an outward current, how would the resulting recordings differ from the one shown here?

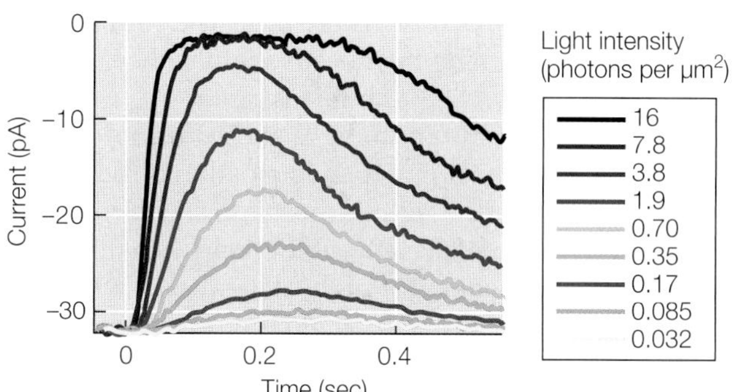


Go to BioPortal for all WORKING WITHDATA exercises

How does the absorption of light by rhodopsin hyperpolarize the rod cell? When rhodopsin is excited by light, it initiates a cascade of events. The dark-adapted rod cell has open $Na^+$ channels, allowing a depolarizing dark current (**Figure 46.18**). Light photons excite rhodopsin, which activates a G protein called transducin. Activated transducin in turn activates a phosphodiesterase (PDE). Activated PDE converts cyclic GMP (cGMP) to GMP, which causes the $Na^+$ channels to close. $Na^+$ is pumped out and the cell hyperpolarizes.

This cascade may seem like a roundabout way of doing business, but its advantage is its enormous amplification ability. Each molecule of light-excited rhodopsin can activate several hundred transducin molecules, thus activating a large number of PDE molecules. The catalytic capacity of PDE is great—one PDE molecule can hydrolyze several hundred molecules of cGMP per second. The bottom line is that a single photon of light can result in the closure of a huge number of $Na^+$ channels.

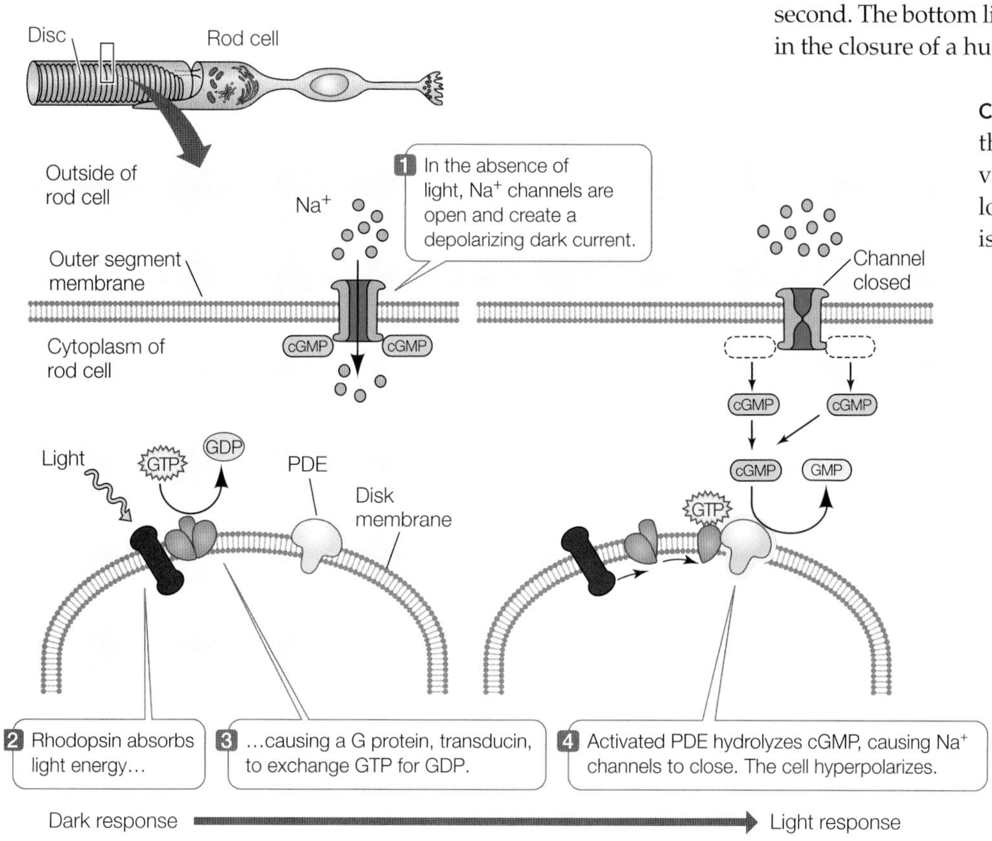

**CONE CELLS** Cone cells are responsible for the high-acuity color vision of day-active vertebrates such as humans. It is therefore logical that the highest density of cone cells is in the area of the retina that receives light

Disc
Rod cell
Outside of rod cell
Outer segment membrane
Cytoplasm of rod cell
Na+
cGMP
1 In the absence of light, Na+ channels are open and create a depolarizing dark current.
Channel closed
Light
GTP
GDP
PDE
Disk membrane
GMP
2 Rhodopsin absorbs light energy…
3 …causing a G protein, transducin, to exchange GTP for GDP.
4 Activated PDE hydrolyzes cGMP, causing Na+ channels to close. The cell hyperpolarizes.
Dark response
Light response

**46.18 Light Absorption Closes Sodium Channels** The absorption of light by rhodopsin initiates a signaling cascade that hyperpolarizes the rod cell. (A) In the dark, $Na^+$ channels in the plasma membrane of the rod cell's outer segment are held open by cGMP, allowing positive charges to enter the cell (upper right of panel). When the rod cell is stimulated by light, it activates transducin (lower portion of panel). (B) Transducin activates a molecule of phosphodiesterase (PDE). (C) Activated PDE catalyzes the breakdown of cGMP to GMP. The depletion of cGMP results in closure of the $Na^+$ channels and hyperpolarization of the cell.

from the center of the visual field, a region called the **fovea**. The human fovea has about 160,000 cones per square millimeter. But humans are not the champions of high-acuity vision; a hawk's fovea has almost twice that number of cones, making the hawk's vision much sharper than ours. Birds also have *two* foveae in each eye; one receives light from straight ahead, the other from a more lateral field of vision. The forward-looking foveae make binocular vision possible, while the lateral-looking foveae provide high-acuity vision. Birds use both sets of foveae by frequently turning their heads slightly; they cannot move their eyes in the sockets as humans can.

Cones have low sensitivity to light and contribute little to night vision. Night vision depends mostly on rod cells, which is why vision in dim light is mostly in shades of gray and acuity is low. You may have trouble seeing a small object at night when you are looking straight at it—that is, when its image is falling on your fovea. If you look a little to the side, so that the image falls on a rod-rich area of your retina, you see the object better. Astronomers looking for faint objects in the sky learned this trick a long time ago. The retinas of nocturnal animals, such as flying squirrels, contain a high percentage of rods. By contrast, some animals that are active only during the day (such as chipmunks) have mostly cones in their retinas.

The human retina has three kinds of cone cells, each containing slightly different opsin molecules that differ in the wavelengths of light they absorb best. Although the same 11-*cis*-retinal group is the light-absorber in all three kinds of cones (see Figure 46.12), its molecular interactions with opsin determine the spectral sensitivity of the cone cell as a whole (**Figure 46.19**). Because different wavelengths of light are differentially absorbed by the different cone cell visual pigments, the brain interprets the relative inputs from the different classes as a full range of color. Color blindness in humans results from the absence or dysfunction of one or more of the three classes of cone cells. Some mammals have only one or two classes of cone cells, whereas birds have four.

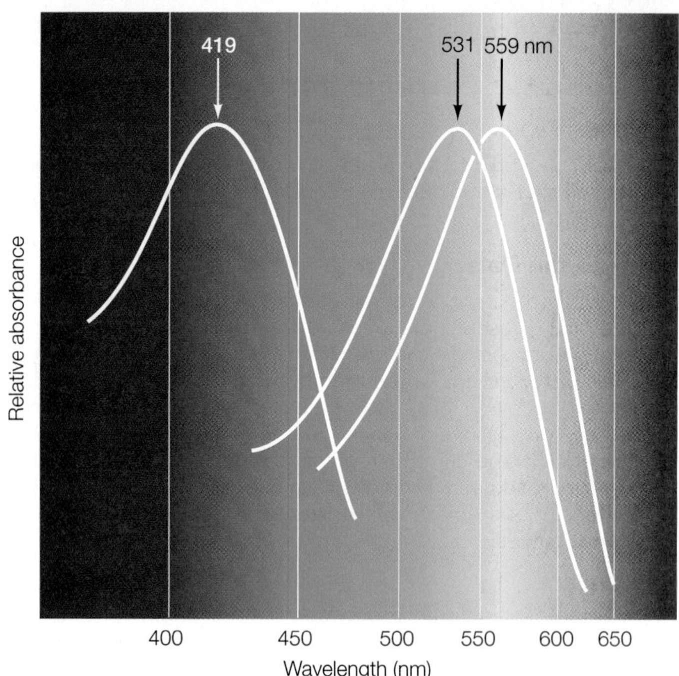


**46.19 Absorption Spectra of Cone Cells** The three kinds of cone cells contain slightly different visual pigments that absorb different wavelengths of light.

## Information flows through layers of neurons in the retina

The human retina is organized into layers of neurons that receive visual information and process it before sending it to the brain (**Figure 46.20**). Closest to the lens (and thus to light

**Go to Media Clip 46.1 Into the Eye Life10e.com/mc46.1**

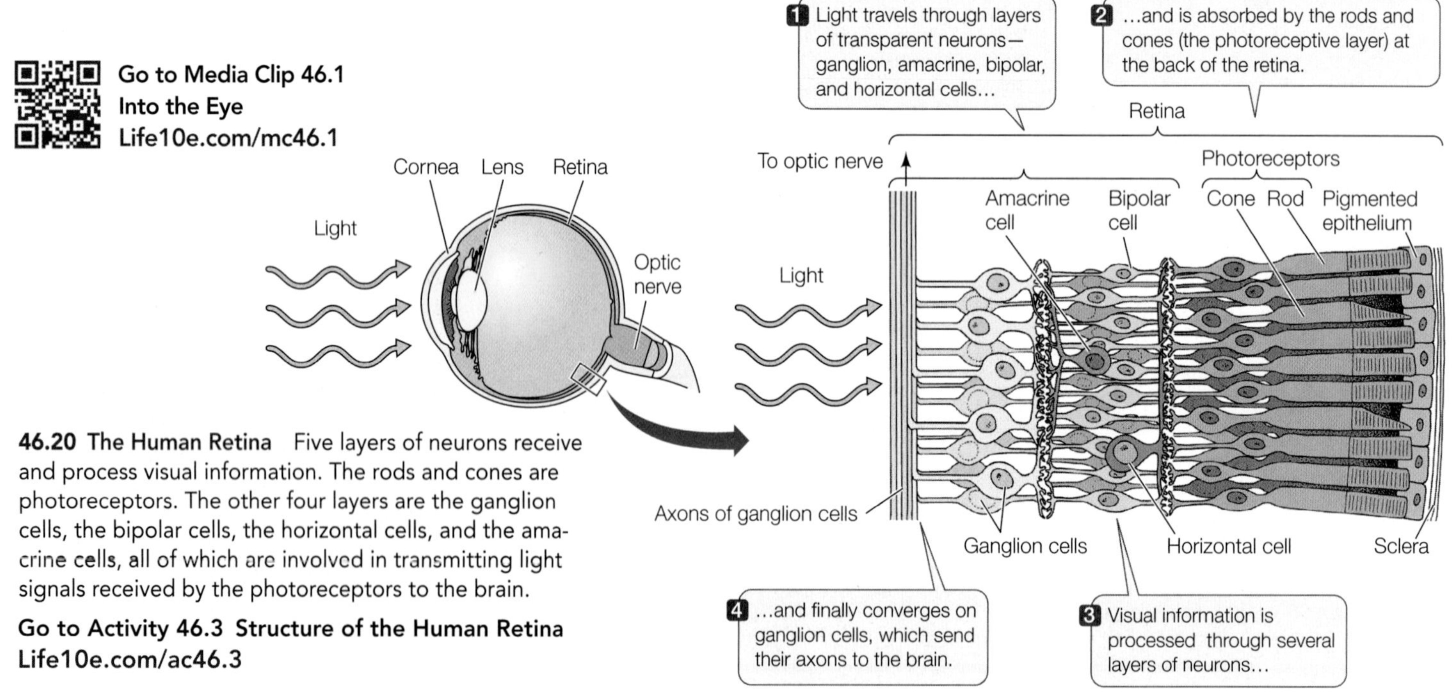


**46.20 The Human Retina** Five layers of neurons receive and process visual information. The rods and cones are photoreceptors. The other four layers are the ganglion cells, the bipolar cells, the horizontal cells, and the amacrine cells, all of which are involved in transmitting light signals received by the photoreceptors to the brain.

**Go to Activity 46.3 Structure of the Human Retina Life10e.com/ac46.3**

input) is a layer of **ganglion cells**; a central layer contains three neuronal types, **bipolar cells**, **horizontal cells**, and **amacrine cells**; and at the "rear" of the retina lie the photoreceptors (rods and cones). The layers of cells between the photoreceptors and the ganglion cells process information about the visual field.

GANGLION AND BIPOLAR CELLS From our discussion of rod cells, we know that the photoreceptor cells at the back of the retina hyperpolarize in response to light and do not generate action potentials. The ganglion cells lie at the front of the retina, and they do fire action potentials. The axons of ganglion cells form the **optic nerve** that travels to the brain.

The ganglion cells are connected to the photoreceptors by bipolar cells. Changes in the membrane potential of rods and cones in response to light alter the rates at which the rods and cones release neurotransmitter at their synapses with the bipolar cells. In response to this neurotransmitter, the membrane potentials of the bipolar cells change, altering the rate at which they release neurotransmitter onto ganglion cells. The rate of neurotransmitter release from the bipolar cells determines the rate at which ganglion cells fire action potentials. Thus the direct flow of information in the retina is from photoreceptor to bipolar cell to ganglion cell. The ganglion cells send the information to the brain via the optic nerve.

Each human eye contains about 1.2 million ganglion cells but more than 100 million rods and cones. Therefore there must be convergence of information as it passes from the photoreceptors to the ganglion cells. A given bipolar cell can receive input from multiple rods or multiple cones, but not from both. The relationship between photoreceptors, bipolar cells, and ganglion cells depends on their location on the retina. In the fovea, a ganglion cell may receive input from as few as five photoreceptors, but in the periphery of the retina, a ganglion cell may receive input from thousands of photoreceptors. Visual acuity is a reflection of these quantitative relationships.

The patch of photoreceptors that communicates with a ganglion cell forms a circular receptive field. When light falls on a receptive field, its ganglion cell can be either excited or inhibited. As mentioned above, each ganglion cell sends an axon to the brain in the optic nerve. Thus the information coming from the retina to the brain is about the pattern of patches of light and dark falling on the retina.

HORIZONTAL AND AMACRINE CELLS The other two cell layers, the horizontal cells and the amacrine cells, consist of interneurons that communicate laterally across the retina. Horizontal cells form synapses with neighboring photoreceptors and bipolar cells. Thus light falling on one photoreceptor can influence the sensitivity of its neighbors to light. This lateral flow of information enables the retina to sharpen the perception of contrast between light and dark patterns. Amacrine cells form local interconnections between bipolar cells and ganglion cells. Some amacrine cells are highly sensitive to changing illumination or to motion. Others assist in adjusting the sensitivity of the eyes according to the overall level of light falling on the retina. When background light levels change, amacrine cell connections to the ganglion cells help the ganglion cells remain sensitive to temporal changes in stimulation. Thus even with large changes in background illumination, the eyes are sensitive to smaller, more rapid changes in the pattern of light falling on the retina.

**RECAP 46.4**

A family of photosensitive visual pigments are responsible for light sensitivity in all animals. Receptor cells, including rod and cone cells in humans, transduce the photosensitivity of visual pigments to light and use it to form images of the environment.

- Explain how a photon of light affects the membrane potential in a rod cell. **See pp. 960–961 and Figures 46.12, 46.17, and 46.18**
- What is the mechanism of color vision? **See pp. 961–962 and Figure 46.19**
- Describe the flow of signals that takes place in the eye in response to light. **See pp. 962–963 and Figure 46.20**

Knowing the path of information from sensory receptor cells to the central nervous system still does not tell us how that information is processed by the brain. What does the eye tell the brain, for example, in response to a pattern of light falling on the retina? In Chapter 47 we will describe how the mammalian brain reassembles sensory information into our perception of the world.

How can bats emit loud pulses of sound and not be deaf to the faint echoes that return within milliseconds?

**ANSWER**

Bat "echolocation" allows these mammals to navigate around objects and to find prey in total darkness. Bats emit sound waves at frequencies well above the range of human hearing; these sound waves bounce off objects, including potential prey. The bat perceives the echoes that bounce back almost immediately and uses these echoes to locate prey and other objects. The loud pulses of sound still being emitted don't "drown out" the weak echoes because small muscles in a bat's ears contract to dampen its hearing sensitivity while the sounds are being emitted, but relax in time for the bat to hear the echo—a truly remarkable ability, considering that the sound pulses are emitted at rates of 20 to 80 per second.

## CHAPTER SUMMARY 46

### 46.1 How Do Sensory Receptor Cells Convert Stimuli into Action Potentials?

- Sensory receptor cells, also known as **sensors** or **receptors**, transduce information about an animal's external and internal environment into action potentials that the brain perceives as different forms of sensory information.
- **Receptor potentials** can spread to regions of the cell's plasma membrane that generate action potentials. Some sensors do not fire action potentials but release neurotransmitter onto sensory neurons that do fire action potentials. **Review Figure 46.1**
- Sensors have **receptor proteins** that cause ion channels to open or close, affecting the receptor cell's membrane potential. Metabotropic receptors act through signal transduction pathways to generate receptor potentials. Mechanoreceptors are ionotropic sensory receptors that open ion channels physically through forces such as pressure or stretch. **Review Figure 46.2**
- The interpretation of action potentials as particular sensations depends on which neurons in the central nervous system receive them.
- **Adaptation** enables the nervous system to ignore irrelevant or continuous stimuli while remaining responsive to relevant or new stimuli.

### 46.2 How Do Sensory Systems Detect Chemical Stimuli?

- **Chemoreceptors** are responsible for **olfaction**, **gustation**, and the sensing of **pheromones**.
- Mammalian **olfactory receptor neurons** (**ORNS**) project directly to the **olfactory bulb** of the brain. ORNs for the same **odorant** project to the same area of the olfactory bulb.
- Each ORN expresses one receptor protein that can bind a specific type of molecule or ion. Binding causes a second messenger to open ion channels, which creates an action potential. **Review Figure 46.3**
- In vertebrates, **taste buds** in the mouth cavity are responsible for gustation. The five basic tastes are sweet, salty, sour, bitter, and umami. **Review Figure 46.5**

### 46.3 How Do Sensory Systems Detect Mechanical Forces?

- The skin contains a variety of ionotropic **mechanoreceptors** that respond to touch and pressure. The density of mechanoreceptors in any skin area determines the sensitivity of that area. **Review Figure 46.6**
- Stretch receptors in **muscle spindles** and in the **Golgi tendon organ** inform the CNS of the positions of and loads on parts of the body. **Review Figure 46.7**
- **Hair cells** are mechanoreceptors of the auditory and vestibular systems. Physical bending of their **stereocilia** alters their receptor proteins and therefore their membrane potentials. **Review Figure 46.8**
- In mammalian auditory systems, ear pinnae collect and direct sound waves to the **tympanic membrane**, which vibrates in response to sound waves. The movements of the tympanic membrane are amplified through a chain of **ossicles** that conduct the vibrations to the **oval window**. Movements of the oval window create pressure waves in the fluid-filled **cochlea**. **Review Figure 46.9, ACTIVITY 46.1**
- The basilar membrane running down the center of the cochlea is distorted by pressure waves at specific locations that depend on the frequency of the wave. These distortions cause hair cells in the **organ of Corti** to bend and to release neurotransmitter, generating action potentials in the cochlear nerve that are transmitted to the auditory cortex of the brain. **Review Figure 46.10, ANIMATED TUTORIAL 46.1**
- Hair cells are also the mechanoreceptors of the organs of equilibrium in the mammalian vestibular system, which include the **semicircular ducts** and the **saccule** and **utricle**. **Review Figure 46.11, ANIMATED TUTORIAL 46.2**

### 46.4 How Do Sensory Systems Detect Light?

- **Photosensitivity** depends on the absorption of photons of light by visual pigment molecules that consist of a protein called **opsin** and a light-absorbing group. Absorption of light is the first step in a cascade of intracellular events leading to a change in the membrane potential of the **photoreceptor cell**. **Review Figure 46.12, ANIMATED TUTORIAL 46.3**
- Visual systems range from the simple **eye cups** of flatworms, which sense the direction of a light source, to the **compound eyes** of arthropods, which detect shapes and patterns, to the image-forming eyes of vertebrates and cephalopods. **Review Figures 46.13, 46.14**
- Vertebrate and cephalopod eyes focus detailed images of the visual field onto dense arrays of photoreceptors that transduce the visual image into neural signals. **Review Figures 46.14, 46.15, ACTIVITY 46.2**
- Vertebrates have two types of photoreceptors, **rod cells** and **cone cells**. Rod cells are more sensitive to light and are responsible for dim light vision. Cone cells are less sensitive to light but are responsible for high-acuity and color vision.
- Photoreceptors do not fire action potentials. When not stimulated by light they release neurotransmitter continuously. Light hyperpolarizes rod cells, and their release of neurotransmitter decreases. **Review Figures 46.17, 46.18**
- **Rhodopsin** is the visual pigment of rod cells. The visual pigments of cone cells have three different opsin components, which gives them different spectral sensitivities. **Review Figure 46.19**
- The vertebrate **retina** consists of layers of neurons lining the back of the eye. The light-absorbing photoreceptor cells are at the rear of the retina. The axons of the **ganglion cells** are bundled together in the **optic nerve**. Between the photoreceptors and the ganglion cells are neurons that process information from the photoreceptors. **Review Figure 46.20, ACTIVITY 46.3**

**Go to the Interactive Summary to review key figures, Animated Tutorials, and Activities Life10e.com/is46**

# CHAPTER**REVIEW**

## REMEMBERING

1. Which statement about sensory systems is *not* true?
   a. Sensory transduction involves the conversion (direct or indirect) of a physical or chemical stimulus into changes in membrane potentials.
   b. In general, a stimulus causes a change in the flow of ions across the plasma membrane of a sensory receptor cell.
   c. The term "adaptation" refers to the process by which a sensory system becomes insensitive to a continuing source of stimulation.
   d. The more intense a stimulus, the greater the magnitude of each action potential fired by a sensory neuron.
   e. Sensory adaptation plays a role in the ability of organisms to discriminate between important and unimportant information.

2. Which statement about olfaction is *not* true?
   a. In general, mammals depend more on vision than on olfaction as their dominant sensory modality.
   b. Olfactory stimuli are recognized by the interaction between odorant molecules and receptor proteins on olfactory hairs.
   c. The more odorant molecules that bind to receptors, the more action potentials are generated.
   d. The greater the number of action potentials generated by an olfactory receptor, the greater the intensity of the perceived smell.
   e. The perception of different smells results from the activation of different combinations of olfactory receptors.

3. The membrane most directly responsible for the ability to discriminate different pitches of sound is the
   a. round window.
   b. oval window.
   c. tympanic membrane.
   d. tectorial membrane.
   e. basilar membrane.

4. Which statement is *not* true?
   a. The transmembrane potential of a rod cell becomes more negative when the rod cell is exposed to light.
   b. A photoreceptor releases the most neurotransmitter when in total darkness.
   c. Whereas in vision the intensity of a stimulus is encoded by the degree of hyperpolarization of photoreceptors, in hearing the intensity of a stimulus is encoded by changes in firing rates of sensory neurons.
   d. Stiffening of the ossicles in the middle ear can lead to deafness.
   e. The interaction among hammer (malleus), anvil (incus), and stirrup (stapes) conducts sound waves across the fluid-filled middle ear.

5. Which of the following statements about information flow in the vertebrate visual system is true?
   a. Action potentials in bipolar cells cause the release of neurotransmitter onto ganglion cells.
   b. Amacrine cells integrate the activity of neighboring rod and cone cells.
   c. When photons of light enter the eye, the first cells they encounter in the retina are ganglion cells.
   d. The highest density of rod cells in the human retina is centrally located in the fovea, resulting in high-acuity dim-light vision.
   e. Pigmented epithelial cells at the back of the retina provide information about the level of ambient light for contrast adjustments.

## UNDERSTANDING & APPLYING

6. What are the similarities and differences in the functioning of olfactory receptors and taste receptors? How do these sensory cells enable the central nervous system to discriminate between an apple and an orange?

7. If you were blindfolded and sitting in a wheeled chair, how would you know if you were being pushed forward or backward?

8. To human ears, sounds are louder underwater than in air. Why is this so?

## ANALYZING & EVALUATING

9. Fish have fluid-filled channels called lateral line canals running down the sides of their bodies. These canals contain hair cells, as illustrated below. Describe what you think the functions of these lateral-line hair cells are.

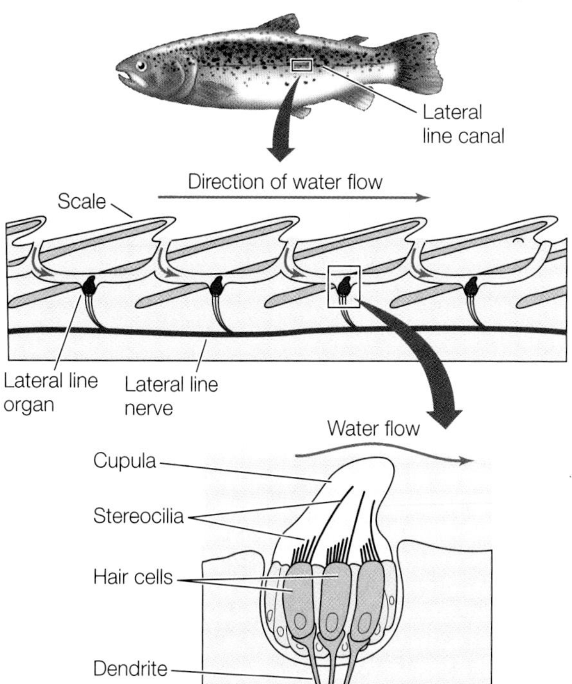

10. An owl can capture a mouse scurrying across a forest floor in total darkness. What sensory information do you think the owl uses, and how does it get directional information from that sensory information?

Go to BioPortal at **yourBioPortal.com** for Animated Tutorials, Activities, LearningCurve Quizzes, Flashcards, and many other study and review resources.

# The Mammalian Nervous System: Structure and Higher Functions

## CHAPTER OUTLINE


**47.1** How Is the Mammalian Nervous System Organized?

**47.2** How Is Information Processed by Neural Networks?

**47.3** Can Higher Functions Be Understood in Cellular Terms?

**A Mind-Expanding Maze** The extraordinary ability of taxicab drivers in London to navigate its maze of streets and byways prompted a study that revealed London cabbies to have larger than normal posterior hippocampi—a brain region implicated in the memory of spatial relationships in the environment.

COMPARE the Google maps of London and New York City at the same scale. In which city do you think it is easier to drive a taxi? In the city with sequentially numbered avenues and streets laid out at right angles to each other, or the city with a maze of arbitrarily named streets going in every direction?

Eleanor Maguire at University College London was so impressed with the navigational abilities of London taxi drivers that she decided to see if there was anything "special" about their brains. Using the brain imaging technique known as MRI (magnetic resonance imaging), Maguire and her colleagues examined the brains of taxi drivers with varying numbers of years of experience and compared them with each other and with the brains of control subjects who were not taxi drivers. The studies revealed significant changes in the anatomy of the hippocampus among taxi drivers.

The hippocampus is an area of the brain involved in learning and memory. The posterior hippocampus in particular is implicated in the memory of spatial relationships among objects in the environment. Maguire found that the posterior hippocampi of taxi drivers was larger than that of the control subjects and that, among the cab drivers themselves, there was a positive correlation between the size of the posterior hippocampus and years of driving experience.

In M. A. Wilson's lab at the Massachusetts Institute of Technology, researchers recorded the activity of hippocampal neurons of rats as they navigated a maze. They located specific neurons, referred to as "place cells," that fire only when the rat is at a particular location in the maze. In a sense these researchers can see what the rat is thinking, because when the animal is not moving through the maze, the neurons will occasionally fire in the same pattern as when it is running the maze—or they will fire in the reverse sequence, representing where the rat has just been. But these replays of firing patterns are much faster; the rat *thinks* about its experience in the maze about 25 times faster than the actual experience. In a maze that had a choice point (go right or go left), the firing pattern seen when the animal was held at the start position predicted which direction it would turn. Similar firing patterns occur when the rats are sleeping. Are they dreaming about the maze? Are they transferring memory of today's experience in the maze into long-term memory?

Does the firing pattern of hippocampal place cells during sleep represent a memory of recent experience being transferred into long-term memory?

See answer on p. 983.

## 47.1 How Is the Mammalian Nervous System Organized?

The organization of the mammalian nervous system can be described anatomically and functionally. In anatomical terms, all vertebrate nervous systems consist of three parts: a brain, a spinal cord, and a set of peripheral nerves that reach all parts of the body. As discussed in Section 45.4, the brain and spinal cord are the **central nervous system**, or **CNS**, and the nerves that connect the CNS to all the tissues and sensors of the body are the **peripheral nervous system**, or **PNS**. An additional division of the nervous system exists in the gut; we will discuss this **enteric nervous system** in Chapter 51.

Recall from Section 45.1 that a neuron is an electrically excitable cell that communicates via an axon. When used in the context of a nervous system, the term **nerve** refers to a bundle of axons that carries information about many things simultaneously. Some axons in a nerve may be carrying information to the CNS while other axons in the same nerve are carrying information from the CNS to the body's organs. A discussion of the functional organization of the nervous system refers to these paths of information flow. In this chapter we will divide the anatomy of the mammalian brain, spinal cord, and peripheral nervous system into smaller, discrete functional units.

### Functional organization is based on flow and type of information

**Figure 47.1** illustrates the major avenues of information flow through the human nervous system. The white boxes represent the four divisions of the peripheral nervous system; two of these bring information from the periphery to the CNS, and two transmit information from the CNS to the periphery.

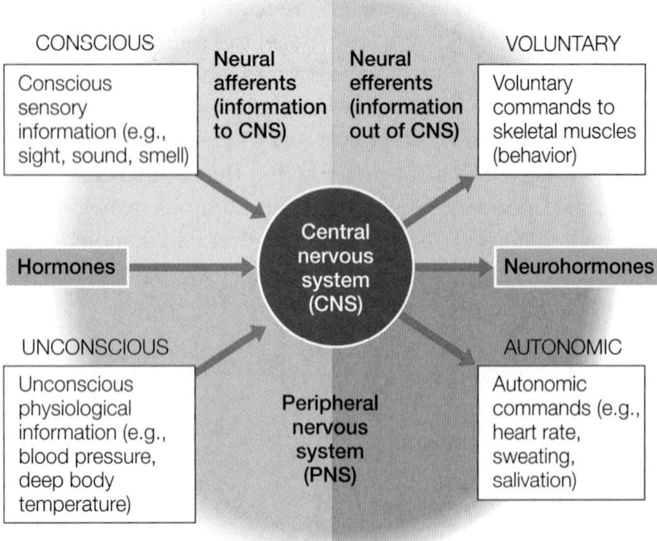


**47.1 Organization of the Nervous System** The peripheral nervous system carries information to (afferent) and from (efferent) the central nervous system (center circle in the diagram). The CNS also receives hormonal inputs and produces hormonal outputs.

- The **afferent** portion of the PNS carries information from sensory receptor cells to the CNS. We are *conscious* of much of this information (e.g., light, sound, skin temperature, limb position), but we are usually *unconscious* of the information involved in physiological regulation (e.g., blood pressure, deep body temperature, blood oxygen levels).
- The **efferent** portion of the PNS carries information from the CNS to the muscles and glands of the body. Efferent pathways are divided into a *voluntary* division that executes our conscious movements; and an *involuntary*, or *autonomic*, division that controls physiological functions.

In addition to the neural information it receives from the PNS, the central nervous system receives chemical information from hormones circulating in the blood. In turn, neurohormones released by neurons enter the circulation and affect neurons and other cells distant from the site of release (see Section 41.1).

### The anatomical organization of the CNS emerges during development

Early in the development of a vertebrate embryo, a tube of neural tissue forms (see Section 44.4). At its anterior end, this neural tube forms three swellings that become the **hindbrain**, **midbrain**, and **forebrain**. The rest of the neural tube becomes the spinal cord. Peripheral nerves sprout from the midbrain and hindbrain (the cranial nerves) and from the spinal cord (the spinal nerves). From these early stages we see the linear axis of information flow in the nervous system. Although the developing brain will fold and become a complex structure, the information flow in the adult nervous system will follow paths that emerge from the simple, linear neural tube.

Each of these three regions of the embryonic brain develops into several structures in the adult brain (**Figure 47.2**). From the embryonic midbrain come structures that integrate information from the different senses and coordinate motor responses. From the hindbrain come the **medulla**, the **pons**, and the **cerebellum**. The medulla is continuous with the spinal cord, the pons is in front of the medulla, and the cerebellum is a dorsal outgrowth of the pons. The medulla and pons contain distinct groups of neurons involved in controlling physiological functions such as breathing, circulation, and basic motor patterns such as swallowing and vomiting. All information traveling between the spinal cord and higher brain areas must pass through the pons, the medulla, and the midbrain, which are collectively known as the **brainstem**.

The cerebellum is involved in coordinating muscle activity and maintaining balance. It is like the director of a movie; the cerebellum receives a "script" of the commands going to the muscles from higher brain areas, and it receives information about the actual performance coming up the spinal cord from the "actors"—the joints and muscles. The cerebellum compares the "script" with the performance and refines motor commands accordingly. Damage to the cerebellum results in loss of fine motor control and coordination.

The embryonic forebrain develops a central region called the **diencephalon** and a surrounding structure called the **telencephalon**. The diencephalon is the core of the forebrain and consists of an upper structure, the thalamus, and a lower

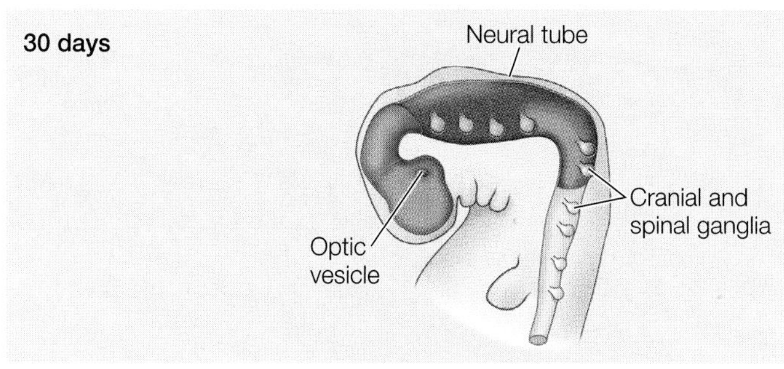


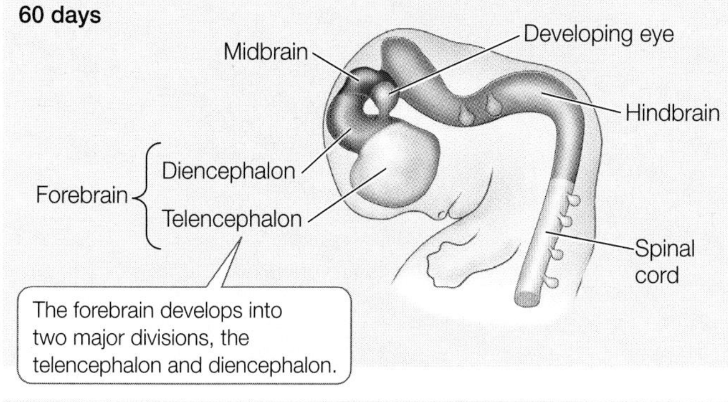


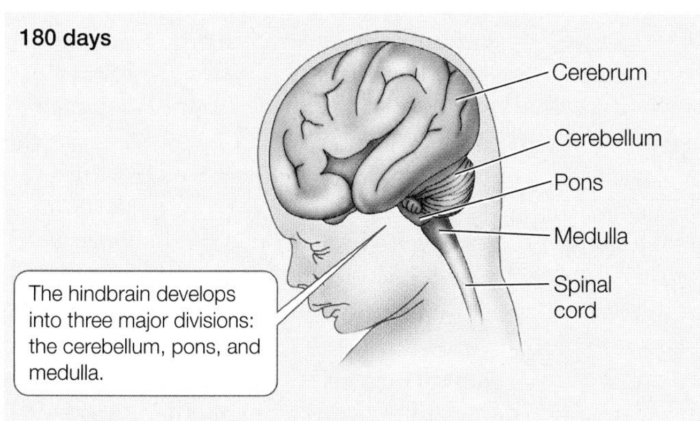


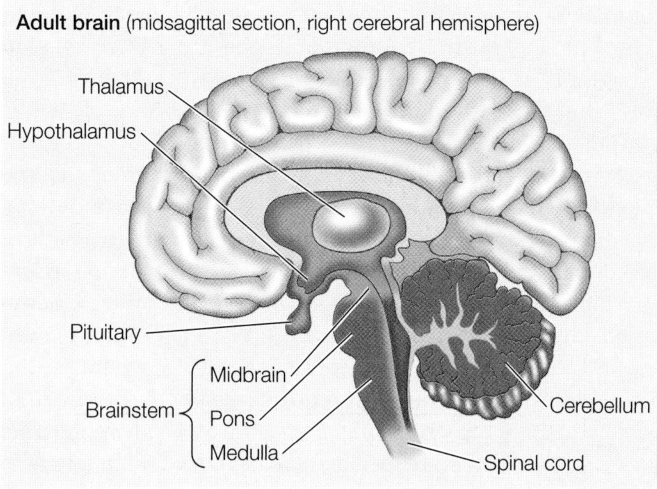


**47.2 Development of the Central Nervous System** In vertebrate embryos, the anterior end of the hollow neural tube differentiates into forebrain, midbrain, and hindbrain. Each of these regions develops into several structures of the adult brain. The remainder of the neural tube becomes the spinal cord.

structure, the hypothalamus. The thalamus is the final relay station for sensory information going to the telencephalon. The hypothalamus receives a lot of physiological information of which we are not conscious, and it uses that information to regulate many physiological functions and biological drives.

The telencephalon—also called the **cerebrum**—consists of the left and right **cerebral hemispheres**. The outer layer of the telencephalon is the **cerebral cortex**, a thin layer rich in cell bodies. If we compare vertebrate groups from fish through amphibians, reptiles, and mammals, the telencephalon increases in size, complexity, and importance—an evolutionary trend called telencephalization (see Figure 45.19). In humans, the telencephalon is by far the largest part of the brain and plays major roles in sensory perception, learning, memory, and conscious behavior.

## The spinal cord transmits and processes information

The spinal cord conveys information to and from the brain. However, the spinal cord is more than an information pipe. As we saw in Section 45.4, the spinal cord carries out integrative functions as well. The knee-jerk reflex (see Figure 45.18) is an example of a circuit between the PNS and the spinal cord that controls a simple behavioral function. That simple circuit, however, can be built on to control more complex behaviors such as the withdrawal reflex, which involves readjusting tension in many muscles on both sides of the body to coordinate movement and maintain balance. Complex motor programs also exist in the spinal cords of many vertebrates. A shark can swim perfectly well after its spinal cord is separated from its brain, and there is the proverbial chicken running around with its head cut off. Central pattern generation in the spinal cord has been demonstrated in mammals by the fact that an experimental animal (usually a cat) with a spinal cord transection that has isolated its hind limbs from its brain can still coordinate its hind limb movements to walk on a treadmill.

## The brainstem carries out many autonomic functions

Swallowing, salivating, breathing, eye movements, blood pressure regulation, and gut activity are only a few of the many autonomic functions that are localized in the medulla, the pons, and the midbrain. To carry out these functions, the brainstem has its own components of the PNS, the 12 paired **cranial nerves**. You encountered the olfactory nerve, the optic nerve, and the auditory nerve (cranial nerves I, II, and VIII) in Chapter 46. Another one, cranial nerve X, is called the vagus ("wandering") nerve because it travels into the body cavity and communicates with many of the organs, including the heart and the gut. We will encounter the vagus nerve in subsequent chapters.

Within the highly complex networks of axons and dendrites in the brainstem, there are many discrete groups of neurons that share a common characteristic such as the neurotransmitter they produce and release. Such an anatomically distinct group of neurons is called a **nucleus** (not to be confused with the nucleus of a single cell). Many of these brainstem nuclei send their axons to various regions of the brain to modulate their activity; for example, brainstem nuclei are involved in keeping the higher brain areas awake or allowing them to sleep. All of the sensory information coming up the neural axis from the spinal cord passes through the brainstem on its way to the forebrain, and many of these ascending neuronal tracts give off collateral branches to the awake-promoting nuclei in the brainstem. Because the neuronal circuitry in this part of the brain is so complicated and because activity in these ascending sensory pathways can promote wakefulness, the core of the brainstem has been termed the **reticular activating system** (reticular means net-like). Damage to the brain or spinal cord below the reticular activating system can result in paralysis but leave sleep–wake cycle behavior normal. Damage above the level of the reticular activating system can result in coma.

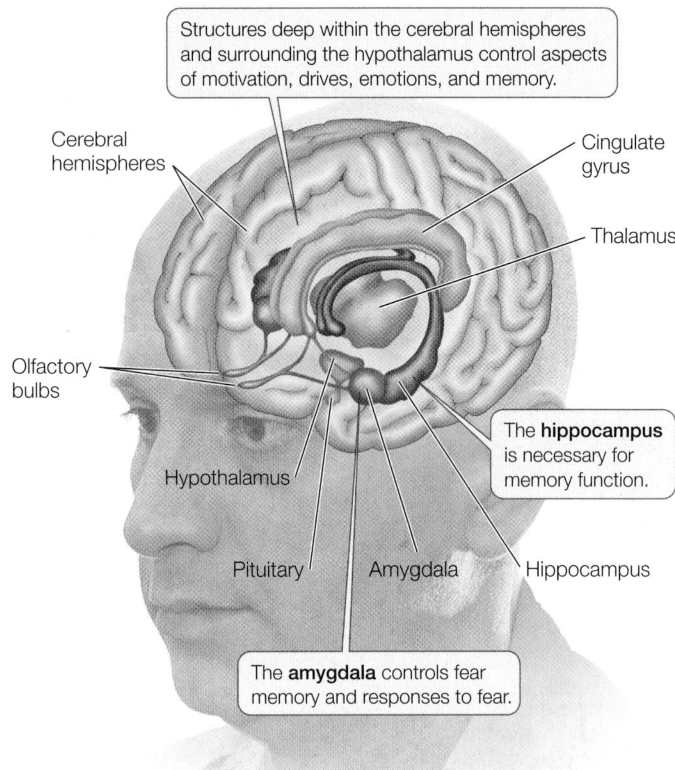


**47.3 The Limbic System** The evolutionarily primitive parts of the telencephalon are referred to as the limbic system. The hippocampus is involved in forming long-term memory. The amygdala triggers fear emotions and fear memories.

## The core of the forebrain controls physiological drives, instincts, and emotions

As mentioned above, the diencephalon consists of the thalamus and the hypothalamus. The thalamus communicates sensory information to the cerebral cortex; the hypothalamus receives information about physiological conditions in the body and regulates many homeostatic functions. Section 40.5 described how the hypothalamus is involved in regulating body temperature, and Section 41.3 discussed the intimate association between the hypothalamus and the pituitary gland in the control of many homeostatic functions.

Surrounding the diencephalon of all vertebrates are phylogenetically older structures of the telencephalon. These structures comprise the **limbic system** (**Figure 47.3**), which is responsible for instincts, long-term memory formation, physiological drives such as hunger and thirst, and emotions such as fear. Within the limbic system are areas that, when stimulated with small electric currents, can cause intense sensations of pleasure, pain, or rage. A rat given the opportunity to stimulate its own pleasure centers by pressing a switch will ignore food and water, pushing the switch until it is exhausted.

Pleasure and pain centers in the limbic system are believed to play roles in learning and in physiological drives. One component of the limbic system—the **amygdala**—is involved in fear and fear memory. If a certain portion of the amygdala is damaged or chemically blocked, an animal cannot learn to be afraid of a stimulus or a situation that would normally induce a strong fear reaction. The amygdala is involved in post-traumatic stress disorder (PTSD). Another part of the limbic system, the **hippocampus**, is necessary in humans for the transfer of certain types of short-term memory to long-term memory, as we will discuss in Section 47.3.

## Regions of the telencephalon interact to control behavior and produce consciousness

The cerebrum is the dominant structure in the mammalian brain. In humans it is so large that it covers all other parts of the brain except the cerebellum (**Figure 47.4A**). The cerebral cortex covering the cerebrum is only about 4 millimeters thick, but it covers a surface area larger than a square meter because it is folded into ridges (**gyri**; singular *gyrus*) and valleys (**sulci**; singular *sulcus*). These foldings, or **convolutions**, enable the large surface of the cortex to fit within the skull.

As we explore the functions of the cerebral cortex and other parts of the brain, we will occasionally mention an individual whose brain was damaged by an accident or other unfortunate event. Until recently the study of such individuals has been the main source of functional information about the human brain, but new imaging technologies such as positron emission tomography (PET) and magnetic resonance imaging (MRI) are providing a wealth of new information and opportunities to study the human brain.

A curious feature of the human nervous system is that the left side of the body is served (in both sensory and motor aspects) mostly by the right side of the brain, and the right side of the body is served mostly by the left side of the brain. Thus sensory input from the right hand goes to the left cerebral hemisphere, and sensory input from the left hand goes to the right cerebral hemisphere. The exception is the head, where the left side is controlled by the left cerebral hemisphere and the right side by the right cerebral hemisphere. The two hemispheres are not symmetrical with respect to all functions. Language abilities, for example, reside predominantly in the left hemisphere.

Different regions of the cerebral cortex have specific functions (**Figure 47.4B**). Some of those functions are easily defined, such as receiving and processing sensory information or generating motor commands, but most of the cortex is involved in

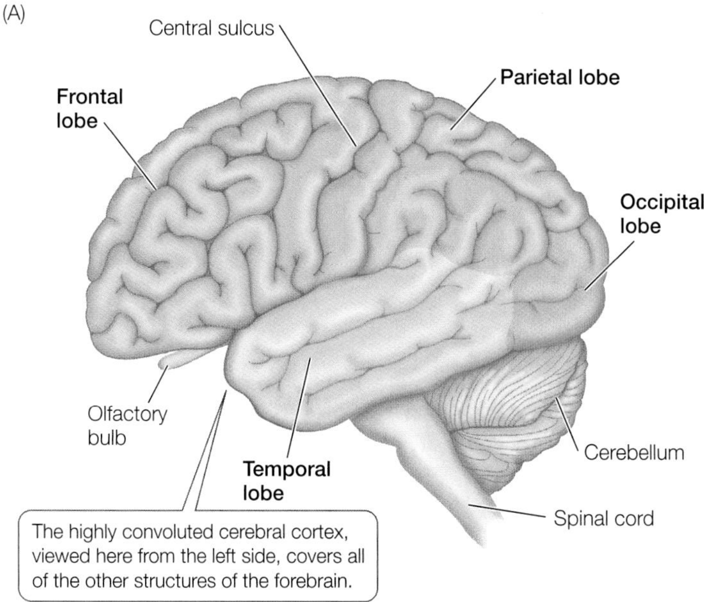


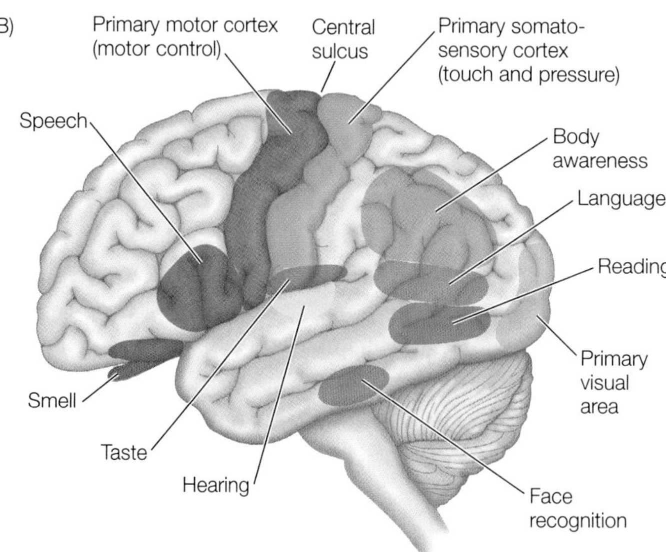


**47.4 The Human Cerebrum** (A) Each cerebral hemisphere is divided into frontal, temporal, parietal, and occipital lobes. (B) Different functions are localized in particular areas of the four cerebral lobes.

**Go to Activity 47.1 The Human Cerebrum**
**Life10e.com/ac47.1**

higher-order information processing that is less easy to define. These latter areas are given the general name of **association cortex**, so named because they integrate, or *associate*, information from different sensory modalities and from memory.

To understand the cerebral cortex, it helps to have an anatomical road map. Viewed from the left side, the left cerebral hemisphere looks like a boxing glove for the right hand with the fingers pointing forward, the thumb pointing out, and the wrist at the rear. The "thumb" area is the **temporal lobe**, the fingers the **frontal lobe**, the back of the hand the **parietal lobe**, and the wrist the **occipital lobe** (see Figure 47.4A). The right cerebral hemisphere shows a mirror image of this arrangement. We will look at each lobe separately.

**THE TEMPORAL LOBE** The upper region of the temporal lobe receives and processes auditory information. The association areas of this lobe are involved in recognizing, identifying, and naming objects. Damage to the temporal lobe results in disorders called agnosias, in which the individual is aware of an object but cannot identify it.

Damage to a certain area of the temporal lobe results in the inability to recognize faces. Even old acquaintances cannot be identified by facial features, although they may be identified by other attributes such as voice, body features, and posture. Using monkeys, it has been possible to record the activity of neurons in this region that respond selectively to faces in general. These neurons do not respond to other stimuli in the visual field, and their responsiveness decreases if some of the facial features are missing or appear in inappropriate locations (**Figure 47.5**). Damage to other association areas of the temporal lobe causes deficits in understanding spoken language, although speaking, reading, and writing abilities may be intact.

**THE FRONTAL LOBE** The frontal and parietal lobes are separated by a deep valley called the central sulcus. A strip of the frontal lobe cortex just in front of the central sulcus is called the **primary motor cortex** (see Figure 47.4B). The neurons in this region control muscles in specific parts of the body; the parts of the body have been mapped onto the primary motor cortex, largely during neurosurgical procedures. As part of these procedures, electrodes were used to stimulate small areas of cortex. In the area just anterior to the central sulcus, stimulation causes specific muscles to contract. Parts of the body with fine motor control, such as the face and hands, have disproportionate representation (**Figure 47.6A**). Stimulation of neurons in the primary motor cortex causes twitches of muscles, not coordinated movements.

The association functions of the frontal lobe are diverse and are best described as having to do with feeling and planning. They contribute significantly to personality. People with

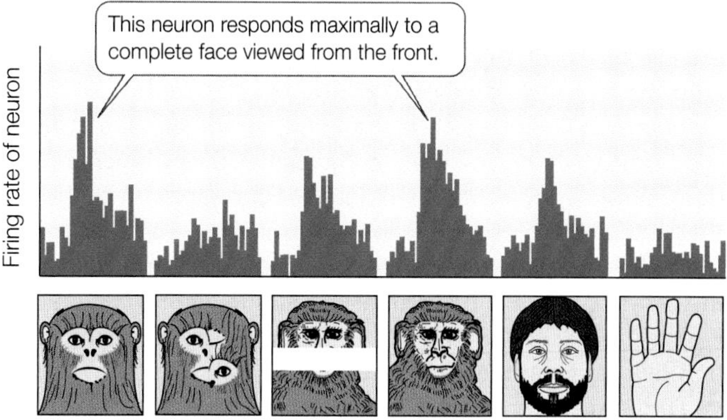


**47.5 "Face Neurons" in One Region of the Temporal Lobe** The electrode traces represent the firing rate of a neuron in the temporal lobe of a monkey in response to the pictures shown below them. Highest firing is stimulated by the appearance of a complete face.

**47.6 The Body Is Represented in Primary Motor and Primary Somatosensory Cortexes** Neurons in the primary motor cortex (A) control muscles in specific parts of the body, while neurons in the primary somatosensory cortex (B) receive information from specific parts of the body. The locations of these neurons within each cortex correspond to "maps" on which regions of the body are represented in proportion to the amount of innervation they receive.



frontal lobe damage have drastic alterations of personality and difficulty planning future events. A dramatic case of frontal lobe damage is the story of Phineas Gage, who in 1848 was an industrious and responsible young railroad construction foreman. Then a blasting accident shot a meter-long, 3-centimeter-wide iron tamping rod through his brain. The rod entered Gage's head below his left eye, passed through his frontal lobe, and exited the top of his head (**Figure 47.7**).

Remarkably, Gage survived, but he was a completely different person. He was quarrelsome, impatient, obstinate, and used profane language, which he had never done before. He lost his railroad job and spent his days as a drifter, earning money by telling his story and exhibiting his scars (and the tamping iron). He died of a seizure in 1860, at the age of 38. If you are in Boston, you can pay him a visit—his skull, death mask, and the tamping iron are on display in the Warren Anatomical Museum of Harvard Medical School.

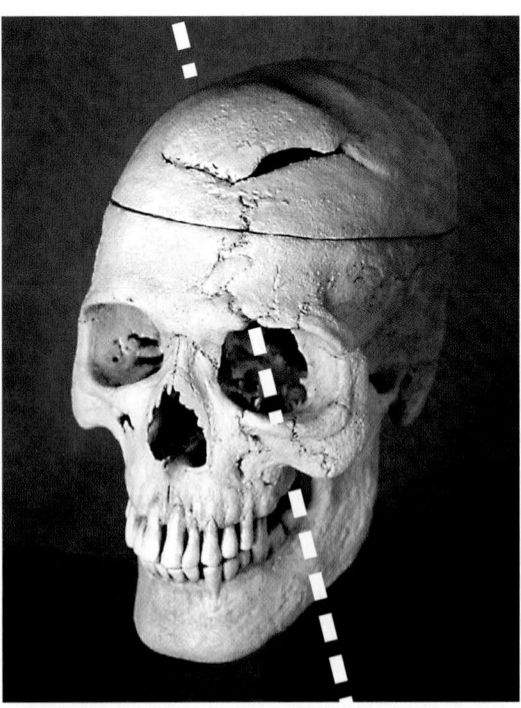

**47.7 A Mind-Altering Experience** Phineas Gage miraculously survived a nineteenth-century railroad construction accident in which an explosion blew an iron rod through his brain. His personality, however, was permanently altered from a responsible foreman to a quarrelsome drifter. The path of the iron rod through Gage's brain is superimposed on this reconstruction of his skull.

**THE PARIETAL LOBE** The strip of parietal lobe cortex just behind the central sulcus is the **primary somatosensory cortex** (see Figure 47.4B). This area receives touch and pressure information relayed from the body through the thalamus.

The entire body surface can be mapped onto the primary somatosensory cortex (**Figure 47.6B**). Areas of the body that have a high density of tactile mechanoreceptors and are capable of making fine discriminations in touch (such as the lips and fingers) have disproportionately large representation. If a very small area of the primary somatosensory cortex is stimulated

**47.8 Evolution of the Human Brain** Brain size scales to body size across a wide range of vertebrates. The higher primates have larger brains than predicted by the correlation, and humans stand outside this relationship with much bigger brains. The increase in brain size in humans is mostly due to an increase in the cerebral cortex. The human brain is also highly convoluted, and more of it is devoted to associative functions.

electrically, the subject reports feeling specific sensations, such as touch, in a localized part of the body.

A major association function of the parietal lobe is attending to complex stimuli. Damage to the right parietal lobe causes a condition called contralateral neglect syndrome, in which the individual tends to ignore stimuli from the left side of the body or the left visual field. Such individuals have difficulty performing complex tasks, such as dressing the left side of the body; an afflicted man may not be able to shave the left side of his face. When asked to copy simple drawings, a person who exhibits this syndrome can do well with the right side of the drawing but not the left.

The parietal cortex is not symmetrical with respect to its role in attention. Damage to the left parietal cortex does not cause the same degree of neglect of the right side of the body. We will see similar asymmetries in cortical function later in the chapter when we discuss language.

**THE OCCIPITAL LOBE** The occipital lobe receives and processes visual information. The association areas of the occipital cortex are essential for making sense of the visual world and translating visual experience into language. Some deficits resulting from damage to these areas are specific. In one case, a woman with limited damage was unable to see motion. Her vision was intact, but she could see a waterfall only as a still image, and an approaching car only as a series of a stationary object at different distances.

## The size of the human brain is off the curve

Humans are sometimes called "big-brain primates," and that is an accurate characterization. Across vertebrate species there is a correlation between body size and brain size (**Figure 47.8**). Higher primates such as chimpanzees, baboons, and gorillas all fall above this regression line, but humans stand out because they are so far above the regression line. Gorillas are much larger than humans, but they have smaller brains. Elephants and whales have large brains, but they fall closer to the regression line. Dolphins and humans stand out as having much larger brains than would be predicted by their body sizes.

The correlation of brain size to body size does not tell the whole story of human brain evolution, however. In Figure 45.19, which compares the brains of four vertebrates, we see that the forebrain is larger than other brain regions, and in mammals this is seen as an elaboration of the cerebral cortex. If we look just at mammals, another feature is the degree of convolution of the cortex. Since the cortex is a layered, two-dimensional array of neurons, the area of cortex is increased by convolutions, which are greatest in humans. And finally, the percent of the cortex that is association cortex (i.e., the brain regions devoted to the integration of information) is by far the greatest in humans. It is these evolutionary changes, primarily in the cortex, that provide the resources for the intellectual capacity of humans—a topic to which we will return at the end of the chapter.

**RECAP 47.1**

The central nervous system communicates with the rest of the body through the peripheral nervous system. We are conscious of some sensory information coming into the CNS, but we are not conscious of other afferent information used in physiological regulation. Different regions of the brain have specific functions. Evolution of the human brain has resulted in a greatly increased cerebral cortex devoted to integration of information.

- Explain how the major functional divisions of the nervous system relate to their origins in the embryonic neural tube. **See pp. 968–969 and Figure 47.2**
- Describe the spatial relations and functions of the major divisions of the telencephalon. **See pp. 970–971 and Figure 47.4**
- What features distinguish the human brain from the brains of other mammals? **See p. 973 and Figure 47.8**

Having briefly described the structure and function of different regions of the nervous system, we will now explore some examples of how information is processed by the neural circuitry in some specific brain regions.

## 47.2 How Is Information Processed by Neural Networks?

Specific functions are localized in specific parts of the nervous system and depend on the neural circuits, or networks, in those structures. A major focus of modern neuroscience is understanding how the various functions of the nervous

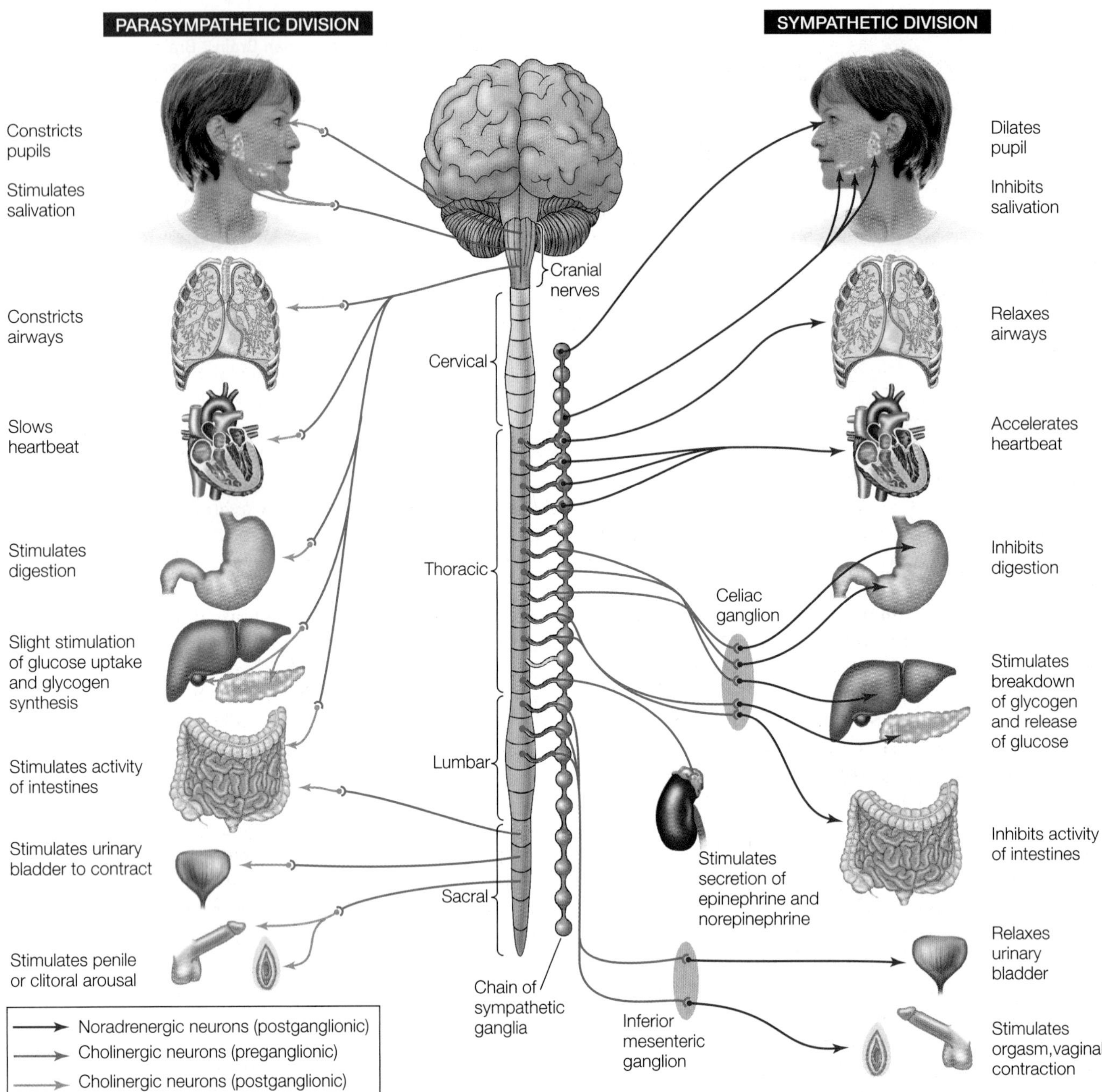


**47.9 The Autonomic Nervous System** The autonomic nervous system is divided into the sympathetic and parasympathetic divisions. The two divisions work in opposition to each other in their effects on most organs; one results in an increase and the other a decrease in activity.

system, ranging from simple reflexes to complex learning and memory, are accomplished by the interactions of neurons in circuits. Two extensively studied examples of how neural networks process information are the autonomic nervous system (an output pathway) and the visual system (an input pathway).

## Pathways of the autonomic nervous system control involuntary physiological functions

The **autonomic nervous system**, or **ANS**, comprises the output pathways of the CNS that control involuntary functions, such as heart rate, blood flow, sweating, and digestive activities. Its control of diverse organs and tissues is crucial to homeostasis. The ANS has two divisions, **sympathetic** and **parasympathetic**, that work in opposition to each other in their effects on most organs: one division causes an increase in an activity and the other a decrease (**Figure 47.9**). The sympathetic and parasympathetic divisions are easily distinguished by their anatomy, neurotransmitters, and actions.

The best-known functions of the ANS are those of the sympathetic division that produce the fight-or-flight response: increasing heart rate, blood pressure, and cardiac output and preparing the body for emergencies (see Figure 41.3). In contrast, the parasympathetic division slows the heart and lowers blood pressure; its actions have been characterized as "rest and digest." It is tempting to think of the sympathetic division as speeding things up and the parasympathetic division as slowing things down, but it is not that simple; for example, the sympathetic division slows down the digestive system whereas the parasympathetic division accelerates it.

Whether sympathetic or parasympathetic, every autonomic efferent pathway begins with a cholinergic neuron (that is, a neuron that uses acetylcholine as its neurotransmitter) that has its cell body in the brainstem or spinal cord. These cells are called preganglionic neurons because the second neuron in the pathway with which they synapse resides in a collection of neurons outside the CNS called a **ganglion** (plural *ganglia*). The second neuron is called a postganglionic neuron because its axon extends out from the ganglion. The axon of the postganglionic neuron synapses with cells in the target organs (see Figure 47.9).

The postganglionic neurons of the sympathetic division are called noradrenergic because they use norepinephrine (also known as noradrenaline) as their neurotransmitter. In contrast, the postganglionic neurons of the parasympathetic division are cholinergic. In organs that receive both sympathetic and parasympathetic input, the target cells respond in an opposite manner to norepinephrine and to acetylcholine. This happens, for example, in a region of the heart called the pacemaker, which generates the heartbeat. Stimulating the sympathetic nerve to the heart or dripping norepinephrine onto pacemaker cells increases their firing rate and causes the heart to beat faster. In contrast, stimulating the parasympathetic nerve to the heart or dripping acetylcholine onto pacemaker cells decreases their firing rate and causes the heart to beat more slowly.

The sympathetic and parasympathetic divisions of the ANS can also be distinguished by anatomy. The preganglionic neurons of the parasympathetic division come from the cranial nerves of the brainstem and the sacral (lowest) region of the spinal cord; those of the sympathetic division come from the thoracic and lumbar regions of the spinal cord (see Figure 47.9). Most of the ganglia of the sympathetic division are lined up in two chains, one on either side of the spinal cord. The parasympathetic ganglia are close to the target organs.

The autonomic nervous system is an important link between the CNS and many physiological functions. Its control of diverse organs and tissues is crucial to homeostasis. Despite the complexity of the ANS, work by neurobiologists and physiologists over many decades has made it possible to understand its functions in terms of neuronal properties and circuits.

## The visual system is an example of information integration by the cerebral cortex

The visual system is one of the most-studied input pathways to the central nervous system. In Section 46.4 we described how light falling on the retina produces signals that are transmitted through the cellular circuits of the retina, resulting in action potentials in the optic nerve. But how does the central nervous system use this information to reconstruct the visual world in the brain? The experiments that have investigated this question are some of the most famous experiments in neurobiology.

**RETINAL RECEPTIVE FIELDS** Section 46.4 described how a retinal ganglion cell collects information from a number of photoreceptors—an example of "convergence of information." Each ganglion cell is communicating to the brain something more than simply the presence or absence or intensity of light falling on a portion of the retina. What information does the retinal ganglion cell extract from the photoreceptors?

This question was addressed in classic experiments by Stephen Kuffler, then at Johns Hopkins University. He used electrodes to record the activity of single ganglion cells of cat eyes while stimulating their retinas with spots of light (**Figure 47.10**). These experiments were the starting point for understanding how the brain assembles information from single cells to create visual images—in other words, of how the brain sees.

Kuffler's experiments revealed that each ganglion cell has a well-defined **receptive field** composed of a group of photoreceptor cells that receive light from a small area of the entire visual field. Stimulating these photoreceptors with light activates the ganglion cell, which sends action potentials to the thalamus and on to the visual cortex (the area of the occipital lobe where visual information is processed; see Figure 47.4). Information from many photoreceptors is therefore communicated to the brain as a single message. Individual photoreceptors may contribute to the receptive fields of multiple ganglion cells, so that receptive fields overlap.

The receptive fields of most ganglion cells are circular, but whether a spot of light falling on a receptive field excites or inhibits its ganglion cell depends both on the nature of the receptive field and on where the spot of light falls on it. Receptive fields have a center and a concentric surround, and can be either "on-center" or "off-center." Light falling on the center of an on-center receptive field excites the ganglion cell, and light falling on the center of an off-center receptive field inhibits the ganglion cell. Light falling on the surround has the opposite effect: the surround for an on-center receptive field inhibits the ganglion cell, and the surround for an off-center field is excitatory. Thus the activity of the ganglion cell reflects how much of the light stimulus is on the center and how much is on the surround of its receptive field (see Figure 47.10).

Center effects are always stronger than surround effects. Thus a small dot of light directly on the center of a receptive field has the maximal effect, and a larger light stimulus illuminating the center and parts of the surround has a smaller effect. A uniform patch of light falling equally on the center and surround has very little effect on the firing rate of the ganglion cell for that receptive field.

**Go to Animated Tutorial 47.1**
**Visual Receptive Fields**
Life10e.com/at47.1

How are cells in the retina connected to each other to create receptive fields? Remember from Section 46.4 that photoreceptors synapse onto bipolar cells and bipolar cells onto ganglion cells. This pattern of connectivity describes the relationship between the photoreceptors in the center of a receptive field. The photoreceptors in the surround area modify communication between the center photoreceptors and their bipolar cells through the lateral connections of horizontal cells and amacrine cells. Thus the receptive field of a ganglion cell results

INVESTIGATING**LIFE**

**47.10 What Does the Eye Tell the Brain?** Stephen Kuffler's experiments recorded the activity of single ganglion cells in the eyes of cats. These groundbreaking experiments revealed the existence of a circular receptive field for each of the retina's ganglion cells. Signals from photoreceptor cells in a receptive field are either excitatory or inhibitory to the ganglion cell, which sends action potentials via the optic nerve to the brain.[a]

**HYPOTHESIS** Retinal ganglion cells are excited or inhibited by light and dark stimuli falling on local areas of the retina.

*Method*
1. Place electrodes next to an individual retinal ganglion cell.
2. Stimulate the retina with different combinations of light and dark stimuli and record the responses of the ganglion cell.
3. Continue recording and move the stimuli around the retina to find the area of sensitivity—the receptive field—for a specific ganglion cell.

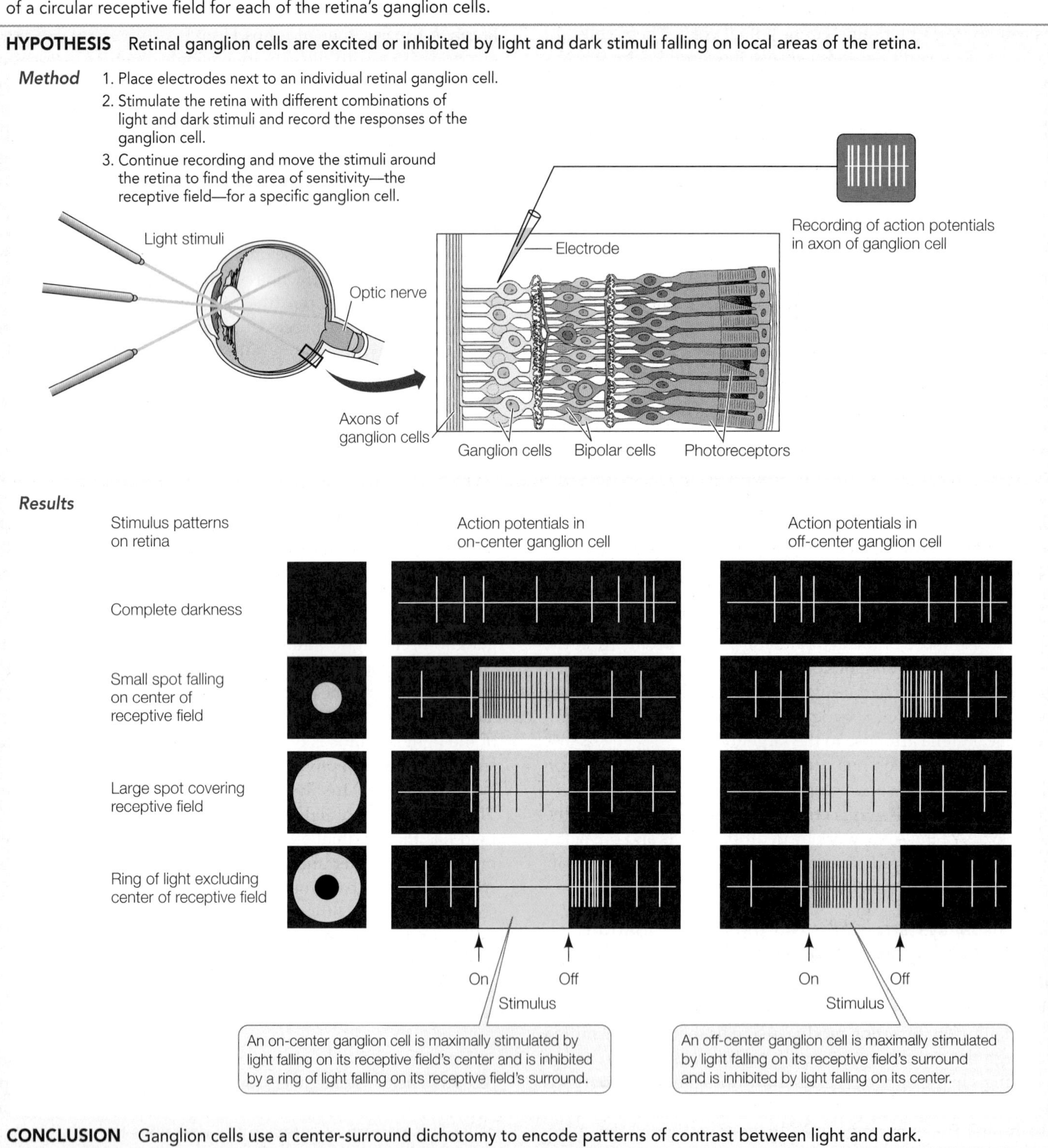


**CONCLUSION** Ganglion cells use a center-surround dichotomy to encode patterns of contrast between light and dark.

Go to **BioPortal** for discussion and relevant links for all INVESTIGATING**LIFE** figures.

[a]Kuffler, S. W. 1953. *Journal of Neurophysiology* 16: 37–68.

from a pattern of synapses between photoreceptors, horizontal cells, amacrine cells, and bipolar cells. A general lesson to learn from this seemingly confusing chain of events is that inhibition can be as important as excitation in neural circuits.

In summary, the neural circuitry of the retina results in the generation of signals in the axons of the optic nerve to the brain that communicate simple information about the contrasting patterns of light and dark falling on different parts of the retina. But once the action potentials in the optic nerve reach their destinations, how does the brain integrate them to construct visual images of the outside world?

**Go to Animated Tutorial 47.2**
**Information Processing in the Retina**
Life10e.com/at47.2

**RECEPTIVE FIELDS OF CELLS IN THE VISUAL CORTEX** The axons of the optic nerves terminate in a region of the thalamus that is a relay station receiving information from both the right and left eyes. From the thalamus, the information encoded in the activity of axons in the optic nerves is relayed to the visual cortex in the occipital lobes at the back of the brain. In the 1960s David Hubel and Torsten Wiesel of Harvard University studied the activity of neurons in the visual cortex by shining spots and bars of light on retinas while recording the activities of single cells in the cortex. They found that neurons in the visual cortex, like retinal ganglion cells, have receptive fields.

Neurons in the visual cortex respond selectively to bars of light of different orientations falling on the retina and in some cases to movement of those bars of light in different directions. The concept that emerges from these experiments is that the brain assembles a mental image of the visual world by analyzing edges in patterns of light falling on the retina. Each retina sends a million axons to the brain, but there are *hundreds of millions* of neurons in the visual cortex. The action potentials from one retinal ganglion cell are received by hundreds of cortical neurons, each responsive to a different combination of orientation, position, color, and movement of contrasting lines in the patterns of light and dark falling on the retina.

## Three-dimensional vision results from cortical cells receiving input from both eyes

How do we perceive objects in three dimensions? The short answer is that a person's front-facing two eyes see overlapping, yet slightly different, visual fields—that is, humans have **binocular vision**. A person who is blind in one eye has difficulty discriminating distances. Animals whose eyes are on the sides of the head rather than facing front have minimal overlap in their fields of vision and, as a result, poor depth vision; however, they can see predators creeping up from all sides.

The story of how the brain integrates information from two eyes begins with the paths of the optic nerves. The two optic nerves run along the underside of the brain, join just under the hypothalamus, and then separate again (**Figure 47.11A**). The place where they join is called the **optic chiasm**. Axons from the half of each retina closest to the nose cross in the optic chiasm and go to the opposite side of the brain. The axons from the

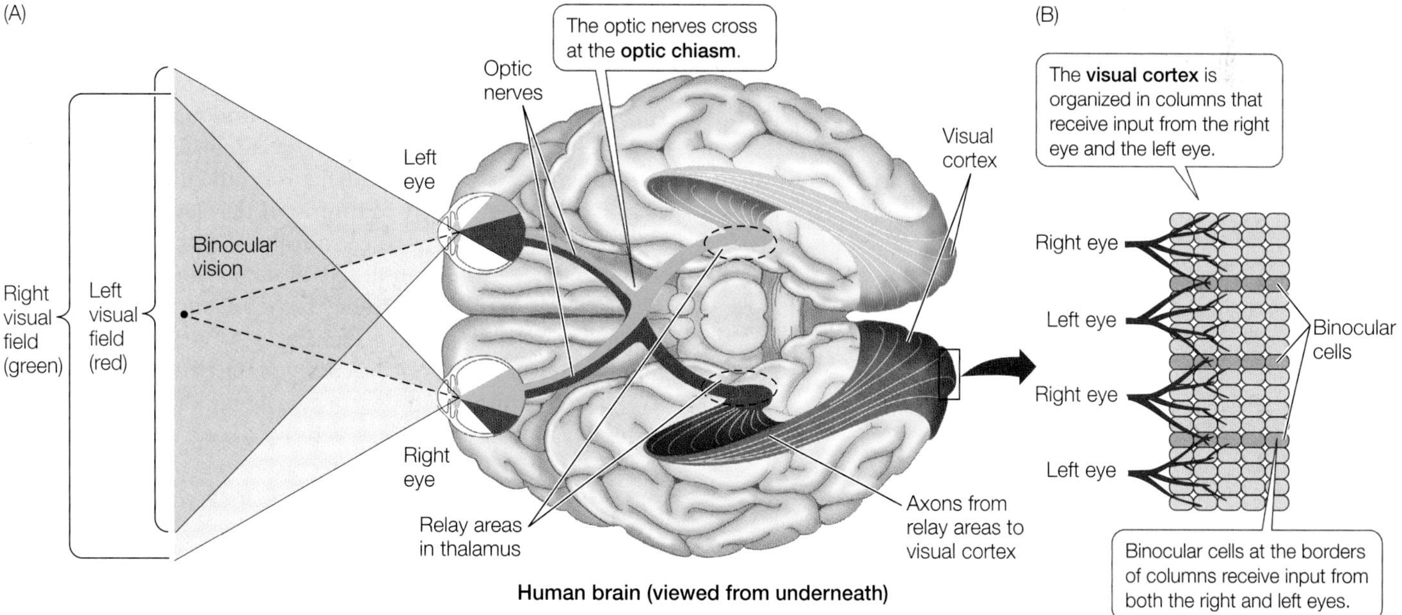


**47.11 Anatomy of Binocular Vision** (A) Each eye transmits information to both sides of the brain; however, the right side of the brain processes all information from the left visual field (red), and the left side of the brain processes all information from the right visual field (green). (B) The visual cortex sorts visual field information according to whether it comes from the right or left eye.

outer half of each retina do not cross over at the optic chiasm; axons from the outer left retina go to the left side of the brain, and vice versa for axons from the outer right retina.

The functional consequence of the optic chiasm is that all of the visual information from the left side of your field of vision when you are looking straight ahead goes to the right side of your brain, and all of the visual information from the right side of your field of vision goes to the left side of your brain. These relationships are shown in red and green in Figure 47.11A.

Cells in the visual cortex are organized in stripes and columns. Stripes refer to the organization across the surface of the cortex, and columns to the organization through the depth of the cortex (**Figure 47.11B**). Stripes and columns alternate according to the source of their input: left eye, right eye, left eye, right eye, and so on. Cells closest to the border between two stripes or columns receive input from both eyes and thus are called binocular cells. Binocular cells interpret distance by measuring the *disparity* between the points at which the same stimulus falls on the two retinas.

What is disparity? Hold your finger out in front of you and look at it closing one eye and then the other. Your finger appears to jump back and forth because its image falls on a different position on each retina. Repeat the exercise with a distant object. It doesn't jump back and forth as much because there is less disparity in the positions of the image on the two retinas. Certain binocular cells respond optimally to a stimulus falling on both retinas with a particular disparity. Which set of binocular cells is stimulated depends on how far away the stimulus is.

When we look at something, we can detect its shape, color, depth, and movement. Where does all this information come together? Is there a single cell that fires only when a red sports car drives by? The answer to that is "no." Specific visual experience comes from simultaneous activity in a large collection of cells. In addition, most visual experiences are enhanced by information from the other senses and from memory, which helps explain why about 75 percent of the cerebral cortex is association cortex.

**RECAP 47.2**

Information in the nervous system is processed by cellular interactions in neural networks. The opposing actions of the sympathetic and parasympathetic divisions of the ANS can be understood in terms of neural pathways consisting of just two neurons. Vision involves a more complex interaction of neurons, organized into receptive fields, to process patterns of light and dark falling on the retina.

- Describe the anatomical and functional differences between the sympathetic and parasympathetic divisions. **See pp. 974–975 and Figure 47.9**
- Explain the cellular basis for the receptive fields of retinal ganglion cells. **See p. 975 and Figure 47.10**
- How do cells in the visual cortex interpret information about how far away an object is? **See pp. 977–978 and Figure 47.11**

By studying the neural circuitry of the visual system and the autonomic nervous system, you have gained some understanding of how information reaches the central nervous system and how the CNS controls various functions of the body. But what about the higher functions of the mammalian CNS—the complex functions between input and output, such as language, learning, memory, and dreams?

## 47.3 Can Higher Functions Be Understood in Cellular Terms?

The higher brain functions discussed in the remaining pages of this chapter are undeniably complex. Nevertheless, neuroscientists, using a wide range of techniques, are making considerable progress in understanding some of the cellular and molecular mechanisms involved in those processes. The following discussion will address several aspects of brain and behavior that present challenges to neuroscientists: sleep and dreaming, learning and memory, language use, and consciousness.

### Sleep and dreaming are reflected in electrical patterns in the cerebral cortex

A dominant feature of behavior is the daily cycle of sleep and waking. All birds and mammals, probably all other vertebrates, and also many invertebrates, sleep. We humans spend one-third of our lives sleeping, yet we do not know why or how. We do know, however, that we need to sleep. Loss of sleep impairs alertness and performance. Many people in our society—certainly most college students—are chronically sleep-deprived. Accidents and serious mistakes that endanger lives can be attributed to impaired alertness caused by lack of sleep. Insomnia (difficulty in falling or staying asleep) is one of the most common medical complaints.

THE ELECTROENCEPHALOGRAM A common tool of sleep researchers is the **electroencephalogram**, or **EEG**. Rather than recording the activity of single neurons, the EEG characterizes activity in huge numbers of neurons. EEG electrodes are much larger than the very fine electrodes used to detect single cell activity. Placed at different locations on the head and scalp (**Figure 47.12A**), EEG electrodes record changes in the electric potential differences between electrodes over time. These differences reflect the electrical activity of the neurons in the brain regions under the electrodes, primarily regions of the cerebral cortex. Usually the electrical activity of one or more skeletal muscles is also recorded; this record is called an electromyogram (EMG). Movements of the eyes are recorded as an electroocculogram (EOG).

EEG, EMG, and EOG patterns reveal the transition from being awake to being asleep. They also reveal that there are different states of sleep. In mammals other than humans, two major sleep states are easily distinguished: **slow-wave sleep** and **rapid eye movement (REM) sleep**. Slow-wave sleep gets its name from the high-amplitude, slow-frequency waves in the EEG. REM sleep gets its name from jerky movements of the eyeballs that occur during this state. In humans, sleep states are characterized as non-REM sleep and REM sleep. Human

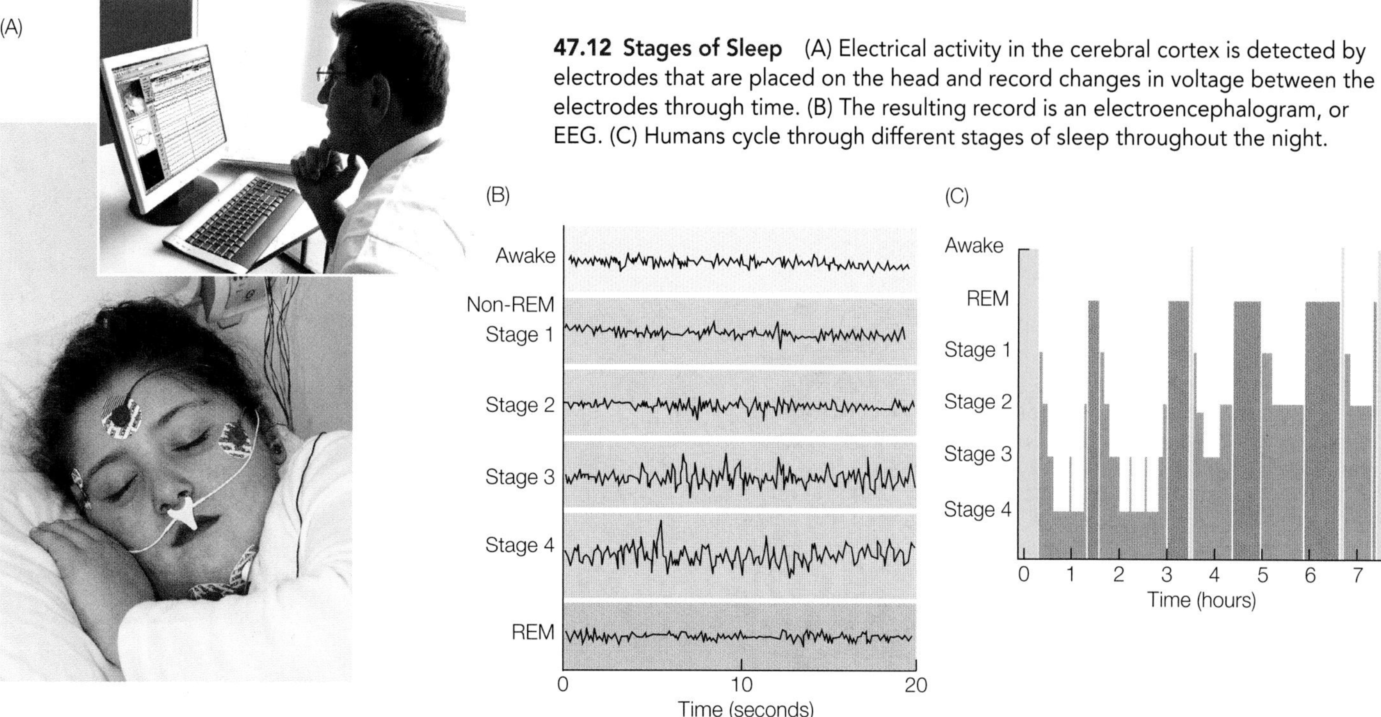


**47.12 Stages of Sleep** (A) Electrical activity in the cerebral cortex is detected by electrodes that are placed on the head and record changes in voltage between the electrodes through time. (B) The resulting record is an electroencephalogram, or EEG. (C) Humans cycle through different stages of sleep throughout the night.

non-REM sleep is divided into four stages, of which the two deepest stages are considered true slow-wave sleep.

When you fall asleep, the first state entered is stage 1 non-REM sleep, which then progresses through stages 2, 3, and 4 (**Figure 47.12B**). Stages 3 and 4 are deep, restorative, slow-wave sleep. This first full cycle of non-REM sleep is followed by an episode of REM sleep. Throughout the night you experience four or five cycles of non-REM and REM sleep (**Figure 47.12C**). About 80 percent of your sleep is non-REM sleep. The most vivid dreams and nightmares occur during the 20 percent of sleep that is REM sleep.

**CELLULAR CHANGES DURING SLEEP** When we are awake, several nuclei in the brainstem reticular formation are continuously active. Axons from neurons in these nuclei extend to the thalamus and throughout the cerebral cortex, where they release depolarizing neurotransmitters (acetylcholine, norepinephrine, and serotonin). These broadly distributed neurotransmitters keep the resting potential of the neurons of the thalamus and cortex close to threshold and sensitive to synaptic inputs, thereby maintaining the responsiveness of the brain that characterizes being awake.

With the onset of sleep, activity in these brainstem nuclei decreases, and their axon terminals release less neurotransmitter. With the withdrawal of the depolarizing neurotransmitters, the resting potentials of the cells of the thalamus and cortex become more negative (hyperpolarized), and the cells are less sensitive to excitatory synaptic input. Their processing of information is inhibited, and consciousness is lost.

An interesting neural event happens as a result of the hyperpolarization with the onset of sleep: cells begin to fire action potentials in bursts. The synchronization of these bursts over broad areas of cerebral cortex results in the EEG slow-wave pattern that characterizes deep non-REM sleep. Studies of neurons of the thalamus and cortex have shown that their hyperpolarization during non-REM sleep is due to increased opening of $K^+$ channels, and that bursting is due to $Ca^{2+}$ channels whose inactivation gates close rapidly and require hyperpolarization to be reopened. We can therefore explain the EEG pattern of non-REM sleep in terms of the properties of neurons and ion channels.

At the transition from non-REM to REM sleep, dramatic changes occur. Some of the brainstem nuclei that were inactive during non-REM sleep become active again, causing a general depolarization of cortical neurons. Thus in REM sleep the synchronized bursts of firing cease, and the EEG resembles that of the awake brain. Because the resting potentials of the neurons return to near threshold levels, the cortex can process information, and vivid dreams occur.

So why don't we act out our dreams? During REM sleep the brain inhibits both afferent (sensory) and efferent (motor) pathways; we are paralyzed during REM sleep. Limb twitches and the jerky eye movements are motor signals breaking through the inhibition. The bizarre nature of dreams may be due to the lack of sensory feedback to the cortex from the body and the outside world. In other words, a functioning cortex is out of touch with reality. The function of muscle paralysis during REM sleep may be to prevent the acting out of dreams.

Knowing the cellular mechanisms of sleep has not yet led to an understanding of its function. Many questions remain. Why do we have two sleep states with very different neurophysiological characteristics? Why does non-REM sleep always occur first? Why do the two states cycle during the rest period? We know sleep is essential for life, but we don't know why. One set of hypotheses is that sleep is necessary for the maintenance and repair of neural connections and for the neural changes

## WORKING WITH**DATA:**

### *Sleep and Learning*

#### Original Paper

Walker, M. P., T. Brakefield, A. Morgan, J. A. Hobson, and R. Stickgold. 2002. Practice with sleep makes perfect: Sleep-dependent motor skill learning. *Neuron* 35: 205–211.

#### Analyze the Data

Does sleep enhance learning and memory? To answer this question, we need to specify the kind of learning and have an accurate means of measuring it. Walker and colleagues at Harvard Medical School investigated procedural (motor skill) memory through a simple task. Subjects used their nondominant hand to type the number sequence 4-1-3-2-4 on a computer keyboard as fast and accurately as they could for 30 seconds. The computer provided scores for speed and accuracy. Training consisted of 12 trials of 30 seconds each, with 30-second rests between trials. Retesting consisted of two similar sessions. Subjects were trained either at 10 A.M. or at 10 P.M. with the first retest after 12 hours and the second retest after 24 hours. Thus, one group had a sleep phase between the two retest events, while the other group had a sleep phase between the training and the first retest event. The results are shown in the bar graphs (right).

Another group of subjects was similarly trained at 10 P.M. and tested the next morning at 10 A.M. However, during their overnight sleep session their EEGs were recorded. Overnight improvement in motor skill scores was then correlated with the quantitative analysis of their sleep stages. The results are given in the table below (remember that most people cycle through 4–5 complete sleep cycles over the course of a night).

**QUESTION 1**

What would you conclude from these data about the role of sleep or wake in the consolidation of a procedural memory?

**QUESTION 2**

Based on the tabular data and the text descriptions of sleep stages, what feature of sleep seems to be most important for procedural memory consolidation?

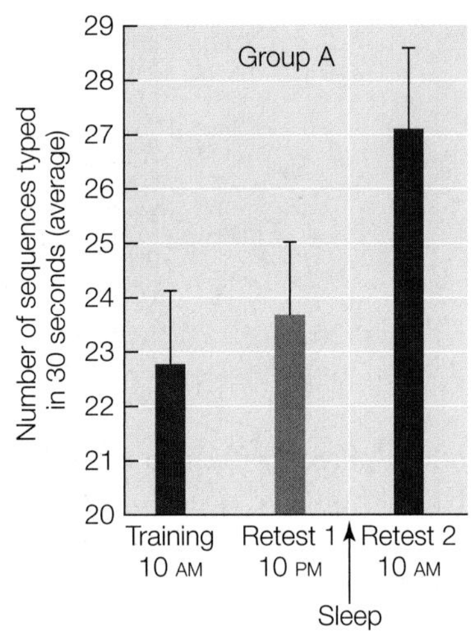


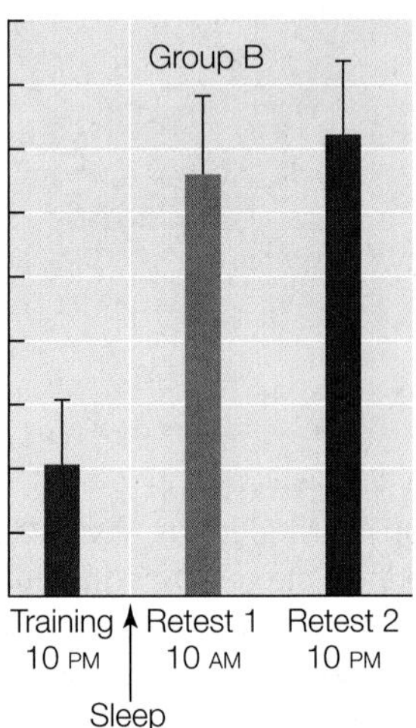


| Group C<br>Stage of sleep cycle | Percent of total time asleep (mean ± standard error) | Correlation with motor skill improvement (correlation coefficient *r*) | Significance (*p* value) |
|---|---|---|---|
| Stage 1 | 3.1 ± 0.68 | 0.41 | 0.17 |
| Stage 2 | 52.1 ± 1.94 | 0.66 | 0.01 |
| Stage 3 (slow-wave) | 10.2 ± 85 | –0.26 | 0.40 |
| Stage 4 (slow-wave) | 11.4 ± 1.51 | –0.17 | 0.59 |
| REM sleep | 18.4 ± 2.05 | –0.32 | 0.30 |

**Go to BioPortal for all** WORKING WITH**DATA exercises**

involved in learning and memory—and possibly forgetting. These hypotheses are supported by many experiments showing that performance of a learned task or recall of declarative information on the day following training is impaired if sleep is prevented, and is best following a good night's sleep.

### Language abilities are localized in the left cerebral hemisphere

No aspect of brain function is as integrally related to human consciousness and intellect as language. Therefore brain mechanisms that underlie the acquisition and use of language are extremely interesting to neuroscientists. A curious observation about language ability is that it resides in one cerebral hemisphere—which in 97 percent of people is the left hemisphere. This phenomenon is referred to as the **lateralization** of language functions.

Fascinating research on this subject was conducted by Roger Sperry and his colleagues at the California Institute of Technology. The two cerebral hemispheres are connected by a tract of white matter called the corpus callosum. In one severe form of epilepsy, bursts of action potentials causing seizures travel between hemispheres via the corpus callosum. Cutting the tract eliminates the problem, and patients function well following surgery. However, these "split-brain" subjects display interesting deficits in language ability.

After the surgery, if an object is shown in the right visual field and the left eye is closed (see Figure 47.11), the patient can describe it verbally and in writing. If the object is shown

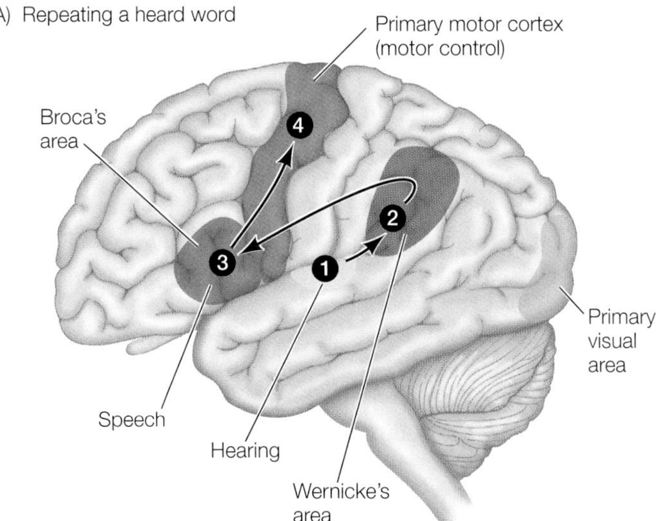


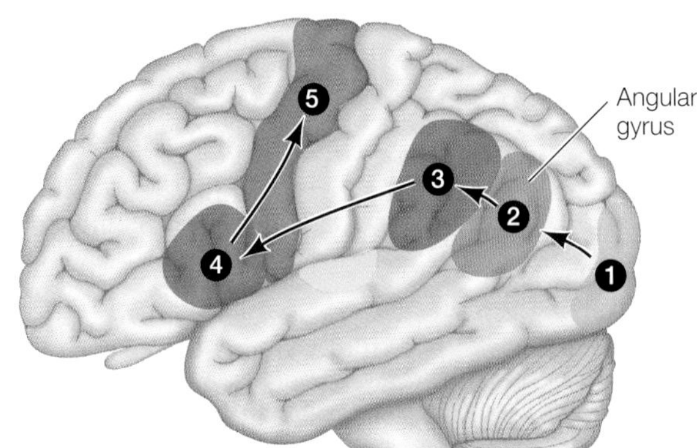


47.13 Language Areas of the Cortex Different regions of the left cerebral cortex participate in the processes of (A) repeating a word that is heard and (B) speaking a written word.

**Go to Activity 47.2 Language Areas of the Cortex Life10e.com/ac47.2**

in the left visual field and the right eye is closed, the patient cannot describe it either verbally or in writing, but can use his or her left hand to point to a picture of the object. Without the connecting tissue between the two hemispheres, knowledge or experience of the right hemisphere can no longer be expressed in language.

Individuals who have suffered damage to the left hemisphere frequently suffer from some form of **aphasia**, a deficit in the ability to use or understand words. Studies of such individuals have identified several language areas in the left hemisphere.

- **Broca's area**, located in the frontal lobe just in front of the primary motor cortex, is essential for speech. Damage to Broca's area results in halting, slow, poorly articulated speech or even complete loss of speech, but the patient can still read and understand language.
- **Wernicke's area**, located in the temporal lobe close to its border with the occipital lobe, is more involved with sensory than with motor aspects of language. Damage to Wernicke's area can cause a person to lose the ability to speak sensibly while retaining the abilities to form the sounds of normal speech and to imitate its cadence. Such a patient cannot understand spoken or written language.
- The **angular gyrus**, located near Wernicke's area, is believed to be essential for integrating spoken and written language.

Normal language ability depends on the flow of information among various areas of the left cerebral cortex. Input from spoken language travels from the auditory cortex to Wernicke's area (**Figure 47.13A**). Input from written language travels from the visual cortex to the angular gyrus to Wernicke's area (**Figure 47.13B**). Commands to speak are formulated in Wernicke's area and travel to Broca's area and from there to the primary motor cortex. Damage to any one of those areas or the pathways between them can result in aphasia. Using modern methods of brain imaging, it is possible to see the metabolic activity in different brain areas when the brain is using language (**Figure 47.14**).

## Some learning and memory can be localized to specific brain areas

*Learning* is the modification of behavior by experience. *Memory* is the ability of the nervous system to retain what is learned and experienced. Even very simple animals can learn and remember, but these two abilities are most highly developed in humans. Consider the amount of information associated with

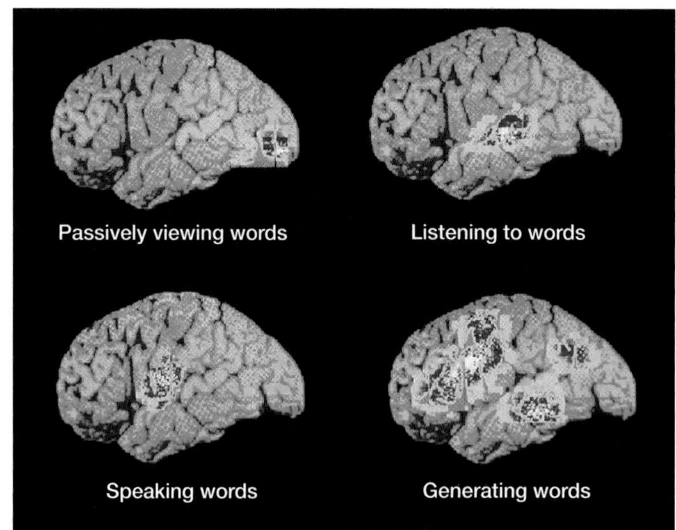


47.14 Imaging Techniques Reveal Active Parts of the Brain Positron emission tomography (PET) scanning reveals the brain regions activated by different aspects of language use. Radioactively labeled glucose is given to the subject. Brain areas take up radioactivity in proportion to their metabolic use of glucose. The PET scan visualizes levels of radioactivity in specific brain regions when a particular activity is performed. The red and white areas are the most active.

learning a language. The capacity of memory and the rate at which memories can be retrieved are remarkable features of the human nervous system.

**LEARNING** Learning that leads to long-term memory and modification of behavior must involve long-lasting synaptic changes. A phenomenon that may explain how long-term synaptic changes might arise is **long-term potentiation**, or **LTP**. LTP results from high-frequency electrical stimulation of certain identifiable circuits that makes these circuits more sensitive to subsequent stimulation. In contrast, continuous, repetitive, low-level stimulation of these same circuits reduces their responsiveness, a phenomenon that has been called **long-term depression** (**LTD**). LTP and LTD may be fundamental cellular or molecular mechanisms involved in learning and memory.

Several kinds of learning exist. A form that is widespread among animal species is **associative learning**, in which two unrelated stimuli become linked to the same response. The simplest example of associative learning was described by the Russian physiologist Ivan Pavlov. Pavlov observed that a dog salivates at the sight or smell of food. He then showed that if a non-food stimulus (such as a sound) was paired with the food, after a number of paired experiences the dog would salivate at the sound even if no food was present (see Figure 53.1). Pavlov called this a **conditioned reflex**.

More complex forms of learning, often referred to as "observational learning," are the foundation of human intelligence. The general pattern of successful observational learning has three elements:

- We pay attention to another person's behavior.
- We retain a memory of what we have observed.
- We try to copy or use that information.

A key to this scheme of learning is the way in which we create and recall memories.

**MEMORY** Some of the first insights into memory processes came from clinical treatment of patients with severe epilepsy, a disorder characterized by uncontrollable, local increases in neural activity. The resulting seizures can endanger these individuals. Serious cases of epilepsy are sometimes treated by destroying the part of the brain from which the surge of activity originates. To find the right area, the surgery is done under local anesthesia, with the patient remaining conscious. As different regions of the brain are electrically stimulated with electrodes, the patient reports the resulting sensations. Stimulation of some regions of the association cortex elicits recall of vivid memories. Such observations provided the first evidence that specific areas in the brain are associated with specific memories and that memory can be attributed to networks of neurons. Destroying a small area of the brain does not completely erase a memory, however, so it is postulated that memory is a function distributed over many brain regions and can be stimulated via many different routes.

You experience several forms of memory everyday. You have **immediate memory** for events that are happening now. Immediate memory is almost perfectly photographic but lasts only seconds. **Short-term memory** contains less information but lasts longer—on the order of 10 to 15 minutes. When you are introduced to a group of several new people, you probably will have forgotten their names in an hour or so if you have not written them down, used them in a conversation, or made a conscious effort to repeat them. Repetition, use, or reinforcement by something that gets your attention (a title such as "President," for example) facilitates the transfer of short-term memory to **long-term memory**, which can last for days, months, years, or a lifetime.

Knowledge about neural mechanisms for the transfer of short-term memory to long-term memory has come from observations of persons who have lost parts of the limbic system, notably the hippocampus. A famous case is that of the man identified as H.M., whose hippocampus on both sides of the brain was removed in 1953 in an effort to control severe epilepsy. After the surgery, H.M. was unable to transfer information to long-term memory. If someone was introduced to him, had a conversation with him, and then left the room for several minutes, when that person returned, H.M did not recognize him—it was as if the conversation had never taken place. Up until his death 55 years later, H.M. remembered events that happened before his surgery but could not remember postsurgery events for more than 10 or 15 minutes.

**Go to Media Clip 47.1**
**The Man with No Short-Term Memory**
**Life10e.com/mc47.1**

Memory of people, places, events, and things is called **declarative memory** because you can consciously recall and describe them. **Procedural memory** cannot be consciously recalled and described; it is the memory of how to perform a motor task. When you learn to ride a bicycle, ski, or use a computer keyboard, you form procedural memories. Although H.M. was incapable of forming declarative memories, he could form procedural memories. When taught a motor task day after day, he could not recall the lessons of the previous day, yet his performance steadily improved. Thus procedural learning and memory must involve mechanisms different from those used in declarative learning and memory.

Memories can have considerable emotional content. As mentioned earlier, the limbic system plays a major role in controlling emotions. The amygdala, a component of the limbic system, is necessary for the emotion of fear and the formation of fear memories. Patients with a damaged amygdala do not associate fear reactions with their declarative memories. Memories can also have positive emotional content, and recalling those memories activates parts of the brain known to be associated with pleasurable sensations and reward, as revealed by brain imaging technologies.

## We still cannot answer the question "What is consciousness?"

This chapter has only scratched the surface of the organization and functions of the human brain. Even with all of our knowledge of the human brain, and with all of the sophisticated new research tools, we still cannot answer the question "What is consciousness?"

The word "consciousness" is used in everyday language to refer to being awake in contrast to being asleep or in a coma. Here we are referring to the deeper meaning of being mentally aware of yourself, your environment, and events going on around you in such a way that you can plan for future events and make decisions based on experience, evidence, value systems, and predicted consequences. Speculations about consciousness have been the realm of philosophers, but we are getting closer to a neurobiological understanding.

The central requirement for conscious experience is a perception of self that can be integrated with information from the physical and social environment and information from past experience. The basis for a perception of self derives from the huge amount of somatosensory and visceral information that comes from all parts of the body. In the CNS of all vertebrates, this information is used for motor control and for homeostatic regulation. It enables animals to find food, seek mates, seek warmth, avoid cold, avoid danger, and so on. This afferent information goes to appropriate control and regulatory systems in the brainstem and forebrain.

In addition, some of this information goes to somatosensory areas of the cerebral cortex, so the animal is aware of certain information in the sense that it responds to it behaviorally. Visceral afferent information goes beyond its regulatory and control centers in the brainstem and hypothalamus to an area deep within the forebrain called the insular cortex, or **insula**. The insula appears to integrate physiological information from all over the body to create a sensation of how the body "feels." Thus when an animal's actions restore homeostasis, it "feels" better, and this is motivation to do the right thing for well-being.

In humans and the great apes, the insula is greatly expanded and has even acquired new types of spindle-shaped neurons not seen in other animals. The circuitry involving the insula has also evolved to communicate with parts of the brain that are involved in planning and decision making. In imaging studies, the insula is seen to be active in a great diversity of situations that involve strong feelings such as pleasure, disgust, humor, pain, lust, craving, humiliation, guilt, or empathy. Damage to the insula causes apathy, loss of ability to enjoy music, loss of sexual response, and even loss of the ability to distinguish good food from spoiled food. Humans and the few other species that have expanded insulas and the new spindle cells are the only species that can recognize themselves in a mirror. Could it be that this very discrete part of our brains and its circuitry are the neurobiological bases for self-awareness and conscious experience?

**RECAP 47.3**

Even complex functions of the nervous system are beginning to be understood in terms of the properties of neurons and neural networks.

- What events in the brain are associated with wakefulness and the stages of sleep? **See p. 979**
- Why do some neurobiologists think that the insula might be involved in conscious experience? **See p. 983**

**Does the firing pattern of hippocampal place cells during sleep represent a memory of recent experience that is being transferred into long-term memory?**

**ANSWER**

Unfortunately we cannot give a rat an exam to ask what it remembers. However, we can look in other brain areas for electrophysiological patterns of activity that correlate with the hippocampal place cell patterns. Wilson and colleagues described the fast replay of the hippocampal place cell patterns as "ripples." They observed that these ripples were tightly coupled to EEG brainwaves that spread to cortical areas of the brain. When they recorded in some of those other areas—the frontal cortex and the visual cortex—they found similar patterns of firing (i.e., ripples) in those areas that were synchronized to the hippocampal ripples. They hypothesized that the ripples did represent memory transcripts that were transferred to and stored in areas of the cortex.

## CHAPTER**SUMMARY** 47

### 47.1 How Is the Mammalian Nervous System Organized?

- The brain and spinal cord make up the **central nervous system (CNS)**; the cranial and spinal nerves make up the **peripheral nervous system (PNS)**.
- The nervous system can be modeled conceptually in terms of the direction of information flow and whether we are conscious of the information. The afferent component carries information from the PNS to the CNS, and the efferent component directs information from the CNS to the peripheral parts of the body. **Review Figure 47.1**
- The vertebrate nervous system develops from a hollow dorsal neural tube. The brain forms from three swellings at the anterior end of the neural tube, which become the **hindbrain**, the **midbrain**, and the **forebrain**. The forebrain develops into the **cerebral hemispheres** (the **telencephalon**, or **cerebrum**) and the underlying thalamus and hypothalamus (which together compose the **diencephalon**). The midbrain and hindbrain develop into the **brainstem** and the **cerebellum**. **Review Figure 47.2**
- The spinal cord communicates information between the brain and the rest of the body.
- The **reticular activating system** is a complex network that directs incoming information to appropriate brainstem **nuclei** that control autonomic functions, and transmits the information to the forebrain that results in conscious sensation. The reticular activating system controls the level of arousal of the nervous system, including sleep and wakefulness.
- The **limbic system** is an evolutionarily primitive part of the telencephalon that is involved in emotions, physiological drives (such as hunger and thirst), instincts, and memory. **Review Figure 47.3**

*continued*

- The cerebral hemispheres are the dominant structures of the human brain. Their surfaces are layers of neurons called the **cerebral cortex**. The cerebral hemispheres can be divided into the **temporal**, **frontal**, **parietal**, and **occipital lobes**. Many motor functions are localized in parts of the frontal lobe. Information from many sensory receptors projects to a region of the parietal lobe. Visual information projects to the occipital lobe, and auditory information projects to a region of the temporal lobe. **Review Figures 47.4, 47.5, 47.6, ACTIVITY 47.1**

## 47.2 How Is Information Processed by Neural Networks?

- The **autonomic nervous system** (**ANS**) consists of efferent pathways that control the physiological function of organs and organ systems. Its **sympathetic** and **parasympathetic** divisions are characterized by their anatomy, neurotransmitters, and effects on target tissues. **Review Figure 47.9**
- The neural network of vision involves patterns of light falling on **receptive fields** in the retina. Receptive fields have a center and a surround, which have opposing effects on ganglion cell firing. **Review Figure 47.10, ANIMATED TUTORIALS 47.1, 47.2**
- Information from retinal ganglion cells is communicated via the optic nerve to the thalamus and then to the visual cortex. The visual cortex seems to assemble an image of the visual world by analyzing edges of patterns of light.
- **Binocular vision** is possible because information from both eyes is communicated to binocular cells in the visual cortex. These cells interpret distance by measuring the disparity between where the same stimulus falls on the two retinas. **Review Figure 47.11**

## 47.3 Can Higher Functions Be Understood in Cellular Terms?

- Humans have a daily cycle of sleep and waking. Sleep can be divided into **rapid eye movement** (**REM**) **sleep** and non-REM sleep. Deep non-REM sleep is known as **slow-wave sleep** because of its characteristic EEG patterns. **Review Figure 47.12**
- Language abilities are localized mostly in the left cerebral hemisphere, a phenomenon known as **lateralization**. Different areas of the left hemisphere—including **Broca's area**, **Wernicke's area**, and the **angular gyrus**—are responsible for different aspects of language. **Review Figures 47.13, 47.14, ACTIVITY 47.2**
- Some learning and memory processes have been localized to specific brain areas. Long-lasting changes in synaptic properties referred to as long-term potentiation (LTP) and **long-term depression** (**LDP**) may be involved in learning and memory.
- Complex memories can be elicited by stimulating small regions of association cortex. Damage to the hippocampus can destroy the ability to form long-term **declarative memory** but not **procedural memory**.
- A sense of the physiological state of the body may be created in the **insula** of the cortex from visceral afferent information. Evolution of this integrative function in higher primates and humans could be the basis for conscious experience.

**See ACTIVITY 47.3 for a concept review of this chapter**

**Go to the Interactive Summary to review key figures, Animated Tutorials, and Activities**
**Life10e.com/is47**

# CHAPTER**REVIEW**

## REMEMBERING

1. Which statement about the limbic system is *not* true?
   a. Damage to one structure in the limbic system makes it impossible to form a fear memory.
   b. The limbic system is involved in basic physiological drives, instincts, and emotions.
   c. The limbic system consists of primitive forebrain structures.
   d. The limbic system contains nuclei that maintain the cortex in an awake state.
   e. In humans, a part of the limbic system is necessary for the transfer of short-term memory to long-term memory.

2. Which of the following represents the largest portion of the human cerebral cortex?
   a. The frontal lobes
   b. The primary somatosensory cortex
   c. The temporal cortex
   d. The association cortex
   e. The occipital cortex

3. Which statement about the autonomic nervous system is true?
   a. The sympathetic division is afferent, and the parasympathetic division is efferent.
   b. The transmitter norepinephrine is always excitatory, and acetylcholine is always inhibitory.
   c. Each pathway in the autonomic nervous system includes two neurons, and the neurotransmitter of the first neuron is acetylcholine.
   d. The cell bodies of many sympathetic preganglionic neurons are in the brainstem.
   e. The cell bodies of most parasympathetic postganglionic neurons are in or near the thoracic and lumbar spinal cord.

4. Which statement is *not* true about some cells in the visual cortex?
   a. They receive input from only the left visual field.
   b. They respond most strongly to bars of light falling at specific locations on the retina.
   c. They receive inputs directly from single retinal ganglion cells.
   d. They receive input from both eyes.
   e. They respond most strongly to an object when it is a certain distance from the eyes.

5. Which of the following conclusions was supported by experiments on split-brain patients?
   a. Language abilities are localized mostly in the left cerebral hemisphere.
   b. Language abilities require both Wernicke's area and Broca's area.
   c. The ability to speak depends on Broca's area.
   d. The ability to read depends on Wernicke's area.
   e. The left hand is served by the left cerebral hemisphere.

## UNDERSTANDING & APPLYING

6. A person receives a stab wound to the left side of his neck. Miraculously, blood vessels are spared. Afterward, however, the man's left pupil remains more constricted than his right pupil, and he drools out of the left side of his mouth. How can you explain these symptoms?

7. The eyes of some animals point in the same direction as do those of humans, but in other animals (such as birds) the eyes are positioned more laterally so they point in different directions. What would be the selective advantages of these different anatomical arrangements, and what kinds of animals would you expect to have one or the other?

8. Sleepwalking occurs in up to 15 percent of the population, although it is more common in children. While sleepwalking, individuals engage in mostly routine activities but are unaware of their actions when they awake. In what state of sleep do you think sleepwalking occurs, and why?

## ANALYZING & EVALUATING

9. Patient X received a gunshot wound that destroyed the right side of his spinal cord at the midthoracic level. He has no conscious motor control over his right leg, but that leg is sensitive to painful stimuli, although not to touch. His right leg shows a reflex response when a painful stimulus is applied to his left foot, but he does not sense that stimulus as pain in his left foot. Considering that different types of information flow in different tracts in the spinal cord and that some types of information cross from one side of the cord to the other, what do the symptoms of this patient tell you about the routes of motor commands, pain sensory information, and mechanosensory information in the spinal cord?

10. High-density EEG recordings are made from arrays of 128 or 256 electrodes evenly spaced over the head. This allows measurements of local changes in EEG activity. When a person was trained in a task to use his left hand to do a difficult motor task, during subsequent sleep the slow-wave activity recorded over the posterior region of his right frontal cortex was greater than that recorded on the opposite side of his brain. Explain these results both in terms of neuroanatomy and possible functions of sleep.

Go to BioPortal at **yourBioPortal.com** for Animated Tutorials, Activities, LearningCurve Quizzes, Flashcards, and many other study and review resources.

# 48 Musculoskeletal Systems

## CHAPTER**OUTLINE**


**48.1 How Do Muscles Contract?**

**48.2 What Determines Skeletal Muscle Performance?**

**48.3 How Do Skeletal Systems and Muscles Work Together?**

**Champion Jumpers** Relative to their size, many animals have more impressive jumping skills than humans. This leopard frog (*Rana pipiens*) can leap distances up to 20 times its body length.

THE OLYMPIC RECORD for the women's long jump is 7.4 meters, set in 1988 by Jackie Joyner-Kersee. Another world-record long jump that still stands was set two years earlier by Rosie the Ribeter, who jumped 6.5 meters. Rosie was a frog competing in the Calaveras County Jumping Frog Contest. In some ways Rosie's jump is more impressive. While Jackie's jump was about 5 times her body length (i.e., her height), Rosie's was about 20 times her body length.

Both jumps were powered by skeletal muscle. Muscle tissue responds to commands from the nervous system. The cellular mechanisms of muscle contraction are essentially the same in the frog and the human, so why is the frog's jump so much more impressive? The answer involves the concept of *leverage*, which depends on the muscles and skeletal elements working together.

Both frog and human jumping muscles pull on bones that are connected at joints to make levers. A lever makes it possible for the same force to move a large mass a small distance or a small mass a large distance. The ratio of a frog's leg length to its body mass is simply greater than that in a human. Thus the frog's legs are better at moving a small mass a long distance than are the human's legs.

Let's add a flea to our interspecies competition. The flea can jump more than *200 times* its body length. This incredible performance is not due to feats of leverage, because no muscle can contract fast enough to explain the take-off velocity of the flea. A different mechanism evolved in the flea—a kind of slingshot action. At the base of the flea's jumping legs is an elastic material that is compressed by muscles while the flea is resting. When a trigger mechanism is released, the elastic material recoils and "fires" the flea into the air.

In a contest of jumping efficiency, the uncontested champion would be the kangaroo. As a human runs faster, the number of strides and the energy expended per minute increase rapidly. Neither is true for the kangaroo. When moving at speeds from about 5 to 25 kilometers per hour, the kangaroo takes the same number of strides per minute and its metabolic rate *does not increase*.

The ability to move about in the environment is a distinguishing feature of most animals. Muscles and skeletons—musculoskeletal systems—enable animals to move.

How can the kangaroo increase its speed fivefold without increasing its metabolic expenditure?

See answer on p. 1003.

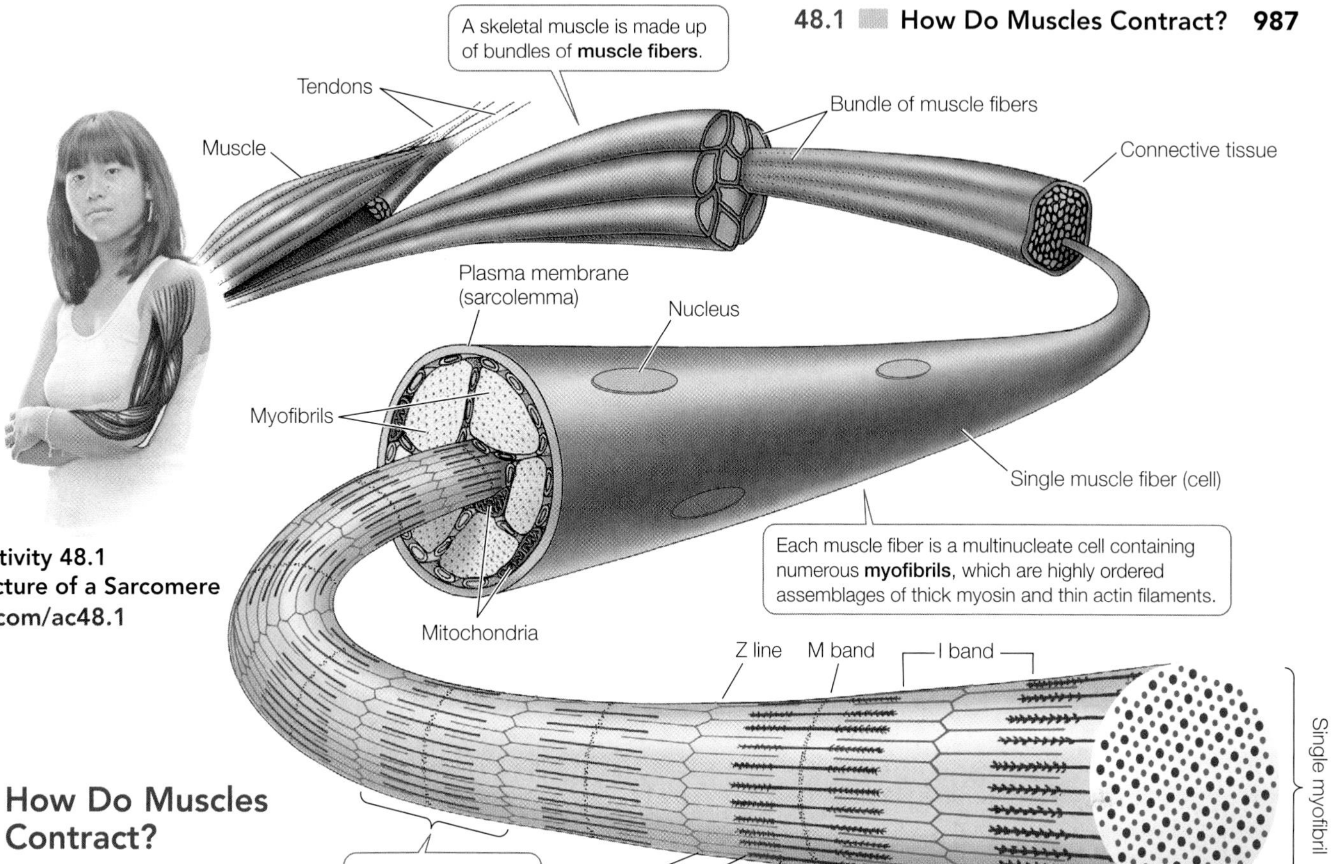

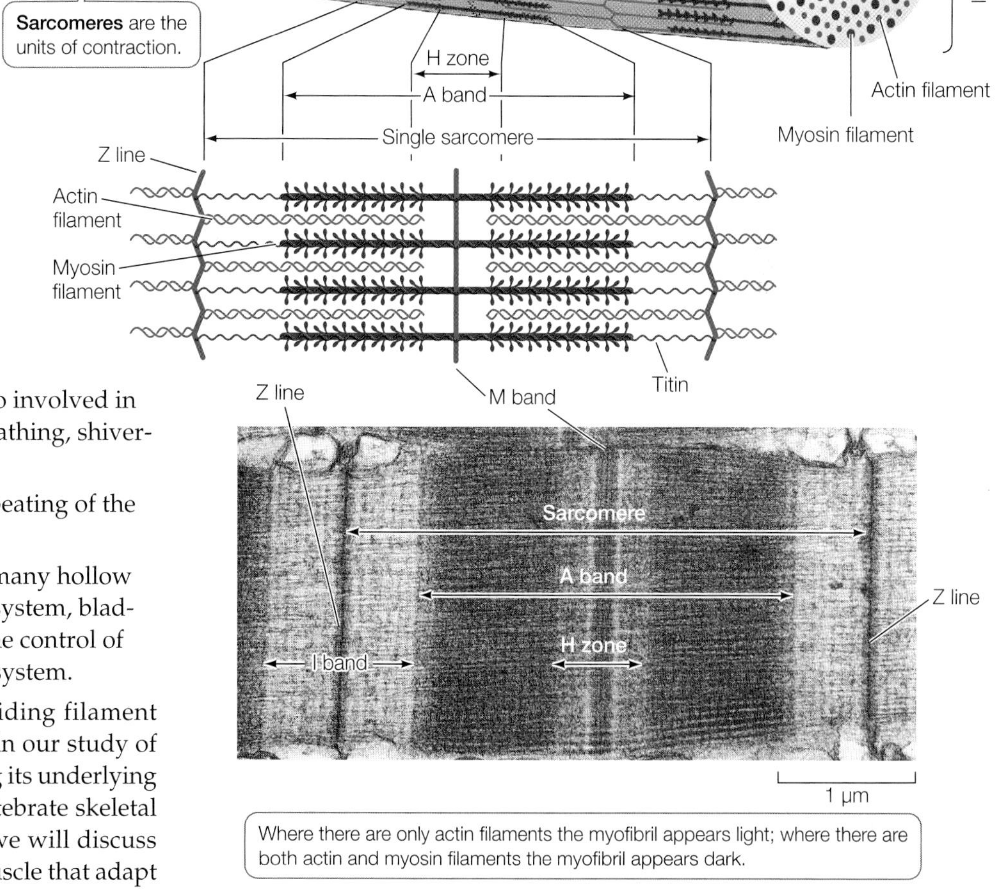

Go to Activity 48.1
The Structure of a Sarcomere
Life10e.com/ac48.1

## 48.1 How Do Muscles Contract?

Most behavior and many physiological actions, such as beating of the heart and moving of food through the digestive tract, depend on muscle contraction. Wherever tissues contract, muscle cells are responsible. As introduced in Section 40.2 and shown in Figure 40.4, there are three types of vertebrate muscle:

- **Skeletal muscle** is responsible for all voluntary movements, such as running or playing a piano. It is also involved in many involuntary actions, such as breathing, shivering, and maintaining posture.
- **Cardiac muscle** is responsible for the beating of the heart.
- **Smooth muscle** creates movement in many hollow internal organs, such as the digestive system, bladder, and blood vessels, and is under the control of the autonomic (involuntary) nervous system.

All three muscle types use the same sliding filament contractile mechanism, and we will begin our study of musculoskeletal movement by describing its underlying molecular mechanisms. We will use vertebrate skeletal muscle as our primary example. Later we will discuss the differences in cardiac and smooth muscle that adapt them to their particular functions.

### Sliding filaments cause skeletal muscle to contract

Skeletal muscle is also called striated muscle because of its striped appearance (**Figure 48.1**; also see Figure 40.4A). Skeletal muscle cells, called **muscle fibers**, are large and have many

**48.1 The Structure of Skeletal Muscle** A skeletal muscle is made up of bundles of muscle fibers. Each muscle fiber is a multinucleate cell containing numerous myofibrils, which are highly ordered assemblages of thick myosin and thin actin filaments. The arrangement of the actin and myosin filaments gives skeletal muscle fibers their characteristic striated appearance.

**48.2 Sliding Filaments**
The banding pattern of the sarcomere changes as it shortens. Observations of electron micrographs such as those on the right led to the sliding filament model of muscle contraction.

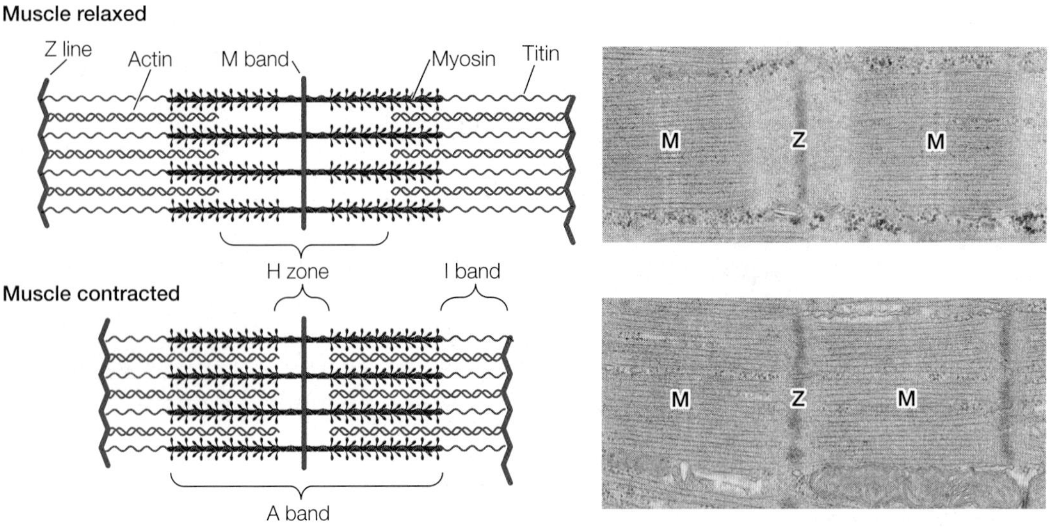

nuclei. These multinucleate cells form in development through the fusion of many individual embryonic muscle cells called myoblasts. A specific muscle such as your biceps (which bends your arm) is composed of hundreds or thousands of muscle fibers bundled together by connective tissue.

Muscle contraction is due to the interaction between the contractile proteins **actin** and **myosin**. Within muscle cells, actin and myosin molecules are organized into filaments. Actin filaments are also called thin filaments, and myosin filaments are thick filaments. The two kinds of filaments lie parallel to each other. When muscle contraction is triggered, the actin and myosin filaments slide past each other in a telescoping fashion.

What is the relationship between a skeletal muscle fiber and the actin and myosin filaments responsible for its contraction? Each muscle fiber (cell) is packed with **myofibrils**—bundles of thin actin and thick myosin filaments arranged in orderly fashion. In most regions of the myofibril, each thick myosin filament is surrounded by six thin actin filaments, and each thin actin filament sits within a triangle of three thick myosin filaments.

A longitudinal view of a myofibril reveals why skeletal muscle appears striated. The myofibril consists of repeating units called **sarcomeres**. Each sarcomere is made of overlapping filaments of actin and myosin, which create a distinct banding pattern (see Figure 48.1). Before the molecular nature of the muscle banding pattern was known, the bands were given names that are still used today. Each sarcomere is bounded by Z lines, which anchor the thin actin filaments. Centered in the sarcomere is the A band, which contains all the myosin filaments. The H zone and the I band, which appear light, are regions where actin and myosin filaments do not overlap in the relaxed muscle. The dark stripe within the H zone is called the M band; it contains proteins that hold the myosin filaments in their regular arrangement.

The bundles of myosin filaments are held in a centered position within the sarcomere by a protein called **titin**. Titin is the largest protein in the body; it runs the full length of the sarcomere from Z line to Z line. Each titin molecule runs right through a myosin bundle. Between the ends of the myosin bundles and the Z lines, titin molecules are very stretchable, like bungee cords. In a relaxed skeletal muscle, resistance to stretch is mostly due to the elasticity of the titin molecules.

As the muscle contracts, the sarcomeres shorten and the band pattern changes. The H zone and the I band become much narrower, and the Z lines move toward the A band as if the actin filaments were sliding into the H zone, the region occupied by the myosin filaments (**Figure 48.2**). In the mid 1950s this observation independently led two teams of British biologists to propose the **sliding filament model** of muscle contraction.

It is not uncommon in science for critical breakthroughs to be made simultaneously in different laboratories, but in this case the coincidences are remarkable. The leaders of the two teams were named Hugh Huxley and Andrew Huxley—but they were not related. Working in separate Cambridge University labs, the two groups proposed the sliding filament model at the same time, and both papers were published in the same issue of the journal *Nature*.

## Actin–myosin interactions cause filaments to slide

To understand how the sliding filament model explains muscle contraction, we must first examine the structures of actin and myosin (**Figure 48.3**). A myosin molecule consists of two long polypeptide chains coiled together, each ending in a large globular head. A myosin filament is made up of many myosin molecules arranged in parallel, with their heads projecting sideways at each end of the filament.

An actin filament consists of actin monomers polymerized into a long molecule that looks like two strands of pearls twisted together. Twisting around the actin chains is another protein, **tropomyosin**, and attached to tropomyosin at intervals are molecules of **troponin**. We'll discuss the latter two proteins in more detail later in this section.

The myosin heads can bind specific sites on actin, to form cross-bridges between the myosin and the actin filaments. Moreover, when a myosin head binds to an actin filament, the head's conformation changes. As the head bends, it exerts a

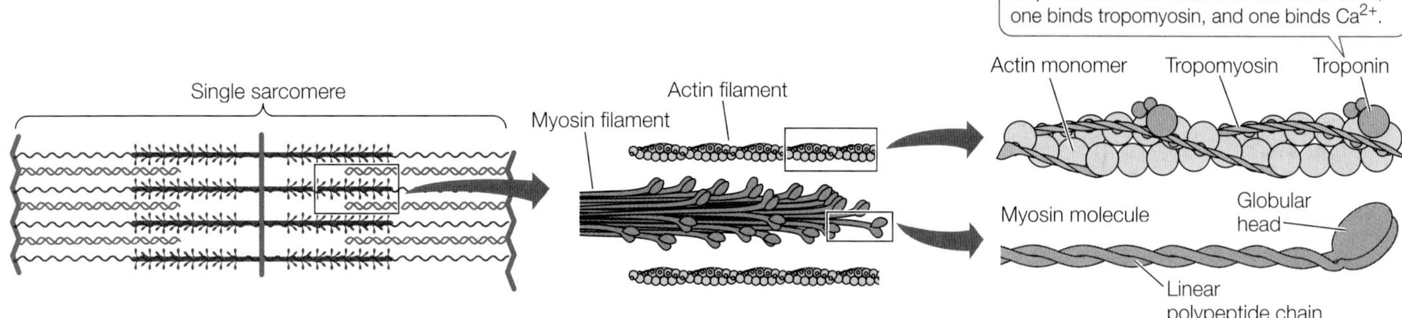


**48.3 Actin and Myosin Filaments Overlap to Form Myofibrils** Myosin filaments are bundles of molecules with globular heads and polypeptide tails; the protein titin holds these filaments centered within the sarcomeres. Actin filaments consist of two chains of actin monomers twisted together. They are wrapped by chains of the polypeptide tropomyosin and are studded at intervals with another protein, troponin.

**Go to Animated Tutorial 48.1**
**Molecular Mechanisms of Muscle Contraction**
Life10e.com/at48.1

tiny force that causes the actin filament to move 5 to 10 nanometers relative to the myosin filament. When the myosin heads are bound to actin, they can bind and hydrolyze ATP. The energy released when this happens changes the conformation of the myosin head, causing it to release the actin and return to its extended position, from which it can bind to actin again.

Together these details help to explain the cycle of events that cause the actin and myosin filaments to slide past each other and shorten the sarcomere. They also explain rigor mortis—the stiffening of muscles soon after death. ATP binding causes myosin to release from actin, so when ATP production stops with death, myosin cannot release and the muscles stay contracted. Eventually, however, the proteins lose their integrity and the muscles soften. The timing of these events helps a medical examiner estimate the time of death.

We have been discussing the cycle of contraction in terms of a single myosin head. Remember that each myosin filament has many myosin heads at both ends and is surrounded by six actin filaments; thus the contraction of the sarcomere involves a great many cycles of interaction between actin and myosin molecules. That is why when a single myosin head breaks its contact with actin, the actin filaments do not slip backward.

## Actin–myosin interactions are controlled by calcium ions

Like neurons, muscle cells are excitable—that is, their plasma membranes can generate and conduct action potentials. In skeletal muscle fibers, action potentials are initiated by motor neurons arriving at a **neuromuscular junction**. The axon terminals of motor neurons are generally highly branched and form synapses with hundreds of muscle fibers (**Figure 48.4**). A motor neuron and all of the fibers with which it forms synapses constitute a **motor unit**. The fibers contract simultaneously when the unit's motor neuron fires. A muscle can consist of many motor units. Thus there are two ways to increase a muscle's strength of contraction—increase the firing rate of an individual motor neuron, or recruit more motor neurons.

When an action potential arrives at a neuromuscular junction, the neurotransmitter acetylcholine is released from the motor neuron terminals, diffuses across the synaptic cleft, binds to receptors in the postsynaptic membrane, and causes ion channels in the motor end plate to open (see Figures 45.13 and 45.14). Most of the ions that flow through these channels are $Na^+$, and therefore the motor end plate is depolarized. The depolarization spreads to the surrounding plasma membrane of the muscle fiber, which contains voltage-gated sodium channels. When threshold is reached, the plasma membrane fires an action potential that is conducted rapidly to all points on the surface of the muscle fiber.

An action potential in a muscle fiber also travels deep within the cell. The plasma membrane is continuous with a system of tubules that descend into the muscle fiber cytoplasm (also called the **sarcoplasm**). The action potential that spreads over the plasma membrane also spreads through this system of transverse tubules, or **T tubules** (**Figure 48.5**).

The T tubules come very close to the endoplasmic reticulum (ER) of the muscle cell. In muscle cells the ER is called the **sarcoplasmic reticulum**, and it is a closed compartment surrounding every myofibril. Calcium pumps in the sarcoplasmic reticulum

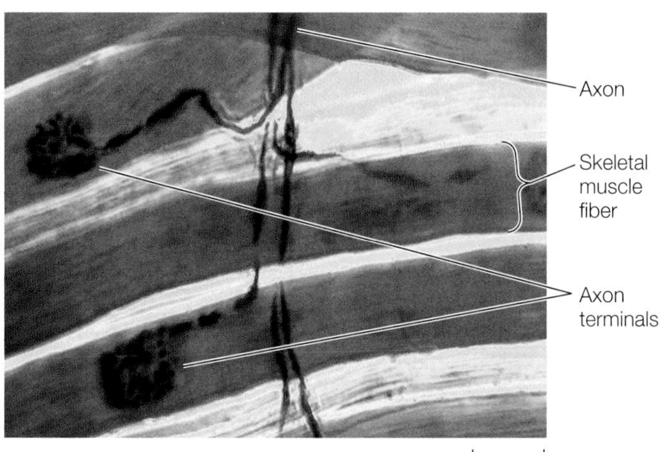


**48.4 The Neuromuscular Junction** Axons branching from a single motor neuron end in terminals that innervate multiple skeletal muscle fibers.

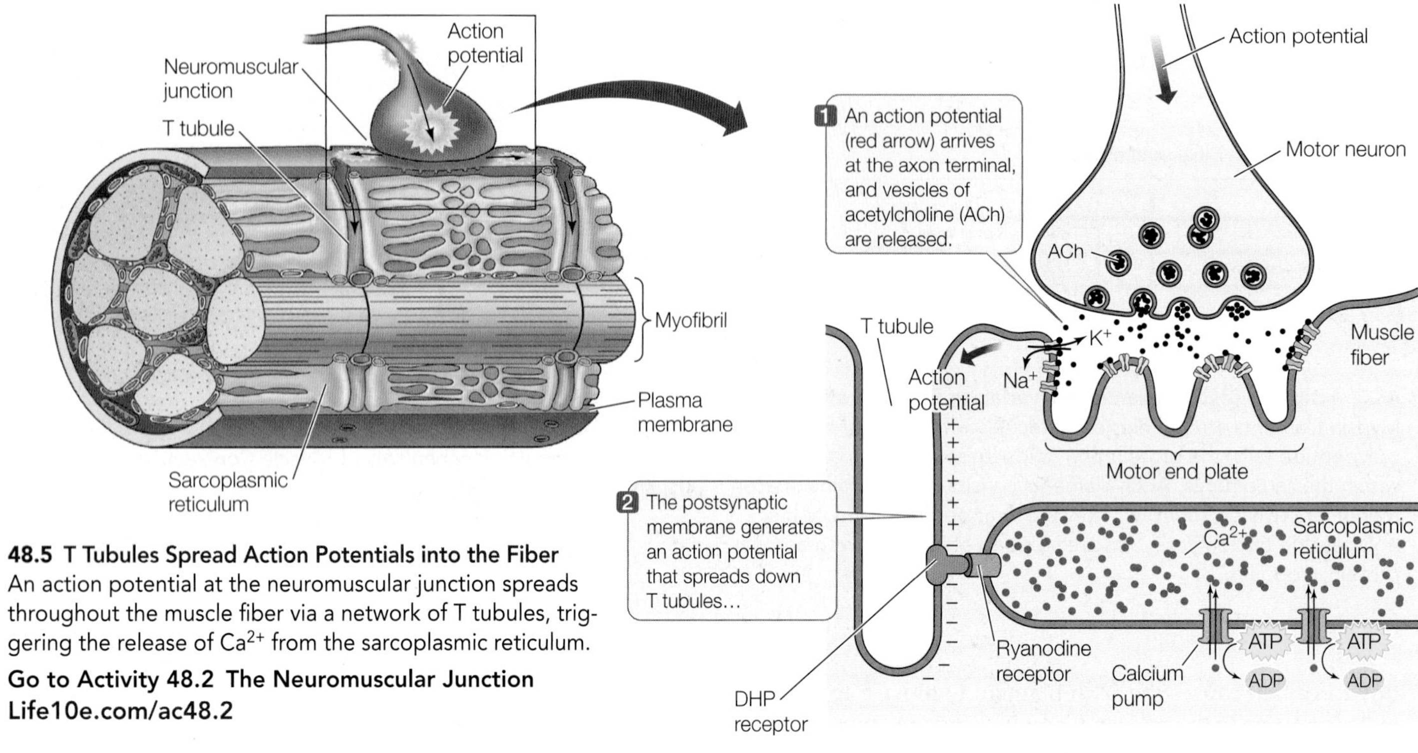


**48.5 T Tubules Spread Action Potentials into the Fiber** An action potential at the neuromuscular junction spreads throughout the muscle fiber via a network of T tubules, triggering the release of $Ca^{2+}$ from the sarcoplasmic reticulum.

**Go to Activity 48.2 The Neuromuscular Junction Life10e.com/ac48.2**

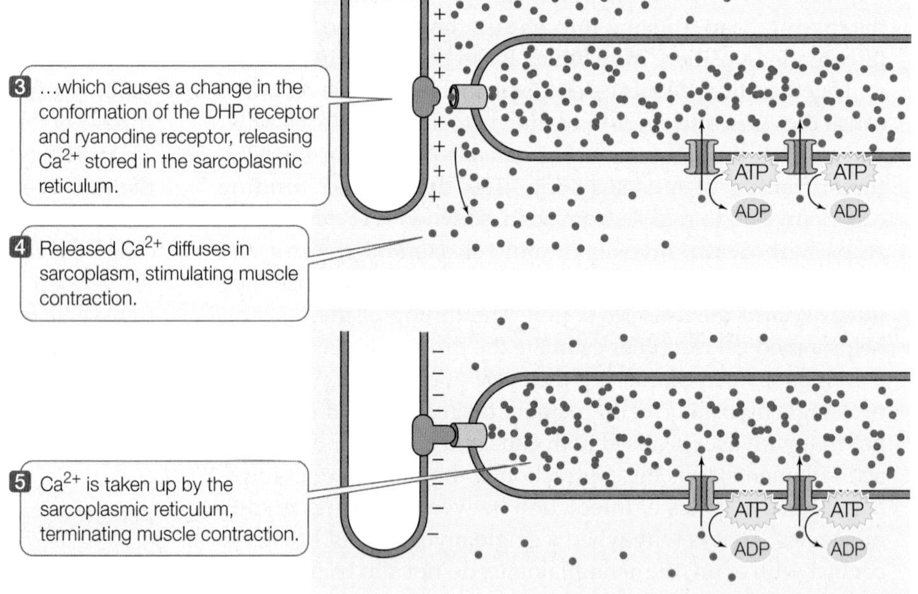


take up $Ca^{2+}$ from the sarcoplasm. Therefore when the muscle fiber is at rest, there is a high concentration of $Ca^{2+}$ in the sarcoplasmic reticulum and a very low concentration of $Ca^{2+}$ in the sarcoplasm.

Spanning the space between the membranes of the T tubules and the membranes of the sarcoplasmic reticulum are two proteins. One protein, the dihydropyridine (DHP) receptor, is located in the T tubule membrane; it is voltage-sensitive and changes its conformation in response to an action potential. The other protein, the ryanodine receptor, is located in the sarcoplasmic reticulum membrane and is a $Ca^{2+}$ channel. These two proteins are physically connected. When the DHP receptor is activated by an action potential, it changes conformation; this allows $Ca^{2+}$ to flow through the ryanodine receptor from the sarcoplasmic reticulum to the sarcoplasm. $Ca^{2+}$ ions diffusing through the sarcoplasm surrounding the actin and myosin filaments trigger the interaction of actin and myosin and the sliding of the filaments. How do the $Ca^{2+}$ ions do this?

An actin filament, as we have seen, is a helical arrangement of actin monomers. Twisted around the actin filament are two strands of the protein tropomyosin (**Figure 48.6**; see also Figure 48.3). At regular intervals, the filament also includes a globular protein, troponin. The troponin molecule has three subunits: one binds actin, one binds tropomyosin, and one binds $Ca^{2+}$.

When the muscle is at rest, the tropomyosin strands are positioned so that they block the sites on the actin filament where myosin heads can bind. When $Ca^{2+}$ is released into the sarcoplasm, it binds to troponin, changing its conformation. Because the troponin is bound to the tropomyosin, this conformational change twists the tropomyosin enough to expose the actin–myosin binding sites. Thus the cycle of making and breaking actin–myosin bonds is initiated, the filaments are pulled past each other, and the muscle fiber contracts. When the calcium pumps in the sarcoplasmic reticulum membranes remove the $Ca^{2+}$ ions from the sarcoplasm, the conformation of the tropomyosin returns to the state in which it blocks the binding of myosin heads to actin, and the muscle fiber returns to its resting condition. Figure 48.6 summarizes this cycle.

**48.6 Release of $Ca^{2+}$ from the Sarcoplasmic Reticulum Triggers Muscle Contraction** When $Ca^{2+}$ binds to troponin, it exposes myosin-binding sites on the actin. As long as binding sites and ATP are available, the cycle of actin and myosin interactions continues and the filaments slide past each other.

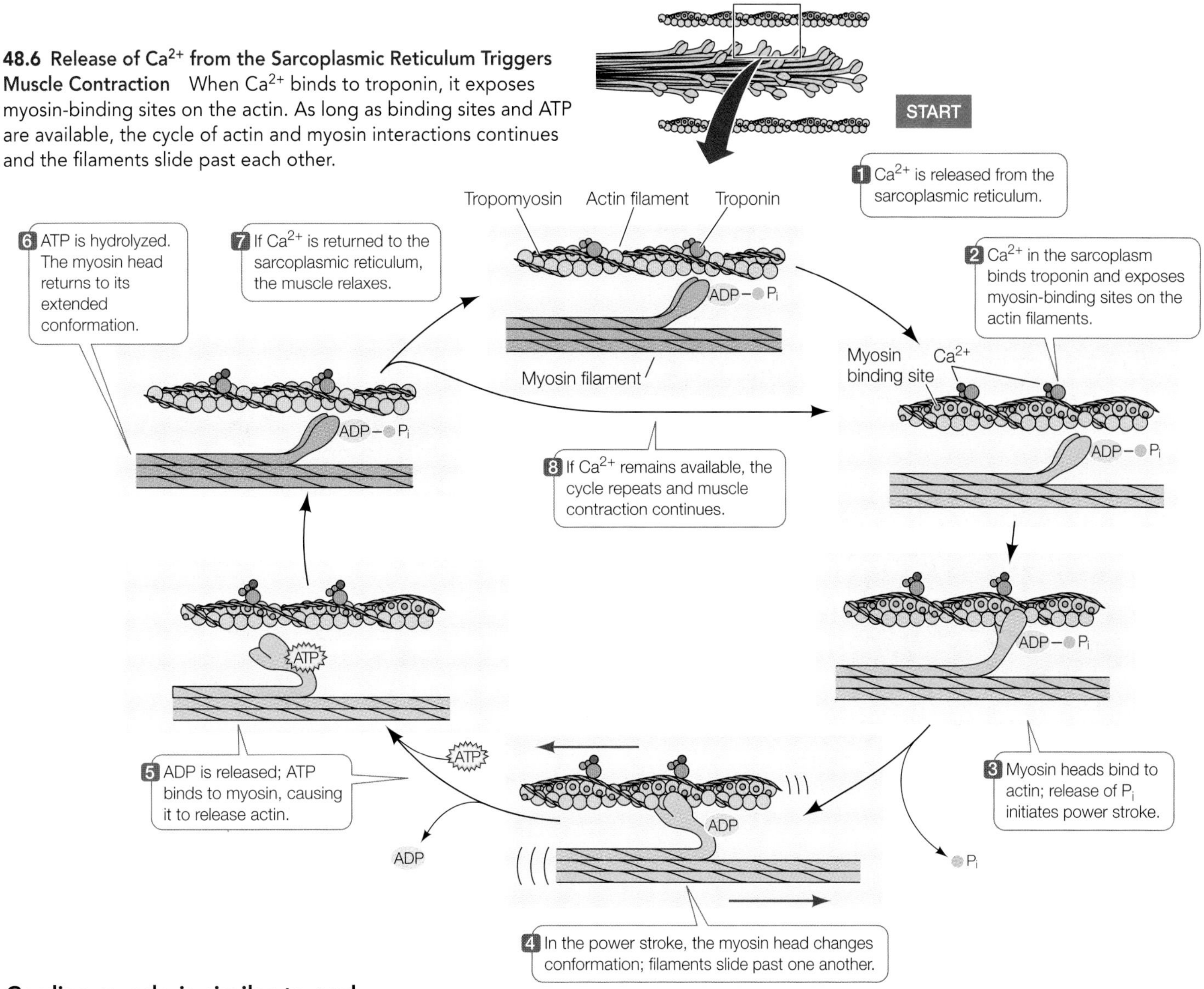

START

1 $Ca^{2+}$ is released from the sarcoplasmic reticulum.

Tropomyosin Actin filament Troponin

ADP – $P_i$

Myosin filament

2 $Ca^{2+}$ in the sarcoplasm binds troponin and exposes myosin-binding sites on the actin filaments.

Myosin binding site $Ca^{2+}$

ADP – $P_i$

ADP – $P_i$

3 Myosin heads bind to actin; release of $P_i$ initiates power stroke.

$P_i$

ADP

4 In the power stroke, the myosin head changes conformation; filaments slide past one another.

ADP

ATP

5 ADP is released; ATP binds to myosin, causing it to release actin.

ATP

6 ATP is hydrolyzed. The myosin head returns to its extended conformation.

ADP – $P_i$

7 If $Ca^{2+}$ is returned to the sarcoplasmic reticulum, the muscle relaxes.

8 If $Ca^{2+}$ remains available, the cycle repeats and muscle contraction continues.

## Cardiac muscle is similar to and different from skeletal muscle

Like skeletal muscle, cardiac muscle appears striated because of the regular arrangement of actin and myosin filaments into sarcomeres (**Figure 48.7**). The difference between cardiac and skeletal muscle is that cardiac muscle cells are much smaller and have only one nucleus each (uninucleate). Cardiac muscle cells branch, and the branches of adjoining cells interdigitate into a meshwork that is resistant to tearing. As a result, the heart walls can withstand high pressures while pumping blood, without the danger of developing leaks. Adding to the strength of cardiac muscle are intercalated discs that provide strong mechanical adhesions between adjacent cells. Gap junctions are an important feature of cardiac muscle. These structures in the intercalated discs allow cytoplasmic continuity between cells (see Figure 7.19A). Because of gap junctions, cardiac muscle cells are electrically coupled. An action potential initiated at one point in a sheet of cardiac muscle spreads rapidly, causing a large number of cardiac muscle cells to contract simultaneously.

Certain cardiac muscle cells are specialized for generating and conducting electric signals. These pacemaker and conducting cells have a low density of actin and myosin filaments, but they initiate and coordinate the rhythmic contractions of the

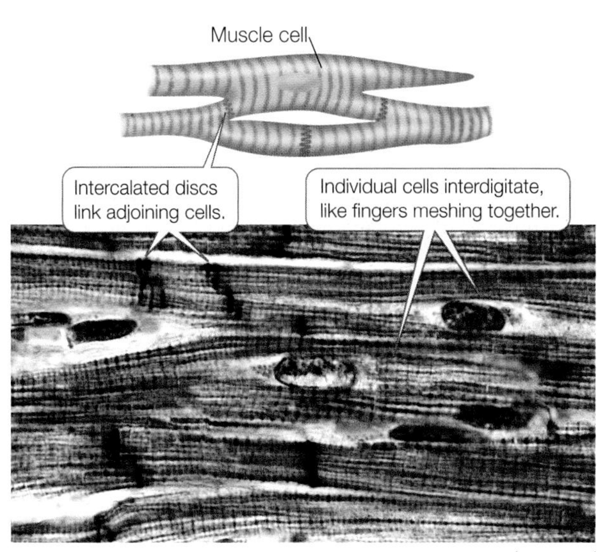


**48.7 Cardiac Muscle Cells Form a Strong Meshwork** Cardiac muscle cells branch and interdigitate, forming a tear-resistant mesh that can withstand the pressure of blood pumping through the heart.

## INVESTIGATING**LIFE**

**48.8 Neurotransmitters Alter the Membrane Potential of Smooth Muscle Cells** Earlier experiments showed that stretching the smooth muscles of the gut (as in the stretch applied by a full stomach) depolarizes the membranes, causing action potentials that activate the contractile mechanism. This follow-up experiment showed that the parasympathetic neurotransmitter acetylcholine and the sympathetic neurotransmitter norepinephrine act antagonistically to alter the membrane potential of smooth muscle.[a]

**HYPOTHESIS** Stimulation from neurotransmitters of the autonomic nervous system (ANS) induce contractions in the smooth muscles of the gut.

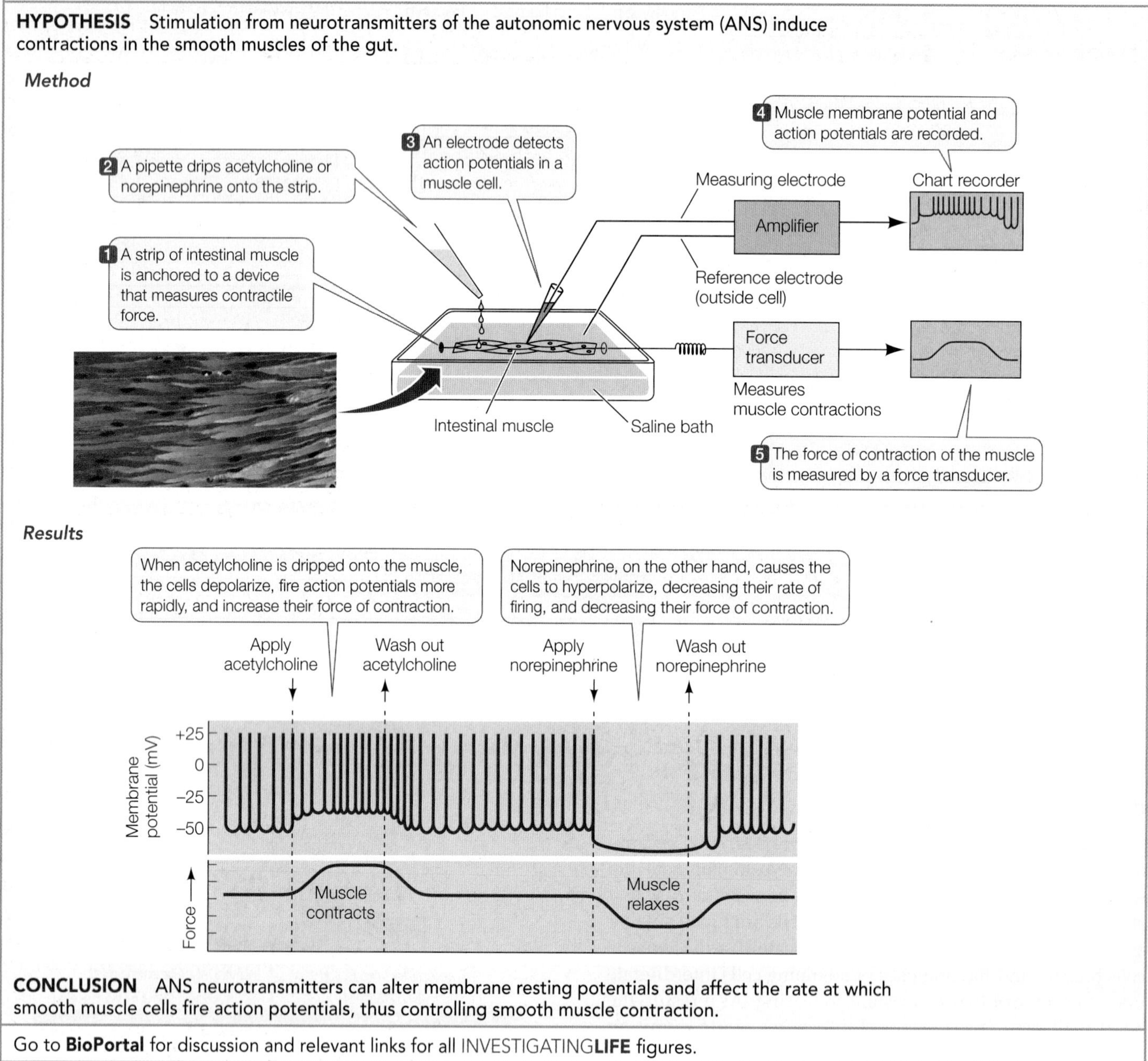


**CONCLUSION** ANS neurotransmitters can alter membrane resting potentials and affect the rate at which smooth muscle cells fire action potentials, thus controlling smooth muscle contraction.

Go to **BioPortal** for discussion and relevant links for all INVESTIGATING**LIFE** figures.

[a]Bolton, T. B. et al. 1999. *Annual Review of Physiology* 61: 85–115.

heart. (The molecular basis for this pacemaking function will be covered in Section 50.3.) Pacemaker cells make the vertebrate heartbeat myogenic, meaning it is generated by the heart muscle itself. A heart removed from a vertebrate can continue to beat with no input from the nervous system. Although input from the autonomic nervous system modifies the *rate* of the pacemaker cells, it is not essential for their continued rhythmic function.

The mechanism of excitation–contraction coupling in cardiac muscle cells is different from that in skeletal muscle cells. The T tubules are larger, and the voltage-sensitive DHP receptor proteins in the T tubules are $Ca^{2+}$ channels. These T tubule proteins are not physically connected with the ryanodine receptors in the sarcoplasmic reticulum. Instead, the ryanodine receptors are ion-gated $Ca^{2+}$ channels. When an action potential spreads down the T tubules, it causes the voltage-gated channels to open, allowing extracellular $Ca^{2+}$ to flow into the sarcoplasm. Increased sarcoplasmic $Ca^{2+}$ concentration results in $Ca^{2+}$ binding to the ryanodine receptors and opens calcium channels in the sarcoplasmic reticulum. The resulting huge rise in sarcoplasmic $Ca^{2+}$ concentration stimulates fiber contraction. This mechanism is called $Ca^{2+}$-induced $Ca^{2+}$ release.

**Go to Media Clip 48.1**
**Be Still My Beating Stem Cell Heart**
Life10e.com/mc48.1

## Smooth muscle causes slow contractions of many internal organs

Smooth muscle provides the contractile force for most of our internal organs, which are under the control of the autonomic nervous system. Smooth muscle moves food through the digestive tract, controls the flow of blood through blood vessels, and empties the urinary bladder. Structurally, smooth muscle cells are the simplest muscle cells. They are smaller than skeletal muscle cells, usually long and spindle-shaped, and each has a single nucleus. They are "smooth" because the actin and myosin filaments are not as regularly arranged as they are in skeletal and cardiac muscle, and so do not produce the striated appearance.

Some smooth muscle tissue, such as that from the wall of the digestive tract, has interesting properties. The cells are arranged in sheets, and individual cells in a sheet are in electrical contact with one another through gap junctions, as they are in cardiac muscle. As a result, an action potential generated in the membrane of one smooth muscle cell can spread to all the cells in the sheet of tissue. Thus the cells in the sheet contract in a coordinated fashion.

The plasma membranes of smooth muscle cells are sensitive to stretch, with important consequences. If the wall of the digestive tract is stretched in one location (as by a mouthful of food passing down the esophagus to the stomach), the membranes of the stretched cells depolarize, reach threshold, and fire action potentials, which cause the cells to contract. Thus smooth muscle contracts after being stretched, and the harder it is stretched, the stronger it contracts. This behavior of smooth muscle is important for moving food through the digestive system.

The walls of blood vessels are mostly smooth muscle. This is especially true on the arterial side where the blood is under higher pressure. Changes in vascular smooth muscle tone are responsible for controlling the distribution of blood in the body.

The neural influences on smooth muscle come from the two divisions of the autonomic nervous system. The neurotransmitters of the sympathetic and parasympathetic postganglionic cells alter the membrane potential of smooth muscle cells. For example, in the digestive tract, acetylcholine causes smooth muscle cells to depolarize, making them more likely to fire action potentials and contract. Antagonistically, norepinephrine causes these muscle cells to hyperpolarize and thus be less likely to fire action potentials and contract (**Figure 48.8**). In contrast, norepinephrine acting through G protein-coupled receptors causes the smooth muscle in arteries serving the gut to contract. Remember that the action of the neurotransmitters depends on the receptors in the target tissues. Sympathetic activity is high in a fight-or-flight situation; in an emergency you don't need to digest your lunch, but you do need to send blood to the tissues critical for survival.

Although smooth muscle cell contraction is not controlled by the troponin–tropomyosin mechanism, calcium still plays a critical role. A $Ca^{2+}$ influx into the sarcoplasm of a smooth muscle cell can be stimulated by action potentials, hormones, or stretching. The $Ca^{2+}$ that enters the sarcoplasm combines with a protein called calmodulin. The calmodulin–$Ca^{2+}$ complex activates an enzyme called myosin kinase that phosphorylates myosin heads. When the myosin heads in smooth muscle are phosphorylated, they undergo cycles of binding and releasing actin, causing muscle contraction. As $Ca^{2+}$ is removed from the sarcoplasm, it dissociates from calmodulin, and the activity of myosin kinase falls. An additional enzyme, myosin phosphatase, dephosphorylates the myosin to help reduce actin–myosin interactions (**Figure 48.9**).

**Go to Animated Tutorial 48.2**
**Smooth Muscle Action**
Life10e.com/at48.2

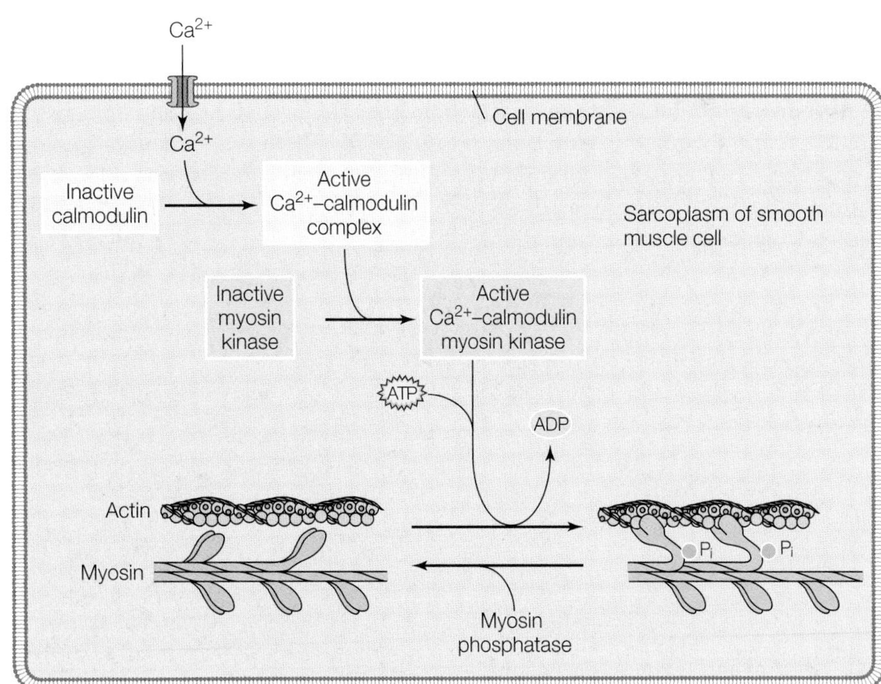


**48.9 The Role of $Ca^{2+}$ in Smooth Muscle Contraction** When a smooth muscle cell is stimulated by neurotransmitter, $Ca^{2+}$ enters the sarcoplasm and binds to calmodulin, which in turn activates an enzyme that phosphorylates the myosin heads, causing them to bind to actin. As long as the myosin remains phosphorylated, actin and myosin go through cycles of binding and release. Thus in smooth muscle the $Ca^{2+}$-mediated change is on myosin, whereas in skeletal and cardiac muscle it is on the actin–tropomyosin filament.

**RECAP 48.1**

The contractile ability of muscle derives from interactions between actin and myosin filaments. The three types of muscle are skeletal, cardiac, and smooth. Contraction in all three depends on control by $Ca^{2+}$ in the sarcoplasm. Tropomyosin and troponin are controlling elements in skeletal and cardiac muscle. Calmodulin is the controlling element in smooth muscle.

- Explain how the cellular and subcellular structures of skeletal muscle relate to the sliding filament theory of muscle contraction. **See pp. 987–989 and Figures 48.1, 48.2, and 48.3**
- What is the role of $Ca^{2+}$ in the contractile mechanism of skeletal, cardiac, and smooth muscle? **See pp. 989–990 and Figures 48.5 and 48.6**
- What roles does ATP play in the actin and myosin interactions that produce contraction? **See Figure 48.6**

Now that we understand how muscles generate force, we can look at what determines the characteristics of a muscle, its performance, and how individual muscles can change their characteristics with regular use and conditioning.

## 48.2 What Determines Skeletal Muscle Performance?

The functions that different muscles perform place different demands on them. Some muscles, such as postural muscles, must sustain a load continuously over long periods of time. Muscles used in locomotion and heavy work must be able to vary their strength of contraction over a wide range. Other muscles, such as those that control your fingers, generally do not have to sustain long, strong contractions, but they must be able to contract quickly. And what is "quick" for humans doesn't begin to compare with insect flight muscles that can contract as fast as 1,000 times per second. How are muscles adapted to specific functions and demands?

### The strength of a muscle contraction depends on how many fibers are contracting and at what rate

In skeletal muscle, the arrival of an action potential at a neuromuscular junction causes an action potential in a muscle fiber. The spread of that action potential through the muscle fiber's T tubule system causes a minimum unit of contraction, called a **twitch**. A twitch can be measured in terms of the tension, or force, it generates (**Figure 48.10A**). A single action potential stimulates a single twitch, but the ultimate force generated by an action potential varies enormously depending on how many muscle fibers it reaches. The level of tension an entire muscle generates depends on two factors: the number of motor units activated, and the frequency at which the motor units fire.

In muscles responsible for fine movements, such as those of the fingers, a motor neuron may innervate only one or a few muscle fibers, but in a muscle that produces large forces, such as the biceps, a motor neuron innervates a large number of muscle fibers.

At the level of a muscle fiber, a single action potential stimulates a single twitch. If action potentials reaching the muscle fiber are adequately separated in time, each twitch is a discrete, all-or-none phenomenon. If action potentials are fired more rapidly, however, new twitches are triggered before the myofibrils have a chance to return to their resting condition. As a result, the twitches sum, and the tension generated by the fiber increases and becomes more sustained. Thus an individual muscle fiber can show a graded response to increased levels of stimulation by its motor neuron.

Twitches sum at high levels of stimulation because the calcium pumps in the sarcoplasmic reticulum (see Figure 48.5) are not able to clear the $Ca^{2+}$ ions from the sarcoplasm between action potentials. Eventually a stimulation frequency can be reached that results in the continuous presence of $Ca^{2+}$ in the sarcoplasm at high enough levels to cause continuous activation of the contractile machinery—a condition known as **tetanus** (**Figure 48.10B**). (Do not confuse this condition with the disease tetanus, which is caused by a bacterial toxin and is characterized by spastic contractions of skeletal muscles.)

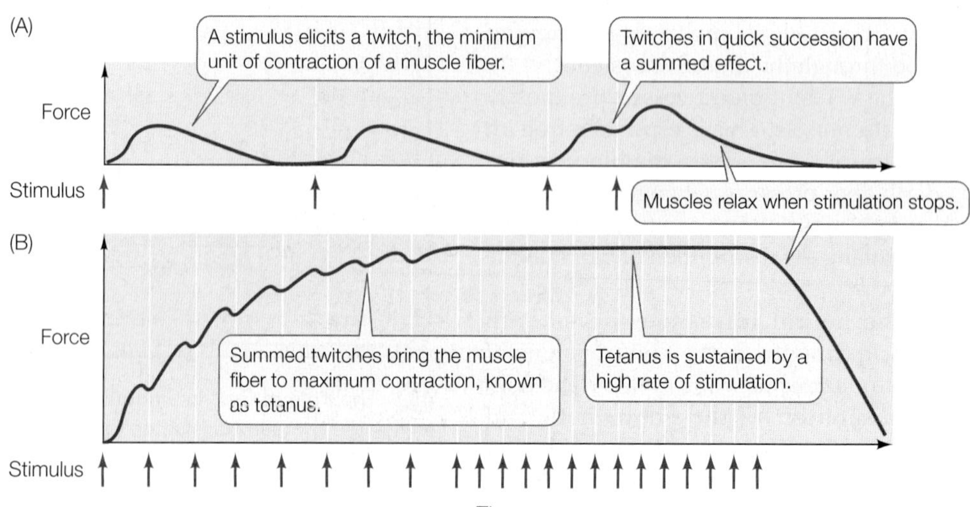


**48.10 Twitches and Tetanus** (A) Action potentials from a motor neuron cause a muscle fiber to twitch. Twitches in quick succession can be summed. (B) Summation of many twitches can bring the muscle fiber to the maximum level of contraction, known as tetanus.

How long a muscle fiber can maintain a tetanic contraction depends on its supply of ATP. Eventually the fiber will become fatigued and be unable to sustain the contraction. It may seem paradoxical that the *lack* of ATP causes fatigue, since the action of ATP is to break actin–myosin bonds. But remember that the energy released from the hydrolysis of ATP "re-cocks" the myosin heads, allowing them to cycle through another power stroke. When a muscle is contracting against a load, the cycle of making and breaking actin–myosin bonds must continue in order to prevent the load from stretching the muscle. The situation is like rowing a boat upstream. You cannot maintain your position relative to the stream bank by just holding the oars out against the current; you have to keep rowing. Likewise, actin–myosin bonds have to keep cycling to maintain tension in the muscle.

Many muscles of the body maintain a low level of tension even when the body is at rest. For example, the muscles of the neck, trunk, and limbs that maintain our posture against the pull of gravity are always working, even when we are standing or sitting still. Muscle tone comes from the activity of a small but changing number of motor units in a muscle; at any one time, some of the muscle's fibers are contracting and others are relaxed. The nervous system is constantly readjusting muscle tone.

## Muscle fiber types determine endurance and strength

Not all skeletal muscle fibers are alike, and a single muscle often contains more than one type of fiber. The two major types of skeletal muscle fibers express different genes for their myosin molecules, and these myosin variants have different rates of ATPase activity. Those with high ATPase activity can recycle their actin–myosin cross-bridges rapidly and are therefore called fast-twitch fibers. Slow-twitch fibers have lower ATPase activity; they develop tension more slowly but can maintain it longer.

**Slow-twitch fibers** are also called oxidative or red muscle because they contain **myoglobin** (an oxygen-binding protein similar to the hemoglobin in red blood cells), have many mitochondria, and are well supplied with blood vessels. These characteristics both increase the fibers' capacity for oxidative metabolism and result in their red appearance. The maximum tension a slow-twitch fiber produces is low and develops slowly but is highly resistant to fatigue. Slow-twitch fibers have substantial reserves of fuel (glycogen and fat), so they can maintain steady, prolonged production of ATP as long as oxygen is available. Muscles with high proportions of slow-twitch fibers are good for long-term aerobic work (that is, work that requires oxygen). Long-distance runners, swimmers, cyclists, and other athletes whose activities require endurance have leg and arm muscles consisting mostly of slow-twitch fibers (**Figure 48.11**).

Some **fast-twitch fibers** are also called glycolytic or white muscle because, compared with slow-twitch fibers, they have few mitochondria, little or no myoglobin, and fewer blood vessels; thus they look pale. Fast-twitch glycolytic fibers can develop maximum tension more rapidly than slow-twitch fibers can, and that maximum tension is greater. However, fast-twitch fibers fatigue rapidly. The myosin of these fibers puts the energy

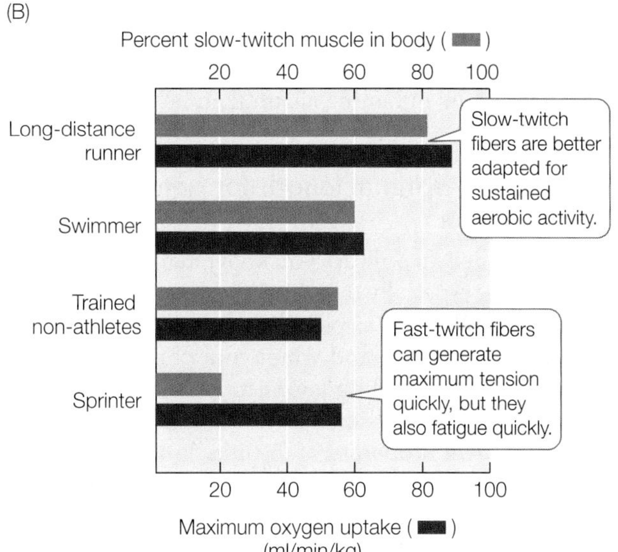


**48.11 Slow- and Fast-Twitch Muscle Fibers** (A) The skeletal muscles in the micrographs were stained with a reagent that shows slow-twitch fibers as dark and fast-twitch muscle as light. (B) Athletes in different sports have different distributions of muscle fiber types.

of ATP to work very rapidly, but the fibers cannot replenish ATP quickly enough to sustain contraction for a long time. Fast-twitch fibers are especially good for short-term work that requires maximum strength. Weight lifters and sprinters have leg and arm muscles with high proportions of fast-twitch fibers.

The types of fibers that make up a muscle influence the performance properties of that muscle, and different muscles have different fiber compositions depending on their function. Postural muscles that maintain continuous contractions are mostly composed of slow-twitch fibers. An example is the soleus muscle that runs up the back of the leg from the heel. Its contraction extends the foot and is therefore used in walking, but its continuous contraction is required for standing. If the soleus muscle fatigued, we would fall forward. A person can walk or stand for a long period of time because the soleus muscle is resistant to fatigue. In contrast, a muscle that is used mostly for short-term work, such as the biceps, has a higher percentage of fast-twitch fibers than does the soleus. We can use our biceps to pick up a heavy weight, but we cannot hold that weight in a given position for a long period of time.

Can you change the fiber composition of your muscles to optimize your performance in a particular activity? To a limited extent, you can alter the properties of your muscle fibers through training. There are fast-twitch fibers that are somewhat oxidative and therefore intermediate in their properties between slow-twitch and fast glycolytic fibers. These intermediate fibers can become more oxidative with endurance training and more glycolytic with strength training. However, the most important determinant of your muscle fiber types is your genetic heritage. There is some truth to the statement that champions are born, not made. A person born with a high proportion of fast-twitch fibers in her legs is unlikely to become a champion marathon runner, and a person born with a high proportion of slow-twitch fibers in her legs is unlikely to become a champion sprinter.

## A muscle has an optimal length for generating maximum tension

If you have ever done a pull-up, you know that two parts of this exercise are especially difficult. When you are hanging from the bar with your arms fully extended, it is hard to get the pull-up started; and when your chin has just about reached the bar, pulling yourself up the last small distance is difficult. Why is this? Part of the explanation comes from the lever properties of the muscle–joint interaction that we will discuss in Section 48.3, and part comes from the structure of the sarcomere.

You can see the relationship between the length of a muscle fiber and its ability to develop tension in **Figure 48.12**. When a muscle is stretched and the sarcomeres are lengthened, there is less overlap between the actin and myosin filaments; therefore fewer cross-bridges can form, and less force can be produced. In fact, if the sarcomeres are stretched too much, actin and myosin do not overlap and no force can be produced. How would a muscle recover from such a difficult situation? Like a bungee cord, titin molecules create enough elastic recoil to pull the actin and myosin fibrils back into an overlapping arrangement.

When the muscle is fully contracted, the actin and myosin filaments overlap so much that the myosin bundles are pressed up against the Z lines. Because they have no place to go, additional shortening is difficult.

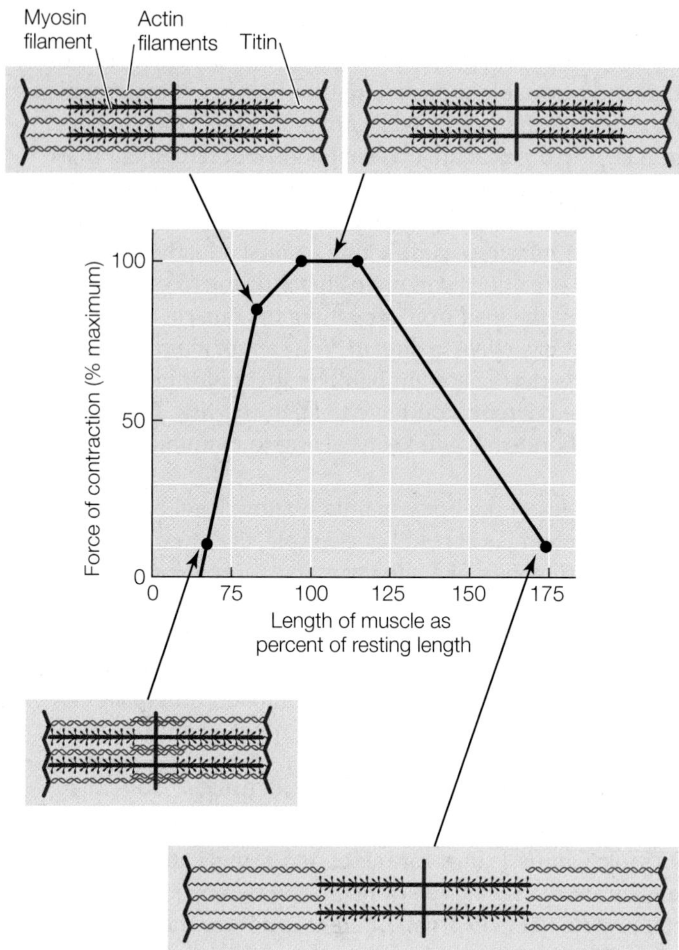


**48.12 Force and Length** The amount of force a sarcomere can generate depends on its resting length. When a muscle is stretched, the sarcomeres lengthen, there is less overlap between the actin and myosin filaments, and less force is produced. Overstretched sarcomeres produce no force because there is no overlap between the actin and myosin.

## Exercise increases muscle strength and endurance

Different types of exercise produce different physical conditioning responses. In general, anaerobic activities, such as weight lifting, increase strength, and aerobic activities, such as jogging, increase endurance. Strength is the maximum force a muscle can exert, and endurance is work capacity or how long a given workload can be sustained. What are the physiological bases for these differences?

Strength is a function of the cross-sectional area of muscles: the more actin and myosin filaments in a muscle fiber, and the more muscle fibers in a muscle, the more tension it can produce. When athletes undertake strength training, they use weights or exercises such as pull-ups to repeatedly contract specific muscles under heavy loads. Repetitions are usually done until the muscle is completely fatigued. Such stress on a muscle does minor tissue damage—hence the soreness the day after a hard workout—but it also induces the formation of new actin and myosin filaments in existing muscle fibers. The muscle fibers, and hence the muscles, get bigger and stronger. In extreme cases, and after serious muscle damage, new muscle fibers can also be produced from stem cells called satellite cells in the muscle. In general, however, the major effect of strength training is to produce bigger, rather than more, muscle fibers.

Aerobic exercise has a completely different effect on muscles: it enhances their oxidative capacity. This effect comes from increases in the number of mitochondria, in enzymes involved in energy use, and in the density of capillaries that deliver oxygen to the muscle. Myoglobin also increases in skeletal muscle cells. Although it is similar to hemoglobin, myoglobin has a higher affinity for oxygen. Therefore myoglobin accepts oxygen from the blood, facilitates the diffusion of oxygen throughout the muscle, and provides a store of oxygen for use when oxygen delivery by the blood is insufficient. By increasing the capacity of muscle to use oxygen to produce ATP, aerobic training increases the length of time that a given workload can be sustained.

## Muscle ATP supply limits performance

Muscles have three systems for supplying the ATP they need for contraction:

1. The *immediate system* uses preformed ATP and creatine phosphate.
2. The *glycolytic system* metabolizes carbohydrates to lactate and pyruvate.
3. The *oxidative system* metabolizes carbohydrates or fats all the way to $H_2O$ and $CO_2$.

The capacity of these three systems and the rates at which they can produce ATP determine both work capacity and endurance (**Figure 48.13**).

ATP is present in muscles in very small amounts. However, muscle fibers also contain a storage compound called creatine phosphate (CP). This molecule stores energy in a phosphate bond, which it can transfer to ADP. The total energy available in all the muscles of your body in the form of ATP and CP—the immediate energy system—is only about 10 kilocalories. When at rest, you metabolize a kilocalorie of energy in less than a minute. Even though the energy available from ATP and CP is limited, it is available immediately, and it enables fast-twitch fibers to generate a lot of force quickly. During burst activity, the immediate system is exhausted in seconds.

The glycolytic system activates within a few seconds to replace the ATP depleted at the onset of muscle activity. The glycolytic enzymes are located in the cytoplasm of the muscle fiber, and therefore the ATP they generate is rapidly available to the myosin filaments. However, as noted in Chapter 9, glycolysis alone is an inefficient way to produce ATP, and it leads to the accumulation of lactic acid, which slows the process. Thus the glycolytic system and the immediate system together can provide most of the energy for active muscles for less than a minute (see Figure 48.13).

Oxidative metabolism becomes fully active in about a minute, producing relatively huge amounts of ATP because it can completely metabolize carbohydrates and fats. However, it requires many reactions (see Section 9.3), and it takes place in the mitochondria, so $O_2$ and substrate must diffuse into the mitochondria, and the formed ATP must diffuse from the mitochondria to the myosin filaments in the muscle. These processes are not instantaneous, so the rate at which oxidative metabolism can make ATP available to do work is slower than the rate at which the other two systems can supply ATP.

The fuel supply available to the muscles influences how long someone can sustain a high level of aerobic exercise.

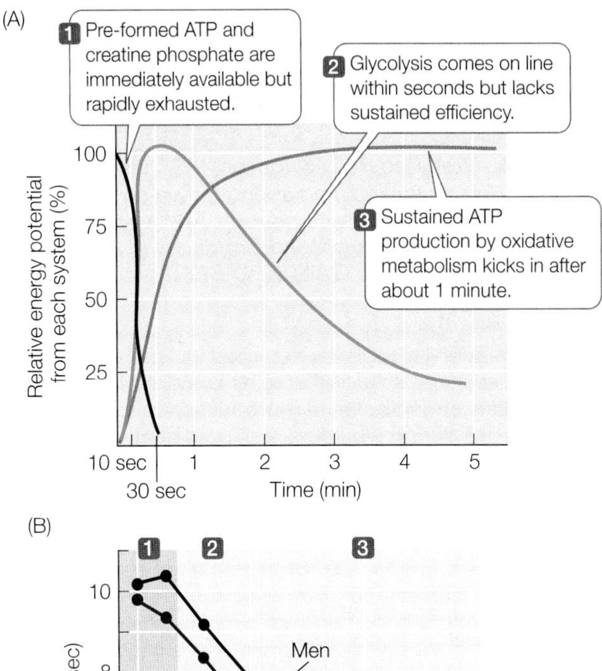


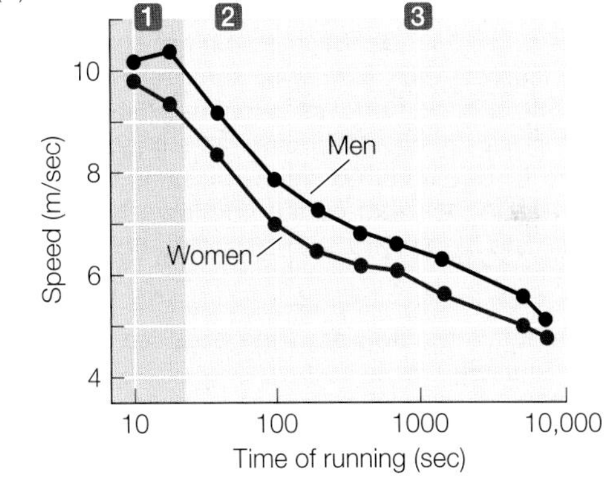


**48.13 Supplying Fuel for High Performance** (A) Muscles have three systems for obtaining the ATP they need for contraction during exertion such as running. (B) Looking at a plot of world-record times for running events of different durations, you can see that the performance of the athletes corresponds to the time courses of the three energy systems.

From the circulating blood, muscle receives glucose and free fatty acids, which it can metabolize to generate ATP. At high levels of aerobic exercise, however, most of the fuel used by muscles to produce ATP comes from the reserve of glycogen stored in the muscle itself. Depletion of muscle glycogen results in fatigue.

The rate at which muscle glycogen is replenished depends on diet: it is high with a high-carbohydrate diet, low with a high-fat diet, and intermediate with a mixed diet. This fact is the basis for a practice called "carbo-loading." For 3 to 5 days, athletes exercise at a level that depletes muscle glycogen. Then, 2 or 3 days before the event, they taper down their level of training and eat a diet rich in complex carbohydrates. The result can be glycogen supercompensation, in which the restoration of muscle glycogen stores "overshoots" and reaches above-normal levels.

## Insect muscle has the greatest rate of cycling

Insect flight muscle can produce a wingbeat frequency of up to 1,000 cycles per second. Since neural action potentials last 1 to 3

## WORKING WITH DATA:

### *Does Heat Cause Muscle Fatigue?*

#### Original Paper

Grahn, D. A., V. H. Cao, C. M. Nguyen, M. T. Lieu, and H. C. Heller. 2012. Work volume and strength training responses to resistive exercise improve with periodic heat extraction from the palm. *Journal of Strength and Conditioning Research*. Epub ahead of print. doi: 10.1519/JSC.0b013e31823f8c1a

#### Analyze the Data

Physical conditioning requires repeated intense physical activity, and the capacity of such workouts is limited by muscle fatigue. Because metabolic heat production raises the temperature of muscles during workouts, it is possible that the rise in temperature contributes to muscle fatigue and limits the capacity of workouts. To test this idea, investigators used the rapid-cooling technology described in the opening of Chapter 40 to extract heat from subjects during 3-minute rests between ten sets of pull-ups twice a week. Each of the ten sets of pull-ups was to muscle failure—the inability to complete an additional pull-up. The control condition was 3-minute rests without cooling. Each subject was his own control; the subjects were randomly assigned to begin with 6 weeks of training with cooling or 6 weeks of training without cooling, followed by a reversal of the treatments. The results are shown in the figure.

**QUESTION 1**

What do these data indicate about the possible role of muscle temperature in muscle fatigue?

**QUESTION 2**

What do these data indicate about the relationship between workout capacity and physical conditioning effects?

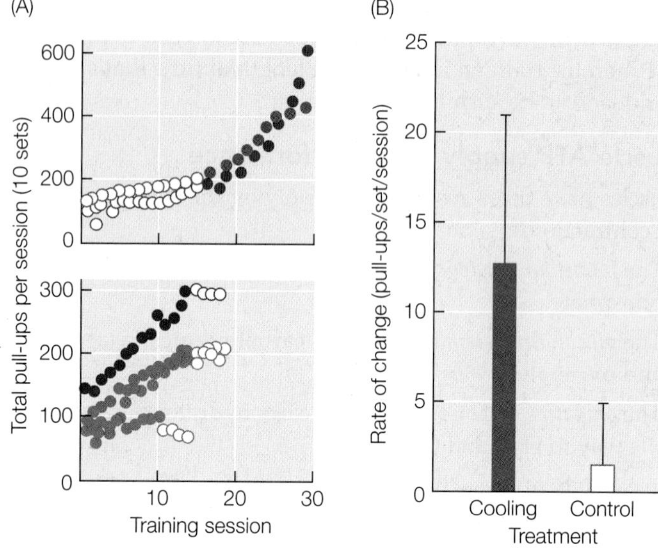


**THE EFFECT OF COOLING ON WORKOUT CAPACITY** (A) The results for individual subjects. Open symbols indicate control treatment, closed symbols indicate cooling treatment; different colors indicate different subjects. (B) Regression analysis of the pull-up data, which shows the rate of increase in workout capacity over the course of the experiment. Mean ± standard deviation of rate of change in pull-ups per set during the two treatment phases ($n = 6$, $P < 0.001$ paired *t*-tests).

**Go to BioPortal for all WORKING WITHDATA exercises**

milliseconds, that number of cycles per second would push the capacity of motor neurons, let alone the mechanism of cycling of striated muscle contraction/relaxation. The extremely fast wingbeat of a hummingbird may only be about 50 cycles per second. How do insects do it?

The mechanism of excitation/contraction coupling is different in insect flight muscle. Vertebrate striated muscle and much of invertebrate striated muscle is called "synchronous" because the cycling of the contractile mechanism is linked to the firing of the motor neurons. This is not true of insect flight muscle, which is therefore called "asynchronous" muscle. The firing of action potentials in the insect flight motor neurons is not particularly fast, but it does cause depolarization of the muscle cell membrane, the spreading of an action potential throughout the membrane, and the release of $Ca^{2+}$ from the sarcoplasmic reticulum. However, once the asynchronous muscle fiber is stimulated, its cycle of contraction/relaxation proceeds at its own characteristic frequency as long as the $Ca^{2+}$ is available to bind to the troponin. The contraction of the muscle fiber deactivates the actin–myosin binding, which in turn permits a stretching of the muscle, which in turn activates the actin–myosin binding. Thus contractile cycling and the resulting wingbeat frequency are not tied to the firing rate of the flight motor neurons.

### RECAP 48.2

Depending on the function a muscle serves, it may need to generate maximum force rapidly, sustain activity for a long period, or contract and relax at a very rapid rate. Properties of muscles can facilitate these types of activities.

- How can a single motor unit exert varying levels of force? **See p. 994 and Figure 48.10**
- Describe the differences between slow-twitch and fast-twitch fibers. **See p. 995 and Figure 48.11**
- How does exercise influence muscle strength and endurance? **See p. 996**
- How do the different sources of ATP influence performance in different types of exercise? **See p. 997 and Figure 48.13**
- Explain how insects can beat their wings ten or more times faster than hummingbirds can. **See p. 998**

Regardless of how much force a muscle can generate, how long it can sustain a workload, or how fast it can contract and relax, a muscle needs something to pull on; otherwise it would just be a lump of pulsating, quivering tissue. Let's look now at how muscle and skeletal elements work together to produce movement.

## 48.3 How Do Skeletal Systems and Muscles Work Together?

Muscles can contract and exert force, or they can relax. To create significant movement, they must have something to pull on and something that stretches the muscle back to a longer position. In some cases muscles pull on each other, as in the trunk of an elephant or the arms of an octopus. In most cases, however, **skeletal systems** are the rigid supports against which muscles pull to create directed movement. In this section we will examine the three types of skeletal systems: hydrostatic skeletons, exoskeletons, and endoskeletons.

### A hydrostatic skeleton consists of fluid in a muscular cavity

Cnidarians, annelids, and other soft-bodied invertebrates have **hydrostatic skeletons** consisting of a volume of fluid enclosed in a body cavity surrounded by muscle (see Section 31.2). When muscles oriented in one direction contract, the fluid-filled body cavity bulges out in a perpendicular direction.

An earthworm uses its hydrostatic skeleton to crawl (**Figure 48.14**). The earthworm's body cavity is divided into many separate segments filled with extracellular fluid. The body wall surrounding each segment has two muscle layers: a circular layer and a longitudinal layer. If the circular muscles in a segment contract, the compartment in that segment narrows and elongates. If the longitudinal muscles in a segment contract, the compartment shortens and bulges outward. Alternating contractions of the earthworm's circular and longitudinal muscles create waves of narrowing and widening, lengthening and shortening, that travel down the body. Bulging, shortened segments serve as anchors as long, narrow segments project forward and longitudinal contractions pull other segments forward. Bristles help the widest parts of the body to hold firm against the substrate, so the body moves forward.

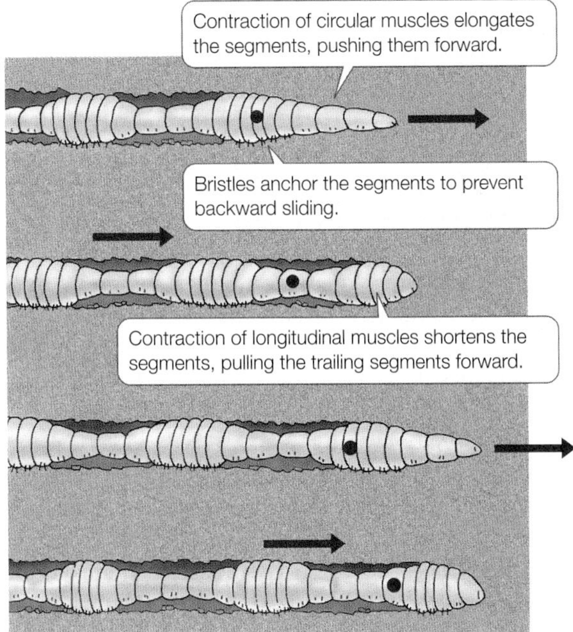


**48.14 A Hydrostatic Skeleton** Alternating waves of muscle contraction move the earthworm through the soil. The red dot enables you to follow the changes in one segment as the worm moves forward.

### Exoskeletons are rigid outer structures

An **exoskeleton** is a hardened, rigid outer surface to which muscles can be attached. Contractions of the muscles cause jointed segments of the exoskeleton to move relative to each other. The simplest example of an exoskeleton is the shell of a mollusk. Some marine mollusks, such as clams, have shells composed of protein strengthened by crystals of calcium carbonate (a rock-hard material). These shells can be massive, affording significant protection against predators. The shells of land mollusks (snails) generally lack the hard mineral component and are much lighter.

The most complex exoskeletons are found among the arthropods. A type of exoskeleton called a **cuticle** covers the outer surfaces of the arthropod body and all its appendages. It is made up of **chitin** secreted by a layer of cells just below the exoskeleton. Chitin stiffens and hardens the cuticle everywhere except at the joints, where flexibility must be retained. Muscles attached to the inner surfaces of the arthropod cuticle move its parts around the joints (see Figure 32.4).

A drawback of the rigid arthropod exoskeleton is that it cannot expand. Therefore, if the animal is to become larger, it must **molt**, shedding its exoskeleton and forming a new, larger one. A molting animal is vulnerable because the new exoskeleton takes time to harden. The animal's body is temporarily unprotected, and without a firm exoskeleton against which its muscles can exert maximum tension, it is unable to move rapidly. Soft-shelled crabs, a gourmet delicacy, are crabs caught while they are molting.

### Vertebrate endoskeletons consist of cartilage and bone

The **endoskeleton** of vertebrates is an internal scaffolding. Muscles are attached to it and pull against it. Endoskeletons are composed of rodlike, platelike, and tubelike bones connected to one another at a variety of joints that allow a wide range of movements. An advantage of endoskeletons over the exoskeletons of arthropods is that bones in the body can grow without the animal shedding its skeleton.

The human skeleton consists of 206 bones, some of which are shown in **Figure 48.15**. It can be divided into an axial skeleton, which includes the skull, vertebral column, sternum, and ribs; and an appendicular skeleton, which includes the pectoral girdle, pelvic girdle, and bones of the arms, legs, hands, and feet.

The vertebrate endoskeleton consists of two kinds of connective tissue, cartilage and bone, which are produced by two kinds of connective tissue cells. Cartilage cells produce an extracellular matrix that is a tough, rubbery mixture of polysaccharides and proteins—mainly fibrous collagen. Collagen fibers run in all directions like reinforcing cords through the gel-like matrix and give it the well-known strength and resiliency of

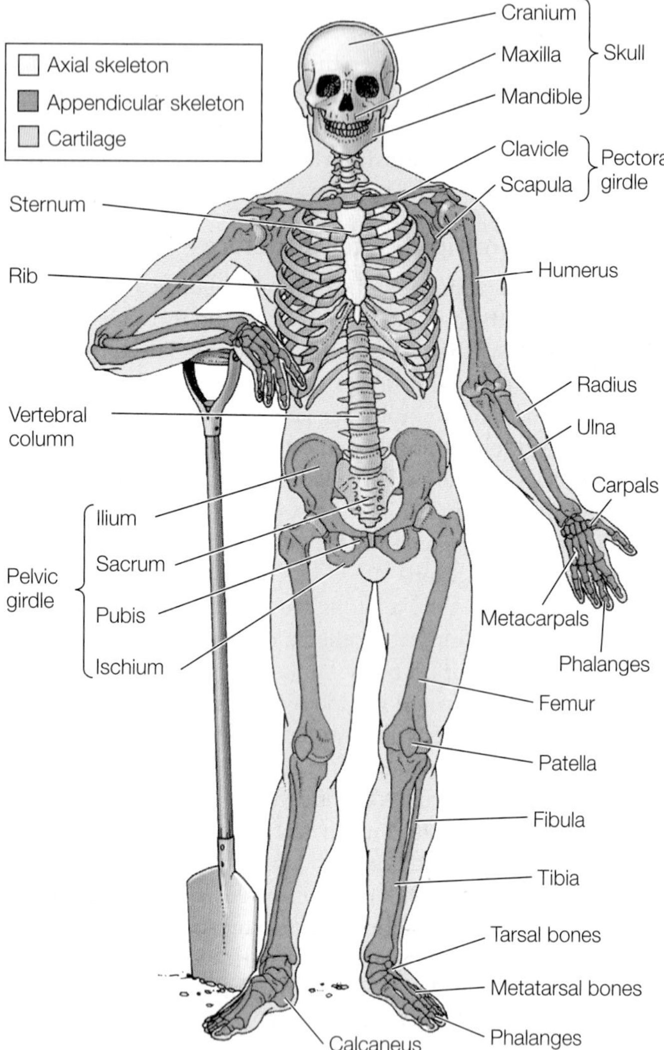


**48.15 The Human Endoskeleton** Cartilage and bone make up the internal skeleton of a human being.

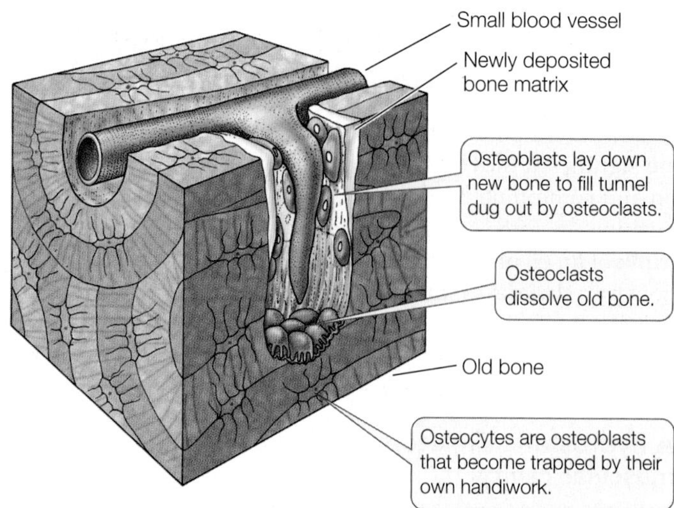


**48.16 Bone Is Living Tissue** Bones are constantly being remodeled by osteoblasts, which lay down bone, and osteoclasts, which resorb bone.

"gristle." This matrix, called **cartilage**, is found in parts of the endoskeleton where both stiffness and resiliency are required, such as on the surfaces of joints where bones move against one another. Cartilage is also the supportive tissue in stiff but flexible structures such as the larynx (voice box), nose, and ear pinnae. Sharks and rays are called cartilaginous fishes because their skeletons are composed entirely of cartilage. In most other vertebrates, cartilage is the principal component of the embryonic skeleton, but during development most of it is gradually replaced by bone.

**Bone** also contains collagen fibers, but it gets its rigidity and hardness from an extracellular matrix of insoluble calcium phosphate crystals. Bone serves as a reservoir of calcium for the rest of the body and is in dynamic equilibrium with soluble calcium in the extracellular fluids of the body. This equilibrium is under the control of calcitonin and parathyroid hormone (see Figure 41.14). If too much calcium is taken from the skeleton, the bones are seriously weakened.

The living cells of bone—osteoblasts, osteocytes, and osteoclasts—are responsible for the constant dynamic remodeling of bone (**Figure 48.16**). **Osteoblasts** lay down new matrix material on bone surfaces. These cells gradually become surrounded by matrix and eventually become enclosed within the bone, at which point they cease laying down matrix but continue to exist within small lacunae (cavities) in the bone. In this state they are called **osteocytes**. Despite the vast amounts of matrix between them, osteocytes remain in contact with one another through long cellular extensions that run through tiny channels in the bone. Communication between osteocytes is important in controlling the activities of the cells that are laying down or removing bone.

The cells that resorb bone are the **osteoclasts**. They are derived from the same cell lineage that produces white blood cells. Osteoclasts erode bone, forming cavities and tunnels. Osteoblasts follow osteoclasts, depositing new bone. Thus the interplay of osteoblasts and osteoclasts constantly replaces and remodels the bones, allowing a bone to recover from damage and adjust to the forces placed on it.

How the activities of the bone cells are coordinated is not understood, but stress placed on bones somehow provides them with information. A remarkable finding in studies of astronauts who spent long periods in zero gravity was that their bones decalcified. Conversely, in athletes, certain bones thicken during training. Both thickening and thinning of bones are experienced by anyone who has had a leg in a cast for a long time: the bones of the uninjured leg carry the person's weight and thicken while the bones of the inactive leg in the cast thin.

Because of the positive effects of physical stress on bone deposition, weight-bearing exercise is effective in preventing and treating osteoporosis, which is the loss of bone density (and hence strength). More than 25 million people in the United States suffer from this debilitating condition. Although osteoporosis is most commonly a problem for postmenopausal women, it can occur in younger people as a result of malnutrition. For example, the condition known as female athlete triad includes eating disorders, cessation of menstrual cycling, and osteoporosis. These are interactive conditions in which the eating disorder and excessive training lead to malnutrition that can result in endocrine disruption and osteoporosis. Excessive training and malnutrition can lead to bone loss in males as well.

## Bones develop from connective tissues

Bones are divided into two types on the basis of how they develop. **Membranous bone** forms on a scaffold of connective tissue membrane. **Cartilage bone** forms first as a cartilaginous structure resembling the future mature bone, then gradually hardens, or ossifies, to become bone. The outer bones of the skull are membranous bones; the bones of the limbs are cartilage bones.

Cartilage bones can grow throughout the ossification process. The long bones of the legs and arms, for example, ossify first at the centers and later at each end (**Figure 48.17**). Growth can continue until these areas of ossification join. The membranous bones forming the skull cap grow until their edges meet.

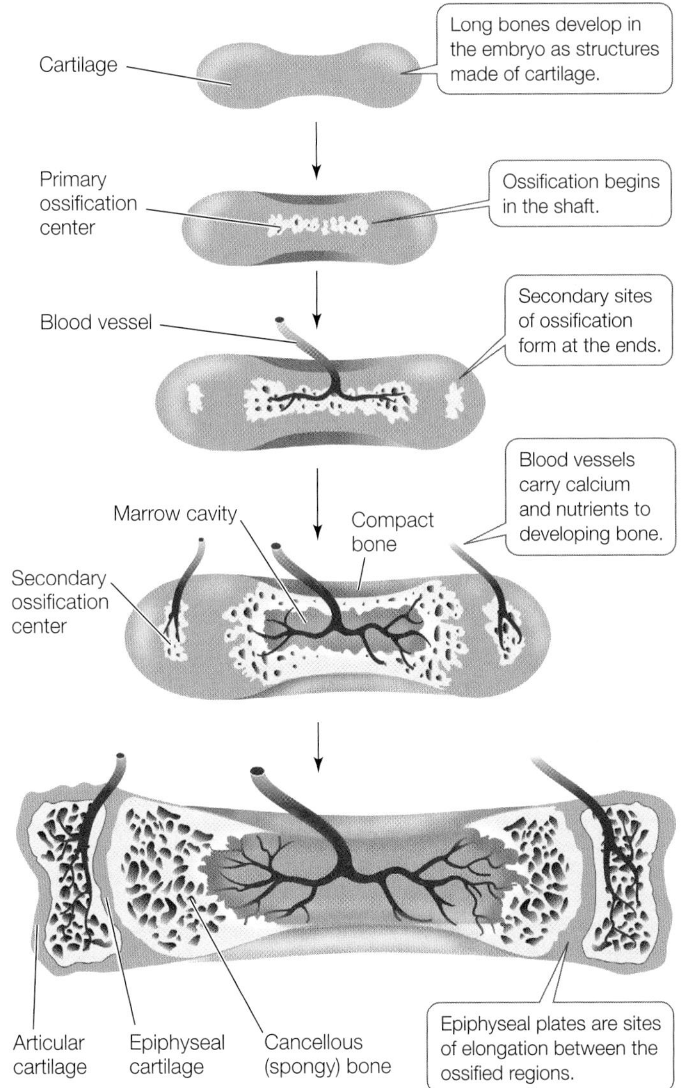


**48.17 The Growth of Long Bones** In the long bones of human limbs, ossification occurs first at the centers and later at each end.

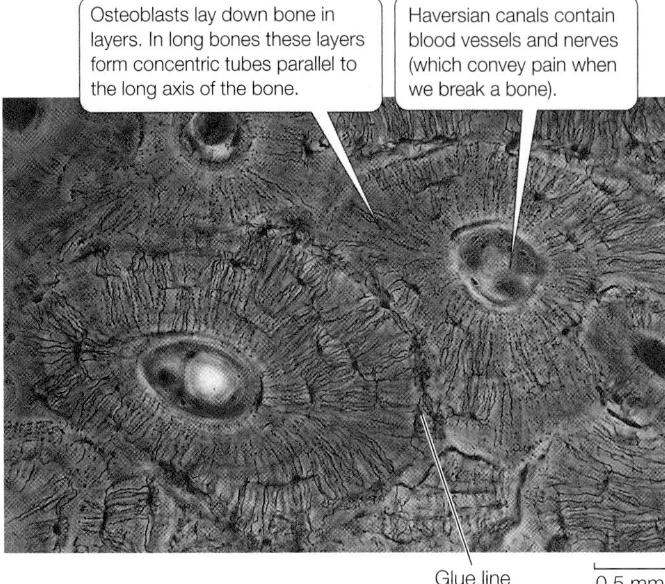


**48.18 Most Compact Bone Is Composed of Haversian Systems** A micrograph of a section of a long bone shows Haversian systems with their central canals. Glue lines separate Haversian systems.

The soft spot on the top of a baby's head (the fontanelle) is the point at which the skull bones have not yet joined.

The structure of bone may be **compact** (solid and hard) or **cancellous** (having numerous internal cavities that make it appear spongy, although it is rigid). The architecture of a specific bone depends on its position and function, but most bones have both compact and cancellous regions. The shafts of the long bones of the limbs, for example, are cylinders of compact bone surrounding central cavities that contain the bone marrow, where the cellular elements of the blood are made. The ends of the long bones are cancellous (see Figure 48.17). Cancellous bone is lightweight because of its numerous cavities, but it is also strong because its internal meshwork constitutes a support system. It can withstand considerable forces of compression. The rigid, tubelike shaft of compact bone can withstand compression and bending forces. Architects and nature alike use hollow tubes as lightweight structural elements.

Most of the compact bone in mammals is called Haversian bone because it is composed of structural units called **Haversian systems** (**Figure 48.18**). Each Haversian system is a set of thin, concentric bony cylinders, between which are the osteocytes in their lacunae. Through the center of each Haversian system runs a narrow canal containing blood vessels and nerves. Adjacent Haversian systems are separated by boundaries called glue lines. Haversian bone is resistant to fracturing because cracks tend to stop at glue lines.

## Bones that have a common joint can work as a lever

Muscles and bones work together around **joints**, where two or more bones come together. Different kinds of joints allow motion in different directions (**Figure 48.19**), but muscles can exert force in only one direction. Therefore muscles create

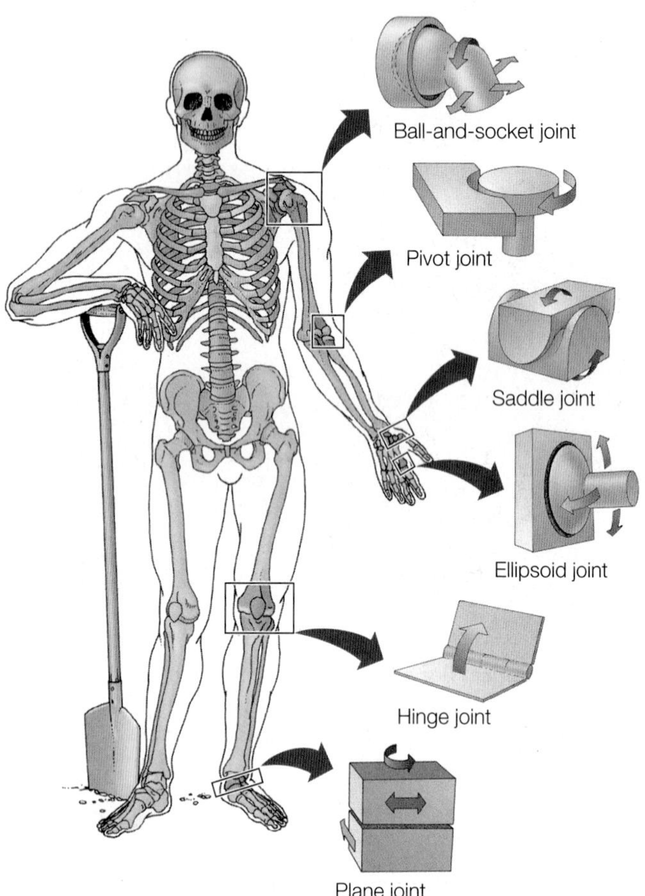


**48.19 Types of Joints** The designs of joints are similar to mechanical counterparts and enable a variety of movements.

**Go to Activity 48.3 Joints**
**Life10e.com/ac48.3**

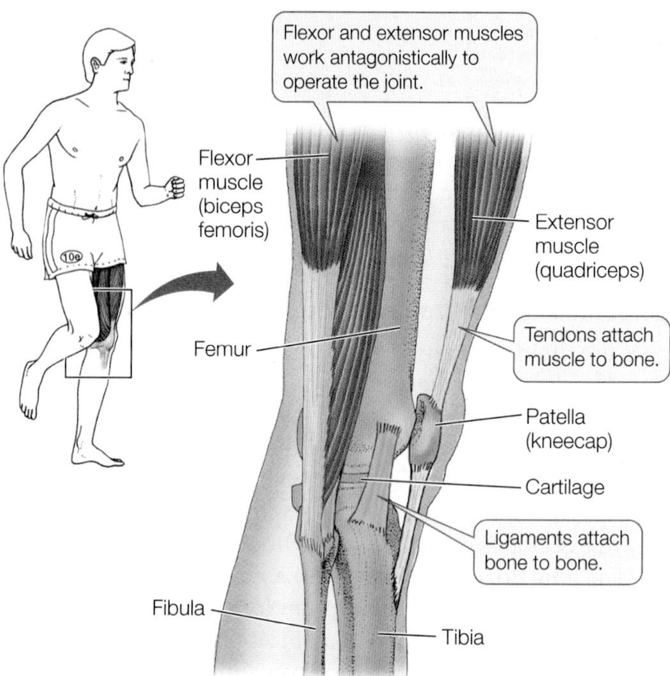


**48.20 Joints, Ligaments, and Tendons** A side view of the knee shows the interactions of muscle, bone, cartilage, ligaments, and tendons at this crucial and vulnerable human joint.

movement around joints by working in antagonistic pairs: when one muscle contracts, the other relaxes. When both contract, the joint becomes rigid (which is important for maintaining posture, for example).

With respect to a particular joint, such as the knee, we refer to the muscle that bends, or flexes, the joint as the **flexor**, and the muscle that straightens, or extends, the joint as the **extensor**. The bones that meet at the joint are held together by **ligaments**, which are flexible bands of connective tissue. Other straps of connective tissue, called **tendons**, attach the muscles to the bones (**Figure 48.20**). In many kinds of joints, only the tendon spans the joint, sometimes moving over the surfaces of the bones like a rope over a pulley. The tendon of the quadriceps muscle traveling over the knee joint is what is tapped to elicit the knee-jerk reflex (see Figure 45.18).

Bones constitute a system of levers that are moved around joints by the muscles. A lever has an effort arm and a load arm that work around a fulcrum (pivot). The length ratio of the two arms determines whether a particular lever can exert a lot of force over a short distance or is better at translating force into large or fast movements. Compare the jaw and knee joints, for example (**Figure 48.21**). The effort arm of the jaw is long relative to the load arm, allowing the jaw to apply great force over a small distance. Think of the powerful jaws of carnivores that can easily crack bones. The effort arm of the lower leg, by contrast, is short relative to the load arm, so you can run fast, jump high, and deliver swift kicks.

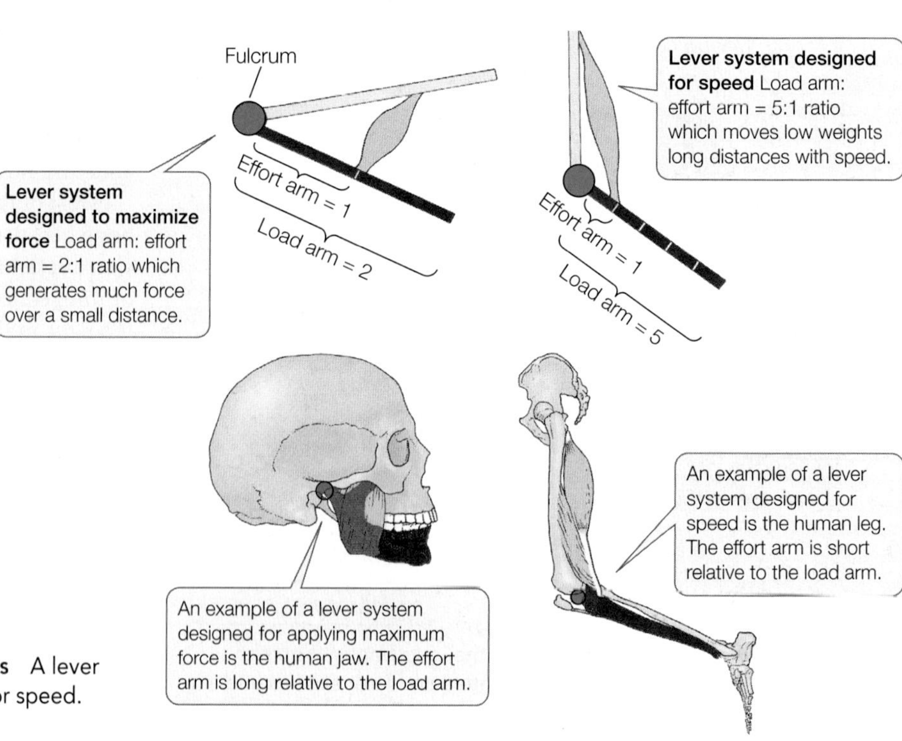


**48.21 Bones and Joints Work Like Systems of Levers** A lever system can be designed for maximizing either force or speed.

## RECAP 48.3

Muscles can only contract and relax; to achieve organized movement, they must pull against rigid structures—other muscles, hydrostatic skeletons, exoskeletons, or endoskeletons.

- How do the muscles and fluid-filled body cavity of an earthworm interact to enable the animal to move? **See p. 999 and Figure 48.14**
- In terms of levers, explain how specific joints can produce maximum force versus maximum speed. **See p. 1002 and Figure 48.21**

How can the kangaroo increase its speed fivefold without increasing its metabolic expenditure?

### ANSWER

In kangaroos, as in other vertebrates, the muscles used to jump are attached to bones by tendons. Tendons are elastic. The kangaroo's tendons stretch when it lands, and their recoil helps power the next jump—similar to the action of a pogo stick. To move faster, the kangaroo simply increases the length of its stride, thereby increasing both the stretch on its tendons each time it lands and the magnitude of the recoil at the initiation of each jump.

# CHAPTER**SUMMARY** 48

## 48.1 How Do Muscles Contract?

- **Skeletal muscle** consists of bundles of **muscle fibers**. Each skeletal muscle fiber is a large cell containing multiple nuclei.
- Skeletal muscles contain numerous **myofibrils**, which are bundles of **actin** and **myosin** filaments. The regular, overlapping arrangement of the actin and myosin filaments into **sarcomeres** gives skeletal muscle its striated appearance. **Review Figure 48.1, ACTIVITY 48.1**
- The changes in the banding patterns of sarcomeres led to the **sliding filament model** of muscle contraction. **Review Figure 48.2**
- The molecular mechanism of muscle contraction involves the binding of the globular heads of myosin molecules to actin. **Review Figures 48.3, 48.6, ANIMATED TUTORIAL 48.1**
- All the fibers activated by a single motor neuron constitute a **motor unit**. Each nerve ending of the motor neuron forms a synapse with the muscle cell membrane. Action potentials spread across the muscle cell membrane and through the **T tubules**, causing $Ca^{2+}$ to be released from the **sarcoplasmic reticulum**. **Review Figure 48.5, ACTIVITY 48.2**
- $Ca^{2+}$ binds to **troponin** and changes its conformation, pulling the **tropomyosin** strands away from the myosin-binding sites on the actin filament. The muscle fiber continues to contract until the $Ca^{2+}$ is returned to the sarcoplasmic reticulum. **Review Figure 48.6**
- **Cardiac muscle** cells are striated, uninucleate, branching, and electrically connected by gap junctions, so that action potentials spread rapidly throughout sheets of cardiac muscle and cause coordinated contractions.
- **Smooth muscle** provides contractile force for internal organs. Smooth muscle cells respond to stretch and to neurotransmitters from the autonomic nervous system. **Review Figure 48.8, ANIMATED TUTORIAL 48.2**

## 48.2 What Determines Muscle Performance?

- In skeletal muscle, a single action potential causes a minimum unit of contraction called a **twitch**. Twitches occurring in rapid succession can be summed to achieve sustained tension, known as **tetanus**. **Review Figure 48.10**
- **Slow-twitch fibers** facilitate extended, aerobic work; **fast-twitch fibers** generate maximum forces for short periods of time. The ratio of slow-twitch to fast-twitch fibers in the muscles of an individual is largely genetically determined. **Review Figure 48.11**
- The force that a muscle fiber can produce depends on its initial state of extension or contraction. **Review Figure 48.12**
- Anaerobic exercise stimulates the enlargement of muscle fibers through production of new microfilaments. Aerobic exercise stimulates greater oxidative capacity of muscle fibers.
- Muscle performance depends on a supply of ATP. **Review Figure 48.13**

## 48.3 How Do Skeletal Systems and Muscles Work Together?

- **Skeletal systems** provide supports against which muscles can pull.
- **Hydrostatic skeletons** are fluid-filled body cavities that can be squeezed by muscles. **Review Figure 48.14**
- **Exoskeletons** are hardened outer surfaces to which internal muscles are attached.
- **Endoskeletons** are internal systems of rigid rodlike, platelike, and tubelike supports, consisting of **bone** and **cartilage** to which muscles are attached. **Review Figure 48.15**
- Bone is continually remodeled by **osteoblasts**, which lay down new bone, and **osteoclasts**, which erode bone. **Review Figure 48.16**
- Bones develop from connective tissue membranes (**membranous bone**) or from cartilage (**cartilage bone**) through ossification. **Review Figure 48.17**
- Bone can be **compact** (solid and hard) or **cancellous** (containing numerous internal spaces). Most of the compact bone of mammals is composed of **Haversian systems**. **Review Figure 48.18**
- **Joints** enable muscles to power movements in different directions. Muscles and bones work together around joints as systems of levers. **Review Figures 48.19, 48.21, ACTIVITY 48.3**
- **Tendons** connect muscles to bones; **ligaments** connect bones to one another. **Review Figure 48.20**

**Go to the Interactive Summary to review key figures, Animated Tutorials, and Activities Life10e.com/is48**

# CHAPTER**REVIEW**

## REMEMBERING

1. The role of $Ca^{2+}$ in the control of muscle contraction is to
   a. cause depolarization of the T tubule system.
   b. change the conformation of troponin, thus exposing myosin-binding sites.
   c. change the conformation of myosin heads, thus causing microfilaments to slide past each other.
   d. bind to tropomyosin and break actin–myosin cross-bridges.
   e. block the ATP-binding site on myosin heads, enabling muscles to relax.

2. Fifteen minutes into a 10-kilometer run, what is the major energy source of the leg muscles?
   a. Preformed ATP
   b. Glycolysis
   c. Oxidative metabolism
   d. Pyruvate and lactate
   e. High-protein drink consumed right before the race

3. Which statement about skeletal muscle contraction is *not* true?
   a. A single action potential at the neuromuscular junction is sufficient to cause a muscle to twitch.
   b. Once maximum muscle tension is achieved, no ATP is required to maintain that level of tension.
   c. An action potential in the muscle cell activates contraction by releasing $Ca^{2+}$ into the sarcoplasm.
   d. Summation of twitches leads to a graded increase in the tension that can be generated by a single muscle fiber.
   e. The tension generated by a muscle can be varied by controlling how many of its motor units are active.

4. Which statement about the structure of skeletal muscle is true?
   a. The light bands of the sarcomere are the regions where actin and myosin filaments overlap.
   b. When a muscle contracts, the A bands of the sarcomere lengthen.
   c. The myosin filaments are anchored in the Z lines.
   d. When a muscle contracts, the H zone of the sarcomere shortens.
   e. The sarcoplasm of the muscle cell is contained within the sarcoplasmic reticulum.

5. Insects can beat their wings at exceptionally high frequencies because
   a. their wing muscles have mostly fast-twitch fibers.
   b. their motor neurons can fire action potentials at a very high frequency.
   c. their wings have exoskeletal supports.
   d. their wing muscles have extensive sarcoplasmic reticulum that cycles $Ca^{2+}$ very fast.
   e. their wing muscles can generate a rapid oscillation of contraction that is asynchronous with motor neuron firing.

## UNDERSTANDING & APPLYING

6. Sarcomeres can shorten only by less than 50 percent of their resting length. Yet muscles cause movements of a wide range of magnitudes—compare the range of movement of a toe and a leg. What are two design features of muscles and skeletons that can maximize a muscle's ability to produce a wide range of movements of an appendage?

7. If an adolescent breaks a leg bone close to the ankle joint, after the break heals, that leg may not grow as long as the other one. Why?

8. Stand with your arms held out at right angles to your body for as long as you can. Which will fatigue first: your shoulders or your legs? Why? What does this tell you about the muscles of your shoulders and legs?

## ANALYZING & EVALUATING

9. A single action potential in a muscle cell lasts about 2 milliseconds, but the single muscle twitch it generates reaches its peak after about 30 msec, and does not return to resting level until about 150 msec after the action potential. Explain three factors that explain the time difference between the action potential and the twitch.

10. Malignant hyperthermia is a rare but frequently lethal condition stimulated by certain anesthetics. Suddenly the individual's muscles become rigid, heart rate shoots up, and body temperature rises rapidly. Those at risk for malignant hyperthermia have a mutation in the gene that codes for the ryanodine receptor. The mutation results in excessive opening of the $Ca^{2+}$ channels in the sarcoplasmic reticulum. Knowing this cause of the condition, how can you explain all of its manifestations—temperature, heart rate, and muscle tension?

Go to BioPortal at **yourBioPortal.com** for Animated Tutorials, Activities, LearningCurve Quizzes, Flashcards, and many other study and review resources.

# 49 Gas Exchange

## CHAPTER OUTLINE


**49.1** What Physical Factors Govern Respiratory Gas Exchange?

**49.2** What Adaptations Maximize Respiratory Gas Exchange?

**49.3** How Do Human Lungs Work?

**49.4** How Does Blood Transport Respiratory Gases?

**49.5** How is Breathing Regulated?

**Proboscis as Snorkel** Elephants can cross deep bodies of water either swimming or by walking on the bottom—whichever allows it to keep its trunk above water.

ELEPHANT TRUNKS HAVE MANY USES. They can pluck leaves, pull up plants, and pick up peanuts and tree trunks; they suck in water to squirt into the mouth for drinking or to spray over the body for cooling; they smell the air around them; they communicate by touch; and mothers can wallop their calves when they misbehave.

Of course, elephants also use their trunks to breathe, and it is a useful appendage indeed when they go into deep water—for example, to cross rivers. Whether the elephant swims or walks on the bottom, it uses its trunk as a snorkel. As long as the tip of the trunk is above water, the elephant can breathe.

You may be familiar with diving using a snorkel about a foot long. When you want to dive deeper, you hold your breath. Water is heavy, and as you swim deeper, water pressure increases and presses on your body. The air in your lungs is compressible, so as you swim deeper, the volume of your lungs decreases.

Think what would happen if you tried to suck air through a long snorkel in deep water. Because the snorkel would be open to the air above, the pressure in your lungs would be the same as at the water surface, but the surrounding water would be pressing on your body. To expand your lungs, you would have to exert enough force to push against the surrounding water. Not so easy! We solve this problem with scuba equipment: by breathing air from a pressurized tank with a regulator valve, we keep the air in our lungs at the same pressure as the surrounding water.

The chest of a snorkeling elephant can be 3 meters below the surface—and for every 3 meters, water pressure increases by 0.3 atmospheres (about 5 pounds per square inch, PSI). Even though the elephant's respiratory muscles are strong enough to work against this pressure, there is still a serious problem for the blood vessels lining the cavity in which the lungs are suspended. The pressure difference across the walls of those blood vessels is the difference between the pressure in the vessels (blood pressure plus the compressive water pressure) and the air pressure at the water surface (where the trunk opens to the air). This difference is at least 5 PSI—a pressure that in other mammals would burst small blood vessels and fill the chest cavity with blood.

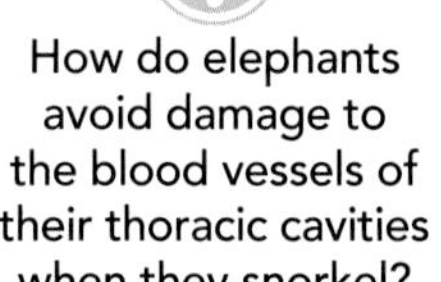

How do elephants avoid damage to the blood vessels of their thoracic cavities when they snorkel?

See answer on p. 1022.

## 49.1 What Physical Factors Govern Respiratory Gas Exchange?

The **respiratory gases** that animals must exchange are oxygen ($O_2$) and carbon dioxide ($CO_2$). Cells need to obtain $O_2$ from the environment to produce an adequate supply of ATP by cellular respiration (see Chapter 9). $CO_2$ is an end product of cellular respiration, and it must be removed from the body to prevent toxic effects. Gas exchange systems of animals consist of (1) specialized body surface areas where these gases can move between the body and the environment, and (2) mechanisms that ventilate the environmental side of those surfaces with air or water and perfuse the internal side of those surfaces with extracellular fluids.

Diffusion is the only means by which respiratory gases are exchanged between an animal's internal body fluids and the outside medium (air or water). There are no active transport mechanisms to move respiratory gases across biological membranes. Because diffusion is a physical process, knowing what physical factors influence rates of diffusion helps us understand the diverse adaptations of gas exchange systems. (You may want to review the discussion about the physical nature of diffusion in Section 6.3.)

### Diffusion of gases is driven by partial pressure differences

Diffusion results from the random motion of molecules, and the net movement of molecules within a medium such as air or water is down their concentration gradient. Concentrations in solutions are simply the amount of solute per volume of solution. Concentrations of gases, however, vary with pressure because gases are compressible. For example, there are twice as many gas molecules in a liter of gas at 2 atmospheres of pressure as there are in a liter of gas at 1 atmosphere of pressure.

Biologists express the concentrations of different gases in a mixture as the **partial pressures** of those gases. To calculate the partial pressure of a gas such as oxygen in a mixture of gases such as air, we have to know the total pressure. In most cases, for an air-breathing animal the total pressure is atmospheric pressure. At sea level, atmospheric pressure is about 760 millimeters of mercury (mm Hg), depending on the weather. Because dry air is 20.9 percent $O_2$, the partial pressure of oxygen ($P_{O_2}$) at sea level is 20.9 percent of 760 mm Hg, or about 159 mm Hg. If two gas mixtures are separated by a membrane permeable to $O_2$, $O_2$ will diffuse from the mixture where its partial pressure is higher to the mixture where its partial pressure is lower.

To calculate the concentration of a gas in a solution, we have to know the solubility of that gas in that particular solvent. The amount of a gas in a liquid depends both on its partial pressure in the gas phase in contact with the liquid *and* on its solubility in that liquid. The bottom line is that diffusion of a gas between the gas phase and the liquid phase is a function of its partial pressures in those two phases; the gas diffuses from the phase with the higher partial pressure to the phase with the lower partial pressure, and at equilibrium the partial pressures in the two phases are equal. However, the *amount* of the gas that can be contained by the liquid depends on the solubility of that gas in that liquid. Furthermore, the solubility of a gas in a particular liquid can vary widely depending on conditions. What follows is a practical illustration of these facts.

Solubility of a gas in a liquid, such as oxygen in water, is a function of temperature—solubility is higher at low temperatures. So if we have similar containers of water in equilibrium with air, but at different temperatures, the concentrations of oxygen in these water containers will be different (less $O_2$ in the warmer ones), but the partial pressures of oxygen will be the same. Thus for water-breathing animals, the warmer the water is, the less $O_2$ there is per liter of water. The important point is that for a gas in solution, its concentration is not the same as its partial pressure, but partial pressures are what drive diffusion. Thus in our continuing discussions of respiratory gas exchange, we will always use partial pressure rather than concentration when referring to the amount of gas in solution.

### Fick's law applies to all systems of gas exchange

Whether in air or water, the diffusion rates of respiratory gases depend on their partial pressure gradients and on other factors that can be described quantitatively with a simple equation called **Fick's law of diffusion**. All environmental variables that limit respiratory gas exchange and all adaptations that maximize respiratory gas exchange are reflected in one or more components of this equation. Fick's law is written as

$$Q = DA\frac{P_1 - P_2}{L}$$

where

- $Q$ is the rate at which a gas such as $O_2$ diffuses between two locations
- $D$ is the diffusion coefficient, which is a characteristic of the diffusing substance, the medium, and the temperature. For example, perfume has a higher $D$ than motor oil vapor, and all substances diffuse faster at higher temperatures and faster in air than in water. Temperature is not expressed explicitly in Fick's law because the diffusion coefficient is usually determined at room temperature (about 20°C)
- $A$ is the area across which the gas is diffusing
- $P_1$ and $P_2$ are the partial pressures of the gas at the two locations
- $L$ is the path length, or distance, between the two locations
- $(P_1 - P_2)/L$ is a partial pressure gradient.

The strict dependence of animals on diffusion for gas exchange with their environments has selected for various adaptations that maximize $Q$, many of which we will describe in this chapter. Animals can maximize $D$ for respiratory gases by using air rather than water as their gas exchange medium whenever possible. All other adaptations for maximizing respiratory gas exchange must influence the surface area ($A$) for gas exchange or the partial pressure gradient across that surface area.

## Air is a better respiratory medium than water

The slow diffusion of $O_2$ molecules in water affects both air- and water-breathing animals. Eukaryotic cells carry out cellular respiration in their mitochondria, which are located in the cytoplasm—an aqueous medium. Cells are bathed in extracellular fluid—also an aqueous medium. In addition, all respiratory surfaces must be protected from drying out by a thin film of fluid through which $O_2$ must diffuse. Even in air-breathing animals, the slow rate of $O_2$ diffusion in water limits the efficiency of $O_2$ distribution from gas exchange surfaces to the sites of cellular respiration.

Diffusion of $O_2$ in water is so slow that even animal cells with low rates of metabolism cannot function more than a few millimeters away from a good source of environmental $O_2$. Therefore there are severe size and shape limits on the many species of invertebrates that lack internal systems for distributing $O_2$. Most of these species are very small, but some have grown larger by evolving a flat, thin body with a large external surface area (**Figure 49.1A**). Others have a very thin body built around a central cavity through which water can circulate (**Figure 49.1B**). A critical factor enabling larger, more complex animal bodies has been the evolution of specialized respiratory systems with large surface areas that are highly permeable to respiratory gases (**Figure 49.1C**).

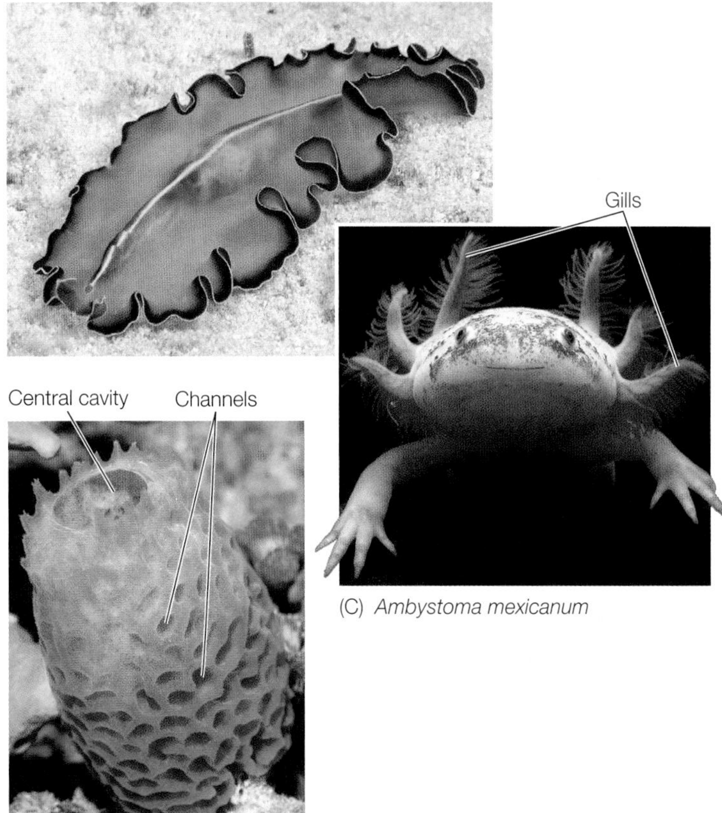


**49.1 Keeping in Touch with the Medium** (A) No cell in the leaf-like body of this marine flatworm is more than a millimeter away from seawater. (B) Sponges have body walls perforated by many channels, which allow water to flow between the outside world and a central cavity. No cell in the sponge is more than a millimeter away from seawater. (C) A feathery fringe of gills on this aquatic salamander provides a large surface area for gas exchange. Blood circulating through the gills comes into close contact with the respiratory medium.

$O_2$ can be obtained more easily from air than from water for several reasons:

- The $O_2$ content of air is much higher than the $O_2$ content of an equal volume of water. The maximum $O_2$ content of a bubbling stream in equilibrium with air is less than 10 milliliters of $O_2$ per liter of water. The $O_2$ content of the air over the stream is about 200 ml of $O_2$ per liter of air.
- $O_2$ diffuses about 8,000 times more rapidly in air than in water. That is why the $O_2$ content of a stagnant pond can be zero only a few millimeters below the surface.
- An animal has to work (expend energy) to ventilate its gas exchange surfaces with water or air. More energy is required to move water than air because water is 800 times denser than air and about 50 times more viscous.

You can appreciate how important these facts were for the evolutionary transition of life to the terrestrial environment, because they meant that there were fewer constraints on the evolution of higher metabolic rates.

## High temperatures create respiratory problems for aquatic animals

Animals that use water for their respiratory exchange medium are in a double bind when environmental temperatures rise. Most water-breathing animals are ectotherms—their body temperatures are closely tied to the temperature of the water around them. As the water temperature rises, an ectotherm's body temperature and metabolic rate rise (see Figure 40.8). Thus water breathers need more $O_2$ as the water gets warmer (**Figure 49.2**). But as mentioned above, warm water holds less dissolved gas than cold water does (think of the gases that escape when you open a warm bottle of soda). In addition, since a water-breathing animal performs work to move water across its gas exchange surfaces, it must expend more energy to breathe as water temperature rises. Therefore, as water temperature goes up, a water-breathing animal must extract more and more $O_2$ from an environment that is increasingly $O_2$ deficient, and a lower percentage of that $O_2$ is available to support activities other than breathing.

## $O_2$ availability decreases with altitude

Just as a rise in water temperature reduces the supply of $O_2$ available to water-breathing animals, an increase in altitude reduces the $O_2$ supply for air breathers. At all altitudes, $O_2$ makes up 20.9 percent of the dry air; however, as you go up in altitude, the total amount of gas per unit of volume decreases, as reflected in the atmospheric pressure. For example, at 5,800 meters, atmospheric pressure is only half what it is at sea level,

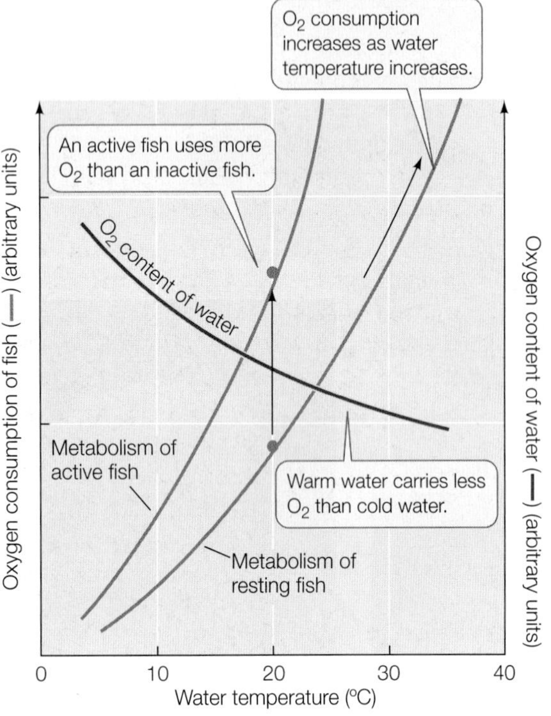


**49.2 The Double Bind of Water Breathers** Fishes need more $O_2$ when the water is warmer, but warm water carries less $O_2$ than cold water.

so the $P_{O_2}$ at that altitude is only about 80 mm Hg. At the summit of Mount Everest (8,850 m), $P_{O_2}$ is only about 50 mm Hg—roughly one-third what it is at sea level.

Because the movement of $O_2$ across respiratory gas exchange surfaces and into the body depends on diffusion, its rate of movement depends on the $P_{O_2}$ difference between the air and the body fluids. Therefore the drastically reduced $P_{O_2}$ in the air at high altitudes constrains $O_2$ uptake. Because of this, mountain climbers attempting the highest peaks usually breathe $O_2$ from pressurized bottles.

### $CO_2$ is lost by diffusion

Respiratory gas exchange is a two-way process: $CO_2$ diffuses out of the body as $O_2$ diffuses in. The direction and rate of diffusion of the respiratory gases across the exchange surfaces depend on the partial pressure gradients of the gases. The partial pressure gradients of $O_2$ and $CO_2$ across these exchange surfaces are quite different. The amount of $CO_2$ in the atmosphere is extremely low (0.03%), so for air-breathing animals there is always a large concentration gradient for diffusion of $CO_2$ from the body to the environment. Whereas the partial pressure gradient for $O_2$ decreases with increasing altitude, the gradient driving $CO_2$ out of the body hardly changes. The partial pressure of carbon dioxide ($P_{CO_2}$) in the atmosphere is close to zero both at sea level and atop Mount Everest.

In general, getting rid of $CO_2$ is not a problem for water-breathing animals because $CO_2$ is much more soluble in water than is $O_2$. Even in stagnant water, where the $P_{CO_2}$ is higher than in moving water, the lack of $O_2$ becomes a problem for an animal long before $CO_2$ exchange difficulties arise.

**RECAP 49.1**

Respiratory gases are exchanged only by diffusion. Air is a better respiratory medium than water because a given volume of air has more $O_2$ than the same volume of water. $O_2$ diffuses faster in air than in water, and less work is required to move air over respiratory exchange surfaces.

- Describe how the variables in Fick's law of diffusion relate to respiratory systems. **See p. 1006**
- Why does a rise in water temperature create a double-bind situation for water-breathing animals? **See p. 1007 and Figure 49.2**
- Explain the concept of partial pressures of gases and how it relates to diffusion rates of $O_2$ and $CO_2$ at different altitudes. **See pp. 1006 and 1007–1008**

Now that we have discussed the physical factors that influence diffusion rates of respiratory gases between animals and their environments, let's look at some of the adaptations that have evolved for maximizing respiratory gas exchange.

## 49.2 What Adaptations Maximize Respiratory Gas Exchange?

As you might expect from the components of Fick's law of diffusion, adaptations to maximize respiratory gas exchange can be categorized as those that:

- Increase the surface area for gas exchange ($A$)
- Maximize the partial pressure difference driving diffusion ($P_1 - P_2$)
- Minimize the diffusion path length ($L$)
- Minimize the diffusion that takes place in an aqueous medium (maximize $D$)

### Respiratory organs have large surface areas

A variety of anatomical adaptations maximize the specialized body surface areas ($A$) for respiratory gas exchange. Water-breathing animals generally have gills, and air-breathing animals have tracheae or lungs. **External gills** are highly branched and folded extensions of the body surface that provide a large surface area for gas exchange (**Figure 49.3A**; see also Figure 49.1C). External gills are found in larval amphibians and in the larvae of many insects. Because they consist of thin, delicate tissues, external gills minimize the path length ($L$) traversed by diffusing molecules of $O_2$ and $CO_2$. External gills are vulnerable to damage, however, and are tempting morsels for predators, so in many animals protective body cavities for gills have evolved. Such **internal gills** are found in most mollusks and arthropods and in all fishes (**Figure 49.3B**).

**Lungs** are internal cavities for respiratory gas exchange with air (**Figure 49.3C**). Their structure is quite different from that of gills. Lungs have a large surface area because they are highly divided; and because they are elastic, they can be inflated with air and deflated.

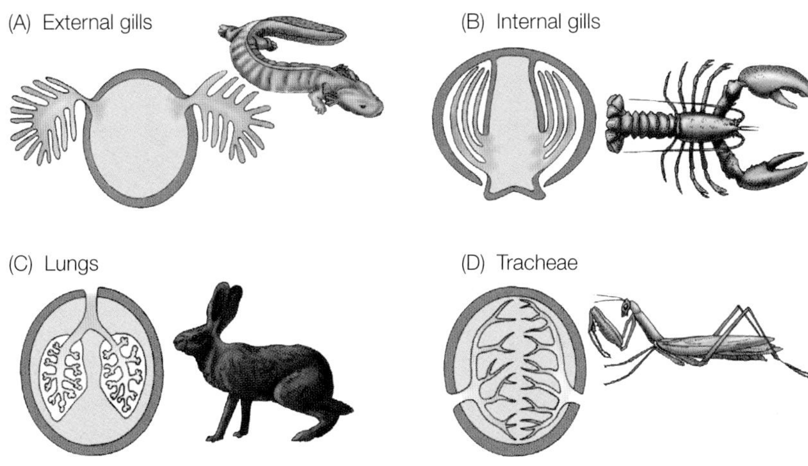


**49.3 Gas Exchange Systems** Large surface areas (blue in these diagrams) for the diffusion of respiratory gases are common features of animals. External (A) and internal (B) gills are adaptations for gas exchange with water. Lungs (C) and tracheae (D) are organs for gas exchange with air.

Insects have a respiratory gas exchange system consisting of a network of air-filled tubes called **tracheae** that branch through all tissues of the insect's body (**Figure 49.3D**). The terminal branches of these tubes are so numerous that they have an enormous surface area compared with the external surface area of the insect's body.

## Ventilation and perfusion of gas exchange surfaces maximize partial pressure gradients

Partial pressure gradients $[(P_1 - P_2)/L]$ drive diffusion across gas exchange surfaces; the larger the gradient, the greater the rate of gas exchange. These gradients can be maximized in several ways:

- *Minimization of path length*: Very thin tissues in gills and lungs reduce the diffusion path length ($L$).
- *Ventilation*: Actively moving the external medium over the gas exchange surfaces (i.e., breathing) regularly exposes those surfaces to fresh respiratory medium containing maximum $O_2$ and minimum $CO_2$ concentrations. This maximizes the partial pressure gradients.
- *Perfusion*: Actively moving the internal medium (e.g., blood) over the internal side of the exchange surfaces transports $CO_2$ to those surfaces and $O_2$ away from them, thus maximizing the partial pressure gradients driving diffusion.

This chapter describes four gas exchange systems. First we will look at the unique gas exchange system of insects. Then we will describe two highly efficient gas exchange systems, fish gills and bird lungs. Lastly we will examine the structure and function of human lungs.

## Insects have airways throughout their bodies

The tracheal system that enables insects to exchange respiratory gases extends to all tissues in the insect body. Thus respiratory gases diffuse through air most of the way to and from every cell. The insect respiratory system communicates with the outside environment through gated openings called spiracles in the sides of the abdomen and thorax (**Figure 49.4A,B**). The spiracles open to allow gas exchange and then close to decrease water loss. They open into tubes called tracheae that branch into even finer tubes, or tracheoles, which end in tiny air capillaries that are the actual gas exchange surfaces (**Figure 49.4C**). In the insect's flight muscles and other highly active tissues, every mitochondrion is close to an air capillary.

## Fish gills use countercurrent flow to maximize gas exchange

The internal gills of fishes are supported by gill arches that lie between the mouth cavity and the protective opercular flaps on

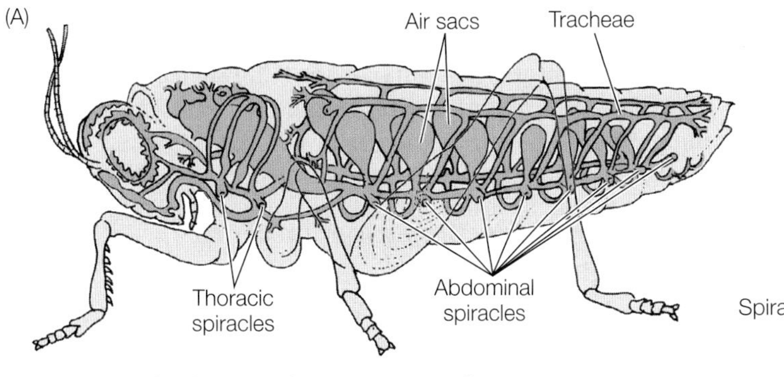


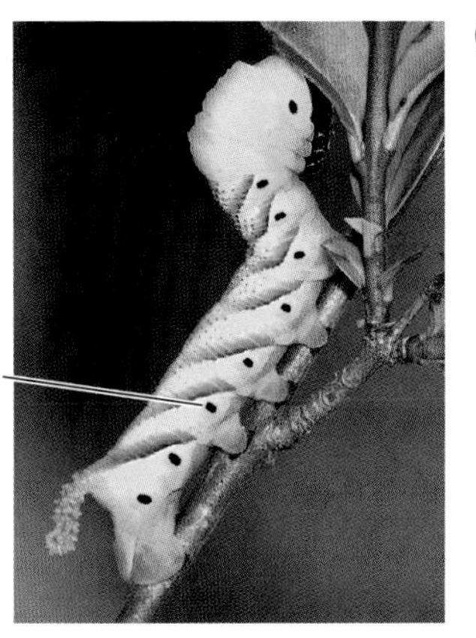

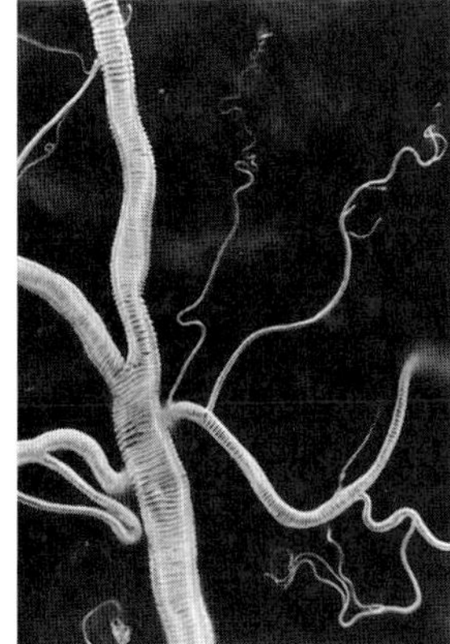


**49.4 The Tracheal Gas Exchange System of Insects** (A) In insects, respiratory gases diffuse through a system of air tubes (tracheae) that open to the external environment through holes called spiracles. (B) The spiracles of a hawkmoth larva run down its sides. (C) A scanning electron micrograph shows an insect trachea dividing into smaller tracheoles and still finer air capillaries.

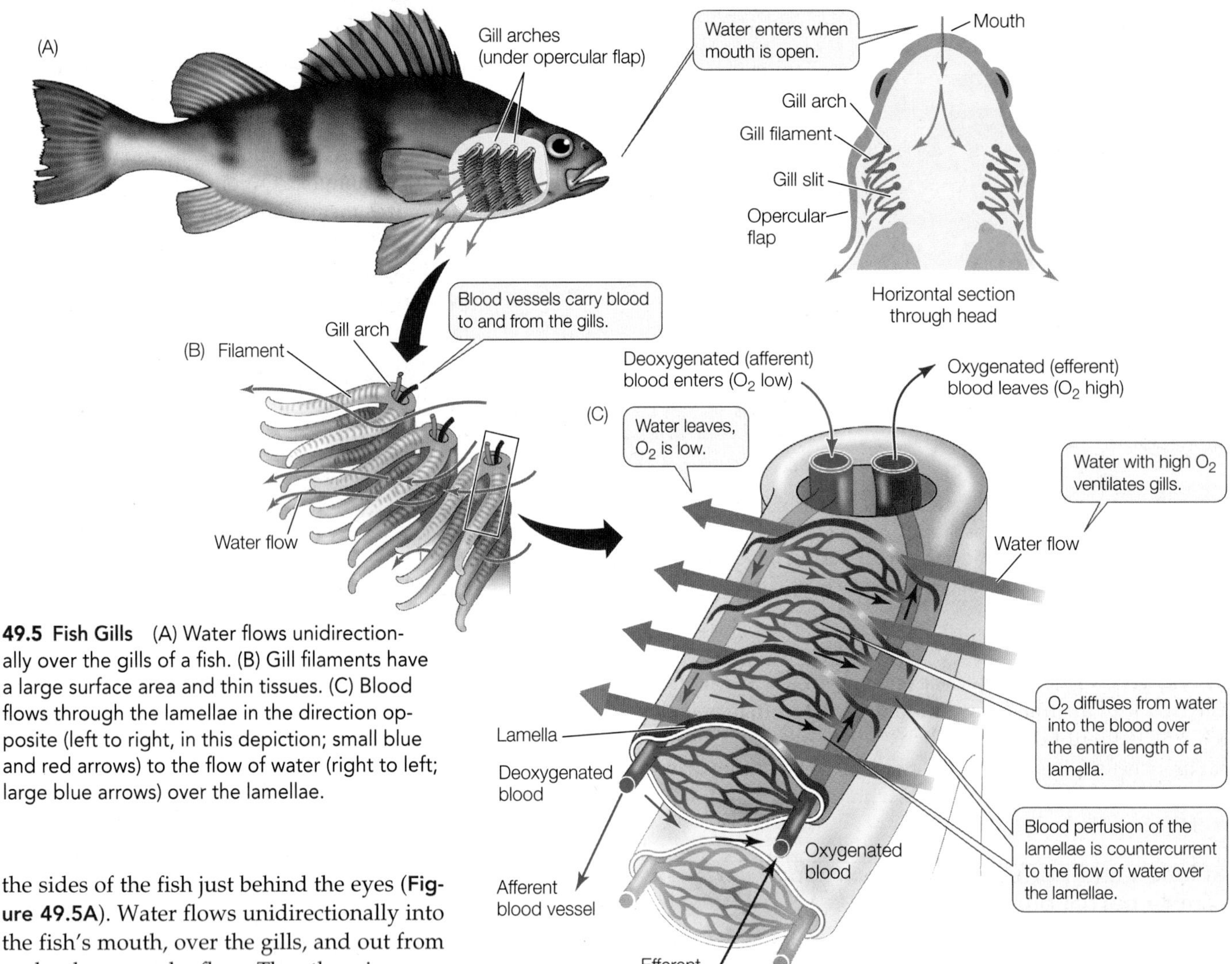


**49.5 Fish Gills** (A) Water flows unidirectionally over the gills of a fish. (B) Gill filaments have a large surface area and thin tissues. (C) Blood flows through the lamellae in the direction opposite (left to right, in this depiction; small blue and red arrows) to the flow of water (right to left; large blue arrows) over the lamellae.

the sides of the fish just behind the eyes (**Figure 49.5A**). Water flows unidirectionally into the fish's mouth, over the gills, and out from under the opercular flaps. Thus there is a constant one-way flow of oxygenated water over the gills, maximizing the $P_{O_2}$ on the external gill surfaces. On the internal side of the gill membranes, the circulation of blood minimizes the $P_{O_2}$ by sweeping $O_2$ away as rapidly as it diffuses across.

Gills have an enormous surface area for gas exchange because they are so highly divided. Each gill consists of hundreds of ribbonlike gill filaments (**Figure 49.5B**). The upper and lower flat surfaces of each gill filament are covered with rows of evenly spaced folds, or lamellae. The lamellae are the actual gas exchange surfaces. Because the lamellae are exceedingly thin, the path length ($L$) for diffusion of gases between blood and water is minimized. The surfaces of the lamellae consist of highly flattened epithelial cells, so the water and the fish's red blood cells are separated by little more than 1 to 2 micrometers.

The flow of blood perfusing the inner surfaces of the lamellae, like the flow of water over the gills, is unidirectional. **Afferent** blood vessels bring deoxygenated blood to the gills, while **efferent** blood vessels take oxygenated blood away from the gills (**Figure 49.5C**). Blood flows through the lamellae in the direction opposite to the flow of water over the lamellae. This **countercurrent flow** optimizes the $P_{O_2}$ gradient between water and blood, making gas exchange more efficient than it would be in a system using concurrent (parallel) flow (**Figure 49.6**).

Some fishes, including anchovies, tuna, and certain sharks, ventilate their gills by swimming almost constantly with their mouths open. Most fishes, however, ventilate their gills by means of a two-pump mechanism. The closing and contracting of the mouth cavity pushes water over the gills, and the expansion of the opercular cavity prior to opening of the opercular flaps pulls water over the gills.

These adaptations for maximizing the surface area ($A$) for diffusion, minimizing the path length ($L$) for diffusion, and maximizing the $P_{O_2}$ gradient allow fishes to extract an adequate supply of $O_2$ from meager environmental sources.

## Birds use unidirectional ventilation to maximize gas exchange

Birds are remarkable for their ability to sustain high levels of activity for a long time—for example, on long-distance flights—even at high altitudes where mammals cannot even survive. The first team to climb Mount Everest (8,850 m) was surprised to see birds flying over the mountain when they themselves could barely move without supplemental $O_2$. Bar-headed geese regularly migrate over Mt. Everest and surrounding peaks, but the highest recorded flight of a bird is from a Ruppell's griffon, a vulture, that was sucked into a jet engine at 11,278 m.

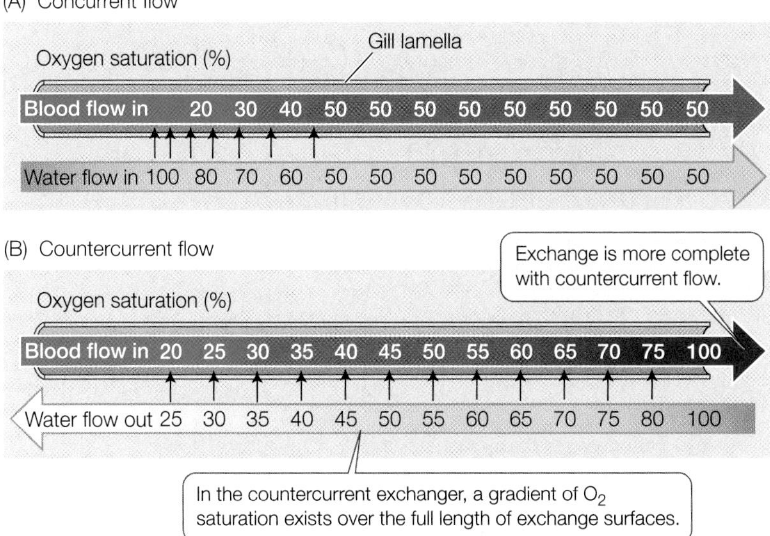


**49.6 Countercurrent Flow Enables More Complete Exchange** In these models of concurrent and countercurrent gas exchange, the numbers represent the $O_2$ saturation percentages of blood and water. (A) In a concurrent exchanger, the saturation percentages of blood and water reach equilibrium halfway across the exchange surface. (B) A countercurrent exchanger allows more complete gas exchange because the water is always more $O_2$-saturated than the blood; thus a gradient of $O_2$ saturation is maintained.

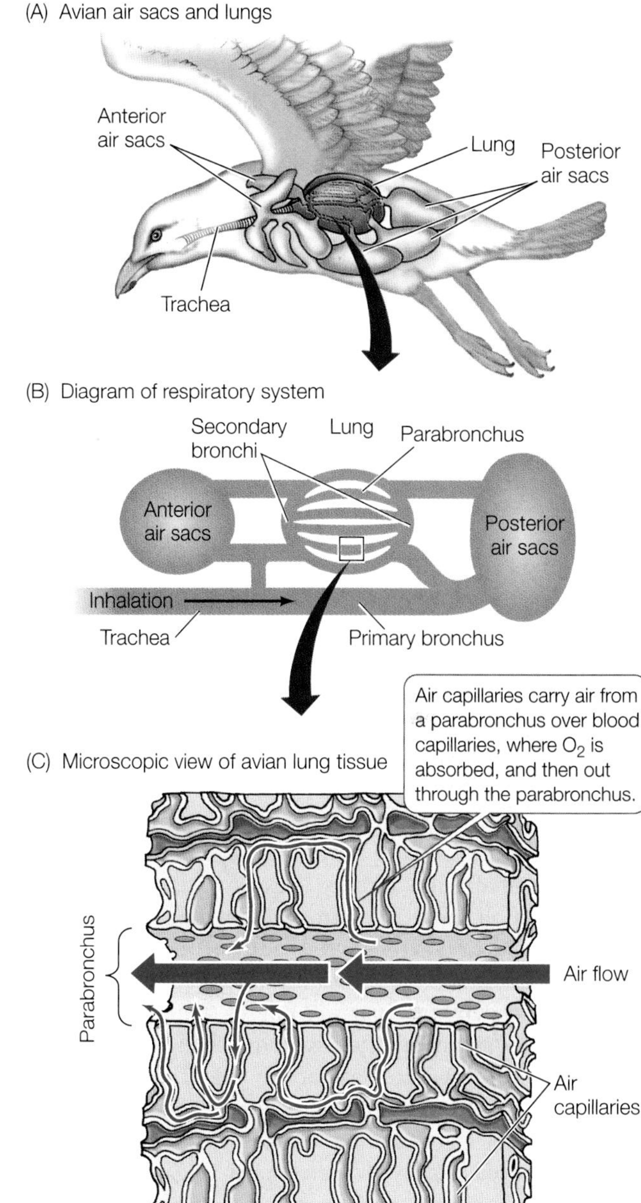


**49.7 The Respiratory System of a Bird** (A) Air sacs and air spaces in the bones are unique to birds. (B) The lung is divided into numerous parabronchi. Primary and secondary bronchi connect the lung to the air sacs. (C) Air flows through bird lungs unidirectionally in parabronchi. Air capillaries, the site of gas exchange, branch off the parabronchi.

Humans cannot survive at such altitudes without supplemental $O_2$. Yet the lungs of a bird are smaller than the lungs of a similar-sized mammal, and bird lungs expand and contract less during a breathing cycle than do mammalian lungs. Furthermore, bird lungs *are compressed* during inhalation and *expand* during exhalation. How do birds accomplish such remarkable feats of respiratory gas exchange?

The structure of bird lungs allows air to flow unidirectionally through the lungs, rather than bidirectionally through all the same airways, as it does in mammals. Because mammalian lungs are never completely emptied of air during exhalation, there is always some lung volume that is not ventilated with fresh air. The air remaining in lungs and airways after exhalation is called **dead space**. Bird lungs, by contrast, have very little dead space, and the fresh incoming air is not mixed with stale air. In this way, a high $P_{O_2}$ gradient is maintained.

An important and unique feature of the avian respiratory system is its **air sacs**, which occupy much of the body cavity of the bird (**Figure 49.7A**). The air sacs are interconnected with each other, with the lungs, and with air spaces in some of the bones. The air sacs receive inhaled air, but they are not gas exchange surfaces; they are integrally involved, however, in the flow of air through the lungs. As in other air-breathing vertebrates, air enters and leaves a bird's gas exchange system through the **trachea** (commonly known as the windpipe, and not to be confused with the air-conducting tracheae of insects) (**Figure 49.7B**). The trachea divides into two smaller airways, the **primary bronchi** (singular *bronchus*). The primary bronchi extend all the way to the posterior air sacs and branch into **secondary bronchi**. The posterior air sacs also have connections to the secondary bronchi. Secondary bronchi divide into tubelike **parabronchi** that run parallel to one another through the lungs.

Branching off the parabronchi are numerous tiny air capillaries (**Figure 49.7C**). Air flows through the parabronchi and diffuses into the air capillaries, which are the gas exchange surfaces. They are so numerous that they provide an enormous surface area for gas exchange. The parabronchi coalesce into larger bronchi that take the air out of the lungs and into anterior air sacs and back to the trachea. Thus the anatomy of a bird's airways allows air to flow unidirectionally through the lungs: trachea, bronchi, posterior air sacs, parabronchi, anterior air sacs, trachea.

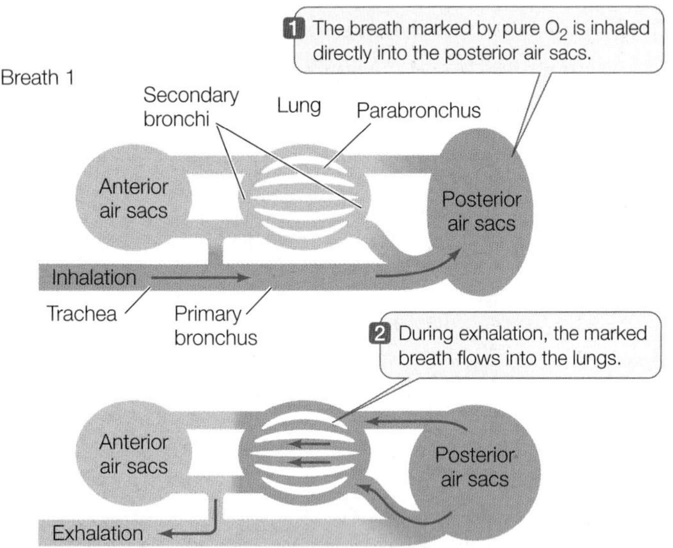


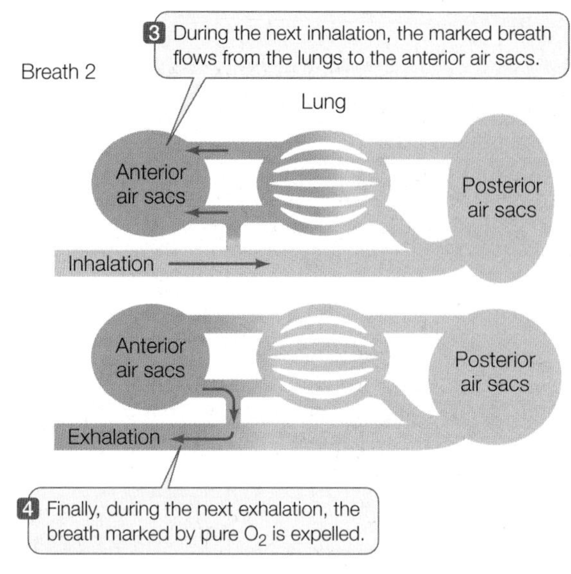


**49.8 The Path of Air Flow through Bird Lungs** The air a bird takes in by breathing (blue) travels through the lungs in one direction, from the posterior to the anterior air sacs. Each breath of air remains in the system for two breathing cycles.

**Go to Animated Tutorial 49.1**
**Airflow in Birds**
Life10e.com/at49.1

The puzzle of how birds breathe was solved by placing small $O_2$ electronic sensors at different locations in birds' air sacs and airways. The birds were then exposed to pure $O_2$ for just a single breath, which made it possible to track that particular inhalation. The experiment demonstrated that a single breath remains in a bird's gas exchange system for two cycles of inhalation and exhalation, and that the air sacs work like bellows; inhalation expands the sacs, and exhalation compresses them to maintain a continuous, unidirectional flow of fresh air through the lungs (**Figure 49.8**).

The advantages of the bird gas exchange system are similar to those of fish gills. The air sacs keep fresh air flowing unidirectionally over the gas exchange surfaces without interruption. Thus a bird can supply its gas exchange surfaces with a continuous flow of fresh air that has a $P_{O_2}$ very close to that of the ambient air. Even when the $P_{O_2}$ of the ambient air is only slightly above that of the blood, $O_2$ can diffuse from air to blood.

## Tidal ventilation produces dead space that limits gas exchange efficiency

Lungs evolved in early lungfish as outpockets of the digestive tract. Although their structure has evolved considerably, lungs remain dead-end sacs in all air-breathing vertebrates except birds (and likely their extinct reptilian ancestors). Because of this, ventilation cannot be constant and unidirectional but must be **tidal**: air flows in and exhaled gases flow out by the same route. Since the lungs and airways can never be completely emptied of air, they always contain dead space. We can easily measure the volumes of air exchanged during breathing, but we have to use an indirect method to measure the dead space contained in the lungs and airways. Measures of dead space are important in assessing lung health and disease.

A spirometer is a device that measures the volume of air breathed in and out (**Figure 49.9**). Using a human as an example, the amount of air that moves in and out per breath when at rest is called the **tidal volume** (**TV**) (about 500 ml for an average human adult). When we breathe in as much as possible,

RESEARCH**TOOLS**

**49.9 Measuring Lung Ventilation** A spirometer is a device that measures the volume of air a person breathes through a mouthpiece. The combined tidal volume, inspiratory reserve volume, and expiratory reserve volume are the lungs' vital capacity.

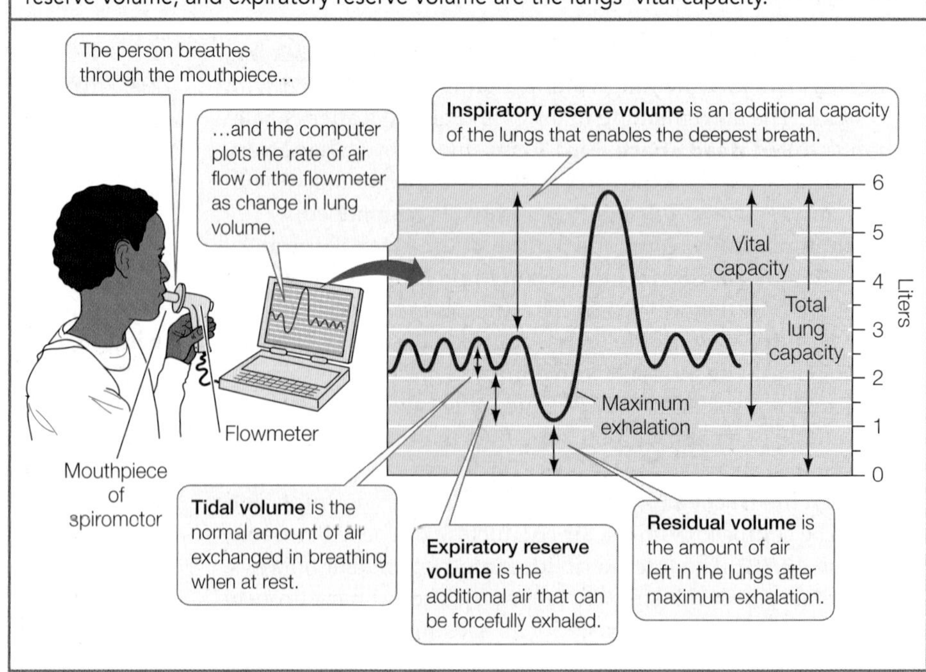

the additional volume is the **inspiratory reserve volume** (**IRV**). Conversely, if we forcefully exhale as much air as possible, the additional amount of air expelled is the **expiratory reserve volume** (**ERV**). The maximum capacity for air exchange in one breath, or the **vital capacity** (**VC**), is the sum of TV + IRV + ERV. The vital capacity of an athlete is generally greater than that of a nonathlete, and vital capacity decreases with age because of stiffening of the lung tissue.

Vital capacity is not the entire lung capacity because of the dead space, also called the **residual volume** (**RV**). We can't measure RV directly with the spirometer, but we can measure it indirectly using the helium dilution method. Briefly, a person breathes from a closed reservoir with a known volume of air containing a known amount of helium (He). The helium is not absorbed from the lungs, so it becomes evenly distributed between the lungs and the reservoir as the subject inhales and exhales. Because the fixed amount of He becomes dispersed in a larger volume of air, its concentration decreases. That decrease in He concentration enables us to calculate the subject's **functional residual volume** (**FRV**), which is the ERV + RV. Since we *can* measure the ERV with the spirometer, we can subtract the ERV from the FRV to obtain the RV.

Why is the RV important? Referring to Figure 49.9, you will see that for a normal person the ERV is about 1,000 ml and the RV is 1,000 ml. Thus the FRV is 2,000 ml, but the tidal volume is only 500 ml; this means that the air that reaches the alveoli (the actual gas exchange surfaces) with each breath consists of only 500 ml of fresh air diluted by 2,000 ml of stale air. The maximum $P_{O_2}$ in this mixed air is much below the $P_{O_2}$ of the outside air, and because of the tidal ventilation pattern, the $P_{O_2}$ in the alveoli is steadily dropping during the breathing cycle. The RV is important because it contributes to the FRC and to the dilution of the $O_2$ in the inhaled air. Any disease or condition that increases the RV (such as emphysema or pulmonary fibrosis) compromises a patient's respiratory ability. Similarly, considering the mixing of fresh air with the FRV, you can understand why reductions in tidal volume can be a problem—and therefore why patients recovering from surgery are encouraged to breathe deeply, even if it hurts.

Offsetting the inefficiencies of tidal breathing, mammalian lungs have some design features to maximize the rate of gas exchange: an enormous surface area and a very short path length for diffusion.

**RECAP 49.2**

The major adaptations that increase animals' efficiency of respiratory gas exchange are a large surface area for exchange and a maximized partial pressure gradient across that surface.

- Describe three different ways that the partial pressure gradient for $O_2$ exchange is maximized across fish gills. **See pp. 1009–1010 and Figures 49.5 and 49.6**
- What respiratory adaptations enable birds to fly at extremely high altitudes? **See pp. 1010–1011 and Figures 49.7 and 49.8**
- Explain why residual volume limits the efficiency of tidal breathing. **See pp. 1012–1013 and Figure 49.9**

Despite their limitations, mammalian lungs serve the respiratory needs of mammals, including humans, well. Next we will look at the human respiratory system as an example.

## 49.3 How Do Human Lungs Work?

Air enters the lungs through the oral cavity or through the nasal passage, which join together in the **pharynx** (**Figure 49.10A**). Below the pharynx, the esophagus conducts food to the stomach, and the trachea conducts air to the lungs. At the beginning of this airway is the **larynx**, or voice box, which houses the vocal cords. The larynx is the lump that you can see or feel on the front of your neck. The trachea is about 2 centimeters in diameter. C-shaped bands of cartilage prevent the thin walls of the trachea from collapsing as air pressure changes during the breathing cycle. If you run your fingers down the front of your neck just below your larynx, you can feel a few of these bands of cartilage.

The trachea branches into two bronchi, one leading to each lung. The bronchi branch repeatedly to generate a treelike structure of progressively smaller airways extending to all regions of the lungs. After four branchings, the cartilage supports disappear, marking the transition to **bronchioles**. After about 16 branchings, the bronchioles are less than a millimeter in diameter, and tiny, thin-walled air sacs called **alveoli** begin to appear. Alveoli are the sites of gas exchange. After the first alveoli there are about six more branchings of the airways that end in clusters of alveoli (**Figure 49.10B**). Because the airways conduct air only to and from the alveoli and do not themselves participate in gas exchange, their volume is dead space.

Human lungs have about 300 million alveoli. Although each alveolus is very small, their combined surface area for diffusion of respiratory gases is about 70 square meters—about one-fourth the size of a basketball court. Each alveolus is made of very thin cells. Between and surrounding the alveoli are networks of capillaries whose walls are also very thin. Thus where capillary meets alveolus, the length of the diffusion path between air and blood is less than 2 micrometers (**Figure 49.10C**).

Diseases of the bronchioles and alveoli are the third leading cause of death in the United States as of 2010. Among these diseases, the most lethal is emphysema, a condition in which inflammation damages and eventually destroys the walls of the alveoli. As a result, the lungs have fewer but larger alveoli, the RV increases, and the lungs lose elasticity. Although genetic factors can contribute to emphysema, the principal cause of the disease is smoking.

### Respiratory tract secretions aid ventilation

Mammalian lungs produce two secretions that do not directly influence their gas exchange but do affect the process of ventilation: mucus and surfactant.

Many cells lining the airways produce sticky mucus that captures bits of dirt and microorganisms that are inhaled. Other cells lining the airways have cilia whose beating continually sweeps the mucus, with its trapped debris, up toward the

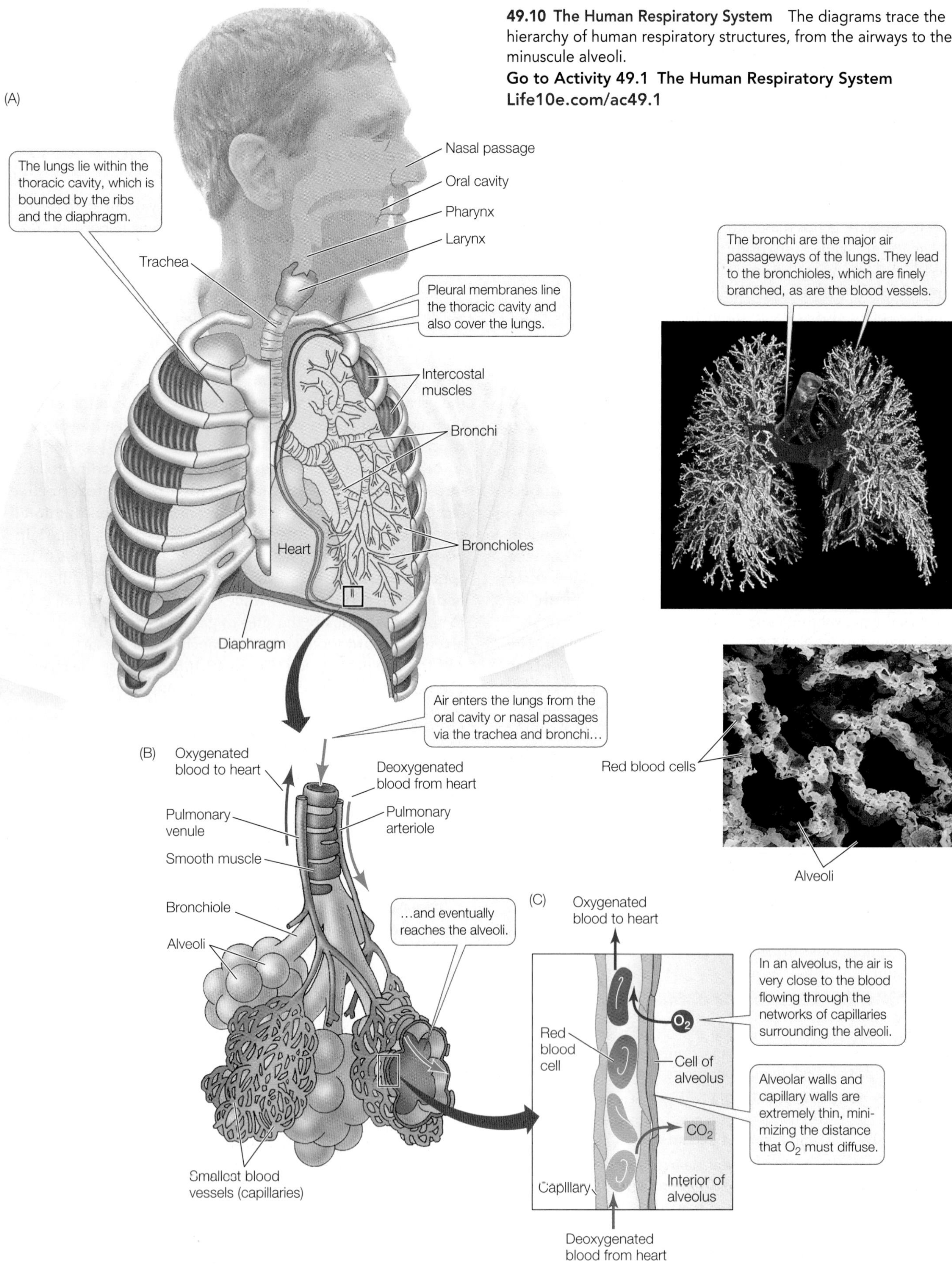


**49.10 The Human Respiratory System** The diagrams trace the hierarchy of human respiratory structures, from the airways to the minuscule alveoli.

**Go to Activity 49.1 The Human Respiratory System**
**Life10e.com/ac49.1**

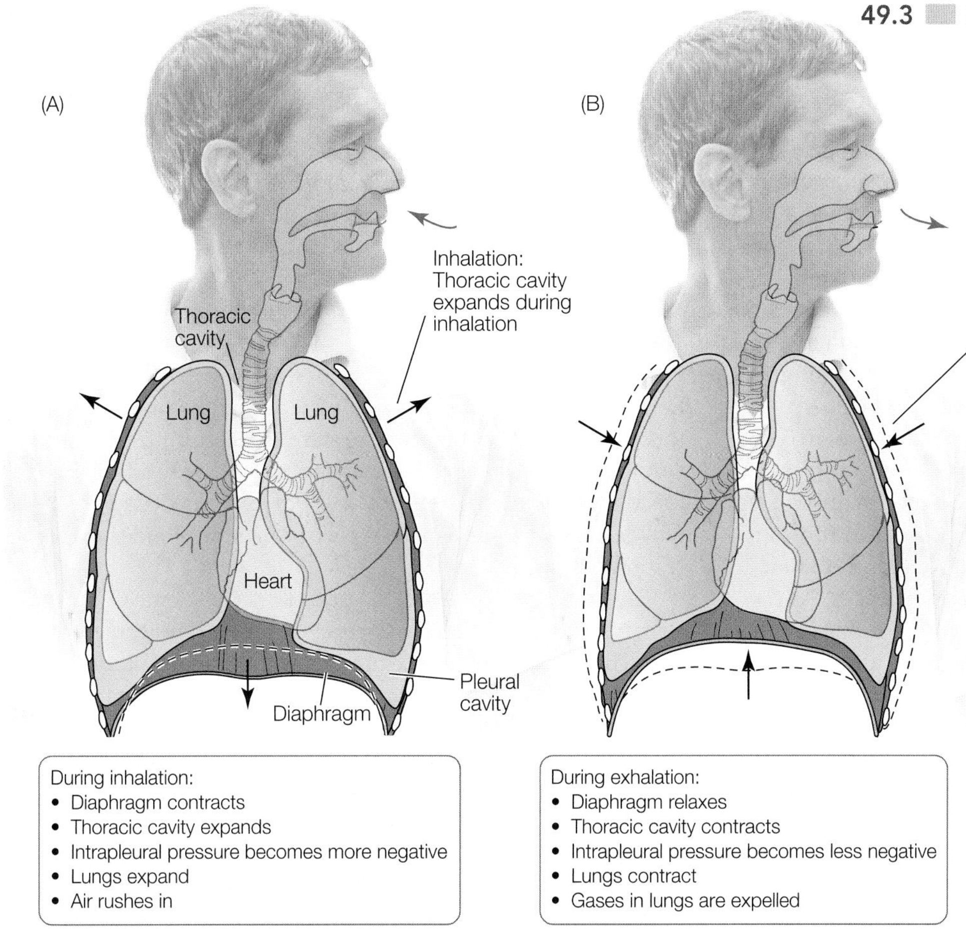

**49.11 Into the Lungs and Out Again** (A) Inhalation is an active process spurred by contraction of the diaphragm. (B) Exhalation generally is a passive process as the diaphragm relaxes.

Go to Animated Tutorial 49.2 Airflow in Mammals
Life10e.com/at49.2

pharynx, where it can be swallowed or spit out. This phenomenon, called the mucus escalator, can be adversely affected by inhaled pollutants. Smoking one cigarette can immobilize the cilia of the airways for hours. A smoker's cough results from the need to clear the obstructing mucus from the airways when the mucus escalator is out of order.

A **surfactant** is a substance that reduces the surface tension of a liquid. **Surface tension** gives the surface of a liquid the properties of an elastic membrane, and it is why certain insects, such as water-striders, can walk on water. As discussed in Section 2.4, surface tension is the result of chemical forces of attraction between water molecules. The attractive forces working on the water molecules at the surface pull from below and from the sides but not from above. This imbalance of forces creates surface tension. The thin film of fluid covering the air-facing surfaces of the alveoli has surface tension that contributes to the lungs' elasticity. To inflate the lungs, enough force has to be generated to overcome both the elasticity of the lung tissue and the surface tension in the alveoli.

Lung surfactant is a fatty, detergent-like substance that is critical for reducing the work necessary to inflate the lungs. Certain cells in the alveoli release surfactant molecules when they are stretched. If a baby is born more than a month prematurely, these cells may not have developed the ability to produce surfactant. A baby with this condition, known as respiratory distress syndrome, will have great difficulty breathing and may die from exhaustion and lack of $O_2$. Common treatments for premature babies have been to put them on respirators to assist their breathing and to give them hormones to speed lung development. A newer approach is to apply surfactant to the lungs via an aerosol.

## Lungs are ventilated by pressure changes in the thoracic cavity

Human lungs are suspended in the **thoracic cavity**, a closed compartment bounded on the bottom by a sheet of muscle called the **diaphragm** (see Figure 49.10A). Each lung is covered by a continuous sheet of tissue called the **pleural membrane** that also lines the thoracic cavity adjacent to that lung. There is no real space between the pleural membranes of the lung and the thoracic cavity, but there is a thin film of fluid. This fluid lubricates the inner surfaces of the pleural membranes so they can slip and slide against each other. Just as we mentioned above in the explanation of surface tension, there are forces of attraction between the molecules of fluid in the pleural membranes. As a result, it is difficult to pull the pleural membranes apart. Think of two wet panes of glass or two wet microscope slides; you can slide them past each other, but it is difficult to separate them. While the inner surfaces of the pleural membranes are "stuck" to each other by surface tension, they can move relative to each other during breathing movements.

Inhalation and exhalation involve changes in the volume of the thoracic cavity (**Figure 49.11**). Because the pleural membranes covering the cavity wall and the lung surface are stuck to each other by surface tension, any attempt to increase the volume of the thoracic cavity increases the tension between the pleural membranes. Even between breaths, there is tension between the pleural membranes because the rib cage is pulling outward and the elasticity of the lung tissue is pulling inward. This slight negative pressure keeps the alveoli partly inflated even at the end of an exhalation. If the thoracic cavity is punctured—by a knife wound, for example—air can leak into the space between the pleural membranes and cause the lung to deflate. If the wound is not sealed, breathing movements pull air in between the pleural membranes rather than into the lung (a "sucking chest wound"), and there is no ventilation of the alveoli in that lung—a condition called "collapsed lung."

At rest, inhalation is initiated by contraction of the muscular diaphragm (see Figure 49.11A). As the domed diaphragm contracts, it pulls down, expanding the thoracic cavity and pulling on the pleural membranes. Since the pleural membranes cannot separate, they pull on the lungs; air rushes in through the trachea from the outside and the lungs expand. Exhalation begins when contraction of the diaphragm ceases. As the diaphragm relaxes, the elastic recoil of the lung tissues pulls the diaphragm up and pushes air out through the airways (see Figure 49.11B). When a person is at rest, inhalation is an active process and exhalation is a passive process.

The diaphragm is not the only muscle that can change the volume of the thoracic cavity. Between the ribs are two sets of **intercostal muscles**. The external intercostal muscles expand the thoracic cavity by lifting the ribs up and outward. The internal intercostal muscles decrease the volume of the thoracic cavity by pulling the ribs down and inward. During strenuous exercise, the external intercostal muscles increase the volume of air inhaled, making use of the inspiratory reserve volume, and the internal intercostal muscles increase the amount of air exhaled, making use of the expiratory reserve volume. The abdominal muscles can also aid in breathing. When they contract, they cause the abdominal contents to push up on the diaphragm and thereby contribute to the expiratory reserve volume.

Remember that ventilation and perfusion work together to maximize the partial pressure gradients across the alveolar membranes. Ventilation delivers $O_2$ to the environmental side of the exchange surface, where it diffuses across and is swept away by the perfusing blood, which carries it to the tissues. The reverse is true for $CO_2$. Perfusion delivers $CO_2$ to the exchange surface, where it diffuses and is swept away by ventilation.

**RECAP 49.3**

The mammalian respiratory system consists of a highly branching system of airways that lead to alveoli—the gas exchange surfaces. Respiratory muscles ventilate the alveoli by creating pressure differences between the lungs and the outside air. $CO_2$ and $O_2$ are exchanged across thin capillary and alveoli walls by diffusion.

- Describe the path that a breath of air takes from the nose to the gas exchange surfaces. **See p. 1013 and Figure 49.10**
- What roles do mucus and surfactant play in maintaining the function of the mammalian respiratory system? **See pp. 1013 and 1015**
- Explain the anatomical and functional relationships between the thoracic cavity, the pleural membranes, and the lungs. **See pp. 1015–1016 and Figure 49.11**

Having discussed how respiratory gases get to and from the environmental side of the gas exchange membranes through ventilation, we will now look at how these gases get to and from the internal side of those membranes through perfusion.

## How Does Blood Transport Respiratory Gases?

Perfusion of the lungs is one of the functions of the circulatory system. The circulatory system uses a pump (the heart) and a network of vessels to transport blood around the body. Circulatory systems are the subject of Chapter 50, so here we will discuss only one aspect of perfusion: how blood transports respiratory gases.

The liquid part of blood, the plasma, carries some $O_2$ in solution, but its ability to transport this nonpolar molecule is limited. The blood plasma of a human can contain in solution only about 0.3 ml of $O_2$ per 100 ml of plasma, which is inadequate to support even basal metabolism. However, the blood of most animals, vertebrate and invertebrate, contains molecules that bind and release $O_2$ and thus augment its transport capacity. These molecules pick up $O_2$ where $P_{O_2}$ is high and release it where $P_{O_2}$ is lower. There are many $O_2$ transport molecules in the animal kingdom, but in vertebrates this role is played by hemoglobin, a protein contained in red blood cells. Hemoglobin increases the capacity of blood to carry 60 times more oxygen than it could carry in solution, making high rates of metabolism possible.

### Hemoglobin combines reversibly with $O_2$

Red blood cells contain enormous numbers of hemoglobin molecules. **Hemoglobin** is a protein consisting of four polypeptide subunits (see Figure 3.11), each of which surrounds a heme group—an iron-containing ring structure that can reversibly bind a molecule of $O_2$. Thus each hemoglobin molecule can bind and release up to four $O_2$ molecules, enabling the blood to carry a large amount of $O_2$ to the body's tissues.

Hemoglobin's ability to pick up or release $O_2$ depends on the $P_{O_2}$ in its environment. When the $P_{O_2}$ of the blood plasma is high, as it usually is in the lung capillaries, each hemoglobin molecule can carry its maximum load of four $O_2$ molecules. As the blood circulates through the rest of the body, it releases some of the $O_2$ it is carrying when it encounters lower $P_{O_2}$ values in the body's tissues.

The relationship between $P_{O_2}$ and the amount of $O_2$ bound to hemoglobin is not linear but S-shaped (sigmoidal). The hemoglobin–oxygen binding curve reflects interactions between the four subunits of the hemoglobin molecule. At low $P_{O_2}$ values, only one subunit will bind an $O_2$ molecule (**Figure 49.12A**). When it does so, the shape of that subunit changes, altering the quaternary structure of the entire hemoglobin molecule. That structural change makes it easier for the other subunits to bind an $O_2$ molecule; that is, their $O_2$ *affinity* is increased. Therefore a smaller increase in $P_{O_2}$ is necessary to get the hemoglobin molecules to bind a second $O_2$ molecule (that is, to become

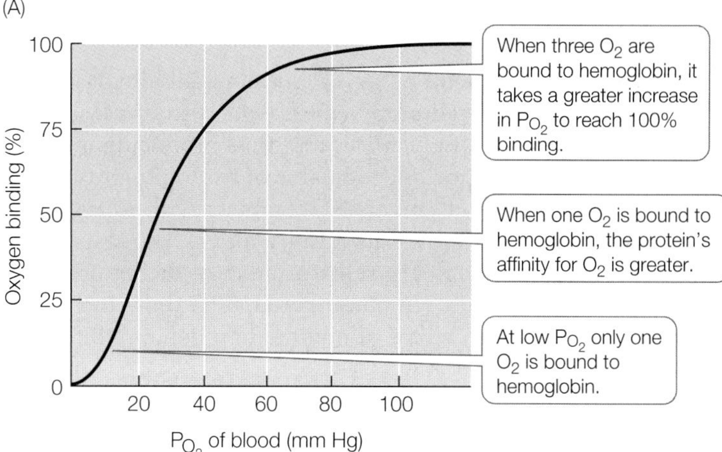


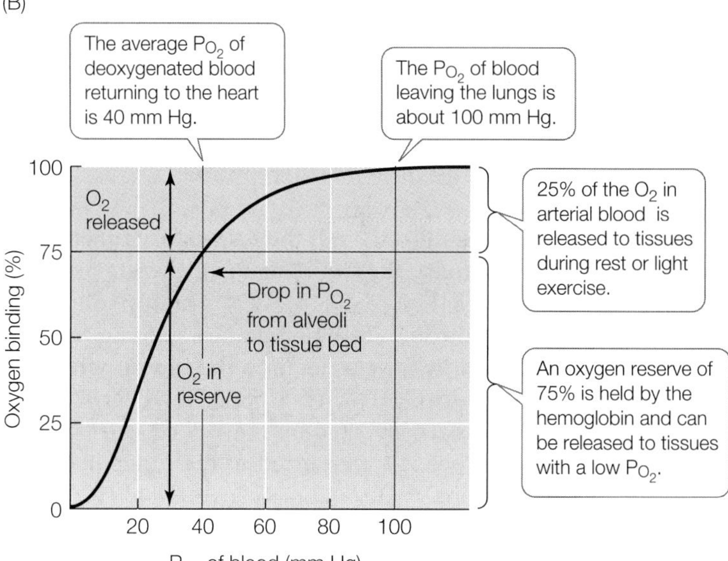


**49.12 Binding of $O_2$ to Hemoglobin Depends on $P_{O_2}$** (A) The sigmoidal shape of hemoglobin's oxygen binding curve reflects positive cooperativity among hemoglobin's subunits. (B) Hemoglobin in blood leaving the lungs is 100 percent saturated (four $O_2$ molecules are bound to each hemoglobin molecule). Most hemoglobin molecules drop only one of their four $O_2$ molecules as they circulate through the body and are still 75 percent saturated when the blood returns to the lungs. The steep portion of this $O_2$-binding curve comes into play when tissue $P_{O_2}$ falls below the normal 40 mm Hg, at which point hemoglobin "unloads" its $O_2$ reserves.

50% saturated) than was necessary to get them to bind one $O_2$ molecule (to become 25% saturated). This change in affinity is reflected in the increased steepness of the $O_2$ binding curve. The influence of $O_2$ binding by one subunit on the $O_2$ affinity of the other subunits is called **positive cooperativity**.

Once the third $O_2$ molecule is bound, the relationship seems to change, as a larger increase in $P_{O_2}$ is required for the hemoglobin to reach 100 percent saturation. This upper bend of the sigmoid curve is due to a probability phenomenon. The closer we get to having all subunits occupied, the less likely it is that any particular $O_2$ molecule will find a place to bind. Therefore it takes a relatively greater $P_{O_2}$ to achieve 100 percent saturation.

The $O_2$-binding/dissociation properties of hemoglobin help get $O_2$ to the tissues that need it most (**Figure 49.12B**). In the lungs, where the $P_{O_2}$ is about 100 mm Hg, hemoglobin is 100 percent saturated. The $P_{O_2}$ in blood returning to the heart from the body (at rest) is usually about 40 mm Hg. You can see that at this $P_{O_2}$ the hemoglobin is still about 75 percent saturated. This means that as the blood circulates around the body, it releases only about one in four of the $O_2$ molecules it carries. This system seems inefficient, but it is really quite adaptive, because the hemoglobin keeps 75 percent of its $O_2$ in reserve to meet peak demands of highly active tissues.

When a tissue becomes starved of $O_2$ and its local $P_{O_2}$ falls below 40 mm Hg, the hemoglobin flowing through that tissue is on the steep portion of its binding/dissociation curve. That means relatively small decreases in $P_{O_2}$ below 40 mm Hg will result in the release of lots of $O_2$ to the tissue. Thus hemoglobin is very effective in making $O_2$ available to tissues precisely when and where it is needed most.

The $O_2$ transport function of hemoglobin can rapidly and tragically be disrupted by a common by-product of incomplete combustion: carbon monoxide (CO). If CO from a faulty heating system, engine exhaust, or burning charcoal accumulates in a closed space, the results can be deadly. Because CO binds to hemoglobin with a 240-fold higher affinity than $O_2$, it prevents hemoglobin from transporting $O_2$. In the United States, more than 5,000 people die each year from CO poisoning.

**Go to Animated Tutorial 49.3**
**Hemoglobin: Loading and Unloading**
Life10e.com/at49.3

## Myoglobin holds an $O_2$ reserve

Muscle cells have their own $O_2$-binding molecule, **myoglobin**. Myoglobin consists of just one polypeptide chain associated with an iron-containing ring structure that can bind one $O_2$ molecule (see Figure 24.11). Myoglobin has a higher affinity for $O_2$ than hemoglobin does, so it picks up and holds $O_2$ at $P_{O_2}$ values at which hemoglobin is releasing its bound $O_2$ (**Figure 49.13**).

Myoglobin facilitates the diffusion of $O_2$ in muscle cells and provides an $O_2$ reserve for times when metabolic demands are high and blood flow is interrupted. Interruption of blood flow in muscles is common because contracting muscles squeeze blood vessels. When tissue $P_{O_2}$ values are low and hemoglobin can no longer supply more $O_2$, myoglobin releases its bound $O_2$. Diving mammals such as seals have high concentrations of myoglobin in their muscles, which is one reason they can stay under water for so long. (We will discuss more adaptations for diving in Chapter 50.) Even in non-diving animals, muscles called on for extended periods of work frequently have more myoglobin than muscles that are used for short, intermittent periods, as noted in Section 48.2.

## Hemoglobin's affinity for $O_2$ is variable

Various factors influence the $O_2$-binding/dissociation properties of hemoglobin, thereby influencing $O_2$ delivery to tissues. Three of these factors are the chemical composition of the hemoglobin, the blood pH, and the presence of 2,3-bisphosphoglyceric acid (BPG) in red blood cells.

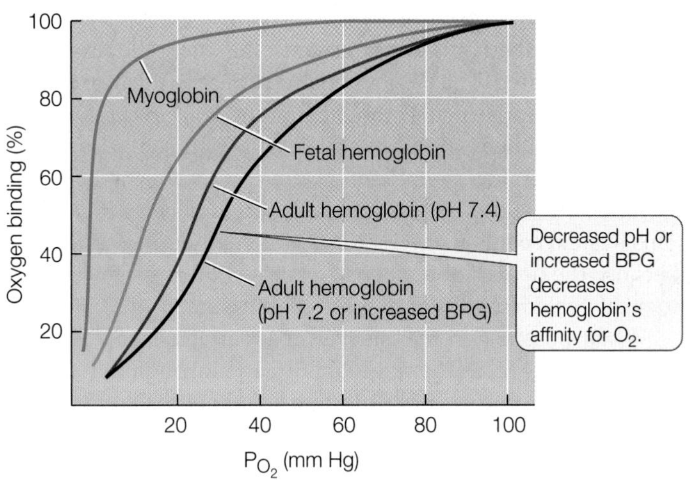


**49.13 Oxygen-Binding Adaptations** Myoglobin and the different hemoglobins have different $O_2$-binding properties adapted to different circumstances. Fetal hemoglobin, for example, has a higher affinity for $O_2$ than does adult hemoglobin, facilitating $O_2$ transfer in the placenta. When high metabolism lowers the pH of the blood, or low $O_2$ increases bisphosphoglyceric acid (BPG), hemoglobin releases more of its $O_2$.

**Go to Activity 49.2 Oxygen-Binding Curves Life10e.com/ac49.2**

HEMOGLOBIN COMPOSITION There is more than one type of hemoglobin, because the chemical composition of the polypeptide chains that form the hemoglobin molecule varies. The normal hemoglobin of adult humans has two each of two kinds of polypeptide chains—two α-globin chains and two β-globin chains. This normal adult hemoglobin has the $O_2$-binding characteristics shown in **Figure 49.13**.

Before birth, the human fetus has a different form of hemoglobin, consisting of two α-globin and two γ-globin chains. The functional difference between fetal and adult hemoglobin is that fetal hemoglobin has a higher affinity for $O_2$. Therefore the fetal hemoglobin–oxygen binding/dissociation curve is shifted to the left compared with the adult curve (see Figure 49.13). You can see from these curves that if both types of hemoglobin are at the same $P_{O_2}$ (as they are in the placenta), fetal hemoglobin will pick up $O_2$ that the adult hemoglobin releases. This difference in $O_2$ affinities enables the efficient transfer of $O_2$ from the mother's blood to the fetus's blood.

HEMOGLOBIN AND PH The $O_2$-binding properties of hemoglobin are also influenced by physiological conditions. The influence of pH (hydrogen ion concentration) on the function of hemoglobin is known as the **Bohr effect**. As blood passes through metabolically active tissue such as exercising muscle, it picks up acidic metabolites such as lactic acid, fatty acids, and $CO_2$. As a result, blood pH falls. The excess $H^+$ binds preferentially to deoxygenated hemoglobin and decreases its affinity for $O_2$ and the $O_2$-binding/dissociation curve of hemoglobin shifts to the right (see Figure 49.13). This shift means the hemoglobin will release more $O_2$ in tissues where pH is low—another way that $O_2$ is supplied where and when it is most needed.

2,3-BISPHOSPHOGLYCERIC ACID BPG is a metabolite of glycolysis. Mammalian red blood cells respond to low $P_{O_2}$ by increasing their rate of glycolysis and thus producing more BPG, which is an important regulator of hemoglobin function. BPG, like excess $H^+$, reversibly combines with *deoxygenated* hemoglobin and lowers its affinity for $O_2$. The result is that at any $P_{O_2}$, hemoglobin releases more of its bound $O_2$ than it otherwise would. In other words, BPG shifts the $O_2$-binding/dissociation curve of mammalian hemoglobin to the right.

When humans go to high altitudes, or when they cease being sedentary and begin to exercise, their red blood cells are exposed to a lower $P_{O_2}$ and their level of BPG goes up, making it easier for hemoglobin to deliver more $O_2$ to tissues. The reason fetal hemoglobin has a left-shifted $O_2$-binding/dissociation curve is that its γ-globin chains have a lower affinity for BPG than do the β-globin chains of adult hemoglobin.

## $CO_2$ is transported as bicarbonate ions in the blood

Delivering $O_2$ to tissues is only half the respiratory function of blood. Blood also must take $CO_2$, a metabolic waste product, away from tissues (**Figure 49.14**). $CO_2$ is highly soluble and readily diffuses through cell membranes, moving from its site of production in the tissues into the blood, where the partial pressure of $CO_2$ ($P_{CO_2}$) is lower. However, very little dissolved $CO_2$ is transported by the blood. Most $CO_2$ produced by the tissues is transported to the lungs in the form of bicarbonate ions, $HCO_3^-$. $CO_2$ is converted to $HCO_3^-$, transported to the lungs, and then converted back to $CO_2$ in several steps.

When $CO_2$ dissolves in water, some of it slowly reacts with the water molecules to form carbonic acid ($H_2CO_3$), some of which then dissociates into a proton ($H^+$) and a bicarbonate ion ($HCO_3^-$). This reversible reaction is expressed as follows:

$$CO_2 + H_2O \rightleftharpoons H_2CO_3 \rightleftharpoons H^+ + HCO_3^-$$

In the extracellular fluid, the reaction between $CO_2$ and $H_2O$ proceeds slowly. But it is a different story in the endothelial cells of the capillaries and in the red blood cells, where the enzyme carbonic anhydrase speeds up the conversion of $CO_2$ to $H_2CO_3$. The newly formed $H_2CO_3$ dissociates, and the resulting bicarbonate ions enter the plasma in exchange for $Cl^-$ (see Figure 49.14). By converting $CO_2$ to $H_2CO_3$, carbonic anhydrase reduces the $P_{CO_2}$ in these cells and in the plasma, facilitating the diffusion of $CO_2$ from tissue cells to endothelial cells, plasma, and red blood cells. Some $CO_2$ is also carried in chemical combination with hemoglobin.

In the lungs, the reactions involving $CO_2$ and bicarbonate ions are reversed. Remember that an enzyme such as carbonic anhydrase only speeds up a reversible reaction; it does not determine its direction. The direction is determined by concentrations of reactants and products. Ventilation keeps the $P_{CO_2}$ in the alveoli low, so $CO_2$ diffuses from the blood plasma into the alveoli, lowering the $P_{CO_2}$ in the blood, which favors the conversion of $HCO_3^-$ into $CO_2$.

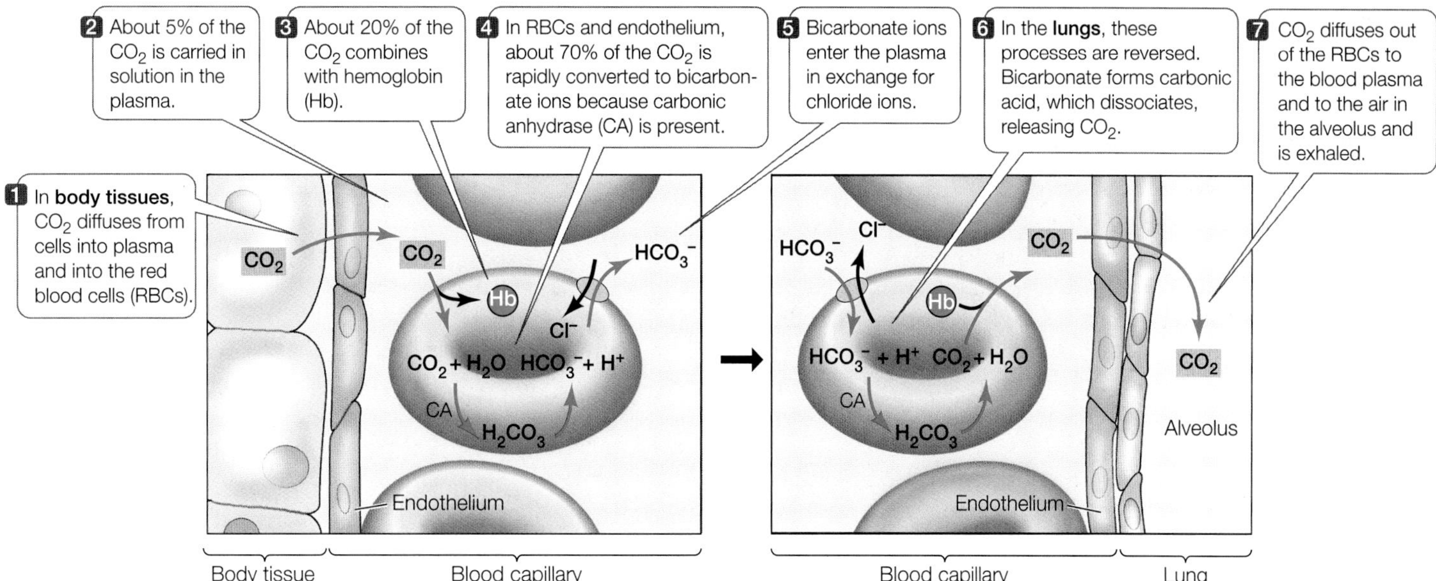


**49.14 Carbon Dioxide Is Transported as Bicarbonate Ions** Carbonic anhydrase (CA) in capillary endothelial cells and in red blood cells facilitates conversion of $CO_2$ produced by tissues into bicarbonate ions carried by the plasma. In the lungs, the process is reversed as $CO_2$ is exhaled.

**RECAP 49.4**

$O_2$ is transported from the lungs to the body's tissues in reversible combination with hemoglobin. Each hemoglobin molecule can reversibly combine with four $O_2$ molecules; the percent saturation of the binding sites is a function of the $P_{O_2}$ in the hemoglobin's environment.

- Explain the advantage of having hemoglobin hold on to three $O_2$ molecules at the usual $P_{O_2}$ of mixed venous blood. **See pp. 1016–1017 and Figure 49.12**
- How is the $O_2$-binding/dissociation curve of hemoglobin influenced by pH? By BPG? By development from fetus to newborn infant? **See p. 1018 and Figure 49.13**
- How is $CO_2$ transported in the blood? **See p. 1018 and Figure 49.14**

We must breathe every minute of our lives, but most of us usually don't worry about it, or even think about it very often. In the next section we will examine how the regular breathing cycle is generated and controlled by the central nervous system.

## 49.5 How Is Breathing Regulated?

Breathing is an involuntary function of the central nervous system. The breathing pattern easily adjusts itself around other activities (such as speech and eating), and breathing rates change to match the metabolic demands of our bodies. How is this accomplished?

### Breathing is controlled in the brainstem

The basic breathing rhythm is an involuntary function driven by a rhythm-generating neural circuit in the brainstem. Breathing is also under voluntary control that enables activities such as talking, singing, sighing, and breath-hold diving. Voluntary control of breathing is eventually and inevitably overridden by feedback information that conveys the body's need for an adequate $O_2$ supply and $CO_2$ elimination.

Breathing ceases if the spinal cord is severed in the neck region, showing that breathing is generated in the brain. If the brainstem is cut just above the medulla (the segment of the brainstem just above the spinal cord), an irregular breathing pattern remains (**Figure 49.15**). A group of respiratory motor neurons in the dorsal medulla increase their firing rates just before an inhalation begins. The axons of these neurons leave the CNS in the neck region to form the phrenic nerve, which innervates the diaphragm. As more and more of these neurons fire—and fire faster and faster—the diaphragm contracts. All of a sudden the neurons stop firing, the diaphragm relaxes, and

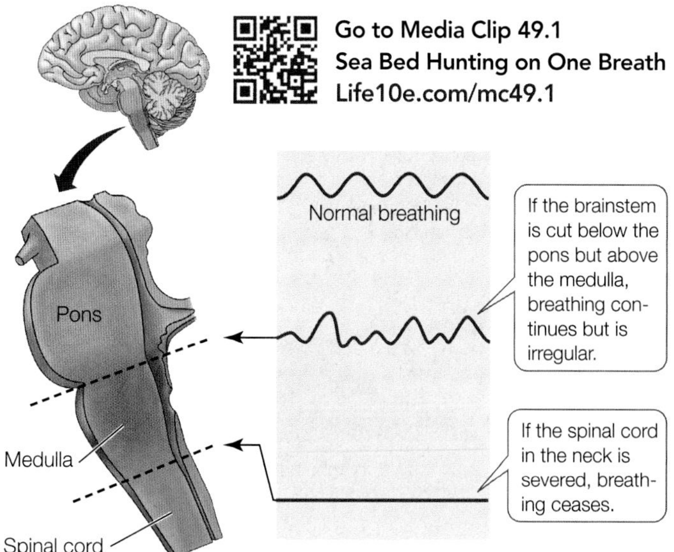


**49.15 Breathing Is Controlled in the Brainstem** Basic breathing rhythm is generated in the medulla and is modified by neurons in or above the pons.

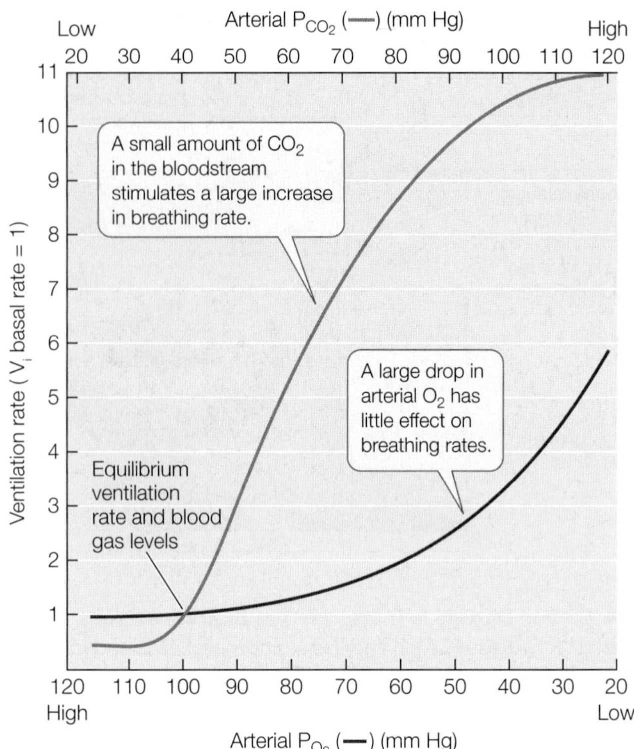


**49.16 Carbon Dioxide Affects Breathing Rate** The breathing mechanism is more sensitive to increased levels of $CO_2$ in arterial blood than to decreased amounts of $O_2$.

exhalation begins. Exhalation is usually a passive process that depends on the elastic recoil of the lung tissues. Another group of respiratory motor neurons in the ventral medulla becomes active when breathing demand is high. These motor neurons communicate through thoracic spinal nerves to the intercostal muscles. By expanding and contracting the rib cage, the intercostal muscles increase both the inhalation and the exhalation volumes.

Neurons in the lower region of the pons help regularize the basic respiratory rhythm. Still higher brain areas modify breathing to accommodate speech, ingestion of food, coughing, and emotional states.

## Regulating breathing requires feedback

When breathing or metabolism changes, it alters the $P_{O_2}$ and $P_{CO_2}$ in the blood. We should therefore expect the blood levels of one or both of these gases to provide feedback information to the breathing rhythm generator in the medulla. Humans and other mammals are remarkably insensitive to falling levels of $O_2$ in arterial blood but are extremely sensitive to increases in $CO_2$. That is, arterial $P_{O_2}$ can deviate considerably from normal without causing much of an increase in ventilation rate, but even a small rise in arterial $P_{CO_2}$ causes a large increase in ventilation (**Figure 49.16**). This relationship is reversed for water-breathing animals, in which $O_2$ is the primary feedback stimulus for gill ventilation.

We might ask whether it is an increase in the $P_{CO_2}$ of the blood that stimulates increased breathing when we exercise. To answer this question, C. R. Bainton observed dogs running on treadmills at different inclines (**Figure 49.17**). As the incline of the treadmill gradually increased, the dogs ran at the same speed but were working harder because they were running uphill. The $P_{CO_2}$ of their blood increased as the incline of the

INVESTIGATING**LIFE**

**49.17 The Respiratory Control System Is Sensitive to $P_{CO_2}$** What is the metabolic feedback signal that controls ventilation rate during exercise? To find out, Cedric Bainton conducted a series of experiments with dogs trained to run on treadmills.[a]

**HYPOTHESIS** Rising levels of blood $CO_2$ during exercise is the feedback signal that stimulates an increase in respiratory rate.

***Method***

1. Dogs are trained to run on a treadmill.
2. The dogs are equipped with instruments that measure respiratory rate and with arterial catheters that enable sampling of blood.
3. As a dog runs, the incline of the treadmill is changed to gradually increase the metabolic workload.
4. Ventilation rate (V; L/min) is plotted as a function of arterial $P_{CO_2}$ (mm Hg).

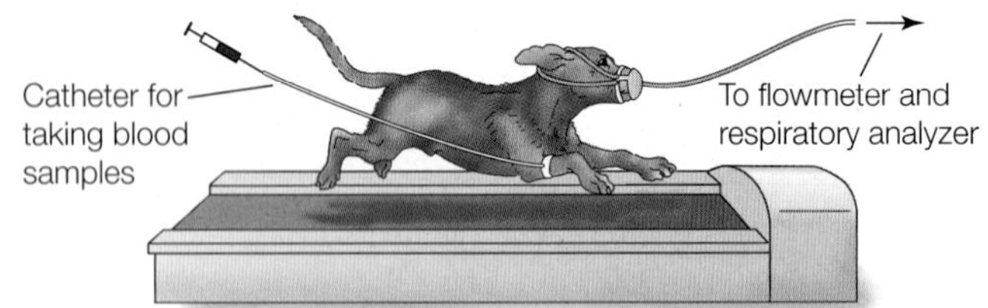


***Results*** When the workload is altered by slowly changing the incline of the treadmill (no change in speed), the ventilation rate is a function of $P_{CO_2}$.

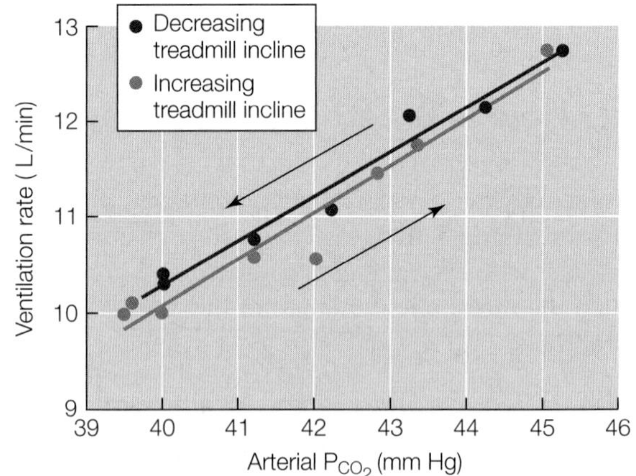


**CONCLUSION** $CO_2$ level in arterial blood is the metabolic feedback signal that regulates respiration in response to workload.

Go to **BioPortal** for discussion and relevant links for all INVESTIGATING**LIFE** figures.

[a]Bainton, C. R. 1972. *Journal of Applied Physiology* 33: 778–787.

## *The Respiratory Control System Is Not Always Regulated by $P_{CO_2}$*

### Original Paper

Bainton, C. R. 1972. Effect of speed vs. grade and shivering on ventilation in dogs during active exercise. *Journal of Applied Physiology* 33: 778–787.

### Analyze the Data

Experiments in which dogs run at different speeds while their $P_{CO_2}$ and V are measured produce results different from those shown in Figure 49.17. When subjects run at different speeds, their $P_{CO_2}$s are the same and yet their V changes. To resolve the differences between these types of experiments, Bainton recorded $P_{CO_2}$ and V in dogs, breath-by-breath immediately after a change in treadmill speed from 3 to 6 mph. The results for one experiment are shown in the table. When this dog was running at 3 mph, its average $P_{CO_2}$ was 39 mm Hg and its V was 9 L/min. After the dog had run at 6 mph for a few minutes its average $P_{CO_2}$ was 41 mm Hg and its V was 15.

**QUESTION 1**

Draw a plot with V on the y-axis and $P_{CO_2}$ on the x-axis that shows the average values at 3 mph and 6 mph and also the breath-by-breath values (in the table) for the transition. Do the data support the hypothesis that $P_{CO_2}$ is the feedback signal controlling ventilation rate? Why or why not?

**QUESTION 2**

How do you explain the average values after the dog had been running at the higher speed for a few minutes?

**QUESTION 3**

Relate these results to those obtained in the experiment in which the incline of the treadmill was gradually raised and lowered. Explain the differences between the results in terms of the $P_{CO_2}$ sensing mechanism (Hint: use the concept of "set point" in your explanation).

| Breath | $P_{CO_2}$ (mm Hg) | V (L/min) |
|---|---|---|
| 1 | 38.0 | 13.0 |
| 2 | 37.0 | 14.0 |
| 3 | 36.5 | 11.2 |
| 4 | 37.2 | 11.2 |
| 5 | 37.2 | 12.0 |
| 6 | 36.8 | 13.0 |
| 7 | 37.2 | 13.0 |

**Go to BioPortal for all** WORKING WITH**DATA** **exercises**

treadmill and respiratory gas exchange rate increased. Bainton observed a similar relationship between blood $P_{CO_2}$ and respiratory rate as he lowered the treadmill back to its original position. These results indicated that blood $P_{CO_2}$ could act as a signal to regulate breathing rate.

Before concluding that blood $P_{CO_2}$ is the *only* signal that controls breathing rate, Bainton changed the experiment. Instead of increasing the incline of the treadmill, he increased its speed (see the Working with Data exercise above). In this experiment the $P_{CO_2}$ of the blood remained constant as treadmill speed increased and as the respiratory gas exchange rate increased. Bainton concluded that blood $P_{CO_2}$ is the primary metabolic feedback information for breathing. However, when an animal starts to run or changes its running speed, additional feedback information from receptors in muscles and joints changes its sensitivity to $CO_2$—an example of feedforward information. As noted in Section 40.1, feedforward information can change the sensitivity or the set point of a regulatory system.

Where are partial pressures of gases in the blood sensed? The major site of $P_{CO_2}$ sensitivity is an area on the ventral surface of the medulla. The primary sensitivity of these chemosensitive cells is not to $CO_2$, however. Rather, they are stimulated by $H^+$ ions. The $H^+$ ion concentration, or pH, in the environment of these cells is a direct reflection of the $P_{CO_2}$ of the blood. When the $P_{CO_2}$ of the blood is higher than that of the extracellular fluid in this area, $CO_2$ diffuses out of the blood. That $CO_2$ interacts with $H_2O$ to form carbonic acid ($H_2CO_3$), which dissociates into $H^+$ ions and $HCO_3^-$ ions (see Figure 49.14). The $H^+$ ions that are produced stimulate the chemosensitive cells that increase respiratory gas exchange. Thus even though we measure blood $P_{CO_2}$ as the stimulus for breathing, the real stimulus is pH.

Sensitivity to blood $P_{O_2}$ resides in nodes of neural tissue on the large blood vessels leaving the heart: the aorta and the carotid arteries (**Figure 49.18**). These **carotid bodies** and

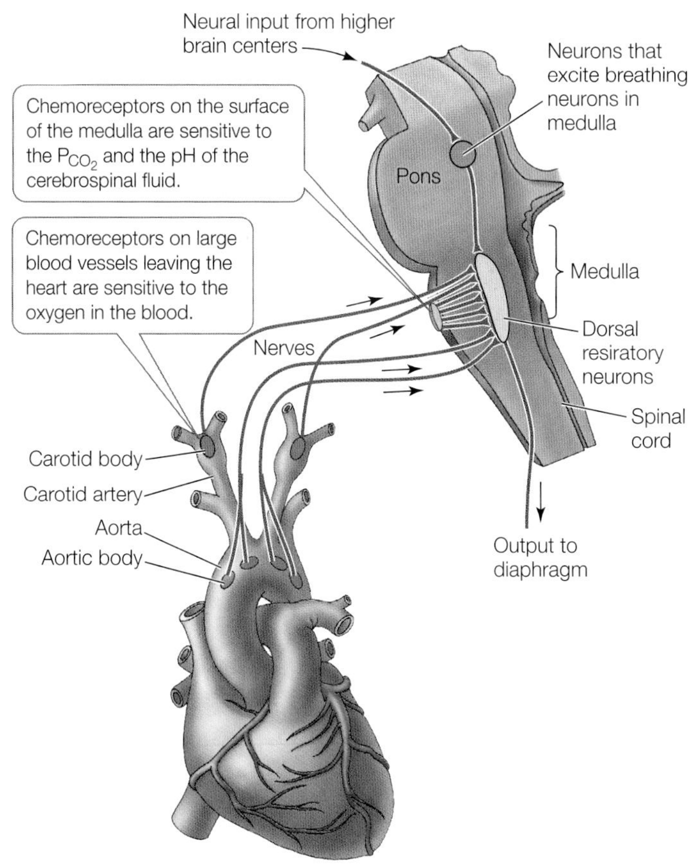


**49.18 Feedback Information Controls Breathing** The body uses feedback information from chemosensors in the heart and brain to match breathing rate to metabolic demand.

**aortic bodies** are chemosensors. If the blood supply to these structures decreases, or if the blood $P_{O_2}$ falls dramatically, the chemosensors are activated and send nerve impulses to the breathing control center. Although we are not very sensitive to changes in blood $P_{O_2}$, the carotid and aortic bodies can stimulate increases in breathing during exposure to very high altitudes or when blood volume or blood pressure is very low. Also, there is a synergism between $CO_2$ and $O_2$ sensing. When blood $P_{CO_2}$ increases, there is an increased sensitivity to low $O_2$, and vice versa.

### RECAP 49.5

The rhythmic contractions of the respiratory muscles that drive breathing are generated by neurons in the brainstem.

- What is the primary chemical stimulus for controlling the respiratory rate, and where is it sensed? **See pp. 1019–1020 and Figures 49.16 and 49.18**
- Explain what feedforward information is and how exercise provides feedforward information in the respiratory control system. **See pp.1020–1021 Figure 49.17**
- What are the functions of the carotid and aortic bodies? **See pp. 1021–1022 and Figure 49.18**

How do elephants avoid damage to the blood vessels of their thoracic cavities when they snorkel?

### ANSWER

In all mammals except elephants, the surfaces of the lungs and the chest cavity move easily against each other. As the lungs inflate and deflate, their surface slips and slides against that of the chest cavity. In elephants, however, dense connective tissue attaches the surface of the lungs to the surface of the chest cavity. This connective tissue acts as reinforcement for the fragile blood vessels on the surface of the chest cavity and prevents them from rupturing and filling the cavity with blood, which would destroy the function of the lungs—exchange of oxygen and carbon dioxide. The elephant is the only mammal that does not have a functional pleural space.

## CHAPTER SUMMARY 49

### 49.1 What Physical Factors Govern Respiratory Gas Exchange?

- Most cells require a constant supply of $O_2$ and continuous removal of $CO_2$. These **respiratory gases** are exchanged between an animal's body fluids and its environment by diffusion.
- **Fick's law of diffusion** shows how various physical factors influence the diffusion rate of gases. Adaptations to maximize respiratory gas exchange influence one or more variables of Fick's law.
- In water-breathing animals, gas exchange is limited by the low diffusion rate and low amount of $O_2$ in water. If water temperature rises, water-breathing animals face a double bind in that the amount of $O_2$ in water decreases, but their metabolism and the amount of work required to move water over the gas exchange surfaces increase. **Review Figure 49.2**
- In air, the **partial pressure** of oxygen ($P_{O_2}$) decreases with altitude.

### 49.2 What Adaptations Maximize Respiratory Gas Exchange?

- Adaptations to maximize gas exchange include increasing the surface area for gas exchange and maximizing partial pressure gradients across those exchange surfaces by ventilating the outer surface with the respiratory medium, and perfusing the inner surface with blood. **Review Figure 49.3**
- Insects distribute air throughout their bodies in a system of **tracheae**, tracheoles, and air capillaries. **Review Figure 49.4**
- The **gills** of fishes have large gas exchange surface areas that are ventilated continuously and unidirectionally with water. The **countercurrent flow** of blood helps increase the efficiency of gas exchange. **Review Figures 49.5, 49.6**
- The gas exchange system of birds includes **air sacs** that communicate with the lungs but are not used for gas exchange. Air flows unidirectionally through bird lungs; gases are exchanged in air capillaries that run between **parabronchi**. **Review Figure 49.7**
- Each breath of air remains in a bird's respiratory system for two breathing cycles. The air sacs work as bellows to supply the air capillaries with a continuous unidirectional flow of fresh air. **Review Figure 49.8, ANIMATED TUTORIAL 49.1**
- In all air-breathing vertebrates except birds, breathing is **tidal**. This is a less efficient form of gas exchange than that of fishes and birds. Although the volume of air exchanged with each breath can vary considerably in tidal breathing, the inhaled air is always mixed with stale air. **Review Figure 49.9**

### 49.3 How Do Human Lungs Work?

- In mammalian lungs, the gas exchange surface area provided by the millions of **alveoli** is enormous, and the diffusion path length between the air and perfusing blood is short. **Surface tension** in the alveoli would make inflation of the lungs difficult if the alveoli did not produce **surfactant**. **Review Figure 49.10, ACTIVITY 49.1**
- Inhalation occurs when contractions of the **diaphragm** increase volume and reduce pressure in the **thoracic cavity**, thereby pulling on the **pleural membranes**. Relaxation of the diaphragm increases pressure in the thoracic cavity and results in exhalation. **Review Figure 49.11, ANIMATED TUTORIAL 49.2**
- During periods of heavy metabolic demands such as strenuous exercise, the **intercostal muscles**, located between the ribs, increase the volume of air inhaled and exhaled.

*continued*

## 49.4 How Does Blood Transport Respiratory Gases?

- $O_2$ is reversibly bound to **hemoglobin** in red blood cells. Each hemoglobin molecule can carry a maximum of four $O_2$ molecules. Because of **positive cooperativity**, hemoglobin's affinity for $O_2$ depends on the $P_{O_2}$ to which the hemoglobin is exposed. Therefore hemoglobin picks up $O_2$ as it flows through respiratory exchange structures and gives up $O_2$ in metabolically active tissues. **Review Figure 49.12, ANIMATED TUTORIAL 49.3**
- **Myoglobin** serves as an $O_2$ reserve in muscle.
- There is more than one type of hemoglobin. Fetal hemoglobin has a higher affinity for $O_2$ than does adult hemoglobin, allowing fetal blood to pick up $O_2$ from the maternal blood in the placenta. **Review Figure 49.13, ACTIVITY 49.2**
- $CO_2$ is transported in the blood principally as bicarbonate ions ($HCO_3^-$). **Review Figure 49.14**

## 49.5 How Is Breathing Regulated?

- The basic breathing rhythm is an involuntary function generated by neurons in the medulla and modulated by higher brain centers. The most important feedback stimulus for breathing is the level of $CO_2$ in the blood. **Review Figures 49.16, 49.17**
- The breathing rhythm is sensitive to feedback from chemoreceptors on the ventral surface of the medulla and in the **carotid** and **aortic bodies** on the large vessels leaving the heart. **Review Figure 49.18**

**See ACTIVITY 49.3 for a concept review of this chapter**

**Go to the Interactive Summary to review key figures, Animated Tutorials, and Activities Life10e.com/is49**

# CHAPTER**REVIEW**

### REMEMBERING

1. Which statement about the gas exchange system of birds is *not* true?
   a. Respiratory gases are not exchanged in the air sacs.
   b. A bird can achieve more complete exchange of $O_2$ from air to blood than humans can.
   c. Air passes through birds' lungs in only one direction.
   d. The gas exchange surfaces in bird lungs are the parabronchi.
   e. A breath of air remains in the system for two breathing cycles.

2. In the human gas exchange system,
   a. the lungs and airways are completely collapsed after a forceful exhalation.
   b. exhalation is driven by contraction of the diaphragm.
   c. the $P_{O_2}$ of the blood leaving the lungs is greater than that of the exhaled air.
   d. the amount of air that is moved per breath during normal, at-rest breathing is termed the total lung capacity.
   e. $P_{CO_2}$ in the air reaching the alveoli during inhalation is close to zero, as it is in the outside air.

3. The hemoglobin of a human fetus
   a. is the same as that of an adult.
   b. has a higher affinity for $O_2$ than adult hemoglobin has.
   c. has only two protein subunits instead of four.
   d. is supplied by the mother's red blood cells.
   e. has a higher affinity for BPG than adult hemoglobin has.

4. Most $CO_2$ in the blood is carried
   a. in the cytoplasm of red blood cells.
   b. as $CO_2$ dissolved in the plasma.
   c. in the plasma as bicarbonate ions.
   d. bound to plasma proteins.
   e. in red blood cells bound to hemoglobin.

5. Myoglobin
   a. binds $O_2$ at $P_{O_2}$ values at which hemoglobin is releasing its bound $O_2$.
   b. has a lower affinity for $O_2$ than hemoglobin does.
   c. consists of four polypeptide chains, just as hemoglobin does.
   d. provides an immediate source of $O_2$ for muscle cells at the onset of activity.
   e. can bind four $O_2$ molecules at once.

### UNDERSTANDING & APPLYING

6. The blood of a certain species of fish that lives in Antarctica has no hemoglobin. What anatomical and behavioral characteristics would you expect to find in this fish, and why is its distribution limited to the waters of Antarctica?

7. In patients with emphysema, the fine structures of alveoli break down, resulting in the formation of larger air cavities in the lungs. Also, the lung tissue becomes less elastic. Give at least two explanations for why such patients have a low tolerance for exercise.

8. At what point in the breathing cycle would the pleural cavity pressure be going down while the alveolar pressure is going up? At what point in the breathing cycle would the alveolar pressure be most positive in relation to atmospheric pressure?

### ANALYZING & EVALUATING

9. Blood banks store whole blood for a much shorter period than they store blood plasma. This is because when blood that has been stored for too long is infused into a patient, it can actually decrease the $O_2$ available to the patient's tissues. Refer to Figure 9.5 on glycolysis and explain why the affinity of hemoglobin for $O_2$ increases with time in storage. Hint: Remember that whole blood is living tissue and requires energy, and that a bag of blood is a closed system. Then think about what fuel can supply energy in a closed system.

10. When you suddenly travel to high altitude, you notice an unusual breathing pattern when you are resting. For a while you stop breathing completely; then suddenly you start breathing rapidly for a short time; then you stop breathing again. This can go on and on in a cyclical pattern called Cheyne–Stokes breathing. Think in terms of the changes in partial pressure gradients when you go to high altitudes and explain how the breathing control system could produce this breathing pattern.

11. South American camelids—llamas, alpacas, guanacos, and vicuñas—are native to the Andes Mountains. In the natural habitat of these mammals, more than 5,000 m above sea level, the $P_{O_2}$ is below 85 mm Hg, and the $P_{O_2}$ in the camelids' lungs is about 50 mm Hg. The hemoglobins of these camelids are different from that of humans; the figure shows the $O_2$-binding/dissociation curves for human and llama hemoglobins.
    a. Compared with human hemoglobin, does llama hemoglobin have a higher or a lower affinity for $O_2$?
    b. Why would llama hemoglobin be advantageous at high altitudes?
    c. How would the characteristics of llama hemoglobin affect the transfer of $O_2$ from the blood to the tissues of the llama?

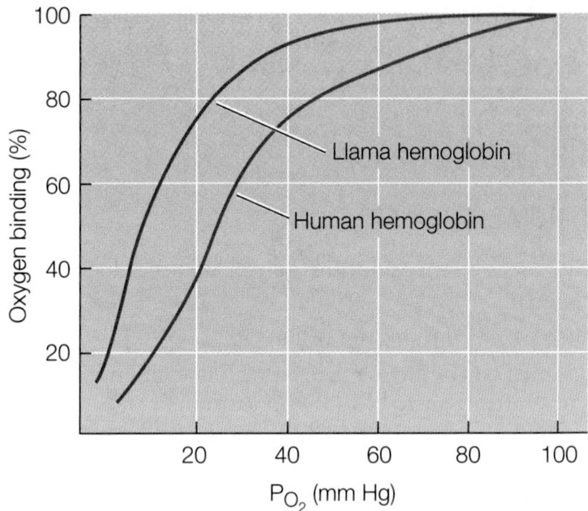


Go to BioPortal at **yourBioPortal.com** for Animated Tutorials, Activities, LearningCurve Quizzes, Flashcards, and many other study and review resources.

# 50 Circulatory Systems

## CHAPTER**OUTLINE**

**50.1** **Why Do Animals Need a Circulatory System?**

**50.2** **How Have Vertebrate Circulatory Systems Evolved?**

**50.3** **How Does the Mammalian Heart Function?**

**50.4** **What Are the Properties of Blood and Blood Vessels?**

**50.5** **How Is the Circulatory System Controlled and Regulated?**

**An Athlete's Heart** Basketball star Reggie Lewis died of heart failure at the age of 27. Although the exact medical situation underlying Lewis's heart problems remains clouded in controversy, great strides have been made in diagnosing a genetic heart condition that can be particularly dangerous for athletes.

ON APRIL 29, 1993, the Boston Celtics met the Charlotte Hornets in an NBA playoff game. The Celtics' star, 27-year-old Reggie Lewis, had just scored 10 points in 3 minutes when he slumped forward and fell to the floor. The team doctor examined him and allowed him to return to the game. But Lewis's legs were "wobbly," and he played only briefly.

Lewis had been experiencing dizzy spells for about a month before the playoff incident. After the playoffs he underwent rigorous testing by cardiologists, who diagnosed him as having a dangerous arrhythmia (irregular heartbeat) indicative of diseased cardiac muscle. Accepting this diagnosis would mean the end of his professional athletic career, and Lewis sought a second opinion. The second medical team concluded that Lewis had undergone a transient irregular heartbeat attributable to normal athletic stresses. The condition was thought to be due to an enlarged heart—a condition common in high-performing athletes. In July 1993, after an hour spent shooting baskets in a pick-up game, Reggie Lewis collapsed and died of heart failure.

Your heart is a muscular pump that, at rest, beats an average of 60 to 70 times per minute. With each beat, it circulates about 70 milliliters of blood through the body. Without taking work or exercise into account, that is 300 liters per hour, 7,200 liters per day, 2.6 million liters per year—no time-outs.

Heart failure is the leading cause of death in the United States, accounting for some one-fourth (about 600,000) of the deaths each year. Heart failure most commonly results from blockage of the vessels that supply the heart muscle with blood, and its risk increases with age. But heart failure is also the leading cause of death among young athletes.

In athletes, the most common cause of heart failure is not blocked vessels but a gene mutation that affects the contractile proteins of the heart. This mutation can lead to a thickening of the walls of the heart. About 0.5 percent of the population has this mutation; most are unaware of it, and most live their entire lives without symptoms.

Diagnosis is improving. In 2008, 33-year-old Cuttino Mobley announced his retirement from basketball after his physical exam revealed arrhythmia and an MRI showed the symptomatic thickening of the heart walls. "Getting the MRI saved my life," Mobley said.

How can the same mutation-based heart condition be fatal to an athlete but innocuous in most other people?

See answer on p. 1046.

## 50.1 Why Do Animals Need a Circulatory System?

A **circulatory system** consists of a muscular pump (the heart), a fluid (blood), and a series of conduits (blood vessels) through which the fluid can be pumped around the body. Heart, blood, and vessels are also known collectively as a **cardiovascular system** (Greek *kardia*, "heart"; Latin *vasculum*, "vessel"). The function of a circulatory system is to transport things around the body. Preceding chapters discussed how circulatory systems transport heat, hormones, respiratory gases, blood cells, platelets, and cells and molecules of the immune system. Succeeding chapters will add nutrients and waste products to that list. In this section we will describe the general types of circulatory systems found in animals.

### Some animals do not have a circulatory system

Single-celled organisms serve all of their needs through direct exchanges with the environment. Such organisms are found mostly in aquatic or very moist terrestrial environments. Similarly, many multicellular aquatic organisms are small or thin enough that all of their cells are close to the external environment. Such species may not have a circulatory system because nutrients, respiratory gases, and wastes can diffuse directly between the cells of their bodies and the environment.

The cells of some larger aquatic multicellular animals without a circulatory system are served by highly branched central cavities called gastrovascular systems that bring the external environment into the animal. All of the cells of a sponge are in contact with, or very close to, the water that surrounds the animal and circulates through its central cavity (see Figure 49.1B). Very small animals without a circulatory system can maintain high levels of metabolism and activity, but larger animals without a circulatory system such as sponges, jellyfishes, and flatworms tend to be inactive, slow, or even sedentary. Large, active animals, however, require a circulatory system.

### Circulatory systems can be open or closed

The cells of large, mobile animals are supported by the extracellular fluid. All nutrients—oxygen, fuel, essential molecules—come from that fluid, and the waste products of cell metabolism go into it. Circulatory systems have muscular chambers, or **hearts**, that move the extracellular fluid around the body. In open circulatory systems, extracellular fluid is the same as the fluid in the circulatory system and is called hemolymph. This fluid leaves the vessels of the circulatory system, percolates between cells and through tissues, and then flows back into the heart or vessels of the circulatory system to be pumped out again. In contrast, closed circulatory systems completely contain the circulating fluid (blood) in a continuous system of vessels. Blood cells and large molecules stay within the system, but water and low-molecular-weight solutes leak out of the smallest vessels, the capillaries, which are highly permeable.

In animals with a closed circulatory system, **extracellular fluid** refers to both the fluid in the circulatory system and the fluid outside it. The fluid in the circulatory system is the blood plasma; the fluid around the cells is the **interstitial fluid** (see Figure 40.1). A 70-kilogram person has a total extracellular fluid volume of about 14 liters. Less than a quarter of it—about 3 liters—is the blood plasma.

### Open circulatory systems move extracellular fluid

**Open circulatory systems** are found in arthropods, mollusks, and some other invertebrate groups. In these systems a muscular pump, or heart, helps move the hemolymph through vessels leading to different regions of the body. The fluid leaves the vessels to filter through the tissues before returning to the heart. In the generalized arthropod shown in **Figure 50.1A**, the fluid returns directly to the heart through openings called ostia. Ostia have valves that allow hemolymph to enter the relaxed heart but prevent it from flowing in the reverse direction when the heart contracts. In mollusks, open vessels collect hemolymph from different regions of the body and return it to the heart (**Figure 50.1B**).

Lest you think that open circulatory systems are inefficient and can support only sluggish lifestyles such as those of mollusks, remember that crabs scuttling along the beach, yellow jackets buzzing around your picnic, and scorpions dashing across the desert all have open circulatory systems.

### Closed circulatory systems circulate blood through a system of blood vessels

In **closed circulatory systems**, a system of vessels keeps circulating blood separate from the interstitial fluid. Blood is pumped through this vascular system by one or more muscular hearts, and some components of the blood never leave the vessels. Closed circulatory systems characterize vertebrates and some invertebrate groups, among them annelids.

A simple example of a closed circulatory system is that of the earthworm (**Figure 50.1C**). One large ventral blood vessel carries blood from the worm's anterior end to its posterior end. Smaller vessels branch off and transport the blood to even smaller vessels serving the tissues in each body segment. In the smallest vessels, respiratory gases, nutrients, and metabolic wastes diffuse between the blood and interstitial fluid. The blood then flows from these vessels into larger vessels that lead into one large dorsal vessel, which carries the blood from the posterior to the anterior end of the body. Five pairs of muscular vessels connect the large dorsal and ventral vessels in the anterior end, thus completing the circuit. The dorsal vessel and the five connecting vessels serve as hearts for the earthworm; their contractions keep the blood circulating. The direction of circulation is determined by one-way valves in the dorsal and connecting vessels.

Closed circulatory systems have several advantages compared with open systems:

- Fluid can flow more rapidly through vessels than through intercellular spaces and can therefore transport nutrients and wastes to and from tissues more rapidly.

(A) Arthropod

Extracellular fluid (hemolymph) of arthropods percolates through tissues and enters the heart through openings called ostia.

Tubular heart

Digestive system

(B) Mollusk

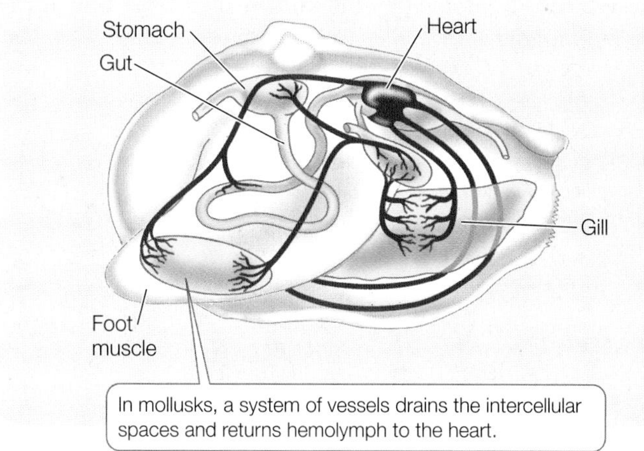


(C) Annelid worm

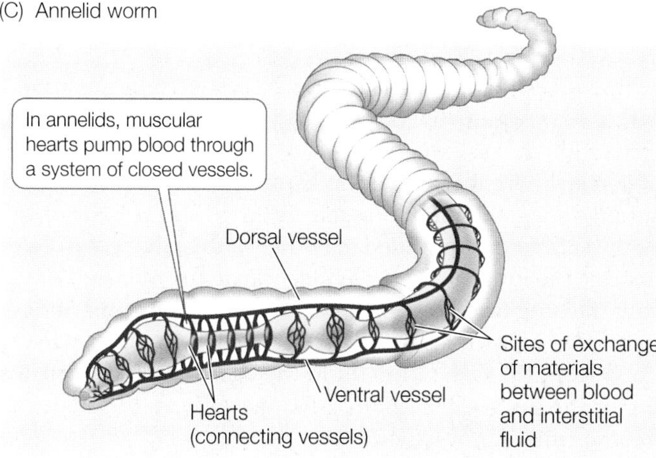


**50.1 Circulatory Systems** Arthropods, illustrated here by an insect (A), and mollusks such as clams (B) have an open circulatory system. Hemolymph is pumped by a tubular heart and directed to different regions of the body through vessels that open into intercellular spaces. (C) Annelids such as earthworms have a closed circulatory system, in which the cellular and macromolecular elements of the blood are confined in a system of vessels and the blood is pumped through those vessels by one or more muscular hearts.

- By changing the diameter (and hence the resistance) of specific vessels, closed systems can control the flow of blood to selective tissues and organs to match their needs.
- Specialized cells and large molecules that aid in transporting hormones and nutrients can be kept in the vessels but can drop their cargo in the tissues where it is needed.

With all of these "advantages" of closed circulatory systems, you might wonder how some species with open circulatory systems can sustain such high levels of activity. In the case of insects, the answer is clear: insects do not depend on their circulatory systems for respiratory gas exchange (see Chapter 49). The open systems of some other species, such as the crab referred to above, involve elaborate systems of vessels that can direct the movement of tissue fluids from one part of the animal's body to another.

**RECAP 50.1**

Circulatory systems consist of a pump and an open or closed set of vessels through which a fluid transports oxygen, nutrients, wastes, and a variety of other substances.

- Circulatory systems are transport systems. What do they transport? **See p. 1026**
- Why are many species with open circulatory systems rather limited with respect to metabolic activity, and why does this limitation not apply to insects? **See p. 1026 and Figure 50.1**
- What are some advantages of a closed circulatory system? **See pp. 1026–1027**

This overview of the open and closed systems found among invertebrates introduced some basic concepts about circulatory systems. Next we will turn to describing the closed circulatory systems of vertebrates.

## 50.2 How Have Vertebrate Circulatory Systems Evolved?

Vertebrates have a closed circulatory system and a heart with two or more chambers. When a heart chamber contracts, it squeezes the blood, putting it under pressure. Blood then flows out of the heart and into vessels, where pressure is lower. Resistance to flow in the vessels dissipates the pressure imparted to the blood by the heart. One-way valves prevent the backflow of blood as the heart cycles between contraction and relaxation.

As we explore the features of the circulatory systems of different classes of vertebrates, a general evolutionary theme will emerge: as circulatory systems become more complex, the blood that flows to the gas exchange organs (gills or lungs; see Chapter 49) becomes increasingly separated from the blood that flows to the rest of the body.

In fishes, the phylogenetically oldest vertebrates, blood is pumped from the heart to the gills and then to the tissues of the body and back to the heart—a single circuit. In birds and mammals, blood is pumped from the heart to the lungs and back to the heart in a **pulmonary circuit**, and then from the heart to the rest of the body and back to the heart in a **systemic circuit**. In all other vertebrates we see various adaptations for separating the blood flow into pulmonary and systemic circuits.

Both pulmonary and systemic circuits begin with vessels called **arteries** that carry blood away from the heart. Arteries branch into smaller vessels called **arterioles** that feed blood into capillary beds. **Capillaries** are the tiny, thin-walled vessels where materials are exchanged between the blood and the tissue fluid. Small vessels called **venules** drain capillary beds. The venules join to form larger vessels called **veins** that ultimately deliver blood back to the heart.

We can trace the evolutionary history of vertebrate circulatory systems by comparing the circulatory systems of ray-finned fish, lungfish, amphibians, non-avian reptiles, birds, and mammals.

## Circulation in fish is a single circuit

The fish heart has four chambers. Blood returning from all parts of the body collects in a **sinus venosus**, which feeds into the muscular **atrium**. When the atrium contracts, it pumps blood into the more muscular chamber, the **ventricle**. Contraction of the ventricle pushes blood into the last part of the fish heart, the **bulbus arteriosus**, which is a highly elastic chamber. The pressure imparted to the blood by the ventricle stretches the bulbus arteriosus, and its elastic recoil dampens the blood pressure oscillations generated by the beating of the heart. The arterial blood leaving the bulbus arteriosus under pressure flows through the gills, where respiratory gases are exchanged. Blood leaving the gills collects in a large dorsal artery, the **aorta**, which distributes blood to smaller arteries and arterioles leading to all the organs and tissues of the body. In the tissues, blood flows through beds of tiny capillaries, collects in venules and veins, and eventually returns to the sinus venosus of the heart. The unidirectional flow of blood in this circuit is enabled by one-way valves between the sinus venosus and the atrium, between the atrium and the ventricle, and between the ventricle and the bulbus arteriosus.

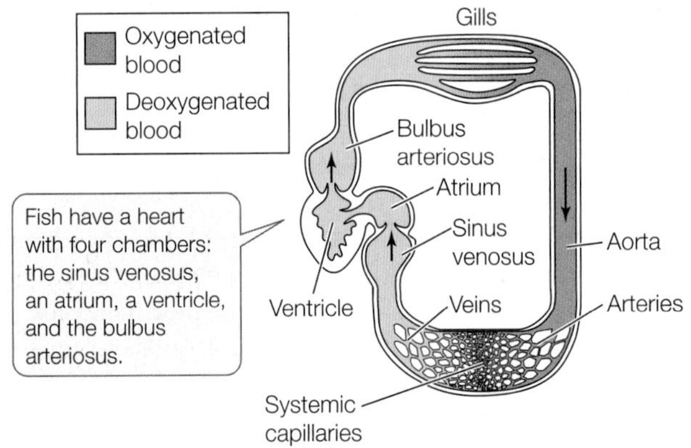


Most of the pressure imparted to the blood by the contraction of the ventricle is dissipated as a result of resistance to flow in the many narrow spaces in the gill lamellae. Therefore blood leaving the gills and entering the aorta is under low pressure, limiting the capacity of the fish circulatory system to supply the tissues with oxygen and nutrients. Yet this limitation on arterial blood pressure does not seem to limit swimming performance. Some species, such as tuna and marlin, can swim at remarkably high rates of speed for long distances.

The evolutionary transition from breathing water to breathing air had important consequences for the vertebrate circulatory system. An example of how the system changed to serve a primitive lung can be seen in the African lungfish.

## Lungfish evolved a gas-breathing organ

Lungfish are periodically exposed to water with low oxygen content or to situations in which their aquatic environment dries up. The adaptation that deals with these conditions is an outpocketing of the gut that serves as a lung. The lung contains many thin-walled blood vessels, so blood flowing through those vessels can pick up oxygen from air gulped into the lung.

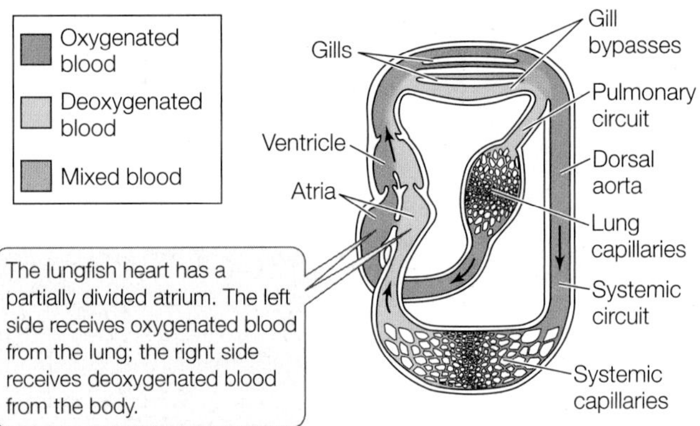


How does the lungfish circulatory system take advantage of this new organ? In fishes, the gills are arranged on supportive gill arches (see Figure 49.5). Blood flows into the gill arch in an afferent arteriole and leaves in an efferent arteriole. In lungfishes, the blood vessels in the posterior pair of gill arteries have been modified to be a low-resistance conduit for blood to the lung, and a new vessel carries oxygenated blood from the lung back to the heart. In addition, two anterior gill arches have lost their gills, and their blood vessels deliver blood from the heart directly to the dorsal aorta. Because a few of the gill arches retain gills, the African lungfish can breathe either air or water.

The lungfish heart partially separates its flow of blood into pulmonary and systemic circuits; it has a partially divided atrium. The left side receives oxygenated blood from the lung and the right side receives deoxygenated blood from the sinus venosus. These two bloodstreams stay mostly separate as they flow through the ventricle and the bulbus arteriosus. As a result, oxygenated blood goes mostly to the anterior gill arteries leading to the dorsal aorta, and deoxygenated blood goes mostly to the other gill arches that have functional gills as well as to the gill arteries that serve the lung.

We can conclude that the lungfish lung evolved as a means of supplementing oxygen uptake from the gills. When the water is oxygenated, the lungfish can obtain oxygen through its gills; but in oxygen-depleted water, it can depend on getting oxygen from its lung. Associated modifications of the lungfish vascular system set the stage for the evolution of separate pulmonary and systemic circulations in higher vertebrates.

### Amphibians have partial separation of systemic and pulmonary circulation

In adult amphibians, a single ventricle pumps blood to the lungs and the rest of the body, but two atria receive blood returning to the heart. The left atrium receives oxygenated blood from the lungs, and the right atrium receives deoxygenated blood from the body.

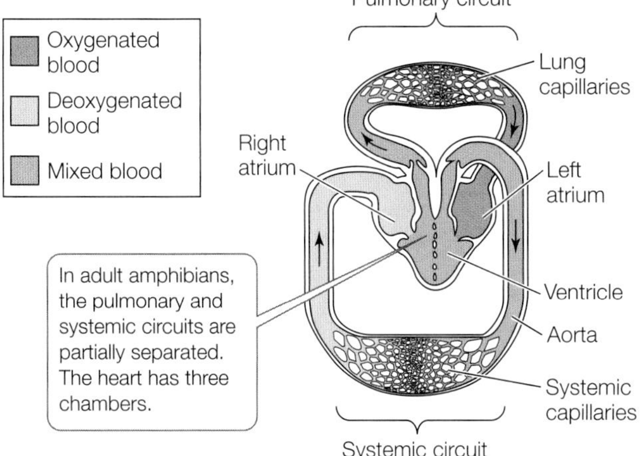


Because both atria deliver blood to the same ventricle, the oxygenated and deoxygenated blood could mix, in which case blood going to the tissues would not carry a full load of oxygen. Mixing is limited, however, because anatomical features of the ventricle direct the flow of deoxygenated blood from the right atrium to the pulmonary circuit and the flow of oxygenated blood from the left atrium to the aorta. Partial separation of pulmonary and systemic circulation has the advantage of allowing blood destined for the tissues to sidestep the large pressure drop that occurs in the gas exchange organ. Blood leaving the amphibian heart for the tissues moves directly to the aorta, and hence to the body, at a higher pressure than if it had first flowed through the lungs.

Amphibians have another adaptation for oxygenating their blood: they can pick up a considerable amount of oxygen in blood flowing through small blood vessels in their skin.

### Reptiles have exquisite control of pulmonary and systemic circulation

As described in Chapter 33, the reptiles include turtles, snakes, lizards, crocodilians, and birds (see Figure 33.20). Crocodilians and birds have cardiovascular systems with two completely separated ventricles, creating a four-chambered heart. All other reptiles have ventricles that are not completely separated into left and right chambers.

Consider the behavior, ecology, and physiology of ectothermic reptiles (i.e., excluding birds). Many are active, powerful, fast animals, but their activity comes in bursts that are interspersed with long periods of inactivity. At these times the animals' metabolic rates are much lower than the resting metabolic rates of the endothermic birds and mammals. So enormous is the range of metabolic demand in ectothermic reptiles that they do not need to breathe continuously. Some species are accomplished divers and spend long periods under water, where they cannot breathe air.

When these animals are not breathing, it would be a waste of energy for them to pump blood through their lungs. Thus they have evolved the capability to send blood to the lungs and the rest of the body when they are breathing, but when they are not breathing, they can bypass the pulmonary circuit and pump all the blood to the body. How do they do this?

In ectothermic reptiles with a three-chambered heart—that is, the turtles, snakes, and lizards—the ventricle is partially divided into left and right halves by a septum. Oxygenated blood from the lungs enters the left side of the ventricle through the left atrium. Deoxygenated blood from the body enters the right side of the ventricle through the right atrium. These species have two aortas, left and right. The left aorta is positioned so that it receives oxygenated blood from the left side of the ventricle. The right aorta, however, is positioned so that it can receive blood from either the right or left side of the ventricle.

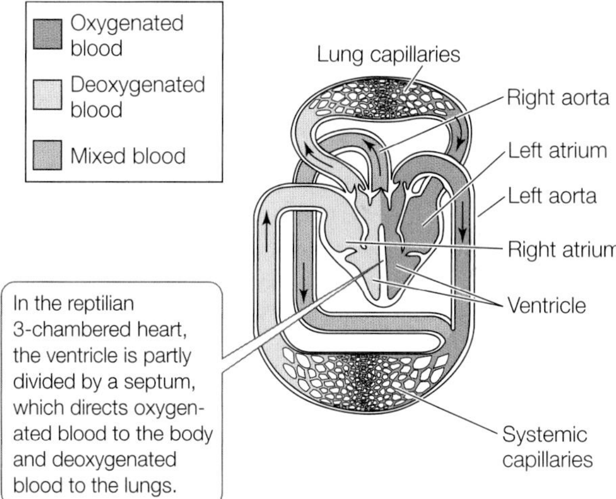


When the animal is breathing air, the resistance in the pulmonary circuit is lower than the resistance in the systemic circuit, so blood from the right side of the ventricle tends to flow into the pulmonary artery rather than the right aorta. When the animal is not breathing, pulmonary vessels constrict, resistance in the pulmonary circuit goes up, and blood from the right side of the ventricle tends to flow into the right aorta. As a result, blood from both sides of the ventricle flows through both aortas to the systemic circuit.

Crocodilians, like birds, have two completely separated ventricles. Unlike birds, they have two aortas, one originating in each ventricle. But there is a connection between the two aortas just as they leave the heart, and this connection enables them to alter the proportions of blood going to their pulmonary and systemic circuits. When a crocodile or alligator is breathing and resistance in the pulmonary circuit is low, backpressure from the stronger left ventricle closes the valve between the right ventricle and the right aorta, forcing all of the blood from the right ventricle to flow into the pulmonary circuit. When the animal stops breathing, pulmonary vessels constrict, resistance in the pulmonary circuit rises, and blood from the right ventricle flows into the right aorta. This ability of all ectothermic reptiles to direct blood to their pulmonary or systemic circuits is highly adaptive for their lifestyle of intermittent breathing.

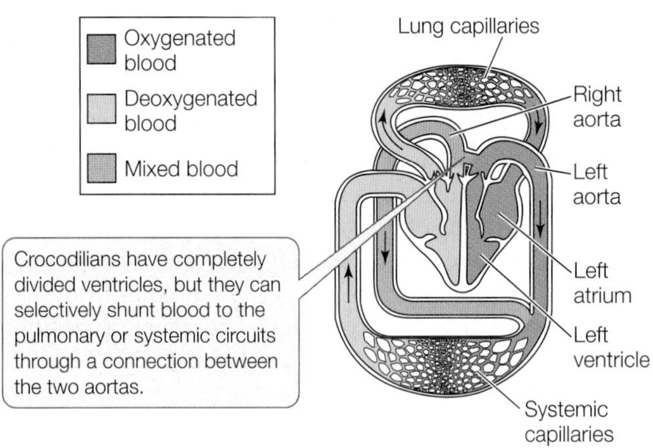


### Birds and mammals have fully separated pulmonary and systemic circuits

The four-chambered hearts of birds and mammals have completely separate pulmonary and systemic circuits. Separate circuits have several advantages for these active animals with continuously high metabolic rates:

- Oxygenated and deoxygenated blood cannot mix; therefore the systemic circuit always receives blood with the highest oxygen content.
- Respiratory gas exchange is maximized because the blood with the lowest oxygen content and highest $CO_2$ content is sent to the lungs.
- Separate systemic and pulmonary circuits can operate at different pressures.

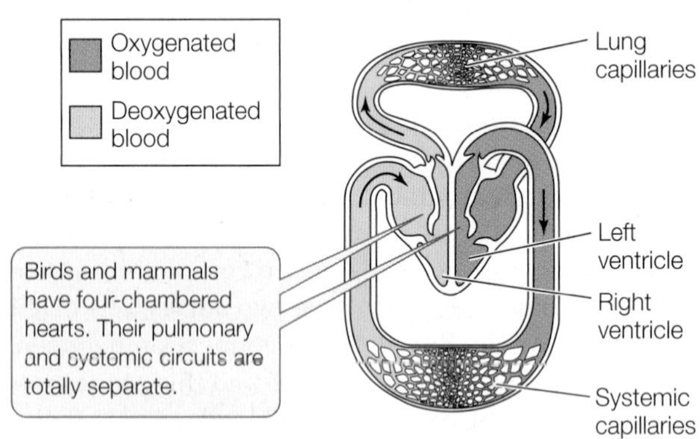


The tissues of birds and mammals have high nutrient demands and thus a very high density of blood vessels, requiring the heart to generate a high blood pressure to perfuse all the vessels of the systemic circuit. The pulmonary circuit of these animals receives a blood flow equal to that of the systemic circuit, but the lungs have far fewer blood vessels. Thus the pulmonary circuit of birds and mammals can function at lower pressures, and the four-chambered heart makes that possible.

**Go to Activity 50.1 Vertebrate Circulatory Systems Life10e.com/ac50.1**

**RECAP 50.2**

The closed circulatory system of vertebrates has evolved from a single circuit system in fishes to separate pulmonary and systemic circuits in birds and mammals.

- Explain why fish cannot supply blood to their tissues at high pressure. **See p. 1028**
- By comparing lungfish and amphibian circulatory systems, explain how a heart with two separate atria could have evolved. **See pp. 1028–1029**
- What are some advantages of separate pulmonary and systemic circuits? **See pp. 1029–1030**

## 50.3 How Does the Mammalian Heart Function?

Mammals have hearts consisting of a right and a left atrium and a right and a left ventricle. We can use the human heart as our example (**Figure 50.2**). The right ventricle pumps blood through the pulmonary circuit, and the left ventricle pumps blood through the systemic circuit.

One-way valves between the atria and ventricles, the **atrioventricular** (**AV**) **valves**, prevent backflow of blood into the atria when the ventricles contract. The **pulmonary valve** and **aortic valve**, one-way valves between the ventricles and the major arteries, prevent backflow of blood into the ventricles when they relax.

In this section we will describe the flow of blood through the heart and body and examine the unique electrical properties of cardiac muscle that result in the rhythmic contractions of the heart.

### Blood flows from right heart to lungs to left heart to body

The heart's right atrium receives deoxygenated blood from the **superior** (upper) **vena cava** and the **inferior** (lower) **vena cava** (see Figure 50.2)—the large veins that collect blood returning to the heart from the upper and lower body, respectively. The veins of the heart itself also drain into the right atrium. From the right atrium, the blood flows through the right AV valve into the right ventricle. Most of the filling of the ventricle results from passive flow while the heart is relaxed between beats. Just at the end of this period of passive ventricular filling, the atrium contracts and adds a little more blood to the

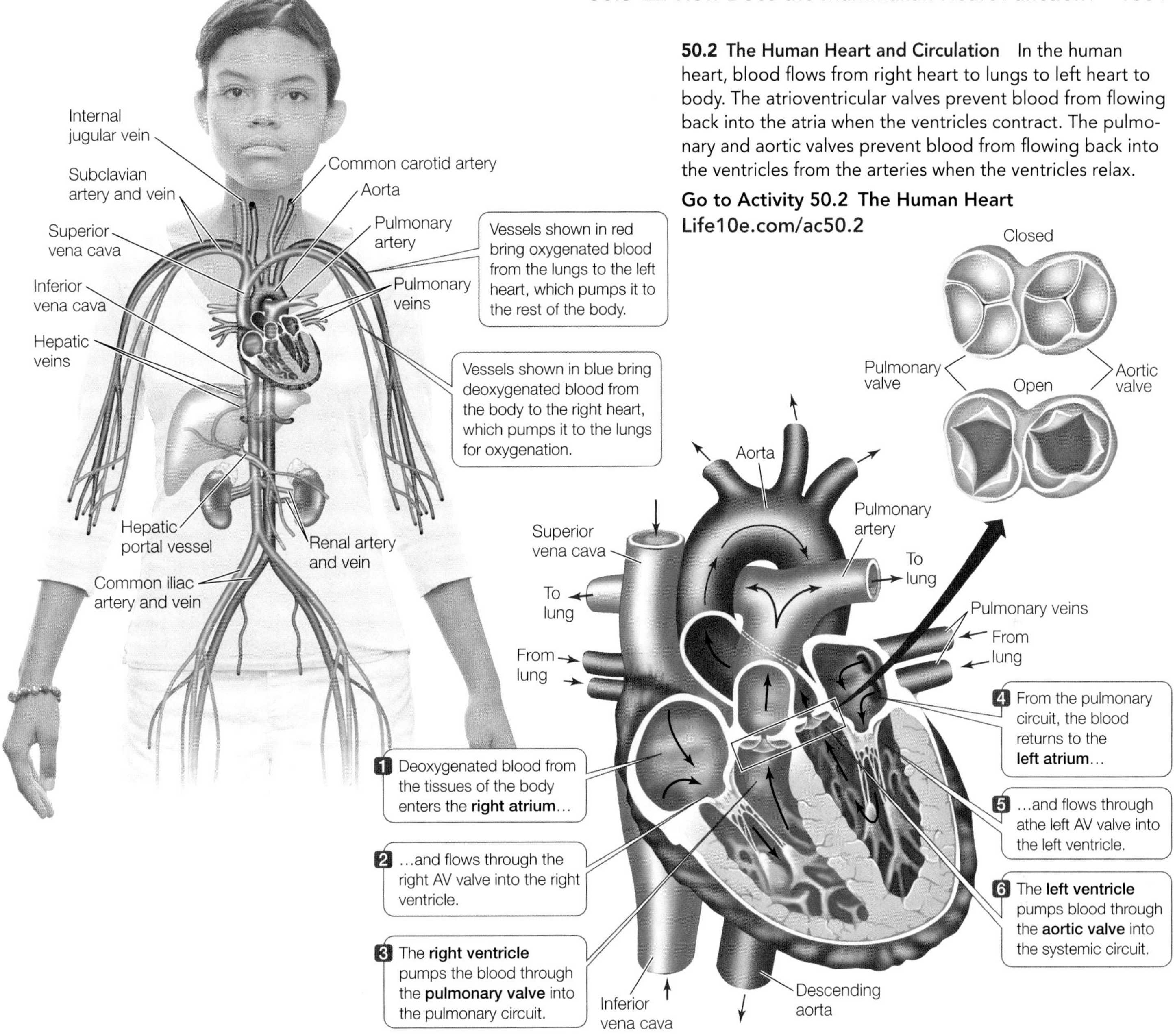

**50.2 The Human Heart and Circulation** In the human heart, blood flows from right heart to lungs to left heart to body. The atrioventricular valves prevent blood from flowing back into the atria when the ventricles contract. The pulmonary and aortic valves prevent blood from flowing back into the ventricles from the arteries when the ventricles relax.

**Go to Activity 50.2 The Human Heart Life10e.com/ac50.2**



ventricular volume. The right ventricle then contracts, causing the AV valve to close and pumping the blood into the **pulmonary artery** leading to the lungs.

**Pulmonary veins** return the oxygenated blood from the lungs to the left atrium, from which the blood enters the left ventricle through the left AV valve. As on the right side of the heart, most left ventricular filling is passive, but ventricular filling reaches a maximum when the atria contract.

The walls of the left ventricle are powerful muscles that contract around the blood with a wringing motion starting from the bottom. When pressure in the left ventricle is high enough to push open the aortic valve, blood rushes into the aorta to begin its circulation throughout the body. In Figure 50.2, observe that the walls of the left ventricle are thicker than those of the right ventricle. The left ventricle has to propel blood through many more kilometers of blood vessels than does the right ventricle, and must therefore push against more resistance, even though both ventricles pump the same volume of blood.

Both sides of the heart contract at the same time. Contraction of the two atria, followed by contraction of the two ventricles and then relaxation, is the **cardiac cycle**. The cardiac cycle is divided into two phases: **systole** (pronounced sís-toll-ee), when the ventricles contract, and **diastole** (die-ás-toll-ee), when the ventricles relax (**Figure 50.3**). At the very end of diastole (step 1 in Figure 50.3), just before the ventricles contract, the atria contract and top off the volume of blood in the ventricles.

The sounds of the cardiac cycle—the "lub-dup" heard through a stethoscope—are created by the heart valves slamming shut. The closing and opening of these valves are simple mechanical events resulting from pressure differences on the two sides of the valves. As the ventricles begin to contract (step 2 in Figure 50.3), the pressure in them rises above the pressure in the atria, so the

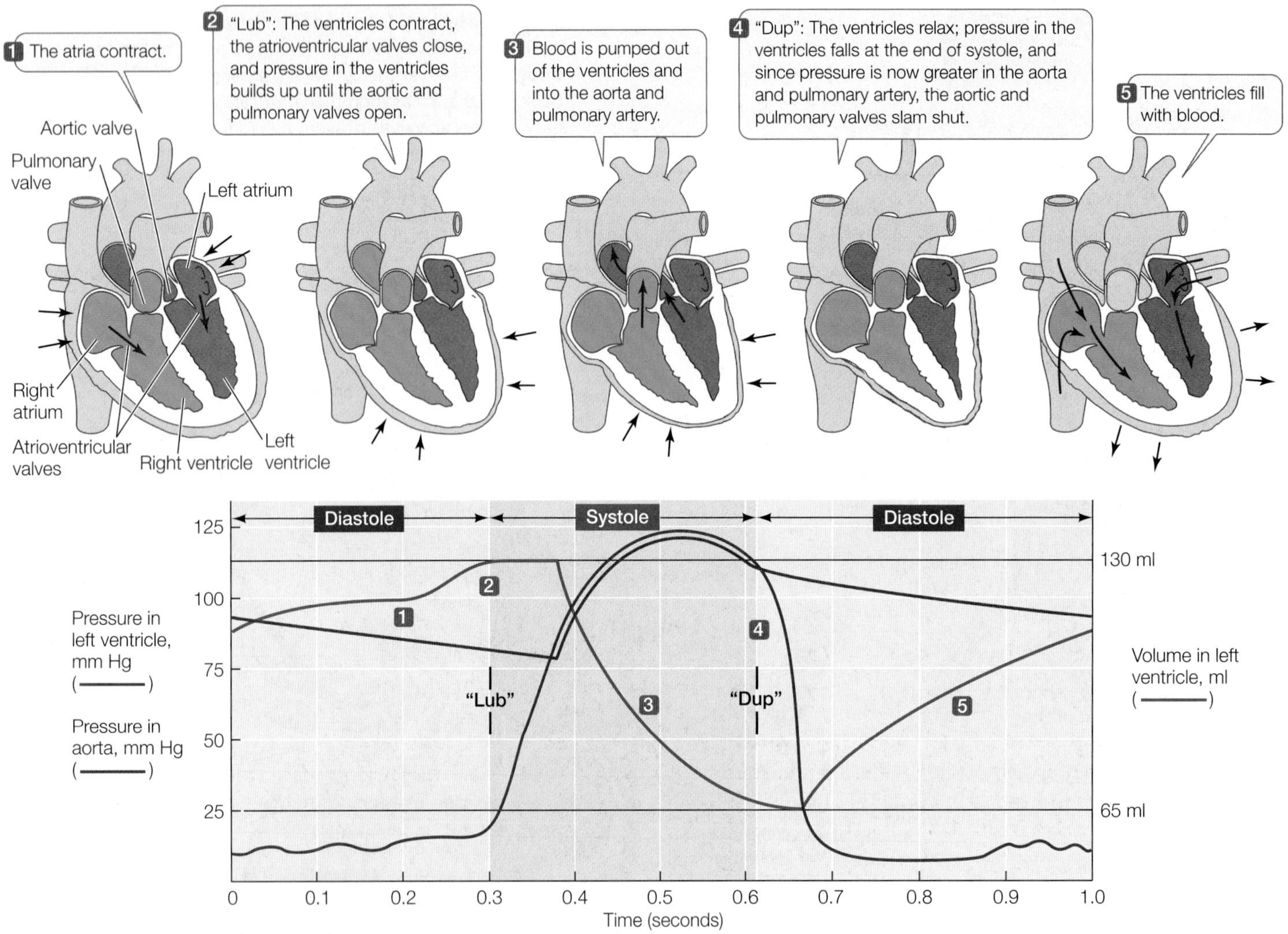

1 The atria contract.

2 "Lub": The ventricles contract, the atrioventricular valves close, and pressure in the ventricles builds up until the aortic and pulmonary valves open.

3 Blood is pumped out of the ventricles and into the aorta and pulmonary artery.

4 "Dup": The ventricles relax; pressure in the ventricles falls at the end of systole, and since pressure is now greater in the aorta and pulmonary artery, the aortic and pulmonary valves slam shut.

5 The ventricles fill with blood.

Aortic valve
Pulmonary valve
Left atrium
Right atrium
Atrioventricular valves
Right ventricle
Left ventricle

Diastole
Systole
Diastole
Pressure in left ventricle, mm Hg (———)
Pressure in aorta, mm Hg (———)
Volume in left ventricle, ml (———)
130 ml
65 ml
"Lub"
"Dup"
125
100
75
50
25
0 0.1 0.2 0.3 0.4 0.5 0.6 0.7 0.8 0.9 1.0
Time (seconds)

**50.3 The Cardiac Cycle** The rhythmic contraction (systole) and relaxation (diastole) of the ventricles is called the cardiac cycle. The representation below shows pressure and volume changes during the cardiac cycle for the left ventricle only.

**Go to Animated Tutorial 50.1**
**The Cardiac Cycle**
Life10e.com/at50.1

AV valves close ("lub"). When the ventricles begin to relax (step 4 in Figure 50.3), the high pressure in the aorta and pulmonary artery closes the aortic and pulmonary valves ("dup").

Defective valves that do not close completely produce turbulent blood flow and the sounds known as heart murmurs. For example, if an AV valve does not close completely, blood will flow back into the atrium with a "whoosh" at the beginning of systole.

Blood pressure changes associated with the cardiac cycle can be measured in the large artery in your arm by using an inflatable pressure cuff and a pressure gauge, together called a sphygmomanometer, and a stethoscope (**Figure 50.4**). This method measures the minimum pressure necessary to compress an artery so blood does not flow through it at all (the systolic value) and the minimum pressure that causes intermittent flow through the artery (the diastolic value). A conventional blood pressure reading is expressed as the systolic value placed over the diastolic value. Healthy values for a young adult might be 120 millimeters of mercury (mm Hg) during systole and 70 mm Hg during diastole, or 120/70.

## The heartbeat originates in the cardiac muscle

Cardiac muscle has unique adaptations that enable it to function as a pump. First, cardiac muscle cells are in electrical contact with one another through gap junctions, which enable action potentials (see Section 45.2) to spread rapidly from cell to cell. Because a spreading action potential stimulates contraction, large groups of cardiac muscle cells contract in unison. This coordinated contraction is essential for pumping blood effectively.

Second, some cardiac muscle cells are **pacemaker cells** that can initiate action potentials without stimulation from the nervous system. When they fire action potentials, they stimulate neighboring cells to contract. The primary pacemaker of the heart is a group of modified cardiac muscle cells, the **sinoatrial node**, located at the junction of the superior vena cava and right atrium (see Figure 50.7). The resting membrane potentials of these cells are less negative than those of other cardiac muscle cells and are not stable; instead they gradually become even less negative until they reach threshold for initiating an

1 The cuff is inflated beyond the point that shuts off all blood flow.

2 Pressure in the cuff is gradually lowered until the sound of a pulsing flow of blood through the constriction in the artery is heard. At this time, pressure in the cuff is just below the peak **systolic pressure** in the artery.

3 Pressure is further lowered until the sound becomes continuous. At this time, the cuff is just below the **diastolic pressure** in the artery. This person's blood pressure is 120/70.

120 200 160 80 240 70 20 300

Pulsing sounds

120 200 160 80 240 70 20 300

Pulsing sound gives way to smooth "whoosh" of blood flow

**50.4 Measuring Blood Pressure** Blood pressure in the major artery of the arm can be measured with a device called a sphygmomanometer, which combines an inflatable cuff and a pressure gauge. A stethoscope is also used to detect sounds created by the blood vessels in response to changes in pressure during the cardiac cycle.

action potential. The action potentials of pacemaker cells are very different from those of neurons and other muscle cells (see Figure 45.10). They are slower to rise; they are broader; and they are slower to return to resting potential (**Figure 50.5A**). These properties of pacemaker cells are due to the ion channels in their membranes.

Pacemaker potentials involve $Na^+$, $Ca^{2+}$, and $K^+$ channels (**Figure 50.5B**). As we discussed in Section 45.2, when $Na^+$ or $Ca^{2+}$ channels open, positive charges flow into the cell and the membrane potential becomes less negative. When $K^+$ channels open, positive charges flow out of the cell and the membrane potential becomes more negative. Because the $Na^+$ channels of pacemaker cells are open more of the time than are those of other cardiac muscle cells, the pacemaker resting potential is less negative. The action potential of pacemaker cells is due to voltage-gated $Ca^{2+}$ channels rather than voltage-gated $Na^+$ channels as in neurons, skeletal muscle, and other cardiac muscle cells. These $Ca^{2+}$ channels open and close more slowly than voltage-gated $Na^+$ channels, explaining the shape of pacemaker action potentials.

The unstable resting potential of pacemaker cells is due to the behavior of cation channels. As in neurons and skeletal muscle cells, there are voltage-gated $K^+$ channels that slowly open on the rising phase of the action potential. The opening of these channels allows $K^+$ ions to leave the cell and thereby restore the negative charge on the cell membrane. That restoration of a negative membrane potential causes the opening of a unique class of voltage-gated cation channels that mostly conduct $Na^+$. At the same time, the voltage-gated $K^+$ channels that opened during the action potential slowly close. The result is that there are more $Na^+$ ions coming into the cell than there are $K^+$ ions leaving, and the cell membrane potential gradually becomes less negative (see Figure 50.5).

The gradual rise in membrane potential closes the channels that allow $Na^+$ to move into the cell, but as the membrane

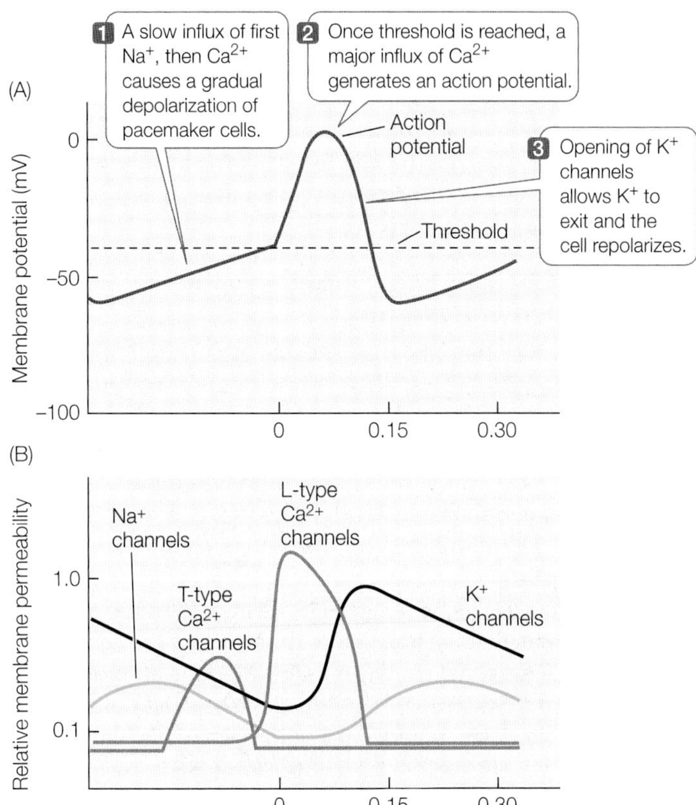


**50.5 The Pacemaker Potential** (A) The resting potential of the pacemaker cells of the sinoatrial node is less negative and gradually drifts upward between action potentials. Action potentials are slow to rise and are broad. These characteristics are due to properties of $Na^+$, $K^+$, and two types of $Ca^{2+}$ channels. (B) At rest, sinoatrial pacemaker cells are more permeable to $Na^+$ than are neurons or other muscle cells. During rest, the cation channels gradually close, but some voltage-gated $Ca^{2+}$ channels open; these are designated T-type channels because their opening is transient. When a threshold is reached, other voltage-gated $Ca^{2+}$ channels open (designated as L-type for long-lasting), generating the action potential.

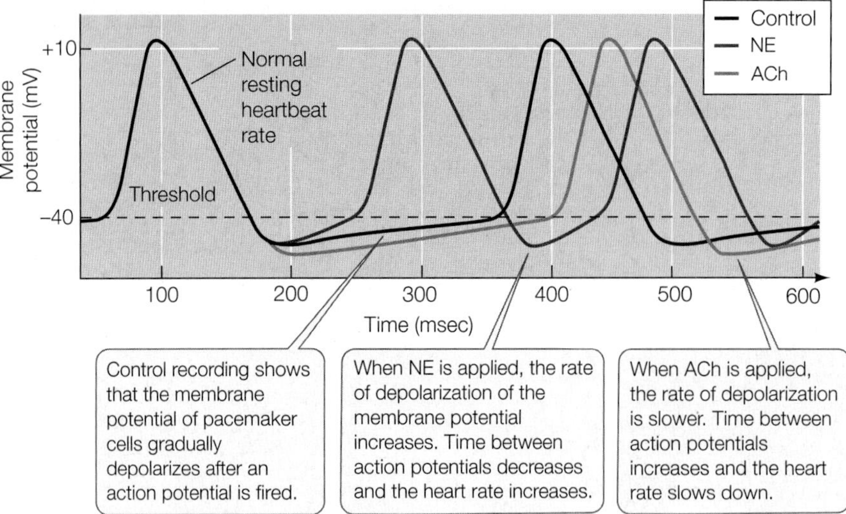


**50.6 The Autonomic Nervous System Controls Heart Rate** Isolated pacemaker cells continue firing action potentials in a culture dish. Neurotransmitter signals from the two divisions of the autonomic nervous system speed up and slow down the rate at which the pacemaker membrane potential drifts upward, thereby controlling the rate at which pacemaker cells fire action potentials.

becomes less negative, some $Ca^{2+}$ channels open, causing the membrane potential to continue its gradual rise. Eventually this rising membrane potential reaches threshold for the major voltage-gated $Ca^{2+}$ channels, and another action potential is generated. The intricate interaction of these ion channels through their effects on membrane potential causes the rhythmic generation of action potentials that characterizes pacemaker cells.

The autonomic nervous system controls the heartbeat (speeds it up or slows it down) by influencing the rate at which the resting potentials of pacemaker cells drift upward (**Figure 50.6**). Norepinephrine (NE) released onto pacemaker cells by sympathetic nerves increases the permeability of the $Na^+$ channels and the $Ca^{2+}$ channels. The result is that the resting potential of the pacemaker cells drifts up more rapidly, the interval between action potentials is decreased, and the heart beats faster. Conversely, the parasympathetic neurotransmitter acetylcholine (ACh) has opposite effects. ACh increases the permeability of $K^+$ channels so that the membrane potential becomes even more negative following an action potential and rises more slowly. ACh also decreases the permeability of the $Ca^{2+}$ channels so that the rate of rise of the membrane potential slows, the interval between pacemaker action potentials lengthens, and the heart slows down.

## A conduction system coordinates the contraction of heart muscle

A normal heartbeat begins with an action potential in the sinoatrial node (**Figure 50.7**). This action potential spreads rapidly throughout the electrically coupled cells of the atria, causing them to contract in unison. Because there are no gap junctions between the cells of the atria and those of the ventricles, the action potential does not spread directly to the ventricles. Therefore the ventricles do not contract in unison with the atria.

How does the action potential move from the atria to the ventricles? Situated at the junction of the atria and the ventricles is a nodule of modified cardiac muscle cells—the **atrioventricular node**—which is stimulated by the depolarization of the atria. With a slight delay, it generates action potentials that are conducted to the ventricles via the **bundle of His**, which consists of modified cardiac muscle fibers that do not contract but do conduct action potentials. These fibers divide into right and left bundle branches that run to the tips of the ventricles and then spread throughout the ventricular muscle mass as **Purkinje fibers**. These conducting fibers ensure that the cardiac action potential spreads rapidly and evenly throughout the ventricular muscle mass, starting at the very bottom of the ventricles. The short delay in the spread of the action potential imposed by the atrioventricular node ensures that the atria contract before the ventricles do, so that the blood passes progressively from the atria to the ventricles to the arteries.

## Electrical properties of ventricular muscles sustain heart contraction

Electrical properties of ventricular muscle fibers allow them to contract for about 300 milliseconds—much longer than those of skeletal muscle fibers. As in neuronal and skeletal muscle action potentials, the rising phase of the ventricular muscle cell action potentials is due to the opening of voltage-gated $Na^+$ channels. Unlike neurons and skeletal muscle fibers, however, ventricular muscle cells remain depolarized for a long time. This extended plateau of the action potential is due to sustained opening of voltage-gated $Ca^{2+}$ channels (**Figure 50.8**). Like other muscle, cardiac muscle is stimulated to contract when $Ca^{2+}$ is available to bind with troponin (see Figure 48.6). As long as $Ca^{2+}$ remains in the sarcoplasm, the ventricular muscle cells continue to contract.

To terminate systole and allow the ventricles to fill again, $Ca^{2+}$ must be rapidly cleared from the sarcoplasm of the ventricular cells. $Ca^{2+}$ pumps in the sarcoplasmic reticulum membrane actively transport $Ca^{2+}$ ions out of the sarcoplasm and into the sarcoplasmic reticulum; there, the ions are sequestered until the next action potential triggers another round of $Ca^{2+}$ release and muscle contraction. Thus the rate of cycling of $Ca^{2+}$ into and out of the sarcoplasmic reticulum puts limits on the heart rate and strength of contraction of the ventricle (**Figure 50.9**).

The drug digitalis has been used since the late 1700s to treat weakened hearts or hearts with irregular patterns of contraction. Digitalis strengthens and slows the heartbeat by slowing the reuptake of $Ca^{2+}$ by the sarcoplasmic reticulum and thereby increasing the concentration of $Ca^{2+}$ in the sarcoplasm. Before being introduced into the practice of medicine, digitalis prepared from the purple foxglove plant (*Digitalis purpurea*) was a folk remedy for heart problems.

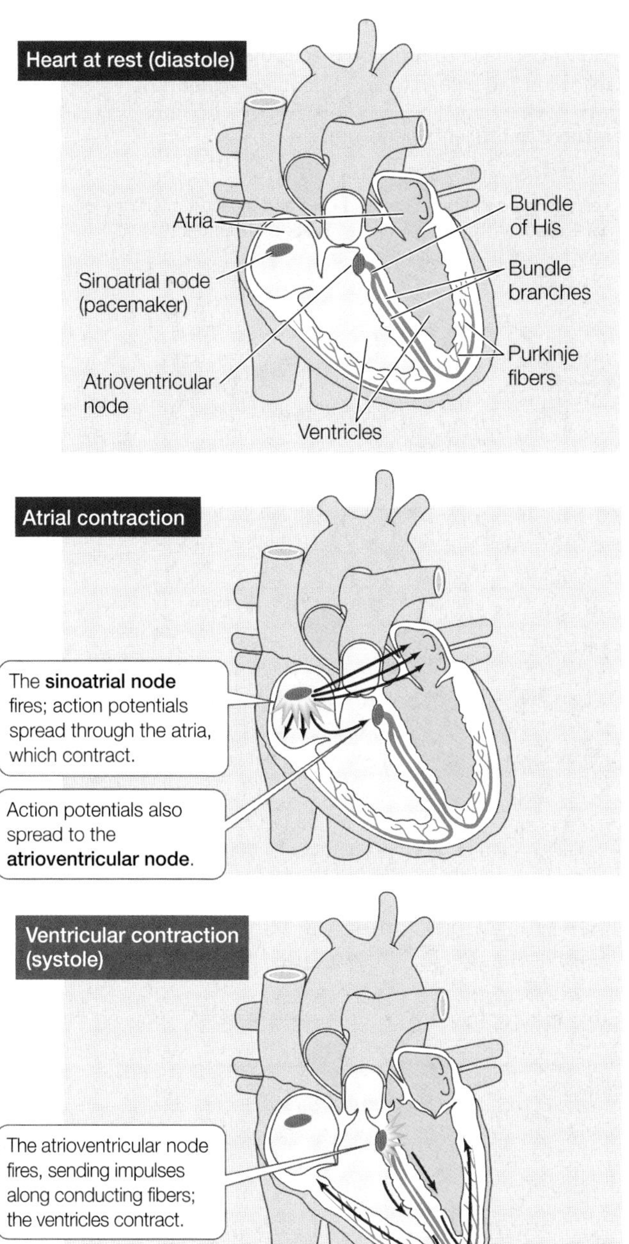


**50.7 The Heartbeat** Pacemaker cells in the sinoatrial node initiate the heartbeat by firing action potentials that spread through the electrically coupled atrial muscle. The atrial action potential eventually spreads to the atrioventricular node which, with a delay, conducts it through the bundle of His and Purkinje fibers to the cells of the ventricles.

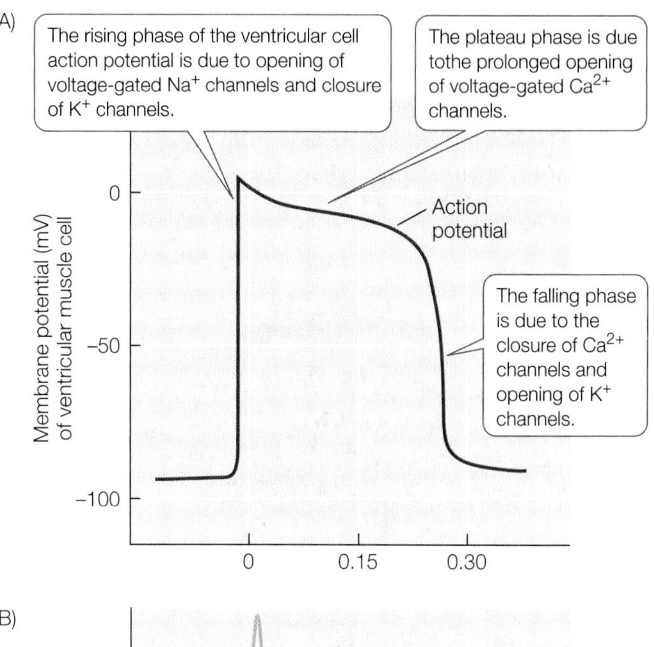


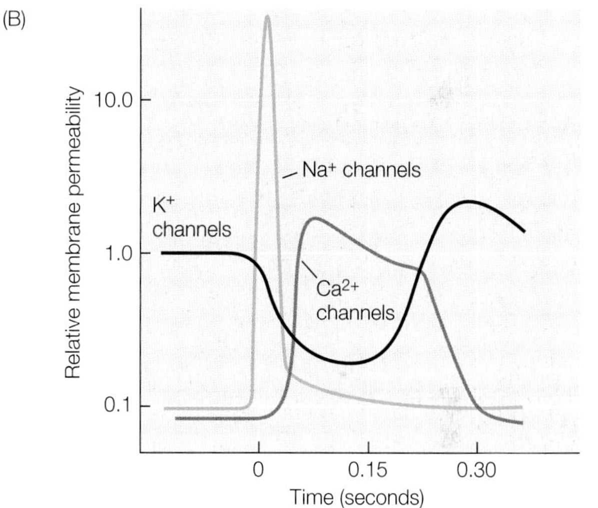


**50.8 The Action Potential of Ventricular Muscle Fibers** (A) The three phases of the action potential of ventricular muscle fibers are due to the opening and closing of voltage-gated channels. (B) At the initiation of the action potential, voltage-gated Na+ channels open rapidly but briefly. At the same time, but more slowly, K+ channels are closing and $Ca^{2+}$ channels are opening. The open $Ca^{2+}$ channels sustain the depolarization. Repolarization occurs when the $Ca^{2+}$ channels close, and the slow opening of the $K^+$ channels also contribute to the repolarization.

## The ECG records the electrical activity of the heart

Electrical events in the cardiac muscle during the cardiac cycle can be recorded by electrodes placed on the surface of the body. Such a recording is an **electrocardiogram**, or **ECG**. **EKG** is also used because German physicians who invented the method used the Greek spelling (*kardia*) and called it the elektrokardiogramm. The ECG is an important tool for diagnosing heart problems (**Figure 50.10A**).

The action potentials that sweep through the muscles of the atria and ventricles before they contract are such massive, localized electrical events that they cause electric currents to flow throughout the body. Electrodes placed at different locations on the skin detect those currents at different times and register a voltage difference between them. The appearance of the ECG depends on the placement of the electrodes. Electrodes placed on the right wrist and left ankle produced the normal ECG shown in **Figure 50.10B**. The wave patterns of the ECG are designated P, Q, R, S, and T, each letter representing a particular event in the cardiac muscle, as shown in the figure.

## INVESTIGATINGLIFE

**50.9 Hot Fish, Cold Heart** In "hot" fish like the bluefin tuna, highly active tissues such as the swimming muscles remain warm while the heart temperature falls (see Figure 40.14). How can a "cold heart" meet the metabolic demands of active muscles? Barbara Block and her colleagues demonstrated that the sarcoplasmic reticulum in the heart muscle cells of a bluefin takes up $Ca^{2+}$ much more rapidly after each contraction than is the case for "cold" fish species, allowing the bluefin heart to beat at a faster rate.[a]

**HYPOTHESIS** The bluefin tuna can maintain a fast heart rate at low temperatures because its heart muscle cells cycle $Ca^{2+}$ more rapidly than in typical fishes.

***Method*** Isolate membrane-enclosed vesicles derived from sarcoplasmic reticulum (SR) from hearts of bluefin (a "hot" fish) and from albacore and yellowfin tuna ("cold" fish). Incubate the SR vesicles in a solution containing $Ca^{2+}$ and measure their rate of $Ca^{2+}$ uptake. Repeat the experiment at a range of temperatures.

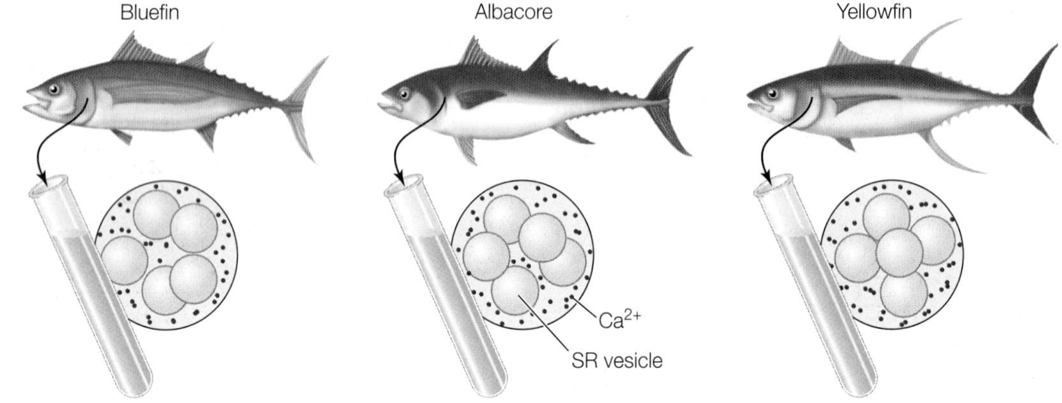


***Results*** Over a wide range of temperatures, SR vesicles from bluefin take up $Ca^{2+}$ more rapidly than do those from the other fishes.

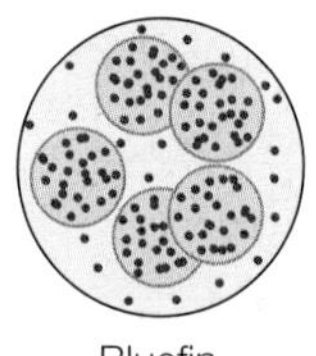


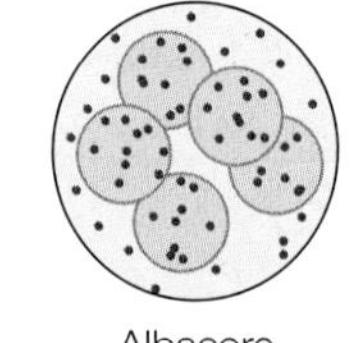


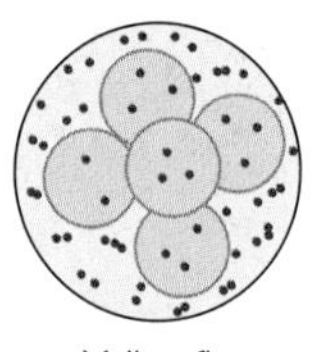


**CONCLUSION** The sarcoplasmic reticulum in bluefin heart muscle cycles $Ca^{2+}$ more rapidly than in other tuna species, allowing the heart to beat faster at cold temperatures.

Go to **BioPortal** for discussion and relevant links for all INVESTIGATING**LIFE** figures.

[a]Landeira-Fernandez, A. et al. 2004. *American Journal of Physiology: Regulatory, Integrative, and Comparative Physiology* 286: R398–R404.

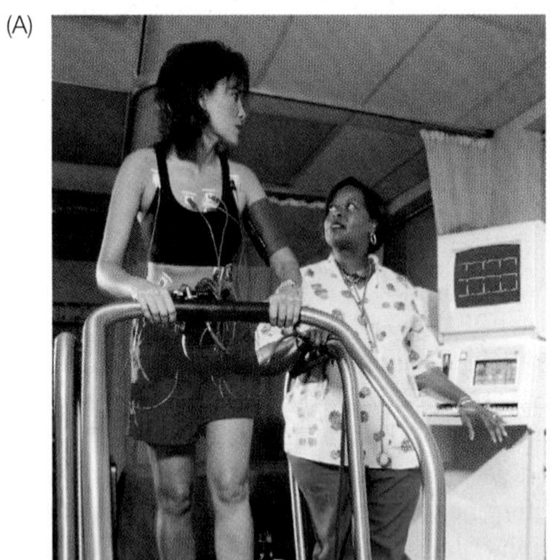


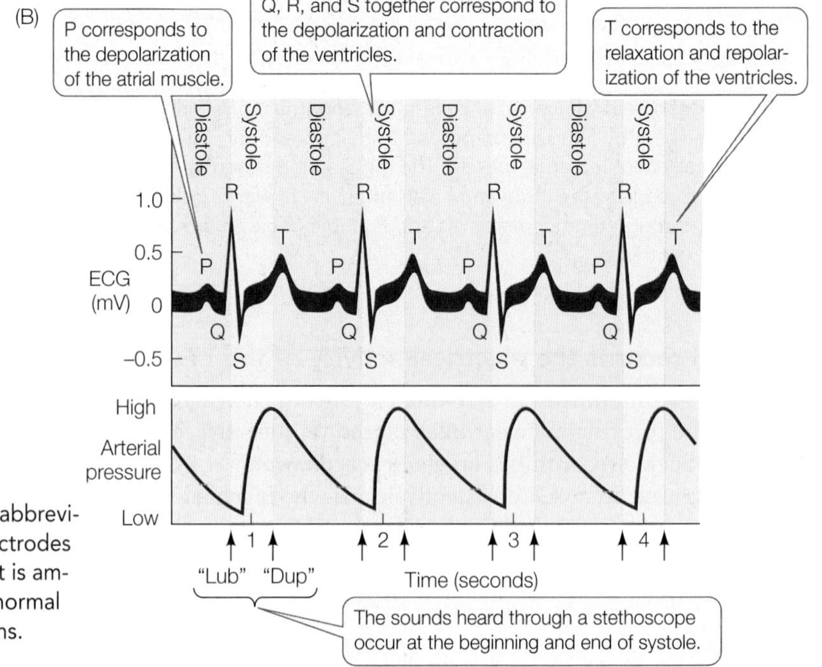


**50.10 The Electrocardiogram** (A) An electrocardiogram (abbreviated as ECG or EKG) is used to monitor heart function. Electrodes attached to the person on the treadmill record an ECG that is amplified and displayed on a monitor. (B) Variations from the normal pattern shown here can be used to diagnose heart problems.

WORKING WITH**DATA:**

## *Warm Fish with Cold Hearts*

### Original paper

Landeira-Fernandez, A., M. M. Morrissette, J. M. Blanc, and B. A. Block. 2004. Temperature dependence of the $Ca^{2+}$-ATPase (SERCA2) in the ventricles of tuna and mackerel. *American Journal of Physiology: Regulatory, Integrative, and Comparative Physiology* 286: R398–R404.

### Analyze the Data

"Hot fish" like the bluefin tuna (see Figure 40.14) face a dilemma when swimming in cold waters. As in all fish, the rate at which their hearts can beat goes down as the heart temperature goes down, which lowers the supply of oxygenated blood to the tissues of the body. But, unlike in "cold fish," the swimming muscles in bluefin tuna remain warm even in cold waters, so their demand for oxygen remains high. How does a bluefin tuna maintain an adequate blood supply to the swimming muscles at cold temperatures? The heart muscle of the bluefin tuna is adapted to cycle $Ca^{2+}$ more rapidly than the heart muscle of "cold fish," allowing the heart to contract more frequently. Barbara Block and her colleagues hypothesized that this adaptation could be due to an increased ability to sequester $Ca^{2+}$ back into the sarcoplasmic reticulum after each contraction (see Figure 50.9). The researchers isolated vesicles derived from the sarcoplasmic reticulum of heart muscle from three different fish: one "hot fish" (bluefin tuna) and two closely related "cold fish" (albacore and yellowfin tuna). The vesicles were incubated in a $Ca^{2+}$ solution and the rate of $Ca^{2+}$ uptake into the vesicles measured at a range of temperatures. The results are shown in the top panel of the figure at the right.

**QUESTION 1**

How do these data support the hypothesis that bluefin tuna have an increased ability to sequester $Ca^{2+}$ in the sarcoplasmic reticulum? How would these data look if the hypothesis weren't true?

**QUESTION 2**

Based on these data, which fish would you expect to be able to achieve the highest cardiac output at 25°C, albacore or yellowfin? Albacore at 25°C or bluefin at 15°C?

**QUESTION 3**

The high rate of $Ca^{2+}$ uptake observed in the bluefin vesicles means that the $Ca^{2+}$/ATPase pumps in the sarcoplasmic reticulum of the bluefin must be either different from those of other fish *or* more numerous. To distinguish between these two possibilities, the researchers plotted their data in a different way; this time they expressed the uptake rates as a percentage of the maximum rate (the rate at 30°C). This plot is shown in the bottom panel of the figure. What does this plot tell you about the properties of the $Ca^{2+}$/ATPase pumps in the different fish? What is the explanation for the higher rate of $Ca^{2+}$ uptake in bluefin tuna?

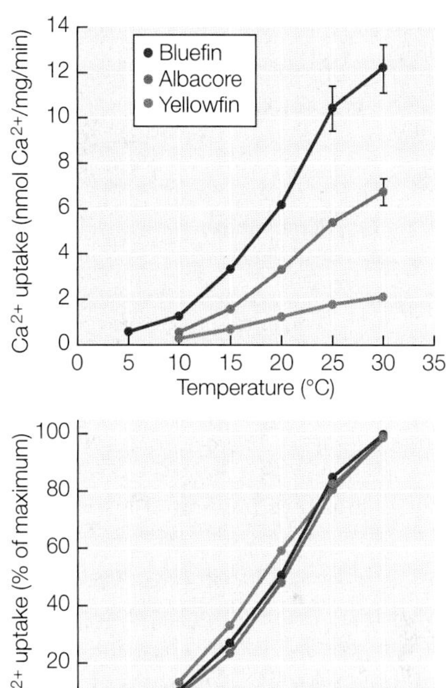


Go to BioPortal for all WORKING WITHDATA exercises

### RECAP 50.3

The mammalian heart has two atria and two ventricles. Modified cardiac muscle tissue in the right atrium functions to spontaneously generate pacemaker action potentials. Other modified cardiac muscle tissue between the atria and ventricles and throughout the ventricles conducts those signals and coordinates the heart contraction. Broad action potentials in ventricular muscle reflect $Ca^{2+}$ cycling in the ventricular muscle cells and make sustained contractions possible.

- Trace the path of blood through both sides of the heart, naming the major blood vessels and heart valves. **See pp. 1030–1031 and Figure 50.2**
- Differentiate systole and diastole and describe the events of the cardiac cycle. **See p. 1031 and Figure 50.3**
- How do cells of the sinoatrial node generate the heartbeat? **See pp. 1032–1034 and Figures 50.5 and 50.7**
- What determines the duration of the contraction of the ventricles during systole? **See p. 1034 and Figure 50.8 and 50.9**

Next we will consider the composition of the blood and the characteristics of the vessels through which blood circulates around the body, illustrating once again how structure serves function. We will also consider the role of the lymphatic vessels that return interstitial fluid to the blood.

## 50.4 What Are the Properties of Blood and Blood Vessels?

Blood is a connective tissue. It consists of cells suspended in an extracellular matrix of complex, yet specific, composition. The unusual feature of blood is that the extracellular matrix is a liquid, so blood is a fluid tissue.

The cells of the blood can be separated from the fluid matrix, called **plasma**, by centrifugation (**Figure 50.11**). If a sample of blood is spun in a centrifuge, all the cells move to the bottom of the tube, leaving the clear, straw-colored plasma on top. The packed-cell volume, or hematocrit, is the percentage of the

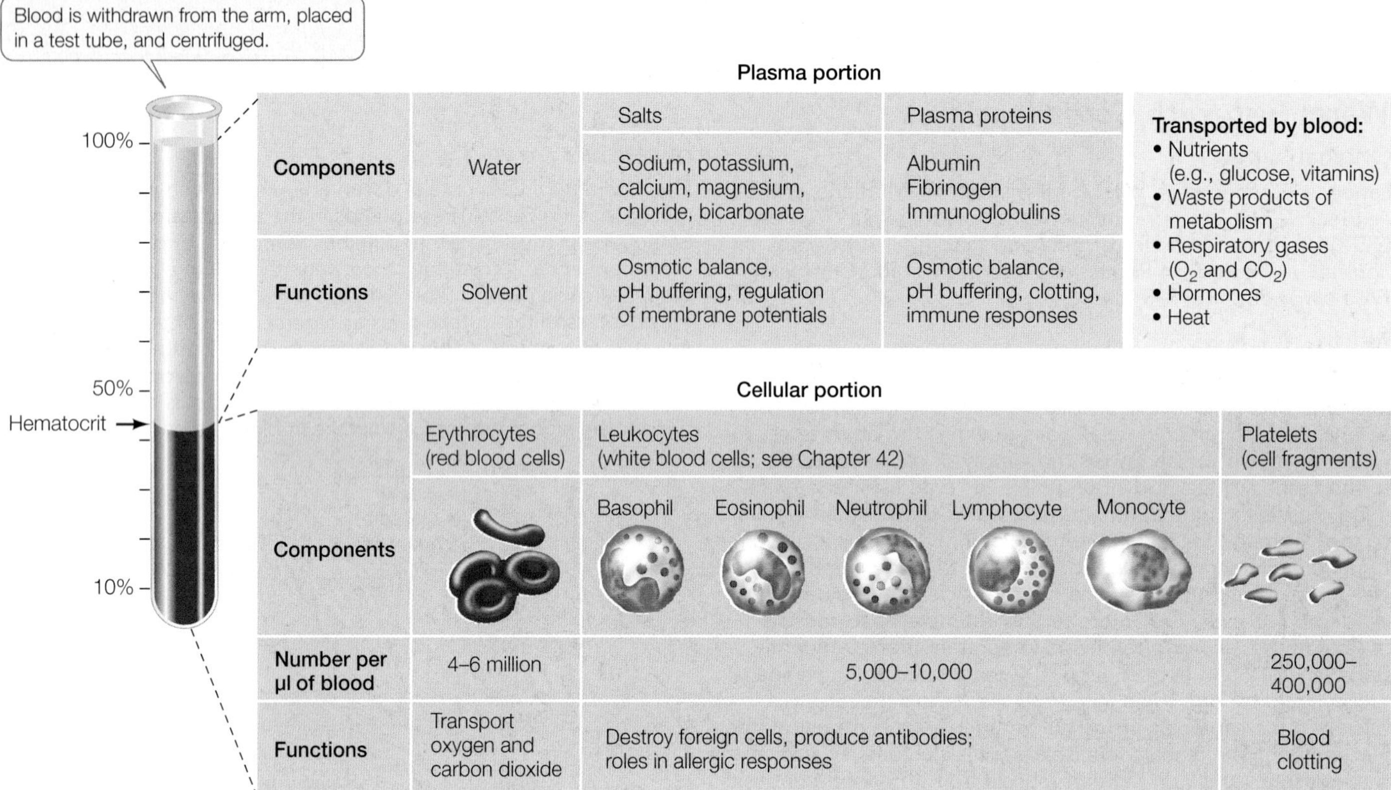


**50.11 The Composition of Blood** Blood consists of a complex aqueous solution (the plasma) and numerous cell types and cell fragments. The hematocrit (arrow) is a measure of the cellular portion as a percentage of total blood volume.

blood volume made up by red blood cells. Normal hematocrit is about 42 percent for women and 46 percent for men, but these values can vary considerably. They are usually higher, for example, in people who live and work at high altitudes, because the low oxygen concentrations there stimulate the production of more red blood cells.

Here we will consider two elements in blood: the red blood cells and the platelets, which are pinched-off fragments of cells. White blood cells, or leukocytes, are the cells of the immune system, discussed in Chapter 42.

## Red blood cells transport respiratory gases

Most of the cells in the blood are **erythrocytes**, or red blood cells. Mature red blood cells are biconcave, flexible discs packed with hemoglobin. Their function is to transport respiratory gases. Their shape gives them a large surface area for gas exchange, and their flexibility enables them to squeeze through narrow capillaries. Men have 4.5 to 6.0 million red blood cells per microliter of blood, and women have 3.5 to 5.0 million.

Red blood cells, as well as all the other cellular components of blood, are generated by stem cells in the bone marrow, particularly in the ribs, breastbone, pelvis, vertebrae, and the long bones of the limbs. Red blood cell production is controlled by a hormone, **erythropoietin**, which is released by cells in the kidneys in response to insufficient oxygen—**hypoxia**. Many tissues respond to hypoxia by expressing a transcription factor called hypoxia-inducible factor 1 (HIF-1). When the kidneys become hypoxic and express HIF-1, one of the actions of the transcription factor is to activate the gene encoding erythropoietin. Increased circulating erythropoietin extends the lives of mature red blood cells and stimulates production of new red blood cells in the bone marrow.

Under normal conditions, your bone marrow produces about 2 million red blood cells every second. Developing red blood cells divide many times while still in the bone marrow, and during this time they produce hemoglobin. When hemoglobin makes up about 30 percent of the immature red blood cell, the nucleus, endoplasmic reticulum, Golgi apparatus, and mitochondria begin to break down. This process is almost complete when the newly mature red blood cell squeezes between the endothelial cells of blood vessels in the bone marrow and enters the circulation. Loss of nuclei from the red blood cells occurs in most mammalian species, but the red cells of a few mammals and of all other vertebrates are nucleated.

Each red blood cell circulates for about 120 days. As it gets older, its membrane becomes less flexible and more fragile, so older red blood cells are more likely to rupture as they bend to fit through narrow capillaries. Red blood cells are particularly squeezed in the **spleen**, an organ that sits near the stomach in the upper left side of the abdominal cavity. The spleen has many sinuses (cavities) that serve as reservoirs for red blood cells. To get into the sinuses, however, the red blood cells must squeeze between spleen cells. When old red blood cells are ruptured by this squeezing, their remnants are taken up and degraded by macrophages (a class of white blood cells that ingest debris and foreign materials).

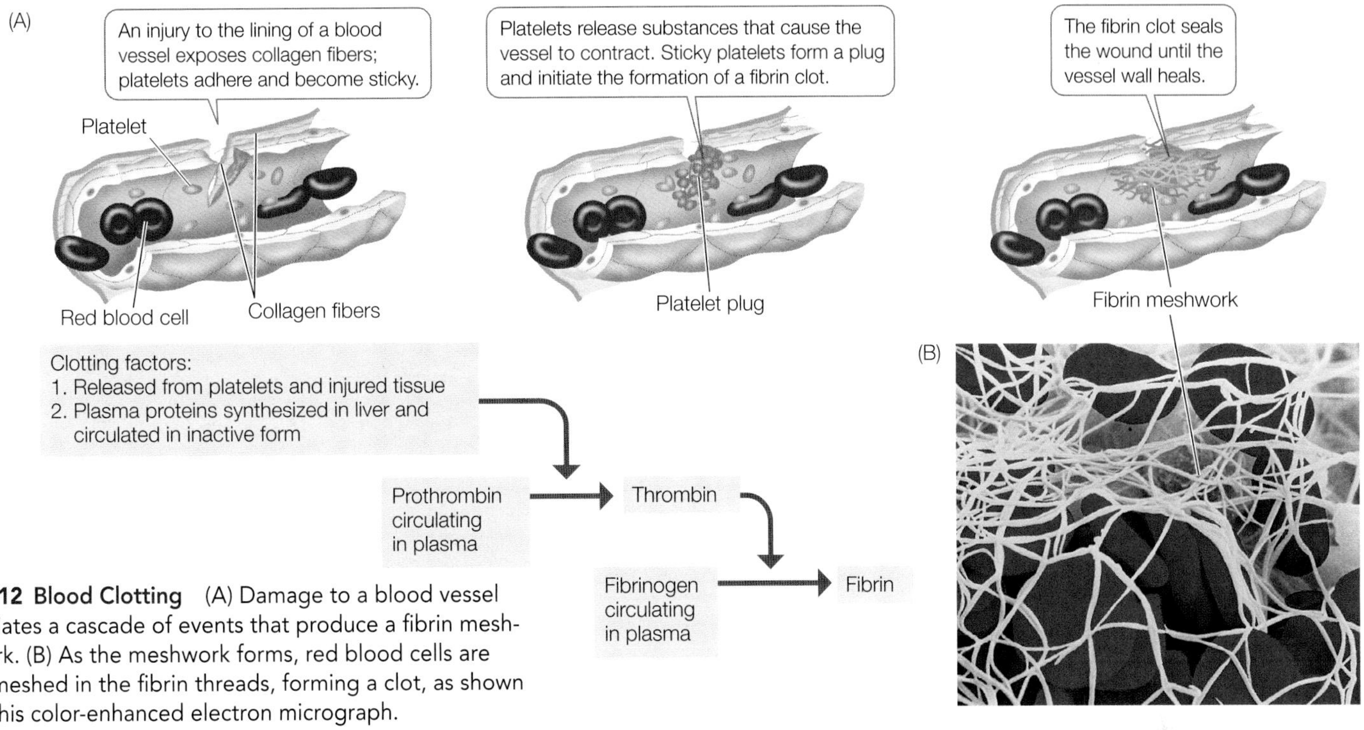

**50.12 Blood Clotting** (A) Damage to a blood vessel initiates a cascade of events that produce a fibrin meshwork. (B) As the meshwork forms, red blood cells are enmeshed in the fibrin threads, forming a clot, as shown in this color-enhanced electron micrograph.

### Platelets are essential for blood clotting

Besides producing erythrocytes and leukocytes, the bone marrow stem cells described in Section 42.1 also produce cells called megakaryocytes. Megakaryocytes are large cells that remain in the bone marrow and release cell fragments called **platelets** into the circulation. A platelet is just a tiny fragment of a cell without cell organelles, but it is packed with enzymes and chemicals necessary for its function: sealing leaks in blood vessels and initiating **blood clotting** (**Figure 50.12**).

Damage to a blood vessel exposes collagen fibers. An encounter with collagen fibers activates a platelet. The platelet swells, becomes irregularly shaped and sticky, and releases chemicals that activate other platelets and initiate the clotting of blood. The sticky platelets also form a plug at the damaged site.

Blood clotting requires many steps and many clotting factors, most of which are circulating in the blood in an inactive form. The absence of any one of these proteins can impair clotting and cause excessive bleeding. Because the liver produces most of the clotting factors, liver diseases such as hepatitis and cirrhosis can result in excessive bleeding. People with hemophilia experience uncontrolled bleeding because of a genetic inability to produce one of the clotting factors.

Blood clotting factors participate in a cascade of chemical activations of other substances circulating in the blood. The cascade begins with blood vessel and other tissue damage that exposes the blood to proteins such as collagen that are normally separated from the blood by endothelial cells lining the blood vessels. This exposure activates platelets and begins the clotting factor cascade. The end result of this cascade is to convert an inactive circulating enzyme, **prothrombin**, to its active form, **thrombin**. Thrombin cleaves molecules of **fibrinogen**, a plasma protein, forming insoluble threads of **fibrin**. The fibrin threads form the meshwork that binds platelets, seals the vessel, and provides a scaffold for the formation of scar tissue (see Figure 50.12).

### Arteries withstand high pressure, arterioles control blood flow

Blood circulates through the vertebrate body in a system of closed vessels, and the properties of those different classes of vessels reflect their functions. The walls of the large arteries have many extracellular collagen and elastin fibers, which enable them to withstand the high blood pressures generated by the heart (**Figure 50.13A**). These elastic tissues have another important function: as we saw with the bulbus arteriosus in fish, they are stretched during systole, and thereby store some of the energy imparted to the blood by the heart. Elastic recoil during diastole returns this energy to the blood by squeezing it and pushing it forward. As a result, even though pressure in the arteries pulsates with the beating of the heart, the flow of blood is smoother than it would be through a system of rigid pipes.

Smooth muscle cells in the walls of the arteries and arterioles constrict or dilate those vessels. When the diameter of the vessels changes, their resistance to blood flow also changes, and the amount of blood flowing through them changes as a result. Neural and hormonal mechanisms act on smooth muscle cells in the walls of the arteries and arterioles, controlling the flow of blood through these vessels. The arterioles are referred to as resistance vessels because their resistance can vary to control the blood flow to specific tissues.

### Materials are exchanged in capillary beds by filtration, osmosis, and diffusion

Beds of capillaries lie between arterioles and venules (**Figure 50.13B**). Few cells are more than a few cell diameters away from a capillary. (Notable exceptions include developing

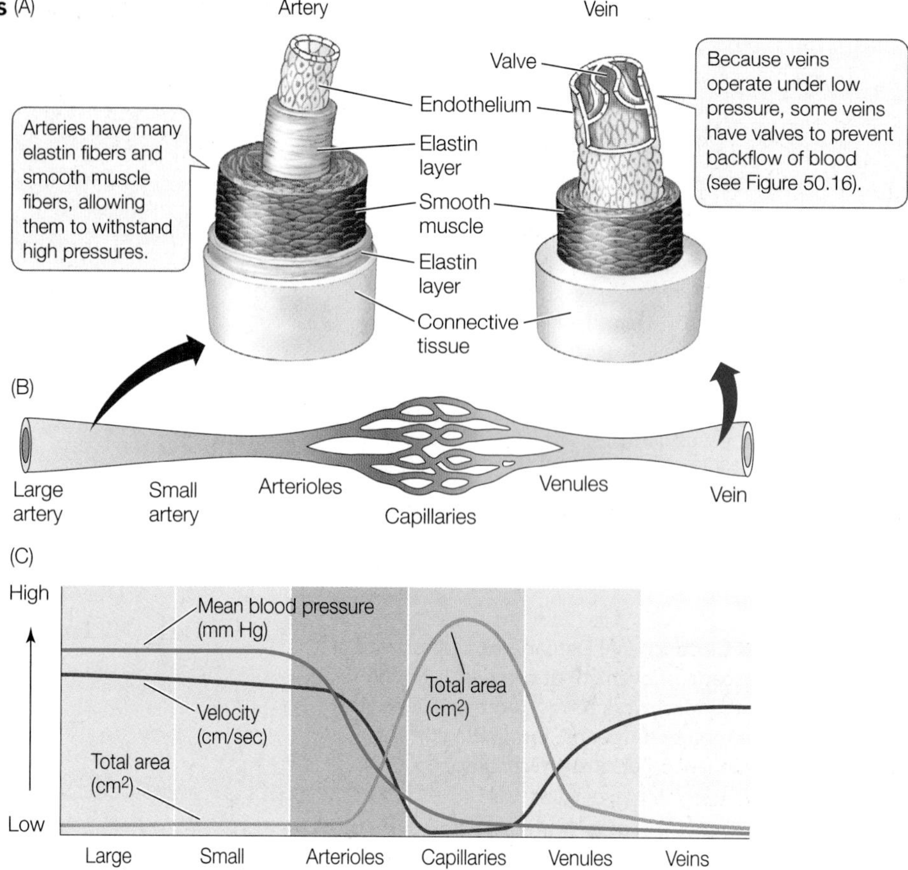


**50.13 Anatomy of Blood Vessels** (A) The different anatomical characteristics of arteries and veins match their functions. (B) Blood from the arterial system feeds into capillary beds, where exchanges with the interstitial fluid occur. The venous system returns the blood to the heart. (C) The area encompassed by each vessel type is graphed along with the pressure and velocity of the blood within them.

**Go to Activity 50.3 Structure of a Blood Vessel Life10e.com/ac50.3**

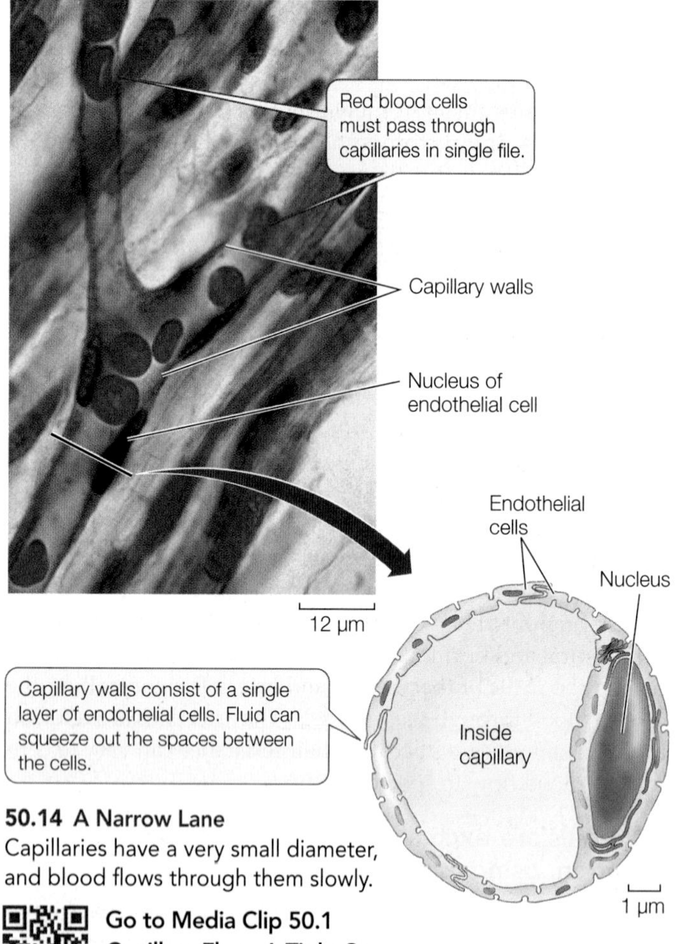


**50.14 A Narrow Lane**
Capillaries have a very small diameter, and blood flows through them slowly.

**Go to Media Clip 50.1 Capillary Flow: A Tight Squeeze Life10e.com/mc50.1**

oocytes and the cells of the lens and cornea.) The cells' needs are served by the exchange of materials between blood and interstitial fluid across the capillary walls. Capillary walls are thin and permeable to water and many solutes. Also, blood flows slowly through capillaries, allowing time for exchange.

It may seem strange that blood flows through the large arteries rapidly at high pressures, but when it reaches the small capillaries the pressure and rate of flow decrease (**Figure 50.13C**). When you place your thumb over the opening of a garden hose, the pressure in the hose increases, which in turn increases the velocity of the water spraying out. But keep in mind that the arteries branch into many arterioles, which give rise to a huge number of capillaries. Even though each capillary has a diameter so small that red blood cells must pass through in single file, there are so many capillaries that their total cross-sectional area is much greater than that of any other class of vessels. As a result, all the capillaries together have a much greater capacity for blood than do the arterioles. Returning to our garden hose analogy, if we connected the hose to large number of lawn sprinklers, the pressure and flow in each sprinkler would be low.

Capillary walls consist of a single layer of endothelial cells (**Figure 50.14**). Capillaries are permeable to water, some ions, and some small molecules but not to large molecules such as proteins. At the arterial (high pressure) end, blood pressure squeezes water and small solutes out through spaces between the cells of the capillary walls into the surrounding intercellular space.

Why don't water and small-molecular-weight solutes collect in the intercellular spaces? How is the blood volume maintained if fluid is continuously leaking out of the capillaries?

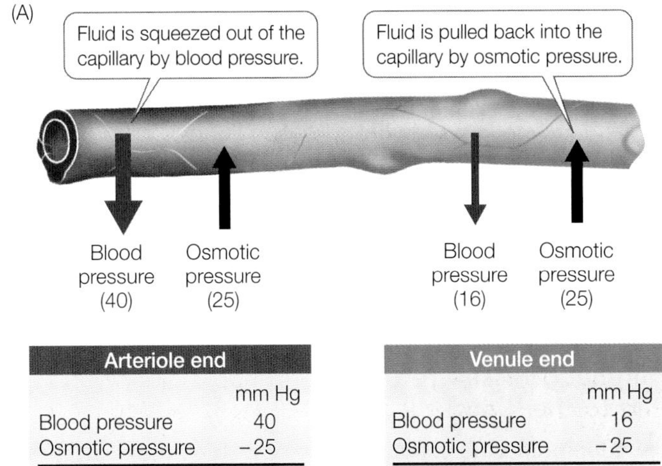


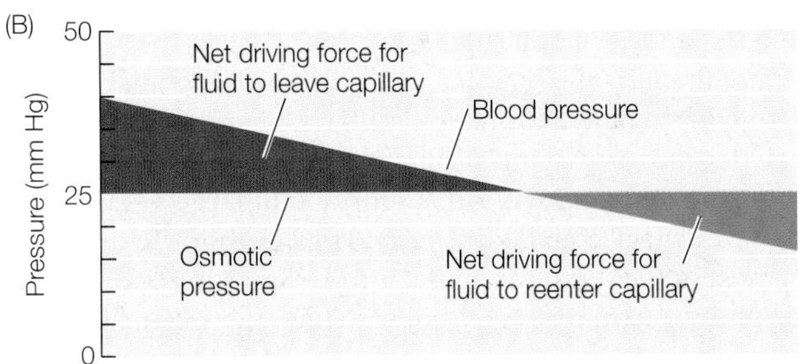


**50.15 Starling's Forces** Starling's model explains how blood volume is maintained in the capillary beds. (A) When blood pressure is greater than osmotic pressure, fluid leaves the capillaries; when blood pressure falls below osmotic pressure, fluid returns to the capillaries. (B) The balance of these two forces changes over the capillary bed as blood pressure falls.

An answer to this question was put forth more than 100 years ago by the physiologist E. H. Starling. Starling suggested that water movement across capillary walls is a result of two opposing forces, which are now known as **Starling's forces**:

- Blood pressure squeezes water and small solutes out of the capillaries.
- Osmotic pressure pulls water back into the capillaries.

Blood pressure is high at the arterial end of a capillary bed and steadily drops as blood moves toward the venous end (**Figure 50.15**). The osmotic pressure is due to the large protein molecules that cannot leave the capillaries, and it is relatively constant along the capillaries. As long as the blood pressure is above the osmotic pressure, fluid leaves the capillaries. At the venule end of most capillaries, blood pressure falls below the osmotic pressure, so fluid returns to the capillaries. The actual numbers for a normal capillary bed in a resting person suggest that there would be a *slight* net loss of fluid to the intercellular spaces. This loss, about 4 liters per day, percolates between cells as the interstitial fluid before it returns to the venous blood via the lymphatic system, which we will discuss later in this chapter.

Several observations support Starling's model. In people with severe liver disease or protein starvation, a fall in blood protein concentration leads to an accumulation of fluid in the extracellular spaces, which results in tissue swelling, or **edema**. Edema is also characteristic of the inflammation response accompanying tissue damage or allergic responses (see Figure 42.5). **Histamine**, a mediator of inflammation released by certain white blood cells, increases capillary permeability and relaxes the smooth muscles of the arterioles, raising blood pressure in the capillaries and leading to fluid leakage into tissues.

A few situations are not explained by Starling's hypothesis. During strenuous exercise, the blood pressure in the arterioles serving the muscles rises substantially but does not result in edema. In birds, the blood pressure in arterioles is much higher than in mammals, and the osmotic pressure is lower. If edema is not a chronic problem in exercising muscles and in birds, what is missing from Starling's model?

Recent research suggests that bicarbonate ions ($HCO_3^-$) in the blood plasma contribute significantly to the osmotic attraction that draws water back into the capillaries. The $CO_2$ produced by cellular metabolism diffuses into the endothelial cells lining the capillaries, where it is converted into $HCO_3^-$ and released into the plasma. When an individual is at rest, the increasing $HCO_3^-$ concentration can cause the osmotic pressure of the blood at the venous end to be 30 mm Hg higher than at the arterial end, and during strenuous exercise this difference can be much higher. Thus it appears that $CO_2$ and $HCO_3^-$ are major factors that pull water back into the capillaries.

All capillaries are permeable to $O_2$, $CO_2$, and small ions. Lipid-soluble substances readily pass through the capillary walls. Water and small solutes pass through spaces in the capillary wall, and in some cases through holes called **fenestrations**. These less selective capillaries are found in the digestive tract, where nutrients are absorbed, and in the kidneys, where wastes are filtered. The capillary walls also contain transporters that can facilitate the passage of specific molecules, such as glucose and lactate. Overall, permeability varies widely in different capillary beds—an important consideration in the design and delivery of drugs.

The capillaries of the brain are a special case, being rather impermeable and wrapped by glia. Not much can pass through them other than lipid-soluble substances (including alcohol and anesthetics). This high selectivity of brain capillaries is known as the **blood–brain barrier**. Even in the brain, however, there are specific regions where the capillaries are more permeable, enabling the brain to detect non-lipid-soluble hormones.

## Blood flows back to the heart through veins

The pressure of the blood flowing from capillaries to venules is extremely low and is insufficient to propel blood back to the heart. The walls of veins are more expandable than the walls of arteries, and blood tends to accumulate in veins. As much as 60 percent of your total blood volume may be in your veins when you are resting. Because of their high capacity to stretch and store blood, veins are called capacitance vessels.

Blood flow through veins that are above the level of the heart is assisted by gravity. Below the level of the heart, however, venous return is against gravity. The most important force propelling blood from these regions is the squeezing of the veins by the contractions of surrounding skeletal muscles.

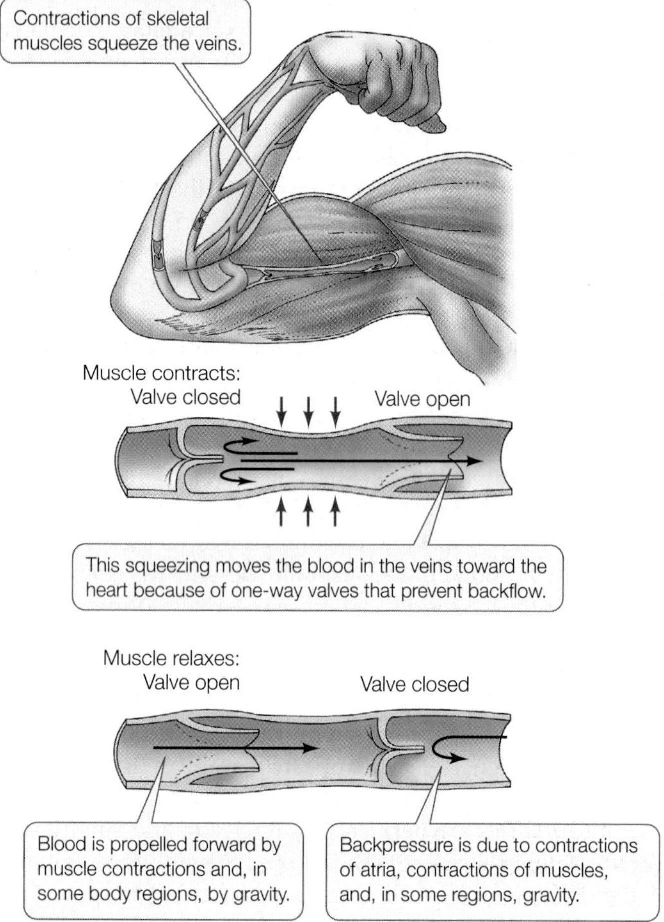


**50.16 One-Way Flow** Veins have valves that prevent blood from flowing backward, and contractions of skeletal muscle help move blood toward the heart.

As muscles contract, the vessels are compressed and blood is squeezed through them. Blood flow may be temporarily obstructed during a prolonged muscle contraction, but when muscles relax, blood is free to move again. One-way valves in the veins of the extremities prevent backflow of blood. Thus whenever a vein is squeezed, blood is propelled forward toward the heart (**Figure 50.16**).

In a resting person, gravity causes blood accumulation in the veins of the lower body and exerts backpressure on the capillary beds. This backpressure shifts the balance between blood pressure and osmotic pressure, causing increased loss of fluid to the intercellular spaces. That is why your feet swell during a long airline flight.

Because of the one-way valves in the veins of the legs, the contractions of leg muscles act as auxiliary vascular pumps when an animal walks or runs and facilitate the return of blood to the heart from the lower body. As a greater volume of blood is returned to the heart, the heart contracts more forcefully and its pumping action is enhanced. The heartbeat gets stronger because of a property of cardiac muscle cells described by the **Frank–Starling law**: if the cardiac muscle cells are stretched, as they are when the volume of returning blood increases, they contract more forcefully.

The actions of breathing also help return venous blood to the heart. The muscles involved in inhalation create negative pressure that pulls air into the lungs (see Figure 49.11), and this negative pressure also pulls blood toward the chest, increasing venous return to the right atrium. In addition, some of the largest veins closest to the heart contain smooth muscle that contracts at the onset of exercise. Contraction of veins can rapidly increase venous return and stimulate the heart in accord with the Frank–Starling law, increasing cardiac output.

## Lymphatic vessels return interstitial fluid to the blood

The interstitial fluid contains water and small molecules, but no red blood cells, and less protein than found in plasma. A separate system of vessels—the **lymphatic system**—returns interstitial fluid to the blood. Each capillary bed contains at least one blind-ended lymph capillary.

Once it enters the lymphatic vessels, the interstitial fluid is called **lymph**. Fine lymphatic capillaries merge into progressively larger vessels and ultimately into two lymphatic vessels—the **thoracic ducts**—that empty into large veins at the base of the neck (see Figure 42.1). The left thoracic duct carries most of the lymph from the lower part of the body and is much larger than the right thoracic duct. Lymphatic vessels, like veins, have one-way valves that keep the lymph flowing toward the thoracic duct. Therefore lymph, like blood, is propelled toward the heart by skeletal muscle contractions and breathing movements. Lymph nodes along the major lymphatic vessels are a major site of lymphocyte production and of the phagocytic action that removes microorganisms and other foreign materials from the circulation (see Section 42.1).

## Vascular disease is a killer

As mentioned at the start of this chapter, cardiovascular disease is responsible for about one-fourth of all deaths each year in the United States, and the same is true for Europe. The immediate cause of most of these deaths is not a defect in heart muscle as in athletes, but is heart attack or stroke—both of which are usually the end result of a disease called **atherosclerosis** ("hardening of the arteries") that begins many years before symptoms are detected.

Healthy arteries have a smooth internal lining of endothelial cells (**Figure 50.17A**) that can be damaged by chronic high blood pressure, smoking, a high-fat diet, or microorganisms. Deposits called **plaque** begin to form at sites of endothelial damage. First, the damaged endothelial cells attract certain white blood cells to the site. These cells are then joined by smooth muscle cells migrating from the deeper layers of the arterial wall. Lipids, especially cholesterol, are deposited in these cells, so that the developing plaque becomes fatty. Fibrous connective tissue made by the invading smooth muscle cells in the plaque, along with deposits of calcium, make the artery wall less elastic—hence "hardening of the arteries." The growing plaque deposit narrows the artery and causes turbulence in the blood flow. Blood platelets stick to the plaque (see Figure 50.12) and initiate formation of an intravascular blood clot, a **thrombus**, which can block the artery (**Figure 50.17B**).

(A)

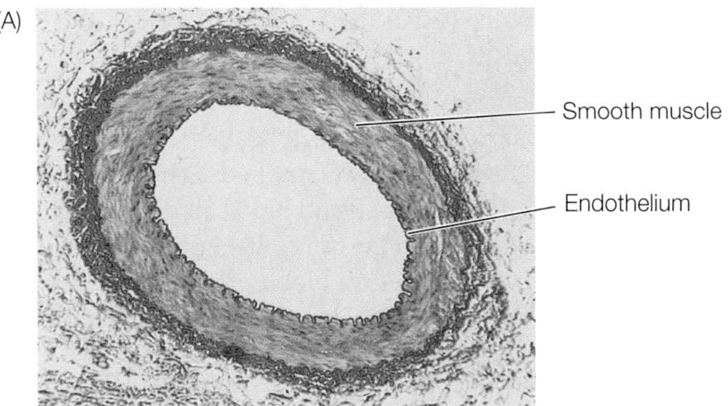


(B)

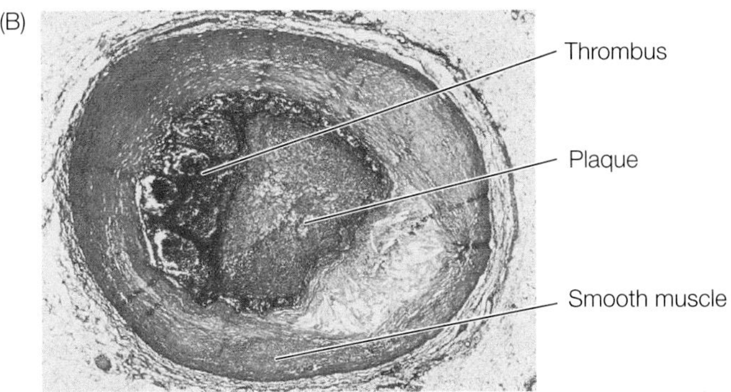


**50.17 Atherosclerotic Plaque** (A) A healthy, clear artery. (B) An atherosclerotic artery, clogged with plaque and a thrombus.

The blood supply to the heart muscle flows through the **coronary arteries**, which are highly susceptible to atherosclerosis. As these arteries narrow, blood flow to the heart muscle decreases, causing the symptoms of chest pain and shortness of breath during mild exertion. A person with atherosclerosis is at high risk of forming a thrombus in a coronary artery. This condition, called **coronary thrombosis**, can totally block the vessel, causing a **myocardial infarction** (heart attack).

A piece of a thrombus that breaks loose, called an **embolus**, is likely to travel to and become lodged in a vessel of smaller diameter, blocking its flow (an **embolism**). Arteries already narrowed by plaque formation are likely places for an embolism. An embolism in an artery in the brain causes the cells fed by that artery to die. This event is a **stroke**. The specific damage resulting from a stroke, such as memory loss, speech impairment, or paralysis, depends on the location of the blocked artery.

Important risk factors for developing atherosclerosis are your genetic predisposition and your age. Environmental risk factors also play a large role, however. These include high-fat and high-cholesterol diets, smoking, and a sedentary lifestyle. Certain untreated medical conditions such as hypertension (high blood pressure), obesity, and diabetes are also risk factors for atherosclerosis. For those who have a genetic predisposition to atherosclerosis, it is even more important to minimize environmental risk factors. Changes in diet and behavior and treatment of predisposing medical conditions can prevent and reverse early atherosclerosis and help fend off this silent killer.

### RECAP 50.4

Blood is a fluid tissue with cellular components that play roles in transport of respiratory gases, immune system function, and blood clotting. The properties of the arteries, arterioles, capillaries, venules, and veins reflect their functions. Exchanges between the blood and interstitial fluids occur in the smallest of those vessels, the capillaries.

- How are the structural differences among the various classes of vessels related to their functions? **See pp. 1039–1040 and Figure 50.13**
- Why are arterioles called resistance vessels and veins called capacitance vessels? **See p. 1039 and p. 1041**
- What factors control the movement of fluids between the vascular and extravascular spaces? **See pp. 1040–1041 and Figures 50.14 and 50.15**
- What propels blood from the lower part of the body back to the heart? **See pp. 1041–1042 and Figure 50.16**

Every tissue in the body requires an adequate flow of oxygen-saturated blood. Blood flow depends on the maintenance of an appropriate blood pressure, and the distribution of blood flow throughout the body depends on control of the resistance in the blood vessels supplying different tissues.

## 50.5 How Is the Circulatory System Controlled and Regulated?

When we investigate how a physiological process is regulated, we start by identifying the critical components of that process, how they can be controlled, and the information used to govern that control. Because blood flow depends on pressure, we can identify the pressure in the aorta as a critical variable of the circulatory system. The pressure in the aorta oscillates between systole and diastole, so we define our variable as the mean arterial pressure (MAP). MAP is determined by the cardiac output (CO) and the resistance to flow in the blood vessels, or total peripheral resistance (TPR):

$$\text{MAP} = \text{CO} \times \text{TPR}$$

Since CO is equal to the heart rate (HR) times how much blood the heart pumps with each beat (stroke volume [SV]), the critical relationships can be expressed as:

$$\text{MAP} = \text{HR} \times \text{SV} \times \text{TPR}$$

HR, SV, and TPR are controlled by neural and hormonal mechanisms at both the local and systemic levels. At the local level, each tissue controls its own blood flow through autoregulatory mechanisms that cause the arterioles supplying that tissue to constrict or dilate.

The collective autoregulatory actions in the capillary beds in all tissues of the body determine TPR and therefore MAP. If many arterioles suddenly dilate, TPR goes down and MAP falls. If many arterioles constrict, TPR goes up and MAP goes up. Changes in MAP provide information about changing needs of the body. In addition, as blood flows through capillary beds, its composition changes—its $CO_2$ content goes up and its $O_2$

content goes down. Thus blood composition also provides information the body uses to regulate the circulatory system.

The nervous and endocrine systems respond to changes in MAP and blood composition by changing breathing rate, heart rate, stroke volume, and peripheral resistance to match the metabolic needs of the body.

**Go to Animated Tutorial 50.2**
**Blood Pressure and Heart Rate Regulation**
Life10e.com/at50.2

## Autoregulation matches local blood flow to local need

The amount of blood that flows through a capillary bed is controlled by the smooth muscle of the arteries and arterioles feeding that bed. **Figure 50.18** illustrates the flow of blood in a typical capillary bed. Blood flows into the bed from an arteriole. Smooth muscle "cuffs," or **precapillary sphincters**, on the arteriole can shut off the supply of blood to the capillary bed. When the precapillary sphincters are relaxed and the arteriole is open, the arterial blood pressure pushes blood into the capillaries.

Autoregulation depends on the sensitivity of the smooth muscle to its local chemical environment. Low $O_2$ concentrations and high $CO_2$ concentrations cause the smooth muscle to relax, thus increasing the supply of blood, which brings in more $O_2$ and carries away $CO_2$—a response known as hyperemia, which means "excess blood." Increases in other by-products of metabolism, such as lactic acid, hydrogen ions, potassium, and adenosine (all of which increase in exercising muscle), also promote hyperemia. Hence activities that increase the metabolism of a tissue also induce hyperemia in that tissue.

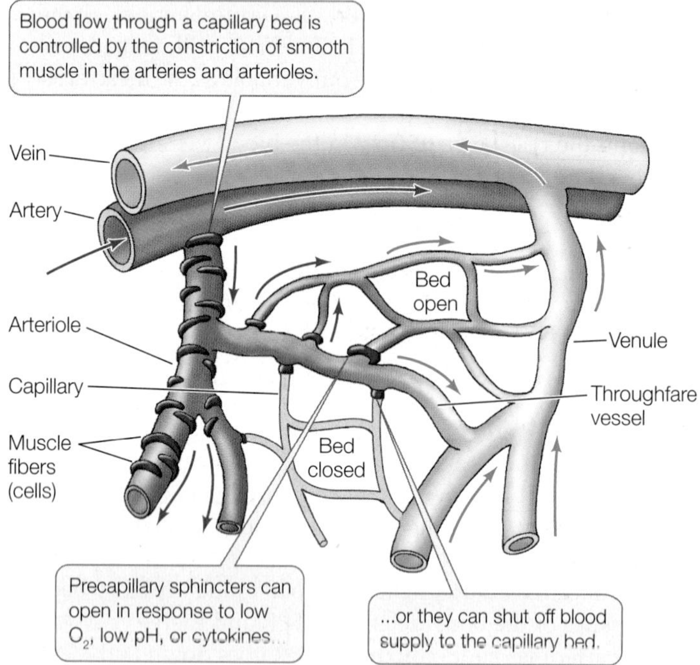


**50.18 Local Control of Blood Flow** Low $O_2$ concentrations or high levels of metabolic by-products cause the smooth muscle of the arteries and arterioles to relax, thus increasing the supply of blood to the capillary bed.

## Arterial pressure is regulated by hormonal and neural mechanisms

Control and regulation of the circulatory system begins with the local autoregulatory mechanisms that alter the resistance of arteries and arterioles feeding capillary beds. The demands of the capillary beds influence MAP and blood composition. Both of these provide information for the control of endocrine and neural responses that act to return blood pressure and composition to normal. Thus circulatory functions are matched to the regional and overall needs of the body.

Arteries and arterioles are innervated by the autonomic nervous system, particularly the sympathetic division. The sympathetic postganglionic neurotransmitter norepinephrine binds to receptors in smooth muscle in blood vessels in the gut and other tissues not essential for "fight or flight" and causes these vessels to constrict, resulting in reduced blood flow through them and an elevation in MAP. As we discussed earlier in this chapter, increased sympathetic activity increases heart rate, and by increasing the strength of the cardiac muscle contraction, it also increases stroke volume.

Hormones also play a role in regulating arterial pressure. Epinephrine has actions similar to those of norepinephrine and is released from the adrenal medulla during massive sympathetic activation stimulated by a fall in arterial pressure or by activation of the fight-or-flight response to a dangerous threat. Another hormone, **angiotensin**, is produced when blood pressure in the kidneys falls (**Figure 50.19**). These hormones influence arterioles located in peripheral tissues (extremities) or in tissues whose functions need not be maintained continuously (such as the digestive system). By reducing blood flow in those arterioles, the hormones increase central blood pressure and blood flow to essential organs such as the heart, brain, and kidneys.

The autonomic nervous system activity that controls heart rate and constriction of blood vessels originates in a cardiovascular control center in the medulla. Many inputs converge on this central integrative network and influence the commands it issues via parasympathetic and sympathetic nerves (**Figure 50.20**). Of special importance is incoming information about changes in blood pressure and composition from both **baroreceptors** (stretch receptors) and chemoreceptors in the walls of the large arteries leading to the brain—the aorta and the carotid arteries.

Increased activity in baroreceptors of the large arteries signals rising blood pressure and inhibits sympathetic nervous system signaling to arteries and arterioles while increasing parasympathetic signaling to the heart's pacemaker. As a result, the heart slows and arterioles in peripheral tissues dilate, reducing blood pressure. If pressure in the large arteries falls, the activity of the baroreceptors decreases, stimulating sympathetic output to the arteries and arterioles while reducing parasympathetic output to the heart's pacemaker. As a result, the heart beats faster and the arterioles in peripheral tissues constrict, increasing blood pressure.

Another hormone that helps stabilize blood pressure is antidiuretic hormone (ADH, also called vasopressin), which is secreted by the posterior pituitary in response to a fall in the activity of the baroreceptors, signaling a fall in arterial pressure. ADH causes the kidneys to resorb more water and thereby

**50.19 Control of Blood Pressure through Local and Systemic Mechanisms** A drop in arterial pressure reduces blood flow to tissues, resulting in local accumulation of metabolic wastes. This change in the extracellular environment stimulates autoregulatory opening of the arteries. A fall in central blood pressure is prevented by negative feedback mechanisms (including the release of antidiuretic hormone, ADH) that constrict arteries in less essential tissues and stimulate maintenance of blood volume and blood pressure.

maintain blood volume and increase blood pressure (see Section 52.6). Increased activity of the baroreceptors inhibits the release of ADH, and as a result the kidneys excrete more water, reducing blood volume and contributing to a fall in arterial pressure (see Figure 50.19).

Other information that causes the cardiovascular control center to increase heart rate and blood pressure comes from chemoreceptors in the medulla, aorta, and the carotid arteries. As we discussed in Section 49.5, the medullary chemosensors are activated by increases in arterial $CO_2$ levels, and the carotid and aortic bodies are activated by falls in arterial $O_2$ levels. Chemosensors send signals to the cardiovascular regulatory center as well as to the respiratory regulatory center.

## RECAP 50.5

The delivery of blood to tissues is controlled locally by autoregulatory mechanisms that dilate or constrict arterioles. These local actions are translated into alterations in central blood pressure and composition that are detected by neural and hormonal mechanisms, which then mediate corrective cardiovascular adjustments.

- How do autoregulatory changes in blood flow to capillary beds result in adjustments to MAP? **See p. 1044 and Figure 50.18**
- What are the roles of hormones in regulating blood pressure? **See p. 1044 and Figure 50.19**
- Describe the role of baroreceptors and chemoreceptors in regulating blood pressure. **See pp. 1044–1045 and Figure 50.20**

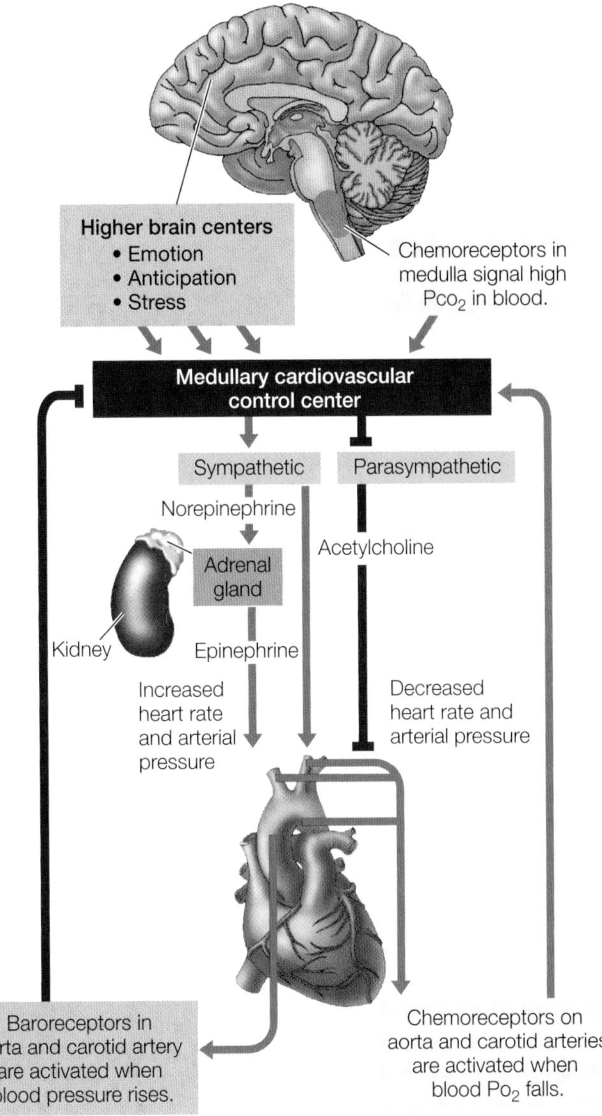


**50.20 Regulating Cardiac Output** The autonomic nervous system controls heart rate in response to information about blood pressure and blood composition originating in baroreceptors and chemoreceptors shown at the bottom of the figure. Information from these sensors goes to the cardiovascular control center in the medulla, where it is integrated with other information. The medullary center generates responses in the sympathetic and parasympathetic nervous systems that control cardiac output.

How can the same mutation-based heart condition be fatal to an athlete but innocuous in most other people?

ANSWER

The mutation responsible for the condition described in the opening story encodes a contractile protein necessary for the pumping action of cardiac muscle. The mutant form of this protein renders the heart less efficient at pumping blood, and the heart compensates by getting larger. In most people, the heart enlarges enough to meet their needs without causing other problems. High-performing athletes, however, regularly engage in heavy exercise that places huge demands on their hearts, and in athletes with this mutation heart enlargement can be excessive. Eventually thickening of the ventricular walls, especially those of the left ventricle, can disrupt the electrical impulses that coordinate contractions of the heart muscle. When heavy demand is placed on such an enlarged heart, muscle fiber contractions can suddenly become uncoordinated or blocked, rendering the heart incapable of pumping blood.

# CHAPTER SUMMARY 50

## 50.1 Why Do Animals Need a Circulatory System?

- The metabolic needs of the cells of many small animals are met by direct exchange of materials with the external medium. The metabolic needs of the cells of larger animals are met by a **circulatory system** that transports nutrients, respiratory gases, and metabolic wastes throughout the body.
- In **open circulatory systems**, extracellular fluid leaves vessels and percolates through tissues. In **closed circulatory systems**, the blood is contained in a system of vessels. Closed circulatory systems have the ability to selectively direct blood, hormones, and nutrients to specific tissues. **Review Figure 50.1**

## 50.2 How Have Vertebrate Circulatory Systems Evolved?

- The circulatory system of vertebrates consists of a heart and a closed system of vessels containing blood that is separate from the interstitial fluid. **Arteries** and **arterioles** carry blood from the heart; **capillaries** are the site of exchange between blood and interstitial fluid; **venules** and **veins** carry blood back to the heart.
- The vertebrate circulatory system evolved from a single circuit in fishes to partially or completely separate pulmonary and systemic circuits in amphibians, reptiles, and mammals.
- In the single-circuit system of fishes, blood flow is unidirectional and is propelled by one-way valves between the **sinus venosus** and the **atrium**, between the atrium and the **ventricle**, and between the ventricle and the **bulbus arteriosus**.
- In birds and mammals, blood circulates through two completely separate circuits. The **pulmonary circuit** transports blood between the heart and lungs, and the **systemic circuit** transports oxygen-rich blood between the heart and tissues. **See ACTIVITY 50.1**

## 50.3 How Does the Mammalian Heart Function?

- The mammalian heart has four chambers. Valves in the heart prevent the backflow of blood. **Review Figure 50.2, ACTIVITY 50.2**
- The **cardiac cycle** has two phases: **systole**, in which the ventricles contract, and **diastole**, in which the ventricles relax. The sequential heart sounds ("lub-dup") are made by the closing of the heart valves. **Review Figure 50.3, ANIMATED TUTORIAL 50.1**
- Blood pressure can be measured using a sphygmomanometer and a stethoscope. **Review Figure 50.4**
- **Pacemaker cells** of the sinoatrial node set the heart rate as a result of the properties of their ion channels. The autonomic nervous system controls heart rate: sympathetic activity increases heart rate, and parasympathetic activity decreases it by altering the rate of depolarization of the pacemaker cell resting membrane potentials following the termination of systole. **Review Figures 50.5, 50.6**
- The **sinoatrial node** controls the cardiac cycle by initiating a wave of depolarization in the atria, which is conducted to the ventricles through a system consisting of the **atrioventricular node**, **bundle of His**, and **Purkinje fibers**. **Review Figure 50.7**
- Sustained contraction of ventricular muscle cells is due to long-duration action potentials that are generated by voltage-gated $Na^+$ and $Ca^{2+}$ channels. **Review Figures 50.8, Figure 50.9**
- An **electrocardiogram** (**ECG** or **EKG**) records electrical events associated with the contraction and relaxation of the cardiac muscles. **Review Figure 50.10**

## 50.4 What Are the Properties of Blood and Blood Vessels?

- Blood consists of a **plasma** portion (water, salts, and proteins) and a cellular portion (**erythrocytes** or red blood cells, platelets, and white blood cells). All of the cellular components are produced from stem cells in the bone marrow. **Review Figure 50.11**
- Erythrocytes transport oxygen. Their production in the bone marrow is stimulated by **erythropoietin**, which is produced in response to **hypoxia** (low oxygen levels) in the tissues.
- **Platelets**, along with circulating proteins, are involved in **blood clotting**, which results in a meshwork of **fibrin** threads that help seal damaged vessels. **Review Figure 50.12**
- Abundant smooth muscle cells allow vessels to change their diameter, altering their resistance and thus blood flow. Arteries have elastic fibers that enable them to withstand high pressures. **Review Figure 50.13, ACTIVITY 50.3**
- Capillary beds are the site of exchange of materials between blood and tissue fluid.
- **Starling's forces** suggest that blood volume is maintained in the capillary beds by an exchange of fluids driven by both blood pressure and osmotic pressure. **Review Figure 50.15**

*continued*

- An accumulation of fluid in the extracellular spaces leads to **edema**. Bicarbonate ions in the blood plasma contribute to the osmotic forces that draw water back into capillaries.
- The ability of a specific molecule to cross a capillary wall depends on the architecture of the capillary, the type of substance, and the concentration gradient between the blood and the tissue fluid.
- Veins have a high capacity for storing blood. Aided by gravity, by contractions of skeletal muscle, and by the actions of breathing, they return blood to the heart. **Review Figure 50.16**
- The **Frank–Starling law** describes forces that increase cardiac output, such as stretch of the cardiac muscles cells caused by increased venous return.
- The **lymphatic system** returns the interstitial fluid to the blood.

## 50.5 How Is the Circulatory System Controlled and Regulated?

- Blood flow through capillary beds is controlled by local **autoregulatory mechanisms**, hormones, and the autonomic nervous system. **Review Figure 50.18, ANIMATED TUTORIAL 50.2**
- Blood pressure is controlled in part by the hormones ADH and **angiotensin**, which stimulate contraction of blood vessels. **Review Figure 50.19**
- Heart rate is controlled by the autonomic nervous system, which responds to information about blood pressure and blood composition that is integrated by regulatory centers in the medulla. **Review Figure 50.20**

**Go to the Interactive Summary to review key figures, Animated Tutorials, and Activities Life10e.com/is50**

# CHAPTER**REVIEW**

### REMEMBERING

1. Which statement about vertebrate circulatory systems is *not* true?
   a. In fish, oxygenated blood from the gills returns to the heart through the left atrium.
   b. In mammals, deoxygenated blood leaves the heart through the pulmonary artery.
   c. In amphibians, deoxygenated blood enters the heart through the right atrium.
   d. In reptiles, the blood in the pulmonary artery has a lower oxygen content than the blood in the aorta.
   e. In birds, the pressure in the aorta is higher than the pressure in the pulmonary artery.

2. Which statement about the human heart is true?
   a. The walls of the right ventricle are thicker than the walls of the left ventricle.
   b. Blood flowing through atrioventricular valves is always deoxygenated blood.
   c. The second heart sound is due to the closing of the aortic valve.
   d. Blood returns to the heart from the lungs in the vena cava.
   e. During systole, the aortic valve is open and the pulmonary valve is closed.

3. The pacemaker action potentials in the heart
   a. are due to opposing actions of norepinephrine and acetylcholine.
   b. are generated by the bundle of His.
   c. depend on the gap junctions between the cells that make up the atria and those that make up the ventricles.
   d. are due to spontaneous depolarization of the plasma membranes of modified cardiac muscle cells.
   e. reflect large depolarizations of membrane potential due to opening of voltage-gated $Na^+$ channels.

4. Blood velocity through capillaries is slow because
   a. much blood volume is lost from the capillaries.
   b. the pressure in venules is high.
   c. the total cross-sectional area of capillaries is larger than that of arterioles.
   d. the osmotic pressure in capillaries is very high.
   e. erythrocytes must pass through in single file.

5. Autoregulation of blood flow to a tissue is due to
   a. sympathetic innervation.
   b. the release of ADH by the hypothalamus.
   c. increased activity of baroreceptors.
   d. chemoreceptors in the aorta and the carotid arteries.
   e. the effect of the local chemical environment on arterioles.

### UNDERSTANDING & APPLYING

6. At the beginning of a race, cardiac output increases immediately before there is any change in blood $O_2$ or $CO_2$ concentrations. Explain two factors that contribute to this effect. Include the Frank–Starling law in your answer.

7. Is there a time in the mammalian cardiac cycle when all four heart valves are open? Explain.

8. A sudden and massive loss of blood results in a decrease in blood pressure. Describe several mechanisms that help return blood pressure to normal and how they do so.

### ANALYZING & EVALUATING

9. The hearts of most mammals stop beating when their temperature falls more than a few degrees below 20°C, but the hearts of hibernating mammals can beat at 0°C. What adaptations of cardiac muscle might explain this capacity of the hearts of hibernators?

10. You can describe the cycle of events in a ventricle of the heart by a graph that plots the pressure in the ventricle on the *y* axis and the volume of blood in the ventricle on the *x* axis. Draw such a graph for a left ventricle using maximum and minimum pressures of 125 and 25 mm Hg and maximum and minimum ventricular volumes of 130 and 60 ml. Where would the heart sounds be on this graph? How would the graph differ for the left and the right ventricles?

Go to BioPortal at **yourBioPortal.com** for Animated Tutorials, Activities, LearningCurve Quizzes, Flashcards, and many other study and review resources.

# 51 Nutrition, Digestion, and Absorption

## CHAPTER**OUTLINE**


**51.1 What Do Animals Require from Food?**

**51.2 How Do Animals Ingest and Digest Food?**

**51.3 How Does the Vertebrate Gastrointestinal System Function?**

**51.4 How Is the Flow of Nutrients Controlled and Regulated?**

**Efficiency Genes** The Pima are an example of a human population that repeatedly experienced periods of severe food deprivation. These historic occurrences may have imposed selection for genes that improve the efficiency of managing the energy obtained from food. With modern diets and lifestyles, these "efficiency genes" can contribute to obesity.

FOR THOUSANDS of years the Pima of southwestern North America were hunters and gatherers who supplemented their diet with subsistence agriculture. Their environment was arid, so they developed sophisticated irrigation systems; even so, they frequently encountered drought and subsequent starvation. Today most individuals of the ethnic Pima population in North America are clinically obese. In fact, as a population they are one of the heaviest in the world.

With obesity come related health problems such as diabetes, high blood pressure, and heart disease. Diabetes incidence in the Pima is seven times the national average; two-thirds of adults over the age of 40 are diabetic. Moreover, diabetes is occurring in younger individuals than ever before. What has caused such a radical health change in an entire population? Two interacting factors are involved: genetics and lifestyle.

Geneticists hypothesize that recurring episodes of starvation produce strong selective pressure for "thrifty genes"—particular alleles of the genes involved in digestion, absorption, and energy storage that result in greater-than-average efficiency in converting food into energy reserves, such as fat. Thrifty genes would carry a strong selective advantage when food is scarce. The Pima display a "thrifty" phenotype. They have low resting metabolic rates and convert food into fat readily. For many Pima, consuming a standard amount of glucose causes their insulin levels to rise three times higher than it does in Americans of European ancestry. Insulin is the hormone that facilitates conversion of dietary sugar into fat.

The other factor in the Pima obesity epidemic is an abrupt change in their traditional lifestyle. Instead of eating their traditional diet, the Pima now eat a high-calorie, high-fat Western diet, and they engage in less physical activity than their ancestors did.

Another population of Pima lives in the Sierra Madre of northern Mexico. Genetically they are the same as the Arizona population. However, they eat traditional foods and live a traditional lifestyle that involves much physical activity. Whereas the Arizona Pima engage in an average of only 2 hours of physical work per week, the Mexico Pima average 23 hours per week. Obesity and diabetes are not prevalent among the Mexico Pima.

A high-calorie diet and sedentary lifestyle affect not just the Pima but contribute to the overall increase in obesity throughout the U.S. population. Researchers are studying the Pima to learn more about the genetics of obesity and related pathologies.

Are there genes that predispose a person to obesity?

See answer on p. 1068.

(A)

(B)

**51.1 Heterotrophs Get Energy from Autotrophs** (A) Herbivores get their energy directly from autotrophs. The large herbivores of the African grasslands must consume huge amounts of plant matter to fulfill their nutritional needs. (B) A carnivore's energy is indirectly obtained from autotrophs, since the energy stored in a prey animal was originally obtained from autotrophs.

## 51.1 What Do Animals Require from Food?

Animals are **heterotrophs**—they derive their nutrition from eating other organisms. In contrast, **autotrophs** (most plants, some bacteria, some archaea, and some protists) can use solar energy or inorganic chemical energy to synthesize all of their components. Directly and indirectly, heterotrophs take advantage of—indeed, depend on—the organic synthesis carried out by autotrophs and have evolved an enormous diversity of adaptations to exploit this resource (**Figure 51.1**). In this section we will describe how animals use food, be it plants or other animals, to obtain energy and building blocks of complex molecules. We will also consider the need for special mineral nutrients and organic molecules and the diseases that result when they are lacking in the diet.

### Energy needs and expenditures can be measured

Energy—the capacity to do work—comes in different forms, including electric, heat, chemical, and nuclear energy. As discussed in Chapter 8, a calorie (note the small *c*) is a unit of heat energy; specifically, it is the amount of heat necessary to raise the temperature of 1 gram of water 1°C. Because this is such a tiny amount of energy, physiologists commonly use the **kilocalorie** (**kcal**) as a unit of measure (1 kcal = 1,000 calories). Nutritionists also use the kilocalorie as a standard unit of energy, but they traditionally refer to it as the **Calorie** (**Cal**), which is capitalized to distinguish it from the single calorie.

Just about any food container you pick up in the U.S. carries the label "Nutrition Facts" that includes the item "Calories." How do Calories (or kcal) relate to the discussion in Chapter 9 about how energy in the chemical bonds of food molecules is transferred to the high-energy phosphate bonds of adenosine triphosphate (ATP) and is then used to do cellular work? And why do we use heat energy as a measure of nutrition?

The reason is found in the laws of thermodynamics, which tell us that energy cannot be created or destroyed, but can be converted from one form to another (see Section 8.1). However, every energy conversion is inefficient and a large portion of the original energy always ends up as heat. Whether we are using the energy in glucose to make ATP or are using that ATP to power muscle contraction or ion transport, most of the available chemical energy is lost as heat. The bottom line is that if an animal is not growing, not doing any external work, and not changing its body temperature, the heat it loses to the environment is a measure of its total energy expenditure, or metabolism.

An animal's energy needs must be met by the ingestion, digestion, and assimilation of food. The basal energy expenditure of a human is 1,300 to 1,500 Cal/day for an adult female and 1,600 to 1,800 Cal/day for an adult male. Physical activity adds to this basal energy requirement. For a person doing sedentary work, about 30 percent of the Calories expended are used for skeletal muscle activity; for a person doing heavy physical labor, more than 95 percent of caloric expenditure is for skeletal muscle activity.

The components of food that provide energy are fats, carbohydrates, and proteins. Fats yield 9.5 Cal/gram, carbohydrates 4.2 Cal/gram, and proteins about 4.1 Cal/gram. **Figure 51.2** shows some equivalencies of food, energy, and energy consumption.

Even though the units calorie, kilocalorie, and Calorie remain in popular use, most scientists now use the International System of Units (ISU). In this system the basic unit of energy is the joule: 1 joule = 0.239 calories, and the measure of energy use is 1 joule/second = 1 watt. You are familiar with light bulb ratings, so think about that when you convert kcal/day into watts. The 1,700 Cal/day energy expenditure of the average man converts to 82 watts (note that a watt includes the time dimension, so it is a rate of energy use).

Thus it is possible to quantify the caloric value of any food an animal eats. It is also possible to quantify the caloric

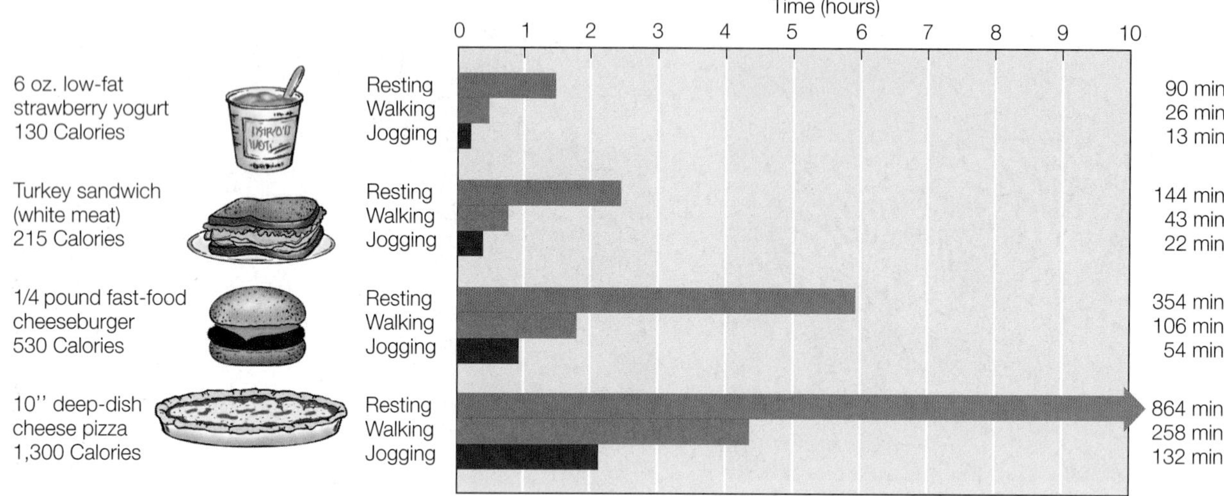


**51.2 Food Energy and How We Use It** The energy contained in several common food items is shown at the left. The graphs indicate about how long it would take a person with a basal energy requirement of about 1,800 Cal/day to use the equivalent amount of energy while resting, walking, or jogging.

expenditure of any activity or behavior an animal performs. By comparing calories consumed with calories expended, we can construct **energy budgets** that allow ecologists and evolutionary biologists to apply a cost–benefit analysis to feeding behavior, as we will explain in Section 53.4.

## Sources of energy can be stored in the body

Although the cells of the body use energy continuously, most animals do not eat continuously and so must store fuel molecules that can be released as needed between meals. Carbohydrates are stored in liver and muscle cells as glycogen, but the total glycogen stored represents only about a day's basal energy requirement. Fat is the most important form of stored energy in the bodies of animals. Not only does fat have more energy per gram than glycogen, but it can be stored with little associated water, making it more compact. Migrating birds store energy as fat to fuel their long flights; if they had to store the same amount of energy as glycogen, they would be too heavy to fly. Proteins are not used as energy storage compounds, although the body's structural protein can be metabolized as an energy source of last resort.

If an animal takes in too little food to meet its energy requirements, it must start metabolizing some of the molecules of its own body. This "self-consumption" begins with the energy storage compounds glycogen and fat. Once fat reserves are seriously depleted, the body increases its metabolism of proteins for energy (**Figure 51.3A**). The first proteins to be sacrificed are those of the blood plasma. The loss of plasma proteins decreases the osmotic concentration of the plasma, resulting in increased loss of fluid from the blood to the interstitial spaces (edema; see Section 50.4). Accumulation of fluid in the extremities and abdomen is the classic sign of kwashiorkor, a disease caused by chronic protein deficiency (**Figure 51.3B**). Continued protein loss damages the body's organs, leading eventually to death.

When an animal consistently takes in more food than it needs to meet its energy requirements, the excess nutrients are stored as increased body mass. First glycogen reserves build

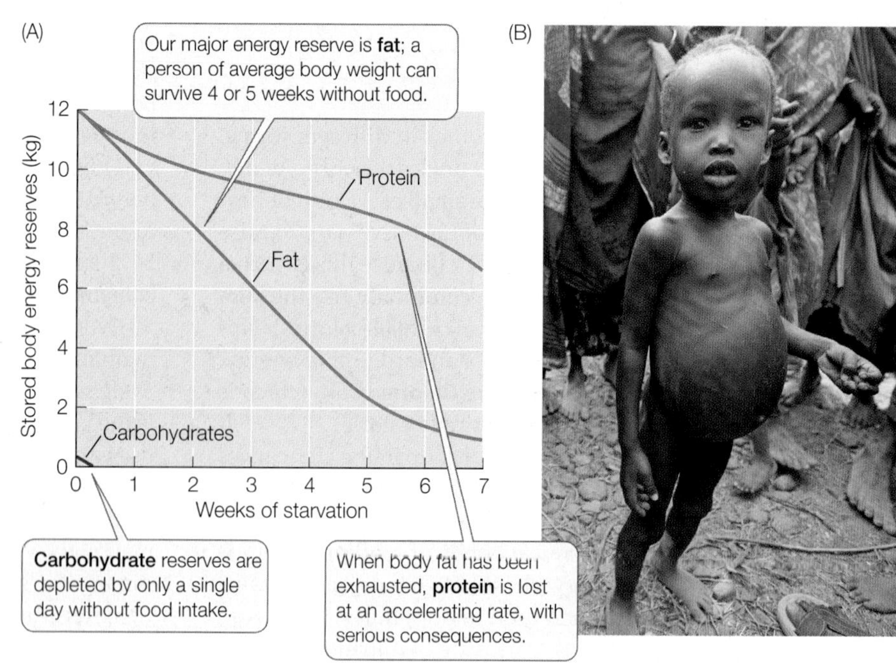


**51.3 The Course of Starvation** (A) In a person subjected to undernutrition, the body's energy reserves are eventually depleted. (B) The swollen abdomen, face, hands, and feet of this girl are due to edema. Along with her spindly limbs, these are symptoms of kwashiorkor, a syndrome resulting from the body breaking down blood proteins and muscle tissue to fuel metabolism.

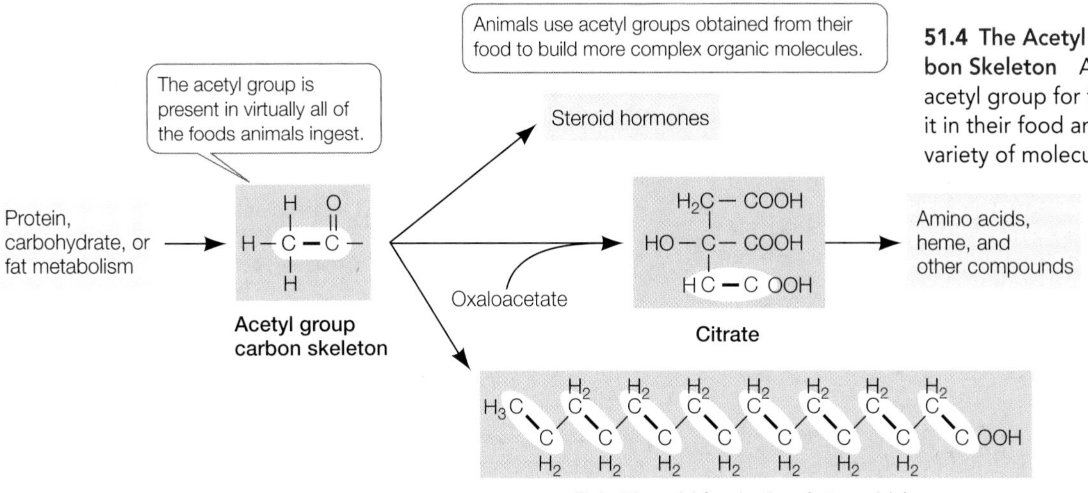


**51.4 The Acetyl Group Is an Acquired Carbon Skeleton** Animals cannot synthesize the acetyl group for themselves, but they ingest it in their food and use it to synthesize a wide variety of molecules.

up; then additional dietary carbohydrates, fats, and proteins are converted to body fat. In some species, such as hibernators, seasonal overnutrition is an important adaptation for surviving periods when food is not available. In humans, however, overnutrition can be a serious health hazard, increasing the risk of high blood pressure, heart attack, diabetes, and other disorders, as seen in the Pima discussed in the chapter opener.

## Food provides carbon skeletons for biosynthesis

Every animal must take in certain organic molecules that it cannot synthesize for itself but needs to form the building blocks of its own complex organic molecules. The acetyl group ($CH_3CO$—) is one such required building block, supplying the **carbon skeleton** of larger organic molecules (**Figure 51.4**). Animals cannot synthesize acetyl groups from carbon, oxygen, and hydrogen molecules but must obtain them from food. Acetyl groups can be derived from the metabolism of almost any food, but they originate in plants.

Acetyl groups are never in short supply for an adequately nourished animal. However, some groups supplying carbon skeletons can be deficient in an animal's diet even if caloric intake is adequate. One such group includes certain amino acids, the building blocks of proteins. Animals can synthesize some of their own amino acids using carbon skeletons from acetyl or other groups and transferring to them amino groups (—$NH_2$) derived from other amino acids. However, most animals cannot synthesize all the amino acids they need and thus must obtain certain **essential amino acids** from food. If an animal does not take in enough of even one of its essential amino acids, its protein synthesis is impaired and its capacity to maintain enzymatic and transport functions is challenged.

Essential amino acids vary by species. Most researchers agree that adult humans must obtain eight essential amino acids from their food: isoleucine, leucine, lysine, methionine, phenylalanine, threonine, tryptophan, and valine. All eight are available in milk, eggs, meat, and soybean products, but most plant foods do not contain adequate quantities of all eight, so a strict vegetarian diet carries a risk of protein malnutrition. A **complementary diet** of plant foods, however, supplies all eight essential amino acids (**Figure 51.5**). In general, grains (such as rice, wheat, and corn) are complemented by legumes (such as beans and peas). Long before the chemical basis for complementarity was understood, societies with little access to meat developed complementary diets. Many Central and South American peoples traditionally eat beans with corn, and the native peoples of North America complemented their beans with squash.

Human infants are thought to require four additional amino acids in their diets: histidine, tyrosine, cysteine, and arginine. Also, some amino acids are required by individuals with certain metabolic disorders who cannot synthesize them adequately. For example, individuals with the genetic disease phenylketonuria lack the enzyme for converting phenylalanine to tyrosine (see Section 15.2) and must obtain tyrosine from their diets. They must keep their dietary intake of phenylalanine low to prevent its accumulation to toxic levels.

Why are dietary proteins completely digested to their constituent amino acids before being used by the body? Wouldn't it be more energy-efficient to reuse some dietary proteins

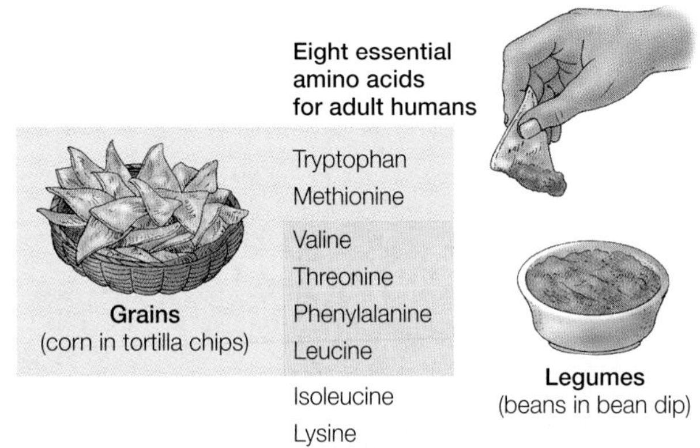


**51.5 A Strategy for Vegetarians** By combining cereal grains with legumes, an adult vegetarian can obtain all eight essential amino acids.

directly? There are several reasons why ingested proteins are not used "as is":

- Macromolecules such as proteins are not readily absorbed by the cells of the gut, but their constituent monomers (such as amino acids) are readily absorbed.
- Protein structure and function are highly species-specific. A protein that functions optimally in one species might not function well in another.
- Foreign proteins entering the body directly from the gut would be recognized as invaders and would be attacked by the immune system.

Humans can synthesize almost all the lipids required by the body using acetyl groups obtained from food (see Figure 51.4), but we must have a dietary source of certain **essential fatty acids**—notably, linoleic acid—that we cannot synthesize. Linoleic acid is needed by mammals to synthesize other unsaturated fatty acids, such as arachidonic acid, which is a component of several signaling molecules, including prostaglandins. Essential fatty acids are also necessary components of membrane phospholipids. A deficiency of linoleic acid can lead to problems such as infertility and impaired lactation, but because it is commonly present in vegetable oils, a deficiency is unlikely in an adequately nourished individual.

### Animals need mineral elements for a variety of functions

**Table 51.1** lists the principal mineral elements that animals require. Elements required in large amounts are called **macronutrients**; those required in only tiny amounts (generally less than 100 mg/day) are called **micronutrients**. Some micronutrients are required in such minute amounts that deficiencies are never observed, but they are nevertheless essential elements.

Calcium is an example of a macronutrient. It is the fifth most abundant element in the body; a 70-kg person contains about 1.2 kg of calcium. Calcium phosphate is the principal structural

TABLE **51.1** Mineral Elements Required by Animals

| Element | Source in Human Diet | Major Functions |
|---|---|---|
| MACRONUTRIENTS | | |
| Calcium (Ca) | Dairy foods, eggs, green leafy vegetables, whole grains, legumes, nuts, meat | Found in bones and teeth; blood clotting; nerve and muscle action; enzyme activation |
| Chlorine (Cl) | Table salt (NaCl), meat, eggs, vegetables, dairy foods | Water balance; digestion (as HCl); principal negative ion in extracellular fluid |
| Magnesium (Mg) | Green vegetables, meat, whole grains, nuts, milk, legumes | Required by many enzymes; found in bones and teeth |
| Phosphorus (P) | Dairy, eggs, meat, whole grains, legumes, nuts | Component of nucleic acids, ATP, and phospholipids; bone formation; buffers; metabolism of sugars |
| Potassium (K) | Meat, whole grains, fruits, vegetables | Nerve and muscle action; protein synthesis; principal positive ion in cells |
| Sodium (Na) | Table salt, dairy foods, meat, eggs | Nerve and muscle action; water balance; principal positive ion in extracellular fluid |
| Sulfur (S) | Meat, eggs, dairy foods, nuts, legumes | Found in proteins and coenzymes; detoxification of harmful substances |
| MICRONUTRIENTS | | |
| Chromium (Cr) | Meat, dairy, whole grains, legumes, yeast | Glucose metabolism |
| Cobalt (Co) | Meat, tap water | Found in vitamin $B_{12}$; formation of red blood cells |
| Copper (Cu) | Liver, meat, fish, shellfish, legumes, whole grains, nuts | Found in active site of many redox enzymes and electron carriers; production of hemoglobin; bone formation |
| Fluorine (F) | Most water supplies | Found in teeth; helps prevent tooth decay |
| Iodine (I) | Fish, shellfish, iodized salt | Found in thyroid hormones |
| Iron (Fe) | Liver, meat, green vegetables, eggs, whole grains, legumes, nuts | Found in active sites of many redox enzymes and electron carriers, hemoglobin, and myoglobin |
| Manganese (Mn) | Organ meats, whole grains, legumes, nuts, tea, coffee | Activates many enzymes |
| Molybdenum (Mo) | Organ meats, dairy, whole grains, green vegetables, legumes | Found in some enzymes |
| Selenium (Se) | Meat, seafood, whole grains, eggs, milk, garlic | Fat metabolism |
| Zinc (Zn) | Liver, fish, shellfish, and many other foods | Found in some enzymes and some transcription factors; insulin physiology |

material in bones and teeth. Muscle contraction, neural function, and many other intracellular functions in animals require calcium ions ($Ca^{2+}$). The turnover of calcium in the extracellular fluid is high, as bones are constantly being remodeled and calcium is constantly entering and leaving cells. Calcium is lost from the body in urine, sweat, and feces, so it must be replaced regularly. Adult humans require 800 to 1,000 mg of calcium per day in their diet.

Iron is an example of a micronutrient. It is found throughout the body because it is the oxygen-binding atom in hemoglobin and myoglobin and is a component of enzymes in the electron transport chain. Nevertheless, the total amount of iron in a 70-kg person is only about 4 grams, and since iron is recycled efficiently in the body and is not lost in the urine, we require only about 15 mg per day in our food. Despite the small amount required, insufficient iron is the most common mineral nutrient deficiency in the world today. Iron deficiency leads to anemia, a condition that renders individuals weak and tired all the time.

**Go to Activity 51.1 Mineral Elements Required by Animals Life10e.com/ac51.1**

## Animals must obtain vitamins from food

Like essential amino acids and fatty acids, **vitamins** are carbon compounds that an animal requires for growth and metabolism but cannot synthesize for itself. They are required in very small amounts compared with the essential amino acids and fatty acids that are incorporated into large body structures. Most vitamins function as coenzymes or parts of coenzymes.

The list of vitamins varies from species to species. Most mammals, for example, can make their own ascorbic acid. Primates (including humans) cannot, so for primates, ascorbic acid is a vitamin—vitamin C. If we do not get vitamin C in our food, we develop scurvy, a disease characterized by bleeding gums, loss of teeth, subcutaneous hemorrhages, and slow wound healing. Scurvy was a frequently fatal problem for sailors on long voyages until late in the eighteenth century, when a Scottish physician, James Lind, discovered that the disease could be prevented if the sailors ate fresh greens and citrus fruit. The British Admiralty made limes standard provisions for its ships (and British sailors have been called "limeys" ever since). When the active ingredient in limes was isolated, it was named ascorbic ("without scurvy") acid.

Humans require 13 vitamins; these are divided into two groups, water-soluble and fat-soluble (**Table 51.2**). When water-soluble vitamins are ingested in excess of bodily needs, they are simply eliminated in the urine. (This is the fate of much of the large doses of vitamin C that people take.) Fat-soluble vitamins, however, can accumulate in body fat and may build up to toxic levels in the liver if taken in excess.

TABLE **51.2**
Vitamins in the Human Diet

| Vitamin | Source | Function | Deficiency Symptoms |
|---|---|---|---|
| WATER-SOLUBLE | | | |
| $B_1$ (thiamin) | Liver, legumes, whole grains | Coenzyme in cellular respiration | Beriberi, loss of appetite, fatigue |
| $B_2$ (riboflavin) | Dairy, meat, eggs, green leafy vegetables | Coenzyme in FAD | Lesions in corners of mouth, eye irritation, skin disorders |
| Niacin | Meat, fowl, liver, yeast | Coenzyme in NAD and NADP | Pellagra, skin disorders, diarrhea, mental disorders |
| $B_6$ (pyridoxine) | Liver, whole grains, dairy foods | Coenzyme in amino acid metabolism | Anemia, slow growth, skin problems, convulsions |
| Pantothenic acid | Liver, eggs, yeast | Found in acetyl CoA | Adrenal problems, reproductive problems |
| Biotin | Liver, yeast, bacteria in gut | Found in coenzymes | Skin problems, loss of hair |
| $B_{12}$ (cobalamin) | Liver, meat, dairy foods, eggs | Formation of nucleic acids, proteins, red blood cells | Pernicious anemia |
| Folic acid | Vegetables, eggs, liver, whole grains | Coenzyme in formation of heme and nucleotides | Anemia |
| C (ascorbic acid) | Citrus fruits, tomatoes, potatoes | Formation of connective tissues; antioxidant | Scurvy, slow healing, poor bone growth |
| FAT-SOLUBLE | | | |
| A (retinol) | Fruits, vegetables, liver, dairy | Found in visual pigments | Night blindness |
| D (calciferol) | Fortified milk, fish oils, sunshine | Absorption of calcium and phosphate | Rickets |
| E (tocopherol) | Meat, dairy foods, whole grains | Muscle maintenance, antioxidant | Anemia |
| K (menadione) | Intestinal bacteria, liver | Blood clotting | Blood clotting problems |

The fat-soluble vitamin D (calciferol), which is essential for absorbing and metabolizing calcium, is a special case because the body can synthesize it. (As noted in Section 41.4, vitamin D is by definition a hormone.) Certain lipids present in the human body can be converted into vitamin D by the action of ultraviolet light on the skin. Thus vitamin D must be obtained in the diet by individuals with inadequate exposure to the sun.

The need for vitamin D may have been an important factor in the evolution of skin color. For humans living in equatorial and low latitudes, dark skin pigmentation is adaptive, as it is a protection against the damaging effects of ultraviolet radiation. These peoples generally expose extensive areas of skin to the sun on a regular basis, so their skin synthesizes adequate amounts of vitamin D. Most races that adapted to life in the higher latitudes lost this dark skin pigmentation, probably because lighter skin facilitates vitamin D production in the relatively small areas of skin exposed to sunlight during the short days of winter. The dark-skinned Inuit peoples of the Arctic are an exception to the correlation between latitude and skin pigmentation, but the Inuit obtain ample vitamin D from the large amounts of animal fat (especially whale blubber) and fish oils in their diet.

**Go to Activity 51.2 Vitamins in the Human Diet**
**Life10e.com/ac51.2**

### Nutrient deficiencies result in diseases

The lack of any essential nutrient in the diet produces a state of **malnutrition**, and chronic malnutrition leads to a characteristic **deficiency disease** (see Table 51.2). We have discussed kwashiorkor (protein deficiency) and scurvy (vitamin C deficiency). Another deficiency disease, beriberi, was directly involved in the discovery of vitamins.

Beriberi, which means "extreme weakness," became prevalent in Asia in the nineteenth century when it became standard practice to mill rice to a white polish and discard the hulls present in brown rice. A critical observation was that chickens and pigeons developed beriberi-like symptoms when they were fed only polished rice. In 1912 Casimir Funk, a Polish scientist working in England, cured pigeons of beriberi by feeding them discarded rice hulls.

At the time of Funk's discovery, all diseases were thought either to be caused by microorganisms or to be inherited. Funk suggested that beriberi and some other diseases are dietary in origin and result from deficiencies in specific substances. Funk coined the term "vitamines" from "vital amines" because he mistakenly thought that all these substances vital for life were compounds with amino groups. In 1926 thiamin (vitamin $B_1$)—the substance lost in the rice milling process—was the first vitamin to be isolated in pure form.

Deficiency diseases can also result from an inability to absorb or process an essential nutrient even if it is present in the diet. Vitamin $B_{12}$ (cobalamin), for example, is present in all foods of animal origin. Since plants neither use nor produce vitamin $B_{12}$, a strictly vegetarian diet (not supplemented with dairy products or vitamin pills) can lead to a $B_{12}$ deficiency disease called pernicious anemia, characterized by a failure of red blood cells to mature. The most common cause of pernicious anemia, however, is not a lack of vitamin $B_{12}$ in the diet but an inability to absorb it. Normally cells in the stomach lining secrete a peptide called intrinsic factor, which binds to vitamin $B_{12}$ and makes it absorbable by the small intestine. Conditions that damage the stomach lining, such as alcoholism or gastritis, can thus lead to pernicious anemia.

Inadequate mineral nutrition can also lead to deficiency diseases. Examples are hypothyroidism and goiter resulting from iodine deficiency (see Section 41.4), and anemia resulting from iron deficiency. Iodine deficiency is almost unheard of in the developed world because we add iodide to salt. However, it is still a major health problem in large segments of the human population.

**RECAP 51.1**

As heterotrophs, animals must obtain the energy and molecular building blocks for biosynthesis from their food. Energy can come from the metabolism of carbohydrates, fats, and proteins. Molecular building blocks include carbon skeletons, vitamins, and minerals.

- Explain the different roles of dietary proteins, carbohydrates, and fats in producing metabolic energy. **See p. 1050 and Figure 51.3**
- Explain the nutritional roles of an essential carbon skeleton, a micronutrient, and a macronutrient. **See pp. 1051–1052, Figure 51.4, and Table 51.1**
- Why should fat-soluble vitamins not be taken in excess? **See p. 1053**

We have surveyed the essential elements of nutrition in animals. Next we will look at various methods and adaptations by which animals obtain the food they need, and the mechanisms they use to extract nutrients from their food.

## 51.2 How Do Animals Ingest and Digest Food?

Heterotrophic organisms can be classified by how they acquire their nutrition. **Saprobes** (also called saprotrophs) are organisms—most of which are either protists or fungi—that absorb nutrients from dead organic matter. **Detritivores** or **decomposers**, such as earthworms and crabs, actively feed on dead organic material. Animals that feed on living organisms are **predators**: **herbivores** prey on plants, **carnivores** prey on animals, and **omnivores** prey on both. **Filter feeders**, such as clams and blue whales, prey on small organisms by filtering them from the aquatic environment. **Fluid feeders** include mosquitoes, aphids, leeches, and hummingbirds. The anatomical adaptations that enable a species to exploit a particular source of nutrition are usually obvious, but physiological and biochemical adaptations are also important, although less obvious.

### The food of herbivores is often low in energy and hard to digest

Most vegetation is coarse, difficult to break down, and has low energy content. Therefore herbivores spend a great deal of time feeding and processing their food. Many have striking adaptations for feeding, such as the trunk (a flexible, gripping nose) of the elephant or the huge bill of the fruit-eating toucan, which can be half as long as its body. Many types of grinding, rasping, cutting, and shredding mouthparts have evolved in invertebrates for ingesting plant material, and the teeth of herbivorous vertebrates have been shaped by selection to tear, crush, and grind coarse plant matter.

The digestive processes of herbivores can also be quite specialized. An example is the koala, which almost exclusively eats leaves of eucalyptus trees. These leaves are very fibrous, low in usable energy and protein, and high in toxic chemicals. The koala has strong jaws for grinding the leaves, a very long gut for fermenting them, enzymes in its liver for detoxifying chemicals in the leaves, and a low metabolic rate (i.e., it expends little energy) to compensate for low energy intake.

## Carnivores must find, capture, and kill prey

The predatory behaviors of many carnivores are legendary—the hunting skills of hawks, wolves, and tigers, for example. Carnivores have evolved stealth, speed, power, large jaws, sharp teeth, and strong gripping appendages. They also have evolved remarkable means of detecting prey. Bats use echolocation, pit vipers sense infrared radiation from the warm bodies of their prey, and certain fish detect electric fields created in the water by their prey. There are many fascinating examples of adaptations for capturing prey, such as the immobilizing venom of many snakes, the long sticky tongues of chameleons, and the webs of spiders.

Some predators digest their prey externally. For example, a spider injects its insect prey with digestive enzymes and then sucks out the liquefied contents, leaving behind the empty exoskeletons frequently seen in old spider webs. The majority of animals, however, digest their food internally. For many, the process of digestion begins with the physical breaking down of the food items by the teeth.

## Vertebrate species have distinctive teeth

Teeth are adapted for the acquisition and initial processing of specific types of foods. Because they are among the hardest structures of the body, an animal's teeth remain in the environment long after it dies. Paleontologists use teeth to identify animals that lived in the distant past and to deduce their feeding behavior.

All mammalian teeth have the same general, three-layered structure (**Figure 51.6A**). An extremely hard material called **enamel**, composed principally of calcium phosphate, covers the crown of the tooth. Both the crown and root contain a layer of bony material called **dentine**, inside of which is a **pulp cavity** containing blood vessels, nerves, and the cells that produce the dentine.

There is a great deal of homology in the dentition of mammals, but the shapes and organization of mammalian teeth are adaptations to different diets (**Figure 51.6B**). In general, incisors are used for cutting, chopping, and gnawing; canines are used for stabbing, gripping, and ripping; and molars and premolars (the cheek teeth) are used for shearing, crushing, and grinding. The highly varied diet of humans is reflected in our multipurpose set of teeth, as is common among omnivores.

Current methods make it possible to analyze residues derived from food in and on tooth enamel of prehistoric animals.

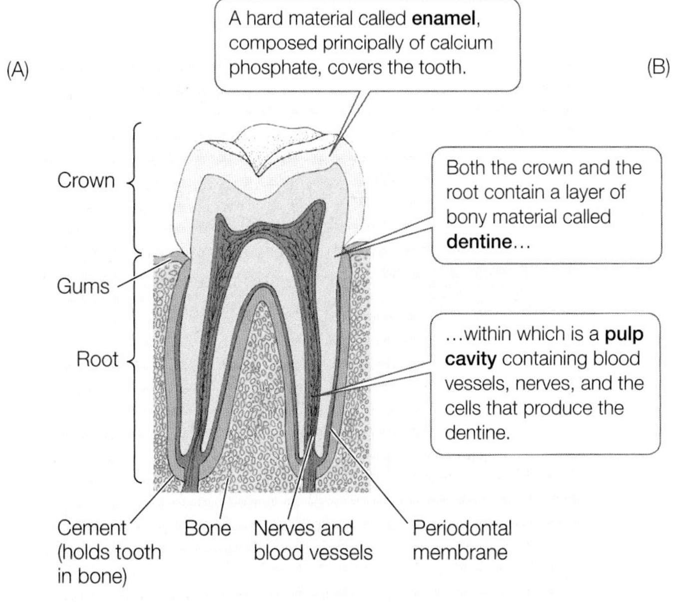


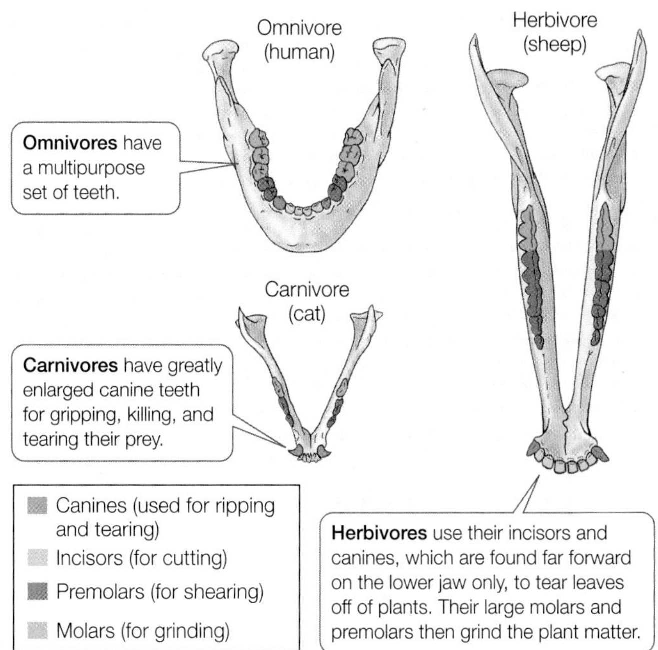


**51.6 Mammalian Teeth** (A) A mammalian tooth has three layers: enamel, dentine, and a pulp cavity. (B) The teeth of different mammalian species are specialized for different diets. This illustration depicts the teeth of the lower jaw, viewed from above.

**Go to Activity 51.3 Mammalian Teeth Life10e.com/ac51.3**

Such analysis of some hominin fossils (*Paranthropus boisei*; see Figure 33.34) from South Africa has revealed that they were predominantly grass eaters. What was the evidence for this conclusion? Tropical grasses are $C_4$ plants (see Section 10.4), whereas leafy trees that produce fruits and nuts are $C_3$ plants. The carbon residues on the *Paranthropus* teeth were those produced by $C_4$ plants, presumably grasses and sedges.

## Digestion usually begins in a body cavity

Animals take food into a body cavity that is continuous with the outside environment. They secrete digestive enzymes into that cavity, and the enzymes break down the food into nutrient molecules that can be absorbed by the cells lining the cavity.

The simplest digestive system is found in the simplest animals, the sponges. Water flows from the environment through the body of the sponge in water channels, and individual cells capture food particles from the water (see Figure 31.2B). A slightly more specialized digestive system is a **gastrovascular cavity**, which connects to the outside world through a single opening. Cnidarians, such as jellyfish, capture prey using stinging nematocysts and use their tentacles to cram the prey into their gastrovascular cavity. Enzymes in the gastrovascular cavity partially digest the prey. Cells lining the cavity take in small food particles by endocytosis. The vesicles created by endocytosis then fuse with lysosomes containing digestive enzymes, and intracellular digestion completes the breakdown of the food. Nutrients are released to the cytoplasm as the vesicles break down.

## Tubular guts have an opening at each end

The guts of most animals are tubular: a **mouth** takes in food; molecules are digested and absorbed throughout the length of the gut; and solid digestive wastes are eliminated through an **anus**. Different regions in the tubular gut are specialized for particular functions (**Figure 51.7**). These functions must be coordinated so they occur in the proper sequence and at rates that maximize the efficiency of digestion and absorption of nutrients.

At the anterior end of the gut is the mouth cavity where food can be fragmented by teeth (in many vertebrates), by a **radula** (in snails), or by **mandibles** (in many arthropods). In most birds, food is ground by small stones in an early, muscular portion of the gut called the **gizzard**. Some animals, such as snakes, simply ingest whole prey with little or no fragmentation. **Stomachs** and **crops** are storage chambers that enable animals to ingest relatively large amounts of food when it is available, and then digest it gradually. In these storage chambers, food may be further fragmented and mixed, and in most vertebrates it is an important site of digestion. Food delivered into the next section of the gut, the **intestine**, is in small particles, well mixed, and usually partially digested.

Most digestion occurs in the intestine, and nutrients, water, and ions are absorbed across its walls. Glands secrete digestive enzymes into the intestine, and other enzymes are produced and secreted by cells lining the intestine. The final segment of the intestine recovers water and ions and stores undigested

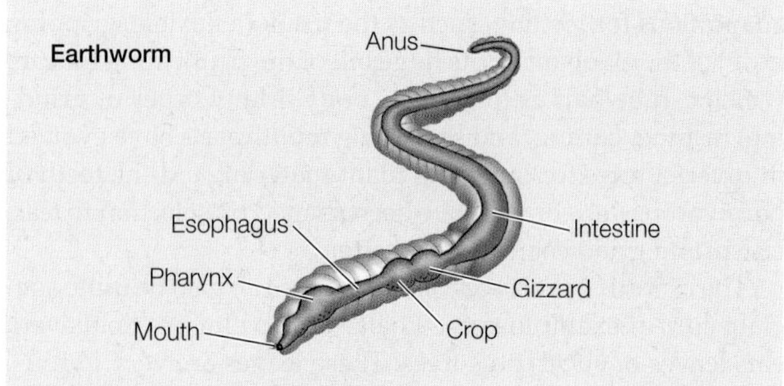


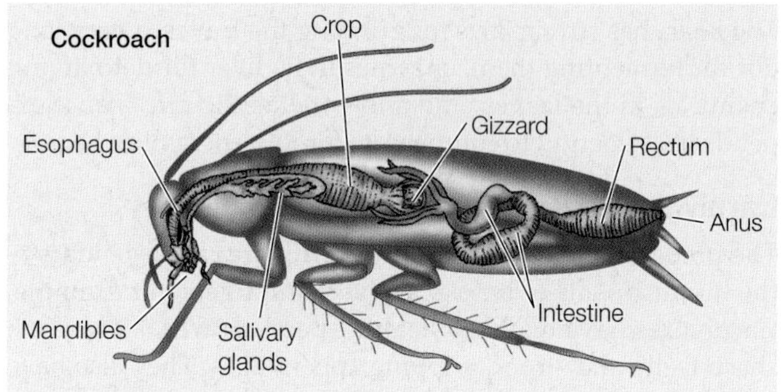


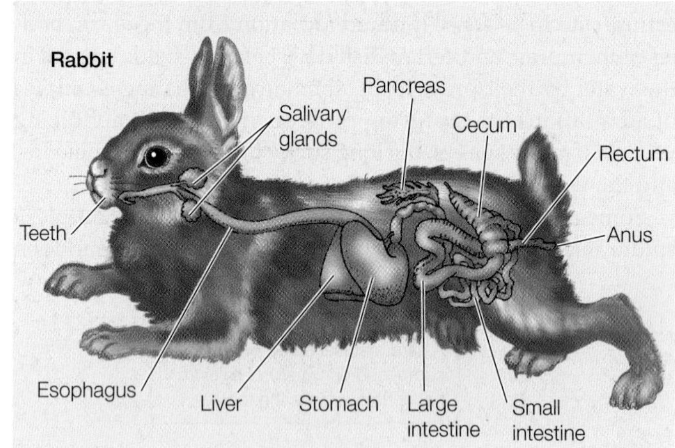


**51.7 Compartments for Digestion and Absorption** Most invertebrates and all vertebrates have a tubular gut that begins with a mouth, which takes in food, and ends in an anus, which eliminates wastes. Between these two structures are specialized regions for digestion and nutrient absorption; the structures in these regions are adapted to different diets and vary from species to species.

wastes, or **feces**, so they can be released to the environment at an appropriate time or place. A muscular **rectum** near the anus assists in expelling feces.

Endosymbiotic bacteria colonize the intestines. These bacteria obtain their nutrition from the food passing through the host's gut while contributing to the host's digestive processes. Members of the leech genus *Hirudo*, for example, produce no enzymes that can digest the proteins in the blood they suck from vertebrates; instead they depend on bacteria to perform this service. The resulting amino acids are subsequently used by both the leech and the bacteria. The microorganisms in

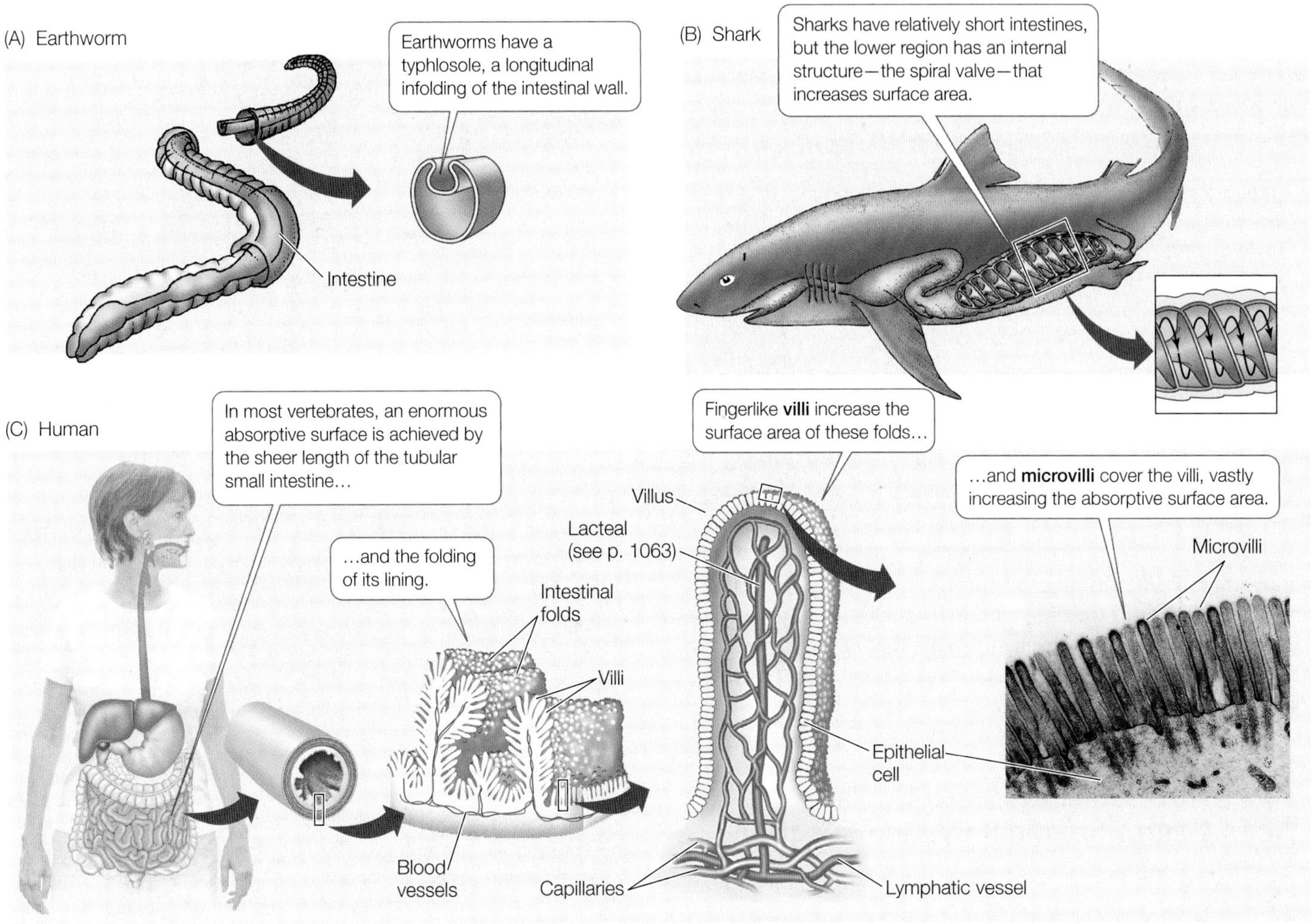

**51.8 Intestinal Surface Area and Nutrient Absorption** Maximizing the surface area of the gut increases an animal's ability to absorb nutrients.

the human gut are called the "forgotten organ" because they provide important services in digestion, prevent the establishment of harmful microorganisms, and even produce some vitamins (vitamin K and biotin). This forgotten organ is huge. It is estimated that the human body consists of $10^{13}$ cells, but our guts contain probably ten times that number of unicellular organisms representing at least 500 different species (see Figure 26.21).

In many animals, the parts of the gut that absorb nutrients have greater surface areas than would be expected of a simple tube. The simplest way to increase the surface area of a tube without changing its diameter is to produce an infolding. Such an infolding of the gut, called a typhlosole, is seen in earthworms (**Figure 51.8A**). Sharks, which have large stomachs but short intestines, have a unique adaptation called a spiral valve (**Figure 51.8B**). The lower region of the intestine is enlarged and has an internal structure that forces the food to pass through it in a spiral fashion (like going down a spiral staircase). The walls of the spiral present a large surface area for absorption of nutrients. This spiral passage, however, will not accommodate large chunks of food or undigestible matter, so food remains in the stomach for a long time to be broken down, and some larger, undigestible items (such as pieces of surfboard) do not leave the stomach except by regurgitation.

In humans, as in most other vertebrates, the wall of the intestine is highly folded, with the individual folds bearing legions of tiny fingerlike projections called **villi** (**Figure 51.8C**). The cells that line the surfaces of the villi, in turn, have microscopic projections called **microvilli**. The microvilli give the intestine an enormous internal surface area for absorbing nutrients.

## Digestive enzymes break down complex food molecules

Protein, carbohydrate, and fat macromolecules are broken down into their simplest monomeric units by hydrolytic enzymes produced at different locations in the digestive tract. Many are secreted into the lumen of the gut, and others remain associated with the membranes of the microvilli. All of these enzymes cleave the chemical bonds of macromolecules through hydrolysis, a reaction that adds a water molecule (see Figure 3.4B). Digestive enzymes are classified according to the substances they hydrolyze: **proteases** break the bonds between adjacent amino acids in proteins; **carbohydrases** hydrolyze carbohydrates; **peptidases** break down peptides; **lipases**, fats; and **nucleases**, nucleic acids.

**RECAP 51.2**

Heterotrophs have diverse adaptations for acquiring food. Once captured or ingested, food is digested extracellularly by secreted enzymes to release nutrients, which are absorbed into the animal's body, usually via a tubular gut.

- Why do herbivores typically spend a great deal of their time feeding? **See p. 1054–1055**
- What characteristics of tooth structure distinguish herbivores, carnivores, and omnivores? **See p. 1055 and Figure 51.6.**
- What are three adaptations of the human gut that increase its surface area for absorption of nutrients? **See p. 1057 and Figure 51.8**

Once ingested by an animal, food may be fragmented and moved into the gut for digestion by hydrolytic enzymes. The processes of digestion release nutrients that are absorbed into the animal's body. Next we will focus on how those processes occur in vertebrates.

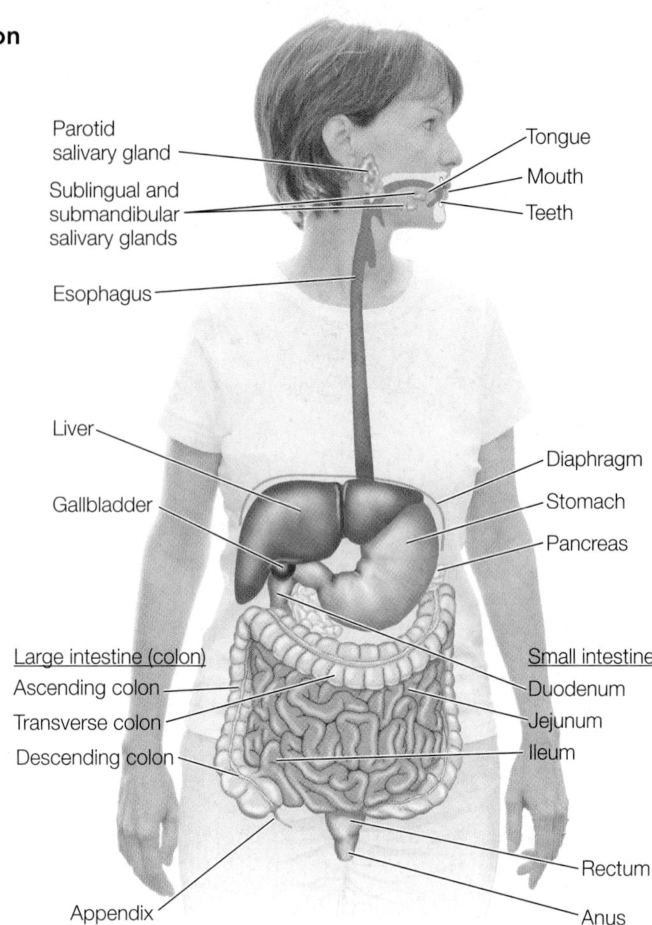


**51.9 The Human Digestive System** Different compartments in the long tubular gut specialize in digesting food, absorbing nutrients, and storing and expelling wastes. Accessory organs contribute secretions containing enzymes and other molecules.

**Go to Activity 51.4 The Human Digestive System Life10e.com/ac51.4**

## 51.3 How Does the Vertebrate Gastrointestinal System Function?

Digestion in vertebrates occurs in the gastrointestinal system, which includes a tubular gut running from mouth to anus and several accessory structures that produce secretions that play important roles in digestion (**Figure 51.9**). In this section we will consider three important processes of this system: the movement of food through it, the sequential steps of digestion, and the absorption of nutrients. We will use as our primary example a typical vertebrate, the human.

### The vertebrate gut consists of concentric tissue layers

The tissues of the vertebrate gut are arranged in concentric layers that have a similar organization throughout its length (**Figure 51.10**). Starting in the internal cavity, or **lumen**, the first layer is the **mucosa**, which consists of delicate epithelial cells with underlying connective tissue. Some cells of this mucosal epithelium secrete mucus to lubricate and protect the walls of the gut; some secrete digestive enzymes; and some secrete hormones. Mucosal epithelial cells in the stomach secrete hydrochloric acid and, as we noted in Section 51.1, some secrete intrinsic factor to aid the absorption of vitamin $B_{12}$. In some regions of the gut, nutrients are absorbed by mucosal epithelial cells. The apical plasma membranes of these absorptive cells have microvilli that increase the surface area over which absorption can take place (see Figure 51.8C).

At the base of the mucosa are smooth muscle cells that move the mucosa to improve contact with gut contents, and just under the mucosa is the submucosal tissue layer. Here we find the blood and lymph vessels that carry absorbed nutrients to the rest of the body. The **submucosa** also contains a network of nerves; the neurons in this network have sensory functions (responsible for stomach aches) and also control various secretory functions of the gut.

External to the submucosa are two layers of smooth muscle responsible for the large movements of the gut. Innermost is the circular muscle layer, with its cells oriented around the gut. Outermost is the longitudinal muscle layer, with its cells oriented along the length of the gut (see Figure 51.10). The circular muscles constrict the gut, and the longitudinal muscles shorten it. Between the two layers of smooth muscle is another nerve network that controls and coordinates the movements of the gut. The coordinated activity of the two smooth muscle layers mixes the content of the gut and moves it continuously toward the rectum. The stomach has a third layer of smooth muscle that is closest to the lumen. The orientation of its fibers is oblique to the longitudinal and circular layers. This third layer is important in generating churning motions of the stomach that mix the food and digestive juices.

**Go to Media Clip 51.1 Following Food from Mouth to Gut Life10e.com/mc51.1**

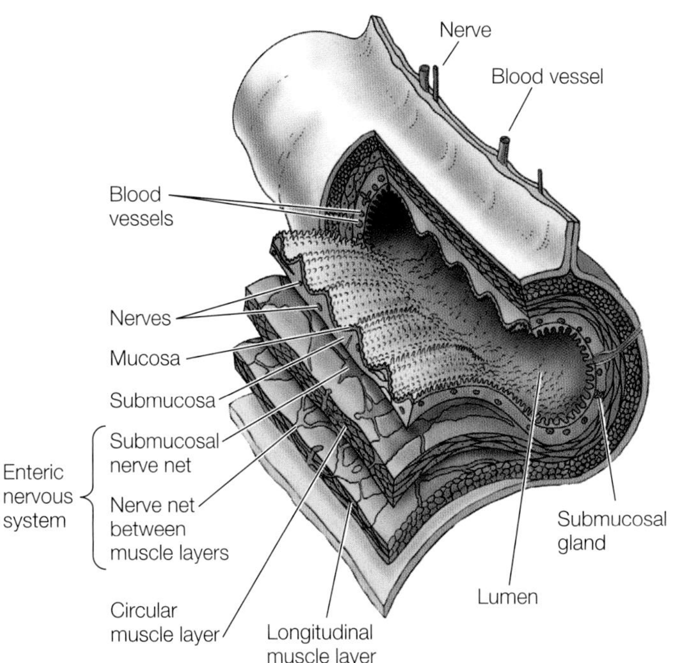


**51.10 Tissue Layers of the Vertebrate Gut** The organization of tissue layers is the same in all compartments of the gut, but specialized adaptations of specific tissues characterize different regions.

The nerve nets in the submucosa and between the smooth muscle layers are called the **enteric nervous system**, and they are unusual. Whereas most neurons of the peripheral nervous system either receive synapses from neurons in the central nervous system (CNS) or contribute synapses to neurons in the CNS, most of the neurons in the enteric nervous system form synapses only with other neurons in their network. Thus they are responsible for communication within the gut. The CNS can influence activity in the enteric nervous system and receive information from it, but the gut truly has "a mind of its own."

A tissue membrane called the **peritoneum** surrounds the gut, as it does all of the organs of the abdominal cavity as well as lining the wall of the cavity. The peritoneum includes connective and epithelial tissues that secrete a fluid that lubricates the organs so they can easily slide against each other in the body cavity.

## Mechanical activity moves food through the gut and aids digestion

In humans and most other mammals, food is chewed in the mouth and mixed with saliva. Periodically the tongue pushes a bolus (mass) of the chewed food toward the throat. By making contact with the soft palate at the back of the mouth cavity, the food bolus initiates swallowing, which is a complex series of reflexes. Swallowing propels the food through the pharynx (where the mouth cavity and nasal passages join) and into the **esophagus** (food tube). To prevent food from entering the trachea (windpipe), the larynx (voice box) closes, and a flap of tissue called the **epiglottis** covers the entrance to the larynx (**Figure 51.11A**).

Once a bolus of food enters the esophagus, it is moved toward the stomach both by the force of gravity and by waves of muscle contraction called **peristalsis** (**Figure 51.11B**). The muscle of the upper region of the esophagus is striated (i.e., skeletal muscle) and is controlled by the central nervous system reflexes of swallowing. The muscles of the rest of the esophagus are smooth muscle controlled by the autonomic and enteric nervous systems.

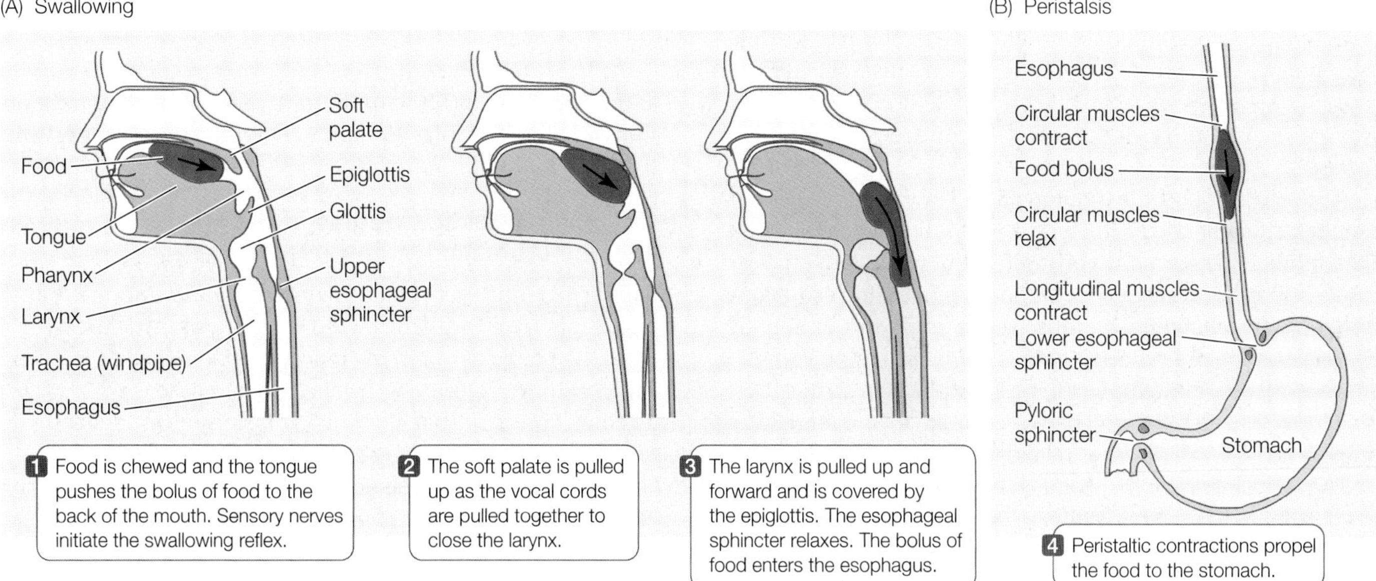


**51.11 Swallowing and Peristalsis** (A) Food pushed to the back of the mouth triggers the swallowing reflex. (B) Once a food bolus enters the esophagus, peristalsis propels it from mouth to anus by coordinated actions of the circular and longitudinal muscle layers of the gut.

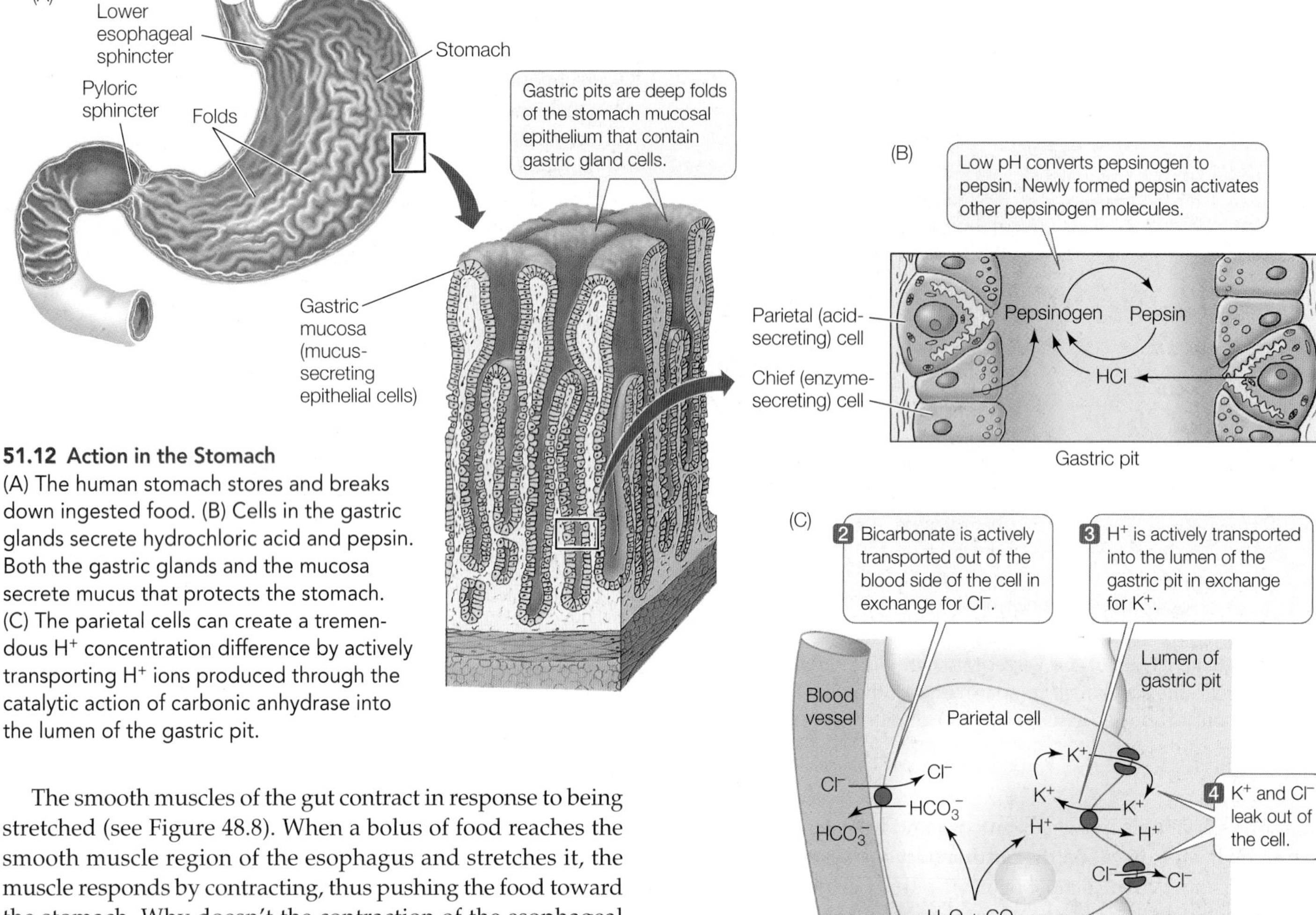


**51.12 Action in the Stomach**
(A) The human stomach stores and breaks down ingested food. (B) Cells in the gastric glands secrete hydrochloric acid and pepsin. Both the gastric glands and the mucosa secrete mucus that protects the stomach. (C) The parietal cells can create a tremendous $H^+$ concentration difference by actively transporting $H^+$ ions produced through the catalytic action of carbonic anhydrase into the lumen of the gastric pit.

The smooth muscles of the gut contract in response to being stretched (see Figure 48.8). When a bolus of food reaches the smooth muscle region of the esophagus and stretches it, the muscle responds by contracting, thus pushing the food toward the stomach. Why doesn't the contraction of the esophageal smooth muscle push the food back toward the mouth? The nerve net between the two smooth muscle layers coordinates the muscles so that contraction is always preceded by an anticipatory wave of relaxation. When a region of the gut smooth muscle contracts, the circular smooth muscle just beyond it relaxes while the longitudinal smooth muscle contracts, pushing the food into that area. The resulting stretch causes that circular smooth muscle to contract while the next region relaxes. In this way peristalsis moves food down the gut from the mouth to the anus.

At the junction of the esophagus and stomach is the lower esophageal **sphincter**, a thick ring of circular smooth muscle. This sphincter is normally constricted, but waves of peristalsis cause it to relax enough to let food pass from the esophagus into the stomach. Sphincter muscles are found throughout the digestive tract: the pyloric sphincter governs the passage of stomach contents into the small intestine; the ileocaecal sphincter controls the flow of food between the small and large intestines; and the anal sphincter relaxes to allow defecation.

The movements of the stomach and small intestine are not as coordinated as the peristaltic movements of the esophagus. They are referred to as **segmentation movements** because segments of the gut periodically contract but do not generate a peristaltic wave of contraction that moves the food in one direction. Segmentation movements move the content of the gut in either direction and thereby mix it with the digestive juices and bring it in contact with the gut walls. Weak peristaltic activity in the small intestine still moves the gut content toward the large intestine.

## Chemical digestion begins in the mouth and the stomach

Salivary glands secrete the enzyme amylase into the mouth where it is mixed with the food being chewed. Amylase hydrolyzes the bonds between the glucose monomers that make up carbohydrate molecules. The action of amylase is what makes a chewed piece of bread or cracker taste slightly sweet if you hold it in your mouth long enough.

A main role of the stomach is to store food so that digestion can occur more slowly than ingestion, but the stomach also secretes digestive enzymes. **Gastric pits** in the stomach walls are lined with three types of secretory cells (**Figure 51.12A**). **Chief cells** secrete a proteolytic enzyme, pepsin, that begins the digestion of protein. **Parietal cells** pump out $H^+$ ions, and the lowered pH kills most ingested microorganisms. This mix of substances could damage the stomach walls, but the epithelial cells of the gastric mucosa secrete mucus that provides a protective coating for the walls of the gastric pits and stomach.

Chief cells actually secrete an inactive digestive enzyme, or **zymogen**, called pepsinogen. The extremely low pH of the stomach juices initiates the conversion of pepsinogen to pepsin by cleaving away a sequence of amino acids that masks the active site of the enzyme. Newly activated pepsin activates other pepsinogen molecules, creating a positive feedback process called **autocatalysis** (**Figure 51.12B**).

The parietal cells of the gastric pits can produce about 2 liters of hydrochloric acid (HCl) per day—enough to bring the pH of the stomach contents below 1, which is the same as battery acid and ten times more acidic than pure lemon juice. This means that across their plasma membranes, gastric pits can create a $H^+$ ion concentration difference of 3 million-fold. Such a feat of transport is not seen anywhere else in the body. How do the gastric pits do it? Enzymes and transporters are involved.

The enzyme carbonic anhydrase in parietal cells catalyzes the hydration of $CO_2$ to $H_2CO_3$, which dissociates into $H^+$ and bicarbonate ion ($HCO_3^-$). An antiporter transport protein (see Figure 6.13) exchanges $HCO_3^-$ for $Cl^-$ on the blood side of the gastric pits, and an antiporter on the gastric pit side exchanges $H^+$ for $K^+$ (**Figure 51.12C**). However, this $K^+$ can leak out again down its concentration gradient. Thus the inward transport of $K^+$ acts like an endless conveyer belt moving $H^+$ out into the stomach lumen. $Cl^-$ also passively leaks out of the gastric lumen side of the parietal cells to maintain electrical neutrality.

## The stomach gradually releases its contents to the small intestine

Contractions of the smooth muscles in the walls of the stomach churn its contents, thoroughly mixing them with the stomach secretions. The acidic, fluid mixture of gastric juice and partly digested food in the stomach is called **chyme**. A few substances can be absorbed across the stomach wall, including alcohol (hence its rapid effects), aspirin, and caffeine, but even these substances are absorbed in rather small quantities from the stomach.

Contractions of the stomach walls push the chyme toward the bottom of the stomach. These waves of contractions cause the pyloric sphincter to relax briefly so that little squirts of the chyme can enter the small intestine. In this manner the human stomach empties itself gradually over a period of approximately 4 hours. This slow introduction of food into the small intestine enables it to work on a little material at a time.

## Most chemical digestion occurs in the small intestine

In the **small intestine**, the digestion of carbohydrates and proteins continues, and the digestion of fats and absorption of nutrients begin. The small intestine takes its name from its diameter; it is in fact a very large organ, about 6 meters long in an adult human. Given its length and the folds, villi, and microvilli of its lining, its inner surface area is roughly the size of a tennis court. Across this surface the small intestine absorbs all the nutrient molecules derived from food.

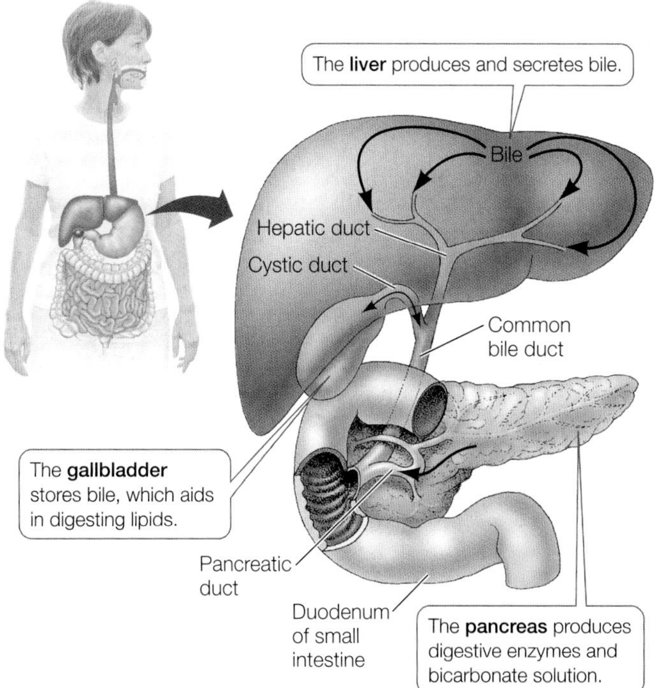


**51.13 Ducts of the Gallbladder and Pancreas** Bile produced in the liver leaves the liver via the hepatic duct. Branching off this duct is the gallbladder, which stores bile. Below the gallbladder, the hepatic duct is called the common bile duct and is joined by the pancreatic duct before entering the duodenum.

The small intestine of humans has three sections. The initial section (about 25 cm long) is called the **duodenum** and is the site of most digestion; the **jejunum** and the **ileum** (together about 600 cm) carry out 90 percent of the absorption of nutrients (see Figure 51.9).

Digestion in the small intestine requires many specialized enzymes, as well as several other secretions. Two accessory organs that are not part of the digestive tract—the liver and the pancreas—produce many of these secretions and deliver them to the lumen of the intestine through ducts.

LIVER The liver synthesizes **bile salts** from cholesterol and secretes them as **bile**. Bile also includes other substances, such as phospholipids and bilirubin (the breakdown product of hemoglobin). Bile flows from the liver through the hepatic duct. A side branch off the hepatic duct called the cystic duct goes to the **gallbladder**, where bile is stored. Below this junction, the hepatic duct is called the common bile duct. Before it reaches the duodenum, the common bile duct is joined by the pancreatic duct (**Figure 51.13**).

Fat entering the duodenum stimulates cells of the duodenal epithelium to release the hormone **cholecystokinin** (**CCK**), which stimulates the walls of the gallbladder to contract rhythmically. As a result, bile is squeezed out of the gallbladder and through the cystic duct to the common bile duct. A small sphincter at the junction of the common bile duct with the duodenum relaxes in response to waves of peristalsis and allows squirts of bile to enter the duodenal lumen.

To understand the role of bile in fat digestion, think of an oil-and-vinegar salad dressing. The oil, which is hydrophobic, tends to aggregate in large globules. For that reason, many salad dressings include an emulsifier—something that prevents oil droplets from aggregating. Mayonnaise, for example, is oil and vinegar with egg yolk added as an emulsifier. Bile salts emulsify fats in the chyme. One end of each bile salt molecule is lipophilic (soluble in fat), and the other end is hydrophilic (soluble in water). The lipophilic ends of bile molecules merge with the fat droplets, leaving their hydrophilic ends sticking out. As a result, bile salts prevent the fat droplets from sticking together and thereby greatly enlarge the surface area of the fats exposed to the lipases—the enzymes that digest fats. The very small fat particles that result are called **micelles** (**Figure 51.14A**).

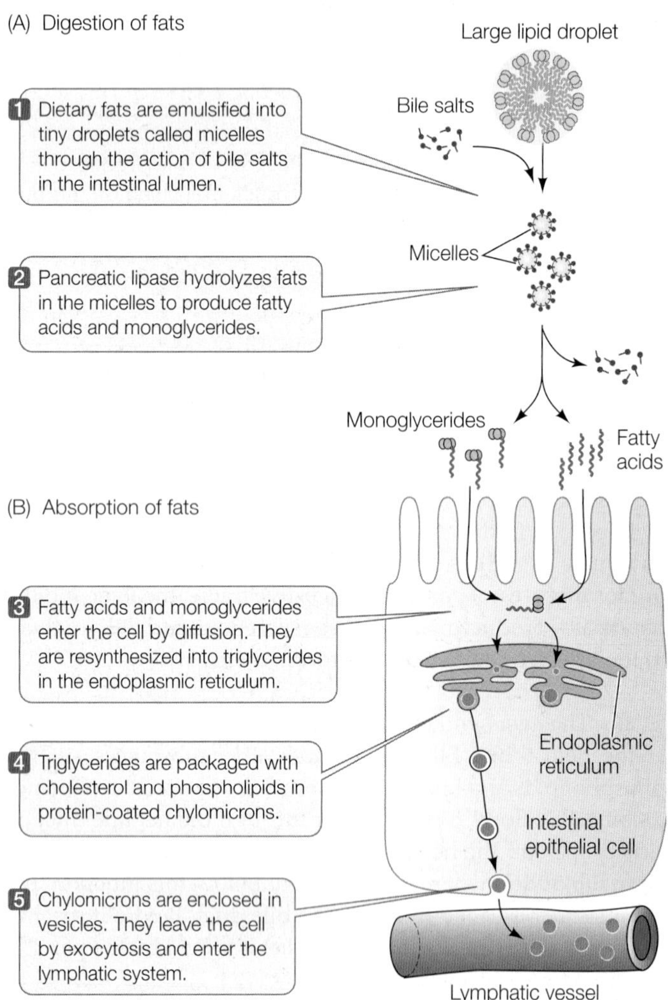


**51.14 Digesting Fats** (A) Dietary fats are broken up by bile into small micelles that present a large surface area to lipases. (B) The products of fat digestion are absorbed by intestinal mucosal cells, where they are resynthesized into triglycerides and exported to lymphatic vessels.

**Go to Animated Tutorial 51.1**
**The Digestion and Absorption of Fats**
Life10e.com/at51.1

PANCREAS The **pancreas** is a large gland that lies just behind and below the stomach (see Figures 51.9 and 51.13). It is both an endocrine gland (secreting hormones into the extracellular fluid; see Section 41.1) and an exocrine gland (secreting digestive juices through the pancreatic duct to the gut lumen). The exocrine tissues of the pancreas produce a host of digestive enzymes, including lipases, amylases, proteases, and nucleases (**Table 51.3**). As in the stomach, the protease enzymes are released as zymogens; if proteases were not in this inactive state, they would digest the pancreas and its ducts before ever reaching the duodenum. Once in the duodenum, the zymogen trypsinogen is activated by the enzyme enterokinase (secreted by cells lining the duodenum) to produce the active protease **trypsin**. Trypsin cleaves other zymogens, releasing other proteases as well as more active trypsin.

The mixture of zymogens produced by the pancreas can be dangerous if the pancreatic duct is blocked or if the pancreas is injured by infection or physical trauma such as a blow to the abdomen. A few activated trypsin molecules can initiate a chain reaction of enzyme activity that digests the tissues of the pancreas (a condition called pancreatitis), destroying both its endocrine and exocrine functions.

The pancreas also produces a secretion rich in bicarbonate ions ($HCO_3^-$). Bicarbonate ions are alkaline (basic) and neutralize the acidic pH of the chyme that enters the duodenum from the stomach. Intestinal enzymes function best at a neutral or slightly alkaline pH.

TABLE **51.3**
Major Digestive Enzymes of Humans

| Source/Enzyme | Action |
|---|---|
| SALIVARY GLANDS | |
| Salivary amylase | Starch → Maltose |
| STOMACH | |
| Pepsin | Proteins → Peptides; autocatalysis |
| PANCREAS | |
| Pancreatic amylase | Starch → Maltose |
| Lipase | Fats → Fatty acids and glycerol |
| Nuclease | Nucleic acids → Nucleotides |
| Trypsin | Proteins → Peptides; zymogen activation |
| Chymotrypsin | Proteins → Peptides |
| Carboxypeptidase | Peptides → Shorter peptides and amino acids |
| SMALL INTESTINE | |
| Aminopeptidase | Peptides → Shorter peptides and amino acids |
| Dipeptidase | Dipeptides → Amino acids |
| Enterokinase | Trypsinogen → Trypsin |
| Nuclease | Nucleic acids → Nucleotides |
| Maltase | Maltose → Glucose |
| Lactase | Lactose → Galactose and glucose |
| Sucrase | Sucrose → Fructose and glucose |

### Nutrients are absorbed in the small intestine

The final step in digesting proteins and carbohydrates and absorbing their components occurs among the microvilli. Mucosal epithelial cells produce peptidases that cleave small peptides into absorbable amino acids. These epithelial cells also produce the enzymes maltase, lactase, and sucrase that cleave the common disaccharides into absorbable monosaccharides—glucose, galactose, and fructose. There is also some lipase activity for fat digestion.

Many humans stop producing the enzyme lactase in childhood and thereafter have difficulty digesting lactose (the sugar in milk). Lactose is a disaccharide and cannot be absorbed without being cleaved into its constituents, glucose and galactose. Unabsorbed lactose is metabolized by bacteria in the large intestine, causing gas, diarrhea, and abdominal cramps.

The mechanisms by which cells of the intestinal epithelium absorb nutrients and inorganic ions are diverse and include diffusion, facilitated diffusion, osmosis, active transport, and co-transport. Many inorganic ions such as sodium, calcium, and iron are actively transported by these cells. For example, active $Na^+$ transporters exist on the basal and lateral sides of the epithelial cells. They maintain a low concentration of $Na^+$ in those cells so that $Na^+$ can diffuse in from the chyme in the intestinal lumen. About 30 grams of $Na^+$ are transported this way every day, and $Cl^-$ follows.

The transport of $Na^+$ and other ions is also important for water absorption because it creates an osmotic concentration gradient. At least 7 to 8 liters of water per day move through the spaces between the epithelial cells in response to this osmotic gradient. Because the water moves through *spaces* between the cells and not through the cells themselves, it can carry with it nutrients that are in solution—a transport mechanism called "solvent drag."

Many different kinds of transport proteins exist in the epithelial cell membranes. Some, such as the transport protein for fructose, only facilitate diffusion, and that requires a concentration gradient. This mechanism works for fructose because once fructose enters the cell it is converted to glucose. Thus the concentration of fructose in the cell is always low and the concentration gradient is maintained. Transport proteins known as symporters (see Figure 6.13) exploit the concentration gradient of $Na^+$ between the inside and outside of the cell that is maintained by the $Na^+/Ka^+$ ATPase common to all cells. Symporters combine the transport of $Na^+$ and another molecule, such as glucose, galactose, or an amino acid. As $Na^+$ moves down its concentration gradient into the cell, the "hitchhiking" molecules are carried along with it.

The absorption of the products of fat digestion is relatively simple. Triglycerides are hydrolyzed to diglycerides, monoglycerides, and fatty acids, all of which are lipid-soluble and thus able to pass through the plasma membranes of the microvilli. In the intestinal epithelial cells, these molecules are resynthesized into triglycerides, combined with cholesterol and phospholipids, and coated with protein to form water-soluble **chylomicrons** (**Figure 51.14B**). Rather than enter the blood directly, chylomicrons pass into blind-ended lymph vessels called **lacteals** that are inside each villus (see Figure 51.8C). They then flow through the lymphatic system, entering the bloodstream through the thoracic ducts at the base of the neck. After a meal rich in fats, chylomicrons can be so abundant in the blood that they give the plasma a milky appearance. Chylomicrons deliver their triglyceride and cholesterol cargo as they circulate through tissues.

The bile salts that emulsify fats are not absorbed along with the monoglycerides, diglycerides, and the fatty acids, but are shuttled back and forth between the gut contents and the microvilli. In the ileum, bile salts are actively reabsorbed and returned to the liver via the bloodstream.

### Absorbed nutrients go to the liver

Blood leaving the digestive tract flows to the liver in the **hepatic portal vein**. This large vein delivers the blood to small spaces called sinusoids between groups of liver cells. These cells absorb the nutrients coming from the digestive tract and either store them or convert them to molecules the body needs. Glucose, sucrose, and fructose are used to synthesize glycogen. Amino acids are used to build proteins. Lipids from the chylomicrons are either stored as triglycerides or used to make lipoproteins, which are released by the liver and carry the triglycerides and cholesterol to other tissues (see Section 51.4).

### Water and ions are absorbed in the large intestine

The motility of the small intestine gradually pushes its contents into the large intestine, or **colon**. Most of the available nutrients have been removed from the chyme that enters the colon, but it contains a lot of water and inorganic ions. Segmentation movements bring the colon's contents into contact with its walls and promote reabsorption of ions and water, producing feces, a semisolid mass of waste products. Absorption of too much water from the colon can cause constipation. The opposite condition, diarrhea, results if too little water is absorbed; in this case, water in the colon is excreted with the feces. Excessive diarrhea caused by diseases such as cholera can produce such rapid loss of water and electrolytes that death can occur in hours.

Fecal matter is stored in the descending colon and in the rectum until defecation occurs. This is usually once a day and is preceded by strong peristaltic activity. The distension of the walls of the rectum by fecal matter initiates a parasympathetic reflex, causing the rectal muscle to contract and the internal anal sphincter to relax. In addition, there is an external anal sphincter that is under conscious control so that defecation is not entirely an involuntary act.

### Herbivores rely on microorganisms to digest cellulose

As the primary component of plant cell walls, cellulose is the principal component of the food of herbivores. Most herbivores, however, cannot produce cellulases, the enzymes that break down cellulose. (Exceptions include earthworms, shipworms, and the silverfish that eat books and stored papers.) From termites to cattle, herbivores rely on microorganisms in their digestive tracts to digest cellulose.

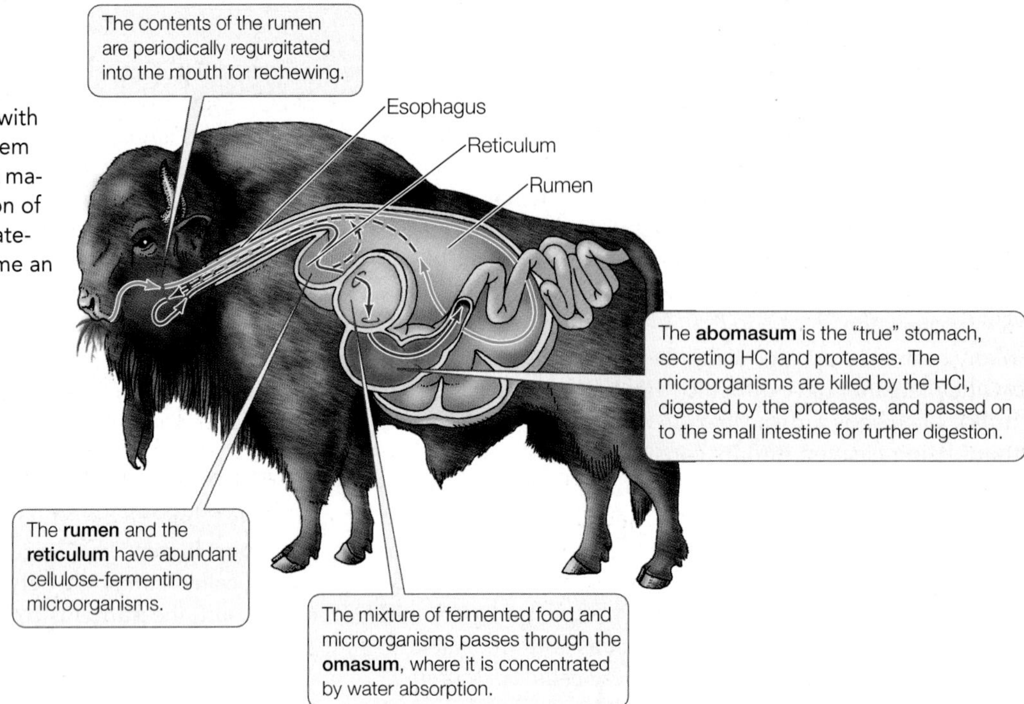


**51.15 A Ruminant's Stomach**
Bison, like their relatives domestic cattle, have a specialized stomach with four compartments that enables them to obtain energy from coarse plant material through bacterial fermentation of the otherwise indigestible plant material. The bacteria themselves become an important source of nutrition.

The stomachs of **ruminants** (cud chewers) such as cattle are large, four-chambered organs that take advantage of their endosymbiotic microorganisms (**Figure 51.15**). The first two chambers, the **rumen** and the **reticulum**, are packed with microorganisms that break down cellulose by fermentation. The ruminant periodically regurgitates the contents of the rumen (the cud) into the mouth for more chewing. When swallowed again, the vegetal fibers present more surface area to the microorganisms. The microorganisms metabolize cellulose and other nutrients to simple fatty acids, which are nutrients for their host.

Enormous numbers of microorganisms leave the rumen along with the partially digested food. This mass is concentrated by water absorption in the **omasum** before it enters the true stomach, the **abomasum**, where the microorganisms are killed by secreted hydrochloric acid, digested by proteases, and passed on to the small intestine for further digestion and absorption. A cow derives more than 100 grams of protein per day from digestion of its endosymbiotic microorganisms. The rate of multiplication of microorganisms in the rumen offsets their loss, so a well-balanced, mutually beneficial relationship is maintained.

Some mammalian herbivores have a microbial fermentation chamber called a **cecum** extending from the large intestine. An example is the rabbit (see Figure 51.7). Since the cecum empties into the large intestine, absorption of some nutrients produced by the microorganisms is inefficient, because of the large intestine's limited surface area. Such species frequently produce two kinds of feces—ones that are pure waste and ones that contain cecal material. In a behavior known as **coprophagy**, these species reingest the cecal feces directly from the anus so they can digest and absorb the nutrients that would otherwise be lost. In humans the cecum is small and ends in the vestigial **appendix**, which serves no digestive function.

**RECAP 51.3**

The vertebrate gastrointestinal system is a tubular gut that is adapted to ingest food, fragment it, digest it, and absorb nutrients. Peristalsis moves food through the gut. Segmentation movements mix the gut contents. Digestion and absorption of nutrients occur mostly in the small intestine; water and ions are absorbed in the large intestine.

- What digestive functions occur in the mouth and stomach? **See p. 1060–1061 and Figure 51.12**
- How do bile salts assist in the digestion of fats? **See p. 1062 and Figure 51.14**
- Describe how symporters drive the absorption of nutrients. **See p. 1063**

The steps included in ingestion and digestion of food—from fragmentation in the mouth to the digestive processes in the gastrointestinal tract—make the nutrients in food available for absorption and ultimately for metabolism. We look next at how the processes of digestion are controlled and how nutrients are handled by the body once food has been digested.

## 51.4 How Is the Flow of Nutrients Controlled and Regulated?

The vertebrate gut is an assembly line in reverse—a *dis*assembly line. As with a standard assembly line, the control and coordination of the sequential processes of digestion are critical. Both neuronal and hormonal controls govern these processes. Once the products of digestion are absorbed, their availability to the cells of the body must also be controlled.

You have certainly experienced salivation at the sight or smell of food. That response is an unconscious reflex, as is swallowing. Many such autonomic reflexes coordinate activity in different regions of the digestive tract. For example, the introduction of food into the stomach stimulates increased activity in the colon that may lead to defecation.

Neuronal messages travel from one region of the digestive tract to another in the enteric nervous system. One function of the gut's nervous system is coordinating the movement of food through the gut. Of course, this intrinsic nervous system communicates information to the CNS and receives input from the CNS, but its most important role is to coordinate actions throughout the digestive tract. In spite of this marvelous intrinsic nervous system, however, much of the control and regulation of the digestive system and nutrient management involves hormonal mechanisms.

## Hormones control many digestive functions

Several hormones control the activities of the digestive tract and its accessory organs (**Figure 51.16**). The first hormone ever discovered came from the duodenum; it was called **secretin** because it causes the pancreas to secrete digestive juices. We now know that secretin is only one of several hormones that control pancreatic secretion; specifically, secretin stimulates the pancreas to secrete a solution rich in bicarbonate ions.

The stimulus that causes the duodenum to release secretin is low pH caused by the arrival of acidic chyme from the stomach. Similarly, the presence of fats and proteins in the chyme stimulates the release of cholecystokinin (CCK), the hormone that stimulates the gallbladder to release bile. CCK also stimulates the pancreas to release digestive enzymes. Both CCK and secretin slow the movements of the stomach, thus slowing the delivery of chyme into the small intestine and allowing more complete digestion in the duodenum.

The presence of food in the stomach stimulates cells in the lower region of the stomach to secrete a hormone called **gastrin**. Gastrin returns to the stomach in the blood and stimulates the secretion of digestive juices and also the movements of the stomach. Gastrin release begins to be inhibited when the pH of the stomach contents falls below 3—an example of negative feedback.

Most animals do not eat continuously, so they can be either in an **absorptive state** (food in the gut) or in a **postabsorptive state** (no food in the gut). Nutrient requirements for energy metabolism and biosynthesis are continuous, however. Thus nutrient traffic must be controlled so that reserves accumulate in the liver, muscle, and adipose (fat) tissue while the animal is in the absorptive state and are then used efficiently during the postabsorptive state.

## The liver directs the traffic of the molecules that fuel metabolism

When fuel molecules are abundant in the blood, the liver stores them in the form of glycogen and fats. The liver also synthesizes blood plasma proteins from circulating amino acids. When levels of fuel molecules in the blood decline, the liver taps its reserves and delivers nutrients into the blood.

The liver has an enormous capacity to interconvert fuel molecules. Liver cells can convert monosaccharides into either

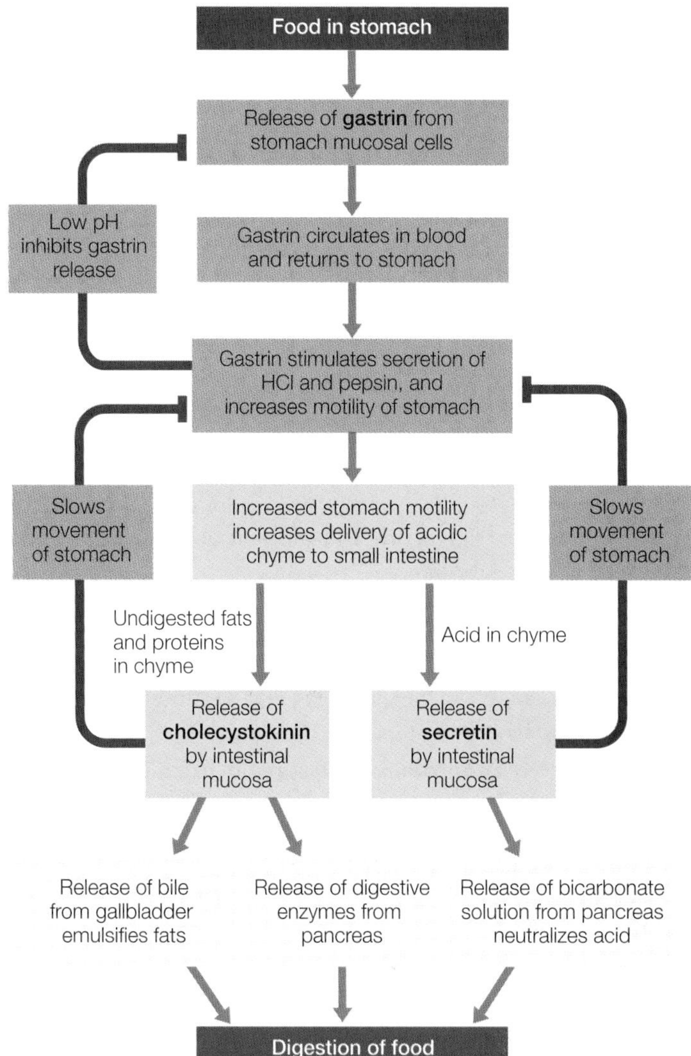


**51.16 Hormones Control Digestion** The hormones gastrin, cholecystokinin, and secretin are involved in feedback loops that control the sequential processing of food in the digestive tract. Red lines indicate inhibitory actions; green lines indicate stimulatory actions.

glycogen or fats, and vice versa. The liver can also convert certain amino acids and some other molecules, such as pyruvate and lactate, into glucose—the process of gluconeogenesis (see Section 9.5). Gluconeogenesis provides an indirect pathway for exercising muscle to contribute to blood glucose levels. At high levels of aerobic activity, muscle cells break down their stores of glycogen to provide metabolic fuel (see Section 48.2). The glucose released from muscle glycogen cannot leave muscle cells as it can leave liver cells. However, when the activity of the muscle becomes anaerobic, pyruvate and lactate build up, leave the muscle cells, and enter the circulation. Circulating pyruvate and lactate are taken up by the liver and converted to glucose that can then move out of the liver cells and into the blood.

The liver is also the major controller of fat metabolism through its production of lipoproteins. A **lipoprotein** is a particle made up of a core of hydrophobic fat and cholesterol with a covering of hydrophilic protein that allows it to be suspended in water.

**LIPOPROTEINS: THE GOOD, THE BAD, AND THE UGLY** Lipoproteins move fats, the most abundant fuel reserve in the body, from sites of absorption or synthesis to sites of storage, and from sites of storage to sites of use. We saw in Section 51.3 how, in the intestine, bile solves the problem of processing hydrophobic fats in an aqueous medium. The transport of fats in the circulatory system presents the same problem, and lipoproteins provide the solution.

The chylomicrons (see p. 1063) produced by the mucosal cells of the intestine are the largest lipoprotein particles in the blood. As the circulation carries chylomicrons through the liver and adipose tissue, lipoprotein lipases begin to break them down, and their triglyceride and cholesterol cargo is absorbed into the liver or fat cells.

Lipoproteins other than chylomicrons are synthesized in the liver. These lipoproteins can be classified according to their density. Fat has a low density (it floats on water) and protein has a high density, so the greater the fat-to-protein ratio in the lipoprotein, the lower its density.

- **High-density lipoproteins** (**HDLs**) remove cholesterol from tissues and carry it to the liver, where it can be used to synthesize bile. HDL consists of about 50% protein, 35% lipids, and 15% cholesterol. These are the "good" lipoproteins, and their levels are higher in people who exercise and are fit.
- **Low-density lipoproteins** (**LDLs**) transport cholesterol around the body for use in biosynthesis and for storage. LDL consists of about 25% protein, 25% lipids, and 50% cholesterol. These are the "bad" lipoproteins associated with a high risk for cardiovascular disease.
- **Very low-density lipoproteins** (**VLDLs**) contain mostly triglyceride fats, which they transport to fat cells in adipose tissues around the body. VLDL consists of about 2% protein, 94% lipids, and 3% cholesterol. These are the "ugly" lipoproteins, as they are associated with excessive fat deposition as well as a high risk for cardiovascular disease.

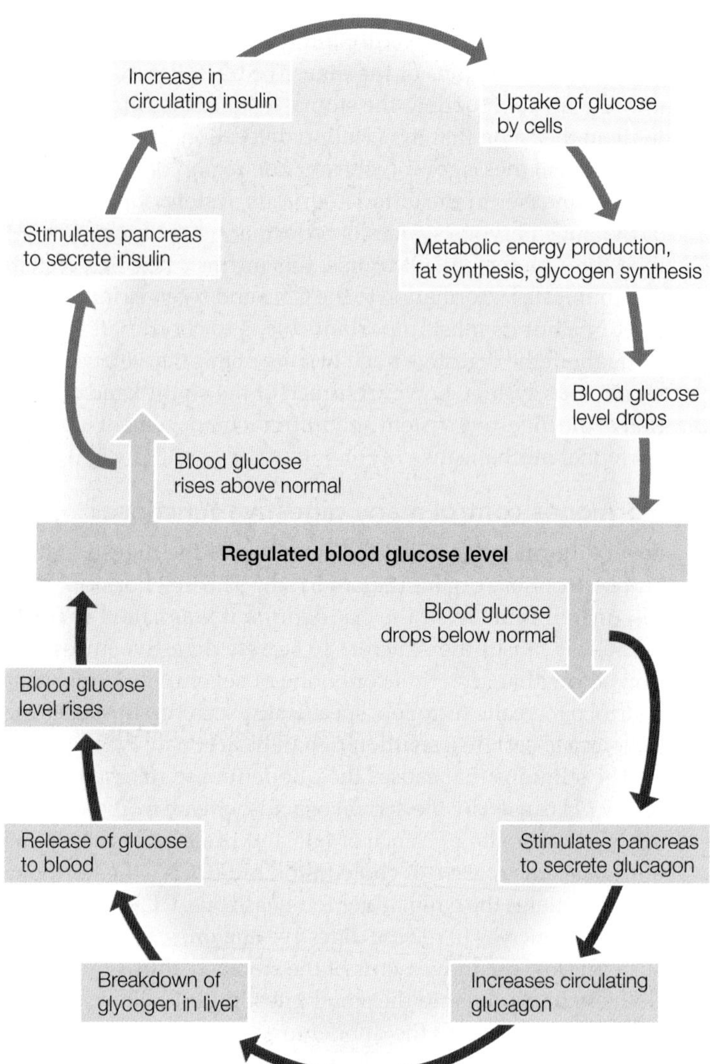


**51.17 Regulating Glucose Levels in the Blood** Insulin (blue) and glucagon (brown) interactions maintain the homeostasis of circulating glucose. It is important for blood glucose to remain stable because it is the essential source of fuel for the nervous system.

**Go to Animated Tutorial 51.2**
**Insulin and Glucose Regulation**
Life10e.com/at51.2

**INSULIN AND GLUCAGON CONTROL FUEL METABOLISM** During the absorptive state, blood glucose levels rise as carbohydrates are digested and absorbed (**Figure 51.17**). During this time, beta cells of the pancreas release the hormone **insulin**, which plays a major role in directing glucose to where it will be used or stored. The actions of insulin vary in different tissues, but they are all aimed at promoting the use of glucose for metabolic fuel and directing the excess glucose into storage as either glycogen or fat.

Glucose enters cells by diffusion. This diffusion is facilitated by transporters, but they are not active transporters—they require a concentration gradient, which is why it is important to regulate blood glucose levels so there is always an adequate glucose concentration gradient across the cell membranes. There are several kinds of glucose transporters, and those in skeletal muscle and adipose tissues are normally sequestered in cytoplasmic vesicles until insulin binds its receptors on the cell surface and triggers the insertion of transporters into the plasma membrane.

Insulin plays many roles in controlling how cells use the glucose they take up from the circulation. In adipose cells, insulin inhibits lipase and promotes fat synthesis from glucose. In the liver, insulin activates an enzyme that phosphorylates glucose as it enters the liver cell so it cannot diffuse back out again, enhancing the overall diffusion of glucose into the cells. Insulin also activates the enzymes in liver cells that catalyze the synthesis of glycogen.

During the postabsorptive state, a fall in blood glucose decreases the release of insulin, and the uptake of glucose by most cells is curtailed (see Figure 51.17). To maintain blood glucose levels, liver cells break down their stored glycogen, releasing glucose into the blood. The liver and adipose tissues break down fats and supply fatty acids to the blood, and most cells preferentially use those fatty acids as their metabolic fuel. *The most important control of fuel metabolism in the postabsorptive state is the lack of insulin.*

One tissue that does not switch fuel sources when an animal is postabsorptive is the nervous system. The cells of the nervous system require a constant supply of glucose and can use other fuels only to a very limited extent. Most neurons do not require insulin to absorb glucose from the blood, but they do need an adequate glucose concentration gradient to drive the facilitated diffusion of glucose across their plasma membranes. Therefore it is critical that blood glucose levels are maintained when an animal is postabsorptive. The overall dependence of neural tissues on glucose, and their requirement for constant blood glucose levels, are the reasons it is so important for other cells of the body to shift to fat metabolism during the postabsorptive state.

The metabolism of fuel molecules during the postabsorptive state is mostly controlled by the lack of insulin, but if blood glucose falls below a certain level, another pancreatic hormone, **glucagon**, is called into play. Glucagon's effect is opposite that of insulin: it stimulates liver cells to break down glycogen and to carry out gluconeogenesis. Thus, under the influence of glucagon, the liver produces glucose and releases it into the blood.

### The brain plays a major role in regulating food intake

Multiple brain areas and signals are involved in the regulation of food intake. Long ago it was discovered that damaging a region in the center of the rat hypothalamus resulted in the rats increasing their food intake and becoming obese. Damage to the lateral hypothalamus, however, led to decreased food intake and the rats became thin. In both cases the rats eventually reached a new equilibrium body weight; thus it appeared that a capacity for regulation remained, but the set point was altered.

We now know that another region of the hypothalamus, the **arcuate nucleus**, plays an important role in integrating a variety of feedback signals that influence food intake and body mass. Cells within the arcuate nucleus send axons to the ventromedial and dorsal hypothalamus, as well as to other brain areas that influence food intake and metabolism. One group of arcuate neurons projects to brain areas that inhibit food intake while the other projects to brain areas that stimulate food intake. But what stimulates or inhibits the activity of the arcuate neurons?

Several factors have been identified that reflect the body's energy balance. Three of these are the proteins insulin, leptin, and ghrelin. Insulin, as detailed earlier, is released when blood glucose levels are high. **Leptin** (Greek *leptos*, "thin") is released by fat cells in proportion to how much lipid they contain. Evidence that leptin is a satiety signal is described in **Figure 51.18**. **Ghrelin** is released by the stomach when it is empty; its levels rise before meals and fall after meals. In the arcuate nucleus, insulin and leptin activate the neurons that inhibit feeding and inhibit the neurons that stimulate feeding. Ghrelin has the opposite effect on these two groups of neurons.

INVESTIGATING**LIFE**

**51.18 A Single-Gene Mutation Leads to Obesity in Mice**
In mice the *Ob* gene codes for the protein leptin, a satiety factor that signals the brain when enough food has been consumed. The recessive *ob* allele is a loss-of-function allele, so *ob/ob* mice do not produce leptin; they do not experience satiety and become obese. The *Db* gene encodes the leptin receptor, so mice homozygous for the recessive loss-of-function allele *db*, even if they produce leptin, cannot use it and so become obese.[a]

**HYPOTHESIS** Mice who cannot produce the satiety signal protein leptin will not become obese if they are able to obtain leptin from an outside source.

***Method***
1. Create two strains of genetically obese laboratory mice, one of which lacks functional leptin (genotype *ob/ob*) and one which lacks the receptor for leptin (genotype *db/db*).
2. Create parabiotic pairs by surgically joining the circulatory systems of a non-obese (wild-type) mouse with a partner from one of the obese strains.
3. Allow mice to feed at will.

Parabiotic pair

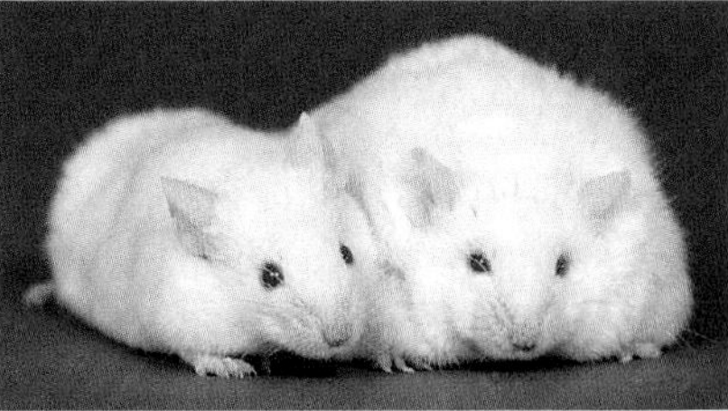

Wild-type mouse (*Ob*/– and *Db*/–)   Genetically obese mouse (either *ob/ob* or *db/db*)

***Results*** Parabiotic *ob/ob* mice obtain leptin from the wild-type partner and lose fat. Parabiotic *db/db* mice remain obese because they lack the leptin receptor and thus the leptin they obtain from their partner has no effect.

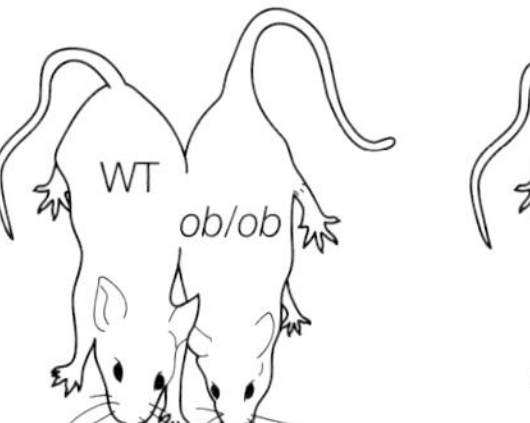


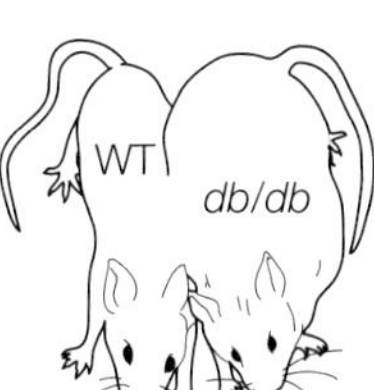


**CONCLUSION** The protein leptin is a satiety signal that acts to prevent overeating and resultant obesity.

Go to **BioPortal** for discussion and relevant links for all INVESTIGATING**LIFE** figures.

[a]Coleman, D. L. and K. P. Hummel. 1969. *American Journal of Physiology* 217: 1298–1304.

**Go to Animated Tutorial 51.3**
**Parabiotic Mice**
Life10e.com/at51.3

## WORKING WITH**DATA**:

### *Is Leptin a Satiety Signal?*

#### Original Papers

Coleman, D. L., and K. P. Hummel. 1969. Effects of parabiosis of normal with genetically diabetic mice. *American Journal of Physiology* 217: 1298–1304.

Coleman, D. L. 1973. Effects of parabiosis of obese with diabetes and normal mice. *Diabetologia* 9: 294–297.

#### Analyze the Data

The study summarized in Figure 51.18 supported the hypothesis that there are two separate genes that signal the brain when enough food has been obtained—one that encodes a satiety signal and another that encodes a receptor protein for the signal. Subsequent studies revealed that the genes in question encode the hormone leptin and its receptor.

In addition to the experiment in Figure 51.18, the researchers purified leptin and injected it into wild-type mice (*Ob/Ob*) and into genetically obese mice that lacked the leptin gene (*ob/ob*). The data in the table below were collected before the injections began (baseline) and 10 days later.

QUESTION 1

Do the data support the hypothesis that leptin is a satiety signal? Explain your answer.

QUESTION 2

What factors might explain the loss of body mass in the leptin-injected *ob/ob* mice?

| | Baseline | | Day 10 | |
|---|---|---|---|---|
| Parameter | *ob/ob* | *Ob/Ob* | *ob/ob* | *Ob/Ob* |
| Food intake (g/day) | 12.0 | 5.5 | 5.0 | 6.0 |
| Body mass (g) | 64 | 35 | 50 | 38 |
| Metabolic rate (ml $O_2$/kg/hr) | 900 | 1,150 | 1,100 | 1,150 |
| Body temperature (°C) | 34.8 | 37.0 | 37.0 | 37.0 |

Go to BioPortal for all WORKING WITHDATA exercises

An integrative signal that could be playing a central role in regulation of feeding is the enzyme AMP-activated protein kinase (AMPK). When most cells are nutrient deprived, they produce AMPK, which stimulates the oxidation of substrates to replenish ATP. Fasting increases AMPK levels in the hypothalamus, and feeding reduces them. Insulin and leptin decrease AMPK activity in the hypothalamus, and ghrelin increases it. Thus AMPK could be a final common pathway for various signals controlling food intake.

#### RECAP 51.4

The major controlling factors of gut function are an intrinsic nervous system and the hormones gastrin, secretin, and cholecystokinin. Insulin is the major hormonal controller of fuel metabolism. The hypothalamus controls food intake by generating sensations of hunger and satiety influenced by feedback from blood glucose and hormones, including insulin, leptin, and ghrelin.

- What are the roles of the three different classes of lipoproteins? **See p. 1066**
- By what actions does insulin promote uptake and storage of energy during the absorptive state? **See pp. 1066–1067 and Figure 51.17**
- What evidence supports the hypothesis that leptin influences satiety? **See pp. 1067–1068 and Figure 51.18**

Are there genes that predispose a person to obesity?

#### ANSWER

Many studies have demonstrated that a propensity for obesity is genetic and therefore heritable. About 30 studies comparing monozygotic and dizygotic twins led to the conclusion that the heritability of body mass is between 64 and 84 percent. Studies of monozygotic twins raised apart concluded that heritability of body mass was around 68 percent.

Single-gene causes of obesity have been observed but are rare. An example is a mutation of the leptin gene which has been found in a very small number of severely obese individuals, and these responded rapidly to leptin replacement therapy. To date the most commonly documented single-gene association with obesity is a mutation of the *MC4R* gene that encodes the melanocortin-4 receptor found on many hypothalamic neurons that inhibit feeding behavior. In a study population of Danish men, 2.5 percent of those who qualified as obese had mutations of the *MC4R* gene.

The Pima community of Arizona was among the first to be involved in a genome-wide survey. Many candidate genes were shown to be associated with obesity, but only weakly so. The conclusion must be that obesity is heritable, but except in a few very rare cases, it is a highly multigenic trait. Body mass is a function of Calories in minus Calories out, and genes influencing either side of this equation may be involved in obesity.

# CHAPTER**SUMMARY** 51

## 51.1 What Do Animals Require from Food?

- Animals are **heterotrophs** that derive their energy and molecular building blocks, directly or indirectly, from **autotrophs**.
- Carbohydrates, fats, and proteins in food supply animals with energy. A measure of the energy content of food is the **kilocalorie (kcal)**. Excess caloric intake is stored as glycogen and fat. **Review Figure 51.2**
- For many animals, food provides essential **carbon skeletons** that they cannot synthesize themselves. **Review Figure 51.4**
- Most researchers consider 8 amino acids to be essential for adult humans; some believe that infants require as many as 12 **essential amino acids** in their diet. **Macronutrients** are mineral elements needed in large quantities; **micronutrients** are needed in small amounts. **Review Figure 51.5, Table 51.1, ACTIVITY 51.1**
- **Vitamins** are organic molecules that must be obtained in food. **Review Table 51.2, ACTIVITY 51.2**
- **Malnutrition** results when any essential nutrient is lacking from the diet. Chronic malnutrition causes **deficiency disease**.

## 51.2 How Do Animals Ingest and Digest Food?

- Animals can be characterized by how they acquire nutrients: **saprobes** and **detritivores**, or **decomposers**, depend on dead organic matter, **filter feeders** strain the aquatic environment for small food items, **herbivores** eat plants, and **carnivores** eat other animals. Behavioral and anatomical adaptations reflect these feeding strategies. **See ACTIVITY 51.3**
- Digestion involves the breakdown of complex food molecules into monomers that can be absorbed and used by cells. In most animals, digestion takes place in a tubular gut. **Review Figure 51.7**
- Absorptive areas of the vertebrate gut are characterized by a large surface area produced by extensive folding and numerous **villi** and **microvilli**. **Review Figure 51.8**
- Hydrolytic enzymes break down proteins, carbohydrates, and fats into their monomeric units.

## 51.3 How Does the Vertebrate Gastrointestinal System Function?

- The vertebrate gut can be divided into several compartments with different functions. **Review Figure 51.9, ACTIVITY 51.4**
- The cells and tissues of the vertebrate gut are organized in the same way throughout its length. The innermost tissue layer, the **mucosa**, is the secretory and absorptive surface. The **submucosa** contains blood and lymph vessels and a nerve network that is sensory and also controls gut secretions. External to the submucosa are two smooth muscle layers. Between the two muscle layers is another nerve network that controls the movements of the gut. **Review Figure 51.10**
- Swallowing is a reflex that pushes a bolus of food into the **esophagus**. **Peristalsis** and **segmentation movements** of the gut move the bolus down the esophagus and through the entire length of the gut. **Sphincters** block the gut at certain locations, but they relax as a wave of peristalsis approaches. **Review Figure 51.11**
- Digestion begins in the **mouth**, where amylase is secreted with the saliva. Digestion of protein begins in the **stomach**, where parietal cells secrete HCl and chief cells secrete pepsinogen, a zymogen that becomes pepsin when activated by low pH and **autocatalysis**. The mucosa also secretes mucus, which protects the tissues of the gut. **Review Figure 51.12**
- In the **duodenum**, pancreatic enzymes carry out most of the digestion of food. **Bile** from the liver and **gallbladder** emulsifies fats into **micelles**. Bicarbonate ions from the **pancreas** neutralize the pH of the **chyme** entering from the stomach to produce an environment conducive to the actions of pancreatic enzymes such as **trypsin**. **Review Figure 51.13, Table 51.3**
- Final enzymatic cleavage of polypeptides and disaccharides occurs among the microvilli of the intestinal mucosa. Amino acids, monosaccharides, and inorganic ions are absorbed by the microvilli. Specific transporter proteins are sometimes involved. Symporters often power the absorption of nutrients.
- Fats broken down by **lipases** are absorbed mostly as monoglycerides and fatty acids and are resynthesized into triglycerides within the gut epithelium. The triglycerides are combined with cholesterol and phospholipids and coated with protein to form **chylomicrons**, which pass out of the mucosal cells and into lymphatic vessels in the submucosa. **Review Figure 51.14, ANIMATED TUTORIAL 51.1**
- Water and ions are absorbed in the large intestine as waste matter and consolidated into **feces**, which are periodically eliminated.
- Microorganisms in some compartments of the gut digest materials that their host cannot. **Review Figure 51.15**

## 51.4 How Is the Flow of Nutrients Controlled and Regulated?

- Autonomic reflexes coordinate activity of the digestive tract, which has an intrinsic nervous system that can act independently of the CNS.
- The actions of the stomach and small intestine are largely controlled by the hormones **gastrin**, **secretin**, and **cholecystokinin**. **Review Figure 51.16**
- The liver plays a central role in directing the traffic of fuel molecules. In the **absorptive state**, the liver takes up and stores fats and carbohydrates, converting monosaccharides to glycogen or fats. The liver also takes up amino acids and uses them to produce blood plasma proteins, and can engage in gluconeogenesis.
- Fat and cholesterol are shipped out of the liver as **low-density lipoproteins**. **High-density lipoproteins** act as acceptors of cholesterol and bring fat and cholesterol back to the liver.
- **Insulin** largely controls fuel metabolism during the absorptive state and promotes glucose uptake as well as glycogen and fat synthesis. In the **postabsorptive state**, lack of insulin blocks the uptake and use of glucose by most cells of the body except neurons. If blood glucose levels fall, **glucagon** secretion increases, stimulating the liver to break down glycogen and release glucose to the blood. **Review Figure 51.17, ANIMATED TUTORIAL 51.2**
- Food intake is governed by sensations of hunger and satiety, which are determined by brain mechanisms responding to feedback signals such as insulin, **leptin**, and **ghrelin**. **Review Figure 51.18, ANIMATED TUTORIAL 51.3**

**Go to the Interactive Summary to review key figures, Animated Tutorials, and Activities**
**Life10e.com/is51**

# CHAPTERREVIEW

## REMEMBERING

1. Which statement about essential amino acids is true?
   a. They are not found in vegetarian diets.
   b. They are stored by the body until they are needed.
   c. Without them, the body is undernourished.
   d. All animals require the same ones.
   e. Humans can acquire all of theirs by eating milk, eggs, and meat.

2. The digestive enzymes of the small intestine
   a. do not function well at a low pH.
   b. are produced and released in response to circulating secretin.
   c. are produced and released under neural control.
   d. are all secreted by the pancreas.
   e. are all activated by an acidic environment.

3. Which statement about nutrient absorption by the intestinal mucosal cells is true?
   a. Carbohydrates are absorbed as disaccharides.
   b. Fats are absorbed as fatty acids and monoglycerides.
   c. Amino acids move across the plasma membrane only by diffusion.
   d. Bile transports fats across the plasma membrane.
   e. Most nutrients are absorbed in the duodenum.

4. Chylomicrons are like the tiny micelles of dietary fat in the lumen of the small intestine in that both
   a. are coated with bile.
   b. are lipid-soluble.
   c. travel through the lymphatic system.
   d. contain triglycerides.
   e. are coated with lipoproteins

5. Which of the following is stimulated by cholecystokinin?
   a. Stomach motility
   b. Release of bile
   c. Secretion of hydrochloric acid
   d. Secretion of bicarbonate ions
   e. Secretion of mucus

## UNDERSTANDING & APPLYING

6. Several popular diet books recommend high fat and protein intake and low carbohydrate intake as a means of losing body mass. What is the rationale for a high-fat and high-protein diet?

7. Explain how insulin is involved in at least five differences between the management of fuel molecules in the absorptive and the postabsorptive state.

8. Trace the history of a fatty acid molecule from a slice of cheese pizza to a plaque on a coronary artery. Into what possible forms and structures might it have been converted as it passed through the body? Describe a direct and an indirect route it could have taken.

## EVALUATING & ANALYZING

9. The pancreatic duct cells and the parietal cells of the stomach both express high levels of carbonic anhydrase. Explain how the actions of this enzyme are similar in both cases, but how the consequences in terms of influences on the pH of the blood are different.

10. Body mass is a function of Calories in minus Calories out. If genome scans of obese individuals show that most of the associated genes play roles in the hypothalamic regulatory pathways, what does that tell us about which side of the energy balance equation is the most important in contributing to obesity?

Go to BioPortal at **yourBioPortal.com** for Animated Tutorials, Activities, LearningCurve Quizzes, Flashcards, and many other study and review resources.

# 52 Salt and Water Balance and Nitrogen Excretion

## CHAPTER OUTLINE


52.1 How Do Excretory Systems Maintain Homeostasis?

52.2 How Do Animals Excrete Nitrogen?

52.3 How Do Invertebrate Excretory Systems Work?

52.4 How Do Vertebrates Maintain Salt and Water Balance?

52.5 How Does the Mammalian Kidney Produce Concentrated Urine?

52.6 How Are Kidney Functions Regulated?

**Blood as Fast Food** The vampire bat *Desmodus rotundus* is able to adjust its excretory physiology rapidly from water-excreting to water-conserving, depending on whether it is ingesting or digesting its blood meal.

BLOOD, SWEAT, AND TEARS taste salty because they have the ionic composition of the extracellular fluid that bathes the cells of the body. The volume and composition of the extracellular fluid must be regulated and kept relatively free of wastes. Maintaining homeostasis of the extracellular fluid is the job of the excretory system. It is a challenging job that sometimes requires getting rid of excess fluids and conserving ions, and at other times requires conserving fluids and excreting excess ions.

The nature of the challenge depends on an animal's environment and lifestyle. Some desert animals rarely if ever encounter open water, so drinking is not an option. Animals that live in fresh water have the opposite challenge: water continuously enters their bodies by osmosis and in their food. Animals that live in salt water face a challenge similar to that of the desert dwellers—they need to conserve water and excrete ions. The physiological mechanisms all animals have to maintain salt and water balance are similar, but they are used in different ways to solve the unique problems of each species. Consider, for example, vampire bats.

Vampire bats feed on the blood of animals such as goats and cattle, using sharp incisor teeth to make a small incision (usually on the legs and ankles of a sleeping victim) and then lapping up the blood. The bat's saliva has an anticoagulant that keeps the blood flowing. Blood contains nutritious protein, but it consists mostly of water. Blood meals may be few and far between, so the bat quickly consumes as much as it can—up to half its body mass. To maximize its protein intake and keep its weight low enough to fly, it rapidly eliminates the water from its meal. Within minutes of starting to feed, the bat is producing copious dilute urine.

Once feeding ends, this high rate of water loss must stop—now the bat is metabolizing protein and must excrete large amounts of nitrogenous wastes while conserving water. Within minutes, the bat's excretory system switches from producing abundant, dilute urine to producing a tiny amount of highly concentrated urine. To conserve water, vampire bats (and some desert rodents) can produce urine that is 15 times more concentrated than their own blood. In one feeding cycle, the vampire bat rapidly transitions from an excretory physiology typical of a mammal living in an environment with abundant fresh water to that of a desert mammal that never sees water.

?

How do desert rodents and vampire bats make highly concentrated urine?

See answer on p. 1090.

# 52.1 How Do Excretory Systems Maintain Homeostasis?

Homeostasis of the extracellular fluid (the blood plasma and interstitial fluid; see Section 40.1) is critical for several reasons:

- The solute concentration of the extracellular fluid determines the water balance of the cells of the body.
- The specific ionic composition of the extracellular fluid influences many functions of various types of cells. Consider, for example, the importance of ion concentration gradients between the extracellular fluid and the cytoplasm of nerve and muscle cells (see Sections 45.2 and 48.1).
- The health of cells requires the elimination of nitrogenous wastes.

The problems that have to be solved to maintain homeostasis of the extracellular fluid depend on the environment in which a species lives (salt water, fresh water, or terrestrial) and its lifestyle, as we saw in the case of the vampire bat in the opening story. Animals depend on **excretory systems** to maintain the volume, concentration, and composition of their extracellular fluids, and to excrete wastes.

## Water enters or leaves cells by osmosis

The volume of cells depends on whether they take up water from or lose water to the extracellular fluid. The movement of water across cell plasma membranes depends on differences in solute concentration on the two sides of the membrane and on the permeability of the membrane. This is the process of osmosis, which we discussed in Sections 6.3 and 35.1. If the solute concentration of the extracellular fluid is less than that of the cytoplasm, water moves into the cells, causing them to swell and possibly burst (see Figure 6.9). If the solute concentration of the extracellular fluid is greater than that of the cytoplasm, the cells lose water and shrink. Thus the solute concentration of the extracellular fluid affects both the volume and the solute concentration of the cells.

Animal physiologists use the term **osmolarity** in discussing osmosis. The osmolarity of a solution is the number of moles of osmotically active solutes per liter of solvent. Thus a 1 molar solution of glucose is also a 1 osmolar (1 osmole per liter) solution, but a 1 molar solution of sodium chloride (NaCl) is a 2 osmolar solution, because each NaCl molecule dissociates into two osmotically active ions.

## Excretory systems control extracellular fluid osmolarity and composition

Excretory systems control the osmolarity and composition of the extracellular fluids by excreting solutes that are present in excess (such as NaCl when we eat lots of salty food) and conserving solutes that are valuable or in short supply (such as glucose and amino acids). Excretory systems also eliminate the toxic waste products of protein metabolism. The output of the excretory system is called **urine**.

Three basic processes are common to a wide variety of animal excretory systems: filtration, secretion, and reabsorption. Filtered extracellular fluid contains no cells or large molecules, such as proteins. In animals with closed circulatory systems, the blood plasma is usually filtered from capillaries into associated tubules. The walls of the capillaries and of the tubules are the filter, and the filtration is driven by blood pressure. As the filtrate flows through the tubules, its composition and concentration are modified through processes of secretion and reabsorption to form the urine that leaves the body.

In all of the discussions that follow about the movement of water across membranes, it is important to remember that there are no mechanisms for the active transport of water. The movement of water is due either to a pressure difference (filtration) or to a difference in solute concentration (osmosis). Water always flows down a pressure gradient or up a solute concentration gradient.

## Aquatic invertebrates can conform to or regulate their osmotic and ionic environments

Most invertebrates that live in seawater conform to the osmotic concentration of their environment over a fairly wide range of salinities and are therefore called **osmoconformers** (**Figure 52.1**). The osmolarity of seawater in the open ocean is about 1,000 milliosmoles/liter (mosm/l), but it can vary quite a bit in estuaries where it is diluted by an influx of fresh water or in evaporating tide pools as the salt gets concentrated. Osmoconformity can result in considerable energetic savings, as it costs metabolic energy to move ions across membranes to achieve

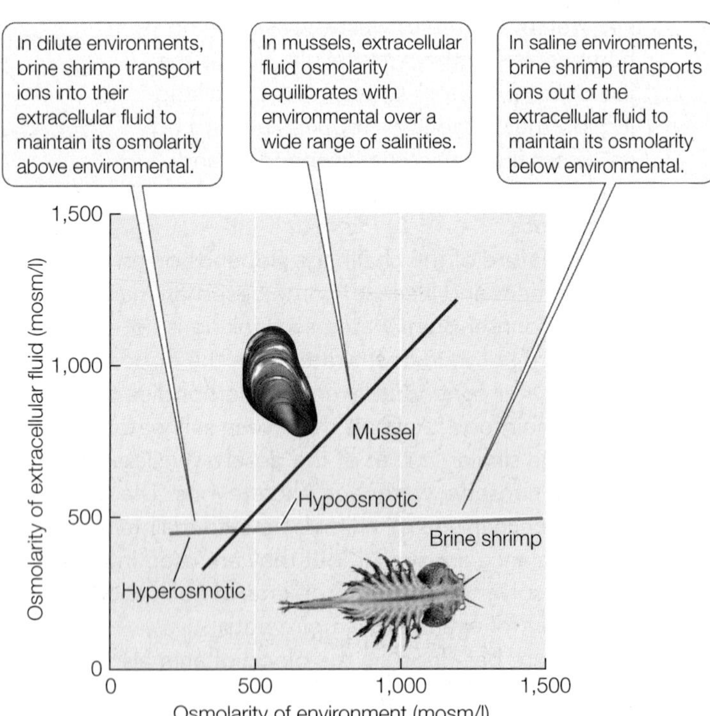


**52.1 Some Marine Invertebrates Osmoregulate** Some aquatic animals, such as mussels, experience an extreme range of salt concentrations in their environment and are osmoconformers over much of that range. Other aquatic animals, such as brine shrimp, are osmoregulators in that they maintain a relatively constant osmolarity of their extracellular fluids as environmental salinity varies.

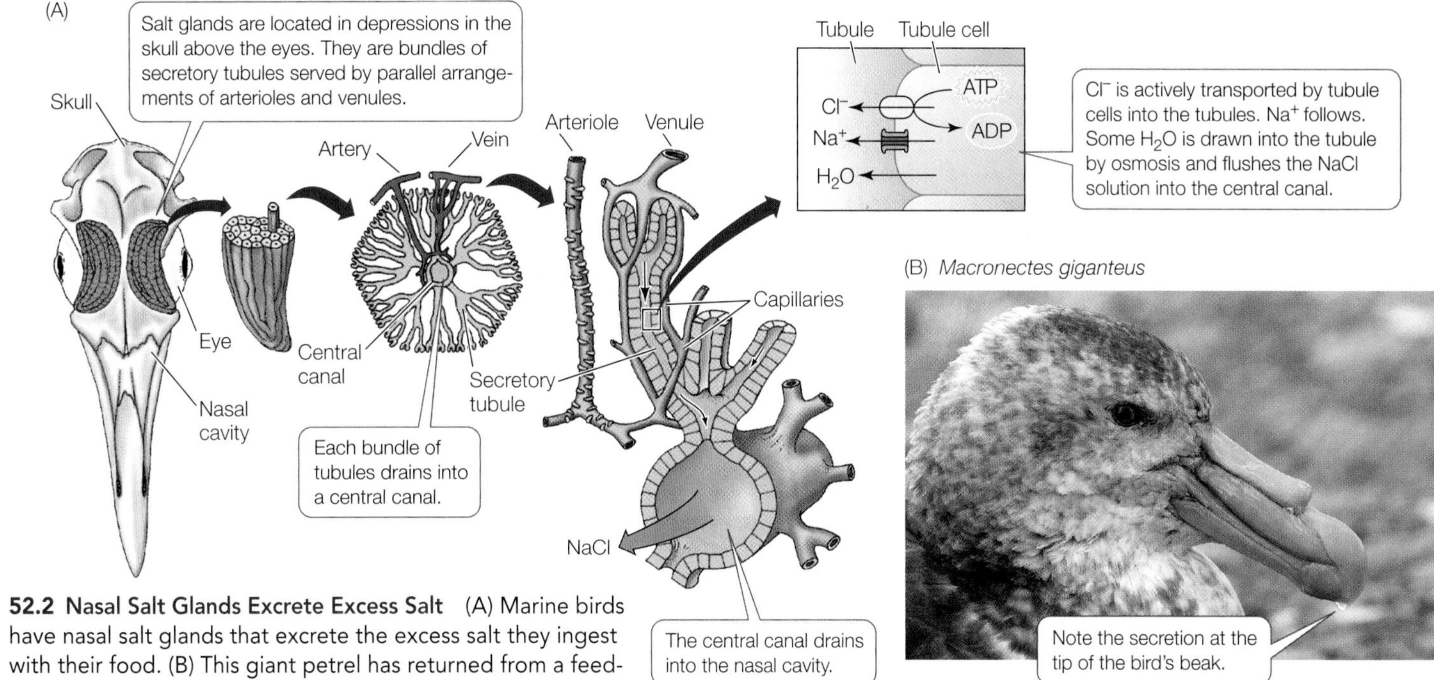


**52.2 Nasal Salt Glands Excrete Excess Salt** (A) Marine birds have nasal salt glands that excrete the excess salt they ingest with their food. (B) This giant petrel has returned from a feeding trip at sea and is secreting salt through its nasal salt gland.

osmotic regulation. However, osmoconformity is not an option for freshwater animals. No cell can do without ions or nutrients, so freshwater animals always have to expend energy to conserve salts and excrete water.

Some marine invertebrates maintain a rather constant osmolarity of their extracellular fluids as the osmolarity of the environment changes, and these animals are therefore called **osmoregulators** (see Figure 52.1). Osmoregulators in the marine environment mostly maintain their osmotic concentration considerably below that of the environment and are therefore engaged in **hypoosmotic regulation**. Occasionally, however, seawater is diluted by an influx of fresh water, as in estuaries, and animals must maintain their osmotic concentrations above that of the environment—a process called **hyperosmotic regulation**.

The brine shrimp *Artemia* illustrated in Figure 52.1 is an osmoregulator with an enormous range of tolerances. *Artemia* are found in huge numbers in the most salty environments known, such as Utah's Great Salt Lake and in coastal evaporation ponds where salt is concentrated for commercial purposes (see Figure 26.17) and can reach an osmolarity of 2,500 mosm/l. No animal could survive with internal osmolarities that high; such a solute concentration would cause proteins to denature. *Artemia* are able to exploit these environments because of their ability to regulate hypoosmotically by actively transporting NaCl from their extracellular fluid out across their gill membranes to the environment. *Artemia* cannot survive in fresh water, but they can live in dilute seawater by reversing the direction of transport of NaCl across their gill membranes to maintain the osmolarity of their extracellular fluids above that of the environment, thus becoming hyperosmotic regulators.

Virtually all marine invertebrates are ionic conformers with respect to certain ions such as $Na^+$ and $Cl^-$ in their extracellular fluids—that is, the extracellular concentrations of these ions are the same as in seawater. Thus these animals avoid the considerable metabolic costs of moving ions across membranes to maintain concentration differences between their extracellular fluids and the seawater. However, most marine animals (with the exception of echinoderms) are ionic regulators with respect to certain ions; they employ active transport mechanisms to maintain these ions in their extracellular fluid at optimal concentrations.

## Vertebrates are osmoregulators and ionic regulators

All aquatic vertebrates, with two exceptions, regulate the osmolarity of their extracellular fluids at 300 mosm/l. In doing so, they are selective in which ions they conserve and which ions they excrete; thus they are ionic regulators as well as osmoregulators. One exception is the hagfish, a primitive jawless fish and a very ancient vertebrate group (see Figure 33.10). Hagfish are osmoconformers as well as ionic conformers for many of the ions found in seawater. The other exception is the chondrichthyans (cartilaginous fishes, the sharks and rays). Chondrichthyans retain in their extracellular fluid two organic solutes, urea and trimethylamine oxide (TMAO), that are products of protein metabolism. As a result, their extracellular fluid is slightly hyperosmotic to the seawater and they gain water by osmosis.

Terrestrial vertebrates obtain their salts mostly from food and regulate the ionic composition of their extracellular fluids by conserving some ions and excreting others. For example, herbivores have to conserve $Na^+$ because the plants they eat have low concentrations of $Na^+$. By contrast, birds that feed on marine animals must excrete the excess sodium they ingest with their food. Such birds, which include penguins and gulls, excrete excess salt through their nasal salt glands. These glands use active transport of $Cl^-$ ions (with $Na^+$ and some water following passively) into a series of canals and ducts to produce a concentrated solution of NaCl that empties into the nasal cavity (**Figure 52.2**). These birds can be seen frequently sneezing or shaking their heads to get rid of the salty droplets excreted from their nasal salt glands.

### RECAP 52.1

Excretory systems control water and salt balance and the excretion of nitrogenous waste products through three mechanisms: filtration of body fluids to form urine, active secretion of substances into the urine, and active reabsorption of substances from the urine.

- Describe the two mechanisms used to move water across membranes. **See p. 1072**
- What different salt and water balance problems might animals encounter in marine, freshwater, and terrestrial environments? What are some of the ways they meet those challenges? **See pp. 1072–1073 and Figures 52.1 and 52.2**

In addition to maintaining salt and water balance, animals must eliminate the waste products of metabolism from their extracellular fluids. The major problem is nitrogen. When nitrogen-containing molecules are broken down by metabolism, the end product can be toxic.

## 52.2 How Do Animals Excrete Nitrogen?

The end products of the metabolism of carbohydrates and fats are water and carbon dioxide, which are not difficult to eliminate. Proteins and nucleic acids, however, contain nitrogen, so their metabolism produces nitrogenous wastes in addition to water and carbon dioxide.

### Animals excrete nitrogen in a number of forms

The most common nitrogenous waste is **ammonia** ($NH_3$). Because it is highly toxic, ammonia is either excreted continuously to prevent its accumulation or is detoxified by conversion into **urea** or **uric acid** (Figure 52.3).

AMMONIA Ammonia is highly soluble in water and diffuses rapidly, so its continuous excretion is relatively simple for many aquatic animals that continuously lose ammonia from their blood to the environment by diffusion across their gill membranes. Animals that excrete ammonia, such as aquatic invertebrates and bony fishes, are **ammonotelic**.

If ammonia builds up in the extracellular fluids, it becomes toxic at rather low levels and is a dangerous metabolite for terrestrial animals and for those aquatic animals that cannot continuously excrete ammonia. These animals must convert ammonia into urea or uric acid.

UREA **Ureotelic** animals, such as mammals, amphibians, chondrichthans (sharks and rays), and hagfish excrete urea as their principal nitrogenous waste product. Urea is quite soluble in water, but its excretion can result in a large loss of water that many animals can ill afford. As we will see later in this chapter, mammals have evolved excretory systems that conserve water by producing urine that has a high concentration of urea. As we mentioned in Section 52.1, sharks, rays, and hagfish retain high concentrations of urea and TMAO in their extracellular fluid, so that it is hyperosmotic to seawater.

URIC ACID Animals that conserve water by excreting nitrogenous wastes mostly as uric acid are **uricotelic**. Insects, reptiles (including birds), and some amphibians are uricotelic. Uric acid is not very soluble in water, so it forms a colloidal suspension in the urine and is excreted as a semisolid (for example, the whitish material in bird droppings). A uricotelic animal loses very little water as it disposes of its nitrogenous wastes.

### Most species produce more than one nitrogenous waste

Humans are ureotelic, but we also excrete uric acid. The uric acid in human urine comes largely from the metabolism of nucleic acids and caffeine. If uric acid levels in the extracellular fluid rise too high, uric acid crystals can precipitate in joints and cause the age-old malady called gout. Because solubility goes down with temperature, uric acid crystals usually precipitate first in the extremities, especially the big toe. Pain in the big toe is a telltale symptom of gout.

Humans can also excrete ammonia, which is an important mechanism for regulating the pH of the extracellular fluids. As

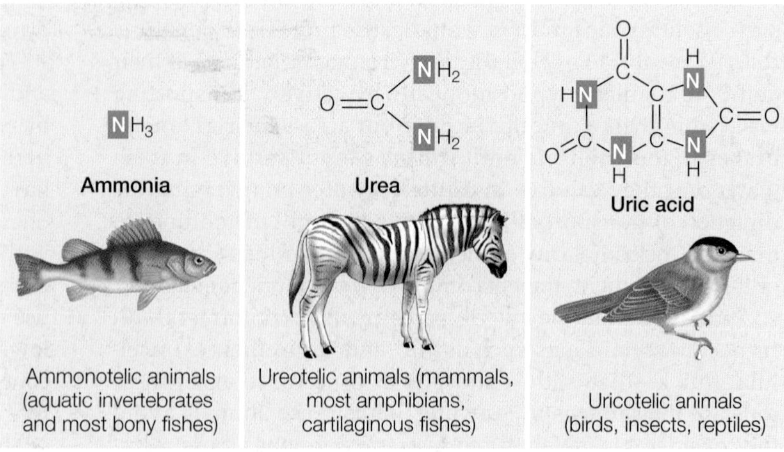


**52.3 Waste Products of Metabolism** The metabolism of proteins and nucleic acids produces nitrogenous wastes. Many aquatic animals, including most fishes, excrete nitrogenous wastes as ammonia, which is highly diffusible and soluble in an aqueous environment. Most terrestrial animals and some aquatic animals excrete either urea or uric acid. Urea is more soluble in water and is the major nitrogenous excretory product for mammals, amphibians, and some fishes. Uric acid is not very soluble in water and is the major nitrogenous excretory product for reptiles, birds, insects, and some amphibians.

we will see later in this chapter, excreted ammonia buffers the urine and enables the excretion of excess hydrogen ions.

Species that live in different habitats at different developmental stages may use more than one mechanism of nitrogen excretion. The tadpoles of frogs and toads, for example, excrete ammonia across their gill membranes, but adult frogs and toads generally excrete urea. Some adult amphibians that live in arid habitats excrete uric acid.

**RECAP 52.2**

Ammonia is a common metabolic waste product of nitrogen-containing molecules. Most aquatic animals excrete ammonia by diffusion into the water. Terrestrial animals and some aquatic animals detoxify ammonia by conversion to urea or uric acid.

- Explain the significance of high concentrations of urea in the blood of sharks. **See p. 1074**
- Why might you expect a species from an arid habitat to use uric acid as its primary nitrogenous waste product? **See p. 1074 and Figure 52.3**

Animals exhibit a variety of adaptations for dealing with the challenges of salt and water balance in different environments. All of these adaptations, however, are based on two basic mechanisms—namely, filtration and tubular processing of the filtrate to conserve some solutes and excrete others.

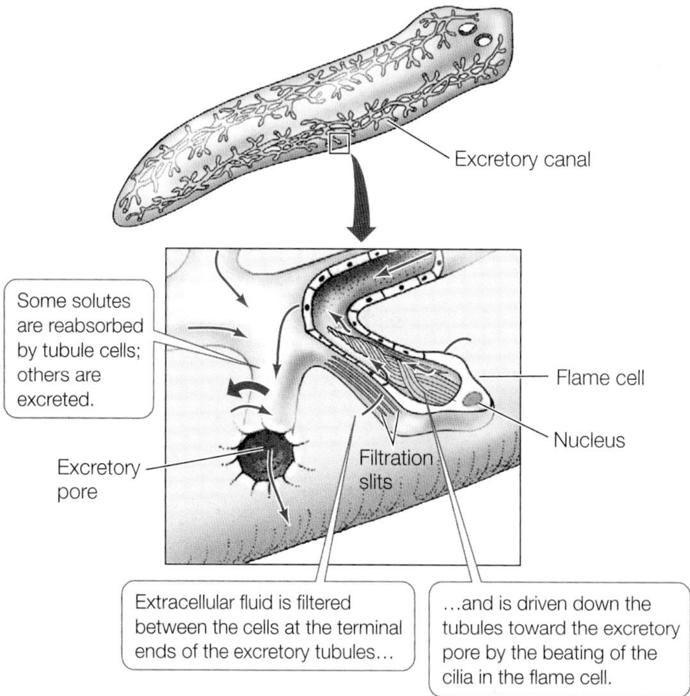


**52.4 Protonephridia in Flatworms** The protonephridia of the flatworm *Planaria* consist of tubules ending in flame cells. In the region of the flame cells, body fluid is filtered between the tubule cells. The composition of the filtrate is modified as it flows down the tubule.

## 52.3 How Do Invertebrate Excretory Systems Work?

Freshwater and terrestrial invertebrates have a wide variety of adaptations for maintaining salt and water balance and excreting nitrogen. In this section we will explore three examples of invertebrate excretory systems: protonephridia, metanephridia, and Malpighian tubules. Each of these systems produces an extract of interstitial fluid lacking large molecules. They then change the solute composition (ions and small molecules) of that fluid to form an excretory product.

### The protonephridia of flatworms excrete water and conserve salts

Many free-living flatworms, such as *Planaria*, live in fresh water. These animals excrete water through an elaborate network of tubules running throughout their bodies. The tubules end in flame cells, so called because each cell has a tuft of cilia projecting into the tubule (**Figure 52.4**). The beating of the cilia gives the appearance of a flickering flame. A flame cell and a tubule together form a **protonephridium** (plural *protonephridia*; Greek *proto*, "before"; *nephros*, "kidney").

Extracellular fluid enters the tubules by filtration. The beating of the cilia causes a slight negative pressure in the tubule, and movements of the animal create positive pressure in the extracellular fluid. This pressure difference causes extracellular fluid to be filtered through tiny spaces between tubule cells. The filtrate flows toward the animal's excretory pore, and along the way the cells of the tubules modify the composition of the fluid by reabsorption and secretion of specific ions and molecules. Because more ions are reabsorbed than are secreted, the urine that leaves the flatworm's body is less concentrated than the extracellular fluid. Thus the protonephridium conserves ions and excretes water and wastes.

### The metanephridia of annelids process coelomic fluid

Filtration of body fluids and modification of urine by tubules are highly developed processes in annelid worms such as the earthworm. Annelids are segmented, and in each segment they have a fluid-filled body cavity called a coelom (see Figure 32.11). Annelids have a closed circulatory system through which blood is pumped under pressure. The pressure causes the blood to be filtered across the thin, permeable capillary walls into the coelom. Some waste products, such as ammonia, diffuse directly from the tissues into the coelom. Where does this coelomic fluid go?

Each segment of the earthworm contains a pair of **metanephridia** (singular *metanephridium*; Greek *meta*, "akin to"). Each metanephridium begins as a ciliated, funnel-like opening called a nephrostome. The nephrostome resides in one segment and continues as a tubule in the next segment. The tubule ends in a pore, called a nephridiopore, that opens to the outside of the animal (**Figure 52.5**). Coelomic fluid is swept into the metanephridia through the ciliated nephrostomes. As the fluid passes through the tubules, their cells actively reabsorb certain molecules from it and actively secrete other molecules into it. What leaves the animal through the nephridiopores is a dilute urine containing nitrogenous wastes and other solutes.

**52.5 Metanephridia in Earthworms** The metanephridia of annelids are arranged segmentally. The cross section at the left end shows a pair of metanephridia. Three longitudinal sections (right) show only one metanephridium of the two in each segment. Coelomic fluid enters the nephrostome and flows through tubules leading to the nephridiopore. A close association of the tubules and blood capillaries facilitates the active exchange of substances between the blood and the tubular fluid.

**Go to Activity 52.1 Annelid Metanephridia Life10e.com/ac52.1**

2 The tubule cells of the metanephridium alter the composition of the fluid as it flows through the tubule…
Capillaries
Bladder
Coelomic cavity
Metanephridium
Collecting tubules
Nephridiopore
Urine
Nephrostome
3 …producing a dilute urine that is excreted through the nephridiopore.
1 Coelomic fluid is swept into the metanephridium by cilia surrounding the nephrostome.

## Malpighian tubules of insects use active transport to excrete wastes

Insects can excrete nitrogenous wastes with very little loss of water and can therefore live in the driest habitats on Earth. The insect excretory system consists of **Malpighian tubules**. An individual insect has from 2 to more than 100 of these blind-ended tubules that open into the gut between the midgut and hindgut (**Figure 52.6**).

Insects have an open circulatory system and therefore cannot use a pressure difference to filter extracellular fluids into the Malpighian tubules. Instead, the cells of the tubules actively transport uric acid, potassium ions, and sodium ions from the extracellular fluid into the tubules. The high concentration of solutes in the tubules causes water to follow osmotically, which flushes the tubule contents toward the gut.

The epithelial cells of the hindgut and rectum actively transport sodium and potassium ions from the gut contents back into the extracellular fluid. This local transport of salts

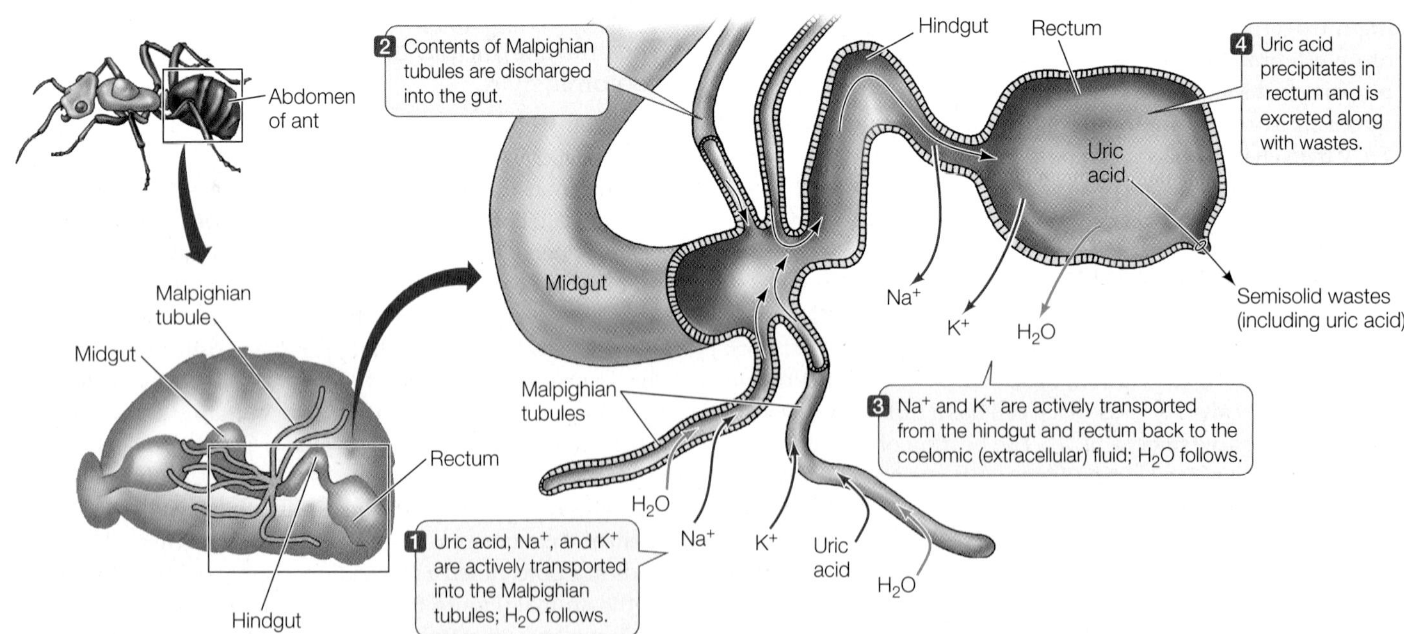


**52.6 Malpighian Tubules in Insects** The blind, thin-walled Malpighian tubules are attached to the junction of the insect's midgut and hindgut and project into the spaces containing extracellular fluid. This system makes it possible to excrete wastes with very little loss of water.

creates an osmotic gradient that pulls water out of the rectal contents. As its concentration increases, the uric acid forms a colloidal suspension, freeing even more water to be reabsorbed. Remaining in the rectum is the uric acid mixed with other wastes; this semisolid matter is what the insect excretes. If you ever park your car under a tree being visited by bees, you will find the little excretory droplets from the bees all over the car. The Malpighian tubule system is a highly effective mechanism for excreting nitrogenous wastes and some salts without giving up much water.

**RECAP 52.3**

Protonephridia and metanephridia work by creating a filtrate of the body fluids that is modified by the secretion and reabsorption of specific substances before being excreted. Insect Malpighian tubules actively secrete uric acid and other solutes into closed tubules.

- Describe how an earthworm filters its blood and produces urine. **See p. 1075 and Figure 52.5**
- How do Malpighian tubules make it possible for some insect species to survive where there is no free water? **See pp. 1076–1077 and Figure 52.6**

Having described how several invertebrate groups handle nitrogen excretion, we will next consider the nephron—the basic unit of the vertebrate excretory system—and how it evolved to be able to respond to a variety of salt and water balance challenges and maintain a relatively constant internal environment.

## 52.4 How Do Vertebrates Maintain Salt and Water Balance?

The main excretory organ of vertebrates is the **kidney**, and the functional unit of the kidney is the **nephron**, which has a blood vessel component and a tubule component. The vascular component begins with a knot of capillaries that are highly permeable and filter the blood into the tubule component. The blood vessels also carry substances to the tubules for secretion and carry away substances that the tubule cells reabsorb. Nephrons can filter large volumes of blood and achieve bulk reabsorption of salts and other valuable molecules such as glucose, making the vertebrate kidney well adapted for the excretion of excess water.

The evolution of vertebrates is thought to have begun with a marine predecessor that moved into a freshwater habitat. The excretory system of this vertebrate ancestor would have evolved to excrete large quantities of water, enabling the osmoregulatory capacities seen in present-day vertebrates. But if the early vertebrates evolved to excrete water, how did subsequent vertebrate lineages adapt to environments where water must be conserved and salts excreted? The answer to this question differs among vertebrate groups. Even among the marine fishes, the excretory adaptations of the bony fishes differ from those of the chondrichthyans. Reptiles, birds, and mammals have excretory systems that conserve water. Reptiles and birds achieve this mainly by being uricotelic and producing a semisolid excretory product that contains little water. Mammals, in contrast, are ammonotelic; they excrete a liquid waste product but have evolved the ability to produce a highly concentrated urine.

### Marine fishes must conserve water

Marine bony fishes osmoregulate their extracellular fluids to maintain them at one-third to one-half the osmolarity of seawater. Their only source of water is the sea around them, so they must conserve water and excrete excess solutes. Marine bony fishes cannot produce urine that is more concentrated than their extracellular fluids, so they minimize water loss by producing very little urine. In contrast, freshwater fishes produce lots of dilute urine.

If marine bony fishes cannot excrete excess solutes in their urine, how do they deal with the large salt loads they ingest with food? Marine bony fishes do not absorb from their guts some of the ions they take in, especially divalent ions such as $Mg^{2+}$ or $SO_4^{2-}$. NaCl, the major salt ingested, is actively excreted across the gill membranes. As we mentioned earlier, bony fishes can lose their nitrogenous waste, ammonia, by diffusion across their gill membranes.

Chondrichthyans (sharks and rays) are osmoconformers but not ionic conformers. As we have discussed, they raise the osmolarity of their body fluids in a unique way. Unlike bony fishes, they retain urea and trimethylamine oxide in their extracellular fluids so that it is hyperosmotic to seawater. These species have adapted to a concentration of urea in the body fluids that would be toxic to other vertebrates. Sharks and rays still have the problem of excreting the large amounts of salts they take in with their food. They solve this problem by having a gland in the rectum that actively secretes NaCl by a mechanism similar to that of the nasal salt glands of seabirds.

### Terrestrial amphibians and reptiles must avoid desiccation

Most amphibians live in or near fresh water, and they stay in humid habitats when they do venture from the water. Like freshwater fishes, most amphibians produce large amounts of dilute urine and conserve salts. Some amphibians, however, have adapted to habitats that require water conservation.

Amphibians living in dry terrestrial environments have skin with a reduced permeability to water. Some secrete a waxy substance over the skin to waterproof it. Several species of frogs that live in arid regions of Australia burrow deep into the ground and remain there during long dry periods. They enter **estivation**, a state of very low metabolic activity and therefore low water turnover. When it rains, the frogs come out of estivation, feed, and reproduce. Their most interesting adaptation is an enormous urinary bladder. Before entering estivation, they fill the bladder with dilute urine, which can amount to one-third of their body weight. This dilute urine serves as a water reservoir that is gradually reabsorbed into the blood during the long period of estivation. Aboriginal peoples have learned to locate these buried frogs and use them as an emergency source of water.

Reptiles occupy habitats ranging from aquatic to extremely hot and dry. In fact, snakes, lizards, and birds are among the most prominent members of many desert faunas. Three major adaptations have freed reptiles from the close association with water that is necessary for most amphibians (see Section 33.4):

- Reptiles are amniotes that do not need fresh water to reproduce because they employ internal fertilization and lay eggs with shells that retard evaporative water loss.
- Reptiles have a dry epidermis (skin) that retards evaporative water loss.
- Reptiles excrete nitrogenous wastes as uric acid semisolids, losing little water in the process.

## Mammals can produce highly concentrated urine

Mammals occupy diverse habitats, many of which present special excretory system challenges. The most challenging environments are those in which water is severely limited. Mammals have a variety of adaptations to conserve water, but chief among them is the ability to produce urine that is more concentrated than their extracellular fluids. They are able to concentrate their urine because of adaptations of their kidneys that we will explore in detail in Section 52.5. To understand how these adaptations work, however, we must first describe the structure and function of the vertebrate nephron.

## The nephron is the functional unit of the vertebrate kidney

Urine formation in vertebrate nephrons involves three main processes (**Figure 52.7**):

- *Filtration*. Each nephron has a dense bed of capillaries called a **glomerulus** (plural *glomeruli*). The glomerulus is highly permeable to water, ions, and small molecules but impermeable to large molecules. Blood pressure drives the movement of water and small-molecular-weight solutes out of the glomerular capillaries.
- *Tubular reabsorption*. The filtrate from the glomerulus flows into the **renal tubule**. Cells in the renal tubule modify the filtrate by reabsorbing specific ions, nutrients, and water, returning these to the blood, and leaving behind and concentrating excess ions and waste products such as urea.
- *Tubular secretion*. The filtrate in the renal tubule is further modified by tubule cells transporting substances into the tubule. These are substances that the body needs to excrete.

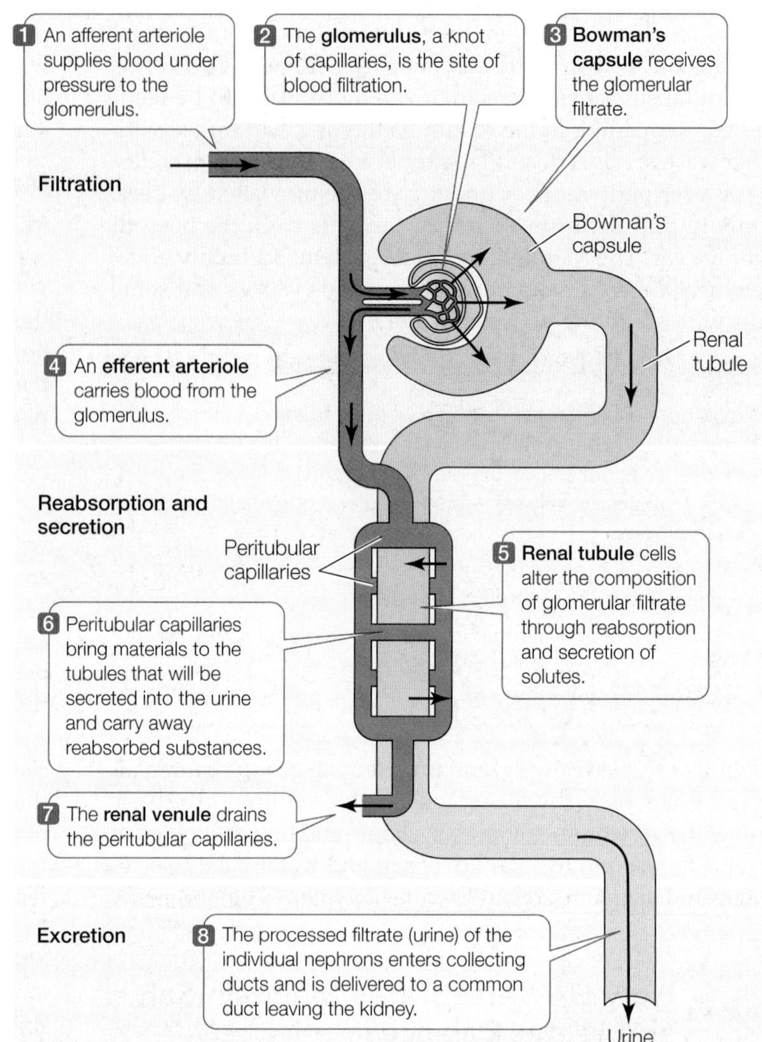


**52.7 The Vertebrate Nephron** The vertebrate nephron consists of a renal tubule closely associated with two capillary beds, the glomerulus and the peritubular capillaries.

**Go to Activity 52.2 The Vertebrate Nephron**
**Life10e.com/ac52.2**

Blood enters the glomerular capillaries via an **afferent arteriole** and leaves the glomerulus in an **efferent arteriole**. This short vessel is called an arteriole because it feeds another capillary bed, called the **peritubular capillaries**, that intimately surrounds the renal tubules. Peritubular capillaries deliver substances to the renal tubule cells that these cells secrete into the urine. The peritubular capillaries also carry away substances that the tubule cells reabsorb from the urine.

## Blood is filtered into Bowman's capsule

The renal tubule begins with **Bowman's capsule** (see Figure 52.7), which encloses the glomerulus (**Figure 52.8A and B**). The glomerulus appears to be pushed into Bowman's capsule much like a fist pushed into an inflated balloon. The cells of the capsule that are in direct contact with the glomerular capillaries are called **podocytes** (**Figure 52.8C**). These highly specialized cells have numerous armlike extensions, each with hundreds of fine, fingerlike processes. The podocytes wrap around the capillaries so that their processes interdigitate and intimately cover the capillaries.

The glomerulus filters the blood to produce a fluid (the renal filtrate) that lacks cells and large molecules. The walls of the capillaries, the basal lamina of the capillary endothelium, and the

(A) Proximal renal tubule

Podocytes Capillaries Bowman's capsule

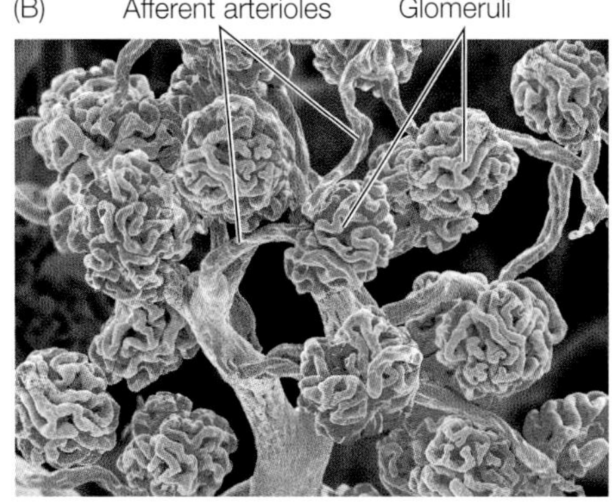


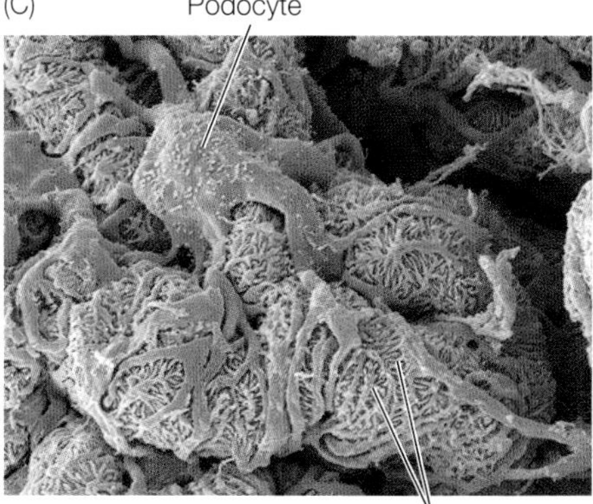


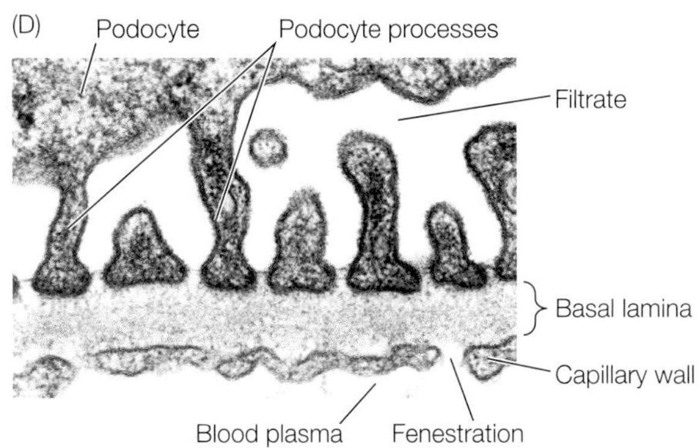


**52.8 A Tour of the Nephron** Scanning electron micrographs illustrate the anatomical basis for blood filtration by the kidneys. (A) This cross section of an intact glomerulus shows the tubule cells that form Bowman's capsule. (B) In a preparation showing only the blood vessels (tubular tissue has been digested away), the glomeruli appear as balls of capillaries served by arterioles. (C) Higher magnification of a glomerulus with the tubule cells intact shows the podocytes that wrap around the glomerular capillaries. (D) The glomerular filter has three layers: the fenestrated endothelial cells of the capillaries, the meshwork of collagen fibers making up the basement membrane, and the filtration slits between the podocyte processes. All three layers have negative charges, which contribute to their ability to prevent the passage of protein molecules.

podocytes of Bowman's capsule all participate in filtration. Fenestrations in the walls of the capillaries (see Section 50.4) allow water and many solute molecules, but not red blood cells, to pass through. The meshwork of the basal lamina and the spaces between the processes of the podocytes are even finer and prevent large molecules from leaving the capillaries (**Figure 52.8D**). The arterial pressure of the blood entering the permeable capillaries causes the filtration of water and small molecules out of the glomerular capillaries and into the Bowman's capsule. The glomerular filtration rate is high because blood pressure in the glomerular capillaries is unusually high, and because the capillaries of the glomerulus, along with their covering of podocytes, are more permeable to water than other capillary beds in the body are.

### The renal tubules convert glomerular filtrate to urine

The composition of the filtrate that enters the renal tubule is similar to that of the blood plasma, with the exception of high-molecular-weight solutes such as proteins. Reabsorption and secretion cause the composition of this fluid to change as it passes down the renal tubule. Cells of the tubule actively reabsorb certain molecules from the tubule fluid (which are returned to the blood flowing through the peritubular capillaries). For example, glucose and amino acids are reabsorbed. Most NaCl is reabsorbed. Other substances in the blood of the peritubular capillaries are actively secreted into the tubule fluid. An example is paraminohippuric acid (PAH), which is produced in the liver from benzoic acid, a common food preservative. Because of the actions of the renal tubules, the excreted urine is very different from the original filtrate.

**RECAP 52.4**

The kidney is the major excretory organ of vertebrates. Its functional unit is the nephron, which includes a glomerulus that filters blood and a renal tubule that secretes and reabsorbs solutes, modifying the filtrate to produce urine. The nephron is a mechanism for excreting excess water while conserving valuable solutes.

- Explain the differences in the osmoregulatory adaptations of freshwater fish, marine bony fish, and chondrichthyans. **See p. 1077**
- What are the functional relationships between the glomerular and peritubular capillaries? **See p. 1078 and Figure 52.7**
- Describe how blood is filtered by the glomerulus. **See pp. 1078–1079**
- How is the composition of the urine made different from the composition of the blood? **See p. 1079**

The adaptations that enable the mammalian kidney to produce urine more concentrated than extracellular fluids were important steps in vertebrate evolution, and they were largely achieved through changes in the structure and regional functions of the renal tubules. These changes converted a kidney that evolved to excrete water into an organ that conserves water.

## How Does the Mammalian Kidney Produce Concentrated Urine?

Mammals have high body temperatures and high metabolic rates, and therefore have the potential for a high rate of water

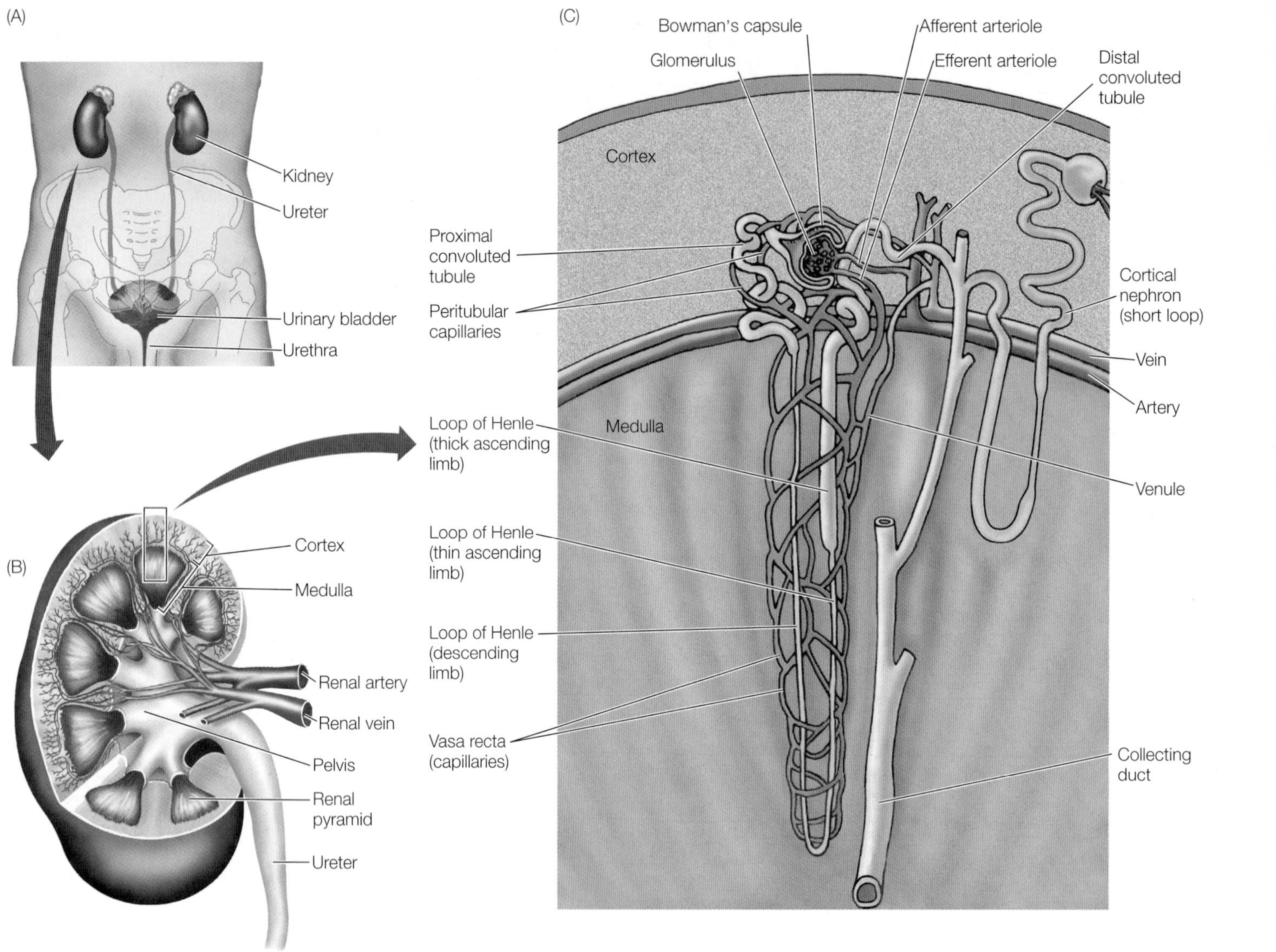

**52.9 The Human Excretory System** (A) The human kidneys lie against the back wall of the abdominal cavity, in the region of the middle back. (B) A highly organized internal structure is the basis for kidney function. Certain parts of the nephrons are in the organ's outer region, called the cortex; other parts are in the internal region, called the medulla. (C) The glomeruli and the proximal and distal convoluted tubules are located in the cortex of the kidney. The loops of Henle run in parallel as straight sections down into the renal medulla and back up to the cortex. Collecting ducts run from the cortex to the inner surface of the medulla, where they open into the ureter. The vasa recta are peritubular capillaries that parallel the loops of Henle.

**Go to Activity 52.3 The Human Excretory System Life10e.com/ac52.3**

loss. Having an excretory system that minimizes water loss made it possible for these highly active species to occupy arid habitats.

**Go to Media Clip 52.1**
**Inside the Bladder**
**Life10e.com/mc52.1**

## Kidneys produce urine and the bladder stores it

Mammalian excretory systems are similar, so we will use that of humans as our example. Humans have two kidneys at the back of the upper region of the abdominal cavity (**Figure 52.9A**). Each kidney filters blood, processes the filtrate into urine, and releases that urine into a duct called the **ureter**. The ureter of each kidney leads to the **urinary bladder**, where the urine is stored until it is excreted through the **urethra**, a short tube that opens to the outside of the body.

Two sphincter muscles surrounding the base of the urethra control urination. One of these sphincters is a smooth muscle and is controlled by the autonomic nervous system. As the bladder fills, stretch receptors in the walls of the bladder trigger a spinal reflex that relaxes this sphincter. This reflex is the only control of urination in infants, hence their frequent "accidents." The other sphincter is a skeletal muscle and is controlled by the voluntary nervous system. When the bladder is *very* full, only deliberate conscious effort prevents urination. Toilet training of children teaches them to control this sphincter.

WORKING WITH**DATA**:

## *What Kidney Characteristics Determine Urine Concentrating Ability?*

### Original Paper

Schmidt-Nielsen, B. and R. O'Dell. 1961. Structure and concentrating mechanism in the mammalian kidney. *American Journal of Physiology* 200: 1119–1124

### Analyze the Data

In the 1950s it was hypothesized that the loops of Henle in mammalian kidneys are a countercurrent multiplier mechanism responsible for the ability of mammals to concentrate their urine. Bodil Schmidt-Nielsen and Roberta O'Dell extended this hypothesis to make a prediction: the longer the loops of Henle, the greater the concentration gradient a mammal could establish. There was a problem with their prediction, however. Since kidney size varies with body size, it presumes that a very small mammal could not have loops of Henle longer than those of a large mammal—yet many mammals that inhabit arid environments (and thus need to conserve water by producing highly concentrated urine) are extremely small. Schmidt-Nielsen and O'Dell refined their prediction to say that the *relative* lengths of the loops of Henle should correlate with ability to concentrate urine. Since the longest loops of Henle can only be as long as the renal medulla is thick, they conceived of a measure they called "relative medullary thickness," or RMT (see the table).

Another possibility existed, however. As can be seen in Figure 52.9C, not all loops of Henle extend all the way to the tip of the medulla—there are short and long loops of Henle. Could the concentrating ability of the kidney be a function of the *proportion* of the loops that are long? Schmidt-Nielsen and O'Dell measured aspects of kidney size and function in several animals from habitats of different aridity and determined the maximum concentration of urine the animals could produce when water-deprived. Some of their data are summarized in the table below.

QUESTION 1

Is the length of the loops of Henle relative to overall kidney size predictive of ability to concentrate urine? (To answer this, plot RMT versus FPD.)

QUESTION 2

Are animals with a higher percentage of long loops of Henle more capable of producing concentrated urine? (Plot percent long loops of Henle versus FPD.)

QUESTION 3

Is percent long loops of Henle or RMT the better predictor of concentration ability?

| Animal | Kidney size (mm)[a] | Relative medullary thickness, RMT[b] | Percent long loops of Henle | Freezing point depression, FPD (°C)[c] |
|---|---|---|---|---|
| Human | 64.0 | 3.0 | 14 | 2.6 |
| Pig | 66.0 | 1.6 | 3 | 2.0 |
| Domestic cat | 24.0 | 4.8 | 100 | 5.80 |
| Beaver | 36.0 | 1.3 | 0 | 0.96 |
| Lab rat | 14.0 | 5.8 | 28 | 4.85 |
| Sand rat | 13.0 | 10.7 | 100 | 9.2 |
| Kangaroo rat | 5.9 | 8.5 | 27 | 10.4 |
| Jerboa | 4.5 | 9.3 | 33 | 12.0 |

[a] Kidney size = cube root of overall length × width × thickness

[b] RMT = thickness of medulla in mm × 10 divided by kidney size.

[c] FPD = the number of degrees below 0°C at which urine freezes. FPD is a measure of solute concentration: the more concentrated the urine, the higher the FPD.

**Go to BioPortal for all** WORKING WITH**DATA** **exercises**

## Nephrons have a regular arrangement in the kidney

The kidney is shaped like a kidney bean; when sliced along its long axis, its key anatomical features are revealed (**Figure 52.9B**). The ureter and the **renal artery** and **renal vein** enter the kidney on its concave (punched-in) side. The ureter extends into the kidney in several branches, the ends of which envelop kidney tissues called **renal pyramids**. The renal pyramids make up the internal core, or **medulla**, of the kidney. The medulla is covered by an outer layer, or **cortex**, that has a granular appearance. Between the cortex and the medulla, the renal artery divides into the many arterioles that serve the nephrons. In this same region, the renal vein collects blood from the many venules that drain the peritubular capillaries.

The organization of the nephrons within the kidney is very regular. All of the glomeruli with their Bowman's capsules are located in the cortex. The initial segment of a renal tubule is called the **proximal convoluted tubule**—"proximal" because it is closest to the glomerulus, and "convoluted" because it is twisted (**Figure 52.9C**). All of the proximal convoluted tubules are also located in the cortex.

At the point at which the renal tubule descends into the medulla, it becomes thin, straightens, and descends directly down into the medulla. In the medulla the tubule makes a hairpin turn and ascends back to the cortex, forming what is called the **loop of Henle**. Some nephrons have longer loops of Henle than others. Some 20 to 30 percent of human nephrons that have glomeruli deep in the cortex (i.e., near the border with the medulla) have long loops of Henle that go deep into the medulla. Nephrons that have glomeruli farther up in the cortex generally have short loops of Henle that descend only a short distance into the medulla. As we will see, the long loops are the critical adaptation of the mammalian nephron that enables the kidney to concentrate the urine.

The ascending limb of the loop of Henle becomes the **distal convoluted tubule** when it reaches the cortex—"distal" because it is farther from the glomerulus. The distal convoluted tubules of many nephrons join a common **collecting duct** in the cortex. The collecting ducts descend back down through the renal pyramid, parallel to and past the tips of the loops of Henle, and empty into a funnel-shaped structure called the pelvis. Divisions of the pelvis that surround each renal pyramid join together to leave the kidney as the ureter (see Figure 52.9B).

The organization of the blood vessels of the kidney closely parallels the organization of the nephrons (see Figure 52.9C). Smaller arteries branch from the renal artery and radiate into the cortex, forming the afferent arterioles that carry blood to each glomerulus. Each glomerulus is drained by an efferent arteriole that gives rise to the peritubular capillaries, most of which surround the proximal and distal convoluted tubules. As we have seen, the intimate associations of the glomerular and peritubular capillaries with the renal tubules permit exchanges between the blood and the specialized regions of the tubules.

Some of the peritubular capillaries run into the medulla in parallel with the loops of Henle and the collecting ducts, forming a vascular network called the **vasa recta**. All of the peritubular capillaries from a nephron join back together into a venule that joins with venules from other nephrons and eventually leads to the renal vein. As we will see, the concentrating ability of the mammalian kidney depends on water reabsorption in the renal medulla, and the vasa recta are the avenue by which that water gets out of the renal medulla and back into the circulation.

## Most of the glomerular filtrate is reabsorbed by the proximal convoluted tubule

Most of the water and solutes filtered by the glomerulus are reabsorbed and do not appear in the urine. We can reach this conclusion by comparing the rate of filtration by the glomeruli with the rate of urine production. The kidneys receive about 1 liter of blood per minute, or about 1,500 liters of blood per day. How much of this huge volume is filtered out of the glomeruli? The answer is about 12 percent. This is still a large volume—180 liters per day! We normally urinate less than 2 liters per day, so about 99 percent of the fluid volume that is filtered out of the glomerulus is returned to the blood. Where and how is this enormous fluid volume reabsorbed?

The proximal convoluted tubule (PCT) is responsible for most of the reabsorption of water and solutes from the glomerular filtrate. The cells of this section of the renal tubule have many microvilli that increase their apical (facing into the tubule) surface area for reabsorption, and they have many mitochondria—an indication that they are metabolically active. PCT cells actively transport $Na^+$ (with $Cl^-$ following) and other solutes, such as glucose and amino acids, out of the tubule fluid.

Almost all glucose and amino acid molecules that are filtered from the blood are actively reabsorbed by PCT cells and transported into the extracellular fluid. The active transport of solutes from the proximal tubule into the interstitial fluid causes water to follow osmotically. The water and solutes moved into the interstitial fluid are taken up by the peritubular capillaries and returned to the venous blood. These processes accomplish the reabsorption of more than 75 percent of the fluid that initially enters the nephron.

Despite the bulk reabsorption of water and solutes by the proximal convoluted tubule, the overall osmolarity of the fluid flowing through the PCT does not change. Thus the process that is occurring in the PCT is called isosmotic reabsorption. The fluid that enters the loop of Henle has the same osmolarity as the blood plasma, although its composition is different. How then does the kidney produce urine that is more concentrated than the blood plasma?

## The loop of Henle creates a concentration gradient in the renal medulla

Humans can produce urine that is four times more concentrated than their blood plasma. The vampire bat we encountered at the beginning of this chapter can produce urine that is 15 times more concentrated than its blood plasma. The concentrating ability of the mammalian kidney arises from a **countercurrent multiplier** mechanism made possible by the anatomical arrangement of the loops of Henle. The term "countercurrent" refers to the opposing directions in which the tubule fluid in the descending and ascending limbs flows. The term "multiplier" refers to the ability of this system to create a solute concentration gradient in the renal medulla.

The loops of Henle do not themselves produce concentrated urine; rather, they increase the osmolarity of the extracellular fluid in the medulla in a graduated way. In humans, for example, the extracellular fluid at the top of the medulla bordering the cortex will be about 300 mosm/L (the concentration of blood plasma). But at the bottom of the medulla, where the loops of Henle make their hairpin turns, the extracellular fluid can be 1,200 mosm/L (see Figure 52.10). How do the loops produce this effect?

The cells that make up the different segments of the loop of Henle differ anatomically and functionally. Cells of the descending limb and the initial cells of the ascending limb are thin, with no microvilli and few mitochondria. They are not specialized for transport. Partway up the ascending limb, the cells become specialized for active transport. These cells are thick and have many mitochondria. Accordingly, the segments of the loop of Henle are named the thin descending limb, the thin ascending limb, and the thick ascending limb (**Figure 52.10**).

The countercurrent multiplier mechanism may be more easily understood by first considering events occurring in the thick ascending limb (Figure 52.10, note 1). The cells of the thick ascending limb reabsorb $Na^+$ and $Cl^-$ from the tubule fluid and move it into the interstitial fluid. (In the following discussion, we will distinguish between the two components of extracellular fluid—the blood plasma and the interstitial fluid.) The thick ascending limb is not permeable to water, so the reabsorption of $Na^+$ and $Cl^-$ from the tubular fluid raises the concentration of those solutes in the surrounding interstitial fluid

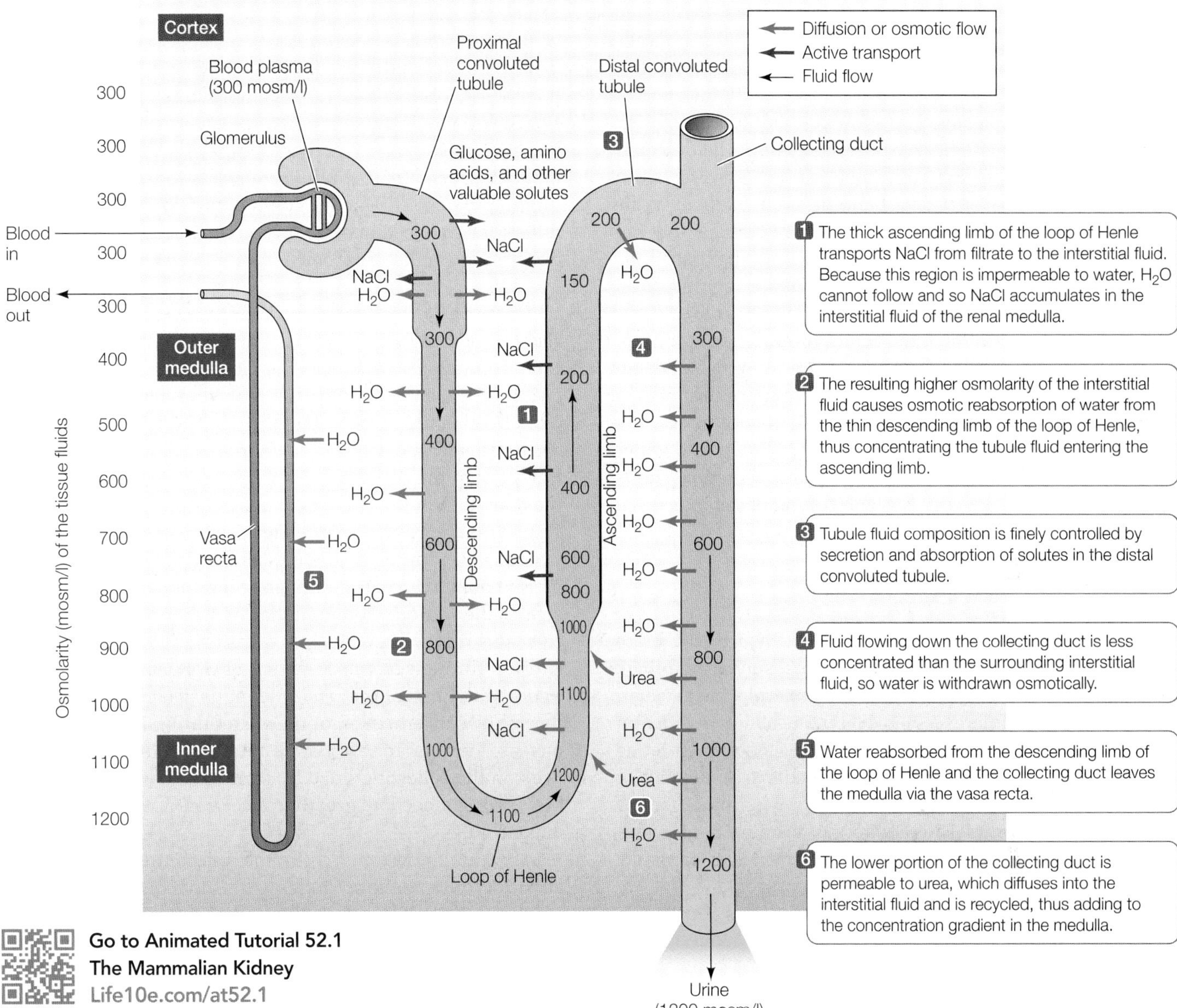


**Go to Animated Tutorial 52.1**
**The Mammalian Kidney**
Life10e.com/at52.1

**52.10 Concentrating the Urine** A countercurrent multiplier mechanism enables the mammalian kidney to produce urine that is far more concentrated than the blood plasma. The composition—but not the concentration—of the filtrate is changed by the proximal convoluted tubule, which reabsorbs valuable molecules (including NaCl). Bulk reabsorption of water follows osmotically. The urine concentration process begins in the thick ascending limb of the loop of Henle, which reabsorbs NaCl but is impermeable to $H_2O$. Some of the reabsorbed NaCl enters the descending limb and is thereby trapped in the renal medulla, creating a concentration gradient in the interstitial fluid. As urine in the collecting duct passes through this concentration gradient, it can lose water osmotically and become highly concentrated.

and decreases the concentration of the tubular fluid entering the distal convoluted tubule.

The thin descending limb, in contrast, is highly permeable to water but not very permeable to $Na^+$ and $Cl^-$. Since the local interstitial fluid has been made more concentrated by the $Na^+$ and $Cl^-$ reabsorbed from the neighboring thick ascending limb, water is withdrawn osmotically from the fluid in the descending limb. Therefore the fluid in the descending limb becomes more concentrated as it flows toward the hairpin turn at the bottom of the renal medulla (Figure 52.10, note 2).

The thin ascending limb, like the thick ascending limb, is not permeable to water. It is, however, permeable to $Na^+$ and $Cl^-$. As the concentrated tubule fluid flows up the thin ascending limb, it is more concentrated than the surrounding interstitial fluid, so $Na^+$ and $Cl^-$ diffuse out. When the tubule fluid reaches the thick ascending limb, active transport continues to move $Na^+$ and $Cl^-$ from the tubule fluid to the interstitial fluid.

As a result of the processes described above, the tubule fluid reaching the distal convoluted tubule is less concentrated than the blood plasma (Figure 52.10, note 3), and the solutes that have been left behind in the renal medulla have created a concentration gradient in the interstitial fluid of the medulla (indicated by the background color gradient in Figure 52.10).

You may wonder why the blood flow through the medulla does not wash out the concentration gradient established by the loops of Henle. The parallel arrangement of the descending and ascending peritubular capillaries—the vasa recta—in the medulla helps preserve the concentration gradient in the medulla. These capillaries are permeable to both salt and water. Therefore, as blood flows down the descending limb of the vasa recta into the increasingly concentrated interstitial fluid of the medulla, it loses water and gains solutes. As blood flows up from the bottom of the medulla in the ascending limb of the vasa recta, the opposite happens (water is gained and solutes are lost) because now the blood is more concentrated than the surrounding interstitial fluid (Figure 52.10, notes 4–6). The dynamics of this countercurrent exchange of salts and water between the blood in the vasa recta and the interstitial fluids result in little net change in the composition of the interstitial fluid in the medulla.

## Water permeability of kidney tubules depends on water channels

We have noted that some tubule regions, such as the PCT, are highly permeable to water whereas others, such as the thick ascending limb of the loop of Henle, are impermeable to water. What causes these differences in water permeability in different regions of the nephron? Aquaporins are a class of membrane proteins that form water channels (see Section 6.3). Regions of the nephron that are highly permeable to water have greater numbers of aquaporins. Thus aquaporins are abundant in kidney PCT cells and in descending limbs of the loops of Henle, but not in the ascending limbs of the loop of Henle.

As an interesting evolutionary note, aquaporins are also important in maintaining water balance in amphibians. When not in an aqueous environment, many amphibians can gain water from a moist substrate because they have aquaporins in the epithelial cells of their belly skin. Thus water can cross their skin into the interstitial fluid by osmosis.

## The distal convoluted tubule fine-tunes the composition of the urine

The first portion of the distal convoluted tubule is similar to the thick ascending limb of the loop of Henle. $Na^+$ and $Cl^-$ are transported out of the tubule fluid, and water cannot follow. As a result, the tubule fluid becomes even more dilute. The later sections of the distal convoluted tubule, however, can be permeable to water, and water can be osmotically drawn from the tubule into the interstitial fluid. As the tubule fluid flows from the distal tubule to the collecting duct, it can be below or equal to the osmolarity of the blood plasma.

An important function of the distal tubule is the fine-tuning of the ionic composition of the urine. Even though bulk reabsorption of substances such as calcium, phosphate, bicarbonate, and potassium occurs in the proximal convoluted tubule, changes in the concentrations of these substances occur in the distal convoluted tubule. In the case of potassium, for example, if a person is potassium depleted, this ion is reabsorbed in the distal convoluted tubule, but if a person has an abundance of potassium, this ion is secreted in the distal convoluted tubule. As we will see below, this exchange of $K^+$ is controlled by the hormone aldosterone. Another example is reabsorption of $Ca^{2+}$ in the distal convoluted tubule, which is controlled by the actions of vitamin D. The fine-tuning of urine composition continues in the collecting duct. As you can imagine, the list of ion transporters in the distal convoluted tubule is large.

## Urine is concentrated in the collecting duct

The tubule fluid entering the collecting duct is at about the same solute *concentration* as the blood plasma, but its solute *composition* is considerably different from that of the plasma. The major solute in the tubular fluid is now urea, since salts were reabsorbed earlier in the nephron. As the tubule fluid flows down the collecting duct, it loses water osmotically to the interstitial fluid, and that water returns to the circulatory system via the vasa recta (see Figure 52.10, note 4).

The concentration gradient established in the renal medulla by the countercurrent multiplier actions of the loops of Henle creates the osmotic potential that withdraws water from the collecting ducts. The collecting ducts begin in the renal cortex and run through the renal medulla before emptying into the ureter at the tips of the renal pyramids. During this journey, the solute concentration of the surrounding interstitial fluid increases, and more and more water can be absorbed from the urine in the collecting duct. By the time it reaches the ureter, the urine can become greatly concentrated, with urea as the major solute.

As water is withdrawn from the collecting duct, some urea also leaks out into the medullary interstitial fluid, adding to its osmotic potential. This urea diffuses back into the loop of Henle and is returned to the collecting duct. The recycling of urea in the renal medulla contributes significantly to the concentration gradient and therefore the ability of the kidney to concentrate the urine in the collecting duct. The ability of a mammal to concentrate its urine is determined by the maximum concentration gradient it can establish in its renal medulla.

## The kidneys help regulate acid–base balance

Besides regulating salt and water balance and excreting nitrogenous wastes, the kidneys have another important role: they regulate the hydrogen ion concentration (the pH) of the extracellular fluids. pH is a critical variable because it influences the structure and function of proteins.

One way to minimize pH changes in a chemical solution is to add a buffer—a substance that can either absorb or release hydrogen ions (see Section 2.4). The major buffer in the blood is bicarbonate ions ($HCO_3^-$; see Figure 49.14) that are formed from the dissociation of carbonic acid, which in turn is formed by the hydration of $CO_2$ according to the following equilibrium reaction:

$$CO_2 + H_2O \rightleftharpoons H_2CO_3 \rightleftharpoons H^+ + HCO_3^-$$

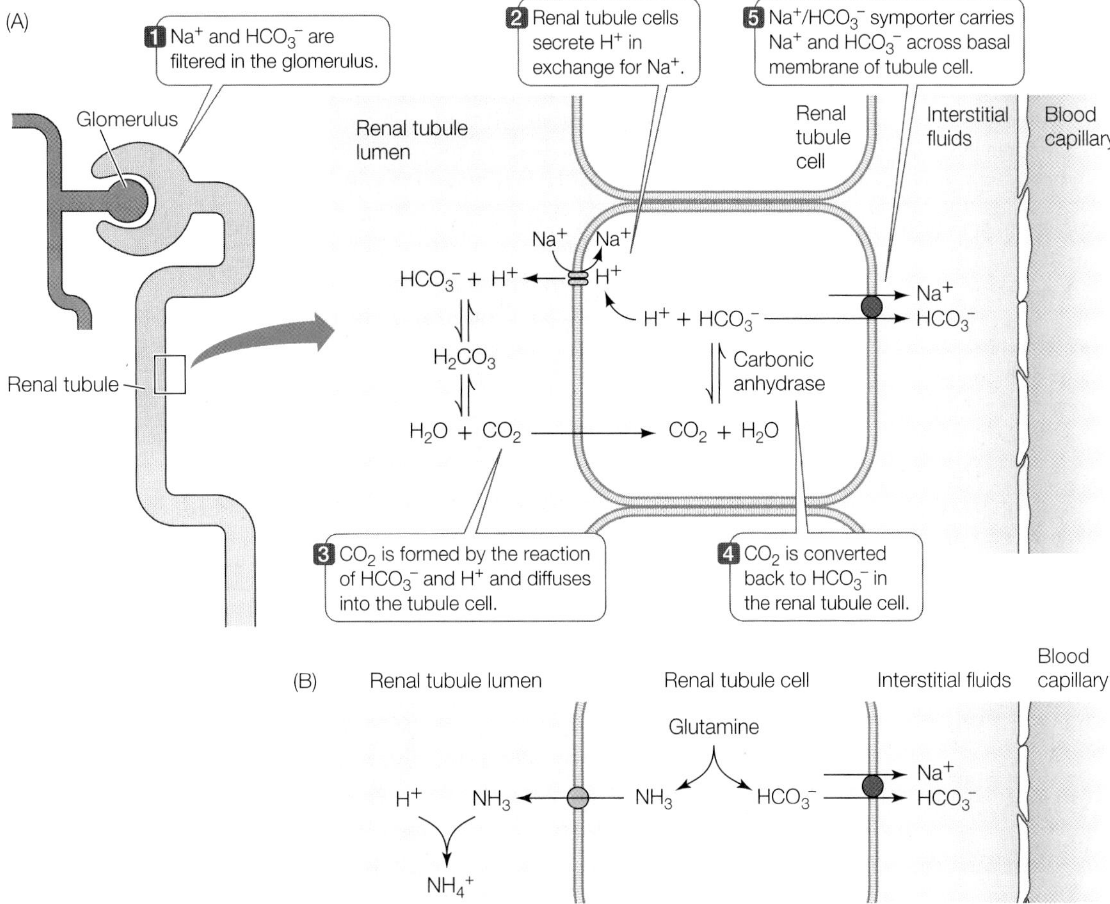


**52.11 The Kidney Excretes Acids and Conserves Bases** (A) Bicarbonate ions are filtered out of the blood at the glomerulus, and renal tubule cells secrete hydrogen ions into the tubule fluid. In the renal tubule, the filtered bicarbonate buffers the secreted hydrogen ions and keeps the urine from becoming too acidic. The $CO_2$ formed by the reaction of bicarbonate and hydrogen ions is converted back to bicarbonate by the renal tubule cells and transported back into the interstitial fluid. (B) Excretion of ammonia ($NH_3$) by renal tube cells is also important for acid–base balance.

From this equation, you can see that if excess hydrogen ions are added to this reaction mixture, the reaction will move to the left and absorb the excess $H^+$. If hydrogen ions are removed from the reaction mixture, however, the reaction will move to the right and supply more $H^+$.

The $HCO_3^-$ buffer system is important for controlling the pH of the blood, and therefore of the interstitial fluids as well, because the reaction can be pushed to the right and pulled to the left physiologically. The lungs control the levels of $CO_2$ in the blood, thus altering the acid portion of the reaction. $CO_2$ is considered the acid portion of the reaction because if you add additional $CO_2$, the reaction shifts to the right, producing more $H^+$ ions. The kidneys control the base portion of the reaction by removing $H^+$ from the blood and returning $HCO_3^-$ to the blood. How does this occur?

$HCO_3^-$ is filtered in the glomerulus and is therefore present in the tubular fluid. Tubule cells transport $H^+$ into the tubule fluid in exchange for $Na^+$. In the tubule, the excreted $H^+$ combines with the filtered $HCO_3^-$ to produce $H_2CO_3$ that then dissociates into $H_2O$ and $CO_2$. The $CO_2$ diffuses into the tubule cells, where in the presence of the enzyme carbonic anhydrase it produces $HCO_3^-$ that is transported out of the basal end of the cell into the interstitial fluid and thence to the blood (**Figure 52.11A**). Thus for each $H^+$ secreted into the tubule fluid, a $HCO_3^-$ ion is released into the blood.

Another mechanism for $H^+$ secretion and $HCO_3^-$ reabsorption involves ammonia ($NH_3$). The metabolism of glutamine in tubule cells produces $NH_3$ and $HCO_3^-$ (**Figure 52.11B**). The $HCO_3^-$ is reabsorbed into the interstitial fluid. The $NH_3$ is transported into the tubule fluid and combines with $H^+$ to form ammonium ($NH_4^+$), which is excreted in the urine. This process results in the net excretion of $H^+$ from the body. The $NH_3$ is transported into tubules by means of an $NH_3$ transporter that has been characterized recently in an effort to identify novel proteins coming from the sequencing of the human and other genomes (**Figure 52.12**).

## Kidney failure is treated with dialysis

Loss of kidney function (renal failure) results in the retention of salts and water (hence high blood pressure), retention of urea (uremic poisoning), and a decreasing pH (acidosis). A person who suffers complete renal failure will die within 2 weeks if not treated. A drastic but highly successful treatment is kidney transplant, but it is usually necessary to sustain a patient for

## INVESTIGATING**LIFE**

**52.12 An Ammonium Transporter in the Renal Tubules?** An important way the kidney excretes hydrogen ions and buffers the blood is to secrete ammonia ($NH_3$) into the renal tubules. It was thought that ammonia simply diffused into the tubules until the function of Rhcg, a protein in the Rh blood antigen family was discovered. This experiment demonstrated that loss of the *Rhcg* gene in mice impairs their ability to buffer their blood pH by excreting excess $H^+$ (in the form of ammonium)[a]

**HYPOTHESIS** The protein Rhcg is an ammonia transporter and is critical for the kidney's role in acid–base balance.

*Method*

1. Create a line of mice in which the gene for the protein Rhcg is knocked out (see Section 18.4).
2. Measure starting blood pH, plasma bicarbonate ($HCO_3^-$) levels, and urine ammonium ($NH_4^+$) levels in experimental and control (wild-type) mice.
3. For 6 days, administer drinking water containing a mild acid to control mice and to Rhcg knockout mice.
4. Measure the three variables (see item 2 above) at days 2 and 6.

*Results*

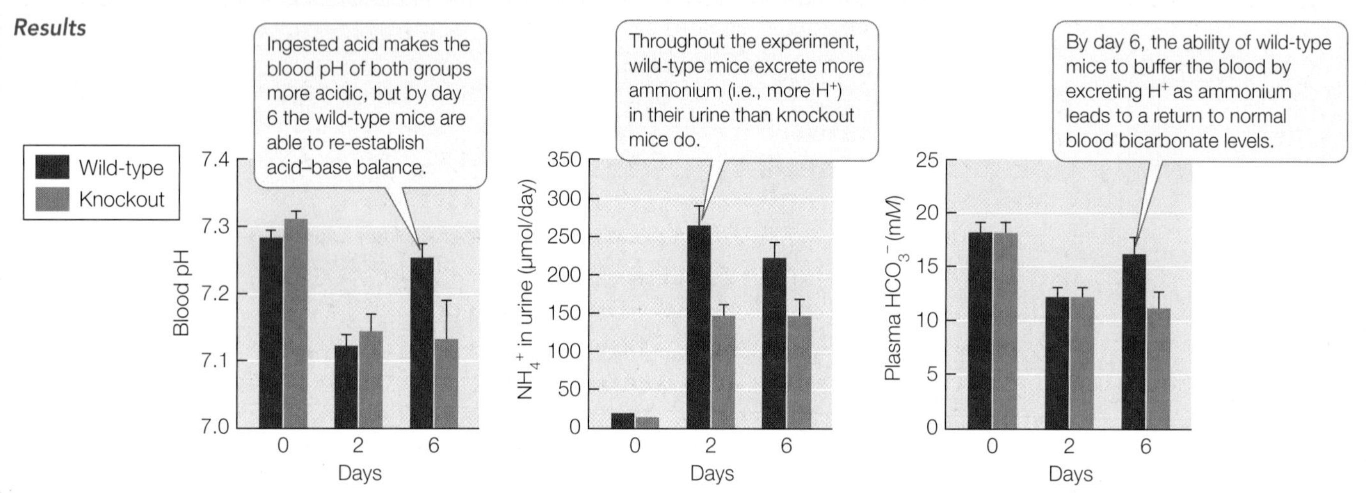


**CONCLUSION** Lack of functional Rhcg protein impairs a mouse's ability to secrete ammonium ions in its urine, thus limiting the capacity to regulate acid–base balance. This protein is probably an ammonia transporter in the renal tubules.

Go to **BioPortal** for discussion and relevant links for all INVESTIGATING**LIFE** figures.

[a]Biver, S. et al. 2008. *Nature* 456: 339–343.

considerable time while waiting for a kidney to become available. Therefore artificial kidneys, or renal dialysis machines, are essential modes of treatment.

In a dialysis machine, the patient's blood flows through many small channels made of semipermeable membranes (**Figure 52.13**). A dialysis solution flows on the other side of these membranes, through which small molecules can diffuse. Molecules and ions diffuse from an area of higher concentration to an area of lower concentration, so the composition of the dialysis fluid is crucial. The concentrations of the molecules or ions that need to be conserved must be at the same concentration in the dialysis fluid as they are in the blood. The concentrations of molecules and ions that need to be removed from the blood are zero in the dialysis fluid. The total osmotic potential of the dialysis fluid must equal that of the plasma.

About 500 ml of the patient's blood is in the dialysis machine at any one time, and the unit processes several hundred milliliters of blood per minute. A patient with no kidney function must be on the dialysis machine for 4 to 6 hours three times a week.

### RECAP 52.5

The anatomical organization of nephrons makes it possible for the mammalian kidney to produce a urine more concentrated than the blood, thereby conserving water to maintain extracellular fluid volume. Bulk reabsorption of salts, other valuable solutes, and water takes place in the proximal convoluted tubule. The loops of Henle act as a countercurrent multiplier, creating a concentration gradient of the interstitial fluids in the renal medulla. Collecting ducts run through the renal medulla and lose water osmotically to the surrounding interstitial fluids, concentrating the urine.

- Explain how the countercurrent multiplier mechanism of the nephron makes it possible for the kidney to form a concentrated urine. **See pp. 1082–1083 and Figure 52.10**
- Why doesn't the blood flow through the kidney wash out the concentration gradient in the medulla? **See p. 1082 and Figure 52.10**
- How does the kidney contribute to acid–base balance? **See pp. 1085–1086 and Figures 52.11 and 52.12**

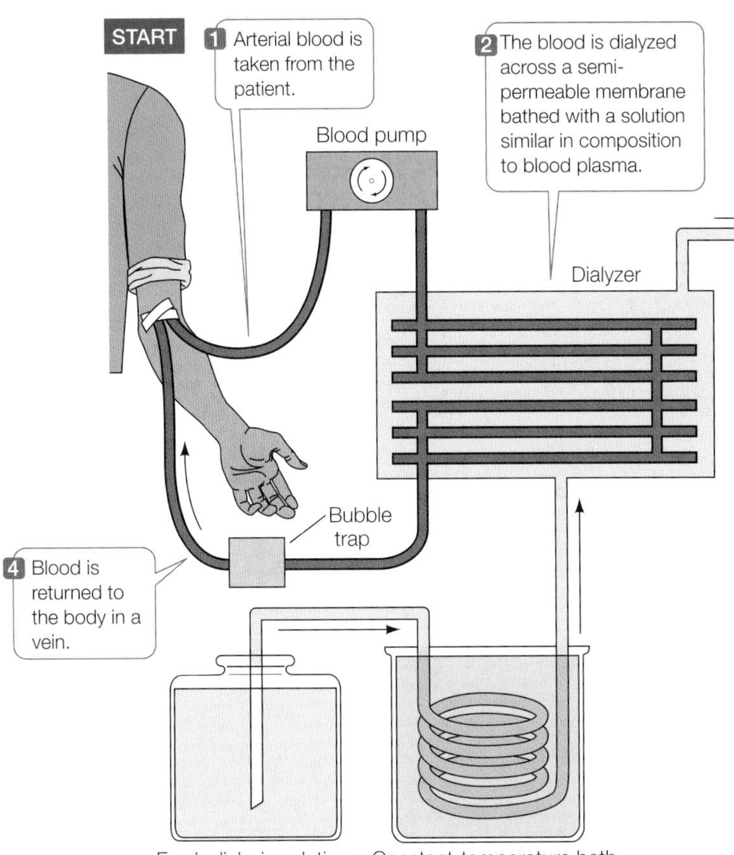


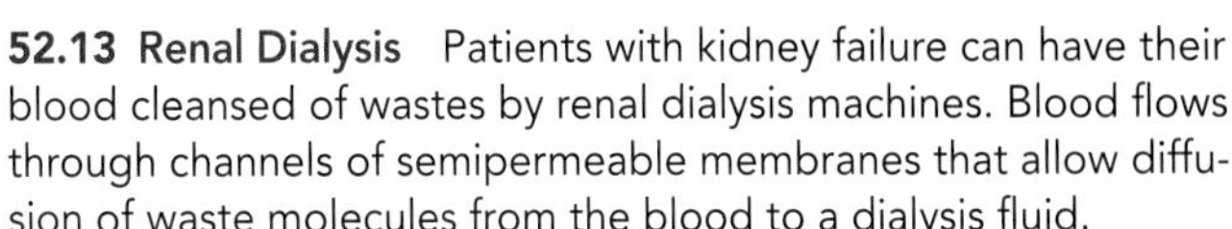

**52.13 Renal Dialysis** Patients with kidney failure can have their blood cleansed of wastes by renal dialysis machines. Blood flows through channels of semipermeable membranes that allow diffusion of waste molecules from the blood to a dialysis fluid.

The kidney contributes to homeostasis in several ways, including regulating extracellular fluid volume, maintaining the osmotic concentration and ionic composition of the extracellular fluid, and regulating pH. As we will see next, the kidneys also play a major role in regulating blood pressure.

## 52.6 How Are Kidney Functions Regulated?

Several regulatory mechanisms act on the kidneys to maintain blood pressure, blood osmolarity, and blood composition. We will discuss these mechanisms separately, but keep in mind that they are always working together.

### Glomerular filtration rate is regulated

If the kidneys stop filtering blood, they cannot accomplish any of their functions. The maintenance of a constant **glomerular filtration rate** (**GFR**) depends on an adequate blood supply to the kidneys at an adequate blood pressure. Renal arteries usually deliver blood to the kidneys at high pressure because they are early branches off the aorta. In addition, autoregulatory mechanisms ensure adequate blood supply and blood pressure for kidney function regardless of what is happening elsewhere in the body. The kidney's autoregulatory adjustments compensate for decreases in cardiac output or decreases in blood pressure so that the GFR remains constant.

One autoregulatory mechanism is the dilation (expansion) of the afferent renal arterioles when blood pressure falls. This dilation decreases the resistance in the arterioles and helps maintain blood pressure in the glomerulus. If arteriole dilation does not keep the GFR from falling, the kidney releases an enzyme, **renin**, into the blood. Renin converts a circulating protein, angiotensinogen, into angiotensin I, which is then acted on by angiotensin-converting enzyme (ACE) to form the active hormone angiotensin II, or simply **angiotensin** (**Figure 52.14**). Angiotensin has several effects that help restore the GFR to normal:

- It constricts the efferent renal arterioles, raising the resistance for blood leaving the glomerulus. Like putting a finger over the end of a garden hose, this restriction of drainage elevates blood pressure in the glomerular capillaries.
- It constricts peripheral blood vessels all over the body, an action that elevates blood pressure.
- It stimulates the adrenal cortex to release the hormone **aldosterone**. Aldosterone stimulates sodium reabsorption by the kidney, making its reabsorption of water more effective. Enhanced water reabsorption helps maintain blood volume and therefore blood pressure.
- It acts on the brain to stimulate thirst. Increased water intake in response to thirst increases blood volume and blood pressure.

Thus the renin-angiotensin-aldosterone system, or RAAS, coordinates many responses to maintain blood pressure and kidney function.

### Regulation of GFR uses feedback information from the distal tubule

A remarkable anatomical feature of the nephron is that where its renal tubule returns to the cortex and becomes the distal

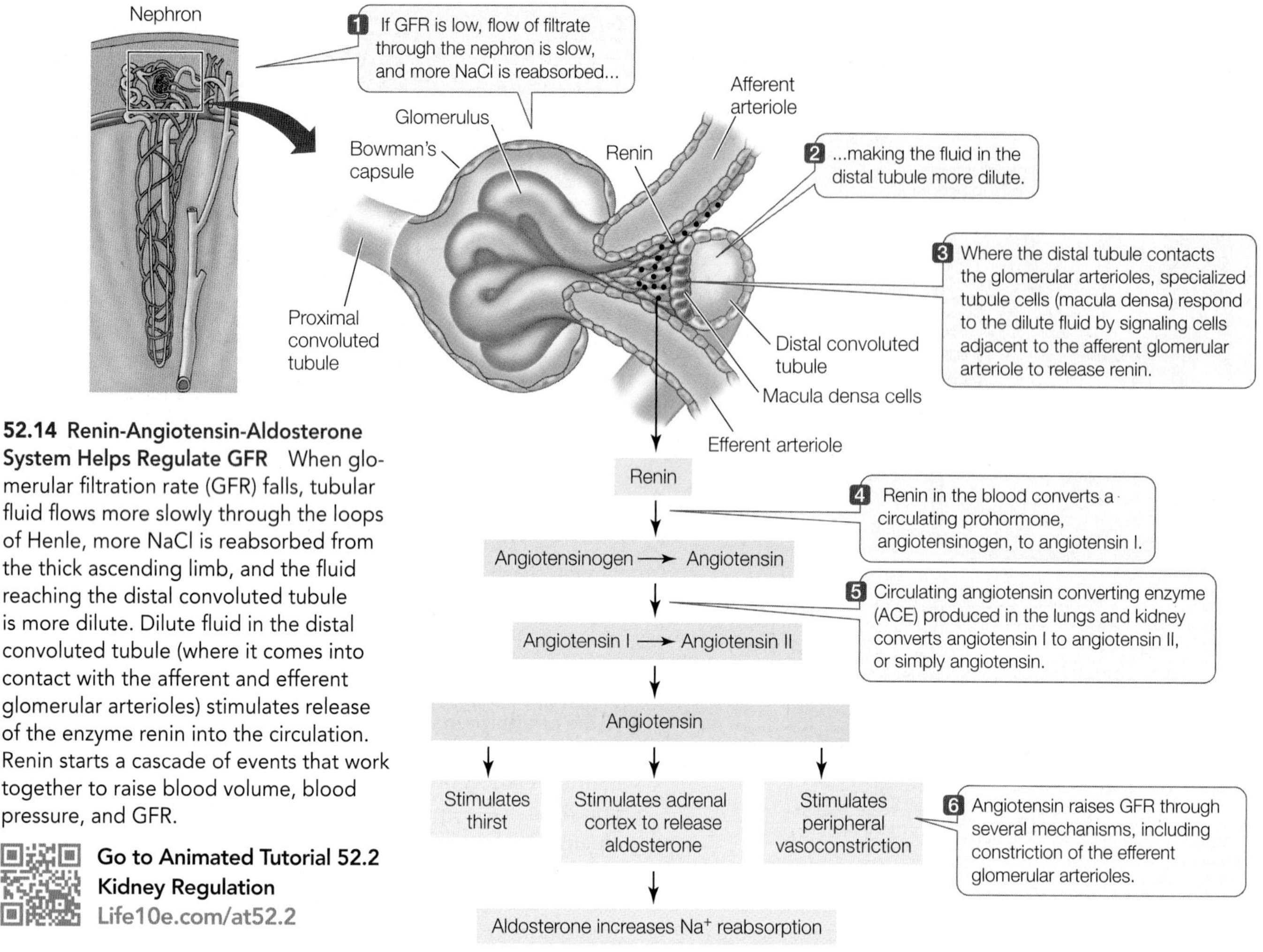


**52.14 Renin-Angiotensin-Aldosterone System Helps Regulate GFR** When glomerular filtration rate (GFR) falls, tubular fluid flows more slowly through the loops of Henle, more NaCl is reabsorbed from the thick ascending limb, and the fluid reaching the distal convoluted tubule is more dilute. Dilute fluid in the distal convoluted tubule (where it comes into contact with the afferent and efferent glomerular arterioles) stimulates release of the enzyme renin into the circulation. Renin starts a cascade of events that work together to raise blood volume, blood pressure, and GFR.

**Go to Animated Tutorial 52.2 Kidney Regulation**
Life10e.com/at52.2

convoluted tubule, it makes contact with the afferent and efferent arterioles of its glomerulus. At this location the cells of the tubule are modified to form a structure called the macula densa (see Figure 52.14), and the arteriole cells are also modified and are called juxtaglomerular cells. The macula densa cells sense the concentration of NaCl in the fluid entering the distal convoluted tubule. If glomerular filtration rate is high, flow through the loop of Henle is high and the cells of the thick ascending limb cannot reabsorb all of the NaCl passing through it. If GFR is low, flow through the loop of Henle is slower and the cells of the thick ascending limb can reabsorb more of the NaCl. If the NaCl level in the distal convoluted tubule drops too low, the macula densa cells signal the juxtaglomerular cells to release renin and trigger the RAAS response. Thus the concentration of NaCl in the fluid passing over the macula densa is a function of the GFR and is information that enables fine control of the RAAS system.

## Blood osmolarity and blood pressure are regulated by ADH

Cells in the hypothalamus can stimulate the release of a hormone called **antidiuretic hormone** (**ADH**, also called vasopressin) from the posterior pituitary. ADH can act on cells of the collecting duct to insert aquaporins (water channels) into their plasma membranes. The aquaporins increase the permeability of these membranes to water, and therefore more water is reabsorbed from the collecting duct fluid into the interstitial spaces of the renal medulla. The higher the circulating levels of ADH, the greater the number of aquaporins. Various factors can stimulate or inhibit the release of ADH. Of key importance to kidney function are osmoreceptors that monitor blood osmolarity and stretch receptors that monitor blood pressure (**Figure 52.15**).

Osmoreceptor neurons in the hypothalamus are activated by a rise in blood osmolarity, and they increase the release of ADH. ADH helps regulate blood osmolarity by controlling water reabsorption. The osmoreceptors also stimulate thirst. The resulting water retention and water intake dilute the blood as they expand blood volume.

Stretch receptors in the walls of the aorta and the carotid arteries (see Figure 50.19) that detect an increase in blood pressure will *inhibit* the release of ADH. With less circulating ADH, less water is reabsorbed, which decreases blood volume and hence acts to lower blood pressure.

If blood pressure falls, as when you lose blood volume through hemorrhage or excessive evaporative water loss,

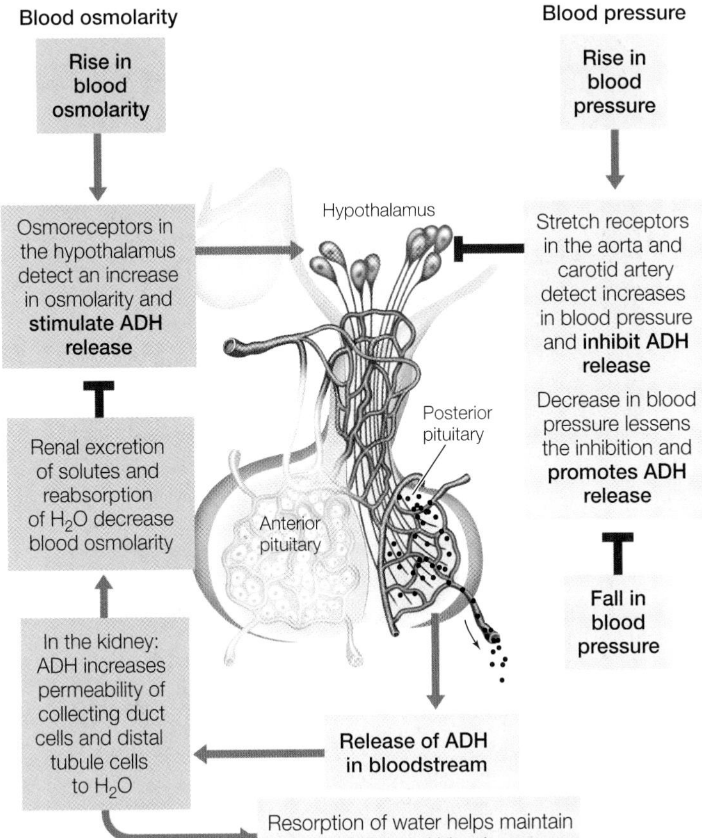


**52.15 Antidiuretic Hormone Increases Blood Pressure and Promotes Water Reabsorption** ADH is produced by neurons in the hypothalamus and released from nerve endings in the posterior pituitary. The release of ADH is stimulated by hypothalamic osmoreceptors and inhibited by stretch receptors in the great arteries. Red lines indicate inhibitory actions; green lines show stimulatory actions.

activity of the stretch receptors in the aorta and carotid arteries decreases. Input via cranial nerves to the hypothalamus from these receptors inhibits the release of ADH, so when the firing rates of these stretch receptors fall, ADH release increases. More ADH results in more efficient water reabsorption and therefore a protection of blood volume and blood pressure.

Alcohol inhibits ADH release, explaining why excessive beer drinking leads to excessive urination and dehydration, which contributes to the symptoms of a hangover.

As mentioned earlier, the presence of aquaporins in plasma membranes determines their water permeability. Aquaporins play an important and unique role in the collecting duct. Several members of the aquaporin family of water channels are found in the plasma membranes of cells in the distal tubules. At least two aquaporins (AQP-3 and AQP-4) are localized in the basolateral membranes (facing the blood vessels). A different aquaporin, called AQP-2, is found in the apical plasma membranes (facing into the tubule). The presence of AQP-2 in these membranes is controlled by ADH, as described in **Figure 52.16**.

INVESTIGATING**LIFE**

**52.16 ADH Induces Insertion of Aquaporins into Plasma Membranes** Aquaporin proteins make some regions of renal tubules permeable to water. One aquaporin, AQP-2, is responsible for the permeability of the collecting duct cells. M. A. Knepper and colleagues did an experiment to find out how antidiuretic hormone acts on these proteins to control the level of permeability in renal cells.[a]

**HYPOTHESIS** Antidiuretic hormone (ADH) controls the location of aquaporin proteins.

*Method*

1. Isolate collecting ducts from rat kidney.
2. Use immunochemical staining to localize the AQP-2 aquaporins in collecting duct cells both with and without the presence of ADH. Also localize the aquaporins after ADH is applied and then washed away.

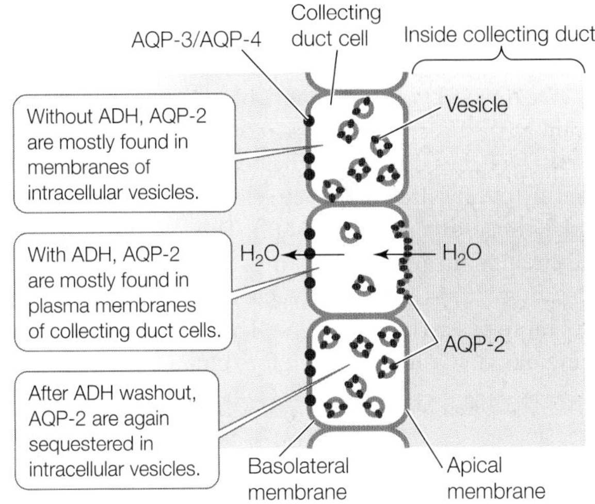


3. Measure the water permeability of the collecting duct cells under the same three conditions.

*Results*

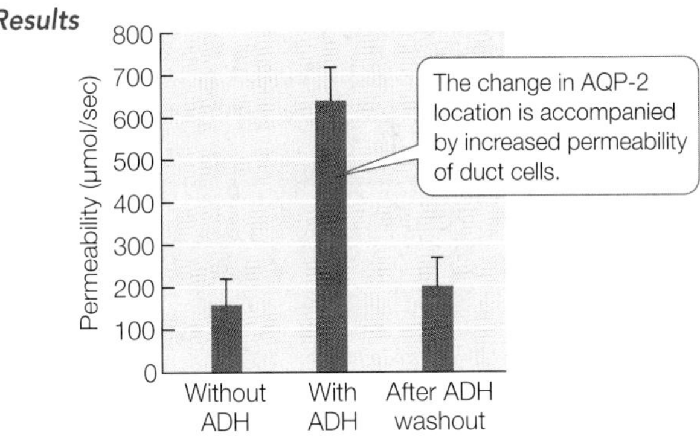


**CONCLUSION** In the absence of ADH, AQP-2 is sequestered intracellularly. When ADH is present, AQP-2 channels are inserted into the plasma membranes, making the cells more permeable to water.

Go to **BioPortal** for discussion and relevant links for all INVESTIGATING**LIFE** figures.

[a]Nielsen, S. et al. 1995. *Proceedings of the National Academy of Sciences USA* 92: 1013–1017.

When ADH levels are low, such as when a person is well hydrated, most of the AQP-2 protein is sequestered in the membranes of intracellular vesicles, and the collecting duct permeability is low. A rise in ADH levels stimulates the insertion of these vesicles along with their AQP-2 channels into the apical plasma membranes. As a result, the membranes become more permeable to water. Water that enters the collecting duct cells passes into the interstitial fluid through the aquaporins in the basolateral membranes. ADH also stimulates the synthesis of new AQP-2 proteins. Thus circulating ADH controls the number of AQP-2 water channels in the plasma membranes of the collecting duct cells, and therefore the permeability of the collecting duct to water.

## The heart produces a hormone that helps lower blood pressure

You may not think of the heart as an endocrine organ, but it is. When blood volume is high, blood pressure is high, putting strain on the heart. Under these conditions, the increased venous return stretches the atria of the heart. When the atrial muscle fibers are overly stretched, they release a peptide hormone called **atrial natriuretic peptide** (**ANP**). This peptide hormone enters the circulation, and in the kidney it decreases the reabsorption of sodium. If less sodium is reabsorbed, less water is reabsorbed, and more passes into the urine. Thus ANP has the effect of lowering blood volume and therefore blood pressure.

### RECAP 52.6

Glomerular filtration is essential for kidney function and is sustained by autoregulatory mechanisms. Sensors that monitor blood pressure and blood osmolarity may stimulate or inhibit the release of hormones that regulate kidney function.

- Explain how falling GFR results in an increase in circulating angiotensin and how angiotensin restores GFR. **See pp. 1087–1088 and Figure 52.14**
- Explain how falling blood pressure or increasing blood osmolarity results in changes in permeability of the collecting ducts. **See Figure 52.15**

(A) Desert gerbil

(B) Laboratory rat

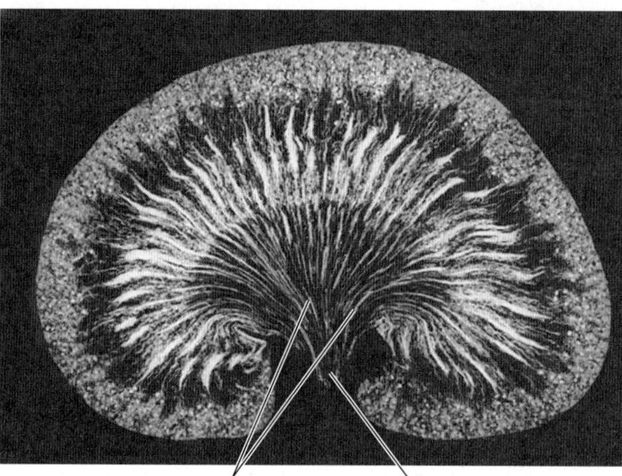


**52.17 The Ability to Concentrate** The ability of the mammalian kidney to concentrate urine depends on the lengths of its loops of Henle relative to the overall size of the kidney. (A) The kidney of a desert gerbil has a single renal pyramid with loops of Henle so long that the pyramid extends far into the ureter (ureter not shown). (B) An ordinary laboratory rat has much shorter loops of Henle.

How do desert rodents and vampire bats make highly concentrated urine?

### ANSWER

As we described in the opening story, some desert rodents and vampire bats conserve water by producing urine that can be as much as 15 times more concentrated than their blood. In fact, the urine is so concentrated that it contains crystals of solute. For comparison, the maximum urine concentration of humans is about four times that of the blood.

Recall from Section 52.5 that the mammalian body's ability to produce concentrated urine depends on the concentration gradient set up in the renal medulla by the loops of Henle. An important adaptation for increasing the concentration gradient is to increase the lengths of the loops of Henle relative to overall kidney size. Some tiny desert gerbils, for example, have such extremely long loops of Henle that the renal pyramid (each of the rodent kidneys has only one, in contrast to humans) extends far out of the concave surface of the kidney and into the ureter (**Figure 52.17**). The large concentration gradient that results draws most of the water out of the urine as it passes down the collecting duct. Desert gerbils are so effective in conserving water that they can survive on the water released by the metabolism of their food. The vampire bat kidney is particularly good at concentrating urea. Within minutes after feeding, the urea concentration in its urine can increase by more than 500-fold.

## 52.1 How Do Excretory Systems Maintain Homeostasis?

- **Excretory systems** maintain the osmolarity and volume of the extracellular fluids and eliminate the waste products of nitrogen metabolism through the processes of filtration, reabsorption, and secretion. **Urine** is the output of excretory systems.
- There is no active transport of water, so water must be moved across membranes by a difference in either **osmolarity** or pressure.
- Water enters and leaves cells by osmosis. To achieve cellular water balance, animals must maintain the osmolarity of their extracellular fluids within an acceptable range.
- Marine animals can be **osmoconformers** or **osmoregulators**. Freshwater animals must be osmoregulators and must continually excrete water and conserve salts. Terrestrial animals are osmoregulators, but the nature of their regulation depends on environment and lifestyle.
- Apart from regulating osmolarity of cells and extracellular fluids, animals must also regulate their ionic composition by conserving some ions and secreting others. Salt glands are adaptations for secretion of NaCl. **Review Figure 52.2**

## 52.2 How Do Animals Excrete Nitrogen?

- Aquatic animals that breathe water can eliminate nitrogenous wastes such as **ammonia** by diffusion across their gill membranes. Terrestrial animals and some aquatic animals must detoxify ammonia by converting it to **urea** or **uric acid** before excretion. **Review Figure 52.3**
- Depending on the form in which they excrete their nitrogenous wastes, animals are classified as **ammonotelic**, **ureotelic**, or **uricotelic**.

## 52.3 How Do Invertebrate Excretory Systems Work?

- The **protonephridia** of flatworms consist of flame cells and excretory tubules. Extracellular fluid is filtered into the tubules, which process the filtrate to produce a dilute urine. **Review Figure 52.4**
- In annelid worms, blood pressure causes filtration of the blood across capillary walls. The filtrate enters the coelomic cavity, where it is taken up by **metanephridia**, which alter the composition of the filtrate by active transport mechanisms. **Review Figure 52.5, ACTIVITY 52.1**
- The **Malpighian tubules** of insects receive ions and nitrogenous wastes by active transport across the tubule cells. Water follows by osmosis. Ions and water are reabsorbed from the rectum, so the insect excretes semisolid wastes. **Review Figure 52.6**

## 52.4 How Do Vertebrates Maintain Salt and Water Balance?

- Marine bony fishes produce little urine. Chondrichthyans retain urea and TMAO, so the osmolarity of their body fluids remains close to that of seawater.
- Reptiles and birds have skin with low water permeability and excrete nitrogenous wastes as uric acid in a semisolid form.
- Mammals produce urine that is more concentrated than their extracellular fluids.
- The **nephron**, the functional unit of the vertebrate **kidney**, consists of a **glomerulus**, in which blood is filtered, a **renal tubule**, which use processes of active secretion and reabsorption to convert the glomerular filtrate into urine to be excreted, and a system of **peritubular capillaries**, which surround the tubule and support its functions of secretion and reabsorption. **Review Figure 52.7, ACTIVITY 52.2**

## 52.5 How Does the Mammalian Kidney Produce Concentrated Urine?

- The concentrating ability of the mammalian kidney is a function of its anatomy, which enables countercurrent exchange. **Review Figure 52.9**
- The glomeruli and the **proximal** and **distal convoluted tubules** are located in the **cortex** of the kidney. Certain molecules are actively reabsorbed from the glomerular filtrate by the tubule cells, and other molecules are actively secreted. Straight sections of renal tubules called **loops of Henle** and **collecting ducts** are arranged in parallel in the **medulla** of the kidney. **Review ACTIVITY 52.3**
- Salts, water, and valuable molecules such as glucose and amino acids are reabsorbed in the proximal convoluted tubule without the renal filtrate becoming more concentrated, although its composition changes.
- The loops of Henle create a concentration gradient in the interstitial fluid of the renal medulla by a **countercurrent multiplier** mechanism. Urine flowing down the collecting ducts to the **ureter** is concentrated by the osmotic reabsorption of water caused by the concentration gradient in the surrounding interstitial fluid. **Review Figure 52.10, ANIMATED TUTORIAL 52.1**
- Hydrogen ions secreted by the renal tubules are buffered in the urine by bicarbonate and other chemical buffering systems. **Review Figures 52.11, 52.12**

## 52.6 How Are Kidney Functions Regulated?

- Kidney function in mammals is controlled by autoregulatory mechanisms that maintain a constant high **glomerular filtration rate** (**GFR**) even if blood pressure varies.
- An important autoregulatory mechanism is the release of **renin** by the kidney when blood pressure falls. Renin activates **angiotensin**, which causes the constriction of efferent glomerular arterioles and peripheral blood vessels, causes the release of **aldosterone** (which enhances water reabsorption), and stimulates thirst. **Review Figure 52.14, ANIMATED TUTORIAL 52.2**
- Changes in blood pressure and osmolarity influence the release of antidiuretic hormone (ADH), which controls the permeability of the collecting duct to water and therefore the amount of water that is reabsorbed from the urine. ADH stimulates the expression of and controls the intracellular location of aquaporins, which serve as water channels in the membranes of collecting duct cells. **Review Figures 52.15, 52.16**
- When the volume of blood returning to the heart increases and stretches the atrial walls, **atrial natriuretic peptide** (**ANP**) is released, which causes increased excretion of salt and water.

**See Activity 52.4 for a review of the major human organ systems.**

**Go to the Interactive Summary to review key figures, Animated Tutorials, and Activities Life10e.com/is52**

# CHAPTER REVIEW

## REMEMBERING

1. Which statement is *true*?
   a. Most marine invertebrates are osmoregulators.
   b. All freshwater invertebrates are osmoconformers.
   c. Marine bony fish and chondrichthyans have similar extracellular osmolarities.
   d. Freshwater fishes are ionic regulators.
   e. Marine mammals gain water osmotically.

2. The excretion of nitrogenous wastes
   a. by humans can be in the form of urea and uric acid.
   b. by mammals is never in the form of uric acid.
   c. by marine fishes is mostly in the form of urea.
   d. does not contribute to the osmolarity of the urine.
   e. requires more water if the waste product is the rather insoluble uric acid.

3. What is the role of renal podocytes?
   a. They prevent red blood cells and large molecules from entering the renal tubules.
   b. They reabsorb most of the glucose that is filtered from the plasma.
   c. They control the glomerular filtration rate by changing the resistance of renal arterioles.
   d. They provide a large surface area for tubular secretion and reabsorption.
   e. They release renin when the glomerular filtration rate falls.

4. Which part of the nephron is responsible for most of the difference in mammals between the glomerular filtration rate and the urine production rate?
   a. The glomerulus
   b. The proximal convoluted tubule
   c. The loops of Henle
   d. The distal convoluted tubule
   e. The collecting duct

5. Which of the following would *not* be a response stimulated by a large drop in blood pressure?
   a. Constriction of afferent renal arterioles
   b. Increased release of renin
   c. Increased release of antidiuretic hormone
   d. Increased thirst
   e. Constriction of efferent renal arterioles

## UNDERSTANDING & APPLYING

6. Persons with uncontrolled diabetes mellitus can have very high levels of glucose in their blood. Why do such individuals have a high level of urine production?

7. Patients with high blood pressure are sometimes treated with ACE inhibitors. Explain the rationale of this treatment.

8. Inulin is a molecule that is filtered out of the glomerulus but is not secreted or reabsorbed by the renal tubules. If you injected inulin into an animal and after a brief time measured the concentration of inulin in the animal's blood and urine, how could you determine the animal's glomerular filtration rate? Assume that the rate of urine production is 1 milliliter per minute.

## ANALYZING & EVALUATING

9. After you did the inulin experiment to measure glomerular filtration rate, how could you use that information to determine whether another substance is secreted or reabsorbed by the renal tubules? Assume you can measure the concentration of that substance in the blood and urine. Urine production is still 1 milliliter per minute.

10. Acetazolamide is a drug used by mountain climbers for short periods of time to help them acclimate to high altitude. The drug is an inhibitor of carbonic anhydrase. Referring back to Chapter 49, how do you think this drug helps in the acclimation process? In the context of this chapter, explain why a side effect of this drug is frequent urination.

Go to BioPortal at **yourBioPortal.com** for Animated Tutorials, Activities, LearningCurve Quizzes, Flashcards, and many other study and review resources.

# 53 Animal Behavior

## CHAPTER**OUTLINE**


**53.1** What Are the Origins of Behavioral Biology?

**53.2** How Do Genes Influence Behavior?

**53.3** How Does Behavior Develop?

**53.4** How Does Behavior Evolve?

**53.5** What Physiological Mechanisms Underlie Behavior?

**53.6** How Does Social Behavior Evolve?

**Big Babies** Chicks hatched from eggs laid by parasitic cowbirds (*Molothrus ater*) are frequently found in the nests of much smaller species such as the yellow warbler (*Dendroica petechia*). The cowbird chick grows rapidly, exceeding the size of the host parent. The cowbird chick's demand for food is so great that the host's own offspring may be ignored and die of neglect.

BROWN-HEADED COWBIRDS (*Molothrus ater*) are abundant and ubiquitous in North America, ranging over an enormous diversity of habitats. They are brood, or nest, parasites, laying their eggs in the nests of other species. A female cowbird may lay 40 eggs a season in as many different nests, but she never incubates her eggs or feeds her young. Cowbird eggs have been found in the nests of at least 220 other species. The host incubates the egg along with its own and feeds the young cowbird—often to the detriment of its own chicks because cowbird chicks are usually larger and more demanding.

Cowbirds have enormous reproductive success and it is not difficult to understand the evolution of brood parasitism from the standpoint of the parasite. But what about the selective pressures on the individuals that are parasitized? Cowbird eggs usually look very different from the host's own eggs, so why haven't host species evolved the ability to recognize an intruder egg and remove it from the nest? Indeed, the European common cuckoo (*Cuculus canorus*) is also a nest parasite, but its eggs closely mimic the eggs of its host species, presumably an adaptive response to an improved ability among host birds to discriminate and remove cuckoo eggs (an example of an evolutionary "arms race").

But in some cases eliminating the parasite's eggs can have negative consequences for the host. Researchers have reported observing what they hypothesize to be a kind of "mafia behavior" by cowbirds, in which the female cowbird periodically returns to inspect the nests she has parasitized. If she finds her egg missing, she destroys the nest along with the host's eggs. Female cowbirds have also been observed to destroy the unparasitized nest and eggs of a potential host. Researchers suggest this "farming behavior" may force the owner of the destroyed nest to rebuild and lay a new clutch of eggs—at which time the cowbird can lay her egg in the new nest.

Students of animal behavior seek to understand how behaviors such as nest parasitism evolve, how they develop, and what their underlying mechanisms are. That quest, as in this cowbird example, frequently requires research that includes the environment and interacting species. Behavioral biology is a highly integrative field that incorporates approaches from virtually all the biological subdisciplines. The study of animal behavior can give us insights into our own behavior.

Could cowbird behaviors create selective pressure for host species *not* to develop egg discrimination behavior?

See answer on p. 1117.

## 53.1 What Are the Origins of Behavioral Biology?

Humans have studied animal behavior since prehistoric times. Understanding the habits of potential prey, as well as those of their predators, was of great value to hunters. Appreciation of behavioral traits led to the domestication of animal species. Accounts of animal behaviors such as seasonal appearances and disappearances, mating displays, aggression, prey capture, parental care, and communication are found throughout recorded history. Yet the scientific study of animal behavior did not truly get under way until the early 1900s.

### Conditioned reflexes are a simple behavioral mechanism

In the late 1800s, the Russian physiologist Ivan Pavlov was studying the neural control of digestive juice secretion when he observed that not only did the dog he was experimenting on salivate when it smelled food, the animal also salivated whenever the technician who routinely fed the dog entered the room—even when no food was present. Following up this observation, Pavlov substituted a sound stimulus for the technician; a metronome ticked while the dog was fed. After several trials, the dog salivated when it heard the metronome, even if no food was offered.

Salivation in response to the sight, smell, or taste of food is a natural reflex response to a stimulus, but salivation in response to a sound was a learned response. The pairing of a sound with the experience of receiving food conditioned the dog's nervous system to generate a response, which Pavlov dubbed the **conditioned reflex** (**Figure 53.1**). Pavlov received a Nobel prize in 1904 for his work showing that a simple behavior controlled by the nervous system could be modified through experience. This work stimulated much new research because Pavlov had developed an experimental model of learning.

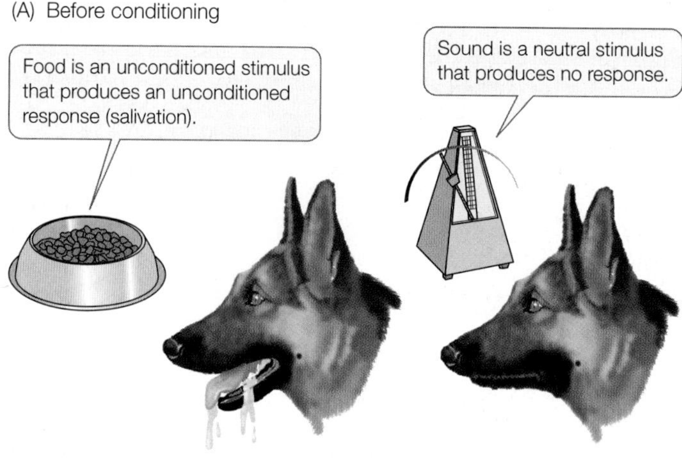

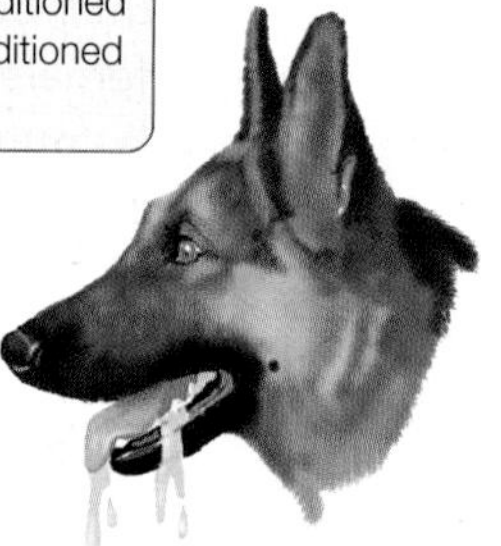


**53.1 The Conditioned Reflex** Ivan Pavlov discovered that when a normal response is paired with an artificial or neutral stimulus, an animal learns to produce the response even when only the artificial stimulus is presented.

Going beyond the conditioning of autonomic reflexes, the psychologist B. F. Skinner showed that any random action of an animal could become a conditioned response to a stimulus if a reward was temporally associated with the action and the stimulus. A rat, for example, could be conditioned to press a lever in response to a stimulus if it got a reward when it behaved as the experimenter desired. Because the animal was conditioned to perform an *operation* on its environment, this experimental protocol was known as **operant conditioning** and was viewed as another model of learning.

The experimental approaches to behavior initiated by Pavlov and Skinner had powerful effects on the nature of research on animal behavior, and the focus of scientists using their approaches was quite specific:

- They focused on laboratory environments rather than natural environments because in the laboratory, the variables in their experiments could be precisely controlled.
- They focused on only a few species (predominantly the albino rat) as model systems rather than studying diverse species from nature.
- They focused on questions of learning and memory, largely to the exclusion of other types of behavior (e.g., mating, feeding, communication).

Thus defined, the field of animal behavior research became known as behaviorism and was largely the domain of psychologists.

### Ethologists focused on the behavior of animals in their natural environment

An alternative approach to the study of animal behavior arose at the same time as behaviorism, but largely in Europe. Scientists there focused on describing the characteristics of animals in their natural environment, an approach that became known as **ethology** (Greek *ethos*, "character"; *logos*, "study"). In contrast to the behaviorists, the ethologists were interested in a wide variety of species, their evolutionary relationships, and

(A) *Larus marinus*

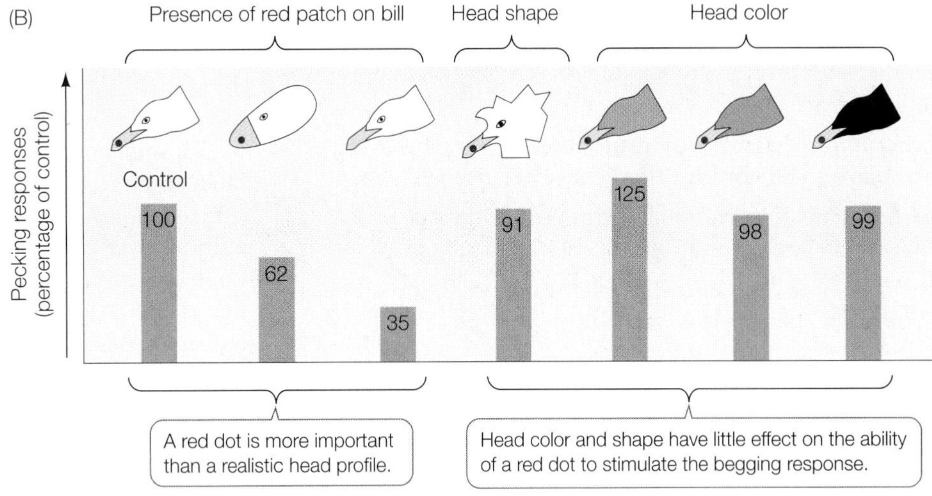


**53.2 Releasing a Fixed Action Pattern** (A) In many gull species, chicks instinctively peck at the red dot on the parent's lower bill, a behavior that induces the parent to regurgitate food into the chick's mouth. (B) Tinbergen's work showed that the red dot on the parent's lower bill is the critical component that releases the pecking response.

the ways in which their behaviors were adapted to their environments. The leaders of the ethology movement were Karl von Frisch, who discovered the dance language of honey bees; Konrad Lorenz, who discovered that the strong bond between parent and offspring develops during a "critical period" following birth; and Niko Tinbergen, who studied inborn patterns of behavior commonly known as instincts. These three scientists shared the Nobel Prize in 1973 for "their discoveries concerning organization and elicitation of individual and social behavior patterns."

The ethologists were mainly interested in species-specific or instinctive behaviors that therefore had to have genetic components. Behaviors were thought to be genetically determined if they:

- are performed without learning
- are stereotypic (that is, they are performed the same way each time)
- cannot be modified by learning

The ethologists called such behaviors **fixed action patterns**.

To demonstrate that a behavior was genetically determined, ethologists performed **deprivation experiments** in which an animal was raised in an environment devoid of opportunities to learn its species-specific behavior. An example of a natural deprivation experiment is the web spinning of a spider. The parents of a young spider die before it hatches, and in a seasonal environment it has no model webs to copy when it spins its first web, which requires thousands of stereotyped sequential movements. Yet a young spider creates a perfect, species-specific web the first and every time it spins a web. Thus the information for the web spinning behavior has to be genetically programmed into its nervous system.

Fixed action patterns are usually responses to specific stimuli. The ethologists carefully characterized such stimuli, which they called **releasers**. In general, releasers are simple subsets of the information available in the environment. For example, Tinbergen studied the begging behavior of gull chicks. Adult gulls have a red dot on their lower bill. When a parent returns to the nest to feed its chicks, the chicks peck on the red dot, which stimulates the parent to regurgitate food (**Figure 53.2A**). Experimenters investigated what stimulated the chicks to peck their parents' bills. Models of gull heads of different shapes and colors were tested (**Figure 53.2B**), as were models of a beak without a head. The results showed that the red dot was necessary for the release of chick pecking behavior. In fact, a pencil with a red eraser elicited a more robust pecking response than an accurate model of a gull head without a red dot.

## Ethologists probed the causes of behavior

The ethologists demonstrated the genetic basis for fixed action patterns by interbreeding closely related species. Konrad Lorenz studied the courtship behaviors of different species of dabbling ducks. Some of these species, such as mallards, teals, pintails, and gadwalls, are closely related and can interbreed, but they rarely do so in nature. Each male duck performs a courtship display consisting of a precise series of movements that is typical of his species. A female is not likely to accept him unless the entire display is successfully and correctly completed.

When Lorenz crossbred these duck species, the hybrid offspring expressed some elements of each parent's courtship display, but in novel combinations. Furthermore, Lorenz observed that hybrids sometimes exhibited display elements that were not in the repertoire of either parent species but were seen in other dabbling duck species. Lorenz's interbreeding studies demonstrated that the stereotypic motor patterns of the courtship displays are inherited. The observation that females were not interested in males performing hybrid displays was evidence that sexual selection had shaped these genetically determined behaviors (see Section 21.2).

The ethologists recognized the importance of development and motivation in behavior, and they laid the foundation for the application of modern biological methods to the study of animal behavior. Tinbergen outlined the challenges for investigators as four questions:

- *Causation:* What is the stimulus for the behavior, and how can the relationship between stimulus and behavior be modified by learning?
- *Development:* What experiences are necessary for a behavior to be displayed, and how does the behavior change with age?
- *Function:* How does the behavior affect the animal's chances for survival and reproduction?
- *Evolution:* How does the behavior compare with similar behaviors in related species, and how might it have evolved?

The first two questions refer to the **proximate causes** of behavior: the immediate genetic, physiological, neurological, and developmental mechanisms that determine how an individual is behaving at a particular time. The third and fourth questions refer to the **ultimate causes** of behavior: the evolutionary processes that produced the animal's capacity and tendency to behave in particular ways. In the sections that follow, we will describe many experiments on animal behavior. For each one, ask yourself which of Tinbergen's four questions it addresses and whether it focuses on proximate or ultimate causes of behavior.

**RECAP 53.1**

Early scientific studies of animal behavior took two approaches. Behaviorists focused on the study of conditioned behavior in a few species of laboratory animals and asked questions about learning. Ethologists studied genetically determined behavior in many species in their natural environments and asked evolutionary questions.

- Describe the difference between conditioned reflexes and operant conditioning. **See p. 1094 and Figure 53.1**
- What is the relationship between a releaser and a fixed action pattern? **See p. 1095 and Figure 53.2**
- Explain the difference between proximate and ultimate causes of behavior and how Tinbergen's questions address these two types of causes. **See p. 1096**

The work of the ethologists left no doubt that behavior can be genetically determined, but how? Genes code for proteins, whereas behaviors are highly complex traits involving sensory input and intricate patterns of control over responses to that input. Is it reasonable to think that a single gene can have a specific effect on a behavior?

## 53.2 How Do Genes Influence Behavior?

Most behaviors are complex traits that depend on many genes. Nevertheless, evidence from multiple approaches shows that alterations in single genes can result in discrete behavioral phenotypes on which natural selection can operate.

### Breeding experiments can produce behavioral phenotypes

Behavioral geneticists identify individuals with unusual behavioral phenotypes and conduct breeding experiments to see if those traits are inherited and to identify the genes involved. The origin of the unusual behavioral phenotypes may come from diversity in natural populations such as the dabbling ducks that Lorenz crossed; from screening animals subjected to mutagenesis; or from artificial selection experiments.

Examples of single genes involved in complex behavior have come from studies of the circadian rhythms of mutagenized animals. Circadian rhythms, which we will discuss in Section 53.5, are rhythms of behavior and physiology that continue to be expressed under constant conditions with a period that is about (but not exactly) 24 hours—in other words, they reflect an internal clock mechanism. The *per* gene (for period) was discovered in fruit flies, and variants of this gene caused the circadian period to be long, short, or eliminated. *Per* genes are now known to be an important element of the circadian clock in a wide variety of species, including humans. Other genes involved in circadian behavior have been discovered in subsequent mutagenesis and screening experiments.

Very few behaviors can be linked to a single gene, however. Most behaviors are complex traits and are influenced by many genes. A technique called quantitative trait analysis (or QTL, for "quantitative trait loci," referring to locations on chromosomes) has been developed to identify multiple genes that influence a given trait. QTL analysis requires animals, such as mice, with completely sequenced and well-mapped genomes that have many identifiable genetic markers (i.e., unique DNA sequences). When individuals from two strains of a species that differ with respect to the trait of interest are crossed, the trait can be quantified in the offspring and the offspring can be genotyped to show which of the genetic markers inherited from each parent correlate with the trait of interest in the offspring. Those markers are assumed to be close to the genes that influence the trait. QTL analysis thus points to regions on chromosomes where there are genes that influence the trait of interest.

### Knockout experiments can reveal the roles of specific genes

Some modern molecular genetic approaches start with identified genes and eliminate or silence them to see what effects their elimination has on a behavioral phenotype (see Section 18.4). As you might expect, knocking out genes involved in sensory pathways can have pronounced effects on behavior. One example is a gene for a specific olfactory receptor in mice.

As we saw in Section 46.2, mice have two olfactory organs: the nasal olfactory epithelium common to all mammals, and a small organ in the nasal passages called the vomeronasal organ, or VNO (**Figure 53.3**). Catherine Dulac at Harvard University discovered that a large number of pheromone receptors were expressed in that organ. (As described in Section 46.2, pheromones are signaling molecules released into the environment.) Dulac hypothesized that when the receptors in the male's VNO bound to sex pheromones produced by female mice, they stimulated mating behavior. To test this hypothesis, Dulac created a genetically engineered male mouse in which a gene for VNO receptor signaling was knocked out. Contrary to the prediction of the hypothesis, the knockout males in fact did pursue and mate with females placed in their cages. However, they also pursued and tried to mate with *males* placed in their cages. Normally a male mouse reacts aggressively to

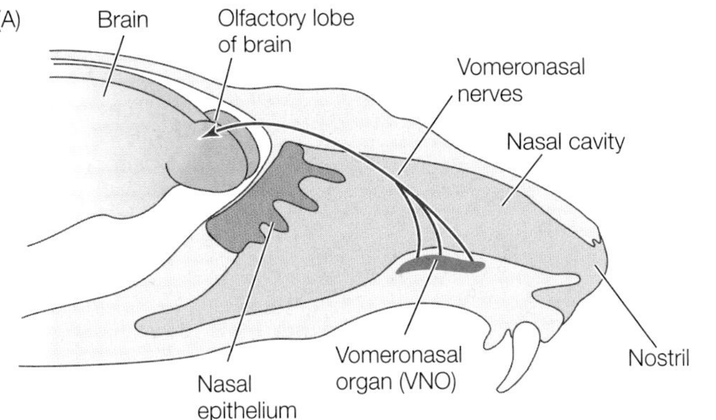

**53.3 The Mouse Vomeronasal Organ Identifies Gender** (A) The mouse VNO is located adjacent to the nasal passages. It contains pheromone receptors whose input travels to a specific region of the olfactory bulb (the accessory olfactory bulb). (B) In male mice, information from the VNO appears to be crucial in identifying gender and thus a potential sexual partner.

a strange male, but the knockout male could not discriminate between males and females placed in his cage. Thus properly functioning VNO receptors appear to be essential not for sexual attraction, but for gender identification. It is possible to imagine how selection working on this one gene could modify the intensity of male–male aggression and lead to changes in social behavior.

## Behaviors are controlled by gene cascades

Male courtship behavior in *Drosophila melanogaster*, the laboratory fruit fly, is stereotypic, species-specific, and requires no learning—a classic fixed action pattern. When a male encounters a potential mate, he follows her, taps her body with his foreleg, extends and vibrates one wing, and licks her genitals (**Figure 53.4A**). The development of this complex male behavior is under the control of a single gene, called *fruitless* (*fru*), just as the development of male anatomy is under the control of another gene, called *doublesex* (*dsx*). In both males and females, these two genes are part of a gene

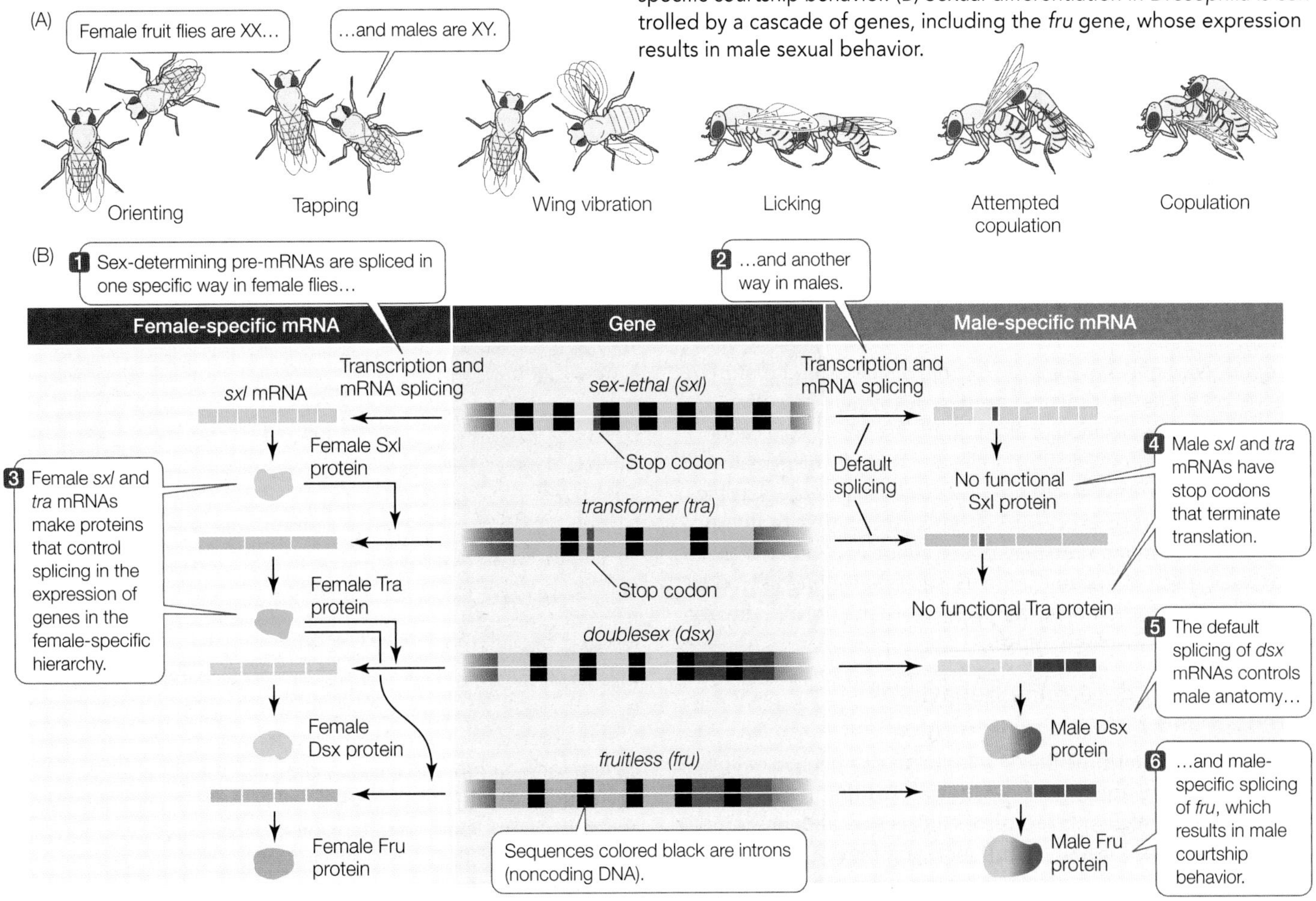

**53.4 The *fruitless* Gene** (A) Male fruit flies display stereotypic, species-specific courtship behavior. (B) Sexual differentiation in *Drosophila* is controlled by a cascade of genes, including the *fru* gene, whose expression results in male sexual behavior.

expression cascade that results in different *dsx* and *fru* gene products in males and females (**Figure 53.4B**). The female version of the Dsx protein controls the development of female anatomy, and the expression of *fru* in the male nervous system results in the organization of the neural circuitry controlling male sexual behavior.

There are two take-home lessons from this example. First, genes that control aspects of behavior, like other genes, are generally embedded in gene cascades that offer multiple opportunities for simple genetic changes that will alter the phenotype of even complex behaviors. Second, certain genes, such as *dsx* and *fru*, influence a range of other genes that contribute to complex behaviors. Modifications in any one of those genes or its expression can alter behavior. Thus even though no behavior is coded for by a single gene, alterations in single genes can influence behavior in ways that affect an animal's fitness.

**RECAP 53.2**

Breeding experiments show that behavioral phenotypes can be inherited and modified by natural selection. Although most behaviors are controlled by complex cascades of genes, molecular genetic methods have shown that a single, identifiable gene in the cascade can influence a complex behavior.

- How can knockout experiments reveal genes controlling behavior? **See p. 1096 and Figure 53.3**
- What is the evidence that the vomeronasal organ in mice is responsible for gender identification? **See pp. 1096–1097**
- Describe some of the ways a gene can control expression of a complex behavior. **See pp. 1097–1098 and Figure 53.4**

How can the genetic cascades that underlie complex behaviors be programmed to respond selectively to specific sets of stimuli? How can their expression be limited to appropriate times in an animal's life? The answers to these questions can by found by studying how behaviors develop over the life span.

## 53.3 How Does Behavior Develop?

The emergence of behavior as an animal develops and matures depends on the development of the nervous system as well as on the growth and maturation of other body systems. A bird cannot fly until its wings grow and its muscles and flight feathers mature. But even with anatomical and physiological competence, specific behaviors may not be expressed. Behaviors that are adaptive at one stage in an animal's life may not be adaptive at other stages. Behaviors typical of juvenile animals, such as begging for food, may disappear and new behavior patterns of a mature individual, such as courtship displays, appear.

### Hormones can determine behavioral potential and timing

Hormones can determine the development of a behavioral potential at an early age and the expression of that behavior at a later age. An excellent example of this is sexual behavior in rats (**Figure 53.5**). Normally, adult male and female rats exhibit different patterns of sexual behavior: females adopt a sexually

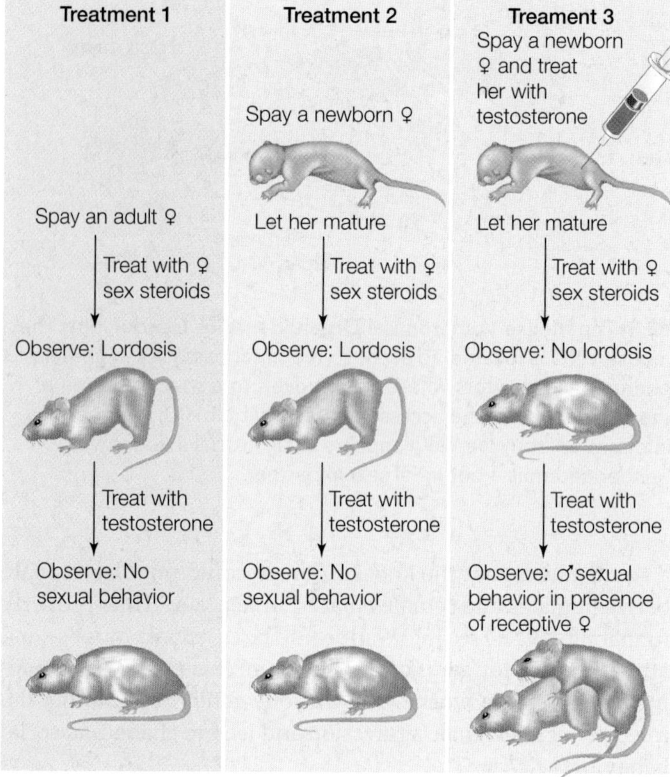


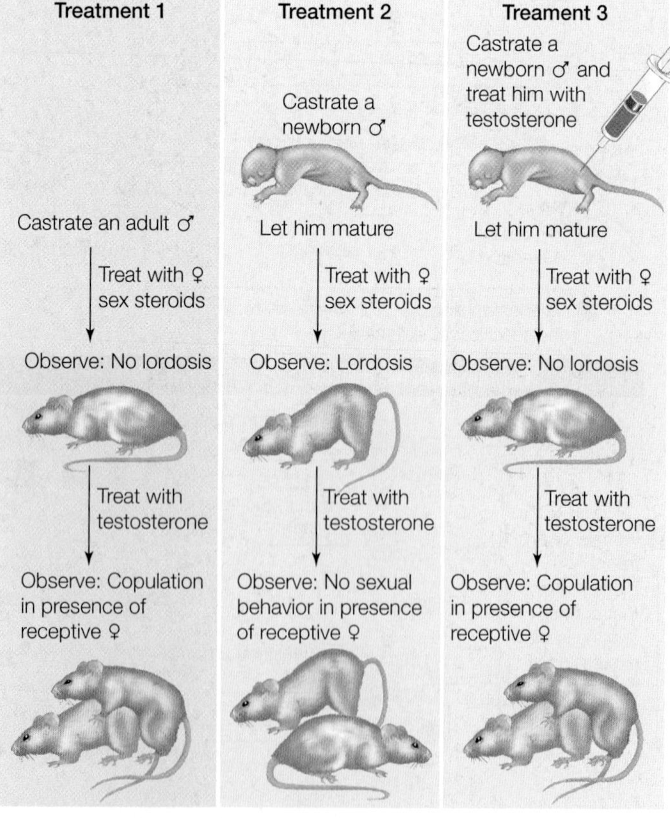


**53.5 Hormonal Control of Sexual Behavior** Experimental hormone treatments of rats demonstrated that the sex steroids present during early development determine what sexual behavior patterns develop, whereas the sex steroids present in adulthood control the expression of those behavior patterns.

receptive posture, called lordosis, in the presence of males, and males copulate with receptive females. Neither sex, however, expresses these behaviors until the animals have reached adulthood. Experiments in which newborn and adult rats were neutered (to remove the influence of sex steroids naturally produced by their gonads) and artificially treated with hormones led to the following conclusions:

- Development of male sexual behavior requires that the brain of the newborn rat be exposed to testosterone, but development of female sexual behavior does not require exposure to estrogen.
- Testosterone masculinizes the nervous systems of both genetic males and genetic females.
- Exposure to sex steroids in adulthood is necessary for the expression of sexual behavior, but testosterone produces male sexual behavior only in adult rats whose brains were masculinized when they were newborns, and estrogen produces female sexual behavior only in adult rats whose brains were not masculinized when they were newborns.

Thus the sex steroids that are present at birth determine which pattern of behavior develops, and the sex steroids that are present in adulthood determine when that pattern is expressed.

## Some behaviors can be acquired only at certain times

Responsiveness to simple releasers is sufficient for certain behaviors such as begging behavior in gull chicks, but more complex information that cannot be genetically programmed is required for other behaviors. An example is parent–offspring recognition. When animals live in close proximity to other individuals, as in a herd or a nesting colony, it is important for parent and offspring to learn each other's identity soon after birth so they will be able to find each other in a crowded situation. In many such cases, a parent–offspring bond is formed by **imprinting**. What characterizes imprinting is that an animal learns a specific set of stimuli during a limited time called a **critical period**, also known as a **sensitive period**.

Konrad Lorenz demonstrated that young graylag geese (*Anser anser*) imprint on their parents between 12 and 16 hours after hatching. By positioning himself to be present during this critical period, Lorenz succeeded in imprinting goslings on himself. The imprinted goslings followed him around as if he were their parent (**Figure 53.6A**). In a subsequent experiment he had his assistants wear boots with different patterns on them. The goslings imprinted on the boots, and even in a situation that mixed different groups of goslings, they always sorted themselves out by following their "parental" boots.

Imprinting requires only a brief exposure, but its effects are strong and long-lasting. Emperor penguins (*Aptenodytes forsteri*) reproduce during the coldest, darkest time of year in Antarctica. The parents walk up to 150 kilometers inland to form a dense colony, where the female lays her egg. She then walks back to the ocean to feed while her mate incubates the egg. By the time she returns, the chick has hatched. She then takes over its care and feeding, and the father walks back to the ocean to feed. Generally he is away so long that the mother must leave

(A) *Anser anser*

(B) *Aptenodytes forsteri*

**53.6 Imprinting Helps Parents and Offspring Recognize Each Other** (A) Greylag geese that imprinted on Konrad Lorenz as hatchlings followed him everywhere he went. (B) Imprinting allows a male emperor penguin to find his own chick among many others.

to find food and to avoid starvation. Thus after being away for weeks, the father must find his chick in a crowded, milling colony of chicks, all calling for their parents (**Figure 53.6B**). Yet he can unerringly locate his own offspring by recognizing its call, which he imprinted on before he left to feed.

The critical or sensitive period for imprinting may be determined by a brief developmental or hormonal state. For example, if a mother goat does not nuzzle and lick her newborn within 10 minutes after its birth, she will not recognize it as her own offspring later. For goats, the sensitive period is associated with peaking levels of the hormone oxytocin in the mother's circulatory system at the time she gives birth and is sensing the olfactory cues emanating from her newborn kid. A female goat rendered incapable of smelling before giving birth is unable to differentiate between her own kid and other kids after giving birth.

## Birdsong learning involves genetics, imprinting, and hormonal timing

Male songbirds use species-specific song to claim and advertise a breeding territory, compete with other males, and declare dominance. They also use song to attract females, which recognize the song of their species even though they do not

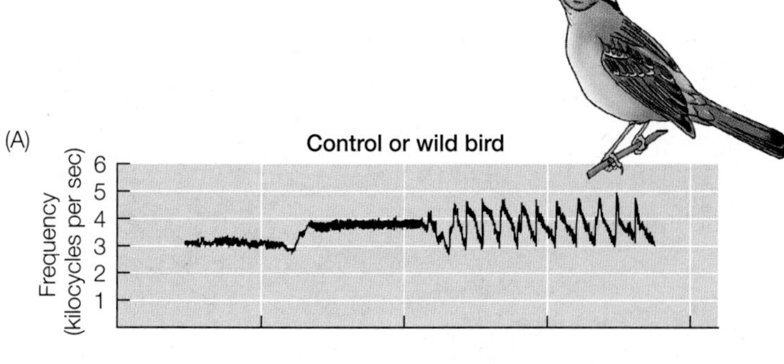


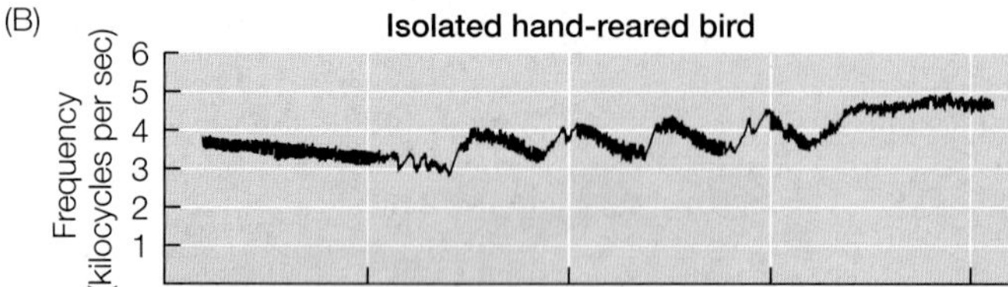


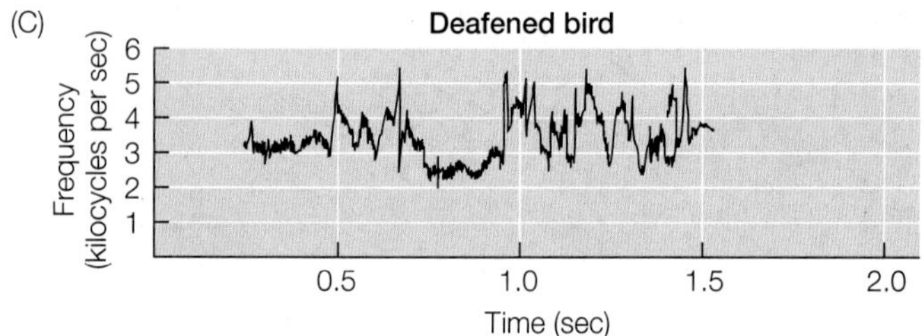


**53.7 Sensitive Periods for Song Learning** (A) Sonogram showing the species-specific song of an adult male white-crowned sparrow (*Zonotrichia leucophrys*). (B) Song of an adult male raised in isolation (never having heard the song as a nestling). (C) Song of an adult male who heard the song as a nestling but was deafened prior to ever singing himself. Marler's experiments showed that the bird must first acquire a song memory by hearing the song as a nestling, and must then be able to hear himself as he attempts to match his singing to that song memory.

sing it. For males of many species, such as the white-crowned sparrow (*Zonotrichia leucophrys*), learning is an essential step in the acquisition of song, but *what* they can learn seems to be influenced by genes, and there is a limited developmental time frame for learning. A hatchling in nature hears his father and other white-crowned sparrows singing. He also hears the songs of many other bird species. But he does not sing until he approaches sexual maturity almost a year later, and when he does, he sings his father's type of song.

Studies of song learning in this species were initiated in the 1960s by Peter Marler at the University of California at Berkeley. Marler incubated eggs of white-crowned sparrows and hand-reared the hatchlings in the laboratory. He was able to expose them to different recorded songs at different times in their development, and discovered that adult male sparrows cannot produce their species-specific song unless they heard it as nestlings in the first 2 months of their lives (**Figure 53.7**). If nestlings hear recordings of white-crowned sparrow song during those first 2 months, they begin to sing (poorly) as they approach sexual maturity. Through trial and error, the maturing birds match their singing to their stored song memory, and from then on they sing their species-specific song. To reach this point, the young bird must be able to hear himself sing. If he is deafened before he begins to sing, he will not be able to match his stored song memory correctly. If he is deafened *after* he has sung his correct species-specific song, however, he will continue to sing normally. We say that at this point the behavior pattern is "crystallized." Thus there are two sensitive periods for song learning, the first in the nestling stage when song memory is imprinted, and the second as the bird approaches sexual maturity, when he learns to match that song memory.

In nature, nestling male white-crowned sparrows hear the songs of many species, so why do they learn only the song of their own species? Marler investigated this question in his isolation experiments by playing recordings of other songs to the hatchlings. The young male sparrows did not learn the songs of other species, even if they heard them many times, but hearing songs of their own species just a few times was sufficient for imprinting. Thus although male sparrows must learn their song, they seem to have a genetic predisposition to learn their own song and not the songs of other species. Marler called this phenomenon "an instinct to learn." There is an important reason why male white-crowned sparrows learn only their species-specific song: female white-crowned sparrows also listen to their fathers' songs while they are nestlings, and when they mature, they choose mates that sing like their fathers did.

More recent investigations have revealed additional complexity in white-crowned sparrow song learning capacity. First, it was demonstrated that each population of white-crowns has its own dialect, and that males from one population can learn the dialect of another population. Second, it was observed that in nature, white-crowns are occasionally heard singing the songs of other species. Could Marler's laboratory experiments have missed a critical natural variable?

Luis Baptista, a curator of birds at the California Academy of Sciences, took up the study of white-crowned sparrows to explore the effects of social interactions on song learning. He discovered that when a hand-reared white-crowned sparrow nestling was exposed to the sight and sound of a related species in an adjacent cage while recordings of his own species' song played in the background, the white-crowned sparrow sang the song of the other species when he matured. Thus social experience has a powerful effect on what the young bird can and will learn.

To ask the question of what the adaptive significance of plasticity in the male singing behavior is, it is necessary to take female choice into consideration. Sarah Woolley and Allison Doupe at the University of California, San Francisco have done that with zebra finches (*Taeniopygia guttata*), a species that has become a valuable model system for studying the neurobiology of birdsong. When male zebra finches sing in isolation, their song is variable ("undirected"), almost as if they are improvising. When they sing in the presence of females, however, their song is very stereotyped ("directed"). Experiments by Woolley and Doupe

**53.8 Practice Makes Perfect** When male zebra finches (*Taeniopygia guttata*) sing alone, they improvise (undirected song), but when they sing in the presence of a female, they sing a stereotyped (directed) song. (A) In a behavioral choice experiment, female zebra finches show which song they prefer by the amount of time they spend in the chambers where those songs are played. (B, C) The graphs show the percent of time the female bird spends on the perch nearest each of the two songs being played. (B) Females prefer their own mate's directed song to his undirected song. (C) Females prefer their mate's directed song to the directed song of an unfamiliar male, indicating that male vocal improvisations may serve to perfect their directed song.

showed that female zebra finches preferred directed song and that the directed song resulted in greater activation of auditory areas in their brains. The strongest effects were seen in mated females hearing the directed song of their mates (**Figure 53.8**). Clearly females are differentially responsive to small variations in the male's song. It appears that the males practice variations in their song but then consolidate the song into a "performance" version that attracts females.

## The timing and expression of birdsong are under hormonal control

As we have seen, both male and female songbirds hear their species-specific song as nestlings, but only the males of most species sing as adults, and most do so only in spring. Hormones underlie both the difference in song expression between male and female songbirds and the timing of song expression.

To determine whether the absence of testosterone is the main reason female songbirds don't sing, investigators injected adult female songbirds with testosterone in spring. In response to these injections, females sang their species-specific song just as males did. Apparently females form a memory of their species-specific song when they are nestlings, and have the physical capacity to sing, but under normal circumstances they simply lack the hormonal stimulation.

How does testosterone cause a songbird to sing? A study by Fernando Nottebohm at Rockefeller University revealed that each spring an increase in circulating testosterone levels causes certain parts of the male's brain necessary for learning and developing song to grow larger. Individual neurons in those regions of the brain increase in size and grow longer extensions, and the number of neurons in those regions increases. Thus hormones can control behavior by changing brain structure as well as brain function, both developmentally and in response to environmental cues.

What triggers the release of testosterone in response to the onset of spring? Takashi Yoshimura conducted a DNA microarray analysis of brain samples from another bird species, the Japanese quail (*Coturnix japonica*), which is a model species for avian genetic analyses. Knowing that photoperiod was the environmental cue, he examined the responses of 38,000 genes to determine which genes were activated by changes in day length. Fourteen hours after dawn on the first day of the critical photoperiod that induces singing behavior, genes in the brain that code for thyroid-stimulating hormone were switched on. This hormone acts on the pituitary gland, which in turn regulates the production of other hormones that stimulate the growth of the testes and the production of testosterone (see Figure 43.10).

RECAP 

Hormones can control what behavior patterns are programmed in the brain long before those behaviors are expressed. Learning and expression of behaviors can also be controlled by hormones, and therefore timed for particular stages in the life cycle.

- What hormonal conditions are necessary for the development of adult sexual behavior in male and female rats? **See pp. 1098–1099 and Figure 53.5**
- What is the adaptive value of imprinting? **See p. 1099**
- Describe the series of events necessary for a male white-crowned sparrow to sing its species-specific song in spring. **See pp. 1100–1101 and Figure 53.7**

Complex behaviors are the product of interactions of genetic, physiological, and environmental factors. Many genes are involved in shaping behavior, and therefore there are multiple opportunities for selection to favor behavioral modifications.

Questions about how changes in behavior adapt animals to environmental conditions are the province of an evolution-based field called **behavioral ecology**.

## 53.4 How Does Behavior Evolve?

Animal behaviors are variable within and among species. We saw at the start of this chapter that the behaviors of North American brown-headed cowbirds and European common cuckoos—two groups of avian brood parasites—have taken subtly different evolutionary pathways, and that these differences have resulted in different responses on the part of the host birds these parasites affect.

Along with the forces of intra- and interspecific interactions on evolution, environmental conditions are also highly variable over both time and space. Behavioral ecologists strive to discover the relationships between behavior and environment with the intent of understanding the evolutionary mechanisms underlying behavior.

WORKING WITH**DATA:**

### *Why Put Up with a Parasite?*

Original Paper

Hoover, J. P. and S. K. Robinson. 2007. Retaliatory mafia behavior by a parasitic cowbird favors host acceptance of parasitic eggs. *Proceedings of the National Academy of Sciences USA* 104: 4479–4483.

Analyze the Data

A nest or brood parasite lays her eggs in the nest of a different species, who then raises the parasite's chicks at the expense of the host's own reproductive success. Brood parasitism has evolved in several bird species, including cuckoos and cowbirds. Why do the host birds accept this behavior, especially if the parasite's eggs and/or hatchlings are clearly distinct from their own? Why don't the host parents just push the parasite's eggs or offspring out of the nest?

In several cases researchers have observed what they call "mafia behavior," in which the parasitic female returns to the host nest and, if her eggs have been removed, destroys the nest completely. Jeffrey Hoover and Scott Robinson studied the interactions of the brown-headed cowbird (*Molothrus ater*), a North American brood parasite, and one of its common host species, the prothonotary warbler (*Protonotaria citrea*). These experiments placed warbler nests in five treatment categories, two of which protected the warbler nests from any approaching cowbird ("cowbird access denied") and three of which allowed the cowbirds access to the nests (see the table). Hoover and Robinson then tracked the incidence of nest destruction by cowbirds and the average number of warbler offspring fledged per warbler nest in each treatment category, as summarized in the two histograms below.

| Treatment | Description |
|---|---|
| 1 | Cowbird egg ejected, cowbird access allowed |
| 2 | Nonparasitized nest, cowbird access allowed |
| 3 | Cowbird egg accepted, cowbird access allowed |
| 4 | Cowbird egg ejected, cowbird access denied |
| 5 | Nonparasitized nest, cowbird access denied |

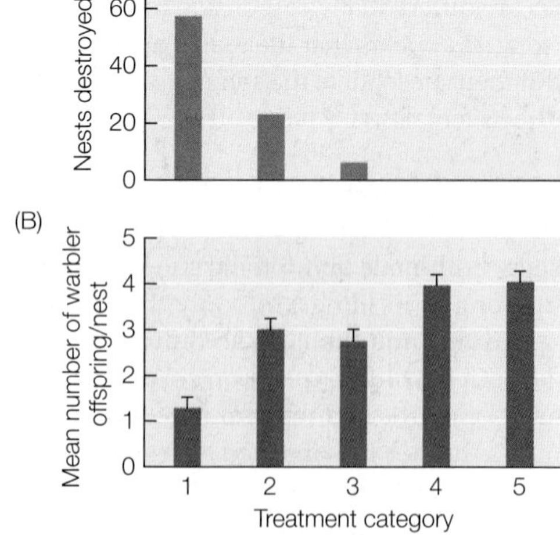


**QUESTION 1**

Under which conditions were warblers most successful? Under which conditions were warbler nests most likely to be destroyed?

**QUESTION 2**

Do these results demonstrate a cost–benefit approach to the evolution of behavior? Explain your answer.

Go to BioPortal for all WORKING WITH**DATA** exercises

## Animals are faced with many choices

Over an animal's lifetime, its behavior is largely a sequence of choices: where and when to move, where to build a nest, what to eat, when to fight and when to flee, with whom to associate, with whom to mate. Making wrong choices reduces fitness. Behavioral ecologists seek to discover what information animals use to make behavioral choices and how that information relates to aspects of the environment that influence their fitness.

Where an animal lives is referred to as its **habitat**. In most cases the habitat provides not only a protected nest site, but also food and access to mates. The environmental cues animals use to select their habitat may be quite simple. For example, seabirds select cliffs or offshore rocks for nesting, and both of those sites offer protection from predators. Animals with very specialized food requirements select habitats where those foods are abundant. The general hypothesis that guides behavioral ecologists is that the cues animals use to select habitats are reliable predictors of conditions suitable for future survival and reproduction.

For many species, the presence of **conspecifics**—other members of the same species—can be a valuable cue. Observing conspecifics can provide animals with information about the quality of a habitat. After all, you can't argue with success. In Europe, collared flycatchers (*Ficedula albicollis*) in the spring breeding season are nosy neighbors, regularly visiting the nests of conspecifics. Researchers hypothesized that this behavior allows the flycatchers to assess the quality of the habitat by seeing how well their neighbors are faring. To test this hypothesis, they created some areas with supersized broods—normally an indication of abundant food—by taking young birds from some nests and adding them to nests in another area. The next year, flycatchers preferentially settled in the areas where broods had been artificially enlarged.

## Behaviors have costs and benefits

Behavioral ecologists often use a cost–benefit approach to investigate the relationship between behavior, environment, and fitness. A **cost–benefit approach** assumes that an animal has only a limited amount of time and energy, and therefore cannot afford to engage in behaviors that cost more to perform than they bring it in benefits. A cost–benefit approach provides a framework that behavioral ecologists can use to make observations, construct hypotheses, and design experiments to investigate why behavior patterns evolve as they do.

The benefits of a behavior can be measured in terms of the enhancement in fitness an animal accrues by performing the behavior. The cost of a behavior typically has three components:

- **Energetic cost** is the difference between the energy the animal expends performing the behavior and not performing it.
- **Risk cost** is the increased chance of being injured or killed as a result of performing the behavior.
- **Opportunity cost** is the benefit the animal forgoes by not being able to perform other behaviors during the same time interval.

## Territorial behavior carries significant costs

Cost–benefit analysis has been used extensively in the study of **territorial behavior**, which is aggressive behavior that actively denies other animals access to a habitat or resource. Optimal habitats and resources are frequently in short supply, so conspecifics must compete for them. Many animals—usually males—defend all-purpose territories that provide a nest site, food, and access to mates. The territory holder stakes out his boundaries by engaging in aggressive interactions with neighbors, and must then patrol those boundaries constantly and respond to trespassers. These aggressive interactions usually consist of highly stereotypic, species-specific displays such as birdsong. Through territorial behavior, the male obtains the resources he needs for reproductive success, but he also pays a price.

Territorial displays require considerable expenditure of energy, they make a male more vulnerable to predation, and they detract from the time he has for feeding or engaging in parental behavior. Michael Moore and Catherine Marler at Arizona State University performed an experiment to estimate the costs incurred by male Yarrow's spiny lizards (*Sceloporus jarrovii*) when defending a territory. These lizards defend territories that include the home ranges of several females. Their territorial behavior is normally most intense during September and October, when the circulating testosterone levels of the males are high and the females are most receptive to mating. The researchers varied the intensity of the lizards' territorial behavior by implanting testosterone capsules in some males in summer, when they are not normally highly territorial (**Figure 53.9**).

Testosterone-treated males spent more time patrolling their territories, performed more displays, and expended about one-third more energy than control males (an energetic cost). As a result, they had less time to feed (an opportunity cost), captured fewer insects, stored less energy, and had a higher death rate (a risk cost). In summer, when females are not normally receptive, these high costs of vigorous territorial defense outweigh the reproductive benefits of territoriality. Thus natural selection has favored seasonal variation in the level of the hormone controlling territorial behavior in this species.

**Go to Animated Tutorial 53.1**
**The Costs of Defending a Territory**
Life10e.com/at53.1

The cost–benefit approach explains the diversity of territorial behaviors seen in different species. Even if a resource is absolutely essential to an animal, if it cannot be defended economically, the animal will not engage in territorial behavior. Food is essential for all animals, but if the food is widely distributed in space or fluctuating in availability, there is no benefit to balance the high costs of trying to defend it. For example,

INVESTIGATING**LIFE**

**53.9 The Costs of Defending a Territory** By using testosterone implants to increase territorial behavior, Michael Moore and Catherine Marler measured the costs to male Yarrow's spiny lizards (*Sceloporus jarrovii*) of defending a territory during the summer, when they do not normally do so.[a]

**HYPOTHESIS** Yarrow's spiny lizards do not defend a territory during summer because the energetic costs of territorial behavior in that season outweigh the benefits.

*Method*

1. During the summer, when female lizards are not sexually receptive, insert testosterone capsules under the skin of some males; leave other males untreated as controls.
2. Observe the patterns of territorial behavior and the survival rate of the two groups of males.

*Results*

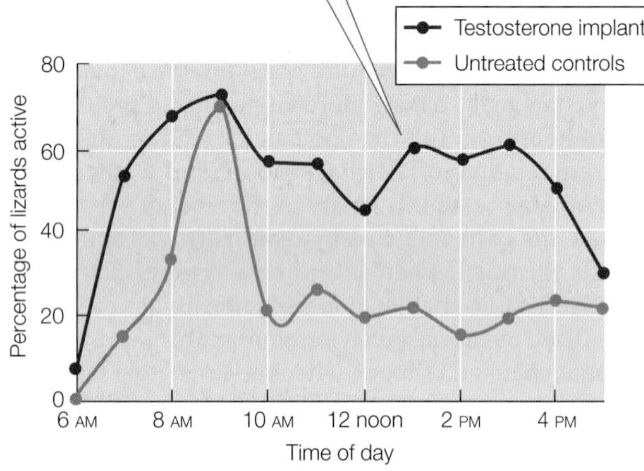


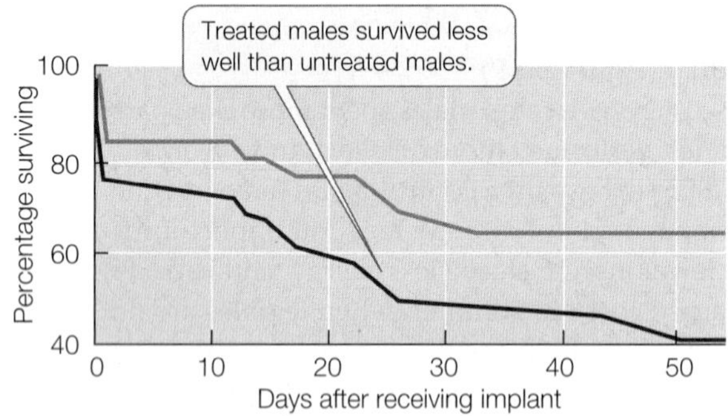


**CONCLUSION** For these lizards, the cost of defending territories during summer significantly reduces their survival rate without increasing their reproductive success.

Go to **BioPortal** for discussion and relevant links for all INVESTIGATING**LIFE** figures.

[a]Marler, C. A. and M. C. Moore. 1988. *Behavioral Ecology and Sociobiology* 23: 21–26.

the open ocean where seabirds feed cannot be defended. But safe nest sites on islands or rocky cliffs are in short supply, and they can be defended. Thus the territories of seabirds may be no larger than the distance the birds can reach while sitting on their nests (**Figure 53.10A**).

In some cases the resource that is defended is the female herself. Elephant seals spend most of their lives at sea, but females come to land at traditional beach sites to give birth to their pups. Male elephant seals arrive at these sites ahead of time and stake out territories through vigorous fighting (**Figure 53.10B**). When the females arrive on the beaches, they enter the territories of the males. As long as the male territory holder can fend off challengers, he will be able to mate with all the females using his piece of the beach.

One unusual form of male territorial behavior arises in situations in which neither food, nest sites, nor females are defended. A **lek** is an area where males gather for the purpose of engaging in intense displays of their territorial prowess aimed at impressing females and winning the opportunity to mate. Even though space is not limited, each male defends a small piece of real estate on which he performs a display (**Figure 53.10C**). Those territories closest to the center of the lek are the prime sites, and males compete intensely for those locations. The females visit into the lek, observe the males, and generally mate with the males holding the prime sites. The benefit of this system to the female is that she is inseminated by a successful competitor, and therefore her offspring will carry the genes that contributed to his success. This is another example of sexual selection (see Section 21.2). The costs of lekking to males are high, as they engage in continuous, intense territorial behavior that precludes eating, drinking, and sleeping until they are displaced. The benefit is the chance to maximize their fitness by mating with many females.

## Cost–benefit analysis can be applied to foraging behavior

When an animal forages (searches for food), among the decisions it must make are how much time to spend in each location before giving up and moving on, what resources at each location are actually edible, and which of the different types of potential food should be eaten and which should be left alone. By applying cost–benefit approaches to feeding behavior, scientists have produced a body of knowledge known as **optimal foraging theory**, which helps them identify the fitness value of feeding choices. The primary benefit of foraging is the nutritional value of the food obtained: the energy, minerals, and vitamins it contains (see Section 51.1). The costs of foraging are similar to those of other behaviors: energy expended, time lost from other activities that could enhance fitness, and the risk of increased exposure to predators.

**Go to Animated Tutorial 53.2**
**Foraging Behavior**
Life10e.com/at53.2

Animals frequently have to make choices among food items that may differ not only in terms of energy content, but also abundance or ease of acquisition and processing. Optimal foraging theory predicts that in such situations, animals will make choices that will maximize the rate at which they obtain energy. The more rapidly a foraging animal satisfies its

(A) *Diomedea melanophris*

(B) *Mirounga angustirostris*

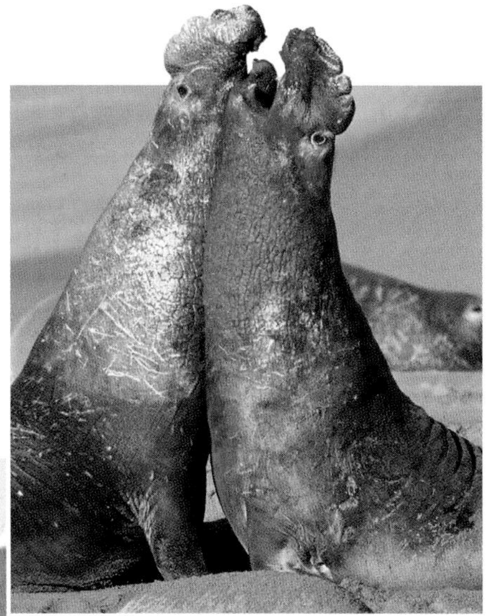

(C) *Centrocercus urophasianus*

**53.10 Animals Defend Territories of Different Sizes** (A) The nesting territories of many seabirds consist of only as much space as they can defend without leaving the nest. (B) Male elephant seals fight vigorously to defend areas of beach where females haul out of the water to give birth to their pups. (C) Male greater sage-grouse gather at a lek in Colorado to perform displays aimed at impressing females and winning the opportunity to mate.

energetic requirements, the lower the opportunity costs and risk costs of foraging.

Earl Werner and Donald Hall of Michigan State University performed laboratory experiments with bluegill sunfish (*Lepomis macrochirus*) to test this energy maximization hypothesis. In preparation for their experiments, they measured the energy content of water fleas (*Daphnia*) of different sizes (the different food types), how much time bluegill sunfish (the foragers) needed to capture and eat those different food types, the energy they spent pursuing and capturing the different food types, and the rates at which they encountered the different food types under different food densities. Werner and Hall then stocked experimental environments with different densities and proportions of large, medium, and small water fleas. They made two predictions from the energy maximization hypothesis: first, that in an environment with abundant large water fleas, the fish would ignore smaller water fleas; and second, that in an environment stocked with low densities of all three sizes of water fleas, the fish would eat every water flea they encountered. The proportions of large, medium, and small water fleas eaten by the fish under different conditions were close to those predicted by the hypothesis (**Figure 53.11**).

The energy maximization hypothesis considers food items in terms of the energy they provide, but animals have nutrient requirements in addition to energy that can play a role in shaping their foraging behavior. Essential minerals, for example, are in short supply in some animals' diets, and those animals may incur large energetic costs and risks to obtain them (**Figure 53.12**). Foods may also have medicinal value. Chimpanzees, for example, have been observed eating the pith of the plant *Vernonia amygdalina*. The pith contains small quantities of a secondary metabolite (vernonioside B1) that is toxic to chimps at high concentrations, but at low concentrations can kill their intestinal parasites. Chimps that consume this pith have fewer parasites.

**RECAP 53.4**

Behavioral ecologists seek to explain relationships between variation in behavior and variation in environmental conditions. They seek to discover what information animals use to make behavioral choices and how that information relates to aspects of the environment that influence their fitness. Cost–benefit analysis has been applied to territorial and foraging behavior.

- What might the presence of conspecifics in a habitat tell an animal about that habitat? **See p. 1103**
- Describe three types of territorial behavior and the costs and benefits of each. **See pp. 1103–1104 and Figure 53.9**
- How can cost–benefit analysis be applied to a behavior? **See pp. 1104–1105 and Figures 53.9 and 53.11**

Whereas behavioral ecologists are interested in understanding how the natural environment influences the fitness value of behavioral choices—the ultimate causes of those behaviors, in Tinbergen's terms—other behavioral biologists focus on the physiological mechanisms and principles that underlie behavior—the proximate causes.

INVESTIGATING**LIFE**

**53.11 Bluegill Sunfish Are Energy Maximizers** Based on energy maximization calculations, Earl Werner and Donald Hall predicted (1) that in an environment with abundant large food items (i.e., the water flea *Daphnia*), bluegill sunfish (*Lepomis macrochirus*) would ignore smaller food items and feed preferentially on larger water fleas; and (2) that in an environment where all sizes of *Daphnia* were scarce, the fish would eat every one they encountered. Such a strategy is in keeping with the cost–benefit hypothesis of foraging behavior.[a]

**HYPOTHESIS** Bluegill food item selection will match the energy maximization predictions of cost–benefit analysis.

*Method*

1. Measure the respective energy content of large, medium, and small water fleas (*Daphnia*).
2. Use cost–benefit energy maximization calculations to predict the rate at which bluegill sunfish will consume the different sized water fleas under different levels of food abundance (i.e., density of water fleas).
3. Provide bluegills with *Daphnia* of different sizes in varying proportions (represented by the different colored bars) and at different densities.

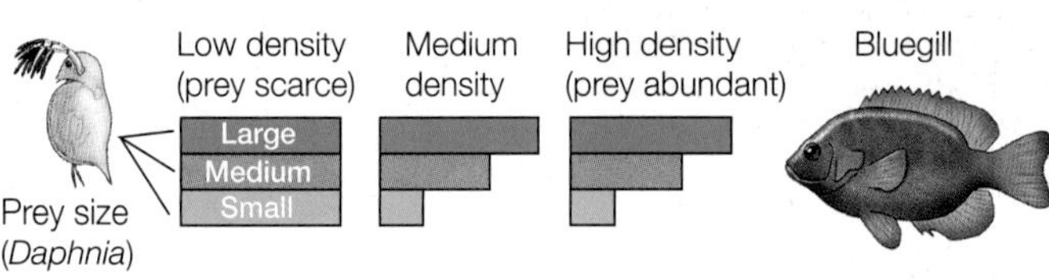


4. Note the proportions of small, medium, and large *Daphnia* actually eaten by the fish under the different conditions and compare these proportions to the predictions of energy maximization.

*Results*

The food choices made by bluegills match the predictions of the energy maximization hypothesis.

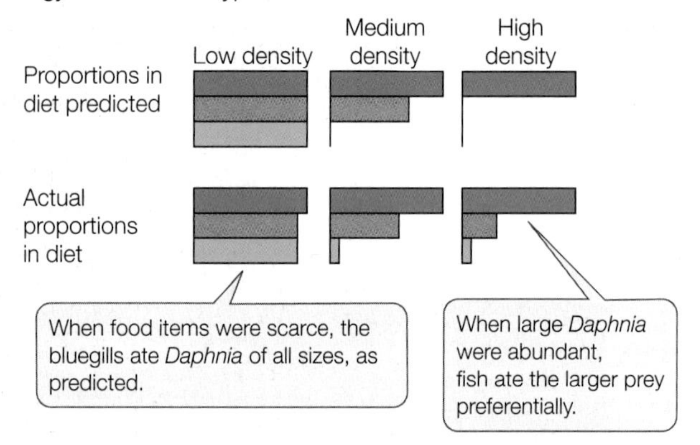


**CONCLUSION** Bluegills select food items in accordance with the predictions of energy maximization calculations.

Go to **BioPortal** for discussion and relevant links for all INVESTIGATING**LIFE** figures.

[a]Werner, E. E. and D. J. Hall. 1974. *Ecology* 55: 1042–1052.

## 53.5 What Physiological Mechanisms Underlie Behavior?

Control of behavior involves the nervous and endocrine systems. Execution of behavior involves the musculoskeletal system as well as other effector mechanisms, such as those that produce secretions, color changes, electrical impulses, sound, and even light. We have already considered many of the physiological systems that are involved in these processes, including hormones, reproductive systems, nervous systems, sensory systems, and feeding mechanisms. The field of behavioral physiology, which encompasses aspects of all of these systems, is enormous, so here we will dig deeper into just three different phenomena studied by behavioral physiologists: the timing of behavior, navigation, and communication.

### Biological rhythms coordinate behavior with environmental cycles

Earth turns on its axis once every 24 hours, generating daily cycles of light and dark, temperature, humidity, and tides. In addition, Earth is tilted on its axis, so the light–dark cycle

(A) *Ara chloropterus*

(B)

**53.12 Herbivores Seek Out Unusual Sources of Minerals** (A) Red-and-green macaws of the Amazon jungle obtain essential minerals by eating dried clay. (B) Pierid butterflies obtain needed salts by drinking secretions from the skin and nostrils of a caiman.

On a cycle of 12 h light/12 h dark, the mouse is mostly active in the dark and has a rest–activity cycle of 24 hours.

In constant dark, the mouse still expresses a daily cycle of rest and activity, but the period of the cycle is less than 24 hours. As a result, the mouse starts its activity and ends its activity earlier each day.

If the mouse is given 20 minutes of light at 24-hour intervals, its rest–activity cycle is entrained to a 24-hour period.

**53.13 Circadian Rhythms Are Entrained by Environmental Cues** The activity–rest cycle of a laboratory mouse (a nocturnal animal) responds to the light–dark cycle under which it is kept. The gray bars indicate times when the mouse is running on an activity wheel. Two days of activity are recorded on each horizontal line; the data for each day are plotted twice—once on the *right* half of each line (hours 24–48) and again on the *left* half of the line below it (hours 0–24). This double plotting is merely to make the pattern easier to see.

**Go to Animated Tutorial 53.3 Circadian Rhythms**
Life10e.com/at53.3

changes as Earth revolves around the sun. These daily and seasonal cycles profoundly influence the physiology and behavior of animals. Animals tend to be active either during the day (diurnal) or at night (nocturnal) and have sensory capabilities appropriate to this distinction. Therefore it is adaptive to organize behavior on a cycle that corresponds with the environmental cycle of light and dark. Similarly, a behavior that is adaptive at one time of year (such as midsummer) may not be adaptive at another times (midwinter). Thus it is important for animals to organize their behavior with respect to time of the day or year and to be able to anticipate those times.

CIRCADIAN RHYTHMS Experimental animals kept in constant darkness and at a constant temperature, with food and water available all the time, still demonstrate daily cycles of activities such as locomotor activity, sleeping, eating, drinking, learning, and just about anything else that can be measured. The persistence of these daily cycles in the absence of environmental time cues suggests that animals have an internal clock. Because these daily cycles are not exactly 24 hours long, they are known as **circadian rhythms** (*circa*, "about"; *dies*, "day").

As described in Section 37.4, any biological rhythm can be viewed as a series of cycles, and the length of one of those cycles is the period of the rhythm. Any point in the cycle is a phase of that cycle. When two rhythms completely match, they are in phase, and if a rhythm is shifted (as in the resetting of a clock), it is phase-advanced or phase-delayed. Because the period of a circadian rhythm is not exactly 24 hours, it must be phase-advanced or phase-delayed each day to remain in phase with the daily cycle of the environment. In other words, the rhythm has to be **entrained** to the cycle of light and dark in the animal's environment.

An animal kept under constant conditions will not be entrained to the light–dark cycle of the environment, and its circadian clock will run according to its natural period—it will be **free-running**. If the period is less than 24 hours, the animal will begin its activity a little earlier each day (**Figure 53.13**). The period of the free-running circadian rhythm is under genetic control. Different species may have different average periods, and within a species, mutations can lead to different period lengths.

Under natural conditions, environmental time cues, such as the onset of light or dark, entrain the free-running rhythm to the light–dark cycle of the environment. In the laboratory it is possible to entrain the circadian rhythms of free-running animals with short pulses of light or dark administered every 24 hours (see the bottom panel of Figure 53.13).

In mammals, the master circadian "clock" consists of two clusters of neurons just above the optic chiasm (the area of the brain where the optic nerves come together). These structures are called the **suprachiasmatic nuclei (SCN)**. If they are destroyed, the animal becomes arrhythmic (loses its circadian rhythm) and is just as likely to eat, drink, sleep, or wake at any time of day.

In 1990, a notable study by Martin Ralph and his colleagues, then at the University of Virginia, showed that the SCN is the source of circadian rhythms. When the SCNs of adult hamsters with typical 24-hour rhythms were destroyed, the animals became arrhythmic. After several weeks of this arrhythmic behavior, the researchers transplanted SCN tissue from hamster fetuses bred for an atypical (mutant) short-day rhythmicity into the original hamsters' brains. The experiment produced two remarkable results (**Figure 53.14**). First, circadian rhythms were restored by the transplanted SCN tissue, demonstrating that the SCN is sufficient to generate circadian rhythms—a unique case of a behavior being restored by a neural transplant. Second, the restored circadian rhythms had the period length of the *donor* strain, demonstrating that the specific phenotype of the behavior was a property of the donor neural tissue, and thus wholly generated by the SCN.

The molecular mechanism of the circadian clock involves negative feedback loops. Although there are a number of genes involved, including the *per* genes discussed in Section 53.2, we can generalize about the mechanism by saying that when certain "clock genes" are expressed in SCN cells, the mRNA enters the cytoplasm, where it is translated. The resulting proteins combine, and the dimer returns to the nucleus as a transcription factor that shuts off the expression of the clock genes. The period of this cycle is about a day. These findings show that it is possible to understand circadian rhythms of behavior at all levels, from the molecular rhythm generators to the environmental stimuli that entrain them to the daily cycle of light and dark.

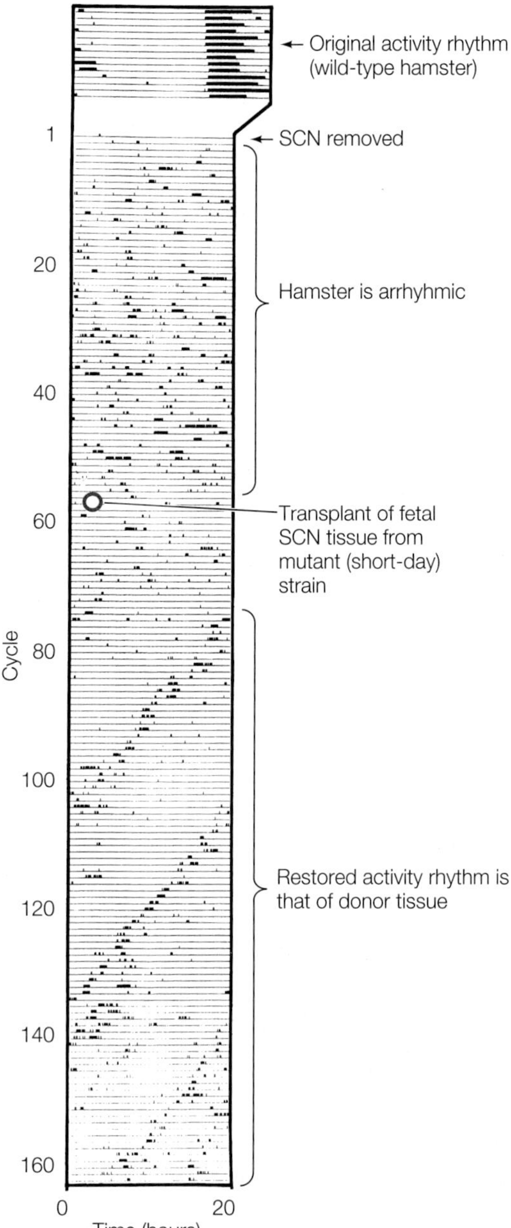


**53.14 The Brain Clock Can Be Transplanted** In this experiment, the activity rhythm of a wild-type (i.e., genetically typical) hamster was measured; this animal had a circadian period of 24.5 hours (top of bar). After its SCN was removed (lesioned), the hamster became arrhythmic. SCN tissue from a fetal "short-day" hamster (a mutant strain with a 19-hour circadian period) was then transplanted into the region where the lesion was made in the wild-type hamster. The transplanted tissues restored circadian rhythm in the lesioned hamster, but the restored rhythm had the period of the donor animal.

**CIRCANNUAL RHYTHMS** Seasonal changes in the environment present challenges to many species. Most animals reproduce most successfully if they time their reproductive behavior to coincide with the most favorable time of year for the survival of their offspring. Many species require considerable advance preparation for reproduction. Migratory animals must arrive on their breeding grounds at the right time, and animals that have specialized structures used in mating displays, such as the antlers of deer, moose, and caribou, must grow these structures before the breeding season arrives.

For many species, a change in day length—the **photoperiod**—is a reliable indicator of seasonal changes to come. For others, however, change in day length is not a reliable seasonal cue. Hibernators, for example, spend long months in dark burrows underground but must be physiologically prepared to breed almost as soon as they emerge in the spring. A bird overwintering near the equator cannot use changes in photoperiod as a cue to time its migration to its temperate-zone breeding grounds. When held under constant laboratory conditions, such animals show endogenous circannual rhythms that keep track of the time of year. Unlike circadian rhythms, the neural basis for circannual rhythms is unknown.

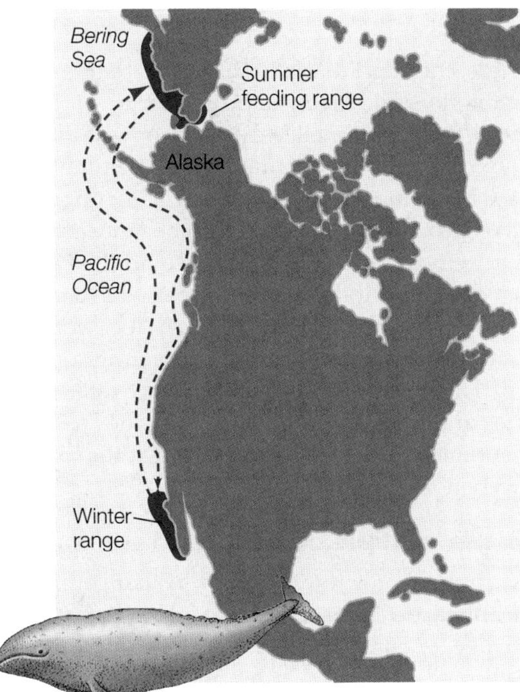


**53.15 Piloting** Gray whales (*Eschrichtius robustus*) migrate south in winter from the Bering Sea to the coast of Baja California by piloting, in part by following the western coast of North America.

## Animals must find their way around their environment

To locate suitable habitats, find food and mates, and avoid predators and bad weather, an animal needs to be able to find its way around its environment. Within its local habitat, an animal can orient to landmarks. But what if its destination is a considerable distance away?

PILOTING: ORIENTATION BY LANDMARKS Most animals find their way by knowing and remembering the structure of their environment. This form of navigation is called **piloting**. Gray whales, for example, migrate seasonally between the Bering Sea and the coastal lagoons of Mexico (**Figure 53.15**). They find their way in part by following the west coast of North America. Coastlines, mountain chains, rivers, water currents, and wind patterns can all serve as piloting cues for the whales. But some remarkable cases of long-distance orientation and movement cannot be explained by piloting.

HOMING: RETURN TO A SPECIFIC LOCATION The ability to return to a nest site, burrow, or other specific location is called **homing**. Homing can be accomplished by piloting in a known environment, but some animals that travel long distances through unfamiliar territory perform much more sophisticated homing. The ability of pigeons to return to their home loft even after being transported to remote sites is well known. How do they find their way home? Experiments have shown that pigeons use the sun as a compass, but they can still find their way home when the sun is not visible. Other experiments have shown that pigeons equipped with frosted contact lenses can find their way home, suggesting that visual cues are not essential. Most amazing has been the demonstration that pigeons can detect Earth's magnetic field and orient to it much as a human orients with a compass. Taken together, the studies of homing by pigeons suggest that they can use multiple, redundant sources of directional information and can switch among those sources depending on the circumstances.

**Go to Animated Tutorial 53.4**
**Homing Simulation**
Life10e.com/at53.4

MIGRATION: NAVIGATION OVER GREAT DISTANCES For as long as humans have inhabited temperate latitudes, they have been aware that entire populations of animals, especially birds, disappear and reappear seasonally. Not until the early nineteenth century, however, were patterns of migration traced by marking individual birds with identification bands around their legs. Only when individuals could be unmistakably identified was it possible to show that the same birds and their offspring returned to the same breeding grounds year after year, and that these same birds could be found during the nonbreeding season at locations hundreds or even thousands of kilometers from their breeding grounds.

Many homing and migrating species take direct routes to their destinations through environments they have never experienced because they use mechanisms of navigation other than piloting. Humans use two major forms of navigation:

- *Distance–direction navigation* requires knowing in what direction and how far away the destination is. With a compass to determine direction and a means of measuring distance, humans can navigate.
- *Bicoordinate navigation*, also known as true navigation, requires knowing the latitude and longitude (the map coordinates) of both the current position and the destination, as well as a compass to determine direction.

Many non-human animals seem to have a "compass sense" that allows them to use environmental cues to determine direction, and some seem to have a "map sense" that allows them to determine their position.

The behavior of many animals suggests that they are capable of bicoordinate navigation. Gray-headed albatrosses (*Thalassarche chrysostoma*), for example, breed on oceanic islands in the Southern Hemisphere. When a young albatross leaves its parents' nest, it flies widely over the southern oceans for 8 or 9 years (appropriately, the genus name *Thalassarche* is Greek for "first on the sea"). At that time, it reaches reproductive maturity and flies back to the island where it was raised, where it mates and builds a nest (**Figure 53.16**). How can the bird find a tiny island in an enormous ocean after years of wandering? A circadian clock probably gives the albatross information about the time of day, and additional information from the position of the sun might allow it to determine its map coordinates—much as sailors did in the days before global positioning satellites.

The ability to locate a position by calculating the angles between celestial objects such as the sun and stars and the horizon

(A)

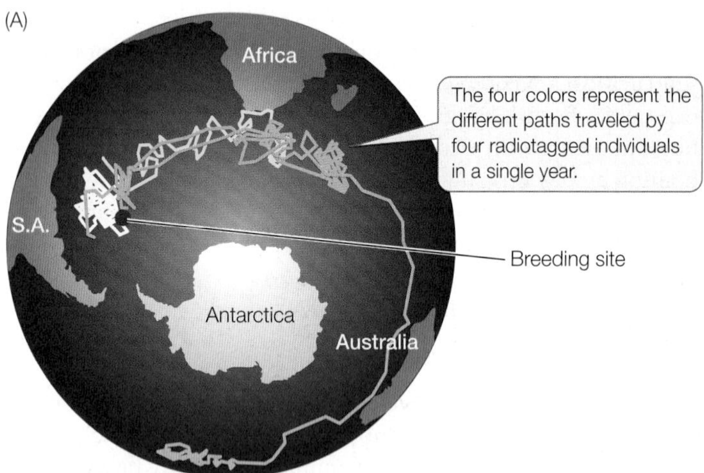


(B) *Thalassarche chrysostoma*

**53.16 Coming Home** (A) Gray-headed albatrosses are born on islands in the subantarctic oceans. Young birds roam widely over the southern oceans for 8 or 9 years. (B) Once they reach maturity, the birds return to the island where they were hatched to mate and raise their own young. A courting couple is shown here.

at specific times of day is called celestial navigation. During the day, the sun can serve as a compass if you know what time it is, and animals can use their circadian clocks for that information. This capacity has been demonstrated by "clock-shifting" experiments such as the one shown in **Figure 53.17**. Similar experiments have shown that many animal species can orient by means of a time-compensated solar compass.

**Go to Animated Tutorial 53.5**
**Time-Compensated Solar Compass**
Life10e.com/at53.5

Many animals are normally nocturnal; in addition, many diurnal bird species migrate at night and thus cannot use the sun to determine direction. The stars offer two sources of information about direction: moving constellations and a fixed point. The positions of constellations (like that of the sun) change because Earth is rotating. With a star map and a clock, direction can be determined using any constellation. But one point that does not change position during the night is the point directly over the axis on which Earth turns. In the Northern Hemisphere, the star Polaris—the "North Star"—lies in that region of the sky and reliably indicates north.

Stephen Emlen at Cornell University showed that birds can learn to use the stars for orientation. As the time of year approaches when young birds would normally migrate to their winter range, young captive birds become more active and orient their activity in the direction they would fly. How do they know that direction? If these birds are raised in a planetarium with a natural star pattern, but one that does not rotate, the birds do not learn to orient, and their premigratory activity is random. However, if the planetarium sky rotates, and even if it rotates around a different point than the North Star, the birds orient their premigratory activity as if the fixed point in the sky were north.

## Animals use multiple modalities to communicate

As individual animals interact, they exchange information; therefore animal behavior can evolve into systems of information exchange, or **communication**. The behaviors of individuals may become elaborated into communication signals, but only if the transmission of information benefits both the sender and the receiver. To understand why these conditions must be met, consider male courtship displays. These displays will be favored if they increase the male's probability of mating and passing on his genes, and sexual selection will occur if the display conveys information to a female (the receiver) about his qualities as a potential father for her offspring.

Animals communicate using a variety of sensory modalities that vary in the nature of the signal produced, the specificity of the information conveyed, the speed and persistence of the signal, and its suitability in different environments. Behavioral physiologists interested in communication must take into consideration the sensory and motor characteristics of their study animals, the physics of the communication modalities they use, and the environment in which the communication takes place.

**CHEMICAL SIGNALS** Because of the diversity of their molecular structures, pheromones can communicate very specific, information-rich messages (see Section 46.2). Pheromones are effective day and night, and they can cover a broad range of transmission distances. Pheromones used in different types of communication vary in their volatility (ease of vaporization) and diffusibility; these chemical properties are functions of the nature and size of the pheromone molecule. Pheromones that act as alarm signals, for example, are highly volatile and diffusible, so their message spreads rapidly but disappears rapidly. Territory-marking and trail-marking pheromones have low volatility and diffusibility and stay effective for a long time, so they can convey directional information. Sex pheromones, such as that of the gypsy moth (see Figure 46.4), are intermediate in these properties, so they can spread a long distance but do not disappear rapidly.

Pheromones are an effective way to exchange species-specific information, and because the recipient must have the proper receptor molecule to detect the pheromone, it is not a signal that is easily intercepted by predators. Pheromonal signals cannot

INVESTIGATING**LIFE**

**53.17 A Time-Compensated Solar Compass** Experiments show that pigeons use the sun to establish directions for navigation and finding food. "Clock-shifting" experiments demonstrate that the birds' circadian clocks factor into their ability to judge direction correctly based on the sun's position.[a]

**HYPOTHESIS** Pigeons determine compass direction from the position of the sun with respect to their internal circadian clocks.

*Method* 1. Place a pigeon in a circular cage from which it can see the sun and sky, but not the horizon or any other visual cue.

2. Surround the cage with multiple food bins but place food only in the southernmost bin, thus training the bird to look for food in the south. (Rotating the cage but always placing the food in the southernmost bin confirms that the bird is navigating to find south.)

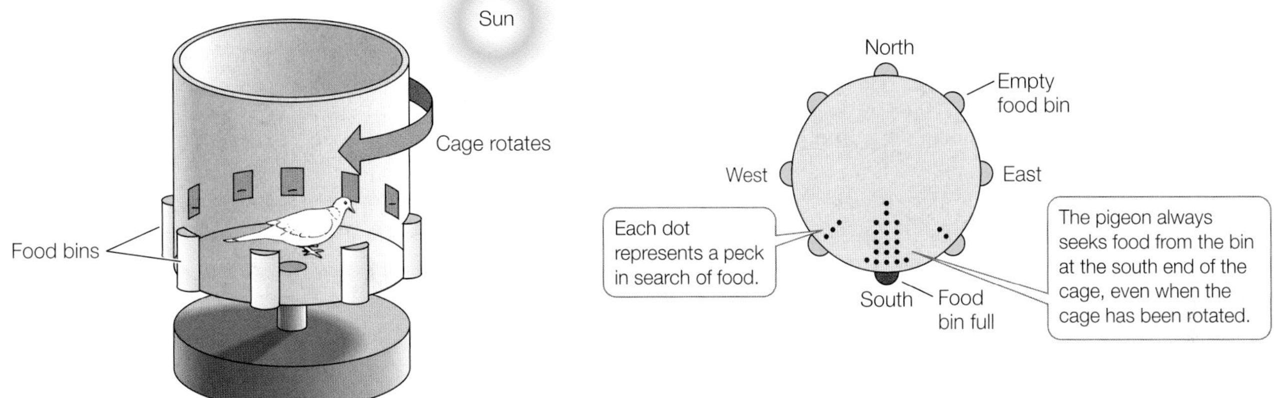


3. Place the trained pigeon in a room with a controlled light cycle for 2 weeks. Turn the lights on at midnight and off at noon to phase-advance its circadian rhythm by 6 hours (i.e., 6 A.M. feels like noon to the bird).

4. Return the pigeon to the circular cage under natural light and observe its food-seeking behavior.

*Results* A 6-hour shift in the circadian clock results in a 90-degree error in the pigeon's orientation.

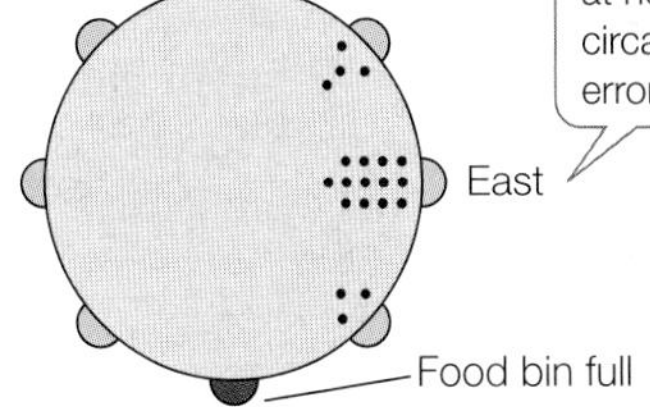


**CONCLUSION** Pigeons have the ability to determine direction using the position of the sun as a compass.

Go to **BioPortal** for discussion and relevant links for all INVESTIGATING**LIFE** figures.

[a]Kramer, G. 1952. *Ibis* 94: 265-285.

be changed rapidly, but they can convey static, complex information. Mammals that mark their territories with pheromones reveal a great deal of information about themselves: species, individual identity, reproductive status, size (indicated by the height of the marking), and how recently the animal has been in the area (indicated by the strength of the scent).

**VISUAL SIGNALS** Visual signals offer the advantage of rapid delivery of information over considerable distances (depending on the environment and the visual acuity of the receiver); they also convey the exact position of the signaler. Signal content can be enhanced by movements (as in a courtship display) or by different postures. Effective visual signals, however, require sufficient light, and the receiver must be looking at the signaler. Thus visual communication is not particularly useful at night or in environments that lack light, such as caves and ocean depths. Some species have overcome this constraint with light-emitting mechanisms. Fireflies, for example, use an enzymatic mechanism to create flashes of light. By emitting flashes in species-specific patterns, fireflies advertise for mates at night.

Another drawback of visual signals is that they can be intercepted by other species. There are predatory firefly species, for example, that mimic the flash pattern of females of other species. A male that approaches the mimicking "female" becomes a meal rather than a mate. Thus deception can be part of animal communication systems, just as it is part of human communication.

**ACOUSTIC SIGNALS** Sound cannot convey complex information as rapidly as visual signals can. But acoustic signals, unlike visual signals, can be used at night and in dark environments. They are not hindered by objects that would interfere with visual signals, so they can be transmitted in complex environments such as forests. They are often better than visual signals at getting the attention of a receiver because the receiver does not have to be looking at the signaler for the message to be received. Sounds are also useful for communicating over long distances. Even though the intensity of a sound decreases with distance from the source, loud sounds can transmit information over much longer distances than visual signals can. The complex songs of humpback whales, when produced at ocean depths of about 1,000 meters, can be heard hundreds of kilometers away, allowing these whales to locate one another across vast expanses of ocean.

The information content of acoustic signals can be increased by varying their frequency, as we can see in the sonograms of the species-specific song of white-crowned sparrows shown in Figure 53.7, and as we practice in our own speech. However, acoustic signals place the signaler at risk for detection by predators. This danger can be minimized by adjustments of frequency and signal structure that decrease the directional information the receiver can extract from the signal. Alarm calls tend to be pure tones (a single frequency) without much temporal structure (starts and stops). It is very difficult to localize such calls. By contrast, territorial calls tend to cover a broad frequency range and have temporal structure. These calls are easy to localize. The frequencies and structures of acoustic signals are also adapted to specific habitats. Different vegetation types, for example, have different sound-absorbing properties: pure tones at lower frequencies carry better in forests, and more complex calls at higher frequencies carry well in open habitats.

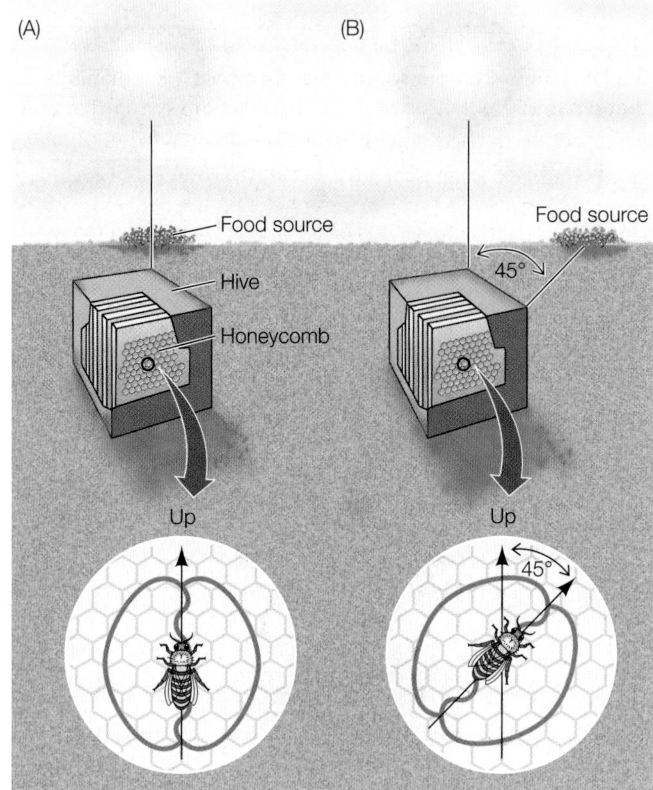


**53.18 The Honey Bee Waggle Dance** (A) A honey bee (*Apis mellifera*) runs straight up on the vertical surface of the honeycomb in the dark hive while wagging her abdomen to tell her hivemates that there is a food source in the direction of the sun. (B) When her waggle runs are at an angle from the vertical, the other bees know that the same angle separates the direction of the food source from the direction of the sun.

**Go to Activity 53.1 Honey Bee Dance Communication Life10e.com/ac53.1**

**MECHANOSENSORY SIGNALS** Animals in close contact with one another can communicate by touch. A classic case of mechanosensory communication is the dance of honey bees (*Apis* spp.), first described by Karl von Frisch. Honey bees have a spectacular ability to navigate and can accurately communicate the location of food sources as far away as 10 kilometers. When a forager bee finds food, she returns to the hive and communicates her discovery to her hivemates by performing a vigorous **waggle dance** in the dark hive on the vertical surface of the honeycomb. Other bees follow the dancer and receive her message.

The waggle dance conveys information about both the distance and the direction of the food source. The dancing bee repeatedly traces out a figure-eight pattern as she runs on the honeycomb. She alternates half-circles to the left and right with vigorous wagging of her abdomen in the short, straight run between turns (**Figure 53.18**). Bees use the sun as their compass, and the angle of the straight run indicates the direction of the food source relative to the position of the sun projected down to the horizon. Even under cloudy conditions, the forager can provide directions to the food source because she can see polarized light. The bee's circadian clock allows the dancer to adjust her dance to take into account the sun's movement during her return flight. The clocks of the recruits enable them to adjust their flight direction to accommodate the sun's movement.

The distance to the food source is communicated by the duration of the waggle portion of the dance. The farther away the food source is, the longer the duration of each waggle run. When food is close to the hive, the waggle portion of the dance becomes so short that it appeared to von Frisch that it was a different dance, which he called a round dance. Recent research, however, has shown that the round and waggle dances are really on a continuum that grades one into the other. Thus the honey bee has one dance language that communicates both the distance and the direction to a food source.

When challenged to prove that the bees were not simply using an odor trail to find the indicated food source, von Frisch responded with a very common sense observation. Bees returning from a new food source fly around barriers such as buildings, but the recruits going out to the food source fly over the barriers in a "beeline," which would be impossible if they were following an odor trail. Careful observation is still one of the best tools for studying behavior.

**COMMUNICATION IN MULTIPLE SENSORY MODALITIES** Avoiding ambiguity is a high priority in any signaling system. Signal specificity is enhanced if multiple sensory modalities are used. Courtship behavior in fruit flies, for example, involves visual, tactile, chemical, and acoustic signals (see Figure 53.4A). The male fruit fly orients toward the female's line of vision (visual signal) and taps her body with his foreleg (tactile signal). Upon detecting pheromones in her cuticle (chemical signal), the male begins to vibrate one wing, producing a species-specific courtship song (acoustic signal). The male then extends his mouthparts to taste the female's genitalia (chemical and tactile signals); if she is receptive, he initiates copulation. If at any point sensory feedback indicates to either the male or the female that their pairing is inappropriate, the courtship abruptly ends.

**RECAP 53.5**

Biological rhythms allow an animal to anticipate changes in its environment. In mammals, a circadian clock located in the suprachiasmatic nuclei generates a rhythm that is entrainable by environmental information. The navigational abilities of animals range from simple piloting by landmarks to distance–direction and bicoordinate navigation. Behaviors may evolve into communication signals if the transmission of information benefits both the sender and the receiver.

- What is meant by a free-running rhythm? Describe how it can be entrained to the 24-hour day. **See pp. 1107–1108 and Figure 53.13**
- Explain the difference between piloting, distance–direction navigation, and bicoordinate navigation. **See pp. 1109–1110**
- Explain how a time-compensated solar compass works. **See pp. 1110 and Figure 53.17**
- Describe an advantage and a disadvantage for each of three modalities of communication. **See pp. 1110–1112**

When behaviors involve multiple individuals—or different species, as is the case in the chapter-opening story—we have to consider how natural selection operates on all of the interacting parties, whether they are sending signals or receiving them. This concern becomes particularly relevant in the case of social behaviors.

## 53.6 How Does Social Behavior Evolve?

The evolution of social behavior became a field of study in its own right in 1975, with the publication of E. O. Wilson's landmark book *Sociobiology*. We begin our consideration of this enormous field with simple interactions that involve a single male and a single female, but already we see diversity. Species differ in their mating systems, which vary from monogamous to promiscuous; in the amount of parental care they give their young; and in the degree to which the male contributes to raising the young. Beyond these relatively simple mating systems, there are associations of larger numbers of reproductive individuals in **polygynous** mating systems, in which a male has more than one mate, or **polyandrous** mating systems, in which a female has more than one mate. Even more complex interactions exist in which extended families participate in raising young, and finally there are societies such as honey bee colonies in which large numbers of nonreproductive individuals assist a single reproductive individual. Sociobiology seeks to understand the evolution of the diversity of social behaviors by asking how the behaviors contribute to the fitness of *all* of the individuals involved.

### Mating systems maximize the fitness of both partners

At the start of Chapter 7 we learned about the mating behavior of two species of voles. Prairie voles (*Microtus ochrogaster*) are monogamous, forming strong pair bonds that can last for life, and both parents participate in rearing the young. In contrast, montane voles (*M. montanus*) are promiscuous: males mate with many females, and the young are raised by the females alone. Behavioral physiologists have explained the proximate mechanisms behind these stark behavioral differences in terms of the release of neurohormones and the distribution of the receptors for those hormones in the brains of the two species. The ultimate question—and the one asked here—is why two such different mating systems evolved in two species that are so closely related.

We begin with the premise that there is an asymmetry in the contributions of male and female animals to their offspring at the time of fertilization. Females produce a limited number of eggs, and each egg is generously stocked with resources. Males produce an almost infinite number of sperm, which contain next to no resources. So the energetic and opportunity costs of reproduction are greater for the female than for the male. In mammals this asymmetry increases throughout gestation as the female bears most of the costs. By the time of birth or hatching, the female's investment in the young is much greater than the male's investment, and the main way for the female to maximize her fitness is to make sure her young are healthy and survive to pass on her genes.

The male has different options for maximizing his fitness. He can simply move on after inseminating the female and seek additional mates as a means of maximizing his reproductive success—as in the case of the meadow vole. Or he can stay with the female he inseminated, protect her, and help care for their young—as in the case of the prairie vole. Which strategy maximizes his fitness depends on a number of factors that are influenced by the species' environment, such as the likelihood that a female and her offspring will survive without a male's help, and a male's likelihood of finding another fertile female. Thus sociobiologists seek to quantify these factors in nature as a means of explaining observed differences in mating systems.

**POLYGYNY** Systems in which a male has more than one mate involve a different asymmetry. In situations in which a male can sequester a group of females from other males, he can increase his fitness by increasing the number of females in his group.

As we saw in Section 53.4, male elephant seals accomplish this by protecting an area of beach where females give birth. Male baboons do so by herding females. Male red-winged blackbirds may acquire more than one mate by defending high-quality nesting territories where females prefer to build their nests. Since sex ratios in all these species are close to 50:50, a large differential in male fitness is established, with some males having high reproductive success while many males have none. Thus selection favors males that are successful in competing with other males to obtain and protect access to many females. In general, bigger, stronger males are the winners, and sexual dimorphism in body size evolves. The elephant seal is an extreme example: males may weigh more than three times as much as females. When species with polygynous mating systems are compared, there is a strong correlation between the number of females a male controls and the degree of sexual dimorphism.

Why do females participate in these polygynous mating systems? Why doesn't a female seek out a nice, kind, noncompetitive male? In some cases she has no choice. If a female elephant seal wants to have her pup on a safe beach, she must enter the territory of a male. If a female red-winged blackbird wants to nest in an optimal territory, she will have to share the attentions of the territory's owner with other females. However, even if the female has a choice of mates, she is likely to maximize her fitness by mating with a male that is strong and dominant enough to control a number of females. Why? If her mate is a dominant male, her male offspring are likely to have their father's traits, become dominant males, and give her more grandchildren. The ultimate result of females selecting males for their prowess and dominance in male–male competition is the lek mating system (see Figure 53.10C), in which the *only* thing a male offers a female is the display of his dominance over other males.

**POLYANDRY** Mating systems in which one female mates with multiple males are relatively rare, but it is seen in some birds and a few mammal species in which paternal care for the young can have a large effect on fitness. An example of a polyandrous species is the golden lion tamarin (*Leontopithecus rosalia*), a primate native to Brazil's tropical rainforests (**Figure 53.19**). Tamarins are very small—adults weigh less than 1 kilogram—and so face high predation pressure. Females usually give birth to twins, and thus newborns constitute a higher percentage of maternal weight than is typical of other primates. The young also grow more rapidly than other primates, so nursing costs are high. For all these reasons, young tamarins cared for by their mother alone are unlikely to survive.

What can a male tamarin do to help guarantee his reproductive success? Watching out for predators is one obvious contribution; gathering food for the female and her young is another. Like other primate parents, tamarins carry their young most of the time, but most other primates have single offspring. When tamarin mothers are carrying twins, they spend 92 percent of the time resting, compared with 58 percent of the time when they are not carrying young. Resting is not compatible with foraging and filling the mother's high energy requirements. When a male is present, however, he carries the young about one-third of the time, so the mother has much more time for foraging and feeding.

*Leontopithecus rosalia*

**53.19 Polyandry in a Small Primate** The endangered golden lion tamarins of Brazil are small primates whose unique life history has given rise to polyandry in some groups, with males playing a major role in rearing the young.

If one male tamarin is helpful in protecting and raising young, then two should be even more helpful. Some females can attract a second mate by being sexually receptive to him. Neither male can be sure that any eventual offspring are his, so it is in the best interest of both to help in their rearing. Of the social groups observed in field studies, only 22 percent had one male and one female, whereas 61 percent had multiple males and one female.

## Fitness can include more than your own offspring

As humans, we readily understand the concept of extended family—brothers, sisters, aunts, uncles, nieces, nephews. Extended families are a common form of social organization in other species as well, and members of these families may cooperate in territory defense, predator avoidance, foraging, and rearing of young. If behavior is favored when it increases the fitness of the individual performing it, then how can we explain the evolution of social behaviors that do not lead to the performer having more offspring and that may even appear to be **altruistic**—benefiting another individual at a cost to the performer?

An individual's fitness is increased by having offspring because those offspring carry the parent's genes into the next generation. Fitness gained by producing offspring is referred to as **direct fitness**. However, an individual's genes are carried into the next generation by more than his or her own offspring. In diploid organisms, two offspring of the same parents share, on average, 50 percent of the same alleles, and an individual is likely to share 25 percent of its alleles with its siblings' offspring (nieces or nephews). Therefore, by helping parents and other relatives raise their offspring, an individual

*Aphelocoma coerulescens*

**53.20 Helpers at the Nest** Young Florida scrub-jays often forego reproduction in their first few years of adulthood to help their parents raise their siblings. These young birds help their parents feed the nestlings, defend the territory, and protect the nest from predators.

increases the transmission of those shared alleles to the next generation. **Inclusive fitness** is the individual's direct fitness plus its **indirect fitness**: the reproductive success of the individual's relatives, to the extent that those relatives share the individual's alleles.

The maximization of inclusive fitness is the mechanism driving **kin selection**, selection for behaviors that increase the reproductive success of relatives even when they come at a cost to the performer. One example is "helping at the nest" behavior, which was studied extensively in Florida scrub-jays (*Aphelocoma coerulescens*) by Glen Woolfenden and John Fitzpatrick. Scrub-jay pairs mate for life and establish large territories, which they defend aggressively. The mating pair may be assisted in rearing their young by three to five helpers (**Figure 53.20**). The helpers guard against predators, feed the young, clean the nest, and fly with fledglings. Why are these birds helping others rather than rearing their own young? Through a long-term study, Woolfenden and Fitzpatrick were able to establish a number of important facts:

- The helpers are prior offspring of the mating pair and are usually 1 to 3 years old.
- Young birds that attempt to breed have almost zero reproductive success.
- Mating pairs with helpers have approximately three times the reproductive success of those without helpers.

These results support the conclusion that helper scrub-jays are maximizing their inclusive fitness by helping their parents raise siblings until they are mature enough to have a reasonable probability of successfully raising their own offspring.

The concept of kin selection was formalized by W. D. Hamilton in what has become known as **Hamilton's rule**. He argued that, for an apparent altruistic behavior to be adaptive, the fitness benefit of that act to the recipient times the degree of relatedness between the performer and the recipient has to be greater than the cost to the performer. This relationship was clearly stated years before by the eminent geneticist J. B. S. Haldane, who said during an argument about altruism that he would not be willing to risk his life to save his brother, but for two brothers or eight cousins, he would consider it.

## Eusociality is the extreme result of kin selection

Hamilton's rule can be applied to explain **eusociality**: social groups that include nonreproductive members (that is, members who as individuals cannot reproduce). The most obvious examples of eusociality occur among the Hymenoptera, an insect group that includes wasps, bees, and ants. In a honey bee colony, for example, the thousands of individuals in the colony are sterile females.

Go to Media Clip 53.1
Social Shrimps
Life10e.com/mc53.1

The key to understanding the evolution of eusociality in hymenopterans is their sex determination mechanism, **haplodiploidy**, in which diploid individuals are female and haploid individuals are male. The queen carries a lifetime supply of sperm obtained during her single mating flight, and she controls whether her eggs are fertilized or not. An unfertilized egg develops into a haploid male; a fertilized egg develops into a diploid female. The queen's daughters share all of their father's genes and, on average, half of their mother's genes. As a result, the sterile female workers in the hive—all sisters—share, on average, 75 percent of their alleles (**Figure 53.21**). Were these females to reproduce, they would share only 50 percent of their alleles with their own female offspring. Thus they potentially increase their inclusive fitness more by raising sisters than by producing and caring for their own offspring.

Eusociality may also arise if it is costly or dangerous to establish new colonies. Nearly all eusocial animals construct elaborate nests or burrow systems within which their offspring are reared. Such a structure represents an enormous investment of resources. Naked mole-rats are eusocial mammals that live in elaborate underground tunnel systems (**Figure 53.22**). A colony includes 70 to 80 individuals but only 1 reproductive female and a few reproductive males. The other colony members are sterile workers that dig and maintain the tunnels, guard against intruders, harvest food (tubers), and use their feces to feed the queen and her offspring. Individuals attempting to found new colonies have a high risk of failing or being captured by predators. When chances of individual reproductive success are practically zero,

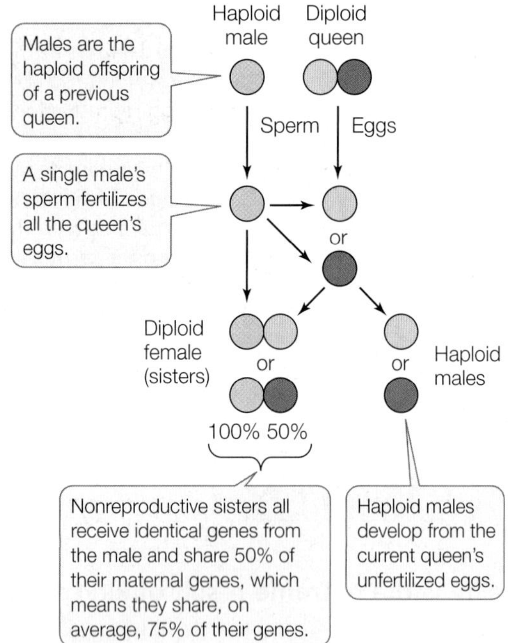


**53.21 Favoring Sisters over Daughters** Female honey bees are diploid and males are haploid. Thus if a female worker bee were to reproduce, she would share approximately 50 percent of her genes with her daughters. However, she shares an average of 75 percent of her genes with her sisters. In terms of inclusive fitness, then, a sister is more valuable than a daughter for this species.

an individual can best maximize its inclusive fitness by staying with and helping maintain the colony.

## Group living has benefits and costs

Apart from their direct influences on reproductive success, social systems can contribute to survival in many ways, but they can also involve costs. Thus the cost–benefit approach of behavioral ecology is relevant to understanding the evolution of social behavior.

**53.22 A Eusocial Mammal** Naked mole-rats live in a large colony with one reproductive female and a few reproductive males. Their home is an elaborate tunnel system excavated by the colony over time.

An obvious example of a benefit of group living is improved foraging efficiency. By hunting in packs, African wild dogs (see Figure 59.17) employ cooperative strategies that enable them to bring down larger prey than could a single dog. The larger the pack, the greater the hunting success rate. Once the prey is killed, the presence of conspecifics also reduces the risk that the wild dogs will lose their prey to larger scavengers, such as hyenas.

Living in a group can also reduce the risk of individuals becoming prey themselves. Many small birds forage in flocks. To test the hypothesis that flocking provides protection against predators, R. E. Kenward released a trained goshawk (*Accipiter gentilis*) near wild common wood-pigeons (*Columba palumbus*) in England. The hawk was most successful when it attacked solitary pigeons. Its success in capturing a pigeon in a flock decreased as the number of pigeons in the flock increased (**Figure 53.23A**). The larger the flock, the sooner some individual in the flock spotted the hawk and flew away. This escape behavior stimulated other individuals in the flock to take flight as well.

Alarm calling is another means of reducing predation risk, but the caller incurs a risk cost by calling attention to itself. Belding's ground squirrels live in large colonies in open meadows. When one squirrel announces the presence of a predator with loud, sharp barks, all the nearby squirrels dive into their burrows (**Figure 53.23B**). Paul Sherman showed that callers double their risk of being preyed on—so why do they do it? Research by Sherman and by others has shown that this altruistic behavior is a product of kin selection. In this polygynous species, males establish large territories in the spring that include the territories of several females, whom they inseminate. The females then drive off the males. Female offspring settle near their mothers, so neighboring females in a colony tend to be sisters, and they defend each other's young. Sherman showed that males are less likely to give alarm calls than females, and that females are more likely to give alarm calls when related individuals are nearby.

Social behavior has many costs as well as benefits. Foraging in a group may reduce the amount of food available to each individual, and the foraging individuals may interfere with one another's foraging activities. Individuals living in groups may face more competition for mates, as well as for food, than solitary individuals would. A large group may actually attract the attention of predators. And living at high population densities can increase the risk of disease transmission. The study of disease transmission in wild animal populations is a relatively new field, but such studies have made it apparent that species living in social groups are more prone to outbreaks of disease than are solitary species.

## Can the concepts of sociobiology be applied to humans?

Ever since the publication of E. O. Wilson's *Sociobiology*, applications of the concepts of evolutionary genetics to human behavior have been hugely controversial. The intensity of the debate may stem from impressions that sociobiological approaches resemble previously discredited pseudoscientific movements such as social Darwinism, biological determinism, and eugenics, all of which have been used as rationales

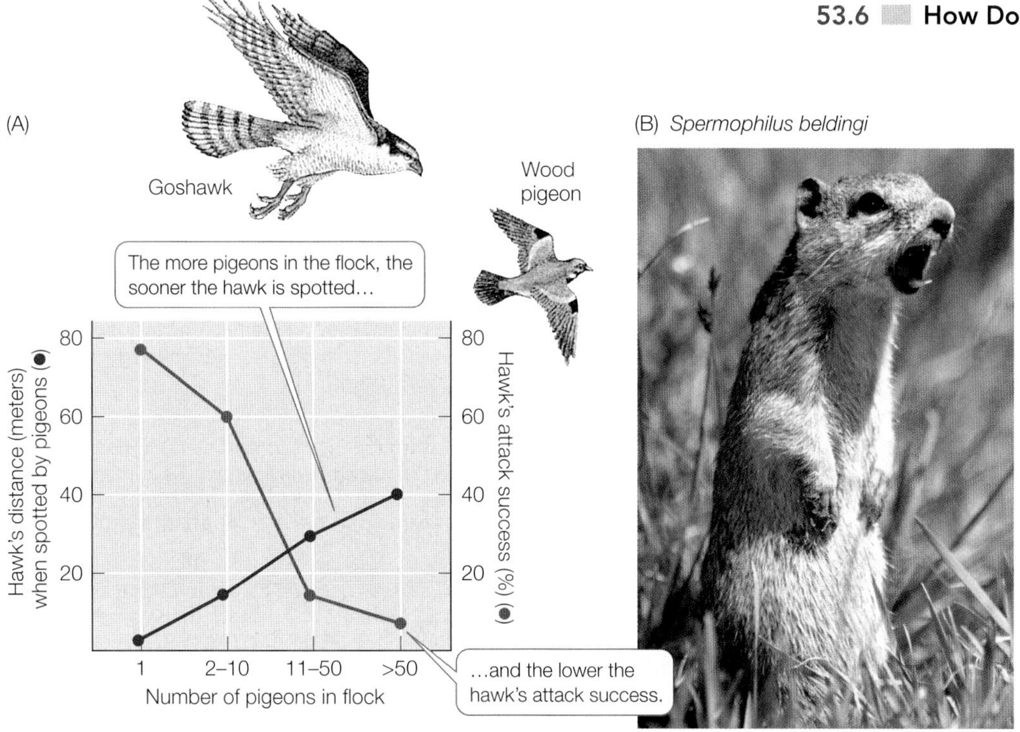


**53.23 Group Living Provides Protection from Predators** Animals that live in groups can spread the cost of looking out for predators. (A) The larger the number of common wood-pigeons in a flock, the greater the chances that one of the pigeons will spot a predatory goshawk before it attacks, and the lower the chances that the hawk will capture one of the pigeons. (B) A male Belding's ground squirrel gives an alarm call upon spotting a predator. Although this behavior increases his individual risk of becoming prey, he increases the survival chances of many of his close relatives.

for racism and discrimination. The proponents of sociobiology maintain that theirs is an objective science and that questions of "what is" should not be conflated with questions of "what ought to be." For example, we can demonstrate the genetic basis of sexual dimorphism in body size and muscle mass in humans, and we can compare this dimorphism with that in other mammals. However, any attempt to use such data as a political or legal defense for polygamy would not be science and should not be confused with science.

The fact that our biochemistry, our cell biology, physiology, and anatomy are shaped by our genes is beyond argument. However, it is also clear that these genetically shaped characteristics are also influenced by factors such as environment, nutrition, social interactions, and culture. Why should it be different for behavior? Studies of identical twins reared apart have produced evidence for inheritance of uncanny similarities in behavioral propensities. Studies of isolated human cultures around the world have also revealed remarkable similarities in social organization. None of these studies, however, would even begin to challenge the dominant role of learning and culture in the shaping of human behavior.

## RECAP 53.6

Social behavior can be understood by asking how it contributes to the fitness of the individuals involved. Asymmetry between the sexes in parental investment is a key factor in the evolution of mating systems. According to the theory of kin selection, an individual can increase its fitness by helping related individuals with whom it shares alleles. In extreme cases, kin selection has given rise to eusociality.

- What environmental conditions can lead to monogamy, polygamy, or polyandry? **See pp. 1113–1114**
- Explain how an individual can increase its fitness by helping its relatives. **See pp. 1114–1115**
- Why is eusociality so common among hymenopterans? **See p. 1115 and Figure 53.21**
- What are some of the costs and benefits of group living? **See p. 1116 and Figure 53.23**

Knowledge of the behavior of particular species—how they use the environment, how they obtain food, how they organize their activities spatially and temporally—is essential for understanding how species interact in nature. These interactions are one focus of the science of ecology, the subject of Part Ten of this book.

Could cowbird behaviors create selective pressure for host species *not* to develop egg discrimination behavior?

### ANSWER

Results of studies by Jeffrey Hoover and Scott Robinson, some of which were described in the Working with Data exercise on p. 1102, showed that because of nest destruction and other behaviors by cowbirds, hosts that tolerated the cowbird nest parasitism may successfully rear more of their own offspring than those that discriminate against cowbird eggs—a selection pressure on host birds not to discriminate. Of course, when the experimenters prevented cowbird access to the host nests, the hosts had even greater reproductive success. Thus the host birds were in effect paying "protection money" in the form of nurturing cowbird eggs (and thus reducing but not totally eliminating their own reproductive success), leading the researchers to dub the cowbirds' actions "mafia behavior."

# CHAPTERSUMMARY 53

## 53.1 What Are the Origins of Behavioral Biology?

- Ivan Pavlov's discovery of **conditioned reflexes** and B. F. Skinner's research on **operant conditioning** as a model for learning led to an approach called behaviorism that mainly carried out laboratory experiments on rats and a few other animal models. **Review Figure 53.1**
- **Ethology** focuses on both the **proximate causes** of behavior (the immediate cause of the behavior, and how the behavior develops) and on the **ultimate causes** (how the behavior affects the animal's evolutionary fitness).
- A major focus of the ethologists was **fixed action patterns** and their **releasers**. They performed **deprivation experiments** as well as breeding experiments to demonstrate that certain behaviors are genetically determined. **Review Figure 53.2.**

## 53.2 How Do Genes Influence Behavior?

- Breeding experiments can reveal whether a behavioral phenotype is inherited. Quantitative trait analysis can reveal candidate genes that influence a behavior. Gene knockout experiments can reveal the roles of specific genes underlying a behavioral phenotype. **Review Figure 53.3**
- Most behaviors are complex traits involving many genes that function in cascades and offer many points for a change in a single gene to influence behavior. **Review Figure 53.4**

## 53.3 How Does Behavior Develop?

- Hormones can determine the pattern of behavior that develops and the timing of its expression. **Review Figure 53.5**
- **Imprinting** is a process by which an animal learns a specific set of stimuli during a limited **critical** or **sensitive period**. That critical period may be determined by hormones.
- The development and expression of song in white-crowned sparrows involves a genetic predisposition to learn the species-specific song, a critical period for imprinting of a song memory, and hormonally controlled timing of song expression. Social interactions may also play a role. **Review Figures 53.7, 53.8**

## 53.4 How Does Behavior Evolve?

- An animal's behavior involves a series of choices that influence its fitness. To make these choices, animals use environmental cues that are reliable predictors of the potential effects of their choice on their fitness.
- The **cost–benefit approach** can be used to investigate the fitness value of specific behaviors. The cost of a behavior typically has three components: **energetic cost**, **risk cost**, and **opportunity cost**. **Review Figure 53.9, ANIMATED TUTORIAL 53.1**
- According to **optimal foraging theory**, animals should practice feeding behaviors that maximize their energetic gain at the least cost. **Review Figure 53.11, ANIMATED TUTORIAL 53.2**

## 53.5 What Physiological Mechanisms Underlie Behavior?

- **Circadian rhythms** control the daily cycle of behavior. Without environmental time cues, circadian rhythms free-run with a period that is genetically programmed. They are normally **entrained** to the light–dark cycle by environmental cues. **Review Figure 53.13, ANIMATED TUTORIAL 53.3**
- Forms of navigation used by animals to find their way in the environment include **piloting** (orienting to landmarks), distance–direction navigation, and bicoordinate navigation. Navigation mechanisms include celestial navigation and a time-compensated solar compass. **Review Figures 53.15–53.17, ANIMATED TUTORIALS 53.4, 53.5**
- The behaviors of individuals may become **communication signals** if the transmission of information benefits both the sender and the receiver. **Review Figure 53.18, ACTIVITY 53.1**
- Chemical communication signals (pheromones) can be highly specific and have different time courses. Visual signals can convey complex messages rapidly, but only if the recipient can see the sender. Acoustic signals can travel over long distances, do not require a focused recipient, and can be modified to reveal or conceal directional information. Tactile signals can convey complex messages when animals are in close proximity.

## 53.6 How Does Social Behavior Evolve?

- **Polygynous** mating systems, in which one male controls and mates with many females, can result in great variation in male reproductive success. **Polyandry**—a female mating with multiple males—can evolve in circumstances in which a male can make a substantial contribution to the survival of his offspring.
- The fitness an individual gains by producing offspring (**direct fitness**) plus the fitness it gains by increasing the reproductive success of relatives with whom it shares alleles (**indirect fitness**) is called **inclusive fitness**. **Kin selection** may favor **altruistic** behavior toward relatives, despite its cost to the performer, if it increases the performer's inclusive fitness.
- As a result of **haplodiploidy**, the sex determination mechanism of hymenopteran insects, nonreproductive female workers (sisters) share more alleles with one another than reproductive females share with their own offspring. **Review Figure 53.21**
- Haplodiploidy has probably facilitated the evolution of **eusocial** behavior in this group through kin selection. Eusociality has also arisen in diploid species in which chances of individual reproductive success are extremely low.
- Group living confers benefits such as greater foraging efficiency and protection from predators, but it also has costs, such as increased competition for food and ease of transmission of diseases.

**See ACTIVITY 53.2 for a concept review of this chapter**

**Go to the Interactive Summary to review key figures, Animated Tutorials, and Activities Life10e.com/is53**

# CHAPTER**REVIEW**

## REMEMBERING

1. Which of the following is *not* true of a fixed action pattern?
   a. Its expression may depend on hormonal conditions.
   b. It is induced by complex, species-specific stimuli.
   c. It is highly stereotypic and species-specific.
   d. It can be expressed even if the animal has never seen it performed.
   e. Its genetic basis can be demonstrated by breeding experiments.

2. Which of the following is *not* a component of the cost of performing a behavior?
   a. Its energetic cost
   b. The risk of being injured
   c. Its opportunity cost
   d. The risk of being attacked by a predator
   e. Its information cost

3. Birds that migrate at night
   a. inherit a star map.
   b. determine direction by knowing the time and the position of a constellation in the rotating night sky.
   c. orient to a fixed point in the rotating night sky.
   d. imprint on one or more key constellations.
   e. determine distance, but not direction, from the stars.

4. If a bird is trained to seek food on the western side of a cage open to the sky, and is then placed in a chamber with a controlled light cycle so that its circadian rhythm becomes phase-delayed by 6 hours (i.e., its circadian rhythm is 6 hours behind real time), when it is returned to the open cage at noon in real time, it will seek food in the
   a. north.
   b. south.
   c. east.
   d. west.

5. Which of the following statements about communication is true?
   a. Complex information cannot be conveyed by pheromones.
   b. Visual signaling is advantageous in complex environments.
   c. Acoustic communication always reveals the location of the signaler.
   d. One advantage of pheromones is that the message can persist over time.
   e. The dance of honey bees is an example of visual signaling.

6. A cost commonly associated with group living is
   a. increased risk of predation.
   b. interference with foraging.
   c. increased exposure to diseases and parasites.
   d. increased competition for mates.
   e. All of the above

7. Altruistic behavior
   a. can increase an individual's inclusive fitness.
   b. depends on haplodiploid sex determination.
   c. is most common among unrelated individuals.
   d. always causes a net decrease in the performer's fitness.
   e. characterizes a monogamous mating system.

8. A group is said to be eusocial if
   a. the group's members interact intensively.
   b. some members produce many more offspring than others do.
   c. a dominance hierarchy exists among group members.
   d. young individuals remain in the group to help their parents rear other offspring.
   e. the group contains nonreproductive helper individuals.

## UNDERSTANDING & APPLYING

9. Adult male dogs lift a hind leg when they urinate, whereas young puppies and adult female dogs squat. If a newborn male puppy receives an injection of estrogen, it will never lift its leg to urinate; for the rest of its life, it will always squat. How might this result be explained?

10. In most vertebrate species with helpers, the helpers are individuals capable of reproducing, and often breed later. Among eusocial insects, sterile castes have evolved repeatedly. What differences between vertebrates and insects might explain the failure of sterile castes to evolve in vertebrates?

11. Studies of birdsong in zebra finches showed that males appear to have "practice" (undirected) songs and "performance" (directed) songs (see Figure 53.8). Field studies show that white-crowned sparrow song has local dialects and that the songs of male birds living only a few kilometers apart are distinguishable. Assuming white-crowned sparrows, like zebra finches, have directed and undirected songs, what could be the adaptive significance of this "practice" versus "performance" behavior?

## ANALYZING & EVALUATING

12. Cowbirds are nest parasites, as seen in the opening story of this chapter. What do you think would characterize the acquisition of song in cowbirds? In a given geographical region, cowbirds tend to parasitize the nests of particular bird species. How do you think female cowbirds learn this behavior? How would you test your hypothesis?

13. Some honey bee hives show hygienic behavior: if pupae die in their cells, some workers uncap the cells while other workers remove the carcasses. Some hives do not show this behavior. In a classical behavior genetics study, crosses were made between hygienic and nonhygienic hives. The results were that about 25% of the resulting hives were hygienic and about 75% were not. Of those that were not, about one-third showed uncapping behavior and about one-third showed carcass removal behavior if cells were uncapped. But there was considerable variability in these data. A recent gene-mapping study (using QTL analysis) revealed seven significant correlations between specific allele frequencies and hygienic behavior. What do the classical data suggest about the genetic basis for hygienic behavior? What do the gene-mapping data suggest? How can you resolve the differences?

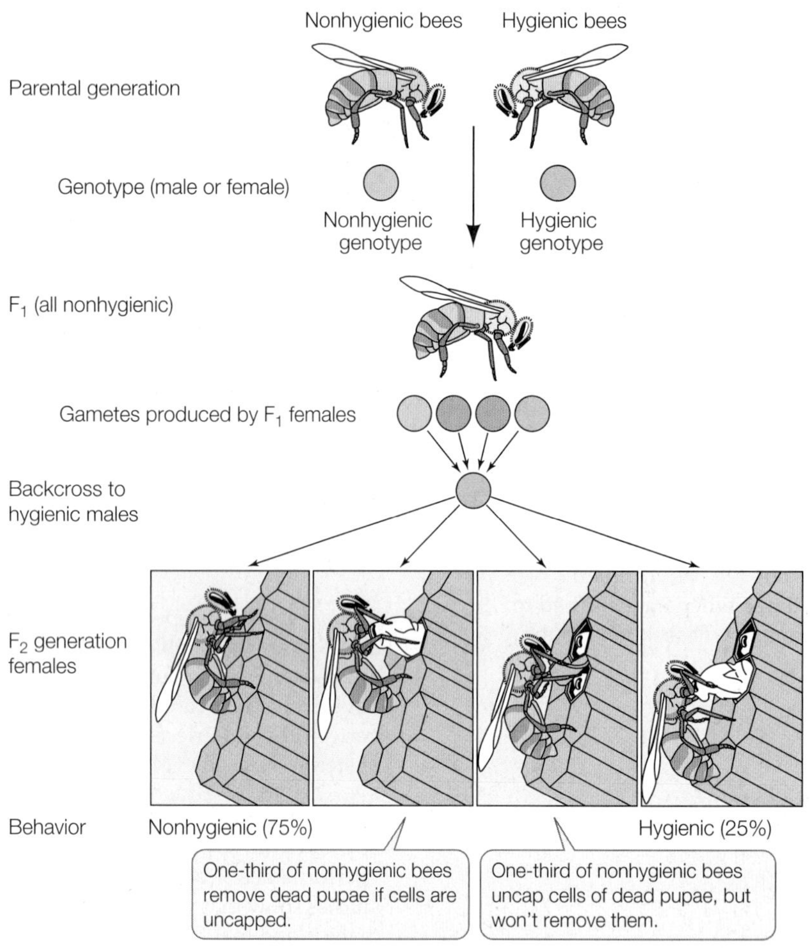


Go to BioPortal at **yourBioPortal.com** for Animated Tutorials, Activities, LearningCurve Quizzes, Flashcards, and many other study and review resources.

# Appendix A The Tree of Life

Phylogeny is the organizing principle of modern biological taxonomy. A guiding principle of modern phylogeny is monophyly. A monophyletic group is considered to be one that contains an ancestral lineage and all of its descendants. Any such group can be extracted from a phylogenetic tree with a single cut.

The tree shown here provides a guide to the relationships among the major groups of extant (living) organisms in the tree of life as we have presented them throughout this book. The position of the branching "splits" indicates the relative branching order of the lineages of life, but the time scale is not meant to be uniform. In addition, the groups appearing at the branch tips do not necessarily carry equal phylogenetic "weight." For example, the ginkgo [75] is indeed at the apex of its lineage; this gymnosperm group consists of a single living species. In contrast, a phylogeny of the eudicots [83] could continue on from this point to fill many more trees the size of this one.

The glossary entries that follow are informal descriptions of some major features of the organisms described in Part Seven of this book. Each entry gives the group's common name, followed by the formal scientific name of the group (in parentheses). Numbers in square brackets reference the location of the respective groups on the tree.

It is sometimes convenient to use an informal name to refer to a collection of organisms that are not monophyletic but nonetheless all share (or all lack) some common attribute. We call these "convenience terms"; such groups are indicated in these entries by quotation marks, and we do not give them formal scientific names. Examples include "prokaryotes," "protists," and "algae." Note that these groups cannot be removed with a single cut; they represent a collection of distantly related groups that appear in different parts of the tree. We also use quotation marks here to designate two groups of fungi that are not believed to be monophyletic.

Go to BioPortal at **yourBioPortal.com** for an interactive version of this tree, with links to photos, distribution maps, species lists, and identification keys.

### – A –

**acorn worms** (*Enteropneusta*) Benthic marine hemichordates [119] with an acorn-shaped proboscis, a short collar (neck), and a long trunk.

**"algae"** Convenience term encompassing various distantly related groups of aquatic, photosynthetic eukaryotes [4].

**alveolates** (*Alveolata*) [5] Unicellular eukaryotes with a layer of flattened vesicles (alveoli) supporting the plasma membrane. Major groups include the dinoflagellates [51], apicomplexans [50], and ciliates [49].

**amborella** (*Amborella*) [78] An understory shrub or small tree found only on the South Pacific island of New Caledonia. Thought to be the sister group of the remaining living angiosperms [15].

**ambulacrarians** (*Ambulacraria*) [29] The echinoderms [118] and hemichordates [119].

**amniotes** (*Amniota*) [36] Mammals, reptiles, and their extinct close relatives. Characterized by many adaptations to terrestrial life, including an amniotic egg (with a unique set of membranes—the amnion, chorion, and allantois), a water-repellant epidermis (with epidermal scales, hair, or feathers), and, in males, a penis that allows internal fertilization.

**amoebozoans** (*Amoebozoa*) [84] A group of eukaryotes [4] that use lobe-shaped pseudopods for locomotion and to engulf food. Major amoebozoan groups include the loboseans, plasmodial slime molds, and cellular slime molds.

**amphibians** (*Amphibia*) [128] Tetrapods [35] with glandular skin that lacks epidermal scales, feathers, or hair. Many amphibian species undergo a complete metamorphosis from an aquatic larval form to a terrestrial adult form, although direct development is also common. Major amphibian groups include frogs and toads (anurans), salamanders, and caecilians.

**amphipods** (*Amphipoda*) Small crustaceans [116] that are abundant in many marine and freshwater habitats. They are important herbivores, scavengers, and micropredators, and are an important food source for many aquatic organisms.

**angiosperms** (*Anthophyta* or *Magnoliophyta*) [15] The flowering plants. Major angiosperm groups include the monocots [82], eudicots [83], and magnoliids [81].

**animals** (*Animalia* or *Metazoa*) [19] Multicellular heterotrophic eukaryotes. The majority of animals are bilaterians [22]. Other groups of animals include the sponges [20], ctenophores [95], placozoans [96], and cnidarians [97]. The closest living relatives of the animals are the choanoflagellates [91].

**annelids** (*Annelida*) [105] Segmented worms, including earthworms, leeches, and polychaetes. One of the major groups of lophotrochozoans [24].

**anthozoans** (*Anthozoa*) One of the major groups of cnidarians [97]. Includes the sea anemones, sea pens, and corals.

**anurans** (*Anura*) Comprising the frogs and toads, this is the largest group of living amphibians [128]. They are tail-less, with a shortened vertebral column and elongate hind legs modified for jumping. Many species have an aquatic larval form known as a tadpole.

**apicomplexans** (*Apicomplexa*) [50] Parasitic alveolates [5] characterized by the possession of an apical complex at some stage in the life cycle.

**arachnids** (*Arachnida*) Chelicerates [114] with a body divided into two parts: a cephalothorax that bears six pairs of appendages (four pairs of which are usually used as legs) and an abdomen that bears the genital opening. Familiar arachnids include spiders, scorpions, mites and ticks, and harvestmen.

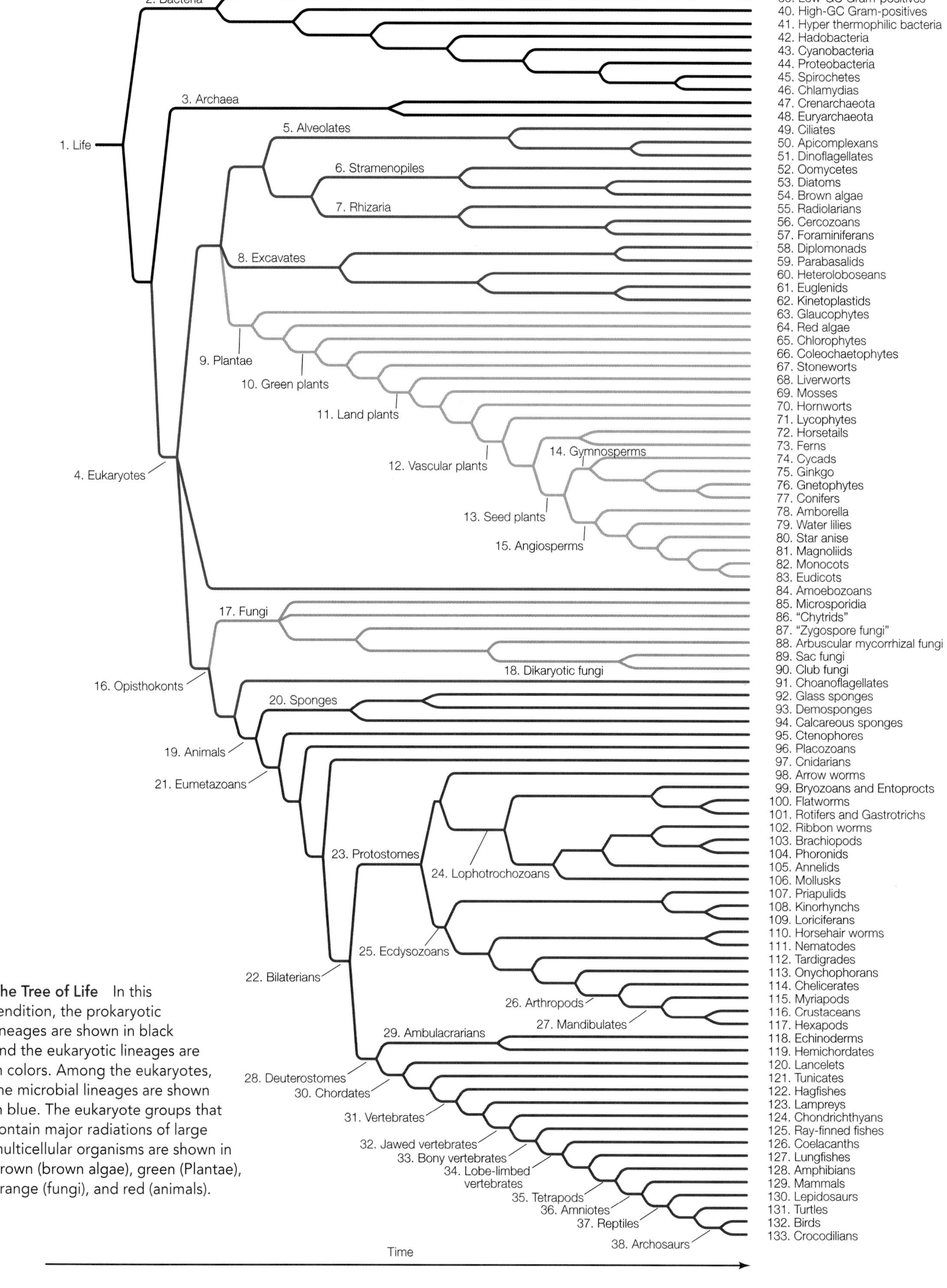


**The Tree of Life** In this rendition, the prokaryotic lineages are shown in black and the eukaryotic lineages are in colors. Among the eukaryotes, the microbial lineages are shown in blue. The eukaryote groups that contain major radiations of large multicellular organisms are shown in brown (brown algae), green (Plantae), orange (fungi), and red (animals).

**arbuscular mycorrhizal fungi** (*Glomeromycota*) [88] A group of fungi [17] that associate with plant roots in a close symbiotic relationship.

**archaeans** (*Archaea*) [3] Unicellular organisms lacking a nucleus and lacking peptidoglycan in the cell wall. Once grouped with the bacteria, archaeans possess distinctive membrane lipids.

**archosaurs** (*Archosauria*) [38] A group of reptiles [37] that includes dinosaurs and crocodilians [133]. Most dinosaur groups became extinct at the end of the Cretaceous; birds [132] are the only surviving dinosaurs.

**arrow worms** (*Chaetognatha*) [98] Small planktonic or benthic predatory marine worms with fins and a pair of hooked, prey-grasping spines on each side of the head.

**arthropods** (*Arthropoda*) The largest group of ecdysozoans [25]. Arthropods are characterized by a stiff exoskeleton, segmented bodies, and jointed appendages. Includes the chelicerates [114], myriapods [115], crustaceans [116], and hexapods (insects and their relatives) [117].

**ascidians** (*Ascidiacea*) "Sea squirts"; the largest group of tunicates [121]. They are sessile (as adults), marine, saclike filter feeders.

## – B –

**bacteria** (*Eubacteria*) [2] Unicellular organisms lacking a nucleus, possessing distinctive ribosomes and initiator tRNA, and generally containing peptidoglycan in the cell wall. Different bacterial groups are distinguished primarily on nucleotide sequence data.

**barnacles** (*Cirripedia*) Crustaceans [116] that undergo two metamorphoses—first from a feeding planktonic larva to a nonfeeding swimming larva, and then to a sessile adult that forms a "shell" composed of four to eight plates cemented to a hard substrate.

**bilaterians** (*Bilateria*) [22] Those animal groups characterized by bilateral symmetry and three distinct tissue types (endoderm, ectoderm, and mesoderm). Includes the protostomes [23] and deuterostomes [28].

**birds** (*Aves*) [132] Feathered, flying (or secondarily flightless) tetrapods [35].

**bivalves** (*Bivalvia*) Major mollusk [106] group; clams and mussels. Bivalves typically have two similar hinged shells that are each asymmetrical across the midline.

**bony vertebrates** (*Osteichthyes*) [33] Vertebrates [31] in which the skeleton is usually ossified to form bone. Includes the ray-finned fishes [125], coelacanths [126], lungfishes [127], and tetrapods [35].

**brachiopods** (*Brachiopoda*) [103] Lophotrochozoans [24] with two similar hinged shells that are each symmetrical across the midline. Superficially resemble bivalve mollusks, except for the shell symmetry.

**brittle stars** (*Ophiuroidea*) Echinoderms [118] with five long, whip-like arms radiating from a distinct central disk that contains the reproductive and digestive organs.

**brown algae** (*Phaeophyta*) [54] Multicellular, almost exclusively marine stramenopiles [6] generally containing the pigment fucoxanthin as well as chlorophylls *a* and *c* in their chloroplasts.

**bryozoans** (*Ectoprocta* or *Bryozoa*) [99] A group of marine and freshwater lophotrochozoans [24] that live in colonies attached to substrates; also known as ectoprocts or moss animals. They are the sister group of entoprocts.

## – C –

**caecilians** (*Gymnophiona*) A group of burrowing or aquatic amphibians [128]. They are elongate, legless, with a short tail (or none at all), reduced eyes covered with skin or bone, and a pair of sensory tentacles on the head.

**calcareous sponges** (*Calcarea*) [94] Filter-feeding marine sponges with spicules composed of calcium carbonate.

**cellular slime molds** (*Dictyostelida*) Amoebozoans [84] in which individual amoebas aggregate under stress to form a multicellular pseudoplasmodium.

**cephalochordates** (*Cephalochordata*) [120] *See* lancelets.

**cephalopods** (*Cephalopoda*) Active, predatory mollusks [106] in which the molluscan foot has been modified into muscular hydrostatic arms or tentacles. Includes octopuses, squids, and nautiluses.

**cercozoans** (*Cercozoa*) [56] Unicellular eukaryotes [4] that feed by means of threadlike pseudopods. Group together with foraminiferans [57] and radiolarians [55] to comprise the rhizaria [7].

**charophytes** (*Charales*) [67] *See* stoneworts.

**chelicerates** (*Chelicerata*) [114] A major group of arthropods [26] with pointed appendages (chelicerae) used to grasp food (as opposed to the chewing mandibles of most other arthropods). Includes the arachnids, horseshoe crabs, pycnogonids, and extinct sea scorpions.

**chimaeras** (*Holocephali*) A group of bottom-dwelling, marine, scaleless chondrichthyan fishes [124] with large, permanent, grinding tooth plates (rather than the replaceable teeth found in other chondrichthyans).

**chitons** (*Polyplacophora*) Flattened, slow-moving mollusks [106] with a dorsal protective calcareous covering made up of eight articulating plates.

**chlamydias** (*Chlamydiae*) [46] A group of very small Gram-negative bacteria; they live as intracellular parasites of other organisms.

**chlorophytes** (*Chlorophyta*) [65] The most abundant and diverse group of green algae, including freshwater, marine, and terrestrial forms; some are unicellular, others colonial, and still others multicellular. Chlorophytes use chlorophylls *a* and *c* in their photosynthesis.

**choanoflagellates** (*Choanozoa*) [91] Unicellular eukaryotes [4] with a single flagellum surrounded by a collar. Most are sessile, some are colonial. The closest living relatives of the animals [19].

**chondrichthyans** (*Chondrichthyes*) [124] One of the two main groups of jawed vertebrates [32]; includes sharks, rays, and chimaeras. They have cartilaginous skeletons and paired fins.

**chordates** (*Chordata*) [30] One of the two major groups of deuterostomes [28], characterized by the presence (at some point in development) of a notochord, a hollow dorsal nerve cord, and a post-anal tail. Includes the lancelets [120], tunicates [121], and vertebrates [31].

**"chytrids"** [90] Convenience term used for a paraphyletic group of mostly aquatic, microscopic fungi [17] with flagellated gametes. Some exhibit alternation of generations.

**ciliates** (*Ciliophora*) [49] Alveolates [5] with numerous cilia and two types of nuclei (micronuclei and macronuclei).

**clitellates** (*Clitellata*) Annelids [105] with gonads contained in a swelling (called a clitellum) toward the head of the animal. Includes earthworms (oligochaetes) and leeches.

**club fungi** (*Basidiomycota*) [90] Fungi [17] that, if multicellular, bear the products of meiosis on club-shaped basidia and possess a long-lasting dikaryotic stage. Some are unicellular.

**club mosses** (*Lycopodiophyta*) [71] Vascular plants [12] characterized by microphylls. *See* lycophytes.

**cnidarians** (*Cnidaria*) [97] Aquatic, mostly marine eumetazoans [21] with specialized stinging organelles (nematocysts) used for prey capture and defense, and a blind gastrovascular cavity. The sister group of the bilaterians [22].

**coelacanths** (*Actinista*) [126] A group of marine lobe-limbed vertebrates [34] that was diverse from the Middle Devonian to the Cretaceous, but is now known from just two living species. The pectoral and anal fins are on fleshy stalks supported by skeletal elements, so they are also called lobe-finned fishes.

**coleochaetophytes** (*Coleochaetales*) [66] Multicellular green algae characterized by flattened growth form composed of thin-walled cells. Thought to be the sister-group to the stoneworts [67] plus land plants [11].

**conifers** (*Pinophyta* or *Coniferophyta*) [77] Cone-bearing, woody seed plants [13].

**copepods** (*Copepoda*) Small, abundant crustaceans [116] found in marine, freshwater, or wet terrestrial habitats. They have a single eye, long antennae, and a body shaped like a teardrop.

**craniates** (*Craniata*) Some biologist exclude the hagfishes [122] from the vertebrates [31], and use the term craniates to refer to the two groups combined.

**crenarchaeotes** (*Crenarchaeota*) [47] A major and diverse group of archaeans [3], defined on the basis of rRNA base sequences. Many are extremophiles (inhabit extreme environments), but the group may also be the most abundant archaeans in the marine environment.

**crinoids** (*Crinoidea*) Echinoderms [118] with a mouth surrounded by feeding arms, and a U-shaped gut with the mouth next to the anus. They attach to the substratum by a stalk or are free-swimming. Crinoids were abundant in the middle and late Paleozoic, but only a few hundred species have survived to the present. Includes the sea lilies and feather stars.

**crocodilians** (*Crocodylia*) [133] A group of large, predatory, aquatic archosaurs [38]. The closest living relatives of birds [132]. Includes alligators, caimans, crocodiles, and gharials.

**crustaceans** (*Crustacea*) [116] Major group of marine, freshwater, and terrestrial arthropods [26] with a head, thorax, and abdomen (although the head and thorax may be fused), covered with a thick exoskeleton, and with two-part appendages. Crustaceans undergo metamorphosis from a nauplius larva. Includes decapods, isopods, krill, barnacles, amphipods, copepods, and ostracods.

**ctenophores** (*Ctenophora*) [95] Radially symmetrical, diploblastic marine animals [19], with a complete gut and eight rows of fused plates of cilia (called ctenes).

**cyanobacteria** (*Cyanobacteria*) [43] A group of unicellular, colonial, or filamentous bacteria that conduct photosynthesis using chlorophyll *a*.

**cycads** (*Cycadophyta*) [74] Palmlike gymnosperms with large, compound leaves.

**cyclostomes** (*Cyclostomata*) This term refers to the possibly monophyletic group of lampreys [123] and hagfishes [122]. Molecular data support this group, but morphological data suggest that lampreys are more closely related to jawed vertebrates [32] than to hagfishes.

## – D –

**decapods** (*Decapoda*) A group of marine, freshwater, and semiterrestrial crustaceans [116] in which five of the eight pairs of thoracic appendages function as legs (the other three pairs, called maxillipeds, function as mouthparts). Includes crabs, lobsters, crayfishes, and shrimps.

**demosponges** (*Demospongiae*) [93] The largest of the three groups of sponges [20], accounting for 90 percent of all sponge species. Demosponges have spicules made of silica, spongin fiber (a protein), or both.

**deuterostomes** (*Deuterostomia*) [28] One of the two major groups of bilaterians [22], in which the mouth forms at the opposite end of the embryo from the blastopore in early development (contrast with protostomes). Includes the ambulacrarians [29] and chordates [30].

**diatoms** (*Bacillariophyta*) [53] Unicellular, photosynthetic stramenopiles [6] with glassy cell walls in two parts.

**dikaryotic fungi** (*Dikarya*) [18] A group of fungi [17] in which two genetically different haploid nuclei coexist and divide within the same hypha; includes club fungi [90] and sac fungi [89].

**dinoflagellates** (*Dinoflagellata*) [51] A group of alveolates [5] usually possessing two flagella, one in an equatorial groove and the other in a longitudinal groove; many are photosynthetic.

**diplomonads** (*Diplomonadida*) [58] A group of eukaryotes [4] lacking mitochondria; most have two nuclei, each with four associated flagella.

## – E –

**ecdysozoans** (*Ecdysozoa*) [25] One of the two major groups of protostomes [23], characterized by periodic molting of their exoskeletons. Nematodes [111] and arthropods [26] are the largest ecdysozoan groups.

**echinoderms** (*Echinodermata*) [118] A major group of marine deuterostomes [28] with five-fold radial symmetry (at some stage of life) and an endoskeleton made of calcified plates and spines. Includes sea stars, crinoids, sea urchins, sea cucumbers, and brittle stars.

**elasmobranchs** (*Elasmobranchii*) The largest group of chondrichthyan fishes [124]. Includes sharks, skates, and rays. In contrast to the other group of living chondrichthyans (the chimaeras), they have replaceable teeth.

**embryophytes** *See* land plants [11].

**entoprocts** (*Entoprocta*) [99] A group of marine and freshwater lophotrochozoans [24] that live as single individuals or in colonies attached to substrates. They are the sister group of bryozoans, from which they differ in having both their mouth and anus inside the lophophore (the anus is outside the lophophore in bryozoans).

**eudicots** (*Eudicotyledones*) [83] A group of angiosperms [15] with pollen grains possessing three openings. Typically with two cotyledons, net-veined leaves, taproots, and floral organs typically in multiples of four or five.

**euglenids** (*Euglenida*) [61] Flagellate excavates characterized by a pellicle composed of spiraling strips of protein under the plasma membrane; the mitochondria have disk-shaped cristae. Some are photosynthetic.

**eukaryotes** (*Eukarya*) [4] Organisms made up of one or more complex cells in which the genetic material is contained in nuclei. Contrast with archaeans [3] and bacteria [2].

**eumetazoans** (*Eumetazoa*) [21] Those animals [19] characterized by body symmetry, a gut, a nervous system, specialized types of cell junctions, and well-organized tissues in distinct cell layers (although there have been secondary losses of some or most of these characteristics in a few eumetazoan lineages).

**euphyllophytes** (*Euphyllophyta*) The group of vascular plants [12] that is sister to the lycophytes [71] and which includes all plants with megaphylls.

**euryarchaeotes** (*Euryachaeota*) [48] A major group of archaeans [3], diagnosed on the basis of rRNA sequences. Includes many methanogens, extreme halophiles, and thermophiles.

**eutherians** (*Eutheria*) A group of viviparous mammals [129], eutherians are well developed at birth (contrast to prototherians and marsupials, the other two groups of mammals). Most familiar mammals outside the Australian and South American regions are eutherians (see Table 33.1).

**excavates** (*Excavata*) [8] Diverse group of unicellular, flagellate eukaryotes, many of which possess a feeding groove; some lack mitochondria.

## – F –

**ferns** Vascular plants [12] usually possessing large, frondlike leaves that unfold from a "fiddlehead." Not a monophyletic group, although most fern species are encompassed in a monophyletic clade, the leptosporangiate ferns [73].

**flatworms** (*Platyhelminthes*) [100] A group of dorsoventrally flattened and generally elongate soft-bodied lophotrochozoans [24]. May be free-living or parasitic, found in marine, freshwater, or damp terrestrial environments. Major flatworm groups include the tapeworms, flukes, monogeneans, and turbellarians.

**flowering plants** *See* angiosperms [15].

**flukes** (*Trematoda*) A group of wormlike parasitic flatworms [100] with complex life cycles that involve several different host species. May be paraphyletic with respect to tapeworms.

**foraminiferans** (*Foraminifera*) [57] Amoeboid organisms with fine, branched pseudopods that form a food-trapping net. Most produce external shells of calcium carbonate.

**fungi** (*Fungi*) [17] Eukaryotic heterotrophs with absorptive nutrition based on extracellular digestion; cell walls contain chitin. Major fungal groups include the microsporidia [85], "chytrids" [86], "zygospore fungi" [87], arbuscular mycorrhizal fungi [88], sac fungi [89], and club fungi [90].

## – G –

**gastropods** (*Gastropoda*) The largest group of mollusks [106]. Gastropods possess a well-defined head with two or four sensory tentacles (often terminating in eyes) and a ventral foot. Most species have a single coiled or spiraled shell. Common in marine, freshwater, and terrestrial environments.

**gastrotrichs** (*Gastrotricha*) [101] Tiny (0.06–3.0 mm), elongate acoelomate lophotrochozoans [24] that are covered in cilia. They live in marine, freshwater, and wet terrestrial habitats. They are simultaneous hermaphrodites.

**ginkgo** (*Ginkgophyta*) [75] A gymnosperm [14] group with only one living species. The ginkgo seed is surrounded by a fleshy tissue not derived from an ovary wall and hence not a fruit.

**glass sponges** (*Hexactinellida*) [92] Sponges [20] with a skeleton composed of four- and/or six-pointed spicules made of silica.

**glaucophytes** (*Glaucophyta*) [63] Unicellular freshwater algae with chloroplasts containing traces of peptidoglycan, the characteristic cell wall material of bacteria.

**gnathostomes** (*Gnathostomata*) *See* jawed vertebrates [32].

**gnetophytes** (*Gnetophyta*) [76] A gymnosperm [14] group with three very different lineages; all have wood with vessels, unlike other gymnosperms.

**green plants** (*Viridiplantae*) [10] Organisms with chlorophylls *a* and *b*, cellulose-containing cell walls, starch as a carbohydrate storage product, and chloroplasts surrounded by two membranes.

**gymnosperms** (*Gymnospermae*) [14] Seed plants [13] with seeds "naked" (i.e., not enclosed in carpels). Probably monophyletic, but status still in doubt. Includes the conifers [77], gnetophytes [76], ginkgo [75], and cycads [74].

## – H –

**hadobacteria** (*Hadobacteria*) [42] A group of extremophilic bacteria [2] that includes the genera *Deinococus* and *Thermus*.

**hagfishes** (*Myxini*) [122] Elongate, slimy-skinned vertebrates [31] with three small accessory hearts, a partial cranium, and no stomach or paired fins. *See also* craniata; cyclostomes.

**hemichordates** (*Hemichordata*) [119] One of the two primary groups of ambulacrarians [29];

marine wormlike organisms with a three-part body plan.

**heteroloboseans** (*Heterolobosea*) [60] Colorless excavates [8] that can transform among amoeboid, flagellate, and encysted stages.

**hexapods** (*Hexapoda*) [117] Major group of arthropods [26] characterized by a reduction (from the ancestral arthropod condition) to six walking appendages, and the consolidation of three body segments to form a thorax. Includes insects and their relatives (see Table 23.2).

**high-GC Gram-positives** (*Actinobacteria*) [40] Gram-positive bacteria with a relatively high (G+C)/(A+T) ratio of their DNA, with a filamentous growth habit.

**hornworts** (*Anthocerophyta*) [70] Nonvascular plants with sporophytes that grow from the base. Cells contain a single large, platelike chloroplast.

**horsehair worms** (*Nematomorpha*) [110] A group of very thin, elongate, wormlike freshwater ecdysozoans [25]. Largely nonfeeding as adults, they are parasites of insects and crayfish as larvae.

**horseshoe crabs** (*Xiphosura*) Marine chelicerates [114] with a large outer shell in three parts: a carapace, an abdomen, and a tail-like telson. There are only five living species, but many additional species are known from fossils.

**horsetails** (*Sphenophyta* or *Equisetophyta*) [72] Vascular plants [12] with reduced megaphylls in whorls.

**hydrozoans** (*Hydrozoa*) A group of cnidarians [97]. Most species go through both polyp and mesuda stages, although one stage or the other is eliminated in some species.

**hyperthermophilic bacteria** [41] A group of thermophilic bacteria [2] that live in volcanic vents, hot springs, and in underground oil reservoirs; includes the genera *Aquifex* and *Thermotoga*.

## – I –

**insects** (*Insecta*) The largest group within the hexapods [117]. Insects are characterized by exposed mouthparts and one pair of antennae containing a sensory receptor called a Johnston's organ. Most have two pairs of wings as adults. There are more described species of insects than all other groups of life [1] combined, and many species remain to be discovered. The major insect groups are described in Table 23.2.

**"invertebrates"** Convenience term encompassing any animal [19] that is not a vertebrate [31].

**isopods** (*Isopoda*) Crustaceans [116] characterized by a compact head, unstalked compound eyes, and mouthparts consisting of four pairs of appendages. Isopods are abundant and widespread in salt, fresh, and brackish water, although some species (the sow bugs) are terrestrial.

## – J –

**jawed vertebrates** (*Gnathostomata*) [32] A major group of vertebrates [31] with jawed mouths. Includes chondrichthyans [124], ray-finned fishes [125], and lobe-limbed vertebrates [34].

## – K –

**kinetoplastids** (*Kinetoplastida*) [62] Unicellular, flagellate organisms characterized by the presence in their single mitochondrion of a kinetoplast (a structure containing multiple, circular DNA molecules).

**kinorhynchs** (*Kinorhyncha*) [108] Small (< 1 mm) marine ecdysozoans [25] with bodies in 13 segments and a retractable proboscis.

**korarchaeotes** (*Korarchaeota*) A group of archaeans [3] known only by evidence from nucleic acids derived from hot springs. Its phylogenetic relationships within the Archaea are unknown.

**krill** (*Euphausiacea*) A group of shrimplike marine crustaceans [116] that are important components of the zooplankton.

## – L –

**lampreys** (*Petromyzontiformes*) [123] Elongate, eel-like vertebrates [31] that often have rasping and sucking disks for mouths.

**lancelets** (*Cephalochordata*) [120] A group of weakly swimming, eel-like benthic marine chordates [30].

**land plants** (*Embryophyta*) [11] Plants with embryos that develop within protective structures; also called embryophytes. Sporophytes and gametophytes are multicellular. Land plants possess a cuticle. Major groups are the liverworts [68], mosses [69], hornworts [70], and vascular plants [12].

**larvaceans** (*Larvacea*) Solitary, planktonic tunicates [121] that retain both notochords and nerve cords throughout their lives.

**lepidosaurs** (*Lepidosauria*) [130] Reptiles [37] with overlapping scales. Includes tuataras and squamates (lizards, snakes, and amphisbaenians).

**leptosporangiate ferns** (*Pteridopsida* or *Polypodiopsida*) [73] Vascular plants [12] usually possessing large, frondlike leaves that unfold from a "fiddlehead," and possessing thin-walled sporangia.

**life** (*Life*) [1] The monophyletic group that includes all known living organisms. Characterized by a nucleic-acid based genetic system (DNA or RNA), metabolism, and cellular structure. Some parasitic forms, such as viruses, have secondarily lost some of these features and rely on the cellular environment of their host.

**liverworts** (*Hepatophyta*) [68] Nonvascular plants lacking stomata; stalk of sporophyte elongates along its entire length.

**lobe-limbed vertebrates** (*Sarcopterygii*) [34] One of the two major groups of bony vertebrates [33], characterized by jointed appendages (paired fins or limbs).

**loboseans** (*Lobosea*) A group of unicellular amoebozoans [84]; includes the most familiar amoebas (e.g., *Amoeba proteus*).

**"lophophorates"** Convenience term used to describe several groups of lophotrochozoans [24] that have a feeding structure called a lophophore (a circular or U-shaped ridge around the mouth that bears one or two rows of ciliated, hollow tentacles). Not a monophyletic group.

**lophotrochozoans** (*Lophotrochozoa*) [24] One of the two main groups of protostomes [23]. This group is morphologically diverse, and is supported primarily on information from gene sequences. Includes bryozoans and entoprocts [99], flatworms [100], rotifers and gastrotrichs [101], ribbon worms [102], brachiopods [103], phoronids [104], annelids [105], and mollusks [106].

**loriciferans** (*Loricifera*) [109] Small (< 1 mm) ecdysozoans [25] with bodies in four parts, covered with six plates.

**low-GC Gram-positives** (*Firmicutes*) [39] A diverse group of bacteria [2] with a relatively low (G+C)/(A+T) ratio of their DNA, often but not always Gram-positive, some producing endospores.

**lungfishes** (*Dipnoi*) [127] A group of aquatic lobe-limbed vertebrates [34] that are the closest living relatives of the tetrapods [35]. They have a modified swim bladder used to absorb oxygen from air, so some species can survive the temporary drying of their habitat.

**lycophytes** (*Lycopodiophyta*) [71] Vascular plants [12] characterized by microphylls; includes club mosses, spike mosses, and quillworts.

## – M –

**magnoliids** (*Magnoliidae*) [81] A major group of angiosperms [15] possessing two cotyledons and pollen grains with a single opening. The group is defined primarily by nucleotide sequence data; it is more closely related to the eudicots and monocots than to three other small angiosperm groups.

**mammals** (*Mammalia*) [129] A group of tetrapods [35] with hair covering all or part of their skin; -females produce milk to feed their developing young. Includes the prototherians, marsupials, and -eutherians.

**mandibulates** (*Mandibulata*) [27] Arthropods [26] that include mandibles as mouth parts. Includes myriapods [115], crustaceans [116], and hexapods [117].

**marsupials** (*Marsupialia*) Mammals [129] in which the female typically has a marsupium (a pouch for rearing young, which are born at an extremely early stage in development). Includes such familiar mammals as opossums, koalas, and kangaroos.

**metazoans** (*Metazoa*) *See* animals [19].

**microbial eukaryotes** *See* "protists."

**microsporidia** (*Microsporidia*) [85] A group of parasitic unicellular fungi [17] that lack mitochondria and have walls that contain chitin.

**mollusks** (*Mollusca*) [106] One of the major groups of lophotrochozoans [24], mollusks have bodies composed of a foot, a mantle (which often secretes a hard, calcareous shell), and a visceral mass. Includes monoplacophorans, chitons, bivalves, gastropods, and cephalopods.

**monilophytes** (*Monilophyta*) A group of vascular plants [12], sister to the seed plants [13], characterized by overtopping and possession of megaphylls; includes the horsetails [72] and ferns [73].

**monocots** (*Monocotyledones*) [82] Angiosperms [15] characterized by possession of a single cotyledon, usually parallel leaf veins, a fibrous root system, pollen grains with a single opening, and floral organs usually in multiples of three.

**monogeneans** (*Monogenea*) A group of ectoparasitic flatworms [100].

**monoplacophorans** (*Monoplacophora*) Mollusks [106] with segmented body parts and a single, thin, flat, rounded, bilateral shell.

**mosses** (*Bryophyta*) [69] Nonvascular plants with true stomata and erect, "leafy" gametophytes; sporophytes elongate by apical cell division.

**moss animals** *See* bryozoans [99].

**myriapods** (*Myriapoda*) [115] Arthropods [26] characterized by an elongate, segmented trunk with many legs. Includes centipedes and millipedes.

## – N –

**nanoarchaeotes** (*Nanoarchaeota*) A group of extremely small, thermophilic archaeans [3] with a much-reduced genome. The only described example can survive only when attached to a host organism.

**nematodes** (*Nematoda*) [111] A very large group of elongate, unsegmented ecdysozoans [25] with thick, multilayer cuticles. They are among the most abundant and diverse animals, although most species have not yet been described. Include free-living predators and scavengers, as well as parasites of most species of land plants [11] and animals [19].

**neognaths** (*Neognathae*) The main group of birds [132], including all living species except the ratites (ostrich, emu, rheas, kiwis, cassowaries) and tinamous. *See* palaeognaths.

## – O –

**oligochaetes** (*Oligochaeta*) Annelid [105] group whose members lack parapodia, eyes, and anterior tentacles, and have few setae. Earthworms are the most familiar oligochaetes.

**onychophorans** (*Onychophora*) [113] Elongate, segmented ecdysozoans [25] with many pairs of soft, unjointed, claw-bearing legs. Also known as velvet worms.

**oomycetes** (*Oomycota*) [52] Water molds and relatives; absorptive heterotrophs with nutrient-absorbing, filamentous hyphae.

**opisthokonts** (*Opisthokonta*) [16] A group of eukaryotes [4] in which the flagellum on motile cells, if present, is posterior. The opisthokonts include the fungi [17], animals [19], and choanoflagellates [91].

**ostracods** (*Ostracoda*) Marine and freshwater crustaceans [116] that are laterally compressed and protected by two clamlike calcareous or chitinous shells.

## – P –

**palaeognaths** (*Palaeognathae*) A group of secondarily flightless or weakly flying birds [132]. Includes the flightless ratites (ostrich, emu, rheas, kiwis, cassowaries) and the weakly flying tinamous.

**parabasalids** (*Parabasalia*) [59] A group of unicellular eukaryotes [4] that lack mitochondria; they possess flagella in clusters near the anterior of the cell.

**phoronids** (*Phoronida*) [104] A small group of sessile, wormlike marine lophotrochozoans [24] that secrete chitinous tubes and feed using a lophophore.

**placoderms** (*Placodermi*) An extinct group of jawed vertebrates [32] that lacked teeth. Placoderms were the dominant predators in Devonian oceans.

**placozoans** (*Placozoa*) [96] A poorly known group of structurally simple, asymmetrical, flattened, transparent animals found in coastal marine tropical and subtropical seas. Most evidence suggests that placozoans are secondarily simplified eumetazoans [21].

**Plantae** [9] The most broadly defined plant group. In most parts of this book, we use the word "plant" as synonymous with "land plant" [11], a more restrictive definition.

**plasmodial slime molds** (*Myxogastrida*) Amoebozoans [84] that in their feeding stage consist of a coenocyte called a plasmodium.

**pogonophorans** (*Pogonophora*) Deep-sea annelids [105] that lack a mouth or digestive tract; they feed by taking up dissolved organic matter, facilitated by endosymbiotic bacteria in a specialized organ (the trophosome).

**polychaetes** (*Polychaeta*) A group of mostly marine annelids [105] with one or more pairs of eyes and one or more pairs of feeding tentacles; parapodia and setae extend from most body segments. May be paraphyletic with respect to the clitellates.

**priapulids** (*Priapulida*) [107] A small group of cylindrical, unsegmented, wormlike marine ecdysozoans [25] that takes its name from its phallic appearance.

**"prokaryotes"** Not a monophyletic group; as commonly used, includes the bacteria [2] and archaeans [3]. A term of convenience encompassing all cellular organisms that are not eukaryotes.

**proteobacteria** (*Proteobacteria*) [44] A large and extremely diverse group of Gram-negative bacteria that includes many pathogens, nitrogen fixers, and photosynthesizers. Includes the alpha, beta, gamma, delta, and epsilon proteobacteria.

**"protists"** This term of convenience is used to encompass a large number of distinct and distantly related groups of eukaryotes, many but far from all of which are microbial and unicellular. Essentially a "catch-all" term for any eukaryote group not contained within the land plants [11], fungi [17], or animals [19].

**protostomes** (*Protostomia*) [23] One of the two major groups of bilaterians [22]. In protostomes, the mouth typically forms from the blastopore (if present) in early development (contrast with deuterostomes). The major protostome groups are the lophotrochozoans [24] and ecdysozoans [25].

**prototherians** (*Prototheria*) A mostly extinct group of mammals [129], common during the Cretaceous and early Cenozoic. The five living species—four echidnas and the duck-billed platypus—are the only extant egg-laying mammals.

**pterobranchs** (*Pterobranchia*) A small group of sedentary marine hemichordates [119] that live in tubes secreted by the proboscis. They have one to nine pairs of arms, each bearing long tentacles that capture prey and function in gas exchange.

**pycnogonids** (*Pycnogonida*) Treated in this book as a group of chelicerates [114], but sometimes considered an independent group of arthropods [26]. Pycnogonids have reduced bodies and very long, slender legs. Also called sea spiders.

## – R –

**radiolarians** (*Radiolaria*) [55] Amoeboid organisms with needlelike pseudopods supported by microtubules. Most have glassy internal skeletons.

**ray-finned fishes** (*Actinopterygii*) [125] A highly diverse group of freshwater and marine bony vertebrates [33]. They have reduced swim bladders that often function as hydrostatic organs and fins supported by soft rays (lepidotrichia). Includes most familiar fishes.

**red algae** (*Rhodophyta*) [64] Mostly multicellular, marine and freshwater algae characterized by the presence of phycoerythrin in their chloroplasts.

**reptiles** (*Reptilia*) [37] One of the two major groups of extant amniotes [36], supported on the basis of similar skull structure and gene sequences. The term "reptiles" traditionally excluded the birds [132], but the resulting group is then clearly paraphyletic. As used in this book, the reptiles include turtles [131], lepidosaurs [130], birds [132], and crocodilians [133].

**rhizaria** (*Rhizaria*) [7] Mostly amoeboid unicellular eukaryotes with pseudopods, many with external or internal shells. Includes the foraminiferans [57], cercozoans [56], and radiolarians [55].

**rhyniophytes** (*Rhyniophyta*) A group of early vascular plants [12] that appeared in the Silurian and became extinct in the Devonian. Possessed dichotomously branching stems with terminal sporangia but no true leaves or roots.

**ribbon worms** (*Nemertea*) [102] A group of unsegmented lophotrochozoans [24] with an eversible proboscis used to capture prey. Mostly marine, but some species live in fresh water or on land.

**rotifers** (*Rotifera*) [101] Tiny (< 0.5 mm) lophotrochozoans [24] with a pseudocoelomic body cavity that functions as a hydrostatic organ, and a ciliated feeding organ called the corona that surrounds the head. Rotifers live in freshwater and wet terrestrial habitats.

**roundworms** (*Nematoda*) [111] *See* nematodes.

## – S –

**sac fungi** (*Ascomycota*) [89] Fungi that bear the products of meiosis within sacs (asci) if the organism is multicellular. Some are unicellular.

**salamanders** (*Caudata*) A group of amphibians [128] with distinct tails in both larvae and adults and limbs set at right angles to the body.

**salps** *See* thaliaceans.

**sarcopterygians** (*Sarcopterygii*) [34] *See* lobe-limbed vertebrates.

**scyphozoans** (*Scyphozoa*) Marine cnidarians [97] in which the medusa stage dominates the life cycle. Commonly known as jellyfish.

**sea cucumbers** (*Holothuroidea*) Echinoderms [118] with an elongate, cucumber-shaped body and leathery skin. They are scavengers on the ocean floor.

**sea spiders** *See* pycnogonids.

**sea squirts** *See* ascidians.

**sea stars** (*Asteroidea*) Echinoderms [118] with five (or more) fleshy "arms" radiating from an indistinct central disk. Also called starfishes.

**sea urchins** (*Echinoidea*) Echinoderms [118] with a test (shell) that is covered in spines. Most are globular in shape, although some groups (such as the sand dollars) are flattened.

**"seed ferns"** A paraphyletic group of loosely related, extinct seed plants that flourished in the Devonian and Carboniferous. Characterized by large, frondlike leaves that bore seeds.

**seed plants** (*Spermatophyta*) [13] Heterosporous vascular plants [12] that produce seeds; most produce wood; branching is axillary (not dichotomous). The major seed plant groups are gymnosperms [14] and angiosperms [15].

**sow bugs** *See* isopods.

**spirochetes** (*Spirochaetes*) [45] Motile, Gram-negative bacteria with a helically coiled structure and characterized by axial filaments.

**sponges** (*Porifera*) [20] A group of relatively asymmetric, filter-feeding animals that lack a gut or nervous system and generally lack differentiated tissues. Includes glass sponges [92], demosponges [93], and calcareous sponges [94].

**springtails** (*Collembola*) Wingless hexapods [117] with springing structures on the third and fourth segments of their bodies. Springtails are extremely abundant in some environments (especially in soil, leaf litter, and vegetation).

**squamates** (*Squamata*) The major group of lepidosaurs [130], characterized by the possession of movable quadrate bones (which allow the upper jaw to move independently of the rest of the skull) and hemipenes (a paired set of eversible penises, or penes) in males. Includes the lizards (a paraphyletic group), snakes, and amphisbaenians.

**star anise** (*Austrobaileyales*) [80] A group of woody angiosperms [15] thought to be the sister-group of the clade of flowering plants that includes eudicots [83], monocots [82], and magnoliids [81].

**starfish** (*Asteroidea*) *See* sea stars.

**stoneworts** (*Charales*) [67] Multicellular green algae with branching, apical growth and plasmodesmata between adjacent cells. The closest living relatives of the land plants [11], they retain the egg in the parent organism.

**stramenopiles** (*Heterokonta* or *Stramenopila*) [6] Organisms having, at some stage in their life cycle, two unequal flagella, the longer possessing rows of tubular hairs. Chloroplasts, when present, surrounded by four membranes. Major stramenopile groups include the brown algae [54], diatoms [53], and oomycetes [52].

## – T –

**tapeworms** (*Cestoda*) Parasitic flatworms [100] that live in the digestive tracts of vertebrates as adults, and usually in various other species of animals as juveniles.

**tardigrades** (*Tardigrada*) [112] Small (< 0.5 mm) ecdysozoans [25] with fleshy, unjointed legs and no circulatory or gas exchange organs. They live in marine sands, in temporary freshwater pools, and on the water films of plants. Also called water bears.

**tetrapods** (*Tetrapoda*) [35] The major group of lobe-limbed vertebrates [34]; includes the amphibians [128] and the amniotes [36]. Named for the presence of four jointed limbs (although limbs have been secondarily reduced or lost completely in several tetrapod groups).

**thaliaceans** (*Thaliacea*) A group of solitary or colonial planktonic marine tunicates [121]. Also called salps.

**therians** (*Theria*) Mammals [129] characterized by viviparity (live birth). Includes eutherians and marsupials.

**theropods** (*Theropoda*) Archosaurs [38] with bipedal stance, hollow bones, a furcula ("wishbone"), elongated metatarsals with three-fingered feet, and a pelvis that points backwards. Includes many well-known extinct dinosaurs (such as *Tyrannosaurus rex*), as well as the living birds [132].

**tracheophytes** *See* vascular plants [12].

**trilobites** (*Trilobita*) An extinct group of arthropods [26] related to the chelicerates [114]. Trilobites flourished from the Cambrian through the Permian.

**tuataras** (*Rhyncocephalia*) A group of lepidosaurs [130] known mostly from fossils; there are only two living tuatara species. The quadrate bone of the upper jaw is fixed firmly to the skull. Sister group of the squamates.

**tunicates** (*Tunicata*) [121] A group of chordates [30] that are mostly saclike filter feeders as adults, with motile larval stages that resemble tadpoles.

**turbellarians** (*Turbellaria*) A group of free-living, generally carnivorous flatworms [100]. Their monophyly is questionable.

**turtles** (*Testudines*) [131] A group of reptiles [37] with a bony carapace (upper shell) and plastron (lower shell) that encase the body in a fashion unique among the vertebrates.

## – U –

**urochordates** (*Tunicata*) [121] *See* tunicates.

## – V –

**vascular plants** (*Tracheophyta*) [12] Plants with xylem and phloem. Major groups include the lycophytes [71] and euphyllophytes.

**vertebrates** (*Vertebrata*) [31] The largest group of chordates [30], characterized by a rigid endoskeleton supported by the vertebral column and an anterior skull encasing a brain. Includes hagfishes [122], lampreys [123], and the jawed vertebrates [32], although some biologists exclude the hagfishes from this group. *See also* craniates.

## – W –

**water bears** *See* tardigrades.

**water lilies** (*Nymphaeaceae*) [79] A group of aquatic, freshwater angiosperms [15] that are rooted in soil in shallow water, with round floating leaves and flowers that extend above the water's surface. They are the sister-group to most of the remaining flowering plants, with the exception of the genus *Amborella* [78].

## – Y –

**"yeasts"** Convenience term for several distantly related groups of unicellular fungi [17].

## – Z –

**"zygospore fungi"** (*Zygomycota*, if monophyletic) [87] A convenience term for a probably paraphyletic group of fungi [17] in which hyphae of differing mating types conjugate to form a zygosporangium.

# Appendix B Statistics Primer

This appendix is designed to help you conduct simple statistical analyses and understand their application and importance. This introduction will help you complete the Apply the Concept and Analyze the Data problems throughout this book. The formulas for a number of statistical tests are presented here, but the presentation is designed primarily to help you understand the purpose and reasoning of the various tests. Once you understand the basis of the analysis, you may wish to use one of many free, online web sites for conducting the tests and calculating relevant test statistics (such as http://faculty.vassar.edu/lowry/VassarStats.html).

## Why Do We Do Statistics?

**ALMOST EVERYTHING VARIES** We live in a variable world, but within the variation we see among biological organisms there are predictable patterns. We use statistics to find and analyze these patterns. Consider any group of common things in nature—all women aged 22, all the cells in your liver, or all the blades of grass in your yard. Although they will have many similar characteristics, they will also have important differences. Men aged 22 tend to be taller than women aged 22, but, of course, not every man will be taller than every woman in this age group.

Natural variation can make it difficult to find general patterns. For example, scientists have determined that smoking increases the risk of getting lung cancer. But we know that not all smokers will develop lung cancer and not all nonsmokers will remain cancer-free. If we compare just one smoker to just one nonsmoker, we may end up drawing the wrong conclusion. So how did scientists discover this general pattern? How many smokers and nonsmokers did they examine before they felt confident about the risk of smoking?

*Statistics helps us to find general patterns, even when nature does not always follow those patterns.*

**AVOIDING FALSE POSITIVES AND FALSE NEGATIVES** When a woman takes a pregnancy test, there is some chance that it will be positive even if she is not pregnant, and there is some chance that it will be negative even if she is pregnant. We call these kinds of mistakes *false positives* and *false negatives*.

Doing science is a bit like taking a medical test. We observe patterns in the world, and we try to draw conclusions about how the world works from those observations. Sometimes our observations lead us to draw the wrong conclusions. We might conclude that a phenomenon occurs, when it actually does not; or we might conclude that a phenomenon does not occur, when it actually does.

For example, the planet Earth has been warming over the past century (see Concept 46.4). Ecologists are interested in whether plant and animal populations have been affected by global warming. If we have long-term information about the locations of species and temperatures in certain areas, we can determine whether species movements coincide with temperature changes. Such information can, however, be very complicated. Without proper statistical methods, one may not be able to detect the true impact of temperature or, instead, may think a pattern exists when it does not.

*Statistics helps us to avoid drawing the wrong conclusions.*

## How Does Statistics Help Us Understand the Natural World?

Statistics is essential to scientific discovery. Most biological studies involve five basic steps, each of which requires statistics:

- **Step 1: Experimental Design**
  Clearly define the scientific question and the methods necessary to tackle the question.
- **Step 2: Data Collection**
  Gather information about the natural world through experiments and field studies.
- **Step 3: Organize and Visualize the Data**
  Use tables, graphs, and other useful representations to gain intuition about the data.
- **Step 4: Summarize the Data**
  Summarize the data with a few key statistical calculations.
- **Step 5: Inferential Statistics**
  Use statistical methods to draw general conclusions from the data about the way the world works.

## Step 1: Experimental Design

We conduct experiments to gain knowledge about the world. Scientists come up with scientific ideas based on prior research and their own observations. These ideas may take the form of a question like "Does smoking cause cancer?," a hypothesis like "Smoking increases the risk of cancer," or a prediction like "If a person smokes, he/she will increase his/her chances of developing cancer." Experiments allow us to test such scientific ideas, but designing a good experiment can be quite challenging.

*We use statistics to guide us in designing experiments so that we end up with the right kinds of data.* Before embarking on an experiment, we use statistics to determine how much data will be required to test our idea, and to prevent extraneous factors from misleading us. For example, suppose we want to conduct an experiment on fertilizers to test the hypothesis that nitrogen increases plant growth. If we include too few plants, we will not be able to determine whether or not nitrogen has an effect on growth, and the experiment will be for naught. If we include too many plants, we will waste valuable time and resources. Furthermore, we should design the experiment so that we can detect differences that are actually caused by nitrogen fertilization rather than by variation, for example, in sunlight or precipitation experienced by the plants.

## Step 2: Data Collection

**TAKING SAMPLES** When biologists gather information about the natural world, they typically collect a few representative pieces of information. For example, when evaluating the efficacy of a candidate drug for medulloblastoma brain cancer, scientists may test the drug on tens or hundreds of patients, and then draw conclusions about its efficacy for all patients with these tumors. Similarly, scientists studying the relationship between body weight and clutch size (number of eggs) for female spiders of a particular species may examine tens to hundreds of spiders to make their conclusions.

We use the expression "sampling from a population" to describe this general method of taking representative pieces of information from the system under investigation (**Figure B1**). The pieces of information in a **sample** are called **observations**. In the cancer therapy example, each observation was the change in a patient's tumor size six months after initiating treatment, and the population of interest was all individuals with medulloblastoma tumors. In the spider example, each observation was a pair of measurements—body size and clutch size—for a single female spider, and the population of interest was all female spiders of this species.

Sampling is a matter of necessity, not laziness. We cannot hope (and would not want) to collect *all* of the female spiders of the species of interest on Earth! Instead, we use statistics to determine how many spiders we must collect in order to confidently infer something about the general population and then use statistics again to make such inferences.

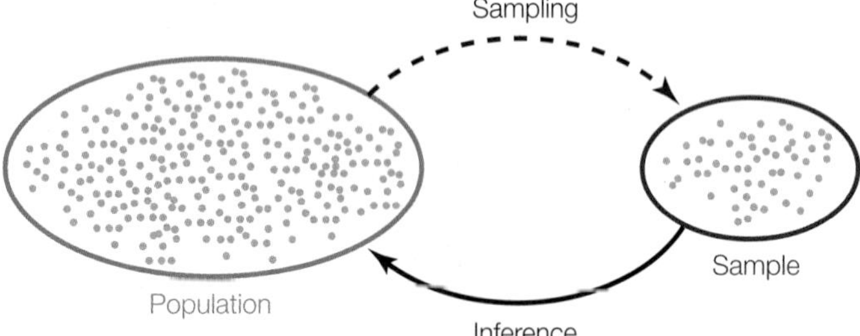


**FIGURE B1 Sampling From a Population** Biologists take representative samples from a population, use descriptive statistics to characterize their samples, and then use inferential statistics to draw conclusions about the original population.

TABLE **B1**
Poinsettia Colors

| Color | Frequency | Proportion |
|---|---|---|
| Red | 108 | 0.59 |
| Pink | 34 | 0.19 |
| White | 40 | 0.22 |
| Total | 182 | 1.00 |

**DATA COME IN ALL SHAPES AND SIZES** In statistics, we use the word *variable* to mean a measurable characteristic of an individual or a system. Some variables are on a numerical scale, like the daily high temperature (a numerical value constrained by the precision of our thermometer), or the clutch size of a spider (a whole number: 0, 1, 2, 3,...). We call these **quantitative variables**. Quantitative variables that only take on whole number values are called **discrete variables**, whereas variables that can also take on any fractional value are called **continuous variables**.

Other variables take categories as values, like a human blood type (A, B, AB, or O) or an ant caste (queen, worker, or male). We call these **categorical variables**. Categorical variables with a natural ordering, like a final grade in Introductory Biology (A, B, C, D, or F), are called **ordinal variables**.

Each class of variables comes with its own set of statistical methods. We will introduce a few common methods in this Appendix that will help you work on the problems presented in this book, but you should consult a biostatistics textbook for more advanced tests and analyses for other data sets and problems.

## Step 3: Organize and Visualize the Data

Tables and graphs can help you gain intuition about your data, design appropriate statistical tests, and anticipate the outcome of your analysis. A **frequency distribution** lists all possible values and the number of occurrences of each value in the sample.

TABLE **B2**
Fish Weights of *Abramis brama* from Lake Laengelmavesi

| Weight (grams) | Frequency | Relative Frequency |
|---|---|---|
| 201–300 | 2 | 0.06 |
| 301–400 | 3 | 0.09 |
| 401–500 | 8 | 0.24 |
| 501–600 | 3 | 0.09 |
| 601–700 | 8 | 0.24 |
| 701–800 | 3 | 0.09 |
| 801–900 | 1 | 0.03 |
| 901–1000 | 6 | 0.18 |
| Total | 34 | 1.00 |

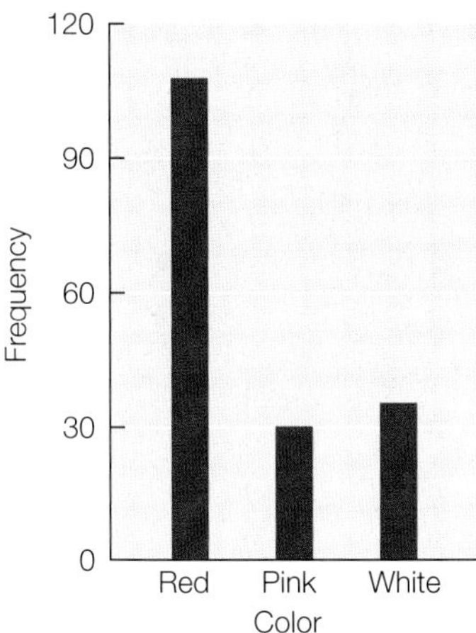


**B2 Bar Charts Compare Categorical Data** This bar chart shows the frequency of three poinsettia colors that result from an experimental cross.

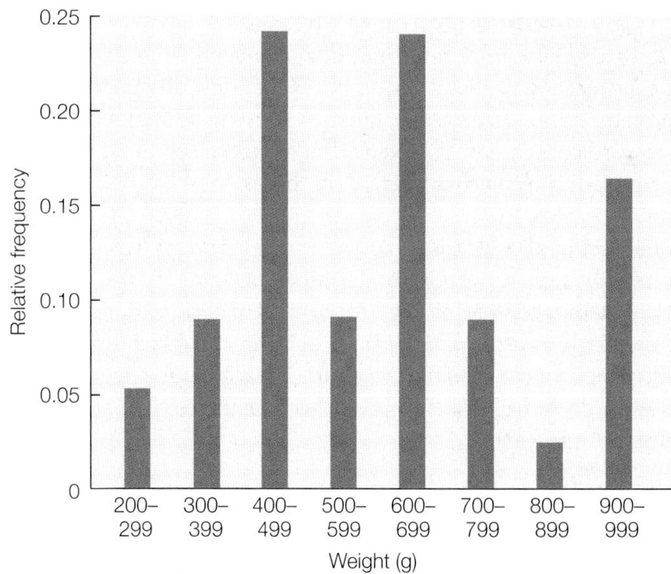


**B4 Histograms Depict Frequency Distributions of Quantitative Data** This histogram shows the relative frequency of different weight-classes of fish (*Abramis brama*).

**Table B1** shows a frequency distribution of the colors of 182 poinsettia plants (red, pink, or white) resulting from an experimental cross between two parent plants. For categorical data like this, we can visualize the frequency distribution by constructing a **bar chart**. The heights of the bars indicate the number of observations in each category (**Figure B2**). Another way to display the same data is in a **pie chart**, which shows the proportion of each category represented like pieces of a pie (**Figure B3**).

For quantitative data, it is often useful to condense your data by grouping (or binning) it into **classes**. In **Table B2**, we see a grouped frequency distribution of fish weights for a sample of 34 fish (*Abramis brama*) caught in Lake Laengelmavesi in Finland. The second column (*Frequency*) gives the number of observations in each class and the third column (*Relative Frequency*) gives the overall proportion of observations falling into each class.

**Histograms** depict frequency distributions for quantitative data. The histogram in **Figure B4** shows the relative frequencies of each weight class in this study. When grouping quantitative data, it is necessary to decide how many classes to include. It is often useful to look at multiple histograms before deciding which grouping offers the best representation of the data.

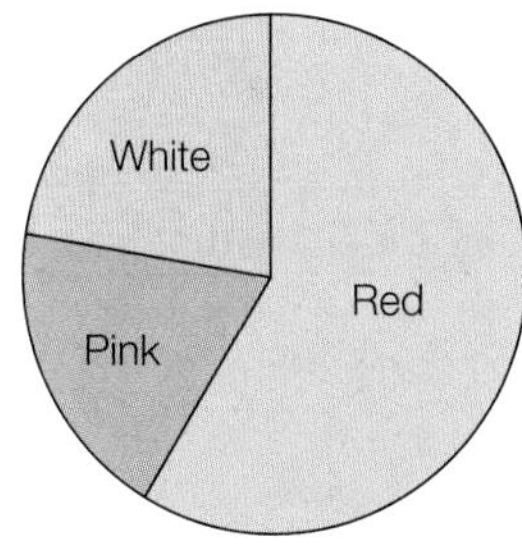


**B3 Pie Charts Show Proportions of Categories** This pie chart shows the proportions of the three poinsettia colors presented in Table B1.

Sometimes we wish to compare two quantitative variables. For example, the researchers at Lake Laengelmavesi investigated the relationship between fish weight and length and thus also measured the length of each fish. We can visualize this relationship using a **scatter plot** in which the weight and length of each fish is represented as a single point (**Figure B5**). We say that these two variables have a **linear relationship** since the points in their scatter plot fall roughly on a straight line.

Tables and graphs are critical to interpreting and communicating data, and thus should be as self-contained and comprehensible as possible. Their content should be easily understood simply by looking at them. Axes, captions, and units should be clearly labeled, statistical terms should be defined, and appropriate groupings should be used when tabulating or graphing quantitative data.

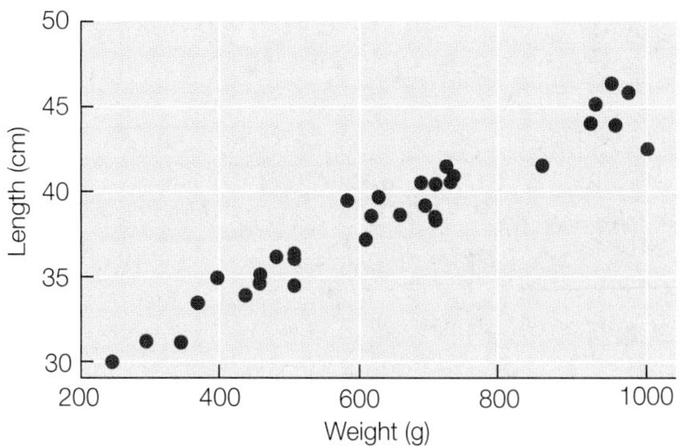


**B5 Scatter Plots Contrast Two Variables** Scatter plot of *Abramis brama* weights and lengths (measured from nose to end of tail). These two variables have a linear relationship since the data points lie close to a straight line.

## Step 4: Summarize the Data

A **statistic** is a numerical quantity calculated from data, while **descriptive statistics** are quantities that describe general patterns in data. Descriptive statistics allow us to make straightforward comparisons between different data sets and concisely communicate basic features of our data.

**DESCRIBING CATEGORICAL DATA** For categorical variables, we typically use proportions to describe our data. That is, we construct tables containing the proportions of observations in each category. For example, the third column in Table B1 provides the proportions of poinsettia plants in each color category, and the pie chart in Figure B3 provides a visual representation of those proportions.

**DESCRIBING QUANTITATIVE DATA** For quantitative data, we often start by calculating the average value or **mean** of our sample. This familiar quantity is simply the sum of all the values in the sample divided by the number of observations in our sample (**Figure B6**). The mean is only one of several quantities that roughly tell us where the *center* of our data lies. We call these quantities **measures of center**. Other commonly used measures of center are the **median**—the value that literally lies in the middle of the sample—and the **mode**—the most frequent value in the sample.

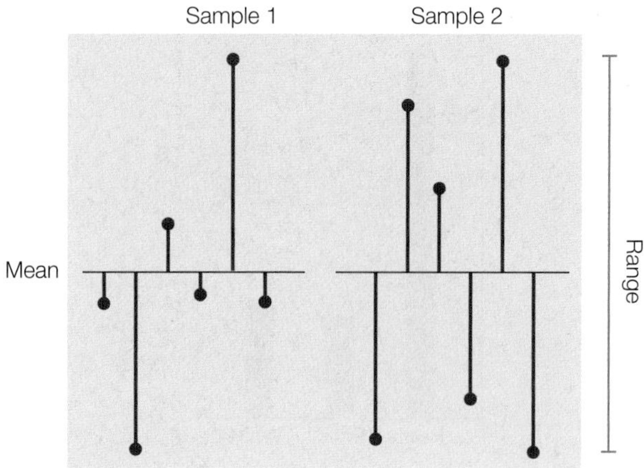


**B7 Measures of Dispersion** Two samples with the same mean (black horizontal lines) and range (blue vertical line). Red lines show the deviations of each observation from the mean. Samples with large deviations have large standard deviations. The left sample has a smaller standard deviation than the right sample.

### RESEARCH**TOOLS**

**B6 Descriptive Statistics for Quantitative Data**

Below are the equations used to calculate the descriptive statistic we discuss in this appendix. You can calculate these statistics yourself, or use free internet resources to help you make your calculations.

*Notation*:

$x_1, x_2, x_3, \ldots x_n$ are the $n$ observations of variable $X$ in the sample.

$\sum_{i=1}^{n} x_i = x_1 + x_2 + x_3, \ldots + x_n$ is the sum of all of the observations. (The Greek letter sigma, $\Sigma$, is used to denote "sum of.")

In regression, the independent variable is $X$, and the dependent variable is $Y$. $b_0$ is the vertical intercept of a regression line. $b_1$ is the slope of a regression line.

*Equations*

1. Mean: $\bar{x} = \dfrac{\sum_{i=1}^{n} x_i}{n}$

2. Standard deviation: $s = \sqrt{\dfrac{\Sigma(x_i - \bar{x})^2}{n-1}}$

3. Correlation coefficient: $r = \dfrac{\Sigma(x_i - \bar{x})(y_i - \bar{y})}{\sqrt{\Sigma(x_i - \bar{x})^2 (y_i - \bar{y})^2}}$

4. Least-squares regression line: $Y = b_0 + b_1 X$

where $b_1 = \dfrac{\Sigma(x_i - \bar{x})(y_i - \bar{y})}{\Sigma(x_i - \bar{x})^2}$ and $b_0 = \bar{y} - b_1 \bar{x}$

It is often just as important to quantify the variation in the data as it is to calculate its center. There are several statistics that tell us how much the values differ from one another. We call these **measures of dispersion**. The easiest to one understand and calculate is the **range**, which is simply the largest value in the sample minus the smallest value. The most commonly used measure of dispersion is the **standard deviation**, which calculates the extent to which the data are spread out from the mean. A deviation is the difference between an observation and the mean of the sample, and the standard deviation is a number that summarizes all of the deviations. Two samples can have the same range, but very different standard deviations if one is clustered closer to the mean than the other. In **Figure B7**, for example, the left sample has a lower standard deviation ($s = 2.6$) than the right sample ($s = 3.6$), even though the two samples have the same means and ranges.

To demonstrate these descriptive statistics, we return to the Lake Laengelmavesi study. The researchers also caught and recorded the weights of six fish in the species *Leusiscus idus*: 270, 270, 306, 540, 800, and 1,000 grams. The mean weight in this sample (equation 1 in Figure B6) is:

$$\bar{x}\frac{\Delta N}{\Delta T} = \frac{(270 + 270 + 306 + 540 + 800 + 1000)}{6} = 531$$

Since there is an even number of observations in the sample, then the median weight is the value halfway between the two middle values:

$$\frac{306 + 540}{2} = 423$$

The mode of the sample is 270, the only value that appears more than once. The standard deviation (equation 2 in Figure B6) is:

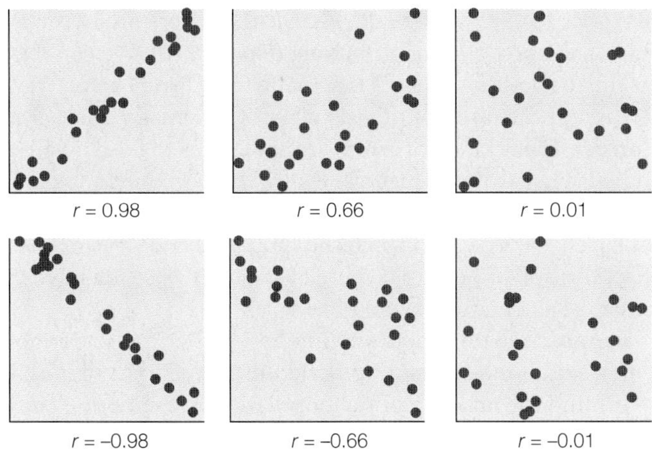


**B8 Correlation Coefficients** The correlation coefficient ($r$) indicates both the strength and the direction of the relationship.

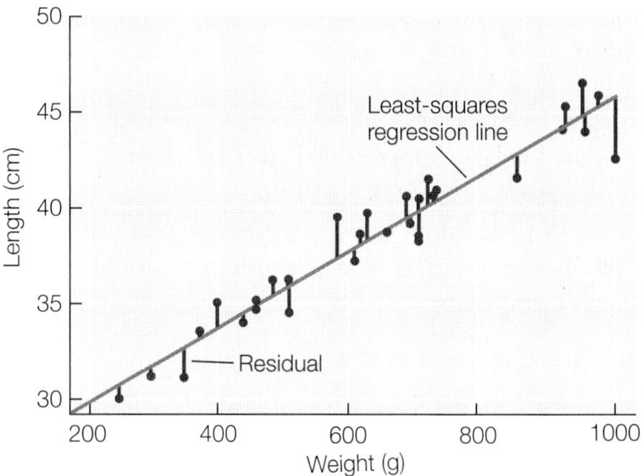


**B9 Linear Regression Estimates the Typical Relationship Between Two Variables** Linear-least squares regression line for *Abramis brama* weights and lengths (measured from nose to end of tail). The regression line (blue line) is given by the equation $Y = 26.1 + 0.02X$. It is the line that minimizes the sum of the squares of the residuals (red lines).

$$s = \sqrt{\frac{(270-531)^2 + (270-531)^2 + (306-531)^2 + (540-531)^2 + (800-531)^2 + (1000-531)^2}{5}} = 309.6$$

and the range is 1000 – 270 = 730.

**DESCRIBING THE RELATIONSHIP BETWEEN TWO QUANTITATIVE VARIABLES** Biologists are often interested in understanding the relationship between two different quantitative variables: How does the height of an organism relate to its weight? How does air pollution relate to the prevalence of asthma? How does lichen abundance relate to levels of air pollution? Recall that scatter plots visually represent such relationships.

We can quantify the strength of the relationship between two quantitative variables using a single value called the Pearson product–moment **correlation coefficient** (equation 3 in Figure B6). This statistic ranges between –1 and 1, and tells us how closely the points in a scatter plot conform to a straight line. A negative correlation coefficient indicates that one variable decreases as the other increases; a positive correlation coefficient indicates that the two variables increase together, and a correlation coefficient of zero indicates that there is no linear relationship between the two variables (**Figure B8**).

One must always keep in mind that *correlation does not mean causation*. Two variables can be closely related without one causing the other. For example, the number of cavities in a child's mouth correlates positively with the size of their feet. Clearly cavities do not enhance foot growth; nor does foot growth cause tooth decay. Instead the correlation exists because both quantities tend to increase with age.

Intuitively, the straight line that tracks the cluster of points on a scatter plot tells us something about the *typical* relationship between the two variables. Statisticians do not, however, simply eyeball the data and draw a line by hand. They often use a method called least-squares **linear regression** to fit a straight line to the data (equation 4 in Figure B6). This method calculates the line that minimizes the overall vertical distances between the points in the scatter plot and the line itself. These distances are called **residuals** (**Figure B9**). Two parameters describe the regression line: $b_0$ (the vertical intercept of the line, or the expected value of variable $Y$ when $X = 0$), and $b_1$ (the slope of the line, or how much values of $Y$ are expected to change with changes in values of $X$).

## Step 5: Inferential Statistics

Data analysis often culminates with statistical inference—an attempt to draw general conclusions about the system under investigation. As depicted in Figure B1, the primary reason we collect data is to gain insight into the larger system from which the data are collected. When we test a new medulloblastoma brain cancer drug on ten patients, we do not simply want to know the fate of those ten individuals; rather, we hope to predict its efficacy on the much larger group of all medulloblastoma patients.

**STATISTICAL HYPOTHESES** When it comes to inferring something about the real world from our data, we often have a "*Whether or not*" question in mind. For example, we would like to know whether or not global warming impacts biodiversity; whether or not the clutch size of a spider increases with body size; or whether or not soil nitrogen increases the growth of a particular plant species.

Before making statistical inferences from data, we must formalize our "*Whether or not*" question into a pair of opposing hypotheses—a **null hypothesis** (denoted $H_0$) and an **alternative hypothesis** (denoted $H_A$). The alternative hypothesis is the "*Whether*"—it is formulated to describe the effect that we expect our data to support; the null hypothesis is the "*or not*"—it is formulated to represent the absence of the effect. In other words, we typically conduct our experiment seeking to demonstrate something new (the alternative hypothesis) and thereby reject the idea that it does not occur (the null hypothesis).

Suppose, for example, we would like to know *whether or not* a new vaccine is more effective than an existing vaccine at immunizing children against influenza. Our hypotheses would be as follows:

$H_0$: The new vaccine is not more effective than the old vaccine.

$H_A$: The new vaccine is more effective than the old vaccine.

If we would like to know whether radiation increases the mutation rate in the bacteria *Escherichia coli*, we would set up the following hypotheses:

$H_0$: Radiation does not increase the mutation rate of *E. coli*.

$H_A$: Radiation does increase the mutation rate of *E. coli*.

**STATISTICAL BURDEN OF PROOF** In the U.S. justice system, people are innocent until proven guilty. In statistics, the world is *null until proven alternative*. Statistics requires overwhelming proof in favor of the alternative hypothesis before rejecting the null hypothesis. In other words, scientists favor existing ideas and resist adopting new ideas until compelling evidence suggests otherwise. This is based on a philosophy that it is worse to accept new claims when they are false than to miss out on discovering some true facts about world.

When testing a new influenza vaccine, the burden of proof is on the new vaccine. Suppose we were to vaccinate three children with the new vaccine (Group A), three with the old vaccine (Group B) and leave three children unvaccinated (Group C). If no children from Group A, one child from Group B, and one child from Group C became infected, would we have enough evidence to conclude that the new vaccine is superior to the old vaccine? No, we would not. If the study were enlarged, and two out of 100 children in group A, seven out of 100 children in group B, and 22 out of 100 children in group C become infected, would we then have sufficient evidence to choose the new vaccine? Perhaps, but we need to use statistics to be sure.

This is the traditional burden of proof in biology and science in general. As a consequence, scientists are more likely to miss out on discovering something new (and true) about the world than they are to make a false discovery. In recent years, scientists have begun to question this approach and develop an alternative statistical approach, called **Bayesian inference**, which makes it easier to favor new hypotheses. In this primer, we discuss only traditional statistical methods, often called **frequentist statistics**.

| Our conclusion | The real world: Null hypothesis true (*not more females*) | The real world: Null hypothesis false (*more females*) |
|---|---|---|
| Null hypothesis true (*not more females*) | √ | Type 2 error (*false negative*) |
| Null hypothesis false (*more females*) | Type 1 error (*false positive*) | √ |

**B10 Two Types of Error** Possible outcomes of a statistical test. Statistical inference can result in correct and incorrect conclusions about the population of interest.

**JUMPING TO THE WRONG CONCLUSIONS** There are two ways that a statistical test can go wrong (**Figure B10**). We can reject the null hypothesis when it is actually true (**Type I error**) or we can accept the null hypothesis when it is actually false (**Type II error**). These kinds of errors are analogous to false positives and false negatives in medical testing, respectively. If we mistakenly reject the null hypothesis when it is actually true, then we falsely endorse the incorrect hypothesis. If we are unable to reject the null hypothesis when it is actually false, then we fail to realize a yet undiscovered truth.

Suppose we would like to know whether there are more females than males in a population of 10,000 individuals. To determine the makeup of the population, we choose 20 individuals randomly and record their sex. Our null hypothesis is that there are *not* more females than males; and our alternative hypothesis is that there are. The following scenarios illustrate the possible mistakes we might make:

- *Scenario 1*: The population actually has 40% females and 60% males. Although our random sample of 20 people is likely to be dominated by males, it is certainly possible that, by chance, we will end up choosing more females than males. If this occurs, and we mistakenly reject the null hypothesis (that there are *not* more females than males), then we make a Type I error.
- *Scenario 2*: The population actually has 60% females and 40% males. If, by chance, we end up with a majority of males in our sample and thus fail reject the null hypothesis, then we make a Type II error.

Fortunately, statistics has been developed precisely to avoid these kinds of errors and inform us about the reliability of our conclusions. The methods are based on calculating the **probabilities** of different possible outcomes. Although you may have heard or even used the word "probability" on multiple occasions, it is important that you understand its mathematical meaning. A probability is a numerical quantity that expresses the likelihood of some event. It ranges between zero and one; zero means that there is no chance the event will occur and one means that the event is guaranteed to occur. This only makes sense if there is an element of chance, that is, if it is possible the event will occur and possible that it will not occur. For example, when we flip a fair coin, it will land on heads with probability 0.5 and land on tails with probability 0.5. When we select individuals randomly from a population with 60% females and 40% males, we will encounter a female with probability 0.6 and a male with probability 0.4.

Probability plays a very important role in statistics. To draw conclusions about the real world (the population) from our sample, we first calculate the probability of obtaining our sample if the null hypothesis is true. Specifically, statistical inference is based on answering the following question:

*Suppose the null hypothesis is true. What is the probability that a random sample would, by chance, differ from the null hypothesis as much as our sample differs from the null hypothesis?*

If our sample is highly improbable under the null hypothesis, then we rule it out in favor of our alternative hypothesis. If, instead, our sample has a reasonable probability of occurring under the null hypothesis, then we conclude that our data are consistent with the null hypothesis and we do not reject it.

Returning to the sex ratio example, we consider two new scenarios:

- *Scenario 3*: Suppose we want to infer whether or not females constitute the majority of the population (our alternative hypothesis) based on a random sample containing 12 females and eight males. We would calculate the probability that a random sample of 20 people includes at least 12 females assuming that the population, in fact, has a 50:50 sex ratio (our null hypothesis). This probability is 0.13, which is too high to rule out the null hypothesis.
- *Scenario 4*: Suppose now that our sample contains 17 females and three males. If our population is truly evenly divided, then this sample is much less likely than the sample in scenario 3. The probability of such an extreme sample is 0.0002, and would lead us to rule out the null hypothesis and conclude that there are more females than males.

This agrees with our intuition. When choosing 20 people randomly from an evenly divided population, we would be surprised if almost all of them were female, but would not be surprised at all if we ended up with a few more females than males (or a few more males than females). Exactly how many females do we need in our sample before we can confidently infer that they make up the majority of the population? And how confident are we when we reach that conclusion? Statistics allows us to answer these questions precisely.

**STATISTICAL SIGNIFICANCE: AVOIDING FALSE POSITIVES** Whenever we test hypotheses, we calculate the probability just discussed, and refer to this value as the **P-value** of our test. Specifically, the *P*-value is the probability of getting data as extreme as our data (just by chance) if the null hypothesis is, in fact, true. In other words, it is the likelihood that chance alone would produce data that differ from the null hypothesis as much as our data differ from the null hypothesis. How we measure the difference between our data and the null hypothesis depends on the kind of data in our sample (categorical or quantitative) and the nature of the null hypothesis (assertions about proportions, single variables, multiple variables, differences between variables, correlations between variables, etc.).

For many statistical tests, *P*-values can be calculated mathematically. One option is to quantify the extent to which the data depart from the null hypothesis and then use look-up tables (available in most statistics textbooks, or on the internet) to find the probability that chance alone would produce a difference of that magnitude. Most scientists, however, find *P*-values primarily by using statistical software rather than hand calculations combined with look-up tables. Regardless of the technology, the most important steps of the statistical analysis are still left to the researcher: constructing appropriate null and alternative hypotheses, choosing the correct statistical test, and drawing correct conclusions.

After we calculate a *P*-value from our data, we have to decide whether it is small enough to conclude that our data are inconsistent with the null hypothesis. This is decided by comparing the *P*-value to a threshold called the **significance level**, which is often chosen even before making any calculations. We reject the null hypothesis only when the *P*-value is less than or equal to the significance level, denoted $\alpha$. This ensures that, if the null hypothesis is true, we have at most a probability $\alpha$ of accidentally rejecting it. Therefore, the lower the value of $\alpha$, the less likely you are to make a Type I error (lower left cell of Figure B10). The most commonly used significance level is $\alpha = 0.05$, which limits the probability of a Type I error to 5%.

If our statistical test yields a *P*-value that is less than our significance level $\alpha$, then we conclude that the effect described by our alternative hypothesis is statistically significant at the level $\alpha$ and we reject the null hypothesis. If our *P*-value is greater than $\alpha$, then we conclude that we are unable to reject the null hypothesis. In this case, we do not actually reject the alternative hypothesis, rather we conclude that we do not yet have enough evidence to support it.

**POWER: AVOIDING FALSE NEGATIVES** The **power** of a statistical test is the probability that we will correctly reject the null hypothesis when it is false (lower right cell of Figure B10). Therefore, the higher the power of the test, the less likely we are to make a Type II error (upper right cell of Figure B10). The power of a test can be calculated, and such calculations can be used to improve your methodology. Generally, there are several steps that can be taken to increase power and thereby avoid false negatives:

- **Decrease the significance level**, $\alpha$. The higher the value of $\alpha$, the harder it is to reject the null hypothesis, even if it is actually false.
- **Increase the sample size**. The more data one has, the more likely one is to find evidence against the null hypothesis, if it is actually false.
- **Decrease variability in the sample**. The more variation there is in the sample, the harder it is to discern a clear effect (the alternative hypothesis) when it actually exists.

It is always a good idea to design your experiment to reduce any variability that may obscure the pattern you seek to detect. For example, it is possible that the chance of a child contracting influenza varies depending on whether he or she lives in a crowded (e.g., urban) environment or one that is less so (e.g., rural). To reduce variability, a scientist might choose to test a new influenza vaccine only on children from one environment or the other. After you have minimized such extraneous variation, you can use power calculations to choose the right combination of $\alpha$ and sample size to reduce the risks of Type I and Type II errors to desirable levels.

There is a trade-off between Type I and Type II errors: As $\alpha$ increases, the risk of a Type I decreases but the risk of a Type II error increases. As discussed above, scientists tend to be more concerned about Type I errors than Type II errors. That is, they believe that it is worse to mistakenly believe a false hypothesis than it is to fail to make a new discovery. Thus, they prefer to use low values of $\alpha$. However, there are many real-world scenarios in which it would be worse to make a Type II error than a Type I error. For example, suppose a new cold medication is being tested for dangerous (life-threatening) side effects. The null hypothesis is that there are no such side effects. A Type II error might lead regulatory agencies to approve a harmful medication that could cost human lives. In contrast, a Type I error would simply mean one less cold medication among the many that already line pharmacy shelves. In such cases, policymakers take steps to avoid a Type II error, even if, in doing so, they increase the risk of a Type I error.

**STATISTICAL INFERENCE WITH QUANTITATIVE DATA** There are many forms of statistical inference for quantitative data. When measuring a single quantitative variable, like birth weight in lambs, calcium concentration in the blood of pregnant women, or migration rate of birds, we often wish to infer the mean value of the population from which we drew the sample. However, the mean of a randomly chosen sample will not necessarily be the same or even close to the population mean. Suppose we wanted to know the average weight of newborn lambs on a particular farm. By chance, we may end up with a random sample that includes an excess of lightweight lambs and therefore a sample mean that is less than the overall mean in the population.

To infer the population mean from the sample data, we can calculate a **confidence interval for the mean**. This is a statistically derived range of values that is centered on the sample mean and is likely to include the population mean. For example, based on the sample of 34 *Abramis brama* weights from Lake Laengelmavesi (see Table B2; Figure B4), the 95% confidence interval for the mean weight ranges from 554 grams to 698 grams. The true average weight for this species of fish is likely, but not guaranteed, to fall within this range.

RESEARCH**TOOLS**

**B11 The t-Test**

What is the t-test? It is a standard method for assessing whether the means of two groups are statistically different from each another.

**Step 1:** State the null and alternative *hypotheses*:

$H_0$: The two populations have the same mean.

$H_A$: The two populations have different means.

**Step 2:** Choose a significance level, $\alpha$, to limit the risk of a Type 1 error.

**Step 3:** Calculate the *test statistic*: $t_s = \dfrac{\bar{y}_1 - \bar{y}_2}{\sqrt{\dfrac{s_1^2}{n_1} + \dfrac{s_2^2}{n_2}}}$

*Notation*: $\bar{y}_1$ and $\bar{y}_2$ are the sample means; $s_1$ and $s_2$ are the sample standard deviations; and $n_1$ and $n_2$ are the sample sizes.

**Step 4:** Use the test statistic to assess whether the data are consistent with the null hypothesis:

Calculate the *P-value* (*P*) using statistical software or by hand using statistical tables.

**Step 5:** Draw conclusions from the test:

If $P \leq \alpha$, then reject $H_0$, and conclude that the population distribution is significantly different.

If $P > \alpha$, then we do not have sufficient evidence to conclude that the means differ.

Biologists frequently wish to compare the mean values in two or more groups; for example, newborn lamb weights on several different farms, calcium concentration in women in early and late stages of pregnancy, or migration rates in birds of different species. Based on the means and standard deviations calculated for each of the samples, they infer whether or not the means in the different populations are statistically different from one another. There are several statistical methods for this, and the correct method depends on the number of groups, the experimental design, and the nature of the data.

**Figure B11** describes the steps of a *t*-test, a simple method for comparing the means in two different groups. To illustrate, we can apply a *t*-test to the Lake Laengelmavesi data to assess whether the two fish species *Abramis brama* and *Leusiscus idus* have significantly different mean weights. We begin by stating our hypotheses and choosing a significance level:

$H_0$: *Abramis brama* and *Leusiscus idus* have the same mean weight.

$H_A$: *Abramis brama* and *Leusiscus idus* have different mean weights.

$\alpha = 0.05$

The test statistic is calculated using the means, standard deviations, and sizes of the two samples:

$$t_s = \frac{626 - 531}{\sqrt{\dfrac{207^2}{34} + \sqrt{\dfrac{310^2}{6}}}} = 0.724$$

We can use statistical software or one of the free statistical sites on the internet to find the *P*-value for this result to be $P = 0.497$. Since *P* is considerably greater than $\alpha$, we fail to reject the null hypothesis and conclude that our study does not provide evidence that the two species have different mean weights.

You may want to consult an introductory statistics textbook to learn more about confidence intervals, *t*-tests, and other basic statistical tests.

**STATISTICAL INFERENCE WITH CATEGORICAL DATA** With categorical data, we often wish to infer the distribution of the different categories within the populations from which our samples are drawn. In the simplest case, we have a single categorical variable with two or more categories. If there are just two categories, we can construct a **confidence interval for the proportion** of the population that belongs to one of the two categories. This is a statistically derived range of values that is centered on the sample proportion and is likely to include the population proportion. If there are three or more categories, we can use a **chi-square goodness-of-fit** test to determine whether the distribution of the different categories in the population is consistent with a specific distribution.

**Figure B12** outlines the steps of a chi-square goodness-of-fit test. As an example, consider the data described in Table B1. Many plant species have simple Mendelian genetic systems in which parent plants produce progeny with three different colors of flowers in a ratio of 2:1:1. However, a botanist believes that these particular poinsettia plants have a different genetic system that does not produce a 2:1:1 ratio of red, pink, and

RESEARCH**TOOLS**

### B12 The Chi-Square Goodness-of-Fit Test

What is the chi-square goodness-of-fit test? It is a standard method for assessing whether a sample came from a population with a specific distribution.

**Step 1:** State the null and alternative *hypotheses*:

$H_0$: The population has the specified distribution.

$H_A$: The population does not have the specified distribution.

**Step 2:** Choose a significance level, $\alpha$, to limit the risk of a Type 1 error.

**Step 3:** Determine the *observed frequency* and *expected frequency* for each category:

The observed frequency of a category is simply the number of observations in the sample of that type.

The expected frequency of a category is the probability of the category specified in $H_0$ multiplied by the overall sample size.

**Step 4:** Calculate the *test statistic*: $\chi_s^2 = \sum_{i=1}^{c} \frac{(O_i - E_i)^2}{E_i}$

*Notation*: $C$ is the total number of categories, $O_i$ is the observed frequency of category *i*, and $E_1$ is the expected frequency of category *i*.

**Step 5:** Use the test statistic to assess whether the data are consistent with the null hypothesis:

Calculate the *P-value* (*P*) using statistical software or by hand using statistical tables.

**Step 6:** Draw conclusions from the test:

If $P \leq \alpha$, then reject $H_0$, and conclude that the population distribution is significantly different than the distribution specified by $H_0$.

If $P > \alpha$, then we do not have sufficient evidence to conclude that population has a different distribution.

white plants. A chi-square goodness-of-fit can be used to assess whether or not the data are consistent with this ratio, and thus whether or not this simple genetic explanation is valid. We start by stating our hypotheses and significance level:

$H_0$: The progeny of this type of cross have the following probabilities of each flower color:

Pr{Red} = .50, Pr{Pink} = .25, Pr{White} = .25

$H_A$: At least one of the probabilities of $H_0$ is incorrect.

$\alpha = 0.05$

We next use the probabilities in $H_0$ and the sample size to calculate the expected frequencies:

| | Red | Pink | White |
|---|---|---|---|
| Observed | 108 | 34 | 40 |
| Expected | (.50)(182) = 91 | (.25)(182) = 45.5 | (.25)(182) = 45.5 |

Based on these quantities, we calculate the chi-square test statistic:

$$\chi_s^2 = \sum_{i=1}^{C} \frac{(O_i - E_i)^2}{E_i} = \frac{(108-91)^2}{91} + \frac{(34-45.5)^2}{45.5} + \frac{(40-45.5)^2}{45.5} = 6.747$$

We find the $P$-value for this result to be $P = 0.0343$ using statistical software. Since $P$ is less than $\alpha$, we reject the null hypothesis and conclude that the botanist is correct: The plant color patterns cannot be explained by the simple Mendelian genetic model under consideration.

This introduction is only meant to provide a brief introduction to the concepts of statistical analysis, with a few example tests. **Figure B13** provides a summary of some of the commonly used statistical tests that you may encounter in biological studies.

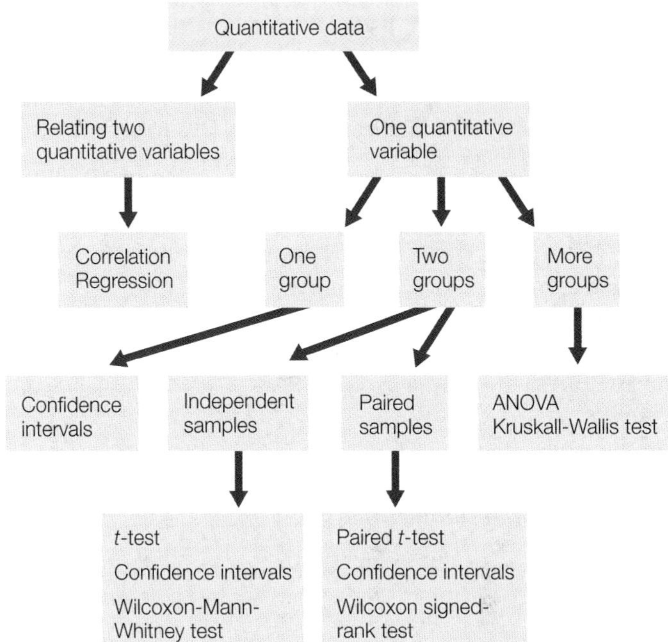


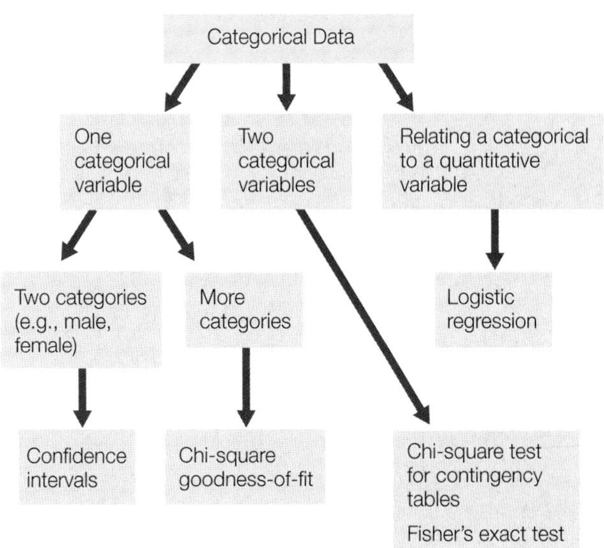


**B13 Some Common Methods of Statistical Inference**
This flow chart shows some of the commonly used methods of statistical inference for different combinations of data. Detailed descriptions of these methods can be found in most introductory biostatistics textbooks.

# Appendix C Some Measurements Used in Biology

| MEASURES OF | UNIT | EQUIVALENTS | METRIC → ENGLISH CONVERSION |
|---|---|---|---|
| Length | meter (m) | base unit | 1 m = 39.37 inches = 3.28 feet = 1.196 yards |
| | kilometer (km) | 1 km = 1000 ($10^3$) m | 1 km = 0.62 miles |
| | centimeter (cm) | 1 cm = 0.01 ($10^{-2}$) m | 1 cm = 0.39 inches |
| | millimeter (mm) | 1 mm = 0.1 cm = $10^{-3}$ m | 1 mm = 0.039 inches |
| | micrometer (µm) | 1 µm = 0.001 mm = $10^{-6}$ m | |
| | nanometer (nm) | 1 nm = 0.001 µm = $10^{-9}$ m | |
| Area | square meter ($m^2$) | base unit | 1 $m^2$ = 1.196 square yards |
| | hectare (ha) | 1 ha = 10,000 $m^2$ | 1 ha = 2.47 acres |
| Volume | liter (L) | base unit | 1 L = 1.06 quarts |
| | milliliter (mL) | 1 mL = 0.001 L = $10^{-3}$ L | 1 mL = 0.034 fluid ounces |
| | microliter (µL) | 1 µL = 0.001 mL = $10^{-6}$ L | |
| Mass | gram (g) | base unit | 1 g = 0.035 ounces |
| | kilogram (kg) | 1 kg = 1000 g | 1 kg = 2.20 pounds |
| | metric ton (mt) | 1 mt = 1000 kg | 1 mt = 2,200 pounds = 1.10 ton |
| | milligram (mg) | 1 mg = 0.001 g = $10^{-3}$ g | |
| | microgram (µg) | 1 µg = 0.001 mg = $10^{-6}$ g | |
| Temperature | degree Celsius (°C) | base unit | °C = (°F – 32)/1.8 |
| | | | 0°C = 32°F (water freezes) |
| | | | 100°C = 212°F (water boils) |
| | | | 20°C = 68°F ("room temperature") |
| | | | 37°C = 98.6°F (human internal body temperature) |
| | Kelvin (K)* | K = °C – 273 | 0 K = –460°F |
| Energy | joule (J) | | 1 J ≈ 0.24 calorie = 0.00024 kilocalorie† |

*0 K (–273°C) is "absolute zero," a temperature at which molecular oscillations approach 0—that is, the point at which motion all but stops.

†A *calorie* is the amount of heat necessary to raise the temperature of 1 gram of water 1°C. The *kilocalorie*, or nutritionist's calorie, is what we commonly think of as a calorie in terms of food.

# Answers to Chapter Review Questions

## CHAPTER 1

1. e 2. b 3. e 4. d 5. e

6. In science, we formulate hypotheses about how the world works, then try to reject those hypotheses with experiments. The experiments must be designed so that we would expect them to uncover problems with our hypothesis. If the experiments are incapable of rejecting a hypothesis, then the experiments are not a rigorous test of the hypothesis.
7. The independent DNA found in mitochondria and chloroplasts is evidence of the origin of these eukaryotic organelles from ancient bacteria that became incorporated in the eukaryotic cell. Since the ancestors of these organelles once existed as independent organisms, they have their own genomes.
8. Controlled experiments, by definition, are able to control many variables in carefully maintained experiments, often in laboratory conditions. Comparative experiments, in contrast, often contain many additional variables that cannot be controlled by the investigator. Comparative experiments often incorporate realistic variation from uncontrolled factors, which accounts for their higher overall variability.
9. If two species share particular changes in the gene we compare, and those changes are not shared by other species we examine, we would expect the two species with the common changes to be more closely related to one another. By comparing many such changes in many genes, we can group species based on their relative evolutionary divergence from one another. For example, we share more changes in our genes with chimpanzees than we do with gorillas. From this, we can deduce that humans and chimpanzees shared a more recent common ancestor than they shared with gorillas.
10. Mitochondrial DNA is often used to follow the history of maternal lineages in a population or species. Nuclear DNA is not used in such cases because it is typically inherited from both parents. This difference can be useful in many circumstances. For example, we might examine a hybrid individual between two species. Equal portions of nuclear DNA from both species could confirm that the individual is a direct hybrid between the two species. If we examine the mitochondrial DNA, however, we can learn which of the two parental species was the female in the cross—and therefore learn by default which was the male.

## CHAPTER 2

1. b 2. d 3. c 4. c 5. a 6. d

7.

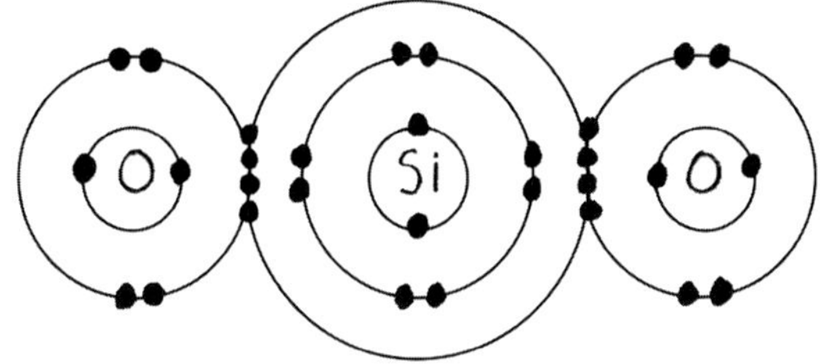


8. An easy way to answer this question is to make a simple table:

| | Covalent H—H | Hydrogen H····O |
|---|---|---|
| Electrons | Shared | Remain with H and O |
| Polarity | Nonpolar | Polar; + at H end |
| Strength | Stronger | Weaker |

9. C—H: nonpolar; hydrophobic
C═O: polar; δ– at O; hydrophilic
O—P: polar; δ– at O; hydrophilic
C—C: nonpolar; hydrophobic
10. This is an example of Van der Waals forces, which act over a short distance and do not involve polarity.
11. The human body has the same elements as Earth's crust but in very different proportions.

## CHAPTER 3

1. e 2. e 3. c 4. a 5. c 6. b

7. The observations support explanation "a." Glycine is small and nonpolar. Glutamic acid and arginine are larger and polar (charged). Serine and alanine are small: the protein retains its shape. But serine is polar (it has –OH as its R group), and that does not affect the structure. Valine is larger and nonpolar, and this affects shape. So the issue is size.
8. Mannose and galactose have the same atomic formula, $C_6H_{12}O_6$, but the arrangement of atoms is different: compare carbons 2 and 4. These sugars have the hydroxyl (–OH) functional group. Its polarity helps the sugars dissolve in water. The –OH group also can participate in bonding the sugar to other molecules through condensation reactions (see Figures 3.4 and 3.17).
9. High temperature disrupts weak interactions such as hydrogen bonds. Heat shock proteins might work by stabilizing the protein so that the weak interactions are not necessary to preserve its structure.
10. A change from lysine is a change in primary structure. The change could affect tertiary structure if the protein folds as a result of electrostatic attractions between charged amino acids (+ to –). In this case, the presence of a negatively charged amino acid (aspartic acid) where there should be a positively charged one (lysine) might prevent correct folding if a negatively charged amino acid elsewhere in the polypeptide chain is involved in folding (it is attracted to a + amino acid). The same forces might be at work in the interaction of separate chains for quaternary structure.
11. See Figure 3.10. Heat breaks hydrogen bonds and other weak interactions that maintain protein shape. Disulfide bonds also required for normal protein shape. Styling and perms partially denature keratin, then renature the protein in a new shape. Your investigation might involve measuring keratin protein structure of hair before and after disrupting hydrogen bonds and disulfide bonds.

## CHAPTER 4

1. c 2. c 3. c 4. c 5. b 6. b

7. The presence of $O_2$ in the atmosphere produces an oxidizing condition that prevents the reduction reactions noted by the Miller–Urey experiment.
8. Oligonucleotides of RNA can fold because of hydrogen bonds forming between bases on the single chain and to a lesser extent because of weak interactions of base stacking when bases come near one another. Short strands of about 20 oligonucleotides are enough to produce uniquely folded RNA.
9. Cells provided concentration and compartmentation chemicals for the reactions needed for life, as well as differential permeability to distinguish life's chemical composition from that of the environment.
10. If microbes survived heat, the initial part of Pasteur's experiment might begin with microbes already present. They would grow in both the open and closed flasks. To get the results he did, Pasteur's flasks must not have contained such microbes. An answer for the proposed experiment on heat-stable microbes might be to inactivate them using reagents, such as mercaptoethanol, that destroy proteins.
11. A suggested experiment might be to dry the samples after the Miller–Urey experiment (allowing condensation reactions—polymerization) and then apply energy in the form of heat. This condition might have existed in volcanic rock in early Earth.

## CHAPTER 5

1. b 2. d 3. e 4. a 5. d 6. b 7. a

8. Four membranes: two in the chloroplast and two in the mitochondrion
Two membranes: the lysosomal membrane and the plasma membrane (via vesicle; the molecules do not themselves cross any membranes)

No membranes: ribosomes do not have membranes. However, if the ribosomes were associated with the endoplasmic reticulum (ER), the answer would be two membranes: into the ER and out of the ER.

9.

| | Animal Cell ECM | Plant Cell Wall |
|---|---|---|
| Composition | Collagen fibers in proteoglycan matrix | Cellulose fibers in polysaccharide and protein matrix |
| Rigidity | Less rigid | More rigid (especially secondary cell walls) |
| Connections | Some specialized proteins and junctions | Plasmodesmata |

10. Microtubules line the long axons of nerve cells, where they act as tracks for vesicles that carry substances down the neuron. Without microtubules, the contents of these vesicles cannot be delivered to their destination, which can result in nerve problems.

    Microtubules are a key part of the mitotic spindle, which is used to move chromosomes during cell division. Depolymerization of microtubules can thus result in loss of dividing cells.

11. For a lysosomal enzyme, the pathway would be ribosome → interior of ER → Golgi → Golgi vesicles → lysosome.

    For an extracellular protein (animal cells), the pathway would be ribosome → interior of ER → Golgi → Golgi vesicles → plasma membrane → extracellular region.

## CHAPTER 6

1. c 2. a 3. d 4. c 5. b 6. e 7. c

8. The pumping of $Ca^{2+}$ requires a lipid bilayer membrane to separate compartments, a protein pump in the membrane, and ATP to provide energy for pumping.

9. Diatom wall components move from the Golgi apparatus to the cell wall by exocytosis.

10. Living in a hypotonic environment (cells hypertonic) results in a tendency for water to enter the organism by osmosis, which can cause swelling and dilute cell contents. Some organisms get around this by using reverse pinocytosis (exocytosis) to remove fluid.

11. Experiments might involve the following:

    To measure membrane fluidity, label a small amount of a lipid or protein with a dye and allow it to incorporate into a cell's membrane. This may make a localized labeled spot on the cell. The localized region will be seen to diffuse over the cell over time. In the cancer cells, this rate of diffusion may be faster.

    To measure cell adhesion, dissociate cancer and normal tissue cells. Incubate for a period of time and determine the rate at which cancer cells and normal lung cells bind to cells from the other tissues besides lung. The cancer cells may bind to a greater extent than normal cells.

## CHAPTER 7

1. d 2. c 3. d 4. a 5. d 6. a 7. d 8. c

9. Different cells can have different target molecules to which cAMP binds, and these target molecules can have different activities and functions. Binding of cAMP changes the structure (e.g., tertiary structure of a protein) and therefore the function of a target molecule. So cAMP can have many effects.

10. Characteristics of direct communication: the size of signal molecules is limited by the size of openings between cells, it is not specific, it is fast, and there can be cytoplasmic connection between cells.

    Characteristics of receptor-mediated communication: the signal molecules can be larger, it is specific, it is slower, and there is no direct cytoplasmic connection.

    Direct communication is useful for a rapid, coordinated response of many cells.

11. See Figure 7.10. A mutation of the *Raf* gene that activates cell division might involve a protein product that does not need binding of Ras to be active. Cell division would occur without activated Ras, thereby eliminating the need for growth factor binding.

    A mutation of the *MAP kinase* gene would stimulate cell division if the resulting MAP kinase protein did not need to be phosphorylated by MEK to be active. No signaling cascade would be needed for the mutant protein to enter the nucleus and stimulate cell division.

12. Experiments might involve applying a solution containing the antibody to the upper part of the *Hydra* body. The antibody would block diffusion of the signal molecule from the apex to the upper body and—if the hypothesis is correct—would allow a bud to form in the upper body. A sham experiment, in which the solution without antibody is applied, would be a control. In this case, a bud would not form in the upper body.

## CHAPTER 8

1. c 2. e 3. c 4. c 5. d 6. d

7. Endergonic reactions are coupled in time and space with exergonic reactions, which release the energy needed for the endergonic reactions.

8. A cytoplasmic enzyme generally has a globular structure with a hydrophilic exterior and an active site for substrate binding. An ion channel generally has a more linear structure with a hydrophobic membrane-spanning region and no active site.

9.

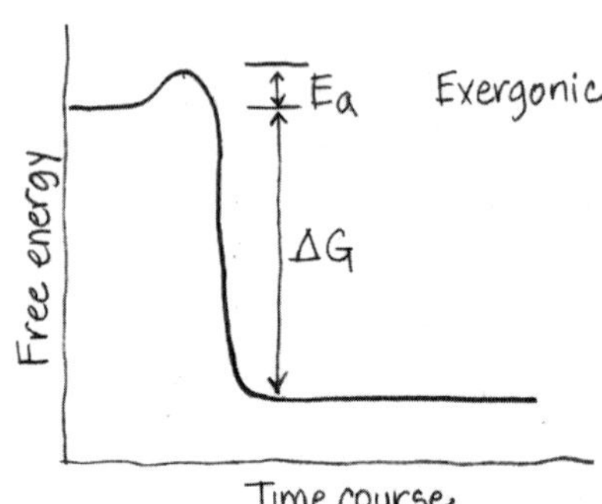


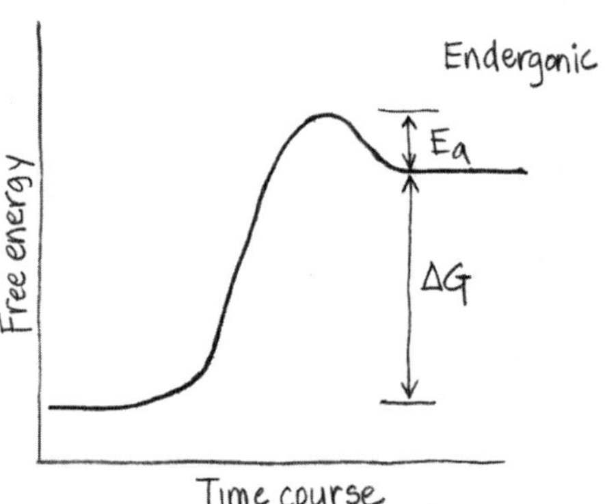


10. (a) The presence of water may prevent $O_2$ from reaching the enzyme. (b) Boiling denatures proteins, so polyphenol oxidase is irreversibly altered by boiling and its active site is destroyed. (c) Proteins have an optimal pH at which ionized R groups are appropriately charged to give the protein its tertiary structure. A pH 3 of may not be that optimal pH for polyphenol oxidase, so the enzyme is denatured and inactive.

11. See Figure 8.17. A competitive inhibitor binds to the active site of the enzyme and shifts the equilibrium to enzyme molecules in the active form.

12. To determine whether catalase has an allosteric or nonallosteric mechanism, perform an experiment with varying amounts of substrate and plot rate of catalase versus substrate concentration. An S-shaped curve will indicate an allosteric mechanism. A hyperbolic curve indicates a nonallosteric enzyme.

    To determine if a pollutant is a competitive or noncompetitive inhibitor, add the pollutant to the catalase to lower the rate of reaction, then add increasing amounts of substrate. A competitive inhibitor will be removed from the active site and the rate of reaction will increase. A noncompetitive inhibitor will not allow the rate to increase as more substrate is added. (There are more sophisticated kinetic experiments that you will learn in a biochemistry course).

## CHAPTER 9

1. d 2. d 3. e 4. c 5. d 6. a

7. If cytochrome *c* remains reduced and cannot accept electrons, the electron transport (respiratory) chain stays reduced and NADH and $FADH_2$ remain reduced. This prevents oxidation reactions in the citric acid cycle and pyruvate oxidation, so pyruvate cannot be converted to acetyl CoA. Instead, pyruvate is converted to lactic acid, regenerating some NAD that can be used so that glycolysis can continue. Because the electron transport chain is not working, there is no proton gradient set up in the mitochondria, and ATP is not made by oxidative phosphorylation.

8. See Figure 9.13. Some amino acids are converted to intermediates of glycolysis. Once they enter glycolysis these intermediates are further metabolized to a

glycolytic intermediate that can be converted to glycerol, which is incorporated into triglycerides. Glycolysis and pyruvate oxidation produce acetyl CoA, which is converted to fatty acids and incorporated into lipids.

Glucose is converted in glycolysis to acetyl CoA, which is then converted to fatty acids as above.

9. (a) Oxidation (removal of H from C2 and C3 of succinate)
   (b) Exergonic (because it is an oxidation)
   (c) It requires the redox coenzyme NAD or FAD.
   (d) The fumarate is converted to other intermediates that regenerate oxaloacetate, the acceptor for the citric acid cycle.
   (e) The reduced coenzyme (NADH or FADH2) is reoxidized in the electron transport chain.
10. Anaerobes use alternate electron acceptors to generate energy, such as sulfur, sulfate, and nitrate. Also, they use substrate-level phosphorylation (direct transfer of phosphate to ADP) to make ATP.
11. The proton gradient in the experiment described in Figure 9.9 was generated artificially from the solution and did not require electron transport (a respiratory chain). The presence of antimycin A thus would have no effect on the experiment.

## CHAPTER 10

1. e 2. b 3. d 4. d 5. d 6. d

7. In the dark, photosynthetic electron transport stops at photosystem II → reduced PQ (plastoquinone). Initially, the chlorophylls in light-harvesting complexes remain reduced, so reaction-center chlorophylls remain reduced and thus photosystem II remains reduced.

   In the dark, the Calvin cycle stops at the reduction phase, which requires NADH. No RuBP is regenerated, so there is no rubisco activity. The initial reactions are no oxidation of photosystem I, and no reduction of NADP to NADPH.
8. These processes can be compared using a table:

| | Cyclic Electron Transport | Noncyclic Electron Transport |
|---|---|---|
| Products | ATP | ATP, NADPH, $O_2$ |
| Source of electrons | Electron transport | Electron transport (photosystem I) or water (photosystem II) |

9. See Figure 10.18. $CO_2$ carbons end up in 3PG, which is converted to pyruvate. Pyruvate goes to the citric acid cycle, where some of the intermediates are converted to amino acids, which are incorporated into protein.

   In the Calvin cycle, some 3PG is converted to G3P, which can enter glycolysis. Some of the intermediates of glycolysis are converted to amino acids, which are incorporated into protein.
10. (a) $O_2$
   (b) NADPH
   (c) 3PG
11. (a) Here is the pathway followed by $^{14}C$: $^{14}CO_2$ → cells → photosynthesis → carbohydrate → combustion → $^{14}CO_2$. Release of $^{14}CO_2$ upon combustion would be evidence of photosynthesis (and life).
   (b) In this case: $^{14}CO_2$ → heat denatured cells, no photosynthesis. If living things were present, $^{14}CO_2$ would be released in experiment (a), but not in experiment (b).

## CHAPTER 11

1. d 2. b 3. d 4. d 5. e 6. d 7. d 8. c

9. See Figure 11.19. In mitotic prophase, there is no pairing of homologous chromosomes, and crossing over is rare. In meiotic prophase I, homologous pairs of chromosomes align, and crossing over is common.

   In mitotic anaphase, sister chromatids separate, with one going to each pole. In meiotic anaphase I, sister chromatids do not separate; homologous pairs of chromosomes separate, with one pair going to each pole.
10. Normally, p53 induces expression of p21, which binds to the G1/S Cdk and prevents cyclin from activating it. Without active Cdk, the cell cycle ceases. If p53 is mutated such that it is nonfunctional, p21 is not induced and the cyclin–Cdk complex can form and stimulate the cell cycle at S phase.
11. Cancers often have multiple mutations in different cells of the tumor. If some of these mutations affect different parts of the cell cycle, targeting the different phases may be a useful therapy.
12. Your proposed experiments should involve isolating the synchronous meiotic cells from the lily anthers and establishing them in the lab. As the cells proceed through the meiotic cell cycle they can be analyzed at different stages for the presence and biochemical activity of various cyclins and Cdks.

## CHAPTER 12

1. e 2. a 3. d 4. d 5. d 6. b 7. b 8. d 9. b

10. $BB \times bb$; $bb \times bb$; $Bb \times bb$; $Bb \times Bb$
11. 1/32
12. (a) Autosomal dominant
   (b) 1/4
13. (a) Males (XY) contain only one allele and will show only one color, black ($X^BY$) or yellow ($X^bY$). Females can be heterozygous ($X^BX^b$).
   (b) $X^bY$, yellow
14. The body color ($G/g$) and wing size ($A/a$) genes are linked; eye color ($R/r$) is unlinked to the other two genes. The distance between the linked genes is 18.5 units.
15. Yellow, blue, and white in a 1:2:1 ratio.
16. $F_1$ will all be wild type, *PpSwsw*. $F_2$ will have phenotypes in the ratio 9:3:3:1; see Figure 12.6 (p. 238) for analogous genotypes.
17. (a) $F_1$ will all be *PpByby* and will have wild-type eye color and wings. The ratio of phenotypes in $F_2$ will be 3:1, *PPByBy* (wild-type eyes and wings) to *ppbyby* (pink eyes and blistery wings.
   (b) $F_1$ will all be *PpbyBy* with wild-type eye color and wings; they will produce just two kinds of gametes (*Pby* and *pBy*). Combine them carefully and see the 1:2:1 phenotypic ratio fall out in the $F_2$: 1 wild-type eyes/blistery wings : 2 wild-type eyes/wild-type wings : 1 pink eyes/wild-type wings.
   (c) Pink–blistery
   (d) See Figures 11.16 and 11.18 (pp. 220–222). Crossing over took place in $F_1$.
18. *Rraa* and *RrAa*
19. (a) $w^+ > w^e > w$
   (b) The parents are $w^ew$ and $w^+Y$. The progeny are $w^+w^e$, $w^+w$, $w^eY$, and $wY$.
20. (a) $BX^a$, $BY$, $bX^a$, $bY$
   (b) The mother is $bbX^AX^a$, the father $BbX^aY$, the son $BbX^aY$, and the daughter $bbX^aX^a$.
21. 75 percent
22. Because the gene is carried on mitochondrial DNA, it is passed through the mother only. Thus if the women does not have the disease but her husband does, their child will not be affected. However, if the woman has the disease but her husband does not, their child will have the disease.
23. The cross *RRYY* × *rryy* produces *RrYy* (round, yellow) $F_1$ offspring. If the seed shape and seed color genes were linked with no recombination between them, the $F_2$ would also be all *RrYy*. A distance of 10 map units between two genes means that on average 10% of the $F_2$ offspring will have recombinant phenotypes, in this case round green (5%) and wrinkled yellow (5%).

   The cross in Figure 12.19 is *BbVgvg* (gray, normal) × *bbvgvg* (black, vestigial). If there were no linkage between the genes, then the gray, normal parent would produce four types of gametes: *BVg*, *bVg*, *Bvg*, and *bvg*. When these combine with the *bvg* gametes produced by the other parent, four types of offspring in a 1:1:1:1 ratio will result: *BbVgvg* (gray, normal), *bbVgvg* (black, normal), *Bbvgvg* (gray, vestigial), and *bbvgvg* (black, vestigial).

## CHAPTER 13

1. a 2. c 3. b 4. b 5. d 6. c 7. d

8. At 3,000 bp per minute in two directions, each origin grows at 6,000 bp per minute. There are 300 minutes in S phase, so the total bp possible for one origin is (300 × 6,000) = 1,800,000. If there are 120 million bp to replicate, then the total number of origins is 120 million/1.8 million = 66 origins. If there are 3 μm of DNA, this means there are about 22 origins per micrometer of DNA.
9. DNA replication adds new nucleotides to the 3′ end of DNA, where there is an —OH group on the sugar at the 3′ position. If there is no —OH group, there cannot be a condensation reaction and formation of a bond to the next nucleotide, so replication stops.
10. After ten rounds there would still have been some DNA (about 1/512th) as hybrid because the original heavy DNA template strands would still have been there. This tiny amount might not have been detectable in the centrifuge, however.
11. The proposed experiments might use S strain pneumococcus and transform R strain as in Figure 13.1. Incubate separate batches of S strain bacteria in $^{32}P$ or $^{35}S$. Make cell-free extracts of the S strains. Incubate with R cells and look for their transformation to the S phenotype. Then check to see if there is $^{32}P$ or $^{35}S$ label in the newly transformed cells. It would be expected that only $^{32}P$ label (DNA) would enter the cells.

## CHAPTER 14

1. b 2. a 3. d 4. b 5. d 6. d 7. d 8. e

9. For 192 amino acids, the triplet genetic code mandates 576 bp of coding sequence. Add the start and stop codons and the total is 582. This is shorter than the actual DNA gene because of promoter and terminator of transcription sequences; introns; and ribosome binding sequences. All except the transcription signals are transcribed into the pre-mRNA. The mature mRNA has the introns removed.

10. Errors in transcription can be tolerated because many copies of each RNA are made; if a few have errors, there are enough perfect ones to overcome any problem. Errors in DNA replication are harmful because DNA is replicated only once in the life of the cell.

11. In the poly CA experiment, threonine is ACA or CAC and histidine is ACA or CAC. In the poly CAA experiment, threonine is CAA, ACA, or AAC. Therefore in the first experiment threonine must be ACA and histidine CAC.

12. Enzymes: 4 → 2 → 3 → 1 → 5

    Compounds: C → F → E → D → G → T

## CHAPTER 15

1. a 2. c 3. b 4. b 5. d 6. b

7. (a) In a loss of function mutation, a phenotype is not present; for example, there may be a loss of enzyme activity. In a gain of function mutation, a new phenotype is present; for example, a new signaling protein may be active.

   (b) In a missense mutation, a single base pair change results in a codon change and thus an amino acid change in a protein. In a nonsense mutation, a single base pair change results in a codon change to a stop codon and thus premature termination of a protein.

   (c) In a spontaneous mutation, DNA changes as a result of unprovoked chemical changes or replication errors. In an induced mutation, DNA changes as a result of outside physical or chemical agents.

8. (a) The mutation that leads to PKU is rare in the human population; most people do not have the harmful allele and the highest probability is that the father is homozygous normal. Because the mother has PKU (she is homozygous mutant), the developing fetus is heterozygous.

   (b) High levels of phenylalanine cause brain damage. If the mother's phenylalanine levels were too high, the baby would be born with brain problems.

   (c) The woman should be on a phenylalanine-restricted diet.

9. Testing for the cystic fibrosis (CF) allele could be done by allele-specific oligonucleotide hybridization with probes for the normal and CF alleles; see Figure 15.18. Or direct DNA sequencing of the CF gene could be done. A person who is a carrier will test positive for both the normal and the mutant alleles.

   To do gene therapy, the normal allele for CF could be inserted into a viral vector that can infect cells in the lung and airway tissues. Then the virus could be sprayed onto these tissues.

10. Early identification of people with multifactorial diseases, even before symptoms appear, could lead to therapeutic interventions to prevent disease development. Ethical issues might include insurability, hiring eligibility, and social stigma.

11. An enzyme test for HEXA would reveal intermediate levels in people who are carriers. This could be done on accessible cells (e.g., blood) if the gene is expressed there. A DNA test could involve sequencing the gene by allele-specific oligonucleotide hybridization (see the answer to Question 9). The advantage of DNA testing is that it can be done on any cells from the body (not just cells that express the enzyme).

    Investigation of the stop codon hypothesis would involve isolating the HEXA protein from patients with Tay-Sachs disease and showing that it is shorter in primary structure than the protein encoded by the normal allele.

12. (a) The amino acid sequence would be Leu-Ile-Ser-Ile-Ala. This is a missense mutation.

    (b) The mutation replaces proline with serine. Proline is a nonpolar amino acid that is usually part of bends or loops in a protein; serine is a polar amino acid with a smaller side chain. The mutation is likely to affect enzyme activity because it is likely to affect protein structure.

    (c) See p. 317. This region of the gene could be amplified by PCR and then digested with *Eco*RV. The mutant DNA will be cut, but the wild-type DNA won't be.

## CHAPTER 16

1. b 2. a 3. e 4. b 5. c 6. d

7. The easiest way to answer this question is to construct a simple table:

| | Lysogenic Bacteriophage | HIV |
|---|---|---|
| (a) Viral entry to host cell | Attachment of viral protein to host cell membrane | Membrane fusion of virus to host cell membrane |
| (b) Virus release from host | Host cell lysis | Budding and exocytotic release |
| (c) Viral genome replication | Host DNA polymerase | Virus reverse transcriptase followed by host RNA polymerase |
| (d) New virus production | Host transcription of virus genes and host-mediated translation of virus proteins | Same as in lysogenic bacteria |

8. In a prokaryotic gene, the promoter is a DNA sequence, there are few transcription factors, and there is one RNA polymerase. In a eukaryotic gene, the promoter is a DNA sequence, there are many transcription factors, and there are several RNA polymerases.

9. Here is the structure of the gene:

   E1 – I1 – E2 – I2 – E3 – I3 – E4

   (E = exon; I = intron). Assuming that initiation of transcription begins at E1, the possible proteins are composed of exons 1234; 134; 124; and 14.

10. To keep a constant, low-level expression of repressor protein, the regulatory gene would have an inefficient promoter, and synthesis of the repressor would be constitutive.

11. You could sequence the relevant genes of colon cancer cells and look for mutations that lead to aberrant function, then isolate the proteins involved and determine that their functions are indeed abnormal. To show epigenetic silencing, you might sequence the promoters of the genes and look for epigenetic changes (e.g., cytosine methylation, which would be increased if there is transcriptional silencing). Then you could examine the tumor cells to see if the active proteins are there but in small amounts.

## CHAPTER 17

1. c 2. b 3. e 4. e 5. b 6. b 7. c 8. c 9. a

10. One gene can produce several proteins by alternative splicing, which makes the proteome highly complex. In addition, many proteins are modified after translation, and this contributes to even more protein diversity. The metabolome is highly variable from cell to cell and from one time to another. It is determined not only genetically but also by responses to environmental conditions.

11. While all of these plants have the same basic genes for "life" as well as "plant" functions (e.g., photosynthesis, cell-wall formation, flowering), there are some genes (and proteins) that are specialized for each plant (e.g., rice genes for growing under water, genes for timing flowering, genes for seed-storage proteins).

12. (a) Extract genomic DNA from the patient's cells and analyze it for SNP polymorphisms. If the SNP that correlates with kidney cancer is present, he has an increased susceptibility.

    (b) Isolate both normal and cancerous kidney cells. Do a metabolomic profile on the kidney cancer cells and the normal kidney cells using chemical analyses for small molecules. By comparing the profiles, generate a metabolomic "signature" for the kidney cancer cells. Next, examine the metabolomic profile of kidney tissue from the patient and compare it with the metabolomic signature for kidney cancer cells.

    (c) For possible drugs involved in kidney cancer treatment, isolate many cancers (or examine stored tissues) and do a SNP analysis, correlating tumor response to the drug with the SNP polymorphism. Then isolate some of the patient's tumor cells and examine the DNA for SNPs that relate to drug response. Use the drug that the patient's genome indicates will be effective.

## CHAPTER 18

1. b 2. c 3. e 4. a 5. e 6. d 7. b 8. c

9. Both PCR and cloning begin with a gene sequence. In PCR, the sequence is amplified in the test tube. In cloning, the sequence is amplified by an organism (typically bacteria). In PCR, amplification is achieved by synthesizing primers that bind to either end of a target DNA sequence and adding nucleotides and DNA polymerase. The doubled DNA is them denatured, and the process is repeated 20 to 40 times.

In cloning, the target DNA is inserted by restriction and ligation into a vector, which has an origin of replication that will function in the organism where amplification will occur. The vector is added to the host cells, which are cultured and divide many times, amplifying the target DNA along with the host chromosome. The vector is then removed from the host cells and cut with a restriction enzyme, releasing amplified, cloned target DNA.

PCR is much simpler and faster but has artifacts where inappropriate fragments of DNA are amplified or sequence errors are introduced by DNA polymerase. Cloning yields the correct DNA without mutations but involves host cell culture and time-consuming DNA purification steps. See Figure 18.12. A simple table can answer this question.

| | Conventional | Recombinant DNA |
|---|---|---|
| (a) Sources of new genes | Other plants of same species | Any organism or synthetic DNA |
| (b) Number of genes transferred | Often many | One |
| (c) How long it takes | At least one growing season, usually many | Weeks |

10. (a) The target gene would be inserted into an expression vector with a promoter such that the gene would be expressed in the developing seed. The vector could be added to cultured wheat cells, and those cells carrying the vector selected (the vector could carry a reporter gene for resistance to an antibiotic). The cells could be induced to form a wheat plantlet, which would be transferred to the field and the seeds examined for the new protein.
    (b) The target gene could be inserted into a sheep expression vector containing the lactoglobulin promoter so that the gene would be expressed in milk glands. The recombinant vector would be inserted into sheep egg cells. After the female offspring grew up, their milk could be tested for the presence of the human enzyme.
11. Public concerns include the artificiality of unnatural interference with nature, the safety of these foods for human consumption, and environmental dangers if non-host plants receive recombinant genes.

## CHAPTER 19

1. c 2. b 3. a 4. e 5. a 6. b 7. c

8. (a) All neuronal precursors might undergo apoptosis and no neurons would form.
   (b) The *p21* gene would be activated and the cell cycle would be blocked; in the presence of other factors, muscle cells would form.
   (c) There might be no gradient of the protein in the developing limb and therefore no differential development of digits—all the digits would be fingers.
   (d) The hunchback protein gradient would not form properly and the embryo would not establish its anterior–posterior axis.
9. (a) No apoptosis would lead to too many cells in developing organs, and the organs would not form properly.
   (b) No gradient of hunchback protein would form, and there would be no posterior end determination in the developing fruit fly.
10. A mutation that caused expression of class A genes instead of class C genes. This would lead to an AB combination instead of AC, and petals would develop instead of stamens.
11. Mechanisms might include cell-cycle inhibition as a result of Cdk blocker; induction of transcription of certain genes; and cytoplasmic segregation, so that when a cell divides only one daughter cell gets a factor important in determination.
12. One could analyze mRNA in egg cells, in the parent differentiated cells, and in the reprogrammed cells. This could be done by reverse transcriptase PCR or by gene expression arrays.

## CHAPTER 20

1. c 2. a 3. a 4. c 5. c

6. If the expression of Gremlin were blocked, this protein could not inhibit BMP4 signaling. The cells in the webbing of the feet would undergo apoptosis, and the duck would be born with unwebbed feet.
7. All of the hatchlings at any temperature would be expected to develop into males. Aromatase is required to convert testosterone into estrogen, which is required for female development.
8. The coexpression of *Hoxc6* and *Hoxc8* appears to be important in the development of thoracic vertebrae (the vertebrae with ribs). This is a short region in mice, and mice have only a small number of thoracic vertebrae, and therefore a short body. In snakes, the coexpression of *Hoxc6* and *Hoxc8* along a much greater length of the embryo results in a much larger number of thoracic vertebrae, and therefore a much-elongated body.
9. The results support the conclusion that higher levels of BMP4 expression result in greater cartilage diameter on the beaks of developing chickens.
10. The observations are consistent with the hypothesis that there has been selection in some human populations for mutations on the enhancer that controls expression of the glycoprotein in red blood cells. This genetic change would be expected to have a selective advantage in human populations that are exposed to malaria at high levels, because the mutation confers greater resistance to malaria in humans that carry it.

## CHAPTER 21

1. d 2. d 3. d 4. e 5. b

6. Humans select traits in domestic plant and animal populations based on our interest in the trait, rather than on how it affects the natural reproductive rate or survivorship of the organisms. Many of the traits artificially selected by humans would not be advantageous in wild populations. For example, humans have selected many cattle breeds for high body fat and high body weight. These traits result in large calves, which in turn result in calving difficulties for cows. Ranchers often have to assist in the birth of such calves, because the calf (and likely its mother) would often die without such assistance. In a natural population, there would be selection for smaller calf size and birth weight, which would increase the successful reproductive rate and survivorship.
7. Behaviors can respond to environmental cues that are predictive of future conditions, and these behaviors can be selected for if they are under genetic control. For example, day length becomes shorter as we move closer to winter, so individual mammals have a survival advantage if they respond to shortening days by going into hibernation. In this case, the environmental cue (day length) is predictive of future environmental conditions (the cold of winter).
8. Natural selection cannot act when there is no effect on the effective reproductive rate of the organism. Diseases such as Alzheimer's usually occur long after the reproductive years have passed. As long as the disease does not affect the relative likelihood of the survival of the affected person's offspring (as a result of reduced parental care, for example), we would not expect natural selection to lead to any reduction in Alzheimer's disease in human populations.
9. (a) Frequency of allele *a*: 0.60; of allele *A*: 0.40
   (b) Frequency of genotype *aa*: 0.40; of genotype *Aa*: 0.40; of genotype *AA*: 0.20
   (c) Expected frequency of genotype *aa*: 0.36; of genotype *Aa*: 0.48; of genotype *AA*: 0.16
   (d) We would expect some level of deviation because the assumptions of Hardy–Weinberg equilibrium are so restrictive. For example, the finite population size, the presence of mutation, any migration of individuals into or out of the population, gene flow from mating with adjacent populations, nonrandom mating within the population, or selection in the population could all lead to deviations from Hardy–Weinberg expectations.
10. The black mice and white mice are highly unlikely to be mating randomly with each other. The combined population is far from Hardy–Weinberg expectations, with far too few heterozygous (*Aa*) individuals. The much higher frequency of the *a* allele among the black mice, and of the *A* allele among the white mice, suggests that the black and white mice are mostly mating within color types, with few between-color-type matings. Another possibility, though, is that the population consists of two subpopulations (one of mostly black mice, the other of mostly white mice) that have only recently come together in the same location. These two hypotheses could be distinguished by following the mice through another generation. If mating is now occurring at random, we would expect the genotype frequencies to be similar to Hardy–Weinberg expectations after one generation of random mating.

## CHAPTER 22

1. e 2. a 3. e 4. a 5. e 6. d

7. The classification is not currently monophyletic. Both genera could be monophyletic if Species 4 were moved from Genus B to Genus A; monophyly could also be achieved if all the species were included in the same genus.
8. Fossils can give us direct evidence of the character states of extinct lineages. For example, all modern birds lack teeth. Is the lack of teeth an ancestral or a derived condition? If we examine extinct species of theropods (the larger group of dinosaurs that includes the living birds), we see that they had teeth. Therefore we know that the lack of teeth is a derived condition in modern birds.
9. The estimated average rate of change is 0.9 amino acid change/500 million years, or 0.0018 amino acid change/million years. If we express this as a percentage rather than as a proportion, we would say there is (on average) 0.18 percent change in amino acid sequences per million years.

10. The West Nile virus in the United States appears to be most closely related to a strain of the virus isolated in Israel. A reasonable hypothesis is that the virus emerged in Africa in the 1930s and subsequently moved into Asia and Europe, probably multiple times. Then in the late 1990s, a strain of the virus from Israel appears to have been transported to New York, perhaps carried by mosquitoes on an airplane or in a cargo shipment. Once in the United States, the virus spread quickly in native bird populations across North America.

## CHAPTER 23

1. e 2. c 3. a 4. e

5. If the only difference between the diverging lineages is at a single locus, then both of the new alleles must be functional when they interact with the products of other gene loci (in both lineages). Any interlocus genetic incompatibility produced by these new alleles would be expected to affect the parental lineages as well. In addition, there are many greater numbers of possible incompatibilities across different gene loci than there are within a single locus. Rather than two deleterious changes at the same locus (one in each lineage), the Dobzhansky–Muller model allows neutral changes at any pair of loci whose products interact. It is the negative interaction of these products in the hybrid between the two lineages that results in genetic compatibility.
6. If two different fusions of chromosomes occur in two different lineages, then the resulting chromosomes cannot pair normally in meiosis in the hybrids. If you attempt to diagram meiosis in the hybrid that would result from a cross of the divergent lineages in Figure 23.4, you will see that homologous pairings require parts of different chromosomes to align with one another. These chromosomes will then be pulled in two different directions as the cell divides in meiosis I, resulting either in a likely failure of the cell to divide, or an uneven distribution of the chromosome arms in the two daughter cells. Production of normal cells with an even distribution of the various chromosomes arms will be limited, so the hybrids will produce few, if any, normal gametes.
7. A likely possibility is that the incompatible alleles have not yet become fixed in the various strains, so only some combinations of crosses result in genetic incompatibility.
8. Species that arise in allopatry initially occur in separate, but usually adjacent, ranges (see Figure 23.6). Therefore we would expect many closely related species to exhibit this same pattern. The ranges of highly mobile species are more likely to change over time, so the pattern should be strongest among relatively sedentary species.
9. There are many possible designs of experiments that might prove informative. Here is an example of one that would examine the effect of flower position on pollinator attraction: Take one species of flower and divide the flowers into two groups. Position each flower to be either upright or pendant, then record the number and type of pollinators that are attracted to flowers in each group. Test to see if the differences between the two groups are statistically significant.
10. (a)

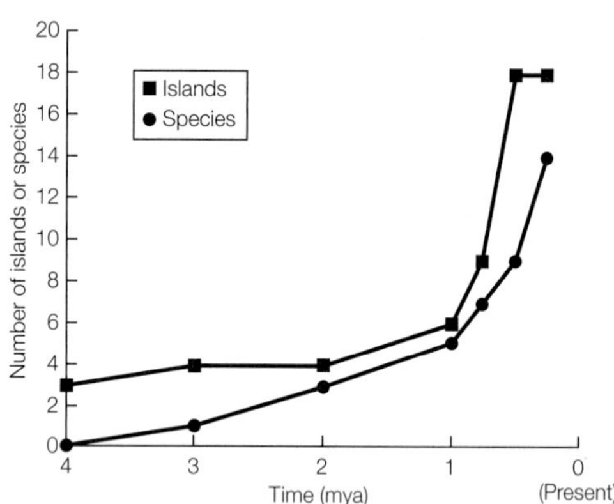


(b) Yes, because the curve for the number of species lags behind the curve for the number of islands, but the two curves exhibit very similar changes in slope through time. As new islands arise, new opportunities for speciation also arise. The number of species at any one time is always just below the number of distinct islands.

(c) There are currently 18 islands in the archipelago and only 14 finch species. This suggests that there are still opportunities for additional speciation by geographic isolation. Based on our graph from Question a, we expect populations of species that occur on two or more islands to diverge into distinct species over time. To test this hypothesis, we could collect samples of each population and examine genetic divergence among the samples. Significant genetic divergence among the populations on different islands suggests that the distance between the islands is a significant barrier to gene flow, so the populations are expected to diverge into distinct species over time.

## CHAPTER 24

1. a 2. a 3. a 4. e 5. b 6. e

7. Molecular clocks work best when they are applied within a group of organisms with similar generation times and populations sizes. Population size makes little difference if all or most changes are neutral, but rates of change among deleterious and beneficial changes are affected by population size. In addition, it is important to make comparisons among homologous genes and proteins, since rates of evolution across different genes are likely to vary widely. When molecular clocks are used to make comparisons across species with very different generation times, it is necessary to account for the different generation times.
8. New mutations are introduced into the experiment shown in Figure 24.14 through the errors made in the PCR amplification step. In other words, the mutation rate is a function of the error rate of the DNA polymerase. Using a different DNA polymerase with a higher error rate would increase the overall mutation rate of the experiment, and that would increase the variation in the population of molecules. Another possible answer is to add a mutagen to the PCR amplification step, which would also increase the mutation rate of the experiment. Any process that increases the mutation rate would be expected to increase the genetic variation present in the pool of molecules prior to the next round of selection.
9. This problem can be investigated by sequencing and comparing the genes for opsins in surface-dwelling (eyed) and cave-dwelling (eyeless) crayfishes. If the genes of the eyeless species are no longer under any selection, we would expect to observe a similar rate of synonymous and nonsynonymous substitutions in the genes. If there has been strong selection for a new function (something other than vision), we would expect a higher rate of nonsynonymous substitutions compared with synonymous substitutions (indicating positive selection). We would compare these rates to the rates seen in the surface-dwelling (eyed) species. In the surface-dwelling species, we would expect to see a higher rate of synonymous compared with nonsynonymous substitutions, which is expected under purifying selection.
10. (a) Codon numbers 12, 15, and 61 are likely to be evolving under positive selection for change because these three codons have each experienced a higher rate of nonsynonymous substitutions (which give rise to amino acid replacements) relative to the rate of synonymous substitutions.

(b) Codon numbers 80, 137, 156, and 226 are likely evolving under purifying selection, as the vast majority of changes at these codons are synonymous substitutions, which do not result in amino acid replacements. Substitutions that result in amino acid changes (nonsynonymous substitutions) undoubtedly occur, but they are usually selected against in the population. Codon number 165 has experienced similar numbers of synonymous and nonsynonymous substitutions. However, since there are approximately three times as many possible nonsynonymous substitutions as there are synonymous substitutions, the number of synonymous substitutions is slightly higher than expected if the rates of each type of substitution are equal. Codon 165 may be evolving under weak purifying selection; it is the codon that is closest to neutral among the codons shown in the table.

## CHAPTER 25

1. b 2. c 3. a 4. c 5. b 6. c

7. There are many possible answers, but four familiar examples include the study of Earth's past atmosphere by examining the chemical composition of rocks; the study of past climates by examining the growth rings of trees; the study of continental drift by examining the geological record; and the study of the origins of the universe (the "Big Bang") by examining the speed at which galaxies are moving apart.
8. Relative dating provides us an order for events; we learn that Event 1 happened before Event 2. But absolute dating provides us with an estimate of the timing of those events. It is important to know not just that Event 1 occurred before Event 2, but also how much time separated the two events.
9. Multicellular organisms require higher concentrations of oxygen, and the levels of oxygen increased throughout the Precambrian. By the end of the Precambrian, atmospheric oxygen levels were sufficiently high to support a variety of multicellular organisms. In addition, the end of widespread glaciation (the "snowball Earth" period) near the end of the Precambrian probably allowed multicellular organisms to flourish.
10. There are many possible experiments that could be devised. For example, the effects of changing oxygen concentrations on other species (besides flying insects, such as the *Drosophila* used in the described experiment) could be tested. An ideal study organism would have a short generation time (so that many generations could be followed in the course of the experiment) and would be easy to raise in the laboratory. For example, guppies could be raised in elevated and reduced oxygen concentrations, and evolution in the size of the swim bladder (a site of oxygen uptake) could be evaluated as a response.

## CHAPTER 26

1. e  2. c  3. e  4. b  5. b  6. d

7. Ribosomal RNA genes are universally present across organisms. They evolve slowly, so they can be compared among the most distantly related species. They are present in multiple copies, so they were relatively easy to isolate and sequence in the earliest days of gene sequencing. Also, since they are required for protein synthesis, and already present in all cellular species, the possibility of lateral gene transfer is greatly reduced. In contrast, different types of metabolism have arisen repeatedly in the history of prokaryotes, so species with similar types of metabolism may not be closely related. Cell structure is useful for identifying some major groups of prokaryotes (Gram-positive versus Gram-negative groups, for example), but the differences are too few to be of great use in classifying most species.
8. A laterally transferred gene does not represent descent from a common ancestor and thus does not reflect a true evolutionary relationship.

Expected tree based on gene x:

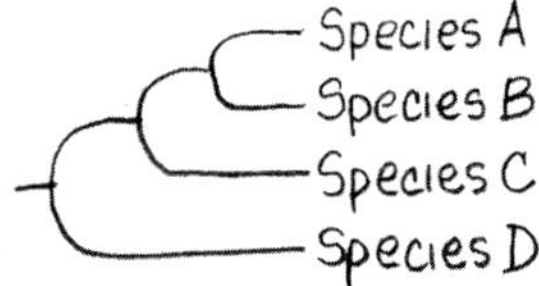


Expected tree based on consensus of non-transferred genes:

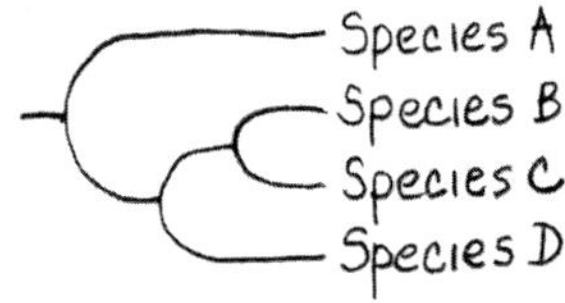


9. Most, but not all, biologists consider viruses to be living organisms. Viruses have their own genomes, and they are composed of proteins much like cellular organisms. They evolved from other living species, and they are clearly a part of life. However, viruses are not composed of cells, and they depend on cellular hosts to carry out many of their biological processes. For these reasons, some biologists consider them to be nonliving components of their cellular hosts rather than distinct living organisms.
10. There are many possible answers, but one widely used approach for detecting new life forms (in any environment, including high-temperature environments) is to directly isolate and amplify conserved gene sequences. The ribosomal RNA genes are often used for such detection because they evolve very slowly and are required for protein synthesis. DNA could be extracted from high-temperature environments, and any ribosomal DNA genes that were present could be amplified and sequenced. The sequences would then be compared with the ribosomal RNA genes of other known species of prokaryotes to classify the organisms living in the extreme environment.

## CHAPTER 27

1. e  2. c  3. e

4. (a) Foraminiferans have external shells of calcium carbonate, whereas radiolarians have long, stiff pseudopods and radial symmetry. The external shells of foraminiferans and the internal skeletons of radiolarians are both important components of ocean sediments and sedimentary rocks.
   (b) Ciliates are covered with numerous hairlike cilia, whereas dinoflagellates generally have two flagella (one in an equatorial groove, and the other in longitudinal groove). Both ciliates and dinoflagellates have sacs, called alveoli, just beneath their plasma membranes, which identify them as alveolates.
   (c) Diatoms are unicellular and are typically composed of two nested plates (like a petri dish). Brown algae are large, multicellular organisms composed of branched elements or leaflike growths. Both diatoms and brown algae are photosynthetic.
   (d) The vegetative unit of a plasmodial slime mold is a plasmodium: a wall-less mass of cytoplasm containing numerous diploid nuclei. The vegetative unit of cellular slime molds consists of separate, single amoeboid cells. In both groups, when environmental conditions become unfavorable, the vegetative units form fruiting structures.
5. The independence of sex and reproduction in ciliates suggests that sex has functions apart from reproduction. Sex is important for recombining genes, which is important for several reasons. Sex allows populations of organisms to avoid the accumulation of deleterious alleles, and it allows the formation of new combinations of beneficial alleles. Thus even organisms that reproduce asexually generally have some other means of achieving sexual recombination of their genomes.
6.

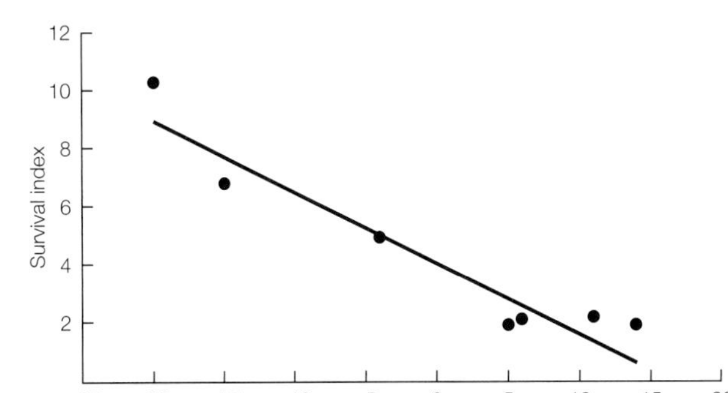


Using the formula for a correlation coefficient shown in Appendix B, $r = -0.948$.

7. The results show that earlier bloom dates are associated with higher survival indices. The relationship between these two measures is very strong and nearly linear, resulting in a correlation coefficient of $r = -0.948$. As noted in the question, larval haddock depend on these blooms for both cover from predation and as a food source. A reasonable hypothesis for this is that earlier blooms provide better cover and more food for the larval haddock, so survivorship of the larval fish is higher in years when the phytoplankton blooms occur earlier. Another (not mutually exclusive) possibility is that the earlier blooms benefit other species that the haddock consume as food, or that the potential predators of haddock target the phytoplankton instead of the haddock.
8. The three rRNA genes of corn are not one another's closest relatives because the nuclear, mitochondrial, and chloroplast genomes have different origins, and the relationships shown in the tree reconstruct the endosymbiotic events that gave rise to mitochondria and chloroplasts.
9. The mitochondrial rRNA gene of corn is more closely related to the rRNA gene of *E. coli* than it is to the nuclear rRNA genes of other eukaryotes because the mitochondria were derived from an endosymbiosis with a proteobacterium. Likewise, the chloroplast rRNA gene of corn is more closely related to the rRNA gene of *Chlorobium* than it is to the nuclear rRNA gene of corn because the chloroplasts were derived from an endosymbiosis with a cyanobacterium.
10. The human and yeast mitochondrial rRNA genes would be expected to cluster on the tree closest to the corn mitochondrial rRNA gene because all of these genes are descended from the same endosymbiotic event (the origin of mitochondria). The human and yeast mitochondrial rRNA genes would be more closely related to each other than either is to the corn mitochondrial rRNA gene because fungi and animals are more closely related to each other than either is to plants (as can be seen in the relationships of the nuclear rRNA genes).

## CHAPTER 28

1. c  2. e  3. b  4. b  5. d

6. Microphylls are usually small and typically have a single vascular strand. In contrast, megaphylls are larger and typically contain branched veins. Microphylls may have originated as sterile sporangia. Megaphylls may have originated from flattening of branching stems, between which photosynthetic tissues developed. Among modern plants, microphylls are found in lycophytes, whereas megaphylls are characteristic of the euphyllophytes (such as ferns and seed plants).
7. One advantage of heterospory is that it allows a greater degree of outcrossing, since there are separate male and female gametophytes.
8. Both mosses and ferns are homosporous, and both alternate between a diploid sporophyte and a haploid gametophyte generation. However, the gametophyte generation is the large, dominant portion of the moss life cycle, whereas the sporophyte generation in the large, dominant portion of the fern life cycle. The sporophyte of a moss is completely dependent on the gametophyte, whereas the sporophyte of a fern becomes independent of the gametophyte.
9. Yes. Heterospory is an example of a trait that appears to have evolved multiple times among different groups of vascular plants.
10. One possibility is to examine leaf size as a function of thermal environment among close relatives of living plant species. We might predict that large leaf size is restricted in hot, dry climates but favored in cooler, wetter climates.

## CHAPTER 29

1. d 2. a 3. d 4. a 5. a

6. To be functional as a reproductive organ, a flower would need to have at least a carpel or a stamen. The petals function largely for pollinator attraction, and so can easily be lost in wind-pollinated species. The sepals function mostly to protect the flower in bud, and so may be lost in species with simplified flowers.
7. The fossil record is not a complete record of life on Earth. Early angiosperms may have been limited in distribution, or may have lived in environments that were not compatible with easy fossil formation. It is likely that the early angiosperms were not very abundant or widespread. They apparently underwent a rapid radiation in the Cretaceous, where they become common in the fossil record.
8. .

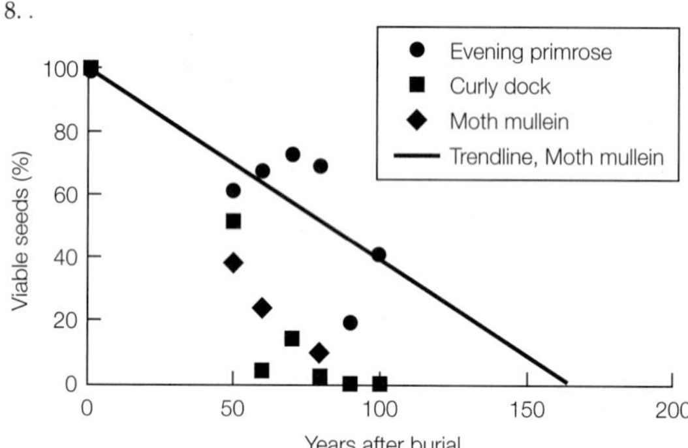


9. One approach to this problem is to calculate a linear trend line for survivorship of moth mullein (*Verbascum blattaria*) seeds by calculating a linear regression line (see Appendix B) and then projecting it forward in time to the point where it intersects zero percent survival. The resulting regression equation is $y = 102.09 - 0.62x$. The graph above shows this approach, which predicts that the last *Verbascum blattaria* seeds would germinate in about Year 165 of the experiment (set $y$ to 0, and solve for $x$ using the regression equation; the result is $x = 164.7$ years). This approach assumes a linear decline in viability of the seeds. It may be more reasonable to assume an exponential decay in seed viability (similar to radioactive decay; see Figure 25.1). If seeds decay exponentially, then we would expect some low level of survivorship of *Verbascum blattaria* seeds well beyond Year 165.
10. At least four factors are related to seed survivorship:
    1. Size of the seed: larger seeds have more food reserves (endosperm).
    2. Density of the seed coat: tougher seed coats provide better protection of the seed.
    3. Level of dormancy of the embryos: deeper dormancy results in longer survivorship.

## CHAPTER 30

1. d 2. b 3. c 4. e 5. c

6. If it is a fungus, we would expect to detect chitin in the cells walls, whereas chitin would be absent if it is a plant. We could also examine the specimen for the presence of chloroplasts, which would be expected to produce the green coloration in a plant but not in a fungus. If it is a vascular plant, we would expect to observe vascular tissues in the sample, which would be absent if it is a fungus. We could also sequence a conserved gene from the sample, such as a ribosomal RNA gene, and compare the sequence phylogenetically with other fungi and plants.
7. It is because the nuclei remain separate in dikaryons, even though the two nuclei are contained within a single cell.
8. Fungi play a critical role in decomposition of plants and parts of animals. If there were no more fungi, there would be enormous accumulation of the remains of dead organisms, especially the cellulose and lignin of plants.
9. Site 5 shows the highest diversity and density of lichens and so is probably farthest from the city center. Site 4 is next, followed by Site 1, then Site 3, and finally Site 2. In addition to distance from the city center and prevailing wind direction, other predictive factors could include distance to point-pollution sources (such as factories or power plants) and distance to major highways (a source of pollution from automobile exhaust). Other answers are also possible; it is important for such studies to control for factors such as the species of tree examined and the exposure of the branches to similar light and humidity conditions.
10. One common source of fungal contaminants in plant samples is symbiotic fungi, including endophytic fungi and mycorrhizal fungi. In either case, it is usually possible to collect specific tissues from the plant that exclude these symbionts. If our hypothesis about the source of the fungal genes is correct, then the fungal sequences should be absent from these symbiont-free tissues.

## CHAPTER 31

1. d 2. b 3. c 4. d 5. d

6. (a) In radial symmetry, body parts are symmetrical across multiple planes that run through a single axis at the body's center. Animals with radial symmetry have no front or rear ends, and they are often sessile or drift freely with currents. If they move under their own power, they can typically move slowly equally well in any direction. In contrast, bilaterally symmetrical animals have mirror-image right and left halves divided by a single plane that runs along an anterior–posterior midline. They have a front end that usually contains a concentration of sensory systems and nervous tissues in a distinct head. Bilateral animals usually move forward in the direction of the head, so the head encounters new environments first.
   (b) Among the bilaterian animals, there are two distinct forms of gastrulation, or the initial indentation of a hollow sphere of cells early in development to form the blastopore. In protostomes, the blastopore eventually develops into the mouth of the animal. In deuterostomes, the blastopore becomes the anus.
   (c) Diploblastic animals have embryos with two cell layers (an outer ectoderm and an inner endoderm). The embryos of triploblastic animals have an additional cell layer between the ectoderm and the endoderm, known as mesoderm.
   (d) Acoelomate animals lack a body cavity enclosed by mesoderm. Pseudocoelomate animals have a body cavity enclosed in mesoderm; this body cavity contains the gut and internal organs composed of endoderm, but these latter organs are not lined with mesoderm. Coelomate animals have a body cavity that is enclosed with mesoderm, and the internal organs are also lined with mesoderm.
7. The answer to this question will depend on the reader's opinion. However, most biologists would answer that the phylogenetic analyses that result from analysis of animal genomes provide the most definitive evidence of animal monophyly.
8. Bilateral organisms have an anterior and a posterior end. As the animal moves through the environment, the anterior end encounters potential food or predators first. It is therefore advantageous for the sensory organs and central nervous system to be concentrated at the anterior end.
9. A slow metabolic rate requires a low energy budget, and hence a low intake of food.
10. Placement of glass microscope slides (or other smooth substrates for placozoan attachment) in warm tropical waters often results in colonization by placozoans. The glass slides can be suspended in water in survey areas, then later retrieved and examined for the presence of placozoans.

## CHAPTER 32

1. e 2. d 3. b 4. d 5. d 6. e

7. Segmentation allows an animal to move different parts of its body independently, which allows for much greater control of movement. However, segmentation tends to constrain the body shape of an organism. Loss of segmentation is often favored in some parasitic and burrowing organisms that live in confined spaces.
8. Several answers are possible, but some examples of key innovations that appear to be associated with major episodes of diversification in protostomes include the evolution of the cuticle in ecdysozoans, the evolution of shells in mollusks, the evolution of jointed limbs in arthropods, and the evolution of wings for flight in insects.
9. Insects have been highly successful in terrestrial environments, in part because flight gives insects greater access to plants. Many insect species are specialists on one or a few plant species, and plant diversity is far greater on land and in freshwater environments than in the oceans. Although some insects live in fresh water for part or all of their life cycles, these freshwater environments are closely associated with surrounding terrestrial environments. Crustaceans have been much more successful in the oceans than have insects, and crustaceans may simply outcompete insects in marine environments.
10. All entomologists agree that many more species of insects remain to be discovered, but many entomologists think that Erwin's estimates were high. Each estimate is highly dependent on how representative *Luehea seemannii* is as a tropical forest tree. If the average tropical forest tree has many fewer host-specific beetle species than does *Luehea seemannii*, then these estimates would be inflated. Likewise, overestimating the number of tropical forest trees, or the percentage of ground-dwelling beetles, or the percentage of all insects that are not beetles, would lead to further inflation of the estimates. In addition, species diversity of beetles may be higher in Panama than in other areas of the tropics. However, any of these estimates could be underestimates as well. Each of Erwin's assumptions is now being tested; these tests require extensive work on additional species of trees, additional groups of insects, and in additional areas of the world.

## CHAPTER 33

1. d 2. a 3. d 4. a 5. e 6. b

7. The four appendages common to most vertebrates are the two pectoral appendages and the two pelvic appendages. In most swimming vertebrates, these appendages function as fins. They are commonly used for propulsion (especially the pectoral fins) but are also used for steering, stabilization, and manipulation of the body position in water. Among tetrapods, the appendages are often modified into limbs used for walking, running, jumping, burrowing, climbing, grasping, and manipulating objects. There have been several reversals to fin-like limbs used by aquatic tetrapods (several times among amphibians, turtles, birds, and mammals, for example). There have also been at least three origins of the pectoral limbs of tetrapods into wings for powered flight (among birds, bats, and the extinct pterosaurs). There have also been several other modifications of the limbs for gliding (in fishes, amphibians, reptiles, and mammals). One or both pairs of appendages have been lost (or greatly reduced) in many groups of fishes, amphibians, reptiles (including birds), and mammals. Some well-known examples of limb reduction or loss include the completely legless caecilians and snakes, the loss of external hind limbs in whales and manatees, and the greatly reduced forelimbs of flightless birds.
8. Amphibians exchange gases and fluids through their permeable skins. This makes them highly vulnerable to many environmental toxins. Many species of amphibians have a biphasic life cycle, so they are vulnerable to habitat degradation and loss of both aquatic and terrestrial environments. Most amphibians do not move long distances, so they do not easily move into new habitats when their local environment is destroyed. For these reasons, they are also sensitive to rapid climate changes. Many species of amphibians have highly specialized habitat requirements and live in very restrictive ranges. Habitat loss or changes within these restricted ranges often result in extinction.
9. Fossil remains of extinct theropod dinosaurs shows that many features once thought to be restricted to birds, such as feathers, actually evolved much earlier among the theropods. Other typical "bird" morphological features, such as air-filled bones and a furcula (wishbone), are also typical of the larger group of theropods. Among living reptiles, DNA sequence analyses clearly unite birds with the crocodilians (the other living archosaurs). The combined evidence from many sources that birds are a surviving group of theropod dinosaurs is now overwhelming.
10. Hair evolved in the ancestor of mammals; feathers evolved among theropod dinosaurs (seen today among the birds). Among the living tetrapods, birds and mammals are endothermic. Hair and feathers provide body insulation for mammals and birds, respectively. Without these forms of insulation, the maintenance of metabolic body heat would be difficult. Fossil evidence shows that many extinct theropod dinosaurs also had feathers, so many paleobiologists predict that they were endothermic as well. Endothermy would also be expected in large, active predators—a description that fits our current view of many theropod dinosaurs.

## CHAPTER 34

1. b 2. e 3. a 4. b 5. b 6. d 7. c 8. c

9. The cell types can be compared using a table.

| Structure/Function | Sclerenchyma | Collenchyma |
|---|---|---|
| Cell walls | Secondary, thickened | Primary, thicker at corners |
| Flexibility | Less flexible | More flexible |
| Cell conditions | Some dead (apoptosis) | Alive |
| Presence | Wood, bark | Petioles, growing areas |

10. Primary growth involves cell division and cell enlargement, and typically results in growth of an organ in length. Secondary growth involves growth of an organ in thickness, by the addition of more cell layers. Only some angiosperms undergo secondary growth. Herbaceous plants such a peonies have only primary growth. Woody plants such as trees have both primary and secondary growth.
11. The initials are still 1.5 meters above the ground today because the plant grows in height at its apex.
12. Some examples might include a larger root apical meristem to produce thicker carrots and reduced internode growth to produce compact heads of cabbage.

## CHAPTER 35

1. c 2. d 3. b 4. b 5. d 6. e

7. Epidermal cells have external walls with a waxy cuticle, which makes them repel water. The epidermal cells of roots might have a thinner (or absent) cuticle, as they take up water; and leaves and stems a thicker cuticle, to conserve water. In addition, the epidermis of leaves, and to a lesser extent stems, has stomata, which regulate gas exchange (including loss of water vapor from the leaf interior).
8. A source is an organ such as a leaf that produces more sugars than it uses. A sink is an organ such as a root that produces less sugars than it needs and so imports sugars from a source. In a deciduous tree, a leaf might be a source in the summer, and roots a sink. But then in spring, the roots might be a source for the buds (newly emerging leaves).
9. To cross the fewest membranes and still get from the soil solution to the atmosphere by way of the stele a water molecule would follow this route: soil to root symplast to stele symplast to stele apoplast to xylem to leaf apoplast to leaf interior air space to stoma to atmosphere. A water molecule could follow this path by crossing as few as two membranes: 1) from soil into a root hair or root cortical cell across a root cell membrane; 2) out of a stele cell into the stele apoplast across a stele cell membrane. Getting from the soil solution to a mesophyll cell in a leaf would require crossing at least three plasma membranes: the two membranes listed for the previous route plus the mesophyll cell membrane (to get from the leaf apoplast into the mesophyll cell).
10. The mutation in the *HARDY* gene might cause increased expression of a gene that inhibits cation accumulation in stomata, thereby keeping them more closed and conserving water. To test this hypothesis, you could look at the response of stomata to light in the leaves of mutant versus wild-type plants. The stomata of wild-type plants should open rapidly in response to light (see Figure 35.9); the stomata of HARDY mutant plants might open more slowly or less wide.
11. (a) Yes. The difference in water potential between the soil and the leaf (1.7 MPa) is enough to overcome gravity and draw water to the top of the tree.
    (b) No. If the soil water potential decreased to –1.0 MPa, it would be more negative than inside the root cells and water would leave the roots (and enter the soil).
    (c) If all the stomata closed, the leaf water potential would not be as negative. This in turn would make the xylem water potential less negative, and so on down to the roots. This would make the difference between the leaf water potential and root water potential insufficient for water to flow from the roots to the leaves (toward a more negative water potential).

## CHAPTER 36

1. d 2. d 3. c 4. a 5. c 6. d

7. The ability of chemists to detect low concentrations of elements is fairly recent. Before then, nutrient solutions thought to be pure often were not.
8. Heavy irrigation after a prolonged dry period may produce runoff of topsoil (the A horizon) and leaching of ions (especially anions) into the subsoil, making fewer nutrients available to plant roots. Converting land use from virgin deciduous forest to crops will change the composition of living organisms in the soil, as many organisms that live in association with tree roots will disappear. The soil structure and texture will also change, because roots will no longer be present to hold the soil together and make air spaces. The soil chemistry will change, because crops take up nutrients from the soils and the nutrients are removed from the system when the crops are harvested.
9. See Figure 36.10, the nitrogen cycle. There are numerous species that fix nitrogen. Loss of one species might allow others to expand and replace it. Loss of all the species would mean that only abiotic methods could be used for nitrogen fixation. This might reduce overall nitrogen in the soil, meaning less would be available for plant growth.
10. The experiment with mutant *Arabidopsis* suggests that *Arabidopsis* uses either its own or exogenous strigolactones for growth regulation and has the appropriate receptor and response mechanisms. This reinforces the idea that an ancient mechanism to attract beneficial microbes also is used for modern plant growth regulation. Or the reverse might be true: the original function of strigolactone might have been as a plant hormone and its role in plant–microbe interactions might have evolved later.
11. Because holoparasitic plants can gain reduced carbon through association with hosts, the genes encoding photosynthesis functions are not under selection pressure, because having them would not confer any survival and reproductive advantage for the parasites. So any mutation that renders such a photosynthesis gene nonfunctional will not be deleterious.

## CHAPTER 37

1. a 2. d 3. b 4. b 5. a 6. b

7. Fire produces ash, which enriches the soil with plant nutrients. A seed that germinated as a result of fire could have an advantage in such a nutrient-rich soil.
8. If a single species has two mechanisms for breaking seed dormancy, then if environmental conditions for Mechanism A are not present, environmental conditions for Mechanism B might be. This enables the plant to respond to a wider array of environmental conditions. In addition, if the cue for, say, Mechanism A turns out to be misleading (not predictive of favorable conditions) and the

seedling dies, there is still a second seed that can germinate at a different time (by Mechanism B), when conditions might be more favorable.

9. The charcoal in the bag absorbs ethylene gas, which is released by ripening fruits. The lack of ethylene prevents over-ripening and decay.
10. To test for the relationship between corn stunt spiroplasma disease and gibberellins, you could measure gibberellins in plants infected with the bacterium and in normal plants; you might expect the spiroplasma-infected plants to exhibit a reduction in gibberellins. Another approach would be to infect normal plants with the spiroplasma and then spray gibberellins on them; you might expect this to reverse the stunt phenotype.
11. (a) See Figure 37.2 Add a mutagen to hundreds of corn seeds and plant them. In a screen, look for plants that are shorter, and propagate these.
    (b) See Figure 37.11. If the transcription factor in the gibberellin signal transduction pathway is inactivated, the plants will be insensitive to gibberellin and be stunted. A mutation that inactivates the gibberellin receptor would have the same effect.
    (c) Other potential effects might include reduced seed germination and reduced seedling growth due to lack of mobilization of stored reserves in the seed (see Figure 37.5). If the mutant is completely gibberellin-insensitive, these effects will *not* be overcome by adding gibberellin to the seeds as they germinate. If, however, the mutant is a dwarf because of reduced amounts of gibberellin in the plant (because of a mutation that affects gibberellin biosynthesis, for example), the germination effects could be reversed with exogenous gibberellin.

## CHAPTER 38

1. b 2. e 3. b 4. e 5. a 6. c

7. In triploid cells undergoing meiosis, there cannot be pairing of homologous chromosomes in meiosis I. So meiosis I is abnormal and functional gametes do not form.
   A fruit is formed from the ovary wall of the flower.
   Seedless grapes are probably propagated by cuttings (vegetative reproduction).
8. Poinsettias are short-day plants; they bloom at a time of year when days are getting shorter (in the Northern Hemisphere).
9. No, it isn't necessary. Just a flash of light during a long night is enough to convert $P_r$ to $P_{fr}$ and to change the photoperiod.
10. (a) The mutation stabilized the CO protein.
    (b) The mutation caused nonfunction of the FD protein.
    (c) The mutation increased expression of the FLC protein.
    (d) The mutation caused constitutive expression of the CO protein.
11. Several approaches might be taken, such as a genetic screen for meiotic cells that do not separate chromosomes at anaphase I, or a search for proteins (and then their genes) that bind to SWII protein.

## CHAPTER 39

1. b 2. c 3. a 4. b 5. c 6. c

7. A plant might make a secondary metabolite that kills an insect pest. Plants making this metabolite would be selected for in evolution. However, the insect might develop resistance to the metabolite. Then the insect population would increase while the plant population decreased—until another defense mechanism evolves. This is coevolution. For more examples, see Chapter 56.
8. *Avr2Avr3* Healthy Healthy Diseased
   *Avr1Avr4* Healthy Healthy Healthy
9. (a) The effects of reduced rainfall could include dehydration and osmotic stress. Genetic responses might include alterations in leaf anatomy, with a thicker cuticle to reduce evaporation; a more extensive root system to obtain water; and accumulation of solutes in the roots, which would reduce root water potential and result in more water uptake in dry soils.
   (b) Flooding reduces the amount of $O_2$ available to the plants and results in reduced respiration. Adaptations might include increased production of pneumatophores or aerenchyma to supply air to submerged plant tissues.
   (c) Wheat rust is a fungal pathogen. Plants can adapt by increasing the ability to seal off infected areas and reduce the spread of the fungus within the plant, by developing specific immunity, and by increasing production of phytoalexin and pathogenesis-related proteins that kill the fungus.
10. You could feed one group of hornworms on normal plants and another group on genetically modified plants. The two groups could then be exposed to the parasite. If nicotine is protective, the hornworms that fed on normal plants should have fewer parasites.

## CHAPTER 40

1. c 2. c 3. a 4. d 5. b

6. Feedforward information makes it possible to anticipate a physiological challenge to homeostasis and to take preemptive action by changing a set point or the sensitivity of a regulatory system. Feedforward information for the regulation of breathing could be the onset of exercise; for blood pressure it could be the fight-or-flight response to a threat; and for secretion of digestive juices it could be the sight, smell, or expectation of food.
7. In the metabolic rate/environmental temperature curve in Figure 40.17, the equivalent of *HL* would be metabolic rate, as long as the animal's temperature is not rising or falling and the animal is not doing external work. *K* would represent the animal's thermal conductance, or how easily it loses heat; $1/K$ would be a measure of the animal's insulation. The curve projects to 0 at an ambient temperature equal to body temperature because this portion of the curve represents the extra metabolic effort necessary to compensate for heat loss to the environment. If body temperature and environmental temperature were the same, there would be no heat loss to the environment.
8. Biological processes proceed more slowly at lower temperatures. Thus the lower the temperature of the heart or skeletal muscle, the slower will be its ability to generate a contractile force. This could pose a physiological challenge for highly active fish such as great white sharks or giant bluefin tuna that depend on fast swimming and endurance to catch prey. An evolutionary adaptation to this challenge can be seen in these fishes' vascular anatomy: blood from their hearts goes to the gills, where it exchanges respiratory gases but also comes into thermal equilibrium with the cold ocean water. Thus these fishes are sending cold blood to their body tissues.
9. Basal metabolic rate (ml $O_2$/hr) vs. body mass (kg):

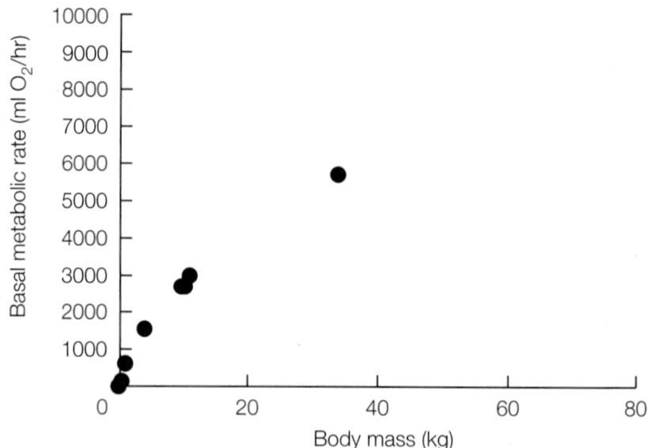


Heart size (g) vs. body mass (kg):

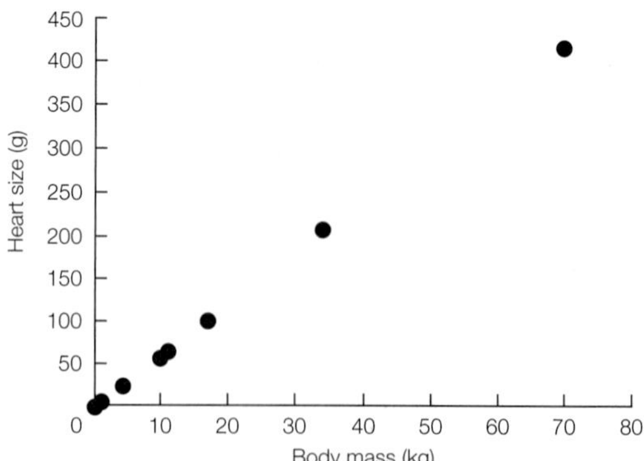


10. The heart has to pump blood against the resistance of the vascular system. Since blood vessels go to all tissues of the body, the total length of blood vessels will be directly proportional to body size, and therefore total peripheral resistance will be directly proportional to body size. The other factor that determines how much blood the heart can pump is the heart rate. The total cardiac output is a function of the size of the heart and the number of times it beats per unit of time.
11. Several factors that determine the heat transfer rate between the iguana and its environment are different in these two conditions. When basking on the lava rocks, the radiation absorbed will be higher and the heat conducted across the skin will be different in comparison to being in the water. Also, when the animal breathes it will be losing heat through evaporation of water in its airways.
    An experiment in which the animal heats up and cools down in the same environment would offer stronger support for the conclusion that blood flow

to the skin is a critical factor. The iguana could be placed in water and in air at two different temperatures (e.g., 20°C and 40°C) to compare the rates of heating and cooling.

## CHAPTER 41

1. b  2. a  3. d  4. e  5. e

6. The time course of a hormone signaling system depends on several factors, including the rate of release of the hormone, its half-life in the blood, and the interactions it has with its receptors. A hormone signaling system that controls a short-term process such as digestion would be expected to have a rapid (e.g., vesicular) release, a short half-life, and rapid action; these are attributes of a peptide hormone. A hormone signaling system that controls a long-term process such as embryonic development would be expected to have continuous release, a long half-life, and to be slow acting, attributes of a steroid hormone.
7. The high levels of synthetic male steroid hormones exert actions through the testosterone receptors that exist in both males and females. The usual effects of promoting muscle hypertrophy and male secondary sexual characteristics (e.g., hair, a deeper voice) will occur. In addition, the high level of negative feedback on the pituitary gonadotropes and on the production of GnRH by the hypothalamus will reduce testicular and ovarian functions in the males and females. Decrease of circulating estrogens in the females will cause reduction of breast tissue.
8. The most common cause of hypothyroidism is lack of iodine in the diet. Thyroglobulin continues to be produced, but there is a lack of functional $T_3$ or $T_4$. As a result, the levels of TRH and TSH rise because of the lack of negative feedback. The elevated TSH induces continued production of thyroglobulin, resulting in goiter. The most common cause of hyperthyroidism is an autoimmune disease called Graves disease. An antibody to the TSH receptor is produced, and the binding of that antibody to the receptor causes activation of the signaling pathway that increases the production of thyroglobulin and the development of goiter. If there is adequate iodine in the diet, however, this condition results in increased secretion of $T_3$ and $T_4$, which produces the symptoms of hyperthyroidism.
9. The large size of the larvae together with the lack of adult moths indicates that cycles of growth and molt of the larvae continued without the induction of pupation. The low temperature probably prevented the usual decline in the production and secretion of juvenile hormone by the corpora allata.
10. Insulin controls the entry of glucose into most cells of the body, but not neurons. When insulin levels fall, as happens during the postabsorptive state (i.e., after ingested nutrients have been fully digested), glucose entry into cells slows down and the cells convert to using other sources of energy. Neurons, however, always require glucose; their lack of insulin means that their access to an adequate glucose supply is protected during the postabsorptive state because there is no decrease in their ability to take up glucose from the blood. Also, the decrease in glucose use by other cells of the body preserves the glucose in the blood for use by the nervous tissue.

## CHAPTER 42

1. a  2. e  3. e  4. c  5. d  6. a

7. See Figure 42.10. The antigen-binding site of an antibody has heavy and light chains in a unique three-dimensional configuration that binds a particular antigenic determinant. This is similar to an enzyme active site that binds a substrate. In both cases, binding is noncovalent. A major difference is in the result of binding: an antigen does not change its covalent structure when it binds to an antibody, whereas a substrate does change covalently when it binds to an active site.
8. Both immunoglobulins and T cell receptors have constant and variable protein regions, bind antigens, and have great variability in primary structure. T cell receptors are only membrane proteins of T cells. Immunoglobulins can be either membrane proteins of B cells or secreted proteins in the blood.
9. The father's haplotypes are A1B7D11 and A3B5D9; the mother's are A4B6D12 and A2B7D11. These two parents could not have the child with the genotype indicated.
10. There are thousands of different enzymes in an individual but potentially millions of different specific antibodies. Every cell in an animal has the genetic information for all enzymes. Each immunoglobulin, however, is derived from a unique gene (produced by DNA rearrangements) in a B cell or a clone.
11. Experiments might involve testing vaccinated people for neutralizing antibodies against HIV (humoral immunity) and looking for T cell activity against HIV-infected cells (cellular immunity).

## CHAPTER 43

1. a  2. d  3. d  4. d  5. d  6. a

7. Leydig and thecal cells have similar functions and characteristics. They are both removed from direct contact with the developing gametes, and they both produce testosterone. Sertoli cells and granulosa cells are both in direct contact with the developing gametes, and they support their development by providing nutrients.
8. Progesterone actions are required to maintain the endometrium in a condition that can support implantation and not degenerate, as occurs during menstruation. By blocking progesterone receptors, RU-486 prevents implantation and the maintenance of the endometrium.
9. Conditions that could favor the evolution of this sexual dimorphism are a sessile existence, dispersed populations, and availability of suitable habitats. If a larva lands on a suitable substrate, it will have high reproductive success if it is a female and can produce lots of eggs; eggs typically have resources that enable them to travel considerable distances if released into the water, and their probability of being fertilized is therefore high. A larva would have a lower probability of success if it developed into a solitary male and produced sperm; sperm have less ability to survive travel over long distances in the water to encounter eggs. However, if a larva lands on a female, it is guaranteed high reproductive success if it can fertilize all of the eggs that female produces. Therefore, by attaching itself to the female and minimizing all of its own physiological processes other than sperm production, a larva can achieve high reproductive success at very little cost.
10. It is likely that the man's offspring would all be daughters. The Y chromosome lacks some essential genes that are on the X chromosome; in the absence of the cytoplasmic bridges, the developing sperm that contain a Y chromosome would lack those gene products. Thus all viable sperm the man produces would contain an X chromosome. This would result in female offspring, since the mother would also contribute an X chromosome.

## CHAPTER 44

1. e  2. a  3. d  4. b  5. b

6. You could inject Disheveled protein into the side of the fertilized egg opposite the gray crescent and see if a secondary organizer formed in that region of the resulting blastula. You could also inject the inhibitor of the Disheveled protein into the region of the gray crescent and see if that prevented the formation of the organizer.
7. You would be destroying the cells that normally migrate from the dorsal blastopore lip to form the notochord, and that also produce the signals that determine the anterior–posterior differentiation of the embryo. Thus you might see defects in the development of the nervous system or abnormal segmental development of the body.
8. The flow of fluid over Henson's node may be asymmetrical and thereby create different physical forces on the primary cilia of cells on either side of the node. These differential forces could influence the expression of *Sonic hedgehog*. Experiments to test this hypothesis could include using gene knockouts that eliminate the motile cilia around Henson's node, or experiments in which early embryos are cultured under conditions in which the flow across Henson's node is opposite that of the normal pattern. The prediction for the first experiments would be that the right–left asymmetry of organs in the embryos would be randomized. The prediction for the second experiment would be that the normal asymmetry of organ development would be reversed.
9. A possible mechanism would be cytoplasmic factors that are not distributed randomly or evenly throughout the cytoplasm of the oogonia or the spermatogonia. Thus when they divide by mitosis, factors that control the fates of the daughter cells could be received by one daughter cell but not the other.
10. At an early stage of blastulation (e.g., the 16- or 32-cell stage,) a few blastomeres could be removed from the embryo and cultured separately to produce a population of stem cells. The embryo could go on to develop normally, and the stem cells could be frozen for later use.

## CHAPTER 45

1. d  2. b  3. c  4. e  5. c

6. When the stimulus occurs at some point along an axon and an action potential is stimulated, depolarizing current will flow in both directions, bringing adjacent areas of the axon to threshold. However, once an action potential is fired, the $Na^+$ channel inactivation gates close and make that section of the axon refractory to further stimulation until they open again. Thus the action potential cannot reverse its direction of propagation, and if the action potential begins at the axon hillock, it cannot reverse its direction of propagation and is unidirectional.
7. Excitatory synapses cause a depolarization of the neuronal membrane, and inhibitory synapses hyperpolarize it. These two influences are summed by virtue of the resulting membrane potential. If it depolarizes enough to reach threshold, an action potential will be fired at the axon hillock.
8. Because the GABA receptor is inhibitory, benzodiazepines would be expected to slow cognitive processes and make a person more likely to fall asleep.

9. The type of information that an action potential transmits depends on the nature of the sensory cell that generated the action potential and on the nature of the cell that receives input as a result of that action potential. Thus photoreceptors transduce light into action potentials, and those action potentials are interpreted as light in the visual circuits that receive those action potentials. Intensity of the stimulus is coded as the frequency of action potentials. Integration is achieved by the summation of excitatory and inhibitory influences on the target cells.

## CHAPTER 46

1. d 2. a 3. e 4. e 5. c

6. Olfactory and taste receptors are both chemosensors that respond to specific molecules in their environment. Olfactory receptors, however, are neurons, whereas taste receptors are epithelial cells that communicate with neurons that are associated with them. Olfactory receptors express a family of genes for olfactory receptor proteins that are then localized on cilia that project out of the olfactory epithelium. All of these olfactory receptor proteins are G protein-linked and are metabotropic. Taste receptors are also located on cilia of the epithelial taste sensor cells. Bitter, sweet, and umami receptors are G protein-linked metabotropic receptors, but salt and sour receptors are ionotropic. The discrimination between an apple and an orange depends on integration of information from both the olfactory and the taste receptors.
7. The sensation of directional motion arises from the vestibular system, which includes the semicircular canals and the vestibule, which contains membranous structures containing a fluid—endolymph. At the base of each semicircular canal is a gelatinous projection, a cupula, that encases a cluster of stereocilia. Movement of the head causes movement of the endolymph, which then exerts force on the cupula and bends the stereocilia, generating action potentials in the vestibular nerves. The vestibule includes two membranous structures called the saccule and the utricle. In these structures, stereocilia tips are in contact with otoliths, which are membranous structures containing crystals of calcium carbonate. When the head is accelerated forward or backward, the momentum of the otoliths causes the stereocilia to bend in a direction that indicates the direction of movement.
8. Underwater, the external ear canals are filled with water. Unlike air, water is not compressible and therefore sound waves are transmitted through water as vibrational movements of the water. These movements exert greater forces on the tympanic membrane than air pressure waves do.
9. As happens in humans in the vestibular system, movements of the fish cause movements of the water in the lateral line canals. The resulting forces are transduced into action potentials by the hair cells and provide information about the movement of the fish through the water. Additionally, vibrations in the water generated by other organisms or physical events will cause movement of the water in the lateral line canal and be transduced into action potentials, providing information to the fish about its environment.
10. The owl depends on auditory stimuli to locate the mouse in total darkness. Directional information comes from the bilateral placement of the ears, which are equally stimulated when the owl is directly facing the source of the sound. The face of the owl is disc-shaped, which helps collect sound waves and direct them to the ears.

## CHAPTER 47

1. d 2. d 3. c 4. c 5. a

6. The stab wound must have severed the sympathetic nerves on the left side of the man's neck. Activity in these nerves causes dilation of the pupil. Severing the sympathetic nerves on the left side would remove all sympathetic activity reaching the pupil on the left side, and therefore it would be more constricted. Similarly, sympathetic activity decreases activity in the salivary glands, whereas parasympathetic activity increases salivation. Thus withdrawal of sympathetic input would release the salivary glands from any inhibition that would counteract even low levels of parasympathetic input.
7. Eyes positioned on the sides of the head enable a wider field of vision. Eyes pointing in the same direction create a narrow field of vision but make depth perception possible. You would expect prey species to benefit from wider fields of vision. You would expect predator species to benefit from depth perception, which would facilitate pursuit and capture of prey.
8. Sleepwalking is more likely to occur in non-REM sleep for two reasons: there is motor inhibition in REM sleep, which renders the individual paralyzed; and the nature of sleepwalking activities does not match with the vivid, bizarre content of REM-sleep dreams.
9. We can break this question down into the different observations. First, the loss of motor control of the right leg indicates that motor commands are ipsilateral—they descend on the same side as the limb that is being controlled. The ability to sense painful stimuli applied to the right leg but not to the left foot indicates that the pain pathways cross over to the opposite, or contralateral, side of the spinal cord before they ascend to the brain. The reflex movement of the right leg to a stimulus applied to the left foot indicates that reflex information is processed at the local level and does not require processing at higher levels of the central nervous system. The conclusion is that motor commands in the spinal cord are ipsilateral to the muscles being controlled but that the pain information ascending to the brain is contralateral. Finally, the different responses to pain and to touch indicate that these two modalities of somatosensory information travel in different tracks: touch ipsilaterally and pain contralaterally.
10. The fact that more slow-wave activity was seen over the right frontal cortex than over the left in response to exercising and training the left hand shows that the right side of the brain controls the left side of the body, and visa versa. This increase in slow-wave activity during sleep following the exercise/training suggests that this sleep slow-wave activity reflects either a restorative process or a learning process.

## CHAPTER 48

1. b 2. c 3. b 4. d 5. c

6. One feature is the length of the muscle; half the length of a long muscle is more than half the length of a short muscle. Another feature is the location of insertion of the muscle on the bone; this determines the relative lengths of the effort arm and the load arm of the lever system created by the muscle, bone, and joint. If the ratio of load to effort arm is small, a large movement that can generate only relatively small forces is possible. If the ratio of load to effort arm is large, only small movements that can exert large force are possible.
7. If the break and healing have damaged the epiphyseal plate, and the primary and secondary areas of ossification fuse, the bone can no longer grow at that end.
8. The shoulders will fatigue first because they are not normally responsible for maintaining posture; they are adapted for rapid movements and sudden applications of large force. Thus shoulder muscles have a higher proportion of fast-twitch fibers. The leg muscles are postural muscles and have a higher proportion of slow-twitch fibers.
9. The action potential is conducted throughout the muscle cell by the system of T tubules. In the T tubules, the action potential causes conformational change of the DHP–ryanodine receptor complex. That change opens $Ca^{2+}$ channels in the sarcoplasmic reticulum, and $Ca^{2+}$ diffuses into and throughout the sarcoplasm. $Ca^{2+}$ binds with the troponin units, causing the tropomyosin to expose the actin–myosin binding sites, cross-bridges to form, and the muscle to contract. When the $Ca^{2+}$ concentration in the sarcoplasm falls as a result of being pumped back into the sarcoplasmic reticulum, the process reverses and actin–myosin binding sites are no longer available. The difference in time course of the contraction versus the action potential is due to the time that it takes for the $Ca^{2+}$ to be released, diffuse throughout the sarcoplasm, and then be sequestered back into the sarcoplasmic reticulum.
10. The increased amount and duration of $Ca^{2+}$ in the sarcoplasm causes increased contraction of the muscles and therefore an increase in muscle tension. The increase in muscle tension requires additional expenditure of ATP, raising metabolism and producing more heat. The increased metabolism causes elevated heart rate. This is in addition to the effect of the increased $Ca^{2+}$ in the cardiac muscle itself.

## CHAPTER 49

1. d 2. e 3. b 4. c 5. a

6. This fish would not be very active. It would move slowly. It would have a larger heart and larger blood vessels than fishes with hemoglobin, to accommodate a high flow of blood at low pressure. Its gill membranes would be well developed. It would have a high blood volume. Its jaws would show adaptations for sit-and-wait capture rather than pursuit. This fish occurs only in Antarctic waters because Antarctic waters are very cold and therefore the solubility of $O_2$ in those waters is high.
7. The total surface area for gas exchange is much smaller in a large air cavity than it is in many smaller cavities (alveoli) that add up to the same total volume. If the lung tissue is less elastic, the vital capacity of the lungs will go down, meaning less air can be exchanged during the breathing cycle. The less elastic lung tissue is also less permeable to respiratory gases.
8. Close to the end of inhalation, the pleural cavity pressure is reaching its maximum negative value. At the same time, the alveolar pressure is rising back up to being the same as the atmospheric pressure. Alveolar pressure would be most positive relative to atmospheric pressure at the midpoint of the exhalation phase.
9. Blood cells require energy. When in storage, their initial energy source is the glucose in the blood plasma, but as that supply gets depleted, blood cells also metabolize the intermediates in the glycolytic pathway. That includes 2,3-BPG. As the 2,3-BPG gets metabolized, there is less of it to bind to deoxygenated hemoglobin and therefore the affinity of the hemoglobin for $O_2$ increases. When

the affinity of the hemoglobin for $O_2$ gets too high, it can lower the $P_{O_2}$ in the plasma to levels that are below the $P_{O_2}$ in the plasma of a patient.

10. When you go up in altitude, the $P_{O_2}$ in the air you breath goes down, but the $P_{CO_2}$ was already low at low altitude and remains low at high altitude. Thus at higher altitude there is less of a concentration gradient driving diffusion of $O_2$ into the blood, but no decrease in the concentration gradient driving $CO_2$ out of the blood. Thus the main stimulus for breathing goes down as the need to increase breathing goes up. As a result, the blood becomes hypoxic and triggers breathing by activating the carotid and aortic chemosensors. The increase in breathing blows off even more $CO_2$, and breathing slows, which causes another bout of hypoxia and even a rise in blood $CO_2$. That triggers another bout of rapid breathing, and this cycle repeats.
11. (a) The llama hemoglobin has a higher affinity for $O_2$.
    (b) Llama hemoglobin would be advantageous at high altitudes because it can become 100 percent saturated at the low $P_{O_2}$ of the high-altitude environment. Therefore the hemoglobin can carry a full load of $O_2$ to the tissues.
    (c) Llama hemoglobin allows the transfer of $O_2$ to occur at lower tissue $P_{O_2}$s.

## CHAPTER 50

1. a   2. c   3. d   4. c   5. e

6. One factor is that at the beginning of a race there is a feedforward signal from the sympathetic nervous system that increases heart rate. Another factor is that the increased heart rate, together with the increased venous return to the heart from the exercising muscles, stretches the ventricles, which then contract with more force as described by the Frank–Starling law. This is due to the fact that a slight stretching of the sarcomeres optimizes the overlap of the actin and myosin fibrils for a maximum contraction. Increased breathing also increases the venous return and induces the Frank–Starling law.
7. There is no time when all four heart valves are open at the same time; if there were, the heart could not pump efficiently. Throughout diastole, the aortic and pulmonary valves are closed and the atrioventricular valves are open. At the beginning of systole, the atrioventricular valves close. There is a brief moment when all four valves are closed, until the aortic and pulmonary valves open, and stay open until the end of systole.
8. There are rapid responses and longer-term responses. A fall in blood pressure lowers the firing rate of baroreceptors in the great arteries. The decrease in baroreceptor input to areas of the brainstem that regulate cardiac function results in increased sympathetic and decreased parasympathetic output to the heart. This increases the heart rate and the force of contraction of the cardiac muscle (the fight-or-flight response). A slower response to blood loss is mediated by the kidney, which responds to the decreased blood pressure by increasing the release of renin, which in turn increases the activation of angiotensin circulating in the blood. Active angiotensin increases blood pressure by constricting peripheral blood vessels and stimulating thirst. Another slow response, stimulated by the fall in baroreceptor activity, is the release of ADH from the posterior pituitary. ADH increases the reabsorption of water by the kidney.
9. The cardiac muscle must be capable of generating and conducting action potentials. This may involve different variants of the ion channels involved in action potential generation and conduction. The cardiac muscle must be able to convert the action potential into the opening of $Ca^{2+}$ channels in the sarcoplasmic reticulum, so there could be adaptive changes in the DHP and ryanodine receptors. Once $Ca^{2+}$ is released into the sarcoplasm, it has to be resequestered into the sarcoplasmic reticulum, and that $Ca^{2+}$ pump is likely to be adapted to operate at lower temperatures in the hibernator.
10. Your graph should look like this:

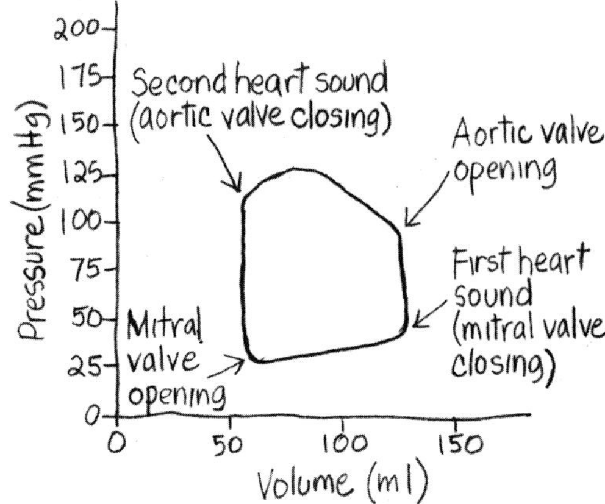


A similar graph for the right ventricle would have the same volumes but lower pressures.

## CHAPTER 51

1. e   2. a   3. b   4. d   5. b

6. The rationale for a high-fat and high-protein diet is that it minimizes the secretion of insulin. Insulin promotes the uptake, metabolism, and storage of glucose by various cells of the body; it also inhibits the action of lipase in the adipose tissue. Thus with low insulin, tissues are more likely to metabolize fats and less likely to store them.
7. The following points might be included in your answer.
    - Insulin stimulates most cells of the body to take up glucose from the blood by stimulating the insertion of glucose transporters into the plasma membranes of those cells in the absorptive state.
    - Insulin inhibits lipase in the adipose tissue, so breakdown of stored lipids is decreased in the absorptive state.
    - Insulin stimulates the synthesis of triglycerides in the adipose tissue.
    - Insulin activates the enzyme that phosphorylates glucose as it enters cells, thereby preventing it from diffusing back out of the cells. This maximizes the uptake of glucose by cells.
    - Insulin activates the liver enzymes that synthesize glycogen.
    - The lack of insulin in the postabsorptive state decreases the uptake of glucose by most cells of the body and activates the enzymes of lipolysis and glycogenolysis.
8. Triglyceride in food is emulsified by bile in the duodenum to form micelles, which are broken down by pancreatic lipase into free fatty acids and monoglycerides. Both are absorbed across the plasma membranes of intestinal epithelial cells. In the intestinal epithelial cells, the free fatty acids and monoglycerides are resynthesized into triglycerides and packaged into chylomicrons, which also contain cholesterol and are coated with lipoproteins. The chylomicrons are secreted from the basal ends of the epithelial cells into the center of the intestinal villi, where they enter the lymphatic vessels and circulate through the lymphatic vessels to the thoracic duct, where they enter the blood.
    A *direct route* the fatty acid could take would be for the chylomicrons circulating in the blood to come into contact with the damaged endothelium of the coronary arteries and be absorbed into the plaque.
    In an *indirect route*, the chylomicrons could be taken up by liver or adipose tissue cells and the triglyceride stored. In the liver, the triglyceride can be repackaged to form low-density lipoproteins or very low-density lipoprotein particles, depending on the amount of cholesterol in the particle. These lipoproteins leave the liver and circulate in the blood. When they come into contact with the damaged endothelial cells of the coronary arteries, they bind to lipoprotein receptors, and their triglyceride and cholesterol are absorbed into the plaque.
9. Carbonic anhydrase catalyzes the hydration of $CO_2$ to produce carbonic acid that dissociates into $H^+$ ions and $HCO_3^+$ ions. In the parietal cells of the stomach, $H^+$ is secreted into gastric pits and then flows into the stomach lumen. The bicarbonate ions are transported out of the basal side of the cells, where they are absorbed into the blood, raising blood pH. In the ducts of the pancreas, bicarbonate is transported into the lumen of the ducts; $H^+$ ions are transported out of the basal sides of the cells, where they are absorbed into the blood, lowering its pH.
10. The hypothalamic regulatory pathways control hunger and satiety, not energy use. This tells us that the side of the energy balance equation that is most important in contributing to obesity is Calories in. Of course, obesity can also result in decreased physical activity, which means that decreased Calories out can be a secondary factor in the causation of obesity.

## CHAPTER 52

1. d   2. a   3. a   4. b   5. a

6. The glucose contributes to the osmotic concentration of the glomerular filtrate and therefore of the tubular fluid. This results in a greater volume of urine flowing through the collecting ducts and being excreted.
7. ACE inhibitors decrease the production of angiotensin II, the active form, from angiontensin I. Decreasing the level of angiotensin in the blood increases the glomerular filtration rate and therefore the production of urine. Losing more water in the urine lowers the blood volume and therefore the blood pressure. Blocking angiotensin also results in dilation of peripheral blood vessels, which lowers blood pressure, and decreases thirst, which helps maintain a lower vascular volume. Angiotensin stimulates the release of aldosterone, which promotes $Na^+$ reabsorption and therefore water retention.
8. The rate at which the inulin is filtered is equal to the concentration of inulin in the blood ($[I_b]$) times the glomerular filtration rate (GFR). The rate at which inulin is excreted is equal to the concentration of inulin in the urine ($[I_u]$) times

the urine flow rate (V = 1 ml/min). Since all inulin that is filtered leaves the body in the urine, the rate at which it is filtered must equal the rate at which it is excreted. Therefore $[I_b] \times GFR = [I_u] \times V$, and $GFR = [I_u] \times V/[I_b]$.

9. Any difference in the rate at which the substance (S) is filtered versus the rate at which it is excreted will be due to tubular reabsorption of tubular secretion. If more of S is excreted than filtered, then S must be secreted by the renal tubules. If less of S is excreted than filtered, then S must be reabsorbed by the tubules. Therefore $[S_b] \times GFR - [S_u] \times V$ = the rate of reabsorption (if negative) or secretion (if positive) of S.
10. At high altitude, the concentration difference driving $O_2$ across the alveolar membranes and into the blood deceases, but the concentration difference driving $CO_2$ across the alveolar membranes and into the expired air does not change. The increased breathing rate driven by hypoxia will therefore blow off too much $CO_2$ and the pH of the blood will increase, which will suppress breathing. Bicarbonate is filtered into the glomerular fluid; as that fluid passes through the renal tubules, $H^+$ ions are excreted, resulting in the formation of $H_2CO_3$, which dissociates into $CO_2$ and $H_2O$. The $CO_2$ is reabsorbed into the tubule cells, where carbonic anhydrase catalyses its hydration to $H^+$ ions that are secreted back into the tubular fluid and bicarbonate ions that are secreted back into the extracellular fluid. Blocking carbonic anhydrase results in fewer $H^+$ ions being secreted into the tubular fluid and more bicarbonate ions remaining in the tubular fluid to be excreted. The retention of $H^+$ and excretion of bicarbonate lower the blood pH, and that stimulates increased breathing. However, the increased bicarbonate in the tubular fluid raises the osmotic concentration of that fluid, causing more water to be excreted.

## CHAPTER 53

1. b 2. e 3. c 4. a 5. d 6. e 7. a 8. e

9. The development of brain circuitry controlling the sexual dimorphism in urination behavior is influenced by the levels of estrogen or testosterone that are circulating in the blood during the early postnatal period. Estrogen must prevent the development of the neuronal patterns of connectivity that are responsible for the male behavior.
10. The major difference between the eusocial insects and vertebrates is the haplodiploid mechanism of sex determination in the insects. This means that a female worker that is a daughter of the queen shares more genes with sisters—the queen's offspring—than she would with her own offspring, which would have a different father. Thus raising a sister contributes more to the female worker's inclusive fitness than raising a daughter would. Haplodiploidy is not found among vertebrates, so this powerful selective force does not operate among them.
11. The variability in the undirected song allows the male to adapt his song to the local variant. The directed song is probably more effective in attracting a female because it more accurately identifies the male as a member of the local population, and is also more effective in competition with neighboring males.
12. Since male cowbirds will not hear the song of his father, his song pattern should be genetically determined. Females probably learn the song of host species and thereby learn to identify and locate potential host nests. You could test the hypothesis about males by raising them in isolation to see what songs develop as they mature. You could test the hypothesis about the females by giving them a choice test such as that done with the zebra finches in testing directed and undirected song preferences (see Figure 53.8). Give each female cowbird a three-chambered cage, and play the songs of host species and non-host species in the opposite end chambers. Place the test bird in the middle chamber. Record the amount of time the female spends in each end chamber as a measure of preference. Another experiment would be to bring a host nest with cowbird eggs into the laboratory and raise the cowbirds in the presence of recordings of another species song. Then repeat the choice experiment with the host's and the other species' songs being played.
13. The classical data suggest that the hygienic behavior is controlled by two genes to give the typical Mendelian ratio of 3 to 1 with 50% hybrid, 25% homozygous dominant, and 25% homozygous recessive. The QTL analysis, however, indicates that more than two genes are involved. The difference in these results could be due to there being two major-effect genes and several modifying genes.

## CHAPTER 54

1. d 2. d 3. a 4. c 5. c

6. Based on its location and its weather conditions, the mattoral should have vegetation typical of that found in other areas with Mediterranean climates. Vegetation should be (and is) tough, shrubby and fire-adapted, with slender, leathery leaves and seeds either stored in fire-safe cones or equipped with elaiosomes and dispersed by ants.
7. The fact that species in the genus are known in Australia and in southern Africa, it is likely that the genus originated before the breakup of Gondwanaland, the supercontinent comprising Antarctica, South America, Africa and Australia. Other species in the genus thus might be found in South America (contemporary conditions are too cold to maintain spider life in Antarctica today but the possibility exists that fossil species in the genus might be found).
8. Examining the *x* axis of the figure suggests that, if average low temperatures shift upward by five degrees C, the tundra biome would experience a substantial decrease in geographic extent (and may cease to exist altogether); boreal forest might also experience a reduction in geographic extent.
9. The extensive radiation of *Drosophila* species in Hawaii suggests that the genus originated here. Hawaii is an isolated island system and its distance from any continents means that dispersal to such remote islands of ancestral species would likely be very rare. Other lines of evidence to suggest that the genus originated in Hawaii would be whether fossil ancestral taxa are found only in Hawaii or anywhere else in the world. If ancestral taxa are not known from any other continent, it's very likely that the genus originated and subsequently diversified in Hawaii (and then dispersed elsewhere).

## CHAPTER 55

1. d 2. c 3. b 4. d 5. a 6. d

7. In both of these cases, human management strategies will be working against the organisms' intrinsic rates of increase. Populations of long-lived organisms with low reproductive rates grow slowly. Such organisms can be categorized as *K*-strategists, which tend to persist at or near the carrying capacity of the environment. They are adapted to predictable environments, and they tend to be more specialized in their resource use, and less tolerant of variation in resource quality, than other organisms. They produce few offspring, but each offspring has a high probability of surviving to adulthood. Recall that the number of births in a population tends to be highest when that population is well below its carrying capacity. For large, long-lived species that we wish to harvest, we should manage the population so that it is far enough below the carrying capacity to have a high birth rate. But some species (such as whales) reproduce so slowly that they cannot sustain any kind of substantial harvest rate. Short-lived organisms with high reproductive rates can be categorized as *r*-strategists. These organisms can generally use a wide variety of resources and tolerate a wide range of conditions, and they can produce large numbers of offspring when conditions are suitable. If we wish to decrease the numbers of a short-lived, rapidly reproducing pest species (such as rats), killing individuals will only increase the birth rate. A better approach is to reduce the species' resources (e.g., clean up garbage) in order to decrease the carrying capacity for the species. In each case, of course, managers need to understand the specific life history and population dynamics of the species they wish to manage.
8. Humans are subject to the same population dynamics that other species are. Resource abundance is a density-dependent population-regulating factor. Like the reindeer population on St. Matthew, the Irish population crashed when its food supply was diminished. Three kinds of changes in the rates of demographic events contributed to the decrease in population size. Recall that $N_1 = N_0 + (B - D) + (I - E)$. First, the emigration rate (*E*) increased. Second, the age at first reproduction, and thus generation time, increased, so the birth rate (*B*) decreased. In other words, the population's life history traits changed with changing environmental conditions. Finally, as a direct effect of the food shortage, the death rate (*D*) increased. A look at the social history of the Irish potato famine might tell us more about the roles of the Irish population's own growth before the famine and whether intraspecific competition (discussed further in Chapter 56) was involved in regulating the food supply—whether other populations monopolized resources and limited the access of the Irish people to those resources.
9. Those who support the view that biological controls should not be used under any circumstances might cite the example of the cane toad in Australia, which not only failed to control the cane beetles it was introduced to control but became a serious pest in its own right. They might note that many species introduced into new regions, where their normal predators and pathogens are absent, reach population densities much higher than those in their native ranges, and they might argue that there is no reason to think this generality would not apply to species introduced as biological control agents. Those who support the view that biological controls can be used safely and effectively might cite the example of the successful control of the cottony-cushion scale in California, which was brought under control within a year by the introduction of a predaceous ladybeetle and a parasitic fly. They might also argue that horror stories like that of the cane toad could be avoided by proper study of the ecology of the proposed biological control agents before they are introduced. Studies in test plots prior to release almost certainly would have revealed that Australian cane beetles stay high on the upper stalks of cane plants, out of reach of the toads, and studies of the toads' life histories might have revealed their generalized and voracious appetites. Strict requirements for extensive testing for specificity and efficacy prior to release can greatly reduce the risk that biological control agents themselves will become pests after they are introduced. But opponents of biological control might respond that, because natural systems are so complex, even careful study might fail to reveal the real risks of introducing a particular species into a new environment.
10. Corridors have to be defined in terms of specific organisms and their dispersal abilities. Corridors consist of habitat between patches through which the organisms of interest can move. An area that serves as a corridor for birds might not be effective as a corridor for small arthropods. On the other hand, the small arthropods would need less habitat area to maintain a viable population. Thus, designing a single study to determine the effects of corridors would be very

difficult. A single experiment might be able to determine effects of corridors on animals that are similar in size and mobility but not if organisms differ widely in those attributes. To understand the effects of corridors in fragmented habitats, it is very important to consider multiple organisms because organisms interact within these habitats. Investigators in the Palenque National Park in Mexico discovered that birds are more likely to be recaptured in home forest patches connected by corridors to the patches where they were released than in home forest patches unconnected to the patches where they were released. However, their ability to navigate these corridors successfully depends on the presence of other species, including predators, and their ability to survive in home patches depends on the presence of other species, including prey species, as well. Designing a single study to determine the effects of corridors would be very difficult.

## CHAPTER 56

1. a 2. c 3. c 4. d 5. e

6. The interactions among ants, cacti, and pollinators described in the Working With Data exercise represent a diversity of types of interactions. By fending off potential herbivores, the pugnacious ant bodyguards act as mutualists of the cactus, as do the bee species that visit the flowers and serve as pollinators. The five ant species all appear to use the extrafloral nectaries on this plant in similar ways and thus may be competitors for extrafloral nectar. Each type of interaction depends on the relative abundance and activities of the interacting species. Cactus plants that grow where there are no herbivorous insects may have no need of pugnacious bodyguards; under those circumstances, the ants might be considered parasites for removing extrafloral nectar without providing any defensive services. The mutualism between ants and plants could also break down if bee pollinators are scarce and the most aggressive ant bodyguard prevents any bees at all from pollinating the flowers.
7. Which pine trees are susceptible to mountain pine beetle attack could be determined by direct observation and by experimentation. Within infested stands, investigators may be able to identify individual pine trees that do not harbor beetle populations and characterize properties that may make them resistant to the beetles (e.g., ability to produce large quantities of resin). Conversely, trees with especially high beetle populations might have properties that make them particularly susceptible (having a history of surviving fire or lightning strike). Experiments can also be conducted under controlled conditions, in which investigators test various species to determine if beetles display a preference for particular species. The fact that this beetle has a symbiotic partner upon which it depends in order to feed on pine trees suggests that a novel strategy for managing the outbreak could be to identify a fungicide that kills the symbiont, thereby rendering the beetle incapable of colonizing and killing the trees. Although there are no such programs currently in use today, many researchers are exploring this dimension of interaction ecology to devise novel methods for pest management.
8. (a) By establishing a microbial population that excludes undesirable species, the poultry industry is applying the principle of competitive exclusion. The principle states that two or more species utilizing a limited resource in similar ways cannot coexist. In the broiler chicks given a culture of three species of bacteria, a microbial community was established in which introduced *Salmonella* could not compete.
   (b) Other ecological outcomes this experiment might have produced include ultimate domination of the gut flora by only one species of bacterium or coexistence of all four bacterial species. Whether these species coexist or whether one or more species goes extinct depends on the availability of resources in the chicken intestines.
   (c) The principle of competitive exclusion might be useful in tackling other problems involving a community of organisms growing under confined conditions. This ecological principle provides part of the rationale for the use of probiotics by humans to improve a variety of conditions. Probiotics are live organisms that are consumed in food for health benefits, which are thought to accrue by altering the microbial balance, inhibiting the growth of deleterious species. Probiotics are being investigation for treatment of intestinal inflammatory diseases, pathogen-related diarrhea, and infections of the urogenital tract.
9. If parasites and their hosts coevolve, then the phylogenetic relationships among parasites should reflect the phylogenetic relationships among the hosts. DNA analysis revealed that flamingoes are actually more closely related to grebes than they are to ducks and geese. This relationship leads to the prediction that lice on flamingoes should be more closely related to the lice on grebes than they are to the lice on ducks and geese. Modern methods of molecular analysis that can be used to determine relationships among bird lice and their hosts include constructing a DNA-based phylogeny of multiple species of waterbirds and their lice and then comparing the phylogenies of the hosts and parasites to see if they are congruent (see Chapter 22). In addition to acquiring parasites by shared ancestry between hosts and parasites, bird species may also acquire parasites by virtue of the fact that they share habitats and come in contact with other bird species, each of which has its own parasite fauna. Because flamingoes, ducks, grebes, and geese are all waterbirds, the possibility exists that flamingoes may have acquired some of its louse parasites by this process of host switching.
10. Among the requirements for a mutualistic pollination system is behavior by the pollinator that ensures it will visit more than one individual of the same plant species. Visiting more than one individual provides a pollinator with the opportunity to carry pollen from one plant individual to the receptive stigmatic surface of another individual of the same species. A pollinator that encounters a feeding deterrent that limits the amount of nectar it can imbibe in a single visit is more likely to continue foraging for nectar on another plant individual. The process of taking a larger number of smaller meals by seeking nectar from flowers of different plant individuals increases the likelihood that the pollinator will carry pollen from one individual to another. Too much nicotine in nectar, however, may reduce the likelihood of pollination if it deters future visits to conspecifics altogether or if it impairs the behavior of its pollinator (nicotine is a neurotoxin). Another factor limiting the amount of nicotine is the cost to the plant of biosynthesizing the compound; investing in increased amounts of nicotine may leave fewer resources to invest in producing flowers, seeds and fruits.

## CHAPTER 57

1. a 2. a 3. b 4. d 5. a 6. c 7. e

8. The diversity of microbes in the human gut can be compared between individuals or across populations by using the same methods employed for comparing diversity of macroscopic communities. Diversity encompasses both the number of different species present, and richness, or abundances of individuals across species. To determine which microbes might be keystone species, selective antibiotics can be used to eliminate particular species and the effects of that elimination on community composition then monitored. Applying the methods for assessing diversity that were developed for macroscopic communities to assessing diversity for microbial communities is limited, however, by our ability to isolate, identify, and quantify all of the microbial species present. Although molecular methods of identifying microbial species has vastly expanded this capacity, there remain challenges in recognizing and categorizing the full expanse of microbial diversity.
9. According to the theory of island biogeography, the number of species on an island represents a balance between the rate at which species immigrate to and colonize the island and the rate at which resident species go locally extinct. With increasing distance from a source pool, the equilibrium number of species on an island decreases; with increasing size of an island, the species number increases. The pattern of hawk moth diversity documented by Beck and Hitching in the 113 islands of Thailand and mainland Malaysia conforms to several of the predictions of island biogeography. The continental source of colonists includes over 180 species. Borneo, a large island close to Thailand, has a larger number of species (between 113 and 135) than does New Guinea (with 46 to 90 species), which is roughly comparable in size and farther away from Thailand. Generally speaking, too, the prediction that larger islands support higher numbers of species is also upheld; Borneo has a larger number of species than the much smaller Philippines (between 46 and 112 species), even though the two places are about equidistant from Thailand.
10. The pattern documented by Marek Sammul, Lauri Oksanen, and M. Magi—that removal of one perennial species from plant communities resulted in an increase in biomass of its competitors in highly productive communities—has been documented in other communities. One hypothesis postulates that interspecific competition becomes more intense when productivity is very high. Goldenrod (*Solidago virgaurea*) is apparently a superior competitor that, when present in a community, can suppress other species. In less productive communities, competition is less intense, so release from competition with goldenrod does not result in increased growth of any remaining species. The results of this study parallel those of a long-term experiment at the Rothamsted Experiment Station in England, in which fertilizer added regularly to selected plots of land to increase their productivity resulted in a decline in the number of plant species compared with the other plots in the study that were unfertilized; in these less productive, unfertilized plots, species diversity remained essentially the same. An alternative hypothesis could be that goldenrod inhibits the growth of co-occurring plant species (as some colonizing species do in early stages of succession). This hypothesis could be tested directly by extracting root exudates of goldenrod and testing their ability to inhibit germination and growth of other species in the community.
11. Whether lampreys should be eliminated as damaging parasites of game fish, or encouraged as ecosystem engineers that create nesting sites that might increase the reproductive success of game fish depends on many factors. Some of the factors are ecological; it is important to quantify the impact of existing sea lamprey populations on survivorship of game fish as well as to estimate the lamprey population size that does not influence survivorship. As well, the beneficial impact of nutrient enrichment and provisioning of nesting habitat should be measured. In addition, designing an ecologically sound lamprey management strategy will also require consideration of local cultural values; assessments of the economic value of the sport fishing industry to the local community and the aesthetic and cultural value placed by the local community on maintaining a more natural assemblage of fish species should be made and factored into management plans.

## CHAPTER 58

1. e 2. d 3. c 4. c 5. a 6. b

7. The rate of turnover would be important to the recovery rate of a lake, and that would depend on its location. In Lake Washington, which is located in a temperate climate, turnover would occur every spring and fall. When sewage was flowing into the lake, the nutrients it contained would have led to eutrophication and thus to oxygen depletion in the bottom water. Once the flow of sewage stopped, however, biomass production would have decreased. There would have been fewer dead organisms to sink to the lake bottom, less accumulation of nutrients on the lake bottom, and less oxygen-consuming decomposition there. Fall and spring turnover would have brought the accumulated nutrients to the lake surface and oxygen to the bottom, improving conditions for organisms that were typical of the lake's preindustrial community. If the lake had been located in a climate where seasonal temperature changes were not great enough to cause turnover, the excess nutrients that had accumulated would have remained on the lake bottom, and eutrophic conditions would have persisted much longer. It's also possible that other conditions in the area might affect the lake's recovery time. Acid precipitation, for example, can affect the viability of freshwater organisms, and, if it were a problem in the region, it might slow the lake community's recovery.

8. A local effect might be nitrogen deposition. Coal—an organic fossil fuel—contains nitrogen, and its combustion would release nitrogen compounds (such as nitrogen dioxide, $NO_2$, and nitrous oxide, $N_2O$) through the smokestacks into the atmosphere. Some of this nitrogen would fall back to land in precipitation or as dry particles. The resulting increase of nitrogen in the soil would favor those plant species that are best adapted to take advantage of high nutrient levels, which would then outcompete other species. Thus the composition of the plant community would change and species diversity would be likely to decrease. Nitrogen deposition might also contribute to eutrophication in lakes, and emission of nitrogen into the atmosphere would contribute to smog.

   A regional effect might be acid precipitation. The combustion of fossil fuels releases $NO_2$ and sulfur dioxide ($SO_2$) into the atmosphere; both compounds react with water molecules in the atmosphere to form nitric acid ($HNO_3$) and sulfuric acid ($H_2SO_4$), respectively. These acids can travel hundreds of kilometers in the atmosphere, so their emission would affect ecosystems far from the smokestacks. Acid precipitation can damage the leaves of plants and reduce their rate of photosynthesis, and it can reduce fish and invertebrate species richness in freshwater lakes.

   A global effect would be climate change. The combustion of fossil fuels releases large amounts of $CO_2$, as well as lesser amounts of $N_2O$—both greenhouse gases. The presence of $N_2O$ in the atmosphere also results in the production of trophospheric ozone; it, too, acts as a greenhouse gas as well as contributing to smog. The increasing concentrations of greenhouse gases in the atmosphere are already resulting in global climate warming. This climate change is having a number of worrisome effects on the global ecosystem, such as the shrinking of Arctic sea ice, rising sea level and potential coastal flooding, and profound changes in the abundances and distributions of species.

   The $SO_2$ released when coal is burned not only produces acid rain, but also contributes to global warming. Scrubbers remove the $SO_2$ but the scrubbing process contributes to pollution in another form—the process generates solid waste byproducts that contain sulfur, which must be deposited in a landfill, along with other solid waste products generated by burning coal.

9. Iron (Fe) is needed by organisms in only small amounts, but it is nevertheless an essential micronutrient. It is scarce in ocean waters because it is insoluble in oxygenated water, so that iron that enters the oceans sinks rapidly to the seafloor. The experiment described in the text demonstrated that iron is a limiting nutrient in the oceans: when the investigators added dissolved iron to surface waters in the equatorial Pacific Ocean, the large phytoplankton bloom that resulted was accompanied by an increase in the uptake of nitrate and carbon dioxide, showing that these nutrients had been available but underused. This experiment showed that adding iron to ocean waters increased photosynthesis, but to better understand the effects of iron fertilization, more such experiments would have to be carried out, still on an ecosystem scale, but over a longer time span. Investigators would have to observe the effects of the iron increase on the entire food web. The fertilized ecosystem would have to be compared with an unfertilized control ecosystem far enough away from the experimental one that the added iron, and its effects, would not reach it.

10. If the "cap and trade" system worked as intended, it could put the brakes on the ongoing increases quickly, holding emissions to the level that prevailed when the law went into effect. It might also be more acceptable to polluters than an outright ban on or regulation of emissions. And it might encourage some companies to invest more in cleaner technology, since doing so might give them credits to sell, or at least spare them having to buy credits. The drawbacks might include the likelihood that the government would have to set up a system to administer and enforce the law. They also include the fact that it is not easy to pass such a law (the United States, for example, has not succeeded in doing so). The biggest polluters might be reluctant to increase their costs of doing business by paying for credits and might thus be likely to lobby against such a law. From a different viewpoint, environmentalists might argue that a cap and trade system is an inadequate response to global warming—that we must not only stop increases in emissions, but decrease them dramatically. They might also argue that there is a moral hazard in allowing anyone to "pay to pollute"—that it might legitimize pollution.

11. No one hurricane—or even several—can be ascribed to global warming. Remember the difference between weather and climate, described in Chapter 54: "Weather is the short-term state of atmospheric conditions at a particular place and time, whereas climate refers to the average atmospheric conditions, and the extent of their variation, at a particular place over a longer time. In other words, climate is what you expect; weather is what you get." But neither does the observation that hurricanes have occurred for many centuries show that the climate has *not* changed. To address this question, we would have to compile temperature and hurricane data over long periods. First, we would have to show that the average temperatures of ocean waters are increasing over time—which has been done. Second, we would have to show that warmer water is correlated with more or stronger hurricanes—that there have been more and stronger hurricanes during years, or longer periods, when the water was warmer than in periods when it was cooler. Such a correlation would supply evidence that warming of the oceans is increasing hurricane frequency and intensity. It is more difficult to demonstrate that the warming of ocean waters is caused by increasing concentrations of greenhouse gases in the atmosphere, but most scientists believe that the evidence supports that claim.

## CHAPTER 59

1. b 2. e 3. e 4. a 5. d 6. c 7. b

8. Conservation biologists and others who wish to preserve biodiversity are usually working with limited resources, so they often face hard choices. They might choose to focus their efforts in biodiversity hotspots and centers of imminent extinction, but within those areas, they would face many other choices. When we discussed the principles of island biogeography, we described the species–area relationship: large islands can support a larger equilibrium number of species than small islands. The same is true of "habitat islands." Protected areas often act as habitat islands, as many are surrounded by habitat that has been made unsuitable for many species by human activities. We must also consider edge effects, keeping in mind that not all of the area we protect will actually remain suitable habitat for communities and species of interest. Therefore, if we wish to preserve natural communities with their full diversity, the larger the preserved area, the better. However, if our concern is focused on one or a few endangered species, we may wish to preserve several separated areas of habitat; that way, if a disturbance or disease should wipe out the population in one area, the entire species will not become extinct. The area or areas we choose, however, must be large enough for the species to maintain a viable population in order to avoid loss of genetic variation. We might favor areas where corridors could be maintained to allow individuals to disperse from the protected area or areas and maintain other populations. But it is rare that a protected area can be designed based on these criteria alone; the plans are also dependent on the willingness of landowners, governments, and area residents to support the preservation of the area.

9. Some might argue that if the sheep constitute only a single population, but there are other populations of pumas, that the puma should be removed from the sheep's range. However, if the puma is a keystone predator in the region, removing it might have unforeseen negative consequences for other species in the community—recall the example of wolves in Yellowstone National Park, described in Chapter 57. If neither the sheep nor the puma is an introduced species—that is, if predator and prey had survived together for a long time before becoming threatened—it might be worth taking a look at what has changed. Have the sheep experienced a loss of habitat or resources, so that their populations are now too small to withstand the rate of predation that they once did? Those observations might suggest an alternative to suppressing the puma population: Could former sheep habitat be restored so that a larger sheep population could be supported, or as a last resort, could the sheep be bred in captivity and then introduced to a new, puma-free area?

10. To some extent, international organizations already have a triage system of sorts in in place. The International Union for the Conservation of Nature (IUCN), e.g., divides species in imminent danger of extinction in all or most of their range as "endangered" or "critically endangered", differentiating them from those who are less likely to go extinct in the near future (and thus are classified as "vulnerable"). To some degree, this classification system can result in prioritization of rescue efforts. One problem with applying the triage system of World War I to species conservation, however, is that medical science is far more successful at predicting the certainty of death than ecological science is at predicting the certainty of extinction. After all, medical science is focused on one species, which has been the subject of intense scrutiny beginning in the earliest days of scientific research. The cost of erring in assigning certainty to extinction is the loss of an entire species—a unique combination of genes that, at least with current technology, can never be reconstructed.

11. Opinion as to the extent to which ethical and moral arguments should enter into discussions of protecting biodiversity varies widely. Your answer might take into consideration a wide range of cultural, historical, and economic factors, as well as the considerations brought to bear by modern scientific knowledge.

# Glossary

## A

**A horizon** *See* topsoil.

**abiotic** (a′ bye ah tick) [Gk. *a*: not + *bios*: life] Nonliving. (Contrast with biotic.)

**abomasum** The true stomach of a ruminant.

**abortion** Any termination of pregnancy, whether induced or natural (in which case it is called a spontaneous abortion), that occurs after a fertilized egg is successfully implanted in the uterus.

**abscisic acid (ABA)** (ab sighs′ ik) A plant growth substance with growth-inhibiting action. Causes stomata to close; involved in a plant's response to salt and drought stress.

**abscission** (ab sizh′ un) [L. *abscissio*: break off] The process by which leaves, petals, and fruits separate from a plant.

**absorption** (1) Of light: complete retention, without reflection or transmission. (2) Of water or other molecules: soaking up (taking in through pores or by diffusion).

**absorption spectrum** A graph of light absorption versus wavelength of light; shows how much light is absorbed at each wavelength.

**absorptive heterotrophs** Organisms (primarily fungi) that feed by **absorptive heterotrophy**, i.e., by secreting digestive enzymes into the environment to break down large food molecules, then absorbing the breakdown products.

**absorptive state** State in which food is in the gut and nutrients are being absorbed. (Contrast with postabsorptive state.)

**abyssal zone** (uh biss′ ul) [Gk. *abyssos*: bottomless] The deepest parts of the ocean.

**accessory pigments** Pigments that absorb light and transfer energy to chlorophylls for photosynthesis.

**accessory sex organs** Anatomical structures that allow transfer of sperm from male to female for internal fertilization. (contrast with primary sex organs.)

**acclimation, acclimatization** Acclimation refers to increased tolerance for environmental extremes (e.g., extreme cold) after prior exposure to them. Acclimatization refers to intrinsic seasonal adjustments in the "set points" of an animal's physiological functioning (e.g., metabolic rate).

**acetyl coenzyme A (acetyl CoA)** A compound that reacts with oxaloacetate to produce citrate at the beginning of the citric acid cycle; a key metabolic intermediate in the formation of many compounds.

**acetylcholine (ACh)** A neurotransmitter that carries information across vertebrate neuromuscular junctions and some other synapses. It is then broken down by the enzyme acetylcholinesterase (AChE).

**acid** [L. *acidus*: sharp, sour] A substance that can release a proton in solution. (Contrast with base.)

**acid growth hypothesis** The hypothesis that auxin increases proton pumping, thereby lowering the pH of the cell wall and activating enzymes that loosen polysaccharides. Proposed to explain auxin-induced cell expansion in plants.

**acid precipitation** Precipitation that has a lower pH than normal as a result of acid-forming precursor molecules introduced into the atmosphere by human activities.

**acidic** Having a pH below 7.0 (i.e., a hydrogen ion concentration greater than $10^{-7}$ molar). (Contrast with basic.)

**acoelomate** An animal that does not have a coelom.

**acrosome** (a′ krow soam) [Gk. *akros*: highest + *soma*: body] The structure at the forward tip of an animal sperm which is the first to fuse with the egg membrane and enter the egg cell.

**ACTH** *See* corticotropin.

**actin** [Gk. *aktis*: ray] A protein that makes up the cytoskeletal microfilaments in eukaryotic cells and is one of the two contractile proteins in muscle. See also myosin.

**action potentials** Generated by neurons, these are electrical signals that transmit information via waves of depolarization or hyperpolarization of the cell membrane.

**action spectrum** A graph of a biological process versus light wavelength; shows which wavelengths are involved in the process.

**activation energy ($E_a$)** The energy barrier that blocks the tendency for a chemical reaction to occur.

**activator** A transcription factor that stimulates transcription when it binds to a gene's promoter. (Contrast with repressor.)

**active site** The region on the surface of an enzyme or ribozyme where the substrate binds, and where catalysis occurs.

**active transport** The energy-dependent transport of a substance across a biological membrane against a concentration gradient—that is, from a region of low concentration (of that substance) to one of high concentration. (*See also* primary active transport, secondary active transport; contrast with facilitated diffusion, passive transport.)

**adaptation** (a dap tay′ shun) (1) In evolutionary biology, a particular structure, physiological process, or behavior that makes an organism better able to survive and reproduce. Also, the evolutionary process that leads to the development or persistence of such a trait. (2) In sensory neurophysiology, a sensory cell's loss of sensitivity as a result of repeated stimulation.

**adaptive defenses** One of the two general types of defenses against pathogens. Involves antibody proteins and other proteins that recognize, bind to, and aid in the destruction of specific viruses and bacteria. Present only in vertebrate animals. (Contrast with innate defenses.)

**adaptive radiation** A series of evolutionary events that results in an array (radiation) of related species that live in a variety of environments, differing in the characteristics each uses to exploit those environments.

**additive growth** Population growth in which a constant number of individuals is added to the population during successive time intervals. (Contrast with multiplicative growth.)

**adenine (A)** (a′ den een) A nitrogen-containing base found in nucleic acids, ATP, NAD, and other compounds.

**adenosine triphosphate** *See* ATP.

**adrenal gland** (a dree′ nal) [L. *ad*: toward + *renes*: kidneys] An endocrine gland located near the kidneys of vertebrates, consisting of two parts, the **adrenal cortex** and **adrenal medulla.**

**adrenaline** *See* epinephrine.

**adrenergic receptors** G protein-linked receptor proteins that bind to the hormones epinephrine and norepinephrine, triggering specific responses in the target cells.

**adrenocorticotropic hormone (ACTH)** *See* corticotropin.

**adsorption** Binding of a gas or a solute to the surface of a solid.

**adventitious roots** (ad ven ti′ shus) [L. *adventitius*: arriving from outside] Roots originating from the stem at ground level or below; typical of the fibrous root system of monocots.

**aerenchyma** In plants, parenchymal tissue containing air spaces.

**aerobic** (air oh′ bic) [Gk. *aer*: air + *bios:* life] In the presence of oxygen; requiring or using oxygen (as in **aerobic metabolism**). (Contrast with anaerobic.)

**afferent** (af′ ur unt) [L. *ad*: toward + *ferre*: to carry] Carrying to, as in neurons that carries impulses to the central nervous system (**afferent neurons**), or a blood vessel that carries blood to a structure. (Contrast with efferent.)

**age structure** The distribution of the individuals in a population across all age groups.

**agonist** A chemical substance (e.g., a neurotransmitter) that elicits a specific response in a cell or tissue. (Contrast with antagonist.)

**air sacs** Structures in the respiratory system of birds that receive inhaled air; they keep fresh air flowing unidirectionally through the lungs, but are not themselves gas exchange surfaces.

**alcoholic fermentation** *See* fermentation.

**aldosterone** (al dohs′ ter own) A steroid hormone produced in the adrenal cortex of mammals. Promotes secretion of potassium and reabsorption of sodium in the kidney.

**aleurone layer** In some seeds, a tissue that lies beneath the seed coat and surrounds the endosperm. Secretes digestive enzymes that break down macromolecules stored in the endosperm.

**allantoic membrane** In animal development, an outgrowth of extraembryonic endoderm plus adjacent mesoderm that forms the allantois, a saclike structure that stores metabolic wastes produced by the embryo.

**allantois** (al′ lun toh is) [Gk. *allant*: sausage] An extraembryonic membrane enclosing a sausage-shaped sac that stores the embryo's nitrogenous wastes.

**allele** (a leel′) [Gk. *allos*: other] The alternate form of a genetic character found at a given locus on a chromosome.

**allele frequency** The relative proportion of a particular allele in a specific population.

**allergic reaction** [Ger. *allergie*: altered] An overreaction of the immune system to amounts of an antigen that do not affect most people; often involves IgE antibodies.

**allopatric speciation** (al′ lo pat′ rick) [Gk. *allos*: other + *patria*: homeland] The formation of two species from one when reproductive isolation occurs because of the interposition of (or crossing of) a physical geographic barrier such as a river. Also called geographic speciation. (Contrast with sympatric speciation.)

**allopolyploidy** The possession of more than two chromosome sets that are derived from more than one species.

**allosteric regulation** (al lo steer′ ik) [Gk. *allos*: other + *stereos*: structure] Regulation of the activity of a protein (usually an enzyme) by the binding of an effector molecule to a site other than the active site.

**α (alpha) helix** A prevalent type of secondary protein structure; a right-handed spiral.

**alternation of generations** The succession of multicellular haploid and diploid phases in some sexually reproducing organisms, notably plants.

**alternative splicing** A process for generating different mature mRNAs from a single gene by splicing together different sets of exons during RNA processing.

**altruistic** Pertaining to behavior that benefits other individuals at a cost to the individual who performs it.

**alveolus** (al ve′ o lus) (plural: alveoli) [L. *alveus*: cavity] A small, baglike cavity, especially the blind sacs of the lung.

**amensalism** (a men′ sul ism) Interaction in which one animal is harmed and the other is unaffected. (Contrast with commensalism, mutualism.)

**amine** An organic compound containing an amino group ($NH_2$).

**amine hormones** Small hormone molecules synthesized from single amino acids (e.g., thyroxine and epinephrine).

**amino acid** An organic compound containing both $NH_2$ and COOH groups. Proteins are polymers of amino acids.

**amino acid replacement** A change in the nucleotide sequence that results in one amino acid being replaced by another.

**ammonia** $NH_3$, the most common nitrogenous waste.

**ammonotelic** (am moan′ o teel′ ic) [Gk. *telos*: end] Pertaining to an organism in which the final product of breakdown of nitrogen-containing compounds (primarily proteins) is **ammonia**. (Contrast with ureotelic, uricotelic.)

**amnion** (am′ nee on) The fluid-filled sac within which the embryos of reptiles (including birds) and mammals develop.

**amniote egg** A shelled egg surrounding four extraembryonic membranes and embryo-nourishing yolk. This evolutionary adaptation permitted mammals and reptiles to live and reproduce in drier environments than can most amphibians.

**amphipathic** (am′ fi path′ ic) [Gk. *amphi*: both + *pathos*: emotion] Of a molecule, having both hydrophilic and hydrophobic regions.

**amplitude** The magnitude of change over the course of a regular cycle.

**amygdala** A component of the limbic system that is involved in fear and the memory of fearful experiences.

**amylase** (am′ ill ase) An enzyme that catalyzes the hydrolysis of starch, usually to maltose or glucose.

**anabolic reaction** (an uh bah′ lik) [Gk. *ana*: upward + *ballein*: to throw] A synthetic reaction in which simple molecules are linked to form more complex ones; requires an input of energy and captures it in the chemical bonds that are formed. (Contrast with catabolic reaction.)

**anaerobic** (an ur row′ bic) [Gk. *an*: not + *aer*: air + *bios*: life] Occurring without the use of molecular oxygen, $O_2$. (Contrast with aerobic.)

**anaphase** (an′ a phase) [Gk. *ana*: upward] The stage in cell nuclear division at which the first separation of sister chromatids (or, in the first meiotic division, of paired homologs) occurs.

**ancestral trait** The trait originally present in the ancestor of a given group; may be retained or changed in the descendants of that ancestor.

**androgen** (an′ dro jen) Any of the several male sex steroids (most notably testosterone).

**aneuploidy** (an′ you ploy dee) A condition in which one or more chromosomes or pieces of chromosomes are either lacking or present in excess.

**angiosperms** Flowering plants; one of the two major groups of living seed plants. (*See also* gymnosperms.)

**angiotensin** (an′ jee oh ten′ sin) A peptide hormone that raises blood pressure by causing peripheral vessels to constrict. Also maintains glomerular filtration by constricting efferent vessels and stimulates thirst and the release of aldosterone.

**angular gyrus** A part of the human brain believed to be essential for integrating spoken and written language.

**animal hemisphere** The metabolically active upper portion of some animal eggs, zygotes, and embryos; does not contain the dense nutrient yolk. (Contrast with vegetal hemisphere.)

**anion** (an′ eye on) [Gk. *ana*: upward] A negatively charged ion. (Contrast with cation.)

**annual** A plant whose life cycle is completed in one growing season. (Contrast with biennial, perennial.)

**antagonist** A biochemical (e.g., a drug) that blocks the normal action of another biochemical substance.

**antagonist interactions** Interactions between two species in which one species benefits and the other is harmed. Includes predation, herbivory, and parasitism.

**antenna system** *See* light-harvesting complex.

**anterior** Toward or pertaining to the tip or headward region of the body axis. (Contrast with posterior.)

**anterior pituitary** The portion of the vertebrate pituitary gland that derives from gut epithelium. Produces trophic hormones.

**anther** (an′ thur) [Gk. *anthos*: flower] A pollen-bearing portion of the stamen of a flower.

**antheridium** (an′ thur id′ ee um) [Gk. *antheros*: blooming] The multicellular structure that produces the sperm in nonvascular land plants and ferns.

**antibody** One of the myriad proteins produced by the immune system that specifically binds to a foreign substance in blood or other tissue fluids and initiates its removal from the body.

**anticodon** The three nucleotides in transfer RNA that pair with a complementary triplet (a codon) in messenger RNA.

**antidiuretic hormone (ADH)** *See* vasopressin

**antigen** (an′ ti jun) Any substance that stimulates the production of an antibody or antibodies in the body of a vertebrate.

**antigen-presenting cell** In cellular immunity, a cell that ingests and digests an antigen, and then exposes fragments of that antigen to the outside of the cell, bound to proteins in the cell's plasma membrane.

**antigenic determinant** The specific region of an antigen that is recognized and bound by a specific antibody. Also called an epitope.

**antiparallel** Pertaining to molecular orientation in which a molecule or parts of a molecule have opposing directions.

**antiporter** A membrane transport protein that moves one substance in one direction and another in the opposite direction. (Contrast with symporter, uniporter.)

**antisense RNA** A single-stranded RNA molecule complementary to, and thus targeted against, an mRNA of interest to block its translation.

**anus** (a′ nus) An opening through which solid digestive wastes are expelled, located at the posterior end of a tubular gut.

**aorta** (a or′ tah) [Gk. *aorte:* aorta] The main trunk of the arteries leading to the systemic (as opposed to the pulmonary) circulation.

**aortic body** A chemosensor in the aorta that senses a decrease in blood supply or a dramatic decrease in partial pressure of oxygen in the blood.

**aortic valve** A one-way valve between the left ventricle of the heart and the aorta that prevents backflow of blood into the ventricle when it relaxes.

**apex** (a′ pecks) The tip or highest point of a structure, as of a growing stem or root.

**aphasia** a deficit in the ability to use or understand words.

**aphotic zone** In bodies of water (lakes and oceans), the region below the reach of light.

**apical dominance** In plants, inhibition by the apical bud of the growth of axillary buds.

**apical hook** A form taken by the stems of many eudicot seedlings that protects the delicate shoot apex while the stem grows through the soil.

**apical meristem** The meristem at the tip of a shoot or root; responsible for a plant's primary growth.

**apomixis** (ap oh mix′ is) [Gk. *apo*: away from + *mixis*: sexual intercourse] The asexual production of seeds.

**apoplast** (ap′ oh plast) In plants, the continuous meshwork of cell walls and extracellular spaces through which material can pass without crossing a plasma membrane. (Contrast with symplast.)

**apoptosis** (ap uh toh′ sis) A series of genetically programmed events leading to cell death.

**aposematism** Warning coloration; bright colors or striking patterns of toxic or toxix-mimic species that act as a warning to predators.

**appendix** In the human digestive system, the vestigial equivalent of the cecum (blind pouc), which serves no digestive function.

**aquaporin** A transport protein in plant and animal cell membranes through which water passes in osmosis.

**aquatic** (a kwa′ tic) [L. *aqua*: water] Pertaining to or living in water. (Contrast with marine, terrestrial.)

**aqueous** (a′ kwee us) Pertaining to water or a watery solution.

**aquifer** A large pool of groundwater.

**archegonium** (ar′ ke go′ nee um) The multicellular structure that produces eggs in nonvascular land plants, ferns, and gymnosperms.

**archenteron** (ark en′ ter on) [Gk. *archos*: first + *enteron*: bowel] The earliest primordial animal digestive tract.

**area phylogeny** Phylogenetic tree n which the names of the taxa are replaced with the names of the places where those taxa live or lived.

**arms race** A series of reciprocal adaptations between species involved in antagonistic interactions, in which adaptations that increase the fitness of a consumer species exert selection pressure on its resource species to counter the consumer's adaptation, and vice versa.

**arteriole** A small blood vessel arising from an artery that feeds blood into a capillary bed.

**artery** A muscular blood vessel carrying oxygenated blood away from the heart to other parts of the body. (Contrast with vein.)

**artificial insemination** An infertility treatment that involves the artificial introduction of sperm into the woman's reproductive tract.

**artificial selection** The selection by human plant and animal breeders of individuals with certain desirable traits.

**ascus** (ass′ cus) (plural: asci) [Gk. *askos*: bladder] In sac fungi, the club-shaped sporangium within which spores (ascospores) are produced by meiosis.

**asexual reproduction** Reproduction without sex.

**assisted reproductive technologies (ARTs)** Any of several procedures that remove unfertilized eggs from the ovary, combine them with sperm outside the body, and then place fertilized eggs or egg–sperm mixtures in the appropriate location in a female's reproductive tract for development.

**association cortex** In the vertebrate brain, the portion of the cortex involved in higher-order information processing, so named because it integrates, or associates, information from different sensory modalities and from memory.

**associative learning** A form of learning in which two unrelated stimuli become linked to the same response.

**asthenosphere** (ass thenn′ o sphere) [Gk. *asthenes*: weak] The viscous, malleable (changeable) layer of Earth's mantle. It is overlain by the solid lithospheric plates.

**astrocyte** [Gk. *astron*: star] A type of glial cell that contributes to the blood–brain barrier by surrounding the smallest, most permeable blood vessels in the brain.

**atherosclerosis** (ath′ er oh sklair oh′ sis) [Gk. *athero*: gruel, porridge + *skleros*: hard] A disease of the lining of the arteries characterized by fatty, cholesterol-rich deposits in the walls of the arteries. When fibroblasts infiltrate these deposits and calcium precipitates in them, the disease become arteriosclerosis, or "hardening of the arteries."

**atom** [Gk. *atomos*: indivisible] The smallest unit of a chemical element. Consists of a nucleus and one or more electrons.

**atomic mass** *See* atomic weight.

**atomic number** The number of protons in the nucleus of an atom; also equals the number of electrons around the neutral atom. Determines the chemical properties of the atom.

**atomic weight** The average of the mass numbers of a representative sample of atoms of an element, with all the isotopes in their normally occurring proportions. Also called atomic mass.

**ATP (adenosine triphosphate)** An energy-storage compound containing adenine, ribose, and three phosphate groups. When it is formed from ADP, useful energy is stored; when it is broken down (to ADP or AMP), energy is released to drive endergonic reactions.

**ATP synthase** An integral membrane protein that couples the transport of protons with the formation of ATP.

**atrial natriuretic peptide** A hormone released by the atrial muscle fibers of the heart when they are overly stretched, which decreases reabsorption of sodium by the kidney and thus blood volume.

**atrioventricular node** A modified node of cardiac muscle that organizes the action potentials that control contraction of the ventricles.

**atrium** (a′ tree um) [L. *atrium*: central hall] An internal chamber. In the hearts of vertebrates, the thin-walled chamber(s) entered by blood on its way to the ventricle(s). Also, the outer ear.

**auditory system** A sensory system that uses mechanoreceptors to convert pressure waves into receptor potentials; includes structures that gather sound waves, direct them to a sensory organ, and amplify their effect on the mechanoreceptors.

**autocatalysis** [Gk. *autos*: self + *kata*: to break down] A positive feedback process in which an activated enzyme acts on other inactive molecules of the same enzyme to activate them.

**autocrine** A chemical signal that binds to and affects the cell that makes it. (Contrast with paracrine.)

**autoimmune diseases** Diseases (e.g., rheumatoid arthritis) that result from failure of the immune system to distinguish between self and nonself, causing it to attack tissues in the organism's own body.

**autoimmunity** An immune response by an organism to its own molecules or cells.

**autonomic nervous system (ANS)** The portion of the peripheral nervous system that controls such involuntary functions as those of guts and glands. Also called the involuntary nervous system.

**autophagy** The programmed destruction of a cell's components.

**autopolyploidy** The possession of more than two entire chromosomes sets that are derived from a single species.

**autoregulatory mechanisms** In mammalian circulatory systems, local control of blood flow through capillary beds by constriction or dilation of incoming arterioles in response to local metabolite concentrations.

**autosome** Any chromosome (in a eukaryote) other than a sex chromosome.

**autotroph** (au' tow trowf') [Gk. *autos:* self + *trophe*: food] An organism that is capable of living exclusively on inorganic materials, water, and some energy source such as sunlight (photoautotrophs) or chemically reduced matter (see chemoautotrophs). (Contrast with heterotroph.)

**auxin** (awk' sin) [Gk. *auxein*: to grow] In plants, a substance (the most common being indoleacetic acid) that regulates growth and various aspects of development.

**avirulence (*Avr*) genes** Genes in a pathogen that may trigger defenses in plants. *See* gene-for-gene resistance.

**Avogadro's number** The number of atoms or molecules in a mole (weighed out in grams) of a substance, calculated to be $6.023 \times 10^{23}$.

**axillary bud** A bud that forms in the angle (axil) where a leaf meets a stem.

**axon** [Gk. axle] The process (branching structure) of a neuron that conducts action potentials away from the cell body. *See also* dendrites.

**axon hillock** The junction between an axon and the neuron's cell body; where action potentials are generated.

**axon terminal** The end portion of an axon, which passes action potentials to another cell. Axon terminals can form synapses and release neurotransmitter.

## B

**B cell** A type of lymphocyte involved in the humoral immune response of vertebrates. Upon recognizing an antigenic determinant, a B cell develops into a plasma cell, which secretes an antibody. (Contrast with T cell.)

**B horizon** *See* subsoil.

**bacillus** (bah sil' us) [L: little rod] Any of various rod-shaped bacteria.

**bacterial conjugation** *See* conjugation.

**bacteriophage** (bak teer' ee o fayj) [Gk. *bakterion*: little rod + *phagein*: to eat] Any of a group of viruses that infect bacteria. Also called phage.

**bacteroids** Nitrogen-fixing organelles that develop from endosymbiotic bacteria.

**bark** All tissues external to the vascular cambium of a plant.

**barometric pressure** Atmospheric pressure; the total pressure of the gas mixture in air.

**baroreceptor** [Gk. *baros*: weight] A pressure-sensing cell or organ. Sometimes called a stress receptor.

**basal metabolic rate (BMR)** The minimum rate of energy turnover in an awake (but resting) bird or mammal that is not expending energy for thermoregulation.

**base** (1) A substance that can accept a hydrogen ion in solution. (Contrast with acid.) (2) In nucleic acids, the purine or pyrimidine that is attached to each sugar in the sugar–phosphate backbone.

**base pair (bp)** In double-stranded DNA, a pair of nucleotides formed by the complementary base pairing of a purine on one strand and a pyrimidine on the other. (*See* complementary base pairing.)

**basic** Having a pH greater than 7.0 (i.e., having a hydrogen ion concentration lower than $10^{-7}$ molar). (Contrast with acidic.)

**basidioma** (plural: basiomata) A fruiting structure produced by club fungi.

**basidium** (bass id' ee yum) In club fungi, the characteristic sporangium in which four **basidiospores** are formed by meiosis and then borne externally before being shed.

**basilar membrane** A membrane in the human inner ear whose flexion in response to sound waves activates hair cells; flexes at different locations in response to different pitch.

**Batesian mimicry** The convergence in appearance of an edible species (mimic) with an unpalatable species (model).

**behavioral ecology** An evolutionary approach to the study of animal behavior that studies how behaviors are adaptive in different environmental conditions.

**behaviorism** One of two classical approaches to the study of proximate causes of animal behavior, derived from the discoveries of Ivan Pavlov and focused on laboratory studies. (Compare with ethology.)

**benefit** Improvement in survival and reproductive success resulting from performing a behavior or having a trait.

**benthic zone** [Gk. *benthos*: bottom] The bottom of the ocean.

**β (beta) pleated sheet** A type of protein secondary structure; results from hydrogen bonding between polypeptide regions running antiparallel to each other.

**biased gene conversion** A mechanism of concerted evolution in which a DNA repair system appears biased in favor of using particular nucleotide sequences as templates for repair, resulting in the rapid spread of the favored sequence across all copies of the gene. (*See* concerted evolution.)

**bicarbonate ion** Ion ($HCO_3^-$) resulting from dissociation of carbonic acid in water; important in pH regulation and carbon dioxide ($CO_2$) transport.

**biennial** A plant whose life cycle includes vegetative growth in the first year and flowering and senescence in the second year. (Contrast with annual, perennial.)

**bilateral symmetry** The condition in which only the right and left sides of an organism, divided by a single plane through the midline, are mirror images of each other.

**bilayer** A structure that is two layers in thickness. In biology, most often refers to the phospholipid bilayer of membranes. (*See* phospholipid bilayer.)

**bile** A secretion of the liver made up of bile salts synthesized from cholesterol, various phospholipids, and bilirubin (the breakdown product of hemoglobin). Emulsifies fats in the small intestine.

**binary fission** Reproduction of a prokaryote by division of a cell into two comparable progeny cells.

**binocular vision** Overlapping visual fields of an animal's two eyes; allows the animal to see in three dimensions.

**binomial nomenclature** A taxonomic naming system in which each species is given a binomial (Gk.: two names), a genus name followed by a species name.

**biodiversity hotspots** Regions identified by conservation biologists as being particularly in need of protection because they harbor great species richness and endemism (i.e., large numbers of species, many of which are found nowhere else).

**biofilm** A community of microorganisms embedded in a polysaccharide matrix, forming a highly resistant coating on almost any moist surface.

**biogeochemical cycle** Movement of inorganic elements such as nitrogen, phosphorus, and carbon through living organisms and the physical environment.

**biogeographic region** One of several defined, continental-scale regions of Earth, each of which has a biota distinct from that of the others. (Contrast with biome.)

**biogeography** The scientific study of the patterns of distribution of populations, species, and ecological communities across Earth.

**bioinformatics** The use of computers and/or mathematics to analyze complex biological information, such as DNA sequences.

**biological control** The use of natural enemies (predators, parasites, or pathogens) to reduce the population density of an economically damaging (pest) species.

**biological species concept** The definition of a species as a group of actually or potentially interbreeding natural populations that are reproductively isolated from other such groups. (Contrast with lineage species concept; morphological species concept.)

**biology** [Gk. *bios*: life + *logos*: study] The scientific study of living things.

**bioluminescence** The production of light by biochemical processes in an organism.

**biome** (bye' ome) A major division of the ecological communities of Earth, characterized primarily by distinctive vegetation. A given biogeographic region contains many different biomes.

**bioremediation** The use by humans of other organisms to remove contaminants from the environment.

**biosphere** (bye' oh sphere) All regions of Earth (terrestrial and aquatic) and Earth's atmosphere in which organisms can live.

**biota** (bye oh' tah) All of the organisms—animals, plants, fungi, and microorganisms—found in a given area. (Contrast with flora, fauna.)

**biotechnology** The use of living cells or organisms to produce materials useful to humans.

**biotic** (bye ah' tick) [Gk. *bios*: life] Alive. (Contrast with abiotic.)

**biotic interchange** The mixing of biotas previously separated by physical, climatic, or other barriers, for example when two formerly separated land masses fuse.

**blastocoel** (blass' toe seal) [Gk. *blastos*: sprout + *koilos*: hollow] The central, hollow cavity of a blastula.

**blastocyst** (blass' toe cist) An early embryo formed by the first divisions of the fertilized egg (zygote). In mammals, a hollow ball of cells.

**blastodisc** (blass' toe disk) An embryo that forms as a disk of cells on the surface of a large yolk mass; comparable to a blastula, but occurring in animals such as birds and reptiles, in which the massive yolk restricts complete cleavage.

**blastomere** Any of the cells produced by the early divisions of a fertilized animal egg.

**blastopore** The opening created by the invagination of the vegetal pole during gastrulation of animal embryos.

**blastula** (blass' chu luh) An early stage of the animal embryo; in many species, a hollow sphere of cells surrounding a central cavity, the blastocoel. (Contrast with blastodisc.)

**block to polyspermy** Any of several responses to entry of a sperm into an egg that prevent more than one sperm from entering the egg.

**blood** A fluid connective tissue that is pumped throughout the body. A component of the circulatory system, blood transports gases such as oxygen and carbon dioxide as well as other essential elements.

**blood clotting** A cascade of events involving platelets and circulating proteins (clotting factors) that seals damaged blood vessels.

**blood–brain barrier** The selective impermeability of blood vessels in the brain that prevents most chemicals from diffusing from the blood into the brain.

**blue-light receptors** Pigments in plants that absorb blue light (400–500 nm). These pigments mediate many plant responses including photo-tropism, stomatal movements, and expression of some genes.

**body plan** The general structure of an animal, the arrangement of its organ systems, and the integrated functioning of its parts.

**Bohr effect** A shift in the $O_2$ binding curve of hemoglobin in response to excess $H^+$ ions such that the hemoglobin releases more $O_2$ in tissues where pH is low.

**Bohr model** A model for atomic structure that depicts the atom as largely empty space, with a central nucleus surrounded by electrons in orbits, or electron shells, at various distances from the nucleus.

**bond** *See* chemical bond.

**bone** A rigid component of vertebrate skeletal systems that contains an extracellular matrix of insoluble calcium phosphate crystals as well as collagen fibers.

**Bowman's capsule** An elaboration of the renal tubule, composed of podocytes, that surrounds and collects the filtrate from the glomerulus.

**brain** The centralized integrative center of a nervous system.

**brainstem** The portion of the vertebrate brain between the spinal cord and the forebrain, made up of the medulla, pons, and midbrain.

**brassinosteroids** Plant steroid hormones that mediate light effects promoting the elongation of stems and pollen tubes.

**Broca's area** A portion of the human brain essential for speech. Located in the frontal lobe just in front of the primary motor cortex.

**bronchioles** The smallest airways in a vertebrate lung, branching off the bronchi.

**bronchus** (plural: bronchi) The major airway(s) branching off the trachea into the vertebrate lung.

**brown fat** In mammals, fat tissue that is specialized to produce heat. It has many mitochondria and capillaries, and a protein that uncouples oxidative phosphorylation.

**budding** Asexual reproduction in which a more or less complete new organism grows from the body of the parent organism, eventually detaching itself.

**buffer** A substance that can transiently accept or release hydrogen ions and thereby resist changes in pH.

**bulbourethral glands** Secretory structures of the human male reproductive system that produce a small volume of an alkaline, mucoid secretion that helps neutralize acidity in the urethra and lubricate it to facilitate the passage of semen.

**bulk flow** The movement of a solution from a region of higher pressure potential to a region of lower pressure potential.

**bundle of His** Fibers of modified cardiac muscle that conduct action potentials from the atria to the ventricular muscle mass.

**bundle sheath cell** Part of a tissue that surrounds the veins of plants.

## C

**C horizon** *See* parent rock.

**$C_3$ plants** Plants that produce 3PG as the first stable product of carbon fixation in photosynthesis and use ribulose bisphosphate as a $CO_2$ receptor.

**$C_4$ plants** Plants that produce oxaloacetate as the first stable product of carbon fixation in photosynthesis and use phosphoenolpyruvate as $CO_2$ acceptor. $C_4$ plants also perform the reactions of $C_3$ photosynthesis.

**calcitonin** Hormone produced by the thyroid gland; lowers blood calcium and promotes bone formation. (Contrast with parathyroid hormone.)

**calcitriol** A hormone derived from vitamin D whose actions include stimulating the cells of the digestive tract to absorb calcium from ingested food.

**calorie** [L. *calor*: heat] The amount of heat required to raise the temperature of 1 gram of water by 1°C. Physiologists commonly use the kilocalorie (kcal) as a unit of measure (1 kcal = 1,000 calories). Nutritionists also use the kilocalorie, but refer to it as the **Calorie** (capital C).

**Calvin cycle** The stage of photosynthesis in which $CO_2$ reacts with RuBP to form 3PG, 3PG is reduced to a sugar, and RuBP is regenerated, while other products are released to the rest of the plant. Also known as the Calvin–Benson cycle.

**calyx** (kay' licks) [Gk. *kalyx*: cup] All of the sepals of a flower, collectively.

**CAM** *See* crassulacean acid metabolism.

**Cambrian explosion** The rapid diversification of multicellular life that took place during the Cambrian period.

**cAMP (cyclic AMP)** A compound formed from ATP that acts as a second messenger.

**cancellous bone** A type of bone with numerous internal cavities that make it appear spongy, although it is rigid. (Contrast with compact bone.)

**canopy** The leaf-bearing part of a tree. Collectively, the aggregate of the leaves and branches of the larger woody plants of an ecological community.

**capillaries** [L. *capillaris*: hair] Very small tubes, especially the smallest blood-carrying vessels of animals between the termination of the arteries and the beginnings of the veins. **Capillary beds** are networks of capillaries where materials are exchanged between the blood and the interstitial fluid.

**capsid** The outer shell of a virus that encloses its nucleic acid.

**carbohydrates** Organic compounds containing carbon, hydrogen, and oxygen in the ratio 1:2:1 (i.e., with the general formula $C_nH_{2n}O_n$). Common examples are sugars, starch, and cellulose.

**carbon skeleton** The chains or rings of carbon atoms that form the structural basis of organic molecules. Other atoms or functional groups are attached to the carbon atoms.

**carbon-fixation reactions** The phase of photosynthesis in which chemical energy captured in the light reactions is used to drive the reduction of $CO_2$ to form carbohydrates.

**carboxylase** An enzyme that catalyzes the addition of carboxyl functional groups (O═C—OH) to a substrate.

**cardiac cycle** Contraction of the two atria of the heart, followed by contraction of the two ventricles and then relaxation.

**cardiac muscle** A type of muscle tissue that makes up, and is responsible for the beating of, the heart. Characterized by branching cells with single nuclei and a striated (striped) appearance. (Contrast with smooth muscle, skeletal muscle.)

**cardiovascular system** [Gk. *kardia*: heart + L. *vasculum*: small vessel] The heart, blood, and vessels are of a circulatory system.

**carnivore** [L. *carn*: flesh + *vorare*: to devour] An organism that eats animal tissues. (Contrast with detritivore, frugivore, herbivore, omnivore.)

**carotenoid** (ka rah' tuh noid) A yellow, orange, or red lipid pigment commonly found as an accessory pigment in photosynthesis; also found in fungi.

**carotid body** A chemosensor in the carotid artery that senses a decrease in blood supply or a dramatic decrease in partial pressure of oxygen in the blood.

**carpel** (kar' pel) [Gk. *karpos*: fruit] The organ of the flower that contains one or more ovules.

**carrier** (1) In facilitated diffusion, a membrane protein that binds a specific molecule and transports it through the membrane. (2) In respiratory and photosynthetic electron transport, a participating substance such as NAD that exists in both oxidized and reduced forms. (3) In genetics, a person heterozygous for a recessive trait.

**carrying capacity (*K*)** The maximum number of individuals in a population (i.e., maximum population size) that can be supported by the resources present in a given environment.

**cartilage** In vertebrates, a tough connective tissue found in joints, the outer ear, and elsewhere. Forms the entire skeleton in some animal groups.

**cartilage bone** A type of bone that begins its development as a cartilaginous structure resembling the future mature bone, then gradually hardens into mature bone. (Contrast with membranous bone.)

**Casparian strip** A band of cell wall containing suberin and lignin, found in the endodermis. Restricts the movement of water across the endodermis.

**caspase** One of a group of proteases that catalyze cleavage of target proteins and are active in apoptosis.

**catabolic reaction** (kat uh bah' lik) [Gk. *kata*: to break down + *ballein*: to throw] A synthetic reaction in which complex molecules are broken down into simpler ones and energy is released. (Contrast with anabolic reaction.)

**catabolite repression** In the presence of abundant glucose, the diminished synthesis of catabolic enzymes for other energy sources.

**catalyst** (kat' a list) [Gk. *kata*: to break down] A chemical substance that accelerates a reaction without itself being consumed in the overall course of the reaction. Catalysts lower the activation energy of a reaction. Enzymes are biological catalysts.

**cation** (cat' eye on) An ion with one or more positive charges. (Contrast with anion.)

**caudal** [L. *cauda*: tail] Pertaining to the tail, or to the posterior part of the body.

**cDNA** *See* complementary DNA.

**cDNA library** A collection of complementary DNAs derived from mRNAs of a particular tissue at a particular time in the life cycle of an organism.

**cecum** (see' cum) [L. blind] A blind branch off the large intestine. In many nonruminant mammals, the cecum contains a colony of microorganisms that contribute to the digestion of food.

**cell** The simplest structural unit of a living organism. In multicellular organisms, many individual cells serve as the building blocks of tissues and organs.

**cell adhesion molecules (CAMs)** Molecules on animal cell surfaces that affect the selective association of cells into tissues during development of the embryo. Also a component of desmosomes.

**cell cycle** The stages through which a cell passes between one mitotic division and the next. Includes all stages of interphase and mitosis. (*See* mitosis.)

**cell cycle checkpoints** Points of transition between different phases of the cell cycle, which are regulated by cyclins and cyclin-dependent kinases (Cdks).

**cell division** The reproduction of a cell to produce two new cells. In eukaryotes, this process involves nuclear division (mitosis) and cytoplasmic division (cytokinesis).

**cell fate** The type of cell that an undifferentiated cell in an embryo will become in the adult.

**cell junctions** Specialized structures associated with the plasma membranes of epithelial cells. Some contribute to cell adhesion, others to intercellular communication.

**cell potency** In multicellular organisms, an undifferentiated cell's potential to become a cell of a specific type. (*See* multipotent; pluripotent; totipotent.)

**cell recognition** Binding of cells to one another mediated by membrane proteins or carbohydrates.

**cell theory** States that cells are the basic structural and physiological units of all living organisms, and that all cells come from preexisting cells.

**cell wall** A relatively rigid structure that encloses cells of plants, fungi, many protists, and most prokaryotes, and which gives these cells their shape and limits their expansion in hypotonic media.

**cellular immune response** Immune system response mediated by T cells and directed against parasites, fungi, intracellular viruses, and foreign tissues (grafts). (Contrast with humoral immune response.)

**cellular respiration** The catabolic pathways by which electrons are removed from various molecules and passed through intermediate electron carriers to $O_2$, generating $H_2O$ and releasing energy.

**cellular specialization** In multicellular organisms, the division of labor such that different cell types become responsible for different functions (e.g., reproduction or digestion) within the organism.

**cellulose** (sell' you lowss) A straight-chain polymer of glucose molecules, used by plants as a structural supporting material.

**central dogma** The premise that information flows from DNA to RNA to polypeptide (protein).

**central nervous system (CNS)** That portion of the nervous system that is the site of most information processing, storage, and retrieval; in vertebrates, the brain and spinal cord. (Contrast with peripheral nervous system.)

**central vacuole** In plant cells, a large organelle that stores the waste products of metabolism and maintains turgor.

**centrifuge** [L. *centrum*: center + *fugere*: to flee] A laboratory device in which a sample is spun around a central axis at high speed. Used to separate suspended materials of different densities.

**centriole** (sen' tree ole) A paired organelle that helps organize the microtubules in animal and protist cells during nuclear division.

**centromere** (sen′ tro meer) [Gk. *centron*: center + *meros*: part] The region where sister chromatids join.

**centrosome** (sen′ tro soam) The major microtubule organizing center of an animal cell.

**cephalization** (sef ah luh zay′ shun) [Gk. *kephale*: head] The evolutionary trend toward increasing concentration of brain and sensory organs at the anterior end of the animal.

**cerebellum** (sair uh bell′ um) [L. diminutive of *cerebrum*, brain] The brain region that controls muscular coordination; located at the anterior end of the hindbrain.

**cerebral cortex** The thin layer of gray matter (neuronal cell bodies) that overlies the cerebrum.

**cerebrum** (su ree′ brum) [L. brain] The dorsal anterior portion of the forebrain, making up the largest part of the brain of mammals; the chief coordination center of the nervous system and the major information-processing areas of the vertebrate brain consists of two **cerebral hemispheres**.

**cervix** (sir′ vix) [L. neck] The opening of the uterus into the vagina.

**cGMP (cyclic guanosine monophosphate)** An intracellular messenger that is part of signal transmission pathways involving G proteins. (*See* G protein.)

**channel protein** An integral membrane protein that forms an aqueous passageway across the membrane in which it is inserted and through which specific solutes may pass.

**chaperone** A protein that guards other proteins by counteracting molecular interactions that threaten their three-dimensional structure.

**character** In genetics, an observable feature, such as eye color. (Contrast with trait.)

**character displacement** An evolutionary phenomenon in which species that compete for the same resources within the same territory tend to diverge in morphology and/or behavior.

**chemical bond** An attractive force stably linking two atoms.

**chemical equilibrium** *See* equilibrium

**chemical evolution** The theory that life originated through the chemical transformation of inanimate substances.

**chemical reaction** The change in the composition or distribution of atoms of a substance with consequent alterations in properties.

**chemical synapse** Neural junction at which neurotransmitter molecules released from a presynaptic cell induce changes in a postsynaptic cell. (Contrast with electrical synapse.)

**chemically gated channel** A type of membrane channel that opens or closes depending on the presence or absence of a specific molecule that binds either to the channel protein itself or to a separate receptor that alters the three-dimensional shape of the channel protein.

**chemiosmosis** Formation of ATP in mitochondria and chloroplasts, resulting from a pumping of protons across a membrane (against a gradient of electrical charge and of pH), followed by the return of the protons through a protein channel with ATP synthase activity.

**chemoautotroph** Organisms that obtain energy by oxidizing inorganic substances, using some of that energy to fix carbon. Also known as chemolithotrophs.

**chemoheterotroph** An organism that must obtain both carbon and energy from organic substances. (Contrast with chemoautotroph, photoautotroph, photoheterotroph.)

**chemoreceptor** A sensory receptor cell that senses specific molecules (such as odorant molecules or pheromones) in the environment.

**chiasma** (kie az′ muh) (plural: chiasmata) [Gk. cross] An X-shaped connection between paired homologous chromosomes in prophase I of meiosis. A chiasma is the visible manifestation of crossing over between homologous chromosomes.

**chief cells** One of three types of secretory cell found in the gastric pits of the stomach wall. Chief cells secrete the protein-digesting enzyme pepsin. (*See* mucosal epithelium; parietal cells.)

**chitin** (kye′ tin) [Gk. *kiton*: tunic] The characteristic tough but flexible organic component of the exoskeleton of arthropods, consisting of a complex, nitrogen-containing polysaccharide. Also found in cell walls of fungi.

**chlorophyll** (klor′ o fill) [Gk. *kloros*: green + *phyllon*: leaf] Any of several green pigments associated with chloroplasts or with certain bacterial membranes; responsible for trapping light energy for photosynthesis.

**chloroplast** [Gk. *kloros*: green + *plast*: a particle] An organelle bounded by a double membrane containing the enzymes and pigments that perform photosynthesis. Chloroplasts occur only in eukaryotes.

**choanocyte** (ko′ an uh site) The collared, flagellated feeding cells of sponges.

**cholecystokinin (CCK)** (ko′ luh sis tuh kai′ nin) A hormone produced and released by the lining of the duodenum when it is stimulated by undigested fats and proteins. It stimulates the gallbladder to release bile and slows stomach activity.

**chorion** (kor′ ee on) [Gk. *khorion*: afterbirth] The outermost of the membranes protecting mammal, bird, and reptile embryos; in mammals it forms part of the placenta.

**chromatid** (kro′ ma tid) A newly replicated chromosome, from the time molecular duplication occurs until the time the centromeres separate (during anaphase of mitosis or of meiosis II).

**chromatin** The nucleic acid–protein complex that makes up eukaryotic chromosomes.

**chromatin remodeling** A mechanism for epigenetic gene regulation by the alteration of chromatin structure.

**chromosomal mutation** Loss of or changes in position/direction of a DNA segment on a chromosome.

**chromosome** (krome′ o sowm) [Gk. *kroma*: color + *soma*: body] In bacteria and viruses, the DNA molecule that contains most or all of the genetic information of the cell or virus. In eukaryotes, a structure composed of DNA and proteins that bears part of the genetic information of the cell.

**chylomicron** (ky low my′ cron) Particles of lipid coated with protein, produced in the gut from dietary fats and secreted into the extracellular fluids.

**chyme** (kime) [Gk. *kymus*: juice] Created in the stomach; a mixture of ingested food with the digestive juices secreted by the salivary glands and the stomach lining.

**cilia** (sil′ ee ah) (singular: cilium) [L. eyelashes] Hairlike organelle used for locomotion by many unicellular organisms and for moving water and mucus by many multicellular organisms. Generally shorter than flagella.

**circadian rhythm** (sir kade′ ee an) [L. *circa*: approximately + *dies*: day] A rhythm of growth or activity that recurs about every 24 hours.

**circannual rhythm** [L. *circa*: + *annus*: year) A rhythm of growth or activity that recurs on a yearly basis.

**circulatory system** A physiological system consisting of a muscular pump (heart), a fluid (blood or hemolymph), and a series of conduits (blood vessels) that transports materials around the body.

**11-*cis*-retinal** The nonprotein, light-absorbing component of the visual pigment rhodopsin. (*See* rhodopsin.)

***cis-trans* isomers** In molecules with a double bond (typically between two carbon items), identifies on which side of the double bond similar atoms or functional groups are found. If they are on the same side, the molecule is a *cis* isomer; in a *trans* isomer, similar atoms are on opposite sides of the double bond. (*See* isomer.)

**citric acid cycle** In cellular respiration, a set of chemical reactions whereby acetyl CoA is oxidized to carbon dioxide and hydrogen atoms are stored as NADH and $FADH_2$. Also called the Krebs cycle.

**clade** [Gk. *klados*: branch] A monophyletic group made up of an ancestor and all of its descendants.

**class I MHC molecules** Cell surface proteins that participate in the cellular immune response directed against virus-infected cells.

**class II MHC molecules** Cell surface proteins that participate in the cell–cell interactions (of T-helper cells, macrophages, and B cells) of the humoral immune response.

**class switching** Occurs when a B cell changes the immunoglobulin class it synthesizes (e.g., a B cell making IgM switches to making IgG).

**cleavage** The first few cell divisions of an animal zygote. *See also* complete cleavage, incomplete cleavage.

**climate** The long-term average atmospheric conditions (temperature, precipitation, humidity, wind direction and velocity) found in a region. (Contrast with weather.)

**climax community** The final stage of succession; a community that is capable of perpetuating itself under local climatic and soil conditions and persists for a relatively long time.

**clinal variation** [Gk. *klinein*: to lean] Gradual change in the phenotype of a species over a geographic gradient.

**cloaca** The opening through which both urinary wastes and digestive wastes are expelled in most amphibians and in reptiles (including birds).

**clonal deletion** Inactivation or destruction of lymphocyte clones that would produce immune reactions against the animal's own body.

**clonal lineages** Asexually reproduced groups of nearly identical organisms.

**clonal selection** Mechanism by which exposure to antigen results in the activation of selected T- or B-cell clones, resulting in an immune response.

**clone** [Gk. *klon*: twig, shoot] (1) Genetically identical cells or organisms produced from a common ancestor by asexual means. (2) To produce many identical copies of a DNA sequence by its introduction into, and subsequent asexual reproduction of, a cell or organism.

**closed circulatory system** Circulatory system in which the circulating fluid is contained within a continuous system of vessels. (Contrast with open circulatory system.)

**clumped dispersion pattern** *See* dispersion

***CO (CONSTANS)*** Gene coding for a transcription factor that activates the synthesis of florigen (FT); involved in the induction of flowering.

**co-repressor** In the regulation of bacterial operons, a molecule that binds to the repressor, causing it to change shape and bind to the operator, thereby inhibiting transcription.

**coastal zone** The marine life zone that extends from the shoreline to the edge of the continental shelf. Characterized by relatively shallow, well-oxygenated water and relatively stable temperatures and salinities.

**coccus** (kock' us) (plural: cocci) [Gk. *kokkos*: berry, pit] Any of various spherical or spheroidal bacteria.

**cochlea** (kock' lee uh) [Gk. *kokhlos*: snail] A spiral tube in the inner ear of vertebrates; it contains the sensory cells involved in hearing.

**codominance** A condition in which two alleles at a locus produce different phenotypic effects and both effects appear in heterozygotes.

**codon** Three nucleotides in messenger RNA that direct the placement of a particular amino acid into a polypeptide chain. (Contrast with anticodon.)

**coelom** (see' loam) [Gk. *koiloma*: cavity] An animal body cavity, enclosed by muscular mesoderm and lined with a mesodermal layer called peritoneum that also surrounds the internal organs.

**coelomate** Possessing a coelom.

**coenocytic** (seen' a sit ik) [Gk. *koinos*: common + *kytos*: container] Referring to the condition, found in some fungal hyphae, of "cells" containing many nuclei but enclosed by a single plasma membrane. Results from nuclear division without cytokinesis.

**coenzyme** A nonprotein organic molecule that plays a role in catalysis by an enzyme.

**coenzyme A (CoA)** A coenzyme used in various biochemical reactions as a carrier of acyl groups.

**coevolution** Evolutionary processes in which an adaptation in one species leads to the evolution of an adaptation in a species with which it interacts; also known as reciprocal adaptation.

**cofactor** An inorganic ion that is weakly bound to an enzyme and required for its activity.

**cohesin** A protein involved in binding chromatids together.

**cohesion** The tendency of molecules (or any substances) to stick together.

**cohort** (co' hort) [L. *cohors*: company of soldiers] A group of similar-aged organisms.

**cold-hardening** A process by which plants can acclimate to cooler temperatures; requires repeated exposure to cool temperatures over many days.

**coleoptile** A sheath that surrounds and protects the shoot apical meristem and young primary leaves of a grass seedling as they move through the soil.

**collagen** [Gk. *kolla*: glue] A fibrous protein found extensively in bone and connective tissue.

**collecting duct** In vertebrates, a tubule that receives urine produced in the nephrons of the kidney and delivers that fluid to the ureter for excretion.

**collenchyma** (cull eng' kyma) [Gk. *kolla*: glue + *enchyma*: infusion] A type of plant cell, living at functional maturity, which lends flexible support by virtue of primary cell walls thickened at the corners. (Contrast with parenchyma, sclerenchyma.)

**colon** [Gk. *kolon*] The portion of the gut between the small intestine and the anus. Also called the large intestine.

**commensalism** [L. *com*: together + *mensa*: table] A type of interaction between species in which one participant benefits while the other is unaffected.

**communication** A signal from one organism (or cell) that alters the functioning or behavior of another organism (or cell).

**community** Any ecologically integrated group of species of microorganisms, plants, and animals inhabiting a given area.

**compact bone** A type of bone with a solid, hard structure. (Contrast with cancellous bone.)

**companion cell** In angiosperms, a specialized cell found adjacent to a sieve tube element.

**comparative experiment** Experimental design in which data from various unmanipulated samples or populations are compared, but in which variables are not controlled or even necessarily identified. (Contrast with controlled experiment.)

**comparative genomics** Computer-aided comparison of DNA sequences between different organisms to reveal genes with related functions.

**competition** In ecology, use of the same resource by two or more species when the resource is present in insufficient supply for the combined needs of the species.

**competitive exclusion** A result of competition between species for resources, in which one species completely eliminates the other from a given habitat.

**competitive inhibitor** A nonsubstrate that binds to the active site of an enzyme and thereby inhibits binding of its substrate. (Contrast with noncompetitive inhibitor.)

**complement system** A group of eleven proteins that play a role in some reactions of the immune system. The complement proteins are not immunoglobulins.

**complementary base pairing** The AT (or AU), TA (or UA), CG, and GC pairing of bases in double-stranded DNA, in transcription, and between tRNA and mRNA.

**complementary DNA (cDNA)** DNA formed by reverse transcriptase acting with an RNA template; essential intermediate in the reproduction of retroviruses; used as a tool in recombinant DNA technology; lacks introns.

**complete cleavage** Pattern of cleavage that occurs in eggs that have little yolk. Early cleavage furrows divide the egg completely and the blastomeres are of similar size. (Contrast with incomplete cleavage.)

**complete metamorphosis** A change of state during the life cycle of an organism in which the body is almost completely rebuilt to produce an individual with a very different body form. Characteristic of insects such as butterflies, moths, beetles, ants, wasps, and flies.

**complex ions** Groups of covalently bonded atoms that carry an electric charge (e.g., $NH_4^+$, the ammonium ion).

**complex life cycle** In reference to parasitic species, a life cycle that requires more than one host to complete.

**composite transposon** Two transposable elements located near one another that transpose together and carry the intervening DNA sequence with them. (*See* transposable element.)

**compound** (1) A substance made up of atoms of more than one element. (2) Made up of many units, as in the **compound eyes** of arthropods.

**concentration gradient** A difference in concentration of an ion or other chemical substance from one location to another, often across a membrane. (*See* active transport; facilitated diffusion.)

**concerted evolution** The common evolution of a family of repeated genes, such that changes in one copy of the gene family are replicated in other copies of the gene family, and thus evolve "in concert." (*See* biased gene conversion; unequal crossing over.)

**condensation reaction** A chemical reaction in which two molecules become connected by a covalent bond and a molecule of water is released ($AH + BOH \rightarrow AB + H_2O$.) (Contrast with hydrolysis reaction.)

**conditional mutation** A mutation that results in a characteristic phenotype only under certain environmental conditions.

**conditioned reflex** A form of associative learning first described by Ivan Pavlov, in which a natural response (such as salivation in response to food) becomes associated with a normally unrelated stimulus (such as the sound of a bell).

**conduction** The transfer of heat from one object to another through direct contact.

**cone** In conifers, a reproductive structure consisting of spore-bearing scales extending from a central axis. (Contrast with strobilus.)

**cone cell** In the vertebrate retina, a type of photoreceptor cell responsible for color vision.

**conidium** (ko nid' ee um) (plural: conidia) [Gk. *konis*: dust] A type of haploid fungal spore borne at the tips of hyphae, not enclosed in sporangia.

**conjugation** (kon ju gay' shun) [L. *conjugare*: yoke together] (1) A process by which DNA is passed from one cell to another through a conjugation tube, as in bacteria. (2) A nonreproductive sexual process by which *Paramecium* and other ciliates exchange genetic material.

**connective tissue** A type of tissue that connects or surrounds other tissues; its cells are embedded in a collagen-containing matrix. One of the four major tissue types in multicellular animals, including cartilage, bone, blood, and fat.

**connexon** In a gap junction, a protein channel linking adjacent animal cells.

**conservation biology** An applied science that carries out investigations with the aim of maintaining the diversity of life on Earth.

**conserved** Pertaining to a gene or trait that has evolved very slowly and is similar or even identical in individuals of highly divergent groups.

**conspecifics** Individuals of the same species.

**constant region** The portion of an immunoglobulin molecule whose amino acid composition determines its class and does not vary among immunoglobulins in that class. (Contrast with variable region.)

**constitutive** Always present; produced continually at a constant rate. (Contrast with inducible.)

**constitutive genes** Genes that are expressed all the time. (Contrast with inducible genes.)

**constitutive proteins** Proteins that an organism produces all the time, and at a relatively constant rate.

**consumer** An organism that eats the tissues of some other organism.

**consumer–resource interactions** Interactions in which organisms gain their nutrition by eating other living organisms or are eaten themselves.

**continental drift** The gradual movements of the world's continents that have occurred over billions of years.

**contraception** Birth control methods that prevent fertilization or implantation (conception).

**contractile vacuole** (kon trak' tul) A specialized vacuole that collects excess water taken in by osmosis, then contracts to expel the water from the cell.

**controlled experiment** An experiment in which a sample is divided into groups whereby experimental groups are exposed to manipulations of an independent variable while one group serves as an untreated control. The data from the various groups are then compared to see if there are changes in a dependent variable as a result of the experimental manipulation. (Contrast with comparative experiment.)

**controlled system** A set of components in a physiological system that is controlled by commands from a regulatory system. (Contrast with regulatory system.)

**convection** The transfer of heat to or from a surface via a moving stream of air or fluid.

**convergent evolution** Independent evolution of similar features from different ancestral traits.

**convolutions** Foldings of the vertebrate brain's cerebral cortex into ridges called gyri (sing. **gyrus**) and valleys called sulci (sing. **sulcus**). The level of cortical convolution increases taxonomically and is especially extensive in humans.

**copulation** Reproductive behavior that results in a male depositing sperm in the reproductive tract of a female.

**cork cambium** [L. *cambiare*: to exchange] In plants, a lateral meristem that produces secondary growth, mainly in the form of waxy-walled protective cells, including some of the cells that become bark.

**cork** In plants, a protective outermost tissue layer composed of cells with thick walls waterproofed with suberin.

**cornea** The clear, transparent tissue that covers the eye and allows light to pass through to the retina.

**corolla** (ko role' lah) [L. *corolla*: a small crown] All of the petals of a flower, collectively.

**corpus luteum** (kor' pus loo' tee um) (plural: corpora lutea) [L. yellow body] A structure formed from a follicle after ovulation; produces hormones important to the maintenance of pregnancy.

**corridor** A connection between habitat patches through which organisms can disperse; plays a critical role in maintaining subpopulations.

**cortex** [L. *cortex*: covering, rind] (1) In plants, the tissue between the epidermis and the vascular tissue of a stem or root. (2) In animals, the outer tissue of certain organs, such as the adrenal gland (adrenal cortex) and the brain (cerebral cortex).

**corticosteroids** Steroid hormones produced and released by the cortex of the adrenal gland.

**corticotropin** A tropic hormone produced by the anterior pituitary hormone that stimulates cortisol release from the adrenal cortex. Also called adrenocorticotropic hormone (ACTH).

**corticotropin-releasing hormone** A hormone produced by the hypothalamus that controls the release of cortisol from the anterior pituitary.

**cortisol** A corticosteroid that mediates stress responses.

**cost** A decrease in fitness resulting from performing a behavior or having a trait.

**cost–benefit approach** An approach to evolutionary studies that assumes an animal has a limited amount of time and energy to devote to each of its activities, and that each activity has fitness costs as well as benefits. (*See also* trade-off.)

**cotyledon** (kot' ul lee' dun) [Gk. *kotyledon*: hollow space] A "seed leaf." An embryonic organ that stores and digests reserve materials; may expand when seed germinates.

**countercurrent flow** An arrangement that promotes the maximum exchange of heat, or of a diffusible substance, between two fluids by having the fluids flow in opposite directions through parallel vessels close together.

**countercurrent heat exchanger** In "hot" fish, an adaptation of the circulatory system such that arterial blood flowing into the muscles is warmed by venous blood flowing out of the muscles, thereby conserving body heat by countercurrent exchange.

**countercurrent multiplier** The mechanism that increases the concentration of the interstitial fluid in the mammalian kidney through countercurrent flow in the loops of Henle and selective permeability and active transport of ions by segments of the loops of Henle.

**covalent bond** Chemical bond based on the sharing of electrons between two atoms.

**CpG islands** DNA regions rich in C residues adjacent to G residues. Especially abundant in promoters, these regions are where methylation of cytosine usually occurs.

**crassulacean acid metabolism (CAM)** A metabolic pathway enabling the plants that possess it to store carbon dioxide at night and then perform photosynthesis during the day with stomata closed.

**critical night length** In the photoperiodic flowering response of short-day plants, the length of night above which flowering occurs and below which the plant remains vegetative. (The reverse applies in the case of long-day plants.)

**critical period** *See* sensitive period.

**crop** A simple food storage sac, the first of two stomachlike organs in many animals (including reptiles, earthworms, and various insects). (*See also* gizzard.)

**cross section** A section taken perpendicular to the longest axis of a structure. Also called a transverse section.

**crossing over** The mechanism by which linked genes undergo recombination. In general, the term refers to the reciprocal exchange of corresponding segments between two homologous chromatids.

**crosstalk** Interactions between different signal transduction pathways.

**crypsis** [Gk. *kryptos*: hidden] The resemblance of an organism to some part of its environment, which helps it to escape detection by enemies.

**cryptochromes** [Gk. *kryptos*: hidden + *kroma*: color] Photoreceptors mediating some blue-light effects in plants and animals.

**ctene** (teen) [Gk. *cteis*: comb] In ctenophores, a comblike row of cilia-bearing plates. Ctenophores move by beating the cilia on their eight ctenes.

**culture** (1) A laboratory association of organisms under controlled conditions. (2) The collection of knowledge, tools, values, and rules that characterize a human society.

**cumulus** A thick gelatinous layer that protects a mammalian ovum.

**cupula** Gelatinous swelling in the semicircular canals of the vestibular system. A cupula encloses hair cell stereocilia that react to shifting fluid in the canal ducts

**currents** Circulation patterns in the surface waters of oceans driven by the prevailing winds.

**cuticle** (1) In plants, a waxy layer on the outer body surface that retards water loss. (2) In ecdysozoans, an outer body covering that provides protection and support and is periodically molted.

**cyclic AMP** *See* cAMP.

**cyclic electron transport** In photosynthetic light reactions, the flow of electrons that produces ATP but no NADPH or $O_2$.

**cyclical succession** Pattern of change in community composition (succession) in which the climax community depends on periodic disturbances (e.g., fire) in order to persist. (Contrast with directional succession.)

**cyclin** A protein that activates a cyclin-dependent kinase, bringing about transitions in the cell cycle.

**cyclin-dependent kinase (Cdk)** A protein kinase whose target proteins are involved in transitions in the cell cycle and which is active only when complexed with additional protein subunits, called cyclins.

**cytokine** A regulatory protein made by immune system cells that affects other target cells in the immune system.

**cytokinesis** (sy′ toe kine ee′ sis) [Gk. *kytos*: container + *kinein*: to move] The division of the cytoplasm of a dividing cell. (Contrast with mitosis.)

**cytokinin** (sy′ toe kine′ in) A member of a class of plant growth substances that plays roles in senescence, cell division, and other phenomena.

**cytoplasm** The contents of the cell, excluding the nucleus.

**cytoplasmic determinants** In animal development, gene products whose spatial distribution may determine such things as embryonic axes.

**cytoplasmic segregation** The asymmetrical distribution of cytoplasmic determinants in a developing animal embryo.

**cytosine (C)** (site′ oh seen) A nitrogen-containing base found in DNA and RNA.

**cytoskeleton** The network of microtubules and microfilaments that gives a eukaryotic cell its shape and its capacity to arrange its organelles and to move.

**cytosol** The fluid portion of the cytoplasm, excluding organelles and other solids.

**cytotoxic T cells ($T_C$)** Cells of the cellular immune system that recognize and directly eliminate virus-infected cells. (Contrast with T-helper cells.)

## D

**DAG** *See* diacylglycerol.

**data** Quantified observations about a system under study.

**daughter chromosomes** During mitosis, the separated chromatids from the beginning of anaphase onward.

**dead space** The lung volume that fails to be ventilated with fresh air (because the lungs are never completely emptied during exhalation).

**dead zones** Regions in aquatic ecosystems that are devoid of aquatic life because eutrophication has resulted in severe oxygen depletion.

**deciduous** [L. *deciduus*: falling off] Pertaining to a woody plant that sheds its leaves but does not die.

**declarative memory** Memory of people, places, events, and things that can be consciously recalled and described. (Contrast with procedural memory.)

**decomposer** An organism that metabolizes organic compounds in debris and dead organisms, releasing inorganic material; found among the bacteria, protists, and fungi. *See also* detritivore, saprobe.

**deductive logic** Logical thought process that starts with a premise believed to be true then predicts what facts would also have to be true to be compatible with that premise. (Contrast with inductive logic.)

**defensin** A type of protein made by phagocytes that kills bacteria and enveloped viruses by insertion into their plasma membranes.

**deficiency disease** A condition (e.g., scurvy and beriberi) caused by chronic lack of any essential nutrient.

**degeneracy** The situation in which a single amino acid may be represented by any of two or more different codons in messenger RNA. Most of the amino acids can be represented by more than one codon.

**deletion** A mutation resulting from the loss of a continuous segment of a gene or chromosome. Such mutations almost never revert to wild type. (Contrast with duplication, point mutation.)

**demethylase** An enzyme that catalyzes the removal of the methyl group from cytosine, reversing DNA methylation.

**demography** The study of population structure and of the processes (**demographic events**, including births and deaths) by which it changes.

**denaturation** Loss of activity of an enzyme or nucleic acid molecule as a result of structural changes induced by heat or other means.

**dendrites** [Gk. *dendron*: tree] Branching fibers (processes) of a neuron. Dendrites are usually relatively short compared with the axon, and commonly carry information to the neuronal cell body.

**denitrification** Metabolic activity by which nitrate and nitrite ions are reduced to form nitrogen gas; carried out by certain soil bacteria.

**denitrifiers** Bacteria that release nitrogen to the atmosphere as nitrogen gas ($N_2$).

**density** *See* population density.

**density-dependent** Pertaining to an effect on population size that increases in proportion to population density.

**density-independent** Pertaining an effect on population size that acts independently of population density.

**deoxyribonucleic acid** *See* DNA.

**deoxyribonucleoside triphosphates (dNTPs)** The raw materials for DNA synthesis: deoxyadenosine triphosphate (dATP), deoxythymidine triphosphate (dTTP), deoxycytidine triphosphate (dCTP), and deoxyguanosine triphosphate (dGTP). Also called deoxyribonucleotides.

**deoxyribose** A five-carbon sugar found in nucleotides and DNA.

**dependent variable** In a scientific experiment, the response that is measured and analyzed as the independent variable is manipulated (See independent variable.)

**depolarization** A change in the resting potential across a membrane so that the inside of the cell becomes less negative, or even positive, compared with the outside of the cell. (Contrast with hyperpolarization.)

**derived trait** A trait that differs from the ancestral trait. (Contrast with synapomorphy.)

**dermal tissue system** The outer covering of a plant, consisting of epidermis in the young plant and periderm in a plant with extensive secondary growth. (Contrast with ground tissue system and vascular tissue system.)

**descent with modification** Darwin's premise that all species share a common ancestor and have diverged from one another gradually over time.

**desmosome** (dez' mo sowm) [Gk. *desmos*: bond + *soma*: body] An adhering junction between animal cells.

**desmotubule** A membrane extension connecting the endoplasmic retituclum of two plant cells that traverses the plasmodesma.

**determinate growth** A growth pattern in which the growth of an organism or organ ceases when an adult state is reached; characteristic of most animals and some plant organs. (Contrast with indeterminate growth.)

**determination** In development, the process whereby the fate of an embryonic cell or group of cells (e.g., to become epidermal cells or neurons) is set (becomes **determined**).

**detritivore** (di try' ti vore) [L. *detritus*: worn away + *vorare*: to devour] An organism that obtains its energy from the dead bodies or waste products (**detritus**) of other organisms.

**development** The process by which a multicellular organism, beginning with a single cell, goes through a series of changes, taking on the successive forms that characterize its life cycle.

**developmental plasticity** The capacity of an organism to alter its pattern of development in response to environmental conditions.

**diacylglycerol (DAG)** In hormone action, the second messenger produced by hydrolytic removal of the head group of certain phospholipids.

**diapause** A period of developmental or reproductive arrest, entered in response to day length, that enables an organism to better survive.

**diaphragm** (dye' uh fram) [Gk. *diaphrassein*: barricade] (1) A sheet of muscle that separates the thoracic and abdominal cavities in mammals; responsible for breathing. (2) A method of birth control in which a sheet of rubber is fitted over the woman's cervix, blocking the entry of sperm.

**diastole** (dye ass' toll ee) [Gk. dilation] The portion of the cardiac cycle when the heart muscle relaxes. (Contrast with systole.)

**dichotomous** (dye cot' oh mus) [Gk. *dichot*: split in two; *tomia*: removed) A branching pattern in which the shoot divides at the apex producing two equivalent branches that subsequently never overlap.

**diencephalon** The portion of the vertebrate forebrain that develops into the thalamus and hypothalamus.

**differential gene expression** The hyposthesis that, given that all cells contain all genes, what makes one cell type different from another is the difference in transcription and translation of those genes.

**differentiation** The process whereby originally similar cells follow different developmental pathways; the actual expression of determination.

**diffuse coevolution** The evolution of similar traits in suites of species experiencing similar selection pressures imposed by other suites of species with which they interact.

**diffusion** Random movement of molecules or other particles, resulting in even distribution of the particles when no barriers are present.

**digestive vacuole** In protists, an organelle specialized for digesting food ingested by endocytosis.

**dihybrid cross** A mating in which the parents differ with respect to the alleles of two loci of interest.

**dikaryon** (di care' ee ahn) [Gk. *di*: two + *karyon*: kernel] A cell or organism carrying two genetically distinguishable nuclei. Common in fungi.

**dioecious** (die eesh' us) [Gk. *di*: two + *oikos*: house] Pertaining to organisms in which the two sexes are "housed" in two different individuals, so that eggs and sperm are not produced in the same individuals. Examples: humans, fruit flies, date palms. (Contrast with monoecious.)

**diploblastic** Having two cell layers. (Contrast with triploblastic.)

**diploid** (dip' loid) [Gk. *diplos*: double] Having a chromosome complement consisting of two copies (homologs) of each chromosome. Designated 2*n*. (Contrast with haploid.)

**direct development** Pattern of development (notably among insects) in which hatchlings look like miniature versions of adults. (Contrast with metamorphosis.)

**direct fitness** That component of fitness resulting from an organism producing its own offspring. (Contrast with inclusive fitness, kin selection.)

**directional selection** Selection in which phenotypes at one extreme of the population distribution are favored. (Contrast with disruptive selection, stabilizing selection.)

**directional succession** Change in community composition after a disturbance (succession) that is characterized by an orderly progression culminating in a persistent state (the climax community). (Contrast with cyclical succession.)

**disaccharide** A carbohydrate made up of two monosaccharides (simple sugars).

**discoidal cleavage** In animal development, a type of incomplete cleavage that is common in fishes, reptiles, and birds, the eggs of which contain a dense yolk mass.

**dispersal** Movement of organisms away from a parent organism or from an existing population.

**dispersion** The distribution of individuals in space within a population. **Clumped dispersion** occurs when individuals tend to occupy the same space; **regular dispersion** is when the presence of one individual decreases the probability of another individual occupying the same space; and **random dispersion** assumes there is equal probability of any individual occupying any given space.

**disruptive selection** Selection in which phenotypes at both extremes of the population distribution are favored. (Contrast with directional selection; stabilizing selection.)

**distal** Away from the point of attachment or other reference point. (Contrast with proximal.)

**distal convoluted tubule** The portion of a renal tubule from where it reaches the renal cortex, just past the loop of Henle to where it joins a collecting duct. (Compare with proximal convoluted tubule.)

**disturbance** A short-term event that disrupts populations, communities, or ecosystems by changing the environment.

**disulfide bridge** The covalent bond between two sulfur atoms (–S—S–) linking two molecules or remote parts of the same molecule.

**DNA (deoxyribonucleic acid)** The fundamental hereditary material of all living organisms. In eukaryotes, stored primarily in the cell nucleus. A nucleic acid using deoxyribose rather than ribose.

**DNA fingerprint** An individual's unique pattern of allele sequences, commonly short tandem repeats and single nucleotide polymorphisms.

**DNA helicase** An enzyme that unwinds the double helix.

**DNA ligase** Enzyme that unites broken DNA strands during replication and recombination.

**DNA methylation** The addition of methyl groups to bases in DNA, usually cytosine or guanine.

**DNA methyltransferase** An enzyme that catalyzes the methylation of DNA.

**DNA microarray** A small glass or plastic square onto which thousands of single-stranded DNA sequences are fixed so that hybridization of cell-derived RNA or DNA to the target sequences can be performed.

**DNA polymerase** Any of a group of enzymes that catalyze the formation of DNA strands from a DNA template.

**DNA replication** The creation of a new strand of DNA in which DNA polymerase catalyzes the exact reproduction of an existing (template) strand of DNA.

**DNA transposons** Mobile genetic elements that move without making an RNA intermediate. (Contrast with retrotransposons.)

**domain** (1) An independent structural element within a protein. Encoded by recognizable nucleotide sequences, a domain often folds separately from the rest of the protein. Similar domains can appear in a variety of different proteins across phylogenetic groups (e.g., "homeobox domain"; "calcium-binding domain"). (2) In phylogenetics, the three monophyletic branches of life (Bacteria, Archaea, and Eukarya).

**dominance** In genetics, the ability of one allelic form of a gene to determine the phenotype of a heterozygous individual in which the homologous chromosomes carry both it and a different (recessive) allele. (Contrast with recessive.)

**dormancy** A condition in which normal activity is suspended, as in some spores, seeds, and buds.

**dorsal** [L. *dorsum*: back] Toward or pertaining to the back or upper surface. (Contrast with ventral.)

**dorsal lip** In amphibian embryos, the dorsal segment of the blastopore. Also called the "organizer," this region directs the development of nearby embryonic regions.

**double fertilization** In angiosperms, a process in which the nuclei of two sperm fertilize one egg. One sperm's nucleus combines with the egg nucleus to produce a zygote, while the other combines with the same egg's two polar nuclei to produce the first cell of the triploid endosperm (the tissue that will nourish the growing plant embryo).

**double helix** Refers to DNA and the (usually right-handed) coil configuration of two complementary, antiparallel strands.

**downregulation** A negative feedback process in which continuous high concentrations of a hormone can decrease the number of its receptors. (Contrast with upregulation.)

**duodenum** (do' uh dee' num) The beginning portion of the vertebrate small intestine. (Contrast with ileum, jejunum.)

**duplication** A mutation in which a segment of a chromosome is duplicated, often by the attachment of a segment lost from its homolog. (Contrast with deletion.)

## E

**ecdysone** (eck die' sone) [Gk. *ek*: out of + *dyo*: to clothe] In insects, a hormone that induces molting.

**ecological economics** Interdisciplinary field that works to assess the economic value of biodiversity.

**ecological efficiency** The overall transfer of energy from one trophic level to the next, expressed as the ratio of consumer production to producer production.

**ecological survivorship curve** *See* survivorship curves

**ecological system** One or more organisms plus the external environment with which they interact.

**ecology** [Gk. *oikos*: house] The scientific study of the interaction of organisms with their living (biotic) and nonliving (abiotic) environments.

**ecosystem** (eek' oh sis tum) The organisms of a particular habitat, such as a pond or forest, together with the physical environment in which they live.

**ecosystem engineer** An organism that builds structures that alter existing habitats or create new habitats.

**ecosystem services** Processes by which ecosystems maintain resources that benefit human society.

**ecotourism** Ecologically responsible travel to natural places.

**ectoderm** [Gk. *ektos*: outside + *derma*: skin] The outermost of the three embryonic germ layers first delineated during gastrulation. Gives rise to the skin, sense organs, and nervous system.

**ectotherm** [Gk. *ektos*: outside + *thermos*: heat] An animal that is dependent on external heat sources for regulating its body temperature (Contrast with endotherm.)

**edema** (i dee' mah) [Gk. *oidema*: swelling] Tissue swelling caused by the accumulation of fluid.

**edge effects** Changes in ecological processes in a community caused by physical and biological factors originating in an adjacent community.

**effector** A component of a physiological system that responds to information by *effecting* changes (making change happen) in the internal environment; examples include muscles and the secretory cells of the digestive tract.

**effector cells** In cellular immunity, B cells and T cells that attack an antigen, either by secreting antibodies that bind to the antigen or by releasing molecules that destroy any cell bearing the antigen.

**effector protein** In cell signaling, a protein responsible for the cellular reponse to a signal transduction pathway.

**efferent** (ef' ur unt) [L. *ex*: out + *ferre*: to bear] Carrying outward or away from, as in neurons that carry impulses outward from the central to the peripheral nervous system (**efferent neurons**), or a blood vessel that carries blood away from a structure. (Contrast with afferent.)

**egg** In all sexually reproducing organisms, the female gamete; in birds, reptiles, and some other vertebrates, a structure within which early embryonic development occurs. *See also* amniote egg, ovum.

**electrical synapse** A type of synapse at which action potentials spread directly from presynaptic cell to postsynaptic cell. (Contrast with chemical synapse.)

**electrocardiogram (ECG or EKG)** A graphic recording of electrical potentials from the heart.

**electrochemical gradient** The concentration gradient of an ion across a membrane plus the voltage difference across that membrane.

**electroencephalogram (EEG)** A graphic recording of electrical potentials from the brain.

**electromagnetic radiation** A self-propagating wave that travels though space and has both electrical and magnetic properties.

**electron** A subatomic particle outside the nucleus carrying a negative charge and very little mass.

**electron shell** The region surrounding the atomic nucleus at a fixed energy level in which electrons orbit.

**electron transport** The passage of electrons through a series of proteins with a release of energy which may be captured in a concentration gradient or in chemical form such as NADH or ATP.

**electronegativity** The tendency of an atom to attract electrons when it occurs as part of a compound.

**electrophoresis** *See* gel electrophoresis.

**element** A substance that cannot be converted to a simpler substance by ordinary chemical means.

**elongation** (1) In molecular biology, the addition of monomers to make a longer RNA or protein during transcription or translation. (2) Growth of a plant axis or cell primarily in the longitudinal direction.

**embolus** (em' buh lus) [Gk. *embolos*: stopper] A circulating blood clot. Blockage of a blood vessel by an embolus or a bubble of gas is called an **embolism**. (Contrast with thrombus.)

**embryo** [Gk. *en*: within + *bryein*: to grow] A young animal, or young plant sporophyte, while it is still contained within a protective structure such as a seed, egg, or uterus.

**embryo sac** In angiosperms, the female gametophyte. Found within the ovule, it consists of eight or fewer cells, membrane bounded, but without cellulose walls between them.

**embryonic stem cell (ESC)** A pluripotent cell in the blastocyst.

**emergent property** A property of a complex system that is not exhibited by its individual component parts.

**emigration** The deliberate and usually oriented departure of an organism from the habitat in which it has been living.

**endemic** (en dem′ ik) [Gk. *endemos*: native] Confined to a particular region, thus often having a comparatively restricted distribution.

**endergonic** A chemical reaction in which the products have higher free energy than the reactants, thereby requiring free energy input to occur. (Contrast with exergonic.)

**endocrine cells** Cells that secrete substances into the extracellular fluid. (*See also* endocrine gland.)

**endocrine gland** (en′ doh krin) [Gk. *endo*: within + *krinein*: to separate] An aggregation of secretory cells that secretes hormones into the blood. The endocrine system consists of all endocrine cells and endocrine glands in the body that produce and release hormones. (Contrast with exocrine gland.)

**endocytosis** A process by which liquids or solid particles are taken up by a cell through invagination of the plasma membrane. (Contrast with exocytosis.)

**endoderm** [Gk. *endo*: within + *derma*: skin] The innermost of the three embryonic germ layers delineated during gastrulation. Gives rise to the digestive and respiratory tracts and structures associated with them.

**endodermis** In plants, a specialized cell layer marking the inside of the cortex in roots and some stems. Frequently a barrier to free diffusion of solutes.

**endomembrane system** A system of intracellular membranes that exchange material with one another, consisting of the Golgi apparatus, endoplasmic reticulum, and lysosomes when present.

**endometrium** The epithelial lining of the uterus.

**endoplasmic reticulum (ER)** [Gk. *endo*: within + L. *reticulum*: net] A system of membranous tubes and flattened sacs found in the cytoplasm of eukaryotes. Exists in two forms: rough ER, studded with ribosomes; and smooth ER, lacking ribosomes.

**endorphins** Molecules in the mammalian brain that act as neurotransmitters in pathways that control pain.

**endoskeleton** [Gk. *endo*: within + *skleros*: hard] An internal skeleton covered by other, soft body tissues. (Contrast with exoskeleton.)

**endosperm** [Gk. *endo*: within + *sperma*: seed] A specialized triploid seed tissue found only in angiosperms; contains stored nutrients for the developing embryo.

**endospore** [Gk. *endo*: within + *spora*: to sow] In some bacteria, a resting structure that can survive harsh environmental conditions.

**endosymbiosis theory** [Gk. *endo*: within + *sym*: together + *bios*: life] The theory that the eukaryotic cell evolved via the engulfing of one prokaryotic cell by another.

**endothelium** The single layer of epithelial cells lining the interior of a blood vessel.

**endotherm** [Gk. *endo*: within + *thermos*: heat] An animal that can control its body temperature by the expenditure of its own metabolic energy. (Contrast with ectotherm.)

**endotoxin** A lipopolysaccharide that forms part of the outer membrane of certain Gram-negative bacteria that is released when the bacteria grow or lyse. (Contrast with exotoxin.)

**energetic cost** The difference between the energy an animal expends in performing a behavior and the energy it would have expended had it rested.

**energy** The capacity to do work or move matter against an opposing force. The capacity to accomplish change in physical and chemical systems.

**energy budget** A quantitative description of all paths of energy exchange between an animal and its environment.

**enhancers** Regulatory DNA sequences that bind transcription factors that either activate or increase the rate of transcription.

**enkephalins** Molecules in the mammalian brain that act as neurotransmitters in pathways that control pain.

**enteric nervous system** The nerve nets in the submucosa and between the smooth muscle layers of the vertebrate gut.

**enthalpy (*H*)** The total energy of a system.

**entrain** To advance or delay an organism's circadian clock each day so that it is in phase with the light-dark cycle of the organism's environment.

**entropy (*S*)** (en′ tro pee) [Gk. *tropein*: to change] A measure of the degree of disorder in any system. Spontaneous reactions in a closed system are always accompanied by an increase in entropy.

**enveloped virus** A virus enclosed within a phospholipid membrane derived from its host cell.

**environment** Whatever surrounds and interacts with or otherwise affects a population, organism, or cell. May be external or internal.

**environmental genomics** Sequencing technique used when biologists are unable to work with the whole genome of a prokaryote species but instead examine individual genes collected from a random sample of the organism's environment.

**environmental resistance** Reduction in a population's growth rate caused by preemption of available resources by other individuals in the population.

**environmentalism** The use of ecological knowledge, along with economics, ethics, and many other considerations, to inform both personal decisions and public policy relating to stewardship of natural resources and ecosystems.

**enzyme** (en′ zime) [Gk. *zyme*: to leaven (as in yeast bread)] A catalytic protein that speeds up a biochemical reaction.

**enzyme–substrate complex (ES)** An intermediate in an enzyme-catalyzed reaction; consists of the enzyme bound to its substrate(s).

**epi-** [Gk. upon, over] A prefix used to designate a structure located on top of another; for example, epidermis, epiphyte.

**epiblast** The upper or overlying portion of the avian blastula which is joined to the hypoblast at the margins of the blastodisc.

**epiboly** The movement of cells over the surface of the blastula toward the forming blastopore.

**epidermis** [Gk. *epi*: over + *derma*: skin] In plants and animals, the outermost cell layers. (Only one cell layer thick in plants.)

**epididymis** (epuh did′ uh mus) [Gk. *epi*: over + *didymos*: testicle] Coiled tubules in the testes that store sperm and conduct sperm from the seminiferous tubules to the vas deferens.

**epigenetics** The scientific study of changes in the expression of a gene or set of genes that occur without change in the DNA sequence.

**epinephrine** (ep i nef′ rin) [Gk. *epi*: over + *nephros*: kidney] The "fight or flight" hormone produced by the medulla of the adrenal gland; it also functions as a neurotransmitter. (Also known as adrenaline.)

**epistasis** Interaction between genes in which the presence of a particular allele of one gene determines whether another gene will be expressed.

**epithelial tissue** A type of animal tissue made up of sheets of cells that lines or covers organs, makes up tubules, and covers the surface of the body; one of the four major tissue types in multicellular animals.

**epitope** *See* antigenic determinant.

**equilibrium** Any state of balanced opposing forces and no net change.

**equilibrium potential** The membrane potential at which an ion is at electrochemical equilibrium, i.e., there is no net flux of the ion across the membrane.

**ER** *See* endoplasmic reticulum.

**error signal** In regulatory systems, any difference between the set point of the system and its current condition.

**erythrocyte** (ur rith′ row site) [Gk. *erythros*: red + *kytos*: container] A red blood cell.

**erythropoietin** A hormone produced by the kidney in response to lack of oxygen that stimulates the production of red blood cells.

**esophagus** (i soff′ i gus) [Gk. *oisophagos*: gullet] That part of the gut between the pharynx and the stomach.

**essential amino acids** Amino acids that an animal cannot synthesize for itself and must obtain from its food.

**essential element** A mineral nutrient required for normal growth and reproduction in plants and animals.

**essential fatty acids** Fatty acids that an animal cannot synthesize for itself and must obtain from its food.

**ester linkage** A condensation (water-releasing) reaction in which the carboxyl group of a fatty acid reacts with the hydroxyl group of an alcohol. Lipids, including most membrane lipids, are formed in this way. (Contrast with ether linkage.)

**estivation** (ess tuh vay' shun) [L. *aestivalis*: summer] A state of dormancy and hypometabolism that occurs during the summer; usually a means of surviving drought and/or intense heat. (Contrast with hibernation.)

**estrogen** Any of several steroid sex hormones; produced chiefly by the ovaries in mammals.

**estrus** (es' trus) [L. *oestrus*: frenzy] The period of heat, or maximum sexual receptivity, in some female mammals. Ordinarily, the estrus is also the time of release of eggs in the female.

**estuary** Aquatic biome in which salt water and fresh water mix, as when a river meets the ocean. Includes such ecosystems as salt marshes and mangrove forests.

**ether linkage** The linkage of two hydrocarbons by an oxygen atom (HC—O—CH). Ether linkages are characteristic of the membrane lipids of the Archaea. (Contrast with ester linkage.)

**ethology** [Gk. *ethos*: character + *logos:* study] An approach to the study of animal behavior that focuses on studying many species in natural environments and addresses questions about the evolution of behavior. (Compare with behaviorism.)

**ethylene** One of the plant growth hormones, the gas $H_2C{=}CH_2$. Involved in fruit ripening and other growth and developmental responses.

**euchromatin** Diffuse, uncondensed chromatin. Contains active genes that will be transcribed into mRNA. (Contrast with heterochromatin.)

**eudicots** Angiosperms with two embryonic cotyledons. (*See also* monocots.)

**Eukarya** One of the three domains of life; organisms made up of one or more eukaryotic cells. (*See also* eukaryotes.)

**eukaryotes** (yew car' ree oats) [Gk. *eu*: true + *karyon*: kernel or nucleus] Organisms whose cells contain their genetic material inside a nucleus. Includes all life other than the viruses, archaea, and bacteria. (Contrast with prokaryotes.)

**eusocial** Pertaining to a social group that includes nonreproductive individuals, as in honey bees.

**eustachian tube** A connection between the middle ear and the throat that allows air pressure to equilibrate between the middle ear and the outside world.

**eutrophication** (yoo trofe' ik ay' shun) [Gk. *eu*: truly + *trephein*: to flourish] The addition of nutrient materials to a body of water, resulting in changes in ecological processes and species composition therein.

**evaporation** The transition of water from the liquid to the gaseous phase.

**evolution** Any gradual change. Most often refers to organic or Darwinian evolution, which is the genetic and resulting phenotypic change in populations of organisms from generation to generation. (*See* macroevolution, microevolution; contrast with speciation.)

**evolutionary developmental biology (evo-devo)** The study of the interplay between evolutionary and developmental processes, with a focus on the genetic changes that give rise to novel morphology. Key concepts of evo-devo include modularity, genetic toolkits, genetic switches, and heterochrony.

**evolutionary radiation** The proliferation of many species within a single evolutionary lineage.

**evolutionary reversal** The reappearance of an ancestral trait in a group that had previously acquired a derived trait.

**evolutionary theory** The understanding and application of the mechanisms of evolutionary change to biological problems.

**excision repair** DNA repair mechanism that removes damaged DNA and replaces it with the appro-priate nucleotide.

**excitable** Capable of generating an action potential.

**excitatory** Input from a neuron that causes depolarization of the recipient cell.

**excited state** The state of an atom or molecule when, after absorbing energy, it has more energy than in its normal, ground state.

**excretion** Release of metabolic wastes by an organism.

**excretory systems** In animals, organs that maintain the volume, solute concentration, and composition of the extracellular fluid by excreting water, solutes, and nitrogenous wastes in the form of urine.

**exergonic** A chemical reaction in which the products of the reaction have lower free energy than the reactants, resulting in a release of free energy. (Contrast with endergonic.)

**exocrine gland** (eks' oh krin) [Gk. *exo*: outside + *krinein*: to separate] Any gland, such as a salivary gland, that secretes to the outside of the body or into the gut. (Contrast with endocrine gland.)

**exocytosis** A process by which a vesicle within a cell fuses with the plasma membrane and releases its contents to the outside. (Contrast with endocytosis.)

**exon** A portion of a DNA molecule, in eukaryotes, that codes for part of a polypeptide. (Contrast with intron.)

**exoskeleton** (eks' oh skel' e ton) [Gk. *exos*: outside + *skleros*: hard] A hard covering on the outside of the body to which muscles are attached. (Contrast with endoskeleton.)

**exotoxin** A highly toxic, usually soluble protein released by living, multiplying bacteria. (Contrast with endotoxin.)

**expanding triplet repeat** A three-base-pair sequence in a human gene that is unstable and can be repeated a few to hundreds of times. Often, the more the repeats, the less the activity of the gene involved. Expanding triplet repeats occur in some human diseases such as Huntington's disease and fragile-X syndrome.

**experiment** A testing process to support or disprove hypotheses and to answer questions. The basis of the scientific method. *See* comparative experiment, controlled experiment.

**expiratory reserve volume** The amount of air that can be forcefully exhaled beyond the normal tidal expiration. (Contrast with inspiratory reserve volume, tidal volume, vital capacity.)

**exploitation competition** Competition in which individuals reduce the quantities of their shared resources. (Contrast with interference competition.)

**exponential growth** Growth, especially in the number of organisms in a population, which is a geometric function of the size of the growing entity: the larger the entity, the faster it grows. (Contrast with logistic growth.)

**expression vector** A DNA vector, such as a plasmid, that carries a DNA sequence for the expression of an inserted gene into mRNA and protein in a host cell.

**expressivity** The degree to which a genotype is expressed in the phenotype; may be affected by the environment.

**extensor** A muscle that extends an appendage. (Contrast with flexor.)

**external fertilization** The release of gametes into the environment; typical of aquatic animals. Also called spawning. (Contrast with internal fertilization.)

**external gills** Highly branched and folded extensions of the body surface that provide a large surface area for gas exchange with water; typical of larval amphibians and many larval insects.

**extinction** The termination of a lineage of organisms.

**extracellular matrix** A material of heterogeneous composition surrounding cells and performing many functions including adhesion of cells.

**extraembryonic membranes** Four membranes that support but are not part of the developing embryos of reptiles, birds, and mammals, defining these groups phylogenetically as amniotes. (*See* amnion, allantois, chorion, yolk sac.)

**extreme halophiles** A group of euryarchaeotes that live exclusively in very salty environments.

**extremophiles** Archaea and bacteria that live and thrive under conditions (e.g., extremely high temperatures) that would kill most organisms.

**eye cups** Photosensory organs in flatworms; components of one of the simplest visual systems in animals.

## F

**5′ end** (5 prime) The end of a DNA or RNA strand that has a free phosphate group at the 5′ carbon of the sugar (deoxyribose or ribose).

**$F_1$** The first filial generation; the immediate progeny of a parental (P) mating.

**$F_2$** The second filial generation; the immediate progeny of a mating between members of the F1 generation.

**facilitated diffusion** Passive movement through a membrane involving a specific carrier protein; does not proceed against a concentration gradient. (Contrast with active transport, diffusion.)

**facilitation** In succession, modification of the environment by a colonizing species in a way that allows colonization by other species. (Contrast with inhibition.)

**facultative anaerobe** A prokaryote that can shift its metabolism between anaerobic and aerobic modes depending on the presence or absence of $O_2$. (Alternatively, facultative aerobe.)

**fast-twitch fibers** Skeletal muscle fibers that can generate high tension rapidly, but fatigue rapidly ("sprinter" fibers). Characterized by an abundance of enzymes of glycolysis. (Compare to slow-twitch fibers.)

**fat** (1) A triglyceride that is solid at room temperature. (Contrast with oil.) (2) Adipose tissue, one type of connective tissue. (See brown fat, white fat.)

**fate map** A diagram of the blastula showing which cells (blastomeres) are "fated " to contribute to specific tissues and organs in the mature body.

**fatty acid** A molecule made up of a long nonpolar hydrocarbon chain and a polar carboxyl group. Found in many lipids.

**fauna** (faw′ nah) All the animals found in a given area. (Contrast with flora.)

***FD* (*FLOWERING LOCUS D*)** Gene coding for a transcription factor in the shoot apical meristem that binds to florigen; involved in the induction of flowering.

**feces** [L. *faeces*: dregs] Waste excreted from the digestive system.

**fecundity** The average number of offspring produced by each female.

**feedback** In regulatory systems, information about the relationship between the set point of the system and its current state (Contrast with feedforward information).

**feedback inhibition** A mechanism for regulating a metabolic pathway in which the end product of the pathway can bind to and inhibit the enzyme that catalyzes the first committed step in the pathway. Also called end-product inhibition.

**feedforward information** In regulatory systems, information that changes the set point of the system. (Contrast with feedback.)

**fermentation** (fur men tay′ shun) [L. *fermentum*: yeast] The anaerobic degradation of a substance such as glucose to smaller molecules such as lactic acid or alcohol with the extraction of energy.

**fertilization** Union of gametes. Also known as syngamy.

**fertilizer** Any of a number of substances added to soil to improve the soil's capacity to support plant growth. May be organic or inorganic.

**fetus** Medical and legal term for the stages of a developing human embryo from about the eighth week of pregnancy (the point at which all major organ systems have formed) to the moment of birth.

**fiber** In angiosperms, an elongated, tapering sclerenchyma cell, usually with a thick cell wall, that serves as a support function in xylem. (*See also* muscle fiber.)

**fibrin** A protein that polymerizes to form long threads that provide structure to a blood clot.

**fibrinogen** A circulating protein that can be stimulated to fall out of solution and provide the structure for a blood clot.

**fibrous root system** A root system typical of monocots composed of numerous thin adventitious roots that are all roughly equal in diameter. (Contrast with taproot system.)

**Fick's law of diffusion** An equation that describes the factors that determine the rate of diffusion of a molecule from an area of higher concentration to an area of lower concentration.

**fight-or-flight response** A rapid physiological response to a sudden threat mediated by the hormone epinephrine.

**filament** In flowers, the part of a stamen that supports the anther.

**filter feeder** An organism that feeds on organisms much smaller than itself that are suspended in water or air by means of a straining device.

**first filial generation** *See* $F_1$.

**first law of thermodynamics** The principle that energy can be neither created nor destroyed.

**fission** *See* binary fission.

**fitness** The contribution of a genotype or phenotype to the genetic composition of subsequent generations, relative to the contribution of other genotypes or phenotypes. (*See* also inclusive fitness.)

**fixed action pattern** In ethology, a genetically determined behavior that is performed without learning, stereotypic (performed the same way each time), and not modifiable by learning.

**flagellum** (fla jell′ um) (plural: flagella) [L. *flagellum*: whip] Long, whiplike appendage that propels cells. Prokaryotic flagella differ sharply from those found in eukaryotes.

**flexor** A muscle that flexes an appendage. (Contrast with extensor.)

**flora** (flore′ ah) All of the plants found in a given area. (Contrast with fauna.)

**floral meristem** In angiosperms, a meristem that forms the floral organs (sepals, petals, stamens, and carpels).

**floral organ identity genes** In angiosperms, genes that determine the fates of floral meristem cells; their expression is triggered by the products of meristem identity genes.

**florigen** A plant hormone involved in the conversion of a vegetative shoot apex to a flower.

**flower** The sexual structure of an angiosperm.

**fluid feeder** An animal that feeds on fluids it extracts from the bodies of other organisms; examples include nectar-feeding birds and blood-sucking insects.

**fluid mosaic model** A molecular model for the structure of biological membranes consisting of a fluid phospholipid bilayer in which suspended proteins are free to move in the plane of the bilayer.

**flux** [L.: flow] In ecology, the flow of an element into or out of a compartment of the biosphere.

**follicle** [L. *folliculus*: little bag] In female mammals, an immature egg surrounded by nutritive cells.

**follicle-stimulating hormone (FSH)** A gonadotropin produced by the anterior pituitary.

**food chain** A portion of a food web, most commonly a simple sequence of prey species and the predators that consume them.

**food web** The complete set of food links between species in a community; a diagram indicating which ones are the eaters and which are eaten.

**forebrain** The region of the vertebrate brain that comprises the cerebrum, thalamus, and hypothalamus.

**fossil** Any recognizable structure originating from an organism, or any impression from such a structure, that has been preserved over geological time.

**fossil fuels** Fuels, including oil, natural gas, coal, and peat, formed over geologic time from organic material buried in anaerobic sediments.

**founder effect** Random changes in allele frequencies resulting from establishment of a population by a very small number of individuals.

**fovea** [L. *fovea*: a small pit] In the vertebrate retina, the area of most distinct vision.

**frame-shift mutation** The addition or deletion of a single or two adjacent nucleotides in a gene's sequence. Results in the misreading of mRNA during translation and the production of a nonfunctional protein. (Contrast with missense mutation, nonsense mutation, silent mutation.)

**Frank–Starling law** The stroke volume of the heart increases with increased return of blood to the heart.

**free energy (*G*)** Energy that is available for doing useful work, after allowance has been made for the increase or decrease of disorder.

**frequency-dependent selection** Selection that changes in intensity with the proportion of individuals in a population having the trait.

**frontal lobe** The largest of the brain lobes in humans; involved with feeling and planning functions; includes the primary motor cortex.

**frugivore** [L. *frugis*; fruit + *vorare*: to devour] An animal that eats fruit.

**fruit** In angiosperms, a ripened and mature ovary (or group of ovaries) containing the seeds. Sometimes applied to reproductive structures of other groups of plants.

***FT* (*FLOWERING LOCUS T*)** Gene that codes for florigen, a small, diffusible protein involved in the induction of flowering.

**fugitive species** A species that leave an otherwise suitable habitat in order to avoid competition with another species.

**full census** A count of every individual in a population. Can only be achieve if individuals are large and distinct enough to be identifiable by the census taker; population sizes are more usually estimated using sampling methods.

**functional genomics** The assignment of functional roles to the proteins encoded by genes identified by sequencing entire genomes.

**functional group** A characteristic combination of atoms that contributes specific properties (such as charge or polarity) when attached to larger molecules (e.g., carboxyl group; amino group).

**fundamental niche** A species' niche as defined by its physiological capabilities. (Contrast with realized niche.)

## G

**G protein** A membrane protein involved in signal transduction; characterized by binding GDP or GTP.

**G protein–linked receptors** A class of receptors that change configuration upon ligand binding such that a G protein binding site is exposed on the cytoplasmic domain of the receptor, initiating a signal transduction pathway.

**G1** In the cell cycle, the gap between the end of mitosis and the onset of the S phase.

**G1-to-S transition** In the cell cycle, the point at which G1 ends and the S phase begins.

**G2** In the cell cycle, the gap between the S (synthesis) phase and the onset of mitosis.

**gain of function mutation** A mutation that results in a protein with a new function. (Contrast with loss of function mutation.)

**gallbladder** In the human digestive system, an organ in which bile is stored.

**gametangium** (gam uh tan' gee um) (plural: gametangia) [Gk. *gamos*: marriage + *angeion*: vessel] Any plant or fungal structure within which a gamete is formed.

**gamete** (gam' eet) [Gk. *gamete*/*gametes*: wife, husband] The mature sexual reproductive cell: the egg or the sperm.

**gametogenesis** (ga meet' oh jen' e sis) The specialized series of cellular divisions that leads to the production of gametes. (*See also* oogenesis, spermatogenesis.)

**gametophyte** (ga meet' oh fyte) In plants and photosynthetic protists with alternation of generations, the multicellular haploid phase that produces the gametes. (Contrast with sporophyte.)

**ganglion** (gang' glee un) (plural: ganglia) [Gk. lump] A cluster of neurons that have similar characteristics or function.

**ganglion cells** Cells at the front of the human retina that transmit information from the bipolar cells to the brain.

**gap genes** In *Drosophila* development, segmentation genes that define broad areas along the anterior–posterior axis of the early embryo. Part of a developmental cascade that includes maternal effect genes, pair rule genes, segment polarity genes, and Hox genes.

**gap junction** A 2.7-nanometer gap between plasma membranes of two animal cells, spanned by protein channels. Gap junctions allow chemical substances or electrical signals to pass from cell to cell.

**gastric pits** Deep infoldings in the walls of the stomach lined with secretory cells.

**gastrin** A hormone secreted by cells in the lower region of the stomach that stimulates the secretion of digestive juices as well as movements of the stomach.

**gastrovascular cavity** Serving for both digestion (gastro) and circulation (vascular); in particular, the central cavity of the body of jellyfish and other cnidarians.

**gastrulation** Development of a blastula into a gastrula. In embryonic development, the process by which a blastula is transformed by massive movements of cells into a *gastrula*, an embryo with three germ layers and distinct body axes.

**gated channel** A membrane protein that changes its three-dimensional shape, and therefore its ion conductance, in response to a stimulus. When open, it allows specific ions to move across the membrane.

**gel electrophoresis** (e lek' tro fo ree' sis) [L. *electrum*: amber + Gk. *phorein*: to bear] A technique for separating molecules (such as DNA fragments) from one another on the basis of their electric charges and molecular weights by applying an electric field to a gel.

**gene** [Gk. *genes*: to produce] A unit of heredity. Used here as the unit of genetic function which carries the information for a polypeptide or RNA.

**gene duplication** The generation of extra copies of a gene in a genome over evolutionary time. A mechanism by which genomes can acquire new functions.

**gene expression** The transcription and translation into a protein of the information (nucleotide sequence) contained in a gene.

**gene family** A set of similar genes derived from a single parent gene; need not be on the same chromosomes. The vertebrate globin genes constitute a classic example of a gene family.

**gene flow** Exchange of genes between populations through migration of individuals or movements of gametes.

**gene pool** All of the different alleles of all of the genes existing in all individuals of a population.

**gene therapy** Treatment of a genetic disease by providing patients with cells containing functioning alleles of the genes that are nonfunctional in their bodies.

**gene tree** A graphic representation of the evolutionary relationships of a single gene in different species or of the members of a gene family.

**gene-for-gene concept** In plants, a mechanism of resistance to pathogens in which resistance is triggered by the specific interaction of the products of a pathogen's *Avr* genes and a plant's *R* genes.

**general transcription factors** In eukaryotes, transcription factors that bind to the promoters of most protein-coding genes and are required for their expression. Distinct from transcription factors that have specific regulatory effects only at certain promoters or classes of promoters.

**genetic code** The set of instructions, in the form of nucleotide triplets, that translate a linear sequence of nucleotides in mRNA into a linear sequence of amino acids in a protein.

**genetic drift** Changes in gene frequencies from generation to generation as a result of random (chance) processes.

**genetic linkage** Association between genes on the same chromosome such that they do not show random assortment and seldom recombine; the closer the genes, the lower the frequency of recombination.

**genetic map** The positions of genes along a chromosome as revealed by recombination frequencies.

**genetic marker** (1) In gene cloning, a gene of identifiable phenotype that indicates the presence of another gene, DNA segment, or chromosome fragment. (2) In general, a DNA sequence such as a single nucleotide polymorphism whose presence is correlated with the presence of other linked genes on that chromosome.

**genetic screen** A technique for identifying genes involved in a biological process of interest. Involves creating a large collection of randomly mutated organisms and identifying those individuals that are likely to have a defect in the pathway of interest. The mutated gene(s) in those individuals can then be isolated for further study.

**genetic structure** The frequencies of different alleles at each locus and the frequencies of different genotypes in a Mendelian population.

**genetic switches** Mechanisms that control how the genetic toolkit is used, such as promoters and the transcription factors that bind them. The signal cascades that converge on and operate these switches determine when and where genes will be turned on and off.

**genetic toolkit** A set of developmental genes and proteins that is common to most animals and is hypothesized to be responsible for the evolution of their differing developmental pathways.

**genetics** The scientific study of the structure, functioning, and inheritance of genes, the units of hereditary information.

**genome** (jee′ nome) The complete DNA sequence for a particular organism or individual.

**genome sequencing** Determination of the nucleotide base sequence of the entire genome of an organism.

**genomic imprinting** The form of a gene's expression is determined by parental source (i.e., whether the gene is inherited from the male or female parent).

**genomic library** All of the cloned DNA fragments generated by the breakdown of genomic DNA into smaller segments.

**genomics** The scientific study of entire sets of genes and their interactions.

**genotype** (jean′ oh type) [Gk. *gen*: to produce + *typos*: impression] An exact description of the genetic constitution of an individual, either with respect to a single trait or with respect to a larger set of traits. (Contrast with phenotype.)

**genotype frequency** The proportion of a genotype among individuals in a population.

**genus** (jean′ us) (plural: genera) [Gk. *genos*: stock, kind] A group of related, similar species recognized by taxonomists with a distinct name used in binomial nomenclature.

**geographic range** The region within which a species occurs.

**germ cell** [L. *germen*: to beget] A reproductive cell or gamete of a multicellular organism. (Contrast with somatic cell.)

**germ layers** The three embryonic layers formed during gastrulation (ectoderm, mesoderm, and endoderm). Also called cell layers or tissue layers.

**germ line mutation** Mutation in a cell that produces gametes (i.e., a germ line cell). (Contrast with somatic mutation.)

**germination** Sprouting of a seed or spore.

**gestation** (jes tay′ shun) [L. *gestare*: to bear] The period during which the embryo of a mammal develops within the uterus. Also known as pregnancy.

**ghrelin** A hormone produced and secreted by cells in the stomach that stimulates appetite.

**gibberellin** (jib er el′ lin) A class of plant growth hormones playing roles in stem elongation, seed germination, flowering of certain plants, etc.

**gill** An organ specialized for gas exchange with water.

**gizzard** (giz′ erd) [L. *gigeria*: cooked chicken parts] The second of two stomachlike organs in birds, other reptiles, earthworms, and various insects, that grinds up food, sometimes with the aid of fragments of stone. (*See also* crop.)

**glia** (glee′ uh) [Gk. *glia*: glue] One of the two classes of neural cells (along with neurons, with which glia interact); glia do not typically conduct action potentials. Types of glia include astrocytes, oligodendrocytes, and Schwann cells.

**global nitrogen cycle** The movement of nitrogen through the biosphere. Steps in the cycle include the fixation of nitrogen gas ($N_2$) to ammonia; nitrification of the fixed nitrogen to nitrate by bacteria; nitrate reduction by plants; and denitrification back to $N_2$ by bacteria.

**glomerular filtration rate (GFR)** The rate at which the blood is filtered in the glomeruli of the kidney.

**glomerulus** (glo mare′ yew lus) [L. *glomus*: ball] Sites in the kidney where blood filtration takes place. Each glomerulus consists of a knot of capillaries served by afferent and efferent arterioles.

**glucagon** Hormone produced by alpha cells of the pancreatic islets of Langerhans. Glucagon stimulates the liver to break down glycogen and release glucose into the circulation.

**gluconeogenesis** The biochemical synthesis of glucose from other substances, such as amino acids, lactate, and glycerol.

**glucose** [Gk. *gleukos*: sugar, sweet wine] The most common monosaccharide; the monomer of the polysaccharides starch, glycogen, and cellulose.

**glyceraldehyde 3-phosphate (G3P)** A phosphorylated three-carbon sugar; an intermediate in glycolysis and photosynthetic carbon fixation.

**glycerol** (gliss′ er ole) A three-carbon alcohol with three hydroxyl groups; a component of phospholipids and triglycerides.

**glycogen** (gly′ ko jen) [Gk. *glyk*: sweet] An energy storage polysaccharide found in animals and fungi; a branched-chain polymer of glucose, similar to starch.

**glycolipid** A lipid to which sugars are attached.

**glycolysis** (gly kol′ li sis) [Gk. *gleukos*: sugar + *lysis*: break apart] The enzymatic breakdown of glucose to pyruvic acid.

**glycoprotein** A protein to which sugars are attached.

**glycosidic linkage** Bond between carbohydrate (sugar) molecules through an intervening oxygen atom (–O–).

**glycosylation** The addition of carbohydrates to another type of molecule, such as a protein.

**glyoxysome** (gly ox′ ee soam) An organelle found in plants, in which stored lipids are converted to carbohydrates.

**Golgi apparatus** (goal′ jee) A system of concentrically folded membranes found in the cytoplasm of eukaryotic cells; functions in secretion from the cell by exocytosis.

**Golgi tendon organ** A mechanoreceptor found in tendons and ligaments; provides information about the force generated by a contracting muscle.

**gonad** (go′ nad) [Gk. *gone*: seed] An organ that produces gametes in animals: either an ovary (female gonad) or testis (male gonad).

**gonadotropin** A trophic hormone that stimulates the gonads.

**gonadotropin-releasing hormone (GnRH)** Hormone produced by the hypothalamus that stimulates the anterior pituitary to secrete gonadotropins.

**Gondwana** The large southern land mass that existed from the Cambrian (540 mya) to the Jurassic (138 mya). Present-day remnants are South America, Africa, India, Australia, and Antarctica.

**graded membrane potential** Small local change in membrane potential caused by opening or closing of ion channels.

**grafting** Artificial transplantation of tissue from one organism to another. In horticulture, the transfer of a bud or stem segment from one plant onto the root of another as a form of asexual reproduction.

**gram stain** A differential purple stain useful in characterizing bacteria. The peptidoglycan-rich cell walls of gram-positive bacteria stain purple; cell walls of gram-negative bacteria generally stain orange.

**gravitropism** [L. *gravitas*: weight, force; Gk. *tropos*: to turn] A directed plant growth response to gravity.

**gray crescent** In frog development, a band of diffusely pigmented cytoplasm on the side of the egg opposite the site of sperm entry. Arises as a result of cytoplasmic rearrangements that establish the anterior–posterior axis of the zygote.

**gray matter** In the nervous system, tissue that is rich in neuronal cell bodies. (Contrast with white matter.)

**greenhouse gases** Gases in the atmosphere, such as carbon dioxide and methane, that are transparent to sunlight, but trap heat radiating from Earth's surface, causing heat to build up at Earth's surface.

**gross primary production** The amount of energy captured by the primary producers in a community.

**gross primary productivity (GPP)** The rate at which the primary producers in a community turn solar energy into stored chemical energy via photosynthesis.

**ground meristem** That part of an apical meristem that gives rise to the ground tissue system of the primary plant body.

**ground tissue system** Those parts of the plant body not included in the dermal or vascular tissue systems. Ground tissues function in storage, photosynthesis, and support.

**growth** An increase in the size of the body and its organs by cell division and cell expansion.

**growth factor** A chemical signal that stimulates cells to divide.

**growth hormone** A peptide hormone released by the anterior pituitary that stimulates many anabolic processes.

**guanine (G)** (gwan' een) A nitrogen-containing base found in DNA, RNA, and GTP.

**guard cells** In plants, specialized, paired epidermal cells that surround and control the opening of a stoma (pore). *See* stoma.

**gustation** The sense of taste.

**gut** An animal's digestive tract.

**gymnosperms** Seed plants that do not produce flowers or fruits; one of the two major groups of living seed plants. (*See also* angiosperms.)

**gyrus** (ji' rus) [Gk. *gyros*: spiral) *See* convolutions.

## H

**habitat** The particular environment in which an organism lives.

**habitat patches** Also called **habitat islands**; areas of suitable habitat for a species that are separated by substantial areas of unsuitable habitat.

**Hadley cells** Patterns of vertical atmospheric circulation that influence surface winds and precipitation patterns according to latitude.

**hair cell** A type of mechanoreceptor in animals. Detects sound waves and other forms of motion in air or water.

**half-life** The time required for half of a sample of a radioactive isotope to decay to its stable, nonradioactive form, or for a drug or other substance to reach half its initial dosage.

**halophyte** (hal' oh fyte) [Gk. *halos*: salt + *phyton*: plant] A plant that grows in a saline (salty) environment.

**Hamilton's rule** The principle that, for an apparent altruistic behavior to be adaptive, the fitness benefit of that act to the recipient times the degree of relatedness of the performer and the recipient must be greater than the cost to the performer.

**haplodiploidy** A sex determination mechanism in which diploid individuals (which develop from fertilized eggs) are female and haploid individuals (which develop from unfertilized eggs) are male; typical of hymenopterans.

**haploid** (hap' loid) [Gk. *haploeides*: single] Having a chromosome complement consisting of just one copy of each chromosome; designated $1n$ or $n$. (Contrast with diploid.)

**haplotype** Linked nucleotide sequences that are usually inherited as a unit (as a "sentence" rather than as individual "words").

**Hardy–Weinberg equililbrium** In a sexually reproducing population, the allele frequency at a given locus that is not being acted on by agents of evolution; the conditions that would result in no evolution in a population.

**haustorium** (haw stor' ee um) (plural: haustoria)[L. *haustus*: draw up] A specialized hypha or other structure by which fungi and some parasitic plants draw nutrients from a host plant.

**Haversian systems** Units of organization in compact bone that reflect the action of intercommunicating osteoblasts.

**heart** In circulatory systems, a muscular pump that moves extracellular fluid around the body.

**heat of vaporization** The energy that must be supplied to convert a molecule from a liquid to a gas at its boiling point.

**heat shock proteins** Chaperone proteins expressed in cells exposed to high or low temperatures or other forms of environmental stress.

**helical** Shaped like a screw or spring (helix); this shape occurs in DNA and proteins.

**helper T cells** *See* T-helper cells.

**hemiparasite** A parasitic plant that can photosynthesize, but derives water and mineral nutrients from the living body of another plant. (Contrast with holoparasite.)

**hemizygous** (hem' ee zie' gus) [Gk. *hemi*: half + *zygotos*: joined] In a diploid organism, having only one allele for a given trait, typically the case for X-linked genes in male mammals and Z-linked genes in female birds. (Contrast with homozygous, heterozygous.)

**hemoglobin** (hee' mo glow bin) [Gk. *heaema*: blood + L. *globus*: globe] Oxygen-transporting protein found in the red blood cells of vertebrates (and found in some invertebrates).

**Hensen's node** In avian embryos, a structure at the anterior end of the primitive groove; determines the fates of cells passing over it during gastrulation.

**hepatic** (heh pat' ik) [Gk. *hepar*: liver] Pertaining to the liver.

**herbivore** (ur' bi vore) [L. *herba*: plant + *vorare*: to devour] An animal that eats plant tissues. (Contrast with carnivore, detritivore, omnivore.)

**heritable trait** A trait that is at least partly determined by genes.

**hermaphroditism** (her maf' row dite ism) The coexistence of both female and male sex organs in the same organism.

**hetero-** [Gk.: *heteros*: other, different] A prefix indicating two or more different conditions, structures, or processes. (Contrast with homo-.)

**heterochromatin** Densely packed, dark-staining chromatin; any genes it contains are usually not transcribed.

**heterochrony** [Gk: different time] Alteration in the timing of developmental events, contributing to the evolution different phenotypes in the adult.

**heterocyst** A large, thick-walled cell type in the filaments of certain cyanobacteria that performs nitrogen fixation.

**heterometry** [Gk: different measure] Alteration in the level of gene expression, and thus in the amount of protein produced, during development, contributing to the evolution of different phenotypes in the adult.

**heteromorphic** (het' er oh more' fik) [Gk.: different form] Having a different form or appearance, as two heteromorphic life stages of a plant. (Contrast with isomorphic.)

**heterosis** The superior fitness of heterozygous offspring as compared with that of their dissimilar homozygous parents. Also called hybrid vigor.

**heterosporous** (het' er os' por us) Producing two types of spores, one of which gives rise to a female megaspore and the other to a male microspore. (Contrast with homosporous.)

**heterotherm** An animal that regulates its body temperature at a constant level at some times but not others, such as a hibernator.

**heterotopy** [Gk: different place] Spatial differences in gene expression during development, controlled by developmental regulatory genes and contributing to the evolution of distinctive adult phenotypes.

**heterotroph** (het' er oh trof) [Gk. *heteros*: different + *trophe*: feed] An organism that requires preformed organic molecules as food. (Contrast with autotroph.)

**heterotrophic succession** Succession in detritus-based communities, which differs from other types of succession in taking place without the participation of plants.

**heterotypy** [Gk.: different kind] Alteration in a developmental regulatory gene itself rather than the expression of the genes it controls. (Contraste with heterochrony; heterometry; heterotopy.)

**heterozygous** (het' er oh zie' gus) [Gk. *heteros*: different + *zygotos*: joined] In diploid organisms, having different alleles of a given gene on the pair of homologs carrying that gene. (Contrast with homozygous.)

**heterozygous carrier** An individual that carries a recessive allele for a phenotype of interest (e.g., a genetic disease); the individual does not show the phenotype, but may have progeny with the phenotype if the other parent also carries the recessive allele.

**hexose** [Gk. *hex*: six] A sugar containing six carbon atoms.

**hibernation** [L. *hibernum*: winter] The state of inactivity of some animals during winter; marked by a drop in body temperature and metabolic rate. (Contrast with estivation)

**high-density lipoproteins (HDLs)** Lipoproteins that remove cholesterol from tissues and carry it to the liver; HDLs are the "good" lipoproteins associated with good cardiovascular health.

**high-throughput sequencing** Rapid DNA sequencing on a micro scale in which many fragments of DNA are sequenced in parallel.

**highly repetitive sequences** Short (less than 100 bp), nontranscribed DNA sequences, repeated thousands of times in tandem arrangements.

**hindbrain** The region of the developing vertebrate brain that gives rise to the medulla, pons, and cerebellum.

**hippocampus** [Gk. sea horse] A part of the forebrain that takes part in long-term memory formation.

**histamine** (hiss' tah meen) A substance released by damaged tissue, or by mast cells in response to allergens. Histamine increases vascular permeability, leading to edema (swelling). (Contrast with histone deacetylase.)

**histone** Any one of a group of proteins forming the core of a nucleosome, the structural unit of a eukaryotic chromosome.

**histone acetyltransferases** Enzymes involved in chromatin remodeling. Add acetyl groups to the tail regions of histone proteins.

**histone deacetylase** In chromatin remodeling, an enzyme that removes acetyl groups from the tails of histone proteins. (Contrast with histone acetyltransferases.)

**HIV** Human immunodeficiency virus, the retrovirus that causes acquired immune deficiency syndrome (AIDS).

**holoparasite** A fully parasitic plant (i.e., one that does not perform photosynthesis).

**homeobox** 180-base-pair segment of DNA found in certain homeotic genes. A specific sequence within the homeobox—the **homeodomain**—regulates the expression of other genes and through this regulation controls large-scale developmental processes. (*See* homeotic genes.)

**homeostasis** (home' ee o sta' sis) [Gk. *homos*: same + *stasis*: position] The maintenance of a steady state, such as a constant temperature, by means of physiological or behavioral feedback responses.

**homeotic genes** Genes that act during development to determine the formation of an organ from a region of the embryo. (Compare with Hox genes.)

**homeotic mutation** Mutation in a homeotic gene that results in the formation of a different organ than that normally made by a region of the embryo.

**homing** In animal navigation, the ability to return to a nest site, burrow, or other specific location.

**hominid** Lineage that includes all modern and extinct Great Apes (i.e., humans, gorillas, chimpanzees, orangutans, and their ancestors.)

**hominin** Lineages that includes modern humans (*Homo sapiens*) and their extinct ancestors (e.g., Australopithecines; *Homo erectus*.)

**homo-** [Gk. *homos*: same] A prefix indicating two or more similar conditions, structures, or processes. (Contrast with hetero-.)

**homolog** (1) In cytogenetics, one of a pair (or larger set) of chromosomes having the same overall genetic composition and sequence. In diploid organisms, each chromosome inherited from one parent is matched by an identical (except for mutational changes) chromosome—its homolog—from the other parent. (2) In evolutionary biology, one of two or more features in different species that are similar by reason of descent from a common ancestor.

**homologous pair** A pair of matching chromosomes made up of a chromosome from each of the two sets of chromosomes in a diploid organism.

**homologous recombination** Exchange of segments between two DNA molecules based on sequence similarity between the two molecules. The similar sequences align and crossover. Used to create knockout mutants in mice and other organisms.

**homology** (ho mol' o jee) [Gk. *homologia*: of one mind; agreement] A similarity between two or more features that is due to inheritance from a common ancestor. The structures are said to be *homologous*, and each is a *homolog* of the others.

**homoplasy** (home' uh play zee) [Gk. *homos*: same + *plastikos*: shape, mold] The presence in multiple groups of a trait that is not inherited from the common ancestor of those groups. Can result from convergent evolution, evolutionary reversal, or parallel evolution.

**homosporous** Producing a single type of spore that gives rise to a single type of gametophyte, bearing both female and male reproductive organs. (Contrast with heterosporous.)

**homotypic** Pertaining to adhesion of cells of the same type. (Contrast with heterotypic.)

**homozygous** (home' oh zie' gus) [Gk. *homos*: same + *zygotos*: joined] In diploid organisms, having identical alleles of a given gene on both homologous chromosomes. An individual may be a homozygote with respect to one gene and a heterozygote with respect to another. (Contrast with heterozygous.)

**horizons** The horizontal layers of a soil profile, including the topsoil (A horizon), subsoil (B horizon) and parent rock or bedrock (C horizon).

**hormone** (hore' mone) [Gk. *hormon*: to excite, stimulate] A chemical signal produced in minute amounts at one site in a multicellular organism and transported to another site where it acts on target cells.

**host** An organism that harbors a parasite or symbiont and provides it with nourishment.

**Hox genes** Conserved homeotic genes found in vertebrates, *Drosophila*, and other animal groups. Hox genes contain the homeobox and specify pattern and axis formation in these animals.

**human chorionic gonadotropin (hCG)** A hormone secreted by the placenta which sustains the corpus luteum and helps maintain pregnancy.

**Human Genome Project** A publicly and privately funded research effort, successfully completed in 2003, to produce a complete DNA sequence for the entire human genome.

**humoral immune response** The response of the immune system mediated by B cells that produces circulating antibodies active against extracellular bacterial and viral infections. (Contrast with cellular immune response.)

**humus** (hew' mus) The partly decomposed remains of plants and animals on the surface of a soil.

**hybrid** (high' brid) [L. *hybrida*: mongrel] (1) The offspring of genetically dissimilar parents. (2) In molecular biology, a double helix formed of nucleic acids from different sources.

**hybrid vigor** *See* heterosis.

**hybrid zone** A region of overlap in the ranges of two closely related species where the species may hybridize.

**hybridize** (1) In genetics, to combine the genetic material of two distinct species or of two distinguishable populations within a species. (2) In molecular biology, to form a double-stranded nucleic acid in which the two strands originate from different sources.

**hydrocarbon** A compound containing only carbon and hydrogen atoms.

**hydrogen bond** A weak electrostatic bond which arises from the attraction between the slight positive charge on a hydrogen atom and a slight negative charge on a nearby oxygen or nitrogen atom.

**hydrologic cycle** The movement of water from the oceans to the atmosphere, to the soil, and back to the oceans.

**hydrolysis reaction** (high drol' uh sis) [Gk. *hydro*: water + *lysis*: break apart] A chemical reaction that breaks a bond by inserting the components of water ($AB + H_2O \rightarrow AH + BOH$). (Contrast with condensation reaction.)

**hydrophilic** (high dro fill' ik) [Gk. *hydro*: water + *philia*: love] Having an affinity for water. (Contrast with hydrophobic.)

**hydrophobic** (high dro foe' bik) [Gk. *hydro*: water + *phobia*: fear] Having no affinity for water. Uncharged and nonpolar groups of atoms are hydrophobic. (Contrast with hydrophilic.)

**hydroponic** Pertaining to a method of growing plants with their roots suspended in nutrient solutions instead of soil.

**hydrostatic pressure** Pressure generated by compression of liquid in a confined space. Generated in plants, fungi, and some protists with cell walls by the osmotic uptake of water. Generated in animals with closed circulatory systems by the beating of a heart.

**hydrostatic skeleton** A fluid-filled body cavity that transfers forces from one part of the body to another when acted on by surrounding muscles.

**hydroxyl group** The —OH group found on alcohols and sugars.

**hyper-** [Gk. *hyper*: above, over] Prefix indicating above, higher, more. (Contrast with hypo-.)

**hyperaccumulators** Plant species that store large quantities of heavy metals such as arsenic, cadmium, nickel, aluminum, and zinc.

**hyperpolarization** A change in the resting potential across a membrane so that the inside of a cell becomes more negative compared with the outside of the cell. (Contrast with depolarization.)

**hypersensitive response** A defensive response of plants to microbial infection in which phytoalexins and pathogenesis-related proteins are produced and the infected tissue undergoes apoptosis to isolate the pathogen from the rest of the plant.

**hypertonic** Having a greater solute concentration. Said of one solution compared with another. (Contrast with hypotonic, isotonic.)

**hypha** (high' fuh) (plural: hyphae) [Gk. *hyphe*: web] In the fungi and oomycetes, any single filament.

**hypo-** [Gk. *hypo*: beneath, under] Prefix indicating underneath, below, less. (Contrast with hyper-.)

**hypoblast** The lower tissue portion of the avian blastula which is joined to the epiblast at the margins of the blastodisc.

**hypothalamus** The part of the brain lying below the thalamus; it coordinates water balance, reproduction, temperature regulation, and metabolism.

**hypothermia** Below-normal body temperature.

**hypothesis** A tentative answer to a question, from which testable predictions can be generated. (Contrast with theory.)

**hypotonic** Having a lesser solute concentration. Said of one solution in comparing it to another. (Contrast with hypertonic, isotonic.)

**hypoxia** A deficiency of oxygen.

## I

**ileum** The final segment of the small intestine. (*See also* duodenum, jejunum.)

**imbibition** Water uptake by a seed; first step in germination.

**immediate hypersensitivity** A rapid, extensive overreaction of the immune system against an allergen, resuting in the release of large amounts of histamine. (Contrast with delayed hypersensitivity.)

**immediate memory** A form of memory for events happening in the present that is almost perfectly photographic, but lasts only seconds.

**immunity** [L. *immunis*: exempt from] In animals, the ability to avoid disease when invaded by a pathogen by deploying various defense mechanisms.

**immunoassay** The use of antibodies to measure the concentration of an antigen in a sample.

**immunoglobulins** A class of proteins containing a tetramer consisting of four polypeptide chains—two identical light chains and two identical heavy chains—held together by disulfide bonds; active as receptors and effectors in the immune system.

**immunological memory** The capacity to more rapidly and massively respond to a second exposure to an antigen than occurred on first exposure.

**imperfect flower** A flower lacking either functional stamens or functional carpels. (Contrast with perfect flower.)

**implantation** The process by which the early mammalian embryo becomes attached to and embedded in the lining of the uterus.

**imprinting** In animal behavior, a rapid form of learning in which an animal learns, during a brief critical period, to make a particular response (which is then maintained for life) to some object or other organism. *See also* genomic imprinting.

**in vitro** [L.: in glass] A biological process occurring outside of the organism, in the laboratory. (Contrast with in vivo.)

**in vitro evolution** A method based on natural molecular evolution that uses artificial selection in the laboratory to rapidly produce molecules with novel enzymatic and binding functions.

**in vivo** [L.: in life] A biological process occurring within a living organism or cell. (Contrast with in vitro.)

**inclusive fitness** The sum of an individual's genetic contribution to subsequent generations both via production of its own offspring and via its influence on the survival of relatives who are not direct descendants. (Contrast with direct fitness)

**incomplete cleavage** A pattern of cleavage that occurs in many eggs that have a lot of yolk, in which the cleavage furrows do not penetrate all of it. (*See also* discoidal cleavage, superficial cleavage; contrast with complete cleavage.)

**incomplete dominance** Condition in which the heterozygous phenotype is intermediate between the two homozygous phenotypes.

**incomplete metamorphosis** Insect development in which changes between instars are gradual. (Contrast with direct development; complete metamorphosis.)

**independent assortment** During meiosis, the random separation of genes carried on nonhomologous chromosomes into gametes so that inheritance of these genes is random. This principle was articulated by Mendel as his second law.

**independent variable** In a scientific experiment, a critical factor that is manipulated while all other factors are held constant. (Contrast with dependent variable.)

**indeterminate growth** A open-ended growth pattern in which an organism or organ continues to grow as long as it lives; characteristic of some animals and of plant shoots and roots. (Contrast with determinate growth.)

**individual fitness** See direct fitness.

**induced fit** A change in the shape of an enzyme caused by binding to its substrate that exposes the active site of the enzyme.

**induced mutation** A mutation resulting from exposure to a mutagen from outside the cell. (Contrast with spontaneous mutation.)

**induced pluripotent stem cells (iPS cells)** Multipotent or pluripotent animal stem cells produced from differentiated cells in vitro by the addition of several genes that are expressed.

**induced responses** Defensive responses that a plant produces only in the presence of a pathogen, in contrast to constitutive defenses, which are always present.

**inducer** (1) A compound that stimulates the synthesis of a protein. (2) In embryonic development, a substance that causes a group of target cells to differentiate in a particular way.

**inducible genes** Genes that are expressed only when their products—**inducible proteins**—are needed. (Contrast with constitutive genes.)

**inducible** Produced only in the presence of a particular compound or under particular circumstances. (Contrast with constitutive.)

**induction** In embryonic development, the process by which a factor produced and secreted by certain cells determines the fates other cells.

**inductive logic** Involves making observations and then formulating one or more possible scenarios—hypotheses—that might explain those observations. (Contrast with deductive logic.)

**inflammation** A nonspecific defense against pathogens; characterized by redness, swelling, pain, and increased temperature.

**inflorescence** A structure composed of several to many flowers.

**inflorescence meristem** A meristem that produces floral meristems as well as other small leafy structures (bracts).

**ingroup** In a phylogenetic study, the group of organisms of primary interest. (Contrast with outgroup.)

**inhibitor** A substance that blocks a biological process.

**inhibitory** Input from a neuron that causes hyperpolarization of the recipient cell.

**initials** Cells that perpetuate plant meristems, comparable to animal stem cells. When an initial divides, one daughter cell develops into another initial, while the other differentiates into a more specialized cell.

**initiation complex** In protein translation, a combination of a small ribosomal subunit, an mRNA molecule, and the tRNA charged with the first amino acid coded for by the mRNA; formed at the onset of translation.

**initiation site** The place within a promoter where transcription begins.

**innate defenses** In animals, one of two general types of defenses against pathogens. Nonspecific and present in most animals. (Contrast with adaptive immunity.)

**inner cell mass** Derived from the mammalian blastula (bastocyst), the inner cell mass that will give rise to the yolk sac (via hypoblast) and embryo (via epiblast).

**inorganic fertilizer** A chemical or combination of chemicals applied to soil or plants to make up for a plant nutrient deficiency. Often contains the macronutrients nitrogen, phosphorus, and potassium (N-P-K).

**inositol trisphosphate ($IP_3$)** An intracellular second messenger derived from membrane phospholipids.

**inspiratory reserve volume** The amount of air that can be inhaled above the normal tidal inspiration. (Contrast with expiratory reserve volume, tidal volume, vital capacity.)

**instar** (in′ star) An immature stage of an insect between molts.

**insula** (in′ su lah) [L. *insula*: island] An area deep within the forebrain that appears to integrate physiological information from all over the body to create a sensation of how the body "feels" and may be involved in human consciousness. Also called the insular cortex.

**insulin** (in′ su lin) [L. *insula*: island] A hormone synthesized in islet cells of the pancreas that promotes the conversion of glucose into the storage material, glycogen.

**integral membrane proteins** Proteins that are at least partially embedded in the plasma membrane. (Contrast with peripheral membrane proteins.)

**integrin** In animals, a transmembrane protein that mediates the attachment of epithelial cells to the extracellular matrix.

**integument** [L. *integumentum*: covering] A protective surface structure. In gymnosperms and angiosperms, a layer of tissue around the ovule which will become the seed coat.

**intercostal muscles** Muscles between the ribs that can augment breathing movements by elevating and suppressing the rib cage.

**interference competition** Competition in which individuals actively interfere with one another's access to resources. (Contrast with exploitation competition.)

**interference RNA (RNAi)** *See* RNA interference.

**interferons** Glycoproteins produced by virus-infected animal cells; interferons increase the resistance of neighboring cells to the virus.

**internal environment** In multicelluar organisms, includes blood plasma and interstitial fluid, i.e., the extracellular fluids that surround the cells.

**internal fertilization** The release of sperm into the female reproductive tract; typical of most terrestrial animals. (Contrast with external fertilization.)

**internal gills** Gills enclosed in protective body cavities; typical of mollusks, arthropods, and fishes.**interneuron** A neuron that communicates information between two other neurons.

**interneuron** A neuron that communicates information between two other neurons.

**internode** The region between two nodes of a plant stem.

**interphase** In the cell cycle, the period between successive nuclear divisions during which the chromosomes are diffuse and the nuclear envelope is intact. During interphase the cell is most active in transcribing and translating genetic information.

**interspecific competition** Competition between members of two or more species. (Contrast with intraspecific competition; see also exploitation competition, interference competition.)

**interstitial fluid** Extracellular fluid that is not contained in the vessels of a circulatory system.

**intertidal zone** A nearshore region of oceans that is periodically exposed to the air as the tides rise and fall.

**intestine** The portion of the gut following the stomach, in which most digestion and absorption occurs.

**intraspecific competition** Competition among members of the same species. (Contrast with interspecific competition.)

**intrinsic rate of increase** The rate at which a population is capable of growing when its density is low and environmental conditions are highly favorable.

**intron** Portion of a of a gene within the coding region that is transcribed into pre-mRNA but is spliced out prior to translation. (Contrast with exon.)

**invasive species** An exotic species that reproduces rapidly, spreads widely, and has negative effects on the native species of the region to which it has been introduced.

**invasiveness** The ability of a pathogen to multiply in a host's body. (Contrast with toxigenicity).

**inversion** A rare 180° reversal of the order of genes within a segment of a chromosome.

**involution** Cell movements that occur during gastrulation of frog embryos, giving rise to the archenteron.

**ion** (eye′ on) [Gk. *ion*: wanderer] An electrically charged particle that forms when an atom gains or loses one or more electrons.

**ion channel** An integral membrane protein that allows ions to diffuse across the membrane in which it is embedded.

**ion exchange** In plants**, a** process by which protons produced by the plant's root displace mineral cations from clay particles in the surrounding soil.

**ionic attraction** An electrostatic attraction between positively and negatively charged ions.

**ionotropic receptors** A receptor that directly alters membrane permeability to a type of ion when it combines with its ligand.

**iris** (eye′ ris) [Gk. *iris*: rainbow] The round, pigmented membrane that surrounds the pupil of the eye and adjusts its aperture to regulate the amount of light entering the eye.

**island biogeography** A theory proposing that the number of species on an island (or in another geographically defined and isolated area) represents a balance, or equilibrium, between the rate at which species immigrate to the island and the rate at which resident species go extinct.

**islets of Langerhans** Clusters of hormone-producing cells in the pancreas.

**iso-** [Gk. *iso*: equal] Prefix used for two separate entities that share some element of identity.

**isomers** Molecules consisting of the same numbers and kinds of atoms, but differing in the bonding patterns by which the atoms are held together.

**isomorphic** (eye so more′ fik) [Gk. *isos*: equal + *morphe*: form] Having the same form or appearance, as when the haploid and diploid life stages of an organism appear identical. (Contrast with heteromorphic.)

**isotonic** Having the same solute concentration; said of two solutions. (Contrast with hypertonic, hypotonic.)

**isotope** (eye′ so tope) [Gk. *isos*: equal + *topos*: place] Isotopes of a given chemical element have the same number of protons in their nuclei (and thus are in the same position on the periodic table), but differ in the number of neutrons.

**isozymes** Enzymes of an organism that have somewhat different amino acid sequences but catalyze the same reaction.

**iteroparous** [L. itero, to repeat + pario, to beget] Reproducing multiple times in a lifetime. (Contrast with semelparous.)

## J

**jasmonate** Also called jasmonic acid, a plant hormone involved in triggering responses to pathogen attack as well as other processes.

**jejunum** (jih jew′ num) The middle division of the small intestine, where most absorption of nutrients occurs. (*See also* duodenum, ileum.)

**joint** In skeletal systems, a junction between two or more bones.

**juvenile hormone** In insects, a hormone maintaining larval growth and preventing maturation or pupation.

## K

***K*-strategist** A species whose life history strategy allows it to persist at or near the carrying capacity ($K$) of its environment. (Contrast with *r*-strategist.)

**karyogamy** The fusion of nuclei of two cells. (Contrast with plasmogamy.)

**karyotype** The number, forms, and types of chromosomes in a cell.

**keystone species** Species that have a dominant influence on the composition of a community.

**kidneys** A pair of excretory organs in vertebrates.

**kilocalorie (kcal)** *See* Calorie.

**kin selection** That component of inclusive fitness resulting from helping the survival of relatives containing the same alleles by descent from a common ancestor. (Contrast with direct fitness.)

**kinase** *See* protein kinase.

**kinetic energy** (kuh-net′ ik) [Gk. *kinetos*: moving] The energy associated with movement. (Contrast with potential energy.)

**kinetochore** (kuh net′ oh core) Specialized structure on a centromere to which microtubules attach.

**knockout** A molecular genetic method in which a single gene of an organism is permanently inactivated.

**Koch's postulates** A set of rules for establishing that a particular microorganism causes a particular disease.

**Krebs cycle** *See* citric acid cycle.

## L

**lagging strand** In DNA replication, the daughter strand that is synthesized in discontinuous stretches. (*See* Okazaki fragments.)

**large intestine** *See* colon.

**larva** (plural: larvae) [L. *lares*: guiding spirits] An immature stage of any animal that differs dramatically in appearance from the adult.

**lateral** [L. *latus*: side] Pertaining to the side.

**lateral gene transfer** The transfer of genes from one species to another, common among bacteria and archaea.

**lateral meristem** Either of the two meristems, the vascular cambium and the cork cambium, that give rise to a plant's secondary growth.

**lateral root** A root extending outward from the taproot in a taproot system; typical of eudicots.

**lateralization** A phenomenon in humans in which language functions come to reside in one cerebral hemisphere, usually the left.

**laticifers** (luh tiss′ uh furs) In some plants, elongated cells containing secondary plant products such as latex.

**Laurasia** The northernmost of the two large continents produced by the breakup of Pangaea.

**law of independent assortment** *See* independent assortment.

**law of segregation** *See* segregation.

**laws of thermodynamics** [Gk. *thermos*: heat + *dynamis*: power] Laws derived from studies of the physical properties of energy and the ways energy interacts with matter. (*See also* first law of thermodynamics, second law of thermodynamics.)

**leaching** In soils, a process by which mineral nutrients in upper soil horizons are dissolved in water and carried to deeper horizons, where they are unavailable to plant roots.

**leading strand** In DNA replication, the daughter strand that is synthesized continuously. (Contrast with lagging strand.)

**leaf** (plural: leaves) In plants, the chief organ of photosynthesis.

**leaf primordium** (plural: primordia) An outgrowth on the side of the shoot apical meristem that will eventually develop into a leaf.

**leghemoglobin** In nitrogen-fixing plants, an oxygen-carrying protein in the cytoplasm of nodule cells that transports enough oxygen to the nitrogen-fixing bacteria to support their respiration, while keeping free oxygen concentrations low enough to protect nitrogenase.

**lek** A display ground within which male animals compete for and defend small display areas as a means of demonstrating their territorial prowess and winning opportunities to mate.

**lens** In the vertebrate eye, a crystalline protein structure that makes fine adjustments in the focus of images falling on the retina.

**leptin** A hormone produced by fat cells that is believed to provide feedback information to the brain about the status of the body's fat reserves.

**leukocyte** *See* white blood cell.

**lichen** (lie′ kun) An organism resulting from the symbiotic association of a fungus and either a cyanobacterium or a unicellular alga.

**life cycle** The entire span of the life of an organism from the moment of fertilization (or asexual generation) to the time it reproduces in turn.

**life history strategy** The way in which an organism partitions its time and energy among growth, maintenance, and reproduction.

**life history** The time course of growth and development, reproduction, and death during an average individual organism's life.

**life table** A summary of information about the progression of individuals in a population through the various stages of their life cycles.

**life zones** In the aquatic (marine and freshwater) biomes, the regions defined by light penetration and water movement such as wave action. Life zones include, e.g., the intertidal, pelagic (open water) and bethic (bottom) zones.

**ligament** A band of connective tissue linking two bones in a joint.

**ligand** (lig′ and) Any molecule that binds to a receptor site of another (usually larger) molecule.

**light reactions** The initial phase of photosynthesis, in which light energy is converted into chemical energy. Followed by the **light-independent reactions** in which the energy captured in the light reactions is used to drive the reduction of $CO_2$ to form carbohydrates.

**light-harvesting complex** In photosynthesis, a group of different molecules that cooperate to absorb light energy and transfer it to a reaction center. Also called *antenna system*.

**lignin** A complex, hydrophobic polyphenolic polymer in plant cell walls that crosslinks other wall polymers, strengthening the walls, especially in wood.

**limbic system** A group of evolutionarily primitive structures in the vertebrate telencephalon that are involved in emotions, drives, instinctive behaviors, learning, and memory.

**limiting resource** The required resource whose supply (or lack thereof) most strongly influences the size of a population.

**limnetic zone** The open-water life zone of a lake

**lineage** A series of populations, species, or genes descended from a single ancestor over evolutionary time.

**lineage species concept** The definition of a species as a branch on the tree of life, which has a history that starts at a speciation event and ends either at extinction or at another speciation event. (Contrast with biological species concept; morphological species concept.)

**linkage** *See* genetic linkage.

**lipase** (lip′ ase; lye′ pase) An enzyme that digests fats.

**lipid** (lip′ id) [Gk. *lipos*: fat] Nonpolar, hydrophobic molecules that include fats, oils, waxes, steroids, and the phospholipids that make up biological membranes.

**lipid bilayer** *See* phospholipid bilayer.

**lipoproteins** Lipids packaged inside a covering of protein so that they can be circulated in the blood.

**lithoosphere** (lith' o sphere) [Gk. *lithos*: strong] The crust of sold rock plates that overlays the viscous mantle of Earth. The movements of the lithosphere are the source of plate tectonics. (Contrast with asthenosphere.)

**littoral zone** The nearshore life zone of a lake that is shallow and is affected by wave action and fluctuations in water level.

**liver** A large digestive gland. In vertebrates, it secretes bile and is involved in the formation of blood.

**loam** A type of soil consisting of a mixture of sand, silt, clay, and organic matter. One of the best soil types for agriculture.

**locus** (low' kus) (plural: loci, low' sigh) In genetics, a specific location on a chromosome. May be considered synonymous with *gene*.

**logistic growth** Growth, especially in the size of an organism or in the number of organisms in a population, that slows steadily as the entity approaches its maximum size. (Contrast with multiplicative growth.)

**long-day plant (LDP)** A plant that requires long days (actually, short nights) in order to flower. (Compare to short-day plant.)

**long-term potentiation (LTP)** A long-lasting increase in the responsiveness of a neuron resulting from a period of intense stimulation.

**loop of Henle** (hen' lee) Long, hairpin loop of the mammalian renal tubule that runs from the cortex down into the medulla and back to the cortex; creates a concentration gradient in the interstitial fluids in the medulla.

**lophophore** A U-shaped fold of the body wall with hollow, ciliated tentacles that encircles the mouth of animals in several different groups. Used for filtering prey from the surrounding water.

**loss of function mutation** A mutation that results in the loss of a functional protein. (Contrast with gain of function mutation.)

**low-density lipoproteins (LDLs)** Lipoproteins that transport cholesterol around the body for use in biosynthesis and for storage; LDLs are the "bad" lipoproteins associated with a high risk of cardiovascular disease.

**lumen** (loo' men) [L. *lumen*: light] The open cavity inside any tubular organ or structure, such as the gut or a renal tubule.

**lung** An internal organ specialized for respiratory gas exchange with air.

**luteinizing hormone (LH)** A gonadotropin produced by the anterior pituitary that stimulates the gonads to produce sex hormones.

**lymph** [L. *lympha*: liquid] A fluid derived from blood and other tissues that accumulates in intercellular spaces throughout the body and is returned to the blood by the lymphatic system.

**lymph node** A specialized structure in the vessels of the lymphatic system. Lymph nodes contain lymphocytes, which encounter and respond to foreign cells and molecules in the lymph as it passes through the vessels.

**lymphatic system** A system of vessels that returns interstitial fluid to the blood.

**lymphocyte** One of the two major classes of white blood cells; includes T cells, B cells, and other cell types important in the immune system.

**lysis** (lie' sis) [Gk. *lysis*: break apart] Bursting of a cell.

**lysogeny** A form of viral replication in which the virus becomes incorporated into the host chromosome and remains inactive. Also called a lysogenic cycle. (Contrast with lytic cycle.)

**lysosome** (lie' so soam) [Gk. *lysis*: break away + *soma*: body] A membrane-enclosed organelle originating from the Golgi apparatus and containing hydrolytic enzymes. (Contrast with secondary lysosome.)

**lysozyme** (lie' so zyme) An enzyme in saliva, tears, and nasal secretions that hydrolyzes bacterial cell walls.

**lytic cycle** A viral reproductive cycle in which the virus takes over a host cell's synthetic machinery to replicate itself, then bursts (lyses) the host cell, releasing the new viruses. (Contrast with lysogeny.)

## M

**M phase** The portion of the cell cycle in which mitosis takes place.

**macroevolution** [Gk. *makros*: large] Evolutionary changes occurring over long time spans and usually involving changes in many traits. (Contrast with microevolution.)

**macromolecule** A giant (molecular weight > 1,000) polymeric molecule. The macromolecules are the proteins, polysaccharides, and nucleic acids.

**macronutrient** In plants, a mineral element required in concentrations of at least 1 milligram per gram of plant dry matter; in animals, a mineral element required in large amounts. (Contrast with micronutrient.)

**macrophage** (mac' roh faj) Phagocyte that engulfs pathogens by endocytosis.

**MADS box** DNA-binding domain in many plant transcription factors that is active in development.

**maintenance methylase** An enzyme that catalyzes the methylation of the new DNA strand when DNA is replicated.

**major histocompatibility complex (MHC)** A complex of linked genes, with multiple alleles, that control a number of cell surface antigens that identify self and can lead to graft rejection.

**malignant** Pertaining to a tumor that can grow indefinitely and/or spread from the original site of growth to other locations in the body. (Contrast with benign.)

**malnutrition** A condition caused by lack of any essential nutrient.

**Malpighian tubule** (mal pee' gy un) A type of protonephridium found in insects.

**mantle** (1) In mollusks, a fold of tissue that covers the organs of the visceral mass and secretes the hard shell that is typical of many mollusks. (2) In geology, the Earth's crust below the solid lithospheric plates.

**map unit** The distance between two genes as calculated from genetic crosses; a recombination frequency.

**marine** [L. *mare*: sea, ocean] Pertaining to or living in the ocean. (Contrast with aquatic, terrestrial.)

**mark–recapture method** A method of estimating population sizes of mobile organisms by capturing, marking, and releasing a sample of individuals, then capturing another sample at a later time.

**mass extinction** A period of evolutionary history during which rates of extinction are much higher than during intervening times.

**mass number** The sum of the number of protons and neutrons in an atom's nucleus.

**mast cells** Cells, typically found in connective tissue, that release histamine in response to tissue damage.

**maternal effect genes** Genes coding for morphogens that determine the polarity of the egg and larva in fruit flies. Part of a developmental cascade that includes gap genes, pair rule genes, segment polarity genes, and Hox genes.

**mating type** A particular strain of a species that is incapable of sexual reproduction with another member of the same strain but capable of sexual reproduction with members of other strains of the same species.

**maturational survivorship curves** *See* survivorship curves

**maximum likelihood** A statistical method of determining which of two or more hypotheses (such as phylogenetic trees) best fit the observed data, given an explicit model of how the data were generated.

**mechanically gated channel** A molecular channel that opens or closes in response to mechanical force applied to the plasma membrane in which it is inserted.

**mechanoreceptor** A cell that is sensitive to physical movement and generates action potentials in response.

**medulla** (meh dull' luh) (1) The inner, core region of an organ, as in the adrenal medulla (adrenal gland) or the renal medulla (kidneys). (2) The portion of the brainstem that connects to the spinal cord.

**medusa** (plural: medusae) In cnidarians, a free-swimming, sexual life cycle stage shaped like a bell or an umbrella.

**megagametophyte** In heterosporous plants, the female gametophyte; produces eggs. (Contrast with microgametophyte.)

**megaphyll** The generally large leaf of a fern, horsetail, or seed plant, with several to many veins. (Contrast with microphyll.)

**megaspore** [Gk. *megas*: large + *spora*: to sow] In plants, a haploid spore that produces a female gametophyte.

**megastrobilus** In conifers, the female (seed-bearing) cone. (Contrast with microstrobilus.)

**meiosis** (my oh' sis) [Gk. *meiosis*: diminution] Division of a diploid nucleus to produce four haploid daughter cells. The process consists of two successive nuclear divisions with only one cycle of chromosome replication. In *meiosis I*, homologous chromosomes separate but retain their chromatids. The second division *meiosis II*, is similar to mitosis, in which chromatids separate.

**melatonin** A hormone released by the pineal gland. Involved in photoperiodicity and circadian rhythms.

**membrane** A phospholipid bylayer forming a barrier that separates the internal contents of a cell from the nonbiological environment, or enclosing the organelles within a cell. The membrane regulates the molecular substances entering or leaving a cell or organelle.

**membrane potential** The difference in electrical charge between the inside and the outside of a cell, caused by a difference in the distribution of ions.

**membranous bone** A type of bone that develops by forming on a scaffold of connective tissue. (Contrast with cartilage bone.)

**memory cells** Long-lived lymphocytes produced after exposure to antigen. They persist in the body and are able to mount a rapid response to subsequent exposures to the antigen.

**Mendel's laws** *See* independent assortment; segregation.

**menopause** In human females, the end of fertility and menstrual cycling.

**menstruation** The process by which the endometrium breaks down, and the sloughed-off tissue, including blood, flows from the body.

**meristem** [Gk. *meristos*: divided] Plant tissue made up of undifferentiated actively dividing cells.

**meristem culture** A method for the asexual propagation of plants, in which pieces of shoot apical meristem are cultured to produce plantlets.

**meristem identity genes** In angiosperms, a group of genes whose expression initiates flower formation, probably by switching meristem cells from a vegetative to a reproductive fate.

**mesenchyme** (mez' en kyme) [Gk. *mesos*: middle + *enchyma*: infusion] Embryonic or unspecialized cells derived from the mesoderm.

**mesoderm** [Gk. *mesos*: middle + *derma*: skin] The middle of the three embryonic germ layers first delineated during gastrulation. Gives rise to the skeleton, circulatory system, muscles, excretory system, and most of the reproductive system.

**mesoglea** (mez' uh glee uh) [Gk. *mesos*: middle + *gloia*, glue] A thick, gelatinous noncellular layer that separates the two cellular tissue layers of ctenophores, cnidarians, and scyphozoans.

**mesophyll** (mez' uh fill) [Gk. *mesos*: middle + *phyllon*: leaf] Chloroplast-containing, photosynthetic cells in the interior of leaves.

**messenger RNA (mRNA)** Transcript of a region of one of the strands of DNA; carries information (as a sequence of codons) for the synthesis of one or more proteins.

**meta-** [Gk.: between, along with, beyond] Prefix denoting a change or a shift to a new form or level; for example, as used in metamorphosis.

**metabolic pathway** A series of enzyme-catalyzed reactions so arranged that the product of one reaction is the substrate of the next.

**metabolism** (meh tab' a lizm) [Gk. *metabole*: change] The sum total of the chemical reactions that occur in an organism, or some subset of that total (as in respiratory metabolism).

**metabolome** The quantitative description of all the small molecules in a cell or organism.

**metabotropic receptor** A receptor that that indirectly alters membrane permeability to a type of ion when it combines with its ligand.

**metagenomics** The practice of analyzing DNA from environmental samples without isolating intact organisms.

**metamorphosis** (met' a mor' fo sis) [Gk. *meta*: between + *morphe*: form, shape] A change occurring between one developmental stage and another, as for example from a tadpole to a frog. (*See* complete metamorphosis, incomplete metamorphosis.)

**metanephridia** The paired excretory organs of annelids.

**metaphase** (met' a phase) The stage in nuclear division at which the centromeres of the highly supercoiled chromosomes are all lying on a plane (the metaphase plane or plate) perpendicular to a line connecting the division poles.

**metapopulation** A population divided into subpopulations, among which there are occasional exchanges of individuals.

**methylation** The addition of a methyl group (—$CH_3$) to a molecule.

**MHC** *See* major histocompatibility complex.

**micelle** A particle of lipid covered with bile salts that is produced in the duodenum and facilitates digestion and absorption of lipids.

**microbiomes** The diverse communities of bacteria that live on or within the body and are essential to bodily function.

**microclimate** A subset of climatic conditions in a small specific area, which generally differ from those in the environment at large, as in an animal's underground burrow.

**microevolution** Evolutionary changes below the species level, affecting allele frequencies. (Contrast with macroevolution.)

**microfibril** Crosslinked cellulose polymers, forming strong aggregates in the plant cell wall.

**microfilament** In eukaryotic cells, a fibrous structure made up of actin monomers. Microfilaments play roles in the cytoskeleton, in cell movement, and in muscle contraction.

**microgametophyte** In heterosporous plants, the male gametophyte; produces sperm. (Contrast with megagametophyte.)

**microglia** Glial cells that act as macrophages and mediators of inflammatory responses in the central nervous system.

**micronutrient** In plants, a mineral element required in concentrations of less than 100 micrograms per gram of plant dry matter; in animals, a mineral element required in concentrations of less than 100 micrograms per day. (Contrast with macronutrient.)

**microphyll** A small leaf with a single vein, found in club mosses and their relatives. (Contrast with megaphyll.)

**micropyle** (mike' roh pile) [Gk. *mikros*: small + *pylon*: gate] Opening in the integument(s) of a seed plant ovule through which pollen grows to reach the female gametophyte within.

**microRNA** A small, noncoding RNA molecule, typically about 21 bases long, that binds to mRNA to inhibit its translation.

**microspore** [Gk. *mikros*: small + *spora*: to sow] In plants, a haploid spore that produces a male gametophyte.

**microstrobilus** In conifers, male pollen-bearing cone. (Contrast with megastrobilus.)

**microtubules** Tubular structures found in centrioles, spindle apparatus, cilia, flagella, and cytoskeleton of eukaryotic cells. These tubules play roles in the motion and maintenance of shape of eukaryotic cells.

**microvilli** (sing.: microvillus) Projections of epithelial cells, such as the cells lining the small intestine, that increase their surface area.

**midbrain** One of the three regions of the vertebrate brain. Part of the brainstem, it serves as a relay station for sensory signals sent to the cerebral hemispheres.

**middle lamella** (la mell' ah) [L. *lamina*: thin sheet] A layer of polysaccharides that separates plant cells; a shared middle lamella lies outside the primary walls of the two cells.

**mineral nutrients** Inorganic ions required by organisms for normal growth and reproduction.

**mismatch repair** A mechanism that scans DNA after it has been replicated and corrects any base-pairing mismatches.

**missense mutation** A change in a gene's sequence that changes the amino acid at that site in the encoded protein. (Contrast with frame-shift mutation, nonsense mutation, silent mutation.)

**mitochondria** (my′ toe kon′ dree uh) (singular: mitochondrion) [Gk. *mitos*: thread + *chondros*: grain] Energy-generating organelles in eukaryotic cells that contain the enzymes of the citric acid cycle, the respiratory chain, and oxidative phosphorylation.

**mitochondrial matrix** The fluid interior of a mitochondrion, enclosed by the inner mitochondrial membrane.

**mitosis** (my toe′ sis) [Gk. *mitos*: thread] Nuclear division in eukaryotes leading to the formation of two daughter nuclei, each with a chromosome complement identical to that of the original nucleus.

**mitosomes** Reduced structures derived from mitochondria found in some organisms.

**model systems** Also known as **model organisms**, these include the small group of species that are the subject of extensive research. They are organisms that adapt well to laboratory situations and findings from experiments on them can apply across a broad range of species. Classic examples include white mice and the fruit fly *Drosophila*.

**moderately repetitive sequences** DNA sequences repeated 10–1,000 times in the eukaryotic genome. They include the genes that code for rRNAs and tRNAs, as well as the DNA in telomeres.

**Modern Synthesis** An understanding of evolutionary biology that emerged in the early twentieth century as the principles of evolution were integrated with the principles of modern genetics.

**modularity** In evolutionary developmental biology, the principle that the molecular pathways that determine different developmental processes operate independently from one another. *See also* developmental module.

**mole** A quantity of a compound whose weight in grams is numerically equal to its molecular weight expressed in atomic mass units. Avogadro's number of molecules: $6.023 \times 10^{23}$ molecules.

**molecular clock** The approximately constant rate of divergence of macromolecules from one another over evolutionary time; used to date past events in evolutionary history.

**molecular evolution** The scientific study of the mechanisms and consequences of the evolution of macromolecules.

**molecular toolkit** *See* genetic toolkit.

**molecular weight** The sum of the atomic weights of the atoms in a molecule.

**molecule** A chemical substance made up of two or more atoms joined by covalent bonds or ionic attractions.

**molting** The process of shedding part or all of an outer covering, as the shedding of feathers by birds or of the entire exoskeleton by arthropods.

**monoclonal antibody** Antibody produced in the laboratory from a clone of hybridoma cells, each of which produces the same specific antibody.

**monocots** Angiosperms with a single embryonic cotyledon; one of the two largest clades of angiosperms. (*See also* eudicots.)

**monoculture** In agriculture, a large-scale planting of a single species of domesticated crop plant.

**monoecious** (mo nee′ shus) [Gk. *mono*: one + *oikos*: house] Pertaining to organisms in which both sexes are "housed" in a single individual that produces both eggs and sperm. (In some plants, these are found in different flowers within the same plant.) Examples include corn, peas, earthworms, hydras. (Contrast with dioecious.)

**monohybrid cross** A mating in which the parents differ with respect to the alleles of only one locus of interest.

**monomer** [Gk. *mono*: one + *meros:* unit] A small molecule, two or more of which can be combined to form oligomers (consisting of a few monomers) or polymers (consisting of many monomers).

**monophyletic** (mon′ oh fih leht′ ik) [Gk. *mono*: one + *phylon*: tribe] Pertaining to a group that consists of an ancestor and all of its descendants. (Contrast with paraphyletic, polyphyletic.)

**monosaccharide** A simple sugar. Oligosaccharides and polysaccharides are made up of monosaccharides.

**monosomic** Pertaining to an organism with one less than the normal diploid number of chromosomes.

**monosynaptic reflex** A neural reflex that begins in a sensory neuron and makes a single synapse before activating a motor neuron.

**morphogen** A diffusible substance whose concentration gradient determines a developmental pattern in embryonic animals and plants.

**morphogenesis** (more′ fo jen′ e sis) [Gk. *morphe*: form + *genesis*: origin] The development of form; the overall consequence of determination, differentiation, and growth.

**morphological species concept** The definition of a species as a group of individuals that look alike. (Contrast with biological species concept; lineage species concept.)

**morphology** (more fol′ o jee) [Gk. *morphe*: form + *logos*: study, discourse] The scientific study of organic form, including both its development and function.

**mortality** Death, or the death rate of a population.

**mosaic development** Pattern of animal embryonic development in which each blastomere contributes a specific part of the adult body. (Contrast with regulative development.)

**motif** *See* structural motif.

**motile** (mo′ tul) Able to move from one place to another. (Contrast with sessile.)

**motor cortex** The region of the cerebral cortex that contains motor neurons that directly stimulate specific muscle fibers to contract.

**motor end plate** The depression in the postsynaptic membrane of the neuromuscular junction where the terminals of the motor neuron sit.

**motor neuron** A neuron carrying information from the central nervous system to a cell that produces movement.

**motor proteins** Specialized proteins that use energy to change shape and move cells or structures within cells.

**motor unit** A motor neuron and the muscle fibers it controls.

**mouth** An opening through which food is taken in, located at the anterior end of a tubular gut.

**mRNA** *See* messenger RNA.

**mucosal epithelium** An epithelial cell layer containing cells that secrete mucus; found in the digestive and respiratory tracts. Also called mucosa.

**mucus** A viscous substance secreted by mucous membranes (e.g., mucosal epithelium). A barrier defense against pathogens in innate immunity in animals and a protective coating in many animal organ systems.

**Muller's ratchet** The accumulation—"ratcheting up"—of deleterious mutations in the nonrecombining genomes of asexual species.

**Müllerian mimicry** Convergence in appearance of two or more unpalatable species.

**multifactorial** The interaction of many genes and proteins with one or more factors in the environment. For example, cancer is a disease with multifactorial causes.

**multipotent** Having the ability to differentiate into a limited number of cell types. (Contrast with pluripotent, totipotent.)

**muscle fiber** A single muscle cell. In the case of skeletal muscle, a syncitial, multinucleate cell.

**muscle tissue** Excitable tissue that can contract through the interactions of actin and myosin; one of the four major tissue types in multicellular animals. There are three types of muscle tissue: skeletal, smooth, and cardiac.

**mutagen** (mute′ ah jen) [L. *mutare*: change + Gk. *genesis*: source] Any agent (e.g., a chemical, radiation) that increases the mutation rate.

**mutation** A change in the genetic material not caused by recombination.

**mutualism** A type of interaction between species that benefits both species.

**mycelium** (my seel′ ee yum) [Gk. *mykes*: fungus] In the fungi, a mass of hyphae.

**mycologists** Scientists who study fungi.

**mycorrhiza** (my′ ko rye′ za) (plural: mycorrhizae) [Gk. *mykes*: fungus + *rhiza*: root] An association of the root of a plant with the mycelium of a fungus.

**myelin** (my′ a lin) Concentric layers of plasma membrane that form a sheath around some axons; myelin provides the axon with electrical insulation and increases the rate of transmission of action potentials.

**myocardial infarction (MI)** Blockage of an artery that carries blood to the heart muscle; a "heart attack."

**MyoD** The protein encoded by the *myo*blast *d*eterming gene. A transcription factor involved in the differentiation of myoblasts (muscle precursor cells).

**myofibril** (my′ oh fy′ bril) [Gk. *mys*: muscle + L. *fibrilla*: small fiber] A polymeric unit of actin or myosin in a muscle.

**myoglobin** (my′ oh globe′ in) [Gk. *mys*: muscle + L. *globus*: sphere] An oxygen-binding molecule found in muscle. Consists of a heme unit and a single globin chain; carries less oxygen than hemoglobin.

**myosin** One of the two contractile proteins of muscle. See also actin.

## N

**natural history** The characteristics of a group of organisms, such as how the organisms get their food, reproduce, behave, regulate their internal environments (their cells, tissues, and organs), and interact with other organisms.

**natural killer cell** A type of lymphocyte that attacks virus-infected cells and some tumor cells as well as antibody-labeled target cells.

**natural selection** The differential contribution of offspring to the next generation by various genetic types belonging to the same population. The mechanism of evolution proposed by Charles Darwin.

**nauplius** (naw′ plee us) [Gk. *nauplios*: shellfish] A bilaterally symmetrical larval form typical of crustaceans.

**necrosis** (nec roh′ sis) [Gk. *nekros*: death] Premature cell death caused by external agents such as toxins.

**negative feedback** In regulatory systems, information that decreases a regulatory response, returning the system to the set point. (Contrast with positive feedback.)

**negative regulation** A type of gene regulation in which a gene is normally transcribed, and the binding of a repressor protein to the promoter prevents transcription. (Contrast with positive regulation.)

**nematocyst** (ne mat′ o sist) [Gk. *nema*: thread + *kystis*: cell] An elaborate, threadlike structure produced by cells of jellyfishes and other cnidarians, used chiefly to paralyze and capture prey.

**neoteny** (knee ot′ enny) [Gk. *neo*: new, recent; *tenein*, to extend] The retention of juvenile or larval traits by the fully developed adult organism.

**nephron** (nef′ ron) [Gk. *nephros*: kidney] The functional unit of the kidney, consisting of a structure for receiving a filtrate of blood and a tubule that reabsorbs selected parts of the filtrate.

**Nernst equation** A mathematical statement that calculates the potential across a membrane permeable to a single type of ion that differs in concentration on the two sides of the membrane.

**nerve** A structure consisting of many neuronal axons and connective tissue.

**nerve nets** Diffuse, loosely connected aggregations of nervous tissues in certain non-bilatarian animals such as cnidarians.

**nervous tissue** Tissue specialized for processing and communicating information; one of the four major tissue types in multicellular animals.

**net primary productivity (NPP)** The rate at which energy captured by photosynthesis is incorporated into the bodies of primary producers through growth and reproduction.

**neural crest cells** During vertebrate neurulation, cells that migrate outward from the neural plate and give rise to connections between the central nervous system and the rest of the body.

**neural network** An organized group of neurons that contains three functional categories of neurons—afferent neurons, interneurons, and efferent neurons—and is capable of processing information.

**neural tube** An early stage in the development of the vertebrate nervous system consisting of a hollow tube created by two opposing folds of the dorsal ectoderm along the anterior–posterior body axis.

**neurohormone** A chemical signal produced and released by neurons that subsequently acts as a hormone.

**neuromuscular junction** Synapse (point of contact) where a motor neuron axon stimulates a muscle fiber cell.

**neuron** (noor′ on) [Gk. *neuron*: nerve] A nervous system cell that can generate and conduct action potentials along an axon to a synapse with another cell.

**neurotransmitter** A substance produced in and released by a neuron (the presynaptic cell) that diffuses across a synapse and excites or inhibits another cell (the postsynaptic cell).

**neurulation** Stage in vertebrate development during which the nervous system begins to form.

**neutral allele** An allele that does not alter the functioning of the proteins for which it codes.

**neutral theory** A view of molecular evolution that postulates that most mutations do not affect the amino acid being coded for, and that such mutations accumulate in a population at rates driven by genetic drift and mutation rates.

**neutron** (new′ tron) One of the three fundamental particles of matter (along with protons and electrons), with mass slightly larger than that of a proton and no electrical charge.

**niche** (nitch) [L. *nidus*: nest] The set of physical and biological conditions a species requires to survive, grow, and reproduce.

**nitrate reduction** The process by which nitrate ($NO_3^-$) is reduced to ammonia ($NH_3$).

**nitric oxide (NO)** An unstable molecule (a gas) that serves as a second messenger causing smooth muscle to relax. In the nervous system it operates as a neurotransmitter.

**nitrification** The oxidation of ammonia ($NH_3$) to nitrate ($NO_3^-$) in soil and seawater, carried out by chemoautotrophic bacteria (nitrifiers).

**nitrogen fixation** Conversion of atmospheric nitrogen gas ($N_2$) into a more reactive and biologically useful form (ammonia), which makes nitrogen available to living things. Carried out by **nitrogen fixers**—bacteria, some of them free-living and others living within plant roots.

**nitrogenase** An enzyme complex found in nitrogen-fixing bacteria that mediates the stepwise reduction of atmospheric $N_2$ to ammonia and which is strongly inhibited by oxygen.

**nitrogenous wastes** The potentially toxic nitrogen-containing end products—ammonia, urea, or uric acid—of protein and nucleic acid catabolism in animals. Eliminated from the body by excretion.

**node** [L. *nodus*: knob, knot] In plants, a (sometimes enlarged) point on a stem where a leaf is or was attached.

**node of Ranvier** A gap in the myelin sheath covering an axon; the point where the axonal membrane can fire action potentials.

**nodule** A specialized structure in the roots of nitrogen-fixing plants that houses nitrogen-fixing bacteria, in which oxygen is maintained at a low level by leghemoglobin.

**non-REM sleep** A state of deep, restorative sleep characterized by high-amplitude slow waves in the EEG. (Contrast with REM sleep.)

**noncompetitive inhibitor** A nonsubstrate that inhibits the activity of an enzyme by binding to a site other than its active site. (Contrast with competitive inhibitor.)

**noncyclic electron transport** In photosynthesis, the flow of electrons that forms ATP, NADPH, and $O_2$.

**nondisjunction** Failure of sister chromatids to separate in meiosis II or mitosis, or failure of homologous chromosomes to separate in meiosis I. Results in aneuploidy.

**nonpolar** Having electric charges that are evenly balanced from one end to the other. (Contrast with polar.)

**nonrandom mating** Selection of mates on the basis of a particular trait or group of traits.

**nonsense mutation** Change in a gene's sequence that prematurely terminates translation by changing one of its codons to a stop codon.

**nonsynonymous substitution** A change in a gene from one nucleotide to another that changes the amino acid specified by the corresponding codon (i.e., AGC → AGA, or serine → arginine). (Contrast with synonymous substitution.)

**norepinephrine** A neurotransmitter found in the central nervous system and also at the postganglionic nerve endings of the sympathetic nervous system. Also called noradrenaline.

**normal flora** Microorganisms that normally live and reproduce on or in the body without causing disease, and which form a nonspecific defense against pathogens by competing with them for space and nutrients. See also microbiota.

**notochord** (no' tow kord) [Gk. *notos*: back + *chorde*: string] A flexible rod of gelatinous material serving as a support in the embryos of all chordates and in the adults of tunicates and lancelets.

**nucleic acid** (new klay' ik) A polymer made up of nucleotides, specialized for the storage, transmission, and expression of genetic information. DNA and RNA are nucleic acids.

**nucleic acid hybridization** A technique in which a single-stranded nucleic acid probe is made that is complementary to, and binds to, a target sequence, either DNA or RNA. The resulting double-stranded molecule is a hybrid.

**nucleoid** (new' klee oid) The region that harbors the chromosomes of a prokaryotic cell. Unlike the eukaryotic nucleus, it is not bounded by a membrane.

**nucleolus** (new klee' oh lus) A small, generally spherical body found within the nucleus of eukaryotic cells. The site of synthesis of ribosomal RNA.

**nucleoside** A nucleotide without the phosphate group; a nitrogenous base attached to a sugar.

**nucleosome** A portion of a eukaryotic chromosome, consisting of part of the DNA molecule wrapped around a group of histone molecules, and held together by another type of histone molecule. The chromosome is made up of many nucleosomes.

**nucleotide** The basic chemical unit in nucleic acids, consisting of a pentose sugar, a phosphate group, and a nitrogen-containing base.

**nucleotide substitution** A change of one base pair to another in a DNA sequence.

**nucleus** (new' klee us) [L. *nux*: kernel or nut] (1) In cells, the centrally located compartment of eukaryotic cells that is bounded by a double membrane and contains the chromosomes. (2) In the brain, an identifiable group of neurons that share common characteristics or functions.

**null hypothesis** In statistics, the premise that any differences observed in an experiment are simply the result of random differences that arise from drawing two finite samples from the same population.

**nutrient** A food substance; or, in the case of mineral nutrients, an inorganic element required for completion of the life cycle of an organism.

## O

**obligate anaerobe** An anaerobic prokaryote that cannot survive exposure to $O_2$.

**occipital lobe** One of the four lobes of the brain's cerebral hemisphere; processes visual information.

**odorant** A molecule that can bind to an olfactory receptor.

**oil** A triglyceride that is liquid at room temperature. (Contrast with fat.)

**Okazaki fragments** Newly formed DNA making up the lagging strand in DNA replication. DNA ligase links Okazaki fragments together to give a continuous strand.

**olfaction** (ole fak' shun) [L. *olfacere*: to smell] The sense of smell.

**olfactory bulb** Structure in the vertebrate forebrain that receives and processes input from olfactory receptor neurons.

**olfactory receptor neurons (ORNs)** Neurons with receptors for different odorants.

**oligodendrocyte** A type of glial cell that myelinates axons in the central nervous system.

**oligophagous** [Gk. *oligo*: few; *phagein*, eat] An animal that feeds on a limited number of foods; generally used of insects that feed on only one or a few plant species.

**oligosaccharide** A polymer containing a small number of monosaccharides.

**omasum** One of the four chambers of the stomach in ruminants; concentrates food by water absorption before it enters the true stomach (abomasum).

**ommatidia** [Gk. *omma*: eye] The units that make up the compound eye of some arthropods.

**omnivore** [L. *omnis*: everything + *vorare*: to devour] An organism that eats both animal and plant material. (Contrast with carnivore, detritivore, herbivore.)

**oncogene** [Gk. *onkos*: mass, tumor + *genes*: born] A gene that codes for a protein product that stimulates cell proliferation. Mutations in oncogenes that result in excessive cell proliferation can give rise to cancer.

**one gene–one polypeptide** The idea, now known to be an oversimplification, that each gene in the genome encodes only a single polypeptide—that there is a one-to-one correspondence between genes and polypeptides.

**oocyte** *See* primary oocyte, secondary oocyte.

**oogenesis** (oh' eh jen e sis) [Gk. *oon*: egg + *genesis*: source] Gametogenesis leading to production of an ovum.

**oogonium** (oh' eh go' nee um) (plural: oogonia) (1) In some algae and fungi, a cell in which an egg is produced. (2) In animals, the diploid progeny of a germ cell in females.

**ootid** In oogenesis, the daughter cell of the second meiotic division that differentiates into the mature ovum.

**open circulatory system** Circulatory system in which extracellular fluid leaves the vessels of the circulatory system, percolates between cells and through tissues, and then flows back into the circulatory system to be pumped out again. (Contrast with closed circulatory system.)

**operator** The region of an operon that acts as the binding site for the repressor.

**operon** A genetic unit of transcription, typically consisting of several structural genes that are transcribed together; the operon contains at least two control regions: the promoter and the operator.

**opportunity cost** The sum of the benefits an animal forfeits by not being able to perform some other behavior during the time when it is performing a given behavior.

**opsin** (op' sin) [Gk. *opsis*: sight] The protein portion of vertebrate visual pigments; associated with the pigment molecule 11-*cis*-retinal. See also rhodopsin.

**optic chiasm** [Gk. *chiasma*: cross] Structure on the lower surface of the vertebrate brain where the two optic nerves come together.

**optic nerve** The nerve that carries information from the retina of the eye to the brain.

**optical isomers** Two molecular isomers that are mirror images of each other.

**optimal foraging theory** The application of a cost–benefit approach to feeding behavior to identify the fitness value of feeding choices.

**oral** [L. *os*: mouth] Pertaining to the mouth, or that part of the body that contains the mouth.

**orbital** A region in space surrounding the atomic nucleus in which an electron is most likely to be found.

**organ** [Gk. *organon*: tool] A body part, such as the heart, liver, brain, root, or leaf. Organs are composed of different tissues integrated to perform a distinct function. Organs, in turn, are integrated into organ systems.

**organ identity genes** In angiosperms, genes that specify the different organs of the flower. (Compare with homeotic genes.)

**organ of Corti** Structure in the inner ear that transforms mechanical forces produced from pressure waves ("sound waves") into action potentials that are sensed as sound.

**organ system** An interrelated and integrated group of tissues and organs that work together in a physiological function.

**organelle** (or gan el′) Any of the membrane-enclosed structures within a eukaryotic cell. Examples include the nucleus, endoplasmic reticulum, and mitochondria.

**organic** (1) Pertaining to any chemical compound that contains carbon. (2) Pertaining to any aspect of living matter, e.g., to its evolution, structure, or chemistry.

**organic fertilizers** Substances added to soil to improve the soil's fertility; derived from partially decomposed plant material (compost) or animal waste (manure).

**organism** Any living entity.

**organizer** Region of the early amphibian embryo that directs early embryonic development. Also known as the primary embryonic organizer.

**organogenesis** The formation of organs and organ systems during development.

**origin of replication (*ori*)** DNA sequence at which helicase unwinds the DNA double helix and DNA polymerase binds to initiate DNA replication.

**orthologs** [Gk. *ortho*: true, direct] Homologous genes whose divergence can be traced to speciation events.

**osmoconformer** An aquatic animal that equilibrates the osmolarity of its extracellular fluid to be the same as that of the external environment. (Contrast with osmoregulator.)

**osmolarity** The concentration of osmotically active particles in a solution.

**osmoregulation** Regulation of the chemical composition of the body fluids of an organism.

**osmoregulator** An aquatic animal that actively regulates the osmolarity of its extracellular fluid. (Contrast with osmoconformer.)

**osmosis** (oz mo′ sis) [Gk. *osmos*: to push] Movement of water across a differentially permeable membrane, from one region to another region where the water potential is more negative.

**ossicle** (oss′ ick ul) [L. *os*: bone] The calcified construction unit of echinoderm skeletons.

**osteoblast** (oss′ tee oh blast) [Gk. *osteon*: bone + *blastos*: sprout] A cell that lays down the protein matrix of bone.

**osteoclast** (oss′ tee oh clast) [Gk. *osteon*: bone + *klastos*: broken] A cell that dissolves bone.

**osteocyte** An osteoblast that has become enclosed in lacunae within the bone it has built.

**outgroup** In phylogenetics, a group of organisms used as a point of reference for comparison with the groups of primary interest (the ingroup).

**oval window** The flexible membrane that, when moved by the bones of the middle ear, produces pressure waves in the inner ear.

**ovarian cycle** In human females, the monthly cycle of events by which eggs and hormones are produced. (Contrast with uterine cycle).

**ovary** (oh′ var ee) [L. *ovum*: egg] Any female organ, in plants or animals, that produces an egg.

**overtopping** Plant growth pattern in which one branch differentiates from and grows beyond the others.

**oviduct** In mammals, the tube serving to transport eggs to the uterus or to the outside of the body.

**oviparity** Reproduction in which eggs are released by the female and development is external to the mother's body. (Contrast with viviparity.)

**ovoviviparity** Pertaining to reproduction in which fertilized eggs develop and hatch within the mother's body but are not attached to the mother by means of a placenta.

**ovulation** Release of an egg from an ovary.

**ovule** (oh′ vule) In plants, a structure comprising the megasporangium and the integument, which develops into a seed after fertilization.

**ovum** (oh′ vum) (plural: ova) [L. egg] The female gamete.

**oxidation** (ox i day′ shun) Relative loss of electrons in a chemical reaction; either outright removal to form an ion, or the sharing of electrons with substances having a greater affinity for them, such as oxygen. Most oxidations, including biological ones, are associated with the liberation of energy. (Contrast with reduction.)

**oxidation–reduction (redox) reaction** A reaction in which one substance transfers one or more electrons to another substance. (*See* oxidation; reduction.)

**oxidative phosphorylation** ATP formation in the mitochondrion, associated with flow of electrons through the respiratory chain.

**oxygenase** An enzyme that catalyzes the addition of oxygen to a substrate from $O_2$.

**oxytocin** A hormone released by the posterior pituitary that promotes social bonding.

**ozone layer** A layer of ozone ($O_3$, a greenhouse gas) in the atmosphere that absorbs a high portion of the sun's potentially mutagenic ultraviology radiation.

## P

**pacemaker cells** Cardiac cells that can initiate action potentials without stimulation from the nervous system, allowing the heart to initiate its own contractions.

**pair rule genes** In *Drosophila* (fruit fly) development, segmentation genes that divide the early embryo into units of two segments each. Part of a developmental cascade that includes maternal effect genes, gap genes, segment polarity genes, and Hox genes.

**paleomagnetic dating** A method for determining the age of rocks based on properties relating to changes in the patterns of Earth's magnetism over time.

**pancreas** (pan′ cree us) A gland located near the stomach of vertebrates that secretes digestive enzymes into the small intestine and releases insulin into the bloodstream.

**Pangaea** (pan jee′ uh) [Gk. *pan*: all, every] The single land mass formed when all the continents came together in the Permian period.

**para-** [Gk. *para*: akin to, beside] Prefix indicating association in being along side or accessory to.

**parabronchi** Passages in the lungs of birds through which air flows.

**paracrine** [Gk. *para*: near] Pertaining to a chemical signal, such as a hormone, that acts locally, near the site of its secretion. (Contrast with autocrine.)

**parallel evolution** The repeated evolution of similar traits, especially among closely related species; facilitated by conserved developmental genes.

**paralogs** Homologous genes whose divergence can be traced to gene duplication events. (Contrast with orthologs.)

**paraphyletic** (par′ a fih leht′ ik) [Gk. *para*: beside + *phylon*: tribe] Pertaining to a group that consists of an ancestor and some, but not all, of its descendants. (Contrast with monophyletic, polyphyletic.)

**parasite** An organism that consumes parts of an organism much larger than itself (known as its host). Parasites sometimes, but not always, kill their host.

**parasympathetic nervous system** The division of the autonomic nervous system that works in opposition to the sympathetic nervous system. (Contrast with sympathetic nervous system.)

**parathyroid glands** Four glands on the posterior surface of the thyroid gland that produce and release parathyroid hormone.

**parathyroid hormone (PTH)** A hormone secreted by the parathyroid glands that stimulates osteoclast activity and raises blood calcium levels. Also called parathormone.

**parenchyma** (pair eng′ kyma) A plant tissue composed of relatively unspecialized cells without secondary walls.

**parent rock** The soil horizon consisting of the rock that is breaking down to form the soil. Also called bedrock, or the C horizon.

**parental (P) generation** The individuals that mate in a genetic cross. Their offspring are the first filial ($F_1$) generation.

**parietal cells** One of three types of secretory cell found in the gastric pits of the stomach wall. Parietal cells produce hydrochloric acid (HCl), creating an acidic environment that destroys many of the harmful microorganisms ingested with food. (See chief cells; mucosal epithelium.)

**parietal lobe** One of four lobes of the cerebral hemisphere; processes complex stimuli and includes the primary somatosensory cortex.

**parsimony** Preferring the simplest among a set of plausible explanations of any phenomenon.

**parthenocarpy** Formation of fruit from a flower without fertilization.

**parthenogenesis** [Gk. *parthenos*: virgin] Production of an organism from an unfertilized egg.

**particulate theory** In genetics, the theory that genes are physical entities that retain their identities after fertilization.

**passive transport** Diffusion across a membrane; may or may not require a channel or carrier protein. (Contrast with active transport.)

**patch clamping** Technique for isolating a tiny patch of membrane to allow the study of ion movement through a particular channel.

**pathogen** (path' o jen) [Gk. *pathos*: suffering + *genesis*: source] An organism that causes disease.

**pattern formation** In animal embryonic development, the organization of differentiated tissues into specific structures such as wings.

**pedigree** The pattern of transmission of a genetic trait within a family.

**pelagic zone** [Gk. *pelagos*: deep sea] The open ocean; a marine life zone.

**penetrance** The proportion of individuals with a particular genotype that show the expected phenotype.

**penis** An accessory sex organ of male animals that enables the male to deposit sperm in the female's reproductive tract.

**pentaradial symmetry** Symmetry in five or multiples of five; a feature of adult echinoderms.

**pentose** [Gk. *penta:* five] A sugar containing five carbon atoms.

**PEP carboxylase** The enzyme that combines carbon dioxide with PEP to form a 4-carbon dicarboxylic acid at the start of $C_4$ photosynthesis or of crassulacean acid metabolism (CAM).

**pepsin** [Gk. *pepsis*: digestion] An enzyme in gastric juice that digests protein.

**pepsinogen** Inactive secretory product that is converted into pepsin by low pH or by enzymatic action.

**peptide hormones** Relatively large hormone molecules made up of amino acids; encoded by genes and produced by translation.

**peptide linkage** The bond between amino acids in a protein; formed between a carboxyl group and amino group (—CO—NH—) with the loss of water molecules.

**peptidoglycan** The cell wall material of many bacteria, consisting of a single enormous molecule that surrounds the entire cell.

**peptidyl transferase** A catalytic function of the large ribosomal subunit that consists of two reactions: breaking the bond between an amino acid and its tRNA in the P site, and forming a peptide bond between that amino acid and the amino acid attached to the tRNA in the A site.

**per capita birth rate (*b*)** In population growth models, the number of offspring that an average individual produces in some time interval.

**per capita death rate (*d*)** In population growth models, the average individual's chance of dying in some time interval.

**per capita growth rate (*r*)** In population models, the average individual's contribution to total population growth rate.

**perennial** (per ren' ee al) [L. *per*: throughout + *annus*: year] A plant that survives from year to year. (Contrast with annual, biennial.)

**perfect flower** A flower with both stamens and carpels; a hermaphroditic flower. (Contrast with imperfect flower.)

**pericycle** [Gk. *peri*: around + *kyklos*: ring or circle] In plant roots, tissue just within the endodermis, but outside of the root vascular tissue. Meristematic activity of pericycle cells produces lateral root primordia.

**periderm** The outer tissue of the secondary plant body, consisting primarily of cork.

**period** (1) A category in the geological time scale. (2) The duration of a single cycle in a cyclical event, such as a circadian rhythm.

**peripheral membrane proteins** Proteins associated with but not embedded within the plasma membrane. (Contrast with integral membrane proteins.)

**peripheral nervous system (PNS)** The portion of the nervous system that transmits information to and from the central nervous system, consisting of neurons that extend or reside outside the brain or spinal cord and their supporting cells. (Contrast with central nervous system.)

**peristalsis** (pair' i stall' sis) Wavelike muscular contractions proceeding along a tubular organ, propelling the contents along the tube.

**peritoneum** The mesodermal lining of the body cavity in coelomate animals.

**peroxisome** An organelle that houses reactions in which toxic peroxides are formed and then converted to water.

**petal** [Gk. *petalon*: spread out] In an angiosperm flower, a sterile modified leaf, nonphotosynthetic, frequently brightly colored, and often serving to attract pollinating insects.

**petiole** (pet' ee ole) [L. *petiolus*: small foot] The stalk of a leaf.

**$P_{fr}$** *See* phytochrome.

**pH** The negative logarithm of the hydrogen ion concentration; a measure of the acidity of a solution. A solution with pH = 7 is said to be neutral; pH values higher than 7 characterize basic solutions, while acidic solutions have pH values less than 7.

**phage** (fayj) *See* bacteriophage.

**phagocyte** [Gk. *phagein*: to eat + *kystos*: sac] One of two major classes of white blood cells; one of the nonspecific defenses of animals; ingests invading microorganisms by **phagocytosis**.

**pharmacogenomics** The study of how an individual's genetic makeup affects his or her response to drugs or other agents, with the goal of predicting the effectiveness of different treatment options.

**pharming** The use of genetically modified animals to produce medically useful products in their milk.

**pharynx** [Gk. throat] The part of the gut between the mouth and the esophagus.

**phenotype** (fee' no type) [Gk. *phanein*: to show] The observable properties of an individual resulting from both genetic and environmental factors. (Contrast with genotype.)

**phenotypic plasticity** *See* developmental plasticity.

**pheromone** (feer' o mone) [Gk. *pheros*: carry + *hormon*: excite, arouse] A chemical substance used in communication between organisms of the same species.

**phloem** (flo' um) [Gk. *phloos*: bark] In vascular plants, the vascular tissue that transports sugars and other solutes from sources to sinks.

**phosphate group** The functional group $—OPO_3H_2$.

**phosphodiester linkage** The connection in a nucleic acid strand, formed by linking two nucleotides.

**phospholipid** A lipid containing a phosphate group; an important constituent of cellular membranes. (*See* lipid.)

**phospholipid bilayer** The basic structural unit of biological membranes; a sheet of phospholipids two molecules thick in which the phospholipids are lined up with their hydrophobic "tails" packed tightly together and their hydrophilic, phosphate-containing "heads" facing outward. Also called lipid bilayer.

**phosphorylation** Addition of a phosphate group.

**photic zone** The life zone in lakes and oceans that is penetrated by light and therefore supports photosynthetic organisms.

**photoautotroph** An organism that obtains energy from light and carbon from carbon dioxide. (Contrast with chemolithotroph, chemoheterotroph, photoheterotroph.)

**photoheterotroph** An organism that obtains energy from light but must obtain its carbon from organic compounds. (Contrast with chemoautotroph, chemoheterotroph, photoautotroph.)

**photomorphogenesis** In plants, a process by which physiological and developmental events are controlled by light.

**photon** (foe' ton) [Gk. *photos*: light] A quantum of visible radiation; a "packet" of light energy.

**photoperiodism** Control of an organism's physiological or behavioral responses by the length of the day or night (the **photoperiod**).

**photophosphorylation** Mechanism for ATP formation in chloroplasts in which electron transport is coupled to the transport of hydrogen ions (protons, $H^+$) across the thylakoid membrane. Compare with chemiosmosis.

**photoreceptors** (1) In plants, pigments that trigger a physiological response when they absorb a photon. (2) In animals, the sensory receptor cells that sense and respond to light energy. (See cone cells; rod cells.)

**photorespiration** Light-driven uptake of oxygen and release of carbon dioxide, the carbon being derived from the early reactions of photosynthesis.

**photosynthesis** (foe tow sin' the sis) [Gk.: creating from light] Metabolic processes carried out by green plants and some microorganisms by which visible light is trapped and the energy used to synthesize compounds such as ATP and glucose.

**photosystem** [Gk. *phos:* light + *systema:* assembly] A light-harvesting complex in the chloroplast thylakoid composed of pigments and proteins. **Photosystem I** absorbs light at 700 nm, passing electrons to ferrodoxin and from there to NADPH. **Photosystem II** absorbs light at 680 nm and passes electrons to the electron transport chain in the chloroplast.

**phototropism** [Gk. *photos*: light + *trope*: turning] A directed plant growth response to light.

**phycobilin** Photosynthetic pigment that absorbs red, yellow, orange, and green light and is found in cyanobacteria and some red algae.

**phylogeny** (fy loj' e nee) [Gk. *phylon*: tribe, race + *genesis*: source] The evolutionary history of a particular group of organisms or their genes. A **phylogenetic tree** is a graphic representation of these lines of evolutionary descent.

**physiological survivorship curves** *See* survivorship curves.

**physiology** (fiz' ee ol' o jee) [Gk. *physis*: natural form] The scientific study of the functions of living organisms and the individual organs, tissues, and cells of which they are composed.

**phytoalexins** Substances toxic to pathogens, produced by plants in response to fungal or bacterial infection.

**phytochrome** (fy' tow krome) [Gk. *phyton*: plant + *chroma*: color] A plant pigment regulating a large number of developmental and other phenomena in plants. It has two isomers: $P_r$, which absorbs red light, and $P_{fr}$, which absorbs far red light. $P_{fr}$ is the active form.

**phytomers** In plants, the repeating modules that compose a shoot, each consisting of one or more leaves, attached to the stem at a node; an internode; and one or more axillary buds.

**phytoplankton** Photosynthetic floating organisms. (*See* plankton.)

**phytoremediation** A form of bioremediation that uses plants to clean up environmental pollution.

**pigment** A substance that absorbs visible light.

**piloting** A form of navigation in which an animal finds its way by remembering landmarks in its environment.

**pineal gland** Gland located between the cerebral hemispheres that secretes melatonin.

**pinocytosis** Endocytosis by a cell of liquid containing dissolved substances.

**pistil** [L. *pistillum*: pestle] The structure of an angiosperm flower within which the ovules are borne. May consist of a single carpel, or of several carpels fused into a single structure. Usually differentiated into ovary, style, and stigma.

**pith** In plants, relatively unspecialized tissue found within a cylinder of vascular tissue.

**pituitary gland** A small gland attached to the base of the brain in vertebrates. Its hormones control the activities of other glands. Also known as the hypophysis.

**placenta** (pla sen' ta) The organ in female mammals that provides for the nourishment of the fetus and elimination of the fetal waste products.

**plankton** Aquatic organisms that float in the water column, dependent on currents and wind for movement. Plankton include many protists, some algae, and larval animals. (See also phytoplankton.)

**planula** (plan' yew la) [L. *planum*: flat] A free-swimming, ciliated larval form typical of the cnidarians.

**plaque** (plack) [Fr.: a metal plate or coin] (1) A circular clearing in a layer (lawn) of bacteria growing on the surface of a nutrient agar gel. (2) An accumulation of prokaryotic organisms on tooth enamel. Acids produced by these microorganisms cause tooth decay. (3) A region of arterial wall invaded by fibroblasts and fatty deposits.

**plasma** (plaz' muh) The liquid portion of blood, in which blood cells and other particulates are suspended.

**plasma cell** An antibody-secreting cell that develops from a B cell; the effector cell of the humoral immune system.

**plasma membrane** The membrane that surrounds the cell, regulating the entry and exit of molecules and ions. Every cell has a plasma membrane, and it is often called the cell membrane.

**plasmid** A DNA molecule distinct from the chromosome(s); that is, an extrachromosomal element; found in many bacteria. May replicate independently of the chromosome.

**plasmodesmata** (singular: plasmodesma) [Gk. *plassein*: to mold + *desmos*: band] Cytoplasmic strands connecting two adjacent plant cells.

**plasmogamy** The fusion of the cytoplasm of two cells. (Contrast with karyogamy.)

**plastid** A class of plant cell organelles that includes the chloroplast, which houses biochemical pathways for photosynthesis.

**plate tectonics** [Gk. *tekton*: builder] The scientific study of the structure and movements of Earth's lithospheric plates, which are the cause of continental drift.

**platelet** A membrane-bounded body without a nucleus, arising as a fragment of a cell in the bone marrow of mammals. Important to blood-clotting action.

**pleiotropy** (plee' a tro pee) [Gk. *pleion*: more] The determination of more than one character by a single gene.

**pleural membrane** [Gk. *pleuras*: rib, side] The membrane lining the outside of the lungs and the walls of the thoracic cavity. Inflammation of these membranes is a condition known as pleurisy.

**pluripotent** [L. *pluri*: many + *potens*: powerful] Having the ability to form all of the cells in the body. (Contrast with multipotent, totipotent.)

**podocytes** Cells of Bowman's capsule of the nephron that cover the capillaries of the glomerulus, forming filtration slits.

**point mutation** A mutation that results from the gain, loss, or substitution of a single nucleotide.

**polar** A molecule with separate and opposite electric charges at two ends, or poles; the water molecule ($H_2O$) is the most prevalent example. (Contrast with nonpolar.)

**polar body** A nonfunctional nucleus produced by meiosis during oogenesis.

**polar covalent bond** A covalent bond in which the electrons are drawn to one nucleus more than the other, resulting in an unequal distribution of charge.

**polar nuclei** In angiosperms, the two nuclei in the central cell of the megagametophyte; following fertilization they give rise to the endosperm.

**polarity** (1) In chemistry, the property of unequal electron sharing in a covalent bond that defines a polar molecule. (2) In development, the difference between one end of an organism or structure and the other.

**pollen** [L. *pollin*: fine flour] In seed plants, microscopic grains that contain the male gametophyte (microgametophyte) and gamete (microspore).

**pollen tube** A structure that develops from a pollen grain through which sperm are released into the megagametophyte.

**pollination** The process of transferring pollen from an anther to the stigma of a pistil in an angiosperm or from a strobilus to an ovule in a gymnosperm.

**poly-** [Gk. *poly*: many] A prefix denoting multiple entities.

**poly A tail** A long sequence of adenine nucleotides (50–250) added after transcription to the 3′ end of most eukaryotic mRNAs.

**polyandry** Mating system in which one female mates with multiple males.

**polygyny** Mating system in which one male mates with multiple females.

**polymer** [Gk. *poly*: many + *meros*: unit] A large molecule made up of similar or identical subunits called monomers. (Contrast with monomer.)

**polymerase chain reaction (PCR)** An enzymatic technique for the rapid production of millions of copies of a particular stretch of DNA where only a small amount of the parent molecule is available.

**polymorphic** (pol′ lee mor′ fik) [Gk. *poly*: many + *morphe*: form, shape] Coexistence in a population of two or more distinct traits.

**polyp** (pah′ lip) [Gk. *poly*: many + *pous*: foot] In cnidarians, a sessile, asexual life cycle stage.

**polypeptide** A large molecule made up of many amino acids joined by peptide linkages. Large polypeptides are called proteins.

**polyphyletic** (pol′ lee fih leht′ ik) [Gk. *poly*: many + *phylon*: tribe] Pertaining to a group that consists of multiple distantly related organisms, and does not include the common ancestor of the group. (Contrast with monophyletic, paraphyletic.)

**polyploid** (pol′ lee ploid ee) Possessing more than two entire sets of chromosomes.

**polyribosome (polysome)** A complex consisting of a threadlike molecule of messenger RNA and several (or many) ribosomes. The ribosomes move along the mRNA, synthesizing polypeptide chains as they proceed.

**polysaccharide** A macromolecule composed of many monosaccharides (simple sugars). Common examples are cellulose and starch.

**pons** [L. *pons*: bridge] Region of the brainstem anterior to the medulla.

**pool** The total amount of an element in a given compartment of the biosphere.

**population** Any group of organisms coexisting at the same time and in the same place and capable of interbreeding with one another.

**population bottleneck** A period during which only a few individuals of a normally large population survive.

**population density** The number of individuals in a population per unit of area or volume.

**population dynamics** The patterns and processes of change in populations.

**population genetics** The study of genetic variation and its causes within populations.

**population size** The total number of individuals in a population.

**positional information** In development, the basis of the spatial sense that induces cells to differentiate as appropriate for their location within the developing organism; often comes in the form of a morphogen gradient.

**positive cooperativity** Occurs when a molecule can bind several ligands and each one that binds alters the conformation of the molecule so that it can bind the next ligand more easily. The binding of four molecules of $O_2$ by hemoglobin is an example of positive cooperativity.

**positive feedback** In regulatory systems, information that amplifies a regulatory response, increasing the deviation of the system from the set point. (Contrast with negative feedback.)

**positive regulation** A form of gene regulation in which a regulatory macromolecule is needed to turn on the transcription of a structural gene; in its absence, transcription will not occur. (Contrast with negative regulation.)

**positive selection** Natural selection that acts to establish a trait that enhances survival in a population. (Contrast with purifying selection.)

**post-** [L. *postere*: behind, following after] Prefix denoting something that comes after.

**postabsorptive state** State in which no food remains in the gut and thus no nutrients are being absorbed. (Contrast with absorptive state.)

**posterior** Toward or pertaining to the rear. (Contrast with anterior.)

**posterior pituitary** A portion of the pituitary gland derived from neural tissue; involved in the storage and release of antidiuretic hormone and oxytocin.

**postsynaptic cell** The cell that receives information from a neuron at a synapse. (Contrast with presynaptic cell.)

**postzygotic isolating mechanisms** Barriers to the reproductive process that occur after the union of the nuclei of two gametes. (Contrast with prezygotic isolating mechanisms.)

**potential energy** Energy not doing work but with the potential to do so, such as the energy stored in chemical bonds. (Contrast with kinetic energy.)

**$P_r$** *See* phytochrome.

**pre-mRNA (precursor mRNA)** Initial gene transcript before it is modified to produce functional mRNA. Also known as the primary transcript.

**Precambrian** The first and longest period of geological time, during which life originated.

**precapillary sphincter** A cuff of smooth muscle that can shut off the blood flow to a capillary bed.

**predator** An organism that kills and eats other organisms.

**pressure flow model** An effective model for phloem transport in angiosperms. It holds that sieve element transport is driven by an osmotically generated pressure gradient between source and sink.

**pressure potential ($\Psi_p$)** The hydrostatic pressure of an enclosed solution in excess of the surrounding atmospheric pressure. (Contrast with solute potential, water potential.)

**presynaptic cell** The neuron that transmits information to another cell at a synapse. (Contrast with postsynaptic cell.)

**prey** [L. *praeda*: booty] An organism consumed by a predator as an energy source.

**prezygotic isolating mechanisms** Barriers to the reproductive process that occur before the union of the nuclei of two gametes (Contrast with postzygotic isolating mechanisms.)

**primary active transport** Active transport in which ATP is hydrolyzed, yielding the energy required to transport an ion or molecule against its concentration gradient. (Contrast with secondary active transport.)

**primary cell wall** In plant cells, a structure that forms at the middle lamella after cytokinesis, made up of cellulose microfibrils, hemicelluloses, and pectins. (Contrast with secondary cell wall.)

**primary consumer** An organism (herbivore) that eats plant tissues.

**primary endosymbiosis** The engulfment of a cyanobacterium by a larger eukaryotic cell that gave rise to the first photosynthetic eukaryotes with chloroplasts.

**primary growth** In plants, growth that is characterized by the lengthening of roots and shoots and by the proliferation of new roots and shoots through branching. (Contrast with secondary growth.)

**primary immune response** The first response of the immune system to an antigen, involving recognition by lymphocytes and the production of effector cells and memory cells. (Contrast with secondary immune response.)

**primary meristem** Meristem that produces the tissues of the primary plant body.

**primary motor cortex** *See* motor cortex

**primary oocyte** (oh′ eh site) [Gk. *oon*: egg + *kytos*: container] The diploid progeny of an oogonium. In many species, a primary oocyte enters prophase of the first meiotic division, then remains in developmental arrest for a long time before resuming meiosis to form a secondary oocyte and a polar body.

**primary plant body** That part of a plant produced by primary growth. Consists of all the *nonwoody* parts of a plant; many herbaceous plants consist entirely of a primary plant body. (Contrast with secondary plant body.)

**primary producer** A photosynthetic or chemosynthetic organism that synthesizes complex organic molecules from simple inorganic ones.

**primary sex determination** Genetic determination of gametic sex, male or female. (Contrast with secondary sex determination.)

**primary somatosensory cortex** *See* somatosensory cortex.

**primary spermatocyte** The diploid progeny of a spermatogonium; undergoes the first meiotic division to form secondary spermatocytes.

**primary structure** The specific sequence of amino acids in a protein.

**primary succession** Succession of ecological communities that begins in an area devoid of life, such as on recently exposed glacial till or lava flows. (Contrast with secondary succession.)

**primase** An enzyme that catalyzes the synthesis of a primer for DNA replication.

**primer** Strand of nucleic acid, usually RNA, that is the necessary starting material for the synthesis of a new DNA strand, which is synthesized from the 3′ end of the primer.

**primordium** (plural: primordia) [L. origin] The most rudimentary stage of an organ or other part.

**pro-** [L.: first, before, favoring] A prefix often used in biology to denote a developmental stage that comes first or an evolutionary form that appeared earlier than another. For example, prokaryote, prophase.

**probe** A segment of single-stranded nucleic acid used to identify DNA molecules containing the complementary sequence.

**procambium** Primary meristem that produces the vascular tissue.

**procedural memory** Memory of motor tasks.; these memories cannot be consciously recalled or described. (Contrast with declarative memory.)

**processive** Pertaining to an enzyme that catalyzes many reactions each time it binds to a substrate, as DNA polymerase does during DNA replication.

**products** The molecules that result from the completion of a chemical reation.

**progesterone** [L. *pro*: favoring + *gestare*: to bear] A female sex hormone that maintains pregnancy.

**prokaryotes** Unicellular organisms that do not have nuclei or other membrane-enclosed organelles. Includes Bacteria and Archaea. (Contrast with eukaryotes.)

**prolactin** A hormone released by the anterior pituitary, one of whose functions is the stimulation of milk production in female mammals.

**prometaphase** The phase of nuclear division that begins with the disintegration of the nuclear envelope.

**promoter** A DNA sequence to which RNA polymerase binds to initiate transcription.

**prop roots** Adventitious roots in some monocots that function as supports for the shoot.

**prophage** (pro′ fayj) The noninfectious units that are linked with the chromosomes of the host bacteria and multiply with them but do not cause dissolution of the cell. Prophage can later enter into the lytic phase to complete the virus life cycle.

**prophase** (pro′ phase) The first stage of nuclear division, during which chromosomes condense from diffuse, threadlike material to discrete, compact bodies.

**prostaglandin** Any one of a group of specialized lipids with hormone-like functions. It is not clear that they act at any considerable distance from the site of their production.

**prostate gland** In male humans, surrounds the urethra at its junction with the vas deferens; supplies an acid-neutralizing fluid to the semen.

**prosthetic group** Any nonprotein portion of an enzyme.

**proteases** Digestive enzymes that digest proteins.

**proteasome** In the eukaryotic cytoplasm, a huge protein structure that binds to and digests cellular proteins that have been tagged by ubiquitin.

**protein** (pro′ teen) [Gk. *protos*: first] Long-chain polymer of amino acids with twenty different common side chains. Occurs with its polymer chain extended in fibrous proteins, or coiled into a compact macromolecule in enzymes and other globular proteins. The component amino acids are encoded in the triplets of messenger RNA, and proteins are the products of genes.

**protein kinase** (kye′ nase) An enzyme that catalyzes the addition of a phosphate group from ATP to a target protein.

**protein kinase cascade** A series of reactions in response to a molecular signal, in which a series of protein kinases activate one another in sequence, amplifying the signal at each step.

**proteoglycan** A glycoprotein containing a protein core with attached long, linear carbohydrate chains.

**proteolysis** [protein + Gk. *lysis*: break apart] An enzymatic digestion of a protein or polypeptide.

**proteome** The complete set of proteins that can be made by an organism. Because of alternative splicing of pre-mRNA, the number of proteins that can be made is usually much larger than the number of protein-coding genes present in the organism's genome.

**prothrombin** The inactive form of thrombin, an enzyme involved in blood clotting.

**protoderm** Primary meristem that gives rise to the plant epidermis.

**proton** (pro′ ton) [Gk. *protos*: first, before] (1) A subatomic particle with a single positive charge. The number of protons in the nucleus of an atom determine its element. (2) A hydrogen ion, $H^+$.

**proton pump** An active transport system that uses ATP energy to move hydrogen ions across a membrane, generating an electric potential.

**proton-motive force** Force generated across a membrane having two components: a chemical potential (difference in proton concentration) plus an electrical potential due to the electrostatic charge on the proton.

**protonephridium** The excretory organ of flatworms, made up of a tubule and a flame cell.

**protoplast** The living contents of a plant cell; the plasma membrane and everything contained within it.

**provirus** Double-stranded DNA made by a virus that is integrated into the host's chromosome and contains promoters that are recognized by the host cell's transcription apparatus.

**proximal convoluted tubule** The initial segment of a renal tubule, closest to the glomerulus. (Compare with distal convoluted tubule.)

**proximal** Near the point of attachment or other reference point. (Contrast with distal.)

**proximate causes** The immediate genetic, physiological, neurological, and developmental mechanisms responsible for a behavior or morphology. (Contrast with ultimate cause.)

**pseudocoelomate** (soo′ do see′ low mate) [Gk. *pseudes*: false + *koiloma*: cavity] Having a body cavity, called a pseudocoel, consisting of a fluid-filled space in which many of the internal organs are suspended, but which is enclosed by mesoderm only on its outside.

**pseudogene** [Gk. *pseudes*: false] A DNA segment that is homologous to a functional gene but is not expressed because of changes to its sequence or changes to its location in the genome.

**pseudopod** (soo′ do pod) [Gk. *pseudes*: false + *podos*: foot] A temporary, soft extension of the cell body that is used in location, attachment to surfaces, or engulfing particles.

**pulmonary** [L. *pulmo*: lung] Pertaining to the lungs.

**pulmonary circuit** The portion of the circulatory system by which blood is pumped from the heart to the lungs or gills for oxygenation and back to the heart for distribution. (Contrast with systemic circuit.)

**pulmonary valve** A one-way valve between the right ventricle of the heart and the pulmonary artery that prevents backflow of blood into the ventricle when it relaxes.

**Punnett square** Method of predicting the results of a genetic cross by arranging the gametes of each parent at the edges of a square.

**pupa** (pew′ pa) [L. *pupa*: doll, puppet] In certain insects (the Holometabola), the encased developmental stage between the larva and the adult.

**pupil** The opening in the vertebrate eye through which light passes.

**purifying selection** The elimination by natural selection of detrimental characters from a population. (Contrast with positive selection.)

**purine** (pure′ een) One of the two types of nitrogenous bases in nucleic acids. Each of the purines—adenine and guanine—pairs with a specific pyrimidine.

**Purkinje fibers** Specialized heart muscle cells that conduct excitation throughout the ventricular muscle.

**pyrimidine** (pe rim' a deen) One of the two types of nitrogenous bases in nucleic acids. Each of the pyrimidines—cytosine, thymine, and uracil—pairs with a specific purine.

**pyrogen** [Gk.: *pry, fire;* ] Molecule that produces a rise in body temperature (fever); may be produced by an invading pathogen or by cells of the immune system in response to infection.

**pyruvate** The ionized form of pyruvic acid, a three-carbon acid; the end product of glycolysis and the raw material for the citric acid cycle.

**pyruvate oxidation** Conversion of pyruvate to acetyl CoA and $CO_2$ that occurs in the mitochondrial matrix in the presence of $O_2$.

## Q

**$Q_{10}$** A value that compares the rate of a biochemical process or reaction over 10°C temperature ranges. A process that is not temperature-sensitive has a $Q_{10}$ of 1; values of 2 or 3 mean the reaction speeds up as temperature increases.

**qualitative** Based on observation of an unmeasured quality of a trait, as in brown vs. blue.

**quantitative** Based on numerical values obtained by measurement, as in quantitative data.

**quantitative trait loci** A set of genes determining a complex character (trait) that exhibits quantitative variation (variation in amount rather than in kind).

**quaternary structure** The specific three-dimensional arrangement of protein subunits. Contrast with primary, secondary, tertiary structure.

**quorum sensing** The use of chemical communication signals to trigger density-linked activities such as biofilm formation in prokaryotes.

## R

**R group** The distinguishing group of atoms of a particular amino acid; also known as a side chain.

***r*-strategist** A species whose life history strategy allows for a high intrinsic rate of population increase (*r*). (Contrast with *K*-strategist.)

**radial symmetry** The condition in which any two halves of a body are mirror images of each other, providing the cut passes through the center; a cylinder cut lengthwise down its center displays this form of symmetry.

**radiation** The transfer of heat from warmer objects to cooler ones via the exchange of infrared radiation. *See also* electromagnetic radiation; evolutionary radiation.

**radicle** An embryonic root.

**radioisotope** A radioactive isotope of an element. Examples are carbon-14 ($^{14}C$) and hydrogen-3, or tritium ($^{3}H$).

**radiometric dating** A method for determining the age of objects such as fossils and rocks based on the decay rates of radioactive isotopes.

**rapid eye movement sleep** *See* REM sleep.

**reactant** A chemical substance that enters into a chemical reaction with another substance.

**reaction center** A group of electron transfer proteins that receive energy from light-absorbing pigments and convert it to chemical energy by redox reactions.

**realized niche** A species' niche as defined by its interactions with other species. (Contrast with fundamental niche.)

**receptive field** The area of visual space that activates a particular cell in the visual system.

**receptor** *See* receptor protein, sensory receptor cell.

**receptor potential** The change in the resting potential of a sensory cell when it is stimulated.

**receptor protein** A protein that can bind to a specific molecule, or detect a specific stimulus, within the cell or in the cell's external environment.

**receptor-mediated endocytosis** Endocytosis initiated by macromolecular binding to a specific membrane receptor.

**recessive** In genetics, an allele that does not determine phenotype in the presence of a dominant allele. (Contrast with dominance.)

**reciprocal crosses** A pair of matings in one of which a female of genotype A mates with a male of genotype B and in the other of which a female of genotype B mates with a male of genotype A.

**recognition sequence** *See* restriction site.

**recombinant** Pertaining to an individual, meiotic product, or chromosome in which genetic materials originally present in two individuals end up in the same haploid complement of genes.

**recombinant DNA** A DNA molecule made in the laboratory that is derived from two or more genetic sources.

**recombinant frequency** The proportion of offspring of a genetic cross that have phenotypes different from the parental phenotypes due to crossing over between linked genes during gamete formation.

**recombination frequency** The proportion of offspring of a genetic cross that have phenotypes different from the parental phenotypes due to crossing over between linked genes during gamete formation.

**reconciliation ecology** The practice of making exploited lands more biodiversity-friendly. Compare with restoration ecology.

**rectum** The terminal portion of the gut, ending at the anus.

**redox reaction** A chemical reaction in which one reactant is oxidized (loses electrons) and the other is reduced (gains electrons). Short for reduction–oxidation reaction.

**reduction** Gain of electrons by a chemical reactant. (Contrast with oxidation.)

**refractory period** The time interval after an action potential during which another action potential cannot be elicited from an excitable membrane.

**regeneration** The development of a complete individual from a fragment of an organism.

**regulative development** A pattern of animal embryonic development in which the fates of the first blastomeres are not absolutely fixed. (Contrast with mosaic development.)

**regulatory gene** A gene that codes for a protein (or RNA) that in turn controls the expression of another gene.

**regulatory sequence** A DNA sequence to which the protein product of a regulatory gene binds.

**regulatory system** A system that uses feedback information to maintain a physiological function or parameter at an optimal level. (Contrast with controlled system.)

**regulatory T cells (Treg)** The class of T cells that mediates tolerance to self antigens.

**reinforcement** The evolution of enhanced reproductive isolation between populations due to natural selection for greater isolation.

**Reissner's membrane** *See* tectonic membrane.

**releaser** Sensory stimulus that triggers performance of a stereotyped behavior pattern.

**REM (rapid-eye-movement) sleep** A sleep state characterized by vivid dreams, skeletal muscle relaxation, and rapid eye movements. (Contrast with non-REM sleep.)

**renal** [L. *renes*: kidneys] Relating to the kidneys.

**renal tubule** A structural unit of the kidney that collects filtrate from the blood, reabsorbs specific ions, nutrients, and water and returns them to the blood, and concentrates excess ions and waste products such as urea for excretion from the body.

**renin** An enzyme released from the kidneys in response to a drop in the glomerular filtration rate. Together with angiotensin converting enzyme, converts an inactive protein in the blood into angiotensin.

**replication** The duplication of genetic material.

**replication complex** The close association of several proteins operating in the replication of DNA.

**replication fork** A point at which a DNA molecule is replicating. The fork forms by the unwinding of the parent molecule.

**replicon** A region of DNA replicated from a single origin of replication.

**reporter gene** A genetic marker included in recombinant DNA to indicate the presence of the recombinant DNA in a host cell.

**repressor** A protein encoded by a regulatory gene that can bind to a promoter and prevent transcription of the associated gene. (Contrast with activator.)

**reproductive isolation** Condition in which two divergent populations are no longer exchanging genes. Can lead to speciation.

**rescue effect** The process by which individuals moving between subpopulations of a metapopulation may prevent declining subpopulations from becoming extinct.

**residence time** The length of time a chemical element (e.g., carbon or nitrogen) remains in a given compartment of the ecosystem (e.g., in an organic body, in soil, in the atmosphere).

**residual volume (RV)** In tidal ventilation, the dead space that remains in the lungs at the end of exhalation.

**resistance (*R*) genes** Plant genes that confer resistance to specific strains of pathogens.

**resource** Something in the environment required by an organism for its maintenance and growth that is consumed in the process of being used.

**resource partitioning** A situation in which selection pressures resulting from interspecific competition cause changes in the ways in which the competing species use the limiting resource, thereby allowing them to coexist.

**respiration** (res pi ra' shun) [L. *spirare*: to breathe] (1) Cellular respiration. (2) Breathing.

**respiratory chain** The terminal reactions of cellular respiration, in which electrons are passed from NAD or FAD, through a series of intermediate carriers, to molecular oxygen, with the concomitant production of ATP.

**respiratory gases** Oxygen ($O_2$) and carbon dioxide ($CO_2$); the gases that an animal must exchange between its internal body fluids and the outside medium (air or water).

**resting potential** The membrane potential of a living cell at rest. In cells at rest, the interior is negative to the exterior. (Contrast with action potential.)

**restoration ecology** The science and practice of restoring damaged or degraded ecosystems.

**restriction enzyme** Any of a type of enzyme that cleaves double-stranded DNA at specific sites; extensively used in recombinant DNA technology. Also called a restriction endonuclease.

**restriction fragment length polymorphism** *See* RFLP.

**restriction point (R)** The specific time during G1 of the cell cycle at which the cell becomes committed to undergo the rest of the cell cycle.

**restriction site** A specific DNA base sequence that is recognized and acted on by a restriction endonuclease.

**reticular activating system** A central region of the vertebrate brainstem that includes complex fiber tracts conveying neural signals between the forebrain and the spinal cord, with collateral fibers to a variety of nuclei that are involved in autonomic functions, including arousal from sleep.

**reticulum** One of the four chambers of the ruminant stomach. Along with the rumen, where food is partially digested with the assistance of gut bacteria.

**retina** (rett' in uh) [L. *rete*: net] The light-sensitive layer of cells in the vertebrate or cephalopod eye.

**retrotransposons** Mobile genetic elements that are reverse transcribed into RNA as part of their transfer mechanism. (Contrast with DNA transposons.)

**retrovirus** An RNA virus that contains reverse transcriptase. Its RNA serves as a template for cDNA production, and the cDNA is integrated into a chromosome of the host cell.

**reverse genetics** Method of genetic analysis in which a phenotype is first related to a DNA variation, then the protein involved is identified.

**reverse transcriptase** An enzyme that catalyzes the production of DNA (cDNA), using RNA as a template; essential to the reproduction of retroviruses.

**reversion mutation** A second- or third-round mutation that reverts the DNA to its original sequence or to a new sequence that results in a non-mutant phenotype.

**RFLP** Restriction fragment length polymorphism, the coexistence of two or more patterns of restriction fragments resulting from underlying differences in DNA sequence.

**rhizoids** (rye' zoids) [Gk. root] Hairlike extensions of cells in mosses, liverworts, and a few vascular plants that serve the same function as roots and root hairs in vascular plants. The term is also applied to branched, rootlike extensions of some fungi and algae.

**rhizome** (rye' zome) An underground stem (as opposed to a root) that runs horizontally beneath the ground.

**rhodopsin** A vertebrate visual pigment involved in transducing photons of light into changes in the membrane potential of certain photoreceptor cells.

**ribonucleic acid** *See* RNA.

**ribose** A five-carbon sugar in nucleotides and RNA.

**ribosomal RNA (rRNA)** Several species of RNA that are incorporated into the ribosome. Involved in peptide bond formation.

**ribosome** A small particle in the cell that is the site of protein synthesis.

**ribozyme** An RNA molecule with catalytic activity.

**ribulose bisphosphate carboxylase/oxygenase** *See* rubisco.

**risk cost** The increased chance of being injured or killed as a result of performing a behavior, compared to resting.

**RNA (ribonucleic acid)** An often single-stranded nucleic acid whose nucleotides use ribose rather than deoxyribose and in which the base uracil replaces thymine found in DNA. Serves as genome from some viruses. (*See* ribosomal RNA, transfer RNA, messenger RNA, and ribozyme.)

**RNA interference (RNAi)** A mechanism for reducing mRNA translation whereby a double-stranded RNA, made by the cell or synthetically, is processed into a small, single-stranded RNA, whose binding to a target mRNA results in the latter's breakdown.

**RNA polymerase** An enzyme that catalyzes the formation of RNA from a DNA template.

**RNA splicing** The last stage of RNA processing in eukaryotes, in which the transcripts of introns are excised through the action of small nuclear ribonucleoprotein particles (snRNP).

**rod cells** Light-sensitive cells in the vertebrate retina; these sensory receptor cells are sensitive in extremely dim light and are responsible for dim light, black and white vision.

**root** The organ responsible for anchoring the plant in the soil, absorbing water and minerals, and producing certain hormones. Some roots are storage organs.

**root apical meristem** Undifferentiated tissue at the apex of the root that gives rise to the organs of the root.

**root cap** A thimble-shaped mass of cells, produced by the root apical meristem, that protects the meristem; the organ that perceives the gravitational stimulus in root gravitropism.

**root hair** A long, thin process from a root epidermal cell that absorbs water and minerals from the soil solution.

**root system** The organ system that anchors a plant in place, absorbs water and dissolved minerals, and may store products of photosynthesis from the shoot system.

**rough endoplasmic reticulum (RER)** The portion of the endoplasmic reticulum whose outer surface has attached ribosomes. (Contrast with smooth endoplasmic reticulum.)

**round window** A flexible membrane at the end of the lower canal of the cochlea in the human ear. (*See also* oval window.)

**rRNA** *See* ribosomal RNA.

**rubisco** Contraction of ribulose bisphosphate carboxylase/oxygenase, the enzyme that combines carbon dioxide or oxygen with ribulose bisphosphate to catalyze the first step of photosynthetic carbon fixation or photorespiration, respectively.

**rumen** One of the four chambers of the ruminant stomach. Along with the reticulum, where food is partially digested with the assistance of gut bacteria.

**ruminant** Herbivorous, cud-chewing mammals such as cows or sheep, characterized by a stomach that consists of four compartments: the rumen, reticulum, omasum, and abomasum.

## S

**S phase** In the cell cycle, the stage of interphase during which DNA is replicated. (Contrast with G1 phase, G2 phase, M phase.)

**salt glands** Glands on the leaves of some halophytic plants that secrete salt, thereby ridding the plants of excess salt.

**saltatory conduction** [L. *saltare*: to jump] The rapid conduction of action potentials in myelinated axons; so called because action potentials appear to "jump" between nodes of Ranvier along the axon.

**saprobe** [Gk. *sapros*: rotten] An organism (usually a bacterium or fungus) that obtains its carbon and energy by absorbing nutrients from dead organic matter.

**sarcomere** (sark' o meer) [Gk. *sark*: flesh + *meros*: unit] The contractile unit of a skeletal muscle.

**sarcoplasm** The cytoplasm of a muscle cell.

**sarcoplasmic reticulum** The endoplasmic reticulum of a muscle cell.

**saturated fatty acid** A fatty acid in which all the bonds between carbon atoms in the hydrocarbon chain are single bonds—that is, all the bonds are saturated with hydrogen atoms. (Contrast with unsaturated fatty acid.)

**Schwann cell** A type of glial cell that myelinates axons in the peripheral nervous system.

**scientific method** A means of gaining knowledge about the natural world by making observations, posing hypotheses, and conducting experiments to test those hypotheses.

**scion** In horticulture, the bud or stem from one plant that is grafted to a root or root-bearing stem of another plant (the stock).

**sclereid** One of the principle types of cells in sclerenchyma.

**sclerenchyma** (skler eng' kyma) [Gk. *skleros*: hard + *kymus*: juice] A plant tissue composed of cells with heavily thickened cell walls. The cells are dead at functional maturity. The principal types of sclerenchyma cells are fibers and sclereids.

**scrotum** In most mammals, a pouch outside the body cavity that contains the testes.

**second filial generation** See $F_2$.

**second law of thermodynamics** The principle that when energy is converted from one form to another, some of that energy becomes unavailable for doing work.

**second messenger** A compound, such as cAMP, that is released within a target cell after a hormone (the first messenger) has bound to a surface receptor on a cell; the second messenger triggers further reactions within the cell.

**second polar body** In oogenesis, the daughter cell of the second meiotic division that subsequently degenerates. (*See also* ootid.)

**secondary active transport** A form of active transport that does not use ATP as an energy source; rather, transport is coupled to ion diffusion down a concentration gradient established by primary active transport.

**secondary cell wall** A thick, cellulosic structure internal to the primary cell wall formed in some plant cells after cell expansion stops (Contrast with primary cell wall.)

**secondary consumer** An organism that eats primary consumers. (Contrast with primary consumer.)

**secondary endosymbiosis** The engulfment of a photosynthetic eukaryote by another eukaryotic cell that gave rise to certain groups of photosynthetic eukaryotes (e.g., euglenids).

**secondary growth** In plants, growth that contributes to an increase in girth. (Contrast with primary growth.)

**secondary immune response** A rapid and intense response to a second or subsequent exposure to an antigen, initiated by memory cells. (Contrast with primary immune response.)

**secondary lysosome** Membrane-enclosed organelle formed by the fusion of a primary lysosome with a phagosome, in which macromolecules taken up by phagocytosis are hydrolyzed into their monomers. (Contrast with lysosome.)

**secondary metabolite** A compound synthesized by a plant that is not needed for basic cellular metabolism. Typically has an antiherbivore or antiparasite function.

**secondary oocyte** In oogenesis, the daughter cell of the first meiotic division that receives almost all the cytoplasm. (*See also* first polar body.)

**secondary plant body** That part of a plant produced by secondary growth; consists of woody tissues. (Contrast with primary plant body.)

**secondary sex determination** Formation of secondary sexual characteristics (i.e., those other than gonads), such as external sex organs and body hair. (Contrast with primary sex determination.)

**secondary spermatocyte** One of the products of the first meiotic division of a primary spermatocyte.

**secondary structure** Of a protein, localized regularities of structure, such as the $\alpha$ helix and the $\beta$ pleated sheet. (Contrast with primary, tertiary, quarternary structure.)

**secondary succession** Succession of ecological communities after a disturbance that did not eliminate all the organisms originally living on the site. (Contrast with primary succession.)

**secretin** (si kreet' in) A peptide hormone secreted by the upper region of the small intestine when acidic chyme is present. Stimulates the pancreatic duct to secrete bicarbonate ions.

**sedimentary rock** Rock formed by the accumulation of sediment grains on the bottom of a body of water. Often contain stratified fossils that allow geologists and biologist to date evolutionary events relative to each other.

**seed** A fertilized, ripened ovule of a gymnosperm or angiosperm. Consists of the embryo, nutritive tissue, and a seed coat.

**seedling** A plant that has just completed the process of germination.

**segment polarity genes** In *Drosophila* (fruit fly) development, segmentation genes that determine the boundaries and anterior–posterior organization of individual segments. Part of a developmental cascade that includes maternal effect genes, gap genes, pair rule genes, and Hox genes.

**segmentation** Division of an animal body into segments.

**segmentation genes** Genes that determine the number and polarity of body segments.

**segregation** In genetics, the separation of alleles, or of homologous chromosomes, from each other during meiosis so that each of the haploid daughter nuclei produced contains one or the other member of the pair found in the diploid parent cell, but never both. This principle was articulated by Mendel as his first law.

**selectable marker** A gene, such as one encoding resistance to an antibiotic, that can be used to identify (select) cells that contain recombinant DNA from among a large population of untransformed cells.

**selective permeability** Allowing certain substances to pass through while other substances are excluded; a characteristic of membranes.

**self-incompatability** In plants, the possession of mechanisms that prevent self-fertilization.

**semelparous** [L. *semel*: once + *pario*: to beget] Reproducing only once in a lifetime. (Contrast with iteroparous.)

**semen** (see' men) [L. *semin*: seed] The thick, whitish liquid produced by the male reproductive system in mammals, containing the sperm.

**semicircular canals** Three canals in the human inner ear that form part of the vestibular system.

**semiconservative replication** The way in which DNA is synthesized. Each of the two partner strands in a double helix acts as a template for a new partner strand. Hence, after replication, each double helix consists of one old and one new strand.

**seminiferous tubules** The tubules within the testes within which sperm production occurs.

**senescence** [L. *senescere*: to grow old] Aging; deteriorative changes with aging; the increased probability of dying with increasing age.

**sensitive period** The life stage during which some particular type of learning must take place, or during which it occurs much more easily than at other times. Typical of song learning among birds. Also known as the critical period.

**sensory receptor cell** Cell that is responsive to a particular type of physical or chemical stimulation. Sometimes referred to as a sensor.

**sensory system** A set of organs and tissues for detecting a stimulus; consists of sensory cells, the associated structures, and the neural networks that process the information.

**sensory transduction** The transformation of environmental stimuli or information into neural signals.

**sepal** (see' pul) [L. *sepalum*: covering] One of the outermost structures of the flower, usually protective in function and enclosing the rest of the flower in the bud stage.

**septate** [L. wall] Divided, as by walls or partitions.

**septum** (plural: septa) (1) A partition or cross-wall appearing in the hyphae of some fungi. (2) The bony structure dividing the nasal passages.

**sequence alignment** A method of identifying homologous positions in DNA or amino acid sequences by pinpointing the locations of deletions and insertions that have occurred since two (or more) organisms diverged from a common ancestor.

**Sertoli cells** Cells in the seminiferous tubules of the testes that nurture the developing sperm.

**sessile** (sess' ul) [L. *sedere*: to sit] Permanently attached; not able to move from one place to another. (Contrast with motile.)

**set point** In a regulatory system, the threshold sensitivity to the feedback stimulus.

**sex chromosome** In organisms with a chromosomal mechanism of sex determination, one of the chromosomes involved in sex determination (in humans and many other animals, these are the X and Y chromosomes).

**sex-linked inheritance** Pattern of inheritance characteristic of genes located on the sex chromosomes of organisms having a chromosomal mechanism for sex determination.

**sex pilus** A thin connection between two bacteria through which genetic material passes during conjugation.

**sexual reproduction** Reproduction involving the union of gametes.

**sexual selection** Selection by one sex of characteristics in individuals of the opposite sex. Also, the favoring of characteristics in one sex as a result of competition among individuals of that sex for mates.

**shared derived trait** *See* synapomorphy.

**shoot apical meristem** Undifferentiated tissue at the apex of the shoot that gives rise to the organs of the shoot.

**shoot system** In plants, the organ system consisting of the leaves, stem(s), and flowers.

**short-day plant (SDP)** A plant that flowers when nights are longer than a critical length specific for that plant's species. (Compare to long-day plant.)

**short tandem repeats (STRs)** Short (1–5 base pairs), moderately repetitive sequences of DNA. The number of copies of an STR at a particular location varies between individuals and is inherited.

**side chain** *See* R group.

**sieve tube element** The characteristic cell of the phloem in angiosperms, which contains cytoplasm but relatively few organelles, and whose end walls (**sieve plates**) contain pores that form connections with neighboring cells.

**sigma factor** In prokaryotes, a protein that binds to RNA polymerase, allowing the complex to bind to and stimulate the transcription of a specific class of genes (e.g., those involved in sporulation).

**signal sequence** The sequence within a protein that directs the protein to a particular organelle.

**signal transduction pathway** The series of biochemical steps whereby a stimulus to a cell (such as a hormone or neurotransmitter binding to a receptor) is translated into a response of the cell.

**silencer** A gene sequence binding transcription factors that repress transcription. (Contrast with promoter.)

**silent mutation** A change in a gene's sequence that has no effect on the amino acid sequence of a protein either because it occurs in noncoding DNA or because it does not change the amino acid specified by the corresponding codon . (Contrast with frame-shift mutation, missense mutation, nonsense mutation.)

**silent substitution** *See* synonymous substitution.

**similarity matrix** A matrix used to compare the degree of divergence among pairs of objects. For molecular sequences, constructed by summing the number or percentage of nucleotides or amino acids that are identical in each pair of sequences.

**simple diffusion** Diffusion that does not involve a direct input of energy or assistance by carrier proteins.

**single nucleotide polymorphisms (SNPs)** Inherited variations in a single nucleotide base in DNA that differ between individuals.

**single-strand binding protein** In DNA replication, a protein that binds to single strands of DNA after they have been separated from each other, keeping the two strands separate for replication.

**sink** In plants, any organ that imports the products of photosynthesis, such as roots, developing fruits, and immature leaves. (Contrast with source.)

**sinoatrial node** (sigh' no ay' tree al) [L. *sinus*: curve + *atrium*: chamber] The pacemaker of the mammalian heart.

**siRNAs (small interfering RNAs)** Short, double-stranded RNA molecules used in RNA interference.

**sister chromatid** Each of a pair of newly replicated chromatids.

**sister species** Two species that are each other's closest relatives.

**skeletal muscle** A type of muscle tissue characterized by multinucleated cells containing highly ordered arrangements of actin and myosin microfilaments. Also called striated muscle. (Contrast with cardiac muscle, smooth muscle.)

**skeletal systems** Organ systems that provide rigid supports—**skeletons**—against which muscles can pull to create directed movements. See also endoskeleton, exoskeleton.

**sliding DNA clamp** Protein complex that keeps DNA polymerase bound to DNA during replication.

**sliding filament model** Mechanism of muscle contraction based on the formation and breaking of crossbridges between actin and myosin filaments, causing the filaments to slide together.

**slow-twitch fibers** Skeletal muscle fibers specialized for sustained aerobic work; contain myoglobin and abundant mitochondria, and are well-supplied with blood vessels. Also called oxidative or red muscle fibers. (Compare to fast-twitch fibers.)

**slow-wave sleep** *See* non-REM sleep.

**small interfering RNAs** *See* siRNAs.

**small intestine** The portion of the gut between the stomach and the colon; consists of the duodenum, the jejunum, and the ileum.

**small nuclear ribonucleoprotein particle (snRNP)** A complex of an enzyme and a small nuclear RNA molecule, functioning in RNA splicing.

**smooth endoplasmic reticulum (SER)** Portion of the endoplasmic reticulum that lacks ribosomes and has a tubular appearance. (Contrast with rough endoplasmic reticulum.)

**smooth muscle** Muscle tissue consisting of sheets of mononucleated cells innervated by the autonomic nervous system. (Contrast with cardiac muscle, skeletal muscle.)

**sodium–potassium ($Na^+$–$K^+$) pump** Antiporter responsible for primary active transport; it pumps sodium ions out of the cell and potassium ions into the cell, both against their concentration gradients. Also called a sodium–potassium ATPase.

**soil horizon** *See* horizons.

**soil solution** The aqueous portion of soil, from which plants take up dissolved mineral nutrients.

**solute** A substance that is dissolved in a liquid (solvent) to form a solution.

**solute potential** ($\Psi_s$) A property of any solution, resulting from its solute contents; it may be zero or have a negative value. The more negative the solute potential, the greater the tendency of the solution to take up water through a differentially permeable membrane. (Contrast with pressure potential, water potential.)

**solution** A liquid (the solvent) and its dissolved solutes.

**solvent** Liquid in which a substance (solute) is dissolved to form a solution.

**somatic cell** [Gk. *soma*: body] All the cells of the body that are not specialized for reproduction. (Contrast with germ cell.)

**somatic mutation** Permament genetic change in a somatic cell (as opposed to a germ cell, the egg or sperm). These mutations affect the individual only; they are not passed on to offspring. (Contrast with germ line mutation.)

**somatosensory cortex** An area of the parietal lobe that receives touch and pressure information from mechanoreceptors throughout the body; neurons in this area are arranged according to the parts of the body with which they communicate.

**somatostatin** Peptide hormone made in the hypothalamus that inhibits the release of other hormones from the pituitary and intestine.

**somite** (so' might) One of the segments into which an embryo becomes divided longitudinally, leading to the eventual segmentation of the animal as illustrated by the spinal column, ribs, and associated muscles.

**source** In plants, any organ that exports the products of photosynthesis in excess of its own needs, such as a mature leaf or storage organ. (Contrast with sink.)

**spatial summation** In the production or inhibition of action potentials in a postsynaptic cell, the interaction of depolarizations and hyperpolarizations produced at different sites on the postsynaptic cell. (Contrast with temporal summation.)

**spawning** *See* external fertilization.

**speciation** (spee' see ay' shun) The process of splitting one population into two populations that are reproductively isolated from one another.

**species** (spee' sees) [L. kind] The base unit of taxonomic classification, consisting of an ancestor–descendant group of populations of evolutionarily closely related, similar organisms. The more narrowly defined "biological species" consists of individuals capable of interbreeding with each other but not with members of other species.

**species composition** The particular mix of species a community contains and the abundances of those species.

**species evenness** A measure of species diversity that reflects the distribution of the species' abundances in a community.

**species richness** The total number of species living in a region.

**species–area relationship** The relationship between the size of an area and the numbers of species it supports.

**specific defenses** Defensive reactions of the vertebrate immune system that are based on the reaction of an antibody to a specific antigen. (Contrast with nonspecific defenses.)

**specific heat** The amount of energy that must be absorbed by a gram of a substance to raise its temperature by one degree centigrade. By convention, water is assigned a specific heat of one.

**sperm** [Gk. *sperma*: seed] The male gamete.

**spermatid** One of the products of the second meiotic division of a primary spermatocyte; four haploid spermatids, which remain connected by cytoplasmic bridges, are produced for each primary spermatocyte that enters meiosis.

**spermatogenesis** (spur mat' oh jen' e sis) [Gk. *sperma*: seed + *genesis*: source] Gametogenesis leading to the production of sperm.

**spermatogonia** In animals, the diploid progeny of a germ cell in males.

**spherical symmetry** The simplest form of symmetry, in which body parts radiate out from a central point such that an infinite number of planes passing through that central point can divide the organism into similar halves.

**sphincter** (sfink' ter) [Gk. *sphinkter*: something that binds tightly] A ring of muscle that can close an orifice, for example, at the anus.

**spicule** [L. arrowhead] A hard, calcareous skeletal element typical of sponges.

**spinal cord** Along with the brain, part of the central nervous system; transmits information between the body and the brain and mediates simple reflexes.

**spinal reflex** The conversion of afferent to efferent information in the spinal cord without participation of the brain.

**spindle apparatus** [O.E. *spindle*, a short stick with tapered ends] Array of microtubules emanating from both poles of a dividing cell during mitosis and playing a role in the movement of chromosomes at nuclear division.

**spleen** Organ that serves as a reservoir for venous blood and eliminates old, damaged red blood cells from the circulation.

**spliceosome** RNA–protein complex that splices out introns from eukaryotic pre-mRNAs.

**splicing** *See* RNA splicing.

**spontaneous mutation** A genetic change caused by internal cellular mechanisms, such as an error in DNA replication. (Contrast with induced mutation.)

**sporangiophore** A stalked reproductive structure produced by zygospore fungi that extends from a hypha and bears one or many sporangia.

**sporangium** (spor an' gee um) (plural: sporangia) [Gk. *spora*: seed + *angeion*: vessel or reservoir] In plants and fungi, any specialized stucture within which one or more spores are formed.

**spore** [Gk. *spora*: seed] (1) Any asexual reproductive cell capable of developing into an adult organism without gametic fusion. In plants, haploid spores develop into gametophytes, diploid spores into sporophytes. (2) In prokaryotes, a resistant cell capable of surviving unfavorable periods.

**sporocyte** Specialized cells of the diploid sporophyte that will divide by meiosis to produce four haploid spores. Germination of these spores produces the haploid gametophyte.

**sporophyte** (spor' o fyte) [Gk. *spora*: seed + *phyton*: plant] In plants and protists with alternation of generations, the diploid phase that produces the spores. (Contrast with gametophyte.)

**stabilizing selection** Selection against the extreme phenotypes in a population, so that the intermediate types are favored. (Contrast with disruptive selection.)

**stamen** (stay' men) [L. *stamen*: thread] A male (pollen-producing) unit of a flower, usually composed of an anther, which bears the pollen, and a filament, which is a stalk supporting the anther.

**starch** [O.E. *stearc*: stiff] A polymer of glucose; used by plants to store energy.

**Starling's forces** The two opposing forces responsible for water movement across capillary walls: blood pressure, which squeezes water and small solutes out of the capillaries, and osmotic pressure, which pulls water back into the capillaries.

**start codon** The mRNA triplet (AUG) that acts as a signal for the beginning of translation at the ribosome. (Contrast with stop codon.)

**stele** (steel) [Gk.: pillar] The central cylinder of vascular tissue in a plant stem.

**stem cell** In animals, an undifferentiated cell that is capable of continuous proliferation. A stem cell generates more stem cells and a large clone of differentiated progeny cells. (*See also* embryonic stem cell.)

**stem** In plants, the organ that holds leaves and/or flowers and transports and distributes materials among the other organs of the plant.

**stereocilia** Fingerlike extensions of hair cell membranes whose bending initiates sound perception. (*See* hair cell.)

**steroid** Any of a family of lipids whose multiple rings share carbons. The steroid cholesterol is an important constituent of membranes and is the base of steroid hormones such as testosterone.

**sticky ends** On a piece of two-stranded DNA, short, complementary, one-stranded regions produced by the action of a restriction endonuclease. Sticky ends facilitate the joining of segments of DNA from different sources.

**stigma** [L. *stigma*: mark, brand] The part of the pistil at the apex of the style that is receptive to pollen, and on which pollen germinates.

**stimulus** [L. *stimulare*: to goad] Something causing a response; something in the environment detected by a receptor.

**stock** In horticulture, the root or root-bearing stem to which a bud or piece of stem from another plant (the scion) is grafted.

**stoma** (plural: stomata) [Gk. *stoma*: mouth, opening] Small opening in the plant epidermis that permits gas exchange; bounded by a pair of guard cells whose osmotic status regulates the size of the opening.

**stomach** An organ that physically (and sometimes enzymatically) breaks down food, preparing it for digestion in the midgut.

**stomatal crypt** In plants, a sunken cavity below the leaf surface in which a stoma is sheltered from the drying effects of air currents.

**stop codon** Any of the three mRNA codons that signal the end of protein translation at the ribosome: UAG, UGA, UAA.

**stratosphere** The upper part of Earth's atmosphere, above the troposphere; extends from approximately 18 kilometers upward to approximately 50 kilometers above Earth's surface.

**stratum** (plural strata) [L. *stratos*: layer] A layer of sedimentary rock laid down at a particular time in the past.

**stretch receptor** A modified muscle cell embedded in the connective tissue of a muscle that acts as a mechanoreceptor in response to stretching of that muscle.

**striated muscle** *See* skeletal muscle.

**strigolactones** Signaling molecules produced by plant roots that attract the hyphae of mycorrhizal fungi.

**strobilus** (plural: strobili) One of several conelike structures in various groups of plants (including club mosses, horsetails, and conifers) associated with the production and dispersal of reproductive products.

**stroke** An embolism in an artery in the brain that causes the cells fed by that artery to die. The specific damage, such as memory loss, speech impairment, or paralysis, depends on the location of the blocked artery.

**stroma** The fluid contents of an organelle such as a chloroplast or mitochondrion.

**structural gene** A gene that encodes the primary structure of a protein not involved in the regulation of gene expression.

**structural isomers** Molecules made up of the same kinds and numbers of atoms, in which the atoms are bonded differently.

**structural motif** A three-dimensional structural element that is part of a larger molecule. For example, there are four common motifs in DNA-binding proteins: helix-turn-helix, zinc finger, leucine zipper, and helix-loop-helix.

**style** [Gk. *stele*: pillar or column] In the angiosperm flower, a column of tissue extending from the tip of the ovary, and bearing the stigma or receptive surface for pollen at its apex.

**sub-** [L. under] A prefix used to designate a structure that lies beneath another or is less than another. For example, subcutaneous (beneath the skin); subspecies

**subduction** In plate tectonics, the movement of one lithospheric plate under another.

**suberin** A waxlike lipid that is a barrier to water and solute movement across the Casparian strip of the endodermis.

**submucosa** (sub mew koe' sah) The tissue layer just under the epithelial lining of the lumen of the digestive tract.

**subsoil** The soil horizon lying below the topsoil and above the parent rock (bedrock); the zone of infiltration and accumulation of materials leached from the topsoil. Also called the B horizon.

**substrate** (sub' strayte) (1) The molecule or molecules on which an enzyme exerts catalytic action. (2) The base material on which a sessile organism lives.

**succession** The gradual, sequential series of changes in the species composition of an ecological community following a disturbance. See also cyclical succession, directional selection, heterotrophic succession.

**succulence** In plants, possession of fleshy, water-storing leaves or stems; an adaptation to dry environments.

**sulcus** (sul' kus; plural sulci) [Gk.: plowed furrow] *See* convolutions.

**superficial cleavage** A variation of incomplete cleavage in which cycles of mitosis occur without cell division, producing a syncytium (a single cell with many nuclei).

**suprachiasmatic nuclei (SCN)** In mammals, two clusters of neurons just above the optic chiasm that act as the master circadian clock.

**surface area-to-volume ratio** For any cell, organism, or geometrical solid, the ratio of surface area to volume; this is an important factor in setting an upper limit on the size a cell or organism can attain.

**surface tension** The attractive intermolecular forces at the surface of liquid; an especially important property of water.

**surfactant** A substance that decreases the surface tension of a liquid. Lung surfactant, secreted by cells of the alveoli, is mostly phospholipid and decreases the amount of work necessary to inflate the lungs.

**survivorship** The fraction of individuals that survive from birth to a given life stage or age.

**survivorship curves** Graphic plot of ages at death of a hypothetical cohort, usually of 1,000 individuals, by plotting the numbers of individuals expected to survive to reach each age category. There are three general shapes. Ecological survivorship is linear: individuals face a constant risk of mortality regardless of their age. Physiological survivorship curves are concave: high survivorship through adulthood with steep declines late in life. Maturational survivorship curves are convex, with high mortality early in life but higher survivorship once individuals reach maturity.

**suspensor** In the embryos of seed plants, the stalk of cells that pushes the embryo into the endosperm and is a source of nutrient transport to the embryo.

**sustainable** Pertaining to the use and management of ecosystems in such a way that humans benefit over the long term from specific ecosystem goods and services without compromising others.

**symbiosis** (sim' bee oh' sis) [Gk. *sym*: together + *bios*: living] The living together of two or more species in a prolonged and intimate relationship.

**symmetry** Pertaining to an attribute of an animal body in which at least one plane can divide the body into similar, mirror-image halves. (*See* bilateral symmetry, radial symmetry.)

**sympathetic nervous system** The division of the autonomic nervous system that works in opposition to the parasympathetic nervous system. (Contrast with parasympathetic nervous system.)

**sympatric speciation** (sim pat' rik) [Gk. *sym*: same + *patria*: homeland] Speciation due to reproductive isolation without any physical separation of the subpopulation. (Contrast with allopatric speciation.)

**symplast** The continuous meshwork of the interiors of living cells in the plant body, resulting from the presence of plasmodesmata. (Contrast with apoplast.)

**symporter** A membrane transport protein that carries two substances in the same direction. (Contrast with antiporter, uniporter.)

**synapomorphy** A trait that arose in the ancestor of a phylogenetic group and is present (sometimes in modified form) in all of its members, thus helping to delimit and identify that group. Also called a shared derived trait.

**synapse** (sin' aps) [Gk. *syn*: together + *haptein*: to fasten] A specialized type of junction where a neuron meets its target cell (which can be another neuron or some other type of cell) and information in the form of neurotransmitter molecules is exchanged across a synaptic cleft.

**synapsis** (sin ap′ sis) The highly specific parallel alignment (pairing) of homologous chromosomes during the first division of meiosis.

**synaptic cleft** The space between the presynaptic cell and the postsynaptic cell in a chemical synapse.

**synergids** [Gk. *syn*: together + *ergos*: work] In angiosperms, the two cells accompanying the egg cell at one end of the megagametophyte.

**syngamy** *See* fertilization.

**synonymous (silent) substitution** A change of one nucleotide in a sequence to another when that change does not affect the amino acid specified (i.e., UUA → UUG, both specifying leucine). (Contrast with nonsynonymous substitution, missense mutation, nonsense mutation.)

**systematics** The scientific study of the diversity and relationships among organisms.

**systemic acquired resistance** A general resistance to many plant pathogens following infection by a single agent.

**systemic circuit** Portion of the circulatory system by which oxygenated blood from the lungs or gills is distributed throughout the rest of the body and returned to the heart. (Contrast with pulmonary circuit.)

**systems biology** The scientific study of an organism as an integrated and interacting system of genes, proteins, and biochemical reactions.

**systole** (sis′ tuh lee) [Gk.: contraction] Contraction of a chamber of the heart, driving blood forward in the circulatory system. (Contrast with diastole.)

## T

**3′ end** (3 prime) The end of a DNA or RNA strand that has a free hydroxyl group at the 3′ carbon of the sugar (deoxyribose or ribose).

**T cell** A type of lymphocyte involved in the cellular immune response. The final stages of its development occur in the thymus gland. (Contrast with B cell; *see also* cytotoxic T cell, T-helper cell.)

**T cell receptor** A protein on the surface of a T cell that recognizes the antigenic determinant for which the cell is specific.

**T tubules** A system of tubules that runs throughout the cytoplasm of a muscle fiber, through which action potentials spread.

**T-helper ($T_H$) cell** Type of T cell that stimulates events in both the cellular and humoral immune responses by binding to the antigen on an antigen-presenting cell; target of the HIV-I virus, the agent of AIDS. (Contrast with cytotoxic T cells.)

**taproot system** A root system typical of eudicots consisting of a primary root (*taproot*) that extends downward by tip growth and outward by initiating lateral roots. (Contrast with fibrous root system.)

**target cell** A cell with the appropriate receptors to bind and respond to a particular hormone or other chemical mediator.

**taste bud** A structure in the epithelium of the tongue that includes a cluster of chemoreceptors innervated by sensory neurons.

**TATA box** An eight-base-pair sequence, found about 25 base pairs before the starting point for transcription in many eukaryotic promoters, that binds a transcription factor and thus helps initiate transcription.

**taxon** (plural: taxa) [Gk. *taxis*: put in order] A biological group (typically a species or a clade) that is given a name.

**$T_C$ cells** *See* cytotoxic T cells.

**tectonic membrane** One of two membranes (the other is the basilar membrane) that extend along the length of the cochlea in the human ear. Also known as Reissner's membrane.

**telencephalon** The outer, surrounding structure of the embryonic vertebrate forebrain, which develops into the cerebrum.

**telomerase** An enzyme that catalyzes the addition of telomeric sequences lost from chromosomes during DNA replication.

**telomeres** (tee′ lo merz) [Gk. *telos*: end + *meros*: units, segments] Repeated DNA sequences at the ends of eukaryotic chromosomes.

**telophase** (tee′ lo phase) [Gk. *telos*: end] The final phase of mitosis or meiosis during which chromosomes become diffuse, nuclear envelopes re-form, and nucleoli begin to reappear in the daughter nuclei.

**template** A molecule or surface on which another molecule is synthesized in complementary fashion, as in the replication of DNA.

**template strand** In double-stranded DNA, the strand that is transcribed to create an RNA transcript that will be processed into a protein. Also refers to a strand of RNA that is used to create a complementary RNA.

**temporal lobe** One of the four lobes of the cerebral hemisphere; receives and processes auditory and visual information; involved in recognizing, identifying, and naming objects.

**temporal summation** In the production or inhibition of action potentials in a postsynaptic cell, the interaction of depolarizations or hyperpolarizations produced by rapidly repeated stimulation of a single point on the postsynaptic cell. (Contrast with spatial summation.)

**tendon** A collagen-containing band of tissue that connects a muscle with a bone.

**tepal** A sterile, modified, nonphotosynthetic leaf of an angiosperm flower that cannot be distinguished as a petal or a sepal.

**termination** In molecular biology, the end of transcription or translation.

**terminator** A sequence at the 3′ end of mRNA that causes the RNA strand to be released from the transcription complex.

**terrestrial** (ter res′ tree al) [L. *terra*: earth] Pertaining to or living on land. (Contrast with aquatic, marine.)

**territorial behavior** Aggressive actions engaged in to defend a habitat or resource such that other animals are denied access.

**tertiary consumers** Carnivores that consume primary carnivores (secondary consumers).

**tertiary endosymbiosis** The mechanism by which some eukaryotes acquired the capacity for photosynthesis; for example, a dinoflagellate that apparently lost its chloroplast became photosynthetic by engulfing another protist that had acquired a chloroplast through secondary endosymbiosis.

**tertiary structure** In reference to a protein, the relative locations in three-dimensional space of all the atoms in the molecule. The overall shape of a protein. (Contrast with primary, secondary, and quaternary structures.)

**test cross** Mating of a dominant-phenotype individual (who may be either heterozygous or homozygous) with a homozygous-recessive individual.

**testis** (tes′ tis) (plural: testes) [L. *testis*: witness] The male gonad; the organ that produces the male gametes.

**tetanus** [Gk. *tetanos*: stretched] (1) A state of sustained maximal muscular contraction caused by rapidly repeated stimulation. (2) In medicine, an often fatal disease ("lockjaw") caused by the bacterium *Clostridium tetani*.

**tetrad** [Gk. *tettares*: four] During prophase I of meiosis, the association of a pair of homologous chromosomes or four chromatids

**thalamus** [Gk. *thalamos*: chamber] A region of the vertebrate forebrain; involved in integration of sensory input.

**theory** [Gk. *theoria*: analysis of facts] A far-reaching explanation of observed facts that is supported by such a wide body of evidence, with no significant contradictory evidence, that it is scientifically accepted as a factual framework. Examples are Newton's theory of gravity and Darwin's theory of evolution. (Contrast with hypothesis.)

**thermoneutral zone (TNZ)** [Gk. *thermos*: temperature] The range of temperatures over which an endotherm does not have to expend extra energy to thermoregulate.

**thermophile** (ther′ muh fyle)[Gk. *thermos*: temperature + *philos*: loving] An organism that lives exclusively in hot environments.

**thoracic cavity** [Gk. *thorax*: breastplate] The portion of the mammalian body cavity bounded by the ribs, shoulders, and diaphragm. Contains the heart and the lungs.

**thoracic duct** The connection between the lymphatic system and the circulatory system.

**threshold** The level of depolarization that causes an electrically excitable membrane to fire an action potential.

**thrombin** An enzyme involved in blood clotting; cleaves fibrinogen to form fibrin.

**thrombus** (throm′ bus) [Gk. *thrombos*: clot] A blood clot that forms within a blood vessel and remains attached to the wall of the vessel.

**thylakoid** (thigh′ la koid) [Gk. *thylakos*: sack or pouch] A flattened sac within a chloroplast. Thylakoid membranes contain all of the chlorophyll in a plant, in addition to the electron carriers of photophosphorylation. Thylakoids stack to form grana.

**thymine (T)** Nitrogen-containing base found in DNA.

**thymus** [Gk. *thymos*: warty] A ductless, glandular lymphoid tissue, involved in development of the immune system of vertebrates. In humans, the thymus degenerates during puberty.

**thyroid gland** [Gk. *thyreos*: door-shaped] A two-lobed gland in vertebrates. Produces the hormone thyroxine.

**thyrotropin** Hormone produced by the anterior pituitary that stimulates the thyroid gland to produce and release thyroxine. Also called thyroid-stimulating hormone (TSH).

**thyrotropin-releasing hormone (TRH)** Hormone produced by the hypothalamus that stimulates the anterior pituitary to release thyrotropin.

**thyroxine** Hormone produced by the thyroid gland; controls many metabolic processes.

**tidal** The bidirectional form of ventilation used by all vertebrates except birds; air enters and leaves the lungs by the same route.

**tight junction** A junction between epithelial cells in which there is no gap between adjacent cells.

**tissue** A group of similar cells organized into a functional unit; usually integrated with other tissues to form part of an organ.

**tissue system** In plants, any of three organized groups of tissues—dermal tissue, vascular tissue, and ground tissue—that are established during embryogenesis and have distinct functions.

**titin** A protein that holds bundles of myosin filaments in a centered position within the sarcomeres of muscle cells. The largest protein in the human body.

**tonoplast** The membrane of the plant central vacuole.

**topsoil** The uppermost soil horizon; contains most of the organic matter of soil, but may be depleted of most mineral nutrients by leaching. Also called the A horizon.

**totipotent** [L. *toto*: whole, entire + *potens*: powerful] Possessing all the genetic information and other capacities necessary to form an entire individual. (Contrast with multipotent, pluripotent.)

**toxigenicity** The ability of some pathogenic bacteria to produce chemical substances that harm the host.

**trachea** (tray′ kee ah) [Gk. *trakhoia*: tube] A tube that carries air to the bronchi of the lungs of vertebrates. When plural (*tracheae*), refers to the major airways of insects.

**tracheary element** Either of two types of xylem cells—tracheids and vessel elements—that undergo apoptosis before assuming their transport function.

**tracheid** (tray′ kee id) A type of tracheary element found in the xylem of nearly all vascular plants, characterized by tapering ends and walls that are pitted but not perforated. (Contrast with vessel element.)

**trade-off** The relationship between the fitness benefits conferred by an adaptation and the fitness costs it imposes. For an adaptation to be favored by natural selection, the benefits must exceed the costs.

**trait** In genetics, a specific form of a character: eye color is a character; brown eyes and blue eyes are traits. (Contrast with character.)

**transcription** The synthesis of RNA using one strand of DNA as a template.

**transcription factors** Proteins that assemble on a eukaryotic chromosome, allowing RNA polymerase II to perform transcription.

**transcription initiation site** The part of a gene's promoter where synthesis of the gene's RNA transcript begins.

**transduction** (1) Transfer of genes from one bacterium to another by a bacteriophage. (2) In sensory cells, the transformation of a stimulus (e.g., light energy, sound pressure waves, chemical or electrical stimulants) into action potentials.

**transfection** Insertion of recombinant DNA into animal cells.

**transfer RNA (tRNA)** A family of folded RNA molecules. Each tRNA carries a specific amino acid and anticodon that will pair with the complementary codon in mRNA during translation.

**transformation** (1) A mechanism for transfer of genetic information in bacteria in which pure DNA from a bacterium of one genotype is taken in through the cell surface of a bacterium of a different genotype and incorporated into the chromosome of the recipient cell. (2) Insertion of recombinant DNA into a host cell.

**transgenic** Containing recombinant DNA incorporated into the genetic material.

**transition state** In an enzyme-catalyzed reaction, the reactive condition of the substrate after there has been sufficient input of energy (activation energy) to initiate the reaction.

**translation** The synthesis of a protein (polypeptide). Takes place on ribosomes, using the information encoded in messenger RNA.

**translational repressor** A protein that blocks translation by binding to mRNAs and preventing their attachment to the ribosome. In mammals, the production of ferritin protein is regulated by a translational repressor.

**translocation** (1) In genetics, a rare mutational event that moves a portion of a chromosome to a new location, generally on a nonhomologous chromosome. (2) In vascular plants, movement of solutes in the phloem.

**transmembrane protein** An integral membrane protein that spans the phospholipid bilayer.

**transpiration** [L. *spirare*: to breathe] The evaporation of water from plant leaves and stem, driven by heat from the sun, and providing the motive force to raise water (plus mineral nutrients) from the roots.

**transpiration–cohesion–tension mechanism** Theoretical basis for water movement in plants: evaporation of water from cells within leaves (transpiration) causes an increase in surface tension, pulling water up through the xylem. Cohesion of water occurs because of hydrogen bonding.

**transposable element (transposon)** A segment of DNA that can move to, or give rise to copies at, another locus on the same or a different chromosome.

**transversion** A mutation that changes a purine to a pyrimidine or vice versa.

**tree of life** A term that encompasses the evolutionary history of all life, or a graphic representation of that history.

**triglyceride** A simple lipid in which three fatty acids are combined with one molecule of glycerol.

**trimesters** The three stages of human pregnancy, approximately 3 months each in length.

**tripartite synapse** The idea that a synapse includes not only the pre- and postsynaptic neurons involved but also encompasses many connections with glial cells called astrocytes.

**triploblastic** Having three cell layers.

**trisomic** Containing three rather than two members of a chromosome pair.

**tRNA** *See* transfer RNA.

**trochophore** (troke′ o fore) [Gk. *trochos*: wheel + *phoreus*: bearer] A radially symmetrical larval form typical of annelids and mollusks, distinguished by a wheel-like band of cilia around the middle.

**trophic cascade** The progression over successively lower trophic levels of the indirect effects of a predator.

**trophic interactions** The consumer–resource relationships among species in a community.

**trophic level** [Gk *trophes*: nourishment] A group of organisms united by obtaining their energy from the same part of the food web of a biological community.

**trophoblast** [Gk *trophes*: nourishment + *blastos*: sprout] At the 32-cell stage of mammalian development, the outer group of cells that will become part of the placenta and thus nourish the growing embryo. (Contrast with inner cell mass.)

**tropic hormones** Hormones produced by the anterior pituitary that control the secretion of hormones by other endocrine glands.

**tropomyosin** [troe poe my′ oh sin] One of the three protein components of an actin filament; controls the interactions of actin and myosin necessary for muscle contraction.

**troponin** One of the three components of an actin filament; binds to actin, tropomyosin, and $Ca^{2+}$.

**true-breeding** A genetic cross in which the same result occurs every time with respect to the trait(s) under consideration, due to homozygous parents.

**trypsin** A protein-digesting enzyme. Secreted by the pancreas in its inactive form (trypsinogen), it becomes active in the duodenum of the small intestine.

**tube feet** A unique feature of echinoderms; extensions of the water vascular system, which functions in gas exchange, locomotion, and feeding.

**tubulin** A protein that polymerizes to form microtubules.

**tumor** [L. *tumor*: a swollen mass] A disorganized mass of cells. Malignant tumors spread to other parts of the body.

**tumor necrosis factor** A family of cytokines (growth factors) that causes cell death and is involved in inflammation.

**tumor suppressor** A gene that codes for a protein product that inhibits cell proliferation; inactive in cancer cells. (Contrast with oncogene.)

**turgor pressure** [L. *turgidus*: swollen] *See* pressure potential.

**turnover** In freshwater ecosystems, vertical movements of water that bring nutrients and dissolved $CO_2$ to the surface and $O_2$ to deeper water.

**twitch** A muscle fiber's minimum unit of contraction, stimulated by a single action potential.

**tympanic membrane** [Gk. *tympanum*: drum] The eardrum.

## U

**ubiquitin** A small protein that is covalently linked to other cellular proteins identified for breakdown by the proteosome.

**ultimate causes** In ethology, the evolutionary processes that produce an animal's capacity and tendency to behave in particular ways. (Contrast with proximate causes.)

**unequal crossing over** When a highly repeated gene sequence becomes displaced in alignment during meiotic crossing over, so that one chromosome receives many copies of the sequence while the second chromosome receives fewer copies. One of the mechanisms of concerted evolution. (See also biased gene conversion.)

**uniporter** [L. *unus*: one + *portal*: doorway] A membrane transport protein that carries a single substance in one direction. (Contrast with antiporter, symporter.)

**unipotent** An undifferentiated cell that is capable of becoming only one type of mature cell. (Contrast with totipotent, multipotent, pluripotent.)

**unsaturated fatty acid** A fatty acid whose hydrocarbon chain contains one or more double bonds. (Contrast with saturated fatty acid.)

**upregulation** A process by which the abundance of receptors for a hormone increases when hormone secretion is suppressed. (Contrast with downregulation.)

**upwelling zones** Areas of the ocean where cool, nutrient-rich water from deeper layers rises to the surface.

**uracil (U)** A pyrimidine base found in nucleotides of RNA.

**urea** A compound that is the main form of nitrogen excreted by many animals, including mammals.

**ureotelic** Pertaining to an organism in which the final product of the breakdown of nitrogen-containing compounds (primarily proteins) is urea. (Contrast with ammonotelic, uricotelic.)

**ureter** (your′ uh tur) Long duct leading from the vertebrate kidney to the urinary bladder or the cloaca.

**urethra** (you ree′ thra) In most mammals, the canal through which urine is discharged from the bladder and which serves as the genital duct in males.

**uric acid** A compound that serves as the main excreted form of nitrogen in some animals, particularly those which must conserve water, such as birds, insects, and reptiles.

**uricotelic** Pertaining to an organism in which the final product of the breakdown of nitrogen-containing compounds (primarily proteins) is uric acid. (Contrast with ammonotelic, ureotelic.)

**urinary bladder** A structure in which urine is stored until it can be excreted to the outside of the body.

**urine** (you′ rin) In vertebrates, the fluid waste product containing the toxic nitrogenous by-products of protein and nucleic acid metabolism.

**uterine cycle** In human females, the monthly cycle of events by which the endometrium is prepared for the arrival of a blastocyst. (Contrast with ovarian cycle).

**uterus** (yoo′ ter us) [L. *utero*: womb] A specialized portion of the female reproductive tract in mammals that receives the fertilized egg and nurtures the embryo in its early development. Also called the womb.

## V

**vaccination** Injection of virus or bacteria or their proteins into the body, to induce immunity. The injected material is usually attenuated (weakened) before injection and is called a *vaccine*.

**vacuole** (vac′ yew ole) Membrane-enclosed organelle in plant cells that can function for storage, water concentration for turgor, or hydrolysis of stored macromolecules.

**vagina** (vuh jine′ uh) [L. sheath] In female animals, the entry to the reproductive tract.

**van der Waals forces** Weak attractions between atoms resulting from the interaction of the electrons of one atom with the nucleus of another. This type of attraction is about one-fourth as strong as a hydrogen bond.

**variable** In a controlled experiment, a factor that is manipulated to test its effect on a phenomenon.

**variable region** The portion of an immunoglobulin molecule or T cell receptor that includes the antigen-binding site and is responsible for its specificity. (Contrast with constant region.)

**vas deferens** (plural: vasa deferentia) Duct that transfers sperm from the epididymis to the urethra.

**vasa recta** Blood vessels that parallel the loops of Henle and the collecting ducts in the renal medulla of the kidney.

**vascular** (vas′ kew lar) [L. *vasculum*: a small vessel] Pertaining to organs and tissues that conduct fluid, such as blood vessels in animals and xylem and phloem in plants.

**vascular bundle** In vascular plants, a strand of vascular tissue, including xylem and phloem as well as thick-walled fibers.

**vascular cambium** (kam′ bee um) [L. *cambiare*: to exchange] In plants, a lateral meristem that gives rise to secondary xylem and phloem.

**vascular tissue system** The transport system of a vascular plant, consisting primarily of xylem and phloem.

**vasopressin** A hormone that promotes water reabsorption by the kidney. Produced by neurons in the hypothalamus and released from nerve terminals in the posterior pituitary. Also called antidiuretic hormone or ADH.

**vector** (1) An agent, such as an insect, that carries a pathogen affecting another species. (2) A plasmid or virus that carries an inserted piece of DNA into a bacterium for cloning purposes in recombinant DNA technology.

**vegetal hemisphere** The lower portion of some animal eggs, zygotes, and embryos, in which the dense nutrient yolk settles. The *vegetal pole* is to the very bottom of the egg or embyro. (Contrast with animal hemisphere.)

**vegetative** Nonreproductive, nonflowering, or asexual.

**vegetative meristem** An apical meristem that produces leaves.

**vegetative reproduction** Asexual reproduction through the modification of stems, leaves, or roots.

**vein** [L. *vena*: channel] A blood vessel that returns blood to the heart. (Contrast with artery.)

**vena cavae** In the circulatory systems of crocodilians, birds, and mammals, large veins that empty into the right atrium of the heart.

**ventral** [L. *venter*: belly, womb] Toward or pertaining to the belly or lower side. (Contrast with dorsal.)

**ventricle** A muscular heart chamber that pumps blood through the lungs or through the body.

**venule** A small blood vessel draining a capillary bed that joins others of its kind to form a vein. (Contrast with arteriole.)

**vernalization** [L. *vernalis*: spring] Events occurring during a required chilling period, leading eventually to flowering.

**vertebral column** [L. *vertere*: to turn] The jointed, dorsal column that is the primary support structure of vertebrates.

**very low-density lipoproteins (VLDLs)** Lipoproteins that consist mainly of triglyceride fats, which they transport to fat cells in adipose tissues throughout the body; associated with excessive fat deposition and high risk for cardiovascular disease.

**vesicle** Within the cytoplasm, a membrane-enclosed compartment that is associated with other organelles; the Golgi complex is one example.

**vessel element** A type of tracheary element with perforated end walls; found only in angiosperms. (Contrast with tracheid.)

**vestibular system** (ves tib′ yew lar) [L. *vestibulum*: an enclosed passage] Structures within the inner ear that sense changes in position or momentum of the head, affecting balance and motor skills.

**vicariant event** (vye care′ ee unt) [L. *vicus*: change] The splitting of a taxon's range by the imposition of some barrier to dispersal.

**villus** (vil′ lus) (plural: villi) [L. *villus*: shaggy hair or beard] A hairlike projection from a membrane; for example, from many gut walls.

**virion** (veer′ e on) The virus particle, the minimum unit capable of infecting a cell.

**virulence** [L. *virus*: poison, slimy liquid] The ability of a pathogen to cause disease and death.

**virus** Any of a group of ultramicroscopic particles constructed of nucleic acid and protein (and, sometimes, lipid) that require living cells in order to reproduce. Viruses evolved multiple times from different cellular species.

**vital capacity (VC)** The maximum capacity for air exchange in one breath; the sum of the tidal volume and the inspiratory and expiratory reserve volumes.

**vitamin** [L. *vita*: life] An organic compound that an organism cannot synthesize, but nevertheless requires in small quantities for normal growth and metabolism.

**vitelline envelope** The inner, proteinaceous protective layer of a sea urchin egg.

**viviparity** (vye vi par′ uh tee) Reproduction in which fertilization of the egg and development of the embryo occur inside the mother's body. (Contrast with oviparity.)

**vivipary** Premature germination in plants.

**voltage** A measure of the difference in electrical charge between two points.

**voltage-gated channel** A type of gated channel that opens or closes when a certain voltage exists across the membrane in which it is inserted.

**vomeronasal organ (VNO)** Chemosensory structure embedded in the nasal epithelium of amphibians, reptiles, and many mammals. Often specialized for detecting pheromones.

## W

**warning coloration** *See* aposematism

**water potential (psi, Ψ)** In osmosis, the tendency for a system (a cell or solution) to take up water from pure water through a differentially permeable membrane. Water flows toward the system with a more negative water potential. (Contrast with solute potential, pressure potential.)

**water vascular system** In echinoderms, a network of water-filled canals that functions in gas exchange, locomotion, and feeding.

**wavelength** The distance between successive peaks of a wave train, such as electromagnetic radiation.

**weather** The state of atmospheric conditions in a particular place at a particular time. (Contrast with climate.)

**weathering** The mechanical and chemical processes by which rocks are broken down into soil particles.

**Wernicke's area** A region in the temporal lobe of the human brain that is involved with the sensory aspects of language.

**white blood cells** Cells in the blood plasma that play defensive roles in the immune system. Also called leukocytes.

**white matter** In the central nervous system, tissue that is rich in axons. (Contrast with gray matter.)

**wild type** Geneticists' term for standard or reference type. Deviants from this standard, even if the deviants are found in the wild, are usually referred to as mutant. (Note that this terminology is not usually applied to human genes.)

**wood** Secondary xylem tissue.

## X-Y-Z

**xerophyte** (zee′ row fyte) [Gk. *xerox*: dry + *phyton*: plant] A plant adapted to an environment with limited water supply.

**xylem** (zy′ lum) [Gk. *xylon*: wood] In vascular plants, the tissue that conducts water and minerals; xylem consists, in various plants, of tracheids, vessel elements, fibers, and other highly specialized cells.

**yolk** [M.E. *yolke*: yellow] The stored food material in animal eggs, rich in protein and lipids.

**yolk sac** In reptiles, birds, and mammals, the extraembryonic membrane that forms from the endoderm of the hypoblast; it encloses and digests the yolk.

**zeaxanthin** A blue-light receptor involved in the opening of plant stomata.

**zona pellucida** A jellylike substance that surrounds the mammalian ovum when it is released from the ovary.

**zone of cell division** The apical and primary meristems of a plant root; the source of all cells of the root's primary tissues.

**zone of cell elongation** The part of a plant root, generally above the zone of cell division, where cells are expanding (growing), primarily in the longitudinal direction.

**zone of maturation** The part of a plant root, generally above the zone of cell elongation, where cells are differentiating.

**zoospore** (zoe′ o spore) [Gk. *zoon*: animal + *spora*: seed] In algae and fungi, any swimming spore. May be diploid or haploid.

**zygospore** Multinucleate, diploid cell that is a resting stage in the life cycle of zygospore fungi.

**zygote** (zye′ gote) [Gk. *zygotos*: yoked] The cell created by the union of two gametes, in which the gamete nuclei are also fused. The earliest stage of the diploid generation.

**zymogen** The inactive precursor of a digestive enzyme; secreted into the lumen of the gut, where a protease cleaves it to form the active enzyme.

# Illustration Credits

**Frontispiece** © Steve Bloom Images/Alamy.

**Table of Contents** Page XXI: © FLPA/Alamy. Page XXII: © Biophoto Associates/Science Source/Photo Researchers, Inc. Page XXIII: Protein data from Sobolevsky et al. 2009. *Nature* 462: 745. Membrane data from Heller et al. 1993. *J. Phys. Chem.* 97: 8343. Page XXIV: © Manabu Kagami/amanaimages/Corbis. Page XXV: © Steve Gschmeissner/SPL/Photo Researchers, Inc. Page XXVI: © zhaoyan/Shutterstock. Page XXVII: © Power and Syred/Science Source/Photo Researchers, Inc. Page XXVIII: © Dr. Fred Hossler/Visuals Unlimited, Inc. Page XXIX: Drawings by Elizabeth Gould, c. 1845. Page XXXI: © SciMAT/Science Source/Photo Researchers, Inc. Page XXXII: David McIntyre. Page XXXIII: © Joe Belanger/Shutterstock. Page XXXIV: © Nigel Cattlin/Alamy. Page XXXV: Courtesy of Andrew D. Sinauer. Page XXXVI: © Martin Harvey/Corbis. Page XXXVIII: © Anthony Bannister/Gallo Images/Corbis. Page XXXIX: © Thomas Deerinck, NCMIR/Science Source/Photo Researchers, Inc. Page XL: David McIntyre. Page XLI: © Cathy Keifer/Shutterstock. Page XLII: © Angelo Gandolfi/Naturepl.com. Page XLIII: © Carol Buchanan/AGE Fotostock. Page XLIV: © Wild Wonders of Europe/O. Haarberg/Naturepl.com.

**Chapter 1** *Opener*: © Pamela S. Turner. 1.1A: © Eye of Science/SPL/Photo Researchers, Inc. 1.1B: © Science Photo Library RF/Photolibrary.com. 1.1C: © Steve Gschmeissner/Photo Researchers, Inc. 1.1D: David McIntyre. 1.1E: © Glen Threlfo/Auscape/Minden Pictures. 1.1F: © Piotr Naskrecki/Minden Pictures. 1.1G: © Tui De Roy/Minden Pictures. 1.3A: © Kwangshin Kim/Photo Researchers, Inc. 1.3B: © Dr. Gopal Murti/Visuals Unlimited, Inc. 1.4A: © Walter Geiersperger/Corbis. 1.4B: © Roger Garwood & Trish Ainslie/Corbis. 1.6A: © Arco Images GmbH/Alamy. 1.6B: © Heather Angel/Natural Visions/Alamy. 1.6C: © Juniors Bildarchiv/Alamy. 1.6D: © Stephen Dalton/Naturepl.com. 1.9A: © A & J Visage/Alamy. 1.9B: © Stefan Huwiler/Rolfnp/Alamy. 1.11: From T. Hayes et al., 2003. *Environ. Health Perspect.* 111: 568. 1.13: Courtesy of Scott Bauer/USDA ARS. 1.14: © Kim Kulish/Corbis. 1.15A: Courtesy of Wayne Whippen. 1.16: Courtesy of the U.S. Geological Survey. 1.17: © Mark Moffett/Minden Pictures/Corbis.

**Chapter 2** *Opener*: © Phil Degginger/Alamy. 2.3: Used with permission of Mayo Foundation for Medical Education and Research, mayoclinic.com. 2.14: © Pablo H Caridad/Shutterstock. 2.15A: © Michael Cole/Corbis. 2.15B: © kawhia/Shutterstock. 2.15C: David McIntyre. Page 38: © Jean Claude Carton/Bruce Coleman USA/AGE Fotostock.

**Chapter 3** *Opener*: © Dennis Kunkel Microscopy, Inc. 3.9: Data from PDB 1IVM. T. Obita, T. Ueda, & T. Imoto, 2003. *Cell. Mol. Life Sci.* 60: 176. 3.11: Data from PDB 2HHB. G. Fermi et al., 1984. *J. Mol. Biol.* 175: 159. 3.18C *left*: © Biophoto Associates/Photo Researchers, Inc. 3.18C *middle*: © Dennis Kunkel Microscopy, Inc. 3.18C *right*: © Don W. Fawcett/Photo Researchers, Inc. 3.19 *Ear*: David McIntyre. 3.19 *Beetle*: © Pan Xunbin/Shutterstock. Page 57: David McIntyre.

**Chapter 4** *Opener*: © Anup Shah/Naturepl.com. 4.7: Courtesy of the Argonne National Laboratory. 4.11B: Courtesy of Janet Iwasa, Szostak group, MGH/Harvard. 4.12: © Stanley M. Awramik/Biological Photo Service. 4.12 *inset*: © Dennis Kunkel Microscopy, Inc.

**Chapter 5** *Opener*: © Roger J. Bick & Brian J. Poindexter/UT-Houston Medical School/Photo Researchers, Inc. 5.1: After N. Campbell, 1990. *Biology*, 2nd Ed., Benjamin Cummings. 5.1 *Protein*: Data from PDB 1IVM. T. Obita, T. Ueda, & T. Imoto, 2003. *Cell. Mol. Life Sci.* 60: 176. 5.1 *T4*: © Dept. of Microbiology, Biozentrum/SPL/Photo Researchers, Inc. 5.1 *Bacterium*: © Jim Biddle/Centers for Disease Control. 5.1 *Plant cells*: © Michael Eichelberger/Visuals Unlimited, Inc. 5.1 *Frog egg*: David McIntyre. 5.1 *Bird*: © Steve Byland/Shutterstock. 5.1 *Baby*: Courtesy of Sebastian Grey Miller. 5.3 *Light microscope*: © Radu Razvan/Shutterstock. 5.3 *Bright-field*: Courtesy of the IST Cell Bank, Genoa. 5.3 *Phase-contrast*: © Michael W. Davidson, Florida State U. 5.3 *DIC*: © Michael W. Davidson, Florida State U. 5.3 *Stained*: © Richard J. Green/SPL/Photo Researchers, Inc. 5.3 *Fluorescence*: © Michael W. Davidson, Florida State U. 5.3 *Confocal*: © Dr. Gopal Murti/SPL/Photo Researchers, Inc. 5.3 *Electron microscope*: © Sinclair Stammers/Photo Researchers, Inc. 5.3 *TEM*: © Dr. Gopal Murti/Visuals Unlimited, Inc. 5.3 *SEM*: © K. R. Porter/SPL/Photo Researchers, Inc. 5.3 *Freeze-fracture*: © D. W. Fawcett/Photo Researchers, Inc. 5.4: © J. J. Cardamone Jr. & B. K. Pugashetti/Biological Photo Service. 5.5A: © Dennis Kunkel Microscopy, Inc. 5.5B: Courtesy of David DeRosier, Brandeis U. 5.6 *Nuclear*: From Y. Mizutani et al., 2001. *J. Cell Sci.* 114: 3727. 5.6 *Mitochondrial*: From L. Argaud et al., 2004. *Cardiovasc Res.* 61: 115. 5.6 *ER*: From Y. Mizutani et al., 2001. *J. Cell Sci.* 114: 3727. 5.7 *Mitochondrion*: © K. Porter, D. Fawcett/Visuals Unlimited, Inc. 5.7 *Cytoskeleton*: © Don Fawcett, John Heuser/Photo Researchers, Inc. 5.7 *Nucleolus*: © Richard Rodewald/Biological Photo Service. 5.7 *Peroxisome*: © E. H. Newcomb & S. E. Frederick/Biological Photo Service. 5.7 *Cell wall*: © Biophoto Associates/Photo Researchers, Inc. 5.7 *Ribosome*: From M. Boublik et al., 1990. *The Ribosome*, p. 177. Courtesy of American Society for Microbiology. 5.7 *Centrioles*: © Barry F. King/Biological Photo Service. 5.7 *Plasma membrane*: Courtesy of J. David Robertson, Duke U. Medical Center. 5.7 *Rough ER*: © Don Fawcett/Science Source/Photo Researchers, Inc. 5.7 *Smooth ER*: © Don Fawcett, D. Friend/Science Source/Photo Researchers, Inc. 5.7 *Chloroplast*: © W. P. Wergin, E. H. Newcomb/Biological Photo Service. 5.7 *Golgi apparatus*: Courtesy of L. Andrew Staehelin, U. Colorado. 5.8A: © Barry King, U. California, Davis/Biological Photo Service. 5.8B: © Biophoto Associates/Science Source/Photo Researchers, Inc. 5.9: © B. Bowers/Photo Researchers, Inc. 5.10: © Sanders/Biological Photo Service. 5.11: © K. Porter, D. Fawcett/Visuals Unlimited, Inc. 5.12: © W. P. Wergin, E. H. Newcomb/Biological Photo Service. 5.13: © Biophoto Associates/Photo Researchers, Inc. 5.14: Courtesy of Vic Small, Austrian Academy of Sciences, Salzburg, Austria. 5.16: Courtesy of N. Hirokawa. 5.17A *upper*: © SPL/Photo Researchers, Inc. 5.17A *lower*, 5.17B: © W. L. Dentler/Biological Photo Service. 5.19: From N. Pollock et al., 1999. *J. Cell Biol.* 147: 493. Courtesy of R. D. Vale. 5.20: © Michael Abbey/Visuals Unlimited, Inc. 5.21: © Biophoto Associates/Photo Researchers, Inc. 5.22 *left*: Courtesy of David Sadava. 5.22 *upper right*: From J. A. Buckwalter & L. Rosenberg, 1983. *Coll. Rel. Res.* 3: 489. Courtesy of L. Rosenberg. 5.22 *lower right*: © J. Gross, Biozentrum/SPL/Photo Researchers, Inc. 5.24: Courtesy of Noriko Okamoto and Isao Inouye. Page 79: Courtesy of Dr. Siobhan Marie O'Connor. Page 92 *Poppy*: David McIntyre. Page 92 *Chromoplast*: © Richard Green/Photo Researchers, Inc. Page 93 *Potatoes*: David McIntyre. Page 93 *Leucoplast*: Courtesy of R. R. Dute.

**Chapter 6** *Opener*: © Mike Franklin/FilmMagic/Getty Images. 6.2: After L. Stryer, 1981. *Biochemistry*, 2nd Ed., W. H. Freeman. 6.4: © D. W. Fawcett/Photo Researchers, Inc. 6.7A: Courtesy of D. S. Friend, U. California, San Francisco. 6.7B: Courtesy of Darcy E.

Kelly, U. Washington. 6.7C: Courtesy of C. Peracchia. 6.9A *top*: © Stanley Flegler/Visuals Unlimited, Inc. 6.9A *bottom*: © Ed Reschke/Getty Images. 6.9B *top*: © David M. Phillips/Photo Researchers, Inc. 6.9B *bottom*: © Ed Reschke/Getty Images. 6.9C *top*: © David M. Phillips/Photo Researchers, Inc. 6.9C *bottom*: © Ed Reschke/Getty Images. 6.11: From G. M. Preston et al., 1992. *Science* 256: 385. 6.17: From M. M. Perry, 1979. *J. Cell Sci.* 39: 26. Page 124: © blickwinkel/Alamy.

**Chapter 7** *Opener*: Courtesy of Todd Ahern. 7.3A: Data from PDB 3EML. V. P. Jaakola et al., 2008. *Science* 322: 1211. 7.3B: © Georgii Dolgykh/istock. 7.14: © Stephen A. Stricker, courtesy of Molecular Probes, Inc. 7.20: Courtesy of David Kirk. Page 143: © Biophoto Associates/Photo Researchers, Inc.

**Chapter 8** *Opener*: © Sinauer Associates. 8.1: Courtesy of Violet Bedell-McIntyre. 8.5B: © Alamy. 8.9: Data from PDB 148L. Kuroki et al., 1993. *Science* 262: 2030. 8.11A: Data from PDB 1AL6. B. Schwartz et al., 1997. 8.11B: Data from PDB 1BB6. V. B. Vollan et al., 1999. *Acta Crystallogr. D. Biol. Crystallogr.* 55: 60. 8.11C: Data from PDB 1AB9. N. H. Yennawar, H. P. Yennawar, & G. K. Farber, 1994. *Biochemistry* 33: 7326. 8.12: Data from PDB 1IG8 (P. R. Kuser et al., 2000. *J. Biol. Chem.* 275: 20814) and 1BDG (A. M. Mulichak et al., 1998 *Nat. Struct. Biol.* 5: 555).

**Chapter 9** *Opener*: © Poulsons Photography/Shutterstock. 9.8: From Y. H. Ko et al., 2003. *J. Biol. Chem.* 278: 12305. Courtesy of P. Pedersen. 9.14: © Ana Abejon/istock.

**Chapter 10** *Opener*: Courtesy of David F. Karnosky. 10.1: © Andrew Syred/SPL/Photo Researchers, Inc. 10.11: Courtesy of Lawrence Berkeley National Laboratory. 10.15A, 10.17A: © E. H. Newcomb & S. E. Frederick/Biological Photo Service. 10.19: © Aflo Foto Agency/Alamy. Table 10.1 *Rice*: © Alan49/Shutterstock. Table 10.1 *Maize*: © piyagoon/Shutterstock. Table 10.1 *Cactus*: © Dan Eckert/istock.

**Chapter 11** *Opener*: © Obstetrics and Gynaecology/Photo Researchers, Inc. 11.1A: © SPL/Photo Researchers, Inc. 11.1B: © Biodisc/Visuals Unlimited, Inc. 11.1C: © Robert Valentic/Naturepl.com. 11.2B: © John J. Cardamone Jr./Biological Photo Service. 11.8 *Chromosome*: © Biophoto Associates/Photo Researchers, Inc. 11.8 *Nucleus*: © D. W. Fawcett/Photo Researchers, Inc. 11.9 *inset*: © Biophoto Associates/Science Source/Photo Researchers, Inc. 11.10: © Nasser Rusan. 11.11B: © Conly L. Rieder/Biological Photo Service. 11.13A: © Robert Brons/Biological Photo Service. 11.13B: © B. A. Palevitz, E. H. Newcomb/Biological Photo Service. 11.14: © Robert E. Ford/Biological Photo Service. 11.15 *left*: © Andrew Syred/SPL/Photo Researchers, Inc. 11.15 *center*: David McIntyre. 11.15 *right*: Courtesy of Andrew D. Sinauer. 11.16: © C. A. Hasenkampf/Biological Photo Service. 11.17: Courtesy of J. Kezer. 11.21: Courtesy of Dr. Thomas Ried and Dr. Evelin Schröck, NIH. 11.22: © Sergey Skleznev/Shutterstock. 11.23A: © Gopal Murti/Photo Researchers, Inc. 11.24: © Dennis Kunkel Microscopy, Inc. Page 231: Courtesy of Paul Schulte.

**Chapter 12** *Opener*: © Gerry Pearce/Alamy. 12.1: © the Mendelianum. 12.10 *Dark*: © Marina Golskaya/istock. 12.10 *Chinchilla*: © purelook/istock. 12.10 *Point*: © Carolyn A. McKeone/Photo Researchers, Inc. 12.10 *Albino*: © ZTS/Shutterstock. 12.13: Courtesy of Madison, Hannah, and Walnut. 12.14: Courtesy of the Plant and Soil Sciences eLibrary (http://plantandsoil.unl.edu); used with permission from the Institute of Agriculture and Natural Resources at the University of Nebraska. 12.15: © Mark Taylor/Naturepl.com. 12.16: © Peter Morenus/U. of Connecticut. 12.23A: © David Scharf/Getty Images. Page 258 *Rose*: © Margo Harrison/Shutterstock. Page 258 *Pea, Walnut, and Single*: David McIntyre.

**Chapter 13** *Opener*: Portrait by Albert Edelfelt, courtesy of the National Library of Medicine. 13.3: © Lee D. Simon/Photo Researchers, Inc. 13.6B: © Science Source/Photo Researchers, Inc. 13.7A: © A. Barrington Brown/Photo Researchers, Inc. 13.7B: Data from S. Arnott & D. W. Hukins, 1972. *Biochem. Biophys. Res. Commun.* 47(6): 1504. 13.14A: Data from PDB 1SKW. Y. Li et al., 2001. *Nat. Struct. Mol. Biol.* 11: 784. 13.19B: © Dr. Peter Lansdorp/Visuals Unlimited, Inc.

**Chapter 14** *Opener*: © CDC/Janice Carr/AGE Fotostock. 14.3: Data from PDB 1MSW. Y. W. Yin & T. A. Steitz, 2002. *Science* 298: 1387. 14.7: From D. C. Tiemeier et al., 1978. *Cell* 14: 237. 14.11: Data from PDB 1EHZ. H. Shi & P. B. Moore, 2000. *RNA* 6: 1091. 14.13: Data from PDB 1GIX and 1G1Y. M. M. Yusupov et al., 2001. *Science* 292: 883. 14.17B: Courtesy of J. E. Edström and *EMBO J.*

**Chapter 15** *Opener*: © Steve Lipofsky/Corbis. 15.3: © Stanley Flegler/Visuals Unlimited, Inc. 15.9: From C. Harrison et al., 1983. *J. Med. Genet.* 20: 280. 15.11B: © David M. Martin, M.D./SPL/Photo Researchers, Inc. 15.13: © Philippe Plailly/Photo Researchers, Inc. 15.14B: U.S. Army photo. 15.16 *Butterfly*: © Bershadsky Yuri/Shutterstock. 15.16 *Bacteria*: Courtesy of Janice Haney Carr/CDC. 15.16 *Fungus*: © Warwick Lister-Kaye/istock. 15.17: © Simon Fraser/Photo Researchers, Inc.

**Chapter 16** *Opener*: © Beyond Fotomedia GmbH/Alamy. 16.12A: © Dennis Kunkel Microscopy, Inc. 16.12B: © Lee D. Simon/Photo Researchers, Inc. 16.21: Courtesy of Irina Solovei, University of Munich (LMU), Germany. Page 336: Data from PDB 2PE5. R. Daber et al., 2007. *J. Mol. Biol.* 370: 609.

**Chapter 17** *Opener*: © moodboard RF/Photolibrary.com. 17.7: Courtesy of Tom Deerinck and Mark Ellisman of the National Center for Microscopy and Imaging Research at the University of California at San Diego. 17.11B: Courtesy of O. L. Miller, Jr. 17.13: Courtesy of Christoph P. E. Zollikofer, Marcia S. Ponce de León, and Elisabeth Daynès. 17.16: From P. H. O'Farrell, 1975. *J. Biol. Chem.* 250: 4007. Courtesy of Patrick H. O'Farrell. 17.18 *left*: © kostudio/Shutterstock. 17.18 *right*: © Bruce Stotesbury/PostMedia News/Zuma Press.

**Chapter 18** *Opener*: U.S. Coast Guard photo by Petty Officer 2nd Class Etta Smith. 18.3: © Dr. Jack Bostrack/Visuals Unlimited, Inc. 18.4: © Stephen Sewell/istock. 18.13: Courtesy of the Golden Rice Humanitarian Board, www.goldenrice.org. 18.14: Courtesy of Eduardo Blumwald. Page 387: © Dr. George Chapman/Visuals Unlimited, Inc.

**Chapter 19** *Opener*: © Frank Franklin II/AP/Corbis. 19.5A: From J. E. Sulston & H. R. Horvitz, 1977. *Dev. Bio.* 56: 100. 19.9A: David McIntyre. 19.12A: From A. Ephrussi and D. St. Johnston, 2004. *Cell* 116: 143. 19.12B: Courtesy of Ruth Lehmann. 19.12C *left*: From E. A. Wimmer, 2012. *Science* 287: 2476. 19.12C *right*: From D. Tautz, 1988. *Nature* 332: 284. 19.13B: Courtesy of C. Rushlow and M. Levine. 19.13C: Courtesy of T. Karr. 19.13D: Courtesy of S. Carroll and S. Paddock. 19.15: Courtesy of F. R. Turner, Indiana U. 19.17: From I. Wilmut et al., 1997. *Nature* 385: 810. Page 402: Courtesy of D. Daily and W. Sullivan.

**Chapter 20** *Opener*: © Theo Allofs/Corbis. 20.1 *Mouse*: © orionmystery@flickr/Shutterstock. 20.1 *Fly*: David McIntyre. 20.1 *Shark*: © Kristian Sekulic/Shutterstock. 20.1 *Squid*: © Gergo Orban/Shutterstock. 20.3: © David M. Phillips/Photo Researchers, Inc. 20.5: © Bone Clones, www.boneclones.com. 20.6: Courtesy of J. Hurle and E. Laufer. 20.7: Courtesy of J. Hurle. 20.8: From M. Kmita and D. Duboule, 2003. *Science* 301: 331. 20.9 *Cladogram*: After R. Galant & S. Carroll, 2002. *Nature* 415: 910. 20.9 *Insect*: © Stockbyte/PictureQuest. 20.9 *Centipede*: © Burke/Triolo/Brand X Pictures/PictureQuest. 20.10: From Wang et al., 2005. *Nature* 436: 714. Courtesy of John Doebley. 20.11: © Neil Hardwick/Alamy. 20.12: © Rob Valentic/ANTPhoto.com. 20.13 *Caterpillars*: © Erick Greene. 20.13 *Adult*: Courtesy of John Gruber. 20.14: © Nigel Cattlin, Holt Studios International/Photo Researchers, Inc. 20.16: Courtesy of Mike Shapiro and David Kingsley.

**Chapter 21** *Opener*: © Pasieka/Photo Researchers, Inc. 21.1 *H.M.S. Beagle*: Painting by Ronald Dean, reproduced by permission of the artist and Richard Johnson, Esquire. 21.1 *Darwin*: © The Art Gallery Collection/Alamy. 21.5A: © Luis César Tejo/Shutterstock. 21.5B: © Duncan Usher/Alamy. 21.5C: © PetStockBoys/Alamy. 21.5D: © Arco Images GmbH/Alamy. 21.8: © Simon G/Shutterstock. 21.14: Courtesy of David Hillis. 21.19A: © Reinhard Dirscherl/Alamy. 21.19B: © Marevision/AGE Fotostock. 21.20A: Courtesy of Edmund D. Brodie, Jr.

**Chapter 22** *Opener*: Courtesy of Misha Matz. 22.6 *Sea squirt larva*: Courtesy of William Jeffery. 22.6 *Sea squirt adult*: © WaterFrame/Alamy. 22.6 *Frog larva*: David McIntyre. 22.6 *Frog adult*: © Mark Kostich/Shutterstock. 22.10: © Alexandra Basolo. 22.11 *L. bicolor*: Courtesy of Steve Matson. 22.11 *L. liniflorus*: Courtesy of

Anthony Valois/National Park Service. 22.14A, B: © Krieger, C./AGE Fotostock. 22.14C: © Photos by Andy/Shutterstock. 22.16: Courtesy of Misha Matz.

**Chapter 23** *Opener*: David McIntyre, courtesy of Exotic Fish and Pet World, Southampton, MA. 23.1A *left*: © Roger K. Burnard/Biological Photo Service. 23.1A *right*: © Richard Codington/Alamy. 23.1B: © Stubblefield Photography/Shutterstock. 23.2A: David McIntyre. 23.2B: © Gerry Bishop/Visuals Unlimited, Inc. 23.4: © Barry Mansell/Naturepl.com. 23.10: © Tim Gainey/Alamy. 23.12 *G. olivacea*: © Phil A. Dotson/Photo Researchers, Inc. 23.12 *G. carolinensis*: © Suzanne L. Collins/Photo Researchers, Inc. 23.13A: © Gustav Verderber/Visuals Unlimited, Inc. 23.13B: © Nathan Derieg. 23.13C: © Bob Gibbons/Alamy. 23.13D: © Daniel L. Geiger/SNAP/Alamy. 23.14: Courtesy of Donald A. Levin. 23.15 *upper*: © Gerhard Schulz/AGE Fotostock. 23.15 *lower*: © Christophe Courteau/Naturepl.com. 23.17A: © blickwinkel/Alamy. 23.17B: © W. Peckover/VIREO. 23.18 *Madia*: © Peter K. Ziminsky/Visuals Unlimited, Inc. 23.18 *Argyroxiphium*: © Ron Dahlquist/Getty Images. 23.18 *Wilkesia*: © Photo Resource Hawaii/Alamy. 23.18 *Dubautia*: © Noble Proctor/The National Audubon Society Collection/Photo Researchers, Inc. 23.19: Courtesy of William R. Rice. Page 484 *left*: © Gustav Verderber/Visuals Unlimited, Inc. Page 484 *right*: © Nathan Derieg.

**Chapter 24** *Opener*: © Jane Burton/Naturepl.com. 24.3 *Rice*: data from PDB 1CCR. H. Ochi et al., 1983. *J. Mol. Biol.* 166: 407. *Tuna*: data from PDB 5CYT. T. Takano, 1984. 24.4: From P. B. Rainey & M. Travisano, 1998. *Nature* 394: 69. © Macmillan Publishers Ltd. 24.7A *Langur*: © blickwinkel/Alamy. 24.7A *Longhorn*: Courtesy of David Hillis.

**Chapter 25** *Opener*: © Graham Cripps/NHMPL. *Opener inset*: © Natasha Litova/istock. 25.3B: © Ric Ergenbright/Corbis. 25.5: Courtesy of Dave Harlow, U.S. Geological Survey. 25.6: © Martin Bond/SPL/Photo Researchers, Inc. 25.7: David McIntyre. 25.9A: © Ted Kinsman/Photo Researchers, Inc. 25.9B: © Georgette Douwma/Photo Researchers, Inc. 25.11: © PjrStudio/Alamy. 25.13A: From S. Xiao et al., 1998. *Nature* 391: 553. © Macmillan Publishers Ltd. 25.13B *top*: Courtesy of Martin Smith. 25.13B *bottom*: © Sinclair Stammers/Photo Researchers, Inc. 25.13C *top*: Courtesy of Chip Clark, National Museum of Natural History. 25.13C *bottom*: © Albert J. Copley/Visuals Unlimited, Inc. 25.14 *Cambrian*: © John Sibbick/NHMPL. 25.14 *Marella*: Courtesy of the Amherst College Museum of Natural History, The Trustees of Amherst College. 25.14 *Ottoia*: © Alan Sirulnikoff/Photo Researchers, Inc. 25.14 *Anomalocaris*: © Kevin Schafer/Alamy. 25.14 *Devonian*: © The Field Museum, #GEO86500_125d. 25.14 *Archaeopteris*: © John Cancalosi/Getty Images. 25.14 *Eusthenopteron*: © Wolfgang Kaehler/Alamy. 25.14 *Permian*: © Karen Carr Studio Inc. 25.14 *Dragonfly*: Image by Roy J. Beckemeyer. 25.14 *Walchia*: © The Natural History Museum, London. 25.14 *Triassic*: © The Natural History Museum (WAC)/Naturepl.com. 25.14 *Ferns*: © Ken Lucas/Visuals Unlimited, Inc. 25.14 *Plateosaurus*: © Koichi Kamoshida/J Press/Zuma Press. 25.14 *Cretaceous*: © Anness Publishing/NHMPL. 25.14 *Chasmosaurus*: © Oleksiy Maksymenko/Alamy. 25.14 *Sapindopsis*: © Barbara J. Miller/Biological Photo Service. 25.14 *Tertiary*: © Publiphoto/Photo Researchers, Inc. 25.14 *Hyracotherium*: Courtesy of the Amherst College Museum of Natural History, The Trustees of Amherst College. 25.14 *Plesiadapis*: © The Natural History Museum, London. 25.15: Courtesy of Conrad C. Labandeira, Department of Paleobiology, National Museum of Natural History, Smithsonian Institution.

**Chapter 26** *Opener*: © Steven Haddock and Steven Miller. 26.2: Courtesy of the Centers for Disease Control. 26.3: © Dennis Kunkel Microscopy, Inc. 26.5: © Dr. Kari Lounatmaa/Photo Researchers, Inc. 26.6: © Dr. Gary Gaugler/Visuals Unlimited, Inc. 26.7: © Don W. Fawcett/Photo Researchers, Inc. 26.8: © David Phillips/Visuals Unlimited, Inc. 26.9A: © Paul W. Johnson/Biological Photo Service. 26.9B: © Dr. Terry Beveridge/Visuals Unlimited, Inc. 26.9C: © RJH Catalog/Alamy. 26.10A: © J. A. Breznak & H. S. Pankratz/Biological Photo Service. 26.10B: © James Cavallini/Photo Researchers, Inc. 26.11: Courtesy of Randall C. Cutlip. 26.12: © Kwangshin Kim/Photo Researchers, Inc. 26.13: © Dean A. Glawe/Biological Photo Service. 26.14: From K. Kashefi & D. R. Lovley, 2003. *Science* 301: 934. Courtesy of Kazem Kashefi. 26.16: © Arco Images GmbH/Alamy. 26.17: © Nancy Nehring/istock. 26.18: From H. Huber et al., 2002. *Nature* 417: 63. © Macmillan Publishers Ltd. Courtesy of Karl O. Stetter. 26.20B: © Science Photo Library RF. 26.22: © Juergen Berger/Photo Researchers, Inc. 26.23A: © Science Photo Library RF/Photolibrary.com. 26.23B, C: © Russell Kightley/Photo Researchers, Inc. 26.23D: © Science Photo Library RF. 26.23E: © animate4.com ltd./Photo Researchers, Inc. 26.23F: © Russell Kightley/Photo Researchers, Inc. 26.24: © Nigel Cattlin/Alamy. 26.26: Courtesy of Margaret McFall-Ngai. Page 548: Courtesy of the Centers for Disease Control.

**Chapter 27** *Opener*: © Don Paulson/PureStock/AGE Fotostock. 27.4: © Dennis Kunkel Microscopy, Inc. 27.5A: © SPL/Photo Researchers, Inc. 27.5B: © Aaron Bell/Visuals Unlimited, Inc. 27.5C: © Steve Gschmeissner/Photo Researchers, Inc. 27.8: © Scenics & Science/Alamy. 27.9A: © Marevision/AGE Fotostock. 27.9B: © Carl W. May/Biological Photo Service. 27.10: David McIntyre. 27.11, 27.12A: © Robert Brons/Biological Photo Service. 27.12B: © Manfred Kage/Photo Researchers, Inc. 27.13A: © J. Paulin/Visuals Unlimited, Inc. 27.13B: © Dr. David M. Phillips/Visuals Unlimited, Inc. 27.15: © Michael Abbey/Photo Researchers, Inc. 27.16: © Wim van Egmond/Visuals Unlimited, Inc. 27.17A: © Matt Meadows/Getty Images. 27.17B: © Ed Reschke/Getty Images. 27.18: Courtesy of R. Blanton and M. Grimson. 27.20B: © London School of Hygiene/SPL/Photo Researchers, Inc. 27.21: Courtesy of M.A. Coffroth and Cindy Lewis, University at Buffalo.

**Chapter 28** *Opener*: Courtesy of the U.S. Coast Guard. 28.2: © Dr. Peter Siver/Visuals Unlimited, Inc. 28.3A: © Wim van Egmond/Visuals Unlimited, Inc. 28.3B: © David Wrobel/Visuals Unlimited, Inc. 28.4A: © Carolina Biological/Visuals Unlimited, Inc. 28.4B: © Marevision/AGE Fotostock. 28.5A: © Larry Mellichamp/Visuals Unlimited, Inc. 28.5B: © Bob Gibbons/Alamy. 28.7 *top*: © Biodisc/Visuals Unlimited, Inc. 28.7 *bottom*: © J. Robert Waaland/Biological Photo Service. 28.8A: David McIntyre. 28.8B: © Dr. Brad Mogen/Visuals Unlimited, Inc. 28.8C: © Verbiesen Henk/AGE Fotostock. 28.9A: © mediacolor's/Alamy. 28.9B: David McIntyre. 28.10A: © Dr. John D. Cunningham/Visuals Unlimited, Inc. 28.10B: © Danilo Donadoni/AGE Fotostock. 28.11: © Daniel Vega/AGE Fotostock. 28.12: © Publiphoto/Photo Researchers, Inc. 28.14A: © Ed Reschke/Getty Images. 28.14B: © Stanislav Sokolov/istock. 28.14C: Courtesy of the Talcott Greenhouse, Mount Holyoke College. 28.14D: © Ted Mead/Getty Images. 28.15 *inset*: © John N. A. Lott/Biological Photo Service. 28.16A: Courtesy of the Biology Department Greenhouses, U. Massachusetts, Amherst. 28.16B: David McIntyre. 28.19: © Patrick Pleul/epa/Corbis.

**Chapter 29** *Opener*: © Mitsuhiko Imamori/Minden Pictures. 29.1 *Cycad*: David McIntyre. 29.1 *Ginkgo*: © hypnotype/Shutterstock. 29.1 *Conifer*: © Irina Tischenko/istock. 29.1 *Magnolia*: © Dole/Shutterstock. 29.3: © Wildlife GmbH/Alamy. 29.4: © Susumu Nishinaga/Photo Researchers, Inc. 29.6A: Courtesy of Jane Sinauer. 29.6B: David McIntyre. 29.6C: © Juan Carlos Muñoz/AGE Fotostock. 29.6D: © Pepbaix/Alamy. 29.7A *left*: © Fritz Poelking/Blickwinkel/AGE Photostock. 29.7A *right*: © Scenics & Science/Alamy. 29.7B *left*: © Gunter Marx/Alamy. 29.7B *right*: © Scenics & Science/Alamy. 29.9: Courtesy of Jim Peaco/National Park Service. 29.10A, B: David McIntyre. 29.10C: © Nigel Cattlin/Alamy. 29.11A: © Phiseksit/Shutterstock. 29.11B: David McIntyre. 29.15: © Ted Kinsman/Photo Researchers, Inc. 29.17A: © Anne Power/Shutterstock. 29.17B: © Klaus Hackenberg/Corbis. 29.17C: © Brian A Jackson/Shutterstock. 29.17D: © Paul Thompson/Corbis. 29.17E: © Arco Images GmbH/Alamy. 29.17F: Courtesy of Keith Weller/USDA ARS. 29.19A: Photo by David McIntyre, courtesy of the U. Massachusetts Biology Department Greenhouses. 29.19B: © Cerealphoto/AGE Fotostock. 29.19C: © Holmes Garden Photos/Alamy. 29.19D: © dora modly-paris/Shutterstock. 29.19E: © Florapix/Alamy. 29.20A: © Floris Slooff/istock. 29.20B: © George Clerk/istock. 29.20C: © Jose B. Ruiz/Naturepl.com. 29.20D: © rotofrank/istock. 29.21A: © BonkersAboutTravel/Alamy. 29.21B: David McIntyre. 29.21C: Courtesy of David Hillis. 29.21D: © A & J Visage/Alamy. 29.22: © Rob Walls/Alamy. 29.23: © Janos Csernoch/Alamy.

**Chapter 30** *Opener*: © Biophoto Associates/Photo Researchers, Inc. 30.2: © Steve Gschmeissner/Science Photo Library/Corbis. 30.3A: © Dr. Jeremy Burgess/Photo Researchers, Inc. 30.4: © Arco Images GmbH/Alamy. 30.5A: © Biophoto Associates/Photo Researchers, Inc. 30.6: © N. Allin & G. L. Barron/Biological Photo Service. 30.7A, C: Courtesy of David Hillis. 30.7B: David McIntyre. 30.9A: © R. L. Peterson/Biological Photo Service. 30.9B: © M. F. Brown/Biological Photo Service. 30.12: © Eye of Science/Photo Researchers, Inc. 30.13: © John Taylor/Visuals Unlimited, Inc. 30.14A: © J. Robert Waaland/Biological Photo Service. 30.14B: © Dr. Jeremy Burgess/Photo Researchers, Inc. 30.15: Photo by David McIntyre; manure courtesy of Myrtle Jackson. 30.16A: © Dr. Cecil H. Fox/Photo Researchers, Inc. 30.16B: © Biophoto Associates/Photo Researchers, Inc. 30.17A: © blickwinkel/Alamy. 30.17B: © Matt Meadows/Getty Images. 30.18: © Dennis Kunkel Microscopy, Inc. 30.19A: David McIntyre. 30.19B: © Mike Norton/Shutterstock. 30.20, 30.21: Courtesy of David Hillis. 30.22: © Dr. Gary Gaugler/Visuals Unlimited, Inc. 30.23: © Biophoto Associates/Photo Researchers, Inc.

**Chapter 31** *Opener*: © Ana Yuri Signorovitch. 31.3: Courtesy of J. B. Morrill. 31.5A: © Ed Robinson/Getty Images. 31.5B: © Steve Gschmeissner/Photo Researchers, Inc. 31.5C: © Konrad Wothe/Minden Pictures/Corbis. 31.6A: © Jurgen Freund/Naturepl.com. 31.6B: © John Bell/istock. 31.6C: © Stockphoto4u/istock. 31.7A: © John A. Anderson/istock. 31.7B: © Kevin Schafer/DigitalVision/Photolibrary.com. 31.7B *inset*: © Mike Rogal/Shutterstock. 31.8A: © Doug Lindstrand/Alaska Stock Images/AGE Fotostock. 31.8B: © blickwinkel/Alamy. 31.9A: © Cathy Keifer/Shutterstock. 31.9B: © Don Johnston/AGE Fotostock. 31.9C: David McIntyre. 31.11 *inset*: © Scott Camazine/Phototake. 31.12: © Gerd Guenther/Photo Researchers, Inc. 31.13A: © First Light/Alamy. 31.13B: © Accent Alaska.com/Alamy. 31.14A: © Helmut Heintges/Corbis. 31.14B: © F1online digitale Bildagentur GmbH/Alamy. 31.15A: © Jurgen Freund/Naturepl.com. 31.15B: David McIntyre. 31.15C: © Robert Brons/Biological Photo Service. 31.16B: © Larry Jon Friesen. 31.17A: Courtesy of Wim van Egmond. 31.18, 31.19: Adapted from F. M. Bayerand & H. B. Owre, 1968. *The Free-Living Lower Invertebrates*, Macmillan Publishing Co. 31.20A: © Charles Wyttenbach/Biological Photo Service. 31.20B: © Georgette Douwma/Naturepl.com. 31.20C, D: © Larry Jon Friesen. 31.21A: © Jurgen Freund/Naturepl.com. 31.21B: © Stephan Kerkhofs/Shutterstock. 31.22: Adapted from F. M. Bayerand & H. B. Owre, 1968. *The Free-Living Lower Invertebrates*, Macmillan Publishing Co.

**Chapter 32** *Opener*: © Mark Moffett/Minden Pictures. 32.2: © blickwinkel/Alamy. 32.3A: From D. C. García-Bellido & D. H. Collins, 2004. *Nature* 429: 40. Courtesy of Diego García-Bellido Capdevila. 32.3B: © Nature's Images/Photo Researchers, Inc. 32.6A: © Larry Jon Friesen. 32.7B: © Roland Birke/Getty Images. 32.7C: Courtesy of David Walter and Heather Proctor. 32.7D: © Michael Abbey/Photo Researchers, Inc. 32.8B: © Larry Jon Friesen. 32.9: © David Wrobel/Visuals Unlimited, Inc. 32.10A: © Fred Bavendam/Minden Pictures. 32.12A: © WaterFrame/Alamy. 32.12B: Courtesy of Cindy Lee Van Dover. 32.12C: © Pakhnyushcha/Shutterstock. 32.12D: © Larry Jon Friesen. 32.13B: © Marevision/AGE Fotostock. 32.13C: © Francesco Tomasinelli/Photo Researchers, Inc. 32.13D: © H. Wes Pratt/Biological Photo Service. 32.13E: © moodboard/Photolibrary.com. 32.14A: © Larry Jon Friesen. 32.14B: © Laura Romin & Larry Dalton/Alamy. 32.14C: © Jeff Rotman/Naturepl.com. 32.15A: Courtesy of Jen Grenier and Sean Carroll, U. Wisconsin. 32.15B: Courtesy of Graham Budd. 32.15C: Courtesy of Reinhardt Møbjerg Kristensen. 32.16B: © Grave/Photo Researchers, Inc. 32.16C: © Steve Gschmeissner/Photo Researchers, Inc. 32.17: © Pascal Goetgheluck/Photo Researchers, Inc. 32.18A: © Steve Gschmeissner/Photo Researchers, Inc. 32.18B: © George Grall/National Geographic Society/Corbis. 32.19: © Gerald & Buff Corsi/Visuals Unlimited, Inc. 32.20A: © David Shale/Naturepl.com. 32.20B: © Joe McDonald/Corbis. 32.21A: © Kelly Swift, www.swiftinverts.com. 32.21B: © Larry Jon Friesen. 32.21C: © Nigel Cattlin/Alamy. 32.21D: SEM by Eric Erbe; colorization by Chris Pooley/USDA ARS. 32.22A: © Rod Williams/Naturepl.com. 32.22B: © John R. MacGregor/Getty Images. 32.23A, B: © Larry Jon Friesen. 32.23C: © Solvin Zankl/Naturepl.com. 32.23D: © Larry Jon Friesen. 32.23E: © Norbert Wu/Minden Pictures. 32.25: © Scenics & Science/Alamy. 32.27A: © Cisca Castelijns/Foto Natura/Minden Pictures/Corbis. 32.27B: © Piotr Naskrecki/Minden Pictures/Corbis. 32.27C: © Pete Oxford/Naturepl.com. 32.27D: David McIntyre. 32.27E: © Papilio/Alamy. 32.27F: © CorbisRF/Photolibrary.com. 32.27G: © Rafael Campillo/AGE Fotostock. 32.27H: © Jean Claude Carton/Bruce Coleman USA/AGE Fotostock.

**Chapter 33** *Opener*: © Michael Tyler/ANTPhoto.com. 33.2: From S. Bengtson, 2000. Teasing fossils out of shales with cameras and computers. *Palaeontologia Electronica* 3(1). 33.3A: © Triarch/Visuals Unlimited, Inc. 33.4: Courtesy of Samuel Chow (CybersamX)/Flickr. 33.5A: © Hal Beral/Visuals Unlimited, Inc. 33.5B, C: © WaterFrame/Alamy. 33.5D: © Marevision/AGE Fotostock. 33.5E: © Robert L. Dunne/Photo Researchers, Inc. 33.6A: © C. R. Wyttenbach/Biological Photo Service. 33.7A: © Stan Elems/Visuals Unlimited, Inc. 33.7B: © Larry Jon Friesen. 33.8A: © Marevision/AGE Fotostock. 33.8B: © Gavin Newman/Alamy. 33.11A: © Ken Lucas/Biological Photo Service. 33.11B *left*: © Marevision/AGE Fotostock. 33.11B *right*: © anne de Haas/istock. 33.12B: © Roger Klocek/Visuals Unlimited, Inc. 33.13A: © Wayne Lynch/AGE Fotostock. 33.13B: © Kelvin Aitken/AGE Fotostock. 33.13C: © Norbert Wu/Minden Pictures. 33.14A: © David Fleetham/Alamy. 33.14B, C: © Larry Jon Friesen. 33.14D: © Norbert Wu/Minden Pictures. 33.15A: © Hoberman Collection/Corbis. 33.15B: © Tom McHugh/Photo Researchers, Inc. 33.15C: © Ted Daeschler/Academy of Natural Sciences/VIREO. 33.18A: © Morley Read/Naturepl.com. 33.18B: © Michael & Patricia Fogden/Minden Pictures. 33.18C: © Jack Goldfarb/Design Pics, Inc./Photolibrary.com. 33.18D: Courtesy of David Hillis. 33.21A: © C. Alan Morgan/Getty Images. 33.21B: © Cathy Keifer/Shutterstock. 33.21C: © Larry Jon Friesen. 33.21D: © Gordon Chambers/Alamy. 33.22A: © Susan Flashman/istock. 33.22B: © Gerry Ellis, DigitalVision/PictureQuest. 33.23A: From X. Xu et al., 2003. *Nature* 421: 335. © Macmillan Publishers Ltd. 33.23B: © Tom & Therisa Stack/Painet, Inc. 33.24: © Melinda Fawver/istock. 33.25A: © Tim Zurowski/All Canada Photos/Getty Images. 33.25B: © Salvador III Manaois/Alamy. 33.25C: © Tom Vezo/Minden Pictures. 33.25D: © Marco Kopp/istock. 33.26A: © John N. A. Lott/Biological Photo Service. 33.26B: © Dave Watts/Visuals Unlimited, Inc. 33.27A: © Ingo Arndt/Naturepl.com. 33.27B: © Greg Harold/Auscape/Minden Pictures/Corbis. 33.27C: © R. Wittek/Arco Images/AGE Fotostock. 33.29A: © Robert McGouey/All Canada Photos/Corbis. 33.29B: © ANT Photo Library/Photo Researchers, Inc. 33.29C: © John E Marriott/All Canada Photos/AGE Fotostock. 33.29D: © Michael S. Nolan/AGE Fotostock. 33.31: © John Warburton-Lee Photography/Alamy. 33.32A: © mike lane/Alamy. 33.32B: © De Agostini Editore/AGE Fotostock. 33.33A: © Steve Bloom Images/Alamy. 33.33B: © Anup Shah/AGE Fotostock. 33.33C: © Lars Christensen/istock. 33.33D: © Anup Shah/Minden Pictures. 33.35A: © Cyril Ruoso/Minden Pictures. 33.35B: Courtesy of Andrew D. Sinauer. 33.35C: © Arco Images GmbH/Alamy. 33.35D: David McIntyre.

**Chapter 34** *Opener*: © Picture Contact BV/Alamy. 34.6A: © Dr. Ken Wagner/Visuals Unlimited, Inc. 34.6B: © Phil Gates/Biological Photo Service. 34.6C: © Biophoto Associates/Photo Researchers, Inc. 34.6D: © Jack M. Bostrack/Visuals Unlimited, Inc. 34.7A: © John D. Cunningham/Visuals Unlimited, Inc. 34.7B: © J. Robert Waaland/Biological Photo Service. 34.7C: © Herve Conge/ISM/Phototake. 34.8 *upper*: © Biodisc/Visuals Unlimited, Inc. 34.8 *lower*: © M. I. Walker/Photo Researchers, Inc. 34.9B: © John N. A. Lott/Biological Photo Service. 34.10A: © Ed Reschke/Getty Images. 34.10B: © Dr. James W. Richardson/Visuals Unlimited, Inc. 34.11: © Larry Jon Friesen. 34.12A: © modesigns58/istock. 34.12B: © Adrian Sherratt/Alamy. 34.12C: © Science Photo Library/Alamy. 34.13A *left*: David McIntyre. 34.13A *right*: © Andrew Syred/Photo Researchers, Inc. 34.13B *left*: © Garry DeLong/Photo Researchers, Inc. 34.13B *right*: © Steve Gschmeissner/Photo Researchers, Inc. 34.14: David McIntyre. 34.15B: Courtesy of Thomas Eisner, Cornell U. 34.15C: © Susumu Nishinaga/Photo Researchers, Inc. 34.18: © Biodisc/Visuals Unlimited, Inc. 34.19: © Phil Gates/Biological Photo Service.

**Chapter 35** *Opener*: © John Carr/Eye Ubiquitous/Corbis. 35.3: © Nigel Cattlin/Alamy. 35.8A: © Susumu Nishinaga/Photo Researchers, Inc. 35.10: © R. Kessel & G. Shih/

Visuals Unlimited, Inc. Page 735 *Aphid and stylet*: © M. H. Zimmermann.

**Chapter 36** *Opener*: © Russ Munn/AgStock Images/Corbis. 36.1: David McIntyre. 36.5: David McIntyre. 36.11A: © J. H. Robinson/ The National Audubon Society Collection/ Photo Researchers, Inc. 36.11B: © Kim Taylor/ Naturepl.com. 36.12: Courtesy of Susan and Edwin McGlew.

**Chapter 37** *Opener*: © Micheline Pelletier/ Sygma/Corbis. 37.2: From J. M. Alonso and J. R. Ecker, 2006. *Nature Reviews Genetics* 7: 524. 37.3A: Courtesy of J. A. D. Zeevaart, Michigan State U. 37.3B: From W. M. Gray, 2004. *PLoS Biol.* 2(9): e311. 37.4: © Sylvan Wittwer/Visuals Unlimited, Inc. 37.9: © Ed Reschke/Getty Images. 37.13, 37.16: David McIntyre. Page 757: © Gerald & Buff Corsi/Visuals Unlimited, Inc. Page 768: David McIntyre. Page 769: Courtesy of Adel A. Kader. Page 771: Courtesy of Eugenia Russinova, VIB Department of Plant Systems Biology, Ghent University, Belgium. Page 777: Clemson University - USDA Cooperative Extension Slide Series, Bugwood.org.

**Chapter 38** *Opener*: © Remco Zwinkels/ Minden Pictures. 38.1A: © kukuruxa/ Shutterstock. 38.1B *left*: © Tish1/Shutterstock. 38.1B *right*: © Pierre BRYE/Alamy. 38.1C: © Bill Beatty/Visuals Unlimited, Inc. 38.3: © Rolf Nussbaumer Photography/Alamy. 38.4: © Christian Guatier/Biosphoto. 38.7A: David McIntyre. 38.7B: © Michael Moreno/ istock. 38.7C: © Scenics & Science/Alamy. 38.9: Courtesy of Richard Amasino. 38.15: Courtesy of Richard Amasino and Colleen Bizzell. 38.17A: © ooyoo/istock. 38.17B: © Nigel Cattlin/Alamy. 38.17C: © Jerome Wexler/Visuals Unlimited, Inc. 38.18: David McIntyre. Page 784 *Thistle*: © John N. A. Lott/ Biological Photo Service. Page 784 *Burrs*: © Scott Camazine/Alamy. Page 795: © yykkaa/ Shutterstock.

**Chapter 39** *Opener*: © Birgit Betzelt/ actionmedeor/hand/dpa/Corbis. 39.4: © Holt Studios International Ltd/Alamy. 39.5A: © Kim Taylor/Naturepl.com. 39.5B: David McIntyre. 39.8: Courtesy of Thomas Eisner, Cornell U. 39.9: © Jon Mark Stewart/Biological Photo Service. 39.10: © Dr. Jack Bostrack/Visuals Unlimited, Inc. 39.11: © TH Foto/Alamy. 39.12: © Simon Fraser/SPL/Photo Researchers, Inc. 39.13: © John N. A. Lott/Biological Photo Service. 39.16: Courtesy of Scott Bauer/USDA. 39.17: © Jurgen Freund/Naturepl.com. 39.18: Courtesy of Ryan Somma.

**Chapter 40** *Opener*: © PCN Black/Alamy. 40.3A: © Gladden Willis/Visuals Unlimited, Inc. 40.3B: From Ross, Pawlina, and Barnash, 2009. *Atlas of Descriptive Histology*. Sinauer Associates: Sunderland, MA. 40.3C: © Ed Reschke/Getty Images. 40.4A: From Ross, Pawlina, and Barnash, 2009. *Atlas of Descriptive Histology*. Sinauer Associates: Sunderland, MA. 40.4B: © Manfred Kage/Photo Researchers, Inc. 40.4C: © SPL/Photo Researchers, Inc. 40.5A: © Chuck Brown/Photo Researchers, Inc. 40.5B: From Ross, Pawlina, and Barnash, 2009. *Atlas of Descriptive Histology*. Sinauer Associates: Sunderland, MA. 40.5C: © Dennis Kunkel Microscopy, Inc. 40.5D: From Ross, Pawlina, and Barnash, 2009. *Atlas of Descriptive Histology*. Sinauer Associates: Sunderland, MA. 40.6A: © James Cavallini/Photo Researchers, Inc. 40.6B: © Innerspace Imaging/SPL/Photo Researchers, Inc. 40.12: © Greg Epperson/ istock. 40.13: © Gerry Ellis/DigitalVision. 40.15: Courtesy of Anton Stabentheiner. 40.18A: © Robert Shantz/Alamy. 40.18B: © Jim Brandenburg/Minden Pictures.

**Chapter 41** *Opener*: © Christian Liepe/ Corbis. 41.2A *Insulin*: Data from PDB 2HIU. Q. X. Hua et al., 1995. *Nat. Struct. Biol.* 2: 129. 41.2A *HGH*: Data from PDB 1HGU. L. Chantalat et al., 1995. *Protein Pept. Lett.* 2: 333. 41.3 *Snake*: © Ameng Wu/istock. 41.4 *Prolactin*: Data from PDB 1RW5. K. Teilum et al., 2005. *J. Mol. Biol.* 351: 810. 41.4 *Fish*: © Alaska Stock LLC/Alamy. 41.4 *Amphibian*: © Gustav W. Verderber/Visuals Unlimited, Inc. 41.4 *Birds*: © Dave Cole/Alamy. 41.4 *Mammals*: © Ale Ventura/PhotoAlto/Photolibrary.com. 41.12A: © Ed Reschke/Getty Images. 41.13: © Scott Camazine/Photo Researchers, Inc. 41.18: Courtesy of Gerhard Heldmaier, Philipps U.

**Chapter 42** *Opener*: © Heritage Images/ Corbis. 42.8: © Steve Gschmeissner/Photo Researchers, Inc. Page 861: © Science Photo Library/Photo Researchers, Inc.

**Chapter 43** *Opener*: David McIntyre. 43.1A: © P&R Photos/AGE Fotostock. 43.1B: © Constantinos Petrinos/Naturepl.com. 43.2A: © Patricia J. Wynne. 43.5: © david gregs/ Alamy. 43.6: © Jane Gould/Alamy. 43.7A: © Morales/AGE Fotostock. 43.7B: © Dave Watts/ Naturepl.com. 43.9B: © Michael Webb/Visuals Unlimited, Inc. 43.12B: © P. Bagavandoss/ Photo Researchers, Inc. 43.15C: © S. I. U. School of Med./Photo Researchers, Inc. 43.17: Courtesy of The Institute for Reproductive Medicine and Science of Saint Barnabas, New Jersey.

**Chapter 44** *Opener*: © Mads Abildgaard/ istock. 44.1: Courtesy of Richard Elinson, U. Toronto. 44.3A *left*: From H. W. Beams and R. G. Kessel, 1976. *American Scientist* 64: 279. 44.3A *center, right*: © Dr. Lloyd M. Beidler/ Photo Researchers, Inc. 44.3B: From H. W. Beams and R. G. Kessel, 1976. *American Scientist* 64: 279. 44.3C: Courtesy of D. Daily and W. Sullivan. 44.4B *left, center*: © Dr. Yorgos Nikas/Science Source/Photo Researchers, Inc. 44.4B *right*: © Petit Format/Science Source/ Photo Researchers, Inc. 44.4C: From J. G. Mulnard, 1967. *Arch. Biol.* (Liege) 78: 107. Courtesy of J. G. Mulnard. 44.14D: Courtesy of K. W. Tosney and G. Schoenwolf. 44.15B: Courtesy of K. W. Tosney. 44.19A: © CNRI/ SPL/Photo Researchers, Inc. 44.19B: © Dr. G. Moscoso/SPL/Photo Researchers, Inc. 44.19C: © Tissuepix/SPL/Photo Researchers, Inc. 44.19D: © Petit Format/Photo Researchers, Inc.

**Chapter 45** *Opener*: © John Birdsall/ AGE Fotostock. 45.3B: © C. Raines/Visuals Unlimited, Inc. 45.4: Courtesy of Philip Haydon. 45.7: From A. L. Hodgkin & R. D. Keynes, 1956. *J. Physiol.* 148: 127.

**Chapter 46** *Opener*: © Casey K. Bishop/ Shutterstock. 46.4: David McIntyre. 46.8A: © Dr. Fred Hossler/Visuals Unlimited, Inc. 46.13A: © Cheryl Power/Photo Researchers, Inc. 46.16: © Omikron/Science Source/Photo Researchers, Inc.

**Chapter 47** *Opener*: © Bluesky International Limited. 47.7: Photo from "Brain: The World Inside Your Head," © Evergreen Exhibitions. 47.12A: Courtesy of Compumedics. 47.14: © Wellcome Dept. of Cognitive Neurology/SPL/ Photo Researchers, Inc.

**Chapter 48** *Opener*: © Oxford Scientific/ Getty Images. 48.1 *Micrograph*: © Frank A. Pepe/Biological Photo Service. 48.2: © Tom Deerinck/Visuals Unlimited, Inc. 48.4: © Kent Wood/Getty Images. 48.7: © Manfred Kage/ Photo Researchers, Inc. 48.8: © SPL/Photo Researchers, Inc. 48.11: Courtesy of Jesper L. Andersen. 48.18: © Robert Brons/Biological Photo Service.

**Chapter 49** *Opener*: © Steve Bloom, stevebloom.com. 49.1A: © Ross Armstrong/ AGE Fotostock. 49.1B: © WaterFrame/Alamy. 49.1C: © Photoshot Holdings Ltd/Alamy. 49.4B: © Andrew Darrington/Alamy. 49.4C: Courtesy of Thomas Eisner, Cornell U. 49.10 *Bronchi*: © SPL/Photo Researchers, Inc. 49.10 *Alveoli*: © P. Motta/Photo Researchers, Inc. 49.17: After C. R. Bainton, 1972. *J. Appl. Physiol.* 33: 775.

**Chapter 50** *Opener*: © NBAE/Getty Images. 50.10A: © Brand X Pictures/Alamy. 50.11: After N. Campbell, 1990. *Biology*, 2nd Ed., Benjamin Cummings. 50.12B: © CNRI/Photo Researchers, Inc. 50.14: © Science Source/ Photo Researchers, Inc. 50.17A: © Chuck Brown/Science Source/Photo Researchers, Inc. 50.17B: © Biophoto Associates/Science Source/ Photo Researchers, Inc.

**Chapter 51** *Opener*: © Marilyn "Angel" Wynn/Nativestock.com. 51.1: Courtesy of Andrew D. Sinauer. 51.3: © Dai Kurokawa/ epa/Corbis. 51.8C *Microvilli*: © Biophoto Associates/Photo Researchers, Inc. 51.18: © Science VU/Jackson/Visuals Unlimited, Inc.

**Chapter 52** *Opener*: © Michael & Patricia Fogden/Corbis. 52.2B: © Morales/AGE Fotostock. 52.8A: © CNRI/SPL/Photo Researchers, Inc. 52.8B: © Susumu Nishinaga/ Photo Researchers, Inc. 52.8C: © Science Photo Library RF. 52.8D: © Dr. Donald Fawcett & D. Friend/Visuals Unlimited, Inc. 52.13: © Hank Morgan/Photo Researchers, Inc. 52.17: From L. Bankir & C. de Rouffignac, 1985. *Am. J. Physiol.* 249: R643-R666. Courtesy of Lise Bankir, INSERM Unit, Hôpital Necker, Paris.

**Chapter 53** *Opener*: © E. R. Degginger/ Photo Researchers, Inc. 53.2A: © FLPA/Alamy. 53.3B: © Maximilian Weinzierl/Alamy. 53.6A: © Nina Leen/Time Life Pictures/Getty Images. 53.6B: © Wayne Lynch/All Canada Photos/ Corbis. 53.10A: © Interfoto/Alamy. 53.10B: ©

Shaun Cunningham/Alamy. 53.10C: © Shattil & Rozinski/Naturepl.com. 53.12A: © Jeremy Woodhouse/Photolibrary.com. 53.12B: © Elvele Images Ltd/Alamy. 53.16B: © Momatiuk - Eastcott/Corbis. 53.19: © Tui De Roy/Minden Pictures. 53.22: © J. Jarvis/Visuals Unlimited, Inc. 53.23B: © Richard R. Hansen/Photo Researchers, Inc.

**Chapter 54** *Opener*: © Bob Gibbons/OSF/Getty Images. 54.6: David McIntyre. *Tundra, left*: © John Schwieder/Alamy. *Tundra, right*: © Imagestate Media Partners Limited - Impact Photos/Alamy. *Ptarmigan*: © Rolf Hicker Photography/Alamy. *Boreal, left*: © John E Marriott/All Canada Photos/AGE Fotostock. *Boreal, right*: © Robert Harding Picture Library Ltd/Alamy. *Boreal owl*: © John Cancalosi/Alamy. *Temperate deciduous, left*: © Carr Clifton/Minden Pictures. *Temperate deciduous, right*: © Jack Milchanowski/AGE Fotostock. *Butterfly*: © Rick & Nora Bowers/Alamy. *Temperate grasslands, left*: © Tim Fitzharris/Minden Pictures. *Temperate grasslands, right*: © Mikhail Yurenkov/Alamy. *Rhea*: © Joe McDonald/Visuals Unlimited, Inc. *Hot desert, left*: © Peter Lilja/AGE Fotostock. *Hot desert, right*: © David M. Schrader/Shutterstock. *Beetle*: © Michal P Fogden/Bruce Coleman USA/AGE Fotostock. *Cold desert, left*: David McIntyre. *Cold desert, right*: © Tui De Roy/Minden Pictures. *Lizard*: Courtesy of Andrew D. Sinauer. *Chaparral*: © Carl W. May/Biological Photo Service. *Maquis*: © José Antonio Jiménez/AGE Fotostock. *Sugarbird*: © Nigel Dennis/AGE Fotostock. *Thorn forest*: © Cyril Ruoso/JH Editorial/Minden Pictures/Corbis. *Savanna*: © Tui De Roy/Minden Pictures. *Termite mound*: Courtesy of Andrew D. Sinauer. *Tropical deciduous, left*: © Staffan Widstrand/Naturepl.com. *Tropical deciduous, right*: © Pete Oxford/Naturepl.com. *Hummingbird*: © Rolf Nussbaumer/Naturepl.com. *Tropical evergreen, left*: © Luciano Candisani/Minden Pictures. *Tropical evergreen, right*: Courtesy of Andrew D. Sinauer. *Sloth*: © Kevin Schafer/AGE Fotostock. 54.9 *Kelp*: © Mauricio Handler/National Geographic Society/Corbis. 54.9 *Fish*: © Bluegreen Pictures/Alamy. 54.9 *Vent*: Courtesy of P. Rona/OAR/NOAA National Undersea Research Program. 54.10 *Reeds*: © Wild Wonders of Europe/Maitland/Naturepl.com. 54.10 *Osprey*: © Robert Bannister/Alamy. 54.13A: © Chris Mattison/Alamy.

**Chapter 55** *Opener*: Courtesy of Dave Klein. 55.1A: Courtesy of Andrew D. Sinauer. 55.1B: © Pete Oxford/Minden Pictures/Corbis. 55.1C: © Heidi & Hans-Juergen Koch/Minden Pictures. 55.3 *inset*: © Kitch Bain/Shutterstock. 55.4A: © Rolf Hicker/All Canada Photos/Corbis. 55.4B: © Ilmarin/Shutterstock. 55.4C: © Ilya Bushuev/istock. 55.6: © M. & C. Photography/Getty Images. 55.10: © Wolfgang Pölzer/Alamy. 55.11 *inset*: © T. W. Davies/California Academy of Sciences. 55.12: David McIntyre. 55.13A: © Rick & Nora Bowers/Alamy. 55.13B: Courtesy of Colin Chapman. 55.15: © Kathie Atkinson.

**Chapter 56** *Opener*: © Mark Moffett/Minden Pictures. 56.1B *upper*: Courtesy of Andrew D. Sinauer. 56.1B *lower*: © Michael Mauro/Minden Pictures. 56.2: © Jodi Jacobson/Shutterstock. 56.3A: © E. A. Kuttapan/Naturepl.com. 56.3B: © Stephen Dalton/Minden Pictures. 56.4A: David McIntyre. 56.4B: © Jason Bazzano/AGE Fotostock. 56.5A: © Cathy Keifer/Shutterstock. 56.5B: © Kelpfish/Shutterstock. 56.5C: © Nicola Vernizzi/Shutterstock. 56.6A *left*: © Georgette Douwma/Naturepl.com. 56.6A *right*: © WaterFrame/Alamy. 56.6B: Courtesy of Andrew Brower. 56.7: David McIntyre. 56.8A: © Eye of Science/Photo Researchers, Inc. 56.8B: © Wolfgang Kaehler/Corbis. 56.9A, C: Courtesy of Erich G. Vallery, USDA Forest Service/Bugwood.org. 56.9B: Courtesy of James Denny Ward, USDA Forest Service/Bugwood.org. 56.10: © Neil Lucas/Naturepl.com. 56.11: © blickwinkel/Alamy. 56.12 *left*: © Trevor Sims/AGE Fotostock. 56.12 *center, right*: Courtesy of Olle Pellmyr. 56.13: © ANT Photo Library/Photo Researchers, Inc. 56.14 *left*: © Krystyna Szulecka Photography/Alamy. 56.14 *right*: © Pete Oxford/Minden Pictures. 56.16: © Mark Moffett/Minden Pictures.

**Chapter 57** *Opener*: © Dr. James L. Castner/Visuals Unlimited, Inc. 57.1A: © Bill Beatty/Visuals Unlimited, Inc. 57.1B: © aaron peterson.net/Alamy. 57.6 *inset*: © Donald M. Jones/Minden Pictures. 57.7: © Chris Howes/Wild Places Photography/Alamy. 57.12: Courtesy of E. O. Wilson. 57.15: Courtesy of the National Park Service.

**Chapter 58** *Opener*: Courtesy of Goddard SVS/NASA. 58.3: From Zhao et al. 2006. *J. Geophys. Res.—Biogeosciences* 111: G01002. 58.4: Image by Robert Simmon, NASA GSFC Earth Observatory, based on data provided by Watson Gregg, NASA GSFC. 58.13: Courtesy of the Scientific Visualization Studio at NASA Goddard Space Flight Center.

**Chapter 59** *Opener*: © Thomas Mangelsen/Minden Pictures. 59.1: © Mark Hallett Paleoart/Photo Researchers, Inc. 59.2B: © Aurora Photos/Alamy. 59.6B: Richard Bierregaard, Courtesy of the Smithsonian Institution, Office of Environmental Awareness. 59.7A: © Xi Zhinong/Minden Pictures/Corbis. 59.7B: © Georgette Douwma/Naturepl.com. 59.8: © John Mitchell/Photo Researchers, Inc. 59.9: © John Worrall. 59.12B: © Craig Hosterman/Shutterstock. 59.12C: © Lee Prince/Shutterstock. 59.13: © Raymond Gehman/Corbis. 59.15: Courtesy of Christopher Baisan and the Laboratory of Tree-Ring Research, U. Arizona, Tucson. 59.16: After S. H. Reichard & C. W. Hamilton, 1997. *Conservation Biology* 11: 193. 59.17: © Peter Johnson/Corbis. 59.18B: Courtesy of the Working for Water Programme. 59.19A: Courtesy of the U.S. Fish and Wildlife Service. 59.19B: © Tom Vezo/Naturepl.com.

# Index

Numbers in *italic* indicate the information will be found in a figure, caption, or table.


## A

A band, *987*, 988
A blood group, 243
A horizon, 745
Aardvarks, *698*
Abalones, 662
Abdul-Jabbar, Kareem, 304
Abiotic ecosystem components
  defined, 1122
  factors influencing biomes, 1126–1127
ABO blood types, 53, 243
Abomasum, 1064
Aboral side, 680
Abortion, 897
Abscisic acid (ABA), 734, *759*, 762, 809
Abscission, 765, 769
Abscission zone, 765
Absolute dating, 506–507
Absorption spectrum, of pigments, 189, 190
Absorptive heterotrophs, 556, 609
Absorptive state, 1065
*Acacia*, 1136
Acacia ants, 1178, *1179*
*Acacia cornigera*, 1178, *1179*
*Acanthamoeba polyphaga* mimivirus (APMV), *543*
*Acanthostega*, *690*
Accessory fruits, 601
Accessory olfactory bulb, 950
Accessory pigments, 190, 571
Accessory sex organs
  defined, 887, 889
  in vertebrate evolution, 888
*Accipiter gentilis*, 1116, *1117*
Acclimation, in plants, 806
Acclimatization, in animals, 821
Accommodation, 958
Acetate, 171
Acetic acid, 34
Acetyl coenzyme A (acetyl CoA)
  allosteric regulation of the citric acid cycle, 182
  anabolic interconversions, 180
  β-oxidation, 179–180
  in the citric acid cycle, 170, 171
  in pyruvate oxidation, 170
Acetyl groups, 1051, 1052
Acetylcholine (ACh)
  actions in neuromuscular junctions, 936–938
  in the autonomic nervous system, 975
  binding to the acetylcholine receptor, 129
  clearing from synapses, 940
  effect on gut muscle, 993
  effect on heartbeat, 1034
  functions of, 939
  in skeletal muscle contraction, 989
  in smooth muscle relaxation, 135–136
Acetylcholine receptors (AChR)
  at neuromuscular junctions, 936–938
  structure and function, 129
  types of, 939–940
Acetylcholinesterase (AChE)
  function at synapses, *937*, 940
  irreversible inhibitor of, 157, *158*
  isozymes in rainbow trout, 162
*N*-Acetylgalactosamine, 243
*N*-Acetylglucosamine, *55*
*Acherontia atropos*, *1009*
Acid–base reactions, 34
Acid growth hypothesis, 766–767
Acid precipitation, 1219–1220
Acid–base balance, regulation by the kidneys, 1084–1085
Acid–base catalysis, 155
Acidification
  of lakes, 1220
  of oceans, 647, 1217
Acidophilic thermophiles, 536
Acids, 34–36, 43
*Acinonyx jubatus*, 62
Acoelomates, 635
Acoels, *632*, 648, 680, *681*
Acorn worms, 682, *683*
Acoustic signals, 1112
Acquired hearing loss, 956
Acquired immune deficiency syndrome (AIDS)
  causal organism, 341
  description of, 876–877
  fungal diseases and, 612
  phylogenetic analyses, 458
  treatments, 301
*Acropora millepora*, *449*
Acrosomal reaction, *885*, 886
Acrosome, 884, *885*, 886, 889, *891*
ACTH. *See* Adrenocorticotropic hormone
Actin
  contractile ring, 216
  in microfilaments, 94–95 (*see also* Microfilaments)
  in muscle contraction, *987*, 988–990, *991*, 993
  in muscle tissues, 817
Actin filaments
  impact of strength training on, 996
  in muscle fiber contraction, *987*, 988–990, *991*
  *See also* Microfilaments
Actin–myosin bonds
  in insect flight muscle contraction, 998
  in skeletal muscle contraction, 988–990, *991*
  tetanic muscle contractions and, 995
Actinobacteria, 532, 1183
Action potentials
  in cardiac muscle, 991
  defined, 925, 927
  generation at neuromuscular junctions, 938
  generation of, 932–934
  membrane potentials and, 927–932
  principles of bioelectricity, 927
  production by sensory receptor cells, 947–948
  properties of, 934–935
  release of neurotransmitter at neuromuscular junctions, 936, *937*
  saltatory conduction, 935
  in smooth muscle, 993
  speed of transmission, 926
  *See also* Cardiac action potentials
Action spectra
  of photosynthesis, 189, 190
  of plant phototropism, 771, *772*
Activation energy, 151–152
Activation gate, 933–934
Activator proteins, 329, 332
Active predators, *638*
Active transport
  defined, 113, 118
  directional nature of, 118
  energy sources for, 118–119
Actuarial tables, 1154
Adaptation (sensory), 948
Adaptations
  to climate, 1122, 1125–1126
  defined, 6, 433
  examples in frogs, *7*
  to reduce heat loss in endotherms, 828
  trade-offs, 445
Adaptations in plants
  defined, 806
  to growth in saturated soils, 808
  to life on land, 574
  to saline environments, 811
  to temperature extremes, 810
  to very dry conditions, 806–808
Adaptive defenses, 857, *858*
Adaptive immunity
  cellular immune response, 871–875
  clonal deletion, 865
  clonal selection, 865, *866*
  discovery of, 862, *863*, 864
  humoral immune response, 867–871
  immunological immunity and vaccinations, 866
  immunological memory and secondary immune response, 865–866
  key features, 862–863
  types of responses in, 863–864
Adaptive radiations, 481–482, 489–490
Adenine
  codons and the genetic code, 288–289
  complementary base pairing, 63–65
  in DNA structure, 264, 265, 266, *267*
  poly A tail, 292
  structure, *63*
Adeno-associated virus, 323, 324
Adenosine, 128
Adenosine 2A receptor, 128
Adenosine diphosphate (ADP)
  in ATP hydrolysis, 150
  in ATP synthesis by ATP synthase, *173*, 174, 175
  as a coenzyme, 167
Adenosine monophosphate (AMP), 150
Adenosine triphosphate (ATP)
  in active transport systems, 118–119
  allosteric regulation of glycolysis, 182
  allosteric regulation of the citric acid cycle, 182
  as an "energy currency," 149–151
  in $C_4$ photosynthesis, 199
  in the Calvin cycle, 194, *196*
  in the charging of tRNAs, 294
  coupled reactions and, 150–151
  effect of ATP supply on muscle performance, 997
  energy released during hydrolysis, 149–150
  functions of, 66, 149
  in protein phosphorylation, 209
  role in catalyzed reactions, *156*

in skeletal muscle contraction, 989, *990, 991*
slow-twitch fibers and, 995
structure, 149
tetanic muscle contractions and, 995
Adenylyl cyclase, 133
*Adiantum, 583*
Adipose cells
effect of insulin on, 1066
fat metabolism in the postabsorptive state, 1067
*See also* Brown fat
Adipose tissues
characteristics of, 819
in the fight-or-flight response, 837
types of, 165
Adrenal cortex, 849–850
Adrenal glands
fight-or-flight response, 837
hormones of, *842*, 849–850
Adrenal medulla, 849, 850
Adrenaline. *See* Epinephrine
Adrenergic receptors, 850, 853
Adrenocorticotropic hormone (ACTH), 849
Adult hemoglobin, 1018
Adventitious roots, 718
*Aegilops speltoides, 226*
*Aegilops tauschii, 226*
*Aegolius funereus, 1129*
*Aegopodium podagraria, 597*
*Aequorea victoria*, 378, 449
Aerenchyma, 808
Aerobic exercise
ATP supply and, 997
impact on muscles, 996–997
slow-twitch fibers and, 995
Aerobic metabolism
atmospheric oxygen levels and, 512
cellular respiration, 166–167
energy yield from cellular respiration and fermentation compared, 178
in the evolution of life, 5
pathways of glucose catabolism, 169–171
in prokaryotes, 537
Aerotolerant anaerobes, 537
Afferent blood vessels, 1010
Afferent nervous system, 968
Afferent neurons
defined, 940
spinal reflexes, 942
Afferent renal arterioles
in the autoregulation of the glomerular filtration rate, 1087, *1088*
in the mammalian kidney, *1080*, 1082
in the vertebrate nephron, 1078
Affinity chromatography, 853
Aflatoxins, 310, 624
Africa
ecotourism, 1242, *1243*
ivory trade, 1240–1241
African ass, *1144*
African clawed frog, 612
African elephants, *1135*
fecundity, 1155
full census counting, 1150
population age structure, 1152
*See also* Elephants
African fish eagle, *218*
African insectivores, *698*
African lions, 1183
African long-tailed widowbird, 435–436
African lungfish, *689*, 1028–1029
African millet beer, 624
African wild dogs, 1116, 1183
Afrosoricida, *698*
After-hyperpolarization, 934
*Agalychnis callidryas, 7*
*AGAMOUS* gene, 786
Agarose, 316
Agave, *1132*
*Agave schottii*, 1183
Age-dependent cohort life tables, 1154
Age structure, 1150, 1151–1152, 1165–1166
Aggregate fruit, 601
Aging, telomerase and, 275
*Aglaophyton major, 580*
Agnosias, 971
Agonists, 940
Agriculture
apomixis studies, 794
applications of biotechnology, 386–388, *389*
biological nitrogen fixation and, 750
dicultures, 1203
effect of levels of atmospheric carbon dioxide on global food production, 202
"Green Revolution," 756
hybrid plants, 778
hybrid vigor, 244–245
impact on ecosystems, 1223
importance of biological research to, 15
important crop plants, 605
improving the nitrogen use efficiency of corn, 740, 753
increasing levels of atmospheric carbon dioxide and, 202
methods to reduce water loss, 738
nitrogen fertilizers and, 740
polyploidy in crops, 225, *226*
problems of monoculture, 1202–1203
secondary succession on agricultural land, 1200
semi-dwarf wheat and rice, 756, 775
soil fertility and fertilizers, 746–747
transgenic rice, 726
use of ethylene in, 769, 770
vegetative reproduction of angiosperms in, 793–794
water demands of, 726
*See also* Crop plants; Transgenic crops
*Agrobacterium tumefaciens*, 377, 534
*Agropyron repens, 597*
AHK protein, 769
AHP protein, 769
AIDS. *See* Acquired immune deficiency syndrome
Air capillaries, 1011
Air quality, lichens as indicators of, 624, *625*
Air sacs, 1011, *1012*, 1013, *1014*
Air temperature
atmospheric circulation patterns and, 1124
causes of seasonal variations, 1123
Airplane metaphor, 1245
*Aix sponsa, 468*
Alanine, *44*
Alarm calls, 1112, 1116, *1117*
Albumin gene, 335
*Alburnus scoranza, 1140*
*Alcedo atthis, 642*
*Alces alces, 1129*
Alcohol dehydrogenase, 155, *177*, 178
Alcoholic beverages, 178
Alcoholic fermentation, 177, 178
Alcoholism, 1054
Aldehyde group, *40*
Alderflies, *672*
Alders, 1200
Aldosterone
actions of, *842*, 849
regulation of blood pressure and blood volume, 1087, *1088*
structure, *836*
*Aleuria aurantia, 622*
Aleurone layer, 761, *762*
*Alexandrium*, 549
Algae
biofuel production and, 585
brown algae, 556
closest relatives of land plants, 572–573
defined, 571
evolution of photosynthesis in, 571–572
red algae, 551, 571–572, 573
*See also* Green algae
Algal blooms, 1219
Alginic acid, 556
Alkaloids, of endophytic fungi, 615
Alkaptonuria, 282
Allantoic membrane, 918
Allantois, 918, 919
Allele frequency
calculating, 436–437
defined, 432
effect of gene flow on, 433–434
effect of nonrandom mating on, 434–436
Hardy–Weinberg equilibrium, 437–438
microevolutionary change, 446
Allele-specific oligonucleotide hybridization, 321
Alleles
codominance, 243
defined, 236, 431
genetic drift, 434
human pedigree analyses, 240–241
incomplete dominance, 242–243
law of independent assortment, 237–239
law of segregation, 236–237
multiple, 242, 243
mutations, 241–242
neutral, 441
pleiotropic, 243–244
using probability to predict inheritance, 239–240
Allergens, 875–876
Allergic reactions, 875–876
Allergies, 862
Alligators, 694, 1029–1030
Allolactose, 331
*Allomyces, 618*
Allopatric speciation, 472–474, *475*
Allopolyploidy, 225, *226*, 473, 475
Allosteric enzymes, 159–160
Allosteric regulation
in cell cycle control, 210–211
characteristics of, 159–160
of glucose catabolism, 181–182
of metabolic pathways, 160–161
Allosteric sites, of enzymes, 159
Allostery, 159
*Alluadia procera, 1135*
*Alnus*, 1200
Alpha (α) helix, 45, *46*
Alpha-lactalbumin, 333
Alpine pennycress, 811
Alpine tundra, 1128–1129
Alternation of generations
in ferns, 582
in land plants, 574–575
overview, *218*
in protists, 562–563
Alternative splicing, 346–347
Altitude, oxygen availability and, 1007–1008
Altman, Sidney, 72
Altricial young, 642
Altruistic behavior, 1114–1115, 1116
Alveolates, 553–555
Alveoli
in alveolates, 553
of ciliates, 554, *555*
diffusion of carbon dioxide from blood, 1018, *1019*
in humans, 1013, *1014*
lung surfactants, 1015
Alzheimer's disease, 336
Amacrine cells, *962*, *963*
Amber, *515*
*Amblyrhynchus cristatus, 824*
*Amborella*, 602–603
Ambros, Victor, 347
Ambulacarians, 680
*Ambystoma mavortium, 691*
*Ambystoma mexicanum, 1007*
Amensalism, *1170*, 1171
American bison, *1131*
American chestnut, 622
American elm, 622
American holly, *780*
American Prairie Foundation, 1238
Amine hormones, 836
Amino acid sequences
of cytochrome *c* from different organisms, *488–489*
genomic information and, 356
identifying evolutionary changes in, 486
Amino acids
binding of tRNA to, 293, 294
catabolic interconversions, *179*, 180
chemical structure, *44*
codons and the genetic code, 288–289

formed in prebiotic synthesis experiments, 70
found on meteorites, 69
need for in animal diets, 1051–1052
optical isomers, 43
peptide linkages, 43–44, *45*
primary structure of proteins, 45, *46*
properties of, 43
side chains, 43, *44*
Amino group
of chemically modified carbohydrates, 55
effect of pH on, 161
peptide linkages in amino acids, 43–44, *45*
properties as a base, 34
properties of, *40*
Amino sugars, 55
Aminoacyl tRNA binding site, 295, 296
Aminoacyl-tRNA synthetases, 294
γ-Aminobutyric acid (GABA), 324, 939
Aminopeptidase, *1062*
Aminopterin, 158
Ammocoetes, 686
Ammonia
in acid–base regulation by the kidneys, 1085, *1086*
as a base, 34
excretion by animals, 1074–1075
in the global nitrogen cycle, 750, *751*
inorganic fertilizers, 747
nitrogen fixation and, 539
oxidation by nitrifiers, 539
Ammonium
in the global nitrogen cycle, 750–751
inorganic fertilizers, 747
organic fertilizers and, 746
sources for plants, 200–202
Ammonium transporters, 744
Ammonotelic animals, 1074
Amniocentesis, 321
Amnion, *907*
in the chicken egg, 918
origins of, 914
in placental mammals, 919
Amniote egg, 692, 888
Amniotes
in the Carboniferous, 520
evolutionary innovations, 692
origin of, 690
phylogeny, *693*
Amniotic cavity, 919
Amniotic fluid, 919
*Amoeba proteus,* 98
Amoebas
amoebozoans, 559–561
heteroloboseans, 558
Amoebozoans, 559–561
AMP-activated protein kinase (AMPK), 1068
Amphibian decline
atrazine and, 1
controlled and comparative experiments in, 12, 13
pathogenic fungi and, 612
Amphibians
actions of prolactin in, *838*
anurans, 690, 692
in the Carboniferous, 520
cell fate determination in the embryo, 394
circulatory system, 1029
diversity in, 690, *691*
effects of atrazine on, 1
egg rearrangements following fertilization, 903–904
environmental requirements, 690
gas exchange through the skin, 1029
gastrulation, 909–913
life cycle, *691*
neurulation, 915–916
nitrogenous wastes, 1075
origin of, 690
recent decline in, 692
salamanders, 692
salt and water balance regulation, 1077
sexual reproduction, 888
social behaviors, 692
*Amphibolurus muricatus,* 421
Amphipathic molecules, 57
Amphipods, 670
*Amphiprion, 888,* 1171
Amphisbaenians, 693
Ampicillin, 527
Amplitude, of biological cycles, 774
Amygdala, 970, 982
Amylases, 144, 1060
Amylopectin, 53
Amylose, 53
*Anabaena, 533,* 749
Anabolic interconversions, 180
Anabolic reactions
endergonic reactions, 147, *148*
energy changes during, 145
integration of anabolism and catabolism, 180–181
linkage to catabolic reactions, 146, 179–180
repressible systems of regulation in prokaryotes, 332
*Anacharis,* 189, 190
Anadromous fishes, 689
Anaerobic metabolism
defined, 5
energy yield from cellular respiration and fermentation compared, 178
fermentation pathways, *166,* 167, 177–178
glycolysis, 166
non-oxygen electron receptors, 176
in prokaryotes, 537
Anal fins, 687
Anal sphincter, 1060, 1063
Anaphase (mitosis)
chromosome separation and movement, 214–216
comparison between mitosis and meiosis, *223*
events in mitosis, 212, *217*
Anaphase I (meiosis), 221
Anaphase II (meiosis), *220*
Anaphase-promoting complex (APC), 214, 215, *216*
Ancestral states, reconstruction, 460
Ancestral traits, 452
Anchor cell, 396, *397*
*Andansonia,* 1136
Andersson, Malte, *435*
Androgens, 850, *851*
*See also* Testosterone
Anemia, 292–293, 1053, 1054
Anemonefish, 887, *888,* 1171
Aneuploidy, 222, 224, 313
Anfinsen, Christian, 48, 49
Angel insects, *672*
Angelman syndrome, 345
Angiosperms
asexual reproduction, 792–794
coevolution with animals, 588, 598–599, 605
defining features, 596
distinguishing characteristics, *574*
double fertilization, 600–601, *781,* 783
endosperm, 591
flowers, 596–598, *599*
fruit development and dispersal, 784–785
fruits, 601
gametophytes, 779–780
life cycle, 600–601
monocots and eudicots, 709–710
monoecious and dioecious, 597
phylogenetic analysis of fertilization mechanisms, 459–460
phylogenetic relationships, 601–603
radiation during the Cretaceous, 521
root systems, 718
seed plants, 589
sexual reproduction, 779–785
shared derived traits, 596
strategies for preventing inbreeding, 782, *783*
in succession, 1201
transition to the flowering state, 785–792
vegetative organs, 709
xylem, 714
Angiotensin, 1044, 1087, *1088*
Angiotensin-converting enzyme, 1087, *1088*
Anglerfishes, *688*
*Angraecum sesquipedale,* 588
Angular gyrus, 981
Animal behavior
altruistic behavior, 1114–1115, 1116
behavioral ecology, 1102
cost–benefit approach, 1103
development of, 1098–1102
evolution of social behavior, 1113–1117
genetic basis of, 1096–1098
nest parasitism, 1093
origins of behavioral biology, 1094–1096
as a sequence of choices, 1103
territorial, 1103–1104, *1105*
underlying physiological mechanisms, 1106–1103
Animal cells
communication through gap junctions, 139–140
cytokinesis, 216
extracellular matrix, 100
insertion of genes into, 376
osmosis, 114, *115*
production of metabolic heat and, 822
relationship between cells, tissues, and organs, 817–820
structure of, *86*
*See also* Eukaryotic cells
Animal communication, 1110–1113
Animal defense systems. *See* Immunology
Animal development
activation by fertilization, 903–904
cleavage, 904–906
determination of polarity, 395
determination of the blastomeres, 906–907
extraembryonic membranes, 918–919
gastrulation, 908–915
germ cell lineage, 908
mosaic and regulative, 907
organogenesis, 915–918
overview, *393*
restriction of cell fate during, 394
*See also* Human development
Animal groups
bilaterians, 643
cnidarians, 645–648
ctenophores, 644–645
eumatozoans, 643
placozoans, 645
sister group to all animals, 648
sponges, 643–644
Animal hemisphere, 903
Animal hormones
of the adrenal gland, 849–850
classes of, 836
compared to plant hormones, *758*
control of digestion, 1065
control of insect molting, 839–841
control of sexual function in human males, 892
detection and measurement with immunoassays, 852–853
discovery of, 838–839
dose–response curves, 853
factors affecting the action of, 837
half-life, 852–853
hormone-mediated signaling cascades, 837
irisin, 834, 838
location and function of receptors, 836–837 (*see also* Hormone receptors)
melatonin, 851
regulation of blood calcium levels, 847–848
regulation of blood glucose concentrations, 848–849
regulation of blood pressure, 1044–1045
role in determining behavioral potential and timing, 1098–1099
secondary sex characteristics and, 250
sex steroids, 850–851
thyroxine, 845–847

Animal nutrition
control and regulation of nutrients in the body, 1064–1068
macronutrients and micronutrients, 1052–1053
nutrient deficiencies, *1053,* 1054
vitamins, 1053–1054
*See also* Food
Animal pole, 395
Animal-pollinated plants, 480
Animal reproduction
budding and regeneration, 881
honey bees, 880
life cycles, 639–640
parthenogenic, 881–882
sexual, 882–889 (*see also* Sexual reproduction)
trade-offs in, 641–642
*See also* Human reproduction
Animals
basic developmental patterns in, 633–634
body plans, 634–637
cloning, 406–407
coevolution with angiosperms, 588, 598–599, 605
common ancestor of, 631, *633*
feeding strategies, 637–639
general characteristics, 630
genomes, *361* (*see also* Eukaryotic genomes)
life cycles, 639–642
major living groups, *632*
monophyly of, 630–631, 633
opisthokonts, 609
pharming, 385–386
phylogenetic tree, *630*
placozoans, 629
plant–pollinator mutualisms, 1180–1181
pollen transport, 780, *782*
smallest genome of, 629
thermoregulation, 822–826
Anions
defined, 29
leaching in soils, 746
Annelids
anatomical characteristics, *652*
body plan, *660*
clitellates, 661–662
closed circulatory system, *1027*
excretory system, 1075, *1076*
hydrostatic skeleton, 999
key features of, 659–660
major subgroups and number of living species, *632*
polychaetes, 660–661
segmentation, 636, 659, *660*
undescribed species, 675
*Annona squamosa,* 785
Annual rings, 722–723
Annuals, 785
*Anomalocaris canadensis, 517*
*Anopheles,* 563, *564*
Anoxygenic photosynthesis, 186, 187
*Anser anser, 642,* 1099
Ant lions, *672*
Antagonistic interactions
arms race analogy, 1172
description of, 1170
impact on communities, 1193–1194
predator–prey interactions, 1172–1175
Antagonistic muscle sets, 942
Antagonists, 128, 940
Antarctica
adaptations to low temperatures, 1125
lichens, 613
Anteaters, *698*
Antelope jackrabbit, *828*
Antenna systems, 190
*See also* Light-harvesting complexes
Antennae, 636
*Antennapedia (Antp)* gene, 404, 415
Anterior, 634, *635*
Anterior pituitary
control by hypothalamic neurohormones, 844
effect of hypothalamic somatostatin on, 849
hormones produced by, *842,* 843
in human puberty, 892
negative feedback loops in the regulation of hormone secretion, 844–845
organization and function of, 842–843
regulation of human spermatogenesis, 892
regulation of the adrenal cortex, 849
regulation of the ovarian and uterine cycles, 893, 894
regulation of thyroxine production, 846
Anterior–posterior axis
body segmentation and Hox genes, 916–918
establishment in the animal zygote, 903
Antheridium, 576
Anthers, *591,* 596, 598, 1180
Anthocerophyta, *574*
*Anthoceros, 579*
Anthocyanin, 93
Anthozoans, 646–647
Anthrax, 531, 542
Anthropoids, 701–702
Anti HIV drugs, 342–343
Anti-inflammatory drugs, 862
Antibiotic resistance
composite transposons and, 359
lateral gene transfer and, 496, 530
methicillin-resistant *S. aureus,* 281, 301
problem of, *626,* 627
Antibiotic resistance genes, 376, 378, 530
Antibiotics
derived from fungi, 608
discovery of penicillin, 608
effects on bacteria, 281, 295, 301, 527
plant phytoalexins, 799–800
targeting bacterial RNA breakdown machinery, 301
Antibodies
classes of, *868*
functions of, 858
in the humoral immune response, 864, *865*
monoclonal, 871
plasma cell production of, 867
specificity of adaptive immunity, 862–863
*See also* Immunoglobulins
Anticancer drugs
antisense drugs, 382
approaches to designing in molecular medicine, 304, 325
histone deacetylase inhibitors, 344
targeting of receptors, 325
taxol, 604
treatment of chronic myelogenous leukemia, 304
Anticoagulants, 661, 662
Anticodons, 293, 294
Antidepressants, 940
Antidiuretic hormone (ADH)
actions of, *842,* 843
regulation of blood pressure and blood osmolarity, 1044–1045, 1088–1089
Antifreezes, 810, 1125
Antigen-presenting cells, 864, *865,* 872, 876
Antigenic determinants, 862, 863
*See also* Antigens
Antigens
ABO blood groups and, 243
in allergic reactions, 875–876
binding of antibodies, 858
binding to T cell receptors, 871–872
exposure in the adaptive immune response, 863–864
immunoglobulin binding sites, 867–868
specificity of adaptive immunity, 862–863
vaccinations and, 866
Antioxidants, 176
Antiparallel strands, in DNA structure, 265, 266
Antipodal cells, 779, *781,* 783
Antiporters, 118, 119
Antique bison, *1229*
Antisense drugs, 382
Antisense RNA, 382
Ants
interference competition, 1183
fungus farming by, 1169, 1178, 1185
mutualisms with plants, 1178–1179
survival of seeds in the fynbos and, 1121, 1135, 1145, 1146
*See also* Hymenoptera
Anurans, 690, 692
Anus, *1058*
in bryozoans and entoprocts, 656
in tubular guts, 1056
Anxiety, 850
Aorta
aortic bodies, *1021,* 1022, 1045
in fish, 1028
in mammals, *1031,* 1032
in reptiles, 1029, 1030
stretch receptors in the regulation of blood pressure, 1088–1089
Aortic bodies, *1021,* 1022, 1045
Aortic valve, 1030, 1031, 1032
Apaf1 protein, 399–400
*APC* gene, *314*
Apes, 701–702
*APETALA1* gene, 786, 792
APETALA1 transcription factor, 786
Aphasia, 981
*Aphelocoma coerulescens,* 1115
Aphids, *672,* 673, 735, 1153
Aphotic zone, *1139,* 1140
Apical buds, 765
Apical complex, 553
Apical dominance, 765
Apical hook, 770
Apical meristems
of deciduous trees, 721–722
origin in plant embryogenesis, 712
in plant growth and development, 710, 715–716, 719–720
Apical–basal axis, in plants, 711
Apicomplexans, 553–554, 563
*Apis mellifera, 1112, 1150*
*See also* Honey bees
Apomixis, 778, 794
Apoplast, 729–730, 737
Apoplastic pathway, 736
Apoptosis
blocking by some cancers, 382
in bone growth, 417
cancer and, 228
in clonal deletion, 865
hypersensitive response in plants, 800–801
in morphogenesis, 394
in pattern formation, 399–400
in plant cells, 226
reasons for, 225–226
signals and pathways in, 226, *227*
Aposematism, 662, 1173, *1174*
Appalachian Mountains, 472
Appendages
of arthropod relatives, 667–668
of arthropods, 655
functions in animals, 636–637
*See also* Jointed appendages
Appendix, *1058,* 1064
Apple maggot fly, 473
Apples, 601
*Aptenodytes forsteri,* 1099
*Aptenodytes patagonicus, 642*
Aquaporins
actions of antidiuretic hormone on, 1088, 1089–1090
of the collecting duct, 1088, 1089–1090
function of, 105, 116
industrial water purification and, 122
in renal tubules, 1084
water movement across membranes and, 728–729
Aquatic biomes
estuaries, 1141
freshwater, 1140
marine, 1139–1140
Aquatic ecosystems
consequences of human alteration, 1223
geographic distribution of net primary production, 1210
Aquatic species
gametic isolation and, 477

invertebrate osmoconformers and osmoregulators, 1072–1073
Aqueous solutions
acids and bases, 34–36
properties of, 33–34
Aquifers, 1215
*Aquifex,* 532
*Aquilegia,* 480
*Aquilegia formosa,* 477
*Aquilegia pubescens,* 477
*Ara chloropterus, 1106*
*Arabidopsis thaliana*
apomixis, 794
camalexin, 801
cytokinin signaling in, 769
dwarfed phenotype, 760, *761*
effect of carbon dioxide on stomatal density, 734
effect of gibberellins on flowering, 791–792
embryogenesis, 711–712
floral organ identity genes, 400–401
genes involved in the signaling response for flowering, 790
genetic screens and the identification of signal transduction pathways, 759, 760
genomic information, *361*
ion transporters, 744
meristem identity genes, 786
metabolone, 370
as a model organism, 14
photoperiodism and flowering, 788
phototropins, 771
phytochromes, 773, 774
vernalization studies, 791
water-use efficiency studies, 726
Arachidonic acid, 1052
Arachnids, 667, 669
Arbuscular mycorrhizae
description of, 614–615, *616,* 619–620
expansion of the plant root system, 748–749
formation, 747, 748
phylogeny of the fungi, *615*
*Arceuthobium americanum,* 752
Archaea
in the evolution of life, 4
examples of, *2*
extremophiles, 532
genomes, 356 (*see also* Prokaryotic genomes)
metabolic pathways, 537–538
nitrifiers, 539
nitrogen fixers, 539
non-oxygen electron receptors, 176
shapes of, 528
in the tree of life, 8
*See also* Prokaryotes
Archaea (domain)
distinguishing characteristics, 81, *527*
habitats, 534
identification through gene sequencing, 534
major groups, 535, 536–537
membrane lipids, 535–536
relationship to Bacteria and Eukarya, *526*
Archaeognatha, *672*
*Archaeopteris, 517*
*Archaeopteryx,* 695
Archean eon, *506–507,* 508
Archegonium, 575–576
Archenteron
in frog gastrulation, 909, 910
in sea urchin gastrulation, 909
Archosaurs, 460, *695*
Arctic
adaptations to low temperatures, 1125
impact of global climate change on sea ice, 1217, *1218*
lichens, 613
Arctic ground squirrel, *10*
Arctic hare, *828*
Arctic tundra, 1128–1129
*Arctostaphylos montaraensis, 1134*
Arcuate nucleus, 1067–1068
Ardipithecines, 702–703
Area phylogenies, 1143, *1144*
*Argema mittrei, 674*
Argentine ant, 1145, 1146
Arginine
in histones, 212
plant canavanine and, 803, 805
structure, *44*
*Argogorytes mystaceus, 1181*
Argon, 1211
Argon-40, 507
*Argopecten irradians, 663*
*Argyroxiphium sandwicense, 481*
Arils, 594
*Ariolimax columbianus, 664*
*Aristolochia littoralis, 602*
*Armadillidium vulgare, 670*
Armadillos, *698*
*Armillaria, 623*
Arms, of cephalopods, 664
Arms race analogy, 1172, 1175–1176
Aromatase, 420, 850, 896
ARR protein, 769
Arrow worms, *630, 632, 652,* 655–656
Arsenic, 1221
*Artemia, 1072,* 1073
*Artemisia annua,* 797
Artemisinin, 797, 812
Arterial blood pressure
control and regulation of blood flow, 1043–1044
in fish, 1028
measuring, 1032, *1033*
regulation of, 1044–1045
Arteries
atherosclerosis, 1042–1043
in blood pressure regulation, 1044
hardening of, 848
structure and function, 1027–1028, 1039, *1040*
Arterioles
in blood pressure regulation, 1043, 1044
functions in vertebrate circulatory systems, 1028
regulation of blood flow, 1039, *1040,* 1044
*Arthrobotrys dactyloides, 613*
Arthropods
anatomical characteristics, *652*
appendages, 636, 655
body plan, *655*
chelicerates, 668–669
compound eyes, 958
ectoparasites, 639
exoskeleton, 999
hemocoel, 652
heterotypy and leg number in, 418, *419*
hormonal control of molting, 839–841
influence of the exoskeleton on evolution, 655
insects, 671–673, *674*
key anatomical features, 667
major living groups, *632,* 667
mandibulates, 669–671
open circulatory system, 1026, *1027*
relatives, 667–668
segmentation, 636
success of, 667
trilobites, 668
Artificial DNA, 380
Artificial insemination, 897
Artificial life studies, 359, *360*
Artificial ribozymes, 72
Artificial selection
in agriculture, 15
Darwin's knowledge of, 6, 433
description of, 432–433
plant domestication, 723–724
*Artocarpus heterophyllus, 601*
Asci, 620, *621,* 622
Ascidians, 684
*Asclepias syriaca,* 806
Ascoma (ascomata), 620, *621,* 622
Ascomycota, 613, *616,* 620–622
*See also* Sac fungi
Ascorbic acid, 1053
Ascospores, 620, *621,* 622
Asexual reproduction
in angiosperms, 792–794
in animals, 881–882
in diatoms, 555
fungi, 616
by mitosis, 217
in protists, 562
Asian apes, 702
Asparagine, *44*
Aspartame, 322
Aspartate, 199
Aspartic acid, *44*
Aspens, 217, 793, 1193–1194
*Aspergillus,* 310, 624
*Aspergillus nidulans,* 624, 626
*Aspergillus niger,* 624
*Aspergillus oryzae,* 624
*Aspergillus tamarii,* 624
*Aspicilia, 613*
Aspirin, 830, 862
Assisted reproductive technologies (ARTs), 897, 899
Association cortex
defined, 970
in human brain evolution, 973
size in humans, 978
Associative learning, 982
Asterozoans, 681
Asthenosphere, 509
Astrocytes, *819,* 926–927
*Astrolithium, 557*
Asymmetry, in animals, 634
"Asynchronous" muscle, 998
Atacama Desert, 808, 1127, 1134
*Ateles geoffroyi, 702*
Atherosclerosis, 1042–1043
Athletes
effect of enhanced cooling on performance, 831
heart failure, 1025, 1046
heat stroke, 815
replenishment of muscle glycogen and, 997
Athlete's foot, 612
Atlantic croaker, 1207
Atmosphere
composition and structure, 1211–1211
global circulation patterns, 1124
as a medium for respiratory gases, 1007
methane and, 536
moderation of Earth's temperature by, 1212
oxygen in the early atmosphere of Earth, 69
prevailing winds, 1124, *1125*
Atmospheric carbon dioxide
biofuels and, 585
from burning fossil fuels, 1216
effect of increasing levels on corals, 647
effect of increasing levels on global food production, 202
effect of increasing levels on photosynthesis, 185, 202
effect of oceanic carbon stores on, 1217
evolution of $C_3$ and $C_4$, 200
evolution of leaves and, 583, *584*
global amounts released by fire, 1214
in the global carbon cycle, 1216–1218
global climate change and, 17, 510, 1217–1218
as a greenhouse gas, 1212
impact on nitrogen fixation by microorganisms, 1221–1223
percent of the total atmosphere, 1211
photosynthetic efficiency of plants and, 201–202
Atmospheric oxygen
body size in insects and, 505, 513, 514, 522
during the Cambrian, 516
during the Carboniferous, 505, 521
cyanobacteria and, 532, 539
impact of life on, 511–513
during the Mesozoic, 521
percent of the total atmosphere, 1211
during the Permian, 505, 522
photosynthesis and the evolution of life, 4–5
during the Proterozoic, 515
Atmospheric pressure, 1006
Atom bomb, 310
Atomic number, 22, *23*
Atomic weight, 23
Atoms
atomic number, 22

bonding to form molecules, 26–31
in chemical reactions, 31
components of, 22
electric charge, 22
electron orbitals and shells, 24–25
electronegativity, 28
elements and, 22
isotopes, 22–24
mass number, 22
molecules and, 24
reactive and stable, 25
role of electrons in chemical reactions, 24, 25
ATP. *See* Adenosine triphosphate
ATP synthases
in ATP synthesis, 171, *173*, 174
in chloroplast photophosphorylation, 192–193
mechanism of operation, 175, *176*
ATP synthesis
in the citric acid cycle, 171
energy cost of NADH shuttle systems and, 178–179
in fermentation pathways, 177, 178
in glucose catabolism, 168, 169
in glucose oxidation, 166
in glycolysis, 169, 170
mitochondria and, 91
in oxidative phosphorylation, 171–176
in photosynthesis, 188, 191, 192–193
Atrazine
experiments on the effects of, 12, 13
impact on sex development, 1, 18
public policy issues, 17
Atria
in amphibians, 1029
in fish, 1028
in lungfish, 1028
in mammals, 1030–1032
in reptiles, 1029
Atrial natriuretic peptide (ANP), 1090
Atrioventricular node, 1034, *1035*
Atrioventricular valve (AV), 1030, 1031, 1032
*Atriplex halimus*, 811
Atropine, *604*
*Atta cephalotes*, *1169*, 1185
*Atta colombica*, *1185*
Attention, the parietal lobe and, 973
Attenuation, 866
Auditory canal, 954
Auditory hair cells
hearing loss and, 956
in sound detection, *954*, 955, 956
stereocilia and mechanoreception, 953–954
Auditory processing, 971
Auditory systems
echolocation in bats, 946, 963
flexion of the basilar membrane, 955–956
hair cell stereocilia and mechanoreception, 953–954
hearing loss, 956
structure of the human ear, 954–955
*Audounia capitata*, 1121
Austin blind salamander, *691*
Australasian region, *1142*
Australia
cane toad populations, 1164, 1235
characteristics of deserts in, 1126
microfossils, 74
Murchison Meteorite, 69
Australian pelican, 412
Australopithecines, 703
*Australopithecus afarensis*, 703
*Austrobaileya*, *598*
Autism, 540
Autocatalysis, 1061
Autocrine signals, 126, 835
Autoimmunity (autoimmune diseases)
causes and examples of, 876
inflammation and, 862
microbiomes and, 540
multiple sclerosis, 926
Autolysis, 1188
Autonomic nervous system (ANS)
in blood pressure regulation, 1044, *1045*
control of heartbeat, 1034
heart rate and, 992
influence on smooth muscle, 993
structure and function, 974–975
Autophagy, 90–91
Autopolyploidy, 224–225, 473
Autosomes, 249
Autotetraploids, 224–225
Autotriploids, 224–225
Autotrophs
in communities, 1189–1190
defined, 196
as food for heterotophs, 1049
Auxins
discovery of, 760, *761*, 762, 763, 764
ethylene and, 769, 770
gravitropism, 765
interactions with cytokinins, 768
lateral transport, 764–765
molecular mechanisms underlying the activity of, 767
phototropism response and, 762, 763, 764, *765*
polar transport, 763–764
structure and typical activities, *759*
Avery, Oswald, 261
Avian malaria, 1236
*Avirulence (Avr)* genes, 800
Avogadro's number, 33
Axial filaments, of spirochetes, 533
Axillary buds
apical dominance and, 765
of Brussels sprouts, 724
in phytomers, 709
stimulation of growth by cytokinins, 768
Axon hillock, 932–933, 938
Axon terminal
defined, 925
mix of synaptic activity impinging on, 938
neuromuscular junctions, 936
synapses and, 926
Axons
conduction of action potentials, 934–935
functions of, 925
generation of action potentials, 932–934
graded membrane potentials, 932
measurement of the resting potential, 928
myelination, 926
nerves, 968
*Azotobacter*, 749

## B

B blood group, 243
B cells
as antigen-presenting cells, 864, 872
clonal deletion, 865
clonal selection, 865, *866*
cytokines, 859
effector B cells, 865
function of, *858*
generation of diversity in immunoglobulins, 868–869, *870*
in the humoral immune response, 863, 864, *865*, 872, *873*
immunoglobulin class switching, 869–871
maturation in bone marrow, *857*
memory B cells, 865
plasma cells, 867
specificity of adaptive immunity and, 862–863
B horizon, 745
Baboons, *1135*
"Baby boom," 1165
*Bacillus*, 531, 539
Bacillus (shape), 528, *529*
*Bacillus anthracis*, 531, 542
*Bacillus subtilis*, 206
*Bacillus thuringiensis* toxin, 387–388, 389
Back substitutions, 488
Bacteria
anoxygenic photosynthesis, 186, 187
antibiotic resistance (*see* Antibiotic resistance)
*Bacillus thuringiensis* toxin in transgenic plants, 387–388
bioluminescence, 525, 534, 546
in bioremediation, 373, 389
cell walls, 527, 528
chlamydias, 533, *534*
conjugation, 253–255
conjugative pili, 84
cyanobacteria, 532, *533*
denitrifiers, 539
DNA replication, 269, 271, 273
effects of antibiotics on, 281, 295, 301, 527
in the evolution of life, 4
experimental studies in molecular evolution, 489–490
expression vectors for, 384
extremophiles, 532
fungal gardens and, 1169
genetic transformation, 260–261
genomes, 356, *361*, 494 (*see also* Prokaryotic genomes)
global nitrogen cycle, 750–751
Gram stains, 527–528
hadobacteria, 532
high-GC Gram-positives, 532
hyperthermophilic, 532
influence on plant uptake of nutrients, 747–748, 748–751
insertion of genes into, 376
lateral gene transfer, 496, 530
low-GC Gram positives, 530–532
metabolic pathways, 537–538
mRNA recycling in, 301
nitrifiers, 539
nitrogen fixers (*see* Nitrogen fixers)
non-oxygen electron receptors, 176
pathogenic, 541–542
proteobacteria, 534
quorum sensing, 525, 539
restriction enzymes, 315
shapes of, 528, *529*
spirochetes, 533
in the tree of life, 8
Bacteria (domain), 81, 526–527
Bacterial conjugation
control of, 254–255
description of, 253–254
in paramecia, 562
Bacteriochlorophyll, 538
Bacteriophage
ecological significance, 546
lysogenic cycle, 340–341
lytic cycle, 339–340, 341
phage λ, 340–341, 377, 379
phage T4, *543*
phage therapy, 545–546
Bacteroids, 748
Bainton, C. R., 1020–1021
Baker's yeast, 620, 623, 624, 626
*Balaenoptera musculus*, 1164
Bald eagles, *638*, 1228
Baldwin, Ian, 803, 804
Bali island, 1141, *1142*
Ball-and-stick models, *27*
Ballast water regulation, 1241
Bamboos, 792, *793*
Banana slugs, *664*
Banded iron formations, 512
Banggai cardinalfish, 1234, *1235*
Banteng, 407
Banting, Frederick, 848
Banyans, 718
Baptista, Luis, 1100
Bar-headed geese, 1010
Bark, 721, 722, *723*
Barklice, *672*
Barley, "malting," 761
Barn owls, *696*
Barnacles, 670–671, 1184
Baroreceptors, 1044, *1045*
Barr, Murray, 346
Barr bodies, 345–346
Barred tiger salamander, *691*
Barrel cactus, *720*

Barrier methods of contraception, *898*
Basal lamina, *100*
Basal metabolic rate (BMR), 826–827
Base pairing
importance to DNA replication and transcription, 65–66, *67*
in nucleic acids, 63–65
Base pairs
in DNA structure, 265, 266, *267*
exposure in DNA grooves, 266, *267*
"hot spots" for mutations, 310
Bases
acid–base reactions, 34
amino acids as, 43
defined, 34
in nucleotides, 63
properties of water as, 34–35
Basidiomata, *621,* 622
Basidiomycota, *615, 616*
*See also* Club fungi
Basidiospores, *622,* 623
Basidium, *621,* 623
Basilar membrane, *954,* 955–956
Basophils, *858*
Batesian mimicry, 1174
*Batrachochytrium dendrobatidis,* 612
Bats
echolocation, 946, 963
excretory physiology of vampire bats, 1071, 1090
number of species, *698*
as pollinators, 599, *1181*
speciation by centric fusion of chromosomes, 470, *471*
success of, 699–700
wing evolution, *423,* 452
Bay checkerspot butterfly, 1161–1162
Bayliss, William, 838
*Bazzania trilobata, 577*
Bcl-2 protein, 399–400
*Bcl2* gene, 382
*BCR–ABL* oncogene, 304, 305
BCR–ABL protein, 304
Bdelloid rotifers, 657
Beach grasses, 793
Beadle, George, 282–283
*Beagle* (HMS), 428, *429*
Beaks
BMP4 expression in birds and, 412, 424
heterometry in development, 416
*Beauveria bassiana,* 627
Beavers, *700,* 1194
Beech–maple forest, 1200
Beef cattle, 245
Beer, 623, 624
Bees
exploitation competition, 1183
as pollinators, 599, 605
*See also* Honey bees; Hymenoptera
Beetles, *674*
estimating the number of living species, 651
fungus farming, 1178
number of living species, *672*
in pollination syndromes, *1181*
Begging behavior, 1095
Behavior. *See* Animal behavior; Complex behaviors; Courtship behavior; Sexual behavior; Social behaviors
Behavioral ecology
defined, 1102
evolution of behavior, 1102–1105, *1106*
Behavioral epigenetics, 328
Behavioral genetics, 1096–1098
Behavioral isolation, 476–477
Behavioral thermoregulation, 822, *823,* 1126
Behavioral traits, 456
Behaviorism, 1094
Behring, Emil von, 862, *863,* 864
Belding's ground squirrel, 1116, *1117*
Belladonna, *604*
Belly button, 897
Belt, Thomas, 1178
Benecke, Mark, 1188
Bengal tiger, *1136*
Benguela Current, *1125*
Benign tumors, 227
Benson, Andrew, 194, 195
Benthic zone, *1139,* 1140
Benzer, Seymour, 294
Benzopyrene, 309, 311
Beriberi, *1053,* 1054
Bermuda bluegrass, 199
Best, Charles, 848
Beta blockers, 850, 853
Beta-carotene, 58
Beta (β) pleated sheet, 45, *46*
*Beta vulgaris,* 718
Beutler, Bruce, 860
Beyer, Peter, 388
Biased gene conversion, 498, *499*
Bicarbonate ions
as a base, 34
blood transport of carbon dioxide and, 1018, *1019*
buffering action in the blood, 1084–1085
in blood, 1041
in buffers, 35
in global carbon cycle, 1216
produced by pancreas, 1062
production of hydrochloric acid and, *1060,* 1061
Biceps, 995
*Bicoid* gene, 402, *403*
Bicoid protein, 402, *403*
Bicoordinate navigation, 1109–1110
Biennials, 785
Bilateral symmetry
in the animal zygote, 903
in animals, 634–635
in bilaterians and cnidarians, 643
in echinoderms and hemichordates, 680, *681*
in flowers, 597
Bilaterians
in animal phylogeny, *630*
functions of the central nervous system, 637
members of, 634
monophyly of, 643
radial cleavage and, 679
Bilayers, lipid, 73–74
*See also* Phospholipid bilayers
Bile, 1061–1062
Bile salts, 1061–1062, 1063
Bin Laden, Kahild, 318
Bin Laden, Osama, 318
Binary fission, 206–207
Bindin, 477, *885,* 886
Bindin receptors, *885,* 886
Binocular cells, 978
Binocular vision, 977–978
Binomial nomenclature, 7, 462, 463–464
BioCassava Plus, 724
Biochemical reactions
energy transformations, 145–149
role of enzymes in, 151–154
Biodiversity
economic value, 1241–1243
human activities threatening, 1232–1237
importance of understanding and appreciating, 17–18
meanings of, 1229
predicting changes in, 1230–1232
species extinctions and, 1229–1230
strategies used to protect, 1237–1245
value to human society, 1230
Biodiversity hotspots, 1237, *1238*
Bioelectric energy, *145,* 927
Biofilms, 281, 539, 540–541
Biofuels, 585
Biogeochemical cycles
carbon cycle, 1216–1218
defined, 1215
hydrologic cycle, 1215
interaction of, 1221–1223
iodine, 1221
iron, 1221
nitrogen, 1218–1219
phosphorus, 1220–1221
sulfur cycle and the burning of fossil fuels, 1219–1220
Biogeographic regions, 1141–1142
Biogeography
area phylogenies, 1143, *1144*
biogeographic regions, 1141–1142
continental drift, 1143
discontinuous distributions from vicariant or dispersal events, 1143–1145
human impact on biogeographic patterns, 1145
origins of, 1141
Bioinformatics, 354
Biological classification, 462–464
Biological clock, 774
*See also* Circadian rhythms
Biological control, 626, 627, 1164
Biological hierarchy, 9
Biological information, 5–6
Biological membranes. *See* Membranes
Biological nitrogen fixation. *See* Nitrogen fixation
Biological research
distinguishing characteristics of science, 14
generalization of discoveries, 14
quantification of data, 11
role of experiments in, 12–13
role of observation in, 11
scientific methods, 11–12
statistics and, 13–14
Biological rhythms
coordination of animal behavior with environmental cycles, 1106–1108
melatonin and, 851
Biological species concept, 468–469
Biological warfare agents, 157
Biology
biological information and genomes, 5–6
defined, 2
evolutionary tree of life, 6–9
impact on public policy, 16–17
importance for understanding ecosystems, 17
importance to modern medicine, 15–16
key characteristics of life and evolution, 2–11
levels of organization in, *8*
modern agriculture and, 15
regulation of the internal environment and, 10
understanding biodiversity and, 17–18
Bioluminescence
ATP and, *149,* 150
by dinoflagellates, 566
green fluorescent protein, 449
by *Vibrio,* 525, 534, 546
Biomass
decrease through time in detritus-based communities, 1202
distribution in communities, 1190–1192
Biomes
abiotic factors influencing, 1126–1127
aquatic, 1139–1141
boreal and temperate evergreen forests, 1129–1130
chaparral, 1134–1135
classification, 1126
cold deserts, 1133–1134
defined, 1126
distribution, 1126, *1127*
hot deserts, 1132–1133
temperate deciduous forest, 1130–1131
temperate grasslands, 1131–1132
thorn forests and tropical savannas, 1135–1136
tropical deciduous forest, 1136–1137
tropical rainforests, 1137–1138
tundra, 1128–1129
"Bioprospecting," 500–501
"Bioreactors," 385–386
Bioremediation
microorganisms and, 373, 389
of oil spills, 373
phytoremediation, 811, *812*
use of fungi in, 625
Biosynthesis
food as a source of carbon skeletons for, 1051–1052
nutrients as the basis for, 10
Biota, 514

Biotechnology
agricultural applications, 386–388
defined, 383
expression vectors, 384
genetically modified organisms in bioremediation, 373
history of, 383–384
medical applications, 384–386
patents and, 373
public concerns, 388–389
*See also* Recombinant DNA technology
Bioterrorism, 531
Biotic ecosystem components, 1122
Biotic potential, 1158
Biotin, *156,* 1057
Bipedal locomotion, 702–703
Bipolar cells, *925, 962,* 963
"Bird flu" virus, *543*
Birds
actions of prolactin in, *838*
advantages of flocking, 1116, *1117*
altricial and precocial young, 642
avian malaria in the Hawaiian Islands, 1236
begging behavior of chicks, 1095
BMP4 and beak development, 412, 424
circulatory system, 1029, 1030
daily torpor, 830
dinosaur origins of, 693, 694
disruptive selection in bill size, 440
extraembryonic membranes, 918
fat as stored energy, 1050
feathers and flight, 695–696
foveae, 962
gastrulation, 913–914
"helping at the nest," 1115
heterometry in beak development, 416
hindlimb development in ducks and chickens, 417
imprinting, 1099
monitoring, *1150,* 1151
nest parasitism, 1093, 1102, 1117
palaeognaths and neognaths, 694–695
as pollinators, 599
salt balance regulation and nasal salt glands, 1073
seed dispersal and, 696
sex determination in, 249
sexual reproduction, 888–889
sexual selection and speciation rates, 480–481
shivering heat production, 827
unidirectional ventilation in gas exchange, 1010–1012
wing evolution, *423,* 452
*See also* Hummingbirds; Songbirds
Birds of paradise, 480–481
Birdsong
factors affecting song acquisition, 1099–1101
hormonal control of song expression, 1101
Birth control. *See* Contraception
Birth control pills, 896, *898*
Birth process
effect of oxytocin on, 843
positive feedback in, 817
the uterus in, 843, 896–897
Bishop's goutweed, *597*
Bison, *1064, 1131,* 1239
*Bison anitquus, 1229*
*Bison bison, 1239*
1,3-Bisphosphoglycerate, 170
2,3-Bisphosphoglyceric acid (BPG), 1018
*Bithorax* gene, 404
Bitter taste, 951
Bivalents, 219
Bivalves, 662, *663*
Black bears, *1130*
Black-bellied seedcracker, 440
Black bread mold, 619
Black-eyed Susans, *599*
Black lace cactus, *603*
Black-legged tick (deer tick), *1151,* 1152
Black rockfish, 1163
Black stem rust, 612
Blackberry, *601*
Bladder, 1080
Bladder cancers, 131, *132*
Blastocoel
in the avian blastula, 914
formation and characteristics of, 904, 906, 907
in frog gastrulation, *910*
in sea urchin gastrulation, 909
Blastocyst
defined, 906
implantation in humans, 892, 893, *907*
pluripotent stem cells, 408–409
secretion of human chorionic gonadotropin, 896
Blastoderm, 905
Blastodisc, 905, 913, *914*
Blastomeres
defined, 904
determination, 906–907
in mammalian cleavage, 906
in sea urchin gastrulation, 909
Blastopore
defined, 633
in deuterostomes, 634, 679
in protostomes, 634, 652
in sea urchin gastrulation, 909
*See also* Dorsal lip of the blastopore
Blastula
defined, 904
determination of blastomeres in, 906–907
gastrulation, 908–915 (*see also* Gastrulation)
role of cleavage in forming, 904–905
Blattodea, *672*
Bleak, 1140
"Blebs," 226
Bleeding shiner, *472*
Blending inheritance, 233, 234
Blenny, *1174*
Block, Barbara, *16, 1036,* 1037
Blocks to polyspermy, *885,* 886–887
Blood
ABO blood groups, 243
carbonic/bicarbonate buffer system, 35, 1084–1085
composition, 1037, *1038*
as connective tissue, 819
countercurrent heat exchange in endotherms, 828
countercurrent heat exchange in "hot" fish, 825
filtration in Bowman's capsule, 1078–1079
hematocrit, 1037–1038
hormonal regulation of calcium levels, 847–848
hormonal regulation of phosphate levels, 848
in the mammalian defense system, 857
platelets, 1039
red blood cells, 1038
regulation of osmolarity by antidiuretic hormone, 1088–1089
thermoregulation through the skin and, 824–825, 828
transport of respiratory gases, 1016–1019
Blood carbon dioxide
in blood pressure regulation, 1045
effect on autoregulation of blood flow, 1044
transport of, 1018–1019
"Blood chamber," 652
Blood clotting, 384–385, 1039
Blood clotting proteins, 323
Blood disorders
hemoglobin C disease, 312
sickle-cell disease, 306, *307,* 312
Blood factor VIII, 323
Blood flow
in arteries and arterioles, 1039, *1040*
control and regulation of, 1043–1044
countercurrent flow in fish gills, 1010, *1011*
through capillary beds, 1040–1041
Blood glucose
hormonal regulation of, 848–849, 1066–1067
maintenance of levels during physical exercise, 180–181
regulatory pathways in glucose catabolism, 181–182
Blood groups, 53, 243
Blood meals, vampire bats and, 1071
Blood pH
effect on hemoglobin's binding of oxygen, 1018
partial pressure of carbon dioxide and, 1021
Blood plasma
composition of, *1038*
defined, 1026
extracellular fluid, 816
filtration in tubule capillaries, 1072
loss of proteins during starvation, 1050
oxygen-carrying capacity, 1016
separation of blood cells from, 1037
synthesis of plasma proteins by the liver, 1065
transport of carbon dioxide, 1018, *1019*
Blood pressure
in capillary beds, 1041
control and regulation of blood flow, 1043–1044
in fish, 1028
measuring, 1032, *1033*
Blood pressure regulation
angiotensin and, 1087
antidiuretic hormone and, 1088–1089
atrial natriuretic peptide and, 1090
hormonal and neural roles in, 1044–1045
vasopressin and, 843
Blood respiratory gases
in blood pressure regulation, 1045
effect on autoregulation of blood flow, 1044
transport, 1016–1019
"Blood sugar." *See* Blood glucose
Blood vessels
anatomy, *1040*
arteries and arterioles, 1039, *1040*
autoregulation of blood flow through, 1044
in blood pressure regulation, 1044, *1045*
capillary beds, 1039–1041
carotid and aortic bodies, 1021–1022
in closed circulatory systems, 1026
clotting and, 1039
constriction in the fight-or-flight response, 837
dilation in penile erection, 890–891
physical stresses in the lungs of snorkeling elephants, 1005, 1022
smooth muscle in (*see* Vascular smooth muscle)
veins, 1041–1042
in vertebrate circulatory systems, 1027–1028
Blood–brain barrier, 926, 1041
"Blooms"
algal, 1219
diatoms, 556
red tides, 549, 564
viral, 546
Blowflies, *674,* 1203
Blubber, 697
"Blue" cheese, *622*
Blue-green bacteria. *See* Cyanobacteria
Blue-light receptors, 771–772, 788
Blue whales, 637, 1164
Bluebells, *462*
Bluebottle flies, 1203
Bluefin tuna, 16–17, 825, *1036,* 1037
Bluegill sunfish, 1105, *1106*
BMP4. *See* Bone morphogenetic protein 4
*BMP4* gene, 417

Boas, 423
Body cavities
  animal movement and, 635–636
  hydrostatic skeletons, 999
  types, 635
Body fat
  buildup during excess food consumption, 1051
  *See also* Adipose tissues
Body plans
  annelids, *660*
  appendages, 636–637
  arrow worms, 655
  arthropods, *655*
  body cavities and movement, 635–636
  chelicerates, 668
  crustaceans, 671
  defined, 634
  insects, 671, *672*
  key features of, 634
  mollusks, *663*
  nematodes, *666*
  nervous systems, 637
  of plants, 709–712
  protostomes, 652
  segmentation, 636
  sponges, 644
  symmetry, 634–635
  of vertebrates, *685*
Body segmentation
  in animal development, 916, *917*
  determination in *Drosophila melanogaster,* 401–405
  regulation by Hox genes, 916–917
Body size
  brain size and, 973
  chromosome number and, 218
  effect of atmospheric oxygen levels on insect body size, 505, 513, 514, 522
  effect on characteristic species density, 1160
  range in birds, 696
  range in mammals, 696–697
  relationship to basal metabolic rate, 826–827
  respiratory gas diffusion and, 1007
  variation between dog breeds, 352
Body temperature
  acclimatization, 821
  classification system, 822
  impact of changes in, 821
  "point restriction" coat patterns and, 245
  regulation (*see* Thermoregulation)
Bogs, 1140
Bohr effect, 1018
Bohr models, *27*
*Boiga irregularis,* 1235
"Bolting," 761
Bombardier beetles, 1173
*Bombina bombina,* 479
*Bombina variegata,* 479
*Bombus lucorum,* 476
*Bombycilla cedrorum, 696*
*Bombyx mori,* 950
Bonding behavior
  factors affecting in voles, 125
  oxytocin and, 843
Bone
  as connective tissue, 819
  development, 1001
  hormonal control of turnover, 847, 848
  interaction with muscle at joints, 1001–1002
  osteoporosis and, 1000–1001
  process of growth, 416–417
  structure of, 1000
  of theropod dinosaurs, 695
Bone marrow
  formation of white blood cells in, 858
  in the lymphatic system, *857*
  red blood cell production, 1038
  types of multipotent stem cells in, 408
Bone morphogenetic protein 4 (BMP4)
  beak development in birds, 412, 424
  loss of foot webbing in nonaquatic birds, 417
Bone morphogenetic proteins (BMPs), 913
Bonner, James, 788, 789
Booklice, *672*
Boreal forest, 1129–1130
Boreal owl, *1129*
Borlaug, Norman, 756
Boron, *741*
Borthwick, Harry, 772
Boston Marathon, 815
*Bothus lunatus, 445*
Botox®, 542, 936
*Botrytis fabae, 801*
Bottle cells, 909, *910*
Botulinum toxin, 936
Botulism, 542
Boulding, Kenneth, 1245
Bowman's capsules, 1078–1079, *1080,* 1081
Boyer, Herbert, 374
Brachiopods
  anatomical characteristics, *652*
  description of, 658–659
  major subgroups and number of living species, *632*
Bracket fungi, 622, 623
*Bradypus variegatus, 1137*
Brain
  alternative splicing and, 347
  blood–brain barrier, 926, 1041
  complexity in vertebrates, 943
  development in mammals, 968–969
  diversity in size and complexity, *943*
  early development, 916
  learning and memory areas, 981–982
  nervous system information flow and, 968
  overinhibition in, 924, 939, 943
  regulation of food intake, 1067–1068
  sleep and dreaming, 978–980
  in small nervous systems, 941
  structure and function in mammals, 969–973
  *See also* Human brain
"Brain coral," *647*
Brain imaging, 970
Brain size
  body size and, 973
  evolution in humans, 704
Brainstem
  changes during sleep, 979
  control of breathing, 1019–1020
  development, 968, *969*
  structure and function, 969–970
Branching, in rhyniophytes, 580
*Branchiostoma lanceolatum, 683*
Branchipods, 670
*Brassica oleracea, 432,* 724
Brassinosteroids, *759,* 771
Braxton Hicks contractions, 896, 897
Brazilian mastiff, *352*
*BRCA1* gene, 245, 246
Bread molds, 282–283, 619
Bread wheat, 225, *226*
Breast cancer
  DNA microarray technology and diagnosis, 383
  expressivity of the *BRCA1* allele and, 246
  human SNP scans, *368*
  oncogene proteins, 228
  penetrance of the *BRCA1* gene and, 245
Breast cancer treatment
  drugs targeting the estrogen receptor, 325
  Herceptin, 229
  immunotherapy, 871
Breast duct cells, 333
Breast milk, 843
Breathing
  effect on venous blood flow, 1042
  regulation of, 1019–1022
  *See also* Respiratory gas exchange; Ventilation
Breeding seasons, 476
Brenner, Sydney, 362
*Brevipalpus phoenicis, 669*
Brewer's yeast, 620, 623, 624, 626
Briggs, Robert, 406
Bright-field microscopy, 80
Brine shrimp, *1072,* 1073
Brines, 536
Bristlecone pine, 593
Brittle stars, 681, *682*
Brno monastery, 233
Broad fish tapeworm, *641*
Broca's area, 981
Broccoli, 724
Brock, Thomas, 277–278
Bromelain, *604*
Bromeliads, 678
Bronchi, 1013, *1014*
Bronchioles, 1013, *1014*
Brood parasitism, 1093, 1102, 1117
Brown, Patrick, 742, 743
Brown algae, 556
Brown fat
  in adults, 834
  characteristics of, *819*
  conversion of white fat into, 834, 839
  heat energy from, 165, 174
  in nonshivering heat production, 827–828
  UCP1 and weight loss, 182
Brown-headed cowbirds, 1093, 1233
Brown molds, 624
Brown tree snake, 1235
Brussels sprouts, 724
Bryophyta, *574*
Bryozoans
  anatomical characteristics, *652*
  colonies, *641*
  description of, 656
  lophophores, *654*
  major subgroups and number of living species, *632*
Bubonic plague, 534, 542
Buckley, Hannah, 1197
Bud scales, *721,* 722
Budding, 562, 620, 881
Budding yeast, 362
  *See also Saccharomyces cerevisiae*
Buds
  defined, 709
  of mosses, *576, 577*
  of twigs, *721,* 722
Buffers
  blood buffer systems, 1084–1085
  properties of, 35–36
*Bufo marinus,* 1164, 1235
*Bufo periglenes, 691*
Bulbourethral glands, 890
Bulbs, 793
Bulbus arteriosus, 1028
Bulk flow
  defined, 727–728
  in phloem, 728, 735–738
  in xylem, 728
Bull, James, 457
Bull trout, 1228
Bullhorn acacia, 1178, *1179*
Bumblebees, 1183
Bundle of His, 1034, *1035*
Bundle sheath cells, *198,* 199, *721*
Burgess Shale, 516
Burs, 784
Bush monkeyflower, 598, *599*
Butane, *41*
*Buteo galapagoensis, 2*
Butterflies, 442–443
  *See also* Lepidopterans; *specific butterflies*

## C

C horizon, 745
C ring, *83*
$C_3$ plants, 198, 199, 200, 202
$C_4$ plants, 198–200, 202
Cabbage plants, 724, 761
Cactaceae, 1133
Cacti
  adaptations to very dry conditions, 807, 808
  number of species, *603*
  stomatal function in, 734
Cactus finches, *474,* 1154–1155
Caddisflies, *672, 674*
Caecilians, 690, *691,* 692
*Caenorhabditis elegans*
  apoptosis studies in development, 399
  genomic information, *361,* 362, *363*
  microRNA, 347
  as model organism, 666

vulval determination, 396, *397*, 398
*See also* Nematodes
Caffeine, 128
Caimans, 693–694, *1106*
Calcareous sponges, *630*, 643–644
Calciferol, 848, *1053*, 1054
Calcitonin
actions of, *842*
influence on blood levels of calcium, 847
thyroid source, 845, *846*
Calcitriol, 848
Calcium
in animal nutrition, 1052–1053
in bone, 1000
hormonal regulation of blood calcium, 847–848
in plant nutrition, *741*
Calcium carbonate
in limestone deposits, 557, 565
in oceanic carbon stores, 1217
otoliths, *956*, 957
in the shells of foraminiferans, 557
Calcium-induced calcium release, 992–993
Calcium ion channels
in cardiac muscle contraction, 992–993, 1033, 1034
in cardiac pacemaker cells, 1033, 1034
in the hyperpolarization of neurons at the onset of sleep, 979
in skeletal muscle contraction, 990
*See also* Voltage-gated calcium channels
Calcium ions
in cardiac muscle contraction, 992–993, 1033, 1034
in cardiac pacemaker cells, 1033, 1034
cycling in cardiac muscle effects heartbeat, 1034, *1036*, 1037
in insect flight muscle contraction, 998
neurotransmitter release at neuromuscular junctions, 936, *937*
roles in animal fertilization, 886, 887
as second messengers, 135, 136
in skeletal muscle contraction, 989–990, *991*
in smooth muscle contraction, 993
in smooth muscle relaxation, 136
tetanic muscle contractions and, 994–995
Calcium phosphate, 21, 1000
Calcium pumps
in sarcoplasmic reticulum of cardiac muscle, 1034
in skeletal muscle contraction, 989–990
California condor, 1244
California plantain, 1161
Callaway, John, *1240*
*Calliarthron*, *572*
Callose, 799
Callus (calli), 405
*Callyspongia plicifera*, *1007*
Calmodulin, 412, 424, 993
*Caloplaca*, *613*
Calorie (Cal), 1049
Calories (cal), 147
Calvin, Melvin, 194, 195
Calvin cycle
in $C_4$ plants, 199
in CAM plants, 200
linkage of photosynthesis and respiration in plants, 201
in photosynthesis, *188*
processes and products, 194—196
radioisotope experiments, 193–194, 195
stimulation by light, 196–197
Calyx, 597
CAM plants, 200
Camalexin, 801
*Camarasaurus*, 21
*Camarhynchus pallidus*, *474*
*Camarhynchus parvulus*, *474*, *1150*
*Camarhynchus pauper*, *474*
*Camarhynchus psittacula*, *474*
Cambrian explosion, 516
Cambrian period, *506–507*, 516, *517*
Camembert cheese, 624
cAMP receptor protein (CRP), 332
*Campanula rotundifolia*, *462*
*Campephilus principalis*, 1230, *1231*
Canary Islands, 440
Canarygrass, 762, 763, 764
Canavanine, 803, 805
Cancellous bone, 1001
Cancer drugs
cisplatin, 259, 268, 278
competitive inhibitors of enzymes, 158
disruptors of microtubule dynamics, 96
targeting the cell cycle, 228–229
Cancer treatment
immunotherapy, 871
Ras inhibitors, 131
targeting the cell cycle, 228–229
use of multipotent stem cells, 408
Cancers
affecting Tasmanian devils, 232
blocking of apoptosis, 382
characteristics of cancer cells, 227
chronic myelogenous leukemia, 304, 305
DNA methylation and, 344
HeLa cells, 205
histone deacetylation and, 344
molecular changes in cancer cells, 228
from mutations to somatic cells, 310–311
normal cell death and, 226
prevalence in the United States, 227
RNA retroviruses and, 544
signal transduction pathways and, 131, *132*
somatic mutations and, 314
telomerase and, 275
unregulated cell division and, 227–229
virus-interrupted breakdown of cyclins and, 349
*See also specific cancers*
Candelabra, *1132*
*Candida albicans*, 612
Cane toad, 1164, 1235
Canines, 1055
*Canis lupus familiaris*, 352
*See also* Dogs
*Canis simensis*, 1242
5′-Cap, 348
Cap-binding complex, 293
Cape sugarbird, *1134*
Capecchi, Mario, 382
Capillary action, 731
Capillary beds
autoregulation of blood flow, 1044
blood–brain barrier, 1041
effect of autoregulatory actions on mean arterial pressure, 1043
structure and function, 1028, 1039–1041
*Capra pyrenaica*, *1134*
Capsid, 340
Capsule, of prokaryotes, *82*, 83
Captive breeding programs, 1244
Carapace, 671
"Carbo-loading," 997
Carbohydrases, 1058
Carbohydrates
biochemical roles, 51
breakdown by digestive enzymes, 1057, 1058
catabolic interconversions, 179
categories of, 51
chemically modified, 55
energy yield, 1049, *1050*
general formula, 51
glycosidic linkages, 53
membrane-associated, *106*, 109
monomer components, 40
monosaccharides, 52
polysaccharides, 53–54
production in photosynthesis, 186, 193–197
proportions in living organisms, *41*
as stored energy in the animal body, 1050
Carbon
covalent bonding capability, *27*
electronegativity, *28*
isotopes, 23
mass number, 22
octet rule for molecule formation, 25
sources for saprobic fungi, 611
Carbon-12, 507
Carbon-14, 507
Carbon cycling
fungi in, 611
global amounts released by fire, 1214
global cycle, 1216–1218
Carbon dioxide
in the bicarbonate buffer system, 1084, 1085
blood transport, 1018–1019
fixation in photosynthesis (*see* Carbon dioxide fixation)
in the global carbon cycle, 1216–1218
impact on plant stomatal density, 734
produced by yeast in food and drink production, 623
produced during glucose catabolism, 169, 170, 171
respiratory exchange by diffusion, 1008
stomatal function in plant uptake of, 732–734
*See also* Atmospheric carbon dioxide
Carbon dioxide fixation
in $C_3$ and $C_4$ plants, 198–200
in CAM plants, 200
evolution of pathways in, 200
in metabolic interactions in plants, 201
in photosynthesis, 186, 193–197
Carbon isotopes, 74
Carbon monoxide, 1017
Carbon skeletons, 1051–1052
Carbonate ions, 1216
Carbonic acid
bicarbonate ion and, 34
in blood transport of carbon dioxide, 1018, *1019*
carbonic/bicarbonate buffering system, 35
in cation exchange by roots, 746
in ocean acidification, 1217
Carbonic anhydrase, 1018, *1019*, *1060*, 1061
Carboniferous period
atmospheric oxygen levels, 512–513
changes on Earth and major events in life, *506–507*
characteristics of life during, 520
decline of fungi in, 611
evolution of leaves, 583
gigantic insects in, 505
vascular plants in, 580
Carboxyl group
acid properties, 34
of chemically modified carbohydrates, 55
effect of pH on, 161
peptide linkages in amino acids, 43–44, *45*
properties of, *40*
Carboxylic acids, 70
Carboxypeptidase, 154, *1062*
*Carcharodon megalodon*, *687*
Carcinogens, 310, 311
Cardiac action potentials
coordination of muscle contraction, 1034, *1035*
electrocardiograms, 1035–1036
pacemaker cells and cardiac muscle contraction, 1032–1034
Cardiac cycle, 1031–1032
Cardiac muscle
atherosclerosis and, 1043
calcium ion cycling in "hot" fish, *1036*, 1037
contraction, 991–993, 1032–1034, *1035*
Frank–Starling law, 1042
functions of, 817, 987

integration of anabolism and catabolism during exercise, 180–181
pacemaker cells, 1032–1034 (*see also* Pacemaker cells)
structure, 991
Cardiac output (CO)
mean arterial pressure and, 1043
regulation of, 1044, *1045*
Cardiovascular disease, 1042–1043
Cardon, *1132*
Caring behavior, 141
Carnivores
defined, 1054
ingestion and digestion of food, 1055
number of species, *698*
teeth, *1055*
Carnivorous plants, 751–752
β-Carotene
accessory pigment in photosynthesis, 190
in genetically modified plants, 388
Carotenoids
accessory pigments in photosynthesis, 190
in brown algae, 556
in light-harvesting complexes, *191*
in red algae, 571
structure and function, 58
Carotid arteries
carotid bodies, 1021–1022, 1045
stretch receptors in the regulation of blood pressure, 1088–1089
Carotid bodies, 1021–1022, 1045
Carpels
anatomy of, *591*
evolution of, 598
function of, *779*
in the structure of flowers, 596–597
Carpenter bees, 1183
*Carpolestes*, 701
Carrier-mediated transport, 117
Carrier proteins
defined, 115
in facilitated diffusion, 117
Carrier viruses, 323, 324
Carrion-feeding beetles, 1203
Carrots, 405–406, 718
Carrying capacity, 1158, 1165
Cartilage
as connective tissue, 819
galactosamine in, 55
in vertebrate skeletons, 999–1000
Cartilage bones, 1001
Cartilaginous fishes. *See* Chondrichthyans
Cartilaginous skeletons, 1000
Cascade mountains, 509
Casparian strip, 730, 744
Caspases, 226, *227*, 399–400
Cassava, 708, 724
Cassowaries, 694
*Castercantha*, 39
*Castor canadensis*, *700*
Catabolic interconversions, 179–180
Catabolic reactions
energy changes during, 146
exergonic reactions, 147, *148*
inducible systems of regulation in prokaryotes, 332
integration of anabolism and catabolism, 180–181
linkage to anabolic reactions, 146, 179–180
*See also* Glucose catabolism
Catabolite repression, 332
Catalase, 156, 176
Catalysts
function of, 151
nonspecific nature of, 152
RNA and the origin of life, 71–72, *73*
*See also* Enzymes
*Catasetum*, 588, 605
*Catasetum macrocarpum*, 605
β-Catenine, 904, 911, 912
Caterpillars, 1176
Catfish, 1140
Cations, 29
Cats
point restriction phenotype, 306
retinal ganglion receptive fields, 975, 976
*Toxoplasma* and, 554
Cattails, 1140
Cattle, 245, 407, 440
Cattle egrets, 1171
Caudal fins, 687
Cauliflower, 724
Cause and effect, 98
*Cavanillesia platanifolia*, *1136*
Cayman crab fly, 1231
CD4 protein, 872
cDNA. *See* Complementary DNA
Cecal material, 1064
Cech, Thomas, 72
Cecum, 1064
CED-3 protein, 399
CED-4 protein, 399, 400
CED-9 protein, 399, 400
Cedar waxwings, *696*
Celestial navigation, 1109–1110, *1111*
Celiac ganglion, *974*
Cell adhesion
defined, 110
to the extracellular matrix, 111–113
homotypic and heterotypic, 111
roles of proteins and carbohydrates in, 111
significance in multicellular organisms, 110–111
types of cell junctions in, 111, *112*
Cell binding. *See* Cell adhesion
Cell body, of neurons, 925
Cell cycle
abnormal regulation in cancer cells, 228
cancer treatments targeting, 228–229
defined, 208
duration of, 208
gain-of-function mutations in tumor suppressor and, 306
internal signals controlling, 208–211
mitosis, 211–217
summary of events, *217*
Cell cycle checkpoints, 210–211, 215
Cell determination. *See* Determination
Cell differentiation. *See* Differentiation
Cell division
in asexual reproduction, 217
cancers and, 227–229
cell differentiation in the embryo and, 399
control of, 208–211
in eukaryotes, 207–211
HeLa cells, 205, 229
important consequences of, *206*
key events in, 206
mitosis, 211–217 (*see also* Mitosis)
in morphogenesis, 394
in prokaryotes, 206–207
in sexual reproduction, 217, 218–219 (*see also* Meiosis)
*See also* Cytokinesis
Cell expansion
auxin-induced expansion in plants, 766–767
in morphogenesis, 394
Cell fate
defined, 394
processes determining, 395–396, *397* (*see also* Determination)
restriction during development, 394
Cell fractionation, 84, 85
Cell fusion experiments, 208–209
Cell junctions, 111, *112*
Cell movement
cilia and flagella, 96–98
integrins and, 112, *113*
microfilaments and, 95, *96*, 98, 99
in morphogenesis, 394
Cell plate, 710
Cell potency, 394
Cell recognition
defined, 110
significance in multicellular organisms, 110–111
Cell theory, 78
Cell-to-cell communication
evolution of multicellularity and, 140–141
gap junctions, 139–140
induction in cell fate determination, 395–396, *397*
plasmodesmata, *139*, 140
*See also* Intercellular signaling; Signal transduction pathways
Cell walls
of bacteria, 527, 528
of diatoms, 555
loss of in the eukaryotic condition, 551
of plant cells (*see* Plant cell walls)
of prokaryotes, *82*, 83
Cellobiose, *53*
Cells
animal cell structure, *86* (*see also* Animal cells)
cell theory, 78
death of, 225–227
diffusion within, 114
effect of osmosis on volume, 1072
effects of signal transduction pathways on, 137–139
experiments on the origin of, 73–74
microscopes, 79, 80–81
plant cell structure, *87* (*see also* Plant cells)
plasma membrane, 79–81
prokaryotic and eukaryotic, 81 (*see also* Eukaryotic cells; Prokaryotes)
regulation of protein longevity in, 348–349
regulation of the internal environment, 10
responses to intercellular signaling, 125, 126–127
surface area-to-volume ratio, 78–79
types of work done by, 10
"Cellular drinking." *See* Pinocytosis
"Cellular eating." *See* Phagocytosis
Cellular immune response
activation and effector phases, *873*, 874
binding of antigens to T cell receptors, 871–872
description of, 863–864, *865*
effector T cells in, 865, 866, 871, 872, *873*, 874
MHC proteins in, 871, 872, *873*, 874
presenting antigens to T cell receptors, 872
suppression by regulatory T cells, 874
in tissue transplants, 874
Cellular respiration
defined, 91
energy yield from cellular respiration and fermentation compared, 178
linkage to photosynthesis in plants, 201
mitochondria and, 91
overview, 166–167, 168
Cellular slime molds, 561
Cellular specialization
in eukaryotes, 4
significance to multicellular life, 9
Cellular therapy, role of transcription factors in, 336, *337*, 338
Cellulases, 1063
Cellulose
digestion in herbivores, 1063–1064
in plant cell walls, 710, 711
plant guard cell function and, 733
structure and function, 54
Cenozoic era, *506–507*, 521–522
"Centers of imminent extinction," 1237, *1238*
Centimorgans (cM), 248
Centipedes, 669–670
"Central dogma," 285
Central nervous system (CNS)
anatomical organization, 968–969

components of, 941
defined, 968
development in humans, 968–969
functions of, 637
glucose as the fuel for, 1067
information flow into and out of, 969
regulation of breathing, 1019–1022
self-perception and, 983
structure and function in mammals, 969–973
Central pattern generators, 969
Central sulcus, 971
Central vacuole, 710
Centric fusion, 470, *471*
Centrioles
in animal cells, *86*
sperm contribution to the zygote, 903
structure and function in cell division, 212, *215*
*Centrocercus urophasianus, 1105*
Centromeres
alignment during meiotic metaphase II, *220*
alignment during mitotic metaphase, *215*
comparison between mitosis and meiosis, *223*
of sister chromatids in mitosis, 212, 213, *216*
Centrosomes
of the animal zygote, 903
structure and function in cell division, 212–213, *214, 215*
Century plants, 785
Cephalization, 634–635
Cephalochordates, 684
Cephalopods
description of, 662, *663,* 664
image-forming eye, 958, *959*
Cephalothorax, 668
*Ceramium, 572*
Cercozoans, 557
Cerebellum, 968, *969*
Cerebral cortex
description of, 969
frontal lobe, 971–972
in human brain evolution, 973
occipital lobe, 973
parietal lobe, 972–973
self-perception and, 983
structure and function, 970–971
temporal lobe, 971
Cerebral hemispheres, 969, 980–981
Cerebrum
cerebral hemispheres, 969
structure and function, 970–973
*Certhidea olivacea, 474*
Cervical caps, *898*
Cervical vertebrae, 418
Cervix
in humans, 892, *893*
in labor and childbirth, 896, *897*
Cesarean section, 540
Cetaceans
blubber, 697
evolution, 700
number of species, *698*
*See also* Whales
Cetartiodactyla, *698*
CFCs. *See* Chlorinated fluorocarbons
*CFTR* gene, 318
*Chaetopleura angulata, 663*
Chagas' disease, *559*
Chain, Ernst, 608
Chakrabarty, Ananda, 373
Chalfie, Martin, 449
Challenger Deep, 557
Chaneton, Enrique, *1162*
Channel proteins, 115–116
"Chaos amoeba," *559*
*Chaos carolinensis, 559*
Chaparral, 1134–1135, 1146
*See also* Fynbos
Chaperone proteins, 51, 131
Chaperonins, 810
*Chara vulgaris, 573*
Character displacement, 1183
Characters
defined, 234
of a phenotype, 431
*Charcharodon charcharis, 688*
Chargaff, Erwin, 264
Chargaff's rule, 264, 265, 266
Chase, Martha, 261–263
*Chasmosaurus belli, 519*
Cheese skipper fly, 1188, 1203
Cheeses, molds and, *622,* 624
Cheetah Conservation Fund, 75
Cheetahs, 62, 75
*Cheirurus ingricus, 668*
Chelicerae, 668, 669
Chelicerates, *632,* 667, 668–669, 675
*Chelonia mydas, 888*
*Chelonoidis nigra, 2*
*Chelonoidis nigra abingdonii, 694*
Chemical bonds
covalent bonds, 26–28
defined, 26
hydrogen bonds, 30
hydrophobic interactions, 30
ionic attractions, 28–29
phosphoric acid anhydride bond, 150
van der Waals forces, 30
Chemical defenses
of prey, 1173
*See also* Plant chemical defenses
Chemical energy
in biological systems, *145*
released during glucose oxidation, 166–169
Chemical equilibrium
defined, 148
enzymes and, 153
free energy and, 148–149
Chemical evolution
hypotheses on the emergence of polymers, 71
in the origin of life, 3
prebiotic synthesis experiments, 69–71
Chemical fertilizers
ecological impact of, 740, 1207, 1219, 1220–1221
impact on the global nitrogen cycle, 1218–1219
Chemical reactions
defined, 31
energy and, 31
energy transformations, 145–149
law of mass action, 35–36
qualitative and quantitative analyses, 33
reactants and products, 31, 145
reversible, 34, 35–36
role of electrons in, 24, 25
role of enzymes in, 151–154
Chemical signaling systems
receptor proteins, 127–131
"responses" to, 125, 126–127
types and sources of signals, 126
*See also* Signal transduction pathways
Chemical signals
in animal communication, 1110–1111
in animals, 835–836 (*see also* Animal hormones)
in slime molds, 838
Chemical synapses, 926, 936–939
*See also* Synapses
Chemical warfare agents, 940
Chemical weathering, 745–746
Chemically gated ion channels, 930, 937
Chemiosmosis
experimental demonstration of, 174, 175
mechanism for ATP synthesis, 174
in oxidative phosphorylation, 171
photophosphorylation and ATP synthesis, 192–193
proton-motive force and, 173–174
Chemoautotrophs, 538, 1189
Chemoheterotrophs, 538, 539
Chemoreceptors
in blood pressure regulation, 1044, 1045
defined, 949
detection of blood levels of respiratory gases, 1021–1022
detection of pheromones, 950
influence on ion channels, 947, *948*
in olfaction, 949–950
in taste buds, 951
in the vomeronasal organ, 950–951
Chemotherapy, 51
Chengjiang, 516
*Chenopodium album,* 1200
Chernobyl nuclear power plant, 811
Chestnut blight, 622
Chewing, 686
Chiasmata, 219, *220, 222*
Chickens
BMP4 and beak development, 424
extraembryonic membranes, 918
gastrulation, 913–914
genomic information, *361*
hindlimb development, 417
Chief cells, 1060
Chihuahua, 352
Childbirth, 896–897
Chimaeras, 687, *688*
Chimpanzees
comparative genomics, 366
foraging behavior for essential minerals, 1105
origin of, 702
skull, 704
China
artemisinin treatment for malaria, 797
stem cell therapy in, 102
Chiroptera, *698*
Chitin
as an elicitor of plant defenses, 799
in exoskeletons, 655, 999
in fungi, 609
structure, *55*
Chitinase, 801
Chitons, 662, *663*
*Chlamydia psittaci, 534*
Chlamydias, 533, *534*
*Chlamydomonas,* 140, 141
*Chlorella,* 14
Chloride ion channels
actions at inhibitory synapses, 938
reduction of overinhibition in the mouse brain, 943
Chloride ions
actions at inhibitory synapses, 938
plant guard cell function and, 733
properties of, 29
transport by nasal salt glands, 1073
uptake by halophytes, 811
Chlorinated fluorocarbons (CFCs), 311, 1211–1212
Chlorine
in animal nutrition, *1052*
electronegativity, *28*
ionic attraction, 28–29
in plant nutrition, *741*
Chlorophyll
absorption of light energy, 188–190
anabolic interconversions and, 180
mutations in plants, 252
Chlorophyll *a*
in cyanobacteria, 532, 538
molecular structure, 190
photochemical changes following light absorption, 191
in photosystems, 190
in red algae, 571, 572
Chlorophyll *b,* 190, *191*
Chlorophytes, 140–141, 572, *573*
Chloroplast DNA (cpDNA), 456
Chloroplasts
carbon dioxide fixation in, 193–197 (*see also* Carbon dioxide fixation)
in cercozoans, 557
endosymbiotic origins, 102, *550,* 551–552
in glaucophytes, 571, *572*
of hornworts, *578*
photophosphorylation and ATP synthesis, 192–193
in photorespiration, 197, *198*
plant cells, *87*
sites of photosynthesis in, *188*
structure and function, 92, *93*

transposition of genes to the nucleus, 366
Choanocytes, *633*, 644
Choanoflagellates, 609, 631, *633*, 644
"Chocolate spot" fungus, *801*
Cholecystokinin (CCK), 1061, 1065
Cholera, 534, 542, 1063
Cholesterol
in the absorption of fats in the small intestine, *1062*, 1063
in atherosclerosis, 1042
biological membranes and, *106*, 107
familial hypercholesterolemia, 121
mechanism of uptake by cells, 121
structure and function, 58
synthesis of calciferol from, 848
in the synthesis of steroid hormones, *836*
Cholinergic neurons, *974*, 975
Chondrichthyans
characteristics of, 687, *688*
excretion of urea, 1074
regulation of ionic composition of extracellular fluid, 1073
salt and water balance regulation, 1077
Chondrocytes, 416–417
Chordamesoderm, *910*, 915–916
Chordates
in animal phylogeny, *630*
derived structures in development, 683–684
evolutionary relationships, *455*
major clades, 683
members of, 679
neurulation, 915–916
tunicates and lancelets, 683, 684
*See also* Vertebrates
Chordin, 916
Chorion
in the chicken egg, 918
in placental mammals, 919
Chorionic villus, *907*
Chorionic villus sampling, 321
Chown, S. L., 1199
Christmas tree worm, *638*
*Chroicocephalus novaehollandiae*, 412
Chromatids
crossing over and genetic exchange, 247–249, *250*
differences between chromatids and chromosomes, 215
events in meiosis, *220–221*
exchanges during meiosis I, 219–220, *222*
*See also* Sister chromatids
Chromatin
defined, 88
remodeling, 344, 791
structure, 211–212, *213*
Chromium, *1052*
*Chromodoris*, *1174*
Chromophores, 773, *774*
Chromoplasts, 92
Chromosomal mutations
description of, 307–308
in human genetic diseases, 312–313
Chromosome 1, 366
Chromosome 8, 366
Chromosome 15, 345
Chromosome 19, 366
Chromosome 21, 224, 308, 924
Chromosome number
haploid and diploid, 218
reduction in meiosis, 219, 221
somatic cells, 218
Chromosomes
defined, 88, 206
differences between chromatids and chromosomes, 215
in eukaryotes, 361
genetic locus, 242
genomic information, *355*
global DNA methylation, 345–346
homologous pairs, 218 (*see also* Homologous chromosomes)
independent segregation during the formation of gametes, 237, 239
karyotype, 224, *225*
linkage of genes, 247, 248, 249–252
meiotic errors, 222, 224
numbers in organisms, 218
origins of replication, 269, 271
replication in prokaryotes, 206, *207*
separation and movement in anaphase, 214–216
in somatic cells, 218
speciation by centric fusion, 470, *471*
telomeres, 275
Chronic myelogenous leukemia (CML), 304, 305
Chronic obstructive pulmonary disease, 862
Chronic protein deficiency, 1050
Chrysanthemums, 761
*Chthamalus stellatus*, 1184
Chylomicrons, *1062*, 1063, 1066
Chyme, 1061
Chymotrypsin, *154*, *1062*
Chytrid fungi
amphibian decline and, 612
description of, 617, 619
life cycle, 617–619
major groups and distinguishing features, *616*
phylogeny of the fungi, 615, *616*
Chytridiomycota, 617
*Chytriomyces hyalinus*, *617*
cI regulatory protein, 340–341
Cicadas, *672*, 673
Cichlids
prezygotic isolating mechanisms, 477
speciation in Lake Malawi, 467 (*see also* Haplochromine cichlids)
*Cicindela campestris*, *1172*
Cigarette smoke
benzopyrene and induced mutations, 309, 311
emphysema and, 1013
public bans on, 311
smoker's cough, 1015
Cilia
cilia motion in nodal cells establishes left–right asymmetry, 902, 914–915, 921
of ciliates, 554, *555*
of ctenophores, 644
of lophophores, 652
structure and function, 96–98
of trophophores, 653
Ciliary muscles, 958–959
Ciliates, 554–555, 562
Cinchona, *604*, 605, 797
Cingulata, *698*
*Ciona*, *683*
Circadian rhythms
in animal behavior, 1107–1108
entrainment by light in plants, 774–775
flowering in angiosperms and, 788
*per* gene and, 1096
Circannual rhythms, 1108
Circular chromosomes
origins of replication, 206, 269, 271
replication, 206–207
Circulatory systems
arteries and arterioles, 1039, *1040*
blood composition, 1037–1039
blood transport of respiratory gases, 1016–1019
capillary beds, 1039–141
components of, 1026
control and regulation of, 1043–1045
countercurrent heat exchange, 825, 828
evolution in vertebrates, 1027–1030
function of, 1026
"heat portals" in the skin, 831
human, *1031*
lymphatic vessels, 1042
mammalian heart function, 1030–1037
open or closed systems, 1026–1027
perfusion of the lungs, 1016
sepsis, 862
thermoregulation through the skin and, 824–825, 828
vascular disease, 1042–1043
veins, 1041–1042
Circumcision, 890
*Cirrhilabrus jordani*, *688*
Cirrhosis, 1039
*cis-trans* Isomers, 41
Cisplatin, 259, 268, 278
Cisternae, of the Golgi apparatus, 90
Citrate, 170, *1051*
Citrate synthase, 154
Citric acid, 624
Citric acid cycle
allosteric regulation, 181–182
anabolic interconversions and, *179*, 180
description of, 170–171
in glucose metabolism, 168
regulation of, 171
relationships among metabolic pathways, *179*
Citrus trees, transgenic, 401
Clades, 451, 463
*Cladonia subtenuis*, 613
Clams, 662, 999
Claspers, 888
Class I MHC proteins, 872, *873*, 874
Class II MHC proteins, 872, *873*
Class switching, 869–871
Classes, 463
Clathrin, 121
*Clavelina dellavallei*, *684*
Clay and clay particles, 71, 745, 746
Clean Air Act of 1990, 1220
Cleaning products, 144
Cleavage
defined, 633, 904
in major animal groups, 633
in mammals, 905–906
spiral, 653
types of, 905
Climate
atmospheric circulation patterns, 1124
defined, 510, 1122
evolutionary adaptations in organisms, 1122, 1125–1126
ocean currents and, 1124–1125
prevailing winds, 1124, *1125*
variation of solar radiation over Earth's surface, 1123
Walter climate diagrams, 1138
Climate change
detection by isotope analysis of water, 36
in Earth's history, 510
photosynthetic efficiency of plants and, 201–202
species extinctions and, 1236–1237
*See also* Global climate change
Climax community, 1200
Clinal variation, 443–444
Clitellates, 661–662
Clitoris, 892, *893*
"Clock genes," 1108
"Clock-shifting" experiments, 1110, *1111*
Clonal deletion, 863, 865, 872
Clonal lineages, 562
Clonal selection, 865, *866*
Clones
asexual reproduction and, 217, 562
of cassava, 708
Cloning
of animals, 406–407
of plants, 405–406
*See also* Molecular cloning
Closed circulatory systems, 1026–1027
*Clostridium*, 531, 936
*Clostridium botulinum*, 542
*Clostridium difficile*, *531*
*Clostridium tetani*, 542
Clothes moths, 1203
Clotting factors, 1039
Clown fish, 887, *888*
Club fungi
description of, 622
distinguishing features, *616*
edible, 624
fruiting structure, *610*
life cycle, *621*, 622–623
phylogeny of the fungi, *615*
*See also* Dikarya
Club mosses, *574*, 580, 581
Clumped dispersion pattern, 1153

*Clunio marinus,* 1152
Cnidarians
  in animal phylogeny, *630*
  anthozoans, 646–647
  bilateral symmetry in, 643
  description of, 645–646
  gastrovascular cavity, 1056
  Hox genes, 413
  hydrostatic skeleton, 999
  hydrozoans, 647–648
  life cycle, 645, *646*
  major subgroups and number of living species, *632*
  nerve net, 940–941
  radial symmetry, 634
  scyphozoans, 647
Cnidocytes, *646*
Co-repressors, 331, 332
Coal deposits, 580
Coastal redwoods, 593, 1130
Coastal sand dunes, 793
Coastal zone, 1139
Coat color
  epistasis in Labrador retrievers, 244
  multiple alleles in rabbits, *242*
Coat color patterns, point restriction phenotype, 245, 306
Coat patterns, 245
Coated pits, 121
Coated vesicles, 121
Cobalamin, *1053*
Cobalt, *1052*
Cobalt-60, 24
Coca plants, 626
Cocaine, 626
Coccolithophores, 521
Coccus, 528, *529*
Cochlea
  anatomy of, *954,* 955
  flexion of the basilar membrane, 955–956
Cochlear canal, *954*
Cochlear fluid, 955, 956
Cochlear nerve, *954,* 956
Cochons Island, 1199
Cockleburs, 788, 789, 790
Cockroaches, *672*
Coconuts, 784
Cocos Island finch, *474*
Cod-liver oil, 848
Codominance, 243
Codons
  binding of tRNA anticodon to, 293
  discovery and description of, 288–289
  frameshift mutations and, 306–307
  nonsense mutations and, 306, *307*
Coelecanths, 689–690
Coelom, 635, 652
Coelomates, 635
Coelomic fluid, 1075, *1076*
*Coelophysis bauri, 518*
Coenocyte, 560
Coenocytic hyphae, 609, *610*
Coenzyme A, *156*
Coenzyme Q10. *See* Ubiquinone
Coenzymes
  description of, 155–156
  in oxidation–reduction reactions, 167–168
  *See also* Acetyl coenzyme A
Coevolution
  diffuse coevolution, 1181–1182
  between herbivores and plants, 1175–1176
  plant–pollinator relationships, 588, 598–599, 605
  resulting from species interactions, 1171–1172
Coffee plantations, 1243
Coffroth, Mary Alice, *565,* 566
Cohen, Stanley, 374
Cohesins
  chromatin structure and, 211–212
  homologous chromosomes and, 219
  meiotic errors and, 224
  removal during mitosis, 212, 213, 214–215, *216*
Cohesion
  of water molecules, 33
  *See also* Transpiration–cohesion–tension mechanism
Cohort life tables, 1154
Cohorts, 1154
Coincident substitutions, 488
Coitus interruptus, 890, *898*
Cold deserts, 1133–1134
"Cold" fish, 825
Cold-hardening, 810
Coleochaetophytes, 572, 573
Coleoptera, *672, 674*
Coleoptiles
  action spectrum of phototropism, 771, *772*
  auxin and phototropism, 762, 763, 764, *765*
  early shoot growth and, 758
*Colias,* 442–443
Collagen
  in arteries, 1039
  in blood clotting, 1039
  in cartilage, 999–1000
  in connective tissues, 818–819
  in extracellular matrix, 100
"Collapsed lung," 1016
Collard lizard, *1133*
Collared flycatchers, 1103
Collecting duct
  aquaporins and water permeability, 1088, 1089–1090
  in the human excretory system, *1080,* 1082
  urine is concentrated in, 1084
Collembola, *672n*
Collenchyma, 713, 719–720
Colon, *1058,* 1063
Colon, Bartolo, 392
Colon cancer, 314
Colonial animals
  bryozoans, *654,* 656
  description of, 640, *641*
  entoprocts, 656
  hydrozoans, 646, 647–648
  sea squirts, 684
Colony-stimulating factor, *385*
Color blindness, *252,* 962
Color vision, 962
Colorectal cancer, 344
Colugos, *698*
*Columba palumbus,* 1116, *1117*
Columbian mammoth, *1229*
Columbines, 477, 480
Columbus, Christopher, 440
Columnar epithelium, *818*
Coma, 970
Comb jellies, 634, 644–645
Combined DNA Index System (CODIS) database, 317–318
Comets, 68–69
Commensalism, *1170,* 1171
Common bile duct, 1061
Common carotid artery, *1031*
Common iliac artery and vein, *1031*
Common names, 462
Common wood-pigeons, 1116, *1117*
Communication. *See* Animal communication; Cell-to-cell communication
Communities
  challenges of identifying boundaries, 1189
  characteristics of, 1189
  correlation between productivity and species diversity, 1192
  defined, 9, 1189
  ecosystems and, 9–10
  energy transfer and biomass relationships between trophic levels, 1190–1192
  food webs, 1190, *1191*
  impact of disturbances on, 1199–1202
  impact of species interactions on, 1193–1195
  impact of species richness on community stability, 1202–1203
  insect colonization of human corpses, 1188
  keystone species, 1194–1195
  microbial, 539, *540*
  patterns of species diversity, 1195–1198, *1199*
  primary producers and gross primary productivity, 1189–1190
Community composition
  boreal and temperate evergreen forest biomes, *1129*
  chapparal biomes, *1134*
  cold desert biomes, *1133*
  hot desert biomes, *1132*
  temperate deciduous forest biome, *1130*
  temperate grassland biome, *1131*
  thorn forest and tropical savanna biomes, *1135*
  tropical deciduous forest biomes, *1136*
  tropical rainforest biomes, *1137*
  tundra biomes, *1128*
Compact bone, 1001
Companion cells, *714,* 715, 735
Comparative experiments, 12, 13
Comparative genomics
  defined, 356
  human genome, 366–367
  myostatin gene, 370
  prokaryotic genomes, 357
"Compass sense," 1109
Competition
  description of, 1170–1171
  exploitation competition, 1182–1183
  importance in determining a species' niche, 1184
  indirect, 1184
  interference competition, 1182, 1183
  types of, 1182–1183
Competitive exclusion, 1182, 1192
Competitive inhibitors, 158
Complement proteins, 860
Complement system, 860
Complementary base pairing
  in DNA, 266
  importance to DNA replication and transcription, 65–66, *67*
  in nucleic acids, 63–65
Complementary diet, 1051
Complementary DNA (cDNA)
  cDNA libraries, 379–380
  creating, 379
  in high-throughput sequencing, 353
  in HIV infection, 341
  RNA retroviruses and, 544
Complementary mRNA, 544
Complete cleavage, 905
Complete flowers, 779
Complete gut, 644
Complete metamorphosis, 672, 841
Complex behaviors
  in deuterostomes, 705
  human evolution and, 704, 705
Complex ions, 29
Complex life cycles
  of eukaryotic viruses, 341
  of parasites, 563, *564,* 640, *641*
  in protostome evolution, 674
Composite transposons, 358–359
Compost, 747
Compound eyes, 958
Compound umbel, *597*
Compounds
  covalent bonding, *27*
  defined, 26
Compulsory vaccinations, 856
Concentrated urine, 1071, 1077, 1082–1084, 1090
Concentration gradients
  effect on diffusion, 114
  passive membrane transport and, 113
  secondary active transport and, 118, 119
Concerted evolution, 498–499
Condensation reactions, 42, 269, *270*
Condensins, 212
Conditional mutations, 306
Conditioned reflexes, 982, 1094
Condoms, *898*
Conducting cells, in cardiac muscle, 991–992
Conduction, in heat exchange, 823, *824*
Conduction deafness, 956
Cone cells, 960, 961–962, 963
Cones, of conifers, 593–594, *595*
Confocal microscopy, 80
Congenital hypothyroidism, 320
Conidia, 616, 622

Coniferophyta, *574*
Conifers
in boreal and tropical evergreen forests, 1129, 1130
cones, 593–594, *595*
distinguishing characteristics, *574*
fire adaptations, 594, *596*
life cycle, 594, *595*
number of species, 593
Conjugation. *See* Bacterial conjugation
Conjugation tube, 253, *254*
Conjugative pili, 84
Connective tissue
characteristics of, 818–819
development of bone from, 1001
endoskeletons, 999–1000
in the gut, 820
Connell, Joseph, 1184
Connexins, 939
Connexons, 139
*Conomyrma bicolor,* 1183
Consciousness, 982–983
Consensus sequences, 292, 332–333
Conservation biology
basic principles of, 1229
defined, 1229
goal of protecting and managing biodiversity, 1229–1230
human activities threatening species persistence, 1232–1237
prediction of changes in biodiversity, 1230–1232
risks of deliberate species introductions, 1228, 1245
strategies used to protect biodiversity, 1237–1245
Conservative replication, 268
Consortium for the Barcode of Life (CBOL), 319
Conspecifics, 1103
*CONSTANS (CO)* gene, 788, 790
CONSTANS (CO) protein, 788, 790
Constant regions, of immunoglobulins, 867, 869–871
Constipation, 1063
Constitutive plant defenses
against herbivores, 801, 802–803
against pathogens, 798, 803
Constitutive proteins, 330
Constricting ring, 612, *613*
Consumers, 1190, *1191*
Continental drift, 509, 1142, 1143
Contraception, 890, 896, 897, *898*
Contractile ring, 216
Contractile vacuoles, 93, 554, *555*
Contractions, in labor and childbirth, 896–897
Contralateral neglect syndrome, 973
"Control" group, 12
Controlled burning, 1240
Controlled experiments, 12–13
Controlled systems, 816
Convection, 823, *824*
Convention on International Trade in Endangered Species (CITES), 1240–1241
Convergent evolution
defined, 452
of eyes, *959*
in lysozyme and foregut fermenters, 492–494
revealed by phylogenetic analyses, 459–460
Convergent extension, 909–910
Convolutions, 970, 973
Cook Strait, 1145
Copepods, 670
Copper
in animal nutrition, *1052*
in plant nutrition, *741*
role in catalyzed reactions, *156*
Coprophagy, 1064
Copulation, 887, 891
"Copy and paste" transposition, 308, *358,* 359, 365
*Coquerelia ventralis, 674*
Coral reefs, 1139
Corals
"bleaching," 565, 566, 1217
description of, 646–647
detection of chemical stimuli, 949
dinoflagellate endosymbionts, 565, 566
endosymbiotic relationship with dinoflagellates, 647
environmental threats to, 647
fluorescent proteins, 449, *464*
*Corixidea major,* 1230
Cork, 722, *723*
Cork cambium, 715, *716,* 721, 722, *723*
Corms, 793
Corn
domestication, 723–724
flowers, *780*
heterotypy and edible kernels, 419
hybrid vigor, 244, 245
improving nitrogen use efficiency, 740, 753
prop roots, 718
volicitin, 804
Corn oil, 57
Cornea, 958, *959*
Corn–sweet potato dicultures, 1203
Corolla, 597
Corona, of rotifers, 657, *658*
Coronary arteries, 1043
Coronary heart disease, *368*
Coronary thrombosis, 1043
Coronavirus, *543*
*Coronosphaera mediterranea, 2*
Corpora allata, 841
Corpse communities, 1188, 1203
Corpus callosum, 980
Corpus cardiacum, *841*
Corpus luteum, 893, *894,* 895
Corridors, 1162, 1233–1234
Cortex
in roots, 717, *718*
in shoots, 719–720
Cortical granules, *885,* 886, 887
Cortical nephron, *1080*
Corticosteroids, *836*
Corticotropin, *842,* 843, 849
Corticotropin-releasing hormone (CRH), 844, 849
Cortisol
actions of, *842*
half-life, 853
negative feedback signaling to the anterior pituitary and hypothalamus, 844
stress response and, 849–850
structure, *836*
Cortisol receptor, 131
*Corynebacterium diphtheriae,* 542
Cost–benefit analysis
applied to animal behavior, 1103
of foraging behavior, 1104–1105, *1106*
of group living, 1116, *1117*
of territorial behavior, 1103–1104, *1105*
Cottony-cushion scale, 1164
*Coturnix japonica,* 1101
Cotyledons
in angiosperms, *600,* 601
in embryogenesis, 711–712
in monocots and eudicots, 602, 709
in seed germination, 758
Countercurrent flow, in fish gills, 1010, *1011*
Countercurrent heat exchange
in the appendages of endotherms, 828
in "hot" fishes, 825
Countercurrent multiplier, 1082–1084
Coupled transporters, 118
Courtship behavior
experiments on the genetic basis of, 1095
gene cascades in *Drosophila,* 1097–1098
impact on speciation, 705
multiple sensory modalities in *Drosophila,* 1113
*See also* Sexual behavior
Covalent bonds
ester linkages, 56
formed in condensation reactions, 42
multiple, 28
orientation, 27–28
overview, 26–27
strength and stability, 27
unequal sharing of electrons, 28
Covalent catalysis, 155
Covalent disulfide bridges, 47
Covalent modifications, of proteins, 50
Cowbirds, 1093, 1102, 1117, 1233
Cows
cloning, 407
*See also* Cattle
Cozumel thrasher, 1232
CpG islands, 343
Crab lice, 1177
Crabs, 670
Cranberry, 615
Cranial nerves
autonomic nervous system and, *974, 975*
development, 968
relationship to the brainstem, 969
"Craniates," 686
Crassulaceae, 200
Crassulacean acid metabolism (CAM) plants, 200, 734
Crayfish, 670, 947
Creatine phosphate, 997
Crenarchaeota, 535, 536
Cretaceous period
changes on Earth and major events in life, *506–507*
characteristics of life during, *519,* 521
mass extinction, 511
Cretinism, 846
Crews, David, 881
Crick, Francis, 264–265, 266, 430–431
Crickets, *672*
Crinoids, 520, 680–681, *682*
Cristae, of mitochondria, *92*
Critical period, in animal behavior, 1095, 1099
Critical temperature, 827
"Critically endangered" species, 1231
*Cro* gene, 341
Cro regulatory protein, 340–341
Crocodiles, 693–694, *695,* 1029–1030
Crocodilians, 693–694, *695,* 1029–1030
*Crocodylus porosus, 695*
Crocuses, 793
*Cromileptes altivelis, 688*
Crop, of the hoatzin, 494
Crop plants
applications of biotechnology, 386–388, *389*
cassava, 708, 724
domestication, 723–724
improvement programs, 724
inoculation of seeds with mycorrhizae, 615
major species, 605
pathogenic fungi, 612
polyploidy in, 225, *226*
*See also* Agriculture; Corn; Rice; Transgenic crops; Wheat
Crop rotation, 750
Crossing experiments, 234
Crossing over
genetic exchange and, 219, 247–249, *250*
during meiosis, 219, *220, 222*
Crosstalk, between signal transduction pathways, 127, 134
*Crotaphytus, 1133*
Crown gall, 534
Crustaceans
description of, 670–671
larval form, 640
major subgroups and number of living species, *632*
undescribed species, 675
Crustose lichens, 613
Crypsis, 1173
Cryptic species, 468, *469*
Crytptochromes, 772
Ctenes, 644
Ctenophores
description of, 644–645
major subgroups and number of living species, *632*

in the phylogeny of animals, *630*, 648
radial symmetry, 634
CTLA4 protein, 876
Cuboidal epithelial cells, *818*
*Cuculus canorus*, 1093
Cud, 1064
"Cuddle hormone," 843
*See also* Oxytocin
Culture, humans and, 705
Cumulus, 886
Cup fungi, 620, *622*
Cupula, *956*, 957
Curare plant, *604*
Currents, 1124–1125
*Cuscuta*, *752*
"Cut and paste" transposition, 308, 359
Cut-flower industry, 770
Cuticle
of ecdysozoans, 654, 665
of exoskeletons, 999
of nonvascular land plants, 575
in plant defenses against herbivory, 1175
in plant evolution, 574
of plants, 713, 732
Cuticular plates, 665
Cutin, 713, 798
Cuttings, of stems, 793
Cuttlefish, 662, *663*
Cyanide
in cassava, 708
clinal variation in white clover, 443–444
in the Miller–Urey experiment, 70
protective storage in plants, 805
Cyanobacteria
atmospheric oxygen and, 532, 539
description of, 532, *533*
endosymbiotic origin of eukaryotic chloroplasts and, 102, *550*, 551, *552*
in eutrophication, 1219
in lichens, 613
microfossils and, 74
nitrogen fixers, 749
photosynthesis in, 532, 538
stromatolites, *5*, 512, *513*
symbiotic relationship with hornworts, 579
Cyanogenic compounds, 708, 724
*Cyathea australis*, *582*
Cycadophyta, *574*
Cycads, *574*, 589, 592, *593*
Cyclic adenosine monophosphate (cAMP)
discovery of, 133, 134
function of, 67
in positive regulation of the *lac* operon, 332, *333*
regulation of, 137
as a second messenger, 133–134
Cyclic electron transport, 192
Cyclic guanosine monophosphate (cGMP)
in penile erection, 890, 891
in smooth muscle relaxation, 136
Cyclical succession, 1201–1202
Cyclin-dependent kinases (Cdk's), 209–211
Cyclin–Cdk complexes, 210–211, 214, *216*
Cyclins, 210–211, 349
*Cyclopoid copepod*, *670*
Cyclosporin, 874
*Cynomys ludovicianus*, *1239*
Cypresses, 808
*Cyprinodon*, 1160
Cysteine, 43, *44*
Cystic duct, 1061
Cystic fibrosis, *313*, 318, 321
Cytochrome *c*
amino acid sequences from different organisms, *488–489*
in the respiratory chain, 172, *173*
Cytochrome *c* oxidase, 172, *173*
Cytochrome oxidase gene, 319
Cytokines
function of, 859
in the humoral immune response, 872, *873*
in inflammation, 861
interferons, 860
Cytokinesis
in animal cells, 216
in cell division, 206
in eukaryotes, 207
in plant cells, 216–217, 710
in prokaryotes, 207
Cytokinins
discovery of, 768
effects on plant growth, 768
signal transduction pathway, 768–769
structure and typical activities, *759*
Cytoplasm
energy pathways in, *168*
of prokaryotes, 82
Cytoplasmic determinants, 395
Cytoplasmic dynein, 98, 216
Cytoplasmic inheritance, 252–253
Cytoplasmic segregation, 395
Cytoplasmic streaming, 560
Cytosine
codons and the genetic code, 288–289
complementary base pairing, 63–65
deamination, 309, 310
in DNA structure, 264, 265, 266, *267*
as a "hot spot" for mutations, 310
methylation, 310, 328, 343 (*see also* DNA methylation)
structure, *63*
Cytoskeleton
animal cells, *86*
asymmetric distribution of cytoplasmic determinants and, 395
attachments to membrane proteins, 109
biological membranes and, *106*
in the evolution of the eukaryotic cell, *550*, 551
functions of, 94
intermediate filaments, 95
microfilaments, 94–95, *96*, 98, 99
microtubules, 95–96
of prokaryotes, 84
Cytosol, 82
Cytotoxic T ($T_C$) cells
in the cellular immune response, 863, *865*, 872, *873*, 874
regulation by Tregs, 874

## D

Daddy longlegs, *669*
Daffodils, *603*
Daily torpor, 830
Daltons, 22
Daly, Herman, 1244–1245
*Danaus plexippus*, *639*, *1174*
Dandelions, *601*, *1153*
*Daphnia*, 1105, *1106*
Dark reactions, 188
*See also* Light-independent reactions
Darters, *472*
Darwin, Charles
beak diversity in finches, 412
concepts of coevolution, 1171–1172
concepts of evolution and natural selection, 6
description of the focus of ecology, 1122
on divergence of character, 1183
on earthworms, 639
on the evolution of flowers, 605
evolutionary theory, 428–430
experiment on phototropism in coleoptiles, 762, 763, *764*
on hybrid vigor, 244
knowledge of artificial selection, 6, 433
on orchids, 588, 605
on sexual selection, 435
Darwin, Francis, 762, 763, *764*
Darwin's black spider, 39
Darwin's finches
*See* Galápagos finches
Darwin's frog, *678*
Darwin's rhea, *1131*
Dasyuromorphia, *698*
Data, importance of quantification in science, 11, 14
Date palms, *603*
Dating methods, 506–508
*Daucus carota*, 718
Daughter chromosomes, 215, 216, *223*
*DAX1* gene, 250
DAX1 protein, 250
Day length
impact on development, 421–422
photoperiodic cues in flowering, 787
*Day of the Dandelion* (Pringle), 778
*Db* gene, *1067*, 1068
*DCC* gene, *314*
Dead Sea, 536
Dead space, 1011, 1012, 1013
Dead zones, 740, 1207, 1214, 1219, 1225
Deafness, 956
Deamination, of cytosine, 309, 310
Death Valley, 1160
Decapods, 670
Declarative memory, 982
Decomposers
characteristics of, 639
corpse communities, 1188, 1203
defined, 1054
in food webs, 1190, *1191*
prokaryotes, 538
saprobic fungi, 611
*See also* Detritivores
Decomposition, 1188
Deductive logic, 12
Deep-sea hydrothermal vent ecosystems
formation of organic polymers and, 71
pogonophorans, 660–661
prokaryotes and, 539
*Deepwater Horizon* (oil well), 373, 569
Deer tick (black-legged tick), *1151*, 1152
Defecation, 1063
Defensins
in animals, 859
in plants, 801
Deficiency diseases, *1053*, 1054
Deforestation, 1232
Dehydration reactions, 42
*Deinococcus*, 532
*Deinococcus radiodurans*, 389
Deiodinase, 845
Delayed hypersensitivity, 876
Deleterious mutations, purging of, 433, 441
Deletions, 308, 312–313
*Delonix regia*, *603*
Demethylase, 343, 345
Demographic events
determination of population size and, 1153–1154
life tables and, 1154–1155
Demography, 1153
Demosponges, *630*, 643
Demyelinating diseases, 926
Denaturation, of proteins, 48, 50
Dendrites, 925–926
Dendritic cells
as antigen-presenting cells, 864, 872
function of, *858*
HIV infection, 876
in innate defenses, 861
pattern recognition receptors, 860
*Dendrobates reticulatus*, *1174*
*Dendroctonus frontalis*, 1178
*Dendroctonus ponderosae*, 1202
*Dendroica petechia*, *1093*
*Dendrosenecio keniensis*, *1128*
Denitrification, 750, *751*
Denitrifiers, 539
Density-dependent population regulation, 1159
Density-independent population regulation, 1159
Dental plaque, 539, *540*
Dentine, 1055
Deoxyribonucleic acid (DNA)
artificial, 380
base sequence reveals evolutionary relationships, 66
chromatin, 88, 211 212, *213*
chromosomes, 88
complementary base pairing, 63–65
distinguishing from RNA, *63*

double helix structure, 65, *66*, 264–265, 266, *267*
effect of cisplatin on, 259, 268, 278
eukaryotic regulatory sequences, 335
evidence for being the genetic material, 260–263
exchange in bacteria, 253–255
genetic information and, 5–6
growth of, 63, *64*
hybridization experiments, 290–291
metagenomics, 357–358
methods of study DNA function, 380–383
mutations (*see* Mutations)
noncoding sequences, 290–291, 494–496
normal, daily damage to in humans, 310
PCR amplification, 277–278
recombinant, 374–375 (*see also* Recombinant DNA technology)
relationship of structure to function, 266
repair mechanisms, 275, 276–277
structure and function, 62, 63–67, 264–267
test tube synthesis, 267–268
transcription, 65
transmission of information by, 65–66
Deoxyribonucleoside monophosphates, 268
Deoxyribonucleoside triphosphates (dNTPs)
in DNA replication, 268–269
in high-throughput sequencing, 353, *354*
in the polymerase chain reaction, 277, *278*
Deoxyribose, *52*, 63, 266
Dependent variable, 13
Depolarization
description of, 930–931, *932*
generation of action potentials and, 932–933, 934
Deprivation experiments, 1095
Derived traits, 452, 453, *454*
Dermal tissue system
description of, 712–713
primary meristem giving rise to, 716
Dermaptera, *672*
Dermoptera, *698*
Descent with modification, 428
*Deschampsia antarctica,* 1125
Desensitization, to allergens, 876
Desert gerbils, 1090
Desert plants
drought avoiders, 806–807
root systems, 808
salt glands, 811
stem modifications, 720
structural adaptations in leaves, 807
succulence, 807
Desert pupfish, 1160
Desert rodents, 1090
Deserts
atmospheric circulation patterns and, 1124
characteristics in Australia, 1126
cold desert biomes, 1133–1134
hot desert biomes, 1132–1133
plant adaptations to, 806–808
*Desmodus rotundus, 1071*
Desmosomes, 95, 111, *112*
Desmotubules, *139*, 140
Detergents, 144, 1221
Determinate growth, 715, 720, 786
Determination
during amphibian gastrulation, 910–913
of blastomeres, 907
of cell fate, 394
by cytoplasmic segregation, 395
defined, 393
in development, 710
differential gene expression and, 393, 397–398
by induction, 395–396, *397*
signal transduction pathways and, 397–398
Detritius-based communities, 1202
Detritivores
characteristics of, 639
commensalisms and, 1171
defined, 637, 1054
in food webs, 1190, *1191*
in succession, 1201
*See also* Decomposers
Detritus, 639, 1190
Deuterium, 22
Deuterostomes
in animal phylogeny, *630*
bilaterians, 634, 643
chordates, 747–679
complex behaviors in, 705
echinoderms and hemichordates, 680–683
fossil ancestors, 679–680
major groups and living species, *632*, 679
pattern of gastrulation in, 634
phylogeny, *679*
shared early developmental patterns, 679
Development
of animal behavior, 1098–1102
defined, 393
developmental constraints on evolution, 444, *445*
DNA methylation and, 344
environmental modulation of, 420–422
evolutionary (*See* Evolutionary developmental biology)
pattern formation, 399–405
phylogenetic analyses and, 455
processes in, 393–394, 710
restriction of cell fates during, 394
role of gene expression in, 397–399
*See also* Animal development; Human development; Plant growth and development
Developmental genes
differences in expression resulting in differences between species, 418–420
discussion of, 413–414
genetic switches, 415
modularity and differences in the patterns of expression, 416–417
Developmental modules
concept of, 415
differences in the patterns of gene expression, 416–417
genetic switches and the genetic toolkit, 415
Developmental plasticity, 420–422
Devonian period
changes on Earth and major events in life, *506–507*
characteristics of life during, *517*, 520
evolution of leaves, 583
evolution of seed plants, 589
vascular plants in, 580
DHAP. *See* Dihydroxyacetone phosphate
d'Herelle, Felix, 545
DHFR (Dihydrofolate reductase), 158
Diabetes mellitus
among the Pima, 1048
human SNP scans, *368*
as a risk factor for atherosclerosis, 1043
Type I, 848, 876
Type II, 848, 853
Diacylglycerol (DAG), 134–135
Diademed sifaka, *701*
*Diadophis punctatus, 694*
Dialysis, 1086, *1087*
Diamondback moth, 1176
Diamondback rattlesnakes, *946*
Diapause, 421
Diaphragm (human respiratory system), *1014*, 1015, 1016, 1019–1020
Diaphragms (in contraception), *898*
Diarrhea, 1063
Diastole, 1031, 1032, 1039
Diastolic pressure, 1032, *1033*
Diatomaceous earth, 565
Diatoms
description of, 555–556
during the Mesozoic, 521
petroleum and natural gas deposits from, 565
as primary producers, 563
red tides and, 564
*Dicaeum hirundinaceum, 1182*
Dichotomous branching, 580
*Dicksonia antarctica, 581*
*Dictyostelium, 98,* 838
*Dictyostelium discoideum, 561*
Didelphimorphia, *698*
*Didelphis virginiana, 699,* 1173
*Didinium nasutum, 554*
Diencephalon, 968–969, 970
Diet
developmental plasticity and, 421, *422*
effect of specialization on speciation, 480
human birth defects and, 916
impact on replenishment of muscle glycogen, 997
manipulation of epigenetic changes through, 349
Differential gene expression
in cell fate determination, 397–398
in development, 393
in differentiation, 398–399
during induction, 396
Differential interference-contrast microscopy, 80
Differentiation
defined, 393
in development, 710
differential gene expression and, 393
differential gene transcription in, 398–399
distinguished from determination, 394
reversibility, 405–409
role of transcription factors in, 336, *337*, 338
significance to multicellular life, 9
Diffuse coevolution, 1181–1182
Diffusion
across membranes, 114–115
within cells and tissues, 114
discussion of, 113–114
facilitated, 115–117
factors affecting, 114
of gases in water and air, 1007
of membrane proteins, 109, 110
of respiratory gases, 1006, 1008
simple diffusion, 114
through plasmodesmata, 140
Digestion
in ciliates, *555*
digestive enzymes, 1057–1058
external, 1055
gastrovascular cavity, 1056
hormonal control of, 1065
in the small intestine, 1061–1062
"thrifty genes" in humans, 1048
tubular guts, 1056–1057
vertebrate teeth and, 1055–1056
Digestive enzymes, 1057–1058, *1062*
Digestive systems
autonomic influence on smooth muscle, 993
control and regulation of the flow of nutrients, 1064–1068
digestive enzymes, 1057–1058
endocrine cells of, 835
gastrointestinal system in vertebrates, 1058–1064
gastrovascular cavity, 1056
gut microbes and, 539, 540–541
pH in the human stomach, 161
tissue composition, 820
tubular guts, 1056–1057
Digestive vacuoles, 555
Digitalis, 1034
*Digitalis purpurea,* 1034
Diglycerides, galactose-substituted, 92
Dihybrid crosses, 238, 239–240
Dihydrofolate, 158
Dihydrofolate reductase (DHFR), 158
Dihydropyridine (DHP) receptor, 990, 992
Dihydroxyacetone phosphate (DHAP), 179
Diisopropyl phosphorofluoridate (DIPF), 157, *158*
Dikarya, *615*

  - club fungi, *621,* 622–623 (*see also* Club fungi)
  - life cycle, 620, *621*
  - sac fungi, 620–622 (*see also* Sac fungi)
  - sister group to arbuscular mycorrhizal fungi, 619
  - *See also* Club fungi; Sac fungi
- Dikaryon, 620, *621*
- Dimethyl sulfide, 1219
- Dimethylsulfoniopropionate (DMSP), 1219
- Dinoflagellates
  - beneficial aspects of, 566
  - bioluminescence, 566
  - description of, 553, *554*
  - endosymbionts in corals, 565, 566, 647
  - endosymbiotic origin of chloroplasts in, 552
  - during the Mesozoic, 521
  - red tides and, 549, 564
- Dinosaurs
  - characteristics of, 694
  - evolution of birds and, 693, 694
  - evolution of feathers, 695
  - migration, 21
- Dioecious plants
  - defined, 597, 779
  - example of, *780*
  - strategies for preventing self-pollination, 782
- Dioecious species, 249, 887
- *Diomedea melanophris, 1105*
- *Dionaea,* 751–752
- *Dionaea muscipula, 752*
- Dipeptidase, *1062*
- DIPF (diisopropyl phosphorofluoridate), 157, *158*
- *Diphyllobothrium latum, 641*
- Diploblastic animals, *630,* 633
- Diploids
  - in alternation of generations life cycle, 562, 563, 574, 575
  - defined, 218, 236
  - in sexual reproduction, 218
- Diplomonads, 558
- Diplontic life cycle, *218*
- *Diploria labyrinthiformis, 647*
- Diplura, *672n*
- Diprotodonts, *698*
- Diptera, 415, *672, 674*
- Direct development, 639
- Direct fitness, 1114
- Directional selection, 439–440
- Directional succession, 1199–1201
- Disaccharides, 51, 53
- Disc flowers, *597*
- Discoidal cleavage, 905
- Disparity, 978
- Dispersal
  - in animal life cycles, 640
  - discontinuous species distributions and, 1143–1145
  - dispersal ability and speciation rates, 481
- Dispersion patterns, 1150, 1152–1153
- Dispersive replication, 268
- Disruptive selection, 439, 440, 473
- Dissociation constant, 128, 153
- Distal convoluted tubule
  - in the mammalian kidney, *1080,* 1082
  - in the production of concentrated urine, *1083,* 1084
  - regulation of the glomerular filtration rate, 1087–1088
- *Distalless (Dll)* gene, 418
- Distance-direction navigation, 1109
- Disturbances
  - defined, 1199
  - impact on species diversity, 1199
  - restoring disturbance patterns in ecosystems, 1239–1240
  - species extinctions and, 1232
  - succession following, 1199–1202
  - variability in the magnitude of effects from, 1199
- Disulfide bonds, 867
- Disulfide bridges, 43, *46,* 47, 196–197
- Diving mammals, 1017
- DNA. *See* Deoxyribonucleic acid
- DNA barcode, 319
- DNA fingerprinting, 317–318
- DNA fragments
  - arranging, 354, *355*
  - genomic libraries, 379
- DNA helicase, 272, 273
- DNA hybridization, 321
- DNA libraries, 379–380
- DNA ligase, 273, *274, 276,* 374
- DNA methylase, 345
- DNA methylation
  - effects of, 343–344
  - gene regulation and, 328
  - genomic imprinting and, 344–345
  - global chromosome changes, 345–346
  - process of, 343
  - in promoters, 328
  - protection from restriction enzymes, 315
- DNA methyltransferase, 343
- DNA microarrays, 368, 382–383
- DNA polymerase I, 273, *274, 276*
- DNA polymerase III, 273, *274, 276*
- DNA polymerases
  - in DNA replication, 269, *270,* 271, *272,* 273–274
  - errors, 276, 309
  - eukaryotic general transcription factors and, 334
  - in high-throughput sequencing, 353, *354*
  - in the PCR reaction, 277–278
  - processive characteristic, 273
  - proofreading function, 276–277
  - sliding DNA clamp, 273–274
  - structure, 271, *272*
  - in test tube synthesis of DNA, 267
  - from *Thermus aquaticus,* 532
- DNA replication
  - in cell division, 206
  - defined, 65
  - DNA structure and, 266
  - errors, 309
  - in eukaryotes, 207
  - importance of base pairing to, 66, *67*
  - lagging strand, 272–273, *274*
  - leading strand, 272–273
  - Okazaki fragments, 273
  - possible replication patterns, 267–268
  - pre-replication complex, 269
  - primers, 63, 271, *272, 274*
  - process of, 268–275
  - in prokaryotes, 206, *207*
  - replication complex, 274–275
  - replication forks, 271, 272–275
  - roles of proteins in, 272, *273*
  - sliding DNA clamp, 273–274
  - teleomeres, 275
- DNA segregation
  - in cell division, 206
  - in eukaryotes, 207
  - in prokaryotes, 207
- DNA sequences
  - alignment, *487*
  - palindromic, 374
  - phylogenetic analyses and, 456
  - recognition by proteins, 335–336
  - using models to calculate evolutionary divergence, 487–489
- DNA sequencing. *See* Genome sequencing
- DNA structure
  - 5′ end and 3′ end, 266
  - antiparallel strands, 265, 266
  - chemical evidence from base composition, 264
  - double helix, 65, *66,* 264–265, 266, *267*
  - importance to DNA function, 266
  - key features of, 265–266, *267*
  - major and minor grooves, 265, 266, *267*
  - physical evidence from X-ray diffraction, 264
  - Watson and Crick's model, 264–265
- DNA technologies
  - DNA barcode, 319
  - DNA fingerprinting, 317–318
  - gel electrophoresis, 316–317
  - restriction enzymes, 315–316
  - reverse genetics, 318
- DNA templates. *See* Templates
- DNA testing, 320–321
- DNA transposons, 365
- DNA viruses
  - as carrier viruses, 323, 324
  - description of, 341, *543,* 544–545
- dNTPs. *See* Deoxyribonucleoside triphosphates
- Dobsonflies, *672*
- Dobzhansky, Theodosius, 470
- Dobzhansky–Muller model, 470, *471*
- Dodders, 752
- Dog Genome Project, 352
- Dogs
  - bleeding and estrus in, 893
  - conditioned reflexes, 1094
  - epistasis and coat color, 244
  - genome, 352
  - myostatin gene, 370
  - olfactory sensitivity, 950
  - partial pressure of carbon dioxide in the regulation of breathing, 1020–1021
  - variation in body size among breeds, 352
- Dolly (cloned sheep), 406–407
- Dolphins, 700, 973
- Domains, of proteins, 291
- Domains (in classification)
  - common ancestor, 527
  - distinguishing characteristics of, *527*
  - lateral gene transfer, 529
  - overview, 81
  - relationships among, *526*
  - shared features of prokaryotes, 526–527
- Domestication, impact on plant form, 723–724
- Dominance, incomplete, 242–243
- Dominance hypothesis, of heterosis, 245
- Dominant trait, 235
- Dopamine, 940
- Dormant seeds. *See* Seed dormancy
- Dorsal, 634, *635*
- Dorsal aorta, 1028
- Dorsal fins, 687
- Dorsal horn, 942
- Dorsal lip of the blastopore
  - amphibian neurulation and, 915–916
  - in embryo formation, 910–911
  - in gastrulation, 909
  - organizer cells and, 912
  - *See also* Primary embryonic organizer
- Dorsal medulla, 1019–1020
- Dorsal spines, 424
- Dorsal–ventral axis, in animal organogenesis, 917
- Dose–response curves, of hormones, 853
- Double fertilization, 600–601, *781,* 783
- Double helix, of DNA, 65, *66,* 264–265, 266, *267*
- Double-stranded DNA viruses, *543,* 544–545
- Double-stranded RNA viruses, 544
- *Doublesex (dsx)* gene, 1097–1098
- Douglas firs, 1160
- Doupe, Allison, 1100–1101
- Doushantuo fossils, *516*
- Down syndrome, 224, 308, 924, 939
- "Down syndrome mouse," 924, 939, 943
- Downregulation, of hormone receptors, 853
- Downy mildews, 556
- Dragline silk, 39, 45
- Dragonflies, 505, 672, *673,* 958
- Dreaming, 979
- *Dreissena polymorpha,* 1160, 1235
- *Drosera,* 751
- *Drosera rotundifolia, 752*
- *Drosophila*
  - allopatric speciation in the Hawaiian Islands, 472, *473*
  - atmospheric oxygen and body size, 513, 514
  - complete cleavage, 905

germ cell lineage, 908
long terminal repeats, 495
as a model organism, 282
reproductive isolation from increasing genetic divergence, *471*
sex-linked inheritance, 251–252
*Drosophila endobranchia*, 1231
*Drosophila melanogaster*
as an *r*-strategist, 1159
artificial selection experiments, 433
determination of body segmentation, 401–405
developmental genes in eye development, 413, *414*
factors involved in characteristic species density, 1160
genetic control of courtship behavior, 1097–1098
genomic information, *361*, 362
homeotic mutations, 404
Hox genes, *414*
laboratory experiments on speciation, 482
multiple sensory modalities in courtship behavior, 1113
wing development, 415
Drought avoiders, 806–807
Drought stress, plant responses to, 809
Drugs
agonists and antagonists of neurotransmitters, 940
anti-inflammatory, 862
determining dosage level, 128
in HIV treatment, 876–877
metabolic inhibitors, 322–323
pharmacogenomics, 368
psychoactive, 135
reduction of overinhibition in the brain, 943
*See also* Cancer drugs; Pharmaceuticals
Drummond, Thomas, 470–471
*Dryas octopetala*, 1200
*Dubautia menziesii*, *481*
Duchenne muscular dystrophy, 313
Duck-billed platypus, 697, 951
Ducks
BMP4 and beak development, 424
hindlimb development, 417
Duckweed, 1140
Dugongs, 700
Dulbecco, Renato, 353
Dung beetles, 1171
Dunman, Paul, 301
Dunn, Casey, 631
Duodenum, 1061, 1065
Duplication-and-divergence hypothesis, 413–414
Duplications, 308
Dutch elm disease, 622, 793
"Dutchman's pipe," *602*
Dwarf mistletoe, 752
Dynein, 97, 98
*Dyscophus guineti*, *7*
Dystrophin, 313

## E

E7 protein, 228
"Ear stones." *See* Otoliths
Eardrum. *See* Tympanic membrane
Earphones, 956
Ears
anatomy of the human ear, 954–955
*See also* Auditory systems; Inner ear
Earth
atmospheric circulation patterns, 1124
magnetic fields and paleomagnetic dating, 508
prevailing winds, 1124, *1125*
variation in received solar radiation, 1123
Earthworms
closed circulatory system, 1026, *1027*
coelomate, *635*
description of, 661
detritivores, 639
excretory system, 1075, *1076*
hydrostatic skeleton, 999
infolding of the gut in, 1057
neural network, 941
rapid climate change and, 1236–1237
simultaneous hermaphroditism, 887
Earwigs, *672*, 673
Easterlies, 1124, *1125*
Eastern gray kangaroo, *699*
eBay, 1241
Ecdysone, 840–841
Ecdysozoans
anatomical characteristics, *652*
in animal phylogeny, *630*
arthropods, 667–673, *674*
cleavage pattern, 633
cuticle and molting, 654
horsehair worms, 666, *667*
major subgroups and number of living species, *632*
nematodes, 666
priapulids, kinorhynchs, and loriciferans, 665–666
Echindnas, 697
*Echiniscus*, *667*
*Echinocereus reichenbachii*, *603*
Echinoderms
in animal phylogeny, *630*
appendages, 636
in deuterostome phylogeny, *679*
features of, 680–682
major groups and living species, *632*, 679
radial symmetry, 634
regeneration in, 881
Echinozoans, 681
"Echo generation," 1165
Echolocation, in bats, 946, 963
Ecological communities, 1189–1193
*See also* Communities
Ecological economics, 1241–1243
Ecological efficiency, 1191
Ecological survivorship curves, 1156
Ecology
defined, 1122
distinguished from environmentalism, 1122
factors affecting climate, 1122–1125
species interactions studied by, 1170–1172 (*see also* Species interactions)
study of the biotic and abiotic components of ecosystems, 1122
terrestrial biomes, 1126–1138 (*see also* Biomes)
Walter climate diagrams, 1138
*Eco*RI restriction enzyme, 315–316, 374
Ecosystem engineers, 1194
Ecosystem services, 1241–1242
Ecosystems
biotic and abiotic components, 1122
causes and impact of dead zones in the Gulf of Mexico, 1207, 1219, 1225
consequences of human alterations, 1223–1224
defined, 9, 1208
examples, 9–10
global climate change and, 17
goods and services provided, 1223–1224
recent metaphors for, 1244–1245
restoration ecology, 1237–1239, *1240*
restoring disturbance patterns, 1239–1240
sustainable management, 1224
*See also* Global ecosystem
Ecotourism, 1242, *1243*
Ectoderm
defined, 908
in diploblastic and triploblastic animals, 633
in extraembryonic membranes, 918
nervous system development in amphibians, 912–913
in neurulation, 916
tissues and organs derived from, *907*, 908
Ectomycorrhizae, 614, 615, 622
Ectoparasites
features of, 639, 1176–1177
leeches, 661
monogeneans, 657
Ectopic pregnancy, 906
*Ectopistes migratorius*, 1230
Ectotherms
control of blood flow to the skin, 824–825
defined, 822
differences from endotherms, 822
energy budget and, 823–824
metabolic heat production, 825–826
reptile circulatory system, 1029–1030
response to changes in environmental temperature, 822, *823*
Edema, 1041, 1050
Edge effects, 1233
Ediacaran fossils, *516*
Eelgrass, 1141
Eels, *688*
Effector B cells. *See* Plasma cells
Effector cells, 865–866
Effector proteins, 130
Effector T cells, 865, 866, 871, 872, *873*, 874
Effector-triggered immunity (ETI), 799
*See also* Specific plant immunity
Effectors
in allosteric regulation, 159
in physiological systems, 816
plant specific immunity and, 799, 800
Efferent blood vessels, in fish gills, 1010
Efferent nervous system, 968
Efferent neurons
defined, 940
spinal reflexes, 942
Efferent pathways, of the autonomic nervous system, 975
Efferent renal arterioles
in the autoregulation of the glomerular filtration rate, 1087, *1088*
in the mammalian kidney, *1080*, 1082
in the vertebrate nephron, 1078
"Efficiency genes," 1048
EGF (epidermal growth factor), 398
Egg cell, 779, *781*, 783
Egg cytoplasm
components of, 903
rearrangement following fertilization, 903–904
Eggs
amniote, 692 (*see also* Amniote egg)
of conifers, *595*
cytoplasmic segregation, 395
fertilization in animals, 884–887 (*see also* Sperm–egg interactions)
fertilization in humans, 892
gametic isolation and, 477
genomic imprinting in mammals, 344–345
ovulation in humans, 892
parthenogenic reproduction, 881–882
production in animals, 882–883, 884
release in spawning, 887
reproductive technologies in humans, 897, 898
reproductive trade-offs in animals, 642
of seed plants, *592*
Ehrlich, Anne, 1245
Ehrlich, Paul, 1161–1162, 1175, 1245
Einkorn wheat, *226*
Ejaculation, 891
Ejaculatory duct, 890
EKG. *See* Electrocardiogram
Elaiosomes, 1135, 1146
Elastin, 819, 1039, *1040*
Elbert, Thomas, 328
Electric charge, of atoms, 22
Electric currents
creation of, 927
ionic, 931–932
Electric eels, 485

Electric organs, 485
Electric signals, employed by fish, 485
Electrical synapses, 926, 936, 939
Electrical work, 10
Electricity. *See* Bioelectricity/Bioelectric energy
Electrocardiogram (ECG), 1035–1036
Electrochemical gradients
  membrane potentials and, 929
  root uptake of mineral ions and, 729
Electrodes, measuring membrane potentials with, 928
Electroencephalogram (EEG), 978–979
Electromagnetic radiation
  photobiology of light, 189–190
  photochemistry of light, 188–189
Electromagnetic spectrum, *189*
Electromyogram (EMG), 978
Electron acceptors
  in denitrifiers, 539
  non-oxygen acceptors, 176
Electron carriers
  in the citric acid cycle, 170–171
  in oxidation–reduction reactions, 167–168
  reoxidation during glucose catabolism, 171
  in the respiratory chain, 172, *173*
Electron donors
  in anoxygenic photosynthesis, 187
  in oxygenic photosynthesis, 187
Electron microscopes, 79, 81, 108
Electron shells, 24–25
Electron transport
  in glucose metabolism, 168
  light-induced, 196–197
  with non-oxygen electron acceptors, 176
  in oxidative phosphorylation, 171
  in photosynthesis, 191–192
  relationships among metabolic pathways, *179*
  toxic intermediates, 175–176
Electronegativity, 28
Electrons
  in atoms, 22
  chemical bonding and, 24, 25
  covalent bonds, 26–28
  shells and orbitals, 24–25
  transport in the respiratory chain, 172, *173*
Electrooculogram (EOG), 978
Electrosensors, 947, *948*
Elegant madtom, *472*
Element cycling
  prokaryotes and, 538–539
  *See also* Biogeochemical cycles
Elementary bodies, 533, *534*
Elements
  atomic number, 22
  atomic weight, 23
  defined, 22
  isotopes, 22–24
  periodic table, 22, *23*
Elephant-nosed fish, *485*
Elephant seals, 1104, *1105*, 1114
Elephant shrews, *698*
Elephantiasis, 666
Elephants, *1135*
  brain size–body size relationship, 973
  fecundity, 1155
  full census counting, 1150
  impact of human overexploitation, 1234
  importance of ending international ivory trade, 1240–1241
  number of species, *698*
  physical challenges of snorkeling, 1005, 1022
  population age structure, 1152
  sound communication, 946
Elicitors
  activation of plant defenses to herbivory, 803–804
  defined, 798
  plant responses to, 798–801
Elk, 1193–1194, 1239
Elliott's milkpea, 1221–1223
Elongation, in transcription, *287*, 288
Embioptera, *672*
Embolism, 1043
Embolus, 1043
Embryo sac, 779, *781*
Embryogenesis
  basic patterns in animals, 633–634
  DNA methylation and, 344
  in mammals, 697
  in plants, 711–712
Embryonic stem cells (ESCs)
  culturing of, 408–409
  in homologous recombination and knockout mice, 381
  in the mammalian embryo, 906
Embryophytes, 573
  *See also* Land plants
Embryos
  of angiosperms, *600*, 601, *781*, 783
  cell fate determination during development, 394
  cleavage of the zygote in animals, 904–906
  of conifers, 594, *595*
  defined, 393
  development in humans, 920
  development in mammals, 697
  development in plants, *393*, 711–712
  developmental modules, 415–418
  features of, 393
  of land plants, 575
  patterns of care and nurture in animals, 889
  in plant evolution, 574
  from in vitro fertilization, 899
Emerging diseases, 501
Emission, 891
Emlen, Stephen, 1110
*Emmenanthe penduliflora*, 757
Emmer wheat, *226*
Emperor penguins, 1099
Emphysema, 1013
Emu, 694
Emulsifiers, 556, 1062
Enamel, 21, 1055
Enceladus (moon of Saturn), 70
*Encephalartos*, *593*
End-product inhibition, of metabolic pathways, 160–161
Endangered species
  captive breeding programs, 1244
  cloning, 407
  defined, 1231
Endangered Species Act, 17
*Endeis*, *668*
Endemism, 1237
Endergonic reactions, 147, *148*, 150–151
Endocrine cells, 835–836
Endocrine glands
  adrenal gland, 849–850
  anterior pituitary, 842–843, 844–845 (*see also* Anterior pituitary)
  defined, 835
  gonads, 850
  in humans, *842*
  pancreas, 848–849
  parathyroid glands, 847–848
  pineal gland, 851
  posterior pituitary, 842, 843
  thyroid gland, 845–847
Endocrine signaling, 835–836
  *See also* Animal hormones
Endocrine system
  in humans, *842*
  interactions with the nervous system, 842–845
  major glands and hormones, 845–852
  types of chemical signaling in, 835–836
  *See also* Animal hormones
Endocytosis
  in ciliates, 555
  defined, 120
  receptor-mediated, 120, 121
  types of, 120, *122*
Endoderm
  in avian gastrulation, 914
  defined, 908
  in diploblastic and triploblastic animals, 633
  in frog gastrulation, 909, 910
  in sea urchin gastrulation, 909
  tissues and organs derived from, *907*, 908
Endodermis, 717, *718*, 729–730
Endogenous retroviruses, 308
Endomembrane system, 88–91
  *See also* Endoplasmic reticulum; Golgi apparatus; Lysosomes
Endometrium
  implantation of the blastocyst, 892, 906, *907*, 919
  in the uterine cycle, 893, *894*
Endoparasites
  features of, 639, 1176
  flatworms, 657
  horsehair worm larvae, 666, *667*
Endophytic fungi, 615
Endoplasmic reticulum (ER)
  evolution in eukaryotic cells, 551
  glycosylation of proteins, 301
  structure and function, 88–90
  *See also* Rough endoplasmic reticulum; Sarcoplasmic reticulum; Smooth endoplasmic reticulum
Endorphins, *842*, 843, 940
Endoskeletons
  connective tissue in, 999–1000
  of humans, 999, *1000*
  interactions with skeletal muscle, 999, 1001–1002
  of radiolarians, *557*, 558
  *See also* Internal skeletons
Endosperm
  in angiosperm seeds, 591, *600*, 601
  formation in angiosperms, 601, *781*, 783
  in seed germination, 758
Endospores, 531
Endosymbionts, 564–565, 566
Endosymbiosis
  chloroplasts and photosynthesis in eukaryotes, 102, *550*, 551–552, 570
  overview, *101*, 102
  primary, 570
  transposons and, 366
Endosymbiotic bacteria
  colonization of the intestines, 1056–1057
  in pogonophorans, 660
Endothermic reactions. *See* Endergonic reactions
Endotherms
  adaptations to cold, 828, 1125
  basal metabolic rate and body size, 826–827
  control of blood flow to the skin, 824–825, 828
  defined, 822
  differences from ectotherms, 822
  dissipation of heat with water and evaporation, 829
  energy budget and, 823–824
  fevers, 830
  heat production in, 822, 827–828
  hibernation, 830, *831*
  hypothermia, 830
  response to changes in environmental temperature, 822, *823*
Endotoxins, 542
Endurance, impact of exercise on, 996–997
*Endymion non-scriptus*, *462*
Energetic costs, in animal behavior, 1103
Energy
  activation energy, 151–152
  ATP and, 149–151
  in chemical reactions, 31
  defined, 31
  from food, 1049–1050
  laws of thermodynamics, 146–147
  measures of, 1049
  nutrients as sources of, 10
  storage in the animal body, 1050
  transfer between trophic levels in communities, 1190–1192
  types of, 145
Energy budgets, 823–824, 1050
Energy maximization hypothesis, of foraging behavior, 1105

Energy transformation, organelles involved in, 91–93
English elm, 793
*Engrailed* gene family, 499–500
Enhancers
  as genetic switches, 415
  of transcription factors, 335
Enkephalins, *842*, 843, 940
Enteric nervous system, 968, 1059
*Enterococcus*, *4*
Enterokinase, 1062, *1062*
Enthalpy, 146, 147
Entomology, forensic, 1188
Entoprocts, *632*, *652*, 656
Entrainment, 1107–1108
Entropy, 146, 147
Enveloped virus, 341
Environment
  carrying capacity, 1158
  gene–environment interactions, 245–246
  modulation of development, 420–422
Environmental cleanup, 373, 389
  *See also* Bioremediation
Environmental cycles, coordination of animal behavior with, 1106–1108
Environmental genomics, 530
Environmental Protection Agency (EPA), 17
Environmental resistance, 1158
Environmentalism, 1122
Enzyme-catalyzed reactions
  effect of pH on, 161
  effect of substrate concentration on reaction rate, 156
  effect of temperature on, 161–162
  lowering of the energy barrier in, 151–154
  mechanisms of enzyme function, 154–156
Enzyme-substrate complex (ES), 152–153
Enzymes
  binding to substrates, 152–153
  as biological catalysts, 151
  chemical equilibrium, 153
  in cleaning aids, 144
  commercial applications, 144, 162
  effect of pH on, 161
  effect of temperature on, 161–162
  effect on the rate of reaction, 153–154
  functions, *42*
  induced fit, 155
  interactions with substrates, 154–155
  isozymes, 162
  loss of function mutations, 311–312
  lowering of the energy barrier in biochemical reactions, 151–154
  mechanisms in catalyzing reactions, 154–156
  modification during signal transduction, 138
  naming convention, 152
  nonprotein chemical "partners," 155–156
  one-gene, one-enzyme hypothesis, 282–283, 284
  regulation of, 156–162
  regulation of signal transduction, 136–137
  relationship of molecular structure to function, 155
  in signal transduction pathways, 126–127
Eons, *506–507*, 508
Eosinophils, *858*
*Ephedra*, *604*
Ephedrine, *604*, *802*
Ephemeroptera, *672*
*Ephestia kuehniella*, 1184
Epiblast, *907*, 913–914, 919
Epidemics, influenza, 427
Epidermal growth factor (EGF), 398
Epidermal growth factor receptor, 325
Epidermis
  of leaves, 720–721
  of plants, 712–713
  of roots, 717, *718*
Epididymis, 890, *891*
Epigenetics
  changes induced by the environment, 344
  defined, 328, 343
  DNA methylation, 343–344 (*see also* DNA methylation)
  effects of "royal jelly" on honey bee development, 899
  *FLC* gene expression in angiosperms, 791
  gene regulation and, 343–346
  histone modifications, 344
  manipulation of epigenetic changes through diet, 349
Epigenomes, 344
Epiglottis, 1059
Epilepsy, 982
Epiloby, 910
Epinephrine (adrenaline)
  actions of, *842*, 850
  activation of glycogen phosphorylase, 132–133
  in blood pressure regulation, 1044, *1045*
  endocrine source, 849
  fight-or-flight response, 837
  half-life, 852–853
  regulation of glucose metabolism in liver cells, 138
  structure, *836*
Epiphytes, 718
Epistasis, 244
Epithelial cells, 112
Epithelial tissues
  characteristics of, 817
  in the gut, 820
  types of, 817–819
Epitopes, 862, 864, 868
  *See also* Antigenic determinants
Epochs, 521, 522
Epstein–Barr virus, 876
Equatorial Countercurrent, 1124, *1125*
Equatorial plate
  in meiosis, *220*, *221*
  in mitosis, *215*
Equidae, 1143, *1144*
Equilibrium, organs of, *956*
  *see also* Vestibular system
*Equisetum*, 581
*Equisetum pratense*, *581*
*Equus*, *1144*
*Equus occidentalis*, *1229*
Erectile dysfunction (ED), 891
*Eremias lugubris*, 1174
Erinaceomorpha, *698*
Erlotinib, 325
Erosion, 1213
Error signals, 816
Erwin, Terry, 651, 673
*Erwinia uredovora*, 388
Erythrocytes. *See* Red blood cells
Erythropoietin, 211, *385*, 1038
*Escherichia*, 542
*Escherichia coli*
  cell division, 206
  circular chromosome, 206
  comparative genomics, 357
  conjugation in, 253–254
  DNA replication, 273
  functional genomics, *357*
  gene cloning, 375–376
  genomic comparison to *S. cerevisiae*, *362*
  genomic information, *361*
  illustration of, *2*
  *lac* operon in, 330, *331*
  LexA protein, 341
  Meselson–Stahl experiment on DNA replication, 268, *269*, 270
  as a model organism, 282
  negative and positive regulation of the *lac* operon, 330–332, *333*
  origins of replication, 269, 271
  pollution of lakes, 1221
  proteobacteria, 534
  recombinant plasmids, 374
  regulation of lactose metabolism, 329–330
  reporter genes, 378
  sigma factors, 333
  strain O157:H7, 357
  uses of genomic information from, 357
  viral infection experiments on DNA, 261–263
*Eschrichtius robustus*, *1109*
*Escovopsis*, 1169
Esophageal sphincter, 1060
Esophagitis, 612
Esophagus, 1059–1060
Essential amino acids, 1051–1052
Essential elements, in plant nutrition, 741–743
Essential fatty acids, 1052
Ester linkages
  in lipids, 535
  in phospholipids, 57, *58*
  in triglycerides, 56
  in waxes, 59
Estivation, 1077
Estradiol, 850
Estrogen receptor, 129, 325
Estrogens
  actions of, *842*
  defined, 850
  in follicle selection for ovulation, 896
  in human pregnancy, 896
  in human puberty, 894
  in labor and childbirth, 896
  in parthenogenic whiptail lizards, 882
  production in human ovaries, 893, *894*
  in regulation of the ovarian and uterine cycles, *894*, 895
  structure, *836*
  temperature-dependent sex determination and, 420
  *See also* Sex steroids
Estrus, 893
Estuaries, 1141
Ethanol
  from alcoholic fermentation, *177*, 178
  produced by yeast, 623
*Etheostoma tetrazonum*, *472*
*Etheostoma variatum*, *472*
Ether linkages, 535
Ethiopian region, *1142*
Ethiopian wolf, 1242
Ethnobotany, 605
Ethology, 1094–1096
Ethylene
  effects on plants, *759*, 769–770
  genetic screen in *Arabidopsis*, 760
  signal transduction pathway in plants, 770–771
  structure, *759*
Etiolated seedlings, 772
*Eublepharis macularius*, *694*
Eucalyptus leaves, 1055
Euchromatin, 345
Eudicots
  characteristics of, 601, 602, 709–710
  early shoot development, 758
  examples of, 603
  leaf anatomy, 720–721
  leaf veins, 720
  root anatomy, *717*
  root systems, 718
  shoot anatomy, 719–720
*Eudocimus ruber*, *696*
*Eudorina*, *140*, 141
*Euglena*, 558, *559*
Euglenids, 551–552, 558–559
Eukarya
  distinguishing characteristics, 81, *527*
  relationship to prokaryotes, *526*, 527
  in the tree of life, 8–9
Eukaryotes
  appearance in the Proterozoic, 515–516
  cellular specialization, 4
  defined, 81
  in the evolution of life, 4
  evolution of multicellularity in, 552–553
  origin of, 550–552
  Precambrian divergence of major groups, 552, *553*
  proteomes, 369, *370*
  protists, 550, 552–561
  relationship to prokaryotes, 527
  shared features with prokaryotes, 526
Eukaryotic cells
  animal cell structure, *86* (*see also* Animal cells)

atmospheric oxygen levels and, 512
cell division, 207–211 (*see also* Meiosis; Mitosis)
cellular locations of energy pathways, *168*
characteristics of gene expression in, *291*
compartmentalization in, 84
cytoskeleton, 94–98
endomembrane system, 88–91
endosymbiosis and, *550,* 551–552
evolution of, 550–552
extracellular structures, 99–100
genes and gene transcription, 290–293
genetic transformation, 263, 376–377
location and functions of RNAs in, *286*
methods of studying and analyzing organelles, 84, 85
nucleus, 85, *86, 87,* 88
organelles that transform energy, 91–93
origin of, 101–102
other types of organelles, 93–94
plant cell structure, *87* (*see also* Plant cells)
posttranscriptional gene regulation, 346–349
ribosomes, 84–85, 88
transcriptional gene regulation, 333–338

Eukaryotic genomes
- *Caenorhabditis elegans,* 362, *363*
- *Drosophila melanogaster,* 362
- features of, 361
- gene families, 363–364
- plants, 363
- repetitive sequences, 364–366
- yeast, 361–362

Eukaryotic viruses, 341
Eumetazoans, *630,* 643
*Eupholus magnificus, 674*
*Euphorbia,* 801
*Euphydryas editha bayensis,* 1161–1162
Euphyllophytes, 583
*Euplectella aspergillum, 643*
*Euplectes progne, 435*
*Euplotes, 554*
*Euprymna scolopes,* 546
Europa (moon of Jupiter), 70
European bee-eaters, *888*
European common cuckoo, 1093
Euryarchaeota, 535, 536–537
*Eurycea waterlooensis, 691*
*Eurylepta californica, 657*
Eusociality, 1115–1116
Eustachian tube, 954–955
*Eusthenopteron, 690*
*Eusthenopteron foordi, 517*

Eutherians
- cleavage in, 905–906
- evolutionary relationships in, 697–698, *699*
- herbivores, 700
- key features, 697
- major living groups and number of species, *698*
- primates, 701–705
- return to an aquatic environment, 700
- rodents and bats, 699–700

Eutrophication, *533,* 1219, 1221
Evans, Martin, 382

Evaporation
- in the global hydrologic cycle, 1215
- heat dissipation in endotherms and, 829
- in heat exchange between animals and their environment, 823, *824*

Evapotranspiration, 1192
Even-toed hoofed mammals, *698*
Evo-devo, 413
- *See also* Evolutionary developmental biology

Evolution
- adaptations, 6, *7*
- biological classification and, 463
- concerted evolution, 498–499
- constraints on, 423–424, 444–446
- defined, 6, 428, 432
- emergence and impact of photosynthesis, 4–5
- evolutionary history and phylogenies, 451
- evolutionary tree of life, 6–9
- genetic basis of, 431
- in genome function, 496–499
- in genome size, 494–496
- maintenance of genetic variation in populations, 441–444
- mechanisms in, 432–436
- methods of measuring, 436–440
- "modern synthesis" with genetics, 430–431
- by natural selection, 428–430
- "opportunistic," 753
- parallel, 423–424
- of populations, 6
- relationship between fact and theory in, 428–431
- role of membranes in, 3–4
- short-term and long-term outcomes, 446
- use of genomes in the study of, 486–491
- using molecular clocks to date evolutionary events, 461–462
- in viruses, 427
- in vitro evolution, 500–501
- *See also* Molecular evolution

Evolutionary developmental biology
- basic principles of, 413
- beak diversity in birds, 412, 424
- developmental genes, 413–414, 423–424
- developmental modules, 415–418
- differences in gene expression resulting in differences between species, 418–420

Evolutionary radiations, 481–482
Evolutionary reversal, 452

Evolutionary theory
- Darwin's and Wallace's concepts, 428–430
- defined, 428
- development following Darwin, 430–431
- practical applications of, 427, 428, 446

Evolutionary tree of life, 6–9
Evolutionary trends, 440
*Ex vivo* gene therapy, 323
Excavates, 558–559
Excision repair, 276, 277
Excitable cells, 925
Excitatory synapses, 938

Excretory systems
- function of, 1071
- of invertebrates, 1075–1077
- in mammals, 1079–1086
- mechanisms to maintain homeostasis, 1072–1074
- nitrogen excretion, 1074–1075
- in vertebrates, 1077–1079
- *See also* Kidneys; Nephrons

Excurrent siphon, 662, *663,* 664

Exercise
- impact on muscle strength and endurance, 996–997
- integration of anabolism and catabolism during, 180–181

Exergonic reactions
- activation energy, 151–152
- ATP and energy-coupling, 150–151
- characteristics of, 147, *148*

Exit site, 295, 296
Exocrine cells, 333
Exocrine glands, 835, 1062

Exocytosis
- description of, *120,* 122
- sweating and, 105

Exons
- alternative splicing, 346–347
- description of, *290,* 291

Exoskeletons
- arthropod evolution and, 655
- of arthropods, 667
- of ecdysozoans, 655
- features of, 999
- in protostome evolution, 674–675

Exothermic reactions. *See* Exergonic reactions
Exotoxins, 542
Expanding triplet repeats, 313–314
Expansins, 710
"Experimental" group, 12
Experiments, types of, 12–13
Expiratory reserve volume (ERV), *1012,* 1013
Exploitation competition, 1182–1183
Exponential population growth, 1157–1158, 1164–1166
Expression vectors, 384
Expressivity, 246
Extensor muscle, 1002
External anal sphincter, 1063
External digestion, 1055
External fertilization, 887, 888
External gills, 1008, *1009*
Extinction. *See* Mass extinctions; Species extinctions
Extinction rates, in island biogeography theory, 1196–1198, *1199*

Extracellular fluid
- in closed circulatory systems, 1026
- in the internal environment of multicellular animals, 816
- in open circulatory systems, 1026
- regulation of homeostasis by excretory systems, 1071, 1072–1074 (*see also* Excretory systems)

Extracellular matrix
- biological membranes and, *106*
- cell adhesion to, 111–113
- collagen fibers in, 818–819
- of ctenophores, 644
- of sponges, 644
- structure and function, 100

Extraembryonic membranes
- in the amniote egg, 692
- in the chicken egg, 918
- functions of, 918
- origins of, 914
- in placental mammals, 919

Extreme halophiles, 536–537
Extremophiles, 532
*Exxon Valdez* (oil tanker), 373

Eye color
- determination in humans, 246
- sex-linked inheritance in *Drosophila,* 251–252

Eye cups, 958
Eye impairments, brain development in humans and, 921
Eye infections, 533
*Eyeless* gene, 413, *414*

Eyes
- of cephalopods, 664
- compound, 958
- developmental genes and, 413, *414*
- image-forming, 958–959
- lens determination in vertebrates, 396
- structure and function of the retina, 959–963

"Eyes," of potatoes, *720*

## F

F-box-containing proteins, 767
Fabaceae, 747–748
"Face neurons," *971*
Facial recognition, 971

Facilitated diffusion
- aquaporins, 116
- carrier proteins and, 117
- channel proteins and, 115–116
- characteristics of, *118*
- saturation of, 117

Facilitation, 1201
Factor VIII, *385*
Facultative anaerobes, 537
Facultative parasites, 611
Fallopian tubes, 892, *893*
Familial hypercholesterolemia, 121, *313*
Family (taxonomic category), 462–463
"Fan worms," *661*
Fanged striped blenny, *1174*

Far-red light
- photomorphogenesis and phytochromes in plants, 772–774
- in photoperiodic control of flowering, 788

Farber, Sidney, 158
"Farming behavior," of cowbirds, 1093
Fas protein, *873,* 874
Fast block to polyspermy, *885,* 886
Fast-twitch fibers, 995
Fat metabolism
  control by the liver, 1065–1066
  in the postabsorptive state, 1067
Fat-soluble vitamins, 1053–1054
Fat tissues. *See* Adipose tissues
Fate maps, of blastomeres, 907
Fats
  absorption in the small intestine, *1062,* 1063
  breakdown by digestive enzymes, 1057, 1058
  digestion in the small intestine, 1061–1062
  energy yield, 1049, *1050*
  storage in the liver, 1065
  as stored energy in the animal body, 1050
  structure and function, 56–57
  triglycerides of animal fats, 57
Fatty acid synthase, 182
Fatty acids
  anabolic interconversions, 180
  catabolic interconversions, 179–180
  formed in prebiotic synthesis experiments, 70
  need for in the human diet, 1051
  in phospholipids, 57, *58,* 107
  significance in the evolution of life, 3–4
  structure and function, 56–57
  in waxes, 59
Fauna, 514
*FD* gene, 790
FD protein, 791
Fear and fear memory, 970, 982
"Feather duster worms," *661*
Feather stars, 520, 680–681
Feathers
  anatomy of, *696*
  evolution of, 695
  flight in birds and, 695–696
  as thermal insulation, 828
Fecal matter, 1063
Feces, 1056, 1063, 1064
Fecundity, 1154–1155, 1156
Federal Bureau of Investigation (FBI), 317–318
Feedback
  in the mechanisms regulating breathing, 1020–1022
  in regulatory systems, 816
Feedback inhibition, of metabolic pathways, 160–161
Feedforward information, 817
Feeding anthers, 1180
Feeding strategies
  detritivores, 639
  filter feeders, 637, *638*
  herbivores, 637–638
  overview, 637
  parasites, 638–639
  predators and omnivores, 638
Feeding structures
  in protostome evolution, 674
  *See also* Teeth
Feigin, Andrew, *323,* 324
Female athlete triad, 1000–1001
Female flowers, 779, *780*
Female reproductive system
  childbirth, 896–897
  components and function of, 892, *893*
  embryo–mother connection in, 889
  follicle selection for ovulation, 895–896
  ovarian and uterine cycles, 893–895
  pregnancy, 896
Females
  genomic imprinting in mammals, 344–345
  length of meiosis in, 220
Feminization, atrazine and, 1, 18
Fenestrations, 1041
Feral cattle, 440
Fermentation
  in beer and wine production, 623–624
  energy yield from cellular respiration and fermentation compared, 178
  in glucose metabolism, 168
  overview, *166,* 167
  pathways in, 177–178
Ferns
  in the Devonian, 520
  distinguishing characteristics, *574,* 582
  life cycle, 582
  monilophytes, 581
  number of chromosomes in, 218
  sexual life cycle, *218*
*Ferocactus wislizeni,* 1179
Ferredoxin, 196, *197*
Ferritin, 348.*23*
Ferrous ion, 187
Fertile Crescent, 388
Fertilization
  activation of development in animals, 903–904
  in animals, 884–887
  defined, 218, 882
  double fertilization in angiosperms, 600–601, *781,* 783
  fruit development in angiosperms and, 785
  in humans, 892
  internal and external, 887
  in mosses, 576
  phylogenetic analysis of mechanisms in angiosperms, 459–460
  in seed plants, 590, *592*
Fertilization anthers, 1180
Fertilization cone, *885,* 886
Fertilizers
  inorganic, 747
  organic, 746–747
  to treat plant deficiencies, 742
  *See also* Chemical fertilizers; Nitrogen fertilizers
Fetal hemoglobin, 1018
Fetscher, Elizabeth, *599*
Fetus
  birth of, 896–897
  defined, 920
  development of, 920
Feulgen, Robert, 260
Fevers, 830, 861–862
Fibers, in plants, 596, *713,* 714
Fibrin, 1039
Fibrinogen, 890, 1039
Fibrinolysin, 890
Fibroblasts, 336, *337,* 338
Fibrous root systems, 718
*Ficedula albicollis,* 1103
Ficke, Henry, 21
Fick's law of diffusion, 1006
Fig trees, 1194–1195
Fight-or-flight response, 837, 850, 852–853, 974
Figs, 601
Filamentous sac fungi, 620, 622
Filaments (of anthers), *591,* 596, 598
Filopodia, 909
Filter feeders, 637, *638,* 1054
Filtration
  of blood plasma in tubule capillaries, 1072
  in the vertebrate nephron, 1078–1079
Fimbriae (fimbria), 84, 892, *893*
Finches, 440
  *See also* Darwin's finches; Galápagos finches
Fins, 686–687, *688*
Fire
  Australian deserts and, 1126
  controlled burning, 1240
  the fynbos and, 1121, 1135, 1145, 1146
  movement of elements through ecosystem compartments and, 1214
Fire adaptations, in pines, 594, *596*
Fire-bellied toad, *479*
Fireflies
  bioluminescence, *149,* 150
  visual signaling, 1111
Firmicutes, 530–532
First filial generation ($F_1$), 234, 235
First law of thermodynamics, 146
First polar body, *883,* 884
Fishes
  acclimatization to seasonal temperatures, 821
  actions of prolactin in, *838*
  calcium ion cycling in the hearts of "hot" fish, *1036,* 1037
  circulatory system, 1027, 1028
  countercurrent heat exchange, 825
  developmental constraints on evolution, 444, *445*
  evolution of electric organs, 485
  external fertilization in, 887, 888
  jawed, 686–689
  jawless, 685–686, *687*
  overharvesting, 16–17, 1163–1164, 1235
  parallel evolution in, 423–424
  respiratory gas exchange in gills, 1009–1010, *1011*
  salt and water balance regulation in marine fish, 1077
  sequential hermaphroditism, 887, *888*
  taste buds, 951
  *See also* Chondrichthyans; Haplochromine cichlids
Fishflies, *672*
Fishhook barrel cactus, 1179
Fishing industry, 16–17, 1163–1164, 1235
Fission yeast, 624, 626
Fitness
  altruistic behavior and, 1114–1115
  defined, 438
  direct, 1114
  inclusive, 1115
  indirect, 1115
  maximization by mating systems, 1113–1114
  sex-specific differences in lizards, 421
Fitzpatrick, John, 1115
5′ Cap, 291–292, 293, 296
Fixed action patterns, 1095
Fixed populations, 437
*Flabellina iodinea, 664*
Flagella
  of chytrids, 617
  of dinoflagellates, 553, *554*
  of euglenids, 558, *559*
  of opisthokonts, 609
  of prokaryotes, *82,* 83–84
  of stramenopiles, 555
  structure and function, 96–98
Flagellin, 799
Flame cells, 1075
Flame fairy wrasse, *688*
"Flame tree," *603*
Flamingos, 637, *638*
Flathead Lake, 1228
Flatworms, 675
  acoelomate, 635
  anatomical characteristics, *652*
  description of, 656–657
  excretory system, 1075
  eye cups, 958
  major subgroups and number of living species, *632*
Flavin adenine dinucleotide (FAD)
  in the citric acid cycle, 170, 171
  in glucose metabolism, 168
  in NADH shuttle systems, 179
  in the respiratory chain, 172, 173
  role in catalyzed reactions, 155, *156*
Flavonoids, 747, *748*
Flax, 800
*FLC* gene, 791
FLC protein, 792
Fleas, 639, *672,* 986
Flegal, A. Russell, 625
Fleming, Alexander, 383, 608
Flesh flies, *1181,* 1203
Flexors, 942, 1002
Flight
  bird physiology and, 695–696
  contraction in insect flight muscles, 997–998
Flight feathers, 695, *696*
Flightless birds, 694, *695*
Flightless weevils, 1144–1145
Flocking, 1116, *1117*
Flooding, 1140
Flor, Harold Henry, 800
Flora, 514

Floral meristems, 400, 715, 785–786
Floral organ identity genes, 400–401, 786
Florey, Howard, 608
Florida scrub-jays, 1115
Florigen, 790
  *See also* FT–FD florigen pathway
Flounders, 444, *445*
Flowering
  categories of, 785
  cues from an "internal clock," 792
  floral organ identity genes, 400–401, 786
  florigen, 790
  the flowering stimulus originates in leaves, 788–790
  induction by gibberellins, 791–792
  induction by temperature, 791
  meristem identity genes, 786
  photoperiodic cues, 787–788
  transition of shoot apical meristems to inflorescence meristems, 785–786
Flowering plants. *See* Angiosperms
Flowers
  anatomy, *591,* 596–597
  in angiosperm sexual reproduction, 779, *781*
  coevolution with animals and, 599
  Darwin on the evolution of, 605
  evolution of, 597–598, *599*
  forms, 596, *597*
  in monocots and eudicots, 710
  monoecious and dioecious plants, 597
  organ determination, 400–401
  parts of, 779, *781*
  perfect and imperfect, 597, 598
  pollination (*see* Pollination)
  pollination syndromes, 1180–1181
  symmetry, 597
  types of, 779
Flu epidemics, 427
Flu vaccines, 427, 446
Fluid feeders, 1054
Fluid mosaic model, 106
Flukes, 657, 675
Fluorescence microscopy, 80
Fluorescent dyes, 353, 354
Fluorescent proteins
  discovery of green fluorescent protein, 449
  evolution in corals, *464*
Fluorine, *1052*
5-Fluorouracil, 228, *229*
Fluxes, 1211
Flying insects. *See* Pterygotes
Flying squirrels, 962
*FMR1* gene, 313–314
FMR1 protein, 314
Fog-basking beetle, *1132*
Folate, 916
Folic acid, 349, 916, *1053*
Foliose lichens, 613
Follicle cells, 895–896
Follicle-stimulating hormone (FSH)
  actions of, *842*
  endocrine source, *842,* 843
  in follicle selection for ovulation, 896
  in puberty, 850–851, 894
  in regulation of human spermatogenesis, 892
  in regulation of the ovarian and uterine cycles, 893–894
Follicles, of the thyroid, 845, *846*
Follicles (human ovaries)
  selection for ovulation, 895–896
  structure and function, 893, *894*
  in the uterine cycle, 894–895
Fontanelle, 1001
Food
  as an energy source, 1049–1050
  brain areas controlling intake, 1067–1068
  effect of food supply on population growth, 1159
  ingestion and digestion in animals, 1054–1058
  movement through the vertebrate gut, 1059–1060
  as a source of carbon skeletons for biosynthesis, 1051–1052
Food acquisition. *See* Foraging
Food allergies, 876
Food chains, 1190
Food webs, 1190, *1191*
Foot
  development in ducks and chickens, 417
  molluscan, 662, *663*
Foraging
  costs of foraging in groups, 1116
  optimal foraging theory, 1104–1105, *1106*
Foraminiferans, 557, 565
Forebrain
  in human brain evolution, 973
  insular cortex, 983
  structure and function, 968–969, 970
Foregut fermenters, 492–494
Forelimbs, loss in snakes, 423
Forensics
  DNA fingerprinting, 317–318
  forensic entomology, 1188
  use of phylogenetic analyses in, 458, *459*
Foreskin, 890
Forest corridors, 1162
Forest fires
  conifer adaptations, 594, *596*
  cyclical succession and, 1201–1202
  *See also* Fire
Forests. *See individual forest types*
Forewings, 672
Forgetting, 980
*Formation of Vegetable Mould Through the Action of Worms, The* (Darwin), 639
Fossil fuel burning
  acid precipitation and, 1219–1220
  global climate change and, 17, 510
  impact on the global carbon cycle, 1216
Fossil fuels
  formation of, 1211
  phytoplankton and, 569
Fossil record
  Cambrian fossil beds, 516
  dating fossils, 506–508
  microfossils, 74
  number of species identified, 514
  phylogenetic analyses and, 455
  reasons for the paucity of, 514–515
  *See also* Paleontology
Founder effect, 434
Founder events, in speciation, 467, 472, *473*
Four-chambered heart
  in birds, 1029, 1030
  in crocodilians, 1029
  in fishes, 1028
  in mammals, 697, 1030
Four o'clock plants, *252*
Fovea, *959,* 962
Foxglove, *604*
*FOXP2* gene, 367
Fraenkel, Gottfried, 1175
Fragile-X syndrome, 313–314
Fragmented habitats, 1233–1234
Frame-shift mutations, 306–307
*Frankenia palmeri,* 811
Franklin, Rosalind, 264
Frank–Starling law, 1042
Free-air concentration enrichment (FACE), 185
Free energy
  ATP and, 149, 150
  chemical equilibrium and, 148–149
  defined, 146–147
  enzymes and, 153
  harvested during glucose oxidation, 166–167
Free radicals, 309
Free-running circadian clock, 1107–1108
Freeze-fracture microscopy, 81, 108
Freeze-tolerant plants, 810
"French flag model," of morphogens, 401
Frequency-dependent selection, 441–442
Freshwater biomes, 1140
Freshwater ecosystems
  consequences of human alteration, 1223
  impact of pollution on, 1233
  transport of elements through ecosystem compartments, 1213–1214
Freshwater protists, 93
Frog eggs
  aquaporins, 116
  complete cleavage, 905
  rearrangements following fertilization, 903–904
Frogs
  adaptations in, *7*
  experiments on the effects of atrazine, 12, 13
  fate map of the blastula, *907*
  features of, 690–691
  feminization induced by atrazine, 1
  gastrulation, 909–910
  germ cell lineage, 908
  jumping ability, 986
  lens determination in the embryo, 396
  mating calls, 705
  nitrogenous wastes excreted, 1075
  nuclear transfer experiments with the embryo, 406
  parental care, 678
  prezygotic isolating mechanisms, 476, *477*
  salt and water balance regulation, 1077
  sexual signals, 435
*Fromia, 881*
Frontal lobe, 971–972
Fructose
  absorption in the small intestine, 1063
  production in photosynthesis, 195
  in seminal fluid, 890
  structure, *52*
Fructose 1,6-bisphosphate, *55, 169*
Frugivores, 1181–1182
Fruit flies
  compound eye, 958
  genomic information, *361,* 362
  as a model organism, 282
  sex determination in, *249*
  sex-linked inheritance, 251–252
  *See also* Drosophila
Fruiting structures
  of fungi, 609–610, 620, *621,* 622
  of slime molds, 561
*Fruitless (fru)* gene, 1097–1098
Fruits
  auxins in fruit development, 766
  defined, 784
  dispersal, 784
  ethylene and fruit ripening, 769–770
  functions of, 784
  gibberellins in fruit growth, 761
  parthenocarpic, 766
  plant–frugivore mutualisms in seed dispersal, 1181–1182
  relationship between seed development and fruit development, 784–785
  seedless, 784–785
  a synapomorphy of angiosperms, 596
Fruticose lichens, 613
FSH receptors, 895–896
*FT* gene, 790
FT protein, 790, 791
FT–FD florigen pathway, 790, 791, 792
Fucoxanthin, 556
Fugitive species, 1183
Fulcrum, 1002
Full census, 1150
Functional genomics, 355–356, *357*
Functional groups
  of chemically modified carbohydrates, 55
  of macromolecules, 40
  properties of, *40*
Functional residual volume (FRV), *1012,* 1013
Fundamental niche, 1184
Fungal mutualisms
  discussion of, 612–615

evolution of land plants and, 574
fungus farming, 1169, 1178, 1185
with nonvascular land plants, 575
*See also* Mycorrhizae
Fungi
antibiotics derived from, 608
arbuscular mycorrhizal fungi, 619–620
in bioremediation, 625
chytrids, 617–619
classification of major groups, *616*
Dikarya, 620–623
distinguishing characteristics, 609
endophytic, 615
in food and drink production, 623–624
global carbon cycle and, 611
hyphae, mycelium, and fruiting structures, 609–610
life cycles, *616, 618, 621*
microsporidia, 617
microtubule organizing centers, 212
as model organisms in lab studies, 625, 626
parasitic, 611–612
pathogenic, 612
phylogenetic relationships, 615–616
predatory, 612, *613*
in reforestation efforts, 626
sexual and asexual reproduction, 616–617
sexual life cycle, *218*
surface area-to-volume ratio of mycelium, 610
tolerance for hypertonic environments, 610–611
tolerance of temperature extremes, 611
unicellular yeasts, 609
used to control diseases and pests, 626, 627
used to study environmental contamination, 624, 625
zygosopore fungi, *618,* 619
*See also* Fungal mutualisms
Fungus farming, 1169, 1178, 1185
Funk, Casimir, 1054
Fur, 697, 825, 828
Furcula, 694
*Fusarium oxysporum,* 626
Fynbos
characteristics of, 1121, 1134–1135
economic benefits of, 1242–1243
fire and, 1121, 1135, 1145, 1146
introduced Argentine ant, 1145
Mediterranean climate, 1126

## G

G protein-linked receptors
in the $IP_3$/DAG pathway, 134, *135*
structure and function, 129–130
G1 phase (of mitosis), 208, *217*
G1-to-S transition
description of, 208
internal signals controlling, 210–211
G2 phase (of mitosis), 208, 211–212, *217*
GABA. *See* γ-Aminobutyric acid
GABA receptors, 943
Gage, Phineas, 972
Gain-of-function mutations, 305, 306, 400–401
*Galactia elliottii,* 1221–1223
Galactosamine, *55*
Galactose, *52*
β-Galactosidase, 329
β-Galactosidase gene, 378
β-Galactoside, 329
β-Galactoside permease, 329
β-Galactoside transacetylase, 329
Galagos, 701
Galápagos finches
adaptive radiation, 481
allopatric speciation, 473, *474*
beak diversity, 412, 424
heterometry in beak development, 416
*See also* Darwin's finches
Galápagos hawk, *2*
Galápagos Islands
Darwin's visit to, 428, *429*
exploitation competition, 1183
Galápagos tortoises, *694*
Gallbladder, *1058,* 1061, 1065
*Gallus gallus, 361*
Gametangia
of mosses, 575, *576*
in plant evolution, 574
of zygospore fungi, *618,* 619
Gametes
in alternation of generations life cycle, 562, 563
genomic imprinting in mammals, 344–345
Mendelian law of segregation and, 236–237, 239
number of chromosomes in, 218
production in animals, 882–884
in sexual reproduction, 218, 219
Gametic isolation, 477
Gametogenesis
in animals, 882–884
defined, 882
Gametophytes
of angiosperms, 596, 600, 779–780
of conifers, 594, *595*
of ferns, 582
in homospory, 584
of hornworts, 578, *579*
of land plants, 575
of liverworts, 577
of mosses, 577–578
of nonvascular land plants, 575–576
relationship to sporophytes in plant evolution, *590*
of seed plants, 589, 590, 591, *592*
γ-Aminobutyric acid (GABA), 324, 939
Ganglia
in annelids, 659–660, 941
of the autonomic nervous system, *974,* 975
Ganglioside, 91
*Ganoderma applanatum,* 623
Gap genes, 403, *404*
Gap junctions
in cardiac muscle, 991
structure and function, 111, *112,* 139–140
Garrod, Archibald, 282
Garter snakes, 445, 889
Gas exchange systems
components of, 1006
fish gills, 1009–1010, *1011*
human lungs, 1013–1016
in insects, 671, 1009
maximization of gas exchange surface area, 1008–1009
maximization of partial pressure gradients, 1009
physical factors affecting, 1006–1008
snorkeling elephants, 1005, 1022
unidirectional ventilation in birds, 1010–1012
Gases
of the atmosphere, 1211
partial pressures and diffusion, 1006
solubility in liquids, 1006
*See also* Carbon dioxide; Hydrogen; Nitrogen; Oxygen; Respiratory gases
*Gasterosteus aculeatus,* 423–424
Gastric brooding, 678
Gastric mucosa, *1060,* 1061
Gastric pits, 1060–1061
Gastrin, 1065
Gastritis, 1054
Gastrointestinal disease, 534
Gastrointestinal disorders, autism and, 540
Gastrointestinal system
absorption of nutrients by the liver, 1063
absorption of nutrients in the small intestine, *1062,* 1063
chemical digestion in the mouth and stomach, 1060–1061
concentric tissue layers in, 1058–1059
control and regulation of the flow of nutrients, 1064–1068
digestion in the small intestine, 1061–1062
digestion of cellulose in herbivores, 1063–1064
enteric nervous system, 1059
large intestine, 1063
movement of food through, 1059–1060
peritoneum, 1059
*Gastrophryne carolinensis, 477*
*Gastrophryne olivacea, 477*
Gastropods, 662, *663, 664*
Gastrotrichs, *632, 652,* 656, 658
Gastrovascular cavity, 645, 648, 1056
Gastrulation
in amphibians, 909–913
basic patterns, 633–634
defined, 908
germ layers formed during, 908
in mammals, 914–915
in reptiles and birds, 913–914
in sea urchins, 908–909
Gated ion channels
chemically gated channels, 930, 937
mechanically gated channels, 930
membrane potentials and, 930–932
as receptors, 129
structure and function, 115, *116*
types of, 930
*See also* Voltage-gated ion channels
Ge Hong, 797
Gehring, Walter, 413
Geiger counter, 24
Gel electrophoresis, 316–317
Gemmae, 577
Gemmae cups, 577
Gemsbok, *1132*
Gender identification, 1096–1097
Gene cascades, in control of animal behavior, 1097–1098
Gene duplications
in the evolution of electric organs, 485
genetic diversity and, 310
as a source of new genome functions, 496–498
transposons and, 365
Gene expression
blocked by antibiotics in bacteria, 281
"central dogma" of molecular biology, 285
changes in plants in response to pathogens, 799, 800
defined, 65
differences between prokaryotes and eukaryotes, *291*
differences in expression resulting in differences between species, 418–420
effect on noncoding sequences on, 494
evidence for proteins as major products of, 282–284
genome characteristics and, 486
genomic information and, *355*
methods of studying, 380–383
overview, 284–285
pattern formation and, 399–405
primary embryonic organizer and, 911–912
in RNA genomes, 285
role in development, 6, 397–399
sequential pattern during determination of fruit fly body segmentation, 401–405
signal sequences and polypeptide movement within the cell, 298–300
transcription, 286–293
translation, 293–297, *298*
visualizing with fluorescent proteins, 449
Gene families
concept of, 497, *498*
concerted evolution and, 498–499
*engrailed* gene family, 499–500
in eukaryotes, 363–364
Gene flow
barriers leading to speciation, 472–475
effect on allele frequencies, 433–434

Gene-for-gene concept, 800
Gene-for-gene resistance, 800
"Gene guns," 376
Gene mutations
  effect on phenotype, 305–306
  molecular medicine and, 304
  point mutations, 306–307
  reversal of, 306
  types in multicellular organisms, 305
  *See also* Mutations
Gene pool, 431
Gene regulation
  DNA methylation and, 328
  effect of stress on during prenatal development, 328
  epigenetic changes, 343–346
  posttranscriptional mechanisms in eukaryotes, 346–349
  potential points for in eukaryotes, *334*
  in prokaryotes, 329–333
  transcriptional regulation in eukaryotes, 333–338
  in viruses, 339–343
Gene regulatory proteins, *42*
Gene sequences
  environmental genomics, 530
  evolutionary relationships in prokaryotes and, 528–529
  identification of the Archaea and, 534
Gene-silencing mechanisms, 347–348
Gene therapy, 323–324
Gene trees
  description of, 497, *498*
  of the *engrailed* gene family, 499–500
  lateral gene transfer in prokaryotes and, 529–530
General plant immunity, 799–800, 801
General transcription factors, 333–334, *335*, *336*
Generative cell, 780, 783
Genes
  alleles, 431
  animal behavior and, 1096–1098
  biological information and, 5, 6
  cloning, 375–376 (*see also* Molecular cloning)
  comparing through sequence alignment, 486–487
  defined, 66
  duplication (*see* Gene duplications)
  epistasis, 244
  evidence for being DNA, 260–263
  gene trees, 499–500
  gene–environment interactions, 245–246
  genetic markers, 318–319
  genome characteristics and, 486 (*see also* Genomes)
  genotype and phenotype, 431
  homologous, 413–414, 499–500
  hybrid vigor, 244–245
  inactivation by homologous recombination, 381–382
  incompatibilities and reproductive isolation, 470, *471*
  inheritance of organelle genes, 252–253
  inserting in cells, 376–377 (*see also* Recombinant DNA technology)
  largest human gene, 291
  lateral gene transfer, 496, 529–530
  law of independent assortment, 237–239
  law of segregation, 236–237, 239
  linkage, 247, 248, 249–252
  in Mendelian inheritance, 236
  methods of studying gene expression, 380–383
  methylated, 343–344 (*see also* DNA methylation)
  multiple alleles, 242
  mutations (*see* Gene mutations; Mutations)
  numbers in the human genome, 366
  one-gene, one-enzyme hypothesis, 282–283, 284
  one-gene, one-polypeptide relationship, 283–284
  phenotype and, 237
  promoters, 286
  pseudogenes, 491, 494–495
  quantitative trait loci, 246
  recombinant frequencies, 248, *250*
  recombination and mapping, 247–249, *250*
  reporter genes, 377–378, *379*
  sizes in the human genome, 366
  therapeutic, 323–324
  transfer in prokaryotes, 253–255
Genetic code
  commonality of, 289
  description of, 288–289
  missense mutations, 306, *307*
  redundancy, 289
"Genetic determinism," 245
Genetic diversity
  gene duplications and, 310
  generated by chromatid exchanges during meiosis I, 219–220, *222*
  generated by independent assortment of homologous chromosomes, 222
  from meiosis and sexual reproduction, 218–219, 245, 882
  mutations and, 310
  *See also* Genetic variation
Genetic drift
  evolution and, 6
  fixation of neutral mutations, 492
  impact on small populations, 434
Genetic markers
  in transformation experiments, 263
  uses of, 318–319, 352
Genetic recombination
  in ciliates, 562
  genetic mapping and, 248–249
  with homologous chromosomes, 247–248, *250*
  inactivation of genes by, 381–382
Genetic screening
  allele-specific oligonucleotide hybridization, 321
  DNA testing, 320–321
  identification of plant signal transduction pathways, 759, 760
  purposes, 320
  screening for disease phenotypes, 320
Genetic structure, 437
Genetic switches, 415
Genetic toolkit, 414, 415
Genetic transformation, 260–261, 263, 376–377
  *See also* Recombinant DNA technology
Genetic variation
  generation by mutation, 432
  geographically distinct populations within species, 443–444
  maintenance by frequency-dependent selection, 441–442
  mechanisms maintaining in populations, 441–444
  phenotypic variation and, 431
  selection on leads to new phenotypes, 432–433
  transposons and, 365–366
  *See also* Genetic diversity
Genetically modified organisms (GMOs)
  in agriculture, 386–388, *389*
  patenting, 373
  public concerns, 388–389
Genetics
  behavioral, 1096–1098
  codominance, 243
  evidence for DNA being the genetic material, 260–263
  incomplete dominance, 242–243
  mechanisms of gene interaction, 244–246
  Mendelian laws of inheritance, 233–241
  model organisms, 282
  "modern synthesis" with evolution, 430–431
  monohybrid crosses, 234–236
  multiple alleles, 242
  mutations, 241–242
  pleiotropy, 243–244
  probability calculations, 239–240
  test crosses, 237, *238*
Genitalia, 887
Genome sequencing
  applications, 353
  defined, 353
  "genetic determinism," 245
  information yield, 355–356
  methods in, 353–354, *355*
  prokaryotes, 356
Genomes
  accumulation of deleterious mutations, 441
  biological information and, 5–6
  characteristics of, 486
  of cheetahs, 62
  defined, 66, 486
  detecting positive and purifying selection in, 492–494
  differential gene expression, 6
  of dogs, 352
  endogenous retroviruses in vertebrate genomes, 545
  environmental genomics, 530
  of eukaryotes, 361–366
  evolution in size, 494–496
  gain of new functions, 496–499
  gene–environment interactions and, 245
  "genetic determinism," 245
  of humans, 366–369 (*see also* Human genome)
  "junk" sequences, 347
  microRNA, 347, 348
  of mimiviruses, 545
  minimal genome studies, 359, 360
  mutations and, 6
  phylogenetic analyses and, 456
  positive-sense, 544
  of prokaryotes, 356–360
  representative organisms, *361*
  small interfering RNAs, 347–348
  smallest animal genome, 629
  in the study of evolution, 486–491
  of *Thermoplasma*, 537
  variation in Tasmanian devils, 232
Genomic imprinting, 344–345
Genomic libraries, 379
Genotype
  defined, 431
  gene–environment interactions, 245–246
  in Mendelian genetics, 236
  relationship of phenotype to, 431
  sexual recombination amplifies the possible number of, 441
Genotype frequency
  calculating, 436–437
  defined, 432
  effect of nonrandom mating on, 434–436
  Hardy–Weinberg equilibrium, 437–438
Genotyping technology, 368
Genus (genera), 7, 462
*Geographical Distribution of Animals, The* (Wallace), 1141
Geology
  geological time scale, *506–507*, 508
  influence on biomes, 1126–1127
Georges Bank, 1163–1164
*Geospiza conirostris*, *474*
*Geospiza difficilis*, *474*
*Geospiza fortis*, *474*
*Geospiza fuliginosa*, *474*, 1183
*Geospiza magnirostris*, *474*
*Geospiza scandens*, *474*, 1154–1155
Germ cell plasm, 908
Germ cells
  in animal gametogenesis, 882–883
  in development, 908
  epigenetic changes and, 344
  in human males, *891*
  mutations and, 310
Germ layers
  formed during gastrulation, 908
  in frog gastrulation, 909–910

in sea urchin gastrulation, 908–909
Germ line gene therapy, 323
Germ line mutations, 305
Germination
of angiosperm pollen, 780, 782
of seeds (*see* Seed germination)
Gestation, 919–920
*See also* Pregnancy
Gey, George and Margaret, 205
GFP. *See* Green fluorescent protein
Gharials, 693–694
Ghost bat, *700*
Ghrelin, 1067, 1068
Giant bluefin tuna, 825
Giant groundsel, *1128*
Giant kelp, 556, *1139*
Giant petrel, *1073*
Giant redwoods, 422
Giant sequoia, 1130, 1160
Giant tortoise, *2*
*Giardia lamblia,* 558
Giardiasis, 558
Gibberellic acid, *759,* 760
Gibberellins
activated by phytochromes in seed germination, 774
discovery of, 760, *761*
effects on plant growth and development, *759,* 760–762
induction of flowering and, 791–792
molecular mechanisms underlying the activity of, 767
semi-dwarf plants and, 775
structure, *759*
Gibbons, 702
*Gigantactis vanhoeffeni, 688*
Gigantism, 843
Gill arches
development from pharyngeal arches, 684
in the evolution of jaws, *687*
of fish gills, 1009–1010
in lungfish, 1028
Gill arteries, 1028
Gill filaments, 1010
Gills
countercurrent heat exchange in fish, 825
in mollusks, 662, *663*
respiratory gas exchange, *1007,* 1009–1010, *1011*
surface area maximization, 1008, *1009*
Gills (of mushrooms), 623
*Ginkgo biloba,* 592, *593*
Ginkgophyta, *574*
Ginkgos, *574,* 589, 592, *593*
Giraffes, 416–417
Girdle, of chitons, 662
Gizzard, 1056
Glacial moraines, 1200
Glaciation
allopatric speciation and, 472
during the Carboniferous, 521
climate change through time and, 510
during the Proterozoic, 515–516
in the Quaternary, 522
sea level and, 509, *510*
Glacier Bay, 1200
Glaciers, impact of climate warming on, *17*
Gladiators, *672*
Gladioli, 793
Glans penis, 890
Glass sponges, *630,* 643
Glaucophytes, 551, *552,* 571, *572, 573*
Gleevec, 304
Glial cells (glia)
characteristics of, 820
neural networks, 940–943
types and functions of, 926–927
*Global Biodiversity Outlook 3,* 1233
Global climate change
atmospheric carbon dioxide levels and, 17, 510, 1217–1218
greenhouse gases and, 1212
methane cycling studies, 357
Global ecosystem
biogeochemical cycles, 1214–1223
energy flow through, 1208–1210
movement of elements through, 1210–1214
Global nitrogen cycle, 750–751
"Globe crab," *670*
α-Globin gene cluster, 364
β-Globin
gene cluster, 364
hemoglobin C disease, 312
missense mutation and sickle-cell disease, 306, 312
nonsense mutation and thalassemia, 306
transcriptional regulation red blood cells, 335
β-Globin gene
differential transcription, 398
transcriptional regulation red blood cells, 335
γ-Globin, 364
Globin gene family, 364, 497, *498*
"Globular" embryo, *393*
Glomeromycota, *615, 616,* 619–620
*See also* Arbuscular mycorrhizae
Glomerular capillaries, 1078–1079, 1082
Glomerular filtration rate (GFR), 1087–1088
Glomeruli, olfactory, *949,* 950
Glomerulus
blood flow to, 1082
function in the vertebrate nephron, 1078–1079
of the mammalian kidney, *1080,* 1081
*Glomus mosseae, 614*
Glucagon, *842,* 848, 849, 1067
β-1,3-Glucan, 799
Glucoamylase, 162
Glucocorticoids, 849
Gluconeogenesis, 180, 1065, 1067
Glucosamine, *55*
Glucose
carrier-mediated transport, 117
catabolic interconversions, 179
energy released during oxidation, 166–169
formation of glucose 6-phosphate, 150, *151*
forms of, 52
as the fuel for the nervous system, 1067
gluconeogenesis, 180
in glycogen, 53, 54
in positive regulation of the *lac* operon, 332
production in photosynthesis, 195
secondary active transport, 119
in starches, 53, *54*
Glucose 1-phosphate, 148, 149
Glucose 6-phosphate, 148, 149, 150, *151*
Glucose catabolism
aerobic pathways, 169–171
allosteric regulation, 181–182
under anaerobic conditions, 177–179
citric acid cycle, 170–171
energy cost of NADH shuttle systems, 178–179
energy released during glucose oxidation, 166–169
energy yield from cellular respiration and fermentation compared, 178
glycolysis, 169–170
oxidative phosphorylation and ATP synthesis, 171–176
pyruvate oxidation, 170
regulation of pyruvate oxidation and the citric acid cycle, 171
Glucose metabolism, liver regulation of, 138, 1065, 1066–1067
Glucose transporters, 117, 1066
Glucuronic acid, 55
Glumes, 419
Glutamate
anabolic interconversions and, 180
catabolic interconversions, 180
as a neurotransmitter, 939
Glutamate decarboxylase, *323,* 324
Glutamic acid, *44*
Glutamine, *44*
Glutamine synthetase (GS), 753
Glyceraldehyde, *52*
Glyceraldehyde 3-phosphate (G3P), *169,* 170, 194, 195, *196,* 201
Glyceraldehyde 3-phosphate dehydrogenase, 170
Glycerate, 197, *198*
Glycerol
catabolic interconversions, 179
in phospholipids, 57, *58*
in triglycerides, 56
Glycerol 3-phosphate, 179
Glycine
as a neurotransmitter, 939
in photorespiration, 197, *198*
side group properties, 43
structure, *44*
Glycogen
buildup during excess food consumption, 1050–1051
in exercising muscles, 997
glucose levels during exercise and, 181
in glucose metabolism, 1066, 1067
regulation by a protein kinase cascade, 138
regulation of blood glucose, 848, 849
storage in the liver, 1065
as stored energy, 1050
structure and function, 53–54
Glycogen phosphorylase, 132–133
Glycogen synthase, 138
Glycogen synthase kinase-3 (GSK-3), 904
Glycolate, 197, *198*
Glycolipids, 109, 111
Glycolysis
allosteric regulation, 181–182
description of, 169–170
energy yield from cellular respiration and fermentation compared, 178
in glucose catabolism, 166, 168
relationships among metabolic pathways, *179*
Glycolytic muscle, 995
Glycolytic system, in skeletal muscle, 997
Glycoproteins
cell adhesion and, 111
formation of, 301
structure and function, 109
Glycosidic linkages, 53, 54
Glycosylation, *300,* 301
Glyoxysomes, 93
Glyphosate, 160, 388, 389
Glyphosate resistance, 389
*Gnathonemus petersi, 485*
Gnathostomes, 686–689
Gnetophytes, *574,* 593
Goats, 1099
Gobi Desert, *1192*
Goiter, 846–847, 1054
Golden feather star, *682*
Golden lion tamarin, 1114, *1137*
Golden toads, *691*
Goldenrod, *752,* 1201
Goldman equation, 929, 931
Golgi, Camillo, 90, 953
Golgi apparatus
in animal cells, *86*
evolution in eukaryotic cells, 551
glycosylation of proteins, 301
Inclusion-cell disease, 300
plant cell plate and, 710
in plant cells, *87*
processing of newly translated polypeptides, *299,* 300
structure and function, *89,* 90
Golgi tendon organs, 953
Gonadotropin-releasing hormone (GnRH)
control of gonadotropins, 851
discovery of, 844
in puberty, 892, 894
in regulation of the ovarian and uterine cycles, *894,* 895
Gonadotropins
defined, 850–851
in pregnancy, 896
in puberty, 894
regulation of the ovarian and uterine cycles, 893–894
Gonads
establishment of the germ cells, 908
gametogenesis in, 882–884
hormones of, *842*

phenotypic determination, 850, *851*
sex steroids produced by, 850–851
Gondwana, 521, 1143
*Gonionemus vertens, 647*
*Gonium,* 140–141
Gonzales, Andrew, *1162*
*Goosecoid* gene, 911, 912
Goosecoid transcription factor, 911, 912
Gooseneck barnacles, *670*
*Gorilla gorilla, 702*
Gorillas
brain size–body size relationship, 973
comparative genomics, 366–367
Goshawks, 1116, *1117*
Gout, 1074
Graded membrane potential, 932, 938, 947
Grafting, 793, *794*
Grains, genetically modified, 388
Gram-negative bacteria, 528, 542
Gram-positive bacteria, 527–528
Gram stains, 527–528
Grant, Peter and Rosemary, 1154–1155
Granulocytes, 858
Granulosa cells, 895, 896
Granum, 92, *93*
Grapevines, 761
Grasses
defenses against herbivory, 1175
endophytic fungi and, 615
spikes, *597*
Grasshoppers, *672, 674*
Grasslands
restoration projects, 1237–1239, *1240*
temperate grassland biome, 1131–1132
Grave's disease, 846–847
Gravitropism, 765
Gravity, vestibular detection of, 956–957
Gray crescent, 903–904, 909–910
Gray-headed albatross, 1109–1110
Gray matter, 941
Gray tree-frogs, *469, 1173*
Gray whales, 1109
Gray wolves, 1193–1194
Graylag geese, 1099
Grazers
trophic cascades in savanna communities, 1194
*See also* Herbivores/Herbivory
Great apes, 983
Great Barrier Reef, 646–647
Great Lakes
eutrophication in, 1221
invasive zebra mussels, 1160
Great Plains, 1131
Great Rift Valley, 509
Great white sharks, 825
Greater bilby, *699*
Greater flamingo, *638*
Greater prairie-chicken, 434
Greater sage-grouse, *1105*
Green algae
biofuel production and, 585
closest relatives of land plants, 572–573
endosymbiotic origin of chloroplasts in, 551
in lichens, 613
"Volvocine line," 140–141
Green fluorescent protein (GFP), 378, *379,* 449
Green molds, 624, 1169
Green plants, 572, 573
"Green revolution," 747
Green sea turtle, *888*
Green sulfur bacteria, 187
Green tiger beetle, *1172*
Greenbottle flies, 1203
Greenhouse gases, 185, 1212
*Gremlin* gene, 417
Gremlin protein, 417
Grévy's zebra, *1144*
Grey goose, *642*
Grooming behavior, 1177
Gross primary production (GPP), 1189–1190
Ground finch, 1183
Ground ivy, *720*
Ground meristem
ground tissue system and, 716
in root growth, 716–717
in shoot growth, 719
Ground squirrels, *831*
Ground tissue system
description of, *712,* 713–714
primary meristem, 716
of shoots, 720
Groundwater, 1213, 1215
Groundwater depletion, 1215
Group living, 1116, *1117*
Growing season, 1138
Growth
defined, 393
determinate, 715, 720, 786
in development, 710
indeterminate, 715–720, 786
processes contributing to, 394
*See also* Plant growth and development
Growth factors
induction of cell division, 211
primary embryonic organizer and, 911, 912
signal transduction pathways and, 131
Growth hormone (GH)
actions of, *842,* 843
production through biotechnology, *385*
Growth hormone deficiency, 386
Growth hormone release-inhibiting hormone, 844
Growth hormone-releasing hormone, 844
Grylloblattodea, *672*
Guam, 1235
Guanine
codons and the genetic code, 288–289
complementary base pairing, 63–65
in DNA structure, 264, 265, 266, *267*
induced mutations from cigarette smoke, 309
mutagens and, 310
structure, *63*
Guanosine diphosphate (GDP)
in the citric acid cycle, 170, 171
G protein-linked receptors and, 130
Guanosine triphosphate (GTP)
5′ cap, 291–292
in the citric acid cycle, 171
function of, 67
G protein-linked receptors and, 130
Guanylyl cyclase, 136
Guard cells, 712, 721, 732, 733–734
"Guide proteins," 559
Guillain–Barre syndrome, 926
Guillemin, Roger, 844
Gulf of Mexico
causes and impact of dead zones, 1207, 1214, 1219, 1225
oil spills, 373, 569
red tides, 549
Gulf Stream, 1124–1125
Gulls, begging behavior of chicks, 1095
Guppies, 889, 1157
Gustation, 951
Gut
defined, 630
of flatworms, 656, *657*
of herbivores, 638
of phoronids, 659, *660*
tissue composition, 820
tubular, 1056–1057
*See also* Gastrointestinal system
Gut hormones, 838
Gut microbes, benefits of, 539, 540–541
Gut muscle
function of, 820
influence of the autonomic nervous system on, 993
layers of, 1058, *1059*
*Gymnogyps californianus, 1244*
Gymnosperms
conifers, 593–596
major groups and distinguishing characteristics, *574,* 592–593
seed development, *592*
swimming sperm, 589
*Gymnothorax meleagris, 688*
Gypsy moth, 1110
Gyres, 1124
Gyri, 970


## H


H zone, *987,* 988
H1N1 influenza virus, *427,* 877
H5N1 influenza virus, *543*
Haber process, 750
Habitat corridors, 1233–1234
Habitat fragmentation, impact on biodiversity, 1233–1234
Habitat islands, 1198
Habitat isolation, 477, 482
Habitat loss
impact on biodiversity, 1233–1234
species extinctions and, 1232
Habitat patches, 1161–1162, 1233
Habitats
animal selection of, 1103
defined, 1103
effect on population dynamics, 1161–1163
impact of interference competition on habitat use, 1183
Haddock, 1163–1164
Hadean eon, *506–507,* 508
Hadobacteria, 532
*Hadrurus arizonensis, 636*
Haeckel, Ernst, 1122
*Haemophilus influenzae*
comparative genomics, 357
functional genomics, 356, *357*
genome sequencing, 356
genomic information, *361*
Hagfish, 685–686, 1073, 1074
Hair
a distinguishing feature of mammals, 697
pigmentation and the *MC1R* gene, 367
Hair cells
mechanism of mechanoreception, 953–954
of the vestibular system, 953–954, *956,* 957
*See also* Auditory hair cells
Hair follicles, 397
Hairgrass, 1125
"Hairy backs." *See* Gastrotrichs
Haldane, J. B. S., 1115
Half-cell reactions, 187
Half-life
of hormones, 852–853
of radioisotopes, 507
*Haliaeetus leucocephalus, 638, 1228*
*Haliaeetus vocifer, 218*
Hall, Donald, 1105, *1106*
Halophiles, 536–537
Haltares, 672
Hamilton, W. D., 1115
Hamilton's rule, 1115
Hamner, Karl, 788, 789
Haplochromine cichlids
prezygotic isolating mechanisms, 476, 477, 482
sexual selection and rates of speciation, 480
speciation, 467
Haplodiploidy, 1115, *1116*
Haploids
in alternation of generations life cycle, 562, 563, 574, 575
defined, 218, 236
generation during meiosis, *220–221, 223* (*see also* Meiosis)
in sexual reproduction, 218, 219
Haplontic life cycle, *218*
Haplotype mapping, 367–368
Haplotypes, 367
Hardy, Godfrey, 437
*HARDY* gene, 726
Hardy–Weinberg equilibrium, 437–438
Hartwell, Leland, 210
Harvestmen, 669
Hashimoto's thyroiditis, 876
*Hatena,* 102
Haustoria, 611–612, 752
Haversian bone, 1001
Haversian systems, 1001
Hawaiian bobtail squid, 546
Hawaiian Islands
adaptive radiations, 481–482
allopatric speciation in *Drosophila,* 472, *473*

avian malaria, 1236
distribution of long-horned beetles, 1145
effect of dispersal ability on speciation rates, 481
Hawks, fovea, 962
Hayes, Tyrone, 1, 12, 13
Hazel, *590*
Head (inflorescence), *597*
Hearing. *See* Auditory systems
Hearing loss, 956
Heart attack
atherosclerosis and, 1042–1043
human SNP scans, *368*
from mutation-based wall thickening, 1025, 1046
treatment with TPA, 385
Heart disease, stem cell therapy, 102
"Heart" embryo, *393*
Heart muscle. *See* Cardiac muscle
Heart pacemaker. *See* Pacemaker cells
Heart stage, 711
Heartbeat
autonomic nervous system control of, 1034
in blood pressure regulation, 1044, *1045*
effect of calcium ion cycling on, 1034, *1036,* 1037
in the fight-or-flight response, 837
Frank–Starling law, 1042
pacemaker cells and cardiac muscle contraction, 1032–1034
Hearts
in amphibians, 1029
calcium ion cycling in "hot" fish, *1036,* 1037
in crocodilians and birds, 1029
in fish, 1028
four-chambered heart of mammals, 697
in lungfish, 1028
mammalian heart function, 1030–1037
mutation-based wall thickening, 1025, 1046
in open circulatory systems, 1026, *1027*
production of atrial natriuretic peptide in blood pressure regulation, 1090
Heat
in biological systems, *145*
from brown fat, 165, 174
muscle fatigue and, 998
Heat of vaporization, 33
Heat-resistant DNA polymerases, 277–278
Heat shock proteins, *51,* 810
Heat shock response, 810
Heavy metal tolerance, in plants, 811–812
Heavy nitrogen, 268, *269,* 270
Hebert, Paul, 319
Hedgehogs, *698*
Heidmann, Thierry, 545
Height, quantitative variation in humans, 246
HeLa cells, 205, 208–209, 229
Helices
alpha helix, 45, *46*
double helix structure of DNA, 65, *66,* 264–265, 266, *267*
left- and right-handed, *47*
*Helicobacter pylori,* 542
*Heliconius, 1174,* 1175
Helium, 22, *22*
*Helix pomatia, 663*
Helix-turn-helix motif, 335–336
"Helping at the nest," 1115
Hematocrit, 1037–1038
Hematopoietic stem cells, 408
Heme, *156*
Hemichordates
in animal phylogeny, *630*
in deuterostome phylogeny, *679*
features of, 680, 682–683
major groups, *632,* 679
Hemiparasites, 752
Hemipenes, 888
Hemipterans, 480, *672,* 673, *674*
*Hemitrichia serpula, 560*
Hemizygous, 251
Hemocoel, 652, 662
Hemoglobin
binding of carbon dioxide, *1019*
binding of carbon monoxide, 1017
binding of oxygen, 1016–1017
factors affecting oxygen affinity, 1017–1018
globin gene family and, 364, 497, *498*
hemoglobin C disease, 312
missense mutation and sickle-cell disease, 306, *307,* 312
in pogonophorans, 660
polymorphism in, *312*
quaternary structure, 48, *49*
in red blood cells, 1038
structure, 1016
β-thalassemia, 292–293, 306
Hemoglobin C disease, 312
Hemolymph, 1026, *1027*
Hemophilia, *313,* 323, 365, 1039
Hendricks, Sterling, 788
Hensen's node, 914
Henslow, John, 428
Hepatic duct, 1061
Hepatic portal vein, 1063
Hepatic portal vessel, *1031*
Hepatic veins, *1031*
Hepatitis, 1039
Hepatophyta, *574*
HER2 receptor, 228, 229
Herbicide resistance, 388, 389
Herbicides, 1, 160
Herbivores/Herbivory
circumvention of plant defenses, 806
defined, 1054, 1170
description of, 1175
digestion of cellulose in, 1063–1064
eutherian, 700
feeding strategy, 637–638
ingestion and digestion of food, 1054–1055
plant defenses against, 801–805, 1175
reciprocal interactions between herbivores and plants, 1175–1176
salt balance in, 1073
teeth, *1055*
trophic cascades in savanna communities, 1194
*See also* Primary consumers
Herbivorous bugs, 480
Herceptin, 229, 871
"Herd immunity," 856
Herelle, Felix d', 545
Heritable traits, 431
Hermaphroditic flowers, 597, 779
Hermaphroditism, in animals, 887–888
*Hermodice carunculata, 636*
Herpes viruses, 341, *543*
Herpesviridae, *543*
Hershey, Alfred, 261–263
Hershey–Chase experiment, 261–263, 339
*Heterocephalus glaber, 1116*
Heterochromatin, 345, 346, 364
Heterochrony, 416–417
Heterocysts, 532, *533*
Heteroloboseans, 558
Heterometry, 416
Heteromorphic alternation of generations, 563
*Heterophrynus batesii, 654*
Heteropods, 662
Heterosis, 244–245
Heterospory
in angiosperms, 600–601
appearance in vascular plants, 584–585
in seed plants, 589–590
Heterotherms, 822
Heterotopy, 417
Heterotrophic succession, 1202
Heterotrophs
absorptive heterotrophy, 609
classified by acquisition of nutrition, 1054
in communities, 1189, 1190
defined, 196
facilitation of succession by, 1201
requirements from food, 1049–1054
Heterotypic cell binding, 111
Heterotypy, 418–419
Heterozygote advantage, 442–443
Heterozygotes
defined, 236
incomplete dominance, 242–243
*Hexacontium, 557*
Hexapods, 667, 671, 673
*See also* Insects
Hexokinase, *151,* 155
Hexose phosphates, 201
Hexoses, 52, 195
Hibernation/Hibernators, 830, *831,* 1108
Hide beetles, 1188, 1203
High-density lipoproteins (HDLs), 1066
High-GC Gram-positives, 532
High-throughput sequencing, 353–354
Highly repetitive sequences, *355,* 364, 366
Hillis, David, 457
Himalayan mountains, 509
*Himanthalia elongata, 556*
Hindbrain, 968, *969*
Hindlimbs
development in ducks and chickens, 417
loss in snakes, 423
Hippocampus
functions of, 970
of London taxi drivers, 967
memory and, 967, 982, 983
"place cells" in rats, 967
regulation of the stress response and, 850
*Hirudo,* 1056
*Hirudo medicinalis,* 661–662
Hispaniola, 440
Histamine, 861, 1041
Histidine, *44*
Histone acetyltransferases, 344
Histone deacetylase inhibitors, 344
Histone deacetylases, 344
Histone methylation, 344
Histone phosphorylation, 344
Histone tails, 344
Histones
in chromatin structure, 212, *213*
modifications in epigenetic gene regulation, 344
HIV-1, 458, 461–462
HIV-2, 458
HIV protease, 301
*HMGA2* gene, 246
Hoatzin, 494
Hodgkin, A. L., 929
Holdfast, 556
Holocene epoch, 522
Holometabolous insects, *672,* 673
Holoparasites, 752, 753
Homeobox, 404, 413, 917
"Homeobox" genes. *See* Hox genes
Homeodomain, 404
Homeosis, 401
Homeostasis
defined, 10
of the internal environment, 816
plasma membrane and, 80
regulation by physiological systems, 816–817
regulation of enzymes and, 156–157
Homeotherms, 822
Homeotic genes
in fruit fly body segmentation, 403–405
regulation of body segmentation, 916–918
Homeotic mutations, 401, 404
Homing, 1109
Hominins
ancestors of humans, 702–705
bipedal locomotion, 702–703
diet of, 1056
Hominoids, 522
*Homo,* early members of, 703–704
*Homo erectus,* 703, 704
*Homo ergaster, 703*
*Homo floresiensis, 703,* 704
*Homo habilis,* 703
*Homo neanderthalensis,* 7, *703,* 704
*Homo sapiens*
evolution of, 522, *703,* 704–705
genome size, 494
meaning of, 7
*See also* Humans

Homogentisic acid, 282
Homogentisic acid oxidase, 282
Homologous chromosomes
  characteristics of, 218
  duplications, 308
  events in meiosis, 219, 220–222, *223*
  genetic recombination, 219, *220*, *222*, 247–249, *250*
  meiotic errors, 222, 224
Homologous features, 452
Homologous genes, 413–414, 499–500
Homologous pairs, 218
  *See also* Homologous chromosomes
Homologous recombination, 381–382
Homologs, 218
Homology
  in genes, 413–414, 499–500
  between macromolecules, 486
Homoplasies, 452
Homospory, 584
Homotypic cell binding, 111
Homozygotes, 236
Honest signals, 435
Honey bees
  epigenetic effects of "royal jelly," 899
  exploitation competition, 1183
  metabolic heat production of colonies, 826
  monitoring, *1150*, 1151
  as pollinators, 1181, *1182*
  sex and reproduction in, 880
  waggle dance, 1112
Honeypot ant, 1183
Hoofed mammals, *698*
Hooke, Robert, 78
Hoover, Jeffrey, 1117
*Hoplostethus atlanticus*, 1235
Horizons, in soils, 745
Horizontal cells, *962*, 963
Horizontal stems, 720
Hormone-based contraceptives, *898*
Hormone receptors
  location and function, 836–837
  multiple receptors for a single hormone, 853
  upregulation and downregulation, 853
  *See also individual hormone receptors*
Hormones
  affinity chromatography, 853
  comparison of animal and plant hormones, *758*
  criteria for defining a molecule as, 853
  defined, 126, 758, 835
  detection and measurement with immunoassays, 852–853
  dose–response curves, 853
  half-life, 852–853
  *See also* Animal hormones; Plant hormones
Hornworts, 573, *574*, 578–579
Horowitz, Norman, 283, 284
Horse family, 1143, *1144*
Horsehair worms, *632*, *652*, 666, *667*
Horseshoe crabs, 668–669
Horsetails, 520, *574*, 580, 581
Hosts
  lytic viral reproductive cycle in, 339–340
  parasites and, 638–639, 1170
Hot deserts, 1132–1133
"Hot" fish, 825, *1036*, 1037
Hot sulfur springs, 536
House flies, 1203
"Housekeeping genes," 332–333
Hox genes
  differences in expression and spine evolution, 418
  duplication-and-divergence hypothesis, 413–414
  in ecdysozoans, 654
  in fruit fly body segmentation, 401, 403–405
  heterotypy and leg number in insects, 418, *419*
  loss of limbs in snakes and, 423
  regulation of body segmentation, 916–918
  wing development in insects, 415
*Hoxc6* gene, 418
*Hoxc8* gene, 418
HSP60, *51*
Hubel, David, 977
Human activities
  impact on ecosystems, 1223–1224
  impact on energy flow through the global ecosystem, 1210
  impact on the global nitrogen cycle, 1218–1219
  impact on the global phosphorus cycle, 1220–1221
  overexploitation of species, 1234–1235
  predicting the effects of humans on biodiversity, 1231–1232
  species extinctions and, 1229–1230
  threatening species persistence, 1232–1237
  *See also* Fossil fuel burning
Human birth defects
  spina bifida, 916
  from thalidomide, 920
Human brain
  brainstem structure and function, 969–970
  complexity of neural networking in, 943
  consciousness, 982–983
  development, 920–921, 968–969
  forebrain structure and function, 970
  language areas, 980–981
  learning and memory areas, 981–982
  size and evolution, 704, 973
  sleep and dreaming, 978–980
  telencephalon structure and function, 970–973
Human chorionic gonadotropin (hCG), 871, 896
Human development
  apoptosis in, 399–400
  left–right asymmetry of internal organs, 902
  regulative, 907
  stages in, 919–921
Human diseases
  fungal, 612
  pathogenic bacteria, 542
  pathogenic protists, 563, *564*
  pathogenic trypanosomes, *559*
  viral agents, 544–545
Human epidermal growth factor (HER2) receptor, 228, 229
Human genetic diseases
  abnormal hemoglobin, 312
  cancer and somatic mutations, 314
  evolutionary studies of sodium channel genes and, 502
  examples of, *313*
  expanding triplet repeats, 313–314
  Inclusion-cell disease, 300
  IPEX, 874
  knockout mouse models, 382
  loss of enzyme function, 311–312
  multifactorial nature of, 314–315
  point mutations, 312
  prevalence, 315
  strategies for treating, 322–325
  β-thalassemia, 292–293
  transposons and, 365
Human genome
  alternative splicing, 346–347
  characteristics of, 366
  comparative genomics, 366–367
  endogenous retroviruses in, 545
  gene duplication in, 497
  genomic library of, 379
  Human Genome Project, 353
  key parameters, *361*
  largest gene, 291
  medical benefits from studying, 367–368
  microRNA, 347
  normal damage to DNA, 310
Human Genome Project, 353
Human growth hormone (hGH), 385–386
Human immunodeficiency virus (HIV)
  as an RNA retrovirus, *543*, 544
  course of infection, 876, *877*
  gene expression in, 285
  gene regulation at the level of transcription elongation, 341–343
  HIV protease, 301
  molecular clock dating of the origin of HIV-1 in humans, 461–462
  phylogenetic analyses, 458, *459*
  treatment, 342–343, 876–877
Human lungs
  anatomy of, 1013, *1014*
  diseases of, 1013
  inhalation and exhalation, 1015–1016
  perfusion by the circulatory system, 1016
  respiratory tract secretions, 1013, 1015
  tidal ventilation, 1012–1013
Human nervous system
  functional and anatomical organization, 968–969
  left–right crossover between brain and body, 970
  *See also* Human brain
Human papillomavirus (HPV), 228, 229, 349
Human reproduction
  contraception, 897, *898*
  female reproductive system, 892–897
  implantation, 906, *907*
  male reproductive system, 889–892
  reproductive technologies to solve infertility, 897, 899
  twins, 907
Humans
  abnormal sex chromosome arrangements, 250
  ABO blood groups, 243
  basal energy expenditure, 1049
  brown fat, 165, 828
  chromosome number, 218
  circulatory system, *1031*
  decomposition and corpse communities, 1188, 1203
  development of language and culture, 705
  digestive enzymes, *1062*
  digestive system, *1058*
  Down syndrome, 224
  ear structure, 954–955
  effects of prenatal stress on child behavior, 328
  embryonic stem cells, 409
  endocrine system overview, *842*
  endosymbiotic bacteria in the intestines, 1057
  essential amino acids, 1051–1052
  evolutionary responses to ectoparasites, 1177
  excretory system, 1080–1086
  exponential population growth, 1164–1166
  extracellular fluid in, 816
  eye, *959*, 962–963
  fevers, 830
  founder effect in, 434
  gene flow in, 434
  genomic imprinting, 345
  global climate change and, 17
  growth hormone deficiency in children, 386
  heart attack from mutation-based wall thickening, 1025, 1046
  heart function, 1030–1037
  heat stroke, 815
  jumping ability, 986
  karyotypes, 224, *225*
  lymphatic system, *857*, 858
  metabolomes, 370
  microbiomes and human health, 539–541
  muscular segmentation, 636
  myostatin gene, 370
  Neanderthal ancestors, 704
  nitrogenous wastes excreted, 1074–1075
  origin of, 702
  overnutrition and obesity, 165, 1048, 1051, 1068
  oxytocin and, 141
  pedigrees, 240–241

primate ancestors, 702–704
proteomes, 369
puberty, 850–851, 894
quantitative variation in, 246
regulation of breathing, 1019–1022
sex-linked inheritance, 252
sex steroids and phenotypic sex determination, 850, *851*
skeleton, 999, *1000*
sociobiology and, 1116–1117
stabilizing selection on birth weight, 439
tactile receptors of the skin, 952
taste buds and taste, 951
trisomies and monosomies, 224
value of biodiversity to, 1230
vitamin requirements, 1053–1054
*See also* Infants
*Humata tyermanii, 218*
Humboldt, Alexander von, 1196
Humerus, 423
Hummingbirds
daily torpor, 830
feeding on nectar, *477*
"nectar corridor" migration, 1137
as pollinators, 598, *599, 1136,* 1181, *1182*
weight of, 696
Humoral immune response
activation and effector phases, 872, *873*
description of, 864, *865*
generation of immunoglobulin diversity, 868–869, *870*
immunoglobulin class switching, 869–871
immunoglobulin classes, 868
immunoglobulin structure and function, 867–868
monoclonal antibodies, 871
overview, 863–864
plasma cells, 867
suppression by regulatory T cells, 874
Humpback whales, 1112
Humus, 745, 746
*Hunchback* gene, 402, *403*
Hunchback protein, 402, *403*
Hungate, Bruce, 1221–1223
Huntington's disease, 314
Hurricane Katrina, 1223–1224
Hussein, Saddam, *317,* 318
Huxley, A. F., 929
Huxley, Thomas, 605
Hyacinths, *603*
*Hyalophora cecropia,* 841
Hybrid seeds, 778
Hybrid vigor, 244–245
Hybrid zones, 478–479
Hybridization
allopolyploidy in wheat and, 225, *226*
lateral gene transfer and, 496
mechanisms preventing, 475–478
Hybrids
hybrid plants in agriculture, 778
postzygotic isolating mechanisms, 478, 479
*Hydra, 881*
Hydrocarbon molecules, 30
Hydrochloric acid
production by the stomach, 1058, *1060,* 1061
properties of, 34
Hydrogen
atomic stability, 25
covalent bonding capability, *27*
electronegativity, *28*
isotopes, 22
Hydrogen bonds
in alpha helices, 45, *46*
in beta pleated sheets, 45, *46*
description of, 30
in DNA, *66,* 266
features of, *26*
in protein binding, 50
in protein quaternary structure, 48
in protein tertiary structure, 47
Hydrogen ions
acids and, 34
bases and, 34
pH of solutions, 35
transfer during oxidation–reduction reactions, 167, 168
Hydrogen peroxide, 93, 175
Hydrogen sulfide
as an electron donor in anoxygenic photosynthesis, 187
as an electron donor in photoautrophs, 538
in the global sulfur cycle, 1219
in pogonophoran metabolism, 660, 661
Hydroid, 578
*Hydrolagus colliei, 688*
Hydrologic cycle, 1215
Hydrolysis
of ATP, 149–150
hydrolysis reactions, 42
Hydrolytic enzymes, 161
Hydronium ion, 34
Hydrophilic molecules, 30
Hydrophilic regions
of integral membrane proteins, 108
of phospholipids, 106
Hydrophobic interactions
description of, 30
features of, *26*
in protein binding, 50
in protein quaternary structure, 48
Hydrophobic molecules, 30
Hydrophobic regions
of integral membrane proteins, 108
of phospholipids, 106
Hydroponics, 742–743
Hydrostatic skeletons, 635–636, 999
Hydrothermal vents. *See* Deep-sea hydrothermal vent ecosystems
Hydroxide ion, 34, 175
Hydroxyl group, *40*
Hydroxyl radical, 175, 176
Hydrozoans, 646, 647–648
*Hyla chrysoscelis, 469*
*Hyla versicolor, 469, 1173*
*Hylobates lar, 702*
Hymen, 892
Hymenoptera
haplodiploidy and eusociality, 1115, *1116*
number of living species, *672*
Hyperaccumulators, 811–812
Hyperemia, 1044
*Hypericum perforatum,* 1175
Hyperosmotic regulation, 1073
Hyperpolarization
description of, 931, *932*
at inhibitory synapses, 938
of neurons at the onset of sleep, 979
of rod cells, 960–961
Hypersensitive response, 226, 799, 800–801
Hypersensitivity, allergic reactions and, 875–876
Hypertension, atherosclerosis and, 1043
Hyperthermophilic bacteria, 532
Hyperthyroidism, 846–847
Hypertonic environments, fungal tolerance of, 610–611
Hypertonic solutions, 114, *115*
Hyphae
characteristics of, 609–610
in lichens, 613
in mycorrhizae, 614
of parasitic fungi, 611–612
of predatory fungi, 612, *613*
Hypoblast
of the human blastocyst, *907*
in the origin of extraembryonic membranes, 918, 919
in yolky eggs, 913, 914
Hypocotyls, 770
Hypoosmotic regulation, 1073
Hypothalamus
body temperature regulation in mammals, 829–830
functions of, 969, 970
hormones of, *842*
in human puberty, 851, 892, 894
interconnections with the pituitary gland, 842, 843
in negative feedback regulation of hormone secretion, 844, *845*
neurohormones of, 843, 844
regulation of blood pressure and blood osmolarity, 1088, 1089
regulation of cortisol release in the stress response, 849–850
regulation of food intake, 1067–1068
regulation of the ovarian and uterine cycles, 894
regulation of thyroxine production, 846
somatostatin production, 849
Hypotheses
defined, 12
hypothesis–prediction methodology, 11–12
significance to scientific inquiry, 14
Hypothyroidism, 846, 847, 1054
Hypotonic solutions, 114, *115*
Hypoxia
dead zones in the Gulf of Mexico, 1207, 1225
release of erythropoietin by the kidneys in response to, 1038
Hypoxia-inducible factor 1 (HIF-1), 1038
Hypoxic conditions, insect body size and, 522
Hyracoidea, *698*
*Hyracotherium leporinum, 519*
Hyraxes, *698*

I

I band, *987,* 988
Ice
effect of ice crystals on plant cells, 810
properties, 32
"Ice ages," 522
Ice caps, impact of global climate change on, 1217, *1218*
Ice-crawlers, *672*
Identical twins, 907
IgA, *868*
IgD, *868*
IgE, *868,* 875, 876
IgG, 868, 876
IgM, *868,* 870
Igneous rocks, 507–508
*Ignicoccus,* 537
Ileocaecal sphincter, 1060
Ileum, 1061
*Ilex opaca, 780*
*Illicium floridanum, 602*
Image-forming eyes
anatomy of, 958, *959*
focusing, 958–959
structure and function of the retina, 959–963
Imatinib, 304
Imbibition, 758
Immediate hypersensitivity, 875–876
Immediate memory, 982
Immigration, in island biogeography theory, 1196–1198, *1199*
Immune system
adaptive immunity (*see* Adaptive immunity)
chemical signaling in, 835
discovery of immunity, 862, *863,* 864
inhibition by cortisol, 849
innate immunity (*see* Innate immunity)
"plant immune system," 799–801
*See also* Immunology
Immune system proteins, types and functions of, 858–859
Immunization, 866
Immunoassays, 852–853, 871
Immunodeficiency viruses, 458
*See also* Human immunodeficiency virus
Immunoglobulin genes, 868–869, *870*
Immunoglobulins
class switching, 869–871
classes of, 868
constant regions, 867, 869–871
generation of diversity in, 868–869, *870*
structure and function, 867–868
variable regions, 867, *869,* 870
*See also* Antibodies

Immunological memory
secondary immune response and, 865–866
vaccinations and, 863, 866
Immunology
adaptive immunity, 862–867
cellular immune response, 871–875
characteristics of innate defenses, 859–862
humoral immune response, 867–871
immune system malfunctions, 875–877
phases in the defensive response, 857
role of immune system proteins, 858–859
roles of blood and lymph in, 857–858
roles of white blood cells in, 858
types of defense systems, 857, *858*
vaccinations, 856
Immunotherapy, 871
Imperfect flowers, 597, 598, 779, *780*
Implantation, 906, *907*, 919, 920
Implantation blockers, *898*
Impotence, 891
Imprinting, 1099
In vitro evolution, 500–501
In vitro fertilization (IVF), 899, 907
*In vivo* gene therapy, 323
Inactivation gate, 933–934
Inbreeding
inbreeding depression, 244–245
self-incompatibility studies in plants, 380–381
strategies for preventing in angiosperms, 782, *783*
Inbreeding depression, 244–245
Incisors, 1055
Inclusion-cell disease, 300
Inclusive fitness, 1115
Incomplete cleavage, 905
Incomplete dominance, 242–243
Incomplete metamorphosis, 672, 841
Incubation temperature, impact on sex determination, 420–421
Incus, *954*, 955
Independent assortment
Mendelian law of, 237–239
separation of homologous chromosomes during meiosis, 220–222
Independent variable, 13
Indeterminate growth, 715–720, 786
Indianmeal moth, 1184
Indirect competition, 1184
Indirect fitness, 1115
Indole-3-acetic acid, *759*, 760, 762, 763
*See also* Auxins
Induced fit, 155
Induced mutations, 309
Induced plant defenses
against herbivores, 803–805
against pathogens, 798–801
Induced pluripotent stem cells (iPS cells), 409
Inducers
in cell fate determination, 395–396, *397*
defined, 330
in negative regulation of the *lac* operon, 331, 332
Inducible promoters, 384
Inducible proteins, 330
Inducible systems, in transcriptional regulation of operons, 330–331, 332
Induction
in cell fate determination, 395–396, *397*
defined, 395
Inductive logic, 12
Industrial nitrogen fixation, 749, 750
Infants
brown fat in, 828
essential amino acids, 1051
genetic screening, 320, 321
microbiomes and, 540
respiratory distress syndrome in premature babies, 1015
Infection thread, 748
Inferior mesenteric ganglion, *974*
Inferior vena cava, 1030, *1031*
Infertility, in humans, 897, 899
Inflammation
anti-inflammatory drugs, 862
description of, 861–862
medical problems, 862
necrosis and, 225
Inflammatory disease therapy, 158
Inflammatory response, 860, 861–862
Inflorescence meristems, 715, 785–786
Inflorescences, 596, *597*, 786
Influenza virus
as an RNA virus, 341
epidemics, 427
strain H5N1, *543*
vaccines, 427, 446
Infrared perception, in rattlesnakes, 946
Ingroup, 453
Inheritance
blending inheritance, 233, 234
human pedigrees, 240–241
Mendelian laws of, 233–241
of organelle genes, 252–253
particulate theory of, 233–234, 236
probability calculations, 239–240
sex-linked, 249, 251–252
Inhibin, 892, 896
Inhibition
in demonstrating cause and effect, 98
of enzymes, 157–159
Inhibitors, 128
Inhibitory synapses, 938
Initials, 715
Initiation complex, in translation, 295, 296
Initiation site, in transcription, 286, *287*
Innate immunity
barrier and local agents, 859–860
cell signaling pathways, 860
inflammation, 861–862
overview, 857, *858*, 859
specialized proteins and cells, 860–861
Inner cell mass, 906, *907*, 914
Inner ear
anatomy of, *954*, 955
flexion of the basilar membrane, 955–956
vestibular system, *954*, 955, 956–957
Inorganic cofactors, of enzymes, 155, *156*
Inorganic fertilizers, 747
Inorganic ions
absorption in the large intestine, 1063
absorption in the small intestine, 1063
Inosine (I), 294
Inositol trisphosphate ($IP_3$), 134–135, 136
Inouye, Isao, 102
Insect wings
development, 415
evolution of, 673, *675*
first appearance of, 520
of pterygotes, 671, 672
success of the insects and, 673
Insecticides
inhibition of acetylcholinesterase, 940
irreversible inhibitors of enzymes, 157
produced by genetically modified plants, 386–388
Insects
appearance of flight in, 520
atmospheric oxygen and body size, 505, 513, 514, 522
chemical defenses, 1173
colonization of human corpses, 1188, 1203
contraction in flight muscles, 997–998
diapause, 421
estimating the number of living species, 651, 673
excretory system, 1076–1077
fossils in amber, *515*
gas exchange in, 671, 1009
gigantic, 505
herbivorous, 1175
heterotypy and leg number in, 418, *419*
hormonal control of molting, 839–841
human impact on the distribution of, 1145
internal fertilization, 887
key features and body plan, 671, *672*
major groups, 671–673, *674*
metabolic heat production, 825–826
as pollinators, 599
superficial cleavage, 905
undescribed species, 675
wingless relatives, 671
Insomnia, 978
Inspiratory reserve volume (IRV), *1012*, 1013
Instars, 671
Insula, 983
Insular cortex, 983
Insulin
actions of, *842*
blood glucose regulation, 848–849, 1066–1067
control of food intake, 1067, 1068
diabetes in the Pima and, 1048
production through biotechnology, *385*
Insulin-dependent diabetes mellitus, 848, 876
Insulin-like growth factor 1 (IGF-1) gene, 352
Insulin receptors, 129, 853
"Insurance population," 255
Integral membrane proteins, *106*, 107–108, 121
Integrase, 544
Integrase inhibitors, 342
Integrin
in cell attachment to the extracellular matrix, 112, *113*
in cell movement, 112, *113*
Integument
in angiosperms, 597
in conifers, 594, *595*
development of seed coat from, 591, *592*, 784
in seed plants, 590
Intercalated discs, 991
Intercellular signaling
bonding behavior in voles and, 125
cell responses to, 125, 126–127
effects on cell function, 137–139
evolution of multicellularity and, 140–141
second messengers, 131–137
signal receptors, 127–131
types and sources of signals, 126
*See also* Signal transduction pathways
Intercostal muscles, *1014*, 1016, 1020
Interference competition, 1182, 1183
Interference RNA. *See* RNA interference
Interferons, 860
Interglacial intervals, 510
Interleukins, 211, 874
Intermediate filaments, 95
Intermediate muscle fibers, 996
Internal anal sphincter, 1063
Internal environment
homeostatic regulation, 816–817
importance of self-regulation, 10
importance to multicellular animals, 816
plasma membrane and, 80
Internal fertilization, 888–889
Internal gills
respiratory gas exchange in fishes, 1009–1010, *1011*
surface area maximization, 1008, *1009*
Internal jugular vein, *1031*
Internal membranes
origins of, 101–102
of prokaryotes, 83
Internal shells, 664

Internal skeletons
of echinoderms, 680, *681*
of glass sponges and demosponges, 643
of humans, 999, *1000*
of vertebrates, 685, 687
*See also* Endoskeletons; Skeletal systems
International animal trade
consequences of, 1234, *1235*
ending, 1240–1241
International Fund for Animal Welfare, 1241
International System of Units (ISU), 1049
International Union for the Conservation of Nature (IUCN), 1231
Interneurons
defined, 940
in the retina, 963
in spinal reflexes, 942–943
Internodes, 709, 719
Interphase
defined, 208
in meiosis, 219
in mitosis, *217*
subphases, 208
Interspecific competition
community productivity and, 1192
defined, 1182
impact on life history traits, 1157
Interstitial fluid, 816, 1026, 1042
Intertidal zone
defined, 1139
impact of competition on barnacle niche determination, 1184
inhibition of succession by green algae, 1201
keystone species, 1194
Intestines
colonization by endosymbiotic bacteria, 1056–1057
digestion in, 1056
epithelial absorption of nutrients and inorganic ions, 1063
infolding of the walls, 1057
*See also* Large intestine; Small intestine
Intracellular receptors, 129, 130–131
Intracytoplasmic sperm injection (ICSI), 899
Intraspecific competition, 1182
Intrauterine device (IUD), *898*
Intrinsic factor, 1054, 1058
Intrinsic rates of increase, 1156, 1159, 1163
Introduced species
high population densities and, 1149, 1160
impact on the fynbos, 1145
population dynamics of reindeer, 1149, 1158, 1166
*See also* Invasive species; Non-native species
Introns
in alternative splicing, 346
in archaea, 356
description of, *290*, 291
splicing to remove, *291*, 292–293
Inuit peoples, 1054
Invasive species
controlling or preventing, 1241, *1242*
"decision tree" for evaluating invasive plants, 1241, *1242*
negative impact of, 1235–1236
*See also* Introduced species; Non-native species
Invasiveness, of pathogens, 542
Inversions, 308
Invertase, 736
Invertebrates
body size and respiratory gas diffusion, 1007
excretory systems, 1075–1077
visual systems, 958
Involution, 909, *910*
Iodine/Iodide, 845, *846*, 847, *1052*, 1221
Iodine/Iodide deficiency, 847, 1054
Ion channel receptors, 129
Ion channels
actions of sensory receptor proteins on, 947, *948*
activation in stereocilia, 954
in cardiac pacemaker cells and heart contraction, 1033–1034
in the generation of membrane potentials, 928, 929
ionotropic receptors, 938–939
opening in response to signals, 137–138
patch clamp studies, 929, *931*
as receptors, 129
root uptake of mineral ions and, 729
specificity of, 115–116
structure and function, 115, *116*
*See also* Gated ion channels; *specific types of ion channels*
Ion pumps, root uptake of mineral ions and, 729
Ion transporters, in the generation of membrane potentials, 928, 929
Ionic attractions
description of, 28–29
features of, *26*
in protein quaternary structure, 48
Ionic electric current, 931–932
Ionic interactions
in protein binding, 50
in protein tertiary structure, 47
Ionic regulators, 1073
Ionization, of water, 34
Ionizing radiation, 309
Ionotropic receptors, 938–939
Ionotropic sensory receptor proteins, 947, *948*
Ions
complex, 29
description of, 29
electrochemical gradients, 929
*See also specific ions*
$IP_3$/DAG pathway, 134–135
IPEX, 874
*Ipomoea batatas*, 718
Iridium, 511
Iris (eye), 958, *959*
Iris (Greek god), 834
Irisin, 834, 838
Iron
in animal nutrition, *1052*, 1053
in catalyzed reactions, *156*
global cycle, 1221
impact on nitrogen fixation, 1222–1223
in plant nutrition, *741*, 742
Iron deficiency, 1053, 1054
Iron oxide, 512
Irreversible inhibition, of enzymes, 157, *158*
Island biogeography theory, 1196–1198, *1199*
Islets of Langerhans, 848–849
Isobutane, *41*
Isocitrate dehydrogenase, 182
Isoleucine, *44*
Isomers, 41
Isomorphic alternation of generations, 563
Isopods, 670
Isoptera, *672*
Isosmotic reabsorption, 1082
Isotonic solutions, 114, *115*
Isotope analysis, of water to detect climate change, 36
Isotopes
description of, 22–24
of oxygen, 21
Isozymes, 162
*Istiophorus albicans*, *1139*
Iteroparous species, 1156
Ivory-billed woodpecker, 1230, *1231*
Ivory trade, 1240–1241
*Ixodes scapularis* (= *Ixodes dammini*), *1151*, 1152

## J

Jackfruit, *601*
Jacky dragons, 421
Jacob, François, 331
Jacobson, Henning, 856
Janzen, Daniel, 1178, *1179*
Japanese macaque, 1177
Japanese mint, *604*
Japanese quail, 1101
Jasmonic acid (jasmonate), 759, 799, 804, 805
Jasper Ridge Biological Preserve, 1161
Jaw worms, 657–658
Jawed fishes, 686–689
Jawless fishes, 685–686, *687*
Jaws
development in vertebrates, 684
as joints, 1002
JAZ protein, 805
Jejunum, 1061
Jellyfish
characteristics of, 646–647
gastrovascular cavity, 1056
green fluorescent protein, 449
life cycle, 646, *647*
*See also* Cnidarians
Johnny jump-ups, *597*
Johnson, R. T., 209
Johnston's organ, 671
Jointed appendages
in animal evolution, 636
arthropods, 667
crustaceans, 671
trilobites, 668
Joints, 1001–1002
Joules (J), 147
Joyner-Kersee, Jackie, 986
Juan de Fuca oceanic plate, 509
Jumping
diversity of ability in animals, 986
efficiency of kangaroos in, 1003
Jumping bristletails, 671, *672*
Junipers, 594, *1133*
*Juniperus*, *1133*
Jurassic period, *506–507*, 521
Juvenile hormone, 841
Juxtaglomerular cells, 1088
Juxtacrine signals, 126

## K

*K*-strategists, 1159, 1201
*Kalanchoe*, 793
Kamen, Martin, 186, 187
Kangaroos
jumping efficiency, 986, 1003
as marsupials, 697, *698*, *699*
Kaposi's sarcoma, 876
*Karenia*, 549
Kartagener's syndrome, 902
Karyogamy, *618*, 619, 620, *621*
Karyotype, 224, *225*
Kashefi, Kazem, 535
Katydid, 1173
Katz, Lawrence, 951
Keeley, John, 757
Kentucky bluegrass, 199
Kenward, R. E., 1116
Keratin, 95, 397
Keratin genes, 66
Keto group, *40*
α-Ketoglutarate, 180
Keystone species, 1194–1195
Kidney stones, 848
Kidney transplants, 1085–1086
Kidneys
in amniote evolution, 692
in blood pressure regulation, 1044–1045
blood vessels of, *1080*, 1082
effect of aldosterone on, 849
effect of antidiuretic hormone on, 843
effect of parathyroid hormone on, 848
in the human excretory system, *1080*
nephron structure in mammals, *1080*, 1081–1082
production of concentrated urine in mammals, 1071, 1077, 1082–1084, 1090
regulation of, 1087–1090
regulation of acid–base balance, 1084–1085
secretion of erythropoietin, 1038
structure and function, 1077, 1078–1079, *1080*
treatments for renal failure, 1085–1086, *1087*
Kilocalorie (kcal), 1049
Kimura, Motoo, 492
Kin selection, 1115–1116
Kinesins
in cilia and flagella, 97–98
in kinetochores and cytokinesis, 216
Kinetic energy, 145
Kinetin, 768

Kinetochore microtubules, 214–216
Kinetochores, 213, 214, *215,* 216
Kinetoplast, 559
Kinetoplastids, 558, 559
King, Thomas, 406
King penguin, *642*
Kingdoms (taxonomic category), 463
Kingfishers, *642*
Kinorhynchs, *632,* 665
"Kiss and run" process, 122
Kitasato, Shibasaburo, 862, *863,* 864
Kiwis, 694
*Klebsiella,* 1169
Klein, David, 1149
Klinefelter syndrome, 250
Klok, C. Jaco, 513
Knee-jerk reflex, 969, 1002
Knee joint, 1002
Knockout model experiments, 381–382, 1096–1097
Koalas, 1055, 1175
Koch, Robert, 541
Koch's postulates, 541–542
Kokanee salmon, 1228
Komodo dragon, 693
Korarchaeota, 535, 537
Krakatau, 510
Krebs cycle. *See* Citric acid cycle
Kruger National Park, 1242
Kuffler, Stephen, 975, 976
Kuroshio Current, *1125*
Kuwait, 373
Kwashiorkor, 1050

## L

L ring, *83*
La Selva Biological Station, 1237
Labia majora, 892, *893*
Labia minora, 892, *893*
Labor, in childbirth, 896–897
Labrador Current, *1125*
Labrador retrievers, 244
*lac* Operon
  description of, 330, *331*
  negative and positive regulation of, 330–332, *333*
*lac* Repressor, 331–332, *333,* 335
Lace lichen, 625
Lacewings, *672*
Lacks, Henrietta, 205
Lactases, *1062,* 1063
Lactate, 177
Lactate dehydrogenase, 177
Lacteals, 1063
Lactic acid
  accumulation in muscle, 997
  formed in prebiotic synthesis experiments, 70
Lactic acid fermentation, 177
Lactoglobulin, 385
Lactose
  in negative regulation of the *lac* operon, 331, *333*
  in the small intestine, 1063
Lactose metabolism, regulation in *E. coli,* 329–332, *333*
*LacZ* gene, 378
*Laetiporus sulphureus, 623*
Lagging strand, in DNA replication, 272–273, *274*
Lagomorpha, *698*
*Lagopus lagopus, 1128*
Lake Erie, 1221
Lake Malawi, 467, 509
Lake Malawi cichlids. *See* Haplochromine cichlids
Lake St. Clair, 1160
Lake Superior, 1189, 1215
Lake trout, 1228
Lakes
  acidification, 1220
  characteristics of, 1140
  effect of nitrate pollution on arsenic levels, 1221
  eutrophication, 1219, 1221
  thermocline, 1214
  turnover, 1213–1214
Lamar Valley, 1193–1194
Lamb's-quarter, 1200
Lamellae, 1010
Lamin proteins, 95
Lampreys, 686
Lampshells, 659
  *See also* Brachiopods
Lancelets, *632, 679,* 683, 684
Land plants
  adaptations to life on land, 574
  alternation of generations, 574–575
  classification, *574*
  closest relatives, 572–573
  colonization of the land, 574–579
  endosymbiotic origin of chloroplasts in, 551
  evolution of, *570,* 571
  major clades, 573, *574*
  nonvascular, 575–579
  vascular, 579–585
Landsteiner, Karl, 243
Language, humans and, 705
Language areas, of the human brain, 980–981
Langurs, 492–494
*Laqueus, 659*
Large cactus finch, *474*
Large ground finch, *474*
Large intestine
  in humans, *1058*
  in ruminants, 1064
Large tree finch, *474*
Larva
  complex life cycles of parasites, 640, *641*
  defined, 639
  dispersal, 640
  in metamorphosis, 639
Larvaceans, 684
Larynx, 1013, *1014,* 1059
Lateral gene transfer, 496, 529–530
Lateral meristems, 715, *716*
Lateral roots, 718, *718*
Lateralization, of language functions, 980–981
Latex, 801, 806
Laticifer-cutting beetles, 1175
Laticifers, 806
*Latimeria chalumnae,* 689
*Latimeria menadoensis,* 689
Latitudinal gradients, in species diversity, 1196, 1197
*Laupala,* 495
Laurasia, 521, 1143
Law of independent assortment, 237–239
Law of mass action, 35–36
Law of segregation, 236–237, 239
Laws of thermodynamics, 146–147
LDL. *See* Low-density lipoproteins
LDL receptors, 121
LEA proteins, 809
Leaching, 745, 746
Leading strand, in DNA replication, 272–273
Leafcutter ants, 1169, 1178, 1185
Leaf primordia, 719, 720
Leafhoppers, *672,* 673, 1175, 1184
*LEAFY* gene, 401, 792
*LEAFY* transcription factor, 786
Learning
  cellular basis of, 982
  defined, 981
  human capacity for, 981–982
  sleep and, 980
Leaves
  abscission, 765, 769
  adaptations to very dry conditions, 807
  aerenchyma, 808
  anatomy, 720–721
  anatomy in $C_3$ and $C_4$ pants, *198,* 199
  of carnivorous plants, 751–752
  defenses against herbivory, 1175
  development, 719, 720
  evolution of, 583, *584*
  function, 709
  hairs, 713, 1175
  invasion by fungal hyphae, *612*
  origin of the flowering stimulus in, 788–790
  salt glands, 811
  senescence, 768, 769–770
  stomatal control, 732–734 (*see also* Stomata)
  transpiration, 732
  trichomes, 713
  vegetative reproduction in angiosperms, 793
  veins, 710, 720, *721*
Lecithin, *58*
Leeches, 661–662, 1056
Left-handed helices, *47*
Left subclavian artery, 858
Leg muscles
  impact on venous blood flow, 1042
  integration of anabolism and catabolism during exercise, 180–181
Leghemoglobin, 750
Legs. *See* Jointed appendages; Limbs
Legumes
  in crop rotation, 750
  cyanide production, 805
  evolution of nitrogen-fixing symbiosis, 521
  root nodule formation, 747–748
*Leiobunum rotundum, 669*
*Leishmania major, 559*
Leishmaniasis, *559*
Leks and lekking, 481, 1104, *1105*
Lemurs, *10,* 701
Lens
  determination in vertebrates, 396
  of the image-forming eye, 958–959
  in ommatidia, 958
Lens placode, 396
Lenticles, 722
*Leontopithecus rosalia,* 1114, *1137*
Leopard frogs
  experiments on the effects of atrazine, 12
  jumping ability, *986*
  temporal isolation, 476
Leopard gecko, *694*
*Lepas pectinata, 670*
*Lepidodermella, 658*
Lepidopterans, *672, 674,* 675
Lepidosaurs, 693
*Lepomis macrochirus,* 1105, *1106*
Leptin, 1067, 1068
Leptin receptor, *1067,* 1068
*Leptosiphon,* 460
*Leptosiphon bicolor,* 460
*Leptosiphon liniflorus, 460*
*Lepus alleni, 828*
*Lepus arcticus, 828*
Lesser long-nosed bats, 1137
Lettuce seeds, 772
Leucine, *44*
Leucoplasts, 92
*Leucospermum cordifolium, 1134*
Leukemia therapy, 158
Leukocytes. *See* White blood cells
Levers, 1002
Levin, Donald, 478, 479
Lewis, Reggie, 1025
Lews, Cynthia, *565,* 566
LexA protein, 341
Leydig cells, 889, *891,* 892
*Libellula quadrimaculata, 674*
Lice, *672,* 1177
Lichens
  description of, 613–614
  edible, 624
  indicators of air pollution, 624, *625*
  in succession on glacial moraines, 1200
Life
  common ancestry of, 3
  common characteristics of, 2
  elements of living organisms, 22
  evolution of, 3–5 (*see also* Evolution)
  origin of, 3 (*see also* Origin of life)
  scale of, *78*
  timeline of, *3, 515*
Life cycles
  amphibians, *691*
  angiosperms, 600–601
  animals, 639–642
  cnidarians, 645, *646*
  complex, 563, *564*
  ferns, 582
  fungi, *616, 618, 621*
  mosses, 575–577
  pine, 594, *595*
  seed plants, 589–590
  of viruses, 339–341
  *See also* Alternation of generations
Life history strategies, 1156–1157, 1159, 1163
Life tables, 1154–1155
Life zones
  in freshwater biomes, 1140
  in marine biomes, 1139–1140

Lifestyle, obesity and, 1048
Ligaments, 953, 1002
Ligand-gated ion channels, 115, *116*
Ligand–receptor complexes, 128
Ligands
  in binding to receptor proteins, 127–128
  chemoreceptors and, 949
  defined, 115, 127
Light
  absorption by pigments, 190–191
  aspects plants are responsive to, 771
  detection by animal sensory systems, 957–963
  entrainment of circadian rhythms in plants, 774–775
  light energy in biological systems, *145*
  photobiology, 189–190
  photochemistry, 188–189
  photomorphogenesis in plants, 772
Light-harvesting complexes, 190–191
Light-independent reactions, 188
Light-induced electron transport, 196–197
Light microscopes, 79, 80
Light reactions, 188
Lignin, 711, 798
Lilies, *598, 780*
*Lilium, 780*
*Lima* (British merchant vessel), 525
Limbic system, 970, 982
Limbs
  evolution in lobe-limbed vertebrates, 690
  evolution of the insect wing, 673, *675*
  heterotypy and leg number in insects, 418, *419*
  loss of forelimbs in snakes, 423
  morphogens and positional information in vertebrate development, 401
  *See also* Hindlimbs; Jointed appendages
Limestone deposits, 557, 565
*Limnephilus, 674*
Limpets, 662
*Limulus polyphemus, 668*
LIN-3 protein, 396, *397,* 398
*Lin*-14 mutations, 347
Lind, James, 1053
Lindley, John, 588
Lineage species concept, 469
Linnaean classification, 462–463
Linnaeus, Carolus, 468
Linnean Society of London, 429
Linoleic acid, *57,* 1052
*Linum usitatissimum,* 800
Lions, exploitation competition, 1183
Lipases, 144, 1058, *1062*
Lipid bilayers, 73–74
Lipid-derived second messengers, 134–135
Lipid monolayers, 535–536
Lipid-soluble hormones, 837
Lipids
  of archaea, 535–536
  in atherosclerosis, 1042
  in biological membranes, 106–107
  carotenoids, 58
  catabolic interconversions, 179–180
  membrane fluidity and, 107
  monomer components, 40
  phospholipids, 57, *58 (see also* Phospholipids)
  properties of, 56
  proportions in living organisms, *41*
  steroids, 58
  thylakoid lipids, 92
  triglycerides, 56–57
  types, 56
  vitamins, 58
  waxes, 58–59
Lipoproteins, 1065–1066
Lithium, 135
Lithosphere, 509
Lithotrophs, 538
Littoral zone, 1139
Liver
  absorption of nutrients by, 1063
  activation of glycogen phosphorylase, 132–133
  control of fat metabolism, 1065–1066
  control of glucose metabolism, 1065, 1066–1067
  effect of insulin on, 1066
  familial hypercholesterolemia, 121
  gluconeogenesis in, 1065, 1067
  in humans, *1058*
  maintenance of blood glucose levels during exercise and, 181
  production of clotting factors, 1039
  reabsorption of bile salts, 1063
  role in digestion, 1061–1062
  uptake of low-density lipoproteins, 121
Liver cells
  in the fight-or-flight response, 837
  regulation of glucose metabolism, 138
Liver diseases, 1039, 1041
Liverworts, 573, *574,* 577
Lizards
  behavioral thermoregulation, 822, *823*
  evolution and characteristics of, 693, *694*
  heat exchange through the skin, 824
  hemipenes, 888
  parthenogenic reproduction, 881–882
  temperature-dependent sex determination and sex-specific fitness differences, 421
  territorial behavior, 1103, *1104*
Loading, of phloem sieve tubes, 736
Loams, 745
Lobe-limbed vertebrates, 687, 689–690
Loboseans, 560
Lobsters, 670
Locomotion
  body cavities and, 635–636
  in ciliates, 554
  in plasmodial slime molds, 560
  in protostome evolution, 675
Locus, 242, 431
*Locusta migratoria, 802*
Lodgepole pines, 594, *596,* 1201–1202, *1236*
Lofenelac, 322
Logic, inductive and deductive, 12
Logistic population growth, 1158
Lombok island, 1141, *1142*
London taxi drivers, 967
Long bones, development, 1001
Long-day plants (LDPs), 787, 788
Long-distance athletes, 995
Long-horned beetles, 1145
Long interspersed elements (LINEs), 364–365
Long-tailed widowbird, 435–436
Long-term depression (LTD), 982
Long-term memory, 982, 983
Long-term potentiation (LTP), 982
Long terminal repeats (LTRs), 364–365, 495
Longhorn cattle, 440
Longwing butterflies, *1174,* 1175, 1176
Loops of Henle
  aquaporins, 1084
  countercurrent multiplier, 1082–1084
  in desert rodents, 1090
  organization of, *1080,* 1081–1082
  regulation of the glomerular filtration rate, 1088
"Lophophorates," 652
Lophophores
  in brachiopods and phoronids, 658, 659, *660*
  in bryozoans, *654,* 656
  description of, 652
  in entoprocts, 656
Lophotrochozoans
  anatomical characteristics, *652*
  in animal phylogeny, *630*
  annelids, 659–662
  brachiopods and phoronids, 658–659, *660*
  bryozoans and entoprocts, 656
  diversity in, 656
  flatworms, rotifers, and gastrotrichs, 656–658
  lophophores and trochophores, 652–653, *654*
  major subgroups and number of living species, *632*
  mollusks, 662–664
  ribbon worms, 658, *659*
  spiral cleavage, 633, 653
  wormlike body forms, 653–654
Lordosis, 1099
Lorenz, Konrad, 1095, 1099
Loriciferans, *632,* 665–666
Lorikeets, *696*
Lorises, 701
Loss-of-function mutations, 305, 311–312, 400, 401
Lovley, Derek, 535
Low-density lipoproteins (LDLs), 121, 1066
Low-GC Gram positives, 530–532
Low temperatures, animal adaptations to, 1125
Lowland gorillas, *702*
*Loxodonta africana, 1135, 1150*
Luciferase, 150
Luciferin, *149,* 150
*Lucilia caesar, 674*
"Lucy," 703
*Luehea seemannii,* 651
*Lumbricus terrestris, 661*
Lumen, of the vertebrate gut, 1058
Lung cancer cells, *227*
Lung surfactants, 1015
Lungfish, *689,* 690, 1028–1029
Lungs
  diffusion of carbon dioxide from blood, 1018, *1019*
  evolution in lungfish, 1028–1029
  human, 1013–1016
  perfusion by the circulatory system, 1016
  physical stresses in snorkeling elephants, 1005, 1022
  surface area maximization, 1008, *1009*
  tidal ventilation, 1012–1013
  unidirectional ventilation in birds, 1010–1012
Luteal phase, 893
Luteinizing hormone (LH)
  actions of, *842*
  endocrine source, *842,* 843
  in follicle selection for ovulation, 895–896
  in puberty, 850–851, 892, 894
  regulation of the ovarian and uterine cycles, 893–894
*Luxilus coccogenis, 472*
*Luxilus zonatus, 472*
*Lycaon pictus,* 1183, 1242, *1243*
*Lycoperdon perlatum, 610*
Lycophytes, *574,* 580, 581, 582–583
Lycopodiophyta, *574*
*Lycopodium, 583*
*Lycopodium annotinum, 581*
Lyell, Charles, 428
Lyme disease, 533, 1152
Lymph, 857, 858, 1042
Lymph capillaries, 858
Lymph ducts, *857*
Lymph nodes, *857,* 858, 1042
Lymphatic system, *857,* 858, 1042
Lymphatic vessels, 1042
Lymphocytes
  cell membranes, *4*
  clonal deletion, 865
  clonal selection, 865, *866*
  diversity of adaptive immunity and, 863
  effector cells and memory cells, 865–866
  specificity of adaptive immunity and, 862–863
  types and functions of, 858
Lymphoma tumors, 876
*Lyperobius huttoni,* 1144–1145
Lysine, *44,* 212
Lysogenic cycle, of bacteriophage, 340–341
Lysosomal storage diseases, 91
Lysosomes
  evolution in eukaryotic cells, 551

Inclusion-cell disease, 300
structure and function, 90–91
Lysozyme
convergent molecular evolution in foregut fermenters, 492–494
in innate defenses, 859
interactions with substrate, *153*, 154
molecular models of, *47*
turnover number, 156
Lytic cycle, of bacteriophage, 339–340, 341

## M

M band, *987*, 988
*Macaca fuscata*, 1177
MacArthur, Robert, 1197
MacKinnon, Roderick, 115
*Macrocystis*, 556, *1139*
*Macroderma gigas*, *700*
Macroevolutionary change, 446
Macromolecules
characteristics of, 39
condensation and hydrolysis reactions, 42
defined, 40
endocytosis and exocytosis, 120–122
functional groups, 40
isomers, 41
molecular evolution, 486
proportions in living organisms, *41*
relationship of structure to function, 41–42
types found in living things, 40
*See also* Carbohydrates; Lipids; Nucleic acids; Proteins
*Macronectes giganteus*, *1073*
Macronucleus, *555*, 562
Macronutrients
in animal nutrition, 1052–1053
in plant nutrition, 741, *742*
Macroparasites, 1176–1177
*Macroperipatus torquatus*, *667*
Macrophages
as antigen-presenting cells, 872
cytokines, 859
degradation of old red blood cells, 1038
digestion of pus, 862
function of, *858*
HIV infection, 876
pattern recognition receptors, 860
*Macropus giganteus*, *699*
Macroscelidea, *698*
*Macrotis lagotis*, *699*
Macula densa, 1088
Macular degeneration, 382
Madagascan shield bug, *674*
Madagascar ocotillo, *1135*
*Madia sativa*, *481*, 482
Madreporite, 680, *681*
MADS box, 400
"Mafia behavior," 1093, 1102, 1117
Magnesium
in animal nutrition, *1052*
in chlorophyll *a*, 190
in plant nutrition, *741*
Magnetic fields
animal navigation and, 1109
paleomagnetic dating and, 508
Magnetic resonance imaging (MRI), 970
Magnolia, *598*
*Magnolia*, *602*
Magnoliids, 602
Maguire, Eleanor, 967
Mahadevan, Lakshminarayanan, 752
Maidenhair tree, 592
*See also* Ginkgos
Maintenance methylase, 343
Major histocompatibility complex (MHC) proteins
antigen-presenting function, 872, *873*
in the cellular immune response, 871, 872, *873*, 874
functions of, 858–859
organ transplant surgery and, 874
role in T cell selection, 872
Malaria
biological control, 627
causative agent, 553, *564*
cause of primary symptoms in, 563
treatments, 605, 797, 812
Malate, 171, 199
Malate dehydrogenase, 171
Malathion, 157, 940
Malawi, 1240–1241
Malay Archipelago, 1141, *1142*
Male flowers, *779*, *780*
Male reproductive system
components of semen, 889, 890
emission and ejaculation, 891
erectile dysfunction, 891
hormonal control of, 892
penile erection, 890–891
spermatogenesis in, 889–890, *891*
Malignant tumors, 227
Maller, James, 209
Malleus, *954*, 955
Malnutrition, 1054
Malpighi, Marcello, 734–735
Malpighian tubules, 1076–1077
Maltase, *1062*
Malthus, Thomas Robert, 1164
"Malting," 761
Maltose, *53*
Mammalian heart
anatomy of, 1030, *1031*
blood flow through, 1030–1031
cardiac cycle, 1031–1032
coordination of muscle contraction, 1034, *1035*
electrocardiograms, 1035–1036
pacemaker cells and cardiac muscle contraction, 1032–1034
sustained contraction of ventricular muscles, 1034, *1035*, *1036*
Mammalian nervous system
brainstem structure and function, 969–970
forebrain structure and function, 969–970
functional and anatomical organization, 968–969
higher functions in cellular terms, 978–983
information processing by neural networks, 973–978
spinal cord functions, 969
Mammalian thermostat, 829–830
Mammals
actions of prolactin in, *838*
basal metabolic rate and body size, 826–827
blocks to polyspermy, 886–887
bone growth in, 416–417
circulatory system, 1030
cleavage in, 905–906
convergent molecular evolution in foregut fermenters, 492–494
defense systems, 857–859 (*see also* Immunology)
dissipation of heat with water and evaporation, 829
distinguishing features, 697
egg-laying, 889
eutherians, 697–705
evolutionary radiation, 696
excretory system, 1079–1086
extraembryonic membranes, 919
gastrulation, 914–915
genomic imprinting during gamete formation, 344–345
heart function, 1030–1037
heat production in, 827–828
heat stroke, 815
hibernation, 830, *831*
Hox genes and body segmentation, 917
hypothalamus as the "thermostat" of, 829–830
length of gestation in, 920
lens of the eye, 958–959
major endocrine glands and hormones, 845–852
major living groups and number of species, *698*
master circadian "clock," 1108
multipotent stem cells, 408
origin of, 692, *693*
production of concentrated urine, 1071, 1077, 1082–1084, 1090
prototherians, 697, *698*
range in body size, 696–697
sex determination in, 249–250
teeth, 697, 1055–1056
therians, 697–700
thermoregulation through the skin, 825
viviparity in, 889
vomeronasal organ, 950, 951
Mammary glands, 697, 843
*Mammuthus columbi*, *1229*
Manatees, 700
Mandibles, 669, 1056
Mandibulates, 667, 669–671
Mandrills, *702*
*Mandrillus sphinx*, *702*
*Manduca sexta*, *802*
Manganese
in animal nutrition, *1052*
in plant nutrition, *741*
Mangold, Hilde, 911, 912
Mangrove forests, 1141
Mangrove island experiment, 1198, *1199*
Mangroves, *808*, *811*, 1141
*Manihot esculenta*, 708
*Manihot glaziovii*, 724
Mannose, *52*
Mantidflies, *672*
Mantids, *672*, 673
Mantle, 662, *663*, 664
Mantodea, *672*
Mantophasmatodea, *672*
Manucodes, *480*, 481
*Manucodia comrii*, *480*
Manure, 747
MAP kinase, *132*
"Map sense," 1109
MAPK (mitogen-activated protein kinase), 139
Mapping, genetic, 248–249
Maquis, *1134*
Marathon runners, 815
*Marchantia*, 577
*Marchantia polymorpha*, *577*
Mariana Islands, 509
Mariana Trench, 509
Marine animals
larval forms and dispersal, 640
undescribed species of annelids, 675
Marine biome, 1139–1140
Marine ecosystems, negative impact of invasive species, 1235
Marine fireworms, *636*
Marine flatworms, *1007*
Marine iguanas, 824
Marine mollusks, 999
Marion Island, 1199
Marker genes, selectable, 376, 378
Mark–recapture method, 1151
Marler, Catherine, 1103, *1104*
Marler, Peter, 1100
*Marrella splendens*, *517*
Mars, 68, 69, 70
Marshall, Barry, *542*
Marshes, 1140
*Marsilea*, *581*
Marsupial moles, *698*
Marsupials, 697, *698*, *699*
*Marthasterias glacialis*, *682*
Maryland Mammoth tobacco, 787
Mass, defined, 22
Mass extinctions
Carboniferous, 521
Cretaceous, 521
Devonian, 520
meteorite-caused, 511, 520, 521
periodic nature of, 508
Permian, 522
sea level drops and, 510
Triassic, 521
*See also* Species extinctions
Mass spectrometry, 369
Mast cells, *858*, 861, *875*
Mastax, 657, *658*
Maternal diet, human birth defects and, 916
Maternal effect genes, 402, *403*
Mating
costs and risks, 882
effect on genotype or allele frequencies, 434–436
heterozygote advantage and polymorphic loci, 442–443
impact of signaling systems on speciation, 705

maximization of the fitness of both partners, 1113–1114
olfactory cues and, 705
types of, 1113
variety in deuterostomes, 705
Mating calls, 476, *477*, 705
Mating seasons, 476
Mating types, of fungi, 617
Matter
atomic structure, 22–25
in chemical reactions, 31
Matthaei, J. H., 288
Maturation promoting factor, 209, 210
Maturational survivorship curves, 1156
Matz, Mikhail, 449, 464
Maxillipeds, *671*
Maximum likelihood methods, 456
Mayflies, 672, *673*
Mayr, Ernst, 468–469
McCulloch, Ernest, 408
McFall-Ngai, Margaret, 546
*MCIR* gene, 367
Mean arterial pressure (MAP), 1043–1044
Measles, 877
Mechanical energy, *145*
Mechanical isolation, 476
Mechanical weathering, 745
Mechanical work, 10
Mechanically gated ion channels, 930
Mechanoreceptors
auditory and vestibular hair cells, 953–954
in auditory systems, 954–956
influence on ion channels, 947, *948*
in muscles, tendons, and ligaments, 952–953
response to physical forces, 952
tactile, 952
vestibular system, 956–957
Mechanosensory signals, in animal communication, 1112
Mecoptera, *672*
Medawar, Sir Peter, 339
Medicinal leeches, 661–662
Medicinal plants, 604–605
Medicine
benefits of human genomics, 367–368
biotechnology and, 384–386
discovery of penicillin, 608
importance of biological research to, 15–16
Koch's postulates, 541–542
medicinal plants, 604–605
phage therapy, 545–546
stem cells and, 77, 392, 408, 409, 410
use of leeches in, 661–662
uses of molecular evolution, 501–502
*See also* Cancer drugs; Cancer treatment; Molecular medicine
Mediterranean climate, 1121, 1126, 1134, 1146
Mediterranean flour moth, 1184
Medium ground finch, *474*
Medium tree finch, *474*
Medulla
in blood pressure regulation, 1044, 1045
in control of breathing, 1019–1020
development in the human brain, 968, *969*
sensitivity to the partial pressure of carbon dioxide in blood, 1021
Medusa
of cnidarians, 645, *646*
of scyphozoans, 647, *648*
Megafauna, 1229
Megagametophyte
in angiosperms, *600*, *779*, *781*
in seed plants, 590, 591, *592*
in vascular plants, 584
Megakaryocyte, 1039
Megaloptera, *672*
*Meganeuropsis permiana*, 505
Megaphylls, 583, *584*
Megasporangia
in angiosperms, 596–597, *600*
in conifers, 594, *595*
in seed plants, 590, 591, *592*
in vascular plants, *584*, 585
Megaspore
in angiosperms, *600*, *779*, *781*
in conifers, *595*
in seed plants, 590, *592*
in vascular plants, 584
Megasporocyte, *595*, *600*
Megastrobilus, 594, *595*
*Megatypus schucherti*, *518*
*Meiacanthus grammistes*, *1174*
Meiosis
chromatid exchanges during, 219–220, *222*
comparison with mitosis, *223*
defined, 207
errors leading to abnormal chromosome structures and numbers, 222, 224
final products, 219, *221*, 222, *223*
during gametogenesis, 883–884
length of, 219–220
nondisjunctions, 309
overall function of, 219
reduction of chromosome number in, 219
segregation of alleles, *237*
separation of homologous chromosomes by independent assortment, 220–222
in sexual life cycles, 217, 218–219
Meiosis I
errors leading to abnormal chromosome structures and numbers, 222, 224
events of, *220–221*
unique features of, 219
Meiosis II
events of, 219, *220–221*
separation of sister chromatids, 222
Meissner's corpuscles, 952
MEK, *132*
*Melampsora lini*, 800
Melanin, 246
Melanocyte-stimulating hormone (MSH), *842*
Melatonin, *842*, 851
Membrane-associated carbohydrates, *106*, 109, 111
Membrane currents, in rod cells, 961
Membrane lipids, of archaea, 535–536
Membrane potential
of cardiac pacemaker cells, 1032–1034
defined, 927
gated ion channels and, 930–932
generation of, 928–929, *930*
generation of action potentials and, 932–934
graded changes, 932
measuring with electrodes, 928
Membrane proteins
cell adhesion and, 111
rapid diffusion of, 109, 110
types of, *106*, 107–109
Membrane receptors, 129
Membrane transport
active, 118–120
function of transporters, *42*
mechanisms in, *118*
passive, 113–117
Membranes
aquaporins and permeability, 116
of archaea, 535–536
diffusion across, 114
dynamic nature of, 109
factors affecting fluidity, 107
fluid mosaic model, 106
fluidity of, 107
membrane-associated carbohydrates, 109
osmosis, 114–115
passive transport, 113–117
pleural membranes, *1014*, 1015–1016
protocells, 73–74
significance in the evolution of life, 3–4
structure of, 106–110
thickness of, 107
*See also* Extraembryonic membranes; Internal membranes; Plasma membranes
Membranous bones, 1001
Memory
cellular basis of, 982
defined, 981
emotional content, 982
fear memory, 970, 982
hippocampus and, 967, 982, 983
human capacity for, 981–982
sleep and, 980
types of, 982
Memory cells, 865–866
Menadione, *1053*
Mendel, Gregor
inheritance experiments with garden peas, 233–236
law of independent assortment, 237–239
law of segregation, 236–237, 239
rediscovery of, 428
test crosses, 237, *238*
Menopause, 893
Menstrual cycle, 892, 893–894
*See also* Uterine cycle
Menthol, *604*
Meristem culture, 793
Meristem identity genes, 786
Meristems
floral, 400, 715, 785–786
hierarchy in plant growth and development, 715, *716*
indeterminate primary growth, 715–720
initials, 715
lateral, 715, *716*
origin in plant embryogenesis, 712
in plant development, 710
secondary, 715, *716*
stem cells in, 408
*See also* Apical meristems; Ground meristem; Primary meristems
Merkel's discs, 952
*Merops apiaster*, *888*
Merozoites, *564*
*Mertensia virginica*, *462*
Meselson, Matthew, 268, *269*, 270
Meselson–Stahl experiment, 268, *269*, 270
Mesenchymal stem cells, 392, 394, 408
Mesenchyme
in acoelomates, 635
in frog gastrulation, *910*
in sea urchin gastrulation, 908, 909
Mesoderm
in avian gastrulation, 914
body cavity types and, 635
body segmentation in vertebrates, 916–918
defined, 908
differentiation of muscle precursor cells, 398–399
in extraembryonic membranes, 918
in frog gastrulation, 909, 910
in protostomes, 652
in sea urchin gastrulation, 908, 909
tissues and organs derived from, *907*, 908
in triploblastic animals, 633
Mesoglea, 644, 646, 647
Mesophyll
in $C_3$ and $C_4$ plants, *198*, 199
in eudicot leaves, 720, *721*
response to water stress, 734
Mesozoic era, *506–507*, 521, 593
Mesquite, 808
Messenger RNA (mRNA)
alternative splicing, 346–347
blocking translation to study gene expression, 382
cDNA libraries and, 379–380
codons and the genetic code, 288–289
DNA microarray technology, 382–383
location and role in eukaryotic cells, *286*
modification of pre-mRNA, 291–293
movement out of the nucleus, 293
nucleic acid hybridization, 290–291

produced by transcription, 286–288
recycling in bacteria, 301
relation to protein abundance, 348
role in gene expression, 285
transcriptional regulation in prokaryotes, 329–330
translation, 293–297, *298*
Metabolic heat, 822, 825
Metabolic inhibitors, 322–323
Metabolic pathways
allosteric regulation, 160–161
cellular locations of energy pathways, *168*
glucose oxidation, 166–169
governing principles, 166
linkages between catabolism and anabolism, 179–180
in prokaryotes, 166, 168
as regulated systems, 181–182
systems biology and, 157
transcriptional regulation in prokaryotes, 329–330
Metabolic pool, 180
Metabolic rate, basal, 826–827
Metabolism
defined, 145
in early prokaryotes, 4
energy transformations, 145–149
linkage of anabolic and catabolic reactions, 146
types of, 145–146
Metabolites, 370
Metabolomes, 370
Metabolomics, 370
Metabotropic receptors, 939
Metabotropic sensory receptor proteins, 947, *948*
Metacarpals, 423
Metagenomics, 357–358
Metal ion catalysis, 155
Metamorphosis
defined, 672
description of, 639–640
holometabolous insects, 673
incomplete, 841
juvenile hormone and, 841
in lampreys, 686
types of, 672
Metanephridia, 1075, *1076*
Metaphase
comparison between mitosis and meiosis, *223*
determination of karyotype during, 224, *225*
in meiosis, *220, 221*
in mitosis, 212, 214, 215, *217*
Metaphase plate
in meiosis, *220, 221*
in mitosis, *215*
Metapopulations, 1161–1162
*Metarhizium anisopliae,* 627
Metastasis, 227
Meteorites
meteorite-caused mass extinctions, 511, 520, 521
origin of life and, 69
Methane
bacteria in methane cycling, 357
covalent bonds in, 27
as a greenhouse gas, 1212
in pogonophoran metabolism, 660, 661
production by methanogens, 536
*Methanococcus,* 357
Methanogens, 536
Methicillin-resistant *S. aureus* (MRSA), 281, 301
Methionine, *44,* 296
Methotrexate, 158
Methyl bromide, 1198
Methylation
of cytosine, 310
of histone, 344
1-Methylcyclopropene, 770
5′-Methylcytosine, 310, 328, 343, 345
Methylglucosinolide, *802*
*Methylococcus,* 357
Meyer, Axel, 328
Mice
albumin gene promoter, 335
"Down syndrome mouse," 924
embryonic stem cells, 408–409
exercise-induced irisin production, 834
homologous recombination and knockout mice, 381–382
Hox genes, *414*
immunoglobulin genes, 868
*inversus viscerum* mutant, 915
knockout experiments in behavior, 1096–1097
vomeronasal organ, 951
Micelles, 1062
Microbial communities, 539, *540*
Microbial eukaryotes, 552
*See also* Protists
Microbial rhodopsin, 537
Microbiomes, 539–541
Microbiothere, *698*
Microclimates, 1126
Microevolutionary change, 446
Microfilaments
asymmetric distribution of cytoplasmic determinants and, 395
cell movement and, 95, *96,* 98, 99
contractile ring, 216
structure and function, 94–95
Microfossils, 74
Microgametophyte
in angiosperms, 600, 779
in seed plants, 589–590
in vascular plants, *584,* 585
*See also* Pollen grains
Microglia, 927
Micromolar solutions, 34
Micronucleus, *555,* 562
Micronutrients
in animal nutrition, 1052, 1053
in plant nutrition, 741–743
Microorganisms
commercial production of proteins, 383–384
digestion of cellulose in herbivores and, 1063–1064
interference competition, 1183
Microparasites, 1176
Microphylls, 581, 583
*Micropogonias undulatus,* 1207
*Micropterus dolomieu,* 1160
Micropyle, *592,* 594, *595*
*Microraptor gui,* 695
MicroRNA (miRNA), *286,* 347, 348, 382
Microscopes, *79,* 80–81, 84
Microsporangium
in angiosperms, 597
in conifers, *595*
in seed plants, 590, 591
in vascular plants, *584,* 585
Microspores
in angiosperms, *600,* 780, *781*
in conifers, *595*
in seed plants, 590
in vascular plants, *584,* 585
Microsporidia, *615, 616,* 617
Microsporocytes, *600, 781*
Microstrobilus, 594, *595*
Microtubule organizing centers, 212
Microtubules
asymmetric distribution of cytoplasmic determinants and, 395
cilia and flagella, 96–98
in the evolution of the eukaryotic cell, *550,* 551
in plant cell cytokinesis, 216–217
in rearrangements of egg cytoplasm following fertilization, 903–904
spindle apparatus, 212, 213–214, *215*
structure and function, 95–96
*Microtus montanus,* 125, 1113
*Microtus ochrogaster,* 125, 1113
Microvilli
in the intestines, 1057, 1063
microfilaments in, 95, *96*
in taste buds, 951
Midbrain, 968, *969*
Middle ear, 954–955, 956
Middle lamella, 713
Mifepristone (RU-486), *898*
Migration
adaptive value of, 1126
of dinosaurs, 21
fat as stored energy in birds, 1050
navigation over great distance, 1109–1110
Milk
breast milk, 843
of prototherians, 697
Milkweed grasshopper, *2*
Milkweeds, 806, 1176, 1200
Miller, Stanley, 70, 71
Millet beer, 624
Millimolar solutions, 34
Millipedes, 669, 670
*Mimetica, 1173*
Mimicry
mechanical reproduction isolation and, *476*
mimicry systems, 1173–1175
Mimiviruses, *543,* 544, 545
*Mimulus aurantiacus,* 598, *599*
Mineral nutrients
plant requirements, 741
transport in xylem, 730–732
uptake by plants, 727–730
Mineralcorticoids, 849
Minty taste, 951
miRNA. *See* MicroRNA
*Mirounga angustirostris, 1105*
Mismatch repair, 276, 277, 310
Missense mutations, 306, *307,* 311, 312
Missense substitutions, 491
*See also* Nonsynonymous substitutions
Missouri saddled darter, *472*
Mistletoebird, 1182
Mistletoes, 752, 1182
Mites, 668, 669
Mitochondria
absence in microsporidia, 617
absence in some excavates, 558
in animal cells, *86*
in animal fertilization, *885,* 886
apoptosis and, 226
β-oxidation in, 179–180
in developing human sperm, 889, *891*
endosymbiotic origin in eukaryotic cells, *550,* 551
energy pathways in, *168*
in exercised muscle fibers, 997
generation of heat in brown fat and, 165
of kinetoplastids, 559
membrane impermeability to NADH, 178–179
oxidative phosphorylation and ATP synthesis, 171–176
in photorespiration, 197, *198*
in plant cells, *87*
structure and function, 91–92
transposition of genes to the nucleus, 366
Mitochondrial DNA (mtDNA)
origins of, 92
phylogenetic analyses and, 456
Mitochondrial genes
inheritance of, 252–253
mutations, 252
Mitochondrial matrix
contents of, 92
energy pathways in, *168*
pyruvate oxidation in, 170
Mitogen-activated protein kinase (MAPK), 139
Mitogens, 139
Mitosis
in asexual reproduction, 217
in the cell cycle, 208
centromeres and the plane of cell division, 212–213
chromatin structure, 211–212, *213*
chromosome separation and movement, 207, 214–216
comparison with meiosis, *223*
cytokinesis, 216–217
determination of karyotype during, 224, *225*
overview of events and phases, 211, 212, *214–215, 217*
spindle apparatus formation, 213–214
Mitosomes, 617
Mitotic center, 213
Mitotic spindles, 905
*See also* Spindle apparatus
Mitter, Charles, 480
*Mnemiopsis, 644*
Mobley, Cuttino, 1025

Model organisms
  apoptosis studies in development, 399
  *Caenorhabditis elegans,* 666
  eukaryotic genome studies, 361–363
  in genetics, 282
  importance in biological research, 14
Moderately repetitive sequences, *355,* 364–366
Modules, developmental, 415–418
Molar solutions, 34
Molars (teeth), 1055
Molds
  in biological control, 626
  bread molds, 282–283, 619
  brown molds, 624
  description of, 622
  green molds, 624, 1169
  water molds, 556, *557*
  *See also* Slime molds
Mole (chemistry), 33–34
Mole (mammal), *698*
Molecular biology
  "central dogma," 285
  using data in phylogenetic analyses, 456
Molecular chaperones, 51
Molecular clocks
  concept of, 492
  defined, 461
  using to date evolutionary events, 461–462
Molecular cloning
  commercial production of proteins, 383–384
  with recombinant DNA technology, 375–379
  sources of DNA used in, 379–380
Molecular evolution
  comparing genes and proteins through sequence alignment, 486–487
  description of, 486
  detecting positive and purifying selection in the genome, 492–494
  experimental studies, 489–491
  in genome size, 494–496
  neutral theory of, 492
  practical applications, 499–502
  synonymous and nonsynonymous substitutions, 491, 492–494
  using models of sequence evolution to calculate evolutionary divergence, 487–489
Molecular medicine
  approaches to gene mutations, 304
  chronic myelogenous leukemia, 304
  DNA microarray technology, 383
  genetic screening, 320–322
  halplotype mapping, 367–368
  knockout mouse models, 382
  reverse genetics, 318
  RNAi-based therapy, 382
  strategies for treating genetic diseases, 322–325
  using genetic markers to find disease-causing genes, 318–319
Molecular mimicry, 876
Molecular weight, 26–27
Molecules
  amphipathic, 57
  chemical bonds, 26–31
  defined, 24
  hydrophilic and hydrophobic, 30
  models for representing, *27*
  molecular weight, 26–27
  octet rule, 25
  *See also* Macromolecules
Mollusks
  anatomical characteristics, 652, *652*
  body plans, *663*
  cephalopod image-forming eye, 958, *959*
  chemical defenses, 1173
  larval form, 640
  major body components, 662, *663*
  major groups, *632,* 662, 664
  monoplacophorans, 662
  open circulatory system, 1026, *1027*
  red tides and, 549
  spiral cleavage, 905
  undescribed species, 675
*Molothrus ater,* 1093, 1233
Molting
  in ecdysozoans, 654
  exoskeletons and, 999
  hormonal control in insects, 839–841
Molybdenum
  impact on nitrogen fixation, 1222–1223
  in plant nutrition, *741*
  in sediments, 69
Momentum, vestibular detection of, 956–957
*Monachanthus,* 588, 605
Monarch butterfly, *639, 1174,* 1176
Monilophytes, *574,* 581–582, 582–583
Monito del monte, 697
Monkeys, 701, *702*
Monoamines, 939–940
Monoclonal antibodies, 871
Monocots
  characteristics of, 601–602, 709–710
  diversity in, 603
  early shoot development, 758
  guard cells, 733
  leaf veins, 720
  root anatomy, *717*
  root systems, 718
  thickening of stems, 723
  vascular bundles, *719*
Monoculture, 1202–1203
Monocytes, *858*
Monod, Jacques, 331
Monoecious plants
  defined, 597, 779
  example of, *780*
  strategies for preventing self-pollination, 782
Monoecious species
  defined, 249, 887
  hermaphroditism in animals, 887–888
Monogeneans, 657
Monohybrid crosses, 234–236, 239
Monomers, 40
  of carbohydrates, 51
  condensation and hydrolysis reactions, 42
Monomorphic populations, 437
Monophyletic groups, 463
Monoplacophorans, 662
Monosaccharides, 51–53
  membrane-associated, 109
Monosodium glutamate (MSG), 951
Monosomics, 224
Monosynaptic reflexes, 942
Monotremes, *698,* 889
Monozygotic twins, 344, 907
Montane voles, 125, 1113
Montreal Protocol, 311
Moore, Michael, 1103, *1104*
Moose, *1129*
Moraines, 1200
Moray eels, *688*
*Morchella esculenta, 622*
Morels, 620, *622*
Morgan, Thomas Hunt, 247, 248, 251, 430
Morphine, *604,* 940
Morphogenesis
  defined, 393
  in development, 710
  impact of mutations in developmental genes, 418–419
  pattern formation, 399–405
  in plants, 710–712
  processes contributing to, 394
Morphogens
  in fruit fly body segmentation, 401–405
  positional information in development and, 401
Morphological species concept, 468, 469
Morphology
  adaptations to low temperature, 1125
  phylogenetic analyses and, 455
Mortality, 1154
*Morus bassanus, 1153*
Mosaic development, 907
Mosaic viruses, 544
Mosquitoes
  biological control, 627
  malaria and, 563, *564*
Mosses
  description of, 577–578
  distinguishing characteristics, *574*
  life cycle, 575–577
  as nonvascular land plants, 573
  in succession on glacial moraines, 1200
Moths
  impact of diet on developmental plasticity, 421, *422*
  in pollination syndromes, 1180–1181
  *See also* Lepidopterans; *individual moths*
Motile animals, 637
Motor end plate, 936, 937, 989, *990*
Motor neurons
  induction by Sonic hedgehog, 917
  neuromuscular junctions, 936–938
  number of muscle fiber innervated, 994
  in skeletal muscle contraction, 989, *990*
  spinal reflexes, 942–943
Motor programs, 969
Motor proteins
  in cilia and flagella, 97–98
  functions of, 94
  microtubules and, 96
  *See also* Myosin
Motor units, 989, 995
Mottle viruses, 544
Mount Everest, 1010
Mount Pinatubo, 510
Mountain avens, 1200
Mountain climbers, 1008
Mountain pine beetle, 1202
Mountain zebra, *1144*
Mountains, rain shadows and, 1126–1127
Mourning cloak butterfly, *1130*
Mouth
  chemical digestion in, 1060
  chewing of food, 1059
  in tubular guts, 1056
Mouthparts
  of chelicerates, 668
  of mandibulates, 669
"Movement proteins," 140
MRSA. *See* Methicillin-resistant *S. aureus*
Mucopolysaccharides, 716
Mucosa, 820, 1058
Mucosal epithelium, 1058
Mucus
  of the human respiratory tract, 1013, 1015
  in innate defenses, 859
  of the nasal epithelium, 949
  in seminal fluid, 890
Mucus escalator, 1013, 1015
Muir Glacier, *17*
Muller, Hermann Joseph, 441, 470
Müllerian mimicry, *1174,* 1175
Muller's ratchet, 441
Mullis, Kary, 278
Multicellular animals
  internal environment and homeostasis, 816–817
  physiological systems (*see* Physiological systems)
  relationship between cells, tissues, and organs, 817–820
Multicellular organisms
  cell communication in, 139–141
  genome and gene expression, 6
  importance of cell adhesion and cell recognition in, 110–111
  importance of cellular specialization and differentiation, 9
  intercellular communication and the evolution of, 140–141
  types of mutations in, 305
Multicellularity
  appearance in the Proterozoic, 515–516

atmospheric oxygen levels and, 512
evolution in eukaryotes, 552–553
nematode genome, 362, *363*
through geological time, 515–522
Multifactorial phenotypes, 314–315
Multiple alleles, 242, 243
Multiple covalent bonds, 28
Multiple fruits, 601
Multiple genes, coordinated regulation, 336–337
Multiple sclerosis, 926
Multiple substitutions, 488, 489
Multipotency, 394
Multipotent stem cells, 408, 410
Multisubunit allosteric enzymes, 159–160
Murchison Meteorite, 69
*Muscari armeniacum, 603*
Muscarinic ACh receptors, 940
Muscle
contraction (*see* Muscle contraction)
differentiation of precursor cells, 398–399
exercise-induced irisin production, 834
integration of anabolism and catabolism during exercise, 180–181
interactions with exoskeletons, 999
interactions with hydrostatic skeletons, 999
jumping ability in animals, 986
myoglobin, 1017
types of, 987
*See also* Cardiac muscle; Skeletal muscle; Smooth muscle
Muscle contraction
in cardiac muscle, 991–993, 1032–1034, *1035*
factors affecting the strength of, 994–995
impact on venous blood flow, 1041–1042
in insect flight muscle, 997–998
sliding filament model, 987–990, *991*
in smooth muscle, *992*, 993
Muscle fatigue
effect of enhanced cooling on, 831
heat and, 998
Muscle fibers
impact of strength training on, 996
intermediate, 996
optimal length for generating maximum tension, 996
role of muscle fiber types in strength and endurance, 995–996
in skeletal muscle, 987–988
sliding filament contractile mechanism, 987–990, *991*
twitches and tetanus, 994–995
Muscle spindles, 952–953
Muscle tissues
characteristics of, 817
lactic acid fermentation in, 177
types of, 817
Muscle tone, 995
Muscular dystrophy, 365, 370
Muscular segmentation, 636
Musculoskeletal systems
factors affecting muscle performance, 994–998
interactions of muscles and skeletal systems, 999–1003
muscle contraction, 987–994
Mushrooms, 609–610, 622, 623, 624
Mussels, 662, 1194
*See also* Zebra mussels
Mustard, 724
Mustard oil glycosides, 1176
Mutagens
defined, 283
induced mutations, 309
natural or artificial, 310
public policy goals regarding, 311
use of transposons in minimal genome studies, 359, 360
Mutant alleles, 241–242
Mutation rates, 432
Mutations
base pairs that are "hot spots," 310
benefits and costs, 310–311
caused by retroviruses and transposons, 308
chromosomal, 307–308
creation of new alleles, 241–242
defined, 283, 305, 432
in demonstrating cause and effect, 98
of DNA, 266
effect on phenotype, 305–306
evolution and, 6
generation of genetic variation, 432
in genomes, 6
morphological impact of mutations in developmental genes, 418–419
Muller's ratchet, 441
in organelle genes, 252
point mutations, 306–307
purging of deleterious mutations, 433, 441
reversal of, 306
spontaneous or induced, 308–309
that lead to human genetic diseases, 311–315
transposons and, 365
types and effects of mutagens, 309, 310
types in multicellular organisms, 305
using to study gene function, 381
*See also* Gene mutations; *specific types of mutations*
Mutualisms
characteristics of, 1177–1178
defined, 612, 1170
food exchange for care or transport, 1178
food exchange for housing or defense, 1178–1179
food exchange for seed transport, 1181–1182
pollination syndromes, 1180–1181
*See also* Fungal mutualisms; Mycorrhizae
*Myanthus*, 588, 605
Mycelium, 609, 610
*Mycobacterium tuberculosis*, 532
Mycologists, 611
*Mycoplasma capricolum*, 359
*Mycoplasma genitalium*
comparative genomics, 357
functional genomics, *357*
genomic information, *361*
minimal genome studies, 359, 360
*Mycoplasma mycoides*, 359
*Mycoplasma mycoides JCV1-syn.1.0.*, 359, *360*
Mycoplasmas, 531–532
Mycorrhizae
arbuscular mycorrhizae, 619–620, 749 (*see also* Arbuscular mycorrhizae)
description of, 614–615
ectomycorrhizae, 614, 615, 622
expansion of the plant root system, 748–749
formation, 747, 748
in plant evolution, 574
in reforestation efforts, 626
Myelin, 926
Myelinated axons, conduction of action potentials, 935
*Myliobatis australis, 688*
Myocardial infarction, 1043
*MyoD* gene, 399
MyoD transcription factors, 399
Myofibrils, *987*, 988
Myogenic heartbeat, 992
Myoglobin
binding of oxygen, 1017
in exercised muscle cells, 996–997
globin gene family and, 497, *498*
in slow-twitch fibers, 995
Myosin
contractile ring, 216
functions of, 94
in muscle tissues, 817
in skeletal muscle contraction, *987*, 988–990, *991*
in smooth muscle contraction, 993
Myosin filaments
impact of strength training on, 996
in muscle fiber contraction, *987*, 988–990, *991*
Myosin kinase, 993
Myosin phosphatase, 993
Myostatin, 370
Myostatin gene, 370
Myotonic dystrophy, 314
Myriapods, *632*, *667*, 669–670, 675
*Myrmecocystus mexicanus*, 1183
*Mysis diluviana*, 1228
*Mytilus californianus*, 1194
Myxamoebas, 561
Myxomyosin, 560
Myxozoans, 646

## N

NADH-Q reductase, 172
*Naegleria*, 558
Naked mole-rats, 1115–1116
*Nanaloricus mysticus, 665*
Nanoarchaeota, 535, 537
*Nanos* gene, 402, *403*
Nanos protein, 402, *403*
*Narcis, 603*
Nasal epithelium, 949, 950
Nasal salt glands, 1073
National Wildlife Federation, 1244
Natural gas deposits, 565
Natural history, 18
Natural killer cells
function of, *858*
in innate defenses, 861
pattern recognition receptors, 860
Natural Resources Defense Council, 17
Natural selection
categories of, 433
Darwin's theory of, 6
direct action on phenotypes, 438
generation of new phenotypes, 433
origin of the concept of, 428–430
possible effects on populations, 439–440
Naupilus, 640
*Nautilus*, 664
Nautiluses, 662, 664
Navigation, 1109–1110, *1111*
Neanderthals, 7, 367, 434, *703*, 704
Nearctic region, *1142*
*Nebela collaris, 560*
Neck, evolution in the giraffe, 416–417
Necrosis, 225
Nectar, 599, 1180
"Nectar corridor," 1137
"Nectar thieves," 1180
Negative feedback
in physiological systems, 816
regulation of hormone secretion, 844–845
Negative gravitropism, 765
Negative regulation
defined, 329
of the *E. coli lac* operon, 330–332, *333*
in eukaryotes, 333
in virus reproductive cycles, 340
Negative selection, 863, 872, 876
Negative-sense single-stranded RNA viruses, *543*, 544
Neher, E., 929
Nematodes
anatomical characteristics, *652*
description of, 666
genomic information, *361*, 362, *363*
major subgroups and number of living species, *632*
predatory fungi, 612, *613*
undescribed species, 675
vulval determination, 396, *397*, 398
*See also Caenorhabditis elegans*
Nemerteans, 658, *659*
*Nemoria arizonaria*, 421, *422*
Nemtaocysts, 645, *646*
Neognaths, 694–695

Neopterans, 672–673
Neoteny, 692, 704
Neotropical region, *1142*
Nephridiopore, 1075, *1076*
Nephrons
production of concentrated urine in mammals, 1082–1084
structure and function in vertebrates, 1077, 1078–1079
structure in mammals, *1080,* 1081–1082
*Nephroselmis,* 102
Nephrostome, 1075, *1076*
Neptune's grass, *603*
Nernst equation, 929, *930*
Neruoptera, *672*
Nerve cells. *See* Neurons
Nerve cord, 659–660, 683
Nerve deafness, 956
Nerve gases, 157, 940
Nerve nets, 637, 940–941
Nerves, 968
Nervous system development
initiation in amphibians, 912–913
neurulation, 915–916
Nervous systems
cell types, 925–927
chemical signaling in, 835, 836
electric signaling in, 927–936 (*see also* Action potentials)
effect of blood calcium levels on, 847
gray matter and white matter, 941
interactions with endocrine system, 842–845
neural networks, 940–943
neurotransmitters, 936–940
overinhibition, 924, 939, 943
synapses, 936–940
types and functions, 637
*See also* Mammalian nervous system
Nervous tissues, 819–820
Nest parasitism, 1093, 1102, 1117
Net primary productivity (NPP), 1190, 1208–1210
Neural crest cells, 916
Neural networks
functional categories of neurons in, 940
information processing in the mammalian nervous system, 973–978
range of complexity in, 940–941
spinal reflexes, 941–943
vertebrate brain, 943
Neural plate, 916
Neural tube, 916, 917, 968
Neuroeconomics, 141
Neurohormones
defined, 836
hypothalamic, 843, 844, 849
Neuromuscular junctions
actions at, 936–938
in skeletal muscle contraction, 989, *990*
Neurons
changes during sleep, 979
characteristics of, 819–820, 925
communication via neurotransmitters, 836
components of, *819*
electroencephalograms, 978
functional categories, 940
generated from fibroblasts, 336, *337,* 338
generation and transmission of electric signals, 927–936 (*see also* Action potentials)
measurement of the resting potential, 928
neural networks, 940–943
number in the human brain, 943
structure and function, 925–926
summation of synaptic input by the postsynaptic cell, 938
synapses and neurotransmitters, 936–940
varied morphologies of, *925*
*Neurospora,* 282–283
*Neurospora crassa,* 624, 626
Neurotoxins, 500
Neurotransmitter receptors
agonists and antagonists, 940
ionotropic and metabotropic, 938–939
multiple types for each neurotransmitter, 940
Neurotransmitters
of the autonomic nervous system, 975
clearing from the synapse, 940
functions of, 836, 926
multiple receptor types for each neurotransmitter, 940
at neuromuscular junctions, 936–938
release from the presynaptic membrane, 936, *937*
response of the postsynaptic membrane to, 936–938
types and properties of, 939–940
Neurulation
in amphibians, 915–916
body segmentation during, 916–918
Neutral alleles
accumulation in populations, 441
genetic drift, 434
Neutral mutations
accumulation in populations, 441
fixation by genetic drift, 492
Neutral theory, of molecular evolution, 492
Neutrons
defined, 22
isotopes and, 22–24
mass number and, 22
Neutrophils, *858,* 862
"New" diseases, 501
New Orleans (LA), 1223–1224
New World monkeys, 701, *702*
New World opossums, 697, *698, 699*
New Zealand, 1144–1145
Newborns. *See* Infants
"Newly rare" species, 1231
Nexin, 97
NF-κB transcription factor, 860
Niacin, *1053*
Niche, 1184
Nickel, *741,* 742, 743
Nicotinamide, 70
Nicotinamide adenine dinucleotide ($NAD^+$/NADH)
in alcoholic fermentation, *177,* 178
allosteric regulation of the citric acid cycle, 182
in catalyzed reactions, *156*
in the citric acid cycle, 170, 171
in glucose catabolism, 169
in glycolysis, 169, 170
impermeability of the mitochondrial membrane to, 178–179
in oxidation–reduction reactions, 167–168
in pyruvate oxidation, 170
in the respiratory chain, 172, 173
Nicotinamide adenine dinucleotide phosphate ($NADP^+$)
in the Calvin cycle, 194, *196*
in photosynthesis, 188, 191, 192
Nicotine, 803, 804
Nicotinic ACh receptors, 940
Night blindness, 58, *1053*
Night length
melatonin release and, 851
photoperiodic cues in flowering, 788
Night vision, 962
"9 + 2" Array, 96, *97*
Nirenberg, Marshall W., 288
Nitrate
as an electron acceptor in denitrifiers, 539
in the global nitrogen cycle, 750, 751
impact on the cycling of arsenic, 1221
inorganic fertilizers and, 747
leaching in soils, 746
organic fertilizers and, 746
production by nitrifiers, 539
Nitrate transporters, 744
Nitric acid, in acid precipitation, 1219–1220
Nitric oxide (NO)
hypersensitive response in plants and, 800–801
as a neurotransmitter, 940
in penile erection, 890, 891
in plant responses to pathogens, 799
as a second messenger, 135–136
Nitrification, 750, *751*
Nitrifiers
in the global nitrogen cycle, 750, *751*
oxidation of ammonia by, 539
Nitrite
in the global nitrogen cycle, 751
as a mutagen, 310
production by nitrifiers, 539
*Nitrobacter,* 539
Nitrogen
in the atmosphere, 1211
covalent bonding capability, *27*
electronegativity, *28*
heavy nitrogen in the Meselson–Stahl experiment, 268, *269,* 270
octet rule for molecule formation, 25
in plant nutrition, *741*
sources for fungi, 611
Nitrogen-14, 507
Nitrogen cycling
global nitrogen cycle, 750–751, 1218–1219
prokaryotes and, 539
Nitrogen deposition, 1219
Nitrogen dioxide, 1219
Nitrogen excretion
forms of nitrogenous waste excreted, 1074–1075
in invertebrates, 1075–1077
*See also* Excretory systems
Nitrogen fertilizers
energy costs, 740, 750
environmental costs, 740
*See also* Chemical fertilizers
Nitrogen fixation
by heterocysts of cyanobacteria, 532, *533*
by humans, 1218–1219
impact of atmospheric carbon dioxide on, 1221–1223
prokaryotes and, 539
by soil bacteria, 749–750
Nitrogen fixers
evolution of symbiosis with legumes, 521
formation of root nodules, 747–748
nitrogen fixation by, 539, 749–750
Nitrogen runoff, 740, 1207, 1219
Nitrogen use efficiency, 740, 753
Nitrogenase, 750
Nitrogenous waste
forms excreted by animals, 1074–1075
*See also* Excretory systems
Nitrosamines, 310
*Nitrosococcus,* 539
*Nitrosomonas,* 539
Nitrous acid, 309
Nitrous oxide
as a greenhouse gas, 1212
increased atmospheric levels from human activity, 1219
NO. *See* Nitric oxide
NO synthase, 136
*Noctiluca,* 549
Nod factors, 747
*Nod* genes, 747
Nodal cells, 902, 914–915, 921
Node, of the mammalian embryo, 902, 914–915, 921
Nodes
of phylogenetic trees, 450
of plants, 709
Nodes of Ranvier, 935
Nodule meristem, 747, *748*
Noggin, 916
Noller, Harry, 296
Non-identical twins, 907
Non-native species
invasives, 1235–1236
regulating the importation of non-native plants, 1241
risks of deliberate introductions, 1228, 1245
*See also* Introduced species; Invasive species
Non-REM sleep, 978–979

Noncoding DNA sequences
in eukaryotic genes, 290–291, 361
genomic information and, 356
significance in genomes, 494–496
Noncompetitive inhibitors, of enzymes, *158,* 159
Noncovalent interactions, in protein binding, 50
Noncyclic electron transport, 191–192
Nondisjunction
abnormal sex chromosome arrangements, 250
in the formation of allopolyploids, 225, *226*
as a mutation, 309
production of aneuploid cells and, 222, 224
Nonpolar covalent bonds, 28
Nonpolar substances, effect on protein structure, 50
Nonrandom mating, 434–436
Nonself, distinguishing from self, 243, 863, 874
Nonsense mutations, 306, *307*
Nonshivering heat production, 827–828
Nonspecific defenses. *See* Innate immunity
Nonsynonymous substitutions, 491, 492–494
Nonvascular land plants
defined, 573
distinguishing characteristics, 575
hornworts, 578–579
liverworts, 577
members and distinguishing characteristics, *574*
mosses, 577–578
sporophyte and gametophyte generations, 575–577
Noradrenaline. *See* Norepinephrine
Noradrenergic neurons, *974,* 975
Norepinephrine (noradrenaline)
actions of, *842,* 850
in the autonomic nervous system, 975
in blood pressure regulation, 1044, *1045*
effect on heartbeat, 1034
effects on gut muscle, 993
endocrine source, 849
as a neurotransmitter, 940
North Atlantic Drift, *1125*
North Pacific Drift, *1125*
Nose
nasal epithelium, 949, 950
olfaction, 949–950
vomeronasal organ, 950–951
*Nostoc,* 749
*Nostoc punctiforme, 533*
*Nothofagus, 1129,* 1130, 1143
Notochord
in chordates, 683–684
development in amphibians, 915–916
dorsal–ventral signaling during organogenesis, 917
evolutionary relationships among chordates and, *455*
in tunicates and lancelets, 684
Notoryctemorphia, *698*
*Notropis, 472*
Nottebohm, Fernando, 1101
*Noturus, 472*
Nuclear bombs, 310
Nuclear envelope
breakdown in meiosis I, *220*
breakdown in mitosis, *214*
in the evolution of the eukaryotic cell, 551
formation during telophase, 216
function of, 85
origins of, 101–102
Nuclear gene sequences, phylogenetic analyses and, 456
Nuclear lamina, *85,* 88
Nuclear localization signal (NLS), 299, *300*
Nuclear pores, 85, 88, 293
Nuclear reactors, 310
Nuclear transfer experiments, 406–407
Nucleases, 1058, *1062*
Nucleic acid hybridization, 290–291, 321, 398
Nucleic acids
anabolic interconversions and, *179,* 180
growth of, 63, *64*
molecular evolution, 486
monomer components, 40
in the origin of life, 3
proportions in living organisms, *41*
in protocells, 74
structure and function, 63–67
types of, 63
*See also* Deoxyribonucleic acid; Ribonucleic acid
Nucleoid, 82
Nucleolus, 85, *87*
Nucleosides, 63
Nucleosomes, 212, *213*
Nucleotide bases
Chargaff's rule, 264, 265, 266
in DNA structure, 264–267
exposure in the grooves of DNA, 266, *267*
formed in prebiotic synthesis experiments, 70
induced mutations, 309
point mutations, 306–307
spontaneous mutations, 308–309
Nucleotide sequences
evolutionary relationships in prokaryotes and, 528–529
identifying evolutionary changes in, 486
phylogenetic analyses and, 456
point mutations, 306–307
Nucleotide substitutions
in gene evolution, 486
synonymous and nonsynonymous substitutions, 491, 492–494
using models of to calculate evolutionary divergence, 487–489
Nucleotides
components of, 63
in DNA replication, 268–269, *270*
genetic information and, 5, 6
other functions of, 66–67
phosphodiester linkages, 63, *64*
Nucleus (atomic), 22
Nucleus (cell)
in animal cells, *86*
in ciliates, 554, *555,* 562
defined, 81
in eukaryotic cells, 85, 88
movement of mature mRNA out of, 293
in plant cells, *87*
transposition of organelle genes to, 366
Nucleus (group of neurons), 969–970
Nudibranchs, 662, *664, 1174*
Null hypothesis, 13
Nutrient deficiencies, *1053,* 1054
*See also* Deficiency diseases
Nutrients
as the basis of cellular biosynthesis, 10
cycling through ecosystems, *1208,* 1214–1223
importance of viruses to nutrient cycling, 546
as sources of energy, 10
*See also* Animal nutrition; Macronutrients; Micronutrients; Mineral nutrients; Plant nutrition
Nutritional categories, of prokaryotes, 537–538
*Nymphaea, 597*
*Nymphalis antiopa, 1130*


## O


O blood group, 243
*Ob* gene, *1067,* 1068
*Obelia, 648*
Obesity, 1048, 1050–1051, 1068
human SNP scans, *368*
recent epidemic in, 165
as a risk factor for atherosclerosis, 1043
single-gene mutations affecting, *1067,* 1068
UCP1 protein and, 165, 182
Obligate aerobes/anaerobes, 536, 537
Obligate parasites, 553–554, 611
Oblimersen, 382
"Observational" learning, 982
Observations, in science, 11–12
Occam's razor, 454
Occipital lobe, 973
Ocean currents, 1124–1125
Oceans
acidification, 647, 1217
biomass distribution, 1191–1192
importance of viruses to ocean ecology, 546
iron in sediments, 1221
marine biome, 1139–1140
osmolarity, 1073
transport of elements through ecosystem compartments, 1214
uptake and release of carbon dioxide, 1217
upwelling zone, 1214
*See also* Sea level
Ochre sea star, 1194
Octet rule, 25
*Octopus macropus, 664*
Octopuses, 662, 664, *959*
Odd-toed hoofed mammals, *698*
Odonata, *672*
Odorant receptors, 138
Odorants, 138, 949, 950
Off-center receptive fields, 975
Oil spills, 569
bioremediation, 373
Oils, 56–57
Okamoto, Noriko, 102
Okazaki fragments, 273
Olcese, James, 895
Old World monkeys, 701, *702*
Olduvai Gorge, 703
Olfaction
description of, 949–950
signal transduction in, 137–138
Olfactory bulb, 949–950
Olfactory cilia, 949
Olfactory cues, in mate selection, 705
Olfactory receptor neurons (ORNs), 949–950
Olfactory receptor proteins, 949–950
Oligochaetes, 661
Oligodendrocytes, 926
Oligonucleotide primers, 380
Oligonucleotide probes, 321
Oligonucleotides, 63, 353, 383
Oligophagous herbivores, 1175, 1176
Oligosaccharides
as antigens on red blood cells, 243
defined, 51
in glycoproteins, 109
glycosidic linkages, 53
Omasum, 1064
Ommatidia, 958
Omnivores
defined, 1054
feeding strategy, 638
in food webs, 1190, *1191*
teeth, *1055*
On-center receptive fields, 975
*On the Origin of Species* (Darwin), 6, 428, 429, 430, 454, 1122, 1171–1172, 1183
Onager, *1144*
Oncogene proteins, 228
Oncogenes
causing chronic myelogenous leukemia, 304
DNA methylation and, 344
gain of function mutations in tumor suppressor and, 306
somatic mutations and cancers, 314
*Oncorhynchus clarkii lewisi,* 1228
*Oncorhynchus mykiss,* 1140
*Oncorhynchus nerka,* 1228
One-gene, one-enzyme hypothesis, 282–283, 284
One-gene, one-polypeptide relationship, 283–284
"One-hour midge," 1152
Onychophorans, *632, 667,* 668
*Onymacris unguicularis, 1132*
Oocytes
in human reproduction, 893, *894*
in oogenesis, *883,* 884

Oogenesis, *883,* 884
Oogonia, 883, 884, 908
Oogpister beetle, 1174
Oomycetes, 556, *557*
Ootid, *883,* 884
Open circulatory systems, 1026, 1027
Open reading frames, 355–356
Operant conditioning, 1094
Operators, 330
Opercular cavity, 1010
Opercular flaps, 1009–1010
Operculum, 687
Operons
  description of, 330, *331*
  negative and positive regulation of, 330–332, *333*
*Ophioglossum reticulatum,* 218
*Ophiopholis aculeata, 682*
*Ophrys apifera, 476*
*Ophrys insectifera, 1181*
Opisthokonts, 559, 609
Opium poppy, *604*
Opossum shrimp, 1228
Opossums, 697, *698, 699,* 1173, 1190
"Opportunistic evolution," 753
Opportunity costs, in animal behavior, 1103
Opsins, 460, 957, 958
Optic chiasm, 977–978
Optic nerve, *962,* 963, 977–978
Optic vesicles, 396
Optical isomers, 41, 43, 52
Optimal foraging theory, 1104–1105, *1106*
*Optix* gene, 1175
Oral hormone contraceptives, 896, *898*
Oral side, 680
Oral–aboral body orientation, 680, *681*
Orange roughy, 1235
Orange trees, transgenic, 401
Orangutans, 366, 702
Orbitals, 24–25
Orcas, *1153*
Orchids
  Darwin and, 588, 605
  mechanical isolation, 476
  mycorrhizae and seed germination, 615
  plant–pollinator mutualisms, 1181
  pollinators, 588, 605
  types of flowers, 605
*Orcinus orca, 1153*
Orders, 463
Ordovician period, *506–507,* 516, 520
Organ identity genes, 400–401
Organ of Corti, *954,* 955, 956
Organ systems
  biological hierarchy concept, 9
  defined, 820
  development in animals, 915–918
  development in humans, 920
Organ transplant surgery
  kidney transplants, 1085–1086
  MHC proteins and, 874
Organelle genes, 252–253
Organelles
  defined, 81
  endomembrane systems, 88–91
  endosymbiotic origin in eukaryotic cells, *550,* 551–552
  in energy transformation, 91–93
  methods of studying and analyzing, 84, 85
  origin of, 101–102
  other types, 93–94
  transposition of genes to the nucleus, 366
  *See also individual organelles*
Organic fertilizers, 746–747
Organismal trees, 529–530
Organizer, 911–913
Organogenesis
  in animal development, 915–918
  in human development, 920
Organophosphates, 387
Organs
  biological hierarchy concept, 9
  development in animals, 915–918
  development in humans, 920
  left–right asymmetry, 902, 914–915, 921
  organization in systems, 820
  in physiological systems, 817
  tissue composition of, 820
Orgasm, 891
*Orgyia antiqua, 1157*
*ori. See* Origin of replication
Oriental region, *1142*
Origin of life
  chemical evolution, 3
  extraterrestrial hypothesis, 69
  importance of water to, 68–69
  prebiotic synthesis experiments, 69–71
  protocell experiments, 73–74
  RNA's catalytic properties and, 71–72, *73*
  spontaneous generation concept, 67–68
  timeline, *74*
Origin of replication *(ori),* 206, 269, 271
*Ornithorhynchus anatinus, 697*
Orotidine monophosphate decarboxylase, 154
Orthologs, 499–500
Orthonectids, *632,* 648
Orthopterans, *672, 674*
*Oryx gazella, 1132*
*Oryza sativa, 361,* 362, 605
  *See also* Rice
Osculum, *633*
Osmoconformers, 1072–1073
Osmolarity, 1072
Osmoreceptors, of the kidney, 1088
Osmoregulators, *1072,* 1073
Osmosis
  effect on cell volume, 1072
  movement of water into root xylem, 730
  osmolarity and, 1072
  water potential and, 727
Osmotic pressure, of blood in capillary beds, 1041
Osprey, *1140*
Ossicles, *954,* 955, 956
Ossification, 1001
Osteoarthritis, 410
Osteoblasts, 847, 1000
Osteoclasts, 847, 1000
Osteocytes, 1000
Osteoporosis, 1000–1001
Ostia, 1026
Ostracods, 670
Ostrander, Elaine, 352
Ostrich, 694, *695,* 696
Otoliths, *956,* 957
Otters, 700
*Ottoia, 517*
Ouachita Mountains, 472
Outer ear, 954
Outer membrane, of prokaryotes, 83
Outgroup, 453
Ova
  fertilization in animals, 884–887 (*see also* Sperm–egg interactions)
  production in animals, 882–883, 884
  *See also* Eggs
Oval window, *954,* 955
Ovarian cycle
  defined, 892
  description of, 893, *894*
  hormonal regulation of, 894–895
Ovaries (in animals)
  hormones of, *842*
  production of eggs in, 882–883, 884
  structure and function in humans, 892, 893, *894*
Ovaries (in plants)
  angiosperms, 597, *600*
  development into fruit, 784
  evolution in flowers, 598
  seed plants, *591*
Ovary wall, in angiosperms, 784
Overdominance hypothesis, of heterosis, 245
Overharvesting
  of fishes, 16–17, 1163–1164, 1235
  species extinctions and, 1234–1235
Overtopping growth, 583
Oviducts
  ectopic pregnancy, 906
  in humans, 892, *893*
Oviparity, 889
Ovoviviparity, 889
Ovulation, in humans, 892, 893, *894,* 895–896
Ovule
  in angiosperms, 597, *600,* 601
  in conifers, 594, *595*
  in seed plants, 590, *591, 592*
Oxaloacetate, 170, 171, 180, 198, 199
Oxidation
  beta-oxidation, 179–180
  defined, 167
  of glucose, 166–169
  of pyruvate, 170, 171
Oxidation–reduction reactions
  coenzyme NAD$^+$ in, 167–168
  glucose oxidation, 166–169
  in glycolysis, 170
  importance of enzymes to, 155
  in photosynthesis, 187
  transfer of electrons during, 167
Oxidative muscle, 995
Oxidative phosphorylation, 171–176
Oxidative system, in skeletal muscle, 997
*Oxycomanthus bennetti, 682*
Oxygen
  as an oxidizing agent, 167
  atomic number, 22
  binding to hemoglobin, 1016–1017
  binding to myoglobin, 1017
  blood transport of, 1016–1018
  chemical reaction with propane, 31
  covalent bonding capability, *27*
  diffusion in water, 1007
  in early atmosphere, 69
  effect of altitude, 1007–1008
  electronegativity, *28*
  factors affecting hemoglobin's affinity for, 1017–1018
  inhibition of nitrogenase, 750
  isotopes, 21
  mass number, 22
  in photorespiration, 197–198
  production during photosynthesis, 186–187
  *See also* Atmospheric oxygen
Oxygenic photosynthesis, 186–188
  *See also* Photosynthesis
Oxytocin
  actions of, *842,* 843
  bonding behavior in voles and, 125
  in labor and childbirth, 896, 897
  sensitive period in parent–offspring recognition and, 1099
Oxytocin receptors, 125, 129
Oysters, 662, 887
Ozark madtom, *472*
Ozark minnow, *472*
Ozark Mountains, 472
Ozone layer, 5, 311, 1211–1212

## P

*P* orbital, 24
P ring, *83*
*p21* gene, 399
p21 protein, 210–211, 399
*p53* gene, *314*
p53 protein, 349
p53 transcription factor, 228
"$P_{680}$" chlorophyll, 191, *192*
"$P_{700}$" chlorophyll, 191, *192*
Pääbo, Svante, 367
Pace, Norman, 357–358
Pacemaker cells
  autonomic nervous system and, 975
  in blood pressure regulation, 1044
  in cardiac muscle contraction, 991–992, 1032–1034, *1035*
Pacific barrel sponges, *643*
Pacific yew, 604
Pacinian corpuscles, 952
Paclitaxel, 228, *229*
*PAH* gene, 311
Paine, Robert, 1194
Pair bonding, 843
Pair rule genes, 403, *404*
Pairwise sequence comparison, *487*

Palaeognaths, 694
Palearctic region, *1142*
Palenque National Park, 1162
Paleomagnetic dating, 508
Paleontology
  dating fossils and rocks, 506–508
  phylogenetic analyses and, 455
  *See also* Fossil record
Paleozoic era, *506–507,* 579
Palindromic DNA sequences, 374
Palisade mesophyll, 720, *721*
Palmitic acid, *57, 1051*
Palms, *603,* 723
PAMP-triggered immunity (PTI), 799–800
Pampas, 1131
PAMPs, 799
*Pan troglodytes, 702*
Pancreas
  blood glucose regulation, 848–849
  hormonal control of, 1065
  hormones produced by, *842*
  in humans, *1058*
  procarboxypeptidase A of exocrine cells, 333
  role in digestion, 1062
  secretion of insulin, 1066
Pancreatic amylase, *1062*
Pancreatic duct, 1061, 1062
Pancreatitis, 1062
Pandanus trees, 718
Pandemics, 427
*Pandion haliaetus, 1140*
*Pandorina, 140,* 141
Pangaea, 520, 521, 1143
Pangolins, *698*
Panther groupers, *688*
*Panthera leo,* 1183
*Panthera tigris, 1172, 1235*
*Panthera tigris tigris, 1136*
Panting, 829
Pantothenic acid, 70, *1053*
  *See also* Vitamin $B_6$
Paper wasps, *674*
Papillomaviruses, 341
*Papio, 1135*
Parabasalids, 558
Parabronchi, 1011, *1012*
Paracrine signals, 126, 835
*Paradisaea minor, 480*
*Paragordius tricuspidatus, 667*
Parallel evolution, 423–424
Parallel substitutions, 488
Paralogs, 499–500
Paralysis, 970
  during REM sleep, 979
Paralytic shellfish poisoning, 549
*Paramecium,* 554, *555,* 562
Paraminohippuric acid (PAH), 1079
*Paranthropus aethiopicus, 703*
*Paranthropus boisei,* 703, 1056
*Paranthropus robustus,* 703
Paraphyletic groups, 463
Parapodia, 660
Parasite–host interactions, pathogenic, 1176–1177
Parasites
  ampicomplexans, 563
  chlamydias, 533, *534*
  complex life cycles, 563, *564*
  defined, 637
  feeding strategy, 638–639
  flatworms, 657
  fungi, 611–612
  kinetoplastids, 559
  life cycles, 640, *641*
  malaria and, 563, *564*
  myxozoans, 646
  nanoarchaeota, 537
  nematodes, 666
  orthonectids, 648
  parasitic fungi, 611–612, 617
  parasitic plants, 752–753, 753
  rhombozoans, 648
Parasitism
  defined, 1170
  pathogenic parasite–host interactions, 1176–1177
  in protostome evolution, 674
Parasitoid wasps
  in indirect competition, 1184
  in sweet potato–corn dicultures, 1203
"Parasol effect," 510
Parasympathetic division
  effect on heartbeat, 1034
  influence on smooth muscle, 993
  structure and function, 974–975
Parathormone, 848
Parathyroid glands, *842,* 847–848
Parathyroid hormone (PTH), *842,* 847–848
Parenchyma, 713, 730
Parent rock, 745
Parental care, in frogs, 678
Parental generation (P), 234
Parent–offspring recognition, 1099
Parietal cells, 1060, 1061
Parietal lobe, 972–973
Parkinson's disease, 323–324
*Parmotrema, 613*
Parotid salivary gland, *1058*
Parsimony principle, 454, 1144–1145
Parthenocarpy, 766
Parthenogenesis, 881–882
Partial pressure gradients, maximization in respiratory gas exchange, 1009
Partial pressure of carbon dioxide
  blood transport of carbon dioxide and, 1018
  regulation of breathing, 1020–1021
  respiratory gas exchange and, 1008
Partial pressure of oxygen
  altitude and, 1008
  binding of oxygen to hemoglobin and, 1016–1017
  binding of oxygen to myoglobin and, 1017
  detection of blood levels by aortic and carotid bodies, 1021–1022
  diffusion of oxygen and, 1006
  gas exchange in birds and, 1012
  gas exchange in fish gills and, 1010
  in lungs with tidal ventilation, 1013
Partial pressures, of gases, 1006
Particulate inheritance, 233–234, 236
Passenger pigeon, 1230
Passeriform birds, *696*
*Passiflora,* 2, *2, 603, 1174,* 1175, 1176
Passionflower, *2, 603, 1174,* 1175, 1176
Passive transport (diffusion), 113–117
  *See also* Diffusion; Facilitated diffusion
Pasteur, Louis, 68, 259
Patch clamping, 929, *931*
Patents, 373
Pathogen associated molecular patterns (PAMPs), 799, 860, 861
Pathogenesis-related (PR) genes, 799, 800
Pathogenic fungi, 612
Pathogens
  animal defense systems and, 857
  bacterial, 357, 541–542
  effect on population growth, 1159
  macroparasites, 1176–1177
  microparasites, 1176
  negative impact as invasives, 1236
Pattern formation
  apoptosis in, 399–400
  body segmentation in *Drosophila melanogaster,* 401–405
  defined, 399
  morphogen gradients and positional information, 401
  plant organ identity genes, 400–401
Pattern recognition receptors (PRRs), 799, 860
Paucituberculata, *698*
Pauling, Linus, 461
Pavlov, Ivan, 838, 982, 1094
*Pax6* gene, 413, *414*
pBR322 plasmid, 377
*Pdm* gene, *675*
Peanut butter, 624
Pears, 601, 714
Peas
  loss of function mutations, 305
  Mendel's experiments with, 233–236
  as a model organism, 282
Peat, 578, 611
Peatlands, 578
Pectin, 713
Pectinase, 162
Pectoral fins, 686–687, *688*
Pedigrees, 240–241
Peforin, *873,* 874
*Pegea, 684*
Pelagic zone, 1139–1140
*Pelecanus conspicillatus,* 412
Pellagra, *1053*
Pellicle, 554, *555*
Pelvic fins, 686–687, *688*
Pelvic spines, 424
Pelvis, 1082
Penetrance, 245
Penicillin, 527, 608
Penicillin resistance, *626*
*Penicillium,* 383, 608, *622,* 624, *626*
Penis
  in birds, 888
  erection, 136, 890–891
  role in internal fertilization, 887
Pentaradial symmetry, 680, *681*
Pentoses, 52, 63
PEP carboxylase, 199, 200
Pepsin, 1060, 1061, *1062*
Pepsinogen, *1060,* 1061
Peptidases, 300, 1058
Peptide bonds. *See* Peptide linkages
Peptide hormones, 836
Peptide linkages, 43–44, *45,* 296, *297*
Peptide neurotransmitters, 936, 940
Peptidoglycans, 527, *528,* 551, *552,* 571
Peptidyl transferase, 296, *297*
Peptidyl tRNA binding site, 295, 296
Per capita growth rate, 1156
*Per* genes, 1096, 1108
Peramelemorphia, *698*
Perching birds, *696*
Pereiopods, *671*
Pereira, Andrew, 726
Perennials, 785
Perfect flowers, 597, 598, 779, *780*
Perfusion
  in the human lung, 1016
  partial pressure gradients, 1009
Perianth, 597
Periarbuscular membrane, 747
Pericycle, 717
Periderm, 712, 722
*Peridinium, 554*
Period, of biological cycles, 774
Periodic table, 22, *23*
Peripheral membrane proteins, *106,* 107, 108
Peripheral nerves, 968
Peripheral nervous system (PNS)
  afferent and efferent portion, 968
  brainstem components, 969
  components of, 941
Periplasmic space, 528
Perissodactyla, *698*
*Perissodus microlepis,* 441–442
Peristalsis, 1059–1060
Peritoneum, 635, 1059
Peritubular capillaries, 1078, *1080,* 1082, 1084
Periwinkle, *604*
Permafrost, 1128
Permian period
  atmospheric oxygen levels, 512–513, 522
  changes on Earth and major events in life, *506–507*
  characteristics of life during, *518,* 520–521
  fungi in, 611
  gigantic insects in, 505
  gymnosperms in, 593
  mass extinction, 522
  vascular plants in, 580
  volcanic activity during, 511
Pernicious anemia, *1053,* 1054
Peroxidase, 443
Peroxides, 93
Peroxisome disorders, 93
Peroxisomes
  in animal cells, *86*

conversion of peroxide to water, 176
in photorespiration, 197, *198*
in plant cells, *87*
structure and function, 93
Personal genomics, 368
Personality, the frontal lobe and, 971–972
Petals, *591*, 597
Petroleum deposits, 565, 569
*Petroscirtes breviceps, 1174*
Pets
cloning, 407
*See also* International animal trade
*Pfiesteria piscicida*, 553
pH
buffers, 35–36
concept of, 35
effect on enzymes, 161
effect on protein structure, 50
generation of low pH in the stomach, 1060–1061
optimal soil pH for plants, 746
regulation by the kidneys, 1084–1085
*See also* Blood pH
*Phaeolus schweinitzii*, 1202
Phage therapy, 545–546
Phagocytes
defensins, 859
functions of, 90, *858*
in inflammation, 861
in innate defenses, 860–861
types of, *858*
Phagocytosis
defined, 120, *122*
in the evolution of the eukaryotic cell, 551
lysosomes and, 90, *91*
Phagosomes, 90, *91*
Phalanges, 423
*Phalaris canariensis*, 762, 763, 764
*Phallus indusiatus, 2*
Phanerozoic eon, 508
Pharmaceuticals
medicinal plants and, 604–605
pharming, 385–386
*See also* Drugs
Pharmacogenomics, 368
Pharming, 385–386
Pharyngeal arches, 684
Pharyngeal basket, 684
Pharyngeal slits, 684
Pharyngotympanic tube, 954
Pharynx
in ecdysozoans, 655
in humans, 1013, *1014*, 1059
in tunicates and lancelets, 684
Phase-contrast microscopy, 80
Phasmida, *672*
Phelloderm, 722
Phenolics, *802*
Phenotype
action of natural selection on, 438
defined, 431
DNA structure and, 266
effects of mutations on, 305–306
epistasis and, 244
fitness and, 438
gene–environment interactions and, 245–246
generation of new phenotypes through selection, 432–433
genes and, 237
incomplete dominance and, 242–243
in Mendelian genetics, 236
modifying in the treatment of genetic diseases, 322–323
multifactorial, 314–315
pleiotropy, 243–244
qualitative and quantitative variation, 246
relationship of genotype to, 431
Phenotypic plasticity, 420–422
Phenotypic sex determination, 850, *851*
Phenotypic variation, genetic variation and, 431
Phenyl butyrate, 899
Phenylalanine
in phenylketonuria, 311, 320, 322, 1051
structure, *44*
Phenylalanine hydroxylase (PAH), 311, 320, 322
Phenylketonuria (PKU)
genetic screening for, 320
knockout mouse model, 382
mutations causing, 311, *312*
pleiotropy in, 243–244
prevalence, *312*
treatment for, 322, 1051
Phenylpyruvic acid, 311
Pheromones
animal communication and, 1110–1111
detection of, 950
*Philodina, 658*
*Philodina roseola, 658*
*Phlebopteris smithii, 518*
Phloem
in angiosperms, 596
bulk flow, 728
function of, 579
in leaves, *721*
in roots, 717
secondary, 721, 722, *723*
in shoots, 719
structure and function, 714–715
translocation in, 734–738
Phloem sap, 734–738
*Phlox cuspidata*, 478, 479
*Phlox drummondii*, 470–471, 478, 479
*Phoenicopterus ruber, 638*
*Phoenix dactylifera, 603*
"Phoenix" virus, 545
Pholidota, *698*
Phoronids, *632*, *652*, 658, 659, *660*
*Phoronis australis, 660*
Phosphatases, 837
Phosphate-based detergents, 1221
Phosphate group
of chemically modified carbohydrates, 55
free energy released by ATP and, 150
in nucleotides, 63
properties of, *40*
protein phosphorylation, *300*, 301
Phosphate ions
hormonal regulation of blood phosphate levels, 848
inorganic fertilizers, 747
Phosphate transporters, 744
Phosphatidylcholine, *58*
Phosphatidylinositol bisphosphate ($PIP_2$), 134, *135*
Phosphodiester linkages
formation during DNA replication, 269, *270*, 273
between nucleotides, 63, *64*
Phosphodiesterase (PDE), 137, 961
Phosphoenolpyruvate (PEP), 199
Phosphofructokinase, 182
Phosphoglucose isomerase (PGI), 442–443
3-Phosphoglycerate (3PG), 170, 194, 197, 198
Phosphoglycerate kinase, 170
Phosphoglycolate, 197, *198*
Phospholipases, 134, 135
Phospholipid bilayers
simple diffusion across, 114
structure, 57, *58*
*See also* Membranes
Phospholipids
in the absorption of fats in the small intestine, *1062*, 1063
in biological membranes, 106–107
fatty acid chain characteristics, 107
hydrophilic and hydrophobic regions, 106
lipid-derived second messengers, 134–135
structure and function, 57, *58*
Phosphoric acid anhydride bond, 150
Phosphorus
in animal nutrition, *1052*
covalent bonding capability, *27*
ecological impact of soil accumulation and runoff, 1207, 1220–1221
electronegativity, *28*
global cycle, 1220–1221
in plant nutrition, *741*
radioactive isotope in the Hershey–Chase experiment, 262–263
Phosphorylase kinase, 138
Phosphorylation
of histones, 344
oxidative, 171–176
photophosphorylation, 192–193
of proteins, 209, *300*, 301
reversible, 161
substrate-level, 170
Photic zone, 1139, 1140
Photoautrophs, 538
Photochemistry, 188–189
Photoheterotrophs, 538
Photomorphogenesis, 772
Photons
absorption by pigments, 188–189
defined, 188
Photoperiod
as an indicator of seasonal change, 1108
flowering and, 787–788
melatonin and, 851
Photophosphorylation, 192–193
Photoreceptor cells (animal)
function of, 957
influence on ion channels, 947, *948*
in ommatidia, 958
receptive fields of ganglion cells and, 963, 975–977
of the retina, 959
rhodopsin and the response to light, 957–958
rod cells and cone cells, 960–962, 963
Photoreceptors (plant), 759, 771–775
Photorespiration, 197–198
Photosensitivity, of visual pigments, 957–958
Photosynthates, phloem translocation, 734–738
Photosynthesis
action spectrum, 189, 190
atmospheric oxygen and, 4–5
in $C_3$ and $C_4$ plants, 198–200
in CAM plants, 200
conversion of light energy into chemical energy, 188–193
in cyanobacteria, 532, 538
defined, 186
effect of increasing levels of atmospheric carbon dioxide on, 185, 202
evolution in plants, 570–573
in the evolution of life, 4–5
general equation for, 186
global consumption of carbon dioxide, 1217
impact on atmospheric oxygen levels, 511–513
interactions with other pathways, 200–202
overview of pathways in, 188
photophosphorylation and ATP synthesis, 192–193
photorespiration and, 197–200
photosynthetic efficiency, 201–202
source of oxygen produced by, 186–187
synthesis of carbohydrates, 193–197
Photosynthetic autotrophs, 1189–1190
Photosynthetic bacteria
impact on atmospheric oxygen levels, 511–512
stromatolites, 512, *513*
Photosynthetic endosymbionts, 553
Photosynthetic lamellae, 532
Photosynthetic pigments
in red algae, 571, 572
*See also* Carotenoids; Chlorophyll
Photosynthetic protists, 252
Photosystem I, 191, 192, *193*
Photosystem II, 191–192, *193*
Photosystems
description of, 191–192
organization of, 190
in photophosphorylation, *193*
Phototropins, 771–772
Phototropism
action spectrum, 771, *772*
coleoptile experiments and the role of auxin in, 762, *763*, 764, *765*

Phthiraptera, *672*
Phycobilins, 190
Phycocyanin, 571
Phycoerythrin, 571, 572
Phyla, 463
*Phyllactinia guttata, 612*
PhyloCode, 464
Phylogenetic trees
  area phylogenies and, 1143, *1144*
  evolutionary history and, 6–9, 451
  gene trees and, 497, 499
  how to read, 450–451
  identifying neutral, purifying, or positive selection in, 492
  lateral gene transfer events and reticulations, 496
  maximum likelihood methods, 456
  methods in construction, 452–458
  monophyletic groups, 463
  parsimony principle, 454
  sources of data for, 454–456
  testing the accuracy of, 456, 457–458
  uses and components of, 450, 458–462
Phylogeny
  defined, 449, 450
  evolutionary history and, 451
  evolutionary perspective on comparing species, 451–452
  evolutionary possibilities of traits, 452
  homologous features, 452
  maximum likelihood methods, 456
  parsimony principle, 454
  relationship to biological classification, 462–464
  sources of data for, 454–456
  testing the accuracy of, 456, 457–458
  using to reconstruct protein sequences from extinct organisms, 464
Phylogeography, 1143, *1144*
*Phymateus morbillosus, 2, 674*
*Physalia physalis, 646*
Physiological survivorship curves, 1155
Physiological systems
  effect of temperature on, 820–821
  heat stroke in mammals, 815
  maintenance of the internal environment, 816
  regulation of homeostasis, 816–817
  relationship between cells, tissues, and organs, 817–820
  thermoregulation (*see* Thermoregulation)
Phytoalexins, 799–800
Phytochromes
  entrainment of circadian rhythms in plants, 774–775
  nuclear localization sequence and protein kinase domain, 774
  in photoperiodic control of flowering, 788
  properties of, 772–773
  stimulation of gene transcription, 773–774
Phytomers, 709, 719
Phytomining, 812
Phytoplankton
  biomass distribution in open oceans and, 1191–1192
  during the Mesozoic, 521
  petroleum deposits and, 569
  as primary producers, 563
Phytoremediation, 811, *812*
Pierid butterflies, *1106*
Pigeons
  artificial selection, 6
  homing, 1109
  time-compensated solar compass in, *1111*
Pigmented epithelium, 959
Pigments
  absorption of photons, 188–189
  absorption spectrum, 189
  accessory pigments, 190, 571
  defined, 189
  photochemical changes, 190–191
  in plant evolution, 574
  in red algae, 571, 572
  in vacuoles, 93
  visual pigments, 957–958
  *See also* Carotenoids; Chlorophyll; Skin pigmentation
Pigs, allantoic sac, 919
Pigweed, 1200
Pijio tree, *1136*
Pili, *82,* 84
*Pilobolus, 619*
Pilosa, *698*
Piloting, 1109
Pima peoples, 1048
*Pinaroloxias inornata, 474*
Pincushion protea, *1134*
Pine bark beetles, 1217–1218
Pineal gland, *842,* 851
Pineapples, *604*
Pines
  cones, 594, *595*
  fire adaptations, 594, *596*
  life cycle, 594, *595*
  pine bark beetle infestations and climate change, 1217–1218
  response to rapid climate change, 1236
Pinnae, 954
Pinocytosis, 120, *122*
*Pinus, 592*
*Pinus contorta, 594,* 1201–1202, *1236*
*Pinus longaeva, 593*
*Pinus ponderosa,* 443, *1241*
Pioneer Hi-Bred seed company, 753
Pioneer species, 793, 1200, 1201
*Piophila casei,* 1188
*Pisaster ochraceus,* 1194
*Pisolithus tinctorius, 614*
Pistils, 597, 598
*Pisum sativum,* 282
  *See also* Peas
Pit organs, 946
Pitcher plants, 751, 1197
Pith, 718, 719, 720
Pith rays, 719, 722
Pits, in xylem, 714
Pituitary gland, 842–845
  *See also* Anterior pituitary; Posterior pituitary
*Pitx1* gene, 424
Pivot, 1002
"Place cells," 967
Placenta
  in childbirth, 897
  development in mammals, 906, *907*
  functions of, 697, 889, 892
  fusion of cells in the outer layer, 545
  hormones produced by, 896
  origin and development, 919
Placental mammals
  characteristics of, 697–700
  cleavage in, 905–906
  extraembryonic membranes, 919
Placoderms, *687*
Placozoans
  in animal phylogeny, *630*
  asymmetry in, 634
  description of, 645
  major subgroups and number of living species, *632*
  structural simplicity of, 629, 648
  *Trichoplax,* 629
Plaice, 444
Plains zebra, *1144*
Plan B®, *898*
*Planaria,* 1075
Plankton, 515
Plant biotechnology, 386–388, *389*
Plant cell walls, *87*
  auxin-induced cell expansion and, 766–767
  in plant development, 710–711
  plasmodesmata, 100
  primary and secondary, 710–711
  responses to pathogenic invasions, 798
  structure and function, 99–100
Plant cells
  apoptosis, 226
  auxin-induced expansion and, 766–767
  cold-hardening and, 810
  communication through plasmodesmata, *139,* 140
  cytokinesis, 216–217, 710
  expansion, 710
  ice crystals and, 810
  methods of transformation, 376, 377
  microtubule organizing centers, 212
  osmosis, 114–115
  structure, *87*
  totipotency and cloning, 405–406
  turgor pressure, 114–115, 727
  vacuoles, 93
  *See also* Eukaryotic cells
Plant chemical defenses
  artemisinin, 797, 812
  constitutive, 798
  lignin, 798
  plant self-protection from, 805
  secondary metabolites, 802–803, 804
Plant defenses
  defensins, 859
  against herbivores, 801–806, 1175
  to pathogens, 798–801
  *See also* Constitutive plant defenses; Induced plant defenses; Specific plant immunity
Plant diseases
  bacterial, 534
  club fungi, 622
  molds, 622
  pathogenic fungi, 612
  powdery mildews, 622
  viral agents, 544
Plant genomes
  features of, 362
  key parameters, *361*
Plant growth and development
  developmental plasticity in response to light, 422
  effect of increasing levels of atmospheric carbon dioxide on, 185
  effects of gibberellins on, 760–762
  embryogenesis, 711–712
  leaf development, 719, 720
  organ identity genes, 400–401
  overview, *393*
  primary and secondary growth defined, 715
  primary indeterminate growth, 715–720
  processes in, 710
  properties affecting, 710–711
  role of auxin in, 762–767
  secondary growth, 721–723
  seed germination and seedling growth, 757–758
Plant growth regulation
  auxin and, 762–767
  brassinosteroids and, 771
  cytokinins and, 768–769
  ethylene and, 769–771
  genetic screens and the identification of signal transduction pathways, 759, *760*
  gibberellins and, 760–762
  hormones and photoreceptors in, 758–759, 771–775
  key factors in, 757
Plant hormones
  auxins, 760, 762–767
  brassinosteroids, 771
  compared to animal hormones, *758*
  cytokinins, 768–769
  ethylene, 769–771
  gibberellins, 760–762, 767
  in plant defenses against herbivory, 804, *805*
  in plant growth and development, 758–759
  production in response to pathogens, 799
  structures of, *759*
"Plant immune system," 799–801
"Plant kingdom," 573
Plant mutualisms
  food exchange for housing or defense, 1178–1179

  food exchange for seed transport, 1181–1182
  pollination syndromes, 1180–1181
  *See also* Mycorrhizae
Plant nutrition
  carnivorous plants, 751–752
  deficiency symptoms, 742
  essential macronutrients and micronutrients, 741–743
  hydroponic experiments, 742–743
  impact of soil on nutrient availability, 744–747
  improving nitrogen use efficiency, 740
  influence of fungi and bacteria on root uptake of nutrients, 747–751
  parasitic plants, 752–753
  plant acquisition of nutrients, 743–744
  plant regulation of nutrient uptake and assimilation, 744
Plant pathogens
  plant defenses against, 798–801
  plant–pathogen signaling, *799*
Plant physiology, 708
Plant reproduction
  asexual reproduction in angiosperms, 792–794
  sexual reproduction in angiosperms, 779–785
Plant signal transduction pathways
  activation in response to pathogens, 799, 800
  ethylene pathway, 770–771
  involving auxins and gibberellins, *767*
  involving cytokinins, 768–769
  in plant defenses against herbivory, 804, *805*
Plant tissue culture, 768
Plant tissue systems
  dermal tissue, 712–713
  ground tissue, *712,* 713–714
  primary meristems in the origin of, 716
  vascular tissue, *712,* 714–715
Plantae, 570–571
Plantlets, 793
Plant–pollinator relationship
  coevolution in, 588, 598–599, 605
  pollination syndromes, 1180–1181
Plants
  adaptations (*see* Adaptations in plants)
  aerenchyma, 808
  agricultural applications of biotechnology, 386–388, *389*
  aspects of light responsive to, 771
  "bolting," 761
  carnivorous, 751–752
  challenges of saline environments, 810–811
  cloning, 405–406
  colonization of the land, 574–579
  cuticle, 732
  cytoplasmic inheritance, *252*
  development in (*see* Plant growth and development)
  effect of pollination strategies on speciation rates, 480
  effects of domestication on, 723–724
  endophytic fungi and, 615
  entrainment of circadian rhythms, 774–775
  environmental stresses on, *806*
  evolution of, *570*
  evolution of photosynthesis in, 570–573
  global nitrogen cycle, 750–751
  heavy metal tolerance, 811–812
  hypersensitive response, 226
  impact of soil structure on, 744–747
  improving nitrogen use efficiency, 740
  interactions with the external environment, 9
  lateral gene transfer, 496
  metabolic interactions involving photosynthesis, 200–202
  metabolomes, 370
  morphogenesis, 710–712
  mycorrhizae and, 614–615
  nitrogen composition, 751
  nutrients and nutrition (*see* Plant nutrition)
  optimal soil pH, 746
  organ identity genes, 400–401
  parasitic, 752–753
  partial reproductive isolation in, 470–471
  photomorphogenesis, 772
  photosynthetic efficiency, 201–202
  phytomining, 812
  phytoremediation, 811, *812*
  plastid gene mutations, 252
  plastid structure and function, 92–93
  polyploidy and agriculture, 225, *226*
  postzygotic isolating mechanisms, 478, 479
  prezygotic isolating mechanisms, 476, 477
  reciprocal interactions with herbivores, 1175–1176
  responses to drought and water stress, 734, 809
  root and shoot systems, 709
  self-incompatibility studies, 380–381
  self-protection from chemical defenses, 805
  starches and starch grains, 53
  stem cells, 408
  stomatal control of water loss and carbon dioxide uptake, 732–734
  stress response, 337
  sympatric speciation through polyploidy, 475
  translocation in phloem, 734–738
  transport systems for ions, 744
  triglycerides of, 57
  uptake of water and solutes, 727–730
  water-use efficiency, 726
  waxes, 59
  xylem transport of water and minerals, 730–732
  *See also* Angiosperms; Crop plants; Gymnosperms; Land plants; Nonvascular land plants; Seed plants; Vascular plants
Planula
  of cnidarians, 645, *646*
  of scyphozoans, 647, *648*
Plaque, in atherosclerosis, 1042
Plasma. *See* Blood plasma
Plasma cells
  development of B cells into, 867
  function of, 865
  in the humoral immune response, 872, *873*
Plasma membranes
  in animal cells, *86*
  in cell adhesion and cell recognition, 110–113
  depolarization and hyperpolarization, 930–931, *932*
  endocytosis, 120–121
  energy pathways on, *168*
  membrane-associated carbohydrates, 109
  membrane proteins, *106,* 107–109
  origin of eukaryotic organelles and, 101–102
  in plant cells, *87*
  principles of bioelectricity, 927
  properties and characteristics of membrane potentials, 927–932
  structure and function, 79–81
Plasmid pBR322, 377
Plasmids
  in bacterial conjugation, 254–255
  in prokaryotes, 356
  recombinant, 374, 376
  reporter genes, 378
  as vectors in transformation, 377
Plasmin, 385
Plasminogen, 385
Plasmodesmata
  blocking in response to pathogenic invasions, 798, *799*
  in phloem, 714, 715
  in plant cells, *87,* 710
  structure and function, 100, *139,* 140
Plasmodial slime molds, 560–561
*Plasmodium,* 553, 554, 563, *564*
*Plasmodium falciparum, 564*
Plasmogamy, *618,* 619, 620, *621*
Plastid genes, inheritance of, 252–253
Plastids
  endosymbiosis theory of, 102
  structure and function, 92–93
Plate tectonics, 509
Platelet-derived growth factor, 211, *385*
Platelets
  in atherosclerosis, 1042
  in blood clotting, *1038,* 1039
  platelet-derived growth factor, 211
Platinum, 259, 278
*Platyspiza crassirostris, 474*
"Playing possum," 1173
Plecoptera, *672*
Pleiotropic alleles, 243–244
Pleistocene epoch, 472, 522
*Pleodorina, 140,* 141
Pleopods, *671*
*Plesiadapis fodinatus, 519*
Pleural membranes, *1014,* 1015–1016
Plimsoll line metaphor, 1244–1245
*Plodia interpunctella,* 1184
*Plumatella repens, 641, 654*
Pluripotency, 394
Pluripotent stem cells, 408–409
Pneumatophores, 808
Pneumococcus, 260–261
*Pneumocystis jirovecii,* 612, 876
Pneumonia, 533
Podocytes, 1078, 1079
*Poecilia reticulata,* 1157
*Poecilotheria metallica, 669*
Pogonophorans, 660–661
Poikilotherms, 822
Point mutations, 306–307, 312, 486
Point restriction phenotype, 245, 306
Poison dart frogs, *1174*
Polar auxin transport, 763–764
Polar bodies, *883,* 884
Polar bonds, 28
Polar covalent bonds, 28
Polar microtubules, 214, *215*
Polar molecules, 30
Polar nuclei, *600, 779, 781,* 783
Polar substances, effect on protein structure, 50
Polar tube, 617
Polarity
  defined, 395
  determination by cytoplasmic segregation, 395
  establishment in the animal zygote, 903
Pole plasm, 908
*Polistes nympha, 674*
Pollen grains
  in angiosperms, 600, 779, 780–782, 783
  in conifers, 594, *595*
  mechanisms of transport, 780, *782*
  in monocots and eudicots, 710
  in seed plants, 590, *591*
Pollen tubes
  in angiosperms, 779, 780–782
  in conifers, 594
  in seed plants, 590, *591*
Pollination
  in angiosperms, 600, 780–782
  evolution of flowers and, 598, *599*
  influence on speciation rates, 480
  in orchids, 588, 605
  in seed plants, 590, *591*
  strategies for preventing inbreeding in angiosperms, 782, *783*
  wind-pollination, 480, 780

*See also* Plant–pollinator relationship; Pollinators
Pollination syndromes, 1180–1181
Pollinators
economic benefits to coffee plantations, 1243
hummingbirds, 598, *599*
of orchids, 588, 605
pollination syndromes, 1180–1181
prezygotic isolating mechanisms in plants, 476, 477
*See also* Plant–pollinator relationship
Pollinia, 605
Poll's stellate barnacle, 1184
Pollution
impact on habitats and biodiversity, 1233
lichens as indicators of air quality, 624, 625
use of fungi to study environmental contamination, 624, 625
Poly A sequence, in expression vectors, 384
Poly A tail, 292
Polyacrylamide, 316
Polyadenylation sequence, 292
Polyandrous mating systems, 1113, 1114
Polychaetes, *638*, 640, 660–661
Polygynous mating systems, 480–481, 1113–1114
Polymerase chain reaction (PCR)
description of, 277–278
in DNA fingerprinting, 317
in DNA testing, 321
in high-throughput sequencing, 353, *354*
in metagenomics, 357–358
RT-PCR, 380
using to create synthetic DNA, 380
in vitro evolution studies, 501
Polymerization, origin of life and, 71
Polymers
condensation and hydrolysis reactions, 42
defined, 40
origin of life and, 71
Polymorphic loci, 242, 442–443
Polymorphisms, 317
*Polymorphus marilis, 658*
Polynucleotides, 63, 74
*Polyorchis penicillatus, 647*
Polypeptide chains, 43, 45, *46*
Polypeptides
modification after translation, 300–301
signal sequences and movement within the cell, 298–300
synthesis in translation, 293–297, *298*
Polyphagous herbivores, 1175–1176
Polyphyletic groups, 463
Polyploidy
description of, 224–225, *226*
gene duplication, 497–498
sympatric speciation and, 473, 475
Polyps
of anthozoans, 646–647
of cnidarians, 645, *646*
of hydrozoans, 647–648
of scyphozoans, 647
Polyribosomes, 297, *298*
Polysaccharides
catabolic interconversions, 179
cellulose, 54
defined, 51
features of, 53
glycogen, 53–54
glycosidic linkages, 53
starch, 53, *54*
Polysomes, 297, *298*
Polyspermy blocks, *885*, 886–887
Polysynaptic reflexes, 942–943
*Polytrichum, 578*
Polyubiquitin, 349
Polyubquitination, 767
*Pombe*, 624
"Pond scum," *533*
Ponderosa pine, 443, *1241*
Ponds, 1140
*Pongo pygmaeus, 702*
Pons, 968, *969*, 1020
Poplar trees, 362
Population bottleneck, 434
Population density
defined, 1150
estimating, 1151
factors limiting, 1157–1161
Population dynamics
defined, 1150
demographic events determine population size, 1153–1154
effect of habitat variation on, 1161–1163
influence on population management, 1163–1164
introduced reindeer populations, 1149, 1158, 1166
life tables, 1154–1155
survivorship curves, 1155–1156
Population growth
effects of density-dependent or density-independent factors on, 1159
exponential, 1157–1158
human population growth, 1164–1166
logistic, 1158
per capita growth rate, 1156
Population management, 1163–1164
Population size, noncoding DNA in the genome and, 496
Populations
age structure, 1150, 1151–1152, 1165–1166
biotic potential, 1158
contribution of genetic variation to phenotypic variation, 431
defined, 6, 432, 1150
dispersion patterns, 1150, 1152–1153
effect of environmental conditions on life histories, 1156–1157
evolution of, 6
fixed, 437
genetic structure, 437
genetic variation in geographically distinct populations, 443–444
Hardy–Weinberg equilibrium, 437–438
interactions between individuals, 9
intrinsic rate of increase, 1156
maintenance of genetic variation by frequency-dependent selection, 441–442
measuring or counting, 1150–1151, 1152
mechanisms of evolution in, 432–436
mechanisms of maintaining genetic variation in, 441–444
possible effects of natural selection on, 439–440
properties of, 1150
*r*-strategists and *K*-strategists, 1159
speciation and, 6
using ecological principles to manage, 1163–1164
*Populus trichocarpa*, 362
Portal blood vessels, 844
Portuguese man-of-war, 640, 646
Portuguese water dogs, 352
*Posidonia oceanica, 603*
Positional information, 401
Positive cooperativity, 1017
Positive feedback, in physiological systems, 816–817
Positive gravitropism, 765
Positive regulation
defined, 329
of the *E. coli lac* operon, 332, *333*
in eukaryotes, 333
in virus reproductive cycles, 339–340
Positive selection, 433, 492–494
Positive-sense genomes, 544
Positive-sense single-stranded RNA viruses, *543*, 544
Positron emission tomography (PET), 970, *981*
Possessions Island, 1199
Post-traumatic stress disorder (PTSD), 970
Postabsorptive state, 1065, 1067
*Postelsia palmiformis, 556*
Posterior, 634, *635*
Posterior hippocampus, 967
Posterior pituitary, 842, 843, 896–897
Posterior–anterior axis, determination in vertebrate limb development, 401
Postganglionic neurons, *974*, 975
Postsynaptic cells
defined, 925
overinhibition of neurons in the brain, 943
summation of excitatory and inhibitory input, 938
Postsynaptic membrane
at electrical synapses, 939
responses to neurotransmitter, 936–938
Posttranscriptional gene regulation, 346–349
Postural muscles, 995
Postzygotic isolating mechanisms, 475, 478, 479
Potassium
in animal nutrition, *1052*
electronegativity, *28*
in plant nutrition, *741*
Potassium-40, 507
Potassium equilibrium potential ($E_K$), 929
Potassium ion channels
in action potentials, 932, 933, 934
in cardiac pacemaker cells and heart contraction, 1033, 1034
in the hyperpolarization of neurons at the onset of sleep, 979
membrane hyperpolarization and, 931, *932*
membrane potentials and, 929
specificity of, 115–116
Potassium ions
in cardiac pacemaker cells and heart contraction, 1033, 1034
generation of action potentials and, 933, 934
inorganic fertilizers, 747
membrane potential and, 927
plant guard cell function and, 733
reabsorption in the kidney, 1084
Potato beetles, 1184
Potatoes
"eyes," 720, 792
sink strength of tubers, 736–737
tubers, 720
vegetative reproduction, 792
Potential energy, 145
Potrykus, Ingo, 388
Powdery mildews, 622
PR genes. *See* Pathogenesis-related genes
PR proteins, 801
Prader-Willi syndrome, 345
Prairie dogs, *1239*
Prairie voles, 125, 1113
Prairies
restoration projects, 1237–1239, *1240*
*See also* Grasslands
Pre-mRNA. *See* Precursor mRNA
Pre-replication complex, 269
Prebiotic synthesis experiments, 69–71
Precambrian, *506–507*, 508, 515–516
Precapillary sphincters, 1044
Precipitation
acid precipitation, 1219–1220
atmospheric circulation patterns and, 1124
boreal and temperate evergreen forest biomes, *1129*
chaparral biomes, *1134*
cold desert biomes, *1133*
effect on terrestrial biomes, 1126–1127
hot desert biomes, *1132*
rain shadows, 1126–1127
temperate deciduous forest biome, *1130*
temperate grassland biome, *1131*

thorn forest and tropical savanna biomes, *1135*
tropical deciduous forest biomes, *1136*
tropical rainforest biomes, *1137*
tundra biomes, *1128*
Walter climate diagrams, 1138
Precocial young, 642
Precursor mRNA (pre-mRNA)
alternative splicing, 346–347
hybridization experiments, 291
processing before translation, 291–293
Predation
defined, 1170
impact on life history traits, 1157
Predation hypothesis, of latitudinal gradients in diversity, 1196
Predator–prey interactions
overview, 1172–1173
prey defenses, 1173–1175
Predators
defined, 637
effect on population growth, 1159
feeding strategy, 638
types of, 1054
Predatory fireflies, 1111
Predatory fungi, 612, *613*
Preganglionic neurons, *974*, 975
Pregnancy
consequences of stress during, 328
effect of exposure to environment factors, 920
in humans, 896
length of, 919
methods of preventing, 897, *898* (*see also* Contraception)
stages of development during, 920
Pregnancy tests, 871, 896
Preimplantation genetic diagnosis (PGD), 899
Preimplantation screening, 321
Premature babies, respiratory distress, 1015
Premolars, 1055
Prenatal screening, 320, 321
Prepenetration apparatus (PPA), 747, *748*
Pressure, detection of, 952
Pressure chambers, 732
Pressure flow model, 735–738
Pressure potential
defined, 727
movement of water and solutes in plants and, 727–728
pressure flow model of phloem transport, 735–738
Presynaptic cell, 925
Presynaptic membrane, 939
"Pretzel mold," *560*
Prevailing winds
description of, 1124, *1125*
ocean currents and, 1124–1125
Prey
defenses, 1173–1175
defined, 638
*See also* Predator–prey interactions
Prezwalski's horse, *1144*
Prezygotic isolating mechanisms, 475, 476–477, 482
Priapulids, *632*, 665
*Priapulus caudatus*, *665*
Priapus, 665
Primary active transport, 118–119
Primary bronchi, 1011, *1012*
Primary cell wall, 710
Primary consumers, 1190, *1191*
Primary embryonic organizer, 911–913
Primary endosymbiosis, 551, *552*, *570*
Primary growth
defined, 715
role of apical meristems in, 715–716
in roots, 716–718
in shoots, 719–720
Primary immune response, 866
Primary lysosomes, 90, *91*
Primary meristems
origins and types of, 715
in root development, 716–718
in shoot growth, 719–720
tissues produced by, 716
Primary metabolites, 370, 802
Primary motor cortex
location, 971
mapping of the body in, 971, *972*
Primary nodule meristem, 747, *748*
Primary oocytes, *883*, 884, 893, *894*
Primary producers, 563, 1189–1190
Primary sex determination, 250
Primary sex organs, 887, 889
*See also* Ovaries; Testes
Primary somatosensory cortex, 972–973
Primary spermatocytes, 883–884, *891*
Primary structure
of proteins, 45, *46*
specifies protein tertiary structure, 48, 49
Primary succession, 1200
Primase, 271, 273, *274*
Primates
anthropoids, 701–702
bipedal locomotion, 702–703
brain size–body size relationship, 973
comparative genomics, 366–367
fossil record, 701
grooming behavior, 1177
hominins, 702–705
number of species, *698*
phylogeny, *701*
prosimians, 701
*See also* Chimpanzees
Primers
in creating synthetic DNA, 380
in DNA replication, 63, 271, *272*, *274*
in the PCR reaction, 277, *278*
Primitive groove, 914
Primitive gut
formation in sea urchins, 909
in frog gastrulation, 909
Primitive streak, 914
"Primordial soup" hypothesis, 71
Probability, 13–14, 239–240
Probes, 290
Proboscidea, *698*
Proboscis
of ecdysozoans, 665
of hemichordates, 682, 683
of ribbon worms, 658, *659*
Procambium
in root growth, 716–718
in shoot growth, 719
vascular tissue system and, 716
Procarboxypeptidase A, 333
Procedural memory, 982
Productivity
defined, 1189–1190
impact of species richness on, 1202, *1203*
species diversity and, 1192
Products, in chemical reactions, 31, 145
Progesterone
actions of, *842*
in human pregnancy, 896
in labor and childbirth, 896
in parthenogenic whiptail lizards, 882
produced in the ovaries, 850, 893, *894*
in regulation of the ovarian and uterine cycles, *894*, 895
Progestin-only pill (Plan B®), *898*
Programmed cell death. *See* Apoptosis
Progymnosperms, 589, 591
Prokaryotes
archaea, 534–537 (*see also* Archaea)
atmospheric oxygen levels and, 511–512
beneficial relationships with eukaryotes, 539
cell division, 206–207
cellular locations of energy pathways, *168*
characteristic features of, 82–84, 526–527
characteristics of gene expression in, *291*
complex communities, 539, *540*
defined, 81
discordant gene trees, 529–530
in element cycling, 538–539
environmental genomics, 530
in the evolution of life, 4
evolutionary relationships from nucleotide sequences, 528–529
gene regulation in, 329–333
gene transfer in, 253–255, 529–530
insertion of genes into, 376–377
major bacterial groups, 530–534
metabolism and metabolic pathways, 4, 166, 168, 537–538
microbiomes and human health, 539–541
nucleic acid hybridization, 291
origins of eukaryotic cells and, 101–102
origins of photosynthesis, 4, *5*
pathogenic, 541–542
phenotypic characteristics used in classification, 527–528, *529*
shared and unique features, 526
success of, 530
Prokaryotic genomes
artificial life studies, 359, *360*
benefits of sequencing, 357
comparative genomics, 357
features of, 356
functional genomics, 356, *357*
metagenomics, 357–358
minimal genome studies, 359, 360
sequencing, 356
transposons, 358–359
Prolactin, *838*, *842*
Prolactin-inhibiting hormone, 844
Prolactin-releasing hormone, 844
Proline, 43, *44*
*Promerops cafer*, *1134*
Prometaphase
in meiosis, *220*
in mitosis, 212, 214, *216*, *217*
Promoters
binding of transcription factors to, 328
consensus sequences, 292, 332–333
eukaryotic general transcription factors and, 333–334, *335*, *336*
in expression vectors, 384
as genetic switches, 415
in transcription, 286, *287*
in the viral lytic reproductive cycle, 340
viral regulatory proteins and, 340–341
Proofreading, 276–277
Prop roots, 718
Propane, 31
Prophage, 340–341
Prophase
centrosome separation, 212
comparison between mitosis and meiosis, *223*
events in mitosis, 212, *214*, *217*
spindle apparatus formation, 213–214
Prophase I (meiosis), 219, 220
Prophase II (meiosis), *220*
*Propithecus diadema*, *701*
*Propithecus verreauxi*, *10*
Propranolol, 940
Prosimians, 701
*Prosopis*, 808
Prostaglandins
in inflammation, 861, 862
production by the prostate gland, 890
Prostate cancer, *368*
Prostate fluid, 890
Prostate gland, 890
Prosthetic groups, of enzymes, 155, *156*
Protease inhibitors, 342, 805
Proteases
function of, 1057–1058
produced by the pancreas, 1062
in proteolysis, 300–301
Proteasomes, 349
Protected areas, 1237, *1238*
Protein hormones, 836
Protein kinase C (PKC), 134–135
Protein kinase cascades, 131–132, 138
Protein kinase receptors, 129, *130*
Protein kinases
in cell cycle control, 209–211

  in hormone-mediated signaling cascades, 837
  in protein phosphorylation, 301
  receptors, 129, *130*
  regulation of glucose metabolism in liver cells, 138
Protein phosphatase, 161
Protein starvation, 1041
Protein synthesis
  inducers, 330
  modifications after translation, 300–301
  polysomes, 297, *298*
  signal sequences and polypeptide movement within the cell, 298–300
  steps in gene expression, 284–285
  transcription, 286–293
  transcriptional regulation in prokaryotes, 329–333
  translation, 293–297, *298*
Proteinases, 300
Proteins
  in animal "self-consumption," 1050
  biological information and, 5, 6
  breakdown by digestive enzymes, 1057–1058
  catabolic interconversions, *179*, 180
  commercial production, 383–384
  comparing through sequence alignment, 486–487
  denaturation, 48, 50
  digestion in animals to constituent amino acids, 1051–1052
  domains, 291
  dysfunctional proteins and human genetic diseases, 311–312
  effect of temperature on, 820
  energy yield, 1049, *1050*
  environmental effects on structure, 50
  fluorescent, 449, 464
  functions, *42*
  gain-of-function mutations, 305
  genomic information and, *355*, 356
  identifying homologous parts, 486
  inducible and constitutive, 330
  loss-of-function mutations, 305, 311–312
  modification after translation, 300–301
  molecular chaperones, 51
  molecular evolution, 486
  monomer components, 40
  in the origin of life, 3
  peptide linkages, 43–44, *45*
  phosphorylation, 209
  point mutations and, 306–307
  primary structure, 45, *46*
  primary structure specifies tertiary structure, 48, 49
  production of medically useful proteins through biotechnology, 384–386
  in prokaryotic gene regulation, 329
  proportions in living organisms, *41*
  protein–DNA interactions, 266, *267*
  proteomes, 369, *370*
  quaternary structure, *46*, 48, *49*
  reconstruction of sequences from extinct organisms, 464
  regulation of longevity in the cell, 348–349
  relation of mRNA abundance to cell protein abundance, 348
  roles in DNA replication, 272, *273*
  secondary structure, 45, *46*
  shape modifications, 50
  specificity of binding, 48, 50
  spider silk, 39
  structural characteristics, 43
  structural motifs and binding to DNA, 335–336
  tertiary structure, 46–48
  use of gene evolution to study protein function, 500
Proteobacteria, 534, 538, *550*, 551
Proteoglycans, 100, 109, 111
Proteolysis, 300–301
Proteomes, 369, *370*
Proteomics, 369, *370*
Proterozoic eon, *506–507*, 508, 515–516
Prothoracicotropic hormone (PTTH), 840, *841*
Prothrombin, 1039
Protists
  alveolates, 553–555
  amoebozoans, 559–561
  ancestor to fungi and animals, 609, 631, *633*
  approaches to classifying, 553
  defined, 550
  endosymbionts, 564–565, 566
  evolution of multicellularity and, 552–553
  excavates, 558–559
  pathogenic, 563, *564*
  primary producers, 563
  rhizaria, 557–558
  sex and reproduction in, 562–563
  stramenopiles, 555–556, *557*
Protocells, 73–74
Protoderm
  dermal tissue system and, 716
  in root growth, 716–717
  in shoot growth, 719
Protohominids, 702–703
Proton gradients, photophosphorylation and ATP synthesis, 192–193
Proton-motive force, 173–174
Proton pumps
  active transport in plants and, 729
  in auxin-induced cell expansion, 766–767
  in polar auxin transport, 763
  in the uptake of ions by roots, 746
Protonema, *576*, 577
Protonephridium, 1075
Protons
  atomic number and, 22
  defined, 22
  mass number and, 22
  proton-motive force and ATP synthesis, 173–176
  transport in the respiratory chain, 172–173
*Protopterus annectens*, *689*
Protostomes
  anatomical characteristics, 652
  in animal phylogeny, *630*
  arrow worms, 655–656
  bilaterians, 634, 643
  defined, 652
  ecdysozoans, 654–655, 665–673, *674*
  key aspects of evolution in, 673–675
  lophotrochozoans, 652–654, 656–664
  major derived traits, 652
  major groups, *632*, 652
  pattern of gastrulation in, 634
  phylogenetic tree, *653*
  undescribed species, 675
Prototherians, 697, *698*, 889
Proturans, 671, *672n*
Province Islands, 1199
Provirus, 341–342, 544
Proximal convoluted tubule (PCT), *1080*, 1081, 1082, *1083*, 1084
Proximate causes, of animal behavior, 1096
Prozac, 940
Przewalski's horse, *1131*
*Pseudobiceros*, *1007*
Pseudocoel, 635, 652
Pseudocoelomates, 635
Pseudogenes, 364, 491, 494–495, 497
*Pseudomonas*, 373, 539
*Pseudomonas aeruginosa*, *82*, *207*
*Pseudomonas fluorescens*, 489–490
*Pseudomyrmex*, 1178, *1179*
*Pseudonocardia*, 1169
Pseudoplasmodium, 561
Pseudopodia, 94
Pseudopods
  of amoebozoans, 559
  of foraminiferans, 557
  of radiolarians, 557
*Pseudotsuga menziesii*, 1160
*Pseudouroctonus minimus*, *669*
Psocoptera, *672*
Psoriasis, 158
Psychoactive drugs, 135
*Pterapogon kauderni*, 1234, *1235*
Pterobranchs, 682–683
*Pteroeides*, *647*
Pterosaur, *423*
Pterygotes
  diversity in, *674*
  instars, 671
  major groups and number of living species, *672*
  mayflies and dragonflies, 672
  metamorphosis, 672
  neopterans, 672–673
  wings and flight, 671, 672, 673
*Pthirus pubis*, *1177*
PTTH. *See* Prothoracicotropic hormone
Puberty
  in females, 894
  in males, 892
  overview, 850–851
Public policy
  importance of biological research to, 16–17
  to reduce the effects of mutagens on human health, 311
*Puccinia graminis*, 612
Puffballs, *610*
Pufferfish, *361*, 500
Pulmonary arteriole, *1014*
Pulmonary artery, 1031, 1032
Pulmonary circuit
  in amphibians, 1029
  in birds and mammals, 1030
  blood vessels of, 1027–1028
  defined, 1027
  in lungfish, 1028
  in reptiles, 1029–1030
Pulmonary valve, 1030, *1031*, 1032
Pulmonary veins, 1031
Pulmonary venule, *1014*
Pulp cavity, 1055
Punnett, Reginald, 236
Punnett square, 236–237
Pupa, *639*
Pupil, 958, *959*
Purifying selection, 433, 492–494
Purines
  anabolic interconversions and, *179*, 180
  found on meteorites, 69
  structure, 63
  transition and transversion mutations, 306
Purkinje cells, *925*
Purkinje fibers, 1034, *1035*
Purple foxglove, 1034
Purple owl's clover, 1161
Purple pitcher plant, 1189
Purple sand crab, *670*
Purple sulfur bacteria, 187, *538*
Pus, 862
Putrefaction, 1188
Pycnogonids, 668
*Pycnophyes kielensis*, *665*
Pygmy tarsiers, 1230–1231
Pyloric sphincter, *1059*, 1060, 1061
Pyramid diagrams, 1191, *1192*
Pyramidal cells, *925*
*Pyrenestes ostrinus*, *440*
Pyrethrin, *802*
Pyridoxine, *1053*
Pyrimidines
  anabolic interconversions and, *179*, 180
  found on meteorites, 69
  structure, 63
  transition and transversion mutations, 306
Pyrogens, 830
Pyrophosphate, 150, 269, *270*
Pyruvate
  in alcoholic fermentation, *177*, 178
  anabolism in the liver during exercise, 181
  in $C_4$ photosynthesis, 199
  in cellular respiration, 166–167
  in glucose catabolism, 168, 169
  in glycolysis, 166
  in lactic acid fermentation, 177
  in metabolic interactions in plants, 201

produced during glycolysis, 169–170
Pyruvate decarboxylase, *177*, 178
Pyruvate dehydrogenase, 171
Pyruvate oxidation
in glucose catabolism, 168, 170
regulation of, 171
relationships among metabolic pathways, *179*
Pyruvic acid. *See* Pyruvate
Pythons, 423

## Q

Q. *See* Ubiquinone
$Q_{10}$, 821
Qiu, Yin-Long, 571
QTL analysis, 1096
Quack grass, *597*
Quadrats, 1151
Quadriceps, 1002
Qualitative analysis, 33
Qualitative traits, 439
Qualitative variation, 246
Quantifiable data, 11, 14
Quantitative analysis, 33
Quantitative trait analysis, 1096
Quantitative trait loci, 246
Quantitative traits
defined, 439
possible actions of natural selection on, 439–440
Quantitative variation, 246
Quaternary period, *506–507*, 510, 522
Quaternary structure, of proteins, *46*, 48, *49*
Queen honey bees, 880, 899
Quiescent center, 716, *717*
Quill, *696*
Quinine, *604*, 605, 797
Quinine-resistant malaria, 797
Quiring, Rebecca, 413
Quorum sensing, 525, 539

## R

*R* genes, 800
R groups
of amino acids, 43
in an alpha helix, 45
of histones, 212
in polypeptide chains, 44
in protein binding, 50
in protein shape changes, 50
in protein tertiary structure, 47
*See also* Side chains
R proteins, 799, 800
*r*-Strategists, 1159, 1201
RAAS system, 1087, *1088*
Rabbits
multiple alleles for coat color, *242*
number of species, *698*
point restriction phenotype, 245, 306
Rachis, *696*
Radcliffe, Paula, 815
Radial axis, in plants, 711
Radial cleavage, 633, 679, 905
Radial symmetry
in animals, 634, *635*
in echinoderms and hemichordates, 680, *681*
in flowers, 597
Radiation
in heat exchange between animals and their environment, 823, *824*
human-made or natural, 310
mutagenic effects, 309
Radiation treatments, 229
Radicle, 718, 758
Radio frequency identification (RFID), 1151
Radioactive contamination, bioremediation of, 389
Radioactive decay, 23
Radioimmunoassays, 852
Radioisotopes
in experiments revealing the Calvin cycle, 193–194, 195
Hershey–Chase experiment on DNA, 262–263
as mutagens, 310
properties of, 23–24
radiometric dating, 507–508
Radiolarians, 557–558, 564
Radiometric dating, 507–508
Radius, 423
Radula, 662, *663*, 1056
*Rafflesia arnoldi*, *603*
Ragweed, 1200
Rain shadows, 1126–1127
Rainbow trout, 162, 1140
Rainey, Paul, 489–490
Ralph, Martin, 1108
*Ramalina menziesii*, 625
*Rana berlandeieri*, *476*
*Rana blairi*, *476*
*Rana pipiens*, 12
*Rana sphenocephala*, *476*
*Rana sylvatica*, *642*, 1125, *1126*
*Randallia ornata*, *670*
Random dispersion pattern, 1153
*Rangifer tarandus*, *700*, 1149
Rao, P. N., 209
Raphidoptera, *672*
"Rapid-cooling" technology, 831
Rapid eye movement (REM) sleep, 978, 979
Ras protein, 131, 132
Ras signaling pathway, 139
Rashes, 876
Ras–MAP signal transduction pathway, 398
Raspberry, 601
Rats
hormonal control of sexual behavior, 1098–1099
mutation affecting obesity, *1067*, 1068
"place cells" of the hippocampus, 967
regulation of food intake by the hypothalamus, 1067
*Toxoplasma* and, 554
Rattlesnakes, 946
Raven, Peter, 1175
Ray-finned fishes, *685*, 687–689
Ray flowers, *597*
Rays
cartilaginous skeleton, 1000
claspers in sexual reproduction, 888
evolution in body morphology, 444, *445*
excretion of urea, 1074
features of and diversity in, 687, *688*
regulation of ionic composition of extracellular fluid, 1073
salt and water balance regulation, 1077
Reabsorption, isosmotic, 1082
Reactants, in chemical reactions, 31, 145
Reaction center, of photosystems, 190, 191, 192
Reactive atoms, 25
Reactive oxygen
hypersensitive response in plants and, 800–801
in plant defenses against herbivory, 804
in plant responses to pathogens, 799
Realized niche, 1184
Recent, the. *See* Holocene epoch
Receptacle (floral), *591*
Receptive fields
of neurons in the visual cortex, 977
of photoreceptors, 963
of retinal ganglion cells, 975–977
Receptor cells
in sensory transduction, 947–948
*See also* Sensory receptor cells
Receptor-mediated endocytosis, 120, 121, *122*
Receptor potential, 947
Receptor proteins
binding of signal ligand to, 127–128
classification by function, 129–130
classification by location, 128–129
functions, *42*
intracellular receptors, 130–131
in receptor-mediated endocytosis, 121
in sensory transduction, 947, *948*
specificity, 127
Recessive traits, 235
Recognition sequence, 315, 316
Recombinant chromatids, 219
Recombinant DNA, 374–375
Recombinant DNA technology
agricultural applications, 386–388, *389*
cloning genes, 375–376
methods of creating recombinant DNA, 374–375
methods of transformation, 376–377
origin of, 374
production of medically useful proteins, 385–386
public concerns, 388–389
reporter genes, 377–378, *379*
sources of DNA used in cloning, 379–380
using to produce vaccines, 866
using to study gene function, 381
Recombinant frequencies, 248, *250*
Reconciliation ecology, 1243–1244
Rectum, 1056, *1058*, 1063
Red algae, 551, 571–572, 573
Red-and-green macaws, *1106*
Red blood cells
ABO blood groups, 243
β-globin gene expression in, 398
induction of cell division in, 211
malaria, 563, *564*
pernicious anemia, 1054
production and elimination, 1038
sickle-cell disease, 306, *307*, 312
transcriptional regulation of β-globin, 335
Red-eared slider turtle, 420
Red fluorescent pigments, 449, 464
Red-green color blindness, *252*
Red light
photomorphogenesis and phytochromes in plants, 772–774
in photoperiodic control of flowering, 788
*See also* Far-red light
Red mangrove, 1198, *1199*
Red muscle, 995
Red tides, 549, 564
Red-winged blackbirds, 1114
Redi, Francesco, 67–68
Redox reactions. *See* Oxidation–reduction reactions
Reduction, 167
Reefs
byrozoan, 656
corals and, 646–647
Reflexes. *See* Spinal reflexes
Reforestation, mycorrhizal fungi and, 626
Refractory period, of voltage-gated sodium channels, 933–934
Regeneration, 881
Regular dispersion pattern, 1153
Regulative development, 907
Regulatory sequences
in eukaryotic genomes, 361
genomic information and, 356
of operons, 330
Regulatory subunits, of enzymes, 159
Regulatory systems
components and functions of, 816–817
maintenance of stability in the internal environment, 10
Regulatory T cells (Tregs), 863, 874
Reindeer, *700*, 1149, 1158, 1166
Reindeer moss, 613
Reinforcement, 475, 478
Relative atomic mass, 23
Release factor, 297, *298*
Releasers, 1095
Religion, 14
REM sleep, 978, 979
Remediation. *See* Bioremediation
Renal artery, *1080*, 1081, 1082
Renal cortex, *1080*, 1081
Renal dialysis, 1086, *1087*
Renal failure, 1085–1086, *1087*
Renal medulla, *1080*, 1081, 1082–1084
Renal pyramids, *1080*, 1081, 1090
Renal tubules
conversion of glomerular filtrate to urine, 1079

in the mammalian kidney, 1081
reabsorption in, 1078, 1082
water channels, 1084
Renal vein, *1080*, 1081
Renal venule, *1078*
Renin, 1087, *1088*
Reoxidation reactions, 171
Repetitive sequences, in eukaryotic genomes, 364–366
Replication. *See* DNA replication
Replication complex, 206, 274–275
Replication forks, 271, 272–275
Replicons, 377
Reporter genes, in recombinant DNA technology, 377–378, *379*
Repressible systems, in transcriptional regulation of operons, 330, 331–332
Repressor proteins
as genetic switches, 415
helix-turn-helix motif, 336
in negative regulation, 329
operator–repressor interactions controlling operon transcription, 330–332, *333*
strategies in the repression of transcription, 336
Reproduction. *See* Animal reproduction; Asexual reproduction; Human reproduction; Plant reproduction; Sexual reproduction
Reproductive capacity, estimating, 1154–1155
Reproductive isolation
biological species concept and, 468–469
defined, 468
hybrid zones, 478–479
importance to speciation, 469
from incompatibilities between genes, 470, *471*
from increasing genetic divergence, 470–471
mechanisms preventing hybridization, 475–478
Reproductive signal
in cell division, 206
in eukaryotic cell division, 207
in prokaryotic cell division, 206
Reproductive success, evolution by natural selection and, 6
Reproductive technologies, 897, 899
Reptiles
circulatory systems, 1029–1030
crocodilians and birds, 693–695
evolution of the amniote egg, 888
gastrulation, 913–914
incomplete cleavage, 633
lepidosaurs, 693
origin of, 692, *693*
radiation during the Triassic, 521
salt and water balance regulation, 1078
temperature-dependent sex determination and sex-specific fitness differences, 421
turtles, 693, *694*
Rescue effect, 1161
Residence time, 1211, 1215
Residual volume (RV), *1012*, 1013
Resistance (*R*) genes, 800
Resolution, of microscopes, 79, 80, 81
Resource partitioning, 1182
Respiration. *See* Cellular respiration
Respiratory chain
allosteric regulation, 181
controlled release of energy by, 172
description of, 172–173
in glucose metabolism, 168
in oxidative phosphorylation, 171
Respiratory distress syndrome, 1015
Respiratory gas exchange
in amphibians, 1029
fish gills, 1009–1010, *1011*
fully separated pulmonary and systemic circuits, 1030
human lungs, 1013–1016
partial pressure gradients, 1009
physical factors governing, 1006–1008
regulation of breathing, 1019–1022
snorkeling elephants, 1005, 1022
surface area of respiratory organs, 1008–1009
unidirectional ventilation in birds, 1010–1012
Respiratory gases
air and water as media for, 1007
blood transport, 1016–1019
defined, 1006
diffusion, 1006
maximization of partial pressure gradients in gas exchange, 1009
Respiratory organs
bird lungs, 1010–1012
fish gills, 1009–1010, *1011*
human lungs, 1013–1016
surface area maximization, 1008–1009
Respiratory tract, anatomy of, 1013, *1014*
Resting membrane potential
of cardiac pacemaker cells, 1032–1033
defined, 927
measurement of, 928
Restoration ecology, 1237–1239, *1240*
Restriction digestion, 315
Restriction endonucleases, 315–316
Restriction enzymes, 315–316, 317, 374
Restriction (R) point, 210
Restriction site, 315, 316
Reticular activating system, 970
Reticular formation, 979
Reticulate bodies, 533, *534*
Reticulations (on phylogenetic trees), 496
Reticulum, 1064
Retina
cone cells, 960, 961–962, 963
in the image-forming eye, 958, *959*
information flow in, 962–963
inputs to the visual cortex, 977–978
receptive fields of ganglion cells, 975–977
rod cells, 960–961, *962*, 963
structure of, 959
Retinal
changes with the absorption of light, 957, 958
in cone cells, 962
role in catalyzed reactions, *156*
vitamin A and, 58
Retinal ganglion cells
information flow through the retina, *962*, 963
receptive fields, 975–977
Retinoblastoma, 349
Retinoblastoma (RB) protein, 210, 228
Retinol, *1053*
Retrotransposons, 364–365, 495
Retroviruses
description of, 341, *543*, 544
endogenous retroviruses in the vertebrate genome, 545
mutations caused by, 308
reverse transcription, 285
Reverse genetics, 318
Reverse transcriptase
in the production of cDNA, 379–380
in retrovirus infections, 341
synthesis of DNA from RNA, 72
Reverse transcriptase inhibitors, 342
Reverse transcription, 285
Reversible chemical reactions, 34, 35–36
Reversible inhibition, of enzymes, 157–159
Reversible phosphorylation, 161
Reversion mutations, 306
Reversions, 488
Reznick, David, 1157
*Rhacophorus nigropalmatus, 7*
*Rhagoletis pomonella*, 473
Rhcg protein, *1086*
*Rhea pennata, 1131*
Rheas, 694
Rhenium, 69
*Rheobatrachus silus, 678*
Rheumatoid arthritis, 158, 862, 876
Rhinoceroses, 1234
Rhizaria, 557–558
Rhizobia, formation of root nodules, 747–748
*Rhizobium*, 357, 534
Rhizoids
of fungi, 609
of nonvascular plants, *576*, *577*
of rhyniophytes, 580
Rhizomes, 580, 792, *793*
*Rhizophora*, 1141
*Rhizophora mangle*, 1198, *1199*
*Rhizopus oligosporus, 218*
*Rhizopus stolonifer, 618*, 619
Rhizosphere, interference competition in, 1183
*Rhodnius prolixus*, 839–840, 841
Rhodopsin, 957–958, 960, 961
*Rhogeessa tumida*, 470, *471*
Rhombozoans, *632*, 648
Rhyniophytes, 580
Rhynocoel, 658, *659*
Rhythm method of contraception, *898*
Ribbon model, of protein tertiary structure, *47*, 48
Ribbon worms, *632*, *652*, 658, *659*
Riboflavin, *1053*
Ribonucleic acid (RNA)
antisense RNA, 382
complementary base pairing, 63–64, *65*
distinguishing from DNA, *63*
DNA transcription and, 65
growth of, 63, *64*
origin of life and, 71–72, *73*
as a primer in DNA replication, 271, *272*, *274*
reverse transcription, 285
roles in gene expression, 285
structure and function, 63–67
translation, 65
types of RNAs produced by transcription, 286
in vitro evolution, 501
Ribonucleoside triphosphates, *287*, 288
Ribose, *52*, 63
Ribosomal RNA (rRNA)
in eukaryotic ribosomes, 84–85
location and role in eukaryotic cells, *286*
produced by transcription, 286
role in translation, 285
*See also* Ribosomes
Ribosomal RNA genes
concerted evolution in, 498
evolutionary relationships in prokaryotes and, 528
Ribosomes
action of antibiotics on prokaryotic ribosomes, 295
in animal cells, *86*
functions in translation, 294–295
interaction with tRNAs in translation, 294
in plant cells, *87*
polyribosomes, 297, *298*
in the process of translation, 295–297, *298*
of prokaryotes, *82*, 83
structure and function in eukaryotes, 84–85, 88
subunits, 295, 364
Riboswitch, 348
Ribozymes
as biological catalysts, 72, *73*, 151
lowering of the energy barrier in biochemical reactions, 151–154
Ribulose 1,5-bisphosphate (RuBP), 194, 195, *196*, 197–198
Ribulose bisphosphate carboxylase/oxygenase (rubisco)
in $C_3$ plants, 198
in $C_4$ plants, 199
in the Calvin cycle, 194
photorespiration and, 197–198
Ribulose monophosphate (RuMP), 194, *196*

Rice
genetically modified, 388
genome, *361*, 362
as a primary human food source, 605
quantitative variation in grain production, 246
sake, 624
semi-dwarf, 756
transgenic improvement of water-use efficiency, 726
water demands of, 726
Rice, William, 482
Rice "paddies," *605*
Richardson, A., *300*
Rickets, *1053*
Ricketts, Taylor, 1243
Ridley–Tree Condor Preservation Act, 1244
*Riftia*, *661*
Riggs Glacier, *17*
Right-handed β-spirals, 46
Right-handed helices, *47*, 265
Rigor mortis, 989
Ring canal, 680, *681*
Ringneck snake, *694*
Ringworm, 612
Risk costs, in animal behavior, 1103
Rivers
freshwater biomes, 1140
in the global hydrologic cycle, 1215
nitrogen runoff and "dead zones," 740, 1207, 1219
RNA-dependent RNA polymerase, 544
RNA genes
genomic information and, *355*, 356
moderately repetitive sequences, 364, *365*
RNA interference (RNAi), 346, 382
RNA polymerases
eukaryotic general transcription factors and, 334, *335*, *336*
roles in transcription, 286, *287*, 288
sigma factors in prokaryotes, 333
structure and function, 286
in transcriptional regulation in prokaryotes, 332–333
in the viral lytic reproductive cycle, 340
RNA retroviruses, *543*, 544, 545
RNA splicing, *291*, 292–293
RNA viruses
description of, 341
plant systemic acquired resistance, 801
types of, *543*, 544
*See also* RNA retroviruses
"RNA world," 71–72, *73*
RNAi. *See* RNA interference
Roaches, 673
Robinson, Scott, 1117
Rock barnacle, 1184
Rocks
dating, 506–508
weathering of, 745–746, 1213
Rod cells, 960–961, *962*, 963
Rodents, *698*, 699
Rohm, Otto, 144
Rooibos, 1242
Root apical meristem, 712, 715–716
Root cap, 716, *717*
Root hairs, 713, *717*
Root nodules
formation, 747–748
nitrogen fixation in, 750
Root systems
adaptations to saturated soils, 808
adaptations to very dry conditions, 808
in monocots and eudicots, 710
primary indeterminate growth, 716–718
root apical meristems, 712, 715–716
secondary growth, 721–723
structure and function, 709
types of, 718
uptake of water and minerals, 718
Roots
auxins in the initiation of, 765
cation exchange with the soil solution, 746
endodermis, 729–730
ethylene and, 770
evolution in vascular plants, 582–583
gravitropism, 765
movement of water and mineral ions across the plasma membrane, 728–729
mycorrhizae, 747, 748–749
of phylogenetic trees, 450
root nodule formation, 747–748
uptake and transport of water and mineral ions, 728–730
vegetative reproduction in angiosperms and, 793
Roquefort cheese, *622*, 624
Rosenberg, Barnett, 259, 278
Rotational cleavage, 906
Rothamsted Experiment Station, 1192
Rotifers
anatomical characteristics, *652*
description of, 656, 657–658
major subgroups and number of living species, *632*
Rough endoplasmic reticulum (RER)
in animal cells, *86*
in plant cells, *87*
processing of newly translated polypeptides, *299*, 300
structure and function, 88–89
Rough-skinned newt, 445
Round window, *954*, 955
Roundup, 160
Roundworms, *635*, 666
*See also* Nematodes
Rove beetles, 1203
"Royal jelly," 899
Royal poinciana, *603*
*RPL21* gene, 497
rRNA. *See* Ribosomal RNA
rRNA genes, 364, *365*
RT-PCR, 380
RU-486 (mifepristone), *898*
Ruben, Samuel, 186, 187
Rubisco. *See* Ribulose bisphosphate carboxylase/oxygenase
*Rudbeckia fulgida*, *599*
Ruffini endings, 952
Rufous hummingbirds, 1137
Rumen, 1064
Ruminants, 492–494, 1064
Runners, 720, 792
Ruppell's griffon, 1010
Rushes, 793
Russian steppe, 1131
Rust fungi, 622, 800
Rusty tussock moth, *1157*
Ryanodine receptor, 990, 992
Rye, 718
Rye mosaic virus, 669

## S

*S*-adenosyl methinione (SAM-e), 349
*S* genes, 381, 782
*S* orbital, 24
S phase
cell fusion experiments on cell cycle control, 208–209
centrosomes in, 212
description of, 208
in meiosis, 219
in mitosis, *214*, *217*
Sac fungi
distinguishing features, *616*
edible, 624
filamentous, 620, 622
in lichens, 613
life cycle, 620, *621*
as model organisms, 625, 626
phylogeny of the fungi, *615*
yeasts, 620 (*see also* Yeasts)
*See also* Dikarya
*Saccharomyces cerevisiae*, *609*
characteristics of, 620
genomic information, *361*, 362
insertion of genes into, 376
as a model organism, 624
in the production of food and drink, 623
*Saccharum*, *603*
*Saccoglossus kowalevskii*, *683*
Saccule, 956, 957
*Sagartia modesta*, *647*
Sailfish, *1139*
Sake, 624
Sakmann, B., 929
Salamanders
experiments on embryo formation, 910–911, 912
features of, 690, *691*
gills and respiratory gas exchange, *1007*
neoteny in, 692
Salamone, Daniel, 386
Salicylic acid, 759, 799, 801
Saline environments
challenges to plants, 810–811
plant adaptations to, 811
Saliva, 1059
Salivary amylase, *1062*
Salivary glands, *1058*, 1060, *1062*
Salivation, 1064–1065
Salmon, 689, 887, 1141
*Salmonella*, 357, 542
*Salmonella typhimurium*, 534
Salps, 684
Salt, George, 482
Salt and water balance regulation
in aquatic invertebrates, 1072–1073
in invertebrates, 1075–1077
by the mammalian kidney, 1082–1084
in vampire bats, 1071
in vertebrates, 1073, 1077–1079
Salt bridges, 47
Salt glands
nasal, 1073
in plants, 811
Salt marshes, 1141
Salt tolerance
in genetically modified plants, 388, *389*
*See also* Halophiles
Saltatory conduction, 935
Salty taste, 951
*Salvelinus confluentus*, 1228
*Salvelinus namaycush*, 1228
*Salvinia*, *581*
SAM-e (*S*-adenosyl methinione), 349
San Andreas Fault, 509
Sand verbena, 793
Sandy soils, 745
Sanger, Frederick, 353
*Sapindopsis belviderensis*, *519*
Saprobes, 556, 611, 1054
*Saprolegnia*, 556, *557*
Saprotrophs. *See* Saprobes
Sarcolemma, *987*
Sarcomeres, *987*, 988, 996
*Sarcophilus harrisii*, 232
Sarcoplasm, 989, 990
Sarcoplasmic reticulum
effect of calcium ion cycling on cardiac muscle contraction, 1034, *1036*, 1037
in skeletal muscle contraction, 989–990, *991*
*Sardinella aurita*, *1139*
Sardines, *1139*
Sargasso Sea, 556
*Sargassum*, 556
Sarin, 157
*Sarracenia*, 751
*Sarracenia purpurea*, 1189, 1197
Satiety factors, 1067–1068
Saturated fatty acids
phospholipids and, 107
structure and function, 56, *57*
Saturated soils, plant adaptations to, 808
Saturation, of facilitated diffusion, 117
Savannas, 1135–1136, 1194
*SBE1* gene, 305
SBE1 protein, 305
Scale of life, *78*
Scales
of lepidosaurs, 693
of ray-finned fishes, 687
Scallops, 662
Scandentia, *698*
Scanning electron microscopy, 81
Scarab beetles, 826
Scarlet ibis, *696*
*Sceloporus jarrovii*, 1103, *1104*
Schally, Andrew, 844
Schindler, David, 1220
Schistosomiasis, 657

*Schizosaccharomyces pombe, 361,* 624, 626
Schomburgk, Robert, 588
Schopf, J. William, 74
Schulze, Franz, 629
Schwann cells, 926
Scientific methods
  distinguishing characteristics of, 14
  experiments, 12–13
  key features of, 11–12
  statistics and, 13–14
  *See also* Biological research
Scientific names, 7, 462, 463–464
Scion, 793, *794*
Sclera, 958, *959*
Sclereids, *713,* 714
Sclerenchyma, 713–714
*Scleria goossensii,* 440
*Scleria verrucosa,* 440
Sclerotium, 560
*Scolopendra hardwicki, 670*
Scorpionflies, *672*
Scorpions, *636,* 669
Scottish deerhound, 352
"Scouring rushes," 581
Scrotum, 889
Scurvy, 1053
Scyphozoans, 647
Sea anemones, 646, *647,* 1171
Sea butterflies, 662
Sea cucumbers, 681, 682
Sea grass, 1141
Sea level
  effect of plate tectonics on, 509
  glaciation and, 509, *510*
  mass extinctions and, 510
Sea lilies, 520, 680–681, 682
Sea lions, 700
Sea palms, *556*
Sea pens, 646, *647*
Sea slugs, 662, *664,* 1173, *1174*
Sea spiders, 668
Sea squirts, *455, 683,* 684
Sea stars, 681, 682, 881
  *See also* Starfish
Sea turtles, 693
Sea urchin eggs
  fertilization, 884–886
  maturation promoting factor, 209, 210
Sea urchins
  blastopore, *634*
  determination of polarity in the embryo, 395
  gametic isolation, 477
  gastrulation, 908–909
  key features of, 681, 682
Seabirds
  dispersion patterns, *1153*
  territory of, 1104, *1105*
Seagrasses, *603*
Seals, 700, 1017
Seasonal temperatures, acclimatization in animals, 821
Seasons
  causes of, 1123
  *See also* Environmental cycles
Seawater evaporating ponds, *536*
*Sebastes melanops,* 1163
*Secale cereale,* 718
Second filial generation ($F_2$), 234–236
Second law of thermodynamics, 146–147
Second messengers
  calcium ions, 135
  defined, 131
  discovery of, 132–133, 134
  in hormone-mediated signaling cascades, 837
  lipid-derived, 134–135
  nitric oxide, 135–136
Second polar body, *883,* 884
Secondary active transport, 118, 119
Secondary bronchi, 1011, *1012*
Secondary cell wall, 711
Secondary consumers, 1190, *1191*
Secondary endosymbiosis, 551–552
Secondary growth, 591–592, 715
Secondary immune response, 866
Secondary lysosomes, 90, *91*
Secondary meristems, 715, *716*
Secondary metabolites
  of the metabolome, 370
  in plant defenses against herbivores, 802–803, 804, 1175
Secondary oocyte, *883,* 884
Secondary phloem, 16, *17,* 721, 722, *723*
Secondary sex characteristics, 250, 889
Secondary spermatocytes, *883,* 884, *891*
Secondary structure, of proteins, 45, *46*
Secondary succession, 1200, 1201
Secondary xylem, 591–592, 721, 722–723
  *See also* Wood
Secretin, 838–839, 1065
Sedges, 440
Sedimentary rocks, 506, 507, 565
Seed coat, 591, *592,* 784
Seed companies, 753
Seed dispersal
  birds and, 696
  conifers, 594
  fruits and, 601
  plant–frugivore mutualisms in, 1181–1182
  seed plants, 591
  strategies in, 784
Seed dormancy, 591, 757, 762, 784
Seed ferns, 589, 591
Seed germination
  cytokinins and, 768
  effects of light on, 772
  overview, 758
  phytochromes and, 774
  role of abscisic acid in, 762
  role of gibberellins in, 761–762
  smoke-induced, 1121
  vacuoles and, 93
"Seed leaves." *See* Cotyledons
Seed plants
  angiosperms, 596–604
  euphyllophytes, 583
  evolution of, 589
  gymnosperms, 592–596
  important crop plants, 605
  life cycle, 589–590
  major groups, *574,* 589
  medicinal, 604–605
  pollination, 590, *591*
  secondary growth, 591–592
  seeds, 590, 591, *592*
Seedless fruits, 784–785
Seedless grapes, 761
Seedless watermelon, 225
Seedlings
  ethylene and the apical hook, 770
  etiolation, 772
  growth of, 758
Seeds
  aleurone layer, 761, *762*
  in angiosperm sexual reproduction, *781*
  conifers, 594, *595*
  development, 590, *592*
  dispersal (*see* Seed dispersal)
  dormancy, 591, 757, 762
  fruit development and, 784–785
  germination (*see* Seed germination)
  hybrid, 778
  inoculation with mycorrhizae, 615
  quiescence, 757
  tissues in, 591
Segment polarity genes, 403, *404*
Segmentation
  in animals, 636
  in annelids, 659, *660*
  in arthropods, 667
  in protostome evolution, 673
  in trilobites, 668
Segmentation genes, 402, 403, *404*
Segmentation movements, 1060, 1063
Segregation
  Mendelian law of, 236–237, 239
  *See also* DNA segregation
Selectable marker genes, 376, 378
Selection. *See* Artificial selection; Natural selection; Sexual selection
Selenium, *1052*
Self, distinguishing from nonself, 863, 874
Self antigens, 874, 876
Self-compatibility, 460
Self-incompatibility
  in angiosperms, 782, *783*
  phylogenetic analysis, 459–460
  studies in plants, 380–381
Self-perception, 983
Self-pollination, 598, 782, *783*
Semelparous species, 1156, *1157*
Semen
  components of, 889, 890
  ejaculation, 891
Semi-dwarf grains, 756, 775
*Semibalanus balanoides,* 1184
Semicircular canals, *954,* 956–957
Semicircular ducts, 956, 957
Semiconservative replication, 268, *269,* 270
  *See also* DNA replication
Seminal fluid, 890
Seminal vesicles, 890
Seminiferous tubules, 889, *891*
Senescence, 768, 769–770
Senescence hormone, 769
  *See also* Ethylene
Sensations
  activation of neurons by action potentials, 947–948
  intensity, 948
Sensitive period, in animal behavior, 1099
Sensors
  in regulatory systems, 816
  in sensory transduction, 947–948
  *See also* Sensory receptor cells
Sensory organs, 948
Sensory receptor cells
  adaptation, 948
  chemoreceptors, 949–952
  conversion of stimuli into action potentials, 947–948
  in sensory transduction, 947–948
Sensory receptor proteins, 947, *948*
Sensory systems
  conversion of stimuli into action potentials by sensory receptor cells, 947–948
  defined, 948
  detection of chemical stimuli, 949–952
  detection of light, 957–963
  detection of mechanical forces, 952–957
  functions of, 637
  infrared perception in rattlesnakes, 946
Sensory transduction, 947–948
Sepals, *591,* 597
Separase, 215, *216*
*Sepia, 663*
Sepsis, 862
Septa, 609, *610*
Septate hyphae, 609, *610*
Sequence alignment, 486–487
Sequential hermaphroditism, 887, *888*
*Sequoiadendron giganteum,* 1160
Serine
  in photorespiration, 197, *198*
  structure, *44*
Serotonin, 940
Sertoli cells, 889, *891,* 892
Sessile animals
  dispersal, 640
  filter feeding, 637, *638*
  radial symmetry in, 634
Set point, 816
Setae, *636,* 660
Seven-transmembrane domain receptors. *See* G protein-linked receptors
Severe acute respiratory syndrome (SARS), 357, *543*
Sex
  advantages and disadvantages of, 441
  in protists, 562
Sex chromosomes
  abnormal arrangements, 250
  of ginkgos, 592
  sex determination by, 249–250
  sex-determining gene, 250
  sex-linked inheritance, 249, 251–252

Sex determination
in animal groups, *249*
haplodiploidy, 1115, *1116*
primary, 250
secondary sex characteristics, 250
by sex chromosomes, 249–250
sex steroids and phenotypic determination, 850, *851*
temperature-dependent, 420–421
Sex development, impact of atrazine on, 1, 18
Sex-linked genetic diseases, 312–313
Sex-linked inheritance, 249, 251–252
Sex pheromones, 1110
Sex pili, 84, 253, *254*
Sex steroids
endocrine sources, 850
functions of, 849
in phenotypic sex determination, 850, *851*
in puberty, 850–851, 894
in regulation of the ovarian and uterine cycles, 894, 895
sexual behavior in rats and, 1099
structures, *836*
temperature-dependent sex determination and, 420
types of, 850
*See also* Estrogens; Testosterone
Sexual behavior
hormonal control in rats, 1098–1099
in parthenogenic whiptail lizards, 882
*See also* Courtship behavior
Sexual dimorphisms, 480–481
Sexual life cycles
meiosis and, 217, 218–219 (*see also* Meiosis)
types of, 218–219
Sexual reproduction
costs and risks, 882
evolution in vertebrates, 888–889
fertilization, 884–887
in flowering plants, 779–785
fundamental steps in, 882
fungi, 616–617
genetic diversity and, 218–219, 245, 882
hermaphroditism, 887–888
honey bees, 880
patterns of embryo care and nurture, 889
in protists, 562–563
spawning, 887
*See also* Human reproduction
Sexual selection
description of, 435–436
in evolution, 6
speciation rates and, 480–481
Sexual stimulation
engorgement of the labia minora and clitoris, 892
oxytocin and, 141
penile erection, 890–891
Sexually selected traits, 435–436, 459
Sexually transmitted diseases, 533, 558
Shark Bay, *513*
Sharks
cartilaginous skeleton, 1000
claspers in sexual reproduction, 888
excretion of urea, 1074
eye control gene, *414*
features of, 687, *688*
motor programs of the spinal cord, 969
regulation of ionic composition of extracellular fluid, 1073
salt and water balance regulation, 1077
spiral valve of the intestines, 1057
Sharp-billed ground finch, *474*
Sheep, cloning, 406–407
Shellfish industry, 549
Shells
of bivalves, 662, *663*
of brachiopods, 659
of cephalopods, 664
as exoskeletons, 999
of gastropods, 662, *663*
in protostome evolution, 674–675
of turtles, 693
Sherman, Paul, 1116
*Shigella*, 357
Shimomura, Osamu, 449
Shindagger agave, 1183
Shine, Rick, 421
Shine–Dalgarno sequence, 295
Shiners, *472*
Shivering heat production, 827
Shoot apical meristem
indeterminate growth, 715, *716*, 786
origin in plant embryogenesis, 712
primary shoot growth, 719–720
transition to inflorescence meristems, 785–786
Shoot system
primary indeterminate growth, 719–720
secondary growth, 721–723
shoot apical meristems, 715, *716*
structure and function, 709
Shoots
apical dominance, 765
auxin-induced root formation with cuttings, 765
early development, 758
gravitropism, 765
vegetative reproduction in angiosperms, 792
Short-beaked echidna, *697*
Short-day plants (SDPs), 787, 788
Short interspersed elements (SINEs), 364–365
Short-tailed shrew, 1172–1173
Short-tandem repeats (STRs), 317–318, 364
Short-term memory, 982
Short-term work, fast-twitch fibers and, 995
Shrew opossums, 697, *698*
Shrews, *698*
Shrimps, 670
Shull, George, 244
Siamese cats, 245, 306
*Siamois* gene, 911–912
Siamois transcription factor, 911–912
Siberian hamsters, *851*
Sickle-cell disease
DNA testing by allele-specific oligonucleotide hybridization, *321*
missense mutation causing, 306, *307*, 312
prevalence among African-Americans, 312
use of reverse genetics to discover the DNA mutation in, 318
Side chains, of amino acids, 43, *44*
Siegelman, William, 788
Sierra Madre Occidental, 1137
Sieve plates, *714*, 715, 735
Sieve tube elements
companion cells, 735
pressure flow model of translocation, 735–738
structure and function, 714–715, 735
Sifaka, *10*
Sigma factors, 286, 333
*Sigmoria trimaculata*, *670*
Signal amplification, 133–134, 138
Signal peptides, 298–300
Signal sequences, 298–300, 384
Signal transduction pathways
cancers and, 131, *132*
in cell fate determination, 397–398
crosstalk, 127, 134
defined, 126
effects on cell function, 137–139
elements of, 126–127
initiation of DNA transcription, 139
in innate defenses, 860
protein kinase cascades, 131–132
receptor proteins, 127–131
regulation of, 136–137
second messengers, 131, 132–136
signal amplification, 133–134, 138
types and sources of signals, 126
*See also* Plant signal transduction pathways
Signals
acoustic signals, 1112
autocrine signals, 126, 835
cell responses to signal molecules, 125
effects on cell function, 137–139
electric signals, 485
error signals, 816
functions of signal proteins, *42*
honest signals, 435
initiation of DNA transcription, 139
juxtracrine signals, 126
mechanosensory signals, 1112
paracrine signals, 126, 835
types and sources of intercellular signals, 126
visual signals, 1111
*See also* Chemical signals; Reproductive signal
Sildenafil (Viagra), 136, 137
Silencers, of transcription factors, 335
Silent mutations, 305, 306, *306*
Silent substitutions, 491
*See also* Synonymous substitutions
Silica, 555, 1175
Silicates, 71
Silicon dioxide, 643
Silkworm moth, 59, 841, 950
Silurian period, *506–507*, 520
Silver gull, 412
Silver salts, 770
Silverfish, 671, *672*
Silverswords, 481–482
Simberloff, Daniel, 1198, *1199*
Similarity matrix, 487
Simple diffusion, 114, *118*
Simple fruit, 601
Simple sugars. *See* Monosaccharides
Simultaneous hermaphroditism, 887–888
Single nucleotide polymorphisms (SNPs)
description of, 317
haplotype mapping, 367–368
in the human genome, 366
human genome scans and diseases, *368*
pharmacogenomics and, 368
using to find disease-causing genes, 318, *319*
Single-strand binding proteins, 272
Sink strength, 736–737
Sinks, in phloem translocation, 714, 734, 736–737
Sinoatrial node, 1032–1033, 1034, *1035*
Sinus venosus, 1028
Siphonaptera, *672*
*Siphonops annulatus*, *691*
Siphons, of cephalopods, 662, *663*, 664
Sirenians, *698*
Sirius Passet, 516
siRNAs. *See* Small interfering RNAs
Sister chromatids
centrosomes, 212–213
chromatin structure, 211–212, *213*
comparison between mitosis and meiosis, *223*
defined, 207
events in meiosis, *220–221*
separation in meiosis II, 219, 222
separation in mitosis, 214–215
Sister clades, 451
Sister species, 451, 472
Sit-and-wait predators, *638*
Situs inversus, 902, 915
Skates, 444, 687, *688*
Skeletal muscle
antagonistic sets, 942
effect of ATP supply on performance, 997
effect of heat on fatigue, 998

factors affecting the strength of muscle contraction, 994–995
functions of, 817, 987
impact of exercise on strength and endurance, 996–997
interaction with bone at joints, 1001–1002
jumping ability in animals, 986
in the knee-jerk reflex, 942
motor units, 989
muscle spindles, 952–953
neuromuscular junctions, 936–938
optimal length for generating maximum tension, 996
paralysis during REM sleep, 979
role of muscle fiber types in strength and endurance, 995–996
sliding filament contractile mechanism, 987–990, *991*
structure of, 987–988

Skeletal systems
interactions with muscle, 999, 1001–1002
types of, 999–1001
*See also* Endoskeletons; Exoskeletons; Internal skeletons

Skin
in amniote evolution, 692
blood flow and thermoregulation, 824–825, 828
epithelial tissues of, 817
gas exchange in amphibians, 1029
"heat portals," 831
of lepidosaurs, 693
tactile receptors, 952

Skin beetles, 1203
Skin cancer, 277, 382
Skin cells, induced pluripotent stem cells from, 409
Skin pigmentation
*MC1R* gene and, 367
vitamin D and, 1054

Skinner, B. F., 1094
Skoog, Folke, 768
Skull cap, 1001
Skulls, in humans and chimpanzees, 704
Sleep, 978–980
Sleeping sickness, *559*
Sliding DNA clamp, 273–274
Sliding filament model, 987–990, *991*
Slime, of roots, 716
Slime molds
cellular, 561
chemical signaling in, 838
plasmodial, 560–561

Sloths, *698*
Slow block to polyspermy, *885*, 886
Slow-twitch fibers, 995
Slow-wave sleep, 978
Slug (cellular slime molds), 561
Slugs, 662
Small ground finch, *474*
Small interfering RNAs (siRNAs)
description of, 347–348
location and role in eukaryotic cells, *286*
in plant systemic acquired resistance to RNA viruses, 801
in RNA interference, 382

Small intestine
absorption of nutrients in, *1062*, 1063
digestion in, 1061–1062
digestive enzymes of, *1062*
in humans, *1058*
movement of stomach contents into, 1061
production of secretin, 1065
sections of, 1061
segmentation movements, 1060

Small nuclear ribonucleoprotein particles (snRNPs), 292
Small nuclear RNA (snRNA), *286*, 292
Small populations
impact of genetic drift on, 434
noncoding DNA in the genome and, 496

Small tree finch, *474*
Smallmouth bass, 1160
Smallpox virus, 856
Smell
description of, 949–950
signal transduction in, 137–138

Smith, Hamilton, 356
Smithies, Oliver, 382
Smoke
breaking of seed dormancy, 757, 1121
*See also* Cigarette smoke

Smoker's cough, 1015
Smooth endoplasmic reticulum (SER)
in animal cells, *86*
conversion of nitrites to nitrosamines, 310
in plant cells, *87*
in plasmodesmata, *139*, 140
structure and function, 89–90

Smooth muscle
acetylcholine-stimulated relaxation, 135–136
contraction, *992*, 993
functions of, 817, 987, 993
in the gut, 820, 993, 1058, *1059*
peristalsis, 1059–1060
structure, 993
vascular (*see* Vascular smooth muscle)

Smut fungi, 622
Snails, 662
Snakeflies, *672*
Snakes
features of, 693, *694*
hemipenes, 888
loss of limbs, 423
ovoviviparity, 889
pit organs and infrared perception, 946
vomeronasal organ, 950–951

"Snowball Earth" hypothesis, 515–516
SNPs. *See* Single nucleotide polymorphisms
snRNA. *See* Small nuclear RNA
snRNPs. *See* Small nuclear ribonucleoprotein particles
Soap, 144
Social behaviors
in amphibians, 692
evolution of, 1113–1117

Social organization, effect on characteristic species density, 1160
Sociobiology, 1116–1117
*Sociobiology* (Wilson), 431
*Socratea exorrhiza*, 1194
Sodium
in animal nutrition, *1052*
electronegativity, *28*
ionic attraction, 28–29
uptake by halophytes, 811

Sodium bicarbonate, 36
Sodium channel genes
evolutionary studies, 485, 500, 502
gene duplication, 498

Sodium chloride, 29
*See also* Salt and water balance regulation

Sodium hydroxide, 34
Sodium ion channels
in action potentials, 932–935
blocking by TTX, 500
in cardiac pacemaker cells and heart contraction, 1033–1034
evolutionary studies, 485, 500, 502
membrane depolarization and, 930–931, *932*
in rod cells, 961
in taste bud sensory cells, 951
TTX resistant, 445, 500
*See also* Acetylcholine receptors; Voltage-gated sodium channels

Sodium ion transporters, 1063
Sodium ions
absorption in the small intestine, 1063
in cardiac pacemaker cells and heart contraction, 1033–1034
generation of action potentials and, 932–934
ionic electric current and, 931–932
properties of, 29
salt tolerance in plants, 388

Sodium tripolyphosphate (STPP), 1221
Sodium–potassium pump
action potentials and, 933
in the generation of membrane potentials, 928, 929
in primary active transport, 118–119

Soft-shelled crabs, 999
Soil bacteria
global nitrogen cycle, 750–751
influence on plant uptake of nutrients, 747–748, 748–751
interference competition, 1183
nitrogen fixation, 749–750

Soil fertility
defined, 745
factors determining, 746

Soil fungi, 612, *613*
Soil solution, 741
Soils
adding fertilizers to, 746–747
availability of nutrients to plants, 746
formation of, 745–746
heavy metals and plants, 811–812
leaching, 745, 746
optimal pH for plants, 746
saline, 810–811
structure, 745
of tundra, 1128

Solar radiation
atmospheric circulation patterns and, 1124
geographic distribution and ecosystems, 1208–1210
impact on community productivity, 1192
impact on development, 421–422
photosynthetic efficiency of plants, 201–202
variation in input across Earth's surface, 1123

Sole, 444
Soleus, 995
*Solidago*, *752*
Solute potential, 727, *728*, 736
Solutes
accumulation in xerophytes, 808
defined, 33
uptake by plants, 727–730

Solutions
buffers, 35–36
concentration of gases in, 1006
osmolarity, 1072
pH, 35
properties of aqueous solutions, 33–34
water potential, 727, *728*

Solvents, 33
Somatic cell gene therapy, 323
Somatic cell nuclear transfer experiments, 406–407
Somatic cells
chromosomes in, 218
cloning animals from, 406–407
harmful consequences of mutations in, 310–311

Somatic mutations, 305, 310–311, 314
Somatosensory cortex, 983
Somatostatin, *842*, 844, 848, 849
Somites, 916, *917*
Songbirds
factors affecting song acquisition, 1099–1101
hormonal control of song expression, 1101

*Sonic hedgehog (Shh)* gene, 423
Sonic hedgehog (Shh) protein, 401, 917
Soredia, 613–614
Sorghum, 805
Sori, 582
Soricomorpha, *698*
Sound
in animal communication, 946, 1112
definition of, 946
perception of (*see* Auditory systems)

Sour taste, 951

Sources, in phloem translocation, 714, 734, 736
South Africa. *See* Fynbos
South Georgia Island, 1149, 1166
Southern beeches, *1129,* 1130, 1143
Southern pine bark beetle, 1178
Sow bugs, 670
Soy sauce, 624
Soybeans, 624
Space-filling models, *27,* 47
Spaceship Earth, 1245
Spanish ibex, *1134*
Spatial heterogeneity hypothesis, of latitudinal gradients in diversity, 1196
Spatial summation, 938
Spawning, 882, 887
Specialization hypothesis, of latitudinal gradients in diversity, 1196
Speciation
  from barriers to gene flow, 472–475
  defined, 468
  divergence of populations, 6
  diversification of mating behaviors and, 705
  factors affecting rates of speciation, 480–482
  genetic basis of, 470–471
  hybrid zones, 478–479
  laboratory experiments with *Drosophila,* 482
  Lake Malawi cichlids, 467
  lineage species concept, 469
  mechanisms preventing hybridization, 475–478
  reproductive isolation and, 468–469
Species
  characteristic densities, 1159–1160
  concepts of, 468–469
  defined, 468
  estimated number of living species, 514, 1230
  evolution of populations, 6
  evolutionary perspective on comparing, 451–452
  evolutionary tree of life, 6–9
  genetic variation in geographically distinct populations, 443–444
  making comparisons between, 7
  mechanisms preventing hybridization, 475–478
Species abundance, role of evolutionary history in, 1160
Species concepts, 468–469
Species diversity
  community productivity and, 1192
  contributions of species richness and species evenness to, 1195–1196
  decrease through time in detritius-based communities, 1202
  impact of disturbances on, 1199
  island biogeography theory, 1196–1198, *1199*
  latitudinal gradient in, 1196, 1197
Species evenness, 1195
Species extinctions
  biodiversity loss and, 1229–1230
  "centers of imminent extinction," 1237, *1238*
  from human activity, 1229–1230
  from invasives, 1235–1236
  from overexploitation, 1234–1235
  predictors of, 1231–1232
  from rapid climate change, 1236–1237
  *See also* Mass extinctions
Species immigration, in island biogeography theory, 1196–1198, *1199*
Species interactions
  categories of, 1170–1171
  coevolution and, 1171–1172
  competition, 1182–1185
  evolution of antagonistic interactions, 1172–1177
  existing as a continuum, 1171
  herbivory, 1175–1176
  impact on communities, 1193–1195
  mutualisms, 117–1182
Species names, 462
Species pool, 1197, 1198
Species richness
  defined, 1195
  habitat loss and, 1232
  impact on community stability, 1202–1203
  species diversity and, 1195–1196
  using as a criterion for protected areas, 1237
  wetlands restorations and, 1239, *1240*
Species trees, 529–530
Species–area relationship, 1197
Specific heat, 32
Specific plant immunity
  defined and described, 799
  gene-for-gene resistance, 800
  hypersensitive response, 800–801
  phytoalexins, 799–800
  systemic acquired resistance, 801
Spemann, Hans, 910–911, 912
Sperm
  of conifers, 594, *595*
  contributions to the zygote, 903
  double fertilization in angiosperms, 600–601, *781,* 783
  fertilization in animals, 884–887
  fertilization in humans, 892
  gametic isolation and, 477
  genomic imprinting in mammals, 344–345
  of mosses, 576
  production in animals, 882–884
  production in humans, 889–890, *891*
  release in spawning, 887
  reproductive technologies in humans, 897, 898
  of seed plants, 589
  in semen, 889
  transfer in internal fertilization, 887
  *See also* Sperm–egg interactions
Sperm cells, in angiosperm sexual reproduction, 780, *781,* 783
Spermatids, *883,* 884, 889, *891*
Spermatocytes, 883–884, 889, *891*
Spermatogenesis
  in animals, 883–884
  hormonal regulation in humans, 892
  in humans, 889–890, *891*
Spermatogonia, 883, 889, *891,* 908
Spermatophores, 887
Spermatozoa, 889, *891*
Sperm–egg interactions
  activation of development in animals, 903–904
  blocks to polyspermy, *885,* 886–887
  specificity in, 884–886
Spermicidal jellies, *898*
*Spermophilus beldingi, 1117*
*Spermophilus parryii, 10*
Sperry, Roger, 980
*Sphaerechinus granularis, 682*
*Sphagnum,* 578
*Sphenodon punctatus, 694*
Spherical symmetry, 634
Sphincter muscles, *1059,* 1060, 1061, 1080
Sphinx moth, *588*
Sphygmomanometer, 1032, *1033*
Spicules, *633,* 643, 644
Spicy/hot taste, 951
Spider beetles, 1201, 1203
Spider monkeys, *702*
Spider silk
  bioengineering of, 59
  properties of, 39
  protein structure, 39, 45, 46
Spiders
  deprivation experiments, 1095
  external digestion of food, 1055
  nervous system, 941
  sperm transfer in spermatophores, 887
  webs, 669
Spiegelman, Bruce, 839
Spike mosses, 581
Spikelets, *597*
Spikes (inflorescence), *597*
Spina bifida, 916
Spinal cord
  development, 968
  early development, 916
  functions of, 969
  nerves of the autonomic nervous system, *974,* 975
  reflexes, 942–943
  structure and function, 941–942
Spinal cord injuries, 970
Spinal cord transection, 969
Spinal nerves, 942, 968
Spinal reflexes, 942–943, 969, 1080
Spindle apparatus, 212, 213–214, *215*
  *See also* Mitotic spindles
Spindle assembly checkpoint, 215
Spindle cells, of the insular cortex, 983
Spines
  of cacti, 807
  in plant defenses against herbivory, 1175
  of sea urchins, 681
Spinner dolphins, *700*
Spiny-headed worms, 657–658
Spiracles, 671, *672,* 1009
Spiral cleavage, 633, 653, 905
Spiral valve, 1057
Spiralians, 633, 653
  *See also* Lophotrochozoans
Spirillum, 528, *529*
*Spirobranchus, 638*
Spirochetes, 533
*Spirographis spallanzanii, 661*
Spirometer, 1012
Spleen, *857,* 1038
Spliceosomes, 292
"Split-brain" studies, 980–981
Sponges
  in animal phylogeny, *630*
  asymmetry in, 634
  cell adhesion and cell recognition in, 110, 111
  description of, 643–644
  digestion in, 1056
  filter feeding, 637
  major subgroups and number of living species, *632*
  respiratory gas exchange, *1007*
  similarity of choanoflagellates to, 631, *633,* 644
  as the sister group to all animals, 648
Spongy mesophyll, 720, *721*
Spontaneous abortion, 897
Spontaneous generation, 67–68
Spontaneous mutations, 308–309
Sporangia
  of club mosses, 581
  of ferns, 582
  of land plants, 575
  of liverworts, 577
  of slime molds, 561
  of zygospore fungi, *618,* 619
Sporangiophores, *618,* 619
Spore walls, in plant evolution, 574
Spores
  in alternation of generations, 575
  of cyanobacteria, 532, *533*
  dispersal in liverworts, 577
  in heterospory, 584–585
  in homospory, 584
  of seed plants, 590
  of sporocytes, 563
  of zygospore fungi, *618,* 619
Sporocytes, 563
Sporophytes
  of angiosperms, 596, 600
  of conifers, *595*
  of ferns, 582
  in homospory, 584
  of hornworts, 578, 579
  of land plants, 575
  of liverworts, 577
  of mosses, 578
  of nonvascular land plants, 575, 576–577
  relationship to gametophytes in plant evolution, *590*
  of seed plants, 590, 591
  of vascular plants, 579
Sporopollenin, 590
Sporozoites, *564*
Sporulation, 562
Spring wheat, 791
Springtails, 671, *672n*

Spruces, 1200
Squamates, 693
Squamous cells, *818*
Squid giant axons, 929, *930*
Squids
  eye control gene, *414*
  features of, 662–663
  sperm transfer in spermatophores, 887
Srb, Adrian, 283, 284
*SRY* gene, 250
SRY protein, 250
St. Johnswort, 1175
St. Matthew Island, 1149, 1166
Stabilizing selection, 439
Stable atoms, 25
Stage-dependent cohort life tables, 1154
Stahl, Franklin, 268, *269*, 270
Stained bright-field microscopy, 80
Stamen, *591*, 596, 598, *779*
Standard free energy
  from glucose oxidation, 149, 166
  from the oxidation of NADH, 168
Stapes, *954*, 955
*Staphylococcus*, 531, 608
*Staphylococcus aureus*, 281, 301, 531
Star anise, 602
Starch branching enzyme 1 (SBE1), 237
Starch grains, 53
Starches
  in cassava, 708
  conversion to ethanol by yeast, 623
  production in photosynthesis, 195–196
  structure and function, 53, *54*
Starfish, 681, 682
  *See also* Sea stars
Starling, Ernest, 838, 1041
Starling's forces, 1041
Stars, navigation by, 1109–1110
Start codon, 289, 295, 296
Starvation, 1050
Statistics, 13–14
Stele, 717, *718*
Stellate barnacles, 1184
Stem cell therapy, 102, 392
Stem cells
  defined, 381, 408
  mesenchymal, 392, 394, 408
  multipotent, 408, 410
  pluripotent, 408–409
  potential medical uses, 77
  *See also* Embryonic stem cells
Stem elongation
  gibberellins and, 761
  inhibition by cytokinins, 768
Stems
  aerenchyma, 808
  annual rings, 722–723
  apical dominance, 765
  cuttings, 793
  effect of ethylene on growth, 770
  function, 709, 720
  modified, 720
  primary indeterminate growth, 719–720
  secondary growth, 721–723
  thickening in monocots, 723
  vegetative reproduction in angiosperms, 792–793
*Stenella longirostris*, *700*
Steno, Nicolaus, 506
Steppe, 1131
Steppuhn, Anke, 804
Stereocilia, 953–954, 955
Sterilization, as a method of contraception, *898*
Sternum, 695
Steroid hormones
  of the adrenal cortex, 849–850
  characteristics of, 836
  ecdysone, 840–841
  in plants, 771
  structure and function, 58
  synthesis pathway, *836*
  *See also* Sex steroids
Steward, Frederick, 405
Stewart, Caro-Beth, 492–493, 494
Stick insects, *672*, 673
Stick model, of protein tertiary structure, 47–48
Sticky ends, 374
Stigmas
  in flower structure, *591*, 597
  germination of pollen grains, 780, 782
  retraction response in bush monkeyflowers, 598, *599*
Stilt palm, 1194
Stingrays, *445*, 687
Stinkhorn mushrooms, *2*
Stock, in grafting, 793, *794*
Stolons, 792
Stomach
  buffering of acid, 36
  chemical digestion in, 1060–1061
  digestive enzymes, *1062*
  function of mucosal epithelial cells, 1058
  in humans, *1058*
  pH in humans, 161
  production of gastrin, 1065
  production of ghrelin, 1067
  release of chyme into the small intestine, 1061
  of ruminants, 1064
  segmentation movements, 1060
  in tubular guts, 1056
Stomata
  in CAM plants, 734
  closure in response to drought stress, 809
  control of water loss and carbon dioxide uptake, 732–734
  dermal tissue system origin, 712
  functions of, 198, 721
  in mosses, 577
  in plant evolution, 574
  plant regulation of number and function, 734
  in xerophytes, 807
Stomatal crypts, 807
Stone cells, 714
Stoneflies, *672*, 673
Stoneworts, 572–573
Stop codons
  in the genetic code, 289
  nonsense mutations, 306, *307*
  in translation, 296–297, *298*
Storage proteins, *42*
Strabismus, 921
Stramatolites, 539
Stramenopiles, 555–556, *557*
Strata, 506
Stratified epithelium, *818*
Stratigraphy, 506
Stratosphere, 1211, *1212*
Strawberries, 601
Strawberry plants, 720, 792
Streams, 1140
Strength, impact of exercise on, 996
Strength training, impact on muscle, 996
Strepsipterans, *672*
*Streptococcus pneumoniae*, 260–261
Streptokinase, 385
*Streptomyces*, 532
Streptophytes, 572, 573
Stress, during pregnancy, implications of, 328
Stress response
  cortisol and, 849–850
  in plants, 337
Stress response element (SRE), 337
Stretch receptors
  in blood pressure regulation, 1044, *1045*
  function in crayfish, 947
  in the knee-jerk reflex, 942
  regulation of blood pressure, 1088–1089
Striated muscle. *See* Skeletal muscle
*Striga*, 626, 753
Strigolactones, 747, *748*, 753
Strobili, 581
Strokes
  atherosclerosis and, 1042–1043
  treatment with TPA, 385
Stroma
  carbohydrate synthesis in, 193–197
  light-independent reactions, *188*
  light-induced pH changes, 196
  structure and function, 92, *93*
Stromatolites, *5*, 512, *513*
STRs. *See* Short-tandem repeats
Structural genes, regulation in prokaryotes, 330–332, *333*
Structural isomers, 41
  of hexoses, 52
Structural motifs, 335–336
Structural proteins, *42*
*Struthio camelus*, *695*
Styles, *591*, 597, 782
Subclavian artery, *1031*
Subclavian vein, *1031*
Subduction, 509
Suberin, 717, 798
Sublingual salivary gland, *1058*
Submandibular salivary gland, *1058*
Submucosa, 1058, 1059
Subpopulations, 1161, 1162
Subsoil, 745
Substance P, 940
Substrate-level phosphorylation, 170
Substrates
  effect on reaction rate, 156
  in enzyme-catalyzed reactions, 152–153
  enzyme interactions with, 154–155
Succession, 1201–1203
Succinate dehydrogenase, 172
Succinic acid, 70
Succulence, 807
Succulents
  absence in Australia, 1126
  adaptations to very dry conditions, 807
  crassulacean acid metabolism in, 200
  in hot desert biomes, *1132*, 1133
  vegetative reproduction, 793
Suckers (of leeches), 661
Suckers (of plants), 793
"Sucking chest wound," 1016
Sucrases, 152, *1062*
Sucrose
  phloem translocation, 734–735
  production in plants, 195, 201
  structure, *53*
Sudden Acute Respiratory Syndrome (SARS), 501
Sugar apples, 785
Sugar beets, 718
Sugar phosphates, *55*
Sugarcane, *603*
Sugars
  formation of glycoproteins, 301
  formed in prebiotic synthesis experiments, 70
  found on meteorites, 69
  phloem translocation, 734–738
Sulci, 970
Sulfate ions
  inorganic fertilizers, 747
  leaching in soils, 746
Sulfated polysaccharides, 111
Sulfhydryl group, *40*
Sulfolipids, 92
*Sulfolobus*, *2*, 536
Sulfur
  in animal nutrition, *1052*
  covalent bonding capability, *27*
  in plant nutrition, *741*
  produced by photoautrophic bacteria, 538
  radioactive isotope in the Hershey–Chase experiment, 262–263
Sulfur cycling
  global cycle, 1219–1220
  prokaryotes and, 539
Sulfur dioxide
  in the global sulfur cycle, 1219
  from volcanoes, 510
Sulfuric acid
  acid precipitation and, 1219–1220
  properties of, 34
Sun, Yuxiang, 182
Sundews, 751, *752*
Sunflowers, 811
Sunlight. *See* Solar radiation
"Superbugs," 357, 373
Superficial cleavage, 905
Superior vena cava, 1030, *1031*
Superoxide, 175, 176
Superoxide dismutase, 176
Suprachiasmatic nuclei (SCN), 1108
Surface area-to-volume ratio, of cells, 78–79

Surface runoff
ecological impact of chemical fertilizers, 740, 1207, 1219, 1221
movement of elements and, 1213
Surface tension
lung surfactants and, 1015
of water, 33
Surfactants, 1015
Survival, evolution by natural selection and, 6
Survivorship, 1154, 1156
Survivorship curves, 1155–1156
Suspension feeders, 637, *638*
Suspensor, 711
Sustainable management, 1224
Sutherland, Earl, 132–133
*Sutterella,* 540
Svedberg unit, 364
Swallowing, 1059
Swamps, 1140
Swarm cells, 561
Sweat glands, 105, 697
Sweating, 105, 829
Sweet potato, 718
Sweet potato–corn dicultures, 1203
Sweet taste, 951
Sweet wormwood, 797
*SWI1* gene, 794
Swim bladders, 687
Swine flu, 877
Swordtails, 459
*Sycon, 643*
Symbiotic interactions
defined, 102, 612
dinoflagellate endosymbionts in corals, 565, 566
evolution of nitrogen-fixation in legumes, 521
hornworts and cyanobacteria, 579
between plants and soil bacteria, 747–748
*Vibrio* and Hawaiian bobtail squid, 546
Symmetry
in animal body plans, 634–635
pentaradial, 680, *681*
*See also* Bilateral symmetry; Radial symmetry
Sympathetic division
in blood pressure regulation, 1044, *1045*
effect on heartbeat, 1034
influence on smooth muscle, 993
structure and function, 974–975
Sympatric speciation, 473, 475
Symplast, 729, 730, 744
Symplastic pathway, 736
Symporters, 118, 1063
Synapomorphies, 452
Synapses
clearing of neurotransmitter, 940
function of, 926
functions of astrocytes at, 926–927
inhibitory or excitatory, 938
ionotropic and metabotropic, 938–939
long-term potentiation and long-term depression, 982
neuromuscular junctions, 936–938
number in the human brain, 943
summation of synaptic input by the postsynaptic cell, 938
tripartite, 927
types of, 926, 936
Synapsis, 219, *220*
Synaptic cleft, 936, *937*
*Synaptula, 682*
"Synchronous" muscle, 998
Syncytium, 905
Synergids
in angiosperm sexual reproduction, 779, *781*
degeneration of, 783
pollen tube growth and, 782
Synonymous substitutions
defined, 491
effect of modes of selection on substitution rates, 492–494
Synthetic cells, 359, *360*
Synthetic DNA, 380
Synthetic hormones, in birth control pills, 896
Syphilis, 533
Systematics, 451
Systemic acquired resistance, in plants, 801
Systemic circuit
in amphibians, 1029
in birds and mammals, 1030
blood vessels of, 1027–1028
defined, 1027
in lungfish, 1028
in reptiles, 1029–1030
Systemic lupus erythematosis (SLE), 876
Systems biology, 157, 181
Systole, 1031–1032, 1039
Systolic pressure, 1032, *1033*
Szostak, Jack, 73

## T

T cell receptors
binding of antigens to, 871–872
in the cellular immune response, 864, *865*
function of, 859
in the humoral immune response, 864, *865*
specificity of adaptive immunity and, 862–863
structure, *871*
T cells
as antigen-presenting cells, 864
binding of T cell receptors to antigens, 871–872
in the cellular immune response, 863, 864, *865*
clonal deletion, 865
clonal selection, 865, *866*
cytokines, 859
in delayed hypersensitivity, 876
effector T cells, 865, 866, 871, 872, *873*, 874
function of, *858*
interaction with antigen-presenting cells, 872
maturation in the thymus, *857*
memory T cells, 865
selection in the thymus, 872
T DNA, 377
T-helper ($T_H$) cells
in the cellular immune response, 864, *865*, 872, *873*
in class switching, 870–871
in delayed hypersensitivity, 876
development of plasma cells and, 867
in HIV infections, 876, *877*
in the humoral immune response, 864, *865*, 872, *873*
regulation by Tregs, 874
T tubules, 989, 990, 992
T2 phage, 261–263
T4 phage, *543*
T7 bacteriophage, 316
*Tachyglossus aculeatus, 697*
Tadpole shrimp, *670*
Tadpoles, 678
*Taeniopygia guttata,* 1100–1101
*Taeniura lymma, 445*
Taiga, 1129–1130
Tamoxifen, 325
Tapeworms, 638, *641*, 657, 675, 888, 1176
Taproots, 710, 718, 808
Tarantulas, *669*
*Taraxacum officinale, 1153*
Tardigrades, *632*, 667–668
*Taricha granulosa, 445*
*Tarsius pumilus,* 1230–1231
Tarweeds, 481–482
Tasmanian Devil Genome Project, 255
Tasmanian devils, 232, 245, 255
Taste, 951
Taste buds, 951
Taste pore, *951*
Tat protein, 342
TATA box, 333–334, *335*
Tatum, Edward, 282–283
Taxi drivers, 967
Taxol, 96, 604
Taxon (taxa)
biological nomenclature, 462, 463–464
defined, 451
Linnaean classification, 462–463
monophyletic, 463
Taxonomy, 462–464
*Taxus brevifolia,* 604
Tay-Sachs disease, 91
Tcf-3 transcription factor, 912
Tectorial membrane, *954*
Teeth
enamel, 21
in mammals, 697
plaque, 539, *540*
structure and function, 1055–1056
*Tegeticula yuccasella, 1181*
Telencephalon, 968, 969, 970–973
Teleomeres, 275
Telomerase, 229, 275
Telophase (mitosis), 212, *215*, 216, *217*
Telophase I (meiosis), *221, 223*
Telophase II (meiosis), *221*
*Teloschistes exilis, 613*
Temperate forests
deciduous, 1126, 1130–1131
global climate change and, 1217–1218
Temperate grasslands, 1131–1132
Temperature
acclimatization to, 821
boreal and temperate evergreen forest biomes, *1129*
chaparral biomes, *1134*
cold desert biomes, *1133*
effect on diffusion, 114
effect on enzymes, 161–162
effect on living organisms, 820–821
effect on protein structure, 50
effect on terrestrial biomes, 1126
highest temperature compatible with life, 535
hot desert biomes, *1132*
impact of solubility of gases in liquids, 1006
impact on respiratory gas exchange for aquatic animals, 1007, *1008*
induction of flowering and, 791
influence on sex determination, 420–421
membrane fluidity and, 107
plant adaptations and responses to temperature extremes, 810
temperate deciduous forest biome, *1130*
temperate grassland biome, *1131*
thorn forest and tropical savanna biomes, *1135*
tolerance of extremes in fungi, 611
tropical deciduous forest biomes, *1136*
tropical rainforest biomes, *1137*
tundra biomes, *1128*
Walter climate diagrams, 1138
Temperature-dependent sex determination, 420–421
Temperature-sensitive mutations, 306
Temperature sensitivity ($Q_{10}$), 821
Templates
in DNA replication, 268, *272*
in the PCR reaction, 277, *278*
telomeric, 275
in test tube synthesis of DNA, 267
Temporal isolation, 476
Temporal lobe, 971
Temporal summation, 938
Tendons, 953, 1002, 1003
Tennessee shiner, *472*
Tension
generated by skeletal muscle fibers, 996
*See also* Transpiration–cohesion–tension mechanism
Tentacles
of cephalopods, 664
of lophophores, 652
Teosinte, 419, 723–724
*Teosinte branched 1 (tb1)* gene, 723–724
*Ter* site, 206
Terminal buds, 709

Termination, of transcription, *287*, 288
Termites, *672*, 673, *1135*
Terpenes, *802*
Terrestrial biomes. *See* Biomes
Territorial behavior, 1103–1104, *1105*
Territorial calls, 1112
Tertiary consumers, 1190, *1191*
Tertiary endosymbiosis, 552
Tertiary period, *506–507*, *519*, 522
Tertiary protein structure, 46–49
Test crosses, 237, *238*
Testate amoebas, 560
Testes
  hormone of, *842*
  in humans, 889, *891*
  spermatogenesis in, 882–884, 889–890, *891*
Testicular cancer, 259
Testosterone
  actions of, *842*
  as an androgen, 850
  in control of song expression in songbirds, 1101
  in follicle selection for ovulation, 895–896
  in human puberty, 892
  sexual behavior in rats and, 1099
  spermatogenesis and, 889, 892
  structure, *836*
  temperature-dependent sex determination and, 420
  territorial behavior in lizards and, 1103, *1104*
  *See also* Sex steroids
Tetanus (disease), 542
Tetanus (muscle contraction), 994–995
Tetanus toxin, 936
Tetracycline, 281, 295
Tetrads, 219, 247–249, *250*
Tetrahydrofolate, 158
Tetraiodothyronine ($T_4$), 845, 846, 847
*Tetraodon nigroviridis*, *361*
Tetraploids, 224–225, *226*
Tetrapods
  limb evolution, 690
  modifications to pharyngeal slits, 684
Tetrodotoxin (TTX), 445, 500
Texas Longhorn cattle, 440
TFIIB protein, 334
TFIID protein, 334, *335*
TFIIE protein, 334
TFIIF protein, 334
TFIIH protein, 334
*Tga1* gene, 419
Thalamus
  functions of, 969, 970
  during sleep, 979
  in visual processing, 975, 977
*Thalassarche chrysostoma*, 1109–1110
β-Thalassemia, 292–293
Thale cress. *See Arabidopsis thaliana*
Thaliaceans, 684
Thalidomide, 920
Thalloid liverworts, 577
Thallus, 613
*Thamnophis sirtalis*, *445*
Thecal cells, 895–896
Therapeutic abortion, 897
Therapeutic genes, 323–324
Therians, 697–700, 889
Thermal insulation, 825, 828
Thermal limits, 820
Thermocline, 1214
Thermodynamics, laws of, 146–147
Thermogenin, 827–828
Thermoneutral zone, 826
Thermophiles, 532, 536
*Thermoplasma*, 537
Thermoreceptors, 947, *948*
Thermoregulation
  behavioral, 822, *823*
  conservation of metabolic heat in "hot" fish, 825
  control of blood flow to the skin, 824–825, 828
  energy budgets and, 823–824
  metabolic heat production in ectotherms, 825–826
  production of metabolic heat, 822
  role of the hypothalamus in mammals, 829–830
  strategies in ectotherms, *824*
  strategies in endotherms, *824*, 826–831
*Thermotoga*, 529, 532
*Thermus aquaticus*, 277–278, 532
Theropods, 694, 695
Thiamin, *1053*, 1054
Thioredoxin, 196–197
Thistles, 784, 1201
*Thlaspi caerulescens*, 811
Thoracic cavity, *1014*, 1015–1016
Thoracic ducts, *857*, 858, 1042, 1063
Thoracic vertebrae, 418
Thorn forest, 1135–1136
Thorns, in plant defenses against herbivory, 1175
"Threatened" species, 1231
Three-chambered hearts, 1029
Three-dimensional vision, 977–978
3′ End, 266
Three-spined sticklebacks, 423–424
Three-toed sloth, *1137*
Threonine, *44*
Threshold membrane potential, 933
"Thrifty genes," 1048
Thrips, *672*, *673*
Thrombin, 1039
Thrombus, 1042, 1043
Thylakoids
  electron transport systems, 191–192
  light reactions, *188*
  photophosphorylation and ATP synthesis, 192–193
  photosystems, 190
  structure and function, 92, *93*
Thymine
  complementary base pairing, 63–65
  dimerization by UV light, 277
  in DNA structure, 264, 265, 266, *267*
  effect of ultraviolet radiation on, 309
  formed from cytosine deamination, 310
  structure, *63*
Thymine dimers, 277
Thymosin, *842*
Thymus
  clonal deletion in, 865, 872
  hormones of, *842*
  in the lymphatic system, *857*
  maturation of regulatory T cells in, 874
  T cells selection in, 872
Thyroglobulin, 845, *846*
Thyroid gland, 320, *842*, 845–847
Thyroid-stimulating hormone (TSH), 846, 847, 1101
Thyrotropin, *842*, 843, 846
  *See also* Thyroid-stimulating hormone
Thyrotropin-releasing hormone (TRH), 844, 846
Thyroxine
  actions of, *842*
  goiter, 846–847
  half-life, 853
  production and regulation, 845–846
  structure, *836*
Thysanoptera, *672*
Thysanura, *672*
Ti plasmid, 377
Ticks, 638, 639, 668, 669
Tidal ventilation, 1012–1013
Tigers, *1172*, 1234, *1235*
Tight junctions, 111, *112*
Tijuana Estuary, 1239, *1240*
*Tiktaalik roseae*, *689*, 690
Till, James, 408
Tilman, David, 1202
Time-compensated solar compass, *1111*
Time hypothesis, of latitudinal gradients in diversity, 1196
Tinamous, 694
Tinbergen, Niko, 1095–1096
Tissue plasminogen activator (TPA), 384–385
Tissue-specific promoters, 384
Tissue systems. *See* Plant tissue systems
Tissue transplants, 874
Tissues
  biological hierarchy concept, 9
  diffusion within, 114
  relationship between cells, tissues, and organs, 817–820
  *See also specific tissue types*
Titin, 291, 988, 996
Toads
  characteristics of, 690–691
  hybrid zones, 479
  nitrogenous wastes excreted, 1075
Tobacco
  flowering cues from an "internal clock," 792
  Maryland Mammoth, 787
  nicotine in flower nectar, 1181
Tobacco hornworm moth, 348
Tobacco mosaic virus, 285
Tocopherol, *1053*
Toll-like receptors, 860
Tomatoes
  diseases of, *798*
  dwarfed phenotype, 760, *761*
  genetically modified for salt tolerance, 388, *389*
*Tomocerus minor*, *671*
Tongue, taste buds in humans, 951
Tonoplast, 710
Topography, impact on biomes, 1126–1127
Topsoil, 745, 1132
"Torpedo" embryo, *393*
Total peripheral resistance (TPR), 1043
Totipotency
  animal cloning, 406–407
  defined, 394
  plant cloning, 405–406
  in plant development, 710
Touch, 952
Toxigenicity, 542
Toxins
  *Bacillus thuringiensis* toxin in transgenic plants, 387–388, 389
  bacterial, 542
  of dinoflagellates, 549
  impact on synaptic proteins, 936
  in plant nectar, 1181
*Toxoplasma*, 554
*Toxostoma guttatum*, 1232
*TP53* gene, 306
TP53 protein, 306
TPA. *See* Tissue plasminogen activator
Trachea
  in birds, 1011, *1012*
  in humans, 1013, *1014*
Tracheae, 671, 1009
Tracheal system, 1009
Tracheary elements, 714
Tracheids
  description of, 714
  in the evolution of vascular plants, 579
  in gymnosperms, 593
  in vascular plants, 573
*Trachemys scripta*, 420
Tracheophytes, 573
  *See also* Vascular plants
Trade-offs
  in animal reproduction, 641–642
  constraints on evolution, 445
Trade winds, 1124, *1125*
Traits
  adaptation, 433
  ancestral and derived, 452
  defined, 234, 431
  dominant and recessive, 235
  in Mendel's monohybrid crosses, 234–236
  qualitative and quantitative, 439
  in sexual selection, 435–436
  sources of data for phylogenetic analyses, 454–456
  using to construct phylogenetic trees, 453, *454*
Transcription
  compared in prokaryotes and eukaryotes, *334*
  components needed for, 286
  differential gene transcription in differentiation, 398–399
  of DNA, 65
  effect of histone modifications on, 344
  error rates, 288

in gene expression, 284–285
gene expression in RNA genomes, 285
genetic code, 288–289
importance of base pairing to, 65, *67*
initiation by signal transduction, 139
initiation in eukaryotes, *335, 336*
noncoding sequences, 290–291
processing of gene transcripts before translation, 291–293
regulation in prokaryotes, 329–333
RNAs produced by, 286
role of mRNA in, 285
signals that start and stop, *297*
steps in, 286–288
stimulation by phytochromes in plants, 773–774
structure and function of RNA polymerases, 286
in the viral lytic reproductive cycle, 340
of X chromosome genes, 345–346

Transcription elongation, HIV gene regulation and, 341–343

Transcription factors
binding to promoters, 328
cell differentiation and, 336, *337*, 338, 399
in cell fate determination, 397–398
coordinated regulation of sets of genes, 336–337
determination of fruit fly body segmentation and, 401–405
enhancers and silencers, 335
as genetic switches, 415
as intracellular receptors, 130–131
plant organ identity genes and, 400
primary embryonic organizer and, 911–912
role in transcription, 286
in signal transduction pathways, 126–127
structural motifs and binding to DNA, 335–336
transcriptional regulation in eukaryotes and, 333–334, *335*

Transcriptional regulation
in eukaryotes, 333–338
in prokaryotes, 329–333

Transducin, 961
Transects, 1151
Transfection, 263, 376
*See also* Transformation

Transfer RNA (tRNA)
binding sites on ribosomes, 295
charging with an amino acid, 293, 294
location and role in eukaryotic cells, *286*
produced by transcription, 286
role in translation, 285, 293–294, 295–297, *298*
specificity in binding to an amino acid, 294
wobble, 294

Transformation
defined, 376
methods, 376–377
*See also* Genetic transformation

Transforming growth factor-β (TGF-β), 912
Transfusions, 243

Transgenic animals
cloning, 407
pharming, 385–386

Transgenic cells, 376

Transgenic crops
water-use efficiency, 726
overview, 386–388, *389*
public concerns, 388–389

Transition mutations, 306
Transition state, 151–152
Transition-state intermediates, 152

Translation
blocking to study gene expression, 382
elongation, 296, *297*
in gene expression, 284–285
initiation, 295–296
overview, 293
polyribosomes, 297, *298*
regulation in eukaryotes, 348.*23*
of RNA, 65
role of ribosomes in, 285, 294–295
role of tRNAs in, 285, 293–294, 295–297, *298*
signals that start and stop, *297*
termination, 296–297, *298*
in the viral lytic reproductive cycle, 340

Translocations, 224, 304, 308
Transmembrane domains, 108
Transmembrane proteins, 108, 112, *113*
Transmission electron microscopy, 81
Transpiration, 732
Transpiration–cohesion–tension mechanism, 731–732

Transport proteins
functions, *42*
in the small intestine, 1063

Transposons (transposable elements)
in eukaryotic genomes, 364–366
in the human genome, 366
as mutagens in minimal genome studies, 359, 360
mutations caused by, 308
as noncoding DNA, 495
in prokaryotic genomes, 358–359
small interfering RNAs and, 348

Transverse tubules. *See* T tubules
Transversion mutations, 306
Trastuzumab, 871
Travisano, Michael, 489–490
Tree ferns, 520, 580, *581*
Tree of life, 6–9, 451, 522
Tree shrews, *698*
Treg cells, 863, 874
*Treponema pallidum, 533*
Triassic period, *506–507, 518,* 521
Tricarboxylic acid cycle. *See* Citric acid cycle
*Trichinella spiralis,* 666
Trichinosis, 666
Trichocysts, 554, *555*
*Trichoglossus haematodus, 696*
Trichomes, 713, 807
*Trichomonas vaginalis,* 558
*Trichoplax adhaerens,* 629
Trichoptera, *672*
*Trifolium repens,* 443–444
Trigger hairs, 751–752

Triglycerides
absorption in the small intestine, *1062,* 1063
structure and function, 56–57
synthesis, 56

Triiodothyronine ($T_3$), 845, 846, 847
Trilobites, 668
Trimesters, of pregnancy, 919–920
Trimethylamine oxide (TMAO), 1073
*Triops longicaudatus, 670*
Triose phosphates, 195
Tripartite synapse, 927
Triploblastic animals, 633
Triploids, 224–225, 473, *475,* 783
Trisomics, 224
*Triticum aestivum, 226*
*Triticum monococcum, 226*
*Triticum turgidum, 226*
Tritium, 22
tRNA. *See* Transfer RNA
tRNA genes, 364
Trochophores, 640, 652, 653
Trophic cascades, 1193–1194

Trophic levels
in communities, 1190, *1191*
energy transfer between, 1190–1192

Trophoblast, 906, *907,* 919
Trophosome, 660
Tropic hormones, 843
Tropical alpine tundra, 1128

Tropical forests
deciduous, 1136–1137
evergreen, *1129,* 1130
keystone species, 1194–1195

Tropical rainforests
current rate of loss, 1232
description of, 1137–1138
estimating the number of insect species in, 651
fragmentation and habitat corridors, 1233–1234

Tropical savannas, 1135–1136

Tropics
atmospheric circulation patterns and, 1124
influence of the dry season on, 1126

*Tropidolaemus wagleri, 638*
Tropomyosin, 988, 990, *991*
Troponin, 988, 990, *991,* 998, 1034
Troposphere, 1211, *1212*
*Trp* operon, 331
True bugs (hemipterans), 480, *672,* 673, *674*
True flies, 672, *672, 674*
True navigation, 1109–1110
Truffles, 620
*Trypanosoma brucei, 559*
*Trypanosoma cruzi, 559*
Trypanosomes, 559
Trypsin, 144, 1062
Trypsinogen, 1062
Tryptophan, *44,* 331
TSH. *See* Thyroid-stimulating hormone
TSH receptors, 846–847
Tsien, Roger, 449
Tsunami of 2004, 1223
TTX-resistant sodium channels, 499, 500
Tuataras, 693, *694*
Tubal ligation, *898*
Tubal pregnancy, 906
Tube cell, 780
Tube feet, 680, 681, 682
Tuberculosis, 532
Tubers, 720, 792
Tubocurarine, *604*
*Tubulanus sexlineatus, 659*
Tubular guts, 1056–1057
Tubular heart, *1027*
Tubular reabsorption, in the vertebrate nephron, 1078, 1079
Tubular secretion, in the vertebrate nephron, 1078, 1079
Tubulidentata, *698*

Tubulin
in microtubules, 95–96 (*see also* Microtubules)
spindle apparatus, 213–214

*Tubulinosema ratisbonensis, 617*
Tulips, *603*
Tumor necrosis factor, 861

Tumor suppressor genes
DNA methylation and, 344
gain of function mutation and, 306
mutations in colon cancer, 314

Tumors
benign and malignant, 227
chaperone proteins and, 51
treatments targeting the cell cycle, 228–229

Tundra, 1128–1129
Tunic, 684
Tunicates, *632, 679,* 683, 684
Turbellarians, 657

Turgor pressure
guard cell function and, 733
mechanism generating, 727
osmosis and, 114–115
in plant cell expansion, 766
in plant growth, 710
in the pressure flow model of phloem transport, 736
vacuoles and, 93

Turner syndrome, 250
Turnover, in lake water, 1213–1214
Turtles, 420, 693, *694*
Twigs, 721–722
Twin studies, on epigenetic changes, 344
Twins, 907
Twitches, 994–995
Twitters, 1183
Two-dimensional gel electrophoresis, 369
"Two-point spatial discrimination test," 952
Two-pronged bristletails, 671, *672*
Tympanic canal, *954,* 955
Tympanic membrane, 954, 955, 956
*Tympanuchus cupido,* 434
Type I diabetes, 848, 876
Type II diabetes, 848, 853
Typhlosole, 1057

Tyrosine, 44
*Tyto alba, 696*

U

Ubiquinone (coenzyme $Q_{10}$), 172, *173*
Ubiquitin, 349
UCP1 (uncoupling 1) protein, 165, 174, 182
*Ulmus procera,* 793
Ulna, 423
Ultimate causes, of animal behavior, 1096
*Ultrabithorax (Ubx)* gene, 415, 418, *419*
Ultrabithorax (Ubx) protein, 418
Ultraviolet (UV) radiation
  evolution of life and, 5
  mutagenic effects, 309, 310
  ozone layer and, 1212
  thymine dimers and skin cancer, 277
*Ulva, 538,* 572, 1201
*Ulva rigida, 573*
Umami, 951
Umbelliferone, *802*
Umbels, *597*
Umbilical cord, 897, 906, 919
Umbilicus, 897
Uncompetitive inhibitors, of enzymes, 158
Underground stems, 792, 793
Undershoot, 934
Unequal crossing over, 498, *499*
Unicellular yeasts, 609
Unidirectional ventilation, 1010–1012
Uniporters, 118
Unipotency, 394
United States
  population age structure, 1165–1166
  population growth, 1165
Unloading, of phloem sieve tubes, 736–737
Unsaturated fatty acids
  phospholipids and, 107
  structure and function, 56, *57*
Upregulation, of hormone receptors, 853
Upwelling zone, 1214
Uracil
  codons and the genetic code, 288–289
  formed by cytosine deamination, 309
  in RNA, 63, 64, *65*
  structure, *63*
Uranium-234, *507*
Uranium-235, *507*
Urea
  effect on protein structure, 50
  excretion as a nitrogenous waste, 1074
  in the extracellular fluid of cartilaginous fish, 1073
  inorganic fertilizers, 747
Ureotelic animals, 1074
Ureter, 1080, 1082
Urethra
  excretion through, 1080
  in the female reproductive system, 892, *893*
  in the male reproductive system, 890, 891
Urey, Harold, 70, 71
Uric acid, 1074, 1076–1077
Uricotelic animals, 1074
Urinary bladder, 1080
Urine
  defined, 1072
  formation in the vertebrate kidney, 1078–1079
  glucose levels with diabetes, 848
  production of concentrated urine in mammals, 1071, 1077, 1082–1084, 1090
  *See also* Excretory systems
*Ursus americanus, 1130*
U.S. Coast Guard, 1241
Uterine cycle
  defined, 892
  description of, 893
  hormonal regulation of, 894–895
Uterus
  function of, 889
  in humans, 892, *893*
  in labor and childbirth, 843, 896–897
  origin and development of the placenta, 919
Utricle, 956, 957

V

Vaccinations
  "drive-through," *16*
  eradication of smallpox, 856
  evolution of viruses and, 427, 446
  "herd immunity," 856
  immunological memory and, 863, 866
  reasons people resist, 877
Vaccine proteins, production through biotechnology, *385*
Vacuoles
  in ciliates, 554, *555*
  functions, 93
  in plant cell expansion, 766
  in plant cells, *87*
  in plant development, 710
  in plant self-protection from chemical defenses, 805
Vagina
  in childbirth, 897
  in humans, 892, *893*
  role in internal fertilization, 887
Vagus nerve (cranial nerve X), 969
Valence shell, 25
Valine, *44*
Vampire bats, 1071, 1090
Van der Waals interactions
  description of, 30
  in DNA, 266
  features of, *26*
  in protein quaternary structure, 48
  in protein tertiary structure, 47
Variable regions, of immunoglobulins, 867, *869,* 870
Variables, 12, 13
Variegated darter, *472*
Vas deferens, 890, 891
Vasa recta, *1080,* 1082, 1084
Vascular bundles, 710, 719
Vascular cambium
  in grafting, 793, *794*
  secondary plant growth and, 715, *716,* 721, 722
Vascular disease, 1042–1043
Vascular plants
  ancient forests, 580
  branching, independent sporophyte, 579
  distinguishing characteristics, 573
  evolution of leaves, 583, *584*
  evolution of roots, 582–583
  evolutionary significance of vascular tissue, 579
  heterospory, 584–585
  horsetails and ferns, 581–582
  lycophytes, 581
  major groups and distinguishing characteristics, *574*
  rhyniophyte relatives, 580
Vascular rays, 722
Vascular smooth muscle
  in arteries and arterioles, 1039, *1040*
  in autoregulation of blood flow, 1044
  calcium ions in the relaxation of, 136
  control of blood distribution in the body, 993
Vascular tissue
  in angiosperms, 596
  evolution of land plants and, 579
  in gymnosperms, 593
  *See also* Phloem; Xylem
Vascular tissue system
  description of, *712,* 714–715
  in leaves, 720, *721*
  primary meristem giving rise to, 716
  in roots, 717, *718*
  in shoots, 719
Vasectomy, *898*
Vasopressin
  actions of, 843
  in blood pressure regulation, 1044–1045
  bonding behavior in voles and, 125
  *See also* Antidiuretic hormone
Vasopressin receptors, 125, 129
Vectors, in transformation, 377
Vegetal hemisphere, 903–904
Vegetal pole, 395
Vegetarian diet, 1051, 1054
Vegetarian finch, *474*
Vegetative cells, of cyanobacteria, 532, *533*
Vegetative meristems, 715
  *See also* Shoot apical meristem
Vegetative reproduction
  in agriculture, 793–794
  disadvantages, 793
  forms of, 792–793
  *See also* Asexual reproduction
Veins (blood vessels)
  anatomy, *1040*
  blood flow through, 1041–1042
  function in vertebrate circulatory systems, 1028
Veins (of leaves), 710, 720, *721*
Veldt, 1131
Velvet worms, *667,* 668
Venter, Craig, 356, 359, 360
Ventilation
  in human lungs, 1015–1016
  maximization of partial pressure gradients, 1009
  respiratory tract secretions and, 1013, 1015
  tidal, 1012–1013
  unidirectional ventilation in birds, 1010–1012
Ventral, 634, *635*
Ventral horn, 942
Ventral medulla, 1020
Ventricles
  in amphibians, 1029
  in crocodilians and birds, 1029
  in fish, 1028
  in lungfish, 1028
  mammalian heart, 1030–1032
  mutation-based wall thickening in humans, 1025, 1046
  in reptiles, 1030
  in three-chambered hearts, 1029
*Venturia canescens,* 1184
Venules, 1028
Venus flytraps, 751–752
Vernalization, 791
*Vernonia amygdalina,* 1105
Vernonioside B1, 1105
Vertebrae, 684
Vertebral column
  characteristic of vertebrates, 684
  evolution of the giraffe neck, 416–417
  evolutionary impact of differences in Hox gene expression, 418
Vertebrate genomes
  endogenous retroviruses in, 545
  gene duplication in, 497–498
  Hox genes, 413–414
Vertebrates
  amniotes, 692, *693*
  amphibians, 690–692
  appendages, 636
  body plan, *685*
  brain, 943
  central nervous system, 941
  characteristic features, 684–685
  circulatory systems, 1027–1030
  in deuterostome phylogeny, *679*
  endoskeleton, 999–1002
  evolution of, 685
  gastrointestinal system, 1058–1064
  Hox gene expression and spine evolution, 418
  jawed fishes, 686–689
  jawless fishes and, 685–686, *687*
  lens determination, 396
  lobe-limbed, 689–690
  major subgroups and number of living species, *632*
  mammals, 696–700
  muscle types, 987
  neurulation and body segmentation, 916–918
  osmoregulation and ionic regulation in, 1073
  phylogeny of living vertebrates, *685*